Before you begin, review your basic study skills.

To get the most out of the *Study Skills Workshops* that begin each chapter, you may choose to review them in the early weeks of your course. Each one includes action items, in addition to simple suggestions that can put you on a clear path to success. Below, we have included a table of contents to aid you in locating these:

Study Skills Workshop
Making Homework a Priority

Attending class and taking notes are important, but they are not enough. The only way that you are really going to learn algebra is by doing your homework.

WHEN TO DO YOUR HOMEWORK: Homework should be started on the day it is assigned, when the material is fresh in your mind. It's best to break your homework sessions into 30-minute periods, allowing for short breaks in between.

HOW TO BEGIN YOUR HOMEWORK: Review your notes and the examples in your text before starting your homework assignment.

GETTING HELP WITH YOUR HOMEWORK: It's normal to have some questions when doing homework. Talk to a tutor, a classmate, or your instructor to get those questions answered.

Now Try This
1. Write a one-page paper that describes *when*, *where*, and *how* you go about completing your algebra homework assignments.
2. For each problem on your next homework assignment, find an example in this book that is similar. Write the example number next to the problem.
3. Make a list of questions that you have while doing your next assignment. Then decide whom you are going to ask to get those questions answered.

CHAPTER 1 Committing to the Course, p. 2
Get started on the right foot by making time for your course, knowing what is expected by your instructor, building a support system.

CHAPTER 2 Preparing to Learn, p. 118
Get ready to learn by discovering your learning style, getting the most out of your textbook: and any accompanying media or supplements, and taking good notes.

CHAPTER 3 Successful Test Taking, p. 222
Improve your chances for success by preparing for the test, using test-taking strategies, and evaluating your performance once your test has been graded.

CHAPTER 4 Making Homework a Priority, p. 320
Commit to completing your homework by starting on the day it is assigned, reviewing examples and notes before getting started, and knowing where to go for help if you have questions.

CHAPTER 5 Attending Class Regularly, p. 394
Recognize the value of class attendance and arrive on time. If you must miss a class, find out what you missed and be sure to study this material.

CHAPTER 6 Reading the Textbook, p. 516
Use your textbook strategically by skimming for an overview before going to class and following up by reviewing the material after class by reading and making notes to deepen your understanding.

CHAPTER 7 Study Groups, p. 642
Form study groups early on in the course to share ideas and notes.

CHAPTER 8 Organizing Your Notebook, p. 748
Organize your notebook by chapter. Then, organize the papers within each section.

CHAPTER 9 Preparing for a Final Exam, p. 826
Gear up for the final exam by getting organized, talking with your instructor, and managing your time.

CHAPTER 10 Preparing for Your Next Math Course, p. 938
Take time to reflect on the study habits that have worked for you over the course and what you would like to do differently next time.

CHAPTER 11 Exploring Careers, p. 990
Seek the advice of a career counselor and visit your campus career center. Then, develop your long-term plan to meet your career goals.

INTERMEDIATE ALGEBRA

FOURTH EDITION

Alan S. Tussy
Citrus College

R. David Gustafson
Rock Valley College

Wolfgang Volz © 1991 Christo

BROOKS/COLE
CENGAGE Learning

Australia • Brazil • Japan • Korea • Mexico • Singapore • Spain • United Kingdom • United States

BROOKS/COLE
CENGAGE Learning™

Intermediate Algebra, Fourth Edition
Alan S. Tussy, R. David Gustafson

Executive Editor:
Charlie Van Wagner

Development Editor:
Danielle Derbenti

Assistant Editor:
Laura Localio

Editorial Assistant:
Lynh Pham

Technology Project Manager:
Ed Costin

Senior Marketing Manager:
Greta Kleinert

Marketing Assistant:
Cassandra Cummings

Marketing Communications Manager:
Darlene Amidon-Brent

Project Manager, Editorial Production:
Cheryll Linthicum

Creative Director:
Rob Hugel

Senior Art Director:
Vernon T. Boes

Print Buyer:
Judy Inouye

Permissions Editor:
Bob Kauser

Production Service:
Graphic World Inc.

Text Designer:
Terri Wright

Photo Researcher:
Terri Wright

Illustrator:
Lori Heckelman

Cover Designer:
Terri Wright

Cover Image:
Wolfgang Volz ©1991 Christo

Cover Printer:
Transcontinental Printing / Interglobe

Compositor:
Graphic World Inc.

For product information and technology assistance, contact us at
Cengage Learning Customer & Sales Support, 1-800-354-9706

For permission to use material from this text or product, submit all requests online at **cengage.com/permissions**
Further permissions questions can be emailed to
permissionrequest@cengage.com

Library of Congress Control Number: 2007940739
ISBN-13: 978-0-495-38973-6
ISBN-10: 0-495-38973-0

Brooks Cole
10 Davis Drive
Belmont, CA 94002-3098
USA

Cengage Learning is a leading provider of customized learning solutions with office locations around the globe, including Singapore, the United Kingdom, Australia, Mexico, Brazil, and Japan. Locate your local office at **international.cengage.com/region.**

Cengage Learning products are represented in Canada by Nelson Education, Ltd.

For your course and learning solutions, visit **academic.cengage.com.**
Purchase any of our products at your local college store or at our preferred online store **www.ichapters.com.**

Printed in Canada
1 2 3 4 5 6 7 12 11 10 09 08

In memory of my mother, Jeanene,
and in honor of my dad, Bill.

—AST

In memory of my teacher and mentor,
Professor John Finch.

—RDG

CONTENTS

1 A REVIEW OF BASIC ALGEBRA 1

1.1 The Language of Algebra 2
1.2 The Real Numbers 10
1.3 Operations with Real Numbers 23
1.4 Simplifying Algebraic Expressions Using Properties of Real Numbers 37
1.5 Solving Linear Equations Using Properties of Equality 49
1.6 Solving Formulas; Geometry 63
1.7 Using Equations to Solve Problems 75
1.8 More about Problem Solving 86
Chapter Summary and Review 102
Chapter Test 113
Group Project 115

2 GRAPHS, EQUATIONS OF LINES, AND FUNCTIONS 117

2.1 The Rectangular Coordinate System 118
2.2 Graphing Linear Equations in Two Variables 132
2.3 Rate of Change and the Slope of a Line 146
2.4 Writing Equations of Lines 161
2.5 An Introduction to Functions 176
2.6 Graphs of Functions 192
Chapter Summary and Review 206
Chapter Test 216
Group Project 218
Cumulative Review 218

3 SYSTEMS OF EQUATIONS 221

3.1 Solving Systems of Equations by Graphing 222
3.2 Solving Systems of Equations Algebraically 235
3.3 Problem Solving Using Systems of Two Equations 246
3.4 Solving Systems of Equations in Three Variables 263
3.5 Problem Solving Using Systems of Three Equations 275
3.6 Solving Systems of Equations Using Matrices 282
3.7 Solving Systems of Equations Using Determinants 294
Chapter Summary and Review 305
Chapter Test 314
Group Project 315
Cumulative Review 316

4 INEQUALITIES   319

4.1   Solving Linear Inequalities in One Variable   320
4.2   Solving Compound Inequalities   335
4.3   Solving Absolute Value Equations and Inequalities   347
4.4   Linear Inequalities in Two Variables   360
4.5   Systems of Linear Inequalities   370
Chapter Summary and Review   380
Chapter Test   387
Group Project   389
Cumulative Review   390

5 EXPONENTS, POLYNOMIALS, AND POLYNOMIAL FUNCTIONS   393

5.1   Exponents   394
5.2   Scientific Notation   408
5.3   Polynomials and Polynomial Functions   417
5.4   Multiplying Polynomials   434
5.5   The Greatest Common Factor and Factoring by Grouping   447
5.6   Factoring Trinomials   459
5.7   The Difference of Two Squares; the Sum and Difference of Two Cubes   473
5.8   Summary of Factoring Techniques   483
5.9   Solving Equations by Factoring   488
Chapter Summary and Review   503
Chapter Test   512
Group Project   514

6 RATIONAL EXPRESSIONS AND EQUATIONS   515

6.1   Rational Functions and Simplifying Rational Expressions   516
6.2   Multiplying and Dividing Rational Expressions   530
6.3   Adding and Subtracting Rational Expressions   542
6.4   Simplifying Complex Fractions   554
6.5   Dividing Polynomials   565
6.6   Synthetic Division   576
6.7   Solving Rational Equations   584
6.8   Problem Solving Using Rational Equations   596
6.9   Proportion and Variation   606
Chapter Summary and Review   622
Chapter Test   634
Group Project   635

7   RADICAL EXPRESSIONS AND EQUATIONS   641

7.1   Radical Expressions and Radical Functions   642
7.2   Rational Exponents   658
7.3   Simplifying and Combining Radical Expressions   671
7.4   Multiplying and Dividing Radical Expressions   683
7.5   Solving Radical Equations   696
7.6   Geometric Applications of Radicals   708
7.7   Complex Numbers   721
Chapter Summary and Review   734
Chapter Test   743
Group Project   745

8   QUADRATIC EQUATIONS, FUNCTIONS, AND INEQUALITIES   747

8.1   The Square Root Property and Completing the Square   748
8.2   The Quadratic Formula   761
8.3   The Discriminant and Equations That Can Be Written in Quadratic Form   773
8.4   Quadratic Functions and Their Graphs   783
8.5   Quadratic and Other Nonlinear Inequalities   799
Chapter Summary and Review   811
Chapter Test   818
Group Project   820
Cumulative Review   821

9   EXPONENTIAL AND LOGARITHMIC FUNCTIONS   825

9.1   Algebra and Composition of Functions   826
9.2   Inverse Functions   838
9.3   Exponential Functions   850
9.4   Base-*e* Exponential Functions   866
9.5   Logarithmic Functions   875
9.6   Base-*e* Logarithmic Functions   889
9.7   Properties of Logarithms   897
9.8   Exponential and Logarithmic Equations   909
Chapter Summary and Review   922
Chapter Test   933
Group Project   934

10 CONIC SECTIONS; MORE GRAPHING   937

10.1   The Circle and the Parabola   938
10.2   The Ellipse   952
10.3   The Hyperbola   962
10.4   Solving Nonlinear Systems of Equations   973
Chapter Summary and Review   981
Chapter Test   986
Group Project   988

11 MISCELLANEOUS TOPICS   989

11.1   The Binomial Theorem   990
11.2   Arithmetic Sequences and Series   1000
11.3   Geometric Sequences and Series   1011
Chapter Summary and Review   1024
Chapter Test   1028
Group Project   1029
Cumulative Review   1030

APPENDIXES   **APPENDIX 1: Roots and Powers   A-1**
APPENDIX 2: Answers to Selected Exercises   A-3

INDEX   I-1

PREFACE

Intermediate Algebra, Fourth Edition, is more than a simple upgrade of the third edition. Substantial changes have been made to the example structure, the Study Sets, and the pedagogy. Throughout the process, the objective has been to ease teaching challenges and meet students' educational needs.

Algebra, for many of today's developmental math students, is like a foreign language. They have difficulty translating the words, their meanings, and how they apply to problem solving. With these needs in mind (and as educational research suggests), the fundamental goal is to have students read, write, think, and speak using the *language of algebra*. Instructional approaches that include vocabulary, practice, and well-defined pedagogy, along with an emphasis on reasoning, modeling, communication, and technology skills have been blended to address this need.

The most common student question as they watch their instructors solve problems and as they read the textbook is . . . *Why?* The new fourth edition addresses this question in a unique way. Experience teaches us that it's not enough to know *how* a problem is solved. Students gain a deeper understanding of algebraic concepts if they know *why* a particular approach is taken. This instructional truth was the motivation for adding a **Strategy** and **Why** explanation to the solution of each worked example. The fourth edition now provides, on a consistent basis, a concise answer to that all-important question: *Why?*

This is just one of several changes in this revision, and we trust that all of them will make the course a better experience for both instructor and student.

NEW TO THIS EDITION

- New Example Structure
- New Chapter Opening Applications
- New *Study Skills Workshops*
- New Chapter Objectives
- New *Guided Practice* and *Try It Yourself* sections in the *Study Sets*
- New End-of-Chapter Organization

Chapter Openers Answering The Question: When Will I Use This?

Have you heard this question before? Instructors are asked this question time and again by students. In response, we have written chapter openers called *From Campus to Careers*. This feature highlights vocations that require various algebraic skills. Designed to inspire career exploration, each includes job outlook, educational requirements, and annual earnings information. Careers presented in the openers are tied to an exercise found later in the *Study Sets*.

Examples That Tell Students Not Just How, But WHY

Why? That question is often asked by students as they watch their instructor solve problems in class and as they are working on problems at home. It's not enough to know how a problem is solved. Students gain a deeper understanding of the algebraic concepts if they know why a particular approach was taken. This instructional truth was the motivation for adding a *Strategy* and *Why* explanation to each worked example.

Examples That Offer Immediate Feedback

Each example includes a *Self Check*. These can be completed by students on their own or as classroom lecture examples, which is how Alan Tussy uses them. Alan asks selected students to read aloud the *Self Check* problems as he writes what the student says on the board. The other students, with their books open to that page, can quickly copy the *Self Check* problem to their notes. This speeds up the note-taking process and encourages student participation in his lectures. It also teaches students how to read mathematical symbols. Each *Self Check* answer is printed adjacent to the corresponding problem in the Annotated Instructor's Edition for easy reference. *Self Check* solutions can be found at the end of each section in the student edition before the *Study Sets* begin.

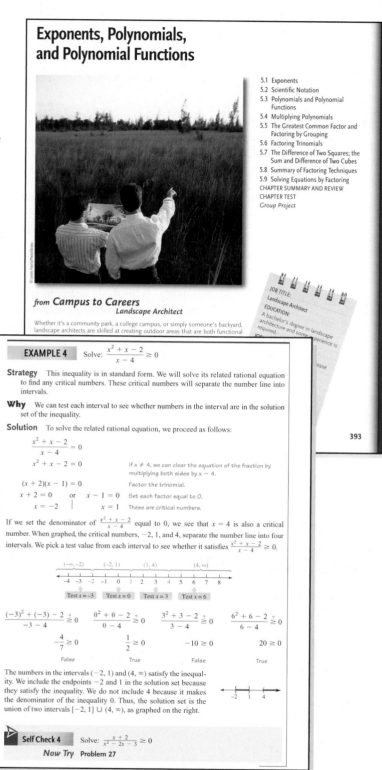

Exponents, Polynomials, and Polynomial Functions

5.1 Exponents
5.2 Scientific Notation
5.3 Polynomials and Polynomial Functions
5.4 Multiplying Polynomials
5.5 The Greatest Common Factor and Factoring by Grouping
5.6 Factoring Trinomials
5.7 The Difference of Two Squares; the Sum and Difference of Two Cubes
5.8 Summary of Factoring Techniques
5.9 Solving Equations by Factoring
CHAPTER SUMMARY AND REVIEW
CHAPTER TEST
Group Project

from **Campus to Careers**
Landscape Architect

Whether it's a community park, a college campus, or simply someone's backyard, landscape architects are skilled at creating outdoor areas that are both functional

JOB TITLE: Landscape Architect
EDUCATION: A bachelor's degree in landscape architecture and some experience is required.

393

EXAMPLE 4 Solve: $\dfrac{x^2 + x - 2}{x - 4} \geq 0$

Strategy This inequality is in standard form. We will solve its related rational equation to find any critical numbers. These critical numbers will separate the number line into intervals.

Why We can test each interval to see whether numbers in the interval are in the solution set of the inequality.

Solution To solve the related rational equation, we proceed as follows:

$$\frac{x^2 + x - 2}{x - 4} = 0$$

$$x^2 + x - 2 = 0 \qquad \text{If } x \neq 4, \text{ we can clear the equation of the fraction by multiplying both sides by } x - 4.$$

$$(x + 2)(x - 1) = 0 \qquad \text{Factor the trinomial.}$$

$$x + 2 = 0 \quad \text{or} \quad x - 1 = 0 \qquad \text{Set each factor equal to 0.}$$

$$x = -2 \qquad \qquad x = 1 \qquad \text{These are critical numbers.}$$

If we set the denominator of $\frac{x^2 + x - 2}{x - 4}$ equal to 0, we see that $x = 4$ is also a critical number. When graphed, the critical numbers, -2, 1, and 4, separate the number line into four intervals. We pick a test value from each interval to see whether it satisfies $\frac{x^2 + x - 2}{x - 4} \geq 0$.

$(-\infty, -2) \quad (-2, 1) \quad (1, 4) \quad (4, \infty)$

Test $x = -3$ Test $x = 0$ Test $x = 3$ Test $x = 6$

$$\frac{(-3)^2 + (-3) - 2}{-3 - 4} \overset{?}{\geq} 0 \qquad \frac{0^2 + 0 - 2}{0 - 4} \overset{?}{\geq} 0 \qquad \frac{3^2 + 3 - 2}{3 - 4} \overset{?}{\geq} 0 \qquad \frac{6^2 + 6 - 2}{6 - 4} \overset{?}{\geq} 0$$

$$-\frac{4}{7} \geq 0 \qquad\qquad \frac{1}{2} \geq 0 \qquad\qquad -10 \geq 0 \qquad\qquad 20 \geq 0$$

False   True   False   True

The numbers in the intervals $(-2, 1)$ and $(4, \infty)$ satisfy the inequality. We include the endpoints -2 and 1 in the solution set because they satisfy the inequality. We do not include 4 because it makes the denominator of the inequality 0. Thus, the solution set is the union of two intervals $[-2, 1] \cup (4, \infty)$, as graphed on the right.

Self Check 4 Solve: $\dfrac{x + 2}{x^2 - 2x - 3} \geq 0$

Now Try Problem 27

Examples That Ask Students To Try

Each example ends with a *Now Try* problem. These are the final step in the learning process. Each one is linked to similar problems found within the *Guided Practice* section of the *Study Sets*.

Emphasis on Study Skills

Each chapter begins with a *Study Skills Work-shop*. Instead of simple suggestions printed in the margins, each workshop contains a *Now Try This* section offering students actionable skills, assignments, and projects that will impact their study habits throughout the course.

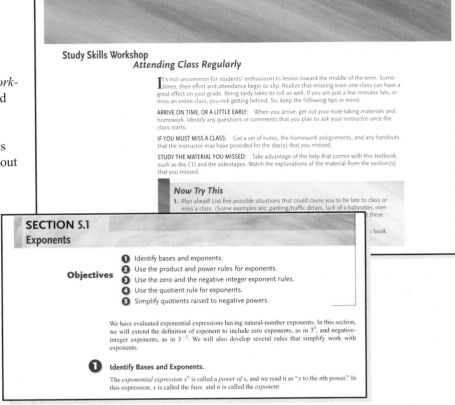

Study Skills Workshop
Attending Class Regularly

It's not uncommon for students' enthusiasm to lessen toward the middle of the term. Sometimes, their effort and attendance begin to slip. Realize that missing even one class can have a great effect on your grade. Being tardy takes its toll as well. If you are just a few minutes late, or miss an entire class, you risk getting behind. So, keep the following tips in mind.

ARRIVE ON TIME, OR A LITTLE EARLY: When you arrive, get out your note-taking materials and homework. Identify any questions or comments that you plan to ask your instructor once the class starts.

IF YOU MUST MISS A CLASS: Get a set of notes, the homework assignments, and any handouts that the instructor may have provided for the day(s) that you missed.

STUDY THE MATERIAL YOU MISSED: Take advantage of the help that comes with this textbook, such as the CD and the videotapes. Watch the explanations of the material from the section(s) that you missed.

Now Try This
1. Plan ahead! List five possible situations that could cause you to be late to class or miss a class. (Some examples are: parking/traffic delays, lack of a babysitter, over-... these
... book.

Useful Objectives Help Keep Students Focused

Objectives are now numbered at the start of each section to focus students' attention on the skills that they will learn as they work through the section. When each objective is introduced, the number and heading will appear again to remind them of the objective at hand.

SECTION 5.1
Exponents

Objectives
1. Identify bases and exponents.
2. Use the product and power rules for exponents.
3. Use the zero and the negative integer exponent rules.
4. Use the quotient rule for exponents.
5. Simplify quotients raised to negative powers.

We have evaluated exponential expressions having natural-number exponents. In this section, we will extend the definition of exponent to include zero exponents, as in 3^0, and negative-integer exponents, as in 3^{-2}. We will also develop several rules that simplify work with exponents.

1. **Identify Bases and Exponents.**

The *exponential expression* x^n is called a *power of x*, and we read it as "x to the nth power." In this expression, x is called the *base*, and n is called the *exponent*.

Heavily Revised Study Sets

The *Study Sets* have been thoroughly revised to ensure every concept is covered even if the instructor traditionally assigns every other problem. Particular attention was paid to developing a gradual level of progression.

Guided Practice

All of the problems in the *Guided Practice* portion of the *Study Sets* are linked to an associated worked example from that section. This feature will promote student success by referring them to the proper example(s) if they encounter difficulties solving homework problems.

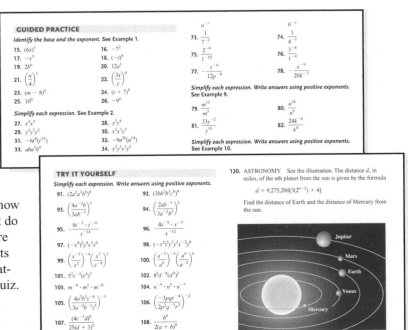

GUIDED PRACTICE

Identify the base and the exponent. See Example 1.

15. $(6x)^3$ 16. -7^2
17. $-x^5$ 18. $(-t)^4$
19. $2b^6$ 20. $12a^2$
21. $\left(\frac{n}{4}\right)^3$ 22. $\left(\frac{3x}{y}\right)^0$
23. $(m-8)^6$ 24. $(t+7)^8$
25. 10^0 26. -9^0

Simplify each expression. See Example 2.

27. x^2x^3 28. y^3y^4
29. $y^3y^7y^2$ 30. $x^2x^3x^5$
31. $-6t^8(t^{15})$ 32. $-9n^{20}(n^{14})$
33. aba^3b^4 34. $x^2y^3x^3y^2$

73. $\frac{1}{7^{-2}}$ 74. $\frac{1}{4^{-3}}$
75. $\frac{2^{-4}}{1^{-10}}$ 76. $\frac{3^{-4}}{1^{-9}}$
77. $-\frac{t^{-6}}{12p^{-8}}$ 78. $-\frac{x^{-9}}{20k^{-7}}$

Simplify each expression. Write answers using positive exponents. See Example 9.

79. $\frac{m^{15}}{m^3}$ 80. $\frac{n^{18}}{n^6}$
81. $\frac{33y^{-2}}{y^{10}}$ 82. $\frac{24k^{-4}}{k^8}$

Simplify each expression. Write answers using positive exponents. See Example 10.

Try It Yourself

To promote problem recognition, some *Study Sets* now include a collection of *Try It Yourself* problems that do not have the example linking. The problem types are thoroughly mixed and are not linked, giving students an opportunity to practice decision making and strategy selection as they would when taking a test or quiz.

TRY IT YOURSELF

Simplify each expression. Write answers using positive exponents.

91. $(2a^2a^3b^0)^4$ 92. $(3bb^2b^3c^0)^4$
93. $\left(\frac{4a^{-2}b}{3ab^{-1}}\right)^3$ 94. $\left(\frac{2ab^{-3}}{3a^{-2}b^2}\right)^2$
95. $\frac{8t^{-3} \cdot t^{-11}}{t^{-14}}$ 96. $\frac{4x^{-9} \cdot x^{-3}}{x^{-12}}$
97. $(-x^8y_2^3y^4x^0)$ 98. $(-x^2)^5y^7y^3x^{-2}y^0$
99. $\left(\frac{x^{-5}}{x^2}\right)^{-4}\left(\frac{x^7}{x^{-8}}\right)^3$ 100. $\left(\frac{a^{-3}}{d^6}\right)^{-3}\left(\frac{a^6}{a^{-2}}\right)^6$
101. $5^2r^{-5}(r^6)^3$ 102. $8^2d^{-8}(d^9)^2$
103. $m^{-4} \cdot m^2 \cdot m^{-8}$ 104. $n^{-9} \cdot n^5 \cdot n^{-7}$
105. $\left(\frac{4a^2b^2z^{-4}}{3a^{-2}b^{-1}z^1}\right)^{-3}$ 106. $\left(\frac{-3pqr^{-4}}{2p^2q^{-3}r^2}\right)^{-2}$
107. $\frac{(4c^{-3}d)^0}{25d(d+3)^0}$ 108. $\frac{b^0}{2(a+b)^0}$

120. **ASTRONOMY** See the illustration. The distance d, in miles, of the nth planet from the sun is given by the formula

$$d = 9,275,200[3(2^{n-2}) + 4]$$

Find the distance of Earth and the distance of Mercury from the sun.

Comprehensive End-of-Chapter Summary with Integrated Chapter Review

The end-of-chapter material has been redesigned to function as a complete study guide for students. New Chapter Summaries that include definitions, concepts, and examples, by section, have been written. Review problems for each section have been placed after each section summary.

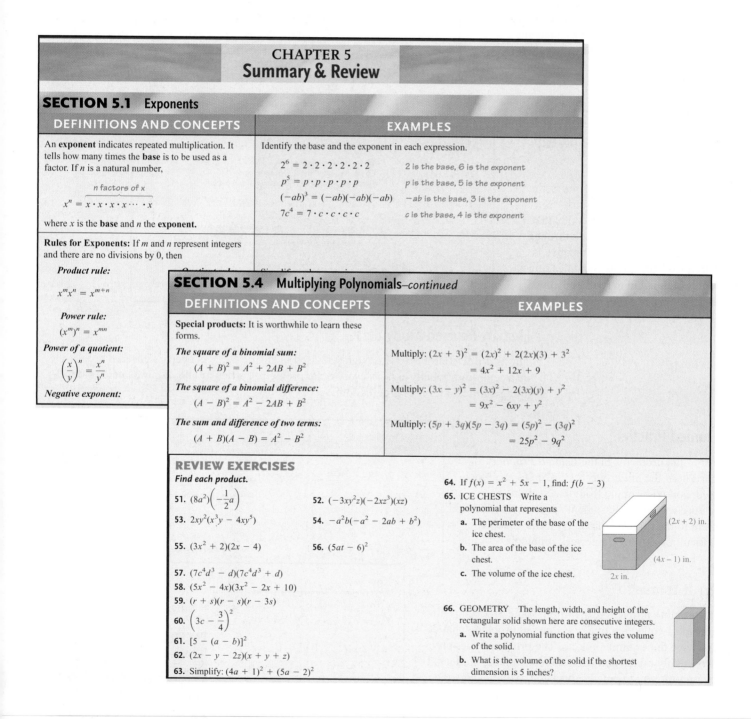

CHAPTER 5
Summary & Review

SECTION 5.1   Exponents

DEFINITIONS AND CONCEPTS	EXAMPLES
An **exponent** indicates repeated multiplication. It tells how many times the **base** is to be used as a factor. If n is a natural number, $$x^n = \overbrace{x \cdot x \cdot x \cdot x \cdots \cdot x}^{n \text{ factors of } x}$$ where x is the **base** and n the **exponent**.	Identify the base and the exponent in each expression. $2^6 = 2 \cdot 2 \cdot 2 \cdot 2 \cdot 2 \cdot 2$ 2 is the base, 6 is the exponent $p^5 = p \cdot p \cdot p \cdot p \cdot p$ p is the base, 5 is the exponent $(-ab)^3 = (-ab)(-ab)(-ab)$ $-ab$ is the base, 3 is the exponent $7c^4 = 7 \cdot c \cdot c \cdot c \cdot c$ c is the base, 4 is the exponent

Rules for Exponents: If m and n represent integers and there are no divisions by 0, then

Product rule:

$$x^m x^n = x^{m+n}$$

Power rule:

$$(x^m)^n = x^{mn}$$

Power of a quotient:

$$\left(\frac{x}{y}\right)^n = \frac{x^n}{y^n}$$

Negative exponent:

SECTION 5.4   Multiplying Polynomials—*continued*

DEFINITIONS AND CONCEPTS	EXAMPLES
Special products: It is worthwhile to learn these forms.	
The square of a binomial sum: $$(A + B)^2 = A^2 + 2AB + B^2$$	Multiply: $(2x + 3)^2 = (2x)^2 + 2(2x)(3) + 3^2$ $\qquad = 4x^2 + 12x + 9$
The square of a binomial difference: $$(A - B)^2 = A^2 - 2AB + B^2$$	Multiply: $(3x - y)^2 = (3x)^2 - 2(3x)(y) + y^2$ $\qquad = 9x^2 - 6xy + y^2$
The sum and difference of two terms: $$(A + B)(A - B) = A^2 - B^2$$	Multiply: $(5p + 3q)(5p - 3q) = (5p)^2 - (3q)^2$ $\qquad = 25p^2 - 9q^2$

REVIEW EXERCISES

Find each product.

51. $(8a^2)\left(-\dfrac{1}{2}a\right)$

52. $(-3xy^2z)(-2xz^3)(xz)$

53. $2xy^2(x^3y - 4xy^5)$

54. $-a^2b(-a^2 - 2ab + b^2)$

55. $(3x^2 + 2)(2x - 4)$

56. $(5at - 6)^2$

57. $(7c^4d^3 - d)(7c^4d^3 + d)$

58. $(5x^2 - 4x)(3x^2 - 2x + 10)$

59. $(r + s)(r - s)(r - 3s)$

60. $\left(3c - \dfrac{3}{4}\right)^2$

61. $[5 - (a - b)]^2$

62. $(2x - y - 2z)(x + y + z)$

63. Simplify: $(4a + 1)^2 + (5a - 2)^2$

64. If $f(x) = x^2 + 5x - 1$, find: $f(b - 3)$

65. ICE CHESTS   Write a polynomial that represents

a. The perimeter of the base of the ice chest.

b. The area of the base of the ice chest.

c. The volume of the ice chest.

$(2x + 2)$ in.
$(4x - 1)$ in.
$2x$ in.

66. GEOMETRY   The length, width, and height of the rectangular solid shown here are consecutive integers.

a. Write a polynomial function that gives the volume of the solid.

b. What is the volume of the solid if the shortest dimension is 5 inches?

TRUSTED FEATURES

- **The Study Sets** found in each section offer a multifaceted approach to practicing and reinforcing the concepts taught in each section. They are designed for students to methodically build their knowledge of the section concepts, from basic recall to increasingly complex problem solving, through reading, writing, and thinking mathematically.

  Vocabulary—Each Study Set begins with the important Vocabulary discussed in that section. The fill-in-the-blank vocabulary problems emphasize the main concepts taught in the chapter and provide the foundation for learning and communicating the language of algebra.

  Concepts—In Concepts, students are asked about the specific subskills and procedures necessary to successfully complete the practice problems that follow.

  Notation—In Notation, the students review the new symbols introduced in a section. Often, they are asked to fill in steps of a sample solution. This helps to strengthen their ability to read and write mathematics and prepares them for the practice problems by modeling solution formats.

  Guided Practice—The problems in Guided Practice are linked to an associated worked example from that section. This feature will promote student success by referring them to the proper examples if they encounter difficulties solving homework problems.

  Try It Yourself—To promote problem recognition, the Try It Yourself problems are thoroughly mixed and are not linked, giving students an opportunity to practice decision-making and strategy selection as they would when taking a test or quiz.

  Applications—The Applications provide students the opportunity to apply their newly acquired algebraic skills to relevant and interesting real-life situations.

  Writing—The Writing problems help students build mathematical communication skills.

  Review—The Review problems consist of randomly selected problems from previous chapters. These problems are designed to keep students' successfully mastered skills fresh and at the forefront of their minds before moving on to the next section.

  Challenge Problems—The Challenge Problems provide students with an opportunity to stretch themselves and develop their skills beyond the basics. Instructors often find these to be useful as extra-credit problems.

- **Detailed Author Notes** guide students along in a step-by-step process continue to be found in the solutions to every example.

- **The Language of Algebra** boxes draw connections between mathematical terms and everyday references to reinforce the language of algebra thread that runs throughout the text.

- **The Notation, Success Tips, Caution,** and **Calculators** boxes offer helpful tips to reinforce correct mathematical notation, improve students' problem-solving abilities, warn students of potential pitfalls and increase clarity, and offer tips on using scientific calculators.

- **Using Your Calculator** (formerly called Accent on Technology) sections are designed for instructors who wish to use calculators as part of the instruction in this course. These sections introduce keystrokes and show how scientific and graphing calculators can be used to solve problems. In the Study Sets, icons are used to denote problems that require a graphing calculator.

- **Strategic use of color** has been implemented within the new design to help the visual learner.

- **Chapter Tests** are available at the end of every chapter as preparation for the class exam.

- **The Cumulative Review** following the end-of-chapter material keeps students' skills sharpened before moving on to the next chapter. Each problem is now linked to the associated section from which the problem came for ease of reference. The final Cumulative Review, found at the end of the last chapter, is often used by instructors as a Final Exam Review.

CHANGES TO THE TABLE OF CONTENTS

Based on feedback from colleagues and users of the third edition, the following changes have been made to the table of contents in an effort to further streamline the text and make it even easier to use.

- In Chapter 1, additional attention has been given to the topics of perimeter, area, and volume with the addition of Section 1.6: *Solving Formulas; Geometry.*

- In Section 2.5: *An Introduction to Functions,* we have inserted a discussion of the concept of *relation* before the formal definition of a function.

- Section 2.6: *Graphs of Functions* now includes more-detailed explanations of how to find function values graphically and how to determine the domain and range of a function from its graph.

- In Chapter 3, greater attention is given to problem solving with the addition of two stand-alone sections entitled Section 3.3: *Problem Solving Using Systems of Two Equations* and Section 3.5: *Problem Solving Using Systems of Three Equations.*

- Section 6.4: *Simplifying Complex Fractions* has been rewritten to better distinguish between Methods 1 and 2 and the problem types for which each method is most appropriate.

- Additional examples of shared-work problems and uniform motion problems are now part of a new section in Chapter 6 entitled Section 6.8: *Problem Solving Using Rational Equations.*

- Section 9.1: *Algebra and Composition of Functions* now includes examples and problems where sum, difference, product, and quotient functions are evaluated graphically.

- There is greater emphasis on *f(x)* function notation in Chapter 9: *Exponential and Logarithmic Functions.*

- Section 11.4: *Permutations and Combinations* and Section 11.5: *Probability,* formerly in the third edition, are now available online from the publisher.

GENERAL REVISIONS AND OVERALL DESIGN

- We have edited the prose so that it is even more clear and concise.

- Strategic use of color has been implemented within the new design to help the visual learner.

- Added color in the solutions highlight strategic steps and improve readability.

- We have updated all data and graphs and have added scaling to all axes in all graphs.

- We have added more real-world applications and deleted some of the more "contrived" problems.
- We have included more problem-specific photographs.

INSTRUCTOR RESOURCES

Print Ancillaries

INSTRUCTOR'S RESOURCE BINDER (0-495-55460-X)

Maria H. Andersen, *Muskegon Community College*

NEW! Offered exclusively with Tussy/Gustafson. Each section of the main text is discussed in uniquely designed Teaching Guides containing instruction tips, examples, activities, worksheets, overheads, assessments, and solutions to all worksheets and activities.

COMPLETE SOLUTIONS MANUAL (0-495-38991-9)

Kristy Hill, *Hinds Community College*

The Complete Solutions Manual provides worked-out solutions to all of the problems in the text.

TEST BANK (0-495-38992-7)

Carol M. Walker & David J. Walker, *Hinds Community College*

Drawing from hundreds of text-specific questions, an instructor can easily create tests that target specific course objectives. The Test Bank includes multiple tests per chapter, as well as final exams. The tests are made up of a combination of multiple-choice, free-response, true/false, and fill-in-the-blank questions.

ANNOTATED INSTRUCTOR'S EDITION (0-495-38988-9)

The Instructor's Edition provides the complete student text with answers next to each respective exercise.

Electronic Ancillaries

WebAssign ENHANCED WEBASSIGN (0-495-38996-X)

Instant feedback and ease of use are just two reasons why WebAssign is the most widely used homework system in higher education. WebAssign's homework delivery system allows you to assign, collect, grade, and record homework assignments via the web. And now, this proven system has been enhanced to include links to textbook sections, video examples, and problem-specific tutorials. Enhanced WebAssign is more than a homework system—it is a complete learning system for math students.

CENGAGENOW™ (0-495-39461-0)

CengageNOW™ is an online teaching and learning resource that gives you more control in less time and delivers the results you want—NOW.

POWERLECTURE™: A 1-STOP MICROSOFT® POWERPOINT® TOOL (0-495-55650-5)

NEW! The ultimate multimedia manager for your course needs. The PowerLecture CD-ROM includes the Complete Solutions Manual, ExamView®, JoinIn™, and custom PowerPoint® lecture slides authored by Richard D. Townsend, North Carolina Central University.

TEXT SPECIFIC DVDs **(0-495-38993-5)**

These text specific DVDs provide additional guidance and support to students when they are preparing for an upcoming quiz or exam.

STUDENT RESOURCES

Print Ancillaries

STUDENT WORKBOOK (0-495-55459-6)
Maria H. Andersen, *Muskegon Community College*

NEW! Get a head start. The Student Workbook contains all of the Assessments, Activities, and Worksheets from the Instructor's Resource Binder for classroom discussions, in-class activities, and group work.

STUDENT SOLUTIONS MANUAL (0-495-38990-0)
Alexander H. Lee, *Hinds Community College*

The Student Solutions Manual provides worked-out solutions to the odd-numbered problems in the text.

Electronic Ancillaries

WebAssign ENHANCED WEBASSIGN (0-495-38996-X)

Get instant feedback on your homework assignments with Enhanced WebAssign (assigned by your instructor). This online homework system is easy to use and includes helpful links to textbook sections, video examples, and problem-specific tutorials.

INSTANT ACCESS CODE, CENGAGENOW™ (0-495-39463-7)

Instant Access gives students without a new copy of Tussy/Gustafson's *Elementary Algebra, Fourth Edition,* one access code to all available technology associated with this text-book. CengageNOW, a powerful and fully integrated teaching and learning system, provides instructors and students with unsurpassed control, variety, and all-in-one utility. CengageNOW ties together the fundamental learning activities: diagnostics, tutorials, homework, personalized study, quizzing, and testing. Personalized Study is a learning companion that helps students gauge their unique study needs and makes the most of their study time by building focused personalized learning plans that reinforce key concepts. Pre-Tests give students an initial assessment of their knowledge. Personalized study plans, based on the students' answers to the Pre-Test questions, outline key elements for review. Post-Tests assess student mastery of core chapter concepts. Results can even be e-mailed to the instructor!

PRINTED ACCESS CARD, CENGAGENOW™ (0-495-39462-9)

This printed access card provides entrance to all the content that accompanies Tussy/Gustafson's *Elementary Algebra, Fourth Edition,* within CengageNOW.

WEBSITE *academic.cengage.com/math/tussy*
Visit us on the web for access to a wealth of free learning resources, including tutorials, final exams, chapter outlines, chapter reviews, web links, videos, flashcards, and more!

ACKNOWLEDGMENTS

We want to express our gratitude to Steve Odrich, Maria H. Andersen, Diane Koenig, Alexander Lee, Ed Kavanaugh, Karl Hunsicker, George Carlson, Jim Cope, Arnold Kondo, John McKeown, Kent Miller, Donna Neff, Eric Robitoy, Maryann Rachford, Chris Scott, Rob Everest, Cathy Gong, Dave Ryba, Terry Damron, Marion Hammond, Lin Humphrey, Doug Keebaugh, Robin Carter, Tanja Rinkel, Bob Billups, Jeff Cleveland, Jo Morrison, Sheila White, Jim McClain, Paul Swatzel, Bill Tussy, Liz Tussy, and the Citrus College Library staff (including Barbara Rugeley) for their help with this project. Your encouragement, suggestions, and insight have been invaluable to us.

We would also like to express our thanks to the Brooks/Cole editorial, marketing, production and design staff for helping us craft this new edition: Charlie Van Wagner, Danielle Derbenti, Greta Kleinert, Laura Localio, Lynh Pham, Cassandra Cummings, Donna Kelley, Sam Subity, Cheryll Linthicum, Vernon Boes, and Graphic World.

Additionally, we would like to say that authoring a textbook is a tremendous undertaking. A revision of this scale would not have been possible without the thoughtful feedback and support from the following colleagues listed below. Their contributions to this edition have shaped this revision in countless ways.

Alan S. Tussy
R. David Gustafson

Advisory Board

Kim Caldwell, Volunteer State Community College

Peter Embalabala, Lincoln Land Community College

John Garlow, Tarrant Community College–Southeast Campus

Becki Huffman, Tyler Junior College

Mary Legner, Riverside Community College

Ann Loving, J. Sargeant Reynolds Community College

Trudy Meyer, El Camino College

Carol Ann Poore, Hinds Community College

Jill Rafael, Sierra College

Pamelyn Reed, Cy-Fair College

Patty Sheeran, McHenry Community College

Valerie Wright, Central Piedmont Community College

Loris Zucca, Kingwood College

Reviewers

Maria Andersen, Muskegon Community College

Scott Barnett, Henry Ford Community College

David Behrman, Somerset Community College

Jeanne Bowman, University of Cincinnati

Carol Cheshire, Macon State College

Suzanne Doviak, Old Dominion University

Peter Embalabala, Lincoln Land Community College

Joan Evans, Texas Southern University

Rita Fielder, University of Central Arkansas

Anissa Florence, Jefferson Community and Technical College

Pat Foard, South Plains College

Tom Fox, Cleveland State Community College

Heng Fu, Thomas Nelson Community College

Kim Gregor, Delaware Technical Community College–Wilmington

Haile Kebede Haile, Minneapolis Community and Technical College

Jennifer Hastings, Northeast Mississippi Community College

Kristy Hill, Hinds Community College

Laura Hoye, Trident Technical College

Becki Huffman, Tyler Junior College

Angela Jahns, North Idaho College

Cynthia Johnson, Heartland Community College

Ann Loving, J. Sargeant Reynolds Community College

Lynette King, Gadsden State Community College

Mike Kirby, Tidewater Community College

Mary Legner, Riverside Community College

Wayne (Paul) Lee, Saint Philip's College

Yixia Lu, South Suburban College

Keith Luoma, Augusta State University

Susan Meshulam, Indiana University/ Purdue University Indianapolis

Trudy Meyer, El Camino College

Molly Misko, Gadsden State Community College

Elsie Newman, Owens Community College

Charlotte Newsom, Tidewater Community College

Randy Nichols, Delta College

Stephen Nicoloff, Paradise Valley Community College

Charles Odion, Houston Community College

Jason Pallett, Longview Community College

Mary Beth Pattengale, Sierra College

Naeemah Payne, Los Angeles Community College

Carol Ann Poore, Hinds Community College

Jill Rafael, Sierra College

Pamela Reed, North Harris Montgomery Community College

Nancy Ressler, Oakton Community College

Emma Sargent, Tennessee State University

Ned Schillow, Lehigh Carbon Community College

Debra Shafer, University of North Carolina

Hazel Shedd, Hinds Community College

Donald Solomon, University of Wisconsin

John Squires, Cleveland State Community College

Robin Steinberg, Pima Community College

Eden Thompson, Utah Valley State College

Carol Walker, Hinds Community College

Diane Williams, Northern Kentucky University

Loris Zucca, Kingwood College

Class Testers

Candace Blazek, Anoka Ramsey Community College

Jennifer Bluth, Anoka Ramsey Community College

Vicki Gearhart, San Antonio College

Megan Goodwin, Anoka Ramsey Community College

Haile Haile, Minneapolis Community and Technical College

Vera Hu-Hyneman, SUNY–Suffolk Community College

Marlene Kutesky, Virginia Commonwealth University

Richard Leedy, Polk Community College

Wendiann Sethi, Seton Hall University

Eleanor Storey, Frontrange Community College

Cindy Thore, Central Piedmont Community College

Gowribalan "Ana" Vamadeva, University of Cincinnati

Cynthia Wallin, Central Virginia Community College

John Ward, Jefferson Community and Technical College

Focus Groups

Khadija Ahmed, Monroe Community College

Maria Andersen, Muskegon Community College

Chad Bemis, Riverside Community College

A. Elena Bogardus, Camden Community College

Carilynn Bouie, Cuyahoga Community College

Kim Brown, Tarrant Community College

Carole Carney, Brookdale Community College

Joe Castillo, Broward Community College

John Close, Salt Lake Community College

Chris Copple, Northwest State Community College

Mary Deas, Johnson County Community College

Maggie Flint, Northeast State

Douglas Furman, SUNY Ulster Community
College

Abel Gage, Skagit Valley College

Amy Hoherz, Johnson County Community
College

Pete Johnson, Eastern Connecticut State
University

Ed Kavanaugh, Schoolcraft College

Leonid Khazanov, Borough of Manhattan
Community College

MC Kim, Suffolk County Community College

Fred Lang, Art Institute of Washington

Hoat Le, San Diego Community College

Richard Leedy, Polk Community College

Daniel Lopez, Brookdale Community
College

Ann Loving, J. Sargeant Reynolds
Community College

Charles Odion, Houston Community
College

Maggie Pasqua Viz, Brookdale Community
College

Fred Peskoff, Borough of Manhattan
Community College

Sheila Pisa, Riverside Community
College–Moreno Valley

Jill Rafael, Sierra College

Christa Solheid, Santa Ana College

Jim Spencer, Santa Rosa Junior College

Teresa Sutcliffe, Los Angeles Valley College

Rose Toering, Kilian Community College

Judith Wood, Central Florida Community
College

Mary Young, Brookdale Community College

Workshops

Andrea Adlman, Ventura College

Rodney Alford, Calhoun Community
College

Maria Andersen, Muskegon Community
College

Hamid Attarzadeh, Jefferson Community
and Technical College

Victoria Baker, University of Houston–
Downtown

Betty Barks, Lansing Community College

Susan Beane, University of Houston–
Downtown

Barbara Blass, Oakland Community College

Charles A. Bower, St. Philip's College

Tony Craig, Paradise Valley Community
College

Patrick Cross, University of Oklahoma

Archie Earl, Norfolk State University

Melody Eldred, State University of New
York at Cobleskill

Joan Evans, Texas Southern University

Mike Everett, Santa Ana College

Betsy Farber, Bucks County Community
College

Nancy Forrest, Grand Rapids Community
College

Radu Georgescu, Prince George's
Community College

Rebecca Giles, Jefferson State Community
College

Thomas Grogan, Cincinnati State

Paula Jean Haigis, Calhoun Community
College

Haile Haile, Minneapolis Community and
Technical College

Kelli Jade Hammer, Broward Community
College

Julia Hassett, Oakton Community College

Alan Hayashi, Oxnard College

Joel Helms, University of Cincinnati

Jim Hodge, Mountain State University

Jeffrey Hughes, Hinds Community
College

Leslie Johnson, John C. Calhoun State
Community College

Cassandra Johnson, Robeson Community
College

Ed Kavanaugh, Schoolcraft College

Alex Kolesnik, Ventura College

Marlene Kustesky, Virginia Commonwealth
University

Lider-Manuel Lamar, Seminole Community
College

Roger Larson, Anoka Ramsey Community
College

Alexander Lee, Hinds Community College,
Rankin Campus

Richard Leedy, Polk Community College

Marcus McGuff, Austin Community College

Owen Mertens, Missouri State University

James Metz, Kapi'olani Community College
Pam Miller, Phoenix College
Tania Munding, Ohlone College
Charlie Naffziger, Central Oregon Community College
Oscar Neal, Grand Rapids Community College
Doug Nelson, Central Oregon Community College
Katrina Nichols, Delta College
Megan Nielsen, St. Cloud State University
Nancy Ressler, Oakton Community College
Elaine Richards, Eastern Michigan University
Harriette Roadman, New River Community College
Lilia Ruvalcaba, Oxnard College
Wendiann Sethi, Seton Hall University
Karen Smith, Nicholls State University
Donald Solomon, University of Wisconsin–Milwaukee
Frankie Solomon, University of Houston–Downtown
Michael Stack, South Suburban College
Kristen Starkey, Rose State College
Kristin Stoley, Blinn College

Eleanor Storey, Front Range Community College–Westminster Campus
Fariheh Towfiq, Palomar College
Gowribalan Vamadeva, University of Cincinnati
Beverly Vredevelt, Spokane Falls Community College
Andreana Walker, Calhoun Community College
Cynthia Wallin, Central Virginia Community College
John Ward, Kentucky Community and Technical College–Jefferson Community College
Richard Watkins, Tidewater Comunity College
Antoinette Willis, St. Philip's College
Nazar Wright, Guilford Technical Community College
Shishen Xie, University of Houston–Downtown
Catalina Yang, Oxnard College
Heidi Young, Bryant and Stratton College
Ghidei Zedingle, Normandale Community College

APPLICATIONS INDEX

Examples that are applications are shown with **boldface page numbers**.
Exercises that are applications are shown with lightface page numbers.

Business and Industry
Advertising, 259, 575
Aluminum foil, 72
Architecture, 501, 782
Assembly lines, 782
Bookstores, 98
Bottled water, 8, **139**
Bottling, 603
Building shelves, 83
Building stairs, **152**
Business, 233
Business expenses, 433
Candy, 258, 482
Carpentry, 10, 74, 670, 707
Catering, **5**
Cereal sales, 82
Choosing furnaces, 260
Clothing stores, 259
Coffee blends, 100
Commuting time, 671
Copiers, 8
Cost of electricity, 73
Cost of water, 73
Costs and revenue, 234
Cubicles, 671
Customer service, 432
Deliveries, 604
Demand equation, 145
Depreciation of a copier, **140**
Directory costs, 529
Discount buying, 721
Drafting, 21, 553, 618
Energy, 97
Entrepreneurs, 99
Furniture equipment, 378
Furniture sales, 378
Heating, ventilation, and air conditioning technician, 345
Hourly pay, 145
Job testing, 99
Landscape architect, 500
Law of supply and demand, 234
Leading employers, **75**
Logging, 670, 1010
Logos, 259
Machining, 83
Making furniture, 83
Making statues, 279
Making tires, 260
Manufacturing, 250, 261

Masonry, 575
Metal fabrication, 772
Mixing candy, 100, 262
Mixing coffee, 262
Mixing nuts, **94**, 281
Mixing solutions, 262
Oil storage, 620
Operating costs, 798
Ordering staircase parts, 9
Packaging, 432, 458
Paper products, 72
Petroleum, 130
Pricing, 575
Production planning, 9, 260, 261
Publishing, 260
Real estate, 332, 620
Salvage values, 175, 862
Scheduling equipment, 333
Scheduling work crews, 529
Selling calculators, 84
Selling seed, 84
Square footage, 144
Steel production, 359
Storage tanks, 432
Supply and demand, 707
Supply equation, 145
Temporary health, 260
Testing steel, **1020**
Tool manufacturing, **275**
Union membership, 144
U.S. employment trends, 773
Utility costs, 529
Water usage, 798
Wiper design, 73
Woodworking, 83
Work schedules, 333

Education
Averaging grades, 333
Bachelor's degrees, 837
Campus parking lots, 130
College costs, 174
Computers, 159
Educational savings plan, **858**
Graduation announcements, **757**
History, 772
Multicultural studies, 131
Police patrol officer, 798
Public education, 333
Real estate sales agent, 1022

Retention study, 529
SAT Scores, 836
School enrollment, 798
School supplies, **163**
Social workers, 873
Wikipedia, 333

Electronics
Computer programming, 333
Computer viruses, 862
Computer-aided drafting, 175
Data analysis, 564
Electronics, 73, 258, **455, 591,** 620, 733
Robotics, **715**

Entertainment and Hobbies
Aquariums, 416, 657, 908
Art history, 720
Bouncing balls, 1023
Broadcast ranges, 951
Broadway shows, 98
Cable tv, 174
Candy, 160
Checkers, 472
Collectibles, 8, 657
Compact discs, 378
Concert touring, 83
Concerts, 260
Craig's List, 175
Crayons, 458
Crossword puzzles, 797
Dances, 772
Deceased celebrities, 280
Digital imaging, 293
Digital photography, 293
Drawing, 618
Drive-ins, 98
Entertainment, **89**
Films, 82
Fine arts, 500
Fireworks, 798
Five-card poker, 416
Framing posters, 500
Graphic arts, 619, 797
Graphic design, 488
Guitars, 863
Halftime performances, 1006
Halloween, 156
Halloween costumes, 837
Hiking, 258
Hockey, 72

Holiday songs, 1010
iMAX screens, 772
Internet, 234
Internet access, **860**
Lighting levels, 849
Magazine sales, 772
Malls, 810
Movie stunts, 482, 760
Music history, 158
Organ pipes, 620
Paintings, 175
Paper airplanes, 719
Parks, 772
Phonograph records, **750**
Photography, 594, **614**
Picture framing, 761
Popcorn, **254**
Puppets, 280
Quilting, 85
Recording companies, 260
Recreation, **496**
Restaurant seating, 369
Roller coasters, 432
Sculpting, 280
Shopping, 618
Spring tours, 83
Swimming pools, 761
Theater productions, 707
Theatre seating, 281
Thrill rides, 603
Ticket sales, 772
Trampolines, 127
TV commercials, 83
TV history, 281
U.S. music sales, 158
Video game systems, 333
Video games, **87**
Video rentals, 128
Winter recreation, 500
World's largest LED screen, 772

Farming
Crop dusting, 605
Farming, 144, 260, 603, 620
Fencing, 85
Fencing pastures, 979
Malthusian model, 872
Milk production, **95**
Ranching, 85, 798

Finance
Accounting, 22, 36, 145, **166,** 595
Bank service charges, **90**
Banking, **68**
Bankruptcy, 671
Break-even point, 501
Checking accounts, 49
Comparing interest rates, 863
Comparing savings plans, 863
Comparison of compounding methods, 872
Compound interest, 863, 919, 920
Computing salaries, 83
Continuous compound interest, 872, 919
Credit cards, 261
Currency exchange, **612**

Declining savings, 1022
Depreciation, 145, 887
Depreciation rates, **704**
Determining initial deposits, 872
Determining the previous balance, 872
Doubling money, 894
Financial presentations, 99
Frequency of compounding, 863
Fundraising, 333
Fundraising letters, 603
Growth of money, 887
Highest rates, 99
Inheritances, 99, **1019**
Installment loans, 1009
Interest income, **92, 251**
Investing, 261, 304, **867,** 887, 979
Investment clubs, 261
Investment in bonds, 36, 73
Investment rates, 773
Investments, 176, 258, 334, 458, 760
IRA accounts, 261
Loans, 261
Market share, 144
Maximizing revenue, 798
Minimizing costs, **793**
Money laundering, 99
National debt, 416
Pension funds, 84
Piggy banks, 281
Portfolio analysis, **78**
Real estate, 174, 175
Retirement, 160, 261
Revenue, 501
Rule of seventy, 920
Salary options, 261
Saving money, 1009
Savings growth, 1022
Social security, **365**
Tax returns, 99
Taxes, 8, 128, 191
Tripling money, 895
Value of IRAs, 84

Geography
Amazon River, 553
Geography, 129, 369, 670, 760
Grand Canyon, 657
Highs and lows, 873
Louisiana Purchase, 864
Maps, 158, 233
New York City, 259
North Star State, 894
Peach State, 894
Silver State, 894
Washington, D.C., 619, 718
Width of rivers, 619

Geometry
Analytic geometry, 695
Angles of quadrilaterals, 85
Area of ellipses, 961
Bracing, 259
Complementary angles, 293
Curve fitting, 281, 282
Dimensions of a rectangle, 761

Dimensions of a triangle, 761
Fractals, 733
Geometric formulas, 457
Geometry, 85, 259, 334, 408, 432, 446, 541, 979
Graphing circles, **943**
Graphing ellipses, **957**
Graphs of systems, 274
Height of triangles, 85
Inscribed squares, 1023
Interior angles, 1010
Parallelograms, **248**
Polygons, 772
Quadrilaterals, 281
Right triangles, 772
Rings, 22
Supplementary angles, 84, 293
Surface area, 74
Triangles, 281, 293
Trigonometry, 695
Vertical angles, 85

Home Management
Baby furniture, 345
Baking, **870**
Bath salts, 262
Blow dryers, 682
Buying appliances, 97
Buying furniture, 98
Carpet cleaning, 9
Clotheslines, 719
Cooking, 500
Decks, 160
Decorating, 603
Designing patios, 1010
Desserts, 258, 279
Disinfectants, 258
Ductwork, 682
Embroidery, 656
Fast foods, 279
Fencing a field, 259
Filling a pool, 529
Fine dining, 603
Fire protection, 174
Floor mats, 72
Gardening, 100, 369
Gourmet cooking, **609**
Health foods, 100
Home construction, 190, **596**
Hot dogs, 280
Housecleaning, 603
Household appliances, 779
Housekeeping, 369
Housepainting, 603
Ice cream, 72
Ironing boards, 719
Kitchen utensils, 564
Kitchens, 500
Landscape design, **958**
Landscaping, **63, 375,** 458
Making JELL-O, 895
Moving expenses, 83
Moving houses, 604
Outdoor cooking, 682
Plumbing, 604
Potpourri, 281

Purchasing pets, 282
Quilts, 447
Roast turkey, 131
Roofing, 603
Room freshener, 262
Salads, 262
Setting up seating, **598**
Spray bottles, 145
Wallpapering, 618
Weatherizing a house, 500

Medicine and Health
Band-aids, **64**
Cardiovascular fitness, 378
Dairy Foods, 100
Decongestants, 190
Dermatology, 262
Dosages, 37
Fitness equipment, 961
Forensic medicine, 501, 895
Hearing tests, 234
Living longer, 144
Medical plans, 334
Medications, 176
Medicine, 874
Nutrition, 280
Nutritional planning, 280
Pediatrics, 37
Pharmacists, 100
Physical fitness, 258
Physical therapy, 293
Pulse rates, 656
Reaction time, **90**
Recommended dosages, 618
Stretching exercises, **713**
Treating fevers, 345
U.S. health care, 346

Miscellaneous
Accidents, 760
Ants, 873
Area codes, 258
Avalanches, 259
Caffeine, 618
Children's height, 887
Cleanup crews, 604
Computers, 97
Confetti, 262
Cosmetology, **186,** 260
Crowd control, 782
Diamonds, 417, 707
Digits problem, 282
Doubling time, **892**
Earthquakes, 129, **884,** 887
Energy, 416
Environmental cleanup, 529
Evening newspapers, **147**
Filling ponds, 603
Fire drill, 603
Firefighting, **709,** 720
Flood damage, 100
Food shortage, **869**
Forestry, 707
Genealogy, 1023
Gift boxes, 447

Groundskeeping, 603
Hurricanes, 129
Ice, 473
Integer problems, 499, 979
Interpersonal relationships, 849
iPAD Building, 84
Kennels, **81**
Leaning Tower of Pisa, 84
Measurement, 260
Newsletters, 604
Number problems, 979
Oceans, 416
Oysters, 604
Packaging, 720
Paper routes, 782
Pest control, 1023
Polls, 346
Population growth, 864, 895, 920
Psychology, 131
Psychology experiments, 176
Radio translators, **944**
Richter scale, 887
Rodent control, 920
Salmon, 261
Search and rescue, 100, 458
Shoelaces, 656
Smoke damage, 603
Stained glass, **495**
Staircases, 159
Statistics, 695
Steep grades, 159
Stepstools, 84
Surveys, 604
Telephone service, 720
Telephones, 618
Timers, 65
Umbrellas, 682
Walkways, 282, 951
Water, 773
Water treatment, **255**
Weather forecasting, 837
Wedding gowns, **88**
World population, 861, 872
Yellow Pages, 446

Politics, Government, and Military
City planning, **869**
Criminology, 174
Flagpoles, 619
Flags, 80, 760
Korean War, 369
Labor statistics, **423**
Law enforcement, 657
Lawyers, **769**
Military history, 798
Political contributions, **329**
Politics, 160
Population of U.S., 872
Statue of Liberty, 83
U.S. postage, 131
World oil reserves/demand, **412**

Science and Engineering
Air pressure, 158
Alpha particles, 971

Astronomy, **120,** 281, 407, 416
Atomic structure, **969**
Atoms, 416
Bacterial cultures, 863, 920
Bacterial growth, 920
Ballistic pendulums, 670
Ballistics, **497,** 501, 798
Big dipper, 417
Biological research, **833**
Biology, 417, 657
Body temperatures, 190
Bronze, 100
Calculating clearance, 961
Carbon-14 dating, 919
Center of gravity, 204
Chemical reactions, 191
Chemistry, 73
Comets, 417, 951
Communications satellites, 682
Converting temperatures, 72
db gain, **884,** 887
Designing an underpass, 961
Diluting solutions, 100
Discharging a battery, 864
Disinfectants, 873
Earth's atmosphere, 190, 280
Engineering, 564, 595, 695
Epidemics, 873
Ergonomics, 618
Error analysis, 359, 360
Evaporation, 101
Finding distance, 620
Fluids, 972
Force of wind, **615**
Free fall, 620, 874
Gas pressure, 620, 621
Global warming, 158, 862
Gravity, 620
Greenhouse gases, 98
Half-life of a drug, 873
Hardware, 719
Hydrogen ion concentration, **905,** 907
Input voltage, 887
Lead decay, 919
Light, 204
Light year, 416
Melting points for metals, 69
Meshing gears, 951
Metallurgy, 100, 837
Microscopes, 407
Newton's law of cooling, 920
Oceanography, 873, 920
Operating temperatures, 359
Optics, 204, 594
Output voltage, 887
Ozone concentrations, 872
Pendulums, **648,** 656, 1023
pH meters, **905**
pH of a solution, 907
pH of pickles, 907
pH scale, 22
Physics, 37, 408, **590**
Physics experiments, 540
Projectiles, 951

Radio communications, 100
Radioactive decay, 864, 919
Rates of speed, 604
Relativity, 670
Rocketry, **420**
Satellite antennas, 952
Satellites, **663**
Solar heating, 85
Sonic boom, 972
Structural engineering, 620, 682
Temperature extremes, 36
Temperature ranges, 359
Tension in a string, 621
Thorium decay, 919
Tolerances, **353**
Tritium decay, 919
Warp speed, 416
Water pressure, 127
Wind power, 707
Windchill, 175
Zoology, 274

Sports
Archery, 191, 979
Area of a track, 962
Badminton, 772
Baseball, 501, 656, 719
Basketball, 97
Bicycle frames, 274
Bicycling, 782
Bike racing, 159
Billiards, 204
Boxing, 604
Bungee jumping, 501
Concessionaires, 190
Cycling, 100
Diving, 862
Downhill skiing, **152**
End zones, 604
Extreme sports, 159
Football, 377
Golf, 129
Ice skating, 293

Jet skiing, 100, 262
Juggling, 431
Long-distance running, 130
Marathons, 100
NBA records, 274
NFL records, 280
Olympics, **277**
Physical fitness, 100
Pool tables, 961
Professional baseball, **856**
Rollerblading, 262
Ski rental packages, **124**
Ski runs, 619
Skiing, 159
Skydiving, 874
Slingshots, 501
Snowmobiling, 261
Soccer, 718
Sport fishing, 98
Sporting goods, 369
Swimming pools, 85
Track and field, 377
Triathlons, **76**
Undersea diving, 174
WNBA Championships, 99

Travel and Transportation
Air traffic control, 100, 234
Airplanes, 128
Airport walkways, 261
Antifreeze, 262
Auto mechanics, 359
Automobile engines, 760
Aviation, 261
Blimps, **253**
Boat depreciation, 1023
Boat sales, 378
Boating, 605
Boating accidents, 97
Bridges, 810
Car depreciation, 145
Comparing travel, 604
Crosswalks, 771

Detailing cars, 604
Driving rates, 979
Flight paths, 619
Flying, 158
Fuel efficiency, 98
Gasoline, 262
Helicopter pads, 446
Highway construction, **616**
Highway design, 706
Improving horsepower, 98
License plates, 407
LORAN, 972
Mass transit, **768**
Navigation, 234
No-fly zones, 378
Parking areas, 48
Parking lots, 717
Railroads, 159
River tours, 605
Riverboat cruises, **601**
Road maps, 128
Road trips, **599**
Shortcuts, **767**
Signaling, 303
Stopping distances, 432
Street intersections, 346
Submarines, 127
Traffic accidents, 501
Traffic engineers, 951
Traffic lights, 281
Traffic signals, 259
Traffic signs, 346
Train travel, 604
Transportation, 564
Transportation engineering, 432
Travel promotions, **77**
Travel time, **93, 99**
Truck deliveries, 604
Trucking, 345
Trucking costs, 620
Vacation mileage costs, 837
Value of cars, 862
Winter travel, 575

CHAPTER 1

A Review of Basic Algebra

© Baerbel Schmidt/Getty Images

1.1 The Language of Algebra
1.2 The Real Numbers
1.3 Operations with Real Numbers
1.4 Simplifying Algebraic Expressions Using Properties of Real Numbers
1.5 Solving Linear Equations Using Properties of Equality
1.6 Solving Formulas; Geometry
1.7 Using Equations to Solve Problems
1.8 More about Problem Solving
CHAPTER SUMMARY AND REVIEW
CHAPTER TEST
Group Project

from **Campus to Careers**
Registered Dietitian

One of the most important things that you can do to protect yourself from cancer, diabetes, heart disease, and stroke is to eat right. No one knows this better than dietitians. They work in hospitals, health care centers, schools, and correctional facilities, where they plan dietary programs and supervise the preparation of healthy meals. The job of dietician requires mathematical skills such as calculating calorie intake, analyzing the nutritional content of food, and budgeting for the purchase of groceries and supplies.

Problem 57 of **Study Set 1.8** shows how a dietician can use algebra to determine how much super-lean hamburger (12% fat) should be added to regular hamburger (30% fat) to obtain a mixture that has a 16% fat content.

JOB TITLE:
Registered Dietitian

EDUCATION:
A bachelor's degree in foods and nutrition (or a related field) and supervised practice

JOB OUTLOOK:
Growth faster than the average, 18%–26% increase

ANNUAL EARNINGS:
The median base salary in 2007 was $47,157.

FOR MORE INFORMATION:
www.bls.gov/oco/ocos077.htm

SECTION 1.1
The Language of Algebra

Objectives

1. Write verbal and mathematical models.
2. Use equations to construct tables of data.
3. Read graphical models.

Algebra is the result of contributions from many cultures over thousands of years. The word *algebra* comes from the title of the book *Al-jabr wa'l muquabalah,* written by the Arabian mathematician al-Khwarizmi around A.D. 800. Using the vocabulary and notation of algebra we can mathematically **model** many situations in the real world. In this section, we will review some of the basic components of the language of algebra.

Write Verbal and Mathematical Models.

In the following rental agreement, we see that two operations need to be performed to calculate the cost of renting the banquet hall.

• First, we must *multiply* the $100-per-hour rental cost by the number of hours that the hall is to be rented.

• To that result, we must *add* the cleanup fee of $200.

Rental Agreement
ROYAL VISTA BANQUET HALL
Wedding Receptions•Dances•Reunions•Fashion Shows

Rented To_____ Date_____
Lessee's Address_____

Rental Charges
• $100 per hour
• Nonrefundable $200 cleanup fee

Terms and conditions
Lessor leases the undersigned lessee the above described property upon the terms and conditions set forth on this page and on the back of this page. Lessee promises to pay rental cost stated herein.

We can describe the process to calculate the rental cost in words using the following **verbal model:**

| The cost of renting the hall | is | 100 | times | the number of hours it is rented | plus | 200. |

The table below lists some key words and phrases that are often used in mathematics to indicate the operations of addition, subtraction, multiplication, and division.

Addition +	Subtraction −	Multiplication ·	Division ÷
added to	subtracted from	multiplied by	divided by
sum	difference	product	quotient
plus	less than	times	ratio
more than	decreased by	percent (or fraction) of	half
increased by	reduced by	twice	into
greater than	minus	triple	per

We can use vocabulary from the table to write a verbal model that describes how to calculate the cost of renting the banquet hall. One such model is:

The cost (in dollars) of renting the hall is the *product* of 100 and the number of hours it is rented, *increased* by 200.

The Language of Algebra
Since the number of hours that the hall is rented can vary, or change, it is represented using a *variable*.

We can also describe the procedure for calculating the rental cost of the banquet hall using *variables* and mathematical symbols. A **variable** is a letter that is used to stand for a number. If we let the letter h represent the number of hours that the hall is rented, the cost to rent the hall can be represented by the notation $100h + 200$. We call $100h + 200$ an **algebraic expression,** or more simply, an **expression.**

Algebraic Expressions

An **algebraic expression** is a combination of variables and/or numbers with the operations of addition, subtraction, multiplication, division, raising to a power, and finding a root.

Here are some more examples of expressions.

$5a - 12$ This expression involves the operations of multiplication and subtraction.

$\dfrac{50 - y}{3y^3}$ This expression involves the operations of subtraction, division, multiplication, and raising to a power.

$\sqrt{a^2 + b^2}$ This expression involves the operations of addition, raising to a power, and finding a root.

In the banquet hall example, if we let the letter c represent the cost to rent the hall, we can translate the verbal model to a **mathematical model.**

The Language of Algebra
The equal symbol = can be represented by verbs such as:

   is   are   gives   yields

The symbol ≠ is read as "is not equal to."

The cost of renting the hall	is	100	times	the number of hours it is rented	plus	200.
c	$=$	100	$\cdot$	h	$+$	200

The statement $c = 100h + 200$ is called an *equation.* An **equation** is a mathematical sentence that contains an = symbol. The = symbol indicates that the expressions on either side of it have the same value.

EXAMPLE 1   Translate each verbal model to a mathematical model.

a. The distance in miles traveled by a vehicle is the product of its average rate of speed in mph and the time in hours it travels at that rate.

b. The sale price of an item is the difference between the regular price and the discount.

Strategy   We will look for key words and phrases that indicate arithmetic operations, and we will represent any unknown quantities using variables.

Why   To translate a verbal (word) model into a mathematical model means to represent it using mathematical symbols.

Solution

a. The word *product* indicates multiplication. If we let d represent the distance traveled in miles, r the vehicle's average rate of speed in mph, and t the length of time traveled in hours, we can write the verbal model in mathematical form as

$$d = rt$$

b. The word *difference* indicates subtraction. If we let s represent the sale price of the item, p the regular price, and d the discount, we have

$$s = p - d$$

Success Tip
The answers to the Self Check problems are given at the end of each section, before each Study Set.

Self Check 1   Express the following relationship as an equation: The simple interest earned by a deposit is the product of the principal, the annual rate of interest, and the time.

Now Try   **Problem 13**

Many applied problems require insight and analysis to determine which mathematical operations to use when writing a verbal or mathematical model.

| EXAMPLE 2 | **Catering.**   It costs $16 per person to have a dinner catered. A $100 discount is given for groups of more than 200 people. Write |

a verbal and a mathematical model that describe the relationship between the catering cost and the number of people being served for groups larger than 200.

Strategy   We will carefully read the problem to identify any phrases that indicate an arithmetic operation and then represent any unknown quantities using variables.

Why   To write a verbal or mathematical model, we must determine what arithmetic operations are involved.

Solution   The phrase *$16 per person* indicates multiplication by 16 and the phrase *a $100 discount* indicates subtraction. Thus, to find the catering cost c (in dollars) for groups larger than 200, we need to *multiply* the number n of people served by $16 and then *subtract* the $100 discount.

A verbal model is:

The catering cost (in dollars) is the *product* of 16 and the number of people served, *decreased by 100.*

Caution   The comma in the verbal model is absolutely essential to convey the correct meaning. Without it, it is unclear what is to be decreased by 100.

In symbols, the mathematical model for groups larger than 200 is:

$$c = 16n - 100$$

| Self Check 2 | After winning a lottery, three friends split the prize equally. Each person had to pay $2,000 in taxes on his or her share. Write a verbal model and a mathematical model that relate the amount of each person's share, after taxes, to the amount of the lottery prize. |

Now Try   **Problem 21**

 2   **Use Equations to Construct Tables of Data.**

In the banquet hall example, the equation $c = 100h + 200$ can be used to determine the cost of renting the banquet hall for *any* number of hours.

| EXAMPLE 3 | Find the cost of renting the banquet hall for 3 hours and for 4 hours. Write the results in a table. |

Strategy   We will substitute 3 (and then 4) for h in the equation $c = 100h + 200$ and evaluate the right side.

Why   The cost c to rent the hall for h hours is given by the equation $c = 100h + 200$.

Solution   First, we construct the table shown on the next page with the appropriate column headings: h for the number of hours the hall is rented and c for the cost (in dollars) to rent the hall. Then we enter the number of hours of each rental time in the left column.

Next, we use the equation $c = 100h + 200$ to find the total rental cost for 3 hours and for 4 hours.

$$c = 100h + 200$$
$$c = 100(3) + 200 \quad \text{Substitute 3 for } h.$$
$$\quad = 300 + 200 \quad\quad \text{Multiply.}$$
$$\quad = 500$$

$$c = 100h + 200$$
$$c = 100(4) + 200 \quad \text{Substitute 4 for } h.$$
$$\quad = 400 + 200 \quad\quad \text{Multiply.}$$
$$\quad = 600$$

Finally, we enter these results in the right column of the table: $500 for a 3-hour rental and $600 for a 4-hour rental.

h	c
3	500
4	600

 Self Check 3   Find the cost of renting the hall for 6 hours and for 7 hours. Write the results in a table.

Now Try   **Problem 25**

3   **Read Graphical Models.**

The cost of renting the banquet hall for various lengths of time can also be presented graphically. The following **bar graph** has a **horizontal axis** labeled "Number of hours the hall is rented." The **vertical axis,** labeled "Cost to rent the hall ($)," is scaled in units of 50 dollars. The bars above each of the times (1, 2, 3, 4, 5, 6, and 7 hours) extend to a height that gives the corresponding cost to rent the hall. For example, if the hall is rented for 5 hours, the bar indicates that the cost is $700.

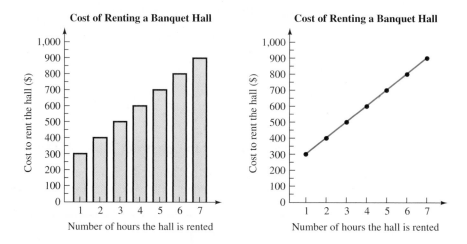

The **line graph** above also shows the rental costs. This type of graph consists of a series of dots drawn at the correct height, connected with line segments. We can use the line graph to find the cost of renting the banquet hall for lengths of time not shown in the bar graph.

EXAMPLE 4   Use the line graph shown above to determine the cost of renting the hall for $4\frac{1}{2}$ hours.

Strategy   Since we know the number of hours the hall is to be rented, we begin on the horizontal axis of the graph and scan up and over to read the answer on the vertical axis.

Why   We scan up and over because the scale on the vertical axis gives the cost of renting the hall.

Solution   In the figure to the right, we locate $4\frac{1}{2}$ on the horizontal axis and draw a vertical line upward to intersect the graph. From the point of intersection with the graph, we draw a horizontal line to the left that intersects the vertical axis. On the vertical axis, we can read that the rental cost is $650 for $4\frac{1}{2}$ hours.

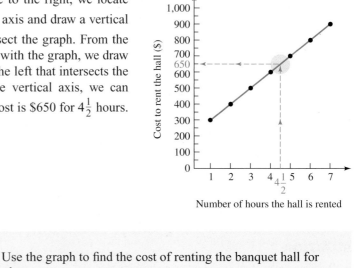

Number of hours the hall is rented

▷ **Self Check 4**   Use the graph to find the cost of renting the banquet hall for $6\frac{1}{2}$ hours.

Now Try   **Problem 30**

▷ **ANSWERS TO SELF CHECKS**   **1.** $I = Prt$ **2.** Each person's share, after taxes, is the quotient of the lottery prize and 3, decreased by 2,000; $S = \frac{p}{3} - 2,000$ **3.**

h	c
6	800
7	900

4. $850

STUDY SET
1.1

VOCABULARY

Fill in the blanks.

1. A _____ is a letter that is used to stand for a number.

2. Variables and/or numbers can be combined with mathematical operations to create algebraic _____.

3. An _____ is a mathematical sentence that contains an = symbol.

4. Words such as *is, was, gives,* and *yields* translate to an ___ symbol.

5. Phrases such as *increased by* and *more than* are used to indicate the operation of _____.

6. Phrases such as *decreased by* and *less than* are used to indicate the operation of _____.

CONCEPTS

7. Classify each of the following as an expression or an equation.

   a. $6x - 5$

   b. $P = a + b + c$

   c. $\frac{s + 9t}{8}$

   d. $\sqrt{2w^2}$

8. What arithmetic operations does the expression $\frac{40 - 8n}{5}$ contain? What variable does it contain?

Use the data in each table to find an equation that mathematically describes the relationship between the two quantities. Then state the relationship in words. (Answers may vary.)

9.

Tower height (ft)	Height of base (ft)
15.5	5.5
22	12
25.25	15.25
45.125	35.125

10.

Seasonal employees	Employees
25	75
50	100
60	110
80	130

NOTATION

11. Translate each verbal model into a mathematical model.

a. 7 times the age of a dog in years gives the dog's equivalent human age.

b. The take-home pay will be $2,500 minus any deductions.

12. Give four verbs that can be represented by an equal symbol $=$.

GUIDED PRACTICE

Translate each verbal model into a mathematical model. **See Example 1.**

13. The cost each semester is the sum of $13 times the number of units taken and a student services fee of $24.

14. The yearly salary is $25,000 plus $75 times the number of years of experience.

15. The quotient of the number of clients and seventy-five gives the number of social workers needed.

16. The difference between 500 and the number of people in a theater gives the number of unsold tickets.

17. Each test score was increased by 15 points to give a new adjusted test score.

18. The weight of a super-size order of French fries is twice that of a regular-size order.

19. The product of the number of boxes of crayons in a case and 12 gives the number of crayons in a case.

20. The perimeter of an equilateral triangle can be found by tripling the length of one of its sides.

Write a verbal and mathematical model for each situation. **See Example 2.**

21. TAXES   A married couple has decided to split the money equally when they receive their federal income tax refund. Furthermore, the husband is going to donate $75 of his share to charity. Describe the relationship between the amount of money that the husband will keep and the amount of the couple's refund.

22. COPIERS   A business is going to rent a copy machine. Under the rental agreement, the company is charged $105 per month and 3¢ for every copy that is made. Describe the relationship between the monthly copier expense and the number of copies made.

23. BOTTLED WATER   A driver left a production plant with 300 five-gallon bottles of drinking water on his truck. His delivery route consisted of office buildings, each of which was to receive 6 bottles of water. Describe the relationship between the number of bottles of water left on his truck and the number of stops that he has made.

24. COLLECTIBLES   A woman inherited 9 antique dolls. She decided to add to her collection by purchasing two more dolls each month. Describe the relationship between the number of antique dolls in her collection and the number of months since she began to purchase them.

Use the given equation to complete each table. **See Example 3.**

25. $c = \dfrac{p}{12}$

Number of packages p	Cartons c
24	
72	
180	

26. $y = 100c$

Number of centuries c	Years y
1	
6	
21	

27. $n = 22.44 - K$

K	n
0	
1.01	
22.44	

28. $y = 2x + 15$

x	y
0	
15	
30	

***Refer to the given graph.* See Example 4.**

29. a. What type of graph is shown?

b. What units are used to scale the horizontal axis? The vertical axis?

c. In what year was the average expenditure on auto insurance the least? Estimate the amount. In what year was it the greatest? Estimate the amount.

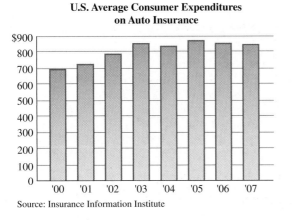

U.S. Average Consumer Expenditures on Auto Insurance

Source: Insurance Information Institute

30. a. What type of graph is shown?

b. What units are used to scale the horizontal axis? The vertical axis?

c. Estimate the height of the candle after it has burned for $3\frac{1}{2}$ hours. For 8 hours.

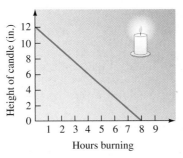

APPLICATIONS

31. PRODUCTION PLANNING Suppose r towel racks are to be manufactured. Complete the four equations that could be used to order the necessary number of oak mounting plates p, bar holders b, chrome bars c, and wood screws s.

$p = \quad b = \quad r \quad c = \quad s = \quad r$

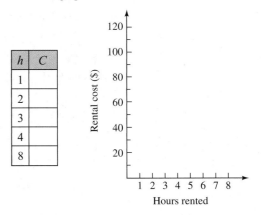

Bar holder
Chrome bar
Oak mounting plate
Wood screws

32. ORDERING STAIRCASE PARTS A builder is going to construct h new homes, each of which will have a staircase as shown. Complete the four equations that could be used to order the necessary number of balusters b, handrails r, posts p, and treads t for the entire project.

$b = \quad h$

$r = \quad h$

$p = \quad h$

$t = \quad h$

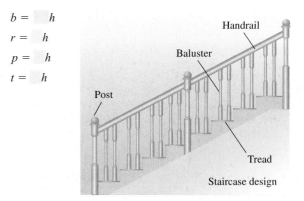

Handrail
Baluster
Post
Tread
Staircase design

33. CARPET CLEANING See the following ad.

Rent the in-home
Carpet Cleaning System
Do it yourself and save!
Safe, effective
Costs only $10 an hour
plus $20 for supplies

a. Write a verbal model that states the relationship between the cost C of renting the carpet-cleaning system and the number of hours h it is rented.

b. Translate the verbal model written in part (a) to a mathematical model.

c. Use your result from part (b) to complete the table, and then draw a line graph.

h	C
1	
2	
3	
4	
8	

Rental cost ($)
Hours rented

34. CARPENTRY   A miter saw can pivot 180° to make angled cuts on molding. The equation that relates the angle measure s on the scrap piece of molding and the angle measure f on the finish piece of molding is $s = 180 - f$. Complete the following table and then draw a line graph.

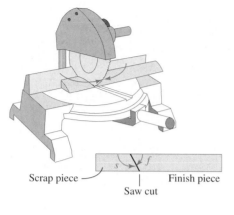

Scrap piece — $s \quad f$ — Finish piece

Saw cut

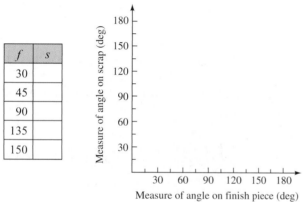

f	s
30	
45	
90	
135	
150	

Measure of angle on scrap (deg)
180 150 120 90 60 30
30 60 90 120 150 180
Measure of angle on finish piece (deg)

35. Explain the difference between an expression and an equation. Give examples.

36. Use each word below in a sentence that indicates a mathematical operation. If you are unsure of the meaning of a word, look it up in a dictionary.

quadrupled	deleted	bisected
confiscated	annexed	docked

CHALLENGE PROBLEMS

37. Use the equation $F = \frac{9}{5}C + 32$ to complete the table.

C	F
5	
	50
15	

38. Fill in the blank: If $T = 16s$ and $s = \dfrac{r}{2}$, then $T = \boxed{}\, r$.

SECTION 1.2
The Real Numbers

Objectives

❶ Define the set of natural numbers, whole number, and integers.

❷ Define the set of rational numbers.

❸ Define the set of irrational numbers.

❹ Classify real numbers.

❺ Graph real numbers.

❻ Order the real numbers.

❼ Find the opposite and the absolute value of a real number.

In this course, we will work with *real numbers*. The set of real numbers is a collection of several other important sets of numbers.

① **Define the Set of Natural Numbers, Whole Numbers, and Integers.**

Natural numbers are the numbers that we use for counting. To write this set, we list its **elements** (or **members**) within **braces** { }.

Natural Numbers	The set of **natural numbers,** denoted by the symbol $\mathbb{N}$, is $\{1, 2, 3, 4, 5, \ldots\}$.
	Read as "the set containing one, two, three, four, five, and so on."

The Language of Algebra
The symbol . . . used in these definitions is called an *ellipsis* and it indicates that the established pattern continues forever.

The symbol $\in$ is used to indicate that an element belongs to a set. For example, we can write

$3 \in \mathbb{N}$ Read as "3 is an element of the set of natural numbers."

We can write $4.5 \notin \mathbb{N}$ to indicate that 4.5 *is not an element* of the set of natural numbers. The natural numbers, together with 0, form the set of **whole numbers.**

Whole Numbers	The set of **whole numbers,** denoted by the symbol $\mathbb{W}$, is $\{0, 1, 2, 3, 4, 5, \ldots\}$.

When all the members of one set are also members of a second set, we say the first set is a **subset** of the second set. Since every natural number is also a whole number, the set of natural numbers is a subset of the set of whole numbers. We can use the symbol $\subseteq$ to indicate this.

$\mathbb{N} \subseteq \mathbb{W}$ Read as "The set of natural numbers is a subset of the set of whole numbers."

Since the set of whole numbers contains an element that the natural numbers do not, namely 0, we can write

$\mathbb{W} \not\subseteq \mathbb{N}$ Read as "The set of whole numbers is not a subset of the set of natural numbers."

Two other important subsets of the whole numbers are the *prime numbers* and the *composite numbers.*

Prime Numbers and Composite Numbers	A **prime number** is a whole number greater than 1 that has only itself and 1 as factors. The first ten prime numbers are 2, 3, 5, 7, 11, 13, 17, 19, 23, and 29.
	A **composite number** is a whole number, greater than 1, that is not prime. The first ten composite numbers are 4, 6, 8, 9, 10, 12, 14, 15, 16, and 18.

Recall from arithmetic that every composite number can be written as the product of prime numbers. For example,

$$6 = 2 \cdot 3, \quad 25 = 5 \cdot 5, \quad \text{and} \quad 168 = 2 \cdot 2 \cdot 2 \cdot 3 \cdot 7$$

Whole numbers are not adequate for describing many real-life situations. For example, if you write a check for more than what is in your account, the account balance will be less than zero.

We can use the **number line** on the next page to visualize numbers less than zero. On the number line, numbers greater than 0 are to the right of 0, and they are called **positive numbers.** Numbers less than 0 are to the left of 0, and they are called **negative numbers.**

For each natural number on the number line, there is a corresponding number, called its *opposite,* to the left of 0. In the diagram, we see that 3 and −3 (negative three) are opposites, as are −5 (negative five) and 5. Note that 0 is its own opposite.

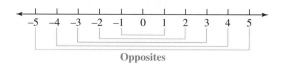

Opposites

| Opposites | Two numbers that are the same distance from 0 on the number line, but on opposite sides of it, are called **opposites.** |

The whole numbers, together with their opposites, form the set of *integers.*

| Integers | The set of **integers,** denoted by the symbol $\mathbb{Z}$, is $\{\ldots, -4, -3, -2, -1, 0, 1, 2, 3, 4, \ldots\}$. |

The Language of Algebra
The positive integers are:

$$1, 2, 3, 4, 5, \ldots$$

The negative integers are:
$-1, -2, -3, -4, -5, \ldots$

Integers that are divisible by 2 are called *even integers,* and integers that are not divisible by 2 are called *odd integers.*

Even integers: $\{\ldots, -6, -4, -2, 0, 2, 4, 6, \ldots\}$
Odd integers: $\{\ldots, -5, -3, -1, 1, 3, 5, \ldots\}$

Since every whole number is also an integer, the set of whole numbers is a subset of the set of integers.

EXAMPLE 1   Determine whether each statement is true or false.
a. $-6 \in \mathbb{W}$ **b.** $\mathbb{Z} \not\subseteq \mathbb{N}$

Solution

a. Since -6 is not an element of the set of whole numbers, the statement $-6 \in \mathbb{W}$ is false.

b. Since the set of integers contains at least one element that does not belong to the set of natural numbers (-1 for example), the set of integers is not a subset of the set of natural numbers. Thus, the statement $\mathbb{Z} \not\subseteq \mathbb{N}$ is true.

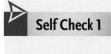

Self Check 1   Determine whether each statement is true or false.
a. $-1.7 \notin \mathbb{Z}$ **b.** $\mathbb{Z} \subseteq \mathbb{W}$

Now Try **Problems 27 and 31**

2 **Define the Set of Rational Numbers.**

In this course, we will work with positive and negative fractions. For example, the slope of a line might be $\frac{7}{12}$ or a tank might drain at a rate of $-\frac{40}{3}$ gallons per minute. We will also work with mixed numbers. For instance, we might speak of $5\frac{7}{8}$ cups of flour or of a river that is $3\frac{1}{2}$ feet below flood stage $\left(-3\frac{1}{2}\ \text{ft}\right)$. These fractions and mixed numbers are examples of *rational numbers.*

Rational Numbers	A **rational number** is any number that can be written in the form $\frac{a}{b}$, where a and b represent integers and $b \neq 0$.

Other examples of rational numbers are

$$\frac{3}{4}, \quad \frac{25}{25}, \quad \text{and} \quad \frac{19}{6}$$

To show that negative fractions are rational numbers, we can use the following fact.

Negative Fractions	Let a and b represent numbers, where $b \neq 0$, $$-\frac{a}{b} = \frac{-a}{b} = \frac{a}{-b}$$

The Language of Algebra
Rational numbers are so named because they can be expressed as the ratio (quotient) of two integers: $\frac{\text{integer}}{\text{integer}}$.

To illustrate this rule, consider $-\frac{40}{3}$. It is a rational number because it can be written as $\frac{-40}{3}$, or as $\frac{40}{-3}$.

Positive and negative mixed numbers such as $5\frac{7}{8}$ and $-3\frac{1}{2}$ are rational numbers because they can be expressed as fractions.

$$5\frac{7}{8} = \frac{47}{8} \quad \text{and} \quad -3\frac{1}{2} = -\frac{7}{2} = \frac{-7}{2}$$

Any natural number, whole number, or integer can be expressed as a fraction with a denominator of 1. For example, $5 = \frac{5}{1}$, $0 = \frac{0}{1}$, and $-3 = \frac{-3}{1}$. Therefore, every natural number, whole number, and integer is also a rational number.

Throughout the book we will also work with decimals. Some examples of uses of decimals are:

- The interest rate of a loan was $11\% = 0.11$.
- In baseball, the distance from home plate to second base is 127.279 feet.
- The third-quarter loss for a business can be represented as -2.7 million dollars.

The Language of Algebra
To *terminate* means to bring to an end. In the movie *The Terminator*, actor Arnold Schwarzenegger plays a heartless machine sent to Earth to bring an end to his enemies.

Terminating decimals such as 0.11, 127.279, and -2.7 are rational numbers because they can be written as fractions with integer numerators and nonzero integer denominators.

$$0.11 = \frac{11}{100} \qquad 127.279 = 127\frac{279}{1,000} = \frac{127,279}{1,000} \qquad -2.7 = -2\frac{7}{10} = \frac{-27}{10}$$

Examples of **repeating decimals** are $0.333 \ldots$ and $4.252525. \ldots$ Any repeating decimal can be expressed as a fraction with an integer numerator and a nonzero integer denominator. For example, $0.333 \ldots = \frac{1}{3}$ and $4.252525 \ldots = 4\frac{25}{99} = \frac{421}{99}$. Since every repeating decimal can be written as a fraction, repeating decimals are also rational numbers.

Rational Numbers	The set of **rational numbers,** denoted by the symbol $\mathbb{Q}$, is the set of all terminating and all repeating decimals.

We can use division to find the *decimal equivalent* of a fraction.

EXAMPLE 2   Find the decimal equivalent of each fraction to determine whether the decimal terminates or repeats.   **a.** $\dfrac{4}{5}$ **b.** $\dfrac{17}{6}$

Strategy   For each fraction, we will divide the numerator by the denominator.

Why   A fraction bar indicates division.

Solution

a. To change $\dfrac{4}{5}$ to a decimal, we divide 4 by 5.

$$
\begin{array}{r}
.8 \\
5\overline{)4.0} \\
\underline{4\ 0} \\
0
\end{array}
$$ *Write a decimal point and a 0 to the right of 4.*

In decimal form, $\dfrac{4}{5}$ is 0.8. This is a terminating decimal.

b. To change $\dfrac{17}{6}$ to a decimal, we divide 17 by 6 and obtain 2.8333. . . . This is a repeating decimal, because the digit 3 repeats forever. It can be written as $2.8\overline{3}$, where the **overbar** indicates that the 3 repeats.

> **Self Check 2**   Find the decimal equivalent of each fraction to determine whether the decimal terminates or repeats.
>
> **a.** $\dfrac{25}{990}$ **b.** $\dfrac{47}{50}$
>
> ***Now Try*** **Problems 35 and 37**

The Language of Algebra
A fraction and its *decimal equivalent* are different notation that represent the same value.

The set of rational numbers is too extensive to list its members in the same way that we listed the members of the natural numbers, whole numbers, and integers. Instead, we will use the following **set-builder** notation to describe it.

Rational Numbers

The set of rational numbers is

$$\left\{\dfrac{a}{b} \,\middle|\, a \text{ and } b \text{ are integers, with } b \neq 0.\right\}$$ *Read as "the set of all numbers of the form $\frac{a}{b}$, such that a and b are integers, with $b \neq 0$."*

The length of a diagonal of the square is $\sqrt{2}$ inches.

The distance around the circle is π inches.

3   **Define the Set of Irrational Numbers.**

Numbers that cannot be expressed as fractions with an integer numerator and a nonzero integer denominator are called **irrational numbers.** One example is $\sqrt{2}$. It can be shown that a square, with sides of length 1 inch, has a diagonal that is $\sqrt{2}$ inches long.

The number represented by the Greek letter π (pi) is another example of an irrational number. It can be shown that a circle, with a 1-inch diameter, has a circumference of π inches. Expressed in decimal form,

$$\sqrt{2} = 1.414213562\ldots \quad \text{and} \quad \pi = 3.141592654\ldots$$

These decimals neither terminate nor repeat.

Irrational Numbers

An **irrational number** is a nonterminating, nonrepeating decimal. An irrational number cannot be expressed as a fraction with an integer numerator and a nonzero integer denominator. The set of irrational numbers is denoted by the symbol ℍ.

Caution

Don't classify a number such as 4.12122122212222 ... as a repeating decimal. Although it exhibits a pattern, no block of digits repeats forever. It is a nonterminating, nonrepeating decimal—an irrational number.

Some other examples of irrational numbers are

$$\sqrt{97} = 9.848857802\ldots$$

$$-\sqrt{7} = -2.64575131\ldots \quad \text{This is a negative irrational number.}$$

$$2\pi = 6.283185307\ldots \quad 2\pi \text{ means } 2 \cdot \pi.$$

Comment   Not all square roots are irrational numbers. When we simplify square roots such as $\sqrt{9}$, $\sqrt{36}$, and $\sqrt{400}$, it is apparent that they are rational numbers: $\sqrt{9} = 3$, $\sqrt{36} = 6$, and $\sqrt{400} = 20$.

Using Your Calculator

Approximating Irrational Numbers

We can approximate the value of irrational numbers with a scientific calculator. To find the value of π, we press the $\boxed{\pi}$ key.

$\boxed{\pi}$ (You may have to use a $\boxed{\text{2nd}}$ or $\boxed{\text{Shift}}$ key first.) $\boxed{3.141592654}$

We see that $\pi \approx 3.141592654$. (Read $\approx$ as "is approximately equal to.") To the nearest thousandth, $\pi \approx 3.142$.

To approximate $\sqrt{2}$, we enter 2 and press the square root key $\boxed{\sqrt{}}$.

$2 \boxed{\sqrt{}}$ $\boxed{1.414213562}$

We see that $\sqrt{2} \approx 1.414213562$. To the nearest hundredth, $\sqrt{2} \approx 1.41$.

To find π and $\sqrt{2}$ with a graphing calculator, we proceed as follows.

$\boxed{\text{2nd}} \boxed{\pi} \boxed{\text{ENTER}}$
$$\begin{array}{l} \pi \\ \qquad\qquad 3.141592654 \end{array}$$

$\boxed{\text{2nd}} \boxed{\sqrt{}} 2 \boxed{)} \boxed{\text{ENTER}}$
$$\begin{array}{l} \sqrt{}(2) \\ \qquad\qquad 1.414213562 \end{array}$$

 Classify Real Numbers.

The set of rational numbers together with the set of irrational numbers form the set of **real numbers.** This means that every real number can be written as either a terminating decimal, a repeating decimal, or a nonterminating, nonrepeating decimal. Thus, the set of real numbers is the set of all decimals.

The Real Numbers

A **real number** is any number that is either a rational or an irrational number. The set of real numbers is denoted by the symbol ℝ.

The figure on the next page shows how the sets of numbers introduced in this section are related; it also gives some specific examples of each type of number. Note that a number can belong to more than one set. For example, -6 is an integer, a rational number, and a real number.

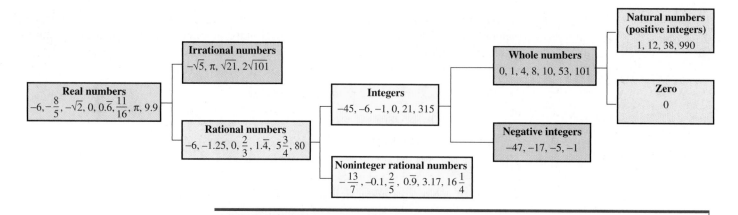

EXAMPLE 3   *Classifying Real Numbers.*   Which numbers in the following set are natural numbers, whole numbers, integers, rational numbers, irrational numbers, and real numbers?

$$\left\{\frac{5}{8},\quad -0.03,\quad 45,\quad -9,\quad \sqrt{7},\quad 5\frac{2}{3},\quad 0,\quad -9.010010001\ldots,\quad 0.\overline{25}\right\}$$

Strategy   We begin by scanning the given set, looking for any natural numbers. Then we scan it five more times, looking for whole numbers, for integers, for rational numbers, for irrational numbers, and, finally, for real numbers.

Why   We need to scan the given set of numbers six times, because numbers in the set can belong to more than one classification.

Solution

Natural numbers: 45   *45 is a member of* {1, 2, 3, 4, 5, . . .}.

Whole numbers: 45, 0   *45 and 0 are members of* {0, 1, 2, 3, 4, 5, . . .}.

Integers: 45, −9, 0   *45, −9, and 0 are members of*
$\qquad\qquad\qquad\qquad${. . . , −3, −2, −1, 0, 1, 2, 3, . . .}.

Rational numbers: $\frac{5}{8}$, 45, −9, $5\frac{2}{3}$, and 0 are rational numbers because each of them can
$\qquad\qquad\qquad\qquad$ be expressed as a fraction: $45 = \frac{45}{1}$, $-9 = \frac{-9}{1}$, $5\frac{2}{3} = \frac{17}{3}$, and $0 = \frac{0}{1}$.
$\qquad\qquad\qquad\qquad$ The terminating decimal $-0.03 = \frac{-3}{100}$ and the repeating decimal
$\qquad\qquad\qquad\qquad$ $0.\overline{25} = \frac{25}{99}$ are also rational numbers.

Irrational numbers: The nonterminating, nonrepeating decimals $\sqrt{7} = 2.645751311\ldots$
$\qquad\qquad\qquad\qquad$ and $-9.010010001\ldots$ are irrational numbers.

Real numbers: $\frac{5}{8}$, -0.03, 45, -9, $\sqrt{7}$, $5\frac{2}{3}$, 0, $-9.010010001\ldots$, $0.\overline{25}$
$\qquad\qquad\qquad\qquad$ *Every natural number, whole number, integer, rational number, and irrational*
$\qquad\qquad\qquad\qquad$ *number is a real number.*

Self Check 3   Use the instructions for Example 3 with the following set:

$$\left\{-\pi,\quad -5,\quad 3.4,\quad \sqrt{19},\quad 1,\quad \frac{16}{5},\quad 9.\overline{7}\right\}$$

Now Try   **Problems 43, 45, and 47**

Graph Real Numbers.

We can illustrate real numbers using a number line. To each real number, there corresponds a point on the line. Furthermore, to each point on the line, there corresponds a real number, called its **coordinate.**

EXAMPLE 4   Graph each number in the set
$$\left\{-\frac{8}{3}, \quad -1.1, \quad 0.\overline{56}, \quad \frac{\pi}{2}, \quad -\sqrt{15}, \quad \text{and} \quad 2\sqrt{2}\right\} \text{ on a}$$
number line.

Strategy   We locate the position of each number on the number line, draw a bold dot, and label it.

Why   To *graph a number* means to make a drawing that represents the number.

Solution   To help locate the graph of each number, we make some observations.

- Expressed as a mixed number, $-\frac{8}{3} = -2\frac{2}{3}$.
- Since -1.1 is less than -1, its graph is to the left of -1.
- $0.\overline{56} \approx 0.6$
- From a calculator, $\frac{\pi}{2} \approx 1.6$.
- From a calculator, $-\sqrt{15} \approx -3.9$.
- $2\sqrt{2}$ means $2 \cdot \sqrt{2}$. From a calculator, $2\sqrt{2} \approx 2.8$.

Self Check 4   Graph the numbers in the set
$$\left\{\pi, \quad -2.\overline{1}, \quad \sqrt{3}, \quad \tfrac{11}{4}, \quad \text{and} \quad -0.9\right\} \text{ on a number line.}$$

Now Try   **Problem 55**

The Language of Algebra
An example of a number that is not on the real number line is $\sqrt{-4}$. It is called an imaginary number. We will discuss such numbers in Chapter 8.

 Order the Real Numbers.

As we move right on the number line, the values of the real numbers increase. As we move left, the values decrease. To compare real numbers, we often use one of the **inequality symbols** shown in the following table.

The Language of Algebra
If a real number x is positive, then $x > 0$. If a real number x is nonnegative, then $x \geq 0$. If a real number x is a negative number, then $x < 0$.

Symbol	Read as	Examples
$\neq$	"is not equal to"	$6 \neq 9$ and $0.33 \neq \frac{3}{5}$
$<$	"is less than"	$\frac{22}{3} < \frac{23}{3}$ and $-7 < -6$
$>$	"is greater than"	$19 > 5$ and $\frac{1}{2} > 0.3$
$\leq$	"is less than or equal to"	$3.5 \leq 3.\overline{5}$ and $1\frac{4}{5} \leq 1.8$
$\geq$	"is greater than or equal to"	$29 \geq 29$ and $-15.2 \geq -16.7$

It is always possible to write an equivalent inequality with the inequality symbol pointing in the opposite direction. For example:

If $-3 < 4$, it is also true that $4 > -3$.

If $5.3 \geq 2.9$, it is also true that $2.9 \leq 5.3$.

EXAMPLE 5   Use one of the symbols $>$ or $<$ to make each statement true.

a. $-24 \quad \boxed{} \quad -25$ **b.** $\dfrac{3}{4} \quad \boxed{} \quad 0.76$

Strategy   To pick the correct inequality symbol to place between a given pair of numbers, we need to determine the position of each on a number line.

Why   For any two numbers on a number line, the number to the *left* is the smaller number and the number to the *right* is the larger number.

Solution

a. Since -24 is to the right of -25 on the number line, $-24 > -25$.

b. If we express the fraction $\frac{3}{4}$ as a decimal, we can easily compare it to 0.76.

Since $\dfrac{3}{4} = 0.75, \quad \dfrac{3}{4} < 0.76.$

Self Check 5   Use one of the symbols $\geq$ or $\leq$ to make each statement true.

a. $\dfrac{2}{3} \underline{\quad} \dfrac{4}{3}$ **b.** $8\dfrac{1}{2} \underline{\quad} 8.4$

Now Try   **Problems 67 and 71**

---

7   **Find the Opposite and the Absolute Value of a Real Number.**

Two numbers that are the same distance from 0 on the number line, but on opposite sides of it, are called **opposites** or **additive inverses**. To write the opposite of a positive number, we simply insert a negative sign $-$ in front of it. For example, the opposite of 10 is -10.

Parentheses are used to express the opposite of a negative number. For example, the opposite of -3 is written as $-(-3)$. Since -3 and 3 are the same distance from zero, the opposite of -3 is 3. In symbols, this can be written as $-(-3) = 3$.

In general, we have the following.

Opposites	The **opposite** of a number a is the number $-a$. If a is a real number, then $-(-a) = a$.

A number line can be used to measure the distance from one number to another. To express the distance that a number is from 0 on a number line, we can use absolute values.

Absolute Value	The **absolute value** of a number is its distance from 0 on the number line.

To indicate the absolute value of a number, we write the number between two vertical bars. From the previous figure, we see that $|-3| = 3$. This is read as "the absolute value of negative 3 is 3" and it tells us that the distance from 0 to -3 is 3 units. It also follows from the figure that $|3| = 3$.

The absolute value of a number can be defined more formally as follows.

Absolute Value	For any real number a, $\begin{cases} \text{If } a \geq 0, \text{ then } |a| = a \\ \text{If } a < 0, \text{ then } |a| = -a \end{cases}$

Caution
The second part of this definition is often misunderstood. Study it carefully. It indicates that the absolute value of a negative number is the opposite (or additive inverse) of the number.

The first part of this definition states that if a is a nonnegative number (that is, if $a \geq 0$), the absolute value of a is a. The second part of the definition states that if a is a negative number (that is, if $a < 0$), the absolute value of a is the opposite of a. For example, if $a = -8$, then

$$|a| = |-8| = \underbrace{-(-8)}_{\text{The opposite of } a} = 8$$

EXAMPLE 6 Find the value of each expression.

a. $|34|$ **b.** $\left|-\dfrac{4}{5}\right|$ **c.** $|0|$ **d.** $-|-1.8|$

Strategy We need to determine the distance that the number within the vertical absolute value bars is from 0.

Why The absolute value of a number is the distance between 0 and the number on a number line.

Solution

a. $|34| = 34$ Because 34 is a distance of 34 from 0 on a number line.

b. $\left|-\dfrac{4}{5}\right| = \dfrac{4}{5}$ Because $-\frac{4}{5}$ is a distance of $\frac{4}{5}$ from 0 on a number line.

c. $|0| = 0$ Because 0 is a distance of 0 from 0 on a number line.

d. The negative sign outside the absolute value bars means to find the opposite of $|-1.8|$.

$$-|-1.8| = -(1.8) \quad \text{Find } |-1.8| \text{ first to get 1.8.}$$
$$= -1.8$$

Self Check 6 Find the value of each expression.
a. $|-9.6|$ **b.** $-|-12|$ **c.** $\left|\dfrac{3}{2}\right|$

Now Try Problems 75 and 79

▷ **ANSWERS TO SELF CHECKS**   **1. a.** True   **b.** False   **2. a.** $\frac{25}{990} = 0.0\overline{25}$, repeating decimal

b. $\frac{47}{50} = 0.94$, terminating decimal   **3.** natural numbers: 1; whole numbers: 1; integers: -5, 1; rational

numbers: -5, 3.4, 1, $\frac{16}{5}$, $9.\overline{7}$; irrational numbers: $-\pi$, $\sqrt{19}$; real numbers; all

4.

$$-2.\overline{1} \quad -0.9 \qquad\qquad \sqrt{3} \quad \frac{11}{4} \quad \pi$$
$$\begin{array}{ccccccccc} | & | & \bullet & \bullet & | & | & | & \bullet & \bullet & | \\ -3 & -2 & -1 & 0 & 1 & 2 & 3 & 4 \end{array}$$

5. a. $\leq$ **b.** $\geq$ **6. a.** 9.6   **b.** -12 **c.** $\frac{3}{2}$

STUDY SET
1.2

VOCABULARY

Fill in the blanks.

1. The set of _____ numbers is {0, 1, 2, 3, 4, 5, . . .}, the set of _____ numbers is {1, 2, 3, 4, 5, . . .}, and the set of _____ is {. . . , $-2, -1, 0, 1, 2, . . .$}.

2. When all the members of one set are members of a second set, we say the first set is a _____ of the second set.

3. A _____ number is a whole number greater than 1 that has only itself and 1 as factors. A _____ number is a whole number greater than 1 that is not prime.

4. A _____ number is any number that can be written as a fraction with an integer numerator and a nonzero integer denominator.

5. _____ numbers are nonterminating, nonrepeating decimals.

6. The set of rational numbers together with the set of irrational numbers form the set of _____ numbers.

7. $>$, $\geq$, $<$, and $\leq$ are called _____ symbols.

8. The _____ _____ of any real number is the distance between the number and zero on a number line.

CONCEPTS

9. Name two numbers that are 6 units away from -2 on the number line.

10. Show that each of the following numbers is a rational number by expressing it as a fraction with an integer numerator and a nonzero integer denominator.

$$7, \quad -7\frac{3}{5}, \quad 0.007, \quad 700.1$$

Determine whether each number is a repeating or a nonrepeating decimal, and whether it is a rational or an irrational number.

11. 0.090090009 . . .

12. $0.0\overline{9}$

13. 5.41414141 . . .

14. 1.414213562 . . .

15. The following diagram can be used to show how the natural numbers, whole numbers, integers, rational numbers, and irrational numbers make up the set of real numbers. If the natural numbers can be represented as shown, label each of the other sets.

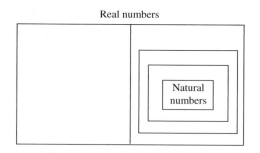

16. Determine whether each statement is true or false.

a. All prime numbers are odd numbers.

b. $6 \geq 6$

c. 0 is neither even nor odd.

d. Every real number is a rational number.

17. Write each statement with the inequality symbol pointing in the opposite direction.

a. $19 > 12$ **b.** $-6 \leq -5$

18. Fill in the blanks:

For any real number a, $\begin{cases} \text{If } a \geq 0, \text{ then } |a| = \rule{1cm}{0.4pt} \\ \text{If } a < 0, \text{ then } |a| = \rule{1cm}{0.4pt} \end{cases}$

NOTATION

Fill in the blanks.

19. The symbol $<$ means "___ ___ ___" and the symbol $\geq$ means "___ ___ ___ ___ ___."

20. $|-2|$ is read as "the _____ value ___ -2."

21. The symbols { } are called _____.

22. The symbol $\in$ is read as "is an _____ of" and the symbol $\subseteq$ is read as "is a _____ of."

23. Describe the set of rational numbers using set-builder notation.

24. What set of numbers does each symbol represent?

   a. $\mathbb{N}$ **b.** $\mathbb{W}$

   c. $\mathbb{Z}$

25. What set of numbers does each symbol represent?

   a. $\mathbb{Q}$ **b.** $\mathbb{H}$

   c. $\mathbb{R}$

26. List two other ways that the fraction $-\frac{2}{3}$ can be written.

GUIDED PRACTICE

Determine whether each statement is true or false. See Example 1.

27. $12 \in \mathbb{N}$ **28.** $9 \in \mathbb{N}$

29. $-5 \notin \mathbb{Z}$ **30.** $-55 \notin \mathbb{H}$

31. $\mathbb{R} \subseteq \mathbb{W}$ **32.** $\mathbb{W} \subseteq \mathbb{R}$

33. $\mathbb{H} \not\subseteq \mathbb{Q}$ **34.** $\mathbb{Q} \not\subseteq \mathbb{Z}$

Find the decimal equivalent of each fraction and determine whether the decimal is terminating or repeating. See Example 2.

35. $\dfrac{3}{5}$ **36.** $\dfrac{21}{50}$

37. $-\dfrac{11}{15}$ **38.** $-\dfrac{7}{30}$

39. $\dfrac{27}{22}$ **40.** $\dfrac{25}{990}$

41. $\dfrac{2}{125}$ **42.** $\dfrac{19}{16}$

List the elements of $\left\{-3,\ -\frac{8}{5},\ 0,\ \frac{2}{3},\ 1,\ \sqrt{3},\ 2,\ \pi,\ 4.75,\ 9,\ 16.\overline{6}\right\}$ that belong to the following sets. See Example 3.

43. Natural numbers

44. Whole numbers

45. Integers

46. Rational numbers

47. Irrational numbers

48. Real numbers

49. Even natural numbers

50. Odd integers

51. Prime numbers

52. Composite numbers

53. Odd composite numbers

54. Odd prime numbers

Graph each set on a number line. See Example 4.

55. $\left\{-\dfrac{5}{2},\ -0.1,\ 2.142765\ldots,\ \dfrac{\pi}{3},\ -\sqrt{11},\ 2\sqrt{3}\right\}$

56. $\left\{2\dfrac{1}{9},\ -3.821134\ldots,\ -\dfrac{\pi}{2},\ \sqrt{15},\ -0.9,\ \dfrac{\sqrt{2}}{2}\right\}$

57. $\left\{3.\overline{15},\ \dfrac{22}{7},\ 3\dfrac{1}{8},\ \pi,\ \sqrt{10},\ 3.1\right\}$

58. $\left\{-0.\overline{331},\ -0.331,\ -\dfrac{1}{3},\ -\sqrt{0.11}\right\}$

59. The set of prime numbers less than 8

60. The set of integers between -7 and 0

61. The set of odd integers between 10 and 18

62. The set of composite numbers less than 10

63. The set of positive odd integers less than 12

64. The set of negative even integers greater than -7

65. The set of even integers from -6 to 6

66. The set of odd natural numbers less than or equal to 5

Insert either a $<$ or a $>$ symbol to make a true statement. See Example 5.

67. -9 -8 **68.** -11 -12

69. $-(-5)$ -10 **70.** $|-3|$ $-(-6)$

71. $6.\overline{1}$ $-(-6)$ **72.** -6.07 $-\dfrac{17}{6}$

73. -7.999 -7.1 **74.** $4\dfrac{1}{2}$ $\dfrac{7}{2}$

Find the value of each expression. See Example 6.

75. $|20|$ **76.** $|-20|$

77. $|-5.9|$ **78.** $-|1.\overline{27}|$

79. $-|-6|$ **80.** $-|-8|$

81. $-\left|\dfrac{9}{4}\right|$ **82.** $-\left|-\dfrac{5}{16}\right|$

APPLICATIONS

83. DRAFTING   Express each dimension in the drawing of a bracket as a four-place decimal. Approximate when necessary.

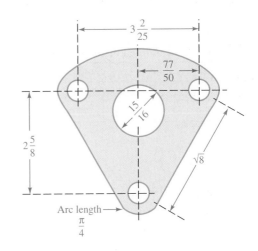

84. pH SCALE The pH scale is used to measure the strength of acids and bases (alkalines) in chemistry. It can be thought of as a number line. On the scale, graph and label each pH measurement given in the table.

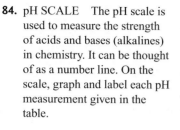

Solution	pH
Seawater	8.5
Cola	2.9
Battery acid	1.0
Milk	6.6
Blood	7.4
Ammonia	11.9
Saliva	6.1
Oven cleaner	13.2
Black coffee	5.0
Toothpaste	9.9
Tomato juice	4.1

85. RINGS The formula $C = \pi D$ gives the circumference C of a circle, where D is the length of its diameter. Find the circumference of the gold wedding band. Give an *exact* answer and then an *approximate* answer, rounded to the nearest hundredth of an inch.

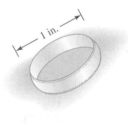

86. ACCOUNTING Business losses are usually written within parentheses on financial statements. Examine the statement below for Delta Air Lines. Rank the years, in order from the greatest loss to the smallest loss.

Delta Air Lines, Annual Income Statement

Net Loss in millions of dollars				
2006	2005	2004	2003	2002
($6,203)	($3,818)	($5,198)	($773)	($1,272)

WRITING

87. Explain why the whole numbers are a subset of the integers.

88. What is a real number? Give examples.

89. Explain why there are no even prime numbers greater than 2.

90. Explain why every integer is a rational number, but not every rational number is an integer.

REVIEW

91. Is $\frac{3x - 4}{2}$ an equation or an expression?

92. Translate into mathematical symbols: The weight of an object in ounces is 16 times its weight in pounds.

Complete each table.

93. $T = x - 1.5$

x	T
3.7	
10	
30.6	

94. $j = 3m$

m	j
0	
15	
300	

CHALLENGE PROBLEMS

95. How many integers have an absolute value that is less than 50?

96. How many odd integers have an absolute value between 20 and 40?

97. The **trichotomy property** of real numbers states that if a and b are real numbers, then $a < b$, $a = b$, or $a > b$. Explain why this is true.

98. Let a and b represent real numbers. Which of the following statements are always true?

a. $|a + b| = |a| + |b|$

b. $|a \cdot b| = |a| \cdot |b|$

c. $|a + b| \leq |a| + |b|$

SECTION 1.3
Operations with Real Numbers

Objectives

 1 Add and subtract real numbers.

2 Multiply and divide real numbers.

3 Find powers and square roots of real numbers.

4 Use the order of operations rule.

5 Evaluate algebraic expressions.

Six operations can be performed with real numbers: addition, subtraction, multiplication, division, raising to a power, and finding a root. In this section, we will review the rules for performing these operations and discuss how to evaluate numerical expressions involving several operations.

1   Add and Subtract Real Numbers.

When two numbers are added, the result is their **sum.** The rules for adding real numbers are as follows:

Adding Two Real Numbers

To **add two positive numbers,** add them in the usual way. The final answer is positive.

To **add two negative numbers,** add their absolute values and make the final answer negative.

To **add a positive number and a negative number,** subtract the smaller absolute value from the larger.

1. If the positive number has the larger absolute value, the final answer is positive.
2. If the negative number has the larger absolute value, make the final answer negative.

EXAMPLE 1   Add: **a.** $-5 + (-3)$ **b.** $8.9 + (-5.1)$

   c. $-\dfrac{13}{15} + \dfrac{3}{5}$ **d.** $6 + (-10) + (-1)$

Strategy   We will use the rules for adding positive and negative real numbers.

Why   Each sum involves signed numbers.

Solution

The Language of Algebra
Positive and negative numbers are often referred to as *signed numbers.*

a. $-5 + (-3) = -8$ Both numbers are negative. Add their absolute values, 5 and 3, to get 8, and make the final answer negative.

b. $8.9 + (-5.1) = 3.8$ One number is positive and the other is negative. Subtract their absolute values, 5.1 from 8.9, to get 3.8. Because 8.9 has the larger absolute value, the final answer is positive.

c. $-\dfrac{13}{15} + \dfrac{3}{5} = -\dfrac{13}{15} + \dfrac{9}{15}$ Express $\frac{3}{5}$ in terms of the lowest common denominator, 15: $\frac{3}{5} = \frac{3 \cdot 3}{5 \cdot 3} = \frac{9}{15}$.

$\qquad\qquad = -\dfrac{4}{15}$ Subtract the absolute values, $\frac{9}{15}$ from $\frac{13}{15}$, to get $\frac{4}{15}$, and make the final answer negative because $-\frac{13}{15}$ has the larger absolute value.

d. To add three or more real numbers, add from left to right.

$$6 + (-10) + (-1) = -4 + (-1)$$
$$= -5$$

▷ **Self Check 1**   Add: **a.** $-34 + 25$ **b.** $-70.4 + (-21.2)$
c. $\dfrac{7}{4} + \left(-\dfrac{3}{2}\right)$ **d.** $-16 + 17 + (-5)$

Now Try **Problems 15 and 21**

When two numbers are subtracted, the result is their **difference.** To find a difference, we can change the subtraction into an equivalent addition. For example, the subtraction $7 - 4$ is equivalent to the addition $7 + (-4)$, because they have the same answer:

$$7 - 4 = 3 \qquad \text{and} \qquad 7 + (-4) = 3$$

This suggests that to subtract two numbers, we can change the sign of the number being subtracted and add.

Subtracting Two Real Numbers

To **subtract two real numbers,** add the first number to the opposite (additive inverse) of the number to be subtracted.

Let a and b represent real numbers,

$$a - b = a + (-b)$$

EXAMPLE 2   Subtract: **a.** $2 - 8$ **b.** $-1.3 - 5.5$ **c.** $-\dfrac{14}{3} - \left(-\dfrac{7}{3}\right)$
d. Subtract 9 from -6 **e.** $-11 - (-1) - 5$

Strategy   To find each difference, we will apply the rule for subtraction: *Add the first number to the opposite of the number to be subtracted.*

Why   It is easy to make an error when subtracting signed numbers. We will probably be more accurate if we write each subtraction as addition of the opposite.

Solution

a.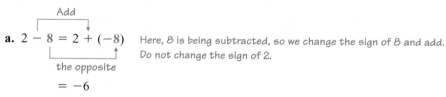

$2 - 8 = 2 + (-8)$ Here, 8 is being subtracted, so we change the sign of 8 and add. Do not change the sign of 2.

$\qquad = -6$

b. $-1.3 - 5.5 = -1.3 + (-5.5)$ Change the sign of 5.5 and add. Do not change the sign of -1.3.

$\qquad\qquad = -6.8$

c. $-\dfrac{14}{3} - \left(-\dfrac{7}{3}\right) = -\dfrac{14}{3} + \dfrac{7}{3}$ Change the sign of $-\frac{7}{3}$ and add.

$$= -\dfrac{7}{3}$$

d. The number to be subtracted is 9. When we translate, we must reverse the order in which 9 and -6 appear in the sentence.

Subtract 9 from -6.

$-6 - 9 = -6 + (-9)$ Add the opposite of 9.
$\qquad\quad = -15$

e. To subtract three or more real numbers, subtract from left to right.

$-11 - (-1) - 5 = -10 - 5$
$\qquad\qquad\qquad = -15$

 Self Check 2 Subtract: **a.** $-15 - 4$ **b.** $-12.1 - (-7.6)$
c. $\dfrac{5}{9} - \dfrac{7}{9}$ **d.** Subtract 1 from -5
e. $5 - 4 - (-15)$

Now Try **Problems 23 and 29**

---

2 **Multiply and Divide Real Numbers.**

When two numbers are multiplied, we call the numbers **factors** and the result is their **product.** The rules for multiplying real numbers are as follows:

Multiplying Two Real Numbers

1. **With unlike signs:** To multiply a positive number and a negative number, multiply their absolute values and make the final answer negative.
2. **With like signs:** To multiply two real numbers with the same sign, multiply their absolute values. The final answer is positive.

EXAMPLE 3 Multiply: **a.** $4(-7)$ **b.** $-5.2(-3)$
c. $-\dfrac{7}{9}\left(\dfrac{3}{16}\right)$ **d.** $8(-2)(-3)$

Strategy We will use the rules for multiplying positive and negative real numbers.

Why Each product involves signed numbers.

Solution

a. $4(-7) = -28$ Multiply the absolute values, 4 and 7, to get 28. Since the signs are unlike, make the final answer negative.

b. $-5.2(-3) = 15.6$ Multiply the absolute values, 5.2 and 3, to get 15.6. Since the signs are like, the final answer is positive.

c. $-\dfrac{7}{9}\left(\dfrac{3}{16}\right) = -\dfrac{7 \cdot 3}{9 \cdot 16}$ Multiply the numerators and multiply the denominators. Since the signs of the factors are unlike, the final answer is negative.

$$= -\dfrac{7 \cdot \overset{1}{\cancel{3}}}{\underset{1}{\cancel{3}} \cdot 3 \cdot 16}$$ Factor 9 as $3 \cdot 3$ and simplify the fraction: $\dfrac{3}{3} = 1$.

$$= -\dfrac{7}{48}$$ Multiply in the numerator and denominator.

d. To multiply three or more real numbers, multiply from left to right.

$$8(-2)(-3) = -16(-3)$$
$$= 48$$

Self Check 3   Multiply: **a.** $(-6)(5)$ **b.** $(-4.1)(-8)$
c. $\left(\dfrac{4}{3}\right)\left(-\dfrac{1}{8}\right)$ **d.** $-4(-9)(-3)$

Now Try   **Problems 31 and 37**

When two numbers are divided, the result is their **quotient.** In the division $\dfrac{x}{y} = q$, the quotient q is a number such that $y \cdot q = x$. We can use this relationship to find rules for dividing real numbers.

$$\dfrac{10}{2} = 5, \text{ because } 2(5) = 10 \qquad\qquad \dfrac{-10}{-2} = 5, \text{ because } -2(5) = -10$$

$$\dfrac{-10}{2} = -5, \text{ because } 2(-5) = -10 \qquad\qquad \dfrac{10}{-2} = -5, \text{ because } -2(-5) = 10$$

These results suggest the following rules for dividing real numbers. Note that they are similar to those for multiplying real numbers.

Dividing Two Real Numbers

To **divide two real numbers,** divide their absolute values.

1. The quotient of two numbers with *like* signs is positive.
2. The quotient of two numbers with *unlike* signs is negative.

EXAMPLE 4   Divide: **a.** $\dfrac{-44}{11}$ **b.** $\dfrac{-2.7}{-9}$

Strategy   We will use the rules for dividing positive and negative real numbers.

Why   Each quotient involves signed numbers.

Solution

a. $\dfrac{-44}{11} = -4$ Divide the absolute values, 44 by 11, to get 4. Since the signs are unlike, make the final answer negative.

b. $\dfrac{-2.7}{-9} = 0.3$ Divide the absolute values, 2.7 by 9, to get 0.3. Since the signs are like, the final answer is positive.

> **Self Check 4** Divide: **a.** $\dfrac{55}{-5}$ **b.** $\dfrac{-7.2}{-6}$
>
> *Now Try* **Problems 39 and 43**

To divide two fractions, we multiply the first fraction by the **reciprocal** of the second fraction. In symbols, if a, b, c, and d are real numbers, and no denominators are 0, then

$$\frac{a}{b} \div \frac{c}{d} = \frac{a}{b} \cdot \frac{d}{c} \qquad \frac{d}{c} \text{ is the reciprocal of } \frac{c}{d}.$$

> **EXAMPLE 5** Divide: **a.** $\dfrac{2}{3} \div \left(-\dfrac{3}{5}\right)$ **b.** $-\dfrac{1}{2} \div (-6)$
>
> **Strategy** We will multiply the first fraction by the reciprocal of the second.
>
> **Why** This is the rule for dividing two fractions.
>
> **Solution**
>
> **a.** $\dfrac{2}{3} \div \left(-\dfrac{3}{5}\right) = \dfrac{2}{3} \cdot \left(-\dfrac{5}{3}\right)$ Multiply by the reciprocal of $-\frac{3}{5}$, which is $-\frac{5}{3}$.
>
> $\qquad\qquad = -\dfrac{10}{9}$ Since the factors have unlike signs, the final answer is negative.
>
> **b.** $-\dfrac{1}{2} \div (-6) = -\dfrac{1}{2} \cdot \left(-\dfrac{1}{6}\right)$ Multiply by the reciprocal of -6, which is $-\frac{1}{6}$.
>
> $\qquad\qquad = \dfrac{1}{12}$ Since the factors have like signs, the final answer is positive.

> **Self Check 5** Divide: **a.** $-\dfrac{7}{8} \div \dfrac{2}{3}$ **b.** $-\dfrac{1}{10} \div (-5)$
>
> *Now Try* **Problem 45**

The Language of Algebra
When we say a division by 0, such as $\frac{4}{0}$, is *undefined*, we mean it is not allowed or it is not defined. That is, $\frac{4}{0}$ does not represent a number.

Students often confuse division problems such as $\frac{0}{4}$ and $\frac{4}{0}$. We know that $\frac{0}{4} = 0$ because $4 \cdot 0 = 0$. However, $\frac{4}{0}$ is undefined, because there is no real number q such that $0 \cdot q = 4$. In general, if $x \neq 0$, $\frac{0}{x} = 0$ and $\frac{x}{0}$ is undefined.

③ Find Powers and Square Roots of Real Numbers.

Exponents indicate repeated multiplication. For example,

$$3^2 = 3 \cdot 3 \qquad \text{Read } 3^2 \text{ as "3 to the second power" or "3 squared."}$$

$$(-9.1)^3 = (-9.1)(-9.1)(-9.1) \qquad \text{Read } (-9.1)^3 \text{ as "−9.1 to the third power" or "−9.1 cubed."}$$

$$\left(\frac{2}{3}\right)^4 = \left(\frac{2}{3}\right)\left(\frac{2}{3}\right)\left(\frac{2}{3}\right)\left(\frac{2}{3}\right) \qquad \text{Read } \left(\frac{2}{3}\right)^4 \text{ as "}\frac{2}{3} \text{ to the fourth power."}$$

These examples suggest the following definition.

| **Natural-Number Exponents** | A natural-number exponent indicates how many times its base is to be used as a factor. For any real number x and any natural number n, |

$$x^n = \overbrace{x \cdot x \cdot x \cdot \ldots \cdot x}^{n \text{ factors of } x}$$

The exponential expression x^n is called a **power of x,** and we read it as "x to the *n*th power." In this expression, x is called the **base,** and n is called the **exponent.** A natural-number exponent indicates how many times the base of an exponential expression is to be used as a factor in a product.

Base $\longrightarrow x^n \longleftarrow$ Exponent

EXAMPLE 6   Find each power: **a.** $(-2)^4$ **b.** $\left(\dfrac{3}{4}\right)^2$ **c.** -0.1 cubed

Strategy   We will write each exponential expression as a product of repeated factors, and then perform the multiplication. This requires that we identify the base and the exponent.

Why   The exponent indicates the number of times the base is to be written as a factor.

Solution

a. $(-2)^4 = (-2)(-2)(-2)(-2) = 16$ *The base is -2. The exponent is 4.*

b. $\left(\dfrac{3}{4}\right)^2 = \dfrac{3}{4}\left(\dfrac{3}{4}\right) = \dfrac{9}{16}$ *The base is $\frac{3}{4}$. The exponent is 2.*

c. -0.1 cubed means $(-0.1)^3$.   *The base is -0.1. The exponent is 3.*

   $(-0.1)^3 = (-0.1)(-0.1)(-0.1) = -0.001$

 Self Check 6   Find each power: **a.** $(-3)^3$ **b.** $(0.8)^2$
   c. 2^4 **d.** $\dfrac{7}{5}$ squared

Now Try   **Problems 49 and 53**

Using Your Calculator   ***The Squaring and Exponential Keys***

A homeowner plans to install a cooking island in her kitchen. (See the figure.) To find the number of square feet of floor space that will be lost, we substitute 3.25 for s in the formula for the area of a square, $A = s^2$. Using the squaring key $\boxed{x^2}$ on a scientific calculator, we can evaluate $(3.25)^2$ as follows:

3.25 $\boxed{x^2}$ $\boxed{10.5625}$

On a graphing calculator, we have:

3.25 $\boxed{x^2}$ $\boxed{\text{ENTER}}$ $\boxed{\begin{array}{l} 3.25^2 \\ \qquad 10.5625 \end{array}}$

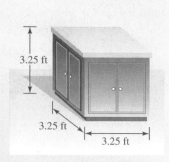

About 10.6 square feet of floor space will be lost.

The number of cubic feet of storage space that the cooking island will add can be found by substituting 3.25 for s in the formula for the volume of a cube, $V = s^3$. Using the exponential key $\boxed{y^x}$ ($\boxed{x^y}$ on some calculators), we can evaluate $(3.25)^3$ on a scientific calculator as follows.

3.25 $\boxed{y^x}$ 3 $\boxed{=}$ $\boxed{34.328125}$

On a graphing calculator, we have:

3.25 $\boxed{\wedge}$ 3 $\boxed{\text{ENTER}}$

$\boxed{\begin{array}{l} 3.25\wedge3 \\ \qquad\qquad 34.328125 \end{array}}$

The cooking island will add about 34.3 cubic feet of storage space.

Although the expressions $(-3)^2$ and -3^2 look alike, they are not. In $(-3)^2$, the base is -3. In -3^2, the base is 3. The $-$ sign in front of 3^2 means the opposite of 3^2. When we evaluate them, we see that the results are different:

$$(-3)^2 = (-3)(-3) \qquad \text{Read as "negative} \qquad\qquad -3^2 = -(3 \cdot 3) \qquad \text{Read as "the opposite of}$$
$$\text{3 squared."} \qquad\qquad\qquad\qquad\qquad\qquad\qquad\qquad \text{the square of 3."}$$

$$= 9 \qquad\qquad\qquad\qquad\qquad\qquad\qquad\qquad\qquad = -9$$

$$\undertwoarrows{\qquad\qquad\qquad \text{Different results} \qquad\qquad\qquad}$$

Using Your Calculator

The Parentheses and Negative Keys

To compute $(-3)^2$ with a scientific calculator, use the *parentheses* keys $\boxed{(}\,\boxed{)}$ and the *negative* key $\boxed{+/-}$. Notice that the negative key is different from the subtraction key $\boxed{-}$. To enter -3, press $\boxed{+/-}$ *after* entering 3.

$\boxed{(}$ 3 $\boxed{+/-}$ $\boxed{)}$ $\boxed{x^2}$ $\boxed{=}$ $\boxed{\qquad\qquad 9}$

If a graphing calculator is used to find $(-3)^2$, press the negative key $\boxed{(-)}$ *before* entering 3.

$\boxed{(}$ $\boxed{(-)}$ 3 $\boxed{)}$ $\boxed{x^2}$ $\boxed{\text{ENTER}}$ $\boxed{\begin{array}{l} (-3)^2 \\ \qquad\qquad 9 \end{array}}$

To compute -3^2 with a scientific calculator, think of the expression as $-1 \cdot 3^2$. First, find 3^2. Then press $\boxed{+/-}$, which is equivalent to multiplying 3^2 by -1.

3 $\boxed{x^2}$ $\boxed{+/-}$ $\boxed{\qquad\qquad -9}$

A graphing calculator recognizes -3^2 as $-1 \cdot 3^2$, so we can find -3^2 by entering the following:

$\boxed{(-)}$ 3 $\boxed{x^2}$ $\boxed{\text{ENTER}}$ $\boxed{\begin{array}{l} -3^2 \\ \qquad\qquad -9 \end{array}}$

Since the product $3 \cdot 3$ can be denoted by the exponential expression 3^2, we say that 3 is squared. The opposite of squaring a number is called finding its **square root.**

All positive numbers have two square roots, one positive and one negative. For example, the two square roots of 9 are 3 and -3. The number 3 is a square root of 9, because $3^2 = 9$, and -3 is a square root of 9, because $(-3)^2 = 9$.

The symbol $\sqrt{}$, called a **radical symbol,** is used to represent the positive (or *principal*) square root of a number.

Principal Square Root	A number b is a square root of a if $b^2 = a$.
	If $a > 0$, the expression $\sqrt{a}$ represents the **principal** (or positive) **square root** of a. The principal square root of 0 is 0: $\sqrt{0} = 0$.

The principal square root of a positive number is always positive. Although 3 and -3 are both square roots of 9, only 3 is the principal square root. The symbol $\sqrt{9}$ represents 3. To represent -3, we place a $-$ sign in front of the radical:

$$\sqrt{9} = 3 \qquad \text{and} \qquad -\sqrt{9} = -3$$

EXAMPLE 7 Find each square root: **a.** $\sqrt{121}$ **b.** $-\sqrt{49}$
c. $\sqrt{\dfrac{1}{4}}$ **d.** $\sqrt{0.09}$

Strategy In each case, we will determine what positive number, when squared, produces the radicand.

Why The symbol $\sqrt{}$ indicates that the positive square root of the number written under it should be found.

Solution

a. $\sqrt{121} = 11$, because $11^2 = 121$. **b.** Since $\sqrt{49} = 7$, $-\sqrt{49} = -7$.

c. $\sqrt{\dfrac{1}{4}} = \dfrac{1}{2}$, because $\left(\dfrac{1}{2}\right)^2 = \dfrac{1}{4}$. **d.** $\sqrt{0.09} = 0.3$, because $(0.3)^2 = 0.09$.

Self Check 7 Find each square root: **a.** $\sqrt{36}$ **b.** $-\sqrt{100}$
c. $\sqrt{\dfrac{4}{25}}$ **d.** $\sqrt{1}$ **e.** $\sqrt{0.81}$ **f.** $-\sqrt{400}$

Now Try Problems 55 and 59

---

4 **Use the Order of Operations Rule.**

We will often have to evaluate expressions involving several operations. For example, consider the expression $3 + 2 \cdot 5$. To evaluate it, we can perform the addition first and then the multiplication. Or we can perform the multiplication first and then the addition. However, we get different results.

Method 1: Add first   *Method 2: Multiply first*

$3 + 2 \cdot 5 = 5 \cdot 5$ Add 3 and 2 first. $3 + 2 \cdot 5 = 3 + 10$ Multiply 2 and 5 first.

$ = 25$ Multiply. $ = 13$ Add.

Different results

This example shows that we need to establish an order of operations. Otherwise, the same expression can have two different values. To guarantee that calculations will have one correct result, we will use the following set of priority rules.

Order of Operations Rule	1. Perform all calculations within parentheses and other grouping symbols, following the order listed in steps 2–4 and working from the innermost pair to the outermost pair.

2. Evaluate all exponential expressions (powers) and roots.

3. Perform all multiplications and divisions as they occur from left to right.

4. Perform all additions and subtractions as they occur from left to right.

When all grouping symbols have been removed, repeat steps 2–4 to complete the calculations.

If a fraction is present, evaluate the expression above the bar (the *numerator*) and the expression below the bar (the *denominator*) separately. Then simplify the fraction, if possible.

To evaluate $3 + 2 \cdot 5$ correctly, we follow steps 2, 3, and 4 of the order of operations rule. Since the expression does not contain any powers or roots, we perform the multiplication first, followed by the addition.

$$3 + 2 \cdot 5 = 3 + 10 \quad \text{Ignore the addition for now and multiply 2 and 5.}$$
$$= 13 \quad \text{Next, perform the addition.}$$

We see that the correct answer is 13.

EXAMPLE 8 Evaluate: **a.** $-5 + 4(-3)^2$ **b.** $-10 \div 5 - 5(3) + 6$

Strategy We will scan the expression to determine what operations need to be performed. Then we will perform those operations, one at a time, following the order of operations rule.

Why If we don't follow the correct order of operations, the expression can have more than one value.

Solution

a. Although the expression contains parentheses, there are no operations to perform within the parentheses. So we proceed with steps 2, 3, and 4 of the order of operations rule.

$$-5 + 4(-3)^2 = -5 + 4(9) \quad \text{First, evaluate the power: } (-3)^2 = 9.$$
$$= -5 + 36 \quad \text{Multiply.}$$
$$= 31 \quad \text{Add.}$$

b. Since the expression does not contain any powers, we perform the multiplications and divisions, working from left to right.

$$-10 \div 5 - 5(3) + 6 = -2 - 5(3) + 6 \quad \text{Divide: } -10 \div 5 = -2.$$
$$= -2 - 15 + 6 \quad \text{Multiply.}$$
$$= -17 + 6 \quad \text{Working from left to right, subtract:}$$
$$ \quad -2 - 15 = -17.$$
$$= -11 \quad \text{Add.}$$

The Language of Algebra
Sometimes, the word *simplify* is used in the place of the word *evaluate*. For instance, Example 8a could read:

Simplify: $-5 + 4(-3)^2$

▷ **Self Check 8** Evaluate: **a.** $-9 + 2(-4)^2$
b. $20 \div (-5) - (-6)(-5) + (-12)$

Now Try **Problems 63 and 67**

Grouping symbols serve as mathematical punctuation marks. They help determine the order in which an expression is evaluated. Examples of grouping symbols are parentheses ( ), brackets [ ], absolute value bars | |, and the fraction bar −.

EXAMPLE 9   Evaluate: **a.** $3 - (4 - 8)^2$ **b.** $2 + 3[-2 - 8(4 - 3^2)]$
c. $|-45 + 30|(2 - 7)$

Strategy   We will perform all calculations within parentheses and other grouping symbols first.

Why   This is the first step of the order of operations rule.

Solution

a. $3 - (4 - 8)^2 = 3 - (-4)^2$ *Perform the subtraction: 4 − 8 = −4.*

$\qquad\qquad\qquad\;\; = 3 - 16$ *Evaluate the power: $(-4)^2 = 16$.*

$\qquad\qquad\qquad\;\; = -13$ *Subtract.*

b. First, we work within the innermost grouping symbols, the parentheses.

$2 + 3[-2 - 8(4 - 3^2)] = 2 + 3[-2 - 8(4 - 9)]$ *Find the power: $3^2 = 9$.*

$\qquad\qquad\qquad\qquad\quad\; = 2 + 3[-2 - 8(-5)]$ *Subtract: 4 − 9 = −5.*

Next, we work within the outermost grouping symbols, the brackets.

$\qquad = 2 + 3[-2 - (-40)]$ *Multiply: 8(−5) = −40.*

$\qquad = 2 + 3(-2 + 40)$

Since only one set of grouping symbols was needed, we wrote $-2 + 40$ within parentheses.

$\qquad = 2 + 3(38)$ *Add: −2 + 40 = 38.*

$\qquad = 2 + 114$ *Multiply.*

$\qquad = 116$ *Add.*

c. Since the absolute value bars are grouping symbols, we perform the operations within the absolute value bars and the parentheses first.

$|-45 + 30|(2 - 7) = |-15|(-5)$

$\qquad\qquad\qquad\quad = 15(-5)$ *Find the absolute value: $|-15| = 15$.*

$\qquad\qquad\qquad\quad = -75$ *Multiply.*

▷ **Self Check 9**   Evaluate: **a.** $(5 - 3)^3 - 40$
b. $-3[-2(5^3 - 3) + 4] - 1$ **c.** $2|-25 - (-6)(3)|$

Now Try   **Problems 71 and 79**

Using Your Calculator   *Order of Operations*

Scientific and graphing calculators are programmed to follow the rules for the order of operations. For example, when finding $3 + 2 \cdot 5$, both types of calculators give the correct answer, 13.

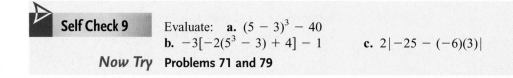

3 [+] 2 [×] 5 [=] | 13

3 [+] 2 [×] 5 [ENTER] | 3+2*5
| 13

Both types of calculators use parentheses keys $\boxed{(}\ \boxed{)}$ when grouping symbols are needed. To evaluate $3 - (4 - 8)^2$, we proceed as follows.

Both types of calculators require that we group the terms in the numerator together and the terms in the denominator together when calculating the value of an expression such as $\frac{200 + 120}{20 - 16}$.

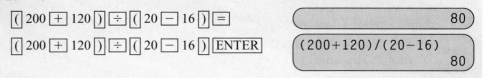

If parentheses aren't used when finding $\frac{200 + 120}{20 - 16}$, you will obtain an incorrect result of 190. That is because the calculator will interpret the entry as $200 + \frac{120}{20} - 16$.

⑤   Evaluate Algebraic Expressions.

Recall that an algebraic expression is a combination of variables and numbers with the operations of arithmetic. To *evaluate* these expressions, we substitute specific numbers for the variables and then apply the order of operations rule.

EXAMPLE 10   If $a = -2$, $b = 9$, and $c = -1$, evaluate

    a. $-\dfrac{1}{2}a^2$ **b.** $\dfrac{-a\sqrt{b} + 3c^3}{c(c - b)}$

Strategy   We will replace each a, b, and c in the expression with the given value of the variable and evaluate the expression using the order of operations rule.

Why   To *evaluate an expression* means to find its numerical value, once we know the value of its variable(s).

Solution

a. We substitute -2 for a and use the order of operations rule.

$$-\frac{1}{2}a^2 = -\frac{1}{2}(-2)^2 \qquad \text{Substitute } -2 \text{ for } a. \text{ Write parentheses around } -2 \text{ so that it is squared.}$$

$$= -\frac{1}{2}(4) \qquad \text{Evaluate the power: } (-2)^2 = 4.$$

$$= -2 \qquad \text{Multiply.}$$

b. $\dfrac{-a\sqrt{b} + 3c^3}{c(c - b)} = \dfrac{-(-2)\sqrt{9} + 3(-1)^3}{-1(-1 - 9)}$ Substitute -2 for a, 9 for b, and -1 for c.

$$= \frac{-(-2)(3) + 3(-1)}{-1(-10)}$$ In the numerator, evaluate the square root and the power: $\sqrt{9} = 3$ and $(-1)^3 = -1$. In the denominator, subtract.

$$= \frac{2(3) + 3(-1)}{-1(-10)}$$ In the numerator, simplify: $-(-2) = 2$.

$$= \frac{6 + (-3)}{10}$$ In the numerator, multiply. In the denominator, multiply.

$$= \frac{3}{10}$$ In the numerator, add.

Self Check 10   If $r = 2$, $s = -5$, and $t = 3$, evaluate:

    a. $-\frac{1}{3}s^3t$ **b.** $\dfrac{\sqrt{-5s}}{(s + t)r^2}$

Now Try   **Problems 95 and 97**

Using Your Calculator

Evaluating Algebraic Expressions

Graphing calculators can evaluate algebraic expressions. For example, to evaluate

$$\frac{-a\sqrt{b} + 3c^3}{c(c - b)}$$

(Example 10, part b) using a TI-84 Plus calculator, we first enter the values of $a = -2$, $b = 9$, and $c = -1$, using the store key $\boxed{\text{STO}}$ and the $\boxed{\text{ALPHA}}$ key. See figure (a).

$\boxed{(-)}$ 2 $\boxed{\text{STO}}$ $\boxed{\text{ALPHA}}$ $\boxed{\text{A}}$ $\boxed{\text{ALPHA}}$ $\boxed{:}$ This enters $a = -2$.

9 $\boxed{\text{STO}}$ $\boxed{\text{ALPHA}}$ $\boxed{\text{B}}$ $\boxed{\text{ALPHA}}$ $\boxed{:}$ This enters $b = 9$.

$\boxed{(-)}$ 1 $\boxed{\text{STO}}$ $\boxed{\text{ALPHA}}$ $\boxed{\text{C}}$ $\boxed{\text{ALPHA}}$ $\boxed{:}$ This enters $c = -1$.

Next, enter the expression as shown in figure (b) and press $\boxed{\text{ENTER}}$ to find that the value of the expression is 0.3. To express the result as a fraction, press $\boxed{\text{MATH}}$, highlight Frac, and then press $\boxed{\text{ENTER}}$ $\boxed{\text{ENTER}}$. See figure (c).

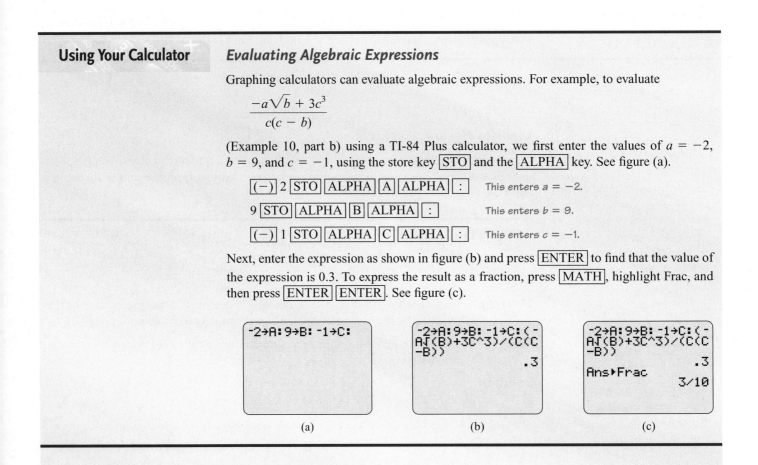

     (a)   (b)   (c)

ANSWERS TO SELF CHECKS   **1. a.** -9 **b.** -91.6 **c.** $\frac{1}{4}$ **d.** -4 **2. a.** -19 **b.** -4.5 **c.** $-\frac{2}{9}$
d. -6 **e.** 16 **3. a.** -30 **b.** 32.8 **c.** $-\frac{1}{6}$ **d.** -108 **4. a.** -11 **b.** 1.2 **5. a.** $-\frac{21}{16}$ **b.** $\frac{1}{50}$
6. a. -27 **b.** 0.64 **c.** 16 **d.** $\frac{49}{25}$ **7. a.** 6 **b.** -10 **c.** $\frac{2}{5}$ **d.** 1 **e.** 0.9 **f.** -20 **8. a.** 23
b. -46 **9. a.** -32 **b.** 719 **c.** 14 **10. a.** 125 **b.** $-\frac{5}{8}$

STUDY SET
1.3

VOCABULARY

Fill in the blanks.

1. When we add two numbers, the result is called the _____. When we subtract two numbers, the result is called the _____.

2. When we multiply two numbers, the result is called the _____. When we divide two numbers, the result is called the _____.

3. The _____ of $\frac{5}{9}$ is $\frac{9}{5}$.

4. In the exponential expression x^2, the _____ is x and 2 is the _____.

5. 6^2 can be read as "six _____" and 6^3 can be read as "six _____."

6. We read $\sqrt{25}$ as the _____ _____ of 25.

7. In the expression $9 + 6[22 - (6 - 1)]$, the _____ are the innermost grouping symbols, and the brackets are the _____ grouping symbols.

8. To _____ an algebraic expression, we substitute values for the variables and then apply the order of operations rule.

CONCEPTS

9. Fill in the blanks.
   a. An exponent indicates repeated _____.
   b. Subtraction is the same as adding the _____ of the number being subtracted.
   c. When multiplying signed numbers, an odd number of negative factors gives a _____ product. An even number of negative factors gives a _____ product.

10. a. What is the related multiplication statement for the division statement $\frac{0}{6} = 0$?
    b. Why isn't there a related multiplication statement for $\frac{6}{0}$?
    c. Fill in the blanks: if $x \neq 0$, $\frac{0}{x} = $ ▢ and $\frac{x}{0}$ is _____.

11. Consider the expression $6 + 3 \cdot 2$.
    a. In what two different ways *might* we evaluate the given expression?
    b. Which result from part (a) is correct and why?

12. In what order should the operations be performed to evaluate $60 - (-9)^2 + 5(-1)$?

NOTATION

13. Translate each expression into symbols, and then evaluate it.
    a. Negative four, squared
    b. The opposite of the square of 4

14. What is the name of the symbol $\sqrt{}$?

GUIDED PRACTICE

Perform the operations. **See Examples 1 and 2.**

15. $-3 + (-5)$ 16. $-2 + (-8)$

17. $-7.1 + 2.8$ 18. $3.1 + (-5.2)$

19. $-9 + (-8) + 4$ 20. $2 + (-6) + (-3)$

21. $\frac{1}{2} + \left(-\frac{1}{3}\right)$ 22. $-\frac{3}{4} + \left(-\frac{1}{5}\right)$

23. $-3 - 4$ 24. $-11 - (-17)$

25. $-3.3 - (-3.3)$ 26. $0.14 - (-0.13)$

27. Subtract $-\frac{3}{5}$ from $\frac{1}{2}$ 28. Subtract $\frac{11}{13}$ from $\frac{1}{26}$

29. $-1 - 5 - (-4)$ 30. $5 - (-3) - 2$

Perform the operations. **See Examples 3–5.**

31. $-2(6)$ 32. $-3(7)$

33. $-0.3(5)$ 34. $-0.4(-0.6)$

35. $-5(6)(-2)$ 36. $-9(-1)(-3)$

37. $\left(-\frac{3}{5}\right)\left(\frac{10}{7}\right)$ 38. $\left(-\frac{6}{7}\right)\left(-\frac{5}{12}\right)$

39. $\frac{-8}{4}$ 40. $\frac{16}{-4}$

41. $\frac{84}{-6}$ 42. $\frac{-78}{6}$

43. $\frac{-10.8}{-1.2}$ 44. $\frac{-13.5}{-1.5}$

45. $-\frac{16}{5} \div \left(-\frac{10}{3}\right)$ 46. $-\frac{5}{24} \div \frac{10}{3}$

Evaluate each expression. **See Example 6.**

47. 6^4 48. 2^5

49. $(-7.9)^2$ 50. $(-4.6)^2$

51. -5^2 52. -8^2

53. $\left(-\frac{3}{5}\right)^3$ 54. $\left(-\frac{4}{3}\right)^3$

Find each square root. **See Example 7.**

55. $\sqrt{64}$ 56. $\sqrt{121}$

57. $-\sqrt{81}$ 58. $-\sqrt{36}$

59. $-\sqrt{\dfrac{9}{16}}$

60. $-\sqrt{\dfrac{81}{49}}$

61. $\sqrt{0.04}$

62. $\sqrt{0.64}$

Evaluate each expression. See Example 8.

63. $3 - 5 \cdot 4$

64. $12 - 2 \cdot 3$

65. $4 \cdot 2^3$

66. $4 \cdot 5^3$

67. $-12 \div 3 \cdot 2$

68. $-18 \div 6 \cdot 3$

69. $7^2 - (-9)^2$

70. $4^2 - (-8)^2$

Evaluate each expression. See Example 9.

71. $(4 + 2 \cdot 3)^4$

72. $|9 - 5(1 - 8)|$

73. $\left(-3 - \sqrt{25}\right)^2$

74. $\left(-1 - \sqrt{144}\right)^2$

75. $-2|4 - 8|$

76. $|\sqrt{49} - 8(4 - 7)|$

77. $2 + 3\left(\dfrac{25}{5}\right) + (-4)$

78. $(-2)^3\left(\dfrac{-6}{-2}\right)(-1)$

79. $30 + 6[-4 - 5(6 - 4)^2]$

80. $7 - 12[7^2 - 4(2 - 5)^2]$

81. $3 - [3^3 + (3 - 1)^3]$

82. $8 - 4|-(3 \cdot 5 - 2 \cdot 6)^2|$

83. $\dfrac{1}{3}\left(\dfrac{1}{6}\right) - \left(-\dfrac{1}{3}\right)^2$

84. $\dfrac{1}{2}\left(\dfrac{1}{8}\right) + \left(-\dfrac{1}{4}\right)^2$

85. $\dfrac{-2 - 5}{-7 + (-7)}$

86. $\dfrac{-3 - (-1)}{-2 + (-2)}$

87. $\dfrac{|-25| - 2(-5)}{2^4 - 9}$

88. $\dfrac{2[-4 - 2(3 - 1)]}{3(3)(2)}$

89. $\dfrac{3[-9 + 2(7 - 3)]}{(8 - 5)(9 - 7)}$

90. $\dfrac{(6 - 5)^4 + 21}{27 - \left(\sqrt{16}\right)^2}$

Evaluate each expression for the given values. See Example 10.

91. $-\dfrac{2}{3}a^2$ for $a = -6$

92. $\left(-\dfrac{2}{3}a\right)^2$ for $a = -6$

93. $\dfrac{y_2 - y_1}{x_2 - x_1}$ for $x_1 = -3, x_2 = 5, y_1 = 12, y_2 = -4$

94. $P_0\left(1 + \dfrac{r}{k}\right)^{kt}$ for $P_0 = 500, r = 4, k = 2, t = 3$

95. $(x + y)(x^2 - xy + y^2)$ for $x = -4, y = 5$

96. $\dfrac{x^2}{a^2} + \dfrac{y^2}{b^2}$ for $x = -3, y = -4, a = 5, b = -5$

97. $\dfrac{-b + \sqrt{b^2 - 4ac}}{2a}$ for $a = 1, b = 2, c = -3$

98. $\dfrac{n}{2}[2a_1 + (n - 1)d]$ for $n = 50, a_1 = -4, d = 5$

99. $\sqrt{(x_2 - x_1)^2 + (y_2 - y_1)^2}$ for $x_1 = -2, x_2 = 4,$ $y_1 = 4,$ and $y_2 = -4$

100. $\dfrac{|Ax_0 + By_0 + C|}{\sqrt{A^2 + B^2}}$ for $A = 3, B = 4, C = -5, x_0 = 2,$ and $y_0 = -1$

101. $-n(4n^2 - 27m^2)^3$ for $m = \dfrac{1}{3}$ and $n = \dfrac{1}{2}$

102. $\dfrac{-s^2 + 1 + 16r^2}{3^2 - 2}$ for $s = -10$ and $r = \dfrac{1}{4}$

APPLICATIONS

103. INVESTMENT IN BONDS   In the following graph, positive numbers represent new cash *inflow* into U.S. bond funds. Negative numbers represent cash *outflow* from bond funds. Was there a net inflow or outflow over the 10-year period? What was it?

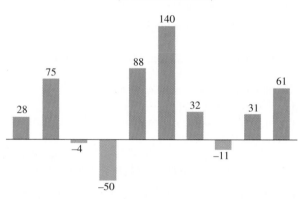

New Net Cash Flow to U.S. Bond Funds (in billions of dollars)

Source: Investment Company Institute

104. ACCOUNTING   On a financial balance sheet, debts (negative numbers) are denoted within parentheses. Assets (positive numbers) are written without parentheses. What is the 2003 fund balance for the preschool whose financial records are shown in the table?

Community Care Preschool Balance Sheet, June 2003	
Fund balances	
Classroom supplies	$ 5,889
Emergency needs	927
Holiday program	(2,928)
Insurance	1,645
Janitorial	(894)
Licensing	715
Maintenance	(6,321)
BALANCE	?

105. TEMPERATURE EXTREMES   The highest and lowest temperatures ever recorded in several cities are shown in the table. List the cities in order, from the smallest to the largest range in temperature extremes.

City	Extreme temperatures	
	Highest	Lowest
Atlanta, Georgia	105	−8
Boise, Idaho	111	−25
Helena, Montana	105	−42
New York, New York	107	−3
Omaha, Nebraska	114	−23

106. PHYSICS Waves are motions that carry energy from one place to another. The illustration shows an example of a wave called a *standing wave*. What is the difference in the height of the crest of the wave and the depth of the trough of the wave? (m stands for meter.)

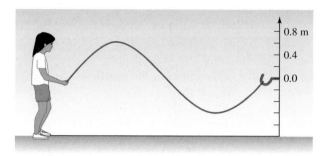

107. PEDIATRICS Young's rule, shown below, is used by some doctors to calculate dosage for infants and children.

$$\frac{\text{Age of child}}{\text{Age of child} + 12}\left(\frac{\text{average}}{\text{adult dose}}\right) = \text{child's dose}$$

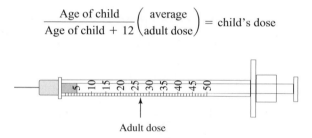

Adult dose

The syringe shows the adult dose of a certain medication. Use Young's rule to determine the dosage for a 6-year-old child. Then use an arrow to locate the dosage on the calibration.

108. DOSAGES The adult dosage of procaine penicillin is 300,000 units daily. Calculate the dosage for a 12-year-old child using Young's rule. (See Exercise 107.)

WRITING

109. Explain what the statement $x - y = x + (-y)$ means.

110. Explain why the order of operations rule is necessary.

REVIEW

111. What two numbers are a distance of 5 away from -2 on the number line?

112. Place the proper symbol ($>$ or $<$) in the blank: -4.6 ___ -4.5

113. Write the set of integers.

114. Translate into mathematical symbols: ten less than twice x.

115. True or false: The real numbers is the set of all decimals.

116. True or false: Irrational numbers are nonterminating, nonrepeating decimals.

CHALLENGE PROBLEMS

117. Insert one pair of parentheses in the expression so that its value is 0.

$$71 - 1 - 2 \cdot 5^2 + 10$$

118. Evaluate:

$$\left(\frac{\dfrac{12 \div 3 \cdot 4}{[-9^2 - 4(-1)^9(20)]^4} - \sqrt{6\sqrt{\left|-\dfrac{3}{2}(24)\right|}}}{\left|-1.5\left|\dfrac{-200}{10^2}\right|\right|^2 + \dfrac{(-2)^3}{-4 - \dfrac{4+2}{1 + \dfrac{8}{2-6}}}}\right)^5$$

SECTION 1.4
Simplifying Algebraic Expressions Using Properties of Real Numbers

Objectives

❶ Identify terms, factors, and coefficients.

❷ Identify and use properties of real numbers.

❸ Simplify products.

❹ Use the distributive property.

❺ Combine like terms.

In algebra, we frequently replace one algebraic expression with another that is equivalent and simpler in form. That process, called *simplifying an expression,* often involves the use of one or more properties of real numbers.

1 **Identify Terms, Factors, and Coefficients.**

Addition signs separate algebraic expressions into parts called *terms.* For example, the expression $3x^2 + x + 4$ has three terms: $3x^2$, x, and 4.

In general, a **term** is a product or quotient of numbers and/or variables. A single number or variable is also a term. Examples of terms are:

$$4, \qquad y, \qquad 6r, \qquad -w^3, \qquad 3.7x^5, \qquad \frac{3}{n}, \qquad -15ab^2$$

Since subtraction can be written as addition of the opposite, the expression $6a - 5b$ can be written in the equivalent form $6a + (-5b)$. We can then see that $6a - 5b$ contains two terms, $6a$ and $-5b$.

A term such as 9, that consists of a single number, is called a **constant term.**

The numerical factor of a term is called the **numerical coefficient** or simply the **coefficient** of the term. The coefficients of the terms of the expression $3x^2 + x + 4$ are 3, 1, and 4, respectively. The coefficients of the terms of $6a - 5b$ are 6 and -5, respectively.

It is important to be able to distinguish between the *terms* of an expression and the *factors* of a term.

EXAMPLE 1   Identify the coefficient of each term of $5n^3 + 10n^2 - n - 2.$

Strategy   First, we will write each subtraction as addition of the opposite so that we can identify the terms of the expression. Then we will identify the coefficient of each term.

Why   Addition symbols separate algebraic expressions into terms.

Solution   If we write $5n^3 + 10n^2 - n - 2$ as $5n^3 + 10n^2 + (-n) + (-2)$, we see that it has four terms: $5n^3$, $10n^2$, $-n$, and -2. The numerical factor of each term is its coefficient.

The coefficient of $5n^3$ is **5** because $5n^3$ means $5 \cdot n^3$.

The coefficient of $10n^2$ is **10** because $10n^2$ means $10 \cdot n^2$.

The coefficient of $-n$ is -1 because $-n$ means $-1 \cdot n$.

The coefficient of the constant term -2 is -2.

Self Check 1   Identify the coefficient of each term of $m^3 - 2m^2 + 3m + 4.$

Now Try   Problem 19

---

2 **Identify and Use Properties of Real Numbers.**

The following properties of real numbers are used to simplify algebraic expressions.

Properties of Real Numbers

If a, b, and c represent real numbers, we have

The commutative properties of addition and multiplication

$$a + b = b + a \qquad ab = ba$$

The associative properties of addition and multiplication

$$(a + b) + c = a + (b + c) \qquad (ab)c = a(bc)$$

The Language of Algebra
Commutative is a form of the word *commute*, meaning to go back and forth. *Commuter* trains take people to and from work.

The *commutative properties* enable us to add or multiply two numbers in either order and obtain the same result. Here are two examples.

$$3 + (-5) = -2 \quad \text{and} \quad -5 + 3 = -2$$
$$-2.6(-8) = 20.8 \quad \text{and} \quad -8(-2.6) = 20.8$$

We can use the commutative properties (and other properties of real numbers) to write *equivalent expressions*. **Equivalent expressions** represent the same number. For example, $x + 3$ and $3 + x$ are equivalent expressions because for each value of x, they represent the same number. For instance, if $x = 6$, both expressions represent 9. If $x = -4$, both expressions represent -1, and so on.

If x = 6	*If x = -4*
$x + 3 = 6 + 3 \qquad 3 + x = 3 + 6$	$x + 3 = -4 + 3 \qquad 3 + x = 3 + (-4)$
$= 9 \qquad\qquad = 9$	$= -1 \qquad\qquad = -1$

Subtraction and division are not commutative, because performing these operations in different orders will give different results. For example,

$$8 - 4 = 4 \quad \text{but} \quad 4 - 8 = -4$$
$$8 \div 4 = 2 \quad \text{but} \quad 4 \div 8 = \frac{1}{2}$$

The Language of Algebra
Associative is a form of the word *associate*, meaning to join a group. For example, the National Basketball *Association* (NBA) is a group of professional basketball teams.

The *associative properties* enable us to group the numbers in an addition or multiplication any way that we wish and get the same result. For example,

$$(19 + 7) + 3 = 26 + 3 = 29 \quad \text{and} \quad 19 + (7 + 3) = 19 + 10 = 29$$
$$(4 \cdot 2)6 = 8 \cdot 6 = 48 \quad \text{and} \quad 4(2 \cdot 6) = 4 \cdot 12 = 48$$

Subtraction and division are not associative, because different groupings give different results. For example,

$$(8 - 4) - 2 = 4 - 2 = 2 \quad \text{but} \quad 8 - (4 - 2) = 8 - 2 = 6$$
$$(8 \div 4) \div 2 = 2 \div 2 = 1 \quad \text{but} \quad 8 \div (4 \div 2) = 8 \div 2 = 4$$

The real numbers 0 and 1 have important special properties.

Properties of 0 and 1

Additive identity: The sum of 0 and any number is the number itself.

$$0 + a = a + 0 = a$$

Multiplicative identity: The product of 1 and any number is the number itself.

$$1 \cdot a = a \cdot 1 = a$$

Multiplication property of 0: The product of any number and 0 is 0.

$$a \cdot 0 = 0 \cdot a = 0$$

For example,

$$7 + 0 = 7, \quad 1(5.4) = 5.4, \quad \left(-\frac{7}{3}\right)1 = -\frac{7}{3}, \quad \text{and} \quad -19(0) = 0$$

If the sum of two numbers is 0, they are called **additive inverses,** or **opposites** of each other. For example, 6 and -6 are additive inverses, because $6 + (-6) = 0$.

| **The Additive Inverse Property** | For every real number a, there exists a real number $-a$ such that $$a + (-a) = -a + a = 0$$ |

If the product of two numbers is 1, the numbers are called **multiplicative inverses** or **reciprocals** of each other.

| **The Multiplicative Inverse Property** | For every nonzero real number a, there exists a real number $\frac{1}{a}$ such that $$a \cdot \frac{1}{a} = \frac{1}{a} \cdot a = 1$$ |

> **Caution**
> The reciprocal of 0 does not exist, because $\frac{1}{0}$ is undefined.

Some examples of reciprocals (multiplicative inverses) are

- 5 and $\frac{1}{5}$ are reciprocals, because $5\left(\frac{1}{5}\right) = 1$.
- $\frac{3}{2}$ and $\frac{2}{3}$ are reciprocals, because $\frac{3}{2}\left(\frac{2}{3}\right) = 1$.
- -0.25 and -4 are reciprocals, because $-0.25(-4) = 1$.

EXAMPLE 2   Complete each statement so that the indicated property is illustrated.

a. $(14 + 92) + 8 =$ _____ (Associative property of addition)

b. $\frac{7}{6} \cdot$ _____ $= \frac{7}{6}$ (Multiplicative identity property)

c. $x \cdot 5 =$ _____ (Commutative property of multiplication)

Strategy   For problems like these, it is important to have memorized the properties of real numbers by name. To fill in each blank, we will determine in what way the indicated property enables us to write an equivalent expression.

Why   We should memorize the properties of real numbers by name because their names remind us how to use them.

Solution

a. To *associate* means to group together. The associative property of addition enables us to group the numbers in a different way. Thus, we have

$$(14 + 92) + 8 = \underline{14 + (92 + 8)} \quad \text{(Associative property of addition)}$$

Note that the order of the numbers on each side of the statement remains the same.

b. The word *identical* means to be exactly the same. The multiplicative identity property indicates that if we multiply $\frac{7}{6}$ by 1, it remains the same. Thus, we have

$$\frac{7}{6} \cdot \underline{1} = \frac{7}{6} \quad \text{(Multiplicative identity property)}$$

c. To *commute* means to go back and forth. The commutative property of multiplication enables us to change the order of the factors. Thus, we have

$$x \cdot 5 = \underline{5 \cdot x} \quad \text{(Commutative property of multiplication)}$$

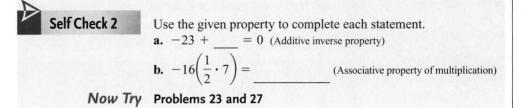

Now Try   **Problems 23 and 27**

Recall that when a number is divided by 1, the result is the number itself, and when a nonzero number is divided by itself, the result is 1.

Division Properties	*Division by 1*: If a represents any real number, then $\frac{a}{1} = a$. *Division of a number by itself*: For any nonzero real number a, $\frac{a}{a} = 1$.

There are three possible cases to consider when discussing division involving 0.

Division with 0	*Division of 0*: For any nonzero real number a, $\frac{0}{a} = 0$. *Division by 0*: For any nonzero real number a, $\frac{a}{0}$ is undefined. *Division of 0 by 0*: $\frac{0}{0}$ is indeterminate.

To show that division of 0 by 0 is indeterminate, we consider $\frac{0}{0} = ?$ and its equivalent multiplication fact $0(?) = 0$.

Multiplication fact

$0(?) = 0$

↑

Any number multiplied by 0 gives 0.

Division fact

$\dfrac{0}{0} = ?$

↑

We cannot determine this—it could be any number.

 Simplify Products.

The commutative and associative properties of multiplication can be used to simplify certain products. For example, let's simplify $6(5x)$.

$$6(5x) = 6 \cdot (5 \cdot x) \quad \text{Rewrite } 5x \text{ as } 5 \cdot x.$$
$$= (6 \cdot 5) \cdot x \quad \text{Use the associative property of multiplication to group 5 with 6.}$$
$$= 30x \quad \text{Multiply within the parentheses.}$$

Since $6(5x) = 30x$, we say that $6(5x)$ simplifies to $30x$.

Success Tip
By the commutative property of multiplication, we can *change* the order of factors. By the associative property of multiplication, we can change the *grouping* of factors.

EXAMPLE 3   Simplify: **a.** $9(10t)$ **b.** $-5.3r(-2s)$ **c.** $-\dfrac{21}{2}a\left(\dfrac{1}{3}\right)$

Strategy   We will use the commutative and associative properties of multiplication to reorder and regroup the factors in each expression.

Why We want to group all of the numerical factors of an expression together so that we can find their product.

Solution

a. $9(10t) = (9 \cdot 10)t$ Use the associative property of multiplication to regroup the factors.

$ = 90t$ Multiply within the parentheses: $9 \cdot 10 = 90$.

b. $-5.3r(-2s) = [-5.3(-2)](r \cdot s)$ Use the commutative and associative properties to group the numbers and group the variables.

$ = 10.6rs$ Multiply within the brackets.

c. $-\dfrac{21}{2}a\left(\dfrac{1}{3}\right) = -\dfrac{21}{2}\left(\dfrac{1}{3}\right)a$ Use the commutative property of multiplication to change the order of the factors a and $\frac{1}{3}$.

$\phantom{-\dfrac{21}{2}a\left(\dfrac{1}{3}\right)} = -\dfrac{7}{2}a$ Multiply: $-\frac{21}{2} \cdot \frac{1}{3} = -\frac{21 \cdot 1}{2 \cdot 3} = -\dfrac{\overset{1}{\cancel{3}} \cdot 7 \cdot 1}{2 \cdot \underset{1}{\cancel{3}}} = -\frac{7}{2}$.

Self Check 3   Simplify: **a.** $14 \cdot 3s$ **b.** $-1.6b(3t)$

            c. $-\frac{2}{3}x(-9)$

Now Try   **Problems 35 and 39**

---

4   **Use the Distributive Property.**

Another property that we can use to simplify algebraic expressions is the **distributive property.** To introduce it, we will evaluate $4(5 + 3)$, in two ways.

The Language of Algebra

To *distribute* means to give from one to several. You have probably *distributed* candy to children coming to your front door on Halloween.

Method 1	**Method 2**
Use the order of operations:	*Distribute the multiplication:*
$4(5 + 3) = 4(8)$	$4(5 + 3) = 4(5) + 4(3)$
$ = 32$	$ = 20 + 12$
	$ = 32$

Each method gives a result of 32. This observation suggests the following property.

The Distributive Property

The distributive property of multiplication over addition

If a, b, and c represent real numbers,

$$a(b + c) = ab + ac$$

EXAMPLE 4   Multiply by using the distributive property to remove parentheses:

         a. $6(a + 9)$ **b.** $-15(4b - 1)$ **c.** $-(-21 - 20m)$

Strategy   We will distribute the multiplication by the factor outside the parentheses over each term within the parentheses.

Why   We cannot simplify the expression within the parentheses. To multiply, we must use the distributive property.

The Language of Algebra
When we use the distributive property to write a product, such as $6(a + 9)$, as the sum, $6a + 54$, we say that we have *removed* or *cleared* parentheses.

Solution

a. $6(a + 9) = 6 \cdot a + 6 \cdot 9$ *Distribute the multiplication by 6.*

$\qquad\qquad = 6a + 54$ *Multiply.*

b. $-15(4b - 1) = -15(4b) - (-15)(1)$ *Distribute the multiplication by −15.*

$\qquad\qquad\quad = -60b + 15$ *Multiply.*

c. To use the distributive property to simplify $-(-21 - 20m)$, we interpret the $-$ symbol as a factor of -1, and proceed as follows.

$$-(-21 - 20m) = -1(-21 - 20m)$$ *Write the − sign in front of the parentheses as −1.*

$$= -1(-21) - (-1)(20m)$$ *Distribute the multiplication by −1.*

$$= 21 + 20m$$ *Multiply.*

$$= 20m + 21$$ *Write the variable term of the answer first.*

Notation
We can use the commutative property of addition to reorder the terms of the result. It is standard practice to write such answers with the variable terms first, followed by the constant term.

> **Self Check 4**   Multiply by using the distributive property to remove parentheses: **a.** $9(r + 4)$ **b.** $-11(-3x - 5)$ **c.** $-(-27k + 15)$
>
> **Now Try**   Problems 43 and 47

A more general form of the distributive property is the **extended distributive property.**

$$a(b + c + d + e + \cdots) = ab + ac + ad + ae + \cdots$$

Since multiplication is commutative, we can write the distributive property in the following forms.

$$(b + c)a = ba + ca, \quad (b - c)a = ba - ca, \quad (b + c + d)a = ba + ca + da$$

> **EXAMPLE 5**   Multiply: **a.** $-0.5(7 - 5y + 6z)$ **b.** $(8x - 3y)\dfrac{3}{2}$

Strategy   We will multiply each term within the parentheses by the factor outside the parentheses.

Why   We cannot simplify the expression within the parentheses. To multiply, we must use an extension of the distributive property.

Solution

a. $-0.5(7 - 5y + 6z)$

$\quad = -0.5(7) - (-0.5)(5y) + (-0.5)(6z)$ *Distribute the multiplication by −0.5.*

$\quad = -3.5 + 2.5y - 3z$ *Multiply.*

$\quad = 2.5y - 3z - 3.5$ *Write the variable terms of the answer first.*

b. $(8x - 3y)\dfrac{3}{2} = (8x)\dfrac{3}{2} - (3y)\dfrac{3}{2}$ Distribute the multiplication by $\frac{3}{2}$.

$$= \dfrac{24x}{2} - \dfrac{9y}{2}$$ Multiply.

$$= 12x - \dfrac{9}{2}y$$ Simplify.

 Self Check 5   Multiply: **a.** $0.8(-6t + 3s - 10)$

b. $(10a + 16b)\dfrac{3}{5}$

***Now Try*   Problems 55 and 59**

---

5   **Combine Like Terms.**

Before we can discuss methods for simplifying algebraic expressions involving addition and subtraction, we must define like and unlike terms.

Like Terms	**Like terms** are terms with exactly the same variables raised to exactly the same powers. Any constant terms in an expression are considered to be like terms. Terms that are not like terms are called **unlike terms.**

Here are some examples of like and unlike terms.

$5x$ and $6x$ are like terms.

$4x$ and $-17y$ are unlike terms, because they have different variables.

$27x^2y^3$ and $-326x^2y^3$ are like terms.

$15x^2y$ and $6xy^2$ are unlike terms, because the variables have different exponents.

If we are to add (or subtract) objects, they must have the same units. For example, we can add dollars to dollars and inches to inches, but we cannot add dollars to inches. The same is true when working with terms of an expression. They can be added or subtracted only when they are like terms.

Simplifying the sum or difference of like terms is called **combining like terms.** To simplify expressions containing like terms, we use the distributive property in reverse. For example,

$$5x + 6x = (5 + 6)x \quad \text{and} \quad 32y - 16y = (32 - 16)y$$
$$= 11x \qquad\qquad\qquad\qquad = 16y$$

These examples suggest the following general rule.

Combining Like Terms	Like terms can be combined by adding or subtracting the coefficients of the terms and keeping the same variables with the same exponents.

EXAMPLE 6   Simplify by combining like terms: **a.** $-8f + 12f$

b. $0.6s^3 - 0.2s^3 - (-0.9s^3)$ **c.** $-\frac{1}{2}ab + \frac{1}{3}ab$

d. $16n + 8n^2 - 42n + 4n^2$

Strategy   We will use the distributive property to add (or subtract) the coefficients of the like terms.

Why   To *combine like terms* means to add or subtract the like terms in an expression.

Solution

a. Since $-8f$ and $12f$ are like terms with the common variable f, we can combine them by adding the coefficients of the like terms and keeping the variable f.

$$-8f + 12f = 4f \quad \text{Think: } (-8 + 12)f = 4f.$$

b. $0.6s^3 - 0.2s^3 - (-0.9s^3) = 0.6s^3 - 0.2s^3 + 0.9s^3$ Add the opposite of $-0.9s^3$.

$$= 1.3s^3 \qquad \text{Think: } (0.6 - 0.2 + 0.9)s^3 = 1.3s^3.$$

c. $-\frac{1}{2}ab + \frac{1}{3}ab = -\frac{1}{2} \cdot \frac{3}{3}ab + \frac{1}{3} \cdot \frac{2}{2}ab$ Build each fraction into an equivalent fraction that has the LCD 6 for its denominator.

$$= -\frac{3}{6}ab + \frac{2}{6}ab \qquad \text{Multiply the numerators. Multiply the denominators.}$$

$$= -\frac{1}{6}ab \qquad \text{Think: } \left(-\frac{3}{6} + \frac{2}{6}\right)ab = -\frac{1}{6}ab.$$

> **Notation**
> We can use the commutative property of addition to reorder the terms of the result. It is standard practice to write such answers in descending powers of the variable.

d. We will combine the n-terms and combine the n^2-terms.

$$16n + 8n^2 - 42n + 4n^2 = -26n + 12n^2 \quad \text{Think: } (16 - 42)n = -26n \text{ and}$$
$$(8 + 4)n^2 = 12n^2.$$

$$= 12n^2 - 26n \quad \text{Write the terms of the result in descending powers of } n.$$

Self Check 6   Simplify by combining like terms: **a.** $5k + 8k$

b. $-600a^2 - (-800a^2) + 100a^2$ **c.** $\frac{2}{3}xy - \frac{3}{4}xy$

d. $c + 32d^2 - 19c - 20d^2$

Now Try   Problems 65, 67, and 71

EXAMPLE 7   Simplify each expression: **a.** $20b^2 - 5(3b^2 + 1) + 8$

b. $6\left(\frac{3}{2}d - \frac{4}{3}\right) + 6\left(\frac{5}{6}d\right)$

Strategy   We will use the distributive property to remove parentheses and then combine any like terms.

Why   Since we cannot simplify the expressions within the parentheses, we will perform the indicated multiplication.

Solution

a. $20b^2 - 5(3b^2 + 1) + 8 = 20b^2 - 15b^2 - 5 + 8$ Distribute the multiplication by -5.

$$= 5b^2 + 3 \qquad \text{Combine like terms.}$$

b. $6\left(\dfrac{3}{2}d - \dfrac{4}{3}\right) + 6\left(\dfrac{5}{6}d\right) = 6\left(\dfrac{3}{2}d\right) - 6\left(\dfrac{4}{3}\right) + 6\left(\dfrac{5}{6}d\right)$ Distribute the multiplication by 6.

$$= \dfrac{18}{2}d - \dfrac{24}{3} + \dfrac{30}{6}d \qquad \text{Multiply.}$$

$$= 9d - 8 + 5d \qquad \text{Simplify each fraction.}$$

$$= 14d - 8 \qquad \text{Combine like terms.}$$

Self Check 7   Simplify: **a.** $44a^3 - 2(10a^3 - a^2) - a^2$
b. $12\left(\dfrac{5}{6}r^2 + \dfrac{1}{4}r\right) + 12\left(\dfrac{2}{3}r\right)$

Now Try   Problems 79 and 85

EXAMPLE 8   Simplify: $3x + 4[6x - 2(7x + 8)]$

Strategy   We will simplify the expression by working from the innermost grouping symbols (the parentheses) to the outermost grouping symbols (the brackets).

Why   To simplify expressions, we follow the order of operations rule.

Solution

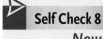

$3x + 4[6x - 2(7x + 8)] = 3x + 4[6x - 14x - 16]$ Remove the innermost parentheses by distributing the multiplication by -2.

$$= 3x + 4[-8x - 16] \qquad \text{Combine like terms within the brackets.}$$

$$= 3x - 32x - 64 \qquad \text{Remove the outermost brackets by distributing the multiplication by 4.}$$

$$= -29x - 64 \qquad \text{Combine like terms.}$$

Self Check 8   Simplify: $8t - 4[10t + 2(2t + 1) - 3]$

Now Try   Problem 87

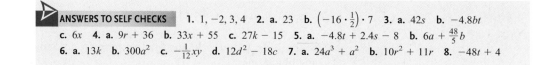

ANSWERS TO SELF CHECKS   **1.** $1, -2, 3, 4$ **2. a.** 23 **b.** $\left(-16 \cdot \frac{1}{2}\right) \cdot 7$ **3. a.** $42s$ **b.** $-4.8bt$
c. $6x$ **4. a.** $9r + 36$ **b.** $33x + 55$ **c.** $27k - 15$ **5. a.** $-4.8t + 2.4s - 8$ **b.** $6a + \frac{48}{5}b$
6. a. $13k$ **b.** $300a^2$ **c.** $-\frac{1}{12}xy$ **d.** $12d^2 - 18c$ **7. a.** $24a^3 + a^2$ **b.** $10r^2 + 11r$ **8.** $-48t + 4$

STUDY SET
1.4

VOCABULARY

Fill in the blanks.

1. A _____ is a product or quotient of numbers and/or variables, such as $6r$, $-t^3$, and $\frac{44}{m}$.

2. The _____ of the term $-8c$ is -8.

3. A term, such as 9, that consists of a single number is called a _____ term.

4. To _____ expressions, we use properties of real numbers to write equivalent expressions in a less complicated form.

5. The _____ properties of real numbers involve changing *order* and the _____ properties of real numbers involve changing *grouping*.

6. We can use the _____ property to remove parentheses in the expression $2(x + 8)$.

7. _____ terms are terms with exactly the same variables raised to exactly the same powers.

8. Simplifying the sum or difference of like terms is called _____ like terms.

CONCEPTS

9. **a.** Using the variables x, y, and z, write the associative property of addition.

   b. Using the variables x and y, write the commutative property of multiplication.

   c. Using the variables r, s, and t, write the distributive property of multiplication over addition.

10. Complete each property of addition. Then give its name.

    a. $a + (-a) = $

    b. $a + 0 = $

    c. $a + b = b + $

    d. $(a + b) + c = a + $

11. Complete each property of multiplication. Then give its name.

    a. $a \cdot b = b \cdot $

    b. $(ab)c = $

    c. $0 \cdot a = $

    d. $1 \cdot a = $

    e. $a\left(\dfrac{1}{a}\right) = $

12. Complete each property of division.

    a. $\dfrac{a}{1} = $ **b.** $\dfrac{a}{a} = $

    c. $\dfrac{0}{a} = $ **d.** $\dfrac{a}{0}$ is

13. **a.** What is the additive identity?

    b. What is the multiplicative identity?

    c. What is the additive inverse (opposite) of x?

    d. What is the multiplicative inverse (reciprocal) of x?

14. What number should be

    a. subtracted from 5 to obtain 0?

    b. added to 5 to obtain 0?

15. By what number should

    a. 5 be divided to obtain 1?

    b. 5 be multiplied to obtain 1?

16. Are the terms listed here like terms? If they are, combine them.

    a. $2x$, $6x$ **b.** $-3x$, $5y$

    c. $-5xy$, $-7yz$ **d.** $24t^2$, $24t^3$

NOTATION

17. In $-(x - 7)$, what does the negative sign in front of the parentheses represent?

18. Does the distributive property apply?

    a. $2(3)(5)$ **b.** $2(3 \cdot 5)$

    c. $2(3x)$ **d.** $2(x - 3)$

GUIDED PRACTICE

What are the terms of the expression? Give the coefficient of each term. **See Example 1.**

19. $3x^3 + 11x^2 - x + 9$

20. $2y^4 - y^3 + 6y + 4$

21. $\dfrac{11}{12}a^4 - \dfrac{3}{4}b^2 + 25b$

22. $0.78m^3 - 1.55n - 0.99$

Complete each statement so that the indicated property is illustrated. **See Example 2.**

23. $3 + 7 = $ _____ (Commutative property of addition)

24. $2(5 \cdot 97) = $ _____ (Associative property of multiplication)

25. $3(2 + d) = $ _____ (Distributive property)

26. $1 \cdot y = $ _____ (Commutative property of multiplication)

27. $c + 0 = $ ___ (Additive identity property)

28. $-4(x - 2) = $ _____ (Distributive property and simplifying)

29. $25 \cdot \dfrac{1}{25} = $ ___ (Multiplicative inverse property)

30. $z + (9 - 27) = $ _____ (Commutative property of addition)

31. $8 + (7 + a) = $ _____ (Associative property of addition)

32. ___ $\cdot 3 = 3$ (Multiplicative identity property)

33. $(x + y)2 =$ _____ (Commutative property of multiplication)

34. $h + (-h) =$ ___ (Additive inverse property)

Multiply. See Example 3.

35. $9(8m)$

36. $12n(4)$

37. $5(-9q)$

38. $-3(2t)$

39. $\dfrac{7}{8}x(-56)$

40. $\dfrac{5}{9}r(-45)$

41. $-4(8r)(-2y)$

42. $-6s(-4t)(-1)$

Multiply. See Example 4.

43. $9(9x + 2)$

44. $7(6y + 1)$

45. $-4(-3t + 3)$

46. $-4(-5y + 3)$

47. $-(24 - d)$

48. $-(19 - w)$

49. $\dfrac{2}{3}(3s^2 - 9)$

50. $\dfrac{1}{5}(5b^3 - 15)$

51. $0.7(m + 2n)$

52. $2.5(6c - 8d)$

53. $100(0.09x + 0.02y)$

54. $100(8.36x - 2.75y)$

Multiply. See Example 5.

55. $5(9t^2 - 12t - 3)$

56. $25(2a^2 - 3a + 1)$

57. $3\left(\dfrac{4}{3}x - \dfrac{5}{3}y + \dfrac{1}{3}\right)$

58. $6\left(-\dfrac{4}{3} + \dfrac{7}{6}s + \dfrac{16}{3}t\right)$

59. $(16t + 24)\dfrac{1}{8}$

60. $(18q + 9)\dfrac{1}{9}$

61. $(y - 2)(-3)$

62. $(2t + 5)(-2)$

Simplify by combining like terms. See Example 6.

63. $3x + 15x$

64. $12y - 17y$

65. $0.7h - 3.8h$

66. $-5.7m + 5.3m$

67. $1.8x^2 - 5.1x^2 + 4.1x^2$

68. $3.7x^2 + 3.3x^2 - 1.1x^2$

69. $-8x + 5x - (-x)$

70. $-20y + 3y - (-6y)$

71. $\dfrac{2}{5}ab - \left(-\dfrac{1}{2}ab\right)$

72. $-\dfrac{3}{4}st - \dfrac{1}{3}st$

73. $\dfrac{3}{5}t + \dfrac{1}{3}t$

74. $\dfrac{3}{16}x - \dfrac{5}{4}x$

75. $-9a + 11ad - 35a + ad$

76. $-7a + 2ab - 7a + 12ab$

77. $4m - t - (-2m) + 3t$

78. $14g + h - (-g) - 8h$

Simplify. See Example 7.

79. $2x^2 + 4(3x - x^2) + 3x$

80. $3p^2 - 6(5p^2 + p) + p^2$

81. $-3(p - 2) + 2(p + 3) - 5(p - 1)$

82. $5(q + 7) - 3(q - 1) - (q + 2)$

83. $36\left(\dfrac{2}{9}x - \dfrac{3}{4}\right) + 36\left(\dfrac{1}{2}\right)$

84. $40\left(\dfrac{3}{8}y - \dfrac{1}{4}\right) + 40\left(\dfrac{4}{5}\right)$

85. $24\left(\dfrac{5}{6}y - \dfrac{9}{8}\right) - 24\left(\dfrac{3}{24}y\right)$

86. $18\left(\dfrac{11}{18}w - \dfrac{7}{2}\right) - 18\left(\dfrac{1}{9}w\right)$

Simplify. See Example 8.

87. $3[2(x + 2)] - 5[3(x - 5)] + 5x$

88. $-5[3(x - 4) - 2(x + 2)] - 7(x - 3)$

89. $2\left[6\left(\dfrac{1}{3}a + 2b\right) - 8\left(\dfrac{1}{4}a - 2b\right) + 3\right]$

90. $10\left[\dfrac{3}{5}(2s + 2t) - \dfrac{4}{5}(s - t) + 1\right]$

TRY IT YOURSELF

Simplify each expression.

91. $-(a + 2A + 1) - (a - A + 2)$

92. $3T - 2(t - T) + t$

93. $8(2cd + 7c) - 2(cd - 3c)$

94. $2tz + 5(tz - 4) - 10(8 - tz)$

95. $6.4a^2 + 11.8a - 9.2a + 5.7$

96. $9.1m^2 - 6.1m + 12.3m - 4.9$

97. $-\dfrac{7}{16}x - \dfrac{3}{4}x$

98. $-\dfrac{5}{9}y - \dfrac{7}{18}y$

99. $-2[4(z - 9) - 6(3z - 7)] - 7(2z - 1)$

100. $9(m^3 + 3) - 5(3 - m^3) - 8(-1 - m^3)$

101. $21\left(\dfrac{6}{7}h^2 - \dfrac{15}{21}h\right) + 21\left(\dfrac{1}{3}h\right)$

102. $\dfrac{1}{12}(y - 12x) - \dfrac{1}{3}(y - 3x)$

103. $4.3(y + 9) - 8.1y$

104. $2.1(4 + 5z) + 0.9z$

105. $3x^2 - (-2x^2) - 5x^2$

106. $8x^3 - x^3 - (-2x^3)$

APPLICATIONS

107. PARKING AREAS   Refer to the illustration on the next page.

　　a. Express the area of the entire parking lot as the product of its length and width.

　　b. Express the area of the entire lot as the sum of the areas of the self-parking space and the valet parking space.

　　c. Write an equation that shows that your answers to parts (a) and (b) are equal. What property of real numbers is illustrated by this example?

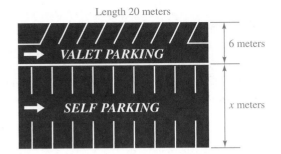

Length 20 meters

6 meters

VALET PARKING

SELF PARKING

x meters

108. CHECKING ACCOUNTS To find the total dollar amount of the checks entered in the register below, we could add the check amounts in the order in which they are written: $39 + $75 + $34 + $25 + $111 + $16. Write an expression with the amounts reordered and grouped in such a way that the addition is easier. Then find the sum. What properties of real numbers did you use?

Number	Date	Description of Transaction	Payment/Debit	
101	3/6	DR. OKAMOTO, DDS	$39	00
102	3/6	UNION OIL CO.	$75	00
103	3/8	STATER BROS.	$34	00
104	3/9	LITTLE LEAGUE	$25	00
105	3/11	NORDSTROM	$111	00
106	3/12	OFFICE MAX	$16	00

WRITING

109. Explain why the distributive property does not apply when simplifying $6(2 \cdot x)$.

110. In each case, explain what you can conclude about one or both of the numbers.
   a. When the two numbers are added, the result is 0.
   b. When the two numbers are subtracted, the result is 0.
   c. When the two numbers are multiplied, the result is 0.
   d. When the two numbers are divided, the result is 0.

111. What are like terms?

112. Use each of the words *commute, associate,* and *distribute* in a sentence in which the context is nonmathematical.

REVIEW

Evaluate each expression.

113. $\left(-\dfrac{3}{2}\right)\left(\dfrac{7}{12}\right)$

114. $\dfrac{1}{2} - \left(-\dfrac{4}{5}\right)$

115. $-3|4 - 8| + (4 + 2 \cdot 3)^3$

116. $\left(\dfrac{-\sqrt{4^3} - 5^2}{2 \cdot 2^2 - (1^9 - 4)}\right)^3$

CHALLENGE PROBLEMS

117. Simplify: $\dfrac{x}{2} + \dfrac{x}{3} + \dfrac{x}{4} + \dfrac{x}{5} + \dfrac{x}{6}$

118. Fill in the blank:

$$(0.005x + 0.02y - 0.0003z) = 50x + 200y - 3z$$

119. What two real numbers are their own reciprocals?

120. Explain how the distributive property can be used to evaluate the expression $52.713(21) + 52.713(79)$ mentally.

SECTION 1.5
Solving Linear Equations Using Properties of Equality

Objectives

1 Determine whether a number is a solution.
2 Use properties of equality to solve equations.
3 Simplify expressions to solve equations.
4 Clear equations of fractions and decimals.
5 Identify identities and contradictions.

One of the most useful concepts in algebra is the equation. Writing and then solving an equation is a powerful problem-solving strategy. In this section, we will review some fundamental properties that are used to solve equations.

① Determine Whether a Number is a Solution.

An **equation** is a statement indicating that two expressions are equal. All equations contain an equal symbol $=$. An example of an equation is $7x - 3 = 4$. The equal symbol separates the equation into two parts: The expression $7x - 3$ is the **left side** and 4 is the **right side.** The letter x is the **variable** (or the **unknown**). Since the sides of an equation can be reversed, we can write $7x - 3 = 4$ or $4 = 7x - 3$.

- An equation can be true: $6 + 3 = 9$.
- An equation can be false: $2 + 4 = 7$.
- An equation can be neither true nor false. For example, $7x - 3 = 4$ is neither true nor false because we don't know what number x represents.

An equation that contains a variable is made true or false by substituting a number for the variable. For example, if $x = 1$, then the equation $7x - 3 = 4$ is true.

The Language of Algebra
It is important to know the difference between an *equation* and an *expression*. An equation contains an $=$ symbol; an expression does not.

$$7x - 3 = 4$$

$$7(1) - 3 \overset{?}{=} 4 \quad \text{Substitute 1 for x. At this stage, we don't know whether the left and right sides of the equation are equal, so we use an "is possibly equal to" symbol } \overset{?}{=} .$$

$$7 - 3 \overset{?}{=} 4$$

$$4 = 4 \quad \text{We obtain a true statement.}$$

A number that makes an equation true when substituted for the variable is called a **solution,** and it is said to *satisfy* the equation. Therefore, 1 is a solution of $7x - 3 = 4$. The **solution set** of an equation is the set of all numbers that make the equation true.

EXAMPLE 1   Determine whether 2 is a solution of $3x + 2 = 2x + 5$.

Strategy   We will substitute 2 for each x in the equation and evaluate the expressions on the left side and the right side separately.

Why   If a true statement results, 2 is a solution of the equation. If we obtain a false statement, 2 is not a solution.

Solution

Evaluate the expression on the left side.

$$3x + 2 = 2x + 5 \quad \text{This is the original equation.}$$

$$3(2) + 2 \overset{?}{=} 2(2) + 5 \quad \text{Substitute 2 for x.}$$

$$6 + 2 \overset{?}{=} 4 + 5$$

$$8 = 9 \quad \text{False}$$

Evaluate the expression on the right side.

Since $8 = 9$ is a false statement, the number 2 does not satisfy the equation. It is not a solution of $3x + 2 = 2x + 5$.

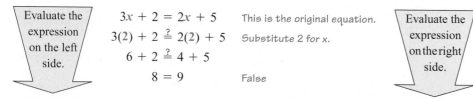

Self Check 1   Is -5 a solution of $2x - 5 = 3x$?
Now Try   Problem 15

2 **Use Properties of Equality to Solve Equations.**

Usually, we do not know the solutions of an equation—we need to find them. In this text, we will discuss how to solve many different types of equations. The easiest equations to solve are *linear equations in one variable.*

Linear Equations	A **linear equation in one variable** can be written in the form
	$$ax + b = c \qquad \text{where } a, b, \text{ and } c \text{ are real numbers, and } a \neq 0.$$

Some examples of linear equations in one variable are

$$2x - 8 = 0, \qquad -\frac{3}{4}y = -7, \qquad \text{and} \qquad 4b - 7 + 2b = 1 + 2b + 8$$

Notice for these linear equations that the highest power on the variable is 1.

When solving linear equations, the objective is to *isolate* the variable on one side of the equation. This is achieved by undoing the operations performed on the variable. As we undo the operations, we produce a series of simpler equations, all having the same solutions. Such equations are called *equivalent equations.*

Equivalent Equations	Equations with the same solutions are called **equivalent equations.**

The solution of the equation $x = 2$ is obviously 2, because replacing x with 2 yields a true statement, $2 = 2$. The equation $x + 4 = 6$ also has a solution of 2. Since $x = 2$ and $x + 4 = 6$ have the same solution, they are equivalent equations.

We can use the following properties to write equivalent equations, in which we will isolate the variable on one side of the equation.

Properties of Equality	Adding the same number to, or subtracting the same number from, both sides of an equation does not change the solution.
	If a, b, and c are real numbers and $a = b$,
	$a + c = b + c$ **Addition property of equality**
	$a - c = b - c$ **Subtraction property of equality**
	Multiplying or dividing both sides of an equation by the same nonzero number does not change the solution.
	If a, b, and c are real numbers with $c \neq 0$, and $a = b$,
	$ca = cb$ **Multiplication property of equality**
	$\dfrac{a}{c} = \dfrac{b}{c}$ **Division property of equality**

EXAMPLE 2   Solve: **a.** $2x - 8 = 0$ **b.** $-35.6 = 77.89 - x$

Strategy   We will use a property of equality to isolate the *variable term* on one side of the equation and then use another property to isolate the *variable.*

Why To solve the original equation, we want to find a simpler equivalent equation of the form $x = $ **a number**, whose solution is obvious.

Solution

a. We note that x is multiplied by 2, and then 8 is subtracted from that product. To isolate x on the left side of the equation, we use the order of operations rule in reverse.

- To undo the subtraction of 8, we add 8 to both sides.

- To undo the multiplication by 2, we divide both sides by 2.

The Language of Algebra
Since division by 2 is the same as multiplication by $\frac{1}{2}$, we can also solve $2x = 8$ using the multiplication property of equality. To isolate x we can multiply both sides by the *multiplicative inverse* of 2, which is $\frac{1}{2}$:

$$2x = 8$$
$$\frac{1}{2} \cdot 2x = \frac{1}{2} \cdot 8$$
$$x = 4$$

$$2x - 8 = 0 \qquad \text{This is the equation to solve.}$$
$$2x - 8 + 8 = 0 + 8 \qquad \text{Use the addition property of equality: Add 8 to both sides to isolate the variable term, } 2x.$$
$$2x = 8 \qquad \text{Simplify both sides of the equation.}$$
$$\frac{2x}{2} = \frac{8}{2} \qquad \text{Use the division property of equality: Divide both sides by 2 to isolate } x.$$
$$x = 4 \qquad \text{Do the divisions.}$$

Check: We substitute 4 for x to verify that it satisfies the original equation.

$$2x - 8 = 0$$
$$2(4) - 8 \stackrel{?}{=} 0 \qquad \text{Substitute 4 for } x.$$
$$8 - 8 \stackrel{?}{=} 0 \qquad \text{Multiply.}$$
$$0 = 0 \qquad \text{True}$$

Since we obtain a true statement, 4 is the solution of $2x - 8 = 0$ and the solution set is $\{4\}$.

The Language of Algebra
Since subtracting 77.89 is the same as adding -77.89, we can also solve the equation using the addition property of equality. To isolate $-x$ we can add the *additive inverse* of 77.89, which is -77.89, to both sides:

$$-35.6 + (-77.89) =$$
$$77.89 - x + (-77.89)$$

b.
$$-35.6 = 77.89 - x \qquad \text{This is the equation to solve.}$$
$$-35.6 - 77.89 = 77.89 - x - 77.89 \qquad \text{Use the subtraction property of equality: Subtract 77.89 from both sides to isolate the variable term, } -x.$$
$$-113.49 = -x \qquad \text{Simplify each side of the equation.}$$

The variable x is not yet isolated, because there is a $-$ sign in front of it. Since the term $-x$ has an understood coefficient of -1, we can write $-x$ as $-1x$. To isolate x, we can either multiply or divide both sides by -1.

$$-113.49 = -1x \qquad \text{Write } -x = -1x.$$
$$\frac{-113.49}{-1} = \frac{-1x}{-1} \qquad \text{Use the division (or multiplication) property of equality: Divide (or multiply) both sides by } -1 \text{ to isolate } x.$$
$$113.49 = x \qquad \text{Simplify each side of the equation.}$$
$$x = 113.49 \qquad \text{Reverse the sides of the equation so that } x \text{ is on the left.}$$

Verify that 113.49 is the solution by checking it in the original equation.

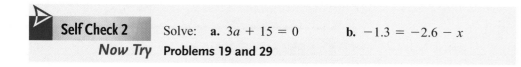

Self Check 2 Solve: **a.** $3a + 15 = 0$ **b.** $-1.3 = -2.6 - x$
Now Try Problems **19** and **29**

EXAMPLE 3   Solve: $\dfrac{3}{4}y = -7$

Strategy   We will isolate y by multiplying both sides of the equation by $\dfrac{4}{3}$.

Why   On the left side, y is multiplied by $\dfrac{3}{4}$. We can undo the multiplication by dividing both sides by $\dfrac{3}{4}$. Since division by $\dfrac{3}{4}$ is equivalent to multiplication by its *reciprocal*, it is easier to isolate y by multiplying both sides by $\dfrac{4}{3}$.

Solution

> **Notation**
> Variable terms with fractional coefficients can be written in two ways. For example, $\frac{3}{4}y = \frac{3y}{4}$ and $\frac{2}{3}b = \frac{2b}{3}$.

$$\dfrac{3}{4}y = -7 \qquad \text{This is the equation to solve.}$$

$$\dfrac{4}{3}\left(\dfrac{3}{4}y\right) = \dfrac{4}{3}(-7) \qquad \text{Use the multiplication property of equality to isolate } y. \text{ Multiply both sides by the reciprocal of } \tfrac{3}{4}, \text{ which is } \tfrac{4}{3}.$$

$$\left(\dfrac{4}{3} \cdot \dfrac{3}{4}\right)y = \dfrac{4}{3}(-7) \qquad \text{Use the associative property of multiplication to regroup.}$$

$$1y = \dfrac{4}{3}(-7) \qquad \text{The product of a number and its reciprocal is 1: } \tfrac{4}{3} \cdot \tfrac{3}{4} = 1.$$

$$y = -\dfrac{28}{3} \qquad \text{On the left side, 1y = y. On the right side, multiply.}$$

Check:

$$\dfrac{3}{4}y = -7 \qquad \text{This is the original equation.}$$

$$\dfrac{3}{4}\left(-\dfrac{28}{3}\right) \stackrel{?}{=} -7 \qquad \text{Substitute } -\tfrac{28}{3} \text{ for } y.$$

$$-\dfrac{\overset{1}{\cancel{3}} \cdot \overset{1}{\cancel{4}} \cdot 7}{\cancel{4} \cdot \cancel{3}} \stackrel{?}{=} -7 \qquad \begin{array}{l}\text{Multiply the numerators and the denominators.}\\ \text{Factor 28 as } 4 \cdot 7 \text{ and simplify.}\end{array}$$

$$-7 = -7 \qquad \text{True}$$

The solution is $-\dfrac{28}{3}$ and the solution set is $\left\{-\dfrac{28}{3}\right\}$.

> **Self Check 3**   Solve: $\frac{2}{3}b - 3 = -15$
> **Now Try**   Problem 35

The equation in Example 3 can be solved using an alternate two-step approach.

$$\dfrac{3}{4}y = -7 \qquad \text{This is the equation to solve.}$$

$$4\left(\dfrac{3}{4}y\right) = 4(-7) \qquad \text{Multiply both sides by 4 to undo the division by 4.}$$

$$3y = -28 \qquad \text{Simplify: } 4\left(\tfrac{3}{4}y\right) = \tfrac{4}{1}\left(\tfrac{3}{4}y\right) = \dfrac{\overset{1}{\cancel{4}} \cdot 3}{1 \cdot \cancel{4}}y = 3y.$$

$$\dfrac{3y}{3} = -\dfrac{28}{3} \qquad \text{To isolate } y, \text{ undo the multiplication by 3 by dividing both sides by 3.}$$

$$y = -\dfrac{28}{3}$$

Simplify Expressions to Solve Equations.

To solve more complicated equations, we often need to combine like terms.

EXAMPLE 4   Solve: $4b - 7 + 2b = 1 + 2b + 8$

Strategy   We will combine like terms on each side of the equation and then eliminate $2b$ from the right side by subtracting $2b$ from both sides.

Why   To solve for b, all the terms containing b must be on the same side of the equation.

Solution

$$4b - 7 + 2b = 1 + 2b + 8 \qquad \text{This is the equation to solve.}$$
$$6b - 7 = 2b + 9 \qquad \text{Combine like terms: } 4b + 2b = 6b \text{ and } 1 + 8 = 9.$$

We note that terms involving b appear on both sides of the equation. To isolate b on the left side, we need to eliminate $2b$ on the right side.

$$6b - 7 = 2b + 9$$
$$6b - 7 - 2b = 2b + 9 - 2b \qquad \text{Subtract } 2b \text{ from both sides.}$$
$$4b - 7 = 9 \qquad \text{Combine like terms on each side: } 6b - 2b = 4b \text{ and } 2b - 2b = 0.$$
$$4b - 7 + 7 = 9 + 7 \qquad \text{To undo the subtraction of 7, add 7 to both sides.}$$
$$4b = 16 \qquad \text{Simplify each side of the equation.}$$
$$b = 4 \qquad \text{To isolate } b, \text{ undo the multiplication by 4 by dividing both sides by 4.}$$

Check:
$$4b - 7 + 2b = 1 + 2b + 8 \qquad \text{This is the original equation.}$$
$$4(4) - 7 + 2(4) \stackrel{?}{=} 1 + 2(4) + 8 \qquad \text{Substitute 4 for } b.$$
$$16 - 7 + 8 \stackrel{?}{=} 1 + 8 + 8$$
$$17 = 17 \qquad \text{True}$$

> **Caution**
> When checking solutions, always use the original equation.

The solution is 4 and the solution set is $\{4\}$.

Self Check 4   Solve: $-6t - 16 + 6t = 1 + 2t - 5$

Now Try   **Problem 49**

EXAMPLE 5   Solve: **a.** $7(a - 2) = 8$ **b.** $d - 3(d - 7) = 2(4d + 10)$
c. $6[x - (2 - x)] = -4(8x + 3)$

Strategy   We will use the distributive property to remove all sets of parentheses (and brackets), simplify each side of the equation by combining like terms, and isolate the variable.

Why   It's best to simplify each side of an equation before isolating the variable.

Solution

a. $7(a - 2) = 8 \qquad \text{This is the equation to solve.}$

$7a - 14 = 8 \qquad \text{Distribute the multiplication by 7.}$

$$7a - 14 + 14 = 8 + 14 \qquad \text{To undo the subtraction of 14, add 14 to both sides.}$$

$$7a = 22$$

$$\frac{7a}{7} = \frac{22}{7} \qquad \text{To isolate } a, \text{ undo the multiplication by 7 by dividing both sides by 7.}$$

$$a = \frac{22}{7}$$

Check:

$$7(a - 2) = 8 \qquad \text{This is the original equation.}$$

$$7\left(\frac{22}{7} - 2\right) \overset{?}{=} 8 \qquad \text{Substitute } \frac{22}{7} \text{ for } a.$$

$$7\left(\frac{22}{7} - \frac{14}{7}\right) \overset{?}{=} 8 \qquad \text{Within the parentheses, express 2 as a fraction that has the LCD for its denominator: } 2 = \frac{14}{7}.$$

$$7\left(\frac{8}{7}\right) \overset{?}{=} 8 \qquad \text{Subtract the fractions.}$$

$$8 = 8 \qquad \text{True}$$

The solution is $\frac{22}{7}$ and the solution set is $\left\{\frac{22}{7}\right\}$.

<table>
<tr><td>

Success Tip

We could have eliminated $8d$ from the right side by subtracting $8d$ from both sides:

$$-2d + 21 = 8d + 20$$
$$-2d + 21 - 8d = 8d + 20 - 8d$$
$$-10d + 21 = 20$$

However, it is usually easier to isolate the variable term on the side that will result in a *positive* coefficient, as we did.

</td></tr>
</table>

b.

$$d - 3(d - 7) = 2(4d + 10) \qquad \text{This is the equation to solve.}$$

$$d - 3d + 21 = 8d + 20 \qquad \text{Distribute the multiplication by } -3 \text{ and by 2.}$$

$$-2d + 21 = 8d + 20 \qquad \text{Combine like terms: } d - 3d = -2d.$$

$$-2d + 21 + 2d = 8d + 20 + 2d \qquad \text{To eliminate the term } -2d \text{ on the left side, add } 2d \text{ to both sides.}$$

$$21 = 10d + 20 \qquad \text{Combine like terms on both sides: } -2d + 2d = 0.$$

$$21 - 20 = 10d + 20 - 20 \qquad \text{To undo the addition of 20, subtract 20 from both sides.}$$

$$1 = 10d \qquad \text{Combine like terms on both sides: } 20 - 20 = 0.$$

$$\frac{1}{10} = \frac{10d}{10} \qquad \text{To isolate } d, \text{ undo the multiplication by 10 by dividing both sides by 10.}$$

$$\frac{1}{10} = d$$

$$d = \frac{1}{10} \qquad \text{Reverse the sides of the equation so that } d \text{ is on the left.}$$

Check: To simplify the computations, we can use the decimal equivalent of $\frac{1}{10}$, which is 0.1, in the check.

$$d - 3(d - 7) = 2(4d + 10) \qquad \text{This is the original equation.}$$

$$0.1 - 3(0.1 - 7) \overset{?}{=} 2[4(0.1) + 10] \qquad \text{Substitute 0.1 for } d.$$

$$0.1 - 3(-6.9) \overset{?}{=} 2[0.4 + 10]$$

$$0.1 + 20.7 \overset{?}{=} 2(10.4)$$

$$20.8 = 20.8 \qquad \text{True}$$

The solution is $\frac{1}{10}$ or 0.1.

c. To simplify the expression on the left side of the equation, we will work from the innermost grouping symbols (the parentheses) to the outermost grouping symbols (the brackets).

$$6[x - (2 - x)] = -4(8x + 3) \qquad \text{This is the equation to solve.}$$

$$6[x - 2 + x] = -32x - 12 \qquad \text{Within the parentheses, distribute the multiplication by } -1. \text{ On the right side, distribute the multiplication by } -4.$$

$$6[2x - 2] = -32x - 12$$ Combine like terms within the brackets.

$$12x - 12 = -32x - 12$$ Distribute the multiplication by 6.

$$12x - 12 + 32x = -32x - 12 + 32x$$ To eliminate the term $-32x$ on the right side, add $32x$ to both sides.

$$44x - 12 = -12$$ Combine like terms.

$$44x - 12 + 12 = -12 + 12$$ To undo the subtraction of 12, add 12 to both sides.

$$44x = 0$$ Simplify each side of the equation.

$$\frac{44x}{44} = \frac{0}{44}$$ To isolate x, undo the multiplication by 44 by dividing both sides by 44.

$$x = 0$$

Verify that 0 is the solution by checking it in the original equation.

▷ **Self Check 5**   Solve: **a.** $-2(x + 3) = 18$
b. $y + 9(y - 5) = 5(4y + 1)$
c. $10[h - (1 - 3h)] = -2(5h + 25)$

Now Try   **Problems 55 and 61**

In general, we will follow these steps to solve linear equations in one variable. Not every step is needed to solve every equation.

Strategy for Solving Linear Equations in One Variable

1. **Clear the equation of fractions or decimals:** Multiply both sides by the LCD to clear fractions or multiply both sides by a power of 10 to clear decimals.
2. **Simplify each side of the equation:** Use the distributive property to remove parentheses and combine like terms on each side.
3. **Isolate the variable term on one side:** Add (or subtract) to get the variable term on one side of the equation and a number on the other using the addition (or subtraction) property of equality.
4. **Isolate the variable:** Multiply (or divide) to isolate the variable using the multiplication (or division) property of equality.
5. **Check the result:** Substitute the proposed solution for the variable in the *original* equation to see if a true statement results.

④   Clear Equations of Fractions and Decimals.

Since equations are often easier to solve when they don't contain fractions, we will use the multiplication property of equality to clear an equation of fractions before we solve it. To do so, we will multiply both sides of the equation by the least common denominator of the fractions contained within the equation.

EXAMPLE 6   Solve: $\frac{1}{3}(2x - 1) = \frac{5}{4}x + \frac{31}{12}$

Strategy   We will follow the steps of the equation-solving strategy.

Why   This is the best way to solve a linear equation in one variable.

Solution

Step 1: We can clear the equation of fractions by multiplying both sides by the least common denominator (LCD) of $\frac{1}{3}$, $\frac{5}{4}$, and $\frac{31}{12}$, which is 12.

$$\frac{1}{3}(2x - 1) = \frac{5}{4}x + \frac{31}{12}$$ *This is the equation to solve.*

$$12\left[\frac{1}{3}(2x - 1)\right] = 12\left[\frac{5}{4}x + \frac{31}{12}\right]$$ *To eliminate the fractions, multiply both sides by the LCD, 12.*

$$4(2x - 1) = 12 \cdot \frac{5}{4}x + 12 \cdot \frac{31}{12}$$ *On the left side, multiply: $12 \cdot \frac{1}{3} = 4$. On the right side, distribute the multiplication by 12.*

$$4(2x - 1) = 15x + 31$$ *Perform the multiplications on the right side.*

> **Success Tip**
>
> Before multiplying both sides of an equation by the LCD, frame the left side and frame the right side with parentheses or brackets.

Step 2: We remove parentheses.

$$8x - 4 = 15x + 31$$ *Distribute the multiplication by 4.*

Step 3: To get the variable term on the right side and the constant on the left side, subtract $8x$ and 31 from both sides.

$$8x - 4 - 8x - 31 = 15x + 31 - 8x - 31$$
$$-35 = 7x$$ *Simplify each side of the equation.*

Step 4: To isolate the variable, undo the multiplication by 7 by dividing both sides by 7.

$$\frac{-35}{7} = \frac{7x}{7}$$ *Divide both sides by 7.*

$$-5 = x$$

$$x = -5$$ *Reverse the sides of the equation so that x is on the left.*

Step 5: We check by substituting -5 for x in the original equation and simplifying:

$$\frac{1}{3}(2x - 1) = \frac{5}{4}x + \frac{31}{12}$$

$$\frac{1}{3}[2(-5) - 1] \stackrel{?}{=} \frac{5}{4}(-5) + \frac{31}{12}$$

$$\frac{1}{3}[-11] \stackrel{?}{=} -\frac{25}{4} + \frac{31}{12}$$

$$-\frac{11}{3} \stackrel{?}{=} -\frac{75}{12} + \frac{31}{12}$$

$$-\frac{11}{3} \stackrel{?}{=} -\frac{44}{12}$$

$$-\frac{11}{3} = -\frac{11}{3}$$ *True*

The solution is -5.

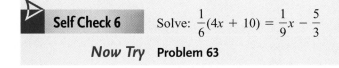

Self Check 6   Solve: $\frac{1}{6}(4x + 10) = \frac{1}{9}x - \frac{5}{3}$

Now Try **Problem 63**

EXAMPLE 7   Solve: $\dfrac{x+2}{5} - 4x = \dfrac{8}{5} - \dfrac{x+9}{2}$

Strategy   We will follow the steps of the equation-solving strategy.

Why   This is the best way to solve a linear equation in one variable.

Solution   Some of the steps used to solve an equation can be done in your head, as you will see in this example.

$$\frac{x+2}{5} - 4x = \frac{8}{5} - \frac{x+9}{2}$$

$$10\left(\frac{x+2}{5} - 4x\right) = 10\left(\frac{8}{5} - \frac{x+9}{2}\right)$$
To clear the equation of the fractions, multiply both sides by the LCD, 10.

$$10\cdot\frac{x+2}{5} - 10\cdot 4x = 10\cdot\frac{8}{5} - 10\cdot\frac{x+9}{2}$$
On each side, distribute the 10.

$$2(x+2) - 40x = 2(8) - 5(x+9)$$
Perform each multiplication by 10.

$$2x + 4 - 40x = 16 - 5x - 45$$
On each side, remove parentheses.

$$-38x + 4 = -5x - 29$$
On each side, combine like terms.

$$33 = 33x$$
Add 38x and 29 to both sides. Since these steps can be done mentally, we don't show them.

$$1 = x$$
Divide both sides by 33. This step is also done mentally.

$$x = 1$$

The solution is 1. Check by substituting it for x in the original equation.

Self Check 7   Solve: $\dfrac{a+3}{2} + 2a = \dfrac{3}{2} - \dfrac{a+27}{5}$

Now Try   **Problem 71**

For more complicated equations involving decimals, we can multiply both sides of the equation by a power of 10 to clear the equation of decimals.

EXAMPLE 8   Solve: $0.04(12) + 0.01x = 0.02(12 + x)$

Strategy   To clear the equation of decimals, we will multiply both sides by a carefully chosen power of 10.

Why   It's easier to solve an equation that involves only integers.

Solution   The equation contains the decimals 0.04, 0.01, and 0.02. Multiplying both sides by $10^2 = 100$ changes the decimals in the equation to integers.

$$0.04(12) + 0.01x = 0.02(12 + x)$$

$$100[0.04(12) + 0.01x] = 100[0.02(12 + x)]$$
To make 0.04, 0.01, and 0.02 integers, multiply both sides by 100.

$$100\cdot 0.04(12) + 100\cdot 0.01x = 100\cdot 0.02(12 + x)$$
On the left side, distribute the multiplication by 100.

Success Tip
When we write the decimals in the equation as fractions, it becomes more apparent why it is helpful to multiply both sides by the LCD, 100.

$$\frac{4}{100}(12) + \frac{1}{100}x = \frac{2}{100}(12 + x)$$

$$4(12) + 1x = 2(12 + x)$$ Perform each multiplication by 100.

$$48 + x = 24 + 2x$$ Remove parentheses.

$$48 + x - 24 - x = 24 + 2x - 24 - x$$ To isolate the variable term on the right side, subtract 24 and x from both sides.

$$24 = x$$ Simplify each side.

$$x = 24$$

Verify that 24 is the solution by substituting it for *x* in the original equation.

 Self Check 8   Solve: $0.08x + 0.07(15,000 - x) = 1,110$

Now Try   **Problem 75**

5 **Identify Identities and Contradictions.**

The equations discussed so far are called **conditional equations.** For these equations, some numbers satisfy the equation and others do not. An **identity** is an equation that is satisfied by every number for which both sides of the equation are defined.

EXAMPLE 9   Solve: $-2(x - 1) - 4 = -4(1 + x) + 2x + 2$

Strategy   We will follow the steps of the equation-solving strategy.

Why   This is the best way to solve a linear equation in one variable.

Solution

$$-2(x - 1) - 4 = -4(1 + x) + 2x + 2$$

$$-2x + 2 - 4 = -4 - 4x + 2x + 2$$ Use the distributive property.

$$-2x - 2 = -2x - 2$$ On each side, combine like terms.

$$-2x - 2 + 2x = -2x - 2 + 2x$$ To attempt to isolate the variable on one side of the equation, add 2x to both sides.

$$-2 = -2$$ True

Success Tip

We know the given equation is an identity because in Step 3 we see that it is equivalent to the equation $-2x - 2 = -2x - 2$, which is true for all values of *x*.

The terms involving *x* drop out. The resulting true statement indicates that the original equation is true for every value of *x*. The solution set is the set of real numbers denoted $\mathbb{R}$. The equation is an identity.

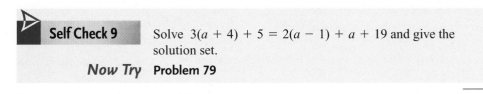 **Self Check 9**   Solve $3(a + 4) + 5 = 2(a - 1) + a + 19$ and give the solution set.

Now Try   **Problem 79**

A **contradiction** is an equation that is never true.

EXAMPLE 10 Solve: $-6.2(-x - 1) - 4 = 4.2x - (-2x)$

Strategy We will follow the steps of the equation-solving strategy.

Why This is the best way to solve a linear equation in one variable.

Solution

$$-6.2(-x - 1) - 4 = 4.2x - (-2x)$$

$$6.2x + 6.2 - 4 = 4.2x + 2x$$ On the left side, remove parentheses. On the right side, write the subtraction as addition of the opposite.

$$6.2x + 2.2 = 6.2x$$ On each side, combine like terms.

$$6.2x + 2.2 - \mathbf{6.2x} = 6.2x - \mathbf{6.2x}$$ To attempt to isolate the variable on one side of the equation, subtract 6.2x from both sides.

$$2.2 = 0$$ False

> **The Language of Algebra**
> *Contradiction* is a form of the word *contradict,* meaning conflicting ideas. During a trial, evidence might be introduced that *contradicts* the testimony of a witness.

The terms involving x drop out. The resulting false statement indicates that no value for x makes the original equation true. The solution set contains no elements and can be denoted as the **empty set** { } or the **null set** $\varnothing$. The equation is a contradiction.

Self Check 10 Solve: $3(a + 4) + 2 = 2(a - 1) + a + 19$

Now Try Problem 81

ANSWERS TO SELF CHECKS **1.** Yes **2. a.** -5 **b.** -1.3 **3.** -18 **4.** -6 **5. a.** -12 **b.** -5 **c.** $-\frac{4}{5}$ **6.** -6 **7.** -2 **8.** 6,000 **9.** All real numbers, $\mathbb{R}$ **10.** No solution, $\varnothing$

STUDY SET
1.5

VOCABULARY

Fill in the blanks.

1. An _____ is a statement that two expressions are equal.

2. $2x + 1 = 4$ and $5(y - 3) = 8$ are examples of _____ equations in one variable.

3. If a number is substituted for a variable in an equation and the equation is true, we say that the number _____ the equation.

4. If two equations have the same solution set, they are called _____ equations.

5. An equation that is true for all values of its variable is called an _____.

6. An equation that is not true for any values of its variable is called a _____.

CONCEPTS

Fill in the blanks.

7. If $a = b$, then $a + c = b + $ ___ and $a - c = b - $ ___. _____ (or subtracting) the same number to (or from) _____ sides of an equation does not change the solution.

8. If $a = b$, then $ca = $ ___ and $\dfrac{a}{c} = \dfrac{b}{c}$. _____ (or dividing) both sides of an equation by the _____ nonzero number does not change the solution.

9. Solve each equation mentally.
   a. $x + 3 = 6$ **b.** $x - 3 = 6$
   c. $3x = 6$ **d.** $\dfrac{x}{3} = 6$

10. a. When solving $\frac{x+1}{3} - \frac{2}{15} = \frac{x-1}{5}$, why would we multiply both sides by 15?

    b. When solving $1.45x - 0.5(1 - x) = 0.7x$, why would we multiply both sides by 100?

11. a. Suppose you solve a linear equation in one variable, the variable drops out, and you obtain $8 = 8$. What is the solution set? What symbol is used to represent the solution set?

    b. Suppose you solve a linear equation in one variable, the variable drops out, and you obtain $8 = 7$. What is the solution set? What symbol is used to represent the solution set?

12. a. Simplify: $5y + 2 - 3y$

    b. Solve: $5y + 2 - 3y = 8$

    c. Evaluate $5y + 2 - 3y$ for $y = 8$.

    d. Check: Is -1 a solution of $5y + 2 - 3y = 8$?

NOTATION

Complete the solution to solve the equation. Then check the result.

13.
$$-2(x + 7) = 20$$
$$\underline{\hspace{1cm}} - 14 = 20$$
$$-2x - 14 + \underline{\hspace{1cm}} = 20 + \underline{\hspace{1cm}}$$
$$-2x = 34$$
$$\frac{-2x}{\underline{\hspace{0.5cm}}} = \frac{34}{\underline{\hspace{0.5cm}}}$$
$$x = -17$$

Check:
$$-2(x + 7) = 20$$
$$-2(\underline{\hspace{1cm}} + 7) \overset{?}{=} 20$$
$$-2(\underline{\hspace{1cm}}) \quad 20$$
$$\underline{\hspace{1cm}} = 20$$

The solution is $\underline{\hspace{0.5cm}}$.

14. Fill in the blanks to make the statements true.

    a. $-x = \underline{\hspace{0.5cm}} x$ **b.** $\frac{2t}{3} = \underline{\hspace{0.5cm}} t$

GUIDED PRACTICE

Determine whether 5 is a solution of each equation. See Example 1.

15. $3x + 2 = 17$ **16.** $7x - 2 = 53 - 5x$

17. $3(2m - 3) = 15$ **18.** $\frac{3}{5}p - 5 = -2$

Solve each equation. Check each result. See Example 2.

19. $2x - 12 = 0$ **20.** $3x - 24 = 0$

21. $8k - 2 = 13$ **22.** $3x + 1 = 3$

23. $\frac{x}{4} - 6 = 1$ **24.** $\frac{m}{3} + 10 = 8$

25. $\frac{y}{6} - 7 = -12$ **26.** $\frac{a}{8} + 1 = -10$

27. $1.6a + (-4) = 0.032$ **28.** $5.51 = 0.05y + (-9)$

29. $0.7 - 4y = 1.7$ **30.** $0.3 - 2x = -0.9$

31. $-x + 12 = -17$ **32.** $6 = -x + 41$

33. $-6 - y = -13$ **34.** $-1 - h = -9$

Solve each equation. Check each result. See Example 3.

35. $\frac{2}{3}c = 10$ **36.** $\frac{9}{7}d = 81$

37. $-\frac{4}{5}s = 2$ **38.** $-\frac{9}{8}s = 3$

39. $-\frac{7}{16}w - 26 = -19$ **40.** $-\frac{5}{8}a - 20 = -10$

41. $\frac{5}{6}k - 7.5 = 7.5$ **42.** $\frac{2}{5}c - 12.2 = 1.8$

Solve each equation. Check each result. See Example 4.

43. $8m + 44 = 4m$ **44.** $9n + 36 = 6n$

45. $60t - 50 = 15t - 5$ **46.** $100s - 75 = 50s + 75$

47. $9.8 - 16r = -15.7 - r$ **48.** $15s + 8.1 - 2s = 8.1 - s$

49. $8b - 2 + b = 5b + 15$ **50.** $w + 7 + 3w = 4 + 10w$

51. $a + 18 = 5a - 3 + a$ **52.** $4a - 21 - a = -2a - 7$

53. $8x = x$ **54.** $-z = 5z$

Solve each equation. Check each result. See Example 5.

55. $3(k - 4) = -36$ **56.** $4(x + 6) = 84$

57. $2(a - 5) - (3a + 1) = 0$

58. $8(3a - 5) - 4(2a + 3) = 12$

59. $9(x - 2) = -6(4 - x) + 18$

60. $3(x + 2) - 2 = -(5 + x) + x$

61. $12 + 3(x - 4) - 21 = 5[5 - 4(4 - x)]$

62. $1 + 3[-2 + 6(4 - 2x)] = -(x + 3)$

Solve each equation. Check each result. See Examples 6 and 7.

63. $\frac{1}{2}(a - 2) = \frac{2}{3}a - 6$ **64.** $\frac{2}{3}(b + 3) = \frac{5}{4}b + \frac{17}{12}$

65. $\frac{1}{2}(3y + 2) - \frac{5}{8} = \frac{3}{4}y$ **66.** $-\frac{3}{4}(4c - 3) + \frac{7}{8}c = \frac{19}{16}$

67. $\frac{3}{4}x - 5 = \frac{2}{3}x + \frac{1}{4}$ **68.** $\frac{3}{5}x + \frac{7}{10} = x - \frac{4}{5}$

69. $\frac{1}{2}b - \frac{19}{6} = \frac{1}{3}b + \frac{5}{6}$ **70.** $\frac{1}{2}w - \frac{7}{6} = \frac{53}{6} - \frac{1}{3}w$

71. $\frac{a + 1}{3} + \frac{a - 1}{5} = \frac{2}{15}$

72. $\frac{2z + 3}{3} + \frac{3z - 4}{6} = \frac{z - 2}{2}$

73. $\dfrac{3 + p}{3} - 4p = 1 - \dfrac{p + 7}{2}$

74. $\dfrac{4 - t}{2} - \dfrac{3t}{5} = 2 + \dfrac{t + 1}{3}$

Solve each equation. Check each result. See Example 8.

75. $0.45 = 16.95 - 0.25(75 - 3x)$

76. $0.02x + 0.0175(15,000 - x) = 277.5$

77. $0.04(12) + 0.01t - 0.02(12 + t) = 0$

78. $0.25(t + 32) = 3.2 + t$

Solve each equation. If the equation is an identity or a contradiction, so indicate. See Examples 9 and 10.

79. $8x + 3(2 - x) = 5x + 6$

80. $4(2 - 3t) + 6t = -6t + 8$

81. $2x - 6 = -2x + 4(x - 2)$

82. $3(x - 4) + 6 = -2(x + 4) + 5x$

83. $2(x - 3) = \dfrac{3}{2}(x - 4) + \dfrac{x}{2}$

84. $y + \dfrac{1}{2} = \dfrac{5}{2}(0.2y + 1) - \dfrac{1}{2}(4 - y)$

85. $-3x = -2x + 1 - (5 + x)$

86. $5(y + 2) + 7 - 3y = 2(y + 9)$

TRY IT YOURSELF

Solve each equation, if possible.

87. $2(2x + 1) = x + 15 + 2x$ **88.** $-2(x + 5) = x + 30 - 2x$

89. $\dfrac{5}{2}a - 12 = \dfrac{1}{3}a + 1$ **90.** $3(x - 2) + 4 = 3x - 2$

91. $\dfrac{4}{5}a = -12$ **92.** $4j + 12.54 = 18.12$

93. $0.06(a + 200) + 0.1a = 172$

94. $0.03x + 0.05(6,000 - x) = 280$

95. $-4[p - (3 - p)] = 3(6p - 2)$

96. $2[5(4 - a) + 2(a - 1)] = 3 - a$

97. $2(x - 2) = \dfrac{2}{3}(3x + 8) - 2$

98. $5 - \dfrac{x + 2}{3} = 7 - x$

99. $13.5y + 16.2 = 0$ **100.** $\dfrac{7}{3}y + 1 = 0$

101. $\dfrac{4}{5}(x + 5) = \dfrac{7}{8}(3x + 23) - 7$

102. $\dfrac{2}{3}(2x + 2) + 4 = \dfrac{1}{6}(5x + 29)$

103. $\dfrac{t - 2}{5} + 5t = \dfrac{7}{5} - \dfrac{t - 2}{2}$

104. $\dfrac{2}{3}(3m - 2) = \dfrac{3}{4}m + \dfrac{11}{12}$

105. $5c - 8 + 3c = 10 + 2c - 3$

106. $6 + 4t - 1 = 6 - 15t + 12t - 8$

WRITING

107. What does it mean to *solve an equation?*

108. Why doesn't the equation $x = x + 1$ have a real-number solution?

109. What is an identity? Give an example.

110. When solving a linear equation in one variable, the objective is to isolate the variable on one side of the equation. What does that mean?

REVIEW

Use variables to state each property of real numbers.

111. **a.** Commutative property of addition
b. Associative property of multiplication
c. Distributive property of multiplication over addition

112. **a.** Additive inverse property
b. Multiplicative inverse property

113. **a.** Additive identity property
b. Multiplicative identity property

114. **a.** Division of 0
b. Division by 0

CHALLENGE PROBLEMS

115. Find the value of k that makes 4 a solution of the following linear equation in x.

$$k + 3x - 6 = 3kx - k + 16$$

116. Solve: $0.75(x - 5) - \dfrac{4}{5} = \dfrac{1}{6}(3x + 1) + 3.2$

SECTION 1.6
Solving Formulas; Geometry

Objectives

❶ Find the perimeter, area, and volume of geometric figures.
❷ Solve for a specified variable.
❸ Solve application problems using formulas.

A **formula** is an equation that states a relationship between two or more variables. Formulas are used in business, science, banking, and many other fields. A large collection of formulas are associated with geometric figures such as squares, rectangles, circles, and cylinders.

❶ **Find the Perimeter, Area, and Volume of Geometric Figures.**

The Language of Algebra
When you hear the word perimeter, think of the distance around the "rim" of a flat figure.

To find the **perimeter** of a plane (two-dimensional, flat) geometric figure, we find the distance around the figure by computing the sum of the lengths of the sides. Perimeter is measured in American units of inches, feet, yards, and in metric units such as millimeters, meters, and kilometers. Several perimeter formulas are shown in the margin.

EXAMPLE 1   *Landscaping.*   Find the number of feet of edging needed to outline a square flowerbed having sides that are 6.5 feet long.

Strategy   We will substitute the length of a side of the flowerbed into the formula for the perimeter of a square, $P = 4s$, and find P.

Why   Since the edging outlines the flowerbed, the concept of perimeter applies.

Solution

$P = 4s$ This is the formula for the perimeter of a square.

$P = 4(6.5)$ Substitute 6.5 for *s*, the length of one side of the square.

$\quad = 26$

26 feet of edging is needed to outline the flowerbed.

Perimeter formulas

$P = 2l + 2w$ (rectangle)
$P = 4s$ (square)
$P = a + b + c$ (triangle)

Turn to the inside back cover for a complete list of geometric formulas.

▷ **Self Check 1**   Find the amount of fencing needed to enclose a rectangular lot that is 205.5 feet long and 165 feet wide.

Now Try   **Problem 11**

The **area** of a plane (two-dimensional, flat) geometric figure is the amount of surface that it encloses. Area is measured in square units, such as square inches, square feet, square yards, and square meters (written as in.^{2}, ft^2, yd^2, and m^2, respectively). Several area formulas are shown in the margin on the next page.

Area formulas

$A = lw$ (rectangle)

$A = s^2$ (square)

$A = \dfrac{1}{2}bh$ (triangle)

$A = \dfrac{1}{2}h(b_1 + b_2)$ (trapezoid)

Turn to the inside back cover for a complete list of geometric formulas.

EXAMPLE 2   ***Band-aids.***   Find the amount of skin covered by the rectangular-shaped bandage.

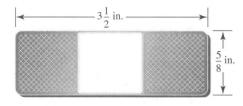

Strategy   We will substitute the length and width of the bandage into the formula for the area of a rectangle, $A = lw$, and find A.

Why   The concept of area is suggested by the phrase *the amount of skin covered.*

Solution

$A = lw$ This is the formula for the area of a rectangle.

$A = 3\dfrac{1}{2}\left(\dfrac{5}{8}\right)$ Substitute $3\frac{1}{2}$ for l, the length of the bandage, and $\frac{5}{8}$ for w, the width.

$A = \dfrac{7}{2}\left(\dfrac{5}{8}\right)$ Write $3\frac{1}{2}$ as a fraction: $3\frac{1}{2} = \frac{7}{2}$.

$A = \dfrac{35}{16}$ Multiply the numerators.
Multiply the denominators.

The bandage covers $\frac{35}{16}$ or $2\frac{3}{16}$ in.^{2} (square inches) of skin.

Caution   When finding area, remember to write the appropriate *square units* in the answer.

Self Check 2   A solar panel is in the shape of a triangle. Its base is 79 centimeters long and the height is 54 centimeters. In square centimeters, how large a surface do the sun's rays strike?

Now Try   **Problem 15**

EXAMPLE 3   **a.** Find the circumference of a circle with diameter 20 feet. Round to the nearest tenth of a foot.   **b.** Find the area of the circle. Round to the nearest tenth of a square foot.

Strategy   We will substitute the given values into the formulas $C = \pi D$ and $A = \pi r^2$ and find C and A.

Why   In the formulas, the variable C represents the circumference of the circle and A represents the area.

Solution

a. Recall that the circumference of a circle is the distance around it. To find the circumference C of a circle with diameter D equal to 20 ft, we proceed as follows.

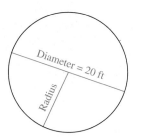

$C = \pi D$ This is the formula for the circumference of a circle. πD means $\pi \cdot D$.

$C = \pi(20)$ Substitute 20 for D.

Circle formulas

$D = 2r$ (diameter)

$r = \dfrac{1}{2}D$ (radius)

$C = 2\pi r = \pi D$ (circumference)

$A = \pi r^2$ (area)

Turn to the inside back cover for a complete list of geometric formulas.

$= 20\pi$ — The exact circumference of the circle is 20π ft.

≈ 62.83185307 — To use a scientific calculator to approximate the circumference, enter π ✕ 20 = . If you do not have a calculator, use 3.14 as an approximation of π. (Answers may vary slightly depending on which approximation of π is used.)

The circumference is exactly 20π ft. Rounded to the nearest tenth, this is 62.8 ft.

b. The radius r of the circle is one-half the diameter, or 10 feet. To find the area A of the circle, we proceed as follows.

$A = \pi r^2$ — This is the formula for the area of a circle. πr^2 means $\pi \cdot r^2$.

$A = \pi(10)^2$ — Substitute 10 for r.

$= 100\pi$ — Evaluate the exponential expression. The exact area is 100π ft^2.

≈ 314.1592654 — To use a scientific calculator to approximate the area, enter 100 ✕ π = .

The area is exactly 100π ft^2. To the nearest tenth, the area is 314.2 ft^2.

Self Check 3   The diameter of a U.S. penny is 0.75 inch. Find the circumference and the area (of one side) of a penny. Round to the nearest hundredth.

Now Try   Problems 19 and 23

The **volume** of a three-dimensional geometric solid is the amount of space it encloses. Volume is measured in cubic units, such as cubic inches, cubic feet, and cubic meters (written as in.^{3}, ft^3, and m^3, respectively). Several volume formulas are shown in the margin.

EXAMPLE 4   *Timers.*   Find the amount of sand in the hourglass.

Strategy   We will substitute the given values into the formula for the volume of a cone, $V = \frac{1}{3}\pi r^2 h$, and find V.

Why   To find the amount of sand in the hourglass, we need to find the amount of space that it occupies by finding a volume.

Volume formulas

$V = lwh$ (rectangular solid)

$V = s^3$ (cube)

$V = \dfrac{4}{3}\pi r^3$ (sphere)

$V = \pi r^2 h$ (cylinder)

$V = \dfrac{1}{3}Bh$* (pyramid)

*B represents the area of the base.
Turn to the inside back cover for a complete list of geometric formulas.

Solution   The sand is in the shape of a cone whose radius is one-half the diameter of the base of the hourglass and whose height is one-half the height of the hourglass. To find the amount of sand, we substitute 1 for r and 2.5 for h in the formula for the volume of a cone.

$V = \dfrac{1}{3}\pi r^2 h$ — This is the formula for the volume of a cone.

$V = \dfrac{1}{3}\pi(1)^2(2.5)$ — Substitute 1 for r, the radius of the circular base, and 2.5 for h, the height of the cone.

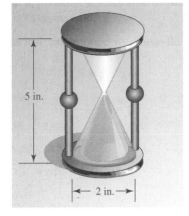
5 in.
2 in.

$$V = \frac{2.5\pi}{3}$$ Simplify. The exact volume of sand is $\frac{2.5\pi}{3}$ in.^{3}.

$$V \approx 2.617993878$$ Use a calculator.

There are exactly $\frac{2.5\pi}{3}$ in.^{3} of sand in the hourglass. Rounded to the nearest tenth, this is 2.6 in.^{3}.

Caution   When finding volume, remember to write the appropriate *cubic units* in your answer.

Self Check 4   Find the volume of a drinking straw that is 250 millimeters long with an inside diameter of 6 millimeters. Round to the nearest cubic millimeter.

Now Try   Problem 27

2   Solve for a Specified Variable.

The Language of Algebra
The word *specified* is a form of the word *specify*, which means to select something for a purpose. Here, we select a variable for the purpose of solving for it.

Real-world applications sometimes call for a formula solved for one variable to be solved for a different variable. To **solve a formula for a specified variable** means to isolate that variable on one side of the equation, with all other variables and constants on the opposite side.

EXAMPLE 5   Solve each formula for the specified variable.

a. $V = lwh$ for w The formula for the volume of a rectangular solid.

b. $A = \frac{1}{2}h(b_1 + b_2)$ for b_1 The formula for the area of a trapezoid.

c. $v_f = v_i + at$ for t A motion formula from physics.

Strategy   To solve for a specified variable, we treat it as if it were the only variable in the equation. To isolate this variable, we will use the same strategy that we used to solve linear equations in one variable. (See page 56 if you need to review the strategy.)

Why   We can solve a formula as if it were an equation in one variable because all the other variables are treated as if they were numbers (constants).

Solution

a.   To solve for w, we will isolate w on this side of the equation.

$$V = lwh$$

$$\frac{V}{lh} = \frac{lwh}{lh}$$ To isolate w, undo the multiplication by l and h by dividing both sides by lh.

$$\frac{V}{lh} = w$$ On the right side, remove the common factors of l and h: $\frac{\overset{1}{\cancel{l}}w\overset{1}{\cancel{h}}}{\underset{1}{\cancel{l}}\underset{1}{\cancel{h}}} = w$.

$$w = \frac{V}{lh}$$ Reverse the sides of the equation so that w is on the left.

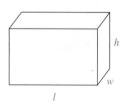
Rectangular solid

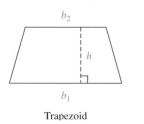

Trapezoid

Notation
The 1 and 2 in b_1 and b_2 are called **subscripts**. This notation allows us to distinguish between the variables b_1 and b_2, while still showing that each represents the length of a base of the trapezoid.

b.

To solve for b_1, we will isolate b_1 on this side of the equation.

$$A = \frac{1}{2}h(b_1 + b_2)$$ Read b_1 as "b-sub-one" and b_2 as "b-sub-two."

$$2 \cdot A = 2 \cdot \frac{1}{2}h(b_1 + b_2)$$ Multiply both sides by 2 to clear the equation of the fraction.

$$2A = h(b_1 + b_2)$$ Simplify each side of the equation.

$$2A = hb_1 + hb_2$$ Distribute the multiplication by h.

$$2A - hb_2 = hb_1$$ Subtract hb_2 from both sides to isolate the variable term hb_1 on the right side. This step is done mentally.

$$\frac{2A - hb_2}{h} = \frac{hb_1}{h}$$ To isolate b_1, undo the multiplication by h by dividing both sides by h.

$$\frac{2A - hb_2}{h} = b_1$$ On the right side, remove the common factor of h: $\frac{\overset{1}{\cancel{h}}b_1}{\cancel{h}}$

$$b_1 = \frac{2A - hb_2}{h}$$ Reverse the sides of the equation so that b_1 is on the left.

When solving formulas for a specified variable, there is often more than one way to express the result. In this case, we could perform the division by h on the right side term-by-term: $b_1 = \frac{2A}{h} - \frac{hb_2}{h}$. After removing the common factor of h in the numerator and denominator of the second fraction, we obtain the following equivalent form of the result: $b_1 = \frac{2A}{h} - b_2$.

Caution Do not try to simplify the result in the following way. It is incorrect because h is not a factor of the entire numerator.

$$b_1 = \frac{2A - \cancel{h}b_2}{\underset{1}{\cancel{h}}}$$

Notation
Variables are also used as subscripts. For example, in physics, the symbol v_i is used to represent initial velocity and v_f final velocity.

c.

To solve for t, we will isolate t on this side of the equation.

$$v_f = v_i + at$$ Read v_f as "v-sub-f and v_i as "v-sub-i."

$$v_f - v_i = at$$ To isolate the term at, subtract v_i from both sides. This step is done mentally.

$$\frac{v_f - v_i}{a} = t$$ To isolate t, undo the multiplication by a by dividing both sides by a. This step is done mentally.

$$t = \frac{v_f - v_i}{a}$$ Reverse the sides of the equation so that t is on the left.

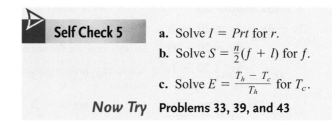

Self Check 5   **a.** Solve $I = Prt$ for r.

b. Solve $S = \frac{n}{2}(f + l)$ for f.

c. Solve $E = \frac{T_h - T_c}{T_h}$ for T_c.

Now Try Problems 33, 39, and 43

Now Try **Problem 83**

STUDY SET
1.6

VOCABULARY

Fill in the blanks.

1. A _____ is an equation that states a relationship between two or more variables.

2. To find the _____ of a plane geometric figure, such as a rectangle or triangle, we calculate the distance around the figure. The _____ of a plane geometric figure is the amount of surface that it encloses.

3. The _____ of a three-dimensional geometric solid is the amount of space it encloses.

4. To _____ a formula for a specified variable means to isolate that variable on one side of the equation, with all other variables and constants on the opposite side.

CONCEPTS

5. Determine which concept (perimeter, circumference, area, or volume) should be used to find each of the following situations. Then determine which unit of measurement, ft, ft^2, or ft^3, would be appropriate.

   a. The amount of ground covered by a lawn

   b. The amount of storage in a safe

   c. The distance traveled by a rider on a Ferris wheel

   d. The distance around a tennis court

6. The area of a circle is exactly 54π ft^2. Approximate the area to the nearest tenth of a square foot.

7. When solving formulas for a specified variable, there can be more than one way to express the result. Fill in the blanks to express this result in an equivalent form:

$$d = \frac{4m - at}{t}$$

$$d = \frac{4m}{t} - \frac{at}{t}$$

$$d = \frac{4m}{t} - \boxed{}$$

8. Fill in the blanks: To solve a formula for a specified variable, we treat it as if it were the _____ variable in the equation. We treat all other variables as if they were _____ (constants).

NOTATION

Complete the solution.

9. Solve $t = ad + bc$ for c.

$$t - \boxed{} = ad + bc - \boxed{}$$

$$t - ad = \boxed{}$$

$$\frac{t - ad}{\boxed{}} = \frac{bc}{\boxed{}}$$

$$\frac{t - ad}{b} = \boxed{}$$

$$c = \boxed{}$$

10. Fill in the blanks: In the notation b_1 and v_f, the number 1 and the variable f are called _____.

GUIDED PRACTICE

Find the perimeter of each figure. **See Example 1.**

11. A square with sides 2 yd long

12. A triangle with sides 1.8, 1.8, and 1.5 cm long

13. A trapezoid with parallel sides 10 in. and 15 in. long and the other two sides each 6 in. long

14. A parallelogram with two adjacent sides 50 m and 100 m long

Find the area of each figure. **See Example 2.**

15. A triangle with a base that is 2.4 ft long and height 8.5 ft

16. A rectangle with sides that measure $8\frac{1}{4}$ ft and $5\frac{1}{2}$ ft

17. A square with sides 17.2 mi long

18. A trapezoid whose parallel sides measure 8 cm and 12 cm and whose height is 10.5 cm.

Find the circumference of each circle to the nearest hundredth. See Example 3. (Answers may vary slightly depending on which approximation of π is used.)

19. A circle with diameter 7.5 in.

20. A circle with diameter $6\frac{1}{4}$ m

21. A circle with radius $2\frac{1}{2}$ ft

22. A circle with radius 12.3 yd

Find the area of each circle to the nearest tenth. See Example 3. (Answers may vary slightly depending on which approximation of π is used.)

23. A circle with radius 5.7 in.

24. A circle with radius $5\frac{3}{4}$ cm

25. A circle with diameter $10\frac{1}{2}$ ft

26. A circle with diameter 12.25 m

Find the volume of each figure to the nearest hundredth. See Example 4. (Answers may vary slightly depending on which approximation of π is used.)

27. A rectangular solid with dimensions 2.51 ft, 3.71 ft, and 10.21 ft

28. A pyramid whose base is a square with each side measuring 2.57 cm and with a height of 12.32 cm

29. A sphere with radius 5.78 meters

30. A cone whose base has a radius of 5.50 in. and whose height is 8.52 in.

Solve each formula for the specified variable. See Example 5.

31. $d = rt$ for t

32. $E = mc^2$ for m

33. $V = lwh$ for h

34. $I = Prt$ for t

35. $V = \frac{1}{3}\pi r^2 h$ for h

36. $A = \frac{1}{2}bh$ for b

37. $T = W + ma$ for W

38. $V = \frac{1}{3}Bh$ for B

39. $h = 48t + \frac{1}{2}at^2$ for a

40. $H = 17 - \frac{A}{2}$ for A

41. $A = \frac{1}{2}h(b_1 + b_2)$ for b_2

42. $\bar{v} = \frac{1}{2}(v + v_0)$ for v_0

43. $l = a + (n - 1)d$ for n

44. $P = 2(l + w)$ for l

45. $P = 2(w + h + l)$ for w

46. $P = 2(w + h + l)$ for h

Solve each formula for the indicated variable. See Example 6.

47. $\lambda = A(x + B)$ for A
   (λ is a letter from the Greek alphabet.)

48. $S = C(1 - r)$ for C

49. $T_f = T_a(1 - F)$ for T_a

50. $S = \frac{n}{2}(f + l)$ for n

51. $l = a + (n - 1)d$ for d

52. $S = \frac{n(a + l)}{2}$ for n

53. $v = \frac{1}{t}(d_1 - d_2)$ for t

54. $A = \frac{1}{2}(b_1 + b_2)h$ for h

Solve each equation for y. See Example 7.

55. $2x - 5y = 20$ **56.** $5x - 6y = 12$

57. $-4x = 12 + 3y$ **58.** $7x - 21 = -3y$

TRY IT YOURSELF

Solve for the specified variable.

59. $y = mx + b$ for x

60. $P = 2l + 2w$ for l

61. $L = 2d + 3.25(r + R)$ for R

62. $l = \frac{a - S + Sr}{r}$ for a

63. $s = \frac{1}{2}gt^2 + vt$ for g

64. $K = \frac{Mv_0^2 + Iw^2}{2}$ for I

65. $y - y_1 = m(x - x_1)$ for x

66. $s = v_o t - 16t^2$ for v_o

67. $G = U - TS + pV$ for S

68. $F = \frac{Gm_1 m_2}{r^2}$ for m_1

69. $PV = nrt$ for r

70. $P = s_1 + s_2 + b$ for s_2

71. $E = IR + Ir$ for R

72. $Ax + By = C$ for x

73. $A = \dfrac{1}{3}(s_1 + s_2 + s_3)$ for s_3

74. $P_1V_1T_2 = P_2V_2T_1$ for V_2

75. $S = \dfrac{n}{2}[2a + (n - 1)d]$ for d

76. $x^2 = 4py$ for p

77. $d = \dfrac{4}{3}\pi h$ for h

78. $I_Q = \dfrac{100M}{C}$ for C

APPLICATIONS

79. FLOOR MATS   What geometric concept applies when finding the length of the plastic trim around the cargo area floor mat? Estimate the amount of trim used.

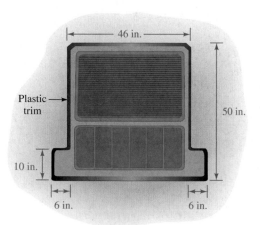

80. ALUMINUM FOIL   Find the number of *square feet* of aluminum foil on a roll if the dimensions printed on the box are $8\frac{1}{3}$ yards $\times$ 12 inches.

81. PAPER PRODUCTS   When folded, the paper sheet shown in the illustration in the next column forms a rectangular-shaped envelope. The formula

$$A = \dfrac{1}{2}h_1(b_1 + b_2) + b_3h_3 + \dfrac{1}{2}b_1h_2 + b_1b_3$$

gives the amount of paper (in square units) used in the design. Explain what each of the four terms in the formula finds. Then evaluate the formula for $b_1 = 6$, $b_2 = 2$, $b_3 = 3$, $h_1 = 2$, $h_2 = 2.5$, and $h_3 = 3$. All dimensions are in inches.

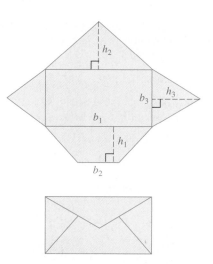

82. HOCKEY   A goal is scored in hockey when the puck, a vulcanized rubber disk 2.5 cm (1 in.) thick and 7.6 cm (3 in.) in diameter, is driven into the opponent's goal. Find the volume of a puck in cubic centimeters and cubic inches. Round to the nearest tenth.

83. CONVERTING TEMPERATURES   In preparing an American almanac for release in Europe, editors need to convert temperature ranges for the planets from degrees Fahrenheit to degrees Celsius. Solve the formula $F = \frac{9}{5}C + 32$ for C. Then use your result to make the conversions for the data shown in the table. Round to the nearest degree.

Planet	High °F	Low °F	High °C	Low °C
Mercury	810	−290		
Earth	136	−129		
Mars	63	−87		

84. ICE CREAM   If the two equal-sized scoops of ice cream melt completely into the cone, will they overflow the cone?

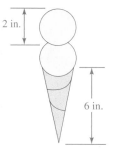

85. WIPER DESIGN   The area cleaned by the windshield wiper assembly shown in the illustration is given by the formula

$$A = \frac{d\pi(r_1^2 - r_2^2)}{360}$$

Engineers have determined the amount of windshield area that needs to be cleaned by the wiper for two different vehicles. Solve the equation for d and use your result to find the number of degrees d the wiper arm must swing in each case. Round to the nearest degree.

Vehicle	Area cleaned	d (deg)
Luxury car	513 in.^{2}	
Sport utility vehicle	586 in.^{2}	

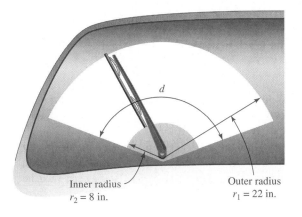

Inner radius
$r_2 = 8$ in.

Outer radius
$r_1 = 22$ in.

86. ELECTRONICS   The illustration is a diagram of a resistor connected to a voltage source of 60 volts. As a result, the resistor dissipates power in the form of heat. The power P lost when a voltage E is placed across a resistance R (in ohms) is given by the formula

$$P = \frac{E^2}{R}$$

Solve for R. If P is 4.8 watts and E is 60 volts, find R.

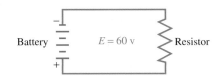

Battery $E = 60$ v   Resistor

87. CHEMISTRY   In chemistry, the ideal gas law equation is $PV = nR(T + 273)$, where P is the pressure, V the volume, T the temperature, and n the number of moles of a gas. R is a constant, 0.082. Solve the equation for n. Then use your result and the data from the student lab notebook in the illustration to find the value of n to the nearest thousandth for trial 1 and trial 2.

Ideal gas law Lab #1			Betsy Kinsell Chem 1 Section A		

Data:	Pressure (Atmosph.)	Volume (Liters)	Temp (°C)
Trial 1	0.900	0.250	90
Trial 2	1.250	1.560	–10

$R = 0.082$ (Constant)

88. INVESTMENTS   An amount P, invested at a simple interest rate r, will grow to an amount A in t years according to the formula $A = P(1 + rt)$. Solve for P. Suppose a man invested some money at 5.5%. If after 5 years, he had \$6,693.75 on deposit, what amount did he originally invest?

89. COST OF ELECTRICITY   The cost of electricity in a city is given by the formula $C = 0.07n + 6.50$, where C is the cost in dollars and n is the number of kilowatt hours used. Solve for n. Then find the number of kilowatt hours used each month by the homeowner whose checks to pay the monthly electric bills are shown in the illustration.

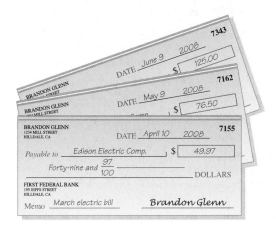

90. COST OF WATER   A monthly water bill in a certain city is calculated by using the formula $n = \frac{5{,}000C - 17{,}500}{6}$, where n is the number of gallons used and C is the monthly cost in dollars. Solve for C and compute the bill for quantities of 500, 1,200, and 2,500 gallons.

91. SURFACE AREA To find the amount of tin needed to make the coffee can shown in the illustration, we use the formula for the surface area of a right circular cylinder,

$$A = 2\pi r^2 + 2\pi rh$$

Solve the formula for h.

92. CARPENTRY A regular polygon has n equal sides and n equal angles. The measure a of an interior angle in degrees is given by $a = 180\left(1 - \frac{2}{n}\right)$. Solve for n. How many sides does the outdoor bandstand shown below have if the performance platform is a regular polygon with interior angles measuring $135°$?

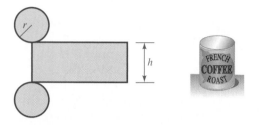

WRITING

93. Explain the difference between what perimeter measures and what area measures.

94. After solving a formula for m, a student compared her answer with that at the back of the textbook. Could this problem have two different-looking answers? Explain why or why not.

Student's answer: $m = \frac{5}{9}ar + 1$

Book's answer: $m = \frac{5ar + 9}{9}$

95. Explain the error made below.

$$T = \frac{\overset{1}{\cancel{adx + y}}}{\underset{1}{\cancel{y}}} = adx + 1$$

96. A student solved $x + 5c = 3c + a$ for c. His answer was $c = \frac{3c + a - x}{5}$. Explain why the equation is not solved for c.

REVIEW

Simplify each expression.

97. $12(2r + 11t + 1) - 11 + 2r$

98. $(16b + 8)\left(\frac{5}{4}\right) - 8b$

99. $-7(a - 3) - 5[3(a - 4) - 2(a + 2)]$

100. $0.9b^3 - 3.81b^3$

101. $-5.7pt - p + 5.1pt + 12p$

102. $\frac{3}{5}t - \frac{2}{3}t$

CHALLENGE PROBLEMS

103. Point C is the center of the largest circle in the figure. Find the area of the shaded region. Round to the nearest tenth.

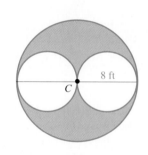

104. Solve $d_1 d_2 = fd_2 + fd_1$ for d_1.

105. Subscripts are used in other disciplines besides mathematics. In what disciplines are the following symbols used?
  a. H_2O and CO_2
  b. C_7 and G_7
  c. B_6 and B_{12}

106. Evaluate $2a_2^2 + 3a_3^3 + 4a_4^4$ for $a_2 = 2$, $a_3 = 3$, and $a_4 = 4$.

SECTION 1.7
Using Equations to Solve Problems

Objectives

1 Apply the steps of a problem-solving strategy.
2 Solve number-value problems.
3 Solve geometry problems.
4 Use formulas to solve problems.

An objective of this course is to improve your problem-solving skills. In the next two sections, you will have the opportunity to do that as we discuss how to use equations to solve many different types of problems.

 Apply the Steps of a Problem-Solving Strategy.

To become a good problem solver, you need a plan to follow, such as the following five-step strategy.

Problem Solving

1. ***Analyze the problem*** by reading it carefully to understand the given facts. What information is given? What are you asked to find? What vocabulary is given? Often a diagram or table will help you visualize the facts of the problem.

2. ***Form an equation*** by picking a variable to represent the quantity to be found. Then express all other unknown quantities as expressions involving the variable. Key words or phrases can be helpful. Finally, translate the words of the problem into an equation.

3. ***Solve the equation.***

4. ***State the conclusion.***

5. ***Check the result*** using the original wording of the problem, not the equation that was formed in step 2.

In order to solve problems, which are usually given in words, we must translate those words into mathematical symbols. In the next example, we use translation to write an equation that mathematically models the situation.

EXAMPLE 1   ***Leading Employers.***   In 2007, Target and Home Depot were two of the nation's top employers. Their combined work forces totaled 658,000 people. If Home Depot employed 46,000 fewer people than Target, how many employees did each company have?

Analyze the Problem

• The phrase *combined work forces totaled* 658,000 suggests that if we add the number of employees of each company, the result will be 658,000.

• The phrase *Home Depot employed 46,000 fewer people than Target* suggests that the number of employees of Home Depot can be found by subtracting 46,000 from the number of employees of Target.

• We are to find the number of employees of each company.

Caution

For this problem, one common mistake is to let

~~x = the number of employees of each company~~

Since Target and Home Depot have different numbers of employees, x cannot represent both unknowns.

Form an Equation   If we let x = the number of employees of Target, then $x - 46,000$ = the number of employees of Home Depot. We can now translate the words of the problem into an equation.

The number of employees of Target	plus	the number of employees of Home Depot	is	658,000.
x	$+$	$x - 46,000$	$=$	658,000

Solve the Equation

$$x + x - 46,000 = 658,000$$
$$2x - 46,000 = 658,000 \quad \text{Combine like terms.}$$
$$2x = 704,000 \quad \text{Add 46,000 to both sides.}$$
$$x = 352,000 \quad \text{To isolate } x, \text{ divide both sides by 2.}$$

Since x represents the number of employees of Target, we can find the number of employees of Home Depot by evaluating $x - 46,000$ for $x = 352,000$.

$$x - 46,000 = \mathbf{352,000} - 46,000$$
$$= 306,000$$

State the Conclusion   In 2007, Target had 352,000 employees and Home Depot had 306,000 employees.

Check the Result   Since $352,000 + 306,000 = 658,000$ and since 306,000 is 46,000 less than 352,000, the answers check.

 Now Try **Problem 13**

When solving problems, diagrams are often helpful, because they allow us to visualize the facts of the problem.

EXAMPLE 2   *Triathlons.* A triathlon includes swimming, long-distance running, and cycling. The long-distance run is 11 times longer than the distance the competitors swim. The distance they cycle is 85.8 miles longer than the run. Overall, the competition covers 140.6 miles. Find the length of each part of the triathlon and round each length to the nearest tenth of a mile.

Analyze the Problem   The entire triathlon course covers a distance of 140.6 miles. We note that the distance the competitors run is related to the distance they swim, and the distance they cycle is related to the distance they run.

Form an Equation   If x = the distance in miles that the competitors swim, then $11x$ = the length of the long-distance run, and $11x + 85.8$ = the distance they cycle. From the diagram on the next page, we can see that the sum of the individual parts of the triathlon must equal the total distance covered.

Swimming	Running	Cycling
x mi	$11x$ mi	$(11x + 85.8)$ mi

|← ——————————— 140.6 m ——————————— →|

We can now form the equation.

The distance they swim	plus	the distance they run	plus	the distance they cycle	equals	the total length of the course.
x	$+$	$11x$	$+$	$11x + 85.8$	$=$	140.6

Solve the Equation

$$x + 11x + 11x + 85.8 = 140.6$$
$$23x + 85.8 = 140.6 \qquad \text{Combine like terms.}$$
$$23x = 54.8 \qquad \text{Subtract 85.8 from both sides.}$$
$$x \approx 2.382608696 \qquad \text{To isolate x, divide both sides by 23.}$$

State the Conclusion   To the nearest tenth, the distance the competitors swim is 2.4 miles. The distance they run is $11x$, or approximately $11(2.382608696) = 26.20869565$ miles. To the nearest tenth, that is 26.2 miles. The distance they cycle is $11x + 85.8$, or approximately $26.20869565 + 85.8 = 112.0086957$ miles. To the nearest tenth, that is 112.0 miles.

Check the Result   If we add the lengths of the three parts of the triathlon and round to the nearest tenth, we get 140.6 miles. The answers check.

 *Now Try*   **Problem 15**

The wording of a problem doesn't always contain key phrases that translate directly to an equation. In such cases, an analysis of the problem will give clues that help us write an equation.

EXAMPLE 3   *Travel Promotions.*   The price of a 7-day Alaskan cruise, normally $2,752 per person, is reduced by $1.75 per person for large groups traveling together. How large a group is needed for the price to be $2,500 per person?

Analyze the Problem   For a group of 20 people, the cost is reduced by $1.75 for each person and the $2,752 price is reduced by $20(\$1.75) = \$35$.

The per-person price of the cruise $= \$2,752 - 20(\$1.75)$

For a group of 30 people, the $2,752 cost is reduced by $30(\$1.75) = \$52.50$.

The per-person price of the cruise $= \$2,752 - 30(\$1.75)$

Form an Equation   If we let x = the group size necessary for the price of the cruise to be $2,500 per person, we can form the following equation:

The price of the cruise	is	$2,752	minus	the number of people in the group	times	$1.75.
2,500	=	2,752	−	x	·	1.75

Solve the Equation

$$2,500 = 2,752 - 1.75x$$
$$2,500 - \mathbf{2,752} = 2,752 - 1.75x - \mathbf{2,752} \qquad \text{Subtract 2,752 from both sides.}$$
$$-252 = -1.75x \qquad \text{Simplify each side.}$$
$$144 = x \qquad \text{To isolate x, divide both sides by } -1.75.$$

State the Conclusion   If 144 people travel together, the price will be $2,500 per person.

Check the Result   For 144 people, the cruise cost of $2,752 will be reduced by 144($1.75) = $252. If we subtract, $2,752 − $252 = $2,500. The answer checks.

 Now Try   Problem 21

2 Solve Number-Value Problems.

Some problems deal with quantities that have a value. In these problems, we must distinguish between the *number of* and the *value of* the unknown quantity. For problems such as these, we will use the relationship

Number · value = total value

EXAMPLE 4   *Portfolio Analysis.*   A college foundation owns stock in Kodak (selling at $26 per share), Coca-Cola (selling at $52 per share), and IBM (selling at $103 per share). The foundation owns an equal number of shares of Kodak and Coca-Cola stock, but five times as many shares of IBM stock. If this portfolio (collection of stocks) is worth $415,100, how many shares of each stock does the foundation own?

Analyze the Problem   The value of the Kodak stock plus the value of the Coca-Cola stock plus the value of the IBM stock must equal $415,100. We need to find the number of shares of each of these stocks held by the foundation.

> **Success Tip**
> We can let x represent the number of shares of Kodak stock *and* the number of shares of Coca-Cola stock because the foundation owns an *equal number* of shares of these stocks.

Form an Equation   If we let x = the number of shares of Kodak stock, then x = the number of shares of Coca-Cola stock. Since the foundation owns five times as many shares of IBM stock as Kodak or Coca-Cola stock, $5x$ = the number of shares of IBM. The value of the shares of each stock is the *product* of the number of shares of that stock and its per-share value. See the table.

Stock	Number of shares ·	Value per share =	Total value of the stock
Kodak	x	26	$26x$
Coca-Cola	x	52	$52x$
IBM	$5x$	103	$103(5x)$
			Total: $415,100

Use the information in this column to form an equation.

We can now form the equation.

The value of Kodak stock	plus	the value of Coca-Cola stock	plus	the value of IBM stock	is	the total value of all of the stock.
$26x$	$+$	$52x$	$+$	$103(5x)$	$=$	$415{,}100$

Solve the Equation

$$26x + 52x + 515x = 415{,}100$$
$$593x = 415{,}100 \quad \text{Combine like terms on the left side.}$$
$$x = 700 \quad \text{To isolate } x, \text{ divide both sides by 593.}$$

State the Conclusion   The foundation owns 700 shares of Kodak, 700 shares of Coca-Cola, and $5(700) = 3{,}500$ shares of IBM.

Check the Result   The value of 700 shares of Kodak stock is $700(\$26) = \$18{,}200$. The value of 700 shares of Coca-Cola is $700(\$52) = \$36{,}400$. The value of 3,500 shares of IBM is $3{,}500(\$103) = \$360{,}500$. The sum is $\$18{,}200 + \$36{,}400 + \$360{,}500 = \$415{,}100$. The answers check.

▷ *Now Try*   **Problem 25**

❸ Solve Geometry Problems.

Sometimes we can use a geometric fact or formula to solve a problem. The following illustrations show two important types of geometric figures: angles and triangles. A **right angle** is an angle whose measure is 90°. A **straight angle** is an angle whose measure is 180°. An **acute angle** is an angle whose measure is greater than 0° and less than 90°. An angle whose measure is greater than 90° and less than 180° is called an **obtuse angle.**

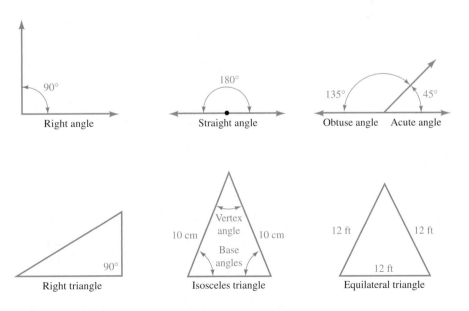

If the sum of two angles equals 90°, the angles are called **complementary,** and each angle is called the **complement** of the other. If the sum of two angles equals 180°, the angles are called **supplementary,** and each angle is the **supplement** of the other.

A **right triangle** is a triangle with one right angle. An **isosceles triangle** is a triangle with two sides of equal measure that meet to form the **vertex angle.** The angles opposite the equal sides, called the **base angles,** are also equal. An **equilateral triangle** is a triangle with three equal sides and three equal angles. It is important to note that *the sum of the angle measures of any triangle is 180°.*

EXAMPLE 5   *Flags.*   The flag of Guyana, a republic on the northern coast of South America, is one isosceles triangle superimposed over another on a field of green, as shown. The measure of a base angle of the larger triangle is 14° more than the measure of a base angle of the smaller triangle. The measure of the vertex angle of the larger triangle is 34°. Find the measure of each base angle of the smaller triangle.

Analyze the Problem   We are working with isosceles triangles. Therefore, the base angles of the smaller triangle have the same measure, and the base angles of the larger triangle have the same measure.

Form an Equation   If we let $x =$ the measure in degrees of one base angle of the smaller isosceles triangle, then the measure of its other base angle is also x. (See the figure.)

     The measure of a base angle of the larger isosceles triangle is $x + 14°$, since its measure is 14° more than the measure of a base angle of the smaller triangle. We are given that the vertex angle of the larger triangle measures 34°.

     The sum of the measures of the angles of any triangle (in this case, the larger triangle) is 180°.

     We can now form the equation.

The measure of one base angle	plus	the measure of the other base angle	plus	the measure of the vertex angle	is	180°.
$x + 14$	$+$	$x + 14$	$+$	34	$=$	180

Solve the Equation

$$x + 14 + x + 14 + 34 = 180$$
$$2x + 62 = 180 \quad \text{Combine like terms.}$$
$$2x = 118 \quad \text{Subtract 62 from both sides.}$$
$$x = 59 \quad \text{To isolate } x, \text{ divide both sides by 2.}$$

State the Conclusion   The measure of each base angle of the smaller triangle is 59°.

Check the Result   If $x = 59$, then $x + 14 = 73$. The sum of the measures of each base angle and the vertex angle of the larger triangle is $73° + 73° + 34° = 180°$. The answer checks.

   Now Try   **Problem 33**

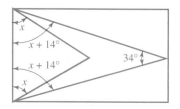

Use Formulas to Solve Problems.

When preparing to write an equation to solve a problem, the given facts of the problem often suggest a formula that we can use to model the situation mathematically.

EXAMPLE 6   *Kennels.*   A man has a 50-foot roll of fencing to make a rectangular kennel. If he wants the kennel to be 6 feet longer than it is wide, find its dimensions.

Analyze the Problem   The perimeter P of the rectangular kennel is 50 feet. Recall that the formula for the perimeter of a rectangle is $P = 2l + 2w$. We need to find its length and width.

Form an Equation   We let w = the width in feet of the kennel shown below. Then the length, which is 6 feet more than the width, is represented by the expression $w + 6$.

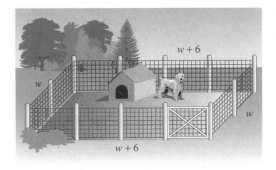

We can now form the equation by substituting 50 for P and $w + 6$ for the length in the formula for the perimeter of a rectangle.

$$P = 2l + 2w$$
$$50 = 2(w + 6) + 2w$$

Solve the Equation

$50 = 2(w + 6) + 2w$	
$50 = 2w + 12 + 2w$	Distribute the multiplication by 2.
$50 = 4w + 12$	Combine like terms.
$38 = 4w$	Subtract 12 from both sides.
$9.5 = w$	To isolate w, divide both sides by 4.

State the Conclusion   The width of the kennel is 9.5 feet. The length is 6 feet more than this, or 15.5 feet.

Check the Result   If a rectangle has a width of 9.5 feet and a length of 15.5 feet, its length is 6 feet more than its width, and the perimeter is 2(9.5) feet + 2(15.5) feet = 50 feet. The answers check.

 Now Try   **Problem 39**

STUDY SET
1.7

VOCABULARY

Fill in the blanks.

1. An _____ angle has a measure of more than 0° and less than 90°.
2. A _____ angle is an angle whose measure is 90°.
3. If the sum of the measures of two angles equals 90°, the angles are called _____ angles.
4. If the sum of the measures of two angles equals 180°, the angles are called _____ angles.
5. If a triangle has a right angle, it is called a _____ triangle.
6. If a triangle has two sides with equal measures, it is called an _____ triangle.
7. The sum of the measures of the _____ of a triangle is 180°.
8. An _____ triangle has three sides of equal length and three angles of equal measure.

CONCEPTS

9. The unit used to measure the intensity of sound is called the *decibel*. In the table, translate the comments in the right column into mathematical symbols to complete the decibels column.

Activity	Decibels	Compared to conversation
Conversation	d	—
Vacuum cleaner		15 decibels more
Circular saw		10 decibels less than twice
Jet takeoff		20 decibels more than twice
Whispering		10 decibels less than half
Rock band		Twice the decibel level

10. INSTRUMENTS   The flute consists of three pieces. Write an algebraic expression that represents
    a. the length of the shortest piece.
    b. the length of the longest piece.
    c. the length of the flute.

11. The following table shows the four types of problems an instructor put on a history test.
    a. Complete the table.
    b. Write an algebraic expression that represents the total number of points on the test.
    c. Write an equation that could be used to find x.

Type of question	Number	·	Value	=	Total value
Multiple choice	x		5		
True/false	$3x$		2		
Essay	$x - 2$		10		
Fill-in	x		5		
					Total: 110 points

12. For each picture shown, what geometric concept studied in this section is illustrated?

APPLICATIONS

13. CEREAL SALES   In 2005, two of the top-selling cereals in the United States were General Mills' Cheerios and Kellogg's Frosted Flakes, with combined sales of $939 million. Frosted Flakes sales were $439 million less than sales of Cheerios. What were the 2005 sales for each brand?

14. FILMS   Denzel Washington's three top domestic grossing films, *Remember the Titans*, *The Pelican Brief*, and *Crimson Tide*, have earned $307.8 million. If *Remember the Titans* earned $14.8 million more than *The Pelican Brief*, and if *The Pelican Brief* earned $9.4 million more than *Crimson Tide*, how much did each film earn as of that date?

15. STATUE OF LIBERTY From the foundation of the large pedestal on which it sits to the top of the torch, the Statue of Liberty National Monument measures 305 feet. The pedestal is 3 feet taller than the statue. Find the height of the pedestal and the height of the statue.

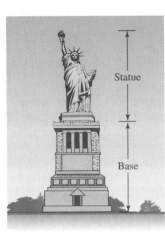

Statue

Base

16. WOODWORKING   The carpenter saws a board that is 22 feet long into two pieces. One piece is to be 1 foot longer than twice the length of the shorter piece. Find the length of each piece.

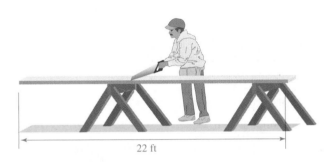

22 ft

17. TV COMMERCIALS   For the typical "one-hour" prime-time television slot, the number of minutes of commercials is $\frac{3}{7}$ of the number of minutes of the actual program. Determine how many minutes of the program are shown in that one hour.

18. CONCERT TOURING   A rock group plans to travel for a total of 23 weeks, visiting three countries. They will be in Germany for 3 weeks longer than they will be in France. Their stay in Great Britain will be 1 week less than that in France. How many weeks will they be in each country?

19. MAKING FURNITURE   A woodworker wants to put two partitions crosswise in a drawer that is 28 inches deep, as shown in the illustration in the next column. He wants to place the partitions so that the spaces created increase by 3 inches from front to back. If the thickness of each partition is $\frac{1}{2}$ inch, how far from the front end should he place the first partition?

28 in.

x in.

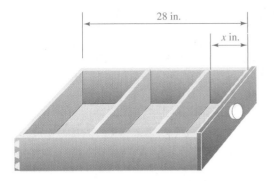

20. BUILDING SHELVES   A carpenter wants to put four shelves on an 8-foot wall so that the five spaces created decrease by 6 inches as we move up the wall. If the thickness of each shelf is $\frac{3}{4}$ inch, how far will the bottom shelf be from the floor? See the illustration.

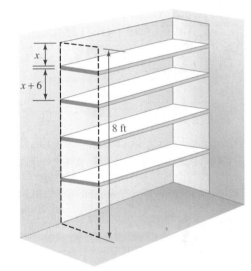

x

x + 6

8 ft

21. SPRING TOURS   A group of junior high students will be touring Washington, D.C. Their chaperons will have the $1,810 cost of the tour reduced by $15.50 for each student they personally supervise. How many students will a chaperon have to supervise so that his or her cost to take the tour will be $1,500?

22. MACHINING   Each pass through a lumber plane shaves off 0.015 inch of thickness from a board. How many times must a board, originally 0.875 inch thick, be run through the planer if a board of thickness 0.74 inch is desired?

23. MOVING EXPENSES   To help move his furniture, a man rents a truck for $41.50 per day plus 35¢ per mile. If he has budgeted $150 for transportation expenses, how many miles will he be able to drive the truck if the move takes 1 day?

24. COMPUTING SALARIES   A student working for a delivery company earns $57.50 per day plus $4.75 for each package she delivers. How many deliveries must she make each day to earn $200 a day?

25. ASSETS OF A PENSION FUND A pension fund owns 2,000 fewer shares in mutual stock funds than mutual bond funds. Currently, the stock funds sell for $12 per share, and the bond funds sell for $15 per share. How many shares of each does the pension fund own if the value of the securities is $165,000?

26. VALUE OF AN IRA In an Individual Retirement Account (IRA) valued at $53,900, a couple has 500 shares of stock, some in Big Bank Corporation and some in Safe Savings and Loan. If Big Bank sells for $115 per share and Safe Savings sells for $97 per share, how many shares of each does the couple own?

27. SELLING CALCULATORS Last month, a bookstore ran the following ad. Sales of $5,370 were generated, with 15 more graphing calculators sold than scientific calculators. How many of each type of calculator did the bookstore sell?

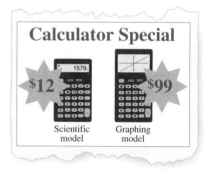

28. SELLING SEED A seed company sells two grades of grass seed. A 100-pound bag of a mixture of rye and Kentucky bluegrass sells for $245, and a 100-pound bag of bluegrass sells for $347. How many bags of each are sold in a week when the receipts for 19 bags are $5,369?

29. NURSING The illustration shows the angle a needle should make with the skin when administering a certain type of injection. Find the measure of both angles labeled.

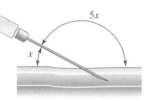

30. SUPPLEMENTARY ANGLES Refer to the illustration and find x.

$2x + 30°$ $2x - 10°$

31. THE LEANING TOWER OF PISA Because of soft soil and a shallow foundation, the Leaning Tower of Pisa in Italy is not vertical. Engineers predict that if the indicated angle gets to be 7°, the walls will not be able to support the structure and it will come crumbling down.

a. How many degrees from vertical is the tower now?

b. How many more degrees of lean must occur to cause the predicted collapse?

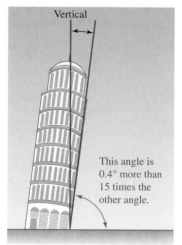

This angle is 0.4° more than 15 times the other angle.

32. THE iPAD BUILDING The sleek design of Apple's iPod has inspired a 23-floor residential and office building in Dubai, called the iPAD. It closely resembles a gigantic iPod sitting at an angle in its charger as shown below. How many degrees from the vertical is the building?

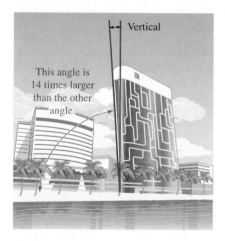

This angle is 14 times larger than the other angle.

33. STEPSTOOLS The sum of the measures of the three angles of any triangle is 180°. In the illustration, the measure of ∠2 (angle 2) is 10° larger than the measure of ∠1. The measure of ∠3 is 10° larger than the measure of ∠2. Find each angle measure.

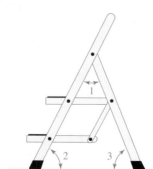

34. ANGLES OF A QUADRILATERAL The sum of the angles of any four-sided figure (called a *quadrilateral*) is 360°. The quadrilateral shown has two equal base angles. Find *x*.

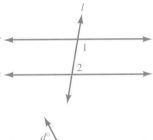

35. GEOMETRY In the illustration, lines *r* and *s* are cut by a third line *l* to form ∠1 (angle 1) and ∠2. When lines *r* and *s* are parallel, ∠1 and ∠2 are supplementary. If ∠1 = *x* + 50°, ∠2 = 2*x* − 20°, and lines *r* and *s* are parallel, find *x*.

36. GEOMETRY In the illustration, *r* ∥ *s* (read as "line *r* is parallel to line *s*"), and *a* = 103. Find *b*, *c*, and *d*. (*Hint:* See Problem 35.)

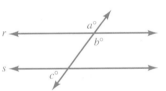

37. VERTICAL ANGLES When two lines intersect, four angles are formed. Angles that are side-by-side, such as ∠1 (angle 1) and ∠2, are called **adjacent angles.** Angles that are nonadjacent, such as ∠1 and ∠3 or ∠2 and ∠4, are called **vertical angles.** From geometry, we know that if two lines intersect, vertical angles have the same measure. If ∠1 = 3*x* + 10° and ∠3 = 5*x* − 10°, find *x*.

38. GEOMETRY In the illustration, *r* ∥ *s* (read as "line *r* is parallel to line *s*"), and *b* = 137. Find *a* and *c*. (*Hint:* See Problems 35 and 37.)

39. SWIMMING POOLS A woman wants to enclose the pool shown and have a walkway of uniform width all the way around. How wide will the walkway be if the woman uses 180 feet of fencing?

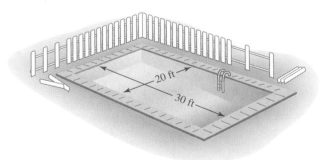

40. QUILTING Throughout history, most artists and designers have felt that *golden rectangles* with a length 1.618 times as long as their width have the most visually attractive shape. A woman is planning to make a quilt in the shape of a golden rectangle. She has exactly 22 feet of a special lace that she plans to sew around the edge of the quilt. What should the length and width of the quilt be? Round both answers up to the nearest hundredth.

41. RANCHING A farmer has 624 feet of fencing to enclose a pasture. Because a river runs along one side, fencing will be needed on only three sides. Find the dimensions of the pasture if its length is double its width.

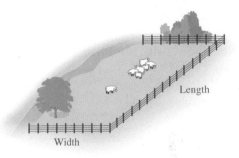

42. FENCING A man has 150 feet of fencing to build the two-part pen shown in the illustration. If one part is a square and the other a rectangle, find the outside dimensions of the pen.

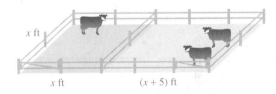

43. SOLAR HEATING One solar panel in the illustration is to be 3 feet wider than the other. To be equally efficient, they must have the same area. Find the width of each.

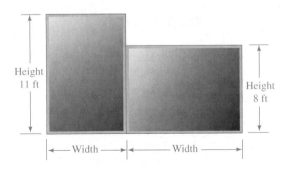

44. HEIGHT OF A TRIANGLE If the height of a triangle with a base of 8 inches is tripled, its area is increased by 96 square inches. Find the height of the triangle.

WRITING

45. Briefly explain what should be accomplished in each of the steps (*analyze, form, solve, state,* and *check*) of the problem-solving strategy used in this section.

46. Write a problem that can be represented by the following verbal model.

The measure of 1st angle	plus	the measure of 2nd angle	plus	the measure of 3rd angle	is	180°.
x	+	$2x$	+	$x + 10$	=	180

REVIEW

47. When expressed as a decimal, is $\frac{7}{9}$ a terminating or repeating decimal?

48. Solve: $x + 20 = 4x - 1 + 2x$

49. Write the set of integers.

50. Solve: $2x + 2 = \frac{2}{3}x - 2$

51. Evaluate $2x^2 + 5x - 3$ for $x = -3$.

52. Solve $T - R = ma$ for R.

CHALLENGE PROBLEMS

A lever will be in balance when the sum of the products of the forces on one side of a fulcrum and their respective distances from the fulcrum is equal to the sum of the products of the forces on the other side of the fulcrum and their respective distances from the fulcrum.

53. MOVING A STONE A woman uses a 10-foot bar to lift a 210-pound stone. If she places another rock 3 feet from the stone to act as the fulcrum, how much force must she exert to move the stone?

54. LIFTING A CAR A 350-pound football player brags that he can lift a 2,500-pound car. If he uses a 12-foot bar with the fulcrum placed 3 feet from the car, will he be able to lift the car?

55. BALANCING A LEVER Forces are applied to a lever as indicated in the illustration. Find x, the distance of the smallest force from the fulcrum.

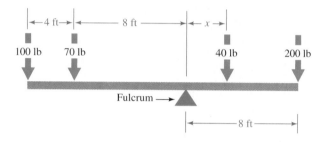

56. BALANCING A SEESAW Jim and Bob sit at opposite ends of an 18-foot seesaw, with the fulcrum at its center. Jim weighs 160 pounds, and Bob weighs 200 pounds. Kim sits 4 feet in front of Jim, and the seesaw balances. How much does Kim weigh?

SECTION 1.8
More about Problem Solving

Objectives

 Solve percent problems.

❷ Find the mean, median, and mode.

❸ Solve investment problems.

❹ Solve uniform motion problems.

❺ Solve mixture problems.

In this section, we will again use equations as we solve a variety of problems.

❶ **Solve Percent Problems.**

Percents are often used to present numeric information. **Percent** means parts per one hundred. One method to solve percent problems is to use the given facts to write a **percent sentence** of the form:

_____ is _____ % of _____ ?

The Language of Algebra
The names of the parts of a percent sentence are:

5	is	50%	of	10.
amount		percent		base

They are related by the formula:

Amount = percent · base

We enter the appropriate numbers in two of the blanks and the word "what" in the remaining blank. Then we translate the sentence to mathematical symbols and solve the resulting equation.

EXAMPLE 1   *Video Games.*   Refer to the following graph. Determine the number of video games sold in the United States in 2005.

BEST-SELLING VIDEO GAMES IN THE UNITED STATES, 2005

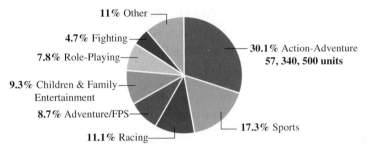

11% Other
4.7% Fighting
7.8% Role-Playing
9.3% Children & Family Entertainment
8.7% Adventure/FPS
11.1% Racing
30.1% Action-Adventure 57, 340, 500 units
17.3% Sports

Source: The NPD Group/Point of Sale Information

Analyze the Problem   In the **circle graph** above, we see that the 57,340,500 action-adventure video games that were sold in the United States in 2005 were 30.1% of the total number of video games sold.

Form an Equation   Let x = the total number of video games sold in the United States in 2005. First, we write a percent sentence using the given data. Then we translate to form an equation.

57,340,500	is	30.1%	of	what?
↓	↓	↓	↓	↓
57,340,500	=	30.1%	·	x

The amount is 57,340,500, the percent is 30.1%, and the base is x.

Solve the Equation   To find the total number of video game sold in the United States in 2005, we solve for x.

$57,340,500 = 0.301x$ *Write 30.1% as a decimal: 30.1% = 0.301.*

$$\frac{57,340,500}{0.301} = \frac{0.301x}{0.301}$$ *To isolate x, divide both sides by 0.301.*

$190,500,000 = x$ *Do the division using a calculator.*

State the Conclusion   There were 190,500,000 video games sold in the United States in 2005.

Check the Result   If a total of 190,500,000 video games were sold, then the 57,340,500 action-adventure video games sold were $\frac{57,340,500}{190,500,000} = 0.301$ or 30.1% of the units sold. The answer checks.

Now Try   **Problem 21**

When the regular price of merchandise is reduced, the amount of reduction is called **markdown** (or discount).

$$\text{Sale price} \quad = \quad \text{regular price} \quad - \quad \text{markdown}$$

Usually, the markdown is expressed as a percent of the regular price.

$$\text{Markdown} \quad = \quad \text{percent of markdown} \quad \cdot \quad \text{regular price}$$

EXAMPLE 2   *Wedding Gowns.*   At a bridal shop, a wedding gown that normally sells for $397.98 is on sale for $265.32. Find the percent of markdown.

Analyze the Problem   In this case, $265.32 is the sale price, $397.98 is the regular price, and the markdown is the *product* of $397.98 and the percent of markdown.

Form an Equation   We let r = the percent of markdown, expressed as a decimal. We then substitute $265.32 for the sale price and $397.98 for the regular price in the formula.

$$\text{Sale price} \quad \text{is} \quad \text{regular price} \quad \text{minus} \quad \text{markdown.}$$
$$265.32 \quad = \quad 397.98 \quad - \quad r \cdot 397.98$$

Markdown = percent of markdown · regular price.

Solve the Equation

$$265.32 = 397.98 - r \cdot 397.98$$
$$265.32 = 397.98 - 397.98r \qquad \text{Rewrite } r \cdot 397.98 \text{ as } 397.98r.$$
$$-132.66 = -397.98r \qquad \text{Subtract 397.98 from both sides.}$$
$$\frac{-132.66}{-397.98} = r \qquad \text{To isolate } r, \text{ divide both sides by } -397.98.$$
$$0.333333\ldots = r \qquad \text{Do the division using a calculator.}$$
$$33.3333\ldots\% = r \qquad \text{Write the decimal as a percent. Multiply}$$

0.333333 ... by 100 by moving the decimal point two places to the right and insert a % sign.

State the Conclusion   The percent of markdown on the wedding gown is 33.3333 . . . % or $33\frac{1}{3}\%$.

Check the Result   The markdown is $33\frac{1}{3}\%$ of $397.98, or $132.66. The sale price is $397.98 − $132.66, or $265.32. The answer checks.

▷ *Now Try*   **Problem 25**

Percents are often used to describe how a quantity has changed. To describe such changes, we use **percent of increase** or **percent of decrease.**

EXAMPLE 3   ***Entertainment.*** Use the following data to determine the percent of increase in the number of indoor movie theater screens in the United States from 1990 to 2005. Round to the nearest one percent.

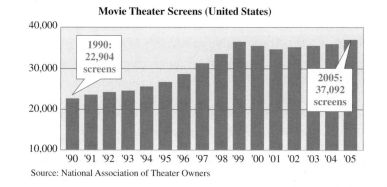

Movie Theater Screens (United States)

1990: 22,904 screens

2005: 37,092 screens

Source: National Association of Theater Owners

Analyze the Problem   To find the percent of increase, we first find the *amount of increase* by subtracting the number of screens in 1990 from the number in 2005.

$$37{,}092 - 22{,}904 = 14{,}188$$

> **Caution**
> Always find the percent of increase (or decrease) with respect to the *original* amount.

Form an Equation   Next, we find what percent of the original 22,904 screens the 14,188 increase represents. We let x = the unknown percent and translate the words into an equation.

14,188	is	what percent	of	22,904?
14,188	=	x	·	22,904

The amount is 14,188, the percent is x, and the base is 22,904.

Solve the Equation

$$14{,}188 = x \cdot 22{,}904$$

$$14{,}188 = 22{,}904x$$

$$\frac{14{,}188}{22{,}904} = \frac{22{,}904x}{22{,}904} \qquad \text{To isolate } x, \text{ divide both sides by 22,904.}$$

$$0.619455117\ldots \approx x \qquad \text{Do the division using a calculator.}$$

$$61.9455117\ldots\% \approx x \qquad \text{Write the decimal as a percent.}$$

$$62\% \approx x \qquad \text{Round to the nearest one percent.}$$

State the Conclusion   There was a 62% increase in the number of movie screens in the United States from 1990 to 2005.

Check the Result   A 50% increase from 22,904 screens would be approximately 11,000 additional screens. It seems reasonable that 14,188 more screens would be a 62% increase.

 Now Try **Problem 31**

2   **Find the Mean, Median, and Mode.**

Statistics is a branch of mathematics that deals with analysis of numerical data. Three types of averages are commonly used in statistics as measures of central tendency of a collection of data: the **mean,** the **median,** and the **mode.**

Mean, Median, and Mode

The **mean** $\bar{x}$ of a collection of values is the sum S of those values divided by the number of values n.

$$\bar{x} = \frac{S}{n} \quad \text{Read } \bar{x} \text{ as "x bar."}$$

The **median** of a collection of values is the middle value. To find the median,

1. Arrange the values in increasing order.
2. If there are an odd number of values, choose the middle value.
3. If there are an even number of values, add the middle two values and divide by 2.

The **mode** of a collection of values is the value that occurs most often.

© Steve Morse photo,
MU Cooperative Media Group

The Language of Algebra
In statistics, the mean, median, and mode are classified as types of *averages*. In daily life, when the word *average* is used, it most often is referring to the mean.

EXAMPLE 4 *Reaction Time.* As a project for a science class, a student measured ten people's reaction times. The times, in seconds, are listed below. Find **a.** the mean **b.** the median **c.** the mode of the collection of data.

0.29, 0.22, 0.19, 0.36, 0.28, 0.23, 0.16, 0.28, 0.33, 0.26

Solution

a. To find the mean, we add the values and divide by the number of values, which is 10.

$$\bar{x} = \frac{0.29 + 0.22 + 0.19 + 0.36 + 0.28 + 0.23 + 0.16 + 0.28 + 0.33 + 0.26}{10}$$

$$= 0.26 \text{ second}$$

b. To find the median, we first arrange the values in increasing order:

0.16, 0.19, 0.22, 0.23,   0.26, 0.28,   0.28, 0.29, 0.33, 0.36

Least   Greatest

Because there is an even number of measurements, the median will be the sum of the middle two values, 0.26 and 0.28, divided by 2.

$$\text{Median} = \frac{0.26 + 0.28}{2} = 0.27 \text{ second}$$

c. Since the time 0.28 second occurs most often, it is the mode.

0.16, 0.19, 0.22, 0.23, 0.26,   0.28,   0.28,   0.29, 0.33, 0.36

▷ *Now Try* **Problem 33**

EXAMPLE 5 *Bank Service Charges.* When the average (mean) daily balance of a customer's checking account falls below $500 in any week, the bank assesses a $15 service charge. What minimum balance will the account shown need to have on Friday to avoid the service charge?

Security Savings
☐ Weekly Statement ☐
Acct: 201-234-002   Type: checking

Day	Date	Daily balance	Comments
Mon	3/11	$730.70	
Tue	3/12	$350.19	
Wed	3/13	−$50.19	overdrawn
Thu	3/14	$275.55	
Fri	3/15		

Analyze the Problem   We can find the average (mean) daily balance for the week by adding the daily balances and dividing by 5. We want the mean to be $500 so that there is no service charge.

Form an Equation   We will let x = the minimum balance needed on Friday. Then we translate the words into mathematical symbols.

Success Tip
The *Form an Equation* step is often the hardest. To help, write a **verbal model** of the situation (as shown on the right) and then translate it into an equation.

The sum of the five daily balances	divided by	5	is	$500.

$$\frac{730.70 + 350.19 + (-50.19) + 275.55 + x}{5} = 500$$

Solve the Equation

$$\frac{730.70 + 350.19 + (-50.19) + 275.55 + x}{5} = 500$$

$$\frac{1{,}306.25 + x}{5} = 500 \qquad \text{Combine like terms in the numerator.}$$

$$5\left(\frac{1{,}306.25 + x}{5}\right) = 5(500) \qquad \text{To clear the equation of the fraction, multiply both sides by 5.}$$

$$1{,}306.25 + x = 2{,}500$$

$$x = 1{,}193.75 \qquad \text{To isolate x, subtract 1,306.25 from both sides.}$$

State the Conclusion   On Friday, the account balance needs to be $1,193.75 to avoid a service charge.

Check the Result   Check the result by adding the five daily balances and dividing by 5.

 Now Try   Problem 35

3   **Solve Investment Problems.**

The money an investment earns is called *interest*. **Simple interest** is computed by the formula $I = Prt$, where I is the interest earned, P is the principal (amount invested), r is the annual interest rate, and t is the length of time the principal is invested.

EXAMPLE 6   *Interest Income.*   To protect against a major loss, a financial analyst suggested the following plan for a client who has $50,000 to invest for 1 year.

1. Alco Development, Inc. Builds mini-malls. High yield: 12% per year. Risky!

2. Certificate of deposit (CD). Insured, safe. Low yield: 4.5% annual interest.

If the client puts some money in each investment and wants to earn $3,600 in interest, how much should be invested at each rate?

Analyze the Problem   In this case, we are working with two investments made at two different rates for 1 year. If we add the interest from the two investments, the sum should equal $3,600.

Form an Equation   If we let $x =$ the number of dollars invested at 12%, the interest earned is $I = Prt = \$x(12\%)(1) = \$0.12x$. If $\$x$ is invested at 12%, there is $\$(50,000 - x)$ to invest at 4.5%, which will earn $\$0.045(50,000 - x)$ in interest. These facts are listed in the table.

	P	$\cdot$ r	$\cdot$ $t =$	I
Alco Development, Inc.	x	0.12	1	$0.12x$
Certificate of deposit	$50,000 - x$	0.045	1	$0.045(50,000 - x)$
			Total: $3,600	

Enter this information first.   Use the information in this column to form an equation.

The sum of the two amounts of interest should equal $3,600. We now translate the words into an equation.

The interest earned at 12%	plus	the interest earned at 4.5%	equals	the total interest earned.
$0.12x$	$+$	$0.045(50,000 - x)$	$=$	$3,600$

Solve the Equation

$$0.12x + 0.045(50,000 - x) = 3,600$$

$$1,000[0.12x + 0.045(50,000 - x)] = 1,000(3,600)$$ To eliminate the decimals, multiply both sides by 1,000.

$$120x + 45(50,000 - x) = 3,600,000$$ Distribute the 1,000 and simplify both sides.

$$120x + 2,250,000 - 45x = 3,600,000$$ Remove parentheses.

$$75x + 2,250,000 = 3,600,000$$ Combine like terms.

$$75x = 1,350,000$$ Subtract 2,250,000 from both sides.

$$x = 18,000$$ To isolate x, divide both sides by 75.

State the Conclusion $18,000 should be invested at 12% and $(50,000 - 18,000) = $32,000 should be invested at 4.5%.

Check the Result   The annual interest on $18,000 is $0.12(\$18,000) = \$2,160$. The interest earned on $32,000 is $0.045(\$32,000) = \$1,440$. The total interest is $\$2,160 + \$1,440 = \$3,600$. The answers check.

Now Try   **Problem 37**

4   Solve Uniform Motion Problems.

Problems that involve an object traveling at a constant rate for a specified period of time over a certain distance are called **uniform motion** problems. To solve these problems, we use the formula $d = rt$, where d is distance, r is rate, and t is time.

EXAMPLE 7   *Travel Time.*   After a stay on her grandparents' farm, a girl is to return home, 385 miles away. To split up the drive, the parents and grandparents start at the same time and drive toward each other, planning to meet somewhere along the way. If the parents travel at an average rate of 60 mph and the grandparents at 50 mph, how long will it take them to meet?

Analyze the Problem   The vehicles are traveling toward each other as shown in the following figure. We know the rates the cars are traveling (60 mph and 50 mph). We also know that they will travel for the same amount of time.

> *Caution*
> When using $d = rt$, make sure the units are consistent. For example, if the rate is given in miles per hour, the time must be expressed in hours.

Form an Equation   We can let $t =$ the time in hours that each vehicle travels. Then the distance traveled by the parents is $60t$ miles, and the distance traveled by the grandparents is $50t$ miles. The total distance traveled by the parents and grandparents is 385 miles. This information is organized in the table.

	r	$\cdot$ $t =$	d
Parents	60	t	$60t$
Grandparents	50	t	$50t$
			Total: 385 mi.

Home   Farm

|— 385 mi —|

Enter this information first.   Use the information in this column to form an equation.

We now translate the words of the problem into an equation.

The distance the parents travel	plus	the distance the grandparents travel	equals	the distance between the child's home and the grandparent's farm.
$60t$	$+$	$50t$	$=$	385

Solve the Equation

$$60t + 50t = 385$$
$$110t = 385 \quad \text{Combine like terms.}$$
$$t = 3.5 \quad \text{To isolate } t\text{, divide both sides by 110.}$$

State the Conclusion   The parents and grandparents will meet in $3\frac{1}{2}$ hours.

Check the Result   The parents travel $3.5(60) = 210$ miles. The grandparents travel $3.5(50) = 175$ miles. The total distance traveled is $210 + 175 = 385$ miles. The answer checks.

 Now Try   **Problem 43**

5 **Solve Mixture Problems.**

We now discuss two types of mixture problems. In the first example, a *dry mixture* of a specified value is created from two differently priced components. The value of each of its ingredients and the value of the dry mixture is given by

$$\text{Amount} \cdot \text{price} = \text{total value}$$

EXAMPLE 8   *Mixing Nuts.* The owner of a produce store notices that 20 pounds of gourmet cashews did not sell because of their high price of $12 per pound. The owner decides to mix peanuts with the cashews to lower the price per pound. If peanuts sell for $3 per pound, how many pounds of peanuts must be mixed with the cashews to make a mixture that could be sold for $6 per pound?

Analyze the Problem   We need to determine how many pounds of peanuts to mix with 20 pounds of cashews to obtain a mixture that could be sold for $6 per pound.

Form an Equation   We can let x = the number of pounds of peanuts to be used. Then $20 + x$ = the number of pounds in the mixture. We enter the known information in the following table

	Amount	$\cdot$ Price	= Total value
Cashews	20	12	240
Peanuts	x	3	$3x$
Mixture	$20 + x$	6	$6(20 + x)$

Enter this information first.

Use the information in this column to form an equation.

We can now form the equation.

The value of the cashews	plus	the value of the peanuts	equals	the value of the mixture.
240	+	$3x$	=	$6(20 + x)$

Solve the Equation

$240 + 3x = 6(20 + x)$

$240 + 3x = 120 + 6x$ Use the distributive property to remove parentheses.

$120 = 3x$ Subtract 3x and 120 from both sides.

$40 = x$ To isolate x, divide both sides by 3.

State the Conclusion   The owner should mix 40 pounds of peanuts with the 20 pounds of cashews.

Check the Result   The cashews are valued at $12(20) = $240, and the peanuts are valued at $3(40) = $120. The mixture is valued at $6(60) = $360. Since the value of the cashews plus the value of the peanuts equals the value of the mixture, the answer checks.

> **Caution**
> Check the result using the original wording of the problem, not by substituting it into the equation. Why? The equation may have been solved correctly, but the danger is that you may have formed it incorrectly.

 Now Try   Problem 51

In the next example, a *liquid mixture* of a desired strength is to be made from two solutions with different strengths (concentrations).

EXAMPLE 9   *Milk Production.*   Owners of a dairy find that milk with a 2% butterfat content is their best seller. Suppose the dairy has large quantities of whole milk having a 4% butterfat content and milk having a 1% butterfat content. How much of each type of milk should be mixed to obtain 120 gallons of milk that is 2% butterfat?

Analyze the Problem   We are to find the amount of 4% milk to mix with 1% milk to get 120 gallons of a milk that has a 2% butterfat content. In the figure, if we let g = the number of gallons of the 4% milk used in the mixture, then $120 - g$ = the number of gallons of the 1% milk needed to obtain the desired concentration. The amount of pure butterfat in each solution is given by

Amount of solution · strength of the solution = amount of pure butterfat

g gallons

High butterfat
4% butterfat

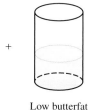

$(120 - g)$ gallons

+

Low butterfat
1% butterfat

120 gallons

=

Mixture
2% butterfat

	Amount ·	Strength =	Amount of pure butterfat
High butterfat	g	0.04	0.04g
Low butterfat	$120 - g$	0.01	0.01(120 − g)
Mixture	120	0.02	0.02(120)

Enter this
information first.

Use the information in this
column to form an equation.

Form an Equation   The amount of butterfat in a tank is the *product* of the percent butterfat and the number of gallons of milk in the tank. In the first tank shown in the figure, 4% of the g gallons, or 0.04g gallons, is butterfat. In the second tank, 1% of the $(120 - g)$ gallons, or 0.01(120 − g) gallons, is butterfat. Upon mixing, the third tank will have 0.02(120) gallons of butterfat in it. These results are recorded in the last column of the table.

We now translate the words of the problem into an equation.

The amount of butterfat in g gallons of 4% milk	plus	the amount of butterfat in $(120 - g)$ gallons of 1% milk	equals	the amount of butterfat in 120 gallons of the mixture.
0.04(g)	+	0.01(120 − g)	=	0.02(120)

Solve the Equation

$$0.04(g) + 0.01(120 - g) = 0.02(120)$$

$$4(g) + 1(120 - g) = 2(120)$$ Multiply both sides by 100 to clear the equation of decimals. This is done mentally.

$$4g + 120 - g = 240$$

$$3g + 120 = 240$$ Combine like terms.

$$3g = 120$$ Subtract 120 from both sides.

$$g = 40$$ To isolate g, divide both sides by 3.

26. BUYING FURNITURE   A bedroom set regularly sells for $983. If it is on sale for $737.25, what is the percent of markdown?

27. FLEA MARKETS   A vendor sells tool chests at a flea market for $65. If she makes a profit of 30% on each unit sold, what does she pay the manufacturer for each tool chest?   (*Hint:* The retail price = the wholesale price + the markup.)

28. BOOKSTORES   A bookstore sells a textbook for $39.20. If the bookstore makes a profit of 40% on each sale, what does the bookstore pay the publisher for each book?   (*Hint:* The retail price = the wholesale price + the markup.)

29. IMPROVING HORSEPOWER   The following graph shows how the installation of a special computer chip increases the horsepower of a truck. Find the percent of increase in horse-power for the engine running at 4,000 revolutions per minute (rpm) and round to the nearest tenth of one percent.

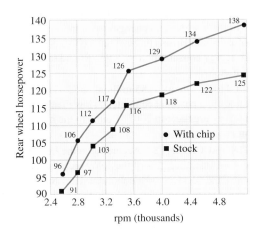

30. GREENHOUSE GASES   The U.S. energy-related carbon dioxide emissions in 2006 were 5,877 million metric tons. In 2005, that figure was 5,955 million metric tons. Find the percent of decrease and round to the nearest tenth of one percent.

31. BROADWAY SHOWS   Complete the table to find the percent of increase or decrease in attendance at Broadway shows for each season compared with the previous season. Round to the nearest tenth of one percent.

Season	Broadway attendance	% of increase or decrease
2003–04	11.61 million	—
2004–05	11.53 million	
2005–06	12.00 million	

Source: LiveBroadway.com

32. DRIVE-INS   The number of drive-in movie theaters in the United States peaked in 1958. Since then, the numbers have steadily declined. Determine the percent of decrease in the number of drive-ins. Round to the nearest one percent.

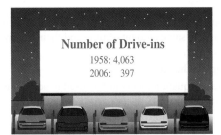

Number of Drive-ins
1958: 4,063
2006:   397

Source: United Drive-in Theater Owners Association

33. FUEL EFFICIENCY   The ten most fuel-efficient cars in 2007, based on manufacturer's estimated city and highway average miles per gallon (mpg), are shown in the table. Find the mean, median, and mode of both sets of data.

Model	mpg city/hwy
Toyota Prius	60/51
Honda Civic Hybrid	49/51
Toyota Camry Hybrid	40/38
Toyota Yaris	34/40
Honda Fit	33/38
Toyota Corolla	32/41
Mini Cooper	32/40
Hyundai Accent	32/35
Honda Civic	30/40
Nissan Versa	30/36

Source: edmonds.com

34. SPORT FISHING   The report shown below lists the fishing conditions at Pyramid Lake for a Saturday in January. Find the median and the mode of the weights of the striped bass caught at the lake.

> **Pyramid Lake**—Some striped bass are biting but are on the small side. Striking jigs and plastic worms. Water is cold: 38°. Weights of fish caught (lb): 6, 9, 4, 7, 4, 3, 3, 5, 6, 9, 4, 5, 8, 13, 4, 5, 4, 6, 9

35. JOB TESTING To be accepted into a police training program, a recruit must have an average score of 85 on a battery of four tests. If a candidate scored 76 on the oral test, 87 on the physical fitness test, and 83 on the psychological test, what is the lowest score she can obtain on the written test and still be accepted into the training program?

36. WNBA CHAMPIONS The results of each 2006 playoff game for the Detroit Shock are shown below. The two Detroit scores that are missing, one in a win and one in a loss, are the same number. If they averaged 74.8 points per game in the playoffs, find the missing scores.

First Round	Conference Finals
Detroit XX, Indiana 56	Detroit 70, Connecticut 59
Detroit 98, Indiana 83	Connecticut 77, Detroit XX
	Detroit 79, Connecticut 59

WNBA Finals
Sacramento 95, Detroit 71
Detroit 73, Sacramento 63
Sacramento 89, Detroit 69
Detroit 72, Sacramento 52
Detroit 80, Sacramento 75

37. HIGHEST RATES Based on the information in the table, a woman invested $12,000, some in an account paying the highest rate and the rest in an account paying the second highest rate. How much was invested in each account if the interest from both investments is $1,060 per year?

First Republic Savings and Loan	
Account	**Rate**
NOW	5.5%
Savings	7.5%
Money market	8.0%
Checking	4.0%
5-year CD	9.0%

38. ENTREPRENEURS Last year, a women's professional organization made two small-business loans totaling $28,000 to young women beginning their own businesses. The money was lent at 7% and 10% simple interest rates. If the annual income the organization received from these loans was $2,560, what was each loan amount?

39. INHERITANCES Paula used some of the money that she received from an inheritance to invest in a certificate of deposit paying 7% annual interest and the rest of the money in a promising biotech company offering an annual return of 10%. She invested twice as much in the 10% investment as she did in the 7% investment. Her combined annual income from the two investments was $4,050.

   a. How much did she invest in each account?

   b. How much did she inherit?

40. TAX RETURNS On a federal income tax form, Schedule B, a taxpayer forgot to write in the amount of interest income he earned for the year. From what is written on the form, determine the amount of interest earned from each investment and the amount he invested in stocks.

Schedule B–Interest and Dividend Income		
Part 1 Interest Income (See pages 12 and B1.)	**Note:** If you had over $400 in taxable income, use this form.	
	1 List name of payer.	**Amount**
	① MONEY MARKET ACCT. DEPOSITED $15,000 @ 3.3%	SAME AMOUNT
	② STOCKS EARNED 5%	FROM EACH

41. MONEY-LAUNDERING Use the evidence compiled by investigators to determine how much money a suspect deposited in the Cayman Islands bank.

   • On 6/1/06, the suspect electronically transferred $300,000 to a Swiss bank account paying an 8% annual yield.

   • That same day, the suspect opened another account in a Cayman Islands bank that offered a 5% annual yield.

   • A document dated 6/3/07 was seized during a raid of the suspect's home. It stated, "The total interest earned in one year from the two overseas accounts was 7.25% of the total amount deposited."

42. FINANCIAL PRESENTATIONS A financial planner showed her client the following investment plan. Find the total amount the client will have to invest to earn $2,700 in interest.

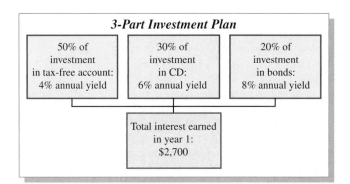

3-Part Investment Plan

| 50% of investment in tax-free account: 4% annual yield | 30% of investment in CD: 6% annual yield | 20% of investment in bonds: 8% annual yield |

Total interest earned in year 1: $2,700

43. TRAVEL TIMES A man called his wife to tell her that they needed to switch vehicles so he could use the family van to pick up some building materials after work. The wife left their home, traveling toward his office in their van at 35 mph. At the same time, the husband left his office in his car, traveling toward their home at 45 mph. If his office is 20 miles from their home, how long will it take them to meet so they can switch vehicles?

44. AIR TRAFFIC CONTROL   An airplane leaves Los Angeles bound for Caracas, Venezuela, flying at an average rate of 500 mph. At the same time, another airplane leaves Caracas bound for Los Angeles, averaging 550 mph. If the airports are 3,675 miles apart, when will the air traffic controllers have to make the pilots aware that the planes are passing each other?

45. CYCLING   A cyclist leaves his training base for a morning workout, riding at the rate of 18 mph. One hour later, his support staff leaves the base in a car going 45 mph in the same direction. How long will it take the support staff to catch up with the cyclist?

46. MARATHONS   Two marathon runners leave the starting gate, one running 12 mph and the other 10 mph. If they maintain the pace, how long will it take for them to be one-quarter of a mile apart?

47. RADIO COMMUNICATIONS   At 2 P.M., two military convoys leave Eagle River, Wisconsin, one headed north and one headed south. The convoy headed north averages 50 mph, and the convoy headed south averages 40 mph. They will lose radio contact when the distance between them is more than 135 miles. When will this occur?

48. SEARCH AND RESCUE   Two search-and-rescue teams leave base camp at the same time, looking for a lost child. The first team, on horseback, heads north at 3 mph, and the other team, on foot, heads south at 1.5 mph. How long will it take them to search a distance of 18 miles between them?

49. JET SKIING   A jet ski can go 12 mph in still water. If a rider goes upstream for 3 hours against a current of 4 mph, how long will it take the rider to return?   (*Hint:* Upstream speed is $(12 - 4)$ mph; how far can the rider go in 3 hours?)

50. PHYSICAL FITNESS   For her workout, Sarah walks north at the rate of 3 mph and returns at the rate of 4 mph. How many miles does she walk if the round trip takes 3.5 hours?

51. MIXING CANDY   How many pounds of red licorice bits that sell for $1.90 per pound should be mixed with 5 pounds of lemon gumdrops that sell for $2.20 per pound to make a candy mixture that could be sold for $2 per pound?

52. COFFEE BLENDS   A store sells regular coffee for $8 a pound and gourmet coffee for $14 a pound. To get rid of 40 pounds of the gourmet coffee, a shopkeeper makes a blend to put on sale for $10 a pound. How many pounds of regular coffee should he use?

53. HEALTH FOODS   A pound of dried pineapple bits sells for $6.19, a pound of dried banana chips sells for $4.19, and a pound of raisins sells for $2.39 a pound. Two pounds of raisins are to be mixed with equal amounts of pineapple and banana to create a trail mix that will sell for $4.19 a pound. How many pounds of pineapple and banana chips should be used?

54. METALLURGY   A 1-ounce commemorative coin is to be made of a combination of pure gold, costing $380 an ounce, and a gold alloy that costs $140 an ounce. If the cost of the coin is to be $200, and 500 are to be minted, how many ounces of gold and gold alloy are needed to make the coins?

55. GARDENING   A wholesaler of premium organic planting mix notices that the retail garden centers are not buying her product because of its high price of $1.57 per cubic foot. She decides to mix sawdust with the planting mix to lower the price per cubic foot. If the wholesaler can buy the sawdust for $0.10 per cubic foot, how many cubic feet of each must be mixed to have 6,000 cubic feet of planting mix that could be sold to retailers for $1.08 per cubic foot?

56. BRONZE   A pound of tin is worth $1 more than a pound of copper. Four pounds of tin are mixed with 6 pounds of copper to make bronze that sells for $3.65 per pound. How much is a pound of tin worth?

57. *from Campus to Careers*
Registered Dietician

Suppose, as registered dietician for a school district, you must make sure that only extra lean ground beef (16% fat) is served in the cafeteria. Further suppose that the kitchen has 8 pounds of regular ground beef (30% fat) on hand. How many pounds of super lean ground beef (12% fat) must be purchased and added to the regular ground beef to obtain a mixture that has the correct fat content?

© Baerbel Schmidt/Getty Images

58. PHARMACISTS   How many liters of a 1% glucose solution should a pharmacist mix with 0.5 liter of a 5% glucose solution to obtain a 2% glucose solution?

59. DAIRY FOODS   Cream is approximately 22% butterfat. How many gallons of cream must be mixed with milk testing at 2% butterfat to get 20 gallons of milk containing 4% butterfat?

60. FLOOD DAMAGE   One website recommends a 6% chlorine bleach-water solution to remove mildew. A chemical lab has 3% and 15% chlorine bleach-water solutions in stock. How many gallons of each should be mixed to obtain 100 gallons of the mildew spray?

61. DILUTING SOLUTIONS   How much water should be added to 20 ounces of a 15% solution of alcohol to dilute it to a 10% alcohol solution?

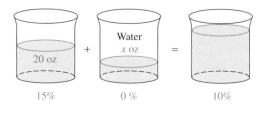

62. EVAPORATION   The beaker shown below contains a 2% saltwater solution.

  a. How much water must be boiled away to increase the concentration of the salt solution from 2% to 3%?

  b. Where on the beaker would the new water level be? (mL means milliliter.)

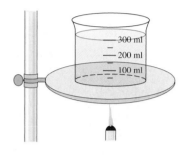

WRITING

63. If a car travels at 60 mph for 30 minutes, explain why the distance traveled is not $60 \cdot 30 = 1,800$ miles.

64. If a mixture is to be made from solutions with concentrations of 12% and 30%, can the mixture have a concentration less than 12%? Can the mixture have a concentration greater than 30%? Explain.

65. Write a mixture problem that can be represented by the following verbal model and equation.

The value of the regular coffee	plus	the value of the gourmet coffee	equals	the value of the blend.
$4x$	$+$	$7(40 - x)$	$=$	$5(40)$

66. Write a uniform motion problem using the facts entered in the table.

	r	$\cdot\ t =$	d
West	8	t	$8t$
East	6	t	$6t$
			Total: 24 mi.

REVIEW

Solve each equation.

67. $9x = 6x$

68. $7a + 2 = 12 - 4(a - 3)$

69. $\dfrac{8(y - 5)}{3} = 2(y - 4)$

70. $\dfrac{t - 1}{3} = \dfrac{t + 2}{6} + 2$

CHALLENGE PROBLEMS

71. Determine a set of 5 values such that the mean is 10, the median is 8, and the mode is 2.

72. Solve the following problem. Then explain why the solution does not make sense.

   Adult tickets cost $4 and student tickets cost $2. Sales of 71 tickets bring in $245. How many of each were sold?

CHAPTER 1
Summary & Review

SECTION 1.1   The Language of Algebra

DEFINITIONS AND CONCEPTS	EXAMPLES

In this course, we will mathematically **model** real-life situations using verbal sentences, mathematical statements (in symbols), tables, and graphs.

For more examples of **tables, bar graphs** and **line graphs** see pages 6–7.

Suppose we want to purchase carpeting for a room and it costs $20 a square yard.

Verbal model: The cost c (in dollars) to carpet the room is $20 times the area a of the room (in square yards).

Mathematical model: $c = 20a$

Table model   *Graphical model*

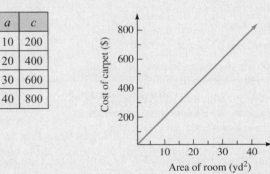

a	c
10	200
20	400
30	600
40	800

A **variable** is a letter (or symbol) that stands for a number.

Algebraic expressions contain numbers, variables, and mathematical operations such as addition, subtraction, multiplication, division, powers, or roots.

An **equation** is a mathematical sentence that contains an = symbol. The = symbol indicates that the expressions on either side of it have the same value.

Variables: x, t, a, and m

Expressions:

$$5y + 7, \quad \frac{12 - x}{5}, \quad \text{and} \quad 8a(b - 3)$$

Equations:

$$3x + 4 = 8, \quad \frac{t}{9} = 12, \quad \text{and} \quad I = Prt$$

REVIEW EXERCISES

1. Translate each verbal model into a mathematical model.

  a. The cost C (in dollars) to rent t tables is $15 more than the product of $2 and t.

  b. A rectangle has an area of 25 in.^{2}. The length of the rectangle is the quotient of its area and its width.

  c. The waiting period for a business license is now 3 weeks less than it used to be.

2. To determine the cooking time for prime rib, a cookbook suggests using the equation $T = 30p$, where T is the cooking time in minutes and p is the weight of the prime rib in pounds. Use this equation to complete the table.

p	T
6.0	
6.5	
7.0	
7.5	
8.0	

3. Use the data from the table in Exercise 2 to draw each type of graph.

  a. Bar graph   **b.** Line graph

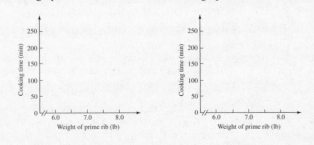

4. Explain the difference between an expression and an equation. Give examples.

SECTION 1.2   The Real Numbers

DEFINITIONS AND CONCEPTS	EXAMPLES
Important sets of numbers	
The set of **natural numbers:**	$\mathbb{N} = \{1, 2, 3, 4, 5, 6, 7, 8, 9, 10, \ldots\}$
The set of **whole numbers:**	$\mathbb{W} = \{0, 1, 2, 3, 4, 5, 6, 7, 8, 9, 10, \ldots\}$
The set of **integers:**	$\mathbb{Z} = \{\ldots, -4, -3, -2, -1, 0, 1, 2, 3, 4, \ldots\}$
The set of **prime numbers:**	$\{2, 3, 5, 7, 11, 13, 17, 19, 23, 29, \ldots\}$
The set of **composite numbers:**	$\{4, 6, 8, 9, 10, 12, 14, 15, 16, 18, \ldots\}$
The set of integers divisible by 2 are **even integers.**	$\{\ldots, -8, -6, -4, -2, 0, 2, 4, 6, 8, \ldots\}$
The set of integers not divisible by 2 are **odd integers.**	$\{\ldots, -7, -5, -3, -1, 1, 3, 5, 7, \ldots\}$
The set of **rational numbers,** denoted by the symbol $\mathbb{Q}$, contains the numbers that can be written as fractions in the form $\frac{a}{b}$, where a and b are integers and $b \neq 0$.	Rational numbers: $$-6, \quad -3.1, \quad 0, \quad \frac{11}{12}, \quad 9\frac{4}{5}, \quad \text{and} \quad 87$$
Terminating and **repeating decimals** can be expressed as fractions and are therefore, rational numbers.	Rational numbers: $$0.25 = \frac{1}{4} \quad \text{and} \quad -0.\overline{6} = -\frac{2}{3}$$
The set of **irrational numbers,** denoted by the symbol $\mathbb{H}$, contains the nonterminating, nonrepeating decimals. An irrational number cannot be expressed as a fraction with an integer numerator and a nonzero integer denominator.	Irrational numbers: $$\sqrt{2}, \quad \pi, \quad -\frac{\sqrt{11}}{4}, \quad 73.050050005\ldots, \quad \text{and} \quad -32\pi$$
The set of **real numbers,** denoted by the symbol $\mathbb{R}$, contains the numbers that are either a rational or an irrational number.	All of the numbers previously listed are real numbers.
Every real number corresponds to a point on the **number line,** and every point on the number line corresponds to exactly one real number.	Graph the set $\left\{-2, -0.75, 1\frac{3}{4}, \pi\right\}$ on a number line.
Symbols used with sets	
The symbol $\in$ is used to indicate that an **element** belongs to a set.	$6 \in \mathbb{N}$ 6 is an element of the set of natural numbers. $3 \notin \mathbb{H}$ 3 is not an element of the set of irrational numbers.
When all the members of one set are also members of a second set, we say the first set is a **subset** of the second set.	
The symbol $\subseteq$ is used to indicate that one set is a **subset** of another set.	$\mathbb{N} \subseteq \mathbb{W}$ The set of natural numbers is a subset of the set of whole numbers. $\mathbb{Q} \nsubseteq \mathbb{Z}$ The set of rational numbers is not a subset of the set of integers.

SECTION 1.2 The Real Numbers—*continued*

DEFINITIONS AND CONCEPTS	EXAMPLES
Inequality symbols $\neq$ means "is not equal to" $<$ means "is less than" $\leq$ means " is less than or equal to" $>$ means "is greater than" $\geq$ means "is greater than or equal to"	$3 \neq 5$ $0.23 < 0.24$ $8 \leq 12$ and $-9 \leq -9$ $\dfrac{3}{4} > \dfrac{2}{3}$ $-5 \geq -6$ and $8 \geq 8$
Two numbers are called **opposites** if they are the same distance from 0 on the number line but are on opposite sides of it. The **opposite of a number** a is $-a$. $\qquad -(-a) = a$	Opposites: 3 and -3 $-(-11) = 11$ and $-(-20) = 20$
The **absolute value** of a number is the distance on the number line between the number and 0. For any real number x: $\begin{cases} \text{If } x \geq 0, \text{ then } \lvert x \rvert = x \\ \text{If } x < 0, \text{ then } \lvert x \rvert = -x \end{cases}$	Find the value of each expression: $\lvert 7.2 \rvert = 7.2 \qquad \lvert -1 \rvert = 1 \qquad -\left\lvert \dfrac{13}{5} \right\rvert = -\dfrac{13}{5}$

REVIEW EXERCISES

List the numbers in $\left\{-5, 0, -\sqrt{3}, 2.4, 7, -\frac{2}{3}, -3.\overline{6}, \pi, \frac{15}{4}, 0.13242368\ldots\right\}$ that belong to the following sets.

5. a. Natural numbers   **b.** Whole numbers

6. a. Integers   **b.** Rational numbers

7. a. Irrational numbers   **b.** Real numbers

8. a. Negative numbers   **b.** Positive numbers

9. a. Prime numbers   **b.** Composite numbers

10. a. Even integers   **b.** Odd integers

11. Graph the prime numbers between 20 and 30 on the number line.

12. Graph the set $\left\{2.75, 2.\overline{3}, \sqrt{7}, \frac{8}{3}, \frac{3\pi}{4}\right\}$ on the number line.

Determine whether each statement is true or false.

13. a. $0 \in \mathbb{N}$ **b.** $\mathbb{W} \subseteq \mathbb{Z}$

14. a. $-5 \notin \mathbb{R}$ **b.** $\mathbb{Q} \not\subseteq \mathbb{H}$

15. Use one of the symbols $>$ or $<$ to make each statement true.

     a. $-16 \quad -17$ **b.** $-(-1.8) \quad 2\frac{1}{2}$

16. Determine whether each statement is true or false.

     a. $23.000001 \geq 23.1$ **b.** $-11 \leq -11$

Find the value of each expression.

17. $\lvert -18 \rvert$ **18.** $-\lvert -6.26 \rvert$

SECTION 1.3 Operations with Real Numbers

DEFINITIONS AND CONCEPTS	EXAMPLES
Adding real numbers: **To add two positive numbers,** add them in the usual way. The final answer is positive.	Add: $3 + 5 = 8$ $5.2 + 7.3 = 12.5$
To add two negative numbers, add their absolute values and make the final answer negative.	Add: $-3 + (-5) = -8$ $-5.2 + (-7.3) = -12.5$
To add a positive number and a negative number, subtract the smaller absolute value from the larger. 1. If the positive number has the larger absolute value, make the final answer positive. 2. If the negative number has the larger absolute value, make the final answer negative.	Add: $5 + (-3) = 2$ $-5.2 + 7 = 1.8$ Add: $6 + (-20) = -14$ $-9 + 3 = -6$
To **subtract two real numbers,** add the first to the opposite of the number to be subtracted. For any real numbers a and b, $$a - b = a + (-b)$$	Subtract: $4 - 7 = 4 + (-7) = -3$ $6 - (-8) = 6 + 8 = 14$ $-1 - (-2) = -1 + 2 = 1$ $-3 - 4 = -3 + (-4) = -7$
Multiplying and dividing real numbers: With **unlike signs,** multiply (or divide) their absolute values and make the final answer negative.	Multiply: $-5(7) = -35$ Divide: $\dfrac{-12}{4} = -3$
With **like signs,** multiply (or divide) their absolute values and make the final answer positive.	Multiply: $-6(-8) = 48$ Divide: $\dfrac{-25}{-5} = 5$
x^n is a *power of x. x* is the **base,** and n is the **exponent.** An exponent represents repeated multiplication. $$x^n = \overbrace{x \cdot x \cdot x \cdot \, \cdots \, \cdot x}^{n \text{ factors of } x}$$	In 4^3, the base is 4 and 3 is the exponent. $4^3 = 4 \cdot 4 \cdot 4$ $(-3)^4 = (-3)(-3)(-3)(-3)$
A number b is a **square root** of a if $b^2 = a$. If $a > 0$, $\sqrt{a}$ represents the **principal** (positive) **square root** of a.	Find each square root: $\sqrt{16} = 4$ Because $4^2 = 16$. $\sqrt{144} = 12$ Because $12^2 = 144$.
Order of operations rule: 1. Perform all calculations within parentheses in the order listed in steps 2–4, working from the innermost pair to the outermost pair. 2. Evaluate all powers and roots. 3. Perform all multiplications and divisions, working from left to right. 4. Perform all additions and subtractions, working from left to right. 5. When all grouping symbols have been removed, repeat steps 2–4 to finish the calculation. If a fraction is present, evaluate the numerator and denominator separately, and then simplify the fraction, if possible.	Evaluate: $3 + 2[-4 - 7(5 - 2)^2]$ $= 3 + 2[-4 - 7(3)^2]$ Work within the parentheses. $= 3 + 2[-4 - 7(9)]$ Evaluate the power. $= 3 + 2(-4 - 63)$ Do the multiplication. $= 3 + 2(-67)$ Do the subtraction with parentheses. $= 3 - 134$ Do the multiplication. $= -131$ Do the subtraction.

SECTION 1.3   Operations with Real Numbers—*continued*

DEFINITIONS AND CONCEPTS	EXAMPLES
To **evaluate an algebraic expression,** substitute the values for the variables and then use the order of operations rule.	Evaluate $\dfrac{a + b}{2(b - a)}$ for $a = 3$ and $b = -5$. $\dfrac{a + b}{2(b - a)} = \dfrac{3 + (-5)}{2(-5 - 3)}$ Substitute 3 for a and -5 for b. $= \dfrac{-2}{2(-8)}$ Evaluate numerator and denominator separately. $= \dfrac{-2}{-16}$ $= \dfrac{1}{8}$ Simplify the fraction.

REVIEW EXERCISES

Perform the operations.

19. $-13 + (-14)$

20. $-70.5 + 80.6$

21. $-\dfrac{1}{2} - \dfrac{1}{4}$

22. $-6 - (-8)$

23. $(-4.2)(-3.0)$

24. $-\dfrac{1}{10} \cdot \dfrac{5}{16}$

25. $\dfrac{-2.2}{-11}$

26. $-\dfrac{9}{8} \div 21$

27. $15 - 25 - 23$

28. $-3.5 + (-7.1) + 4.9$

29. $-3(-5)(-8)$

30. $-1(-1)(-1)(-1)$

Evaluate each expression.

31. $(-3)^5$

32. $\left(-\dfrac{2}{9}\right)^2$

33. 0.4 cubed

34. -5^2

Evaluate each expression.

35. $\sqrt{4}$

36. $-\sqrt{100}$

37. $\sqrt{\dfrac{9}{25}}$

38. $\sqrt{0.64}$

Evaluate each expression.

39. $-6 + 2(-5)^2$

40. $\dfrac{-20}{4} - (-3)(-2)\left(-\sqrt{1}\right)$

41. $4 - (5 - 9)^2$

42. $4 + 6[-1 - 5(25 - 3^3)]$

43. $2|-1.3 + (-2.7)|$

44. $\dfrac{(7 - 6)^4 + 32}{36 - (\sqrt{16} + 1)^2}$

45. $(-10)^3 \left(\dfrac{-6}{-2}\right)(-1)$

46. $-(-2 \cdot 4)^2 \div 8 \cdot 2$

Evaluate the algebraic expression for the given values of the variables.

47. $(x + y)(x^2 - xy + y^2)$ for $x = -2$ and $y = 4$

48. $\dfrac{-b - \sqrt{b^2 - 4ac}}{2a}$ for $a = 2$, $b = -3$, and $c = -2$

SECTION 1.4   Simplifying Algebraic Expressions Using Properties of Real Numbers

DEFINITIONS AND CONCEPTS	EXAMPLES
A **term** is a product or quotient of numbers and/or variables. A single number or variable is also a term. A term that is a number called is a **constant term.**	Terms: z, $-7t$, $3.7x^7$, $\dfrac{3}{5}y$, $25ab^2c^3$, and 6 (a constant)
The numerical factor of a term is called its **coefficient.**	The coefficients of the previous terms are 1, -7, 3.7, $\dfrac{3}{5}$, 25, and 6

SECTION 1.4 Simplifying Algebraic Expressions Using Properties of Real Numbers–*continued*

DEFINITIONS AND CONCEPTS	EXAMPLES

Properties of real numbers

The **commutative properties** enable us to add or multiply two numbers in either order and obtain the same result.

$a + b = b + a$ Commutative property of addition

$3 + (-5) = -5 + 3$ Reorder the addends.

$ab = ba$ Commutative property of multiplication

$3(-5) = -5(3)$ Reorder the factors.

The **associative properties** enable us to group the numbers in an addition or multiplication any way that we wish and get the same result.

$(a + b) + c = a + (b + c)$ Associative property of addition

$(-5 + 7) + 4 = -5 + (7 + 4)$ Regroup the addends.

$(ab)c = a(bc)$ Associative property of multiplication

$(-5 \cdot 7)4 = -5(7 \cdot 4)$ Regroup the factors.

---

0 is the **additive identity**:

$a + 0 = a$ and $0 + a = a$

$5 + 0 = 5$ The sum of a real number and 0 is the number.

1 is the **multiplicative identity**:

$1 \cdot a = a$ and $a \cdot 1 = a$

$1 \cdot 22 = 22$ The product of any real number and 1 is the number.

Multiplication property of 0:

$a \cdot 0 = 0$ and $0 \cdot a = 0$

$-1.78 \cdot 0 = 0$ The product of any real number and 0 is 0.

The **additive inverse property**:

$a + (-a) = 0$ and $-a + a = 0$

$9 + (-9) = 0$ The sum of a real number and its opposite is 0.

The **multiplicative inverse property**:

$a \cdot \dfrac{1}{a} = 1$ and $\dfrac{1}{a} \cdot a = 1$ $(a \neq 0)$

$7 \cdot \dfrac{1}{7} = 1$ The product of a real number and its reciprocal is 1.

Division properties of real numbers:

$\dfrac{a}{1} = a \qquad \dfrac{a}{a} = 1 \qquad \dfrac{0}{a} = 0$

$\dfrac{a}{0}$ is undefined $\dfrac{0}{0}$ is indeterminate

Divide, if possible:

$\dfrac{6.8}{1} = 6.8 \qquad \dfrac{3}{3} = 1$

$\dfrac{0}{15} = 0 \qquad \dfrac{-2.03}{0}$ is undefined

The **distributive property** and its related forms can be used to remove parentheses.

$a(b + c) = ab + ac \qquad a(b - c) = ab - ac$

$(b + c)a = ba + ca \qquad (b - c)a = ba - ca$

The extended distributive property:

$a(b + c + d + e + \cdots) =$
$\qquad ab + ac + ad + ae + \cdots$

Multiply:

$3(a + 4) = 3a + 3 \cdot 4 \qquad (x - 5)x = x \cdot x - 5 \cdot x$
$\qquad\quad = 3a + 12 \qquad\qquad\qquad = x^2 - 5x$

$-4(2a + b - 7) = -4(2a) + (-4)b - (-4)(7)$
$\qquad\qquad\quad = -8a - 4b + 28$

79. $0.0035 = 0.25g$

80. $0 = x + 4$

81. $11 - 5x = -1$

82. $-3x - 7 + x = 6x + 20 - 5x$

83. $-4(y - 1) + (-3) = 25$

84. $5 + 3[2 - 13(x - 1)] = 17 - 18x$

85. $\dfrac{8}{3}(x - 5) = \dfrac{2}{5}(x - 4)$

86. $\dfrac{3y}{4} - 14 = -\dfrac{y}{3} - 1$

87. $-k = -0.06$

88. $\dfrac{5}{4}p = -10$

89. $\dfrac{4t + 1}{3} - \dfrac{t + 5}{6} = \dfrac{t - 3}{6}$

90. $33.9 - 0.5(75 - 3x) = 0.9$

Solve each equation. If the equation is an identity or a contradiction, so state.

91. $2(x - 6) = 10 + 2x$

92. $-5x + 2x - 1 = -(3x + 1)$

SECTION 1.6   Solving Formulas; Geometry

DEFINITIONS AND CONCEPTS	EXAMPLES
A **formula** is an equation that states a relationship between two or more variables.	Turn to the inside back cover for a complete list of geometric formulas.
The **perimeter** of a plane geometric figure is the distance around it. The **area** of a plane geometric figure is the amount of surface that it encloses. The **volume** of a three-dimensional geometric figure is the amount of space it encloses.	Find the volume of a cone whose base has a radius of 6 cm and whose height is 8 cm. $V = \dfrac{1}{3}\pi r^2 h$ This is the formula for the volume of a cone. $= \dfrac{1}{3}\pi(6)^2(8)$ Substitute 6 for r and 8 for h. $= \dfrac{1}{3}\pi(288)$ Evaluate $(6)^2(8)$. $= 96\pi$ Multiply $\frac{1}{3}$ and 288. The exact volume is 96π cm³. Rounded to the nearest tenth of a cubic centimeter, the approximate volume is 301.6 cm³.
To **solve a formula for a specified variable** means to isolate that variable on one side of the equation, with all other variables and constants on the opposite side.	Solve $F = \dfrac{mMg}{r^2}$ for M. $Fr^2 = mMg$ Multiply both sides by r^2. $\dfrac{Fr^2}{mg} = M$ To isolate M, divide both sides by mg. $M = \dfrac{Fr^2}{mg}$ Write M on the left side.

REVIEW EXERCISES

93. Find the perimeter of a trapezoid whose parallel sides measure 10.5 feet and 12.5 feet and whose nonparallel sides measure 3.5 feet and 4.5 feet.

94. Find the circumference and the area of a circle with a diameter of 17 centimeters. Round to the nearest hundredth.

95. Find the volume of a sphere with a radius of 7.5 meters. Round to the nearest hundredth.

96. HIGHWAY SAFTEY CONES

a. Find the area covered by the square rubber base if its sides are 10 inches long.

b. The safety cone is centered atop the base, as shown. Give its exact volume and its approximate volume, rounded to the nearest tenth.

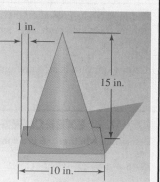

Solve each formula for the specified variable.

97. $V = \dfrac{1}{3}\pi r^2 h$ for h

98. $K = \dfrac{Mv_0^2 + Iw^2}{2}$ for M

99. $l = a + (n - 1)d$ for d

100. $9x - 5y = 35$ for y

SECTION 1.7 Using Equations to Solve Problems

DEFINITIONS AND CONCEPTS	EXAMPLES

Problem-solving strategy:

1. Analyze the problem.

2. Form an equation.

3. Solve the equation.

4. State the conclusion.

5. Check the result.

Some problem-solving hints

Diagrams are often helpful in solving application problems. See page 77.

For problems that deal with quantities that have a value (see page 78), use the relationship:

Number · value = total value

It is often helpful to list the facts of a number-value problem in a **table**.

Sometimes a **geometric fact** or formula can be used to solve a problem. See pages 79–80.

The given facts of a problem often suggest a **formula** that can be used to model the situation mathematically. See page 81.

BILLIONAIRES In 2007, Forbes Magazine ranked Bill Gates and Warren Buffet as the two richest Americans. Their combined net worth was estimated to be $108.4 billion, with Gates the wealthier by $3.6 billion. Find the net worth of each man.

Analyze the problem   The phrase *combined net worth* suggests addition. If Gates was the wealthier, then his net worth was $3.6 billion *more than* Buffet's.

Form an equation   Let $x =$ Buffet's net worth in billions of dollars. Since Gates is the wealthier, $x + 3.6 =$ Gates's net worth. We can use the words of the problem to form an equation.

Buffet's net worth	plus	Gates's net worth	equals	$108.4 billion.
x	$+$	$x + 3.6$	$=$	108.4

Solve the equation

$2x + 3.6 = 108.4$ Combine like terms.

$\quad 2x = 104.8$ Subtract 3.6 from both sides.

$\quad\; x = 52.4$ Divide both sides by 2.

To find Gates's net worth we evaluate $x + 3.6$ for $x = 52.4$.

$x + 3.6 = 52.4 + 3.6 = 56$

State the conclusion   In 2007, Warren Buffet's net worth was $52.4 billion and Bill Gates's net worth was $56 billion.

Check the result   The sum of $52.4 billion and $56 billion is $108.4 billion, and $56 billion is $3.6 billion more than $52.4 billion. The answers check.

REVIEW EXERCISES

101. AIRPORTS   The world's two busiest airports are Hartsfield Atlanta International and Chicago O'Hare International. Together they served 161 million passengers in 2006, with Atlanta handling 8.6 million more than O'Hare. How many passengers did each airport serve?

102. TUITION   A private school reduces the monthly tuition cost of $245 by $5 per child if a family has more than one child attending the school. Write an algebraic expression that gives the monthly tuition cost per child for a family having c children.

103. WAREHOUSING   A large warehouse stores 150 more computers than printers. The monthly storage cost for a computer is $2.50 and a printer is $1.50. If storage for the computers and printers is $2,775 per month, how many printers are in the warehouse?

104. CABLE TV   A 186-foot television cable is to be cut into four pieces. Find the length of each piece if each successive piece is 3 feet longer than the previous one.

105. TOOLING   The illustration shows the angle at which a drill is to be held when drilling a hole into a piece of aluminum. Find the measures of both labeled angles.

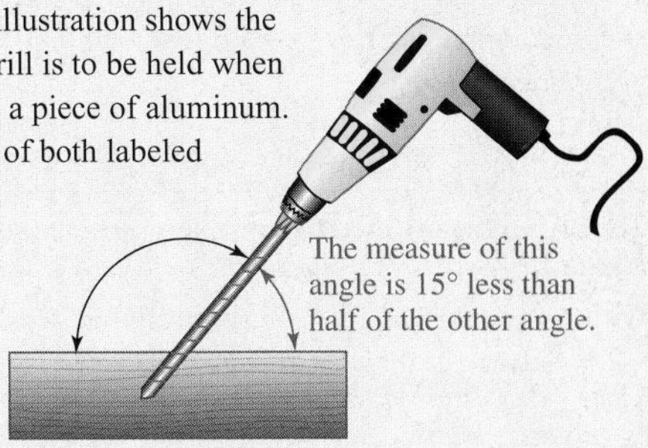

The measure of this angle is 15° less than half of the other angle.

106. COLLECTIBLES   In North America, most new movie releases are advertised using a poster size commonly referred to as a *one-sheet*. A one-sheet movie poster is rectangular, has a perimeter of 134 inches, and its length is 13 inches longer than its width. Find the dimensions of a one-sheet.

SECTION 1.8   More about Problem Solving

DEFINITIONS AND CONCEPTS	EXAMPLES
Percent means parts per one hundred.	$7\% = \dfrac{7}{100} = 0.07$ $125\% = \dfrac{125}{100} = 1.25$
One method to solve percent problems is to use the given facts to write a **percent sentence** of the form: ☐ is ☐ % of ☐ ? Then we translate the sentence to mathematical symbols and solve the resulting equation. Always find the **percent of increase** (or decrease) with respect to the *original* amount.	TAXES In Texas, \$31.25 in state sales tax is charged on a \$500 purchase. What is the Texas state sales tax rate? 31.25 is what percent of 500? ↓ ↓ ↓ ↓ ↓ 31.25 = x · 500 $31.25 = 500x$ $\dfrac{31.25}{500} = x$ To isolate x, divide both sides by 500. $0.0625 = x$ Do the division. $6.25\% = x$ Change the decimal to a percent. The Texas state sales tax rate is 6.25%.
Three types of averages are commonly used in statistics: $\textbf{Mean} = \dfrac{\text{sum of the values}}{\text{number of values}}$ The **median** is the middle value after the values have been arranged in increasing order. The **mode** is the value that occurs most often.	Consider the values: 3, 5, 5, 6, 7, 8, 8, 8, 9, 10. The mean is $\dfrac{3 + 5 + 5 + 6 + 7 + 8 + 8 + 8 + 9 + 10}{10} = 6.9$ The median is the average of the middle terms: $\dfrac{7 + 8}{2} = \dfrac{15}{2} = 7.5$ Since 8 occurs most often, 8 is the mode.
To solve **investment** problems involving simple interest, use the formula $I = Prt$ Interest = principal · rate · time	See page 92 for an example.
To solve **uniform motion** problems, use the formula $d = rt$ Distance = rate · time	See page 93 for an example.
To solve problems where a **dry mixture** of a specified value is created from two differently priced components, use Amount · price = total value	See page 94 for an example.
To solve a **liquid mixture** problem, where a desired strength solution is to be made from two solutions with different strengths (concentrations), use $\dfrac{\text{Amount of}}{\text{solution}} \cdot \dfrac{\text{strength of}}{\text{the solution}} = \dfrac{\text{amount of}}{\text{pure ingredient}}$	See page 95 for an example.

REVIEW EXERCISES

107. GROUNDHOG DAY   According to groundhog.org, the weather-predicting groundhog has emerged from his burrow and seen his shadow 96 times, which is 80% of the years on record. How many groundhog days does this website have on record?

108. EARLY REGISTRATION   An early bird discount lowers the registration fee for a financial seminar from $550 to $375. Find the percent of markdown. Round to the nearest percent.

109. CAR SALES

   a. Determine the percent of increase in the number of Toyota Camrys sold in 2006 compared with 2005. Round to the nearest tenth of one percent.

   b. Determine the percent of decrease in the number of Honda Accords sold in 2006 compared with 2005. Round to the nearest tenth of one percent.

The Two Top-Selling Passenger Cars in the U.S.	
2006	
1. Toyota Camry	448,445
2. Honda Accord	354,441
2005	
1. Toyota Camry	433,703
2. Honda Accord	369,293

Source: *MSN Autos*

110. HURRICANES   The following table gives the number of hurricanes that made landfall in the United States for each of the years 1993–2006. Find the mean, median, and mode.

1993	1994	1995	1996	1997	1998	1999	2000	2001	2002	2003	2004	2005	2006
1	0	3	2	1	3	2	0	0	1	2	6	7	0

Source: Insurance Information Institute

111. INVESTMENTS   Sally has $25,000 to invest. She invests some money at 10% interest and the rest at 9%. If her total annual income from these two investments is $2,430, how much does she invest at each rate?

112. PAPARAZZI   A celebrity leaves a nightclub in his car and travels at 1 mile per minute (60 mph) trying to avoid a tabloid photographer. One minute later, the photographer leaves the nightclub on his motorcycle, traveling at 1.5 miles per minute (90 mph) in pursuit of the celebrity. How long will it take the photographer to catch up with the celebrity?

113. PEST CONTROL   How much of a 4% pesticide solution must be added to 20 gallons of a 12% pesticide solution to dilute it to a 10% solution?

114. COFFEE   Mild coffee that sells for $7.50 per pound is to be mixed with a robust coffee that sells for $8.40 per pound to make 90 pounds of a mixture that will be sold for $7.90 per pound. How many pounds of each type of coffee should be used?

CHAPTER 1
Test

1. Fill in the blanks.

   a. For any nonzero real number a, $\frac{a}{0}$ is _____.

   b. $>$, $\geq$, $<$, and $\leq$ are called _____ symbols.

   c. $9x^2$ and $7x^2$ are _____ _____ because they have the same variable raised to exactly the same power.

   d. To _____ an equation means to find all of the values of the variable that make the equation true.

   e. The _____ property of _____ says that adding the same number to both sides of an equation does not change the solution.

2. Translate each verbal model into a mathematical model.

   a. Each test score T was increased by 10 points to give a new adjusted test score s.

   b. The area A of a triangle is the product of one-half the length of the base b and the height h.

3. COUNTING CALORIES   Refer to the graph on the next page that gives the number of calories in a given number of glazed donuts.

   a. What units are used to scale the vertical axis?

   b. How many calories are in one-half dozen glazed donuts?

   c. How many glazed donuts did a person eat if his calorie intake from the donuts was 700?

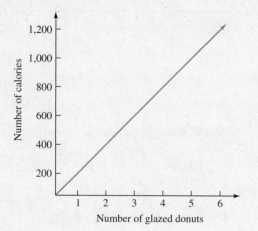

4. Consider the set: $\left\{-2, \pi, 0, -3\frac{3}{4}, 9.2, \frac{14}{5}, 5, -\sqrt{7}\right\}$

   a. Which numbers are integers?

   b. Which numbers are rational numbers?

   c. Which numbers are irrational numbers?

   d. Which numbers are real numbers?

5. Determine whether each statement is true or false.

   a. $-6 \in \mathbb{Z}$ **b.** $76 \notin \mathbb{N}$

   c. $\mathbb{W} \subseteq \mathbb{R}$ **d.** $\mathbb{H} \not\subseteq \mathbb{Q}$

Graph each set on the number line.

6. $\left\{\dfrac{7}{6}, \dfrac{\pi}{2}, 1.8234503\ldots, \sqrt{3}, 1.\overline{91}\right\}$

7. The set of prime numbers less than 12

8. Determine whether each statement is true or false.

   a. $-|-2.78| > -(-2.71)$ **b.** $(-3)^2 \le -3^2$

Evaluate each expression.

9. $-\dfrac{5}{3} \div \left(-\dfrac{25}{4}\right)$ **10.** $\dfrac{2|-4 - 2(3-1)|}{-3\left(\sqrt{9}\right)(-2)}$

11. $10 - 3[5^2 - 6(-1-1)^3]$

12. Evaluate the expression for $a = 2$, $b = -3$, and $c = 4$.

$$\frac{(-3b + c)^2 - 17a}{-b + a^2bc}$$

13. PEDIATRICS   Some doctors use Young's rule in calculating dosage for infants and children.

$$\frac{\text{Age of child}}{\text{Age of child} + 12}\left(\begin{array}{c}\text{average} \\ \text{adult dose}\end{array}\right) = \text{child's dose}$$

The adult dose of Achromycin is 250 milligrams (mg). What is the dose for an 8-year-old child?

Determine which property of real numbers justifies each statement.

14. a. $3 + 5 = 5 + 3$

   b. $x(yz) = (xy)z$

   c. $-17 + 17 = 0$

   d. $\dfrac{1}{2} \cdot 1 = \dfrac{1}{2}$

Simplify each expression.

15. $11.1n^2 - 7.3n + 15.1n - 9.8$

16. $-5(9s)(-2t)$

17. $-7(c - 4) - 5[3(c - 4) - 2(c + 2)]$

18. $\dfrac{2}{9}(xy + 45x) - \dfrac{1}{4}(xy - 24x)$

Solve each equation.

19. $9(x + 4) + 4 - 8x = 4(x - 5) + x$

20. $\dfrac{m - 1}{5} = \dfrac{2m - 3}{3} - 2$

21. $6 - (x - 3) - 5x = 3[1 - 2(x + 2)]$

22. $\dfrac{1}{2}r - \dfrac{7}{6} = -\dfrac{1}{3}r + \dfrac{53}{6}$

23. Use a check to determine whether 6.7 is a solution of $1.6y + (-3) = y + 1.02$.

24. Solve $P = L + \dfrac{s}{f}i$ for i.

25. Solve $y - y_1 = m(x - x_1)$ for x_1.

26. CROP CIRCLES   In 1992, two Hungarian high school students were charged for the damage that they caused in creating a 36-meter diameter crop circle in a wheat field. Find the area covered by the crop circle. Round to the nearest square meter.

27. HAND TOOLS   With each pass that a craftsman makes with a sander over a piece of fiberglass, he removes 0.03125 inch of thickness. If the fiberglass was originally 0.9375 inch thick, how many passes are needed to obtain the desired thickness of 0.6875 inch?

28. RENTALS   The owners of an apartment building rent equal numbers of 1- and 2-bedroom units. The monthly rent for a 1-bedroom is $950, and a 2-bedroom is $1,200. If the total monthly income is $53,750, how many of each type of unit are there?

29. ISOSCELES TRIANGLES   The measure of a base angle of an isosceles triangle is 5° more than eight times the measure of the vertex angle. Find the measure of each angle of the triangle.

30. CALCULATORS   The viewing window of a calculator has a perimeter of 26 centimeters and is 5 centimeters longer than it is wide. Find the dimensions of the window.

31. FUEL EFFICIENCY   Use the data below to determine the percent of increase in U.S. sales of hybrid vehicles from 2005 to 2006. Round to the nearest percent.

U.S. Gas–Electric Hybrid Vehicle Sales
2005: 199,148
2006: 254,545

Source: MSNBC.com

32. AIRLINE ACCIDENTS   Refer to the data in the table.

   a. Find the mean. Round to the nearest tenth.

   b. Find the median.

   c. Find the mode.

Number of Major Accidents for U.S. Air Carriers (1997–2006)				
1997	1998	1999	2000	2001
2	0	2	3	5
2002	2003	2004	2005	2006
1	2	4	2	1

Source: National Transportation Safety Board

33. INVESTING   An investment club invested part of $10,000 at 9% annual interest and the rest at 8%. If the annual income from these investments was $860, how much was invested at 8%?

34. RENTAL CARS   While waiting for his car to be repaired, a man rents a car for $17 per day and 33 cents per mile. His insurance company will pay up to $200 of the rental fee. If he needs the car for four days, how many miles of driving will his policy cover?

35. MIXING ALLOYS   How many ounces of a 40% copper alloy must be mixed with 10 ounces of a 10% copper alloy to obtain an alloy that is 25% copper?

36. MEN'S COLOGNE   How many ounces of *Skin Soother* men's cologne (unit price: $2.40 per ounce) must be mixed with *Cool Sport* men's cologne (unit price: $1.60 per ounce) to make 8 ounces of a mixture having a unit price of $1.90 per ounce?

GROUP PROJECT

WRITING FRACTIONS AS DECIMALS AND AS PERCENTS

Overview: This is a good activity to try at the beginning of the course. You can become acquainted with other students in your class while you review some important arithmetic skills.

Instructions: Form groups of 5 students. Select one person from your group to record the group's responses on the questionnaire. Express the results in fraction form, decimal form, and as percents.

What fraction, decimal, and percent of the students in your group . . .	Fraction	Decimal	Percent
• have the letter *a* in their first names?			
• have a birthday in January or February?			
• say that vanilla is their favorite flavor of ice cream?			
• have ever been on television?			
• live more than 20 miles from campus?			
• say they enjoy rainy days?			
• work full-time or part-time?			

CHAPTER 2

Graphs, Equations of Lines, and Functions

2.1 The Rectangular Coordinate System
2.2 Graphing Linear Equations in Two Variables
2.3 Rate of Change and the Slope of a Line
2.4 Writing Equations of Lines
2.5 An Introduction to Functions
2.6 Graphs of Functions
CHAPTER SUMMARY AND REVIEW
CHAPTER TEST
Group Project
CUMULATIVE REVIEW

from *Campus to Careers*
Certified Fitness Instructor

Because of our busy schedules, many of us have difficulty making exercise a part of our daily routine. A certified fitness instructor can often provide the motivation, discipline, and instruction that a person needs to get and stay in shape. Fitness instructors plan and lead classes, weigh and measure clients, analyze records and graphs, and perform assessment and testing related to weight training and cardiovascular exercise.

Fitness instructors stress the importance of healthy eating and regular exercise. In **problem 89** of **Study Set 2.4,** you will see how a graph can be used to show the relationship between the time spent exercising and the number of calories burned.

JOB TITLE:
Certified Fitness Instructor

EDUCATION:
An increasing number of employers are requiring a bachelor's degree in a health-related field. Also, some level of training certification is often required.

JOB OUTLOOK:
Good because of rapid growth in the fitness industry.

ANNUAL EARNINGS:
$44,088 is the median annual salary.

FOR MORE INFORMATION:
www.bls.gov/oco/home.htm

Study Skills Workshop
Preparing to Learn

Many students feel that there are two types of people—those who are good at math and those who are not—and that this cannot be changed. This isn't true! Here are some suggestions that can increase your chances for success in algebra.

DISCOVER YOUR LEARNING STYLE: Are you a visual, verbal, or audio learner? Knowing this will help you determine how best to study.

GET THE MOST OUT OF THE TEXTBOOK: This book and the software that comes with it contain many student support features. Are you taking advantage of them?

TAKE GOOD NOTES: Are your class notes complete so that they are helpful when doing your homework and studying for tests?

Now Try This

1. To determine what type of learner you are, take the *Learning Style Survey* found online at http://www.metamath.com/multiple/multiple_choice_questions.html. Then, write a one-page paper explaining what you learned from the survey results and how you will use the information to help you succeed in the class.

2. To learn more about the student support features of this book, take the *Textbook Tour* found online at: http://academic.cengage.com/math/tussy/.

3. Rewrite a set of your class notes to make them more readable and to clarify the concepts and examples covered. Be sure to write your notes in outline form. Fill in any information you didn't have time to copy down in class and complete any phrases or sentence fragments.

SECTION 2.1
The Rectangular Coordinate System

Objectives

 1 Plot ordered pairs and determine the coordinates of a point.

2 Graph paired data.

3 Read graphs.

4 Find the midpoint of a line segment.

In this section, we will show how numerical relationships can be described by mathematical pictures called *graphs*. We will draw the graphs on a *rectangular coordinate system*.

1   **Plot Ordered Pairs and Determine the Coordinates of a Point.**

Many cities are laid out on a rectangular grid, as shown on the next page. For example, on the east side of Rockford, Illinois, all streets run north and south, and all avenues run east and west. If we agree to list the street numbers first, every address can be identified by using an ordered pair of numbers. If Jose Montoya lives on the corner of Third Street and Sixth Avenue, his address is given by the ordered pair (3, 6).

This is the street. ──────┐   ┌────── This is the avenue.

(3, 6)

If Lisa Kumar has an address of (6, 3), we know that she lives on the corner of Sixth Street and Third Avenue. From the figure, we can see that

- Bob Anderson's address is (4, 1).
- Rosa Vang's address is (7, 5).
- The address of the store is (8, 2).

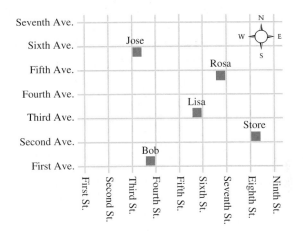

The idea of associating an ordered pair of numbers with points on a grid is attributed to the 17th-century French mathematician René Descartes. Such a grid is called a **rectangular coordinate system,** or **Cartesian coordinate system** after its inventor.

In general, a rectangular coordinate system is formed by two intersecting perpendicular number lines, as shown in the figure. The horizontal number line is usually called the ***x*-axis.** The vertical number line is usually called the ***y*-axis.**

The Language of Algebra

The prefix *quad* means four, as in quadrilateral (4 sides), quadra-phonic sound (4 speakers), and quadruple (4 times).

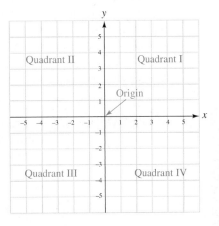

The positive direction on the *x*-axis is to the right, and the positive direction on the *y*-axis is upward. If no scale is indicated on the axes, we assume that the axes are scaled in units of 1.

The point where the axes intersect is called the **origin.** This is the 0 point on each axis. The two axes form a **coordinate plane** and divide it into four regions called **quadrants,** which are numbered using Roman numerals.

Every point on a coordinate plane can be identified by an **ordered pair** of real numbers *x* and *y*, written as (*x*, *y*). The first number in the pair is the ***x*-coordinate,** and the second number is the ***y*-coordinate.** The numbers are called the **coordinates** of the point. Some examples are $(6, -4)$, $(2, 3)$, and $(-4, 6)$.

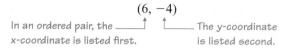

In an ordered pair, the ——┐   ┌—— The y-coordinate
x-coordinate is listed first.   is listed second.

The process of locating a point in the coordinate plane is called **graphing** or **plotting** the point. Below, we use red arrows to graph the point $(6, -4)$. Since the *x*-coordinate, 6, is positive, we begin at the origin and move 6 units to the *right* along the *x*-axis. Since the *y*-coordinate, -4, is negative, we then move *down* 4 units, and draw a dot. This locates the point $(6, -4)$, which lies in quadrant IV.

In the figure, blue arrows are used to show how to plot $(-4, 6)$. We start at the origin, move 4 units to the *left* along the *x*-axis, and then 6 units *up* and draw a dot. This locates the point $(-4, 6)$, which lies in quadrant II.

> **Caution**
> Note that the point $(-4, 6)$ is not the same as the point $(6, -4)$. This illustrates that the order of the coordinates of a point is important.

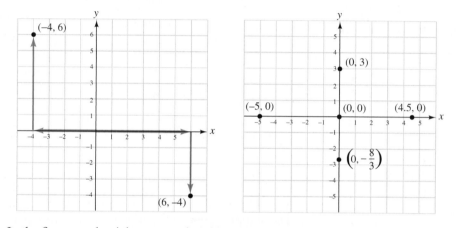

In the figure on the right, we see that the points $(-5, 0)$, $(0, 0)$, and $(4.5, 0)$ all lie on the *x*-axis. In fact, every point with a *y*-coordinate of 0 will lie on the *x*-axis. We also see that the points $\left(0, -\frac{8}{3}\right)$, $(0, 0)$, and $(0, 3)$ all lie on the *y*-axis. In fact, every point with an *x*-coordinate of 0 will lie on the *y*-axis. Note that the coordinates of the origin are $(0, 0)$.

A point may lie in one of the four quadrants or it may lie on one of the axes, in which case the point is not considered to be in any quadrant. For points in quadrant I, the *x*- and *y*-coordinates are positive. Points in quadrant II have a negative *x*-coordinate and a positive *y*-coordinate. In quadrant III, both coordinates are negative. In quadrant IV, the *x*-coordinate is positive and the *y*-coordinate is negative.

EXAMPLE 1   ***Astronomy.***   Halley's comet passes Earth every 76 years as it travels in an orbit about the sun. Use the graph to determine the comet's position for the years 1912, 1930, 1948, 1966, and 1978.

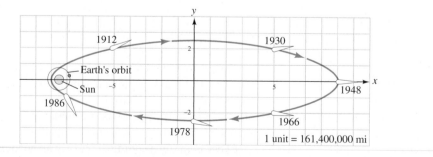

Strategy We will start at the origin and count to the left or right on the *x*-axis, and then up, or down to reach the point.

Why The movement left or right gives the *x*-coordinate of the ordered pair and the movement up or down gives the *y*-coordinate.

Solution To find the coordinates of each position, we start at the origin and move left or right along the *x*-axis to find the *x*-coordinate and then up or down to find the *y*-coordinate.

Year	Position of comet on graph	Coordinates
1912	5 units to the *left*, then 2 units *up*	$(-5, 2)$
1930	5 units to the *right*, then 2 units *up*	$(5, 2)$
1948	9 units to the *right*, no units *up* or *down*	$(9, 0)$
1966	5 units to the *right*, then 2 units *down*	$(5, -2)$
1978	No units *left* or *right*, then 2.5 units *down*	$(0, -2.5)$

Self Check 1 Find the position of the comet in 1986.

Now Try **Problems 27 and 29**

❷ Graph Paired Data.

Every day, we deal with quantities that are related.

- The distance we travel depends on how fast we are going.
- Your test score depends on the amount of time you study.
- The height of a toy rocket depends on the time since it was launched.

Graphs are often used to show relationships between two quantities. For example, suppose we know the height of a toy rocket at 1-second intervals from 0 to 6 seconds after it is launched. We can list this information in a table and write each data pair as an ordered pair of the form (time, height).

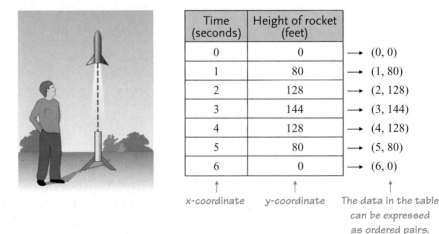

Time (seconds)	Height of rocket (feet)	
0	0	⟶ (0, 0)
1	80	⟶ (1, 80)
2	128	⟶ (2, 128)
3	144	⟶ (3, 144)
4	128	⟶ (4, 128)
5	80	⟶ (5, 80)
6	0	⟶ (6, 0)

↑ *x*-coordinate   ↑ *y*-coordinate   ↑ The data in the table can be expressed as ordered pairs.

The ordered pairs in the table can then be plotted on a rectangular coordinate system and a smooth curve drawn through the points.

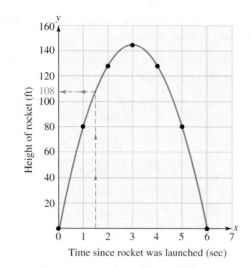

This graph shows the height of the rocket in relation to the time since it was launched. It does not show the path of the rocket.

From the graph, we can see that the height of the rocket increases as the time increases from 0 second to 3 seconds. Then the height decreases until the rocket hits the ground in 6 seconds. We can also use the graph to make observations about the height of the rocket at other times. For example, the dashed blue lines on the graph show that in 1.5 seconds, the height of the rocket will be approximately 108 feet.

3   **Read Graphs.**

Since graphs are becoming an increasingly popular way to present information, the ability to read and interpret them is becoming ever more important.

EXAMPLE 2   *Water Management.*   The graph below shows the water level of a reservoir before, during, and after a storm. On the *x*-axis, 0 represents the day the storm began. On the *y*-axis, 0 represents the normal water level that operators try to maintain.

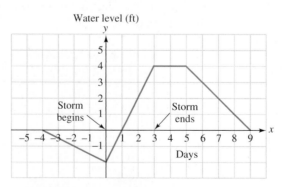

a. In anticipation of the storm, operators released water to lower the level of the reservoir. By how many feet was the water lowered prior to the storm?

b. After the storm ended, on what day did the water level begin to fall?

c. When was the water level 2 feet above normal?

Strategy   We will use ordered pairs to describe the situations mentioned in parts (a), (b), and (c).

Why   The coordinates of specific points on the graph can be used to answer the given questions.

Solution

a. The graph starts at the point $(-4, 0)$. This means that 4 days before the storm began, the water level was at the normal level. If we look below 0 on the y-axis, we see that the point $(0, -2)$ is on the graph. So the day the storm began, the water level had been lowered 2 feet.

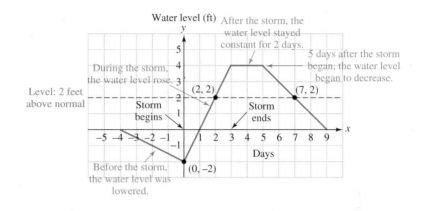

b. If we look at the x-axis, we see that the storm lasted 3 days. From the third to the fifth day, the water level remained constant, 4 feet above normal, as indicated by the horizontal line segment between $(3, 4)$ and $(5, 4)$. The graph does not begin to decrease until day 5.

c. We can draw a horizontal line passing through 2 on the y-axis. This line intersects the graph in two places—at the points $(2, 2)$ and $(7, 2)$. This means that 2 days and 7 days after the storm began, the water level was 2 feet above normal.

Self Check 2   Refer to the graph.   **a.** When was the water at the normal level?

b. By how many feet did the water level rise during the storm?

c. After the storm ended and the water level began to fall, how long did it take for the water level to return to normal?

Now Try   **Problem 37**

Notation

Variables other than x and y can be used to label the horizontal and vertical axes of a rectangular coordinate graph.

The graph on the right shows the cost of renting ski equipment for different periods of time. The horizontal axis is labeled with the variable d. This reinforces the fact that it is associated with the number of *days* the equipment is rented. The vertical axis is labeled with the variable c, for *cost*. In this case, ordered pairs on the graph have the form (d, c). For example, the point $(3, 100)$ on the graph tells us that the cost of renting the equipment for 3 days is $100. The cost of renting the equipment for more than 4 and up to 5 days is $140. We call this type of graph a *step graph*.

The point at the end of each step indicates the rental cost for 1, 2, 3, 4, 5, and 6 days. Each open circle indicates that that point is not on the graph.

Cost of Renting a Ski Equipment Package

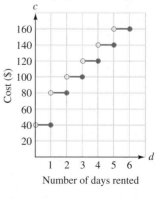

EXAMPLE 3   ***Ski Rental Packages.***   Use the following graph to answer the questions about the rental cost of a ski equipment package (skis, boots, and poles).   **a.** Find the cost of renting the equipment for 2 days.   **b.** Find the cost of renting the equipment for $5\frac{1}{2}$ days.   **c.** How long can you rent the equipment if you have budgeted $120 for the rental?   **d.** Is the cost of renting the equipment the same each day?

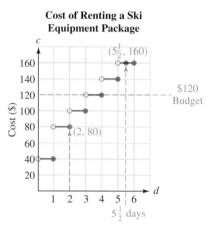

Cost of Renting a Ski Equipment Package

Strategy   To answer questions about the rental costs, we will scan from the horizontal axis, up and over, to the vertical axis. To answer questions about length of time the equipment can be rented, we will scan from the vertical axis, over and down, to the horizontal axis.

Why   The scale on the vertical axis gives the cost to rent the equipment. The scale on the horizontal axis gives the length of time the equipment is rented.

Solution

a. We locate 2 on the *d*-axis and move up to locate the point on the graph directly above the 2. Since that point has coordinates (2, 80), a 2-day rental costs $80.

b. We locate $5\frac{1}{2}$ on the *d*-axis and move straight up to locate the point on the graph with coordinates $\left(5\frac{1}{2}, 160\right)$, which indicates that a $5\frac{1}{2}$-day rental would cost $160.

c. We draw a horizontal line through the point labeled 120 on the *c*-axis. Since this line intersects one of the steps of the graph, we can look down to the *d*-axis to find the *d*-values that correspond to a *c*-value of 120. We see that the equipment can be rented for more than 3 and up to 4 days for $120.

d. The cost each day is not the same. If we look at how the *c*-coordinates change, we see that the first-day rental fee is $40. The second day, the cost jumps another $40. The third day, and all subsequent days, the cost jumps $20.

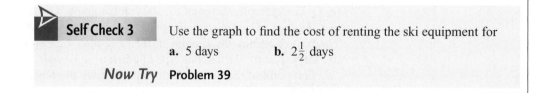

Self Check 3   Use the graph to find the cost of renting the ski equipment for

    a. 5 days   **b.** $2\frac{1}{2}$ days

Now Try   **Problem 39**

4 Find the Midpoint of a Line Segment.

If point M in the following figure lies midway between point P and point Q, it is called the **midpoint** of line segment PQ. We call the points P and Q, the **endpoints** of the segment.

To distinguish between the coordinates of the endpoints of a line segment, we can use *subscript notation*. In the figure, the point P with coordinates (x_1, y_1) is read as "point P with coordinates x sub 1 and y sub 1," and the point Q with coordinates (x_2, y_2) is read as "point Q with coordinates x sub 2 and y sub 2."

To find the coordinates of point M, we find the average of the x-coordinates and the average of the y-coordinates of points P and Q. Using subscript notation, we can write the midpoint formula in the following way.

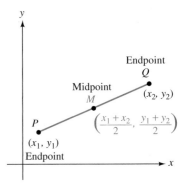

The Midpoint Formula

The **midpoint** of a line segment with endpoints (x_1, y_1) and (x_2, y_2) is the point with coordinates

$$\left(\frac{x_1 + x_2}{2}, \frac{y_1 + y_2}{2} \right)$$

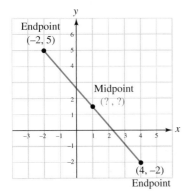

EXAMPLE 4 Find the midpoint of the line segment with endpoints $(-2, 5)$ and $(4, -2)$.

Strategy To find the coordinates of the midpoint, we find the average of the x-coordinates and the average of the y-coordinates of the endpoints.

Why This is what is called for by the expressions $\frac{x_1 + x_2}{2}$ and $\frac{y_1 + y_2}{2}$ of the midpoint formula.

Solution We can let $(x_1, y_1) = (-2, 5)$ and $(x_2, y_2) = (4, -2)$. After substituting these values into the expressions for the x- and y-coordinates in the midpoint formula, we will evaluate each expression to find the coordinates of the midpoint.

$$\frac{x_1 + x_2}{2} = \frac{-2 + 4}{2} \qquad \frac{y_1 + y_2}{2} = \frac{5 + (-2)}{2}$$

$$= \frac{2}{2} \qquad\qquad = \frac{3}{2}$$

$$= 1$$

Thus, the midpoint is $\left(1, \frac{3}{2} \right)$.

Self Check 4 Find the midpoint of the line segment with endpoints $(-1, 8)$ and $(5, 2)$.

Now Try Problem 41

EXAMPLE 5 The midpoint of the line segment joining $(-5, -3)$ and a point Q is the point $(-1, 2)$. Find the coordinates of point Q.

Strategy As in Example 4, we will use the midpoint formula to find the unknown coordinates. However, this time, we need to find x_2 and y_2.

Why We want to find the coordinates of one of the endpoints.

Solution We can let $(x_1, y_1) = (-5, -3)$ and $(x_M, y_M) = (-1, 2)$, where x_M represents the x-coordinate and y_M represents the y-coordinate of the midpoint. To find the coordinates of point Q, we substitute for x_1, x_M, y_1, and y_M in the expressions for the coordinates in the midpoint formula and solve the resulting equations for x_2 and y_2.

$x_M = \dfrac{x_1 + x_2}{2}$	$y_M = \dfrac{y_1 + y_2}{2}$ Read x_M as "x sub M" and y_M as "y sub M."
$-1 = \dfrac{-5 + x_2}{2}$	$2 = \dfrac{-3 + y_2}{2}$
$-2 = -5 + x_2$ Multiply both sides by 2.	$4 = -3 + y_2$ Multiply both sides by 2.
$3 = x_2$ Add 5 to both sides.	$7 = y_2$ Add 3 to both sides.

Since $x_2 = 3$ and $y_2 = 7$, the coordinates of point Q are $(3, 7)$.

Self Check 5 The midpoint of the line segment joining $(-5, -3)$ and a point P is the point $(-2, 5)$. Find the coordinates of point P.

Now Try Problem 49

ANSWERS TO SELF CHECKS **1.** $(-8, -1)$ **2. a.** 4 days before the storm began, 1 day and 9 days after the storm began, **b.** 6 ft, **c.** 4 days **3. a.** $140, **b.** $100 **4.** $(2, 5)$ **5.** $(1, 13)$

STUDY SET
2.1

VOCABULARY

Fill in the blanks.

1. The pair of numbers $(6, -2)$ is called an _____ pair.
2. In the ordered pair $(22, 29)$, the ___-coordinate is 29.
3. The point $(0, 0)$ is the _____.
4. The x- and y-axes divide the coordinate plane into four regions called _____.
5. Ordered pairs of numbers can be graphed on a _____ coordinate system.

6. The process of locating a point on a coordinate plane is called _____ the point.
7. If a point is midway between two points P and Q, it is called the _____ of segment PQ.
8. If a line segment joins points P and Q, points P and Q are called _____ of the segment.

CONCEPTS

Fill in the blanks.

9. To plot $(6, -3.5)$, we start at the _____ and move 6 units to the _____ and then 3.5 units _____.
10. To plot $(-6, 2)$, we start at the _____ and move 6 units to the _____ and then 2 units _____.

11. In which quadrant do points with a negative *x*-coordinate and a positive *y*-coordinate lie?

12. In which quadrant do points with a positive *x*-coordinate and a negative *y*-coordinate lie?

NOTATION

13. Do these ordered pairs name the same point?

$$\left(5.25, -\tfrac{3}{2}\right), \left(5\tfrac{1}{4}, -1.5\right), \left(\tfrac{21}{4}, -1\tfrac{1}{2}\right)$$

14. For the ordered pair (t, d), which variable is associated with the horizontal axis?

15. What type of letter is used to label points?

16. Fill in the blank: The expression x_1 is read as ___ ____ ___.

17. Explain the difference between x^2 and x_2.

18. Fill in the blanks: The *x*-coordinate of the midpoint of the line segment joining (x_1, y_1) and (x_2, y_2) is ____ and the *y*-coordinate is ____ .

GUIDED PRACTICE

Plot each point on a rectangular coordinate system and name the quadrant or axis in which the point lies. **See Objective 1.**

19. $(4, 3)$

20. $(-2, 1)$

21. $(3.5, -2)$

22. $(-2.5, -3)$

23. $(5, 0)$

24. $(-4, 0)$

25. $\left(0, -\tfrac{8}{3}\right)$

26. $\left(0, \tfrac{10}{3}\right)$

Give the coordinates of each point. **See Example 1.**

27. *A*

28. *B*

29. *C*

30. *D*

31. *E*

32. *F*

33. *G*

34. *H*

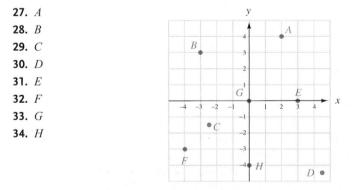

Use the information in the table to draw each graph. **See Objective 2.**

35. WATER PRESSURE   A tub was filled with water from a faucet. The table in the next column shows the number of gallons of water in the tub at 1-minute intervals. Plot the ordered pairs in the table on a rectangular coordinate system and then draw a line through the points.

Time (min)	Water in tub (gal)
0	0
1	8
2	16
3	24
4	32

36. TRAMPOLINES   The table shows the distance a girl is from the ground (in relation to time) as she bounds into the air and back down to the trampoline. Plot the ordered pairs in the table on a rectangular coordinate system and then draw a smooth curve through the points.

Time (sec)	Height (ft)
0	2
0.25	9
0.5	14
1.0	18
1.5	14
1.75	9
2.0	2

Use the graph to answer each question. **See Example 2.**

37. SUBMARINES   The graph in the following illustration shows the depths of a submarine at certain times.

 a. Where is the sub 2 hours after launch?

 b. What is the sub doing as *t* increases from 2 to 3?

 c. For how long does the submarine travel at a depth of 1,000 feet?

 d. How large an ascent does the sub begin to make 6 hours after launch?

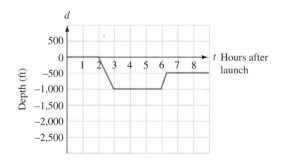

38. AIRPLANES The following graph shows the altitudes of a plane at certain times.

  a. Where is the plane when $t = 0$?

  b. What is the plane doing as t increases from 1 to 2?

  c. For how long does the airplane travel at an altitude of 5,000 feet?

  d. How much of a descent does the plane begin to make 4 hours after take-off?

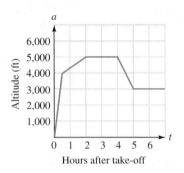

Use each graph to answer each question. **See Example 3.**

39. VIDEO RENTALS The charges for renting a video are shown in the following graph.

  a. Find the 1-day rental charge.

  b. Find the charge if the video is kept for 1 week.

  c. How long can you rent the video if you have budgeted $5 for the rental?

  d. Is the cost of renting the video the same each day?

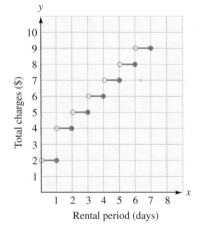

40. TAXIS The following graph gives the fares charged for rides up to 1 mile in length by a taxicab company. In the graph, the symbol $\doteqdot$ indicates a break in the labeling of the vertical axis. The break enables us to omit a large portion of the grid that would not be used.

  a. What is the fare for a $\frac{1}{2}$-mile ride?

  b. Is the fare the same for each $\frac{1}{8}$ mile traveled?

  c. What is a fare for a $\frac{7}{10}$- mile long ride?

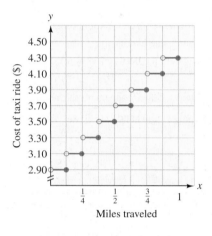

Find the midpoint of line segment with the given endpoints. **See Example 4.**

41. $(0, 0), (6, 8)$

42. $(10, 12), (0, 0)$

43. $(6, 8), (12, 16)$

44. $(10, 4), (2, -2)$

45. $(-2, -8), (3, -8)$

46. $(-5, -2), (7, 3)$

47. $(7, 1), (-10, 4)$

48. $(-4, -3), (4, -8)$

Solve each problem. **See Example 5.**

49. If $(-2, 3)$ is the midpoint of segment PQ and the coordinates of P are $(-8, 5)$, find the coordinates of Q.

50. If $(6, -5)$ is the midpoint of segment PQ and the coordinates of Q are $(-5, -8)$, find the coordinates of P.

51. If $(-7, -3)$ is the midpoint of segment QP and the coordinates of Q are $(6, -3)$, find the coordinates of P.

52. If $\left(\frac{1}{2}, -2\right)$ is the midpoint of segment QP and the coordinates of P are $\left(-\frac{5}{2}, 5\right)$, find the coordinates of Q.

APPLICATIONS

53. ROAD MAPS Maps have a built-in coordinate system to help locate cities. Use the map on the next page to find the coordinates of these cities in South Carolina: Jonesville, Easley, Hodges, and Union. Express each answer in the form (number, letter).

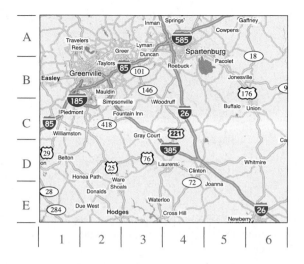

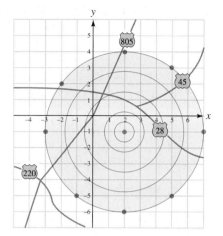

54. HURRICANES   A coordinate system that designates the location of places on the surface of Earth uses a series of latitude and longitude lines, as shown in the illustration.

   a. In 2005, Hurricane Katrina devastated the city of New Orleans. If we agree to list longitude first, what are the coordinates of New Orleans, expressed as an ordered pair?

   b. In August 1992, Hurricane Andrew destroyed Homestead, Florida. Estimate the coordinates of Homestead.

   c. Estimate the coordinates of where Hurricane Andrew hit Louisiana.

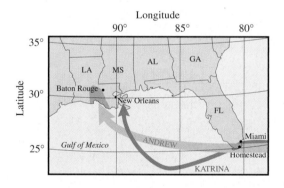

55. EARTHQUAKES   The graph in the next column shows the area damaged by an earthquake.

   a. Find the coordinates of the *epicenter* (the source of the quake).

   b. Was damage done at the point $(4, 5)$?

   c. Was damage done at the point $(-1, -4)$?

56. GEOGRAPHY   The following illustration shows a cross-sectional profile of the Sierra Nevada mountain range in California.

   a. Estimate the coordinates of blue oak, sagebrush scrub, and tundra using an ordered pair of the form (distance, elevation).

   b. The *tree line* is the highest elevation at which trees grow. Estimate the tree line for this mountain range.

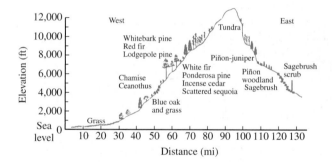

57. GOLF   Refer to the graph on the next page. Tiger Woods came back in the final 18 holes of the 2000 AT&T Pebble Beach National Pro-Am golf tournament to overtake the leader, Matt Gogel. See the graph. (In golf, the player with the score that is the farthest *under* par is the winner.)

   a. At the beginning of the final round, by how many strokes did Gogel lead Woods?

   b. What was the largest lead that Gogel had over Woods in the final round?

   c. On what hole did Woods tie up the match?

   d. On what hole did Woods take the lead?

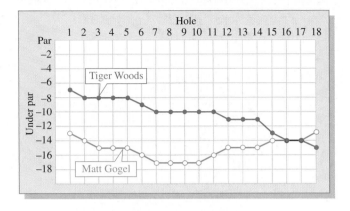

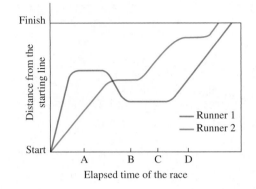

58. PETROLEUM Refer to the following graph.

a. When did the United States net petroleum imports first surpass production?

b. Estimate the difference in U.S. petroleum net imports and production for 2005.

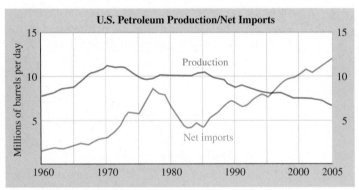

Source: Energy Information Administration

59. LONG-DISTANCE RUNNING Refer to the graph in the next column that describes a two-person race.

a. Which runner ran faster at the start of the race?

b. Which runner stopped to rest first?

c. Which runner dropped his watch and had to go back and get it?

d. At which of these elapsed times (A, B, C, D) was runner 1 stopped and runner 2 running?

e. Describe what was happening at time D.

f. Which runner won the race?

60. CAMPUS PARKING LOT Match each daily parking description with the graph that best illustrates it. The parking lot holds 500 cars.

a. On Mondays, the parking lot is full by noon. It's impossible to find a parking space until late in the afternoon.

b. On Tuesdays, the lot never gets more than half-full.

c. On Wednesdays, the lot fills up quickly. It empties out around lunchtime but then it fills up fast when the evening classes begin.

d. On Thursdays, it is easy to park in the lot until the evening classes begin. Then it is almost impossible to find a parking space.

e. On Fridays, it is very difficult to find a parking space, unless you arrive early. However, the lot clears out by noon.

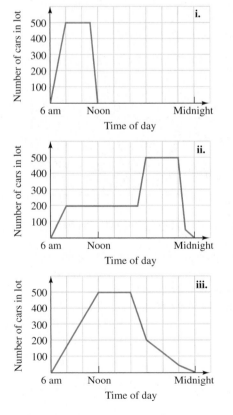

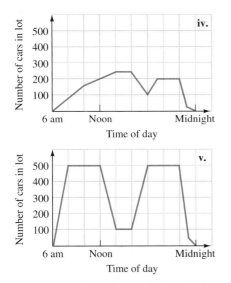

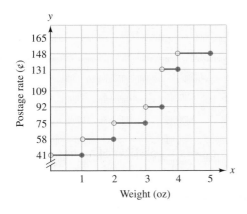

61. U.S. POSTAGE The graph shown below gives the first-class postage rates in 2007 for mailing letters weighing up to 5 ounces. In the graph, the symbol ⋜ indicates a break in the labeling of the vertical axis. The break enables us to omit a large portion of the grid that would not be used.

a. Find the cost of postage to mail a 3-oz letter.

b. Find the difference in cost for a 1.75-oz letter and a 4.75-oz letter.

c. What is the heaviest letter that can be mailed first class for $1?

62. ROAST TURKEY Guidelines that appear on the label of a frozen turkey are listed in the table. Draw a step graph that illustrates these instructions.

Size	Time thawing in refrigerator
10 lb to just under 18 lb	3 days
18 lb to just under 22 lb	4 days
22 lb to just under 24 lb	5 days
24 lb to just under 30 lb	6 days

63. MULTICULTURAL STUDIES Social scientists use the following diagram to classify cultures. The amount of group/family loyalty in a culture is measured on the horizontal *group* axis. The amount of social mobility is measured on the vertical *social grid* axis. In the diagram, four cultures are classified. In which culture, R, S, T, or U, would you expect that

a. anyone can grow up to be president, and parents expect their children to get out on their own as soon as possible?

b. only the upper class attends college, and people must marry within their own social class?

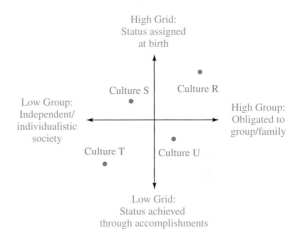

64. PSYCHOLOGY The results of a personal profile test taken by an employee are plotted as an ordered pair on the grid in the illustration. The test shows whether the employee is more task oriented or people oriented. From the results, would you expect the employee to agree or disagree with each of the following statements?

a. Completing a project is almost an obsession with me, and I cannot be content until I am finished.

b. Even if I'm in a hurry while running errands, I will stop to talk with a friend.

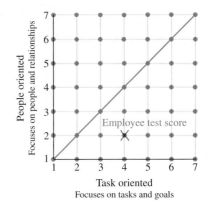

WRITING

65. Explain how to plot the point $(-2, 5)$.

66. Explain why the coordinates of the origin are $(0, 0)$.

67. Explain why the point $(-1, 6)$ is not the same as the point $(6, -1)$.

68. Explain this diagram.

$$
\begin{array}{c|c}
\text{II} & \text{I} \\
(-, +) & (+, +) \\
\hline
\text{III} & \text{IV} \\
(-, -) & (+, -)
\end{array}
$$

REVIEW

Evaluate each expression.

69. $-5^2 - 5 - 5(-5)$

70. $\dfrac{1}{3}\left(\dfrac{1}{6}\right) - \left(-\dfrac{1}{3}\right)^2$

71. $\dfrac{|-25| - 2(-5)}{2^4 - 9}$

72. $\dfrac{3[-9 + 2(7 - 3)]}{(8 - 5)(9 - 7)}$

73. Solve $P = 2l + 2w$ for w.

74. Solve $T_f = T_a(1 - F)$ for T_a.

CHALLENGE PROBLEMS

75. Find the coordinates of the three points that divide the line segment joining (a, b) and (c, d) into four equal parts?

76. AIRPLANES   When designing an airplane, engineers use a coordinate system with 3 axes, as shown. Any point on the airplane can be described by an *ordered triple* of the form (x, y, z). The coordinates of three points on the plane are $(0, 181, 56)$, $(-46, 48, 19)$, and $(84, 94, 24)$. Which highlighted part of the plane corresponds with which ordered triple?

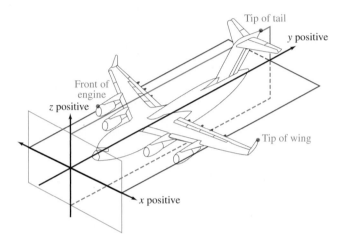

SECTION 2.2
Graphing Linear Equations in Two Variables

Objectives

❶ Determine whether an ordered pair is a solution of an equation.

❷ Find a solution of an equation in two variables.

❸ Graph linear equations by plotting points.

❹ Graph linear equations by finding intercepts.

❺ Graph horizontal and vertical lines.

❻ Use linear models to solve applied problems.

In this section, we will discuss equations that contain two variables. These equations are used to describe algebraic relationships between two quantities.

❶   **Determine Whether an Ordered Pair is a Solution of an Equation.**

We will now extend our equation-solving skills to find solutions of equations in two variables. To begin, let's consider $y = -\frac{1}{2}x + 3$, an equation in x and y. In general, a solution of an equation in two variables is an ordered pair of numbers that make a true statement when substituted into the equation.

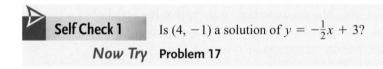

EXAMPLE 1 Is $(-4, 5)$ a solution of $y = -\dfrac{1}{2}x + 3$?

Strategy We will substitute -4 for x and 5 for y and see whether the resulting equation is true.

Why An ordered pair is a solution of $y = -\dfrac{1}{2}x + 3$ if replacing the variables with the values of the ordered pair results in a true statement.

Solution

$$y = -\frac{1}{2}x + 3$$

$$5 \stackrel{?}{=} -\frac{1}{2}(-4) + 3 \qquad \text{Substitute 5 for y and } -4 \text{ for x.}$$

$$5 \stackrel{?}{=} 2 + 3$$

$$5 = 5 \qquad\qquad \text{True}$$

Since the result is a true statement, $(-4, 5)$ is a solution of $y = -\dfrac{1}{2}x + 3$. We say that $(-4, 5)$ *satisfies* the equation.

 Self Check 1 Is $(4, -1)$ a solution of $y = -\dfrac{1}{2}x + 3$?

Now Try **Problem 17**

2 **Find a Solution of an Equation in Two Variables.**

To find a solution of an equation in two variables, we can select a number for one of the variables and find the corresponding value of the other variable. For example, to find a solution of $y = -\dfrac{1}{2}x + 3$, we can select a value for x, say 6, and find the corresponding value of y.

$$y = -\frac{1}{2}x + 3$$

$$y = -\frac{1}{2}(6) + 3 \qquad \text{Substitute 6 for x.}$$

$$y = -3 + 3 \qquad \text{Evaluate the right side.}$$

$$y = 0 \qquad\qquad \text{This is the y-coordinate of the solution.}$$

Thus, $(6, 0)$ is a solution of $y = -\dfrac{1}{2}x + 3$.

Since we can choose any real number for x, and since any choice for x will give a corresponding value of y, the equation $y = -\dfrac{1}{2}x + 3$ has infinitely many solutions. It would be impossible to list all of the solutions. Instead, we can draw a mathematical picture of the solutions, called the *graph of the equation*.

3 **Graph Linear Equations by Plotting Points.**

Equations in two variables can be graphed in several ways. If an equation in x and y is solved for y, we can graph it by selecting values for x and calculating the corresponding values of y.

> **EXAMPLE 2**   Graph: $y = -\dfrac{1}{2}x + 3$

Strategy   We will find three solutions of the equation, plot them on a rectangular coordinate system, and then draw a straight line passing through the points.

Why   To *graph* an equation in two variables means to make a drawing that represents all of its solutions.

Solution   To find three solutions of this equation, we select three values for x that will make the computations easy. Then we find each corresponding value of y. For example, if x is -2, we have

$$y = -\frac{1}{2}x + 3$$

$$y = -\frac{1}{2}(-2) + 3 \qquad \text{Substitute } -2 \text{ for } x.$$

$$y = 1 + 3 \qquad\qquad \text{Evaluate the right side.}$$

$$y = 4$$

Thus, $(-2, 4)$ is a solution. In a similar manner, we find corresponding y-values for x-values of 0 and 2 and enter the solutions in the table below.

When we plot the ordered-pair solutions on a rectangular coordinate system, we see that they lie in a straight line. Using a straight edge or ruler, we then draw a straight line through the points because the graph of any solution of $y = -\frac{1}{2}x + 3$ will lie on this line. Furthermore, every point of this line represents a solution. We call the line the graph of the equation. It represents all of the solutions of $y = -\frac{1}{2}x + 3$.

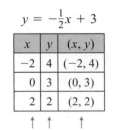

$$y = -\frac{1}{2}x + 3$$

x	y	(x, y)
-2	4	$(-2, 4)$
0	3	$(0, 3)$
2	2	$(2, 2)$

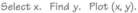

Select x. Find y.   Plot (x, y).

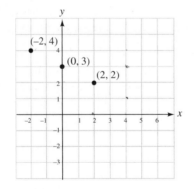

Plot the ordered pairs.

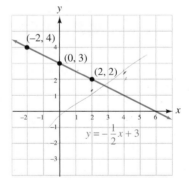

Draw a straight line through the points.
This is the *graph of the equation.*

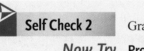

> **Self Check 2**   Graph: $y = \frac{1}{3}x + 1$
>
> **Now Try**   **Problems 21 and 31**

The equation from Example 2, $y = -\frac{1}{2}x + 3$, is called a *linear equation in two variables.*

| Standard (General) Form of a Linear Equation | A **linear equation in two variables** is an equation that can be written in the form $$Ax + By = C$$ where A, B, and C are real numbers and A and B are not both 0. |

Some other examples of linear equations are

$$y = 4x - 7, \qquad 2x - 5y = 10, \qquad y = 3, \qquad \text{and} \qquad x = 2$$

The graph of every linear equation in two variables is a line. We can use the following method to graph linear equations solved for y.

| Graphing Linear Equations Solved for y by Plotting Points | 1. Find three ordered pairs that are solutions of the equation by selecting three values for x and calculating the corresponding values of y.
 2. Plot the solutions on a rectangular coordinate system.
 3. Draw a straight line passing through the points. If the points do not lie on a line, check your computations. |

 Graph Linear Equations by Finding Intercepts.

In Example 2, the graph intersected the y-axis at the point $(0, 3)$, called the y-intercept, and it intersected the x-axis at the point $(6, 0)$ called the x-intercept. In general, we have the following definitions.

| Intercepts of a Line | The **y-intercept** of a line is the point $(0, b)$, where the line intersects the y-axis. To find b, substitute 0 for x in the equation of the line and solve for y.

 The **x-intercept** of a line is the point $(a, 0)$, where the line intersects the x-axis. To find a, substitute 0 for y in the equation of the line and solve for x. |

Plotting the x- and y-intercepts of a graph and drawing a line through them is called the **intercept method of graphing a line.** This method is useful when graphing linear equations written in the standard (general) form $Ax + By = C$.

EXAMPLE 3   Graph $2x - 5y = 10$ by finding the intercepts.

Strategy   We will let $x = 0$ to find the y-intercept of the graph and then let $y = 0$ to find the x-intercept.

Why   Since two points determine a line, the y-intercept and x-intercept are enough information to graph this linear equation.

Solution   To find the y-intercept, we substitute 0 for x and solve for y:

$$2x - 5y = 10 \qquad \text{This is the equation to graph.}$$
$$2(0) - 5y = 10 \qquad \text{Substitute 0 for x.}$$
$$-5y = 10$$
$$y = -2 \qquad \text{To isolate y, divide both sides by } -5.$$

The y-intercept is the point $(0, -2)$. To find the x-intercept, we substitute 0 for y and solve for x:

$$2x - 5y = 10 \quad \text{This is the equation to graph.}$$
$$2x - 5(0) = 10 \quad \text{Substitute 0 for y.}$$
$$2x = 10$$
$$x = 5 \quad \text{To isolate x, divide both sides by 2.}$$

The x-intercept is the point $(5, 0)$.

Although two points provide enough information to draw the graph of the equation, it is a good idea to find and plot a third point as a check. If the three points do not lie on a line, then at least one of them is in error.

To find the coordinates of a third point, we can substitute any convenient number (such as -5) for x and solve for y:

$$2x - 5y = 10 \quad \text{This is the equation to graph.}$$
$$2(-5) - 5y = 10 \quad \text{Substitute } -5 \text{ for x.}$$
$$-10 - 5y = 10$$
$$-5y = 20 \quad \text{Add 10 to both sides.}$$
$$y = -4 \quad \text{To isolate y, divide both sides by } -5.$$

The line will also pass through the point $(-5, -4)$. We plot the intercepts and the check point, draw a straight line through them, and label the line as $2x - 5y = 10$.

The Language of Algebra
For any two points, exactly one line passes through them. We say two points *determine* a line.

The Language of Algebra
Be careful with pronunciation: The point where a line *intersects* the x- or y-axis is called an *intercept*.

$2x - 5y = 10$

x	y	(x, y)	
0	-2	$(0, -2)$	← y-intercept
5	0	$(5, 0)$	← x-intercept
-5	-4	$(-5, -4)$	← Check point

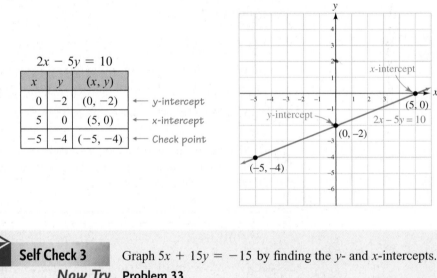

Self Check 3   Graph $5x + 15y = -15$ by finding the y- and x-intercepts.

Now Try   Problem 33

Using Your Calculator   *Generating Tables of Solutions*

If an equation in x and y is solved for y, we can use a graphing calculator to generate a table of solutions. The instructions in this discussion are for a TI-84 or a TI-84 Plus graphing calculator. For specific details about other brands, please consult the owner's manual.

To construct a table of solutions for $2x - 5y = 10$, we first solve for y.

$$2x - 5y = 10$$

$$-5y = -2x + 10 \qquad \text{Subtract 2x from both sides.}$$

$$y = \frac{2}{5}x - 2 \qquad \text{Divide both sides by } -5 \text{ and simplify.}$$

To enter $y = \frac{2}{5}x - 2$, we press $\boxed{\text{Y} =}$ and enter (2/5)x − 2, as shown in figure (a). (Ignore the subscript 1 on y; it is not relevant at this time.)

To enter the x-values that are to appear in the table, we press $\boxed{\text{2nd}}$ $\boxed{\text{TBLSET}}$ and enter the first value for x on the line labeled TblStart =. In figure (b), −5 has been entered on this line. Other values for x that are to appear in the table are determined by setting an increment value on the line labeled ΔTbl =. Figure (b) shows that an increment of 1 was entered. This means that each x-value in the table will be 1 unit larger than the previous x-value.

The final step is to press the keys $\boxed{\text{2nd}}$ $\boxed{\text{TABLE}}$. This displays a table of solutions, as shown in figure (c).

(a)   (b)   (c)

5 **Graph Horizontal and Vertical Lines.**

Equations such as $y = 3$ and $x = -2$ are linear equations, because they can be written in the general form $Ax + By = C$.

| $y = 3$ | is equivalent to | $0x + 1y = 3$ |
| $x = -2$ | is equivalent to | $1x + 0y = -2$ |

EXAMPLE 4   Graph: **a.** $y = 4$ **b.** $x = -2$

Strategy   To find three ordered-pair solutions of $y = 4$, we will select three values for x and use 4 for y each time. To find three ordered-pair solutions of $x = -2$, we will select three values for y and use −2 for x each time.

Why   The first equation requires that $y = 4$ and the second equation requires that $x = -2$.

Solution

a. Since the equation $y = 4$ does not contain x, the numbers chosen for x have no effect on y. The value of y is always 4.

After plotting the ordered pairs shown in the table, and drawing a straight line through them, we see that the graph is a horizontal line, parallel to the x-axis, with a y-intercept of $(0, 4)$. The line has no x-intercept.

b. Since the equation $x = -2$ does not contain y, the value of y can be any number.

After plotting the ordered pairs shown in the table, and drawing a straight line through them, we see that the graph is a vertical line, parallel to the y-axis, with an x-intercept of $(-2, 0)$. The line has no y-intercept.

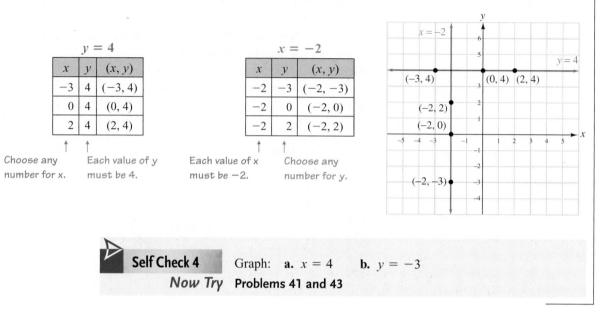

$y = 4$

x	y	(x, y)
-3	4	$(-3, 4)$
0	4	$(0, 4)$
2	4	$(2, 4)$

   ↑ ↑

Choose any   Each value of y
number for x.   must be 4.

$x = -2$

x	y	(x, y)
-2	-3	$(-2, -3)$
-2	0	$(-2, 0)$
-2	2	$(-2, 2)$

   ↑ ↑

Each value of x Choose any
must be -2.   number for y.

> **Self Check 4**   Graph: **a.** $x = 4$ **b.** $y = -3$
>
> **Now Try** Problems 41 and 43

The results of Example 4 suggest the following facts.

Equations of Horizontal and Vertical Lines	The equation $y = b$ represents the horizontal line that intersects the y-axis at $(0, b)$. The equation $x = a$ represents the vertical line that intersects the x-axis at $(a, 0)$.

The graph of the equation $y = 0$ has special significance; it is the x-axis. Similarly, the graph of the equation $x = 0$ is the y-axis.

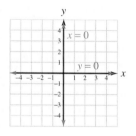

6   **Use Linear Models to Solve Applied Problems.**

In the next examples, we will see how linear equations can model real-life situations. **Linear models,** as they are called, are often written in variables other than x and y. We must make the appropriate changes when labeling the table of solutions and the graph of the equation. We can use linear models to make observations about what has occurred in the past and what might occur in the future.

EXAMPLE 5   ***Bottled Water.*** The increasing popularity of bottled water in the United States can be modeled by the linear equation $w = 1.8t + 17.3$, where t represents the number of years after 2000 and w represents the annual per capita consumption in gallons. (Source: Beverage Marketing Corporation) **a.** Graph the equation.   **b.** Suppose the current trend continues. Use the graph to estimate what the annual per capita consumption of bottled water will be in the year 2012.

Strategy   We will find three solutions of the equation, plot them on a rectangular coordinate system, and then draw a straight line passing through the points.

Why   To *graph* a linear equation in two variables means to make a drawing that represents all of its solutions.

The Language of Algebra
The phrase *per capita* means the *average per person*. For example, in 2007, Connecticut had the highest *per capita* income of any state: $55,536.

Solution

a. The variables t and w are used in the equation. If we associate t with the horizontal axis and w with the vertical axis, then the ordered pairs have the form (t, w).

To graph the equation, we pick three values for t, substitute them into the equation, and find each corresponding value of w. Since t represents the number of years *after* 2000, we will not select any negative values for t. The results are listed in the following table.

For t = 0   (The year 2000)
$w = 1.8t + 17.3$
$w = 1.8(0) + 17.3$
$w = 17.3$

For t = 2   (The year 2002)
$w = 1.8t + 17.3$
$w = 1.8(2) + 17.3$
$w = 3.6 + 17.3$
$w = 20.9$

For t = 5   (The year 2005)
$w = 1.8t + 17.3$
$w = 1.8(5) + 17.3$
$w = 9 + 17.3$
$w = 26.3$

The pairs $(0, 17.3)$, $(2, 20.9)$, and $(5, 26.3)$ are plotted, and a straight line is drawn through them to give the graph of the equation.

Notation
In the graph, the symbol ⧧ indicates a break in the labeling of the vertical axis. The break enables us to omit a large portion of the grid that would not be used.

$w = 1.8t + 17.3$

t	w	(t, w)
0	17.3	(0, 17.3)
2	20.9	(2, 20.9)
5	26.3	(5, 26.3)

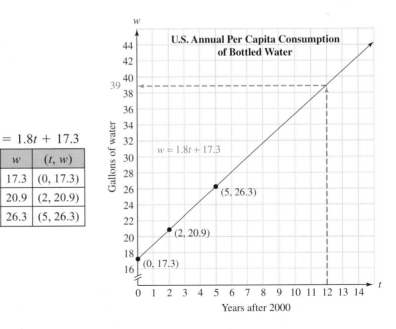

b. To estimate the per capita consumption in 2012 (which is 12 years after 2000), we locate 12 on the horizontal axis. Then we move upward and over (as shown in blue) to estimate a reading of 39 on the vertical axis. This means that if the current trend continues, in the year 2012, the annual per capita consumption of bottled water in the United States will be approximately 39 gallons.

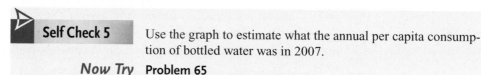

Self Check 5   Use the graph to estimate what the annual per capita consumption of bottled water was in 2007.

Now Try   **Problem 65**

For tax purposes, many businesses use the equation of a line to find the declining value of aging equipment. This method is called **straight-line depreciation.**

EXAMPLE 6   *Depreciation of a Copier.*   A copy machine that was purchased for $6,750 is expected to depreciate according to the straight-line depreciation equation $y = -950x + 6,750$, where y is the value of the copier after x years of use. When will the copier have no value?

Strategy   To find when the copier will have no value, we will substitute 0 for y in the equation $y = -950x + 6,750$ and solve for x.

Why   The variable y represents the value of the computer. When the copier has no value, y will be equal to 0.

Solution

$y = -950x + 6,750$	This is the straight-line depreciation model.
$0 = -950x + 6,750$	Substitute 0 for y.
$-6,750 = -950x$	Subtract 6,750 from both sides.
$7.105263158 \approx x$	To isolate x, divide both sides by −950.

The copier will have no value after it has been in use for approximately 7.1 years.

The equation $y = -950x + 6,750$ is graphed below. Important information can be obtained from the intercepts of the graph.

> **The Language of Algebra**
> *Depreciation* is a form of the word *depreciate*, meaning to lose value due to wear and tear, decay, or declining price. You've probably heard that the minute you drive a new car off the lot, it has *depreciated*.

> **Success Tip**
> When the copier is new, it has been in use **0** years. In that case, x is 0. When the copier has no value, y is 0.

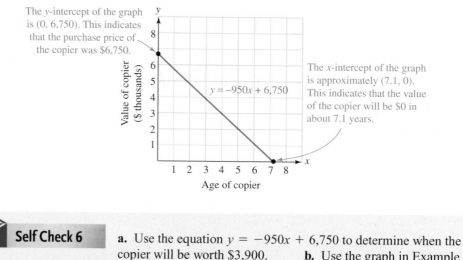

Self Check 6   **a.** Use the equation $y = -950x + 6,750$ to determine when the copier will be worth $3,900.   **b.** Use the graph in Example 6 to determine when the copier will be worth $2,000.

Now Try   **Problem 71**

Using Your Calculator

Graphing Lines

We have graphed linear equations by finding solutions, plotting points, and drawing lines through those points. Graphing is often easier using a graphing calculator.

Window settings

Graphing calculators have a window to display graphs. To see the proper picture of a graph, we must decide on the minimum and maximum values for the x- and y-coordinates. A window with standard settings of

$$\text{Xmin} = -10 \qquad \text{Xmax} = 10 \qquad \text{Ymin} = -10 \qquad \text{Ymax} = 10$$

will produce a graph where the values of x and the values of y are between -10 and 10, inclusive. We can use the notation $[-10, 10]$ to describe such intervals.

Graphing lines

To graph $5x - 2y = 4$, we must first solve the equation for y.

$$y = \frac{5}{2}x - 2 \qquad \text{Subtract 5x from both sides and then divide both sides by } -2.$$

Next, we press $\boxed{Y =}$ and enter the right side of the equation after the symbol $Y_1 =$. See figure (a). We then press the $\boxed{\text{GRAPH}}$ key to get the graph shown in figure (b). To show more detail, we can change the window settings to $[-2, 5]$ for x and $[-4, 5]$ for y by pressing $\boxed{\text{WINDOW}}$ and entering -2 for Xmin, 5 for Xmax, -4 for Ymin, and 5 for Ymax. See figure (c).

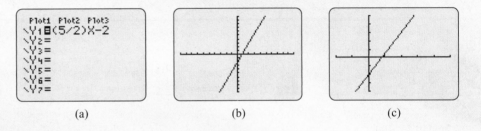

(a)   (b)   (c)

Finding the coordinates of a point on the graph

If we reenter the standard window settings of $[-10, 10]$ for x and for y, press $\boxed{\text{GRAPH}}$, and press the $\boxed{\text{TRACE}}$ key, we get the display shown in figure (d). The y-intercept of the graph is highlighted by the flashing cursor, and the x- and y-coordinates of that point are given at the bottom of the screen. We can use the $\boxed{\blacktriangleright}$ and $\boxed{\blacktriangleleft}$ keys to move the cursor along the line to find the coordinates of any point on the line. After pressing the $\boxed{\blacktriangleright}$ key 12 times, we will get the display in figure (e).

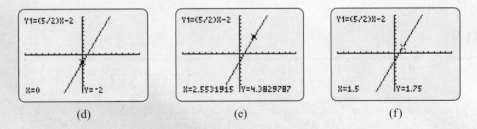

(d)   (e)   (f)

To find the y-coordinate of any point on the line, given its x-coordinate, we press $\boxed{\text{2nd}}$ $\boxed{\text{CALC}}$ and select the value option. We enter the x-coordinate of the point and press $\boxed{\text{ENTER}}$. The y-coordinate is then displayed. In figure (f), 1.5 was entered for the x-coordinate, and its corresponding y-coordinate, 1.75, was found.

The table feature, discussed on pages 136–137, gives us a third way of finding the coordinates of a point on the line.

**Determining the
x-intercepts of a graph**

To determine the x-intercept of the graph of $y = \frac{5}{2}x - 2$, we can use the zero option, found under the CALC menu. (Be sure to reenter the standard window settings for x and y before using CALC.) After we enter left and right bounds and a guess, as shown in figure (g), the cursor automatically moves to the x-intercept of the graph when we press $\boxed{\text{ENTER}}$. Figure (h) shows how the coordinates of the x-intercept are then displayed at the bottom of the screen.

We can also use the trace and zoom features to determine the x-intercept of the graph of $y = \frac{5}{2}x - 2$. After graphing the equation using the standard window settings, we press $\boxed{\text{TRACE}}$. Then we move the cursor along the line toward the x-intercept until we arrive at a point with the coordinates shown in figure (i). To get better results, we press $\boxed{\text{ZOOM}}$, select the zoom in option, and press $\boxed{\text{ENTER}}$ to get a magnified picture. We press $\boxed{\text{TRACE}}$ again and move the cursor to the point with coordinates shown in figure (j). Since the y-coordinate is nearly 0, this point is nearly the x-intercept. We can achieve better results with more zooms and traces.

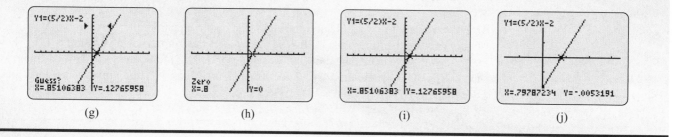

(g)   (h)   (i)   (j)

▷ **ANSWERS TO SELF CHECKS**   **1.** No

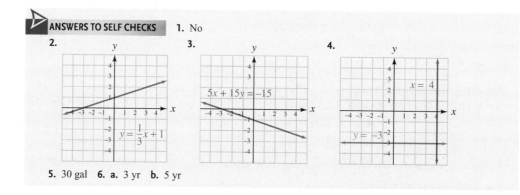

2.   **3.**   **4.**

5. 30 gal   **6. a.** 3 yr   **b.** 5 yr

STUDY SET
2.2

VOCABULARY

Fill in the blanks.

1. A solution of an equation in two variables is an _____ _____ of numbers that make a true statement when substituted into the equation.

2. The graph of $y = 2x + 1$ is the graph of all points (x, y) on the rectangular coordinate system whose coordinates _____ the equation.

3. The equation $y = -6x - 3$ is a _____ equation in two variables.

4. The _____ form of a linear equation in two variables is $Ax + By = C$, where not both A and B are 0.

5. The point where the graph of a linear equation intersects the *y*-axis is called the _____ and the point where it intersects the *x*-axis is called the _____.

6. The graph of any equation of the form $x = a$ is a _____ line. The graph of any equation of the form $y = b$ is a _____ line.

CONCEPTS

7. Use the following graph to determine three solutions of $2x + 3y = 9$.

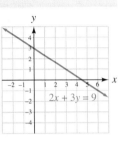

8. Consider the equation $2x + 4y = 8$. How many variables does it contain? How many solutions does it have?

9. Fill in the blanks: The exponent on each variable of a linear equation is an understood ____. For example, $4x + 7y = 3$ can be thought of as $4x\ \ + 7y\ \ = 3$.

10. A table of solutions for a linear equation is given on the right. From the table, determine the *x*-intercept and the *y*-intercept of the graph of the equation.

x	y	(x, y)
-6	0	$(-6, 0)$
-2	2	$(-2, 2)$
0	3	$(0, 3)$

11. Refer to the graph.

    a. What is the *x*-intercept and what is the *y*-intercept of the line?

    b. If the coordinates of point *M* are substituted into the equation of the line that is graphed here, will a true or a false statement result?

12. Consider the linear equation $6x - 4y = -12$.

    a. Find the *x*-intercept of its graph.

    b. Find the *y*-intercept of its graph.

    c. Does its graph pass through $(2, 6)$?

13. A graphing calculator display is shown below. It is a table of solutions for which one of the following linear equations?

$$y = -2x - 1, \quad y = -3x - 1, \quad \text{or} \quad y = -4x - 1$$

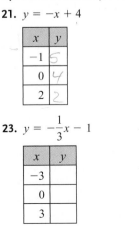

14. The graphing calculator displays below show the graph of $y = -2x - \frac{5}{4}$.

    a. In figure (a), what important feature of the line is highlighted by the cursor?

    b. In figure (b), what important feature of the line is highlighted by the cursor?

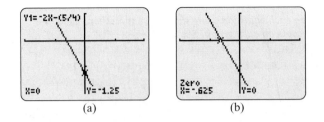

(a)   (b)

NOTATION

15. a. The graph of the equation $x = 0$ is which axis?

    b. The graph of the equation $y = 0$ is which axis?

16. A linear equation in two variables is an equation that can be written in the standard (general) form $Ax + By = C$. The equation $x - 5y = 4$ is written in standard form. Determine A, B, and C.

GUIDED PRACTICE

Determine whether each ordered pair is a solution of the given equation. **See Example 1.**

17. $y = -5x - 2$

    a. $(-1, 3)$ **b.** $(3, -13)$

18. $2x - 5y = 9$

    a. $(-4, 2)$ **b.** $(2, -1)$

19. $y - 6x = 10$

    a. $\left(\frac{1}{6}, 11\right)$ **b.** $(-2.1, -0.6)$

20. $3y - 5x = 30$

    a. $(-2.3, -7.5)$ **b.** $\left(-\frac{23}{5}, \frac{7}{3}\right)$

Complete each table of solutions and use the results to graph the equation. **See Example 2.**

21. $y = -x + 4$

x	y
-1	6
0	4
2	2

22. $y = x - 2$

x	y
-2	
0	
4	

23. $y = -\frac{1}{3}x - 1$

x	y
-3	
0	
3	

24. $y = -\frac{1}{2}x + \frac{5}{2}$

x	y
-1	
3	
5	

Graph each equation. See Example 2.

25. $y = x$

26. $y = -2x$

27. $y = -3x + 2$

28. $y = 2x - 3$

29. $y = 3 - x$

30. $y = 5 - x$

31. $y = \dfrac{x}{4} - 1$

32. $y = -\dfrac{x}{4} + 2$

Graph each equation using the intercept method. Label the intercepts on each graph. See Example 3.

33. $3x + 4y = 12$

34. $4x - 3y = 12$

35. $3y = 6x - 9$

36. $2x = 4y - 10$

37. $3x + 4y - 8 = 0$

38. $-2y - 3x + 9 = 0$

39. $3x = 4y - 11$

40. $-5x + 3y = 11$

Graph each equation. See Example 4.

41. $x = 3$

42. $y = -4$

43. $y = -\dfrac{1}{2}$

44. $x = \dfrac{4}{3}$

Write each equation in $y = b$ or $x = a$ form and then graph it. See Example 4.

45. $y - 2 = 0$

46. $x + 1 = 0$

47. $-2x + 3 = 11$

48. $-3y + 2 = 5$

Use a graphing calculator to graph each equation, and then find the x-coordinate of the x-intercept to the nearest hundredth. See Using Your Calculator: Determining the x-intercepts of a Graph.

49. $y = 3.7x - 4.5$

50. $y = \dfrac{3}{5}x + \dfrac{5}{4}$

51. $1.5x - 3y = 7$

52. $0.3x + y = 7.5$

TRY IT YOURSELF

Graph each equation.

53. $5x - 4y = 13$

54. $3x - 4y = 11$

55. $y = \dfrac{5}{6}x - 5$

56. $y = \dfrac{2}{3}x - 2$

57. $x + 2y = -2$

58. $2x + 4y = -8$

59. $y = \dfrac{5}{2}$

60. $x = 0$

61. $y = 4x$

62. $y = x$

63. $x = 50 - 5y$

64. $3x = -150 - 5y$

APPLICATIONS

65. SQUARE FOOTAGE   The equation $s = 31t + 1{,}920$ approximates the median number of square feet s in a new single-family home, t years after 1995. (Source: U.S. Census Bureau)
   a. Graph the equation.
   b. What information can be obtained from the s-intercept of the graph?

c. Suppose the current trend continues. From the graph, estimate the median number of square feet s that there will be in a new single-family home in 2025.

66. MARKET SHARE   The automobile industry classifies light-duty vehicles as cars, vans, SUVs , and pickup trucks. For sales of light-duty vehicles in the United States, the percent market share m of cars only is approximated by the equation $m = -\dfrac{4}{3}t + 69$, where t is the number of years after 1990. (Source: Environmental Protection Agency)
   a. Graph the equation.
   b. What information can be obtained from the m-intercept of the graph?

   c. Suppose the current trend continues. From the graph, estimate the market share m (in percent) of the sales of cars in 2020.

67. UNION MEMBERSHIP   The equation $p = -0.26t + 11.9$ approximates the percent p of those employed in the private sector who were union members, t years after 1990. (Source: Data360)
   a. Graph the equation.
   b. What information can be obtained from the p-intercept of the graph?

   c. Suppose the current trend continues. From the graph, estimate the percent of people working in the private sector who will be union members in 2015.

68. FARMING   The equation $a = -2.2t + 945$ gives the approximate number of acres a of farmland (in millions) in the United States, t years after 2000. (Source: U.S. Department of Agriculture)
   a. Graph the equation.
   b. What information can be obtained from the a-intercept of the graph?

   c. Suppose the current trend continues. From the graph, estimate the number of acres of farmland in the year 2020.

69. LIVING LONGER   The equation $a = 0.16t + 73.7$ is a linear model that approximates the average life expectancy a (in years) in the United States, t years after 1980. (Source: U.S. Census Bureau)
   a. Graph the equation.
   b. What information can be obtained from the a-intercept of the graph?

   c. Suppose the current trend continues. From the graph, estimate the average life expectancy in the United States in 2030.

70. HOURLY PAY The equation $p = 0.45t + 11$ approximates the average hourly pay p (in dollars) of a U.S. production worker, t years after 1994. (Source: U.S. Department of Labor)

a. Graph the equation.

b. What information can be obtained from the p-intercept of the graph?

c. Suppose the current trend continues. From the graph, estimate the average hourly pay for production workers in 2012.

71. DEPRECIATION The graph on the right shows how the value of a computer decreased over the age of the computer. What information can be obtained from the x-intercept? The y-intercept?

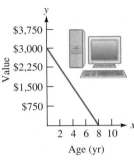

72. SPRAY BOTTLES The following graph shows the amount A of sore throat medication (in ounces) that remains in the bottle after the spray trigger has been pushed down a total of n times. What information can be obtained from the n-intercept? The A-intercept?

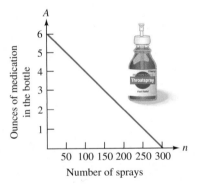

Number of sprays

73. CAR DEPRECIATION A car purchased for $17,000 is expected to lose value according to the straight-line depreciation model $y = -1,360x + 17,000$. When will the car have no value?

74. ACCOUNTING A carpet company purchased a new loom for $124,000. For income tax purposes, company accountants will use the straight-line depreciation equation $y = -15,500x + 124,000$ to describe the declining value of the loom.

a. When will the value of the loom be one-half of its purchase price?

b. When will the loom have no value?

75. DEMAND EQUATION The number of microwave ovens that consumers buy depends on price. The higher the price, the fewer microwaves people will buy. The equation that relates price to the number of microwaves sold at that price is called a **demand equation.** If the demand equation for a 2-cubic-foot countertop microwave is approximated by $p = -\frac{1}{10}q + 170$, where p is the price and q is the number of microwaves sold at that price, how many of that model microwave will be sold at a price of $150?

76. SUPPLY EQUATION The number of microwave ovens that manufacturers produce depends on price. The higher the price, the more microwaves manufacturers will produce. The equation that relates price to the number of microwaves produced at that price is called a **supply equation.** If the supply equation for a 2-cubic-foot countertop microwave is $p = \frac{1}{10}q + 130$, where p is the price and q is the number of microwaves produced for sale at that price, how many of that model microwave will be produced if the price is $150?

WRITING

77. Explain how to graph a line using the intercept method.

78. When graphing a line by plotting points, why is it a good practice to find three solutions instead of two?

79. What does it mean when we say that a linear equation in two variables has infinitely many solutions? Give an example.

80. On a quiz, a student was asked to graph the lines $x = -3$ and $y = 2$. His answer is shown below. Explain his error.

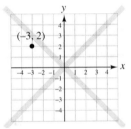

REVIEW

81. Write the set whose elements are the prime numbers between 10 and 30.

82. Write the set whose elements are the first ten composite numbers.

83. In what quadrant does the point $(-2, -3)$ lie?

84. What is the formula that gives the area of a circle?

85. Simplify: $-4(-20s)(-6)$

86. Approximate π to the nearest hundredth.

87. Simplify: $-4[-2(-3x - 8)]$

88. Simplify: $\frac{1}{3}b + \frac{1}{3}b + \frac{1}{3}b$

CHALLENGE PROBLEMS

Graph each equation.

89. $\frac{1}{5}x = 6 - \frac{3}{10}y$

90. $\frac{x}{2} - \frac{y}{3} - 4 = 0$

SECTION 2.3
Rate of Change and the Slope of a Line

Objectives

❶ Calculate an average rate of change.

❷ Find the slope of a line from its graph.

❸ Find the slope of a line given two points.

❹ Find the slope of horizontal and vertical lines.

❺ Solve applications of slope.

❻ Determine whether lines are parallel or perpendicular using slope.

Our world is one of constant change. In this section, we will describe the amount of change in one quantity with respect to the amount of change in another by finding an *average rate of change.*

❶ Calculate an Average Rate of Change.

The following line graphs model the approximate number of morning and evening newspapers published in the United States for the years 1990–2005. We see that the number of morning newspapers increased and the number of evening newspapers decreased over this time span.

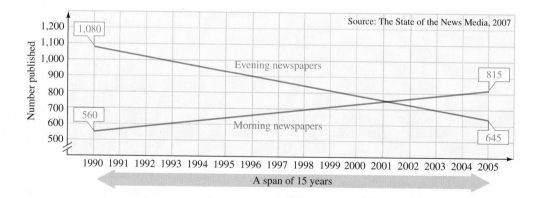

If we want to know the rate at which the number of morning newspapers increased or the rate at which the number of evening newspapers decreased, we must find an average rate of change. To find an average rate of change, we find the *ratio* of the change in the number of newspapers to the length of time in which that change took place and attach the appropriate units.

Ratios and Rates

A **ratio** is a comparison of two numbers using a quotient. In symbols, if a and b are two numbers, the ratio of a to b is $\frac{a}{b}$. Ratios that are used to compare quantities with different units are called **rates.**

The Language of Algebra
Ratios are used in many settings. Mortgage companies calculate *debt-to-income ratios* for loan applicants. Bicyclists choose *gear ratios* that produce a smooth ride uphill and downhill.

In the graph, we see that in 1990, the number of morning newspapers published was 560. In 2005, the number grew to 815. This is a change of $815 - 560$ or 255 over a 15-year time span. So we have

$$\begin{aligned}
\text{Average rate} \atop \text{of change} &= \frac{\text{change in number of morning newspapers}}{\text{change in time}} \\[6pt]
&= \frac{255 \text{ newspapers}}{15 \text{ years}} \\[6pt]
&= \frac{\overset{1}{\cancel{15}} \cdot 17 \text{ newspapers}}{\underset{1}{\cancel{15}} \text{ years}} \\[6pt]
&= \frac{17 \text{ newspapers}}{1 \text{ year}}
\end{aligned}$$

A rate of change is a ratio that includes units.

Factor 255 as $15 \cdot 17$ and simplify: $\frac{15}{15} = 1$. We could also just simply divide: $255 \div 15 = 17$.

Success Tip
Note that the numerator of the rate of change ratio contains units associated with the vertical axis of the graph. The denominator contains units associated with the horizontal axis.

The number of morning newspapers published in the United States increased, on average, at a rate of 17 newspapers per year from 1990 through 2005. This can be written as 17 newspapers/year.

EXAMPLE 1   *Evening Newspapers.*   Refer to the previous graph. Find the average rate at which the number of evening newspapers published in the United States decreased from 1990 through 2005.

Strategy   We will write the ratio of the change in the number of evening newspapers to the change in time and attach the appropriate units. Then we will simplify the result, if possible.

Why   An average rate of change compares the amount of change in one quantity with respect to the amount of change in another quantity using a ratio with units.

Solution   From the graph, we see that in 1990 the number of evening newspapers published was 1,080. In 2005, the number fell to 645. To find the change, we subtract: $645 - 1{,}080 = -435$. The negative result indicates a decline in the number of evening newspapers over the 15-year time span. So we have

Success Tip
In general, to find the change in a quantity, we subtract the earlier value from the later value.

$$\begin{aligned}
\text{Average rate} \atop \text{of change} &= \frac{\text{change in number of evening newspapers}}{\text{change in time}} \\[6pt]
&= \frac{-435 \text{ newspapers}}{15 \text{ years}} \\[6pt]
&= \frac{\overset{-1}{\cancel{-15}} \cdot 29 \text{ newspapers}}{\underset{1}{\cancel{15}} \text{ years}} \\[6pt]
&= \frac{-29 \text{ newspapers}}{1 \text{ year}}
\end{aligned}$$

A rate of change is a ratio that includes units.

Factor -435 as $-15 \cdot 29$ and simplify: $\frac{-15}{15} = -1$. We could also just simply divide: $-435 \div 15 = -29$.

The number of evening newspapers being published changed at a rate of -29 newspapers/year. That is, on average, there were 29 fewer per year, every year, from 1990 through 2005.

Self Check 1   In 1992, there were approximately 888 Sunday edition newspapers being published in the United States. By 2005, that number had risen to 914. Find the average rate at which the number of Sunday edition newspapers increased from 1992 through 2005.

Now Try   **Problem 53**

2   **Find the Slope of a Line from Its Graph.**

In the newspaper example, we measured the steepness of the lines in the graph to determine the average rates of change. In doing so, we found the *slope* of each line. The **slope of a line** is a ratio that compares the vertical change to the corresponding horizontal change as we move along the line from one point to another.

To determine the slope of a line (usually denoted by the letter *m*) from its graph, we first pick two points on the line. Then we write the ratio of the vertical change, called the **rise,** to the corresponding horizontal change, called the **run,** as we move from one point to the other.

$$m = \frac{\text{vertical change}}{\text{horizontal change}} = \frac{\text{rise}}{\text{run}}$$

EXAMPLE 2   Find the slope of the line graphed in Figure (a) below.

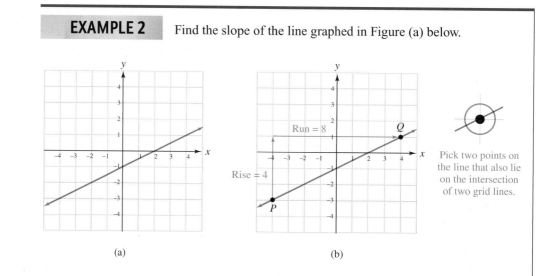

Pick two points on the line that also lie on the intersection of two grid lines.

(a)   (b)

Strategy   We will pick two points on the line, construct a slope triangle, and find the rise and run. Then we will write the ratio of rise to run and simplify the result, if possible.

Why   The slope of a line is the ratio of the rise to the run.

Solution   We begin by choosing two points on the line, *P* and *Q*, as shown in figure (b). One way to move from *P* to *Q* is to start at *P*, move upward, a rise of 4 grid squares, and then to the right, a run of 8 grid squares, to reach *Q*. These steps create a right triangle called a **slope triangle.**

$$m = \frac{\text{rise}}{\text{run}} = \frac{4}{8} = \frac{1}{2}$$ Simplify the fraction. The result is positive.

The slope of the line is $\frac{1}{2}$.

Caution

Slopes are normally written as fractions, sometimes as decimals, but never as mixed numbers.

As with any fractional answer, always express slope in simplified form (lowest terms).

The two-step process to move from P to Q can be reversed. Starting at P, we can move to the right, a run of 8; and then upward, a rise of 4, to reach Q. With this approach, the slope triangle is below the line. When we form the ratio to find the slope, we get the same result as before:

$$m = \frac{\text{rise}}{\text{run}} = \frac{4}{8} = \frac{1}{2}$$

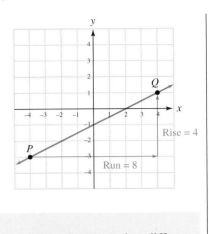

▷ **Self Check 2**   Find the slope of the line shown above using two points different from those used in the solution of Example 2.

Now Try   **Problems 17 and 21**

The identical answers from Example 2 and its Self Check illustrate that *the same value will be obtained no matter which two points on a line are used to find its slope.*

③ Find the Slope of a Line Given Two Points.

We can use the graphic method for finding slope to develop a slope formula. To begin, we select points P and Q on the line shown in the figure on the right. To distinguish between the coordinates of these points, we use **subscript notation.** Point P has coordinates (x_1, y_1) and point Q has coordinates (x_2, y_2).

As we move from point P to point Q, the rise is the difference of the y-coordinates: $y_2 - y_1$. The run is the difference of the x-coordinates: $x_2 - x_1$. Since the slope is the ratio $\frac{\text{rise}}{\text{run}}$, we have the following formula for calculating slope.

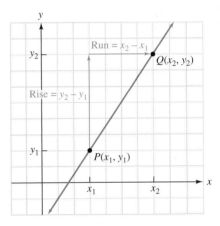

Slope of a Line

The **slope** of a line passing through points (x_1, y_1) and (x_2, y_2) is

$$m = \frac{\text{vertical change}}{\text{horizontal change}} = \frac{\text{rise}}{\text{run}} = \frac{\text{change in } y}{\text{change in } x} = \frac{y_2 - y_1}{x_2 - x_1} \quad \text{if } x_2 \neq x_1$$

Another notation that we use to define slope involves the symbol Δ, which is the letter *delta* from the Greek alphabet. If the change in y is represented by Δy (read as "delta y") and the change in x is represented by Δx (read as "delta x"), then:

$$m = \frac{\Delta y}{\Delta x} \quad \text{where } \Delta x \neq 0$$

The graph on the right shows all of the notation associated with the concept of slope of a line.

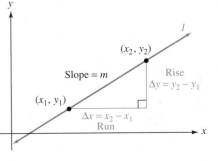

EXAMPLE 3   Find the slope of the line passing through $(-2, 4)$ and $(3, -4)$.

Strategy   We will use the slope formula to find the slope.

Why   We know the coordinates of two points on the line.

Solution   We can let $(x_1, y_1) = (-2, 4)$ and $(x_2, y_2) = (3, -4)$. Then we have

$$m = \frac{y_2 - y_1}{x_2 - x_1} \qquad \text{This is the slope formula.}$$

$$= \frac{-4 - 4}{3 - (-2)} \qquad \text{Substitute } -4 \text{ for } y_2, 4 \text{ for } y_1, 3 \text{ for } x_2, \text{ and } -2 \text{ for } x_1.$$

$$= -\frac{8}{5} \qquad \text{Write } \frac{-8}{5} \text{ with the } - \text{ sign in front of the fraction. The result is negative.}$$

> **Success Tip**
> The slope formula is a valuable tool because it enables us to calculate the slope of a line without having to view its graph.

The slope of the line is $-\frac{8}{5}$.

The graph of the line passing through $(-2, 4)$ and $(3, -4)$ is shown on the right. Notice that we obtain the same result when the slope of the line is found graphically.

$$m = \frac{\text{rise}}{\text{run}} = \frac{-8}{5} = -\frac{8}{5}$$

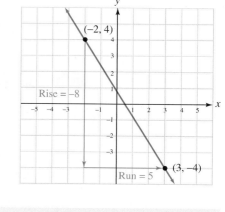

Self Check 3   Find the slope of the line passing through $(-3, 6)$ and $(4, -8)$.
Now Try   **Problems 27 and 31**

When calculating slope, it doesn't matter which point we call (x_1, y_1) and which point we call (x_2, y_2). We will obtain the same result in Example 3 if we let $(x_1, y_1) = (3, -4)$ and $(x_2, y_2) = (-2, 4)$.

$$m = \frac{y_2 - y_1}{x_2 - x_1} = \frac{4 - (-4)}{-2 - 3} = \frac{8}{-5} = -\frac{8}{5}$$

Caution   When using the slope formula, we must be careful to subtract the y-coordinates and the x-coordinates in the same order. For instance, in Example 3 with $(x_1, y_1) = (-2, 4)$ and $(x_2, y_2) = (3, -4)$, it would be incorrect to write either of the following:

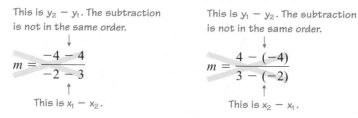

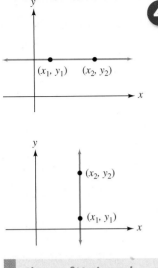

4 Find the Slope of Horizontal and Vertical Lines.

If (x_1, y_1) and (x_2, y_2) are distinct points on the horizontal line shown on the left, then $y_1 = y_2$, and the numerator of the fraction

$$\frac{y_2 - y_1}{x_2 - x_1} \quad \text{On a horizontal line, } x_2 \neq x_1.$$

is 0. Thus, the value of the fraction is 0, and the slope of the horizontal line is 0.

If (x_1, y_1) and (x_2, y_2) are distinct points on the vertical line to the left, then $x_1 = x_2$, and the denominator of the fraction

$$\frac{y_2 - y_1}{x_2 - x_1} \quad \text{On a vertical line, } y_2 \neq y_1.$$

is 0. Since the denominator of a fraction cannot be 0, a vertical line has no defined slope.

Slopes of Horizontal and Vertical Lines	Horizontal lines (lines with an equation of the form $y = b$) have a slope of 0. Vertical lines (lines with equations of the form $x = a$) have no defined slope.

To classify the slope of a line as positive or negative, follow the line from left to right. If a line rises, its slope is positive. If a line drops, its slope is negative. If a line is horizontal, its slope is 0. If a line is vertical, it has undefined slope.

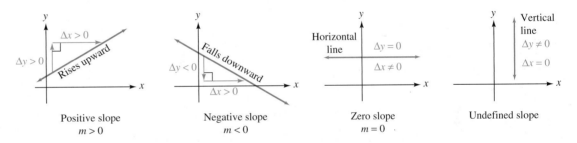

Positive slope $m > 0$	Negative slope $m < 0$	Zero slope $m = 0$	Undefined slope

Caution   Undefined and 0 do not mean the same thing. A horizontal line has a defined slope; it is 0. A vertical line does not have a defined slope; we say its slope is undefined.

5 Solve Applications of Slope.

The concept of slope has many applications. For example, architects use slope when designing ramps and determining the pitch of roofs. Truckers must be aware of the slope, or grade, of a road. Mountain resorts rate the difficulty level of ski runs by the degree of steepness.

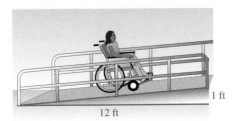

The maximum slope for a wheelchair ramp is 1 foot of rise for every 12 feet of run: $m = \frac{1}{12}$.

A 6% grade means a vertical change of 6 feet for every horizontal change of 100 feet: $m = \frac{6}{100}$.

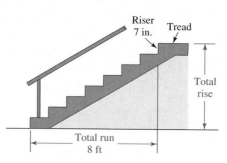

EXAMPLE 4   *Building Stairs.*   The slope of a staircase is defined to be the ratio of the total rise to the total run, as shown in the illustration. Find the slope of the staircase.

Strategy   We will express the total rise and the total run in terms of the same units and form their ratio.

Why   The slope is a ratio that compares two quantities with the same units.

Solution   Since the design has eight 7-inch risers, the total rise is $8 \cdot 7 = 56$ inches. The total run is 8 feet, or 96 inches. With these quantities expressed in the same units, we can form their ratio.

$$m = \frac{\text{total rise}}{\text{total run}}$$

$$= \frac{56}{96}$$

$$= \frac{7}{12}$$ Simplify the fraction: $\frac{56}{96} = \frac{7 \cdot \cancel{8}}{\cancel{8} \cdot 12} = \frac{7}{12}$.

The slope of the staircase is $\frac{7}{12}$.

Self Check 4   Find the slope of the staircase if the riser height is changed to 6.5 inches.

Now Try   **Problem 57**

EXAMPLE 5   *Downhill Skiing.*   It takes a skier 25 minutes to complete the course shown in the illustration on the next page. Find her average rate of descent in feet per minute.

Strategy   We will describe the skier's positions on the course as ordered pairs of the form (time, elevation). Then we will use the slope formula to calculate the average rate of descent.

Why   We use the slope formula because we know the coordinates of the skier at the beginning and at the end of the course.

Solution   The skier's position can be described as an ordered pair of the form (time, elevation), or more simply, (t, E). To find the average rate of descent, we will calculate the slope of the line passing through the points $(t_1, E_1) = (0, 12{,}000)$ and $(t_2, E_2) = (25, 8{,}500)$ and attach the appropriate units to the numerator and denominator.

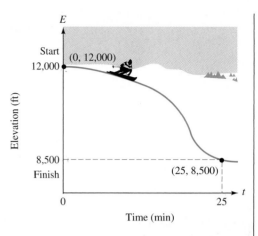

$$\text{Average rate of descent} = \frac{(E_2 - E_1)\text{ feet}}{(t_2 - t_1)\text{ minutes}}$$

This is the slope formula adapted to a coordinate system with ordered pairs of the form (t, E).

$$= \frac{(8{,}500 - 12{,}000)\text{ feet}}{(25 - 0)\text{ minutes}}$$

Substitute 8,500 for E_2, 12,000 for E_1, 25 for t_2, and 0 for t_1.

$$= \frac{-3{,}500\text{ feet}}{25\text{ minutes}}$$

Do the subtractions.

$$= -140 \text{ ft per min}$$

Do the division. The negative symbol indicates a loss of elevation.

The skier's average rate of descent was 140 feet per minute. (The word *descent* itself implies a loss of elevation during the ski run. Therefore, the negative symbol − need not be written in front of 140.)

Self Check 5   Find the average rate of descent if the skier completes the course in 20 minutes.

Now Try   **Problem 63**

6   **Determine Whether Lines Are Parallel or Perpendicular Using Slope.**

To see a relationship between parallel lines and their slopes, we refer to the parallel lines l_1 and l_2 shown on the next page, with slopes of m_1 and m_2, respectively. Because right triangles *ABC* and *DEF* are similar, it follows that

$$m_1 = \frac{\Delta y \text{ of } l_1}{\Delta x \text{ of } l_1}$$

Read l_1 as "line l sub 1."

$$= \frac{\Delta y \text{ of } l_2}{\Delta x \text{ of } l_2}$$

Since the triangles are similar, corresponding sides of $\triangle ABC$ and $\triangle DEF$ are proportional: $\frac{CB}{BA} = \frac{FE}{ED}$.

$$= m_2$$

Thus, if two nonvertical lines are parallel, they have the same slope. It is also true that when two different lines have the same slope, they are parallel.

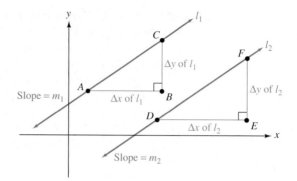

Slopes of Parallel Lines	Nonvertical parallel lines have the same slope, and different lines having the same slope are parallel.

EXAMPLE 6   Determine whether the line that passes through the points $(-6, 2)$ and $(3, -1)$ is parallel to a line with a slope of $-\dfrac{1}{3}$.

Strategy   We will compare the slopes of the lines.

Why   If the slopes are equal, the lines are parallel. If the slopes are not equal, the lines are not parallel.

Solution   We can use the slope formula to find the slope of the line that passes through $(-6, 2)$ and $(3, -1)$.

$$m = \frac{y_2 - y_1}{x_2 - x_1}$$

$$m = \frac{-1 - 2}{3 - (-6)} \qquad \text{Substitute } -1 \text{ for } y_2, \ 2 \text{ for } y_1, \ 3 \text{ for } x_2, \text{ and } -6 \text{ for } x_1.$$

$$= \frac{-3}{9}$$

$$= -\frac{1}{3} \qquad \text{Simplify the fraction. Write the } - \text{ symbol in front of the fraction.}$$

Both lines have slope $-\frac{1}{3}$, and therefore they are parallel.

Self Check 6   Determine whether the line that passes through the points $(4, -8)$ and $(1, -2)$ is parallel to a line with slope 2.

Now Try   **Problems 41 and 49**

The two lines shown in the figure on the next page meet at right angles and are called **perpendicular lines.** Each of the four angles that are formed has a measure of 90°.

The Language of Algebra
The words *perpendicular* and
parallel are used in many settings.
For example, the gymnast on the
parallel bars shown below is a
position that is *perpendicular* to
the floor.

The product of the slopes of two (nonvertical) perpendicular lines is -1. For example, the perpendicular
lines shown in the figure have slopes of $\frac{3}{2}$ and $-\frac{2}{3}$. If we
find the product of their slopes, we have

$$\frac{3}{2}\left(-\frac{2}{3}\right) = -\frac{6}{6} = -1$$

Two numbers whose product is -1, such as $\frac{3}{2}$ and
$-\frac{2}{3}$, are called **negative reciprocals.** The term *negative
reciprocal* can be used to relate perpendicular lines and
their slopes.

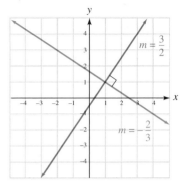

Slopes of Perpendicular Lines	If two nonvertical lines are perpendicular, their slopes are negative reciprocals.
	If the slopes of two lines are negative reciprocals, the lines are perpendicular.

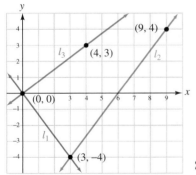

We can also state the fact given above symbolically: If the slopes of two nonvertical lines
are m_1 and m_2, then the lines are perpendicular if

$$m_1 \cdot m_2 = -1 \qquad \text{or} \qquad m_2 = -\frac{1}{m_1}$$

Because a horizontal line is perpendicular to a vertical line, a line with a slope of 0 is perpendicular to a line with no defined slope.

EXAMPLE 7   Are the lines l_1 and l_2 shown in the figure perpendicular?

Strategy   We will compare the slopes of the two lines.

Why   If the slopes are negative reciprocals, the lines are perpendicular. If the slopes are
not negative reciprocals, the lines are not perpendicular.

Solution   We find the slope of each line and see whether they are negative reciprocals.

$$m = \frac{y_2 - y_1}{x_2 - x_1}$$

$$= \frac{-4 - 0}{3 - 0} \qquad \text{Use } (3, -4) \text{ and } (0, 0) \text{ on } l_1.$$

$$= -\frac{4}{3} \qquad \text{This is the slope of } l_1.$$

$$m = \frac{y_2 - y_1}{x_2 - x_1}$$

$$= \frac{4 - (-4)}{9 - 3} \qquad \text{Use } (3, -4) \text{ and } (9, 4) \text{ on } l_2.$$

$$= \frac{8}{6}$$

$$= \frac{4}{3} \qquad \text{This is the slope of } l_2.$$

Since their slopes are not negative reciprocals $\left(-\frac{4}{3} \cdot \frac{4}{3} \neq -1\right)$, the lines are not perpendicular.

Self Check 7   Is l_1 perpendicular to l_3?
Now Try   Problems 43 and 51

STUDY SET
2.3

VOCABULARY

Fill in the blanks.

1. A _____ is a comparison of two numbers using a quotient. In symbols, if *a* and *b* are two numbers, the _____ of *a* to *b* is $\frac{a}{b}$.

2. Ratios that are used to compare quantities with different units are called _____.

3. An average _____ of _____ describes how much one quantity changes with respect to another.

4. _____ is defined as the change in *y* divided by the change in *x*.

5. The _____ in *x* (written Δx) is the horizontal run of the line between two points on the line, and the change in *y* (written Δy) is the vertical _____ of the line between two points on the line.

6. $\frac{7}{8}$ and $-\frac{8}{7}$ are _____ reciprocals.

CONCEPTS

7. HALLOWEEN A couple kept a graphical record of the number of trick-or-treaters who came to their door on Halloween night for the years 1990–2006. Fill in the blanks.

a. Number of trick-or-treaters in 1990:

Number of trick-or-treaters in 2006:

b. Change in the number of trick-or-treaters from 1990–2006:

Time span: _____ years

c. Rate of change $= \dfrac{\text{trick-or-treaters}}{\text{years}}$

$=$ _____ trick-or-treaters/year

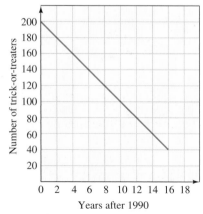

8. Fill in the blanks.

a. $m = \dfrac{\rule{2cm}{0.4pt}}{\text{horizontal change}} = \dfrac{\text{rise}}{\rule{1.5cm}{0.4pt}} = \dfrac{\text{change in } y}{\rule{1.5cm}{0.4pt}} = \dfrac{\rule{1.5cm}{0.4pt}}{\Delta x}$

b. $m = \dfrac{y_2 - y_1}{\rule{1.5cm}{0.4pt}}$

9. Refer to the slope triangle shown on the graph.

a. Find the rise.

b. Find the run.

c. Find the value of $\frac{\text{rise}}{\text{run}}$.

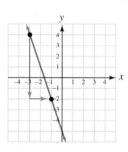

10. Refer to the slope triangle shown on the graph.

a. Find Δy.

b. Find Δx.

c. Find $\frac{\Delta y}{\Delta x}$.

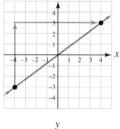

11. Refer to the graph.

a. Which line is horizontal? Find its slope.

b. Which line is vertical? Find its slope.

c. Which line has a positive slope? What is it?

d. Which line has a negative slope? What is it?

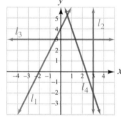

12. Fill in the blanks: _____ lines have the same slope and the slopes of _____ lines are negative reciprocals.

13. Refer to the graph on the next page.

a. Find the slopes of lines l_1 and l_2. Are they parallel?

b. Find the slopes of lines l_2 and l_3. Are they perpendicular?

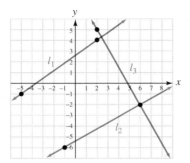

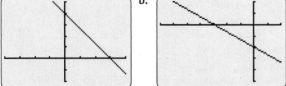

14. Determine the slope of each line.

a. | b.

NOTATION

15. The rate of change of 10 ft/yr is read as "10 feet _____ year."

16. Fill in the blanks: The symbol Δ is the letter _____ from the _____ alphabet.

GUIDED PRACTICE

Find the slope of each line. See Example 2.

17.

18.

19.

20.

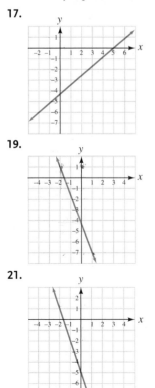

21.

22.

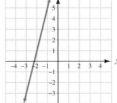

23.

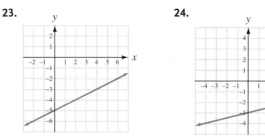

24.

Find the slope of the line that passes through the given points, if possible. See Example 3.

25. $(0, 0), (3, 9)$ **26.** $(-5, -8), (3, 8)$

27. $(-1, 8), (6, 1)$ **28.** $(3, 4)$ and $(2, 7)$

29. $(3, -1), (-6, 2)$ **30.** $(0, -8), (-5, 0)$

31. $(28, 50), (7, 17)$ **32.** $(-7, -2), (70, 40)$

33. $(7, 5), (-9, 5)$ **34.** $(2, -8), (3, -8)$

35. $(-7, -5), (-7, -2)$

36. $(3, -5), (3, 14)$

37. $\left(\frac{1}{4}, \frac{9}{2}\right), \left(-\frac{3}{4}, 0\right)$

38. $\left(\frac{1}{8}, \frac{3}{4}\right), \left(\frac{3}{8}, -\frac{1}{4}\right)$

39. $(0.7, -0.6), (-0.9, 0.2)$

40. $(-1.2, 8.6), (-1.1, 7.6)$

Determine whether the line that passes through the two given points is parallel or perpendicular (or neither) to a line with a slope of -2. See Examples 6 and 7.

41. $(3, 4), (4, 2)$

42. $(6, 4), (8, 5)$

43. $(-2, 1), (6, 5)$

44. $(3, 4), (-3, -5)$

45. $(5, 4), (6, 6)$

46. $(-2, 3), (4, -9)$

47. $(3.2, 12.3), (6.2, 6.3)$

48. $\left(\frac{2}{3}, \frac{3}{4}\right), \left(\frac{1}{2}, \frac{2}{3}\right)$

Determine whether the lines are parallel, perpendicular, or neither. See Examples 6 and 7.

49. A line passing though $(-3, -2)$ and $(4, 5)$
A line passing through $(-1, 0)$ and $(6, 7)$

50. A line passing through $(-3, -1)$ and $(-3, 4)$
A line passing through $(-4, -2)$ and $(3, -2)$

51. A line passing through $(-2, -2)$ and $(-5, 4)$
A line passing through $(-6, 2)$ and $(-2, 4)$

52. A line passing through $(-5, 4)$ and $(5, 4)$
A line passing through $(-6, -2)$ and $(3, -2)$

APPLICATIONS

53. U.S. MUSIC SALES   The following line graph models the approximate number of CDs that were shipped for sale in the United States from 1980 through 2005.

   a. Find the rate of increase in the number of CDs shipped from 1990–2000.

   b. Find the rate of decrease in the number of CDs shipped from 2000–2005.

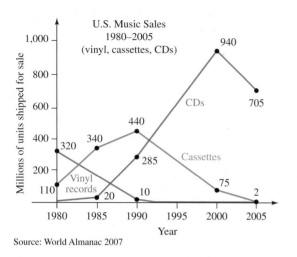

Source: World Almanac 2007

54. MUSIC HISTORY   The line graphs in problem 53 model the approximate number of vinyl records and cassettes that were shipped for sale in the United States for the years 1980 through 2005.

   a. Find the rate of decrease in the number of vinyl records shipped from 1980–1990.

   b. Find the rate of decrease in the number of cassettes shipped from 1990–2000.

55. GLOBAL WARMING   The following graph models the prediction by the National Center for Atmospheric Research of a future average global temperature rise caused by greenhouse gas emissions. Find the predicted average rate of change of temperature for this time span.

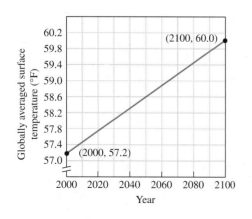

56. AIR PRESSURE   Air pressure, measured in units called **Pascals** (Pa), decreases with altitude. Find the rate of change in Pascals for the fastest and the slowest decreasing steps of the following graph. (km stands for kilometers.)

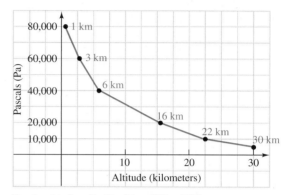

Based on data from *The Blue Planet* (Wiley, 1995)

57. FLYING   A jet descends in a stairstep pattern, as shown in the illustration. (The symbol ′ stands for feet.) The required elevations of the plane's path are given. Find the slope of the descent in each of the three parts of its landing that are labeled. Which part is the steepest?

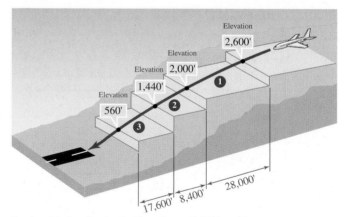

Based on data from *Los Angeles Times* (August 7, 1997), p. A8

58. MAPS   Topographic maps have contour lines that connect points of equal elevation on a mountain. The vertical distance between contour lines in the illustration on the next page is 50 feet. Find the slope of the west face and the slope of the east face of the mountain peak.

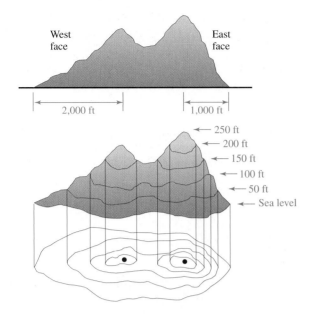

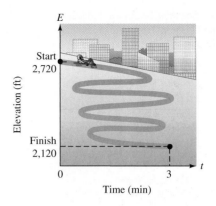

59. STEEP GRADES   Find the grade of the road shown in the illustration.   (*Hint:* 1 mi = 5,280 ft.)

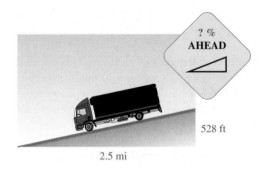

60. RAILROADS   The Saluda Grade in Polk County, North Carolina, is the steepest standard-gauge mainline railway grade in the United States. At one point, the grade reaches 5.1% between the towns of Melrose and Saluda. Explain what a 5.1% grade means.

61. STAIRCASES   Common practice among American architects for interior staircases has been to make the unit rise about $7\frac{1}{2}$ inches and the unit run 9 inches. Write the ratio of rise to run as a fraction in simplest form.

62. EXTREME SPORTS   See the illustration in the next column. Find the average rate of descent of a street luge "pilot" if he completes the course in 3 minutes.

63. BIKE RACING   Find the average rate of descent of mountain biker if she completes the course in 6 minutes.

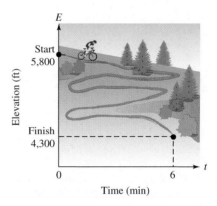

64. COMPUTERS   The price of computers has been dropping for the past ten years. If a desktop PC cost $5,700 10 years ago, and the same computing power cost $400 2 years ago, find the rate of decrease per year. (Assume a straight-line model.)

65. SKIING   The men's giant slalom course shown in the illustration is longer than the women's course. Does this mean that the men's course is steeper? Use the concept of the slope of a line to explain.

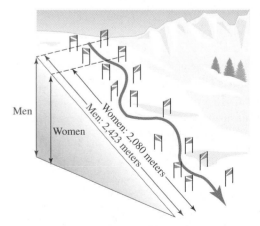

66. DECKS   See the illustration. Find the slopes of the cross-brace and the supports. Is the cross-brace perpendicular to either support?

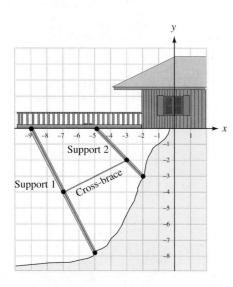

67. POLITICS   The following illustration shows how federal Medicare spending would have continued if the Republican-sponsored Balanced Budget Act hadn't become law in 1997. Explain why Democrats could argue that the budget act "cut spending." Then explain why Republicans could respond by saying, "There was no cut in spending—only a reduction in the rate of growth of spending."

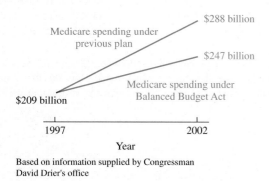

Based on information supplied by Congressman David Drier's office

68. a. When is a rate of change negative? Give an example.

   b. When is a rate of change 0? Give an example.

69. Explain why a vertical line has no defined slope.

70. Explain how to determine from their slopes whether two lines are parallel, perpendicular, or neither.

71. CANDY   A candy maker wants to make a 60-pound mixture of two candies to sell for $2 per pound. If black licorice bits sell for $1.90 per pound and orange gumdrops sell for $2.20 per pound, how many pounds of each should be used?

72. RETIREMENT   A nurse wants to supplement her pension with investment interest. If she invests $28,000 at 6% interest, how much would she have to invest at 7% to achieve a goal of $3,500 per year in supplemental income?

73. The two lines graphed in the illustration are parallel. Find x and y.

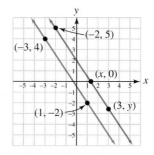

74. The line passing through $(1, 3)$ and $(-2, 7)$ is perpendicular to the line passing through points $(4, b)$ and $(8, -1)$. Without graphing, find b.

75. Find the slope of the line that passes through (a, b) and $(-b, -a)$. Assume $a \neq 0$ and $b \neq 0$.

76. A table of solutions for a linear equation is shown here. Find the slope of the graph of the equation.

SECTION 2.4
Writing Equations of Lines

Objectives

1. Use slope–intercept form to write the equation of a line.
2. Use the point–slope form to write the equation of a line.
3. Use slope as an aid when graphing.
4. Recognize parallel and perpendicular lines.
5. Write a linear equation model to fit a collection of data.

We have seen that linear relationships can be presented in graphs. In this section, we will write equations to model linear relationships.

1 **Use the Slope–Intercept Form to Write the Equation of a Line.**

To develop one form of the equation of a line, consider line l in the figure with y-intercept $(0, b)$ and let (x, y) be another point on the line. If we let $(x_1, y_1) = (0, b)$ and $(x_2, y_2) = (x, y)$, the slope of line l is given by the slope formula:

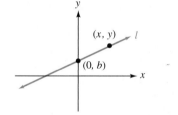

$$m = \frac{y_2 - y_1}{x_2 - x_1}$$ This is the slope formula.

$$m = \frac{y - b}{x - 0}$$ Substitute y for y_2, b for y_1, x for x_2, and 0 for x_1.

$$m = \frac{y - b}{x}$$ Simplify the denominator.

To solve the equation for y, we clear it of the fraction by multiplying both sides by x.

$$m \cdot x = \frac{y - b}{x} \cdot x$$

$$mx = y - b$$ Simplify the right side: $\dfrac{y - b}{\overset{}{\underset{1}{\cancel{x}}}} \cdot \overset{1}{\cancel{x}}$.

$$mx + b = y$$ To isolate y, add b to both sides.

$$y = mx + b$$ Write the equation with y on the left side.

Because this equation displays the slope m and the y-coordinate b of the y-intercept, it is called the **slope–intercept form** of the equation of a line.

Slope–Intercept Form

The equation of the line with slope m and y-intercept $(0, b)$ is

$$y = mx + b$$

When an equation of a line is written in slope–intercept form, the coefficient of the x-term is the line's slope and the constant term gives the y-coordinate of the y-intercept.

$$y = mx + b$$

Slope ↑   ↑ y-intercept: $(0, b)$

Linear equation	Equation written in slope–intercept form	Slope	y-intercept
$y = 4x - 3$	$y = 4x + (-3)$ or $y = 4x - 3$	4	$(0, -3)$
$y = -\dfrac{5}{6}x$	$y = -\dfrac{5}{6}x + 0$	$-\dfrac{5}{6}$	$(0, 0)$

EXAMPLE 1

a. Write an equation of the line with slope -1 and y-intercept $(0, 7)$.

b. Write an equation of the line graphed on the left.

Strategy In each case, we will use the slope–intercept form, $y = mx + b$, to write an equation of the line.

Why We are given (or we can easily find) the slope and the y-intercept of each line.

Solution

a. If the slope is -1 and the y-intercept is $(0, 7)$, then $m = -1$ and $b = 7$.

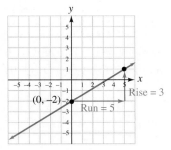

$$y = mx + b \qquad \text{This is the slope–intercept form.}$$
$$y = -1x + 7 \qquad \text{Substitute } -1 \text{ for } m \text{ and } 7 \text{ for } b.$$
$$y = -x + 7 \qquad \text{Simplify: } -1x = -x.$$

The equation of the line with slope -1 and y-intercept $(0, 7)$ is $y = -x + 7$.

b. In the figure on the left, we see that the y-intercept of the line is $(0, -2)$ and the slope of the line is $\frac{3}{5}$. When we substitute $\frac{3}{5}$ for m and -2 for b into the slope–intercept form $y = mx + b$, we obtain an equation of the line: $y = \frac{3}{5}x - 2$.

Self Check 1

a. Write an equation of the line with slope 1 and y-intercept $(0, -12)$.

b. Write an equation of the line whose graph is shown.

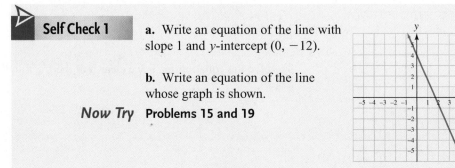

Now Try **Problems 15 and 19**

EXAMPLE 2 Use the slope–intercept form to write an equation of the line that has slope 4 and passes through $(5, 9)$.

Strategy Since the slope of the line is given as $m = 4$, the slope–intercept equation of the line has the form $y = 4x + b$. To find b, we will substitute 5 for x and 9 for y in $y = 4x + b$ and solve for b.

Why If the point $(5, 9)$ lies on the line, it is a solution of the equation and its coordinates must satisfy the equation of the line.

Solution

$$y = mx + b \qquad \text{This is the slope–intercept form.}$$
$$y = 4x + b \qquad \text{Substitute 4 for } m. \text{ We do not know } b.$$

To find b, we use the fact that the point $(5, 9)$ lies on the line and its coordinates, therefore, satisfy the equation.

$9 = 4(5) + b$ Substitute 9 for y and 5 for x.

$9 = 20 + b$ Do the multiplication.

$-11 = b$ To isolate b, subtract 20 from both sides.

Because $m = 4$ and $b = -11$, the equation is $y = 4x - 11$.

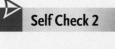

Self Check 2   Use the slope–intercept form to write an equation of the line that has slope -2 and passes through $(-2, 8)$.

Now Try   **Problem 23**

The slope–intercept form of the equation of a line can be used to model linear relationships. However, we will often use variables other than x and y to describe the quantities involved.

EXAMPLE 3   **School Supplies.**   Each turn of the handle of a pencil sharpener shaves off 0.05 inch from a 7.25-inch-long pencil.

a. Write a linear equation that gives the new length L of the pencil after the sharpener handle has been turned t times.

b. How long is the pencil after the sharpener handle has been turned 20 times?

Strategy   We will read the problem with the hope that we can find information about the slope and the y-intercept of the graph of the equation.

Why   If we can determine the slope and y-intercept of the graph of the line, we can use the slope–intercept form to write an equation to model the situation.

Solution

a. Since the length L of the pencil depends on the number of turns t of the handle, the linear equation that models this situation will have the form $L = mt + b$. We need to determine m and b.

- The length of the pencil decreases as the handle is turned. This rate of change, -0.05 inch per turn, is the slope of the graph of the equation. Thus, $m = -0.05$.
- Before any turns of the handle are made (when $t = 0$), the length of the pencil is 7.25 inches. Written as an ordered pair of the form (t, L), we have $(0, 7.25)$. When graphed, this would be the L-intercept of the graph. Thus, $b = 7.25$.

Original length
7.25 in.

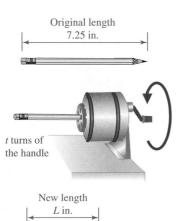

t turns of the handle

New length
L in.

Substituting for m and b, we have the linear equation that models this situation.

$$L = -0.05t + 7.25$$ This is the slope–intercept form using the variables L and t.

The slope is the rate of change of the length of the pencil.   The intercept is the original length of the pencil.

b. To find the pencil's length after the handle is turned 20 times, we proceed as follows:

$L = -0.05t + 7.25$ This is the linear model.

$L = -0.05(20) + 7.25$ Substitute 20 for t, the number of turns.

$L = -1 + 7.25$

$ = 6.25$

If the sharpener handle is turned 20 times, the pencil will be 6.25 inches long.

Now Try **Problem 87**

②   **Use the Point–Slope Form to Write the Equation of a Line.**

To develop another form of the equation of a line, we consider line l in the figure that has slope m and passes through (x_1, y_1). If (x, y) is a second point on the line, the slope of the line is given by the slope formula:

$$m = \frac{y - y_1}{x - x_1} \quad \text{Substitute } x \text{ for } x_2 \text{ and } y \text{ for } y_2.$$

If we multiply both sides by $x - x_1$, and reverse the sides of the equation, we have

$$y - y_1 = m(x - x_1)$$

Because this equation displays the coordinates of the point (x_1, y_1) on the line and the slope m of the line, it is called the **point–slope form** of the equation of a line.

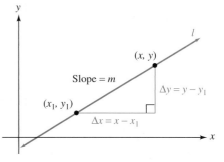

Point–Slope Form	The equation of the line passing through (x_1, y_1) and with slope m is $$y - y_1 = m(x - x_1)$$

EXAMPLE 4   Find an equation of the line that has slope $-\dfrac{2}{3}$ and passes through $(-4, 5)$. Write the equation in slope–intercept form.

Strategy   We will use the point–slope form, $y - y_1 = m(x - x_1)$, to write the equation of the line.

Why   We are given the slope of the line and the coordinates of a point that it passes through.

Success Tip
This is the same type of problem as that in Example 2. Here, it is solved using a different approach. We begin with point–slope form, and then write the result in slope–intercept form.

Solution   We substitute $-\dfrac{2}{3}$ for m, -4 for x_1, and 5 for y_1 into the point–slope form and simplify.

$$y - y_1 = m(x - x_1) \qquad \text{This is the point–slope form.}$$

$$y - 5 = -\frac{2}{3}[x - (-4)] \qquad \text{Substitute } -\tfrac{2}{3} \text{ for } m, -4 \text{ for } x_1, \text{ and 5 for } y_1.$$

$$y - 5 = -\frac{2}{3}(x + 4) \qquad \text{Simplify the expression within the brackets. This equation is in point–slope form.}$$

To write this equation in slope–intercept form, we solve for y.

$$y - 5 = -\frac{2}{3}x - \frac{8}{3} \qquad \text{To remove the parentheses, distribute } -\tfrac{2}{3}.$$

$$y = -\frac{2}{3}x + \frac{7}{3} \qquad \begin{array}{l} \text{To isolate } y, \text{ add 5 in the form of } \tfrac{15}{3} \text{ to both sides and simplify:} \\ -\tfrac{8}{3} + \tfrac{15}{3} = \tfrac{7}{3}. \end{array}$$

The slope–intercept form for the equation of the line is $y = -\dfrac{2}{3}x + \dfrac{7}{3}$.

> **Self Check 4**   Find an equation of the line that has slope $\frac{5}{4}$ and passes through $(2, -6)$. Write the equation in slope–intercept form.
>
>   *Now Try*   **Problems 27 and 31**

EXAMPLE 5   Find an equation of the line passing through $(-5, 4)$ and $(8, -6)$. Write the equation in slope–intercept form.

Strategy   We will use the point–slope form, $y - y_1 = m(x - x_1)$, to write an equation of the line.

Why   Since we know the coordinates of two points on the line, we can calculate the slope of the line and solve the problem using the method of Example 4.

Solution   First we find the slope of the line.

$$m = \frac{y_2 - y_1}{x_2 - x_1} \qquad \text{This is the slope formula.}$$

$$= \frac{-6 - 4}{8 - (-5)} \qquad \text{Substitute } -6 \text{ for } y_2, \ 4 \text{ for } y_1, \ 8 \text{ for } x_2, \text{ and } -5 \text{ for } x_1.$$

$$= -\frac{10}{13} \qquad \text{This is the slope of the line.}$$

Since the line passes through $(-5, 4)$ and $(8, -6)$, we can choose either point and substitute its coordinates into the point–slope form. If we select $(-5, 4)$, we substitute -5 for x_1, 4 for y_1, and $-\frac{10}{13}$ for m and proceed as follows.

> **Success Tip**
>
> Here, either of the given points can be used as (x_1, y_1) when writing the point–slope equation. Looking ahead, we usually choose the point whose coordinates will make the computations the easiest.

$$y - y_1 = m(x - x_1) \qquad \text{This is the point–slope form.}$$

$$y - 4 = -\frac{10}{13}[x - (-5)] \qquad \text{Substitute } -\frac{10}{13} \text{ for } m, \ -5 \text{ for } x_1, \text{ and } 4 \text{ for } y_1.$$

$$y - 4 = -\frac{10}{13}(x + 5) \qquad \begin{array}{l}\text{Simplify the expression within the brackets. This equation is in}\\ \text{point–slope form.}\end{array}$$

To write this equation in slope–intercept form, we solve for y.

$$y - 4 = -\frac{10}{13}x - \frac{50}{13} \qquad \text{To remove the parentheses, distribute } -\frac{10}{13}.$$

$$y = -\frac{10}{13}x + \frac{2}{13} \qquad \begin{array}{l}\text{To isolate } y, \text{ add 4 in the form of } \frac{52}{13} \text{ to both sides and simplify:}\\ -\frac{50}{13} + \frac{52}{13} = \frac{2}{13}.\end{array}$$

The equation of the line in slope–intercept form is $y = -\frac{10}{13}x + \frac{2}{13}$.

> **Self Check 5**   Find an equation of the line passing through $(-2, 5)$ and $(4, -3)$. Write the equation in slope–intercept form.
>
>   *Now Try*   **Problem 35**

Linear models can be used to describe certain types of financial gain or loss. For example, straight-line depreciation is used when aging equipment declines in value and straight-line appreciation is used when property or collectibles increase in value.

EXAMPLE 6   ***Accounting.***   After purchasing a new drill press, a machine shop owner had his accountant prepare a depreciation worksheet for tax purposes. See the illustration.

a. Assuming straight-line depreciation, write an equation that gives the value v of the drill press after x years of use.

b. Find the value of the drill press after $2\frac{1}{2}$ years of use.

c. What is the economic meaning of the v-intercept of the line?

d. What is the economic meaning of the slope of the line?

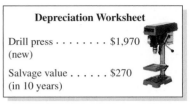

Depreciation Worksheet
Drill press . . . . $1,970 (new)
Salvage value . . . . . . $270 (in 10 years)

Strategy   We will read the problem with the hope that we can find information about the slope of the line and the coordinates of a point (or points) that lie on the line.

Why   If we know the slope of the line and the coordinates of a point (or points) that lie on the line, we can write its equation using slope–intercept form or point–slope form.

Solution

a. The facts presented in the worksheet can be expressed as ordered pairs of the form

$$(x, v)$$

Number of years of use ⤴   ⤴ value of the drill press

• When purchased, the new $1,970 drill press had been used 0 years: (0, 1,970).

• After 10 years of use, the value of the drill press will be $270: (10, 270).

A sketch showing these ordered pairs and the line of depreciation is helpful in visualizing the situation.

Since we know two points that lie on the line, we can write its equation. As we saw in Example 5, the first step is to find the slope of the line.

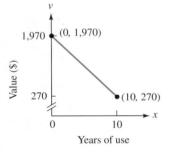

$$m = \frac{v_2 - v_1}{x_2 - x_1} \qquad \text{This is the slope formula using the variables } x \text{ and } v.$$

$$= \frac{270 - 1,970}{10 - 0} \qquad \text{Let } (x_1, v_1) = (0, 1,970) \text{ and } (x_2, v_2) = (10, 270) \text{ and substitute.}$$

$$= \frac{-1,700}{10} \qquad \text{Simplify.}$$

$$= -170 \qquad \text{This is the slope of the line.}$$

To find the equation of the line, we substitute -170 for m, 0 for x_1, and 1,970 for v_1 in the point–slope form and simplify.

$$v - v_1 = m(x - x_1) \qquad \text{This is the point–slope form using the variables } x \text{ and } v.$$

$$v - 1,970 = -170(x - 0) \qquad \text{Substitute.}$$

$$v = -170x + 1,970 \qquad \text{To isolate } v, \text{ add 1,970 to both sides.}$$

The value v of the drill press after x years of use is given by the straight-line depreciation model $v = -170x + 1,970$.

b. To find the value of the drill press after $2\frac{1}{2}$ years of use, we substitute 2.5 for x in the depreciation equation and find v.

$$v = -170x + 1,970 \qquad \text{This is the straight-line depreciation equation.}$$

$$= -170(2.5) + 1,970 \qquad \text{Substitute.}$$

The Language of Algebra
In this problem, the value of the drill press *depreciates*. The value of an item can also *appreciate*, which means to increase in value over time. Certain types of art, antiques, and jewelry *appreciate* quickly.

$$= -425 + 1,970$$
$$= 1,545$$

In $2\frac{1}{2}$ years, the drill press will be worth $1,545.

c. From the sketch, we see that the *v*-intercept of the graph of the depreciation line is (0, 1,970). This gives the original cost of the drill press, $1,970.

d. Each year, the value of the drill press decreases by $170, because the slope of the line is -170. The slope of the line is the annual depreciation rate.

 Now Try   **Problem 93**

③   **Use Slope as an Aid When Graphing.**

If we know the slope and the *y*-intercept of a line, we can graph the line without constructing a table of solutions.

 EXAMPLE 7   Find the slope and the *y*-intercept of the line with the equation $2x + 3y = -9$ and graph the line.

Strategy   We will write the equation in slope–intercept form ($y = mx + b$), plot the *y*-intercept, and use the slope to determine a second point on the line.

Why   Once we locate two points on the line, we can draw the graph of the line.

Solution

$$2x + 3y = -9 \qquad \text{The given equation is in standard (general) form.}$$

$$3y = -2x - 9 \qquad \text{To isolate the term 3y, subtract 2x from both sides.}$$

$$\frac{3y}{3} = \frac{-2x}{3} - \frac{9}{3} \qquad \text{To isolate y, divide both sides by 3.}$$

$$y = -\frac{2}{3}x - 3 \qquad \text{Simplify both sides. We see that } m = -\frac{2}{3} \text{ and } b = -3.$$

Caution
When using the *y*-intercept and the slope to graph a line, remember to draw the slope triangle from the *y*-intercept, *not* from the origin.

Success Tip
To increase accuracy, slope triangles in a staircase pattern can be drawn to locate several points on the line.

The slope of the line is $-\frac{2}{3}$, which can be expressed as $\frac{-2}{3}$. After plotting the *y*-intercept, $(0, -3)$, we move 2 units downward (rise) and then 3 units to the right (run). This locates a second point on the line, $(3, -5)$. From this point, we move another 2 units downward and 3 units to the right to locate a third point on the line, $(6, -7)$. Then we draw a line through the points to obtain the graph shown in the figure.

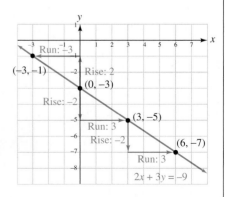

An alternate way to find another point on the line is to write the slope in the form $\frac{2}{-3}$. As before, we begin at the *y*-intercept $(0, -3)$. Since the rise is positive, we move 2 units *upward,* and since the run is negative, we then move 3 units to the *left.* We arrive at $(-3, -1)$, another point on the graph of $y = -\frac{2}{3}x - 3$.

Self Check 7   Find the slope and the y-intercept of the line with the equation $3x - 2y = -4$ and graph the line.

Now Try   **Problems 41 and 43**

④ **Recognize Parallel and Perpendicular Lines.**

Recall these two facts from Section 2.3:

- Different lines having the same slope are parallel.
- If the slopes of two lines are negative reciprocals, the lines are perpendicular.

EXAMPLE 8   **a.** Show that the lines represented by $2y = -3x + 12$ and $6x + 4y = 7$ are parallel.

b. Show that the lines represented by $3(x - 3y) = 10$ and $-3x = y + 5$ are perpendicular.

Strategy   We will write each equation in slope–intercept form and compare their slopes.

Why   If the slopes are equal, the lines are parallel. If the slopes are negative reciprocals, the lines are perpendicular.

Solution

a. We solve each equation for y to see whether the lines are distinct (different) and whether their slopes are equal.

$$2y = -3x + 12$$
$$y = -\frac{3}{2}x + 6 \quad m = -\frac{3}{2} \text{ and } b = 6$$

$$6x + 4y = 7$$
$$4y = -6x + 7$$
$$y = -\frac{3}{2}x + \frac{7}{4} \quad m = -\frac{3}{2} \text{ and } b = \frac{7}{4}$$

Since the values of b in these equations are different $\left(6 \text{ and } \frac{7}{4}\right)$, the lines have different y-intercepts and are distinct. Since the slope of each line is $-\frac{3}{2}$, they are parallel.

b. We solve each equation for y to see whether the slopes of their straight-line graphs are negative reciprocals.

$$3(x - 3y) = 10$$
$$3x - 9y = 10$$
$$-9y = -3x + 10$$
$$y = \frac{1}{3}x - \frac{10}{9} \quad m = \frac{1}{3} \text{ and } b = -\frac{10}{9}$$

$$-3x = y + 5$$
$$-3x - 5 = y$$
$$y = -3x - 5 \quad m = -3 \text{ and } b = -5$$

Since the slopes are negative reciprocals $\left(\frac{1}{3} \text{ and } -3\right)$, the lines are perpendicular.

Self Check 8   **a.** Are the lines represented by $3x - 2y = 4$ and $2x = 5(y + 1)$ parallel?

b. Are the lines represented by $3x + 2y = 6$ and $2x - 3y = 6$ perpendicular?

Now Try   **Problems 51 and 53**

EXAMPLE 9 Find an equation of the line that passes through $(-2, 5)$ and is parallel to the line $y = 8x - 3$. Write the equation in slope–intercept form.

Strategy We will use the point–slope form to write an equation of the line.

Why We know that the line passes through $(-2, 5)$. We can use the fact that the lines are parallel to determine the unknown slope of the desired line.

Solution Since the slope of the line represented by $y = 8x - 3$ is the coefficient of x, the slope is 8. Since the desired equation is to have a graph that is parallel to the graph of $y = 8x - 3$, its slope must also be 8.

We substitute -2 for x_1, 5 for y_1, and 8 for m in the point–slope form and simplify.

$$y - y_1 = m(x - x_1)$$
$$y - 5 = 8[x - (-2)] \qquad \text{Substitute 5 for } y_1, \text{ 8 for } m, \text{ and } -2 \text{ for } x_1.$$
$$y - 5 = 8(x + 2) \qquad \text{Simplify within the brackets. This equation is in point–slope form.}$$
$$y - 5 = 8x + 16 \qquad \text{To remove parentheses, distribute 8.}$$
$$y = 8x + 21 \qquad \text{To isolate } y, \text{ add 5 to both sides.}$$

The equation in slope–intercept form is $y = 8x + 21$.

> **Success Tip**
> If this problem had called for the equation of a line that is perpendicular to $y = 8x - 3$, the slope of the desired line would be the negative reciprocal of 8, which is $-\frac{1}{8}$.

Self Check 9 Write an equation of the line that is parallel to the line $y = 8x - 3$ and passes through the $(0, 0)$.

Now Try **Problems 59 and 63**

When you are asked to *write the equation of a line,* determine what you know about the graph of the line: its slope, its y-intercept, points it passes through, and so on. Then substitute the appropriate numbers into one of the following forms of a linear equation.

Forms of a Linear Equation			
Standard (General) form	$Ax + By = C$	A and B cannot both be 0.
Slope–intercept form	$y = mx + b$	The slope is m, and the y-intercept is $(0, b)$.
Point–slope form	$y - y_1 = m(x - x_1)$	The slope is m, and the line passes through (x_1, y_1).
A horizontal line	$y = b$	The slope is 0, and the y-intercept is $(0, b)$.
A vertical line	$x = a$	There is no defined slope, and the x-intercept is $(a, 0)$.

5 **Write a Linear Equation Model to Fit a Collection of Data.**

In statistics, the process of using one variable to predict another is called **regression.** For example, if we know a man's height, we can usually make a good prediction about his weight because taller men tend to weigh more than shorter men.

The table on the next page shows the results of sampling twelve men at random and recording the height h and weight w of each. In figure (a), the ordered pairs (h, w) from the table are plotted to form a scatter diagram. Notice that the data points fall more or less along an imaginary straight line, indicating a linear relationship between h and w.

Height h in.	Weight w lb
66	145
67	150
68	150
68	165
70	180
70	165
71	175
72	200
73	190
74	190
75	205
75	215

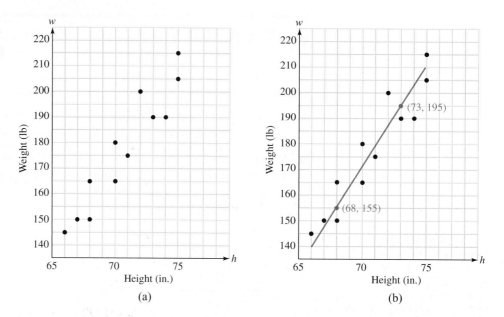

(a)   (b)

To write a prediction equation (sometimes called a **regression equation**) that relates height and weight, we must find the equation of the line that comes closer to all of the data points in the scatter diagram than any other possible line. In statistics, there are exact methods to find this equation; however, they are beyond the scope of this book. In this course, we will draw "by eye" a line that we feel best fits the data points.

In figure (b), a straight edge was placed on the scatter diagram and a line was drawn that seemed to best fit all of the data points. Note that it passes through (68, 155) and (73, 195). To write the equation of that line, we first need to find its slope.

$$m = \frac{w_2 - w_1}{h_2 - h_1} \qquad \text{This is the slope formula using the variables } h \text{ and } w.$$

$$= \frac{195 - 155}{73 - 68} \qquad \text{Choose } (h_1, w_1) = (68, 155) \text{ and } (h_2, w_2) = (73, 195) \text{ and substitute.}$$

$$= \frac{40}{5} \qquad \text{Simplify.}$$

$$= 8$$

> **Caution**
> When drawing a line through the data points of a scatter diagram by eye, the results could vary from person to person. Graphing calculators have a program that finds the line of best fit for a collection of data. Look in the owner's manual under linear regression.

We then use the point–slope form to find the equation of the line. Since the line passes through (68, 155) and (73, 195), we can use either ordered pair to write its equation.

$$w - w_1 = m(h - h_1) \qquad \text{This is the point–slope form using the variables } h \text{ and } w.$$

$$w - 155 = 8(h - 68) \qquad \text{Choose } (68, 155) \text{ for } (h_1, w_1).$$

$$w - 155 = 8h - 544 \qquad \text{To remove the parentheses, distribute 8.}$$

$$w = 8h - 389 \qquad \text{To isolate } w, \text{ add 155 to both sides.}$$

The equation of the line that was drawn through the data points in the scatter diagram is $w = 8h - 389$. We can use this equation to predict the weight of a man who is 72 inches tall.

$$w = 8h - 389 \qquad \text{This is the linear equation model.}$$

$$w = 8(72) - 389 \qquad \text{Substitute 72 for } h.$$

$$w = 576 - 389$$

$$w = 187$$

We predict that a 72-inch-tall man chosen at random will weigh about 187 pounds.

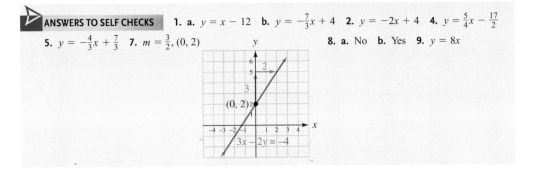

1. a. $y = x - 12$ **b.** $y = -\frac{7}{3}x + 4$ **2.** $y = -2x + 4$ **4.** $y = \frac{5}{4}x - \frac{17}{2}$
5. $y = -\frac{4}{3}x + \frac{7}{3}$ **7.** $m = \frac{3}{2}, (0, 2)$ **8. a.** No **b.** Yes **9.** $y = 8x$

STUDY SET
2.4

VOCABULARY

Fill in the blanks.

1. The _____ form of the equation of a line is
   $y = mx + b$.

2. The point–slope form of the equation of a line is
   _____.

CONCEPTS

3. **a.** Suppose you know the slope of a line. Is that enough
   information about the line to write its equation?

   b. Suppose you know the coordinates of a point on a line. Is
   that enough information about the line to write its
   equation?

4. Find the slope and y-intercept of the graph of the equation
   $y = -\frac{2}{3}x + 1$.

5. Find the slope of the graph of each equation, if possible.
   a. $y = -x$ **b.** $x = -3$

6. Find the y-intercept of the graph of each equation.
   a. $y = 2x$ **b.** $x = -3$

7. For the line in the illustration, find its slope and y-intercept. Then
   write the equation of the line in slope–intercept form by filling in
   the blanks: $y = \boxed{}\, x + \boxed{}$.

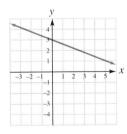

8. **FARMING** The graph below decribes the number of bushels
   of a crop that will be produced on a plot of land for a given
   amount of rainfall.

   a. Find the y-intercept. What information does the y-intercept
   give?

   b. Find the slope of the line. What information does the slope
   give?

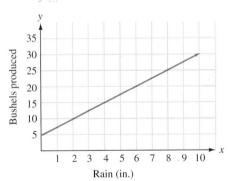

9. Find the slope of the graph of the equation $y - 3 = -\frac{2}{3}(x + 1)$.
   What point does the equation indicate the line will pass
   through?

10. Find the slope of the line in the
    illustration that passes through the
    point $(-2, -3)$ and write its equa-
    tion in point–slope form by filling
    in the blanks: $y + \boxed{} = \boxed{}(x + \boxed{})$.

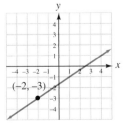

11. Do the equations $y - 2 = 3(x - 2)$ and $y = 3x - 4$ describe
    the same line?

12. a. The graphs of two lines and their equations are shown below. Are the lines perpendicular? Explain.

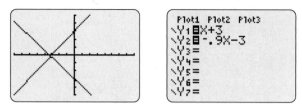

b. The graphs of two lines and their equations are shown below. Are the lines parallel? Explain.

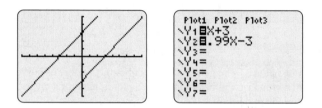

NOTATION

Complete each solution.

13. Write $y + 2 = \frac{1}{3}(x + 3)$ in slope–intercept form.

$$y + 2 = \frac{1}{3}(x + 3)$$

$$y + 2 = \boxed{} + 1$$

$$y + 2 - \boxed{} = \frac{1}{3}x + 1 - \boxed{}$$

$$y = \frac{1}{3}x - \boxed{}$$

$$m = \boxed{} \text{ and } b = \boxed{}$$

14. Write an equation of the line that has slope -2 and passes through the point $(3, 1)$.

$$y - y_1 = m(x - x_1)$$

$$y - \boxed{} = -2(x - \boxed{})$$

$$y - 1 = \boxed{} + 6$$

$$y = -2x + \boxed{}$$

GUIDED PRACTICE

Use the slope–intercept form to write an equation of the line with the given slope and y-intercept. See Example 1.

15. Slope 3; y-intercept $(0, 6)$

16. Slope -2; y-intercept $(0, 11)$

17. Slope $-\frac{2}{3}$; y-intercept $\left(0, -\frac{7}{3}\right)$

18. Slope $\frac{5}{7}$; y-intercept $\left(0, \frac{1}{4}\right)$

Find the slope and y-intercept of each line graphed below. Then use the slope–intercept form to write an equation of the line. See Example 1.

19.
20.
21.
22.

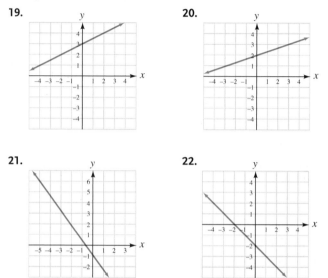

Use the slope–intercept form to write an equation of the line that has the given slope and passes through the given point. See Example 2.

23. Slope 7; passes through $(-7, 5)$

24. Slope 3; passes through $(-2, -5)$

25. Slope -9; passes through $(2, -4)$

26. Slope -10; passes through the origin

Use the point–slope form to write an equation of the line with the given properties or the given graph. Leave the answer in point–slope form. See Example 4.

27. Slope 10; passes through $(1, 7)$

28. Slope -8; passes through $(3, -2)$

29.
30.

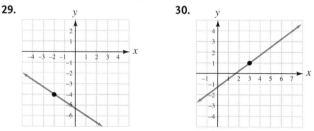

Use the point–slope form to write an equation of the line with the given properties. Then write each equation in slope–intercept form. See Example 4.

31. Slope 5; passes through $(4, -5)$

32. Slope -7; passes through $(-3, -7)$

33. Slope -9; passes through $(-3.5, 2.7)$

34. Slope 4; passes through $(7.2, -3.7)$

Use the point–slope form to write an equation of the line passing through the two given points. Then write each equation in slope–intercept form. See Example 5.

35. $(6, 8)$ and $(2, 10)$

36. $(-4, 5)$ and $(2, -6)$

37.

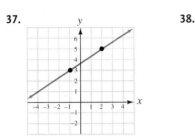

38.

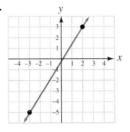

Find the slope and y-intercept and use them to draw the graph of the line. See Example 7.

39. $y = x - 1$

40. $y = -x + 2$

41. $y = -\dfrac{5}{4}x - 3$

42. $y = \dfrac{2}{3}x + 2$

43. $4y - 3 = -3x - 11$

44. $-2x + 4y = 12$

45. $5y - 8x - 30 = 0$

46. $7x + 3y + 15 = 0$

Write each equation in slope–intercept form. Then find the slope and the y-intercept of the line determined by the equation. See Example 8.

47. $3x - 2y = 8$

48. $-2x - 4y = -12$

49. $-2(x + 3y) = 5$

50. $5(2x - 3y) = 4$

Determine whether the graphs of each pair of equations are parallel, perpendicular, or neither. See Example 8.

51. $y = 3x + 4, \ y = 3x - 7$

52. $y = 4x - 13, \ y = \dfrac{1}{4}x + 13$

53. $x + y = 2, \ y = x + 5$

54. $x = y + 2, \ y = x + 3$

55. $3x + 6y = 1, \ y = \dfrac{1}{2}x$

56. $2x + 3y = 9, \ 3x - 2y = 5$

57. $y = 3, \ x = 4$

58. $y = -3, \ y = -7$

Find an equation of the line that passes through the given point and is parallel to the given line. Write the equation in slope–intercept form. See Example 9.

59. $(2, 5), \ y = 4x + 8$

60. $(-6, 3), \ y = -3x - 12$

61. $\left(\dfrac{2}{3}, \dfrac{1}{4}\right), \ y = 3x - 2$

62. $\left(\dfrac{4}{5}, -\dfrac{2}{3}\right), \ y = -5x - \dfrac{1}{2}$

Find an equation of the line that passes through the given point and is perpendicular to the given line. Write the equation in slope–intercept form. See Example 9.

63. $(0, 0), \ y = 4x - 7$

64. $(0, 0), \ y = -\dfrac{1}{3}x + 4$

65. $(4, -2), \ 4x = 5y - 8$

66. $(1, -5), \ 4x = -3y + 20$

TRY IT YOURSELF

Write an equation in slope–intercept form of the line with the given properties or given graph.

67. Passes through

x	y
3	4
-3	-10

68. Passes through

x	y
2	8
6	-8

69. Slope $\dfrac{4}{3}$, passes through $(5, 9)$

70. Slope $-\dfrac{7}{5}$, passes through $(-6, 0)$

71. Passes through $(2, 5)$, perpendicular to $4x - y = 7$

72. Passes through $(-6, 3)$, perpendicular to $y + 3x = -12$

73.

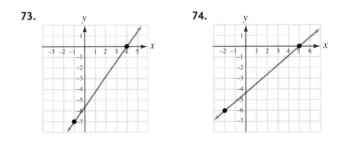

74.

75. Passes through $(-1, -1)$ and $(4, 4)$

76. Passes through $(-5, 5)$ and $(9, -9)$

77. Passes through $(0, 0)$, parallel to $y = 4x - 7$

78. Passes through $(0, 0)$, parallel to $y = -\dfrac{1}{3}x + 4$

79. Slope -3, passes through $(4, -6)$

80. Slope 4, passes through $(-1, 16)$

81. Passes through $\left(\dfrac{2}{3}, \dfrac{1}{4}\right)$, perpendicular to $y = 3x - 2$

82. Passes through $\left(\dfrac{4}{5}, -\dfrac{2}{3}\right)$, perpendicular to $y = -5x - \dfrac{1}{2}$

83.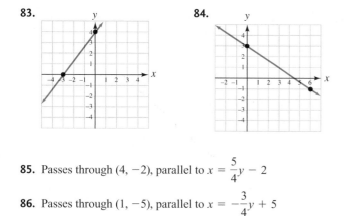

84.

85. Passes through $(4, -2)$, parallel to $x = \dfrac{5}{4}y - 2$

86. Passes through $(1, -5)$, parallel to $x = -\dfrac{3}{4}y + 5$

APPLICATIONS

87. CABLE TV   Since 1990, when the average monthly price for basic cable TV programming in the United States was approximately $16, the cost has risen by about $1.58 a year. (Source: Kagan Research)

   a. Write an equation in slope–intercept form to predict cable TV costs in the future. Use t to represent time in years after 1990 and C to represent the average basic monthly cost.

   b. If the equation in part (a) were graphed, what would be the meaning of the C-intercept and the slope of the line?

   c. Use the model to predict the average monthly price for cable TV in the year 2020.

88. UNDERSEA DIVING   The illustration shows that the pressure p that divers experience is related to the depth d of the dive. A linear model can be used to describe this relationship.

   a. Write the linear model in slope–intercept form.

   b. Pearl and sponge divers often reach depths of 100 feet. What pressure do they experience? Round to the nearest tenth.

   c. Scuba divers can safely dive to depths of 250 feet. What pressure do they experience? Round to the nearest tenth.

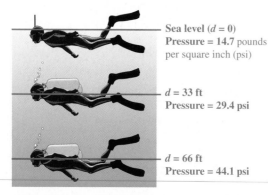

Sea level ($d = 0$)
Pressure = 14.7 pounds per square inch (psi)

$d = 33$ ft
Pressure = 29.4 psi

$d = 66$ ft
Pressure = 44.1 psi

89. *from Campus to Careers*
    Certified Fitness Instructor

Suppose you are a fitness instructor and want to determine the number of calories a client burns during a workout. From exercise tables, you find that during the first part of the workout (aerobics) she will burn 220 calories. During the optional second part of the workout (swimming), she will burn 7.8 calories per minute.

   a. Write a linear model in slope–intercept form that gives the total number of calories c that the client burns if she concludes a workout with m minutes of swimming.

   b. Many fitness instructors recommend that a client burn 300 calories per exercise session to lose weight. How many minutes of swimming should the client perform to satisfy this requirement?

90. COLLEGE COSTS   According to the *College Board,* in 1980, the average tuition and fees at a private college were $8,850 a year. Since then, the annual cost has increased by about $514 per year.

   a. Write a linear model in slope–intercept form that gives the cost c to attend a private college t years after 1980.

   b. Use the model to predict what the average tuition and fees at a private college will be in the year 2050.

91. CRIMINOLOGY   City growth and the number of burglaries for a certain city are related by a linear equation. Records show that 575 burglaries were reported in a year when the local population was 77,000 and that the rate of increase in the number of burglaries was 1 for every 100 new residents.

   a. Using the variables p for population and B for burglaries, write an equation (in slope–intercept form) that police can use to predict future burglary statistics.

   b. How many burglaries can be expected when the population reaches 110,000?

92. FIRE PROTECTION   City growth and the number of fires for a certain city are related by a linear equation. Records show that 113 fires occurred in a year when the local population was 150,000 and that the rate of increase in the number of fires was 1 for every 1,000 new residents.

   a. Using the variables p for population and F for fires, write an equation (in slope–intercept form) that the fire department can use to predict future fire statistics.

   b. How many fires can be expected when the population reaches 200,000?

93. REAL ESTATE

   a. Use the information given in the description on the next page of the property to write a straight-line appreciation equation for the house.

   b. What will be the predicted value of the home when it is 25 years old?

Vacation Home
$254,000
Only 2 years old

• Great investment property!
• Expected to appreciate $4,000/yr

Sq ft: 1,635	Fam rm: yes	Den: no
Bdrm: 3	Ba: 1.5	Gar: enclosed
A/C: yes	Firepl: yes	Kit: built-ins

94. WINDCHILL A combination of cold and wind makes a person feel colder than the actual temperature. The table shows what temperatures of 35°F and 15°F feel like when a 15-mph wind is blowing. The relationship between the actual temperature and the windchill temperature can be modeled with a linear equation.

a. Write the equation that models this relationship. Answer in slope–intercept form.

b. What information is given by the y-intercept of the graph of the equation found in part (a)?

15-mph wind	
Actual temperature x	Windchill temperature y
35°F	25°F
15°F	0°F

95. CRAIGSLIST Find the straight-line depreciation equation for the TV in the following ad found on Craigslist (an online website featuring free classified advertisements).

46 inch SHARP high definition LCDTV - $1,000

Reply to: sale-001234@saleslist.org
Date: 2007-10-30, 7:20PM EST

3-year-old TV with surround sound & vision, remote.
New $3,295. Asking $1,000.

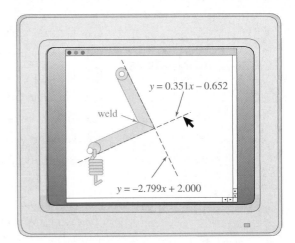

96. SALVAGE VALUES A truck was purchased for $19,984. Its salvage value at the end of 8 years is expected to be $1,600. Find the straight-line depreciation equation.

97. PAINTINGS According to Wikipedia, the estimated value of the *Mona Lisa* was about $100 million in 1960. By the year 2000, its value had increased to approximately $600 million.

a. Write a straight-line appreciation equation for the painting, where y represents the approximate value of the painting in millions of dollars and x represents the number of years after 1960.

b. When does the straight-line appreciation equation predict that the painting will be worth one billion dollars?

© Dennis Hallinan/Alamy

98. REAL ESTATE A house purchased for $275,000 is expected to be worth $282,000 in 2 years. Find the straight-line appreciation equation for the house if it continues to appreciate at the same rate. What will the house be worth in 10 years?

99. COMPUTER-AIDED DRAFTING The illustration shows a computer-generated drawing of an airplane part. When the designer clicks the mouse on a line on the drawing, the computer finds the equation of the line. Use a calculator to determine whether the angle where the weld is to be made is a right angle.

$y = 0.351x - 0.652$

weld

$y = -2.799x + 2.000$

100. PSYCHOLOGY EXPERIMENTS The following scatter-gram shows the performance of a rat in a maze.

   a. Draw a line through $(1, 10)$ and $(19, 1)$. Write its equation using the variables t and E. In psychology, this equation is called the learning curve for the rat.

   b. What does the slope of the line tell us?

   c. What information does the t-intercept of the graph give?

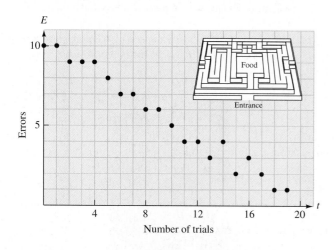

WRITING

101. Explain how to find the equation of a line passing through two given points.

102. Explain what m, x_1, and y_1 represent in the point–slope form of the equation of a line.

103. A student was asked to determine the slope of the graph of the line $y = 6x - 4$. If his answer is $m = 6x$, explain his error.

104. Linear relationships between two quantities can be described by an equation or a graph. Which do you think is the more informative? Why?

REVIEW

105. INVESTMENTS Equal amounts are invested at 6%, 7%, and 8% annual interest. The three investments yield a total of $2,037 annual interest. Find the total amount of money invested.

106. MEDICATIONS A doctor prescribes an ointment that is 2% hydrocortisone. A pharmacist has 1% and 5% concentrations in stock. How many ounces of each should the pharmacist use to make a 1-ounce tube?

CHALLENGE PROBLEMS

Investigate the properties of the slope and the y-intercept by completing the following problems.

107. a. Graph $y = mx + 2$ for several positive values of m. What do you notice?

   b. Graph $y = mx + 2$ for several negative values of m. What do you notice?

108. a. Graph $y = 2x + b$ for several increasing positive values of b. What do you notice?

   b. Graph $y = 2x + b$ for several decreasing negative values of b. What do you notice?

109. If the graph of $y = mx + b$ passes through quadrants I, II, and IV, what do we know about the constants m and b?

110. The graph of $Ax + By = C$ passes only through quadrants I and IV. What do we know about the constants A, B, and C?

SECTION 2.5
An Introduction to Functions

Objectives

  1 Define relation, domain, and range.

  2 Identify functions.

  3 Use function notation.

  4 Find the domain of a function.

  5 Graph linear functions.

  6 Use the vertical line test.

  7 Solve applications involving functions.

The concept of a *function* is one of the most important ideas in all of mathematics. To introduce this topic, we will begin with a table that might be seen on television or printed in a newspaper.

1   **Define Relation, Domain, and Range.**

The following table shows the number of women serving in the U.S. House of Representatives for several recent sessions of Congress.

Women in the U.S. House of Representatives							
Session of Congress	104th	105th	106th	107th	108th	109th	110th
Number of Women Representatives	48	54	56	59	59	68	71

We can display the data in the table as a set of ordered pairs, where the **first component** represents the session of Congress and the **second component** represents the number of women representatives serving during that session:

$$\{(104, 48),\quad (105, 54),\quad (106, 56),\quad (107, 59),\quad (108, 59),\quad (109, 68),\quad (110, 71)\}$$

Sets of ordered pairs like this are called **relations.** The set of all first components is called the **domain of the relation,** and the set of all second components is called the **range of the relation.** A relation may consist of a finite number of ordered pairs or an infinite number of ordered pairs.

EXAMPLE 1   Find the domain and range of the relation:
$$\{(3, 2), (5, -7), (-8, 2), (9, 0)\}$$

Strategy   We will identify the first components and the second components of the ordered pairs.

Why   The set of all first components is the domain of the relation, and the set of all second components is the range.

Solution   The first components of the ordered pairs are highlighted in red, and the second components are highlighted in blue:

$$\{(3, 2), (5, -7), (-8, 2), (9, 0)\}$$

The domain of the relation is $\{-8, 3, 5, 9\}$.

The range of the relation is $\{-7, 0, 2\}$.

The elements of the domain and range are usually listed in increasing order, and if a value is repeated, it is only listed once.

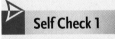

Self Check 1   Find the domain and range of the relation:
$$\{(5, 6), (-12, 4), (8, 6), (-6, -6), (5, 4)\}$$

Now Try   **Problem 17**

---

2   **Identify Functions.**

The relation in Example 1 was defined by a set of ordered pairs. Relations can also be defined using an **arrow** or **mapping diagram.** The data from the U.S. House of Representatives example is presented on the next page in that form.

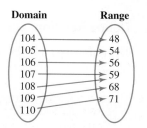

Notice that to each session of Congress, there corresponds exactly one number of women representatives. That is, to each member of the domain there corresponds exactly one member of the range. Relations that have this characteristic are called *functions*.

Function	A **function** is a set of ordered pairs (a relation) in which to each first component there corresponds exactly one second component. The set of first components is called the **domain of the function,** and the set of second components is called the **range of the function.**

Since we will often work with sets of ordered pairs of the form (x, y), it is helpful to define a function using the variables x and y.

y Is a Function of x	Given a relation in x and y, if to each value of x in the domain there corresponds exactly one value of y in the range, then y is said to be a function of x.

In the previous definition, since y depends on x, we call x the **independent variable** and y the **dependent variable.** The set of all possible values that can be used for the independent variable is the **domain** of the function, and the set of all values of the dependent variable is the **range** of the function.

EXAMPLE 2   In each case, determine whether the relation defines y to be a function of x.

a. x y

      5 → 4
      7 → 6
      11 → 10

b.

x	y
8	2
1	4
8	3
9	9

c. $\{(-2, 3), (-1, 3), (0, 3), (1, 3)\}$

Strategy   In each case, we will determine whether there is more than one value of y that corresponds to a single value of x.

Why   If to any x-value there corresponds more than one y-value, then y is not a function of x.

Solution

a. The arrow diagram defines a function because to each value of x there corresponds exactly one value of y.

- 5→4   To the x-value 5, there corresponds exactly one y-value, 4.

- 7→6   To the x-value 7, there corresponds exactly one y-value, 6.

- 11→10   To the x-value 11, there corresponds exactly one y-value, 10.

b. The table does not define a function, because to the *x*-value 8 there corresponds to more than one *y*-value.

- In the first row, to the *x*-value 8, there corresponds the *y*-value 2.
- In the third row, to the same *x*-value 8, there corresponds a different *y*-value, 3.

When the correspondence in the table is written as a set of ordered pairs, it is apparent that the relation does not define a function

The same x-value
$$\{(8, 2), (1, 4), (8, 3), (9, 9)\}\qquad \text{This is not a function.}$$
Different y-values

c. Since to each value of *x*, there corresponds exactly one value of *y*, the set of ordered pairs defines *y* to be a function of *x*.

- $(-2, 3)$ To the *x*-value -2, there corresponds exactly one *y*-value, 3.
- $(-1, 3)$ To the *x*-value -1, there corresponds exactly one *y*-value, 3.
- $(0, 3)$ To the *x*-value 0, there corresponds exactly one *y*-value, 3.
- $(1, 3)$ To the *x*-value 1, there corresponds exactly one *y*-value, 3.

In this case, the same *y*-value, 3, corresponds to each *x*-value.

The results from parts (b) and (c) illustrate an important fact: *Two different ordered pairs of a function can have the same y-value, but they cannot have the same x-value.*

Self Check 2   In each case, determine whether the relation defines *y* to be a function of *x*.

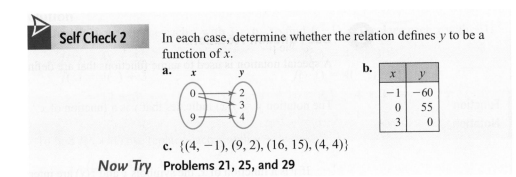

a.

b.

x	y
-1	-60
0	55
3	0

c. $\{(4, -1), (9, 2), (16, 15), (4, 4)\}$

Now Try   **Problems 21, 25, and 29**

A function can also be defined by an equation. For example, $y = \frac{1}{2}x + 3$ sets up a rule in which to each value of *x* there corresponds exactly one value of *y*. To find the *y*-value (called an **output**) that corresponds to the *x*-value 4 (called an **input**), we substitute 4 for *x* and evaluate the right side of the equation.

$$y = \frac{1}{2}x + 3$$

$$= \frac{1}{2}(4) + 3 \qquad \text{Substitute 4 for x. The input is 4.}$$

$$= 2 + 3$$

$$= 5 \qquad\qquad \text{This is the output.}$$

In the function $y = \frac{1}{2}x + 3$, a *y*-value of 5 corresponds to an *x*-value of 4.

Not all equations define functions, as we will see in the next example.

In the next example, we are asked to find the input of a function when we are given the corresponding output.

EXAMPLE 5

Let $f(x) = \frac{1}{3}x + 4$. For what value of x is $f(x) = 2$?

Strategy We will substitute 2 for $f(x)$ and solve for x.

Why In the equation, there are two unknowns, x and $f(x)$. If we replace $f(x)$ with 2, we can use equation-solving techniques to find x.

Solution

$$f(x) = \frac{1}{3}x + 4 \qquad \text{This is the given function.}$$

$$2 = \frac{1}{3}x + 4 \qquad \text{Substitute 2 for } f(x).$$

$$-2 = \frac{1}{3}x \qquad \text{Subtract 4 from both sides.}$$

$$-6 = x \qquad \text{To isolate } x, \text{ multiply both sides by 3.}$$

We have found that $f(x) = 2$ when $x = -6$. To check this result, we can substitute -6 for x and verify that $f(-6) = 2$.

$$f(x) = \frac{1}{3}x + 4$$

$$f(-6) = \frac{1}{3}(-6) + 4 \qquad \text{Substitute } -6 \text{ for } x.$$

$$= -2 + 4 \qquad \text{Evaluate the right side.}$$

$$= 2$$

Self Check 5 For what value(s) of x is $f(x) = -5$?

Now Try Problem 77

---

4 Find the Domain of a Function.

We can think of a function as a machine that takes some input x and turns it into some output $f(x)$, as shown in figure (a). The machine shown in figure (b) turns the input -6 into the output -11. The set of numbers that we put into the machine is the domain of the function, and the set of numbers that comes out is the range.

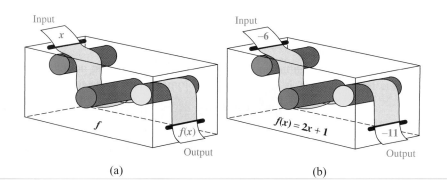

(a) (b)

EXAMPLE 6   Find the domain of each function:

a. $f(x) = 3x + 1$ **b.** $f(x) = \dfrac{1}{x - 2}$

Strategy   We will ask, "What values of x are acceptable replacements for x in $3x + 1$ and $\dfrac{1}{x - 2}$?"

Why   These values of x form the domain of the function.

The Language of Algebra
The last two letters in the word
domain help us remember that it
is the set of all *inputs* of a
function.

Solution

a. We will be able to evaluate $3x + 1$ for any value of x that is a real number. Thus, the domain of the function is *the set of real numbers,* which can be represented by the symbol $\mathbb{R}$.

b. To find the domain of $f(x) = \dfrac{1}{x - 2}$, we must exclude any real number that is not an acceptable replacement for x in $\dfrac{1}{x - 2}$. The number 2 cannot be substituted for x, because that would make the denominator equal to 0. Since any real number except 2 can be substituted for x, the domain is *the set of all real numbers except 2.*

Self Check 6   Find the domain of each function:
a. $f(x) = 2x - 6$
b. $f(x) = \dfrac{2}{x + 3}$

Now Try   Problem 81

5   **Graph Linear Functions.**

We have seen that in a function, a single value of $f(x)$ corresponds to each value of x in the domain. The "input-output" pairs that a function generates can be plotted on a rectangular coordinate system to get the graph of the function.

EXAMPLE 7   Graph the function: $f(x) = \dfrac{1}{2}x + 3$

Strategy   To graph the function, we can think of $f(x)$ as y and use the same methods that we used to graph linear equations in Section 2.2.

Why   The notation $f(x) = \frac{1}{2}x + 3$ is another way to write $y = \frac{1}{2}x + 3$.

Solution   We begin by constructing a table of function values. To make a table, we select several values for x and find the corresponding values of $f(x)$. If $x = -2$, we have

$$f(x) = \frac{1}{2}x + 3 \qquad \text{This is the function to graph.}$$

$$f(-2) = \frac{1}{2}(-2) + 3 \quad \text{Substitute } -2 \text{ for each } x.$$

$$= -1 + 3 \qquad \text{Evaluate the right side.}$$

$$= 2$$

Thus, $f(-2) = 2$ and the ordered pair $(-2, 2)$ lies on the graph of f.

Solve Applications Involving Functions.

We can use functions to describe many relationships where one quantity depends upon another.

EXAMPLE 9   ***Cosmetology.***   A cosmetologist rents a station from the owner of a beauty salon for $18 a day. She expects to make $12 profit from each customer she serves. Write a linear function that describes her daily income if she serves c customers per day. Graph the function.

Strategy   To write a function that describes her daily income, we must write an expression that represents the total daily profit that she makes from serving c customers.

Why   Her daily income will be the profit that she makes from serving c customers less the $18 daily station rental fee.

Solution   The cosmetologist makes a profit of $12 per customer, so if she serves c customers a day, she will make $12c$. To find her income, we must *subtract* the $18 rental fee from the profit. Therefore, the income function is $I(c) = 12c - 18$.

The graph of this linear function is a line with slope 12 and intercept $(0, -18)$. Since the cosmetologist cannot have a negative number of customers, we do not extend the line into quadrant III.

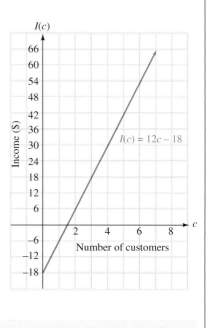

Now Try   **Problem 103**

Using Your Calculator   ***Evaluating Functions***

We can use a graphing calculator to find function values.

For example, to find the income earned by the cosmetologist in Example 9 for different numbers of customers, we first graph the income function $I(c) = 12c - 18$ as $y = 12x - 18$, using window settings of $[0, 10]$ for x and $[0, 100]$ for y to obtain figure (a). To find her income when she serves seven customers, we trace and move the cursor until the x-coordinate on the screen is nearly 7, as in figure (b). From the screen, we see that her income is about $66.25.

To find her income when she serves nine customers, we trace and move the cursor until the x-coordinate is nearly 9, as in figure (c). From the screen, we see that her income is about $90.51.

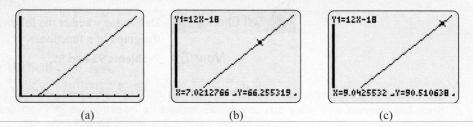

(a)   (b)   (c)

With some graphing calculator models, we can evaluate a function by entering function notation. To find $I(15)$, the income earned by the cosmetologist of Example 9 if she serves 15 customers, we use the following steps on a TI-83 Plus calculator.

With $I(c) = 12c - 18$ entered as $Y_1 = 12x - 18$, we call up the home screen by pressing $\boxed{\text{2nd}}$ $\boxed{\text{QUIT}}$. Then we enter $\boxed{\text{VARS}}$ $\boxed{\blacktriangleright}$ $\boxed{1}$ $\boxed{\text{ENTER}}$. The symbolism Y_1 will be displayed. See figure (a). Next, we enter the input value 15 within parentheses, as shown in figure (b), and press $\boxed{\text{ENTER}}$. In figure (c) we see that $Y_1(15) = 162$. That is, $I(15) = 162$. The cosmetologist will earn \$162 if she serves 15 customers in one day.

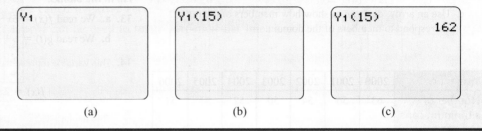

(a)	(b)	(c)

▷ **ANSWERS TO SELF CHECKS**   **1.** D: $\{-12, -6, 5, 8\}$; R: $\{-6, 4, 6\}$ **2. a.** No; $(0, 2), (0, 3)$ **b.** Yes   **c.** No; $(4, -1), (4, 4)$ **3. a.** Yes   **b.** No; $(1, 2), (1, -2)$ **4. a.** -5 **b.** 5   **c.** $2t - 1$ **d.** 0.42   **5.** -27 **6. a.** The set of real numbers   **b.** The set of all real numbers except -3 **7.**   **8.** Not a function

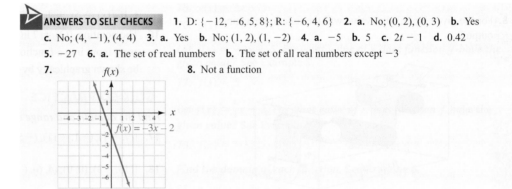

$f(x) = -3x - 2$

STUDY SET
2.5

VOCABULARY

Fill in the blanks.

1. A set of ordered pairs is called a _____. The set of all first components of the ordered pairs is called the _____ and the set of all second components is called the _____.

2. A _____ is a set of ordered pairs (a relation) in which to each first component there corresponds exactly one second component.

3. Given a relation in x and y, if to each value of x in the domain there corresponds exactly one value of y in the range, y is said to be a _____ of x. We call x the independent _____ and y the _____ variable.

4. For a function, the set of all possible values that can be used for the independent variable is called the _____. The set of all values of the dependent variable is called the _____.

5. We call $f(x) = 2x + 1$ a _____ function.

6. We call $f(x) = x$ the _____ function because it assigns each real number to itself. We call $f(x) = 8$ a _____ function, because for any input x, the output is always 8.

Determine whether each graph is the graph of a function. If it is not, find two ordered pairs where more than one value of y corresponds to a single value of x. **See Example 8.**

93.

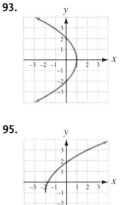

94.

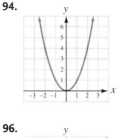

95.

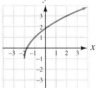

96.

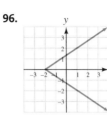

97.

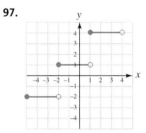

98.

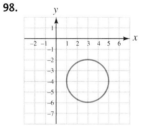

99.

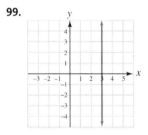

100.

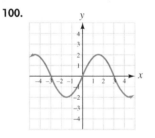

APPLICATIONS

101. DECONGESTANTS   The temperature in degrees Celsius that is equivalent to a temperature in degrees Fahrenheit is given by the linear function

$C(F) = \frac{5}{9}(F - 32)$. Use this function to find the low and high temperature extremes, in degrees Celsius, in which a bottle of decongestant should be stored.

DIRECTIONS: Adults and children 12 years of age and over: Two teaspoons every 4 hours. DO NOT EXCEED 6 DOSES IN A 24-HOUR PERIOD. Store at a controlled room temperature between 68°F and 77°F.

102. BODY TEMPERATURES   The temperature in degrees Fahrenheit that is equivalent to a temperature in degrees Celsius is given by the linear function $F(C) = \frac{9}{5}C + 32$. Convert each temperature in the following excerpt from *The Good House-keeping Family Health and Medical Guide* to degrees Fahrenheit. (Round to the nearest degree.)

> *In disease, the temperature of the human body may vary from about 32.2°C to 43.3°C for a time, but there is grave danger to life should it drop and remain below 35°C or rise and remain at or above 41°C.*

103. CONCESSIONAIRES   A baseball club pays a vendor $125 per game for selling bags of peanuts for $4.75 each.
   a. Write a linear function that describes the income the vendor makes for the baseball club during a game if she sells *b* bags of peanuts.
   b. Find the income the baseball club will make if the vendor sells 110 bags of peanuts during a game.

104. HOME CONSTRUCTION   In a proposal to some clients, a housing contractor listed the following costs:

Fees, permits, miscellaneous	$12,000
Construction, per square foot	$95

   a. Write a linear function that the clients could use to determine the cost of building a home having *f* square feet.

   b. Find the cost to build a home having 1,950 square feet.

105. EARTH'S ATMOSPHERE   The illustration shows a graph of the temperatures of the atmosphere at various altitudes above Earth's surface. The temperature is expressed in degrees Kelvin, a scale widely used in scientific work.
   a. Estimate the coordinates of three points on the graph that have an *x*-coordinate of 200.
   b. Explain why this is not the graph of a function.

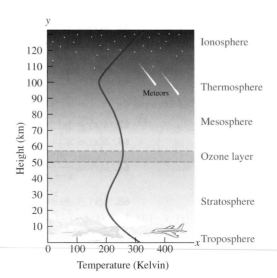

Temperature (Kelvin)

106. CHEMICAL REACTIONS   When students mixed solutions of acetone and chloroform, they found that heat was generated. However, as time passed, the mixture cooled down. The graph shows data points of the form (time, temperature) taken by the students.

a. The linear function $T(t) = -\frac{t}{240} + 30$ models the relationship between the elapsed time t since the solutions were combined and the temperature $T(t)$ of the mixture. Graph the function.

b. Predict the temperature of the mixture immediately after the two solutions are combined.

c. Is $T(180)$ more or less than the temperature recorded by the students for $t = 300$?

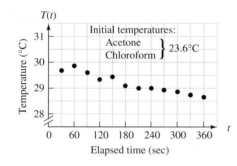

107. TAXES   The function

$$T(a) = 755 + 0.15(a - 7,550)$$

(where a is adjusted gross income) is a model of the instructions given on the first line of the following tax rate Schedule X.

a. Find $T(25,000)$ and interpret the result.

b. Write a function that models the second line on Schedule X.

Schedule X–Use if your filing status is **Single**		20**06**	
If your adjusted gross income is: Over —	But not over —	Your tax is	of the amount over —
$ 7,550	$30,650	$ 755 + 15%	$ 7,550
$30,650	$74,200	$4,220 + 25%	$30,650

108. ARCHERY   The area of a circle with a diameter of length d is given by the function $A(d) = \pi\left(\frac{d}{2}\right)^2$.

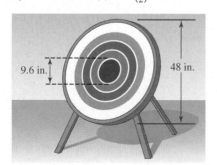

a. Find the area of the archery target to the nearest tenth of a square inch.

b. Find the area of the bull's eye to the nearest tenth of a square inch.

WRITING

109. Explain why we can think of a function as a machine.

110. Consider the function defined by $y = 6x + 4$. Why do you think x is called the *independent* variable and y the *dependent* variable?

111. A website selling nutritional supplements contains the following sentence: "Health is a *function* of proper nutrition." Explain what this statement means.

112. A student was asked to determine whether the graph shown below is the graph of a function. What is wrong with the following reasoning?

When I draw a vertical line through the graph, it intersects the graph only once. By the vertical line test, this is the graph of a function.

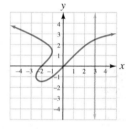

REVIEW

Solve each equation. If the equation is an identity or a contradiction, so indicate.

113. $-2(t + 4) + 5t + 1 = 3(t - 4) + 7$

114. $\dfrac{3}{2}(a - 4) = 2(a - 3) - \dfrac{a}{2}$

CHALLENGE PROBLEMS

115. Let $f(x) = 4x + 6$, function g be defined by $\{(4, 6), (6, 8), (8, 4)\}$, and function h be defined by the arrow diagram below. Find $\dfrac{f(8) + g(8)}{h(8)}$.

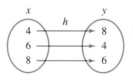

116. Find the domain of $f(x) = \dfrac{1}{5[9(x - 2) - 6(x - 3) + 3]}$.

SECTION 2.6
Graphs of Functions

Objectives

① Find function values graphically.
② Find the domain and range of a function graphically.
③ Graph nonlinear functions.
④ Translate graphs of functions.
⑤ Reflect graphs of functions.

Since a graph is often the best way to describe a function, we need to know how to construct and interpret their graphs.

 Find Function Values Graphically.

From the graph of a function, we can determine function values. In general, the value of $f(a)$ is given by the y-coordinate of a point on the graph of f with x-coordinate a.

EXAMPLE 1 Refer to the graph of function f in figure (a). **a.** Find $f(-3)$. **b.** Find the value of x for which $f(x) = -2$.

Strategy In each case, we will use the information provided by the function notation to locate a specific point on the graph and determine its x- and y-coordinates.

Why Once we locate the specific point, one of its coordinates will equal the value that we are asked to find.

Solution

a. To find $f(-3)$, we need to find the y-coordinate of the point on the graph of f whose x-coordinate is -3. If we draw a vertical line through -3 on the x-axis, as shown in figure (b), the line intersects the graph of f at $(-3, 5)$. Therefore, 5 corresponds to -3, and it follows that $f(-3) = 5$.

b. To find the input value x that has an output value $f(x) = -2$, we draw a horizontal line through -2 on the y-axis, as shown in figure (c) and note that it intersects the graph of f at $(4, -2)$. Since -2 corresponds to 4, it follows that $f(x) = -2$ if $x = 4$.

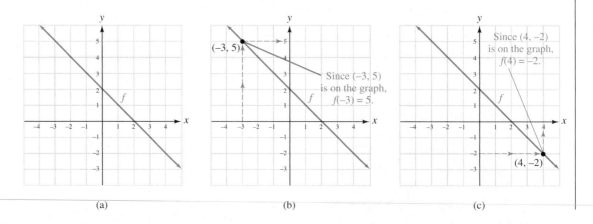

(a)   (b)   (c)

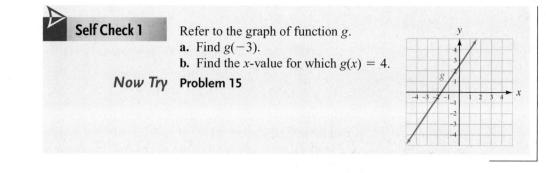

Now Try Problem 15

Self Check 1 Refer to the graph of function g.
a. Find $g(-3)$.
b. Find the x-value for which $g(x) = 4$.

2 Find the Domain and Range of a Function Graphically.

We can find the domain and range of a function from its graph. For example, to find the domain of the linear function graphed in figure (a), we *project* the graph onto the x-axis. Because the graph of the function extends indefinitely to the left and to the right, the projection includes all the real numbers. Therefore, the domain of the function is the set of real numbers.

   To find the range of the same linear function, we project the graph onto the y-axis, as shown in figure (b). Because the graph of the function extends indefinitely upward and downward, the projection includes all the real numbers. Therefore, the range of the function is the set of real numbers.

The Language of Algebra
Think of the *projection* of a graph on an axis as the "shadow" that the graph makes on the axis.

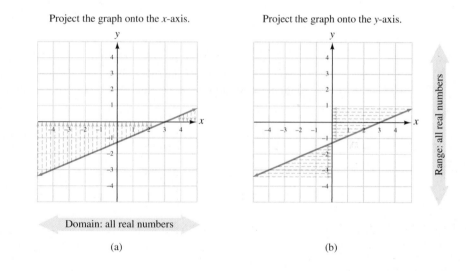

Project the graph onto the x-axis.

Project the graph onto the y-axis.

Range: all real numbers

Domain: all real numbers

(a)   (b)

3 Graph Nonlinear Functions.

We have seen that the graph of a linear function is a line. We will now consider several examples of **nonlinear functions** whose graphs are not lines. We will begin with $f(x) = x^2$, called the **squaring function.**

EXAMPLE 2   Graph $f(x) = x^2$ and find its domain and range.

Strategy   We will graph the function by creating a table of function values and plotting the corresponding ordered pairs.

Why   After drawing a smooth curve though the plotted points, we will have the graph.

Solution   To graph the function, we select several x-values and find the corresponding values of $f(x)$. For example, if we select -3 for x, we have

$$f(x) = x^2 \qquad \text{This is the function to graph.}$$
$$f(-3) = (-3)^2 \qquad \text{Substitute } -3 \text{ for each } x.$$
$$= 9$$

Since $f(-3) = 9$, the ordered pair $(-3, 9)$ lies on the graph of f. In a similar manner, we find the corresponding values of $f(x)$ for six other x-values and list the ordered pairs in the table of values. Then we plot the points and draw a smooth curve through them to get the graph, called a **parabola.**

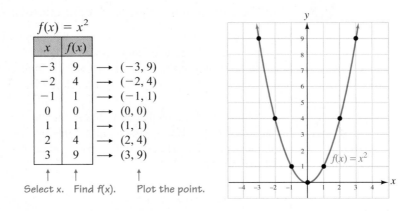

$f(x) = x^2$

x	$f(x)$	
-3	9	$\rightarrow (-3, 9)$
-2	4	$\rightarrow (-2, 4)$
-1	1	$\rightarrow (-1, 1)$
0	0	$\rightarrow (0, 0)$
1	1	$\rightarrow (1, 1)$
2	4	$\rightarrow (2, 4)$
3	9	$\rightarrow (3, 9)$

Select x.   Find $f(x)$.   Plot the point.

Because the graph extends indefinitely to the left and to the right, the projection of the graph onto the x-axis includes all the real numbers. See figure (a). This means that the domain of the squaring function is the set of real numbers.

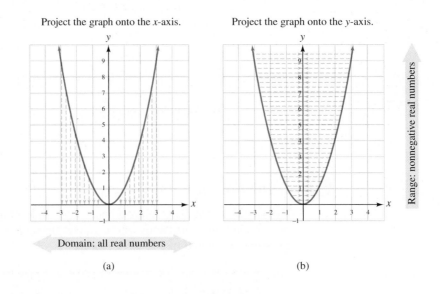

Project the graph onto the x-axis.   Project the graph onto the y-axis.

Domain: all real numbers

Range: nonnegative real numbers

(a)   (b)

Because the graph extends upward indefinitely from the point $(0, 0)$, the projection of the graph on the y-axis includes only positive real numbers and 0. See figure (b) above. This means that the range of the squaring function is the set of nonnegative real numbers.

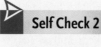 **Self Check 2** Graph $g(x) = x^2 - 2$ and find its domain and range. Compare the graph to the graph of $f(x) = x^2$.

Now Try **Problem 27**

Another nonlinear function is $f(x) = x^3$, called the **cubing function.**

EXAMPLE 3 Graph $f(x) = x^3$ and find its domain and range.

Strategy We will graph the function by creating a table of function values and plotting the corresponding ordered pairs.

Why After drawing a smooth curve though the plotted points, we will have the graph.

Solution To graph the function, we select several values for x and find the corresponding values of $f(x)$. For example, if we select -2 for x, we have

$$f(x) = x^3$$
$$f(-2) = (-2)^3 \quad \text{Substitute } -2 \text{ for each } x.$$
$$= -8$$

Since $f(-2) = -8$, the ordered pair $(-2, -8)$ lies on the graph of f. In a similar manner, we find the corresponding values of $f(x)$ for four other x-values and list the ordered pairs in the table. Then we plot the points and draw a smooth curve through them to get the graph.

Success Tip

To graph a linear function, it is recommended that you find three points on the line to draw its graph. Because the graphs of nonlinear functions are more complicated, more work is required. You need to find a sufficient number of points on the graph so that its entire shape is revealed.

$$f(x) = x^3$$

x	$f(x)$	
-2	-8	$\longrightarrow (-2, -8)$
-1	-1	$\longrightarrow (-1, -1)$
0	0	$\longrightarrow (0, 0)$
1	1	$\longrightarrow (1, 1)$
2	8	$\longrightarrow (2, 8)$

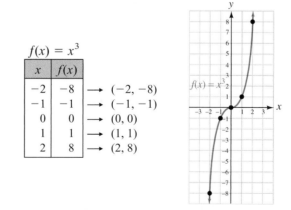

Because the graph of the function extends indefinitely to the left and to the right, the projection includes all the real numbers. Therefore, the domain of the cubing function is the set of real numbers.

Because the graph of the function extends indefinitely upward and downward, the projection includes all the real numbers. Therefore, the range of the cubing function is the set of real numbers.

▷ **Self Check 3**   Graph $g(x) = x^3 + 1$ and find its domain and range. Compare the graph to the graph of $f(x) = x^3$.

Now Try   **Problem 29**

A third nonlinear function is $f(x) = |x|$, called the **absolute value function.**

EXAMPLE 4   Graph $f(x) = |x|$ and find its domain and range.

Strategy   We will graph the function by creating a table of function values and plotting the corresponding ordered pairs.

Why   After drawing lines though the plotted points, we will have the graph.

Solution   To graph the function, we select several *x*-values and find the corresponding values for $f(x)$. For example, if we choose -3 for *x*, we have

$$f(x) = |x|$$
$$f(-3) = |-3| \quad \text{Substitute } -3 \text{ for each } x.$$
$$= 3$$

Since $f(-3) = 3$, the ordered pair $(-3, 3)$ lies on the graph of f. In a similar manner, we find the corresponding values of $f(x)$ for six other *x*-values and list the ordered pairs in the table. Then we plot the points and connect them to get the following V-shaped graph.

Success Tip
To determine the entire shape of the graph, several positive and negative values, along with 0, were selected as *x*-values when constructing the table of values.

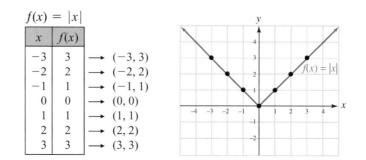

$$f(x) = |x|$$

x	f(x)	
-3	3	→ (-3, 3)
-2	2	→ (-2, 2)
-1	1	→ (-1, 1)
0	0	→ (0, 0)
1	1	→ (1, 1)
2	2	→ (2, 2)
3	3	→ (3, 3)

Because the graph extends indefinitely to the left and to the right, the projection of the graph onto the *x*-axis includes all the real numbers. Thus, the domain of the absolute value function is the set of real numbers.

Because the graph extends upward indefinitely from the point $(0, 0)$, the projection of the graph on the *y*-axis includes only positive real numbers and 0. Thus, the range of the absolute value function is the set of nonnegative real numbers.

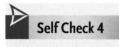 **Self Check 4**   Graph $g(x) = |x - 2|$ and find its domain and range. Compare the graph to the graph of $f(x) = |x|$.

Now Try   **Problem 31**

Using Your Calculator

Graphing Functions

We can graph nonlinear functions with a graphing calculator. For example, to graph $f(x) = x^2$ in a standard window of $[-10, 10]$ for x and $[-10, 10]$ for y, we first press $\boxed{Y =}$. Then we enter the function by typing $x \char`\^ 2$ (or x followed by $\boxed{x^2}$), and press the $\boxed{\text{GRAPH}}$ key. We will obtain the graph shown in figure (a).

To graph $f(x) = x^3$, we enter the function by typing $x \char`\^ 3$ and then press the $\boxed{\text{GRAPH}}$ key to obtain the graph in figure (b). To graph $f(x) = |x|$, we enter the function by selecting abs from the NUM option within the MATH menu, typing x, and pressing the $\boxed{\text{GRAPH}}$ key to obtain the graph in figure (c).

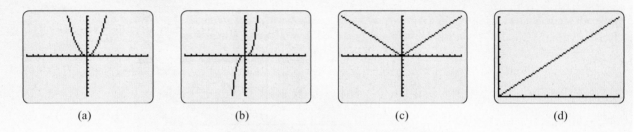

| (a) | (b) | (c) | (d) |

When using a graphing calculator, we must be sure that the viewing window does not show a misleading graph. For example, if we graph $f(x) = |x|$ in the window $[0, 10]$ for x and $[0, 10]$ for y, we will obtain a misleading graph that looks like a line. See figure (d). This is not correct. The proper graph is the V-shaped graph shown in figure (c). One of the challenges of using graphing calculators is finding an appropriate viewing window.

 Translate Graphs of Functions.

Examples 2, 3, and 4 and their Self Checks suggest that the graphs of different functions may be identical except for their positions in the coordinate plane. For example, the figure on the right shows the graph of $f(x) = x^2 + k$ for three different values of k. If $k = 0$, we get the graph of $f(x) = x^2$. If $k = 3$, we get the graph of $f(x) = x^2 + 3$, which is identical to the graph of $f(x) = x^2$ except that it is shifted 3 units upward. If $k = -4$, we get the graph of $f(x) = x^2 - 4$, which is identical to the graph of $f(x) = x^2$ except that it is shifted 4 units downward. These shifts are called **vertical translations.**

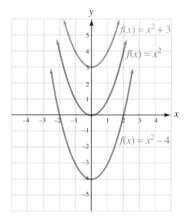

In general, we can make these observations.

Vertical Translations

If f is a function and k represents a positive number, then

- The graph of $y = f(x) + k$ is identical to the graph of $y = f(x)$ except that it is translated k units upward.
- The graph of $y = f(x) - k$ is identical to the graph of $y = f(x)$ except that it is translated k units downward.

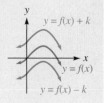

EXAMPLE 5   Graph: $g(x) = |x| + 2$

Strategy   We will graph $g(x) = |x| + 2$ by translating (shifting) the graph of $f(x) = |x|$ upward 2 units.

Why   The addition of 2 in $g(x) = |x| + 2$ causes a vertical shift of the graph of the absolute value function 2 units upward.

Solution   Each point used to graph $f(x) = |x|$, which is shown in gray, is shifted 2 units upward to obtain the graph of $g(x) = |x| + 2$, which is shown in red.

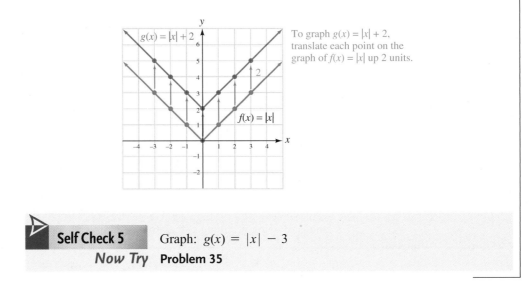

To graph $g(x) = |x| + 2$, translate each point on the graph of $f(x) = |x|$ up 2 units.

▶ **Self Check 5**   Graph: $g(x) = |x| - 3$
   Now Try   **Problem 35**

The figure on the right shows the graph of $f(x) = (x + h)^2$ for three different values of h. If $h = 0$, we get the graph of $f(x) = x^2$. The graph of $f(x) = (x - 3)^2$ is identical to the graph of $f(x) = x^2$ except that it is shifted 3 units to the right. The graph of $f(x) = (x + 2)^2$ is identical to the graph of $f(x) = x^2$ except that it is shifted 2 units to the left. These shifts are called *horizontal translations*.

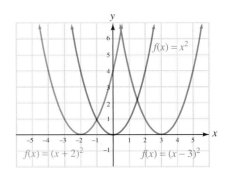

In general, we can make these observations.

Horizontal Translations

If f is a function and h is a positive number, then

- The graph of $y = f(x - h)$ is identical to the graph of $y = f(x)$ except that it is translated h units to the right.
- The graph of $y = f(x + h)$ is identical to the graph of $y = f(x)$ except that it is translated h units to the left.

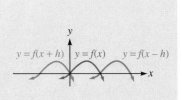

EXAMPLE 6   Graph: $g(x) = (x + 3)^3$

Strategy   We will graph $g(x) = (x + 3)^3$ by translating (shifting) the graph of $f(x) = x^3$ to the left 3 units.

Why   The addition of 3 to x in $g(x) = (x + 3)^3$ causes a horizontal shift of the graph of the cubing function 3 units to the left.

Solution   Each point used to graph $f(x) = x^3$, which is shown in gray, is shifted 3 units to the left to obtain the graph of $g(x) = (x + 3)^3$, which is shown in red.

To graph $g(x) = (x + 3)^3$, translate each point on the graph of $f(x) = x^3$ to the left 3 units.

Self Check 6   Graph: $g(x) = (x - 2)^2$

Now Try   **Problem 45**

The graphs of some functions involve horizontal and vertical translations.

EXAMPLE 7   Graph: $g(x) = (x - 5)^2 - 2$

Strategy   To graph $g(x) = (x - 5)^2 - 2$, we will perform two translations by shifting the graph of $f(x) = x^2$ to the right 5 units and then 2 units downward.

Why   The subtraction of 5 from x in $g(x) = (x - 5)^2 - 2$ causes a horizontal shift of the graph of the squaring function 5 units to the right and the subtraction of 2 causes a vertical shift of the graph 2 units downward.

Solution   Each point used to graph $f(x) = x^2$, which is shown in gray, is shifted 5 units to the right and 2 units downward to obtain the graph of $g(x) = (x - 5)^2 - 2$, which is shown in red.

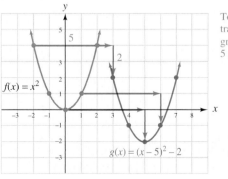

To graph $g(x) = (x - 5)^2 - 2$, translate each point on the graph of $f(x) = x^2$ to the right 5 units and then 2 units downward.

Self Check 7   Graph: $g(x) = |x + 2| - 3$

Now Try   **Problem 51**

5   Reflect Graphs of Functions.

The following figure shows a table of values for $f(x) = x^2$ and for $g(x) = -x^2$. We note that for a given value of x, the corresponding y-value in the tables are opposites. When graphed, we see that the $-$ sign in $g(x) = -x^2$ has the effect of flipping the graph of $f(x) = x^2$ over the x-axis so that the parabola opens downward. We say that the graph of $g(x) = -x^2$ is a **reflection** of the graph of $f(x) = x^2$ about the x-axis.

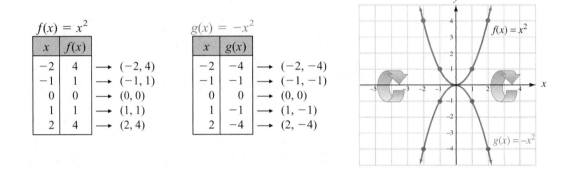

$f(x) = x^2$

x	$f(x)$	
-2	4	$\rightarrow (-2, 4)$
-1	1	$\rightarrow (-1, 1)$
0	0	$\rightarrow (0, 0)$
1	1	$\rightarrow (1, 1)$
2	4	$\rightarrow (2, 4)$

$g(x) = -x^2$

x	$g(x)$	
-2	-4	$\rightarrow (-2, -4)$
-1	-1	$\rightarrow (-1, -1)$
0	0	$\rightarrow (0, 0)$
1	-1	$\rightarrow (1, -1)$
2	-4	$\rightarrow (2, -4)$

Reflection of a Graph	The graph of $y = -f(x)$ is the graph of $y = f(x)$ reflected about the x-axis.

EXAMPLE 8   Graph: $g(x) = -x^3$

Strategy   We will graph $g(x) = -x^3$ by reflecting the graph of $f(x) = x^3$ about the x-axis.

Why   Because of the $-$ sign in $g(x) = -x^3$, the y-coordinate of each point on the graph of function g is the opposite of the y-coordinate of the corresponding point on the graph $f(x) = x^3$.

Solution   To graph $g(x) = -x^3$, we use the graph of $f(x) = x^3$ from Example 3. First, we reflect the portion of the graph of $f(x) = x^3$ in quadrant I to quadrant IV, as shown. Then we reflect the portion of the graph of $f(x) = x^3$ in quadrant III to quadrant II.

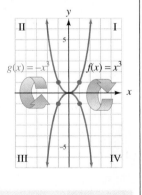

Self Check 8   Graph: $g(x) = -|x|$

Now Try   **Problem 61**

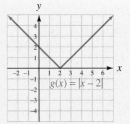

ANSWERS TO SELF CHECKS 1. a. −2 b. 1

2. D: the set of real numbers, R: the set of all real numbers greater than or equal to −2; the graph has the same shape, but is 2 units lower.

3. D: the set of real numbers, R: the set of all real numbers; the graph has the same shape, but is 1 unit higher.

4. D: the set of real numbers, R: the set of nonnegative real numbers; the graph has the same shape, but is 2 units to the right.

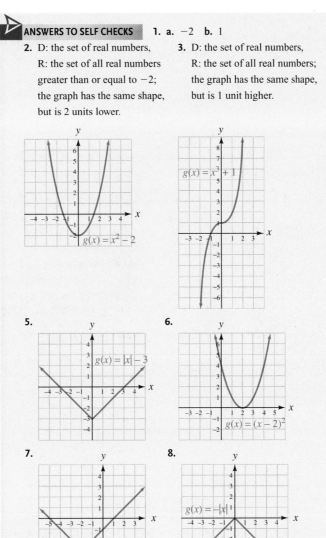

STUDY SET
2.6

VOCABULARY

Fill in the blanks.

1. Functions whose graphs are not lines are called _____ functions.

2. The graph of $f(x) = x^2$ is a cuplike shape called a _____.

3. The set of _____ real numbers is the set of real numbers greater than or equal to 0.

4. A shift of the graph of a function upward or downward is called a vertical _____.

CONCEPTS

5. Sketch the graph of each of these basic functions and give its name.

   a. $f(x) = x^2$

   b. $f(x) = x^3$

   c. $f(x) = |x|$

6. Complete each sentence about finding function values graphically.

   a. To find $f(-3)$, we find the y-coordinate of the point on the graph whose x-coordinate is ____ .

   b. To find the value of x for which $f(x) = -2$, we find the x-coordinate of the point(s) on the graph whose y-coordinate is ____ .

7. Suppose for a function f that $f(5) = 9$. What corresponding ordered pair will be on the graph of function f?

8. Fill in the blanks. The illustration shows the projection of the graph of function f on the _____. We see that the _____ of f is the set of real numbers less than or equal to 0.

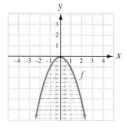

9. Consider the graph of the function f.

   a. Label each arrow in the illustration with the appropriate term: *domain* or *range*.

   b. Give the domain and range of f.

10. The graph of $f(x) = x^2 + k$ for three values of k is shown on the right. Find the three values.

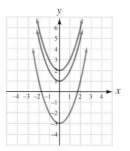

11. The graph of $f(x) = |x + h|$ for three values of h is shown below. Find the three values.

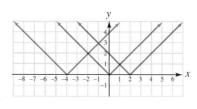

12. a. Translate each point plotted on the graph to the left 5 units and then up 1 unit.

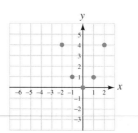

b. Translate each point plotted on the graph to the right 4 units and then down 3 units.

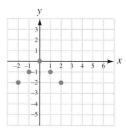

Fill in the blanks.

13. a. The graph of $f(x) = (x + 4)^3$ is the same as the graph of $f(x) = x^3$ except that it is shifted ___ units to the ____.

   b. The graph of $f(x) = x^3 + 4$ is the same as the graph of $f(x) = x^3$ except that it is shifted ___ units ____.

14. a. The graph of $f(x) = |x| - 5$ is the same as the graph of $f(x) = |x|$ except that it is shifted ___ units _____.

   b. The graph of $f(x) = |x - 5|$ is the same as the graph of $f(x) = |x|$ except that it is shifted ___ units to the _____.

GUIDED PRACTICE

Refer to the given graph to find each value. **See Example 1.**

15. a. $f(-2)$

   b. $f(0)$

   c. The value of x for which $f(x) = 4$.

   d. The value of x for which $f(x) = -2$.

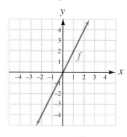

16. a. $g(-2)$

   b. $g(0)$

   c. The value of x for which $g(x) = 3$.

   d. The values of x for which $g(x) = -1$.

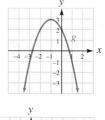

17. a. $s(-3)$

   b. $s(3)$

   c. The values of x for which $s(x) = 0$.

   d. The values of x for which $s(x) = 3$.

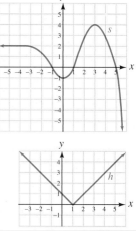

18. a. $h(-3)$

   b. $h(4)$

   c. The values of x for which $h(x) = 1$.

   d. The value of x for which $h(x) = 0$.

Find the domain and range of each function. See Objective 2 and Example 2.

19.

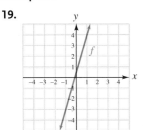

20.

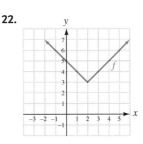

21.

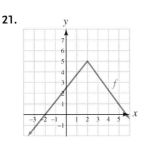

22.

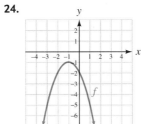

23.

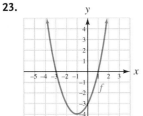

24.

25.

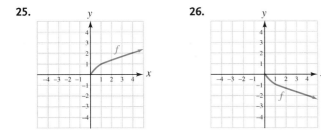

26.

Graph each function by creating a table of function values and plotting points. Give the domain and range of the function. See Examples 2, 3, and 4.

27. $f(x) = x^2 + 2$

28. $f(x) = x^2 - 4$

29. $f(x) = x^3 - 3$

30. $f(x) = x^3 + 2$

31. $f(x) = |x - 1|$

32. $f(x) = |x + 4|$

33. $f(x) = (x + 4)^2$

34. $f(x) = (x - 1)^3$

For each of the following functions, first sketch the graph of its associated function, $f(x) = x^2$, $f(x) = x^3$, or $f(x) = |x|$. Then draw the graph of function g using a translation and give its domain and range. See Examples 5 and 6.

35. $g(x) = |x| - 2$

36. $g(x) = |x + 2|$

37. $g(x) = (x + 1)^3$

38. $g(x) = x^3 + 5$

39. $g(x) = x^2 - 3$

40. $g(x) = (x - 6)^2$

41. $g(x) = (x - 4)^3$

42. $g(x) = |x| + 1$

43. $g(x) = x^3 + 4$

44. $g(x) = x^2 - 5$

45. $g(x) = (x + 4)^2$

46. $g(x) = (x - 1)^3$

For each of the following functions, first sketch the graph of its associated function, $f(x) = x^2$, $f(x) = x^3$, or $f(x) = |x|$. Then draw the graph of function g using translations and/or a reflection. See Examples 7 and 8.

47. $g(x) = |x - 2| - 1$

48. $g(x) = (x + 2)^2 - 1$

49. $g(x) = (x + 1)^3 - 2$

50. $g(x) = |x + 4| + 3$

51. $g(x) = (x - 2)^2 + 4$

52. $g(x) = (x - 4)^2 + 3$

53. $g(x) = |x + 3| + 5$

54. $g(x) = (x - 3)^2 - 2$

55. $g(x) = -x^3$

56. $g(x) = -|x|$

57. $g(x) = -x^2$

58. $g(x) = -(x + 1)^2$

59. $g(x) = -|x + 5|$

60. $g(x) = -(x + 4)^3$

61. $g(x) = -x^2 + 3$

62. $g(x) = -|x| - 4$

Graph each function using window settings of [−4, 4] for x and [−4, 4] for y. The graph is not what it appears to be. Pick a better viewing window and find a better representation of the true graph. See Using Your Calculator: Graphing Functions.

63. $f(x) = x^2 + 8$

64. $f(x) = x^3 - 8$

65. $f(x) = |x + 5|$

66. $f(x) = |x - 5|$

67. $f(x) = (x - 6)^2$

68. $f(x) = (x + 9)^2$

69. $f(x) = x^3 + 8$

70. $f(x) = x^3 - 12$

APPLICATIONS

71. OPTICS See the illustration. The **law of reflection** states that the angle of reflection is equal to the angle of incidence. What function studied in this section models the path of the reflected light beam with an angle of incidence measuring 45°?

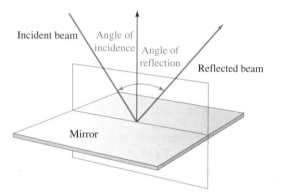

72. BILLIARDS In the illustration, a rectangular coordinate system has been superimposed over a billiard table. Write a function that models the path of the ball that is shown banking off of the cushion.

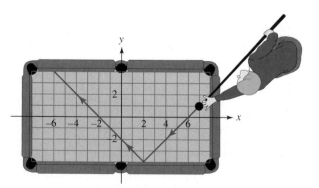

73. CENTER OF GRAVITY See the illustration. As a diver performs a $1\frac{1}{2}$-somersault in the tuck position, her center of gravity follows a path that can be described by a graph shape studied in this section. What graph shape is that?

74. LIGHT Light beams coming from a bulb are reflected outward by a parabolic mirror as parallel rays.

   a. The cross-sectional view of a parabolic mirror is given by the function $f(x) = x^2$ for the following values of x: -0.7, $-0.6, -0.5, -0.4, -0.3, -0.2, -0.1, 0, 0.1, 0.2, 0.3, 0.4$, $0.5, 0.6, 0.7$. Sketch the parabolic mirror using the following graph.

   b. From the lightbulb filament at $(0, 0.25)$, draw a line segment representing a beam of light that strikes the mirror at $(-0.4, 0.16)$ and then reflects outward, parallel to the y-axis.

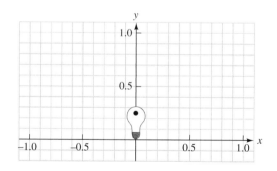

WRITING

75. Explain how to graph a function by plotting points.

76. Explain how to *project* the graph of a function onto the x-axis. Give an example.

77. What does it mean to vertically translate a graph? What does it mean to horizontally translate a graph?

78. What does it mean to reflect the graph of a function about the x-axis?

REVIEW

Solve each formula for the indicated variable.

79. $T - W = ma$ for W

80. $a + (n - 1)d = l$ for n

81. $s = \dfrac{1}{2}gt^2 + vt$ for g

82. $e = mc^2$ for m

CHALLENGE PROBLEMS

Graph each function.

83. $f(x) = \begin{cases} |x| & \text{for } x \geq 0 \\ x^3 & \text{for } x < 0 \end{cases}$

84. $f(x) = \begin{cases} x^2 & \text{for } x \geq 0 \\ |x| & \text{for } x < 0 \end{cases}$

Find the domain and range of each function.

85.

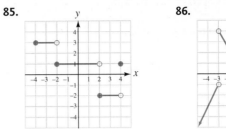

86.

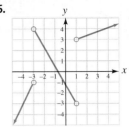

CHAPTER 2
Summary & Review

SECTION 2.1   The Rectangular Coordinate System

DEFINITIONS AND CONCEPTS	EXAMPLES
A **rectangular coordinate** system is formed by two intersecting perpendicular number lines called the **x-axis** and the **y-axis**. The x- and y-axes divide the coordinate plane into four quadrants. The process of locating a point in the coordinate plane is called **plotting** or **graphing** that point.	Plot the point $(-2, 4)$. We begin at the origin, and move 2 units to the left along the x-axis and then 4 units up, and draw a dot.
Graphs can be used to visualize relationships between two quantities.	Refer to the **line graph** on page 122. We see that on day 6, the water level was 3 feet above normal. Refer to the **step graph** on page 123. We see that the cost to rent the ski equipment for 5 days is $140.
The **midpoint** of a line segment with endpoints (x_1, y_1) and (x_2, y_2) is the point with coordinates $$\left(\frac{x_1 + x_2}{2}, \frac{y_1 + y_2}{2} \right)$$	Find the midpoint of the segment joining $(-3, 7)$ and $(5, -8)$. We let $(x_1, y_1) = (-3, 7)$ and $(x_2, y_2) = (5, -8)$ and substitute the coordinates into the midpoint formula. $$\left(\frac{x_1 + x_2}{2}, \frac{y_1 + y_2}{2} \right) = \left(\frac{-3 + 5}{2}, \frac{7 + (-8)}{2} \right)$$ $$= \left(\frac{2}{2}, -\frac{1}{2} \right)$$ $$= \left(1, -\frac{1}{2} \right) \quad \text{This is the midpoint.}$$

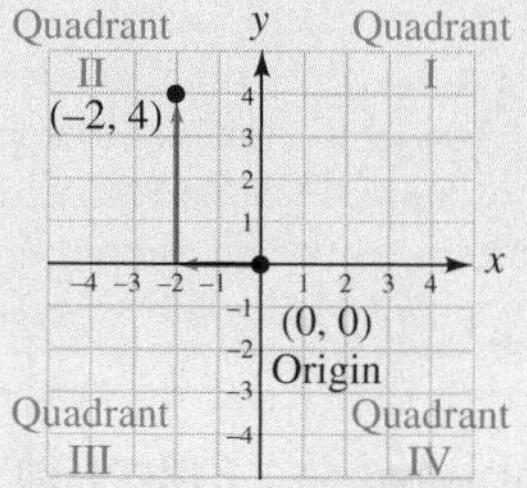

REVIEW EXERCISES

Plot each point on a rectangular coordinate system.

1. a. $(0, 3)$ **b.** $(-2, -4)$

    c. $\left(\frac{5}{2}, -1.75 \right)$ **d.** the origin

    e. $(2.5, 0)$

2. The given graph shows how the height of the water in a flood control channel changed over a 7-day period.

    a. Describe the height of the water at the beginning of day 2.

    b. By how much did the water level increase or decrease from day 4 to day 5?

    c. During what time period did the water level stay the same?

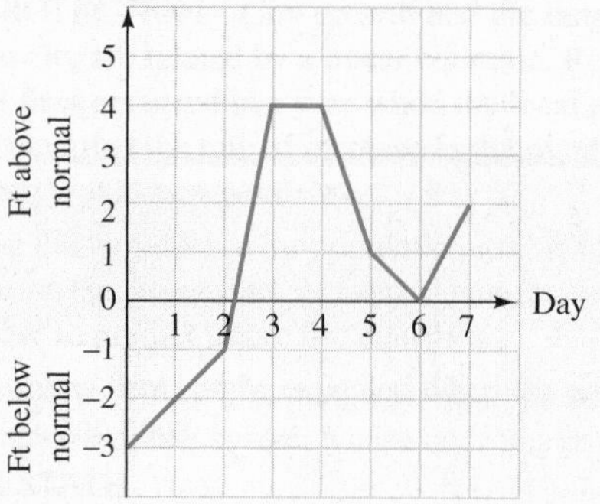

3. AUCTIONS   The dollar increments used by an auctioneer during the bidding process depend on what initial price the auctioneer began with for the item. See the step graph.

   a. What increments are used by the auctioneer if the bidding on an item began at $150?

   b. If the first bid on an item being auctioned is $750, what will be the next price asked for by the auctioneer?

4. Find the midpoint of the segment joining $(8, -2)$ and $(6, -4)$.

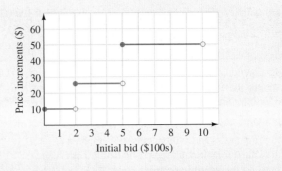

SECTION 2.2   Graphing Linear Equations in Two Variables

DEFINITIONS AND CONCEPTS	EXAMPLES
A **solution of an equation in two variables** is an ordered pair of numbers that makes the equation a true statement when the pair is substituted for the variables.	Determine whether $(1, -5)$ is a solution of $3x - y = 8$. We substitute the coordinates into the equation. $$3x - y = 8$$ $$3(1) - (-5) \overset{?}{=} 8 \quad \text{Substitute 1 for } x \text{ and } -5 \text{ for } y.$$ $$3 + 5 \overset{?}{=} 8 \quad \text{Evaluate the left side.}$$ $$8 = 8 \quad \text{True}$$ Since the result is true, $(1, -5)$ is a solution of the equation.
The **standard** or **general form** of a **linear equation** in two variables is $Ax + By = C$, where A, B, and C are real numbers and A and B are not both zero. The graph of a linear equation in two variables is a line.	Linear equations: $$y = \frac{2}{3}x - 5, \quad 3x + 4y = -8, \quad y = -\frac{5}{2}, \quad \text{and} \quad x = 3$$
Graphing Linear Equations Solved for y by Plotting Points: 1. Find three ordered pairs that are solutions of the equation by selecting three values for x and calculating the corresponding values of y. 2. Plot the solutions on a rectangular coordinate system. 3. Draw a straight line passing through the points. If the points do not lie on a line, check your computations.	Graph: $y = \dfrac{2}{3}x - 5$ We construct a table of solutions, plot the points, and draw the line. $$y = \frac{2}{3}x - 5$$ | x	y	(x, y)
---	---	---
-3	-7	$(-3, -7)$
0	-5	$(0, -5)$
6	-1	$(6, -1)$	 Select x.　Find y.　Plot (x, y).

SECTION 2.2 Graphing Linear Equations In Two Variables—*continued*

DEFINITIONS AND CONCEPTS	EXAMPLES

To find the **y-intercept** of a line, substitute 0 for x in the equation and solve for y. To find the **x-intercept** of a line, substitute 0 for y in the equation and solve for x.

Plotting the x- and y-intercepts of a graph and drawing a line through them is called the **intercept method for graphing a line.**

Use the y- and x-intercepts to graph $3x + 4y = -8$.

y-intercept: let x = 0
$$3x + 4y = -8$$
$$3(0) + 4y = -8$$
$$4y = -8$$
$$y = -2$$

x-intercept: let y = 0
$$3x + 4y = -8$$
$$3x + 4(0) = -8$$
$$3x = -8$$
$$x = -\frac{8}{3}$$

The y-intercept is $(0, -2)$ and the x-intercept is $\left(-\frac{8}{3}, 0\right)$.

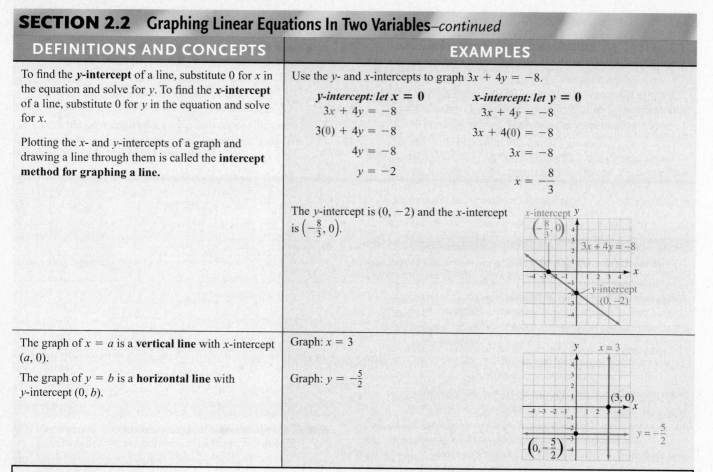

The graph of $x = a$ is a **vertical line** with x-intercept $(a, 0)$.

The graph of $y = b$ is a **horizontal line** with y-intercept $(0, b)$.

Graph: $x = 3$

Graph: $y = -\frac{5}{2}$

REVIEW EXERCISES

5. Is $(3, -6)$ a solution of $y = -5x + 9$?

6. The graph of a linear equation is shown in the illustration.

    a. If the coordinates of point A are substituted in the equation, will a true or false statement result?

    b. If the coordinates of point B are substituted in the equation, will a true or false statement result?

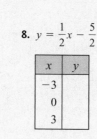

Complete each table.

7. $y = -3x$

x	y
-3	
0	
3	

8. $y = \frac{1}{2}x - \frac{5}{2}$

x	y
-3	
0	
3	

Graph each equation.

9. $y = 3x + 4$

10. $y = -\frac{1}{3}x - 1$

Graph each equation using the intercept method.

11. $2x + y = 4$

12. $3x - 4y - 8 = 0$

Graph each equation.

13. $y = 4$

14. $x = -2$

15. ACCOUNTING   A recording studio purchased a new electronic console for \$82,800. For income tax purposes, company accountants will use the straight-line depreciation equation $y = -13,800x + 82,800$ to describe the declining value of the console.

    a. What will be the value of the console in 2 years?

    b. When will the console have no value?

16. RECYCLING   It takes more aluminum cans to weigh one pound than it used to because manufacturers continue to use thinner materials. The equation $n = 0.42t + 24.08$ gives the approximate number n of empty aluminum cans needed to weigh one pound, where t is the number of years since 1980. Graph the equation. (Source: The Aluminum Association)

    a. What information can be obtained from the n-intercept of the graph?

    b. From the graph, estimate the number of cans it took to weigh one pound in 2006.

SECTION 2.3   Rate of Change and the Slope of a Line

DEFINITIONS AND CONCEPTS	EXAMPLES
The **slope** m of a line is a ratio that compares the vertical and horizontal change as we move along the line from one point to another. $$m = \frac{\text{vertical change}}{\text{horizontal change}} = \frac{\text{rise}}{\text{run}} = \frac{\Delta y}{\Delta x}$$	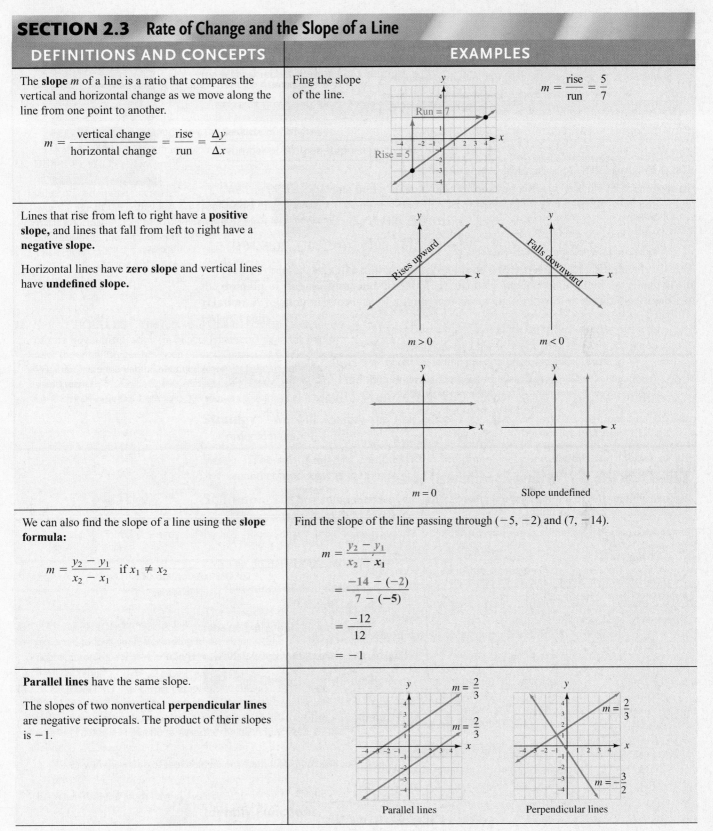 Fing the slope of the line. $m = \dfrac{\text{rise}}{\text{run}} = \dfrac{5}{7}$
Lines that rise from left to right have a **positive slope,** and lines that fall from left to right have a **negative slope.** Horizontal lines have **zero slope** and vertical lines have **undefined slope.**	
We can also find the slope of a line using the **slope formula:** $$m = \frac{y_2 - y_1}{x_2 - x_1} \quad \text{if } x_1 \neq x_2$$	Find the slope of the line passing through $(-5, -2)$ and $(7, -14)$. $m = \dfrac{y_2 - y_1}{x_2 - x_1}$ $\quad = \dfrac{-14 - (-2)}{7 - (-5)}$ $\quad = \dfrac{-12}{12}$ $\quad = -1$
Parallel lines have the same slope. The slopes of two nonvertical **perpendicular lines** are negative reciprocals. The product of their slopes is -1.	

REVIEW EXERCISES

17. Find the slope of lines l_1 and l_2.

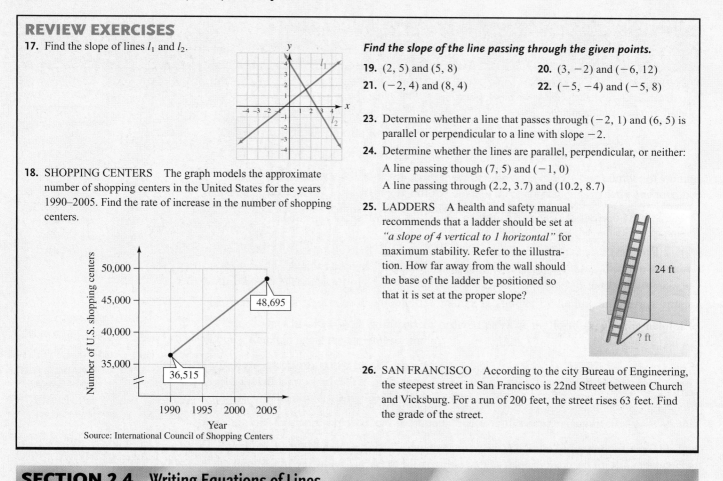

18. SHOPPING CENTERS   The graph models the approximate number of shopping centers in the United States for the years 1990–2005. Find the rate of increase in the number of shopping centers.

Find the slope of the line passing through the given points.

19. $(2, 5)$ and $(5, 8)$

20. $(3, -2)$ and $(-6, 12)$

21. $(-2, 4)$ and $(8, 4)$

22. $(-5, -4)$ and $(-5, 8)$

23. Determine whether a line that passes through $(-2, 1)$ and $(6, 5)$ is parallel or perpendicular to a line with slope -2.

24. Determine whether the lines are parallel, perpendicular, or neither:
A line passing though $(7, 5)$ and $(-1, 0)$
A line passing through $(2.2, 3.7)$ and $(10.2, 8.7)$

25. LADDERS   A health and safety manual recommends that a ladder should be set at *"a slope of 4 vertical to 1 horizontal"* for maximum stability. Refer to the illustration. How far away from the wall should the base of the ladder be positioned so that it is set at the proper slope?

26. SAN FRANCISCO   According to the city Bureau of Engineering, the steepest street in San Francisco is 22nd Street between Church and Vicksburg. For a run of 200 feet, the street rises 63 feet. Find the grade of the street.

SECTION 2.4   Writing Equations of Lines

DEFINITIONS AND CONCEPTS	EXAMPLES
If a linear equation is written in **slope–intercept** form $$y = mx + b$$ the graph of the equation is a line with slope m and y-intercept $(0, b)$.	The equation of the line with slope $\frac{2}{3}$ and y-intercept $(0, 7)$ is: $$y = \frac{2}{3}x + 7$$
Slope can be used as an aid in graphing.	Find the slope and y-intercept of the line whose equation is $4x + 3y = 6$. Then graph the line. To find the slope and y-intercept, we solve the equation for y. $$4x + 3y = 6$$ $$3y = -4x + 6 \quad \text{Subtract 4x from both sides.}$$ $$y = -\frac{4}{3}x + 2 \quad \begin{array}{l}\text{To isolate } y, \text{ divide both sides by 3.}\\ m = -\frac{4}{3} \text{ and } b = 2.\end{array}$$ The slope of the line is $-\frac{4}{3}$ and the y-intercept is $(0, 2)$. To graph the line, plot the y-intercept and then use the rise and run components of the slope to locate other points on the graph.

SECTION 2.4   Writing Equations of Lines–*continued*

DEFINITIONS AND CONCEPTS	EXAMPLES
If a line with slope m passes through the point with coordinates (x_1, y_1), the equation of the line in **point–slope form** is $$y - y_1 = m(x - x_1)$$	Find an equation of the line with slope -5 that passes through $(-1, 3)$. Write the equation in slope–intercept form. We substitute the slope and the coordinates of the point into the point–slope form. $\begin{aligned} y - y_1 &= m(x - x_1) &&\text{This is point–slope form.} \\ y - 3 &= -5[x - (-1)] &&\text{Substitute.} \\ y - 3 &= -5(x + 1) &&\text{Simplify within the brackets.} \\ y - 3 &= -5x - 5 &&\text{Distribute.} \\ y &= -5x - 2 &&\text{To isolate } y, \text{ add 3 to both sides.} \\ &&&\text{This is slope–intercept form.} \end{aligned}$
Slopes can be used to identify **parallel** and **perpendicular lines**.	The graphs of the lines $y = 7x + 5$ and $y = 7x - 3$ are parallel because each line has slope 7. The graphs of the lines $y = 6x + 1$ and $y = -\frac{1}{6}x$ are perpendicular because their slopes are 6 and $-\frac{1}{6}$, which are negative reciprocals.

REVIEW EXERCISES

Find an equation of the line with the given properties. Write the equation in slope–intercept form.

27. Slope 3; passes through $(-8, 5)$

28. Passes through $(-2, 4)$ and $(6, -9)$

29. Passes through $(-3, -5)$; parallel to the graph of $3x - 2y = 7$

30. Passes through $(-3, -5)$; perpendicular to the graph of $3x - 2y = 7$

31. Write $3x + 4y = -12$ in slope–intercept form. Give the slope and y-intercept of the graph of the equation. Then use this information to graph the line.

Write the equation of each line.

32. a. The x-axis   **b.** The y-axis

33. Determine whether the graphs of $y = x + 15$ and $x + y = 4$ are parallel, perpendicular, or neither.

34. Find the slope and the y-intercept of the line shown. Then write an equation of the line.

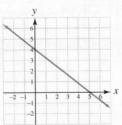

35. BUSINESS GROWTH   City growth and the number of business licenses issued by a certain city are related by a linear equation. Records show that 250 licenses had been issued when the local population was 21,000, and that the rate of increase in the number of licenses issued was 1 for every 150 new residents. Use the variables p for population and L for the number of business licenses to write an equation (in slope–intercept form) that city officials can use to predict future business growth.

36. DEPRECIATION   A manufacturing company purchased a new diamond-tipped saw blade for $8,700 and will depreciate it on a straight-line basis over the next 5 years. At the end of its useful life, it will be sold for scrap for $100.

a. Write a depreciation equation for the saw blade using the variables x and y.

b. If the depreciation equation is graphed, explain the significance of the y-intercept.

GROUP PROJECT

MEASURING SLOPE

Overview: This hands-on activity will give you a better understanding of slope.

Instructions: Form groups of 2 or 3 students. Use a ruler and a level to find the slopes of ramps or inclines on campus by measuring $\frac{rise}{run}$, as shown in the illustration. Record your results in a table, listing the slopes in order from smallest to largest.

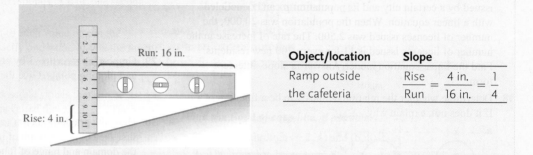

Object/location	Slope		
Ramp outside the cafeteria	$\dfrac{\text{Rise}}{\text{Run}}$ =	$\dfrac{4 \text{ in.}}{16 \text{ in.}}$ =	$\dfrac{1}{4}$

CUMULATIVE REVIEW
Chapters 1–2

1. List the elements of

$$\left\{-2,\ 0,\ 1,\ 2,\ \tfrac{13}{12},\ 6,\ 7,\ \sqrt{5},\ \pi\right\}$$

that belong to the following sets. [Section 1.2]

   a. Natural numbers

   b. Whole numbers

   c. Rational numbers

   d. Irrational numbers

   e. Negative numbers

   f. Real numbers

   g. Prime numbers

   h. Even numbers

2. Graph each element of the following set of numbers on a number line. [Section 1.2]

$$\left\{-1\tfrac{1}{2},\ -\pi,\ -\tfrac{35}{8},\ -0.333\ldots,\ 3,\ 4.25,\ \sqrt{2}\right\}$$

Evaluate each expression.

3. $\dfrac{|-5| + |-3|}{-|4|}$ [Section 1.3]

4. $7 - 12[7^2 - 4(2 - 5)^2]$ [Section 1.3]

5. $-\dfrac{16}{5} \div \left(-\dfrac{10}{3}\right)$ [Section 1.3]

6. $\dfrac{(9 - 8)^4 + 21}{3^3 - \left(\sqrt{16}\right)^2}$ [Section 1.3]

Evaluate each expression for x = 2 and y = −3.

7. $-y - 5xy$ [Section 1.3]

8. $\dfrac{x^2 - y^2}{2x + y}$ [Section 1.3]

9. Determine which property of real numbers justifies each statement. [Section 1.4]

   a. $(a + b) + c = a + (b + c)$

   b. $3(x + y) = 3x + 3y$

   c. $(a + b) + c = c + (a + b)$

   d. $(ab)c = a(bc)$

10. a. What is the additive inverse (opposite) of 6? [Section 1.4]

   b. What is the multiplicative inverse (reciprocal) of $-\tfrac{1}{8}$? [Section 1.4]

Simplify each expression.

11. $-7s(-4t)(-1)$ [Section 1.4]

12. $40\left(\dfrac{3}{8}y - \dfrac{1}{4}\right) + 40\left(\dfrac{4}{5}\right)$ [Section 1.4]

13. $-\dfrac{3}{4}s - \dfrac{1}{3}s$ [Section 1.4]

14. $-5[3(x - 4) - 2(x + 2)] - 7(x - 3)$ [Section 1.4]

Solve each equation, if possible.

15. $-\dfrac{9}{8}s = 3$ [Section 1.5]

16. $4(y - 3) + 4 = -3(y + 5)$ [Section 1.5]

17. $2x - \dfrac{3(x - 2)}{2} = 7 - \dfrac{x - 3}{3}$ [Section 1.5]

18. $0.04(24) + 0.02x = 0.04(12 + x)$ [Section 1.5]

19. $-3x = -2x + 1 - (5 + x)$ [Section 1.5]

20. $2[5(4 - a) + 2(a - 1)] = 3 - a$ [Section 1.5]

Solve each formula for the specified variable.

21. $-Tx + 3By = c$ for B [Section 1.6]

22. $A = \dfrac{1}{2}h(b_1 + b_2)$ for h [Section 1.6]

23. COLLECTING SIGNATURES A student working for a political campaign earns \$45 per day plus \$1.25 for each signature on a petition she obtains. How many signatures must she collect to earn \$250 a day? [Section 1.7]

24. ISOSCELES TRIANGLES The measure of each base angle of an isosceles triangle is 10° less than twice the measure of the vertex angle. Find the measure of each angle of the triangle. [Section 1.7]

25. INVESTMENTS A woman invested part of \$20,000 at 6% and the rest at 7%. If the annual interest earned was \$1,260, how much did she invest at 6%? [Section 1.8]

26. DRIVING RATES John drove to a distant city in 5 hours. When he returned, there was less traffic, and the trip took only 3 hours. If he drove 26 mph faster on the return trip, how fast did he drive each way? [Section 1.8]

27. Find the midpoint of a line segment with endpoints $(-5, -2), (7, 3)$. [Section 2.1]

28. Determine whether $\left(-1, \dfrac{25}{3}\right)$ is a solution of $3y - 5x = 30$. [Section 2.2]

29. Graph: $7x - 3y = 6$ [Section 2.2]

30. Find the slope of the line graphed below. [Section 2.3]

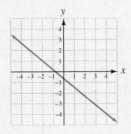

31. Find an equation of the line passing through $(-2, 5)$ and $(8, -9)$. Write the equation in slope–intercept form. [Section 2.4]

32. Find an equation of the line passing through $(-2, 3)$ and parallel to the graph of $3x + y = 8$. Write the equation in slope–intercept form. [Section 2.4]

33. See the illustration in the next column. Explain why there is not a linear relationship between the height of the antenna and its maximum range of reception. [Section 2.4]

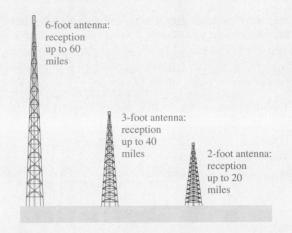

6-foot antenna: reception up to 60 miles

3-foot antenna: reception up to 40 miles

2-foot antenna: reception up to 20 miles

34. TRAFFIC SIGNALS City growth and the number of traffic signals for a certain city are related by a linear equation. Records show that there were 50 traffic signals when the local population was 25,000 and that the rate of increase in the number of traffic signals was 1 for every 1,000 new residents.

   a. Using the variables p for population and T for traffic signals, write an equation (in slope–intercept form) that the transportation department can use to predict future traffic signal needs. [Section 2.4]

   b. How many traffic signals can be expected when the population reaches 35,000? [Section 2.4]

Refer to the following graph of function f.

35. Find $f(1)$. [Section 2.5]

36. Find the value of x for which $f(x) = 1$. [Section 2.5]

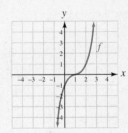

37. Determine whether the graph represents a function. [Section 2.5]

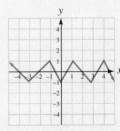

38. Find the domain and range of the relation $\{(1, -12), (-6, 7), (5, 8), (0, 7), (0, 4)\}$. Is it a function? [Section 2.5]

39. Does $y^2 = x$ define y as a function of x? If it does not, find two ordered pairs where more than one value of y corresponds to a single value of x. [Section 2.5]

40. Let $h(x) = -\dfrac{1}{5}x - 12$. For what value of x is $h(x) = 0$? [Section 2.5]

Let $f(x) = 3x^2 + 2$ **and** $g(x) = -2x - 1$. **Find each function value.**

41. $f(-1)$ [Section 2.5]

42. $g(0)$ [Section 2.5]

43. $g(-2)$ [Section 2.5]

44. $f(-r)$ [Section 2.5]

45. Find the domain of the function $f(x) = \frac{5}{x+1}$. [Section 2.5]

46. What is the slope of the graph of the linear funtion $f(x) = 6x + 15$? [Section 2.5]

47. Graph $f(x) = -x^2 + 1$ by plotting points. Then give the domain and range of the function. [Section 2.6]

48. First sketch the graph of the basic function associated with $g(x) = |x - 3| - 4$. Then draw the graph of function g using a translation. Give the domain and range of function g. [Section 2.6]

CHAPTER 3
Systems of Equations

© Radius Images/Alamy

3.1 Solving Systems of Equations by Graphing

3.2 Solving Systems of Equations Algebraically

3.3 Problem Solving Using Systems of Two Equations

3.4 Solving Systems of Equations in Three Variables

3.5 Problem Solving Using Systems of Three Equations

3.6 Solving Systems of Equations Using Matrices

3.7 Solving Systems of Equations Using Determinants

CHAPTER SUMMARY AND REVIEW
CHAPTER TEST
Group Project
CUMULATIVE REVIEW

from *Campus to Careers*
Fashion Designer

Fashion designers help create the billions of clothing articles, shoes, and accessories purchased every year by consumers. Fashion design relies heavily on mathematical skills, including knowledge of lines, angles, curves, and measurement. Designers also use mathematics in the manufacturing and marketing parts of the industry as they calculate labor costs and determine the markups and markdowns involved in retail pricing.

In **Problem 11** of **Study Set 3.5,** we will examine the production side of fashion design as we determine the number of coats, shirts, and slacks that can be made with the available labor.

JOB TITLE:
Fashion Designer

EDUCATION:
Many community colleges and vocational training schools provide training for fashion industry jobs.

JOB OUTLOOK:
The best opportunities will be in designing clothing sold in department stores and retail chains.

ANNUAL EARNINGS:
$69,270

FOR MORE INFORMATION:
www.collegeboard.com/csearch/majors_careers/profiles/

Study Skills Workshop
Successful Test Taking

Taking a math test doesn't have to be an unpleasant experience. Here are some suggestions that can make it more enjoyable and also improve your score.

PREPARING FOR THE TEST: Begin studying several days before the test rather than cramming your studying into one marathon session the night before.

TAKING THE TEST: Follow a test-taking strategy so that you can maximize your score by using the testing time wisely.

EVALUATING YOUR PERFORMANCE: After your graded test is returned, classify the type of errors that you made on the test so that you do not make them again.

Now Try This

1. Write a study session plan that explains how you will prepare on each of the four days before the test, as well as on test day. For some suggestions, see *Preparing for a Test**.

2. Develop your own test-taking strategy by answering the survey questions found in *How to Take a Math Test**.

3. Use the outline found In *Analyzing Your Test Results** to classify the errors that you made on your most recent test.

* Found online at: academic/cengage.com/math/tussy

SECTION 3.1
Solving Systems of Equations by Graphing

Objectives

1. Determine whether an ordered pair is a solution of a system.
2. Solve systems of linear equations by graphing.
3. Use graphing to identify inconsistent systems and dependent equations.
4. Solve equations graphically.

The red line in the graph on the next page shows the cost for a company to produce a given number of skateboards. The blue line shows the revenue the company will receive for selling a given number of those skateboards. The graph offers the company important financial information.

- The production costs exceed the revenue earned if fewer than 400 skateboards are sold. In this case, the company loses money.

- The revenue earned exceeds the production costs if more than 400 skateboards are sold. In this case, the company makes a profit.

- Production costs equal revenue earned if exactly 400 skateboards are sold. This fact is indicated by the point of intersection of the two lines, (400, 20,000), which is called the **break-even point.**

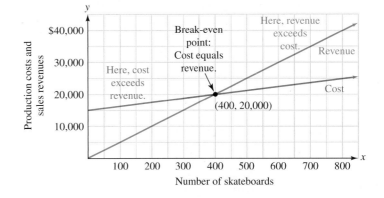

From Chapter 2, we know that the lines in the graph that show the cost and the revenue can be modeled by linear equations in two variables. Together, such a set of equations is called a *system of equations.*

In general, when two equations with the same variables are considered simultaneously (at the same time), we say that they form a **system of equations.** We will use a left brace { when writing a system of equations. An example is

$$\begin{cases} 2x + 5y = -1 \\ x - y = -4 \end{cases}$$ Read as "the system of equations $2x + 5y = -1$ and $x - y = -4$."

① Determine Whether an Ordered Pair Is a Solution of a System.

A **solution of a system** of equations in two variables is an ordered pair that satisfies both equations of the system.

EXAMPLE 1   Determine whether $(-3, 1)$ is a solution of each system of equations.

a. $\begin{cases} 2x + 5y = -1 \\ x - y = -4 \end{cases}$ **b.** $\begin{cases} 5y = 2 - x \\ y = 3x \end{cases}$

Strategy   We will substitute the x- and y-coordinates of $(-3, 1)$ for the corresponding variables in both equations of the system.

Why   If both equations are satisfied (made true) by the x- and y-coordinates, the ordered pair is a solution of the system.

Solution
a. To determine whether $(-3, 1)$ is a solution, we substitute -3 for x and 1 for y in each equation.

Check: $2x + 5y = -1$ First equation. $x - y = -4$ Second equation.

$2(-3) + 5(1) \overset{?}{=} -1$ $-3 - 1 \overset{?}{=} -4$

$-6 + 5 \overset{?}{=} -1$ $-4 = -4$ True

$-1 = -1$ True

Since $(-3, 1)$ satisfies both equations, it is a solution of the system.

b. Again, we substitute -3 for x and 1 for y in each equation.

Check:

$5y = 2 - x$	First equation.	$y = 3x$	Second equation.
$5(1) \overset{2}{=} 2 - (-3)$		$1 \overset{2}{=} 3(-3)$	
$5 \overset{2}{=} 2 + 3$		$1 = -9$	False
$5 = 5$	True		

Although $(-3, 1)$ satisfies the first equation, it does not satisfy the second. Because it does not satisfy both equations, $(-3, 1)$ is not a solution of the system.

Self Check 1   Determine whether $(6, -2)$ is a solution of $\begin{cases} x - 2y = 10 \\ y \overset{\perp}{=} 3x - 20 \end{cases}$.

Now Try   Problem 13

---

2 **Solve Systems of Linear Equations by Graphing.**

To **solve a system** of equations means to find all of the solutions of the system. One way to solve a system of linear equations in two variables is to graph each equation and find where the graphs intersect.

The Graphing Method

1. Graph each equation on the same rectangular coordinate system.
2. Determine the coordinates of the point of intersection of the graphs. That ordered pair is the solution of the system.
3. If the graphs have no point in common, the system has no solution.
4. Check the proposed solution in each equation of the original system.

A system of two linear equations can have exactly one solution, no solution, or infinitely many solutions. When a system of equations (as in Example 2) has at least one solution, the system is called a **consistent system.**

EXAMPLE 2   Solve the system by graphing: $\begin{cases} x + 2y = 4 \\ 2x - y = 3 \end{cases}$

Strategy   We will graph both equations on the same coordinate system.

Why   The graph of a linear equation is a picture of its solutions. If both equations are graphed on the same coordinate system, we can see whether they have any common solutions.

Solution   The intercept method is a convenient way to graph equations such as $x + 2y = 4$ and $2x - y = 3$, because they are in standard $Ax + By = C$ form.

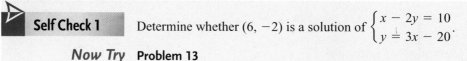

Success Tip

Since accuracy is crucial when using the graphing method to solve a system:
- Use graph paper.
- Use a sharp pencil.
- Use a straightedge.

$x + 2y = 4$

x	y	(x, y)
4	0	$(4, 0)$
0	2	$(0, 2)$
-2	3	$(-2, 3)$

$2x - y = 3$

x	y	(x, y)
$\frac{3}{2}$	0	$\left(\frac{3}{2}, 0\right)$
0	-3	$(0, -3)$
-1	-5	$(-1, -5)$

Although infinitely many ordered pairs (x, y) satisfy $x + 2y = 4$, and infinitely many ordered pairs (x, y) satisfy $2x - y = 3$, only the coordinates of the point where the graphs intersect satisfy both equations. From the graph, it appears that the intersection point has coordinates $(2, 1)$. To verify that it is the solution, we substitute 2 for x and 1 for y in both equations and show that $(2, 1)$ satisfies each one.

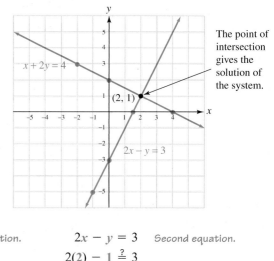

The point of intersection gives the solution of the system.

Check:

$x + 2y = 4$	First equation.	$2x - y = 3$	Second equation.
$2 + 2(1) \stackrel{?}{=} 4$		$2(2) - 1 \stackrel{?}{=} 3$	
$2 + 2 \stackrel{?}{=} 4$		$4 - 1 \stackrel{?}{=} 3$	
$4 = 4$	True	$3 = 3$	True

Since $(2, 1)$ makes both equations true, it is the solution of the system. The solution set is $\{(2, 1)\}$.

Self Check 2 Solve the system by graphing: $\begin{cases} x - 3y = -5 \\ 2x + y = 4 \end{cases}$

Now Try **Problem 21**

③ Use Graphing to Identify Inconsistent Systems and Dependent Equations.

When a system has no solution (as in Example 3), it is called an **inconsistent system.**

EXAMPLE 3 Solve the system $\begin{cases} 2x + 3y = 6 \\ 4x + 6y = 24 \end{cases}$ by graphing, if possible.

Strategy We will graph both equations on the same coordinate system.

Why If both equations are graphed on the same coordinate system, we can see whether they have any common solutions.

Solution Using the intercept method, we graph both equations on one set of coordinate axes, as shown on the right.

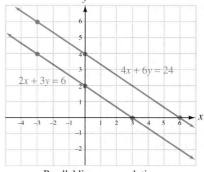

$2x + 3y = 6$

x	y	(x, y)
3	0	$(3, 0)$
0	2	$(0, 2)$
-3	4	$(-3, 4)$

$4x + 6y = 24$

x	y	(x, y)
6	0	$(6, 0)$
0	4	$(0, 4)$
-3	6	$(-3, 6)$

Parallel lines—no solution

In this example, the graphs are parallel, because the slopes of the two lines are equal and they have different y-intercepts. We can see that the slope of each line is $-\frac{2}{3}$ by writing each equation in slope–intercept form.

$2x + 3y = 6$	First equation.	$4x + 6y = 24$	Second equation.
$3y = -2x + 6$		$6y = -4x + 24$	
$y = -\dfrac{2}{3}x + 2$	Divide both sides by 3 and simplify.	$y = -\dfrac{2}{3}x + 4$	Divide both sides by 6 and simplify.

Because the lines are parallel, there is no point of intersection. Such a system has *no solution* and it is called an **inconsistent system.** The solution set is the empty set, which is written $\varnothing$.

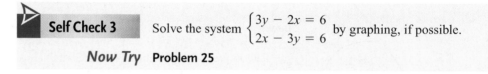

Self Check 3   Solve the system $\begin{cases} 3y - 2x = 6 \\ 2x - 3y = 6 \end{cases}$ by graphing, if possible.

Now Try   **Problem 25**

When the equations of a system have different graphs (as in Examples 2 and 3), the equations are called **independent equations.**

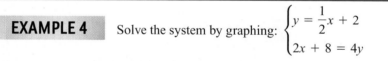

EXAMPLE 4   Solve the system by graphing: $\begin{cases} y = \dfrac{1}{2}x + 2 \\ 2x + 8 = 4y \end{cases}$

Strategy   We will graph both equations on the same coordinate system.

Why   If both equations are graphed on the same coordinate system, we can see whether they have any common solutions.

Solution   We graph each equation on one set of coordinate axes, as shown below.

Graph by using the slope and y-intercept.

$$y = \frac{1}{2}x + 2$$

$$m = \frac{1}{2} \qquad b = 2$$

Slope $= \dfrac{1}{2}$ y-intercept: $(0, 2)$

Graph by using the intercept method.

$$2x + 8 = 4y$$

x	y	(x, y)
-4	0	$(-4, 0)$
0	2	$(0, 2)$
2	3	$(2, 3)$

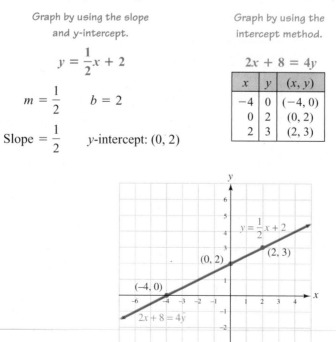

The same line — infinitely many solutions

The graphs appear to be identical. We can verify this by writing the second equation in slope–intercept form and observing that it is the same as the first equation.

$$y = \frac{1}{2}x + 2 \quad \textit{First equation.} \qquad\qquad 2x + 8 = 4y \quad \textit{Second equation.}$$

$$\frac{2x}{4} + \frac{8}{4} = \frac{4y}{4} \quad \textit{Divide both sides by 4.}$$

$$\frac{1}{2}x + 2 = y$$

We see that the equations of the system are equivalent. Because they are different forms of the same equation, they are called **dependent equations.**

Since the graphs are the same line, they have infinitely many points in common. The coordinates of each of those points satisfy both equations of the system. In cases like this, we say that there are *infinitely many solutions.* The solution set can be written using set-builder notation as

$$\left\{ (x, y) \,\middle|\, y = \frac{1}{2}x + 2 \right\} \quad \textit{Read as, "the set of all ordered pairs } (x, y)\textit{, such that } y = \tfrac{1}{2}x + 2\textit{."}$$

We can also express the solution set using the second equation of the system in the set-builder notation: $\{(x, y) \mid 2x + 8 = 4y\}$.

Some instructors prefer that the set-builder notation use an equation in standard form with coefficients that are integers having no common factor other than 1. Such an equation that is equivalent to $y = \frac{1}{2}x + 2$ and $2x + 8 = 4y$ is $x - 2y = -4$. The set-builder notation solution for this example could, therefore, be written as $\{(x, y) \mid x - 2y = -4\}$.

From the graph, it appears that three of the infinitely many solutions are $(-4, 0)$, $(0, 2)$, and $(2, 3)$. Check each of them to verify that both equations of the system are satisfied.

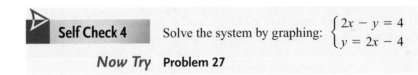

Self Check 4   Solve the system by graphing: $\begin{cases} 2x - y = 4 \\ y = 2x - 4 \end{cases}$

Now Try   **Problem 27**

We now summarize the possibilities that can occur when two linear equations, each with two variables, are graphed.

Solving a System of Equations by the Graphing Method

If the lines are different and intersect, the equations are independent, and the system is consistent. **One solution exists.** It is the point of intersection.

If the lines are different and parallel, the equations are independent, and the system is inconsistent. **No solution exists.**

If the lines coincide, the equations are dependent, and the system is consistent. **Infinitely many solutions exist.** Any point on the line is a solution.

If each equation in one system is equivalent to a corresponding equation in another system, the systems are called **equivalent systems.**

EXAMPLE 5   Solve the system by graphing: $\begin{cases} \dfrac{3}{2}x - y = \dfrac{5}{2} \\ x + \dfrac{1}{2}y = 4 \end{cases}$

Strategy   We will use the multiplication property of equality to clear both equations of fractions and solve the resulting equivalent system by graphing.

Why   It is usually easier to solve systems that do not contain fractions.

Solution   We multiply both sides of $\frac{3}{2}x - y = \frac{5}{2}$ by 2 to eliminate the fractions and obtain the equation $3x - 2y = 5$. We multiply both sides of $x + \frac{1}{2}y = 4$ by 2 to eliminate the fraction and obtain the equation $2x + y = 8$.

The original system   *An equivalent system*

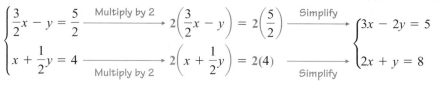

Since the new system is equivalent to the original system, they have the same solution. If we graph the equations of the new system, it appears that the point where the lines intersect is $(3, 2)$.

$3x - 2y = 5$

x	y	(x, y)
$\frac{5}{3}$	0	$\left(\frac{5}{3}, 0\right)$
0	$-\frac{5}{2}$	$\left(0, -\frac{5}{2}\right)$
1	-1	$(1, -1)$

$2x + y = 8$

x	y	(x, y)
4	0	$(4, 0)$
0	8	$(0, 8)$
1	6	$(1, 6)$

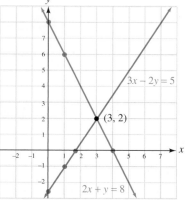

To verify that $(3, 2)$ is the solution, we substitute 3 for x and 2 for y in each equation of the original system.

Check:

$\dfrac{3}{2}x - y = \dfrac{5}{2}$ *First equation.*

$\dfrac{3}{2}(3) - 2 \overset{?}{=} \dfrac{5}{2}$

$\dfrac{9}{2} - \dfrac{4}{2} \overset{?}{=} \dfrac{5}{2}$

$\dfrac{5}{2} = \dfrac{5}{2}$ *True*

$x + \dfrac{1}{2}y = 4$ *Second equation.*

$3 + \dfrac{1}{2}(2) \overset{?}{=} 4$

$3 + 1 \overset{?}{=} 4$

$4 = 4$ *True*

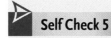

Self Check 5   Solve the system by graphing:
$$\begin{cases} \dfrac{1}{2}x + \dfrac{1}{2}y = -1 \\ \dfrac{1}{3}x - \dfrac{1}{2}y = -4 \end{cases}$$

Now Try   **Problem 29**

Using Your Calculator

Solving Systems by Graphing

The graphing method is limited to equations with two variables. Systems with three or more variables cannot be solved graphically. Also, it is often difficult to find exact solutions graphically. However, the TRACE and ZOOM capabilities of graphing calculators enable us to get very good approximations of such solutions.

To solve the system $\begin{cases} 3x + 2y = 12 \\ 2x - 3y = 12 \end{cases}$

with a graphing calculator, we must first solve each equation for y so that we can enter the equations into the calculator. After solving for y, we obtain the following equivalent system:

$$\begin{cases} y = -\dfrac{3}{2}x + 6 \\ y = \dfrac{2}{3}x - 4 \end{cases}$$

If we use window settings of $[-10, 10]$ for x and for y, the graphs of the equations will look like those in figure (a). If we zoom in on the intersection point of the two lines and trace, we will get an approximate solution like the one shown in figure (b). To get better results, we can do more zooms. We would then find that, to the nearest hundredth, the solution is $(4.63, -0.94)$. Verify that this is reasonable.

We can also find the intersection of two lines by using the INTERSECT feature found on most graphing calculators. To locate INTERSECT, press 2nd , CALC , 5, followed by ENTER . After graphing the lines and using INTERSECT, we obtain a graph similar to figure (c). The display shows the approximate coordinates of the point of intersection.

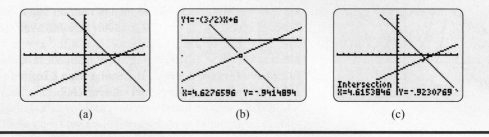

(a)   (b)   (c)

 Solve Equations Graphically.

The graphing method discussed in this section can be used to solve equations in one variable.

EXAMPLE 6   Solve the equation $2x + 4 = -2$ graphically.

Strategy   To find the solution of $2x + 4 = -2$, we will set y equal to the left side and y equal to the right side of the equation and solve the system $\begin{cases} y = 2x + 4 \\ y = -2 \end{cases}$ by graphing.

Why To solve $2x + 4 = -2$, we need to find the value of x that makes $2x + 4$ equal -2. That x-value will be equal to the x-coordinate of the point of intersection of the graphs of $y = 2x + 4$ and $y = -2$.

Solution The graphs of $y = 2x + 4$ and $y = -2$ are shown below. The point of intersection of the graphs is $(-3, -2)$. This indicates that if $x = -3$, the expression $2x + 4$ equals -2. So the solution of $2x + 4 = -2$ is -3. Verify that this result satisfies $2x + 4 = -2$.

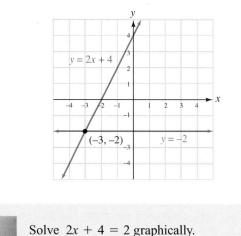

Self Check 6 Solve $2x + 4 = 2$ graphically.

Now Try **Problem 41**

Using Your Calculator | ***Solving Equations Graphically***

To solve $2(x - 3) + 3 = 7$ with a graphing calculator, we graph the left side and the right side of the equation in the same window by entering

$$Y_1 = 2(x - 3) + 3$$
$$Y_2 = 7$$

Figure (a) shows the graphs, generated using settings of $[-10, 10]$ for x and for y.

The coordinates of the point of intersection of the graphs can be determined using the INTERSECT feature found on most graphing calculators. With this feature, the cursor automatically highlights the intersection point, and the x- and y-coordinates are displayed.

In figure (b), we see that the point of intersection is $(5, 7)$, which indicates that 5 is a solution of $2(x - 3) + 3 = 7$.

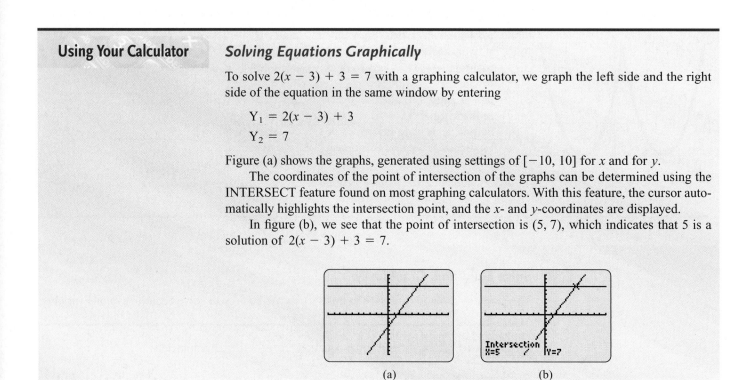

(a)   (b)

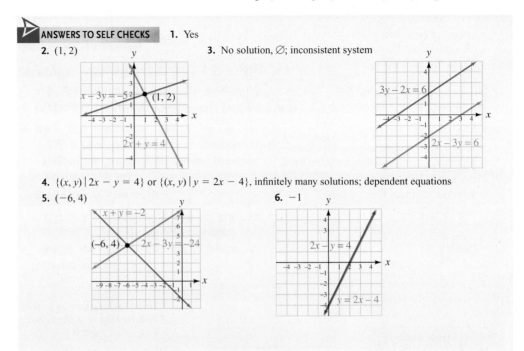

ANSWERS TO SELF CHECKS   **1.** Yes

2. $(1, 2)$

3. No solution, $\emptyset$; inconsistent system

4. $\{(x, y) \mid 2x - y = 4\}$ or $\{(x, y) \mid y = 2x - 4\}$, infinitely many solutions; dependent equations

5. $(-6, 4)$

6. -1

STUDY SET
3.1

VOCABULARY

Fill in the blanks.

1. $\begin{cases} x - 2y = 4 \\ 2x - y = 3 \end{cases}$ is called a _____ of linear equations.

2. When a system of equations has at least one solution, it is called a _____ system. If a system has no solutions, it is called an _____ system.

3. If two equations have different graphs, they are called _____ equations. Two equations with the same graph are called _____ equations.

4. When solving a system of two linear equations by the graphing method, we look for the point of _____ of the two lines.

CONCEPTS

5. Refer to the illustration. Decide whether a true or a false statement would be obtained

   a. when the coordinates of Point A are substituted into the equation for line l_1.

   b. when the coordinates of Point B are substituted into the equation for line l_1.

   c. when the coordinates of Point C are substituted into the equation for line l_1.

   d. when the coordinates of Point C are substituted into the equation for line l_2.

6. Refer to the illustration.

   a. How many ordered pairs satisfy the equation $3x + y = 3$? Name three.

   b. How many ordered pairs satisfy the equation $\frac{2}{3}x - y = -3$? Name three.

   c. How many ordered pairs satisfy both equations? Name it or them.

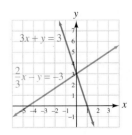

7. a. How many solutions does the
system of equations graphed on
the right have? Are the equations
dependent or independent?

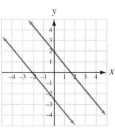

b. How many solutions does the
system of equations graphed on
the right have? Give three of the
solutions. Is the system consistent or inconsistent?

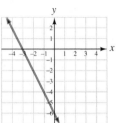

8. Estimate the solution of the system of linear equations
shown in the following display.

Use the graphs in the illustration to solve each equation.

9. $-3x + 2 = x - 2$ **10.** $-x - 4 = x - 2$

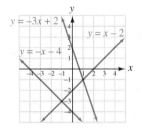

NOTATION

Fill in the blanks.

11. The symbol { is called a left _____. It is used when writing a
system of equations.

12. We read the set-builder notation $\{(x, y)\,|\,3x - 5y = 1\}$ as "the
____ of all ordered pairs (x, y) _____ that $3x - 5y = 1$."

GUIDED PRACTICE

*Determine whether the ordered pair is a solution of the system of
equations. See Example 1.*

13. $(-4, 3)$; $\begin{cases} 4x - y = -19 \\ 3x + 2y = -6 \end{cases}$ **14.** $(-1, 2)$; $\begin{cases} 3x - y = -5 \\ x - y = -4 \end{cases}$

15. $(2, -3)$; $\begin{cases} y + 2 = \dfrac{1}{2}x \\ 3x + 2y = 0 \end{cases}$ **16.** $(1, 2)$; $\begin{cases} 2x - y = 0 \\ y = \dfrac{1}{2}x + \dfrac{3}{2} \end{cases}$

17. $\left(\dfrac{1}{2}, \dfrac{1}{3}\right)$; $\begin{cases} 2x + 3y = 2 \\ 4x - 9y = 1 \end{cases}$ **18.** $\left(-\dfrac{3}{4}, \dfrac{2}{3}\right)$; $\begin{cases} 4x + 3y = -1 \\ 4x - 3y = -5 \end{cases}$

19. $(-0.2, 0.5)$; $\begin{cases} 2x + 5y = 2.1 \\ 5x + y = -0.5 \end{cases}$ **20.** $(2.1, -3.2)$; $\begin{cases} x + y = -1.1 \\ 2x - 3y = 13.8 \end{cases}$

Solve each system by graphing. See Example 2.

21. $\begin{cases} x + y = 6 \\ x - y = 2 \end{cases}$ **22.** $\begin{cases} x - y = 4 \\ 2x + y = 5 \end{cases}$

23. $\begin{cases} y = -2x + 1 \\ x - 2y = -7 \end{cases}$ **24.** $\begin{cases} 3x - y = -3 \\ y = -2x - 7 \end{cases}$

*Solve each system by graphing, if possible. If a system is
inconsistent or if the equations are dependent, state this. See
Examples 3 and 4.*

25. $\begin{cases} 3x - 3y = 4 \\ x - y = 4 \end{cases}$ **26.** $\begin{cases} 5x + 2y = 6 \\ -10x - 4y = -12 \end{cases}$

27. $\begin{cases} x = 3 - 2y \\ 2x + 4y = 6 \end{cases}$ **28.** $\begin{cases} 3x = 5 - 2y \\ 3x + 2y = 7 \end{cases}$

Solve each system by graphing. See Example 5.

29. $\begin{cases} \dfrac{1}{6}x = \dfrac{1}{3}y + \dfrac{1}{2} \\ y = x \end{cases}$ **30.** $\begin{cases} x = y + 3 \\ \dfrac{1}{4}x - \dfrac{1}{6}y = \dfrac{1}{3} \end{cases}$

31. $\begin{cases} \dfrac{1}{3}x - \dfrac{7}{6}y = \dfrac{1}{2} \\ \dfrac{1}{5}y = \dfrac{1}{3}x + \dfrac{7}{15} \end{cases}$ **32.** $\begin{cases} \dfrac{3}{5}x + \dfrac{1}{4}y = -\dfrac{11}{10} \\ \dfrac{1}{8}x = \dfrac{13}{24} + \dfrac{1}{3}y \end{cases}$

Solve each equation graphically. See Example 6.

33. $2x + 1 = 5$ **34.** $3x + 5 = -4$

35. $-2x + 8 = 3x - 7$ **36.** $2x - 3 = 3x - 3$

*Use a graphing calculator to solve each system. Give all
answers to the nearest hundredth. See Using Your
Calculator: Solving Systems by Graphing.*

37. $\begin{cases} y = 3.2x - 1.5 \\ y = -2.7x - 3.7 \end{cases}$ **38.** $\begin{cases} y = -0.45x + 5 \\ y = 5.55x - 13.7 \end{cases}$

39. $\begin{cases} 1.7x + 2.3y = 3.2 \\ y = 0.25x + 8.95 \end{cases}$

40. $\begin{cases} 2.75x = 12.9y - 3.79 \\ 7.1x - y = 35.76 \end{cases}$

61. $\begin{cases} x + 3y = 6 \\ y = -\dfrac{1}{3}x + 2 \end{cases}$

62. $\begin{cases} 2x - y = -4 \\ 2y = 4x - 6 \end{cases}$

Use a graphing calculator to solve each equation.

41. $4x - 4 = 3x$

42. $4(x - 3) - x = x - 6$

43. $11x + 6(3 - x) = 3$

44. $2x + 4 = 12 - 2x$

63. $\begin{cases} x = -\dfrac{3}{2}y \\ 2x = 3y - 4 \end{cases}$

64. $\begin{cases} 4x = 3y - 1 \\ 3y = 4 - 8x \end{cases}$

TRY IT YOURSELF

Solve each system by graphing, if possible. If a system is inconsistent or if the equations are dependent, state this.   **(Hint:** *Several coordinates of points of intersection are fractions.)*

45. $\begin{cases} y = -\dfrac{5}{2}x + \dfrac{1}{2} \\ 2x - \dfrac{3}{2}y = 5 \end{cases}$

46. $\begin{cases} \dfrac{5}{2}x + 3y = 6 \\ y = -\dfrac{5}{6}x + 2 \end{cases}$

47. $\begin{cases} x + y = 0 \\ y = 2x - 6 \end{cases}$

48. $\begin{cases} 3x + y = 3 \\ 3x + 2y = 0 \end{cases}$

49. $\begin{cases} y = 3 \\ x = 2 \end{cases}$

50. $\begin{cases} 2x + 3y = -15 \\ 2x + y = -9 \end{cases}$

51. $\begin{cases} x = \dfrac{11 - 2y}{3} \\ y = \dfrac{11 - 6x}{4} \end{cases}$

52. $\begin{cases} x = \dfrac{1 - 3y}{4} \\ y = \dfrac{12 + 3x}{2} \end{cases}$

53. $\begin{cases} 4x - 3y = 5 \\ y = -2x \end{cases}$

54. $\begin{cases} 2x + 2y = -1 \\ 3x + 4y = 0 \end{cases}$

55. $\begin{cases} x = 13 - 4y \\ 3x = 4 + 2y \end{cases}$

56. $\begin{cases} 3x = 7 - 2y \\ 2x = 2 + 4y \end{cases}$

57. $\begin{cases} x = 2 \\ y = -\dfrac{1}{2}x + 2 \end{cases}$

58. $\begin{cases} y = -2 \\ y = \dfrac{2}{3}x - \dfrac{4}{3} \end{cases}$

59. $\begin{cases} x = \dfrac{5}{2}y - 2 \\ x - \dfrac{5}{3}y + \dfrac{1}{3} = 0 \end{cases}$

60. $\begin{cases} 2x = 5y - 11 \\ 3x = 2y \end{cases}$

APPLICATIONS

65. MAPS

   a. In the following map, what New Mexico city lies on the intersection of Interstate 25 and Interstate 40?

   b. What New Mexico city lies on the intersection of Interstate 25 and Interstate 10?

66. BUSINESS   Estimate the break-even point (where cost = revenue) on the graph in the illustration. Explain why it is called the *break-even point*.

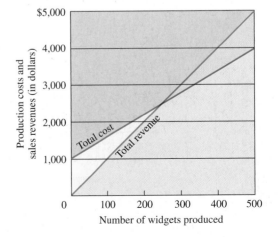

67. HEARING TESTS See the illustration. At what frequency and decibel level were the hearing test results the same for the left and right ear? Write your answer as an ordered pair.

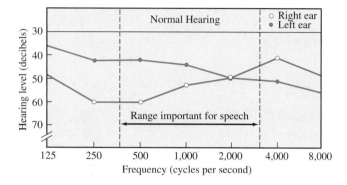

68. LAW OF SUPPLY AND DEMAND The demand function, graphed below, describes the relationship between the price x of a certain camera and the demand for the camera.

a. The supply function, $S(x) = \frac{25}{4}x - 525$, describes the relationship between the price x of the camera and the number of cameras the manufacturer is willing to supply. Graph this function in the illustration.

b. For what price will the supply of cameras equal the demand?

c. As the price of the camera is increased, what happens to supply and what happens to demand?

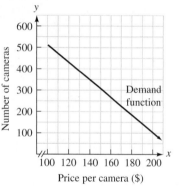

69. THE INTERNET

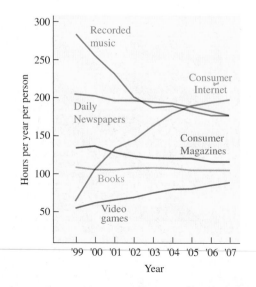

The graph in the previous column shows the growing importance of the Internet in the daily lives of Americans. Determine when the time spent on the following activities was the same. Approximately how many hours per year were spent on each?

a. Internet and reading magazines

b. Internet and reading newspapers

c. Internet and reading books

d. Reading newspapers and listening to recorded music

70. COST AND REVENUE The function $C(x) = 200x + 400$ gives the cost for a college to offer x sections of an introductory class in CPR (cardiopulmonary resuscitation). The function $R(x) = 280x$ gives the amount of revenue the college brings in when offering x sections of CPR.

a. Find the break-even point (where cost = revenue) by graphing each function on the same coordinate system.

b. How many sections does the college need to offer to make a profit on the CPR training course?

71. NAVIGATION The paths of two ships are tracked on the same coordinate system. One ship is following a path described by the equation $2x + 3y = 6$, and the other is following a path described by the equation $y = \frac{2}{3}x - 3$.

a. Graph both equations on the same coordinate system. Is there a possibility of a collision?

b. Is a collision a certainty?

72. AIR TRAFFIC CONTROL Two airplanes, flying at the same altitude, are tracked using the same coordinate system on a radar screen. One plane is following a path described by the equation $y = 0.4x - 2$, and the other is following a path described by the equation $2x = 5y + 7$. Graph both equations on the same coordinate system. Is there a possibility of a collision?

WRITING

73. Suppose the solution of a system of two linear equations is $\left(\frac{14}{5}, -\frac{8}{3}\right)$. Knowing this, explain any drawbacks you might encounter when solving the system by the graphing method.

74. Can a system of two linear equations have exactly two solutions? Why or why not?

75. Suppose the graphs of the two linear equations of a system are the same line. What is wrong with this statement? *The system has infinitely many solutions. Any ordered pair is a solution of the system.*

76. a. Without graphing, how can you tell that the graphs of $y = 2x + 1$ and $y = 3x + 2$ intersect?

b. Without graphing, how can you tell that the graphs of $y = 2x + 1$ and $y = 2x + 2$ do not intersect?

c. Without graphing, how can you tell that the graphs of $y = 2x + 3$ and $2y = 4x + 6$ are the same line?

REVIEW

Let $f(x) = -x^3 + 2x - 2$ *and* $g(x) = \frac{2-x}{9+x}$ *and find each value.*

77. $f(-1)$ **78.** $f(10)$

79. $g(2)$ **80.** $g(-20)$

81. $f(t)$ **82.** $g(s+1)$

83. Find the domain of the function $f(x) = \dfrac{1}{3x+6}$.

84. Find the domain of the function $f(x) = |x|$.

CHALLENGE PROBLEMS

85. Write a dependent system of equations with a solution of $(-5, 2)$.

86. Write a system of *three* independent linear equations in two variables that has the solution $(-5, 2)$.

SECTION 3.2
Solving Systems of Equations Algebraically

Objectives

① Solve systems of linear equations by substitution.

② Solve systems of linear equations by the elimination (addition) method.

③ Use substitution and elimination (addition) to identify inconsistent systems and dependent equations.

④ Determine the most efficient method to use to solve a linear system.

The graphing method enables us to visualize the process of solving systems of equations. However, it can be difficult to determine the exact coordinates of the point of intersection. We now discuss two other methods we can use to find the exact solutions of systems of equations.

① **Solve Systems of Linear Equations by Substitution.**

The substitution method works well for solving systems where one equation is solved, or can be easily solved, for one of the variables. To solve a system of two linear equations in x and y by the substitution method, we can follow these steps.

The Substitution Method	1. Solve one of the equations for either x or y—preferably a variable with a coefficient of 1 or -1. If this is already done, go to step 2. (We call this equation the **substitution equation.**)
	2. Substitute the expression for x or for y obtained in step 1 into the other equation and solve that equation.
	3. Substitute the value of the variable found in step 2 into the substitution equation to find the value of the remaining variable.
	4. Check the proposed solution in each equation of the original system. Write the solution as an ordered pair.

EXAMPLE 1   Solve the system by substitution: $\begin{cases} 4x + y = 13 \\ -2x + 3y = -17 \end{cases}$

Strategy   We will use the substitution method. Since the system does not contain an equation solved for x or y, we must choose an equation and solve it for x or y. It is easiest to solve for y in the first equation, because y has a coefficient of 1.

Why   Solving $4x + y = 13$ for x or solving $-2x + 3y = -17$ for x or y would involve working with cumbersome fractions.

Solution

Step 1: We solve the first equation for y, because y has a coefficient of 1.

$$4x + y = 13$$

$$y = -4x + 13 \qquad \text{To isolate } y \text{, subtract } 4x \text{ from both sides.}$$
$$\text{This is the substitution equation.}$$

The Language of Algebra
Since substitution involves algebra and not graphing, it is called an *algebraic* method for solving a system.

Because y and $-4x + 13$ are equal, we can substitute $-4x + 13$ for y in the second equation of the system.

$$y = \underset{\bigcirc}{(-4x + 13)} \qquad -2x + 3y = -17$$

Success Tip
With this method, the objective is to use an appropriate substitution to obtain *one* equation in *one* variable.

Step 2: We then substitute $-4x + 13$ for y in the second equation to eliminate the variable y from that equation. The result will be an equation containing only one variable, x.

$$-2x + 3y = -17 \qquad \text{This is the second equation of the system.}$$

$$-2x + 3(-4x + 13) = -17 \qquad \text{Substitute } -4x + 13 \text{ for } y.$$
$$\text{Write the parentheses so that the multiplication by 3 is}$$
$$\text{distributed over both terms of } -4x + 13.$$

$$-2x - 12x + 39 = -17 \qquad \text{Distribute the multiplication by 3.}$$

$$-14x + 39 = -17 \qquad \text{Combine like terms.}$$

$$-14x = -56 \qquad \text{Subtract 39 from both sides.}$$

$$x = 4 \qquad \text{To solve for } x \text{, divide both sides by } -14.$$
$$\text{This is the } x\text{-value of the solution.}$$

The Language of Algebra
The phrase *back-substitute* can also be used to describe step 3 of the substitution method. To find y, we *back-substitute* 4 for x in the equation $y = -4x + 13$.

Step 3: To find y, we substitute 4 for x in the substitution equation and evaluate the right side.

$$y = -4x + 13 \qquad \text{This is the substitution equation.}$$

$$= -4(4) + 13 \qquad \text{Substitute 4 for } x.$$

$$= -16 + 13$$

$$= -3 \qquad \text{This is the } y\text{-value of the solution.}$$

Step 4: To verify that $(4, -3)$ satisfies both equations, we substitute 4 for x and -3 for y into each equation of the original system and simplify.

Check:

$4x + y = 13$ First equation.	$-2x + 3y = -17$ Second equation.
$4(4) + (-3) \stackrel{?}{=} 13$	$-2(4) + 3(-3) \stackrel{?}{=} -17$
$16 - 3 \stackrel{?}{=} 13$	$-8 - 9 \stackrel{?}{=} -17$
$13 = 13$ True	$-17 = -17$ True

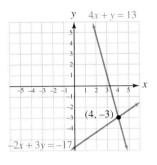

$4x + y = 13$

$(4, -3)$

$-2x + 3y = -17$

Since $(4, -3)$ satisfies both equations of the system, it is the solution of the system. The solution set is $\{(4, -3)\}$. The graphs of the equations of the system help to verify this—they appear to intersect at $(4, -3)$, as shown on the left.

Self Check 1   Solve the system by substitution: $\begin{cases} x + 3y = 9 \\ 2x - y = -10 \end{cases}$

Now Try   **Problem 17**

EXAMPLE 2   Solve the system by substitution: $\begin{cases} \dfrac{2}{9}x - \dfrac{2}{9}y = \dfrac{2}{3} \\ 0.1x = 0.2 - 0.1y \end{cases}$

Strategy   We will find an equivalent system without fractions or decimals and use the substitution method to solve it.

Why   It's usually easier to solve a system of equations that involves only integers.

Solution   To clear the first equation of fractions, we multiply both sides by 9, which is the least common denominator of the fractions in the equation. To clear the second equation of the decimals, we multiply both sides by 10.

The original system   *An equivalent system*

$\begin{cases} \dfrac{2}{9}x - \dfrac{2}{9}y = \dfrac{2}{3} \\ 0.1x = 0.2 - 0.1y \end{cases}$ $\xrightarrow[\text{Multiply by 10}]{\text{Multiply by 9}}$ $9\left(\dfrac{2}{9}x - \dfrac{2}{9}y\right) = 9\left(\dfrac{2}{3}\right)$ $\xrightarrow[\text{Simplify}]{\text{Simplify}}$ $\begin{cases} 2x - 2y = 6 \\ x = 2 - y \end{cases}$

$10(0.1x) = 10(0.2 - 0.1y)$

Notation

We number the equations (1) and (2) to help describe how the system is solved using the substitution method.

These results form the following equivalent system, which has the same solution as the original one.

(1) $\begin{cases} 2x - 2y = 6 \\ x = 2 - y \end{cases}$

(2)   *This is the substitution equation.*

Since the variable x is isolated in equation 2, we will substitute $2 - y$ for x in equation 1. This step will eliminate x from equation 1, leaving an equation containing only one variable, y. We then solve for y.

$2x - 2y = 6$ *This is equation 1 of the equivalent system.*

$2(2 - y) - 2y = 6$ *Substitute $2 - y$ for x.*

$4 - 2y - 2y = 6$ *Distribute the multiplication by 2.*

$-4y = 2$ *Combine like terms and subtract 4 from both sides.*

$y = -\dfrac{1}{2}$ *To solve for y, divide both sides by -4 and simplify the fraction $-\dfrac{2}{4}$. This is the y-value of the solution.*

Caution

Always use the *original* equations when checking a solution. Do not use a substitution equation that you found algebraically. If an error was made, a proposed solution that would not satisfy the original system might appear to be correct.

We can find x by substituting $-\dfrac{1}{2}$ for y in equation 2 and simplifying:

$x = 2 - y$ *This is the substitution equation.*

$x = 2 - \left(-\dfrac{1}{2}\right)$ *Substitute $-\dfrac{1}{2}$ for y.*

$$= 2 + \frac{1}{2}$$

$$= \frac{5}{2} \qquad \text{Do the addition: } 2 + \frac{1}{2} = \frac{4}{2} + \frac{1}{2} = \frac{5}{2}. \text{ This is the x-value of the solution.}$$

The solution is $\left(\frac{5}{2}, -\frac{1}{2}\right)$ and the solution set is $\left\{\left(\frac{5}{2}, -\frac{1}{2}\right)\right\}$. Verify that this result satisfies both equations of the original system.

 Self Check 2   Solve the system by substitution: $\begin{cases} \dfrac{x}{8} + \dfrac{y}{4} = \dfrac{1}{2} \\ 0.01y = -0.02x + 0.04 \end{cases}$

Now Try   **Problem 19**

②   Solve Systems of Linear Equations by the Elimination (Addition) Method.

The substitution method for solving a system of equations can be difficult to use if none of the variables has a coefficient of 1 or -1. This is the case for the system

$$\begin{cases} 5x + 3y = 2 \\ 2x - 3y = -16 \end{cases}$$

Solving either equation for x or y involves working with cumbersome fractions. Fortunately, we can solve systems like this one using an easier method called the **elimination** or the **addition method.**

The elimination (addition) method for solving a system is based on the **addition property of equality:** *When equal quantities are added to both sides of an equation, the results are equal.* In symbols, if $A = B$ and $C = D$, then adding the left sides and the right sides of these equations, we have $A + C = B + D$. This procedure is called *adding the equations.*

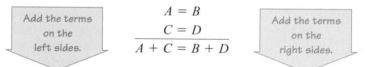

With the elimination method, we add the equations in a way that will eliminate the terms involving one of the variables.

To solve a system of linear equations in x and y by the elimination (addition) method, we can use the following steps.

The Elimination (Addition) Method

1. Write both equations of the system in standard (general) form: $Ax + By = C$.

2. If necessary, multiply one or both of the equations by a nonzero number chosen to make the coefficients of x (or the coefficients of y) opposites.

3. Add the equations to eliminate the terms involving x (or y).

4. Solve the equation resulting from step 3.

5. Find the value of the remaining variable by substituting the solution found in step 4 into any equation containing both variables. Or, repeat steps 2–4 to eliminate the other variable.

6. Check the proposed solution in each equation of the original system. Write the solution as an ordered pair.

EXAMPLE 3   Solve the system by elimination: $\begin{cases} 5x + 3y = 2 \\ 2x - 3y = -16 \end{cases}$

Strategy   Since the coefficients of the y-terms are opposites, we will add the left and right sides of the given equations to eliminate y.

Why   When we add the equations in this way, the result will be an equation that contains only one variable, x.

Solution

Step 1: Since both equations are already written in standard $Ax + By = C$ form, we can move to the next step.

Step 2: Because the coefficients of the terms $3y$ in the first equation and $-3y$ in the second equation are opposites, we can move to the next step.

Step 3:   We can add the left sides and add the right sides of the given equations as shown below.

$5x + 3y = 2$

$\underline{2x - 3y = -16}$ To add the equations, add the like

$7x \quad\quad = -14$ terms, column by column.

$2 + (-16) = -14$

$3y + (-3y) = 0$

$5x + 2x = 7x$

Step 4: Because the sum of the terms $3y$ and $-3y$ is 0, we say that the variable y has been *eliminated*. Since the resulting equation has only one variable, we can solve it for x.

$$7x = -14$$

$$x = -2 \quad \text{To solve for x, divide both sides by 7.}$$

This is the x-value of the solution.

Step 5: To find the y-value of the solution, substitute -2 for x in either equation of the original system.

$$5x + 3y = 2 \quad \text{This is the first equation of the system.}$$

$$5(-2) + 3y = 2 \quad \text{Substitute } -2 \text{ for x.}$$

$$-10 + 3y = 2 \quad \text{Multiply.}$$

$$3y = 12 \quad \text{Add 10 to both sides.}$$

$$y = 4 \quad \text{To solve for y, divide both sides by 3.}$$

This is the y-value of the solution.

Step 6: Now we check the proposed solution $(-2, 4)$ in the equations of the original system.

Check: $5x + 3y = 2$ First equation. $2x - 3y = -16$ Second equation.

$5(-2) + 3(4) \stackrel{?}{=} 2$ $2(-2) - 3(4) \stackrel{?}{=} -16$

$-10 + 12 \stackrel{?}{=} 2$ $-4 - 12 \stackrel{?}{=} -16$

$2 = 2$ True $-16 = -16$ True

Since $(-2, 4)$ satisfies both equations, it is the solution. The solution set is written $\{(-2, 4)\}$. The graphs of the equations of the system help to verify this—they appear to intersect at $(-2, 4)$, as shown on the left.

Notation

When a linear equation is written in standard $Ax + By = C$ form, the *variable terms* are on the left side of the equal symbol and the *constant* term is on the right side.

The Language of Algebra

In the equations $5x + 3y = 2$ and $2x - 3y = -16$, the coefficients of the y-terms, 3 and -3, are *opposites* or *additive inverses*.

Caution

When using the substitution or elimination method to solve a system of equations, a common error is to find the value of one of the variables, say x, and forget to find the value of the other. Remember that a solution of a linear system of two equations is an ordered pair (x, y).

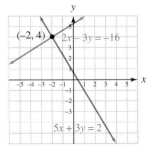

▷ **Self Check 3**   Solve the system by elimination: $\begin{cases} 6x - 7y = 8 \\ 2x + 7y = -16 \end{cases}$

Now Try   **Problem 23**

For many systems, we cannot immediately eliminate a variable by adding. In such cases, we will use the multiplication property of equality to create coefficients that are opposites.

EXAMPLE 4   Solve the system by elimination: $\begin{cases} 9a + 5b = 16 \\ 3a - 4b = -6 \end{cases}$

Strategy   To use the elimination method to solve this system, we will multiply both sides of the second equation by -3 and add the equations to eliminate the terms involving the variable a.

Why   This will give one equation involving only the variable b.

Solution

Step 1: Since both equations are written in standard (general) form, we can move to the next step.

Step 2: The coefficient of a in the first equation is 9. If we multiply both sides of the second equation by -3, the coefficient of a in that equation will be -9. Then the coefficients of a will differ only in sign.

$$\begin{cases} 9a + 5b = 16 \\ 3a - 4b = -6 \end{cases} \xrightarrow[\text{Multiply by } -3]{\text{Unchanged}} \begin{array}{l} 9a + 5b = 16 \\ -3(3a - 4b) = -3(-6) \end{array} \xrightarrow[\text{Simplify}]{\text{Unchanged}} \begin{cases} 9a + 5b = 16 \\ -9a + 12b = 18 \end{cases}$$

Step 3: When the resulting equations are added, the terms involving a drop out (or are eliminated), and we get an equation that contains only the variable b. We then proceed by solving for b.

$$\begin{array}{r} 9a + 5b = 16 \\ \underline{-9a + 12b = 18} \\ 17b = 34 \end{array}$$

Add the like terms, column by column:
$9a + (-9a) = 0$, $5b + 12b = 17b$, and $16 + 18 = 34$.

Step 4: Since the resulting equation has only one variable, we can solve it for b.

$$17b = 34$$
$$b = 2$$

To solve for b, divide both sides by 17.
This is the b-value of the solution.

Step 5: To find a, we can substitute 2 for b in either of the equations of the original system, or in $-9a + 12b = 18$. The computations will be simple if we use $9a + 5b = 16$.

$$9a + 5b = 16$$ This is the first equation of the original system.
$$9a + 5(2) = 16$$ Substitute 2 for b.
$$9a + 10 = 16$$
$$9a = 6$$ To isolate the variable term, subtract 10 from both sides.
$$a = \frac{2}{3}$$ To solve for a, divide both sides by 9 and simplify the fraction $\frac{6}{9}$.
This is the a-value of the solution.

Success Tip
We choose to eliminate a because the coefficient 9 of the term $9a$ is a *multiple* of the coefficient 3 of the term $3a$. The same cannot be said for 5 and -4, the coefficients of b.

Success Tip
It doesn't matter which variable is eliminated first. We don't have to find the values of the variables in alphabetical order. Choose the one that is the easier to eliminate. The basic objective of this method is to obtain two equations whose sum will be *one* equation in *one* variable.

Step 6: Verify that the solution is $\left(\frac{2}{3}, 2\right)$ by substituting $\frac{2}{3}$ for a and 2 for b in the original equations.

▷ **Self Check 4**   Solve the system by elimination: $\begin{cases} 15a - 2b = -9 \\ 5a - 5b = -29 \end{cases}$

Now Try **Problems 25**

EXAMPLE 5   Solve the system by elimination: $\begin{cases} 4x = 3(2 + y) \\ 3(x - 10) = -2y \end{cases}$

Strategy   We will write each equation in standard (general) form $Ax + By = C$ and use the elimination (addition) method to solve the resulting equivalent system.

Why   In their current form, the equations do not contain terms with coefficients that are opposites.

Solution   To write each equation in standard (general) form, we use the distributive property to remove the parentheses and then rearrange the terms.

The first equation	**The second equation**
$4x = 3(2 + y)$	$3(x - 10) = -2y$
$4x = 6 + 3y$	$3x - 30 = -2y$
$4x - 3y = 6$	$3x + 2y = 30$ This is $Ax + By = C$ form.

We proceed by solving the following equivalent system

(1) $\begin{cases} 4x - 3y = 6 \\ \end{cases}$
(2) $\begin{cases} 3x + 2y = 30 \end{cases}$

Since the coefficients of y have *opposite signs,* we choose to eliminate y. To make the y-terms drop out when we add the equations, we multiply both sides of equation 1 by 2 and both sides of equation 2 by 3.

$$\begin{cases} 4x - 3y = 6 \\ 3x + 2y = 30 \end{cases} \xrightarrow[\text{Multiply by 3}]{\text{Multiply by 2}} \begin{cases} 2(4x - 3y) = 2(6) \\ 3(3x + 2y) = 3(30) \end{cases} \xrightarrow{\text{Simplify}} \begin{cases} 8x - 6y = 12 \\ 9x + 6y = 90 \end{cases}$$

When the results are added, the y-terms drop out and we can solve for x.

$$\begin{array}{rl} 8x - 6y = 12 & \text{Add like terms, column by column:} \\ \underline{9x + 6y = 90} & 8x + 9x = 17x, \ -6y + 6y = 0, \text{ and } 12 + 90 = 102. \\ 17x \qquad = 102 & \\ x = 6 & \text{To solve for } x, \text{ divide both sides by 17.} \end{array}$$

To find y, we can substitute 6 for x in any equation involving both variables. If we substitute 6 for x in equation 2, we get

$$\begin{array}{rl} 3x + 2y = 30 & \\ 3(6) + 2y = 30 & \text{Substitute 6 for } x. \\ 18 + 2y = 30 & \text{Perform the multiplication.} \\ 2y = 12 & \text{Subtract 18 from both sides.} \\ y = 6 & \text{Divide both sides by 2.} \end{array}$$

The solution is (6, 6). Check this result by substituting 6 for x and 6 for y in the equations of the original system.

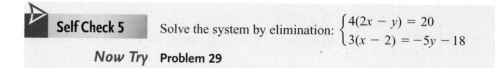

Self Check 5   Solve the system by elimination: $\begin{cases} 4(2x - y) = 20 \\ 3(x - 2) = -5y - 18 \end{cases}$

Now Try   **Problem 29**

3 **Use Substitution and Elimination (Addition) to Identify Inconsistent Systems and Dependent Equations.**

We have solved inconsistent systems and systems of dependent equations by graphing. We can also solve these systems using the substitution and elimination methods.

EXAMPLE 6   Solve the system $\begin{cases} y = 2x + 4 \\ 8x - 4y = 7 \end{cases}$, if possible.

Strategy   We will use the substitution method to solve this system.

Why   The substitution method works well when one of the equations of the system (in this case, $y = 2x + 4$) is solved for a variable.

Solution   Since the first equation is solved for y, we will use the substitution method.

$$y = 2x + 4 \qquad \text{This is the substitution equation.}$$
$$8x - 4y = 7 \qquad \text{This is the second equation of the system.}$$
$$8x - 4(2x + 4) = 7 \qquad \text{Substitute } 2x + 4 \text{ for } y.$$

We can now try to solve this equation for x:

$$8x - 8x - 16 = 7 \qquad \text{Distribute the multiplication by } -4.$$
$$-16 = 7 \qquad \text{Simplify the left side: } 8x - 8x = 0.$$

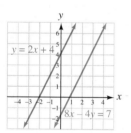

Here, the terms involving x drop out, and we get $-16 = 7$. This false statement indicates that the system has *no solution* and is, therefore, inconsistent. The solution set is $\varnothing$. The graphs of the equations of the system help to verify this—they appear to be parallel lines, as shown on the left.

Self Check 6   Solve the system, if possible: $\begin{cases} x = -2.5y + 8 \\ y = -0.4x + 2 \end{cases}$

Now Try   **Problem 33**

EXAMPLE 7   Solve the system: $\begin{cases} 4x + 6y = 12 \\ -2x - 3y = -6 \end{cases}$

Strategy   Since both equations are written in standard form, we will use the elimination method.

Why   Since no variable has a coefficient 1 or −1, it would be difficult to solve this system using substitution.

Solution We can copy the first equation and multiply both sides of the second equation by 2 to get

$$\begin{array}{rcl} 4x + 6y &=& 12 \\ -4x - 6y &=& -12 \\ \hline 0 &=& 0 \end{array}$$

When we add like terms, column by column, the result is $0x + 0y = 0$, which simplifies to $0 = 0$.

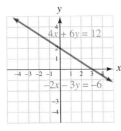

Here, both the x- and y-terms drop out. The resulting true statement $0 = 0$ indicates that the equations are dependent and that the system has *infinitely many solutions*. The solution set is written using set-builder notation as $\{(x, y) \mid 4x + 6y = 12\}$ and is read as "the set of all ordered pairs (x, y) such that $4x + 6y = 12$."

Note that the equations of the system are dependent equations, because when the second equation is multiplied by -2, it becomes the first equation. The graphs of these equations are, therefore, the same line. To find some of the infinitely many solutions of the system, we can substitute 0, 3, and -3 for x in either equation to obtain $(0, 2)$, $(3, 0)$, and $(-3, 4)$.

 Self Check 7 Solve the system: $\begin{cases} 2x - 5y = 19 \\ -\dfrac{2}{5}x + y = -\dfrac{19}{5} \end{cases}$

Now Try **Problem 35**

Examples 6 and 7 illustrate the following facts.

Inconsistent Systems and Dependent Equations	When solving a system of two linear equations in two variables using substitution or elimination (addition):

1. If the variables drop out and a true statement (identity) is obtained, the system has an infinite number of solutions. The equations are dependent and the system is consistent.

2. If the variables drop out and a false statement (contradiction) is obtained, the system has no solution and is inconsistent.

4 **Determine the Most Efficient Method to Use to Solve a Linear System.**

If no method is specified for solving a particular linear system, the following guidelines can be helpful in determining whether to use graphing, substitution, or elimination.

1. If you want to show trends and see the point that two graphs have in common, use the **graphing method.** However, this method can be lengthy and is not exact.

2. If one of the equations is solved for one of the variables, or easily solved for one of the variables, use the **substitution method.**

3. If both equations are in standard (general) $Ax + By = C$ form, and no variable has a coefficient of 1 or -1, use the **elimination (addition) method.**

4. If the coefficient of one of the variables is 1 or -1, you have a choice. You can write each equation in standard form $(Ax + By = C)$ and use elimination, or you can solve for the variable with coefficient 1 or -1 and use substitution.

Here are some examples of suggested approaches:

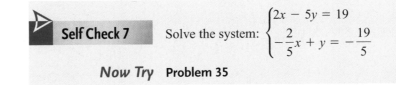

$\begin{cases} 5x + 6y = 1 \\ y = x - 4 \end{cases}$	$\begin{cases} a - 16 = 5b \\ 4a - 8b = -11 \end{cases}$	$\begin{cases} 3x + 2y = 17 \\ 7x + 5y = -33 \end{cases}$	$\begin{cases} 3x - y = -1 \\ 9x + 8y = 2 \end{cases}$
Substitution	Substitution	Elimination	Elimination or substitution

STUDY SET
3.2

VOCABULARY

Fill in the blanks.

1. $Ax + By = C$ is the _____ form of a linear equation.

2. In the equation $x + 3y = -1$, the x-term has an understood _____ of 1.

3. When we add the two equations of the system $\begin{cases} x + y = 5 \\ x - y = -3 \end{cases}$, the y-terms are _____.

4. To solve $\begin{cases} y = 3x \\ x + y = 4 \end{cases}$, we can _____ $3x$ for y in the second equation.

CONCEPTS

5. If the system $\begin{cases} 4x - 3y = 7 \\ 3x - y = 6 \end{cases}$ is to be solved using the substitution method, what variable in what equation would it be easier to solve for?

6. Given the equation $3x + y = -4$,
   a. solve for x.
   b. solve for y.
   c. Which variable was easier to solve for? Explain why.

7. If the system $\begin{cases} 4x - 3y = 7 \\ 3x - 2y = 6 \end{cases}$ is to be solved using the elimination (addition) method, by what number should each equation be multiplied if
   a. the x-terms are to drop out?
   b. the y-terms are to drop out?

8. Consider the system: $\begin{cases} \frac{2}{3}x - \frac{y}{6} = \frac{16}{9} \\ 0.03x + 0.02y = 0.03 \end{cases}$
   a. What algebraic step should be performed to clear the first equation of fractions?
   b. What algebraic step should be performed to clear the second equation of decimals?

9. Can the system $\begin{cases} 2x + 5y = 7 \\ 4x - 3y = 16 \end{cases}$ be solved more easily by the substitution or the elimination method?

10. The substitution method was used to solve three systems of linear equations. The results after y was eliminated and the remaining equation was solved for x are listed below. Match each result with one of the possible graphs shown.
   a. $-2 = 3$ **b.** $x = 3$ **c.** $3 = 3$

Possible graphs

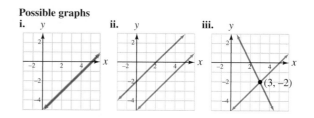

NOTATION

11. For the following system, write each equation in standard $Ax + By = C$ form.
   a. $\begin{cases} 4y = 8 - 7x \\ 3x - y = 2(x + 4) \end{cases} \longrightarrow \begin{cases} \\ \end{cases}$
   b. For the following system, clear the equations of any fractions or decimals.
$$\begin{cases} \frac{x}{5} + \frac{y}{10} = \frac{6}{5} \\ 0.3x - 0.9y = 17 \end{cases} \longrightarrow \begin{cases} \\ \end{cases}$$

12. Fill in the blanks: We read $\{(x, y) \mid x - 5y = 9\}$ as "the set of all _____ pairs (x, y), _____ _____ $x - 5y = 9$."

GUIDED PRACTICE

Solve each system by substitution. **See Examples 1 and 2.**

13. $\begin{cases} y = 3x \\ x + y = 8 \end{cases}$ **14.** $\begin{cases} y = x + 2 \\ x + 2y = 16 \end{cases}$

15. $\begin{cases} x = 2 + y \\ 2x + y = 13 \end{cases}$ **16.** $\begin{cases} x = -5 + y \\ 3x - 2y = -7 \end{cases}$

17. $\begin{cases} x + 2y = 6 \\ 3x - y = -10 \end{cases}$ **18.** $\begin{cases} 2x - y = -21 \\ 4x + 5y = 7 \end{cases}$

19. $\begin{cases} 0.3a + 0.1b = 0.5 \\ \frac{4}{3}a + \frac{1}{3}b = 3 \end{cases}$ **20.** $\begin{cases} 0.9p + 0.2q = 1.2 \\ \frac{2}{3}p + \frac{1}{9}q = 1 \end{cases}$

Solve each system by elimination. See Examples 3–4.

21. $\begin{cases} x - y = 7 \\ x + y = 11 \end{cases}$
22. $\begin{cases} a + b = 5 \\ a - b = 11 \end{cases}$

23. $\begin{cases} 2s + 3t = -8 \\ 2s - 3t = -8 \end{cases}$
24. $\begin{cases} x + 2y = -21 \\ x - 2y = 11 \end{cases}$

25. $\begin{cases} 5x + 2y = 11 \\ 7x + 6y = 9 \end{cases}$
26. $\begin{cases} 3x + 4y = -24 \\ 5x + 12y = -72 \end{cases}$

27. $\begin{cases} 5x + 3y = 72 \\ 3x + 5y = 56 \end{cases}$
28. $\begin{cases} 2x + 3y = 31 \\ 3x + 2y = 39 \end{cases}$

Solve each system by elimination. See Example 5.

29. $\begin{cases} 2(a + b) = 94 \\ 4(a - 9) = 3b - 23 \end{cases}$
30. $\begin{cases} c - 2d = 29 \\ 2(c - 5) = d - 21 \end{cases}$

31. $\begin{cases} 2(x + y) + 1 = 0 \\ 3x + 4y = 0 \end{cases}$
32. $\begin{cases} 5x + 3y = -7 \\ 3(x - y) - 7 = 0 \end{cases}$

Solve each system by any method, if possible. If a system is inconsistent or if the equations are dependent, state this. See Examples 6 and 7.

33. $\begin{cases} 2(a + b) = a + 12 \\ a = 14 - 2b \end{cases}$
34. $\begin{cases} 3c = 2(6 - d) \\ 2d = 3(4 - c) \end{cases}$

35. $\begin{cases} 2x - \dfrac{5}{2} = y \\ 0.04x - 0.02y = 0.05 \end{cases}$
36. $\begin{cases} \dfrac{3}{2}x + 2 = y \\ 0.6x - 0.4y = -0.4 \end{cases}$

TRY IT YOURSELF

Solve each system by any method, if possible. If a system is inconsistent or if the equations are dependent, state this.

37. $\begin{cases} 3x - 4y = 9 \\ x + 2y = 8 \end{cases}$
38. $\begin{cases} 3x - 2y = -10 \\ 6x + 5y = 25 \end{cases}$

39. $\begin{cases} 4(x - 2) = -9y \\ 2(x - 3y) = -3 \end{cases}$
40. $\begin{cases} 2(2x + 3y) = 5 \\ 8x = 3(1 + 3y) \end{cases}$

41. $\begin{cases} 0.16x - 0.08y = 0.32 \\ 2x - 4 = y \end{cases}$
42. $\begin{cases} x = \dfrac{3}{2}y + 5 \\ 2x - 3y = 8 \end{cases}$

43. $\begin{cases} x = \dfrac{2}{3}y \\ y = 4x + 50 \end{cases}$
44. $\begin{cases} y = -2x - 165 \\ x = \dfrac{2}{3}y + 5 \end{cases}$

45. $\begin{cases} \dfrac{m - n}{5} + \dfrac{m + n}{2} = 6 \\ \dfrac{m - n}{2} - \dfrac{m + n}{4} = 3 \end{cases}$
46. $\begin{cases} \dfrac{r - 2}{5} + \dfrac{s + 3}{2} = 5 \\ \dfrac{r + 3}{2} + \dfrac{s - 2}{3} = 6 \end{cases}$

47. $\begin{cases} 0.5x + 0.5y = 6 \\ \dfrac{x}{2} - \dfrac{y}{2} = -2 \end{cases}$
48. $\begin{cases} \dfrac{x}{2} - \dfrac{y}{3} = -4 \\ 0.009x + 0.002y = 0 \end{cases}$

49. $\begin{cases} \dfrac{3}{4}x + \dfrac{2}{3}y = 7 \\ \dfrac{3}{5}x - \dfrac{1}{2}y = 18 \end{cases}$
50. $\begin{cases} \dfrac{2}{3}x - \dfrac{1}{4}y = -8 \\ \dfrac{1}{2}x - \dfrac{3}{8}y = -9 \end{cases}$

51. $\begin{cases} \dfrac{x}{4} = 1 + \dfrac{y}{5} \\ x = \dfrac{4}{5}(y + 10) \end{cases}$
52. $\begin{cases} \dfrac{5}{3}x + 2 = 2(y + 6) \\ 3y + 5 = \dfrac{5}{2}(x - 4) \end{cases}$

53. $\begin{cases} \dfrac{3}{2}x - \dfrac{2}{3}y = 0 \\ \dfrac{3}{4}x + \dfrac{4}{3}y = \dfrac{5}{2} \end{cases}$
54. $\begin{cases} \dfrac{3}{5}x + \dfrac{5}{3}y = 2 \\ \dfrac{6}{5}x - \dfrac{5}{3}y = 1 \end{cases}$

55. $\begin{cases} 12x - 5y - 21 = 0 \\ \dfrac{3}{4}x + \dfrac{2}{3}y = -\dfrac{13}{8} \end{cases}$
56. $\begin{cases} 4y + 5x - 7 = 0 \\ \dfrac{10}{7}x - \dfrac{4}{9}y = \dfrac{17}{21} \end{cases}$

Solve each system. To do so, substitute a for $\frac{1}{x}$ and b for $\frac{1}{y}$ and solve for a and b. Then find x and y using the fact that $a = \frac{1}{x}$ and $b = \frac{1}{y}$.

57. $\begin{cases} \dfrac{1}{x} + \dfrac{1}{y} = \dfrac{5}{6} \\ \dfrac{1}{x} - \dfrac{1}{y} = \dfrac{1}{6} \end{cases}$
58. $\begin{cases} \dfrac{1}{x} + \dfrac{1}{y} = \dfrac{9}{20} \\ \dfrac{1}{x} - \dfrac{1}{y} = \dfrac{1}{20} \end{cases}$

59. $\begin{cases} \dfrac{1}{x} + \dfrac{2}{y} = -1 \\ \dfrac{2}{x} - \dfrac{1}{y} = -7 \end{cases}$
60. $\begin{cases} \dfrac{3}{x} - \dfrac{2}{y} = -30 \\ \dfrac{2}{x} - \dfrac{3}{y} = -30 \end{cases}$

WRITING

61. Which method would you use to solve the system
$\begin{cases} 4x + 6y = 5 \\ 8x - 3y = 3 \end{cases}$? Explain.

62. Which method would you use to solve the system
$\begin{cases} x - 2y = 2 \\ 2x + 3y = 11 \end{cases}$? Explain.

63. Why is the method for solving systems that is discussed in this section called the *elimination method*?

64. When using the elimination (addition) method, how can you tell whether

  a. a system of linear equations has no solution?

  b. a system of linear equations has infinitely many solutions?

65. If the elimination (addition) method is to be used to solve this system, what is wrong with the form in which it is written?

$$\begin{cases} 3x - 5y = 10 \\ 8 + 5x = 7y \end{cases}$$

66. Construct a table like that shown below, listing one advantage and one disadvantage for each of the methods that can be used to solve a system of two linear equations in two variables.

Method	Advantage	Disadvantage
Graphing		
Substitution		
Elimination		

REVIEW

Find the slope of each line.

67.
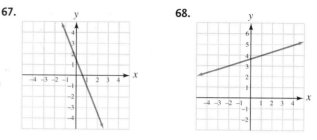

68.

69. The line that passes through $(0, -8)$ and $(-5, 0)$

70. The line wituation $y = -3x + 4$

71. The line with equation $4x - 3y = -3$

72. The line with equation $y = 3$

CHALLENGE PROBLEMS

73. If the solution of the system $\begin{cases} Ax + By = -2 \\ Bx - Ay = -26 \end{cases}$ is $(-3, 5)$, what are the values of the constants A and B?

74. Solve: $\begin{cases} 2ab - 3cd = 1 \\ 3ab - 2cd = 1 \end{cases}$ for a and c. Assume that b and d are constants.

SECTION 3.3
Problem Solving Using Systems of Two Equations

Objectives

1 Assign variables to two unknowns.

2 Use systems to solve geometry problems.

3 Use systems to solve number-value problems.

4 Use systems to find the break point.

5 Use systems to solve interest, uniform motion, and mixture problems.

In Chapter 1, we solved applied problems involving two unknown quantities by modeling the situation with an equation in one variable. It's often easier to solve such problems using a two-variable approach. We write two equations in two variables to model the situation, and then we use the methods of this chapter to solve the system formed by the pair of equations.

 Assign Variables to Two Unknowns.

The following steps are helpful when solving problems involving two unknown quantities.

Problem-Solving Strategy	1. **Analyze the problem** by reading it carefully to understand the given facts. Often a diagram or table will help you visualize the facts of the problem.
	2. Pick different variables to represent two unknown quantities. Translate the words of the problem to **form two equations** involving each of the two variables.
	3. **Solve the system** of equations using graphing, substitution, or elimination.
	4. **State the conclusion.**
	5. **Check the results** in the words of the problem.

EXAMPLE 1   *Pets.*   In 2007, there were 163 million dogs and cats owned in the United States. If the number of dogs was 13 million less than the number of cats, how many dogs and how many cats were owned in the United States that year? (Source: Humane Society)

Analyze the Problem

- In 2007, the total number of dogs and cats was 163 million.
- The number of dogs was 13 million less than the number of cats.
- Find the number of dogs and the number of cats that year.

Form Two Equations   We will let x = the number of dogs (in millions) and y = the number of cats (in millions). We can translate the words of the problem into two equations, each involving x and y.

Caution
If two variables are used to represent two unknown quantities, we must form a system of two equations to find the unknowns.

The number of dogs	plus	the number of cats	was	163 million.
x	$+$	y	$=$	163

The number of dogs	was	13 million	less than	the number of cats.
x	$=$	y	$-$	13

The resulting system is $\begin{cases} x + y = 163 \\ x = y - 13 \end{cases}$

Solve the System   Since the second equation is solved for x, we will use substitution to solve the system.

$$x + y = 163 \qquad \text{This is the first equation of the system.}$$
$$y - 13 + y = 163 \qquad \text{Substitute } y - 13 \text{ for } x.$$
$$2y - 13 = 163 \qquad \text{Combine like terms.}$$
$$2y = 176 \qquad \text{Add 13 to both sides.}$$
$$y = 88 \qquad \text{Divide both sides by 2. This is the number of cats (in millions).}$$

To find x, we substitute 88 for y in the second equation of the system.

$$x = y - 13$$
$$= 88 - 13 \qquad \text{Substitute 88 for } y.$$
$$= 75 \qquad \text{This is the number of dogs (in millions).}$$

State the Conclusion   In 2007, the number of dogs owned in the United States was 75 million and the number of cats was 88 million.

Check the Results   Since 75 million + 88 million = 163 million and 75 million is 13 million less than 88 million, the results check.

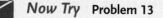 *Now Try*   **Problem 13**

2   **Use Systems to Solve Geometry Problems.**

Sometimes we can use geometric facts or formulas to solve application problems.

EXAMPLE 2   *Parallelograms.*   Refer to parallelogram *ABCD*. Find the unknown degree measures represented by *x* and *y*.

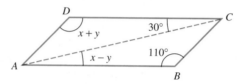

Analyze the Problem   To solve this problem, we will use two facts about parallelograms.

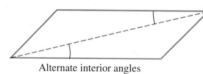

Alternate interior angles

When a diagonal intersects two parallel sides of a parallelogram, pairs of alternate interior angles have the same measure.

Opposite angles

Opposite angles of a parallelogram have the same measure.

Form Two Equations   In the figure, $\angle BAC$ and $\angle DCA$ are alternate interior angles and therefore have the same measure. Thus, $x - y = 30$. Since $\angle B$ and $\angle D$ in the figure are opposite angles of the parallelogram, $x + y = 110$.

Solve the System   To find *x* and *y*, we solve the following system:

$$\begin{cases} x - y = 30 \\ x + y = 110 \end{cases}$$

Since the coefficients of the terms $-y$ and y are opposites, we will use elimination to solve the system.

$$\begin{array}{r} x - y = 30 \\ \underline{x + y = 110} \\ 2x \quad\;\; = 140 \\ x = 70 \end{array}$$

  Add the equations. The y-terms drop out.

  To solve for x, divide both sides by 2.

We can substitute 70 for *x* in the second equation and solve for *y*.

$$x + y = 110$$
$$70 + y = 110 \qquad \text{Substitute 70 for } x.$$
$$y = 40 \qquad \text{To solve for } y, \text{ subtract 70 from both sides.}$$

State the Conclusion   Thus, $x = 70°$ and $y = 40°$.

Check the Results   Since $70° - 40° = 30°$ and $70° + 40° = 110°$, the results check.

 Now Try   **Problem 19**

3   **Use Systems to Solve Number-Value Problems.**

EXAMPLE 3   ***Wedding Pictures.***   A professional photographer offers two different packages for wedding pictures. Use the information in the figure to determine the cost of one 8×10-inch photograph and the cost of one 5×7-inch photograph.

Analyze the Problem

- Eight 8×10 and twelve 5×7 pictures cost $133.

- Six 8×10 and twenty-two 5×7 pictures cost $168.

- Find the cost of one 8×10 photograph and the cost of one 5×7 photograph.

Form Two Equation   Let $x =$ the cost of one 8×10 photograph (in dollars), and let $y =$ the cost of one 5×7 photograph (in dollars). We can use the fact that **Number · value = total value** to write equations that model the cost of each package. For the first package, the cost of eight 8×10 photos is $8 \cdot \$x = \$8x$, and the cost of twelve 5×7 photos is $12 \cdot \$y = \$12y$. For the second package, the cost of six 8×10 photos is $\$6x$, and the cost of twenty-two 5×7 photos is $\$22y$. To find x and y, we must write and solve two equations.

The cost of eight 8×10 photographs	plus	the cost of twelve 5×7 photographs	is	the cost of the first package.
$8x$	$+$	$12y$	$=$	133

The cost of six 8×10 photographs	plus	the cost of twenty-two 5×7 photographs	is	the cost of the second package.
$6x$	$+$	$22y$	$=$	168

Solve the System   To find the cost of one 8×10 and one 5×7 photograph, we must solve the following system:

(1) $\begin{cases} 8x + 12y = 133 \\ \end{cases}$
(2) $\begin{cases} 6x + 22y = 168 \end{cases}$

We will use elimination to solve this system. To make the x-terms drop out, we multiply both sides of equation 1 by 3. Then we multiply both sides of equation 2 by -4, add the resulting equations, and solve for y:

$$24x + 36y = 399 \qquad \text{This is } 3(8x + 12y) = 3(133).$$
$$\underline{-24x - 88y = -672} \qquad \text{This is } -4(6x + 22y) = -4(168).$$
$$-52y = -273 \qquad \text{Add the terms, column by column. The x-terms drop out.}$$
$$y = 5.25 \qquad \text{Divide both sides by } -52. \text{ This is the cost of one } 5 \times 7 \text{ photograph.}$$

To find x, we substitute 5.25 for y in equation 1 and solve for x:

$$8x + 12y = 133$$
$$8x + 12(5.25) = 133 \qquad \text{Substitute 5.25 for y.}$$
$$8x + 63 = 133 \qquad \text{Do the multiplication.}$$
$$8x = 70 \qquad \text{Subtract 63 from both sides.}$$
$$x = 8.75 \qquad \text{Divide both sides by 8. This is the cost of one 8 × 10 photograph.}$$

> ### Caution
> In this problem we are to find two unknowns, the cost of an 8 × 10 photo and the cost of a 5 × 7 photo. Remember to give both in the *State the Conclusion* step of the solution.

State the Conclusion   The cost of one 8 × 10 photo is $8.75, and the cost of one 5 × 7 photo is $5.25.

Check the Results   If the first package contains eight 8 × 10 and twelve 5 × 7 photographs, the value of the package is 8($8.75) + 12($5.25) = $70 + $63 = $133. If the second package contains six 8 × 10 and twenty-two 5 × 7 photographs, the value of the package is 6($8.75) + 22($5.25) = $52.50 + $115.50 = $168. The results check.

 Now Try   **Problem 25**

---

4   ## Use Systems to Find the Break Point.

Running a machine involves both *setup costs* and *unit costs*. Setup costs include the cost of preparing a machine to do a certain job. The costs to make one item are unit costs. They depend on the number of items to be manufactured, including costs of raw materials and labor.

EXAMPLE 4   *Manufacturing.*   The setup cost of a machine that mills brass plates is $750. After setup, it costs $0.25 to mill each plate. Management is considering the purchase of a larger machine that can produce a plate at a cost of $0.20 per plate. If the setup cost of the larger machine is $1,200, how many plates would the company have to produce to make the purchase worthwhile?

Analyze the Problem   We need to find the number of plates (called the **break point**) that will cost equal amounts to produce on either machine.

Form Two Equations   We can let c = the cost (in dollars) of milling p plates. If we call the machine currently in use machine 1 and the new, larger one machine 2, we can form two equations.

The cost of making p plates using machine 1	equals	the setup cost of machine 1	plus	the cost per plate of machine 1	times	the number of plates p to be made.
c	=	750	+	0.25	·	p

The cost of making p plates using machine 2	equals	the setup cost of machine 2	plus	the cost per plate of machine 2	times	the number of plates p to be made.
c	=	1,200	+	0.20	·	p

To find the break point, we must solve the system $\begin{cases} c = 750 + 0.25p \\ c = 1,200 + 0.20p \end{cases}$

Solve the System   Since the costs are equal for the break point, we can use substitution to solve the system.

$$\begin{cases} c = 750 + 0.25p \\ c = 1{,}200 + 0.20p \end{cases}$$

$750 + 0.25p = 1{,}200 + 0.20p$	Substitute 750 + 0.25p for c in the second equation.
$0.25p = 450 + 0.20p$	Subtract 750 from both sides.
$0.05p = 450$	Subtract 0.20p from both sides.
$p = 9{,}000$	Divide both sides by 0.05.

State the Conclusion   Since the cost will be the same on either machine when 9,000 plates are milled, 9,000 is the break point.

Check the Results   We can check the result by substituting 9,000 for p in each equation of the system and verifying that 3,000 is the value of c in both cases.
   If we graph the two equations, we can illustrate the break point.

Machine 1
$c = 750 + 0.25p$

p	c
0	750
1,000	1,000
3,000	1,500

Machine 2
$c = 1{,}200 + 0.20p$

p	c
0	1,200
4,000	2,000
12,000	3,600

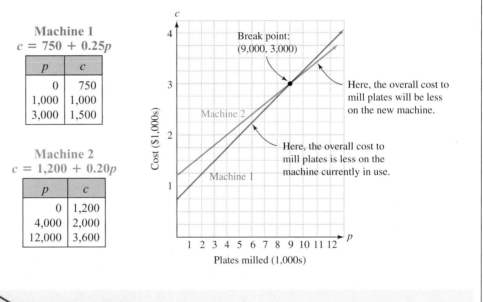

 Now Try   **Problem 31**

5   **Use Systems to Solve Interest, Uniform Motion, and Mixture Problems.**

To compare one-variable and two-variable approaches, we will solve the investment problem of Example 6 in Chapter 1 using two variables to find the unknown investment amounts.

EXAMPLE 5   *Interest Income.*   To protect against a major loss, a financial analyst suggested the following plan for a client who has $50,000 to invest for 1 year.

1. Alco Development, Inc. Builds mini-malls. High yield: 12% per year. Risky!
2. Certificate of deposit (CD). Insured, safe. Low yield: 4.5% annual interest.

If the client puts some money in each investment and wants to earn $3,600 in interest, how much should be invested at each rate?

Analyze the Problem   A total of $50,000 is invested at two different rates for 1 year. The total interest earned is $3,600.

Form Two Equations   Let $x =$ the number of dollars invested at 12% and $y =$ the number of dollars invested at 4.5%. We will use the formula $I = Prt$ to determine that $x invested at 12% for 1 year will earn $0.12x$ and $y invested at 4.5% for 1 year will earn $0.045y$. This information is shown in the table.

Success Tip
With a one-variable approach, we let $x =$ the number of dollars invested at 12% and $50,000 - x =$ the number of dollars invested at 4.5%. With a two-variable approach, again we let $x =$ the number of dollars invested at 12%, but then we let $y =$ the number of dollars invested at 4.5%.

	P	$\cdot$ r	$\cdot t=$	I
Alco	x	0.12	1	$0.12x$
CD	y	0.045	1	$0.045y$
Total	50,000			3,600

↑ One equation comes from this column.   ↑ The other equation comes from this column.

The facts of the problem give the following two equations.

The amount invested in Alco	plus	the amount invested in the CD	is	$50,000.
x	$+$	y	$=$	50,000

The interest earned at 12%	plus	the interest earned at 4.5%	equals	the total interest earned.
$0.12x$	$+$	$0.045y$	$=$	3,600

Solve the System   To find out how much was invested at each rate, we solve the following system:

$$\textbf{(1)}\quad \begin{cases} x + y = 50,000 \\ \textbf{(2)}\quad 0.12x + 0.045y = 3,600 \end{cases}$$

To solve this system by substitution, we can solve equation 1 for y:

$$x + y = 50,000$$
$$y = 50,000 - x \quad \text{This is the substitution equation.}$$

Then we substitute $50,000 - x$ for y in equation 2 and solve for x.

Success Tip
This equation ⟶ is the same as the one that we obtained in Chapter 1 using a one-variable approach.

$$0.12x + 0.045y = 3,600$$

$$0.12x + 0.045(\mathbf{50,000} - x) = 3,600 \qquad \text{Substitute } 50,000 - x \text{ for } y.$$

$$120x + 45(50,000 - x) = 3,600,000 \qquad \text{To clear the equation of the decimals, multiply both sides by 1,000.}$$

$$120x + 2,250,000 - 45x = 3,600,000 \qquad \text{Distribute the multiplication by 45.}$$

$$75x = 1,350,000 \qquad \text{Combine like terms and subtract 2,250,000 from both sides.}$$

$$x = 18,000 \qquad \text{Divide both sides by 75. This is the amount that should be invested at 12\%.}$$

To find y, we can substitute 18,000 for x in the substitution equation:

$$y = 50,000 - x$$

$$= 50,000 - \mathbf{18,000} \qquad \text{Substitute 18,000 for } x.$$

$$= 32,000 \qquad \text{This is the amount that should be invested at 4.5\%.}$$

State the Conclusion $18,000 should be invested at 12% and $32,000 should be invested at 4.5%.

Check the Results   We note that $18,000 plus $32,000 equals the required amount of money invested. The annual interest on $18,000 is $0.12(\$18,000) = \$2,160$. The interest earned on $32,000 is $0.045(\$32,000) = \$1,440$. The total interest is $\$2,160 + \$1,440 = \$3,600$. The results check.

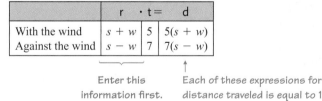 *Now Try*   **Problem 39**

EXAMPLE 6   ***Blimps.***   *The Spirit of America,* one of the Goodyear blimps, flew 175 miles in 5 hours with the wind. The return trip took 7 hours flying against the wind. Find the speed of the blimp in still air and the speed of the wind.

Analyze the Problem   Traveling with the wind, the speed of the blimp will be faster than it would be in still air. Traveling against the wind, the speed of the blimp will be slower than it would be in still air.

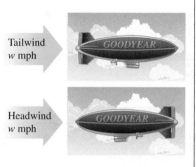

Tailwind
w mph

Headwind
w mph

Form Two Equations   Let s = the speed of the blimp (in mph) in still air and w = the speed of the wind (in mph). Then the speed of the blimp flying with the wind is $s + w$ and the speed of the blimp flying against the wind is $s - w$.

Using the formula $d = rt$, we find that $5(s + w)$ represents the distance traveled with the wind and $7(s - w)$ represents the distance traveled against the wind. This information is shown in the table.

	r	· t =	d
With the wind	$s + w$	5	$5(s + w)$
Against the wind	$s - w$	7	$7(s - w)$

Enter this information first.

Each of these expressions for distance traveled is equal to 175.

Since each trip is 175 miles long, the information in the Distance column of the table can be used to form two equations in two variables. To write each equation in standard form, we use the distributive property.

$$\begin{cases} 5(s + w) = 175 \\ 7(s - w) = 175 \end{cases} \xrightarrow[\text{Distribute}]{\text{Distribute}} \begin{cases} 5s + 5w = 175 \quad \textbf{(1)} \\ 7s - 7w = 175 \quad \textbf{(2)} \end{cases}$$

Solve the System   Since the coefficients of w have opposite signs, we will eliminate w. To do this, we will create terms of $35w$ and $-35w$ by multiplying both sides of the equation 1 by 7 and both sides of the equation 2 by 5.

$$35s + 35w = 1{,}225 \qquad \text{This is } 7(5s + 5w) = 7(175).$$
$$\underline{35s - 35w = 875} \qquad \text{This is } 5(7s - 7w) = 5(175).$$
$$70s \qquad\quad = 2{,}100$$
$$s = 30 \qquad \text{Divide both sides by 70. This is the speed of the blimp in still air.}$$

To find w, we will substitute 30 for s in equation 1.

$$5s + 5w = 175$$
$$5(30) + 5w = 175 \qquad \text{Substitute 30 for } s.$$
$$150 + 5w = 175 \qquad \text{Do the multiplication.}$$
$$5w = 25 \qquad \text{Subtract 150 from both sides.}$$
$$w = 5 \qquad \text{Divide both sides by 5. This is the speed of the wind.}$$

State the Conclusion   The speed of the blimp in still air is 30 mph and the speed of the wind is 5 mph.

Check the Results   The blimp's speed traveling with a 5-mph wind will be $30 + 5 = 35$ mph. In 5 hours, it will travel $35 \cdot 5 = 175$ miles. The blimp's speed traveling against a 5-mph wind will be $30 - 5 = 25$ mph. In 7 hours, it will travel $25 \cdot 7 = 175$ miles. The results check.

Now Try   **Problem 45**

© Kim Karpeles/Alamy

EXAMPLE 7   *Popcorn.*   A tin of jalapeño-flavored popcorn sells for $36, while the same size tin of cheddar cheese–flavored popcorn sells for $24. How many tins of each type of popcorn should be used to create 10 tins of a jalapeño–cheddar mix that can be sold for $27 per tin?

Analyze the Problem   We will use a two-variable approach to solve this dry mixture problem.

Form Two Equations   Let $x =$ the number of tins of jalapeño popcorn and $y =$ the number of tins of cheddar cheese popcorn that should be mixed. The value of the mixture and the value of each of its components are given by

$$\textbf{Amount} \cdot \textbf{price} = \textbf{total value}$$

Thus, the value of x tins of jalapeño popcorn is $36x$ and the value of y tins of cheddar cheese popcorn is $24y$. The sum of these values is also equal to the total value of the final mixture that is $10 \cdot \$27$ or $270. This information is shown in the table.

	Amount ·	Price =	Total value
Jalapeño popcorn	x	36	$36x$
Cheddar cheese popcorn	y	24	$24y$
Mixture	10	27	$10(27)$

The facts in the table give the following equations:

The number of tins of jalapeño popcorn	plus	the number of tins of cheddar cheese popcorn	equals	10.
x	$+$	y	$=$	10

The value of the jalapeño popcorn	plus	the value of the cheddar cheese popcorn	equals	the value of the mixture.
$36x$	$+$	$24y$	$=$	$10(27)$

Success Tip
Determine the similarities and the differences of the one-variable and two-variable approaches by reviewing the dry mixture problem of Example 8, Chapter 1, on page 94.

Solve the System   To find how many tins of each popcorn are needed, we solve the following system:

$$\begin{cases} x + y = 10 \\ 36x + 24y = 270 \end{cases} \quad \text{Multiply: 10(27) = 270.}$$

We will use elimination to solve this system. To make the y-terms drop out, we multiply both sides of the first equation of the system by -24 and add the resulting equations to solve for x:

$$\begin{array}{rcl} -24x - 24y &=& -240 \\ \underline{36x + 24y} &=& \underline{270} \\ 12x &=& 30 \end{array}$$

This is $-24(x + y) = -24(10)$.

Add like terms, column by column.

$$x = \frac{30}{12} = 2.5$$

Divide both sides by 12 and simplify. This is the number of tins of jalapeño popcorn needed.

To find y, we substitute 2.5 for x in the first equation of the system and solve for y:

$$\begin{aligned} x + y &= 10 \\ 2.5 + y &= 10 \qquad \text{Substitute 2.5 for x.} \\ y &= 7.5 \qquad \text{Subtract 2.5 from both sides. This is the number of tins of cheddar cheese popcorn needed.} \end{aligned}$$

State the Conclusion   To obtain 10 tins of jalapeño–cheddar popcorn, 2.5 tins of japaleño and 7.5 tins of cheddar cheese popcorn should be combined.

Check the Results   When 2.5 tins and 7.5 tins are combined, the result is 10 tins. The 2.5 tins of jalapeño popcorn are valued at 2.5($36) = $90 and the 7.5 tins of cheddar cheese popcorn are valued at 7.5($24) = $180. The sum of those values, $90 + $180 = $270, is the same as the value of the mixture, 10($27) = $270. The results check.

▷ *Now Try*   **Problem 51**

EXAMPLE 8   *Water Treatment.* A technician determines that 100 fluid ounces of a 15% muriatic acid solution needs to be added to the water in a swimming pool to kill a growth of algae. If the technician has 5% and 20% muriatic solutions on hand, how many ounces of each must be combined to create the 15% solution?

Analyze the Problem   We will use a two-variable approach to solve this liquid mixture problem. We need to find the number of ounces of a 5% solution and the number of ounces of a 20% solution that must be combined to obtain 100 ounces of a 15% solution.

Form Two Equations   Let $x =$ the number of ounces of the 5% solution and let $y =$ the number of ounces of the 20% solution that are to be mixed. The amount of pure muriatic acid in each solution is given by

Amount of solution · strength = amount pure muriatic acid

Thus, the amount of muriatic acid in the 5% solution is 0.05x ounces, and the amount of muriatic acid in the 20% solution is 0.20y ounces. The sum of these amounts is also the amount of muriatic acid in the final mixture, which is 15% of 100 ounces or 0.15(100). This information is shown in the table.

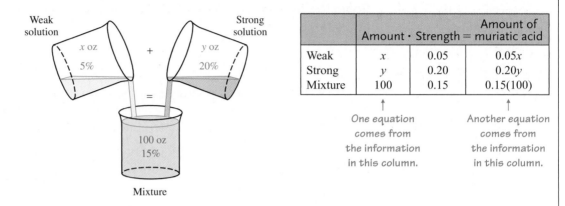

	Amount · Strength =		Amount of muriatic acid
Weak	x	0.05	0.05x
Strong	y	0.20	0.20y
Mixture	100	0.15	0.15(100)

↑
One equation comes from the information in this column.

↑
Another equation comes from the information in this column.

Success Tip

Determine the similarities and the differences of the one-variable and two-variable approaches by reviewing the liquid mixture problem of Example 9, Chapter 1, on page 95.

The facts in the table give the following equations:

The number of ounces of 5% solution	plus	the number of ounces of 20% solution	equals	the total number of ounces in the 15% mixture.
x	$+$	y	$=$	100

The number of ounces of acid in the 5% solution	plus	the number of ounces of acid in the 20% solution	is	the number of ounces of acid in the 15% mixture.
$0.05x$	$+$	$0.20y$	$=$	$0.15(100)$

Solve the System   To find out how many ounces of each are needed, we solve the following system:

$$\textbf{(1)} \quad \begin{cases} x + y = 100 \\ \textbf{(2)} \quad 0.05x + 0.20y = 15 \quad \text{Multiply: 0.15(100) = 15.} \end{cases}$$

To solve this system by substitution, we can solve the first equation for y:

$$x + y = 100$$
$$y = 100 - x \quad \text{This is the substitution equation.}$$

Then we substitute $100 - x$ for y in equation 2 and solve for x.

$$0.05x + 0.20y = 15$$
$$0.05x + 0.20(\textbf{100} - \textbf{x}) = 15 \qquad \text{Substitute } 100 - x \text{ for } y.$$
$$5x + 20(100 - x) = 1,500 \qquad \text{To clear the equation of decimals, multiply both sides by 100.}$$
$$5x + 2,000 - 20x = 1,500 \qquad \text{Distribute the multiplication by 20.}$$
$$-15x = -500 \qquad \text{Combine like terms and subtract 2,000 from both sides.}$$
$$x = \frac{-500}{-15} \qquad \text{To solve for } x, \text{ divide both sides by } -15.$$
$$x = \frac{100}{3} \qquad \text{Simplify. This is the number of ounces of the 5% acid solution that is needed.}$$

To find y, we can substitute $\frac{100}{3}$ for x in the substitution equation:

$$y = 100 - x$$

$$= 100 - \frac{100}{3} \qquad \text{Substitute } \tfrac{100}{3} \text{ for } x.$$

$$= \frac{200}{3} \qquad \text{To subtract, think: } \tfrac{300}{3} - \tfrac{100}{3}. \text{ This is the number of ounces of the 20\% acid}$$
$$\text{solution that is needed.}$$

State the Conclusion   To obtain 100 ounces of a 15% solution, the technician must mix $\frac{100}{3}$ or $33\frac{1}{3}$ ounces of the 5% solution with $\frac{200}{3}$ or $66\frac{2}{3}$ ounces of the 20% solution.

Check the Results   We note that $33\frac{1}{3}$ ounces of solution plus $66\frac{2}{3}$ ounces of solution equals the required 100 ounces of solution. The $33\frac{1}{3} = \frac{100}{3}$ ounces of 5% solution contains $0.05\left(\frac{100}{3}\right) = \frac{5}{3}$ ounces of muriatic acid, and the $66\frac{2}{3} = \frac{200}{3}$ ounces of 20% solution contains $0.20\left(\frac{200}{3}\right) = \frac{40}{3}$ ounces of muriatic acid—a total of $\frac{5}{3} + \frac{40}{3} = \frac{45}{3}$ or 15 ounces of muriatic acid. The 100 ounces of the 15% mixture contains $0.15(100) = 15$ ounces of acid. The results check.

> **Notation**
> For application problems, it is often more meaningful to present the results as mixed numbers $\left(\text{such as } 33\frac{1}{3} \text{ and } 66\frac{2}{3}\right)$ than as improper fractions $\left(\frac{100}{3} \text{ and } \frac{200}{3}\right)$.

▷ **Now Try   Problem 57**

STUDY SET
3.3

VOCABULARY

Fill in the blanks.

1. A parallelogram is a four-sided figure with two pairs of _____ sides.

2. Suppose a hammer can be manufactured in two different ways. The number of hammers that will cost equal amounts to produce either way is called the _____ point.

CONCEPTS

3. **a.** Refer to the parallelogram below. Find the measure of $\angle SRU$ and $\angle TSU$.

   b. Fill in the blank: $\angle SUR$ and $\angle TSU$ are called _____ _____ angles.

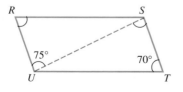

4. A company charges a $75 setup fee plus $5.25 per shirt to silkscreen a design on specialty t-shirts. Write an equation which gives the cost of purchasing x shirts.

5. In still water, a person swims at the rate of x mph. Find the speed of the swimmer for each of the following situations.

   a. With the current

   b. Against the current

6. **a.** Write an expression that represents the total value of x ounces of ginseng tea that costs $32 per pound.

   b. Write an expression that represents the amount of hydrochloric acid in x gallons of a 3% hydrochloric acid solution.

NOTATION

7. Write each percent as a decimal.

   a. 6%   **b.** 4.8%   **c.** $13\frac{1}{2}$%

8. What is the formula that finds
   a. Simple interest
   b. Distance traveled

GUIDED PRACTICE

Complete each table and write a system of two equations that can be used to find x and y. DO NOT SOLVE THE SYSTEM. **See Examples 5–8.**

9. INVESTMENTS   A total of $25,000 was invested in two accounts for 1 year and earned a total of $1,050 in interest.

	Principal ·	Rate ·	Time =	Interest
Township Bank	x	0.05	1	
Ameritech Savings	y	0.04	1	
Total				

10. PHYSICAL FITNESS   A jogger and cyclist started at the same point and traveled for 2 hours in opposite directions until they were 42 miles apart. The cyclist traveled 10 mph faster than the jogger.

	Rate ·	Time =	Distance
Jogger	x	2	
Cyclist	y	2	
Total			

11. CANDY   A company combines dark chocolate (selling for $13.90 per pound) with white chocolate (selling for $5.10 per pound) to get 6 pounds of a mixture that will sell for $10.25 per pound.

	Amount ·	Value =	Total value
Dark chocolate	x	13.90	
White chocolate	y	5.10	
Mixture	6	10.25	

12. DISINFECTANTS   A 1% bleach solution is to be mixed with a 5% bleach solution to obtain 15 ounces of a 3% bleach solution.

	Amount ·	Strength =	Amount of bleach
Weak	x	0.01	
Strong	y	0.05	
Mix	15	0.03	

APPLICATIONS

Write a system of two equations in two variables to solve each problem.

13. ELECTRONICS   In the illustration, two resistors in the voltage divider circuit have a total resistance of 1,375 ohms. To provide the required voltage, R_1 must be 125 ohms greater than R_2. Find both resistances.

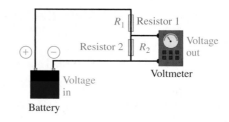

14. DESSERTS   A slice of Mrs. Smith's apple pie and one scoop of Häagen-Dazs vanilla bean ice cream totals 600 calories. The pie has 20 more calories than the ice cream. Find the number of calories in each.

15. AREA CODES   The entire state of Montana has just one telephone area code. The same is true for Idaho. The sum of their area codes is 614 and the difference is 198, and Montana has the numerically larger one. Find the area code of each of these states.

16. HIKING   The Pacific Crest Trail runs from the U.S. border with Mexico to its border with Canada. The Appalachian Trail extends between Springer Mountain in Georgia and Mount Katahdin in Maine. The sum of the lengths of the trails is 4,824 miles, the difference is 476 miles, and the Pacific Crest Trail is the longer. Find the length of each trail.

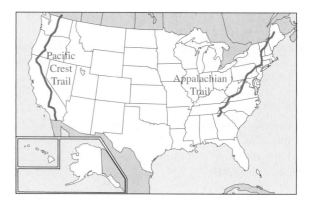

17. AVALANCHES For the 2005–2006 snow season, the total number of avalanche fatalities in the United States and Canada was 32. If the number in the United States was three times greater than the number in Canada, how many avalanche fatalities were there in each country? (Source: Avalanche.org)

18. CLOTHING STORES During the years 2005 and 2006, the Abercrombie and Fitch Company opened a total of 167 new stores. The number of new stores opened in 2006 was 22 less than twice the number opened in 2005. Find the number of new stores that the company opened each of those years. (Source: International Council of Shopping Centers)

19. BRACING The bracing of a basketball backboard forms a parallelogram. Find the unknown degree measures represented by x and y.

20. LOGOS Part of a Chevrolet logo is formed by a parallelogram. Find the unknown degree measures represented by x and y.

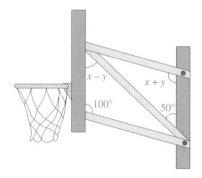

CHEVROLET®

21. TRAFFIC SIGNALS In the illustration, braces A and B are perpendicular. Find the values of x and y.

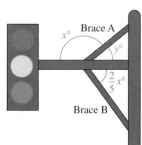

22. GEOMETRY An acute angle is an angle with measure less than 90°. In a right triangle, the measure of one acute angle is 15° greater than two times the measure of the other acute angle. Find the measure of each acute angle.

23. FENCING A FIELD The perimeter of a rectangular field is surrounded by 72 meters of fencing. If the field is partitioned into two parts as shown, a total of 88 meters of fencing is required. Find the dimensions of the field.

24. NEW YORK CITY The triangular-shaped Flatiron Building in Manhattan has a perimeter of 499 feet at its base. It is bordered on each side by a street. The 5th Avenue front of the building is 198 feet long. The Broadway front is 43 feet more than twice as long as the East 22nd Street front. Find the length of the Broadway front and East 22nd Street front. (Source: New York Public Library)

25. ADVERTISING Use the information in the ad to find the cost of a 15-second and a 30-second radio commercial on radio station KLIZ.

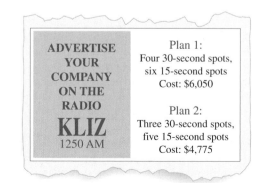

26. TEMPORARY HELP   A law firm hired several workers to help finish a large project. From the following billing records, determine the daily fee charged by the employment agency for a clerk-typist and for a computer programmer.

TEMPORARY EMPLOYMENT, INC.
We meet your employment needs!

Billed to: _Archer Law Offices_ Attn: _B. Kinsell_

Day	Position/Employee Name	Total cost
Mon. 3/22	*Clerk-typists:* K. Amad, B. Tran, S. Smith *Programmers:* T. Lee, C. Knox	$685
Tues. 3/23	*Clerk-typists:* K. Amad, B. Tran, S. Smith, W. Morada *Programmers:* T. Lee, C. Knox, B. Morales	$975

27. PRODUCTION PLANNING   A manufacturer builds racing bikes and mountain bikes, with the per-unit manufacturing costs shown in the table. The company has budgeted $26,150 for materials and $31,800 for labor. How many bicycles of each type can be built?

Model	Cost of materials	Cost of labor
Racing	$110	$120
Mountain	$140	$180

28. FARMING   A farmer keeps some animals on a strict diet. Each animal is to receive 15 grams of protein and 7.5 grams of carbohydrates. The farmer uses two food mixes, with nutrients as shown in the table. How many grams of each mix should be used to provide the correct nutrients for each animal?

Mix	Protein	Carbohydrates
Mix *A*	12%	9%
Mix *B*	15%	5%

29. CONCERTS   According to *StubHub.com,* in 2006, two tickets to a Rolling Stones concert and two tickets to a Jimmy Buffet concert cost, on average, a total of $792. At those prices, four tickets to see the Stones and two tickets to see Jimmy Buffet cost $1,320. What was the average cost of a Rolling Stones ticket and a Jimmy Buffet ticket in 2006?

30. MEASUREMENT   A *furlong* is a measure of distance and is used to express the length of certain horse races. A *fathom* is a measure of distance and is used to express depths of water. Four furlongs and five fathoms is a total of 2,670 feet. Five furlongs and four fathoms is a total of 3,324 feet. Find the length of a furlong and a fathom.

31. MAKING TIRES   A company has two molds to form tires. One mold has a setup cost of $1,000 and the other has a setup cost of $3,000. The cost to make each tire with the first mold is $15, and the cost to make each tire with the second mold is $10.

   a. Find the break point.

   b. Check your result by graphing both equations on the coordinate system below.

   c. If a production run of 500 tires is planned, determine which mold should be used.

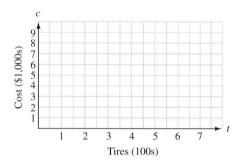

32. CHOOSING A FURNACE   A high-efficiency 90+ furnace can be purchased for $2,250 and costs an average of $824 per year to operate in Chicago, Illinois. An 80+ furnace can be purchased for only $1,710, but it costs $932 per year to operate.

   a. Find the break point.

   b. If you intended to live in Chicago for 4 years, which furnace would you choose?

33. PUBLISHING   A printer has two presses. The older press has a setup cost of $210 and can print the pages of a certain book for $5.98. The newer press has a setup cost of $350 and can print the pages of the same book for $5.95.

   a. Find the break point.

   b. If the publisher has advanced orders for 5,100 copies of the book, which press should be used?

34. COSMETOLOGY   A beauty shop specializing in permanents has fixed costs of $2,101.20 per month. The owner estimates that the cost for each permanent is $23.60, which covers labor, chemicals, and electricity. If her shop can give as many permanents as she wants at a price of $44 each, how many must be given each month to break even?

35. RECORDING COMPANIES   Three people invest a total of $105,000 to start a recording company that will produce reissues of classic jazz. Each release will be a set of 3 CDs that will retail for $45 per set. If each set can be produced for $18.95, how many sets must be sold for the investors to make a profit?

36. PRODUCTION PLANNING   A paint manufacturer can choose between two processes for manufacturing house paint, with monthly costs as shown in the table. Assume that the paint sells for $18 per gallon.

Process	Fixed costs	Unit cost
A	$12,390	$29
B	$20,460	$17

a. Find the **break even point** where production costs equal revenue earned for process A.

b. Find the break even point where production costs equal revenue earned for process B.

c. If expected sales are 7,000 gallons per month, which process should the company use?

37. MANUFACTURING   A manufacturer of automobile water pumps is considering retooling for one of two manufacturing processes, with monthly fixed costs and unit costs as indicated in the table. Each water pump can be sold for $50.

Process	Fixed costs	Unit cost (per gallon)
A	$32,500	$13
B	$80,600	$5

a. Find the break even point where production costs equal revenue earned for process A.

b. Find the break even point where production costs equal revenue earned for process B.

c. If expected sales are 550 per month, which process should be used?

38. SALARY OPTIONS   A sales clerk can choose from two salary plans:

1. a straight 7% commission

2. $150 + 2% commission

How much would the clerk have to sell for each plan to produce the same monthly paycheck?

39. INVESTMENT CLUBS   Part of $8,000 was invested by an investment club at 10% interest and the rest at 12%. If the annual income from these investments is $900, how much was invested at each rate?

40. RETIREMENT INCOME   A retired couple invested part of $12,000 at 6% interest and the rest at 7.5%. If their annual income from these investments is $810, how much was invested at each rate?

41. INVESTING   A woman invested some money in a credit union paying 5% annual simple interest and three times as much in a money market account paying 4.25% annual simple interest. If she earned $1,420 interest in one year, how much did she invest in each account?

42. IRA ACCOUNTS   A teacher started an Individual Retirement Account (IRA) by investing a total of $30,500 in two municipal bond funds, one paying 5% annual interest and the other paying $5\frac{1}{2}$% annual interest. At the end of the year, the funds earned a total of $1,615 in interest. How much did the teacher invest in each fund?

43. CREDIT CARDS   A couple stopped using their VISA credit card charging 1.5% per month interest and their Robinsons-May credit card charging 1.75% per month interest because they had built up a combined debt of $16,500 on the two cards. For 1 month, they made no purchases or payments on the accounts. If the total amount of interest the credit cards accumulated during the month was $259.25, what amount did they owe on each card when they stopped using them?

44. LOANS   A student had a car loan charging 0.75% interest per month and a tuition loan charging 0.5% interest per month. How much did he owe on each account if he paid a total of $95.50 monthly interest on a total debt of $14,200?

45. AVIATION   The jet stream is a wind current that flows across the United States from west to east. Flying with the jet stream, an airplane flew 2,700 miles in 4.5 hours. Against the same wind, the return trip took 6 hours. Find the speed of the plane in still air and the speed of the jet stream.

46. SALMON   It takes a salmon 40 minutes to swim 10,000 feet upstream and 8 minutes to swim that same portion of a river downstream. Find the speed of the salmon in still water and the speed of the current.

47. AIRPORT WALKWAYS   A man walks at a steady pace as he steps onto a moving walkway. It takes him 40 seconds to reach the end, 320 feet away. If he walks at the same rate against the flow of the walkway, it would take him 80 seconds to reach the end. Find his rate of walking and the rate of the moving walkway.

48. SNOWMOBILING   A man rode a snowmobile at the rate of 20 mph and then skied cross country at the rate of 4 mph. During the 6-hour trip, he traveled 48 miles. How long did he snowmobile, and how long did he ski?

49. JET SKIS   A Jet Ski rider can travel 10 miles against the current of the lower Mississippi River in $\frac{1}{2}$ hour and make the return trip with the current in $\frac{1}{3}$ hour. Find the speed of the Jet Ski in still water and the speed of the current.

50. ROLLERBLADING   An in-line skater headed west at the rate of 6 mph. One hour later, a moped rider left the same spot and headed west on the same road at 30 mph. How long will it take the moped rider to catch the skater?

51. MIXING CANDY   How many pounds of each candy shown in the illustration must be mixed to obtain 60 pounds of candy that would be worth $4 per pound?

Gummy Bears
$3.50/lb

Jelly Beans
$5.50/lb

52. GASOLINE   A truck owner drove his pickup to a service station to fill the nearly empty 24-gallon gas tank. If the truck runs on 89-octane gasoline, but the station only sells 87-octane and 93-octane gas, how many gallons of each should be pumped to fill the tank with an 89-octane blend?

53. MIXING COFFEE   How many pounds of regular coffee (selling for $4 per pound) and how many pounds of Kona coffee (selling for $11.50 per pound) must be combined to get 20 pounds of a mixture worth $6 per pound?

54. ROOM FRESHENER   A florist sells mixtures of dried, fragrant plant material that provides a gentle natural scent for houses. She wants to mix lilac (that sells for $18.25 a pound) with lavender (that sells for $12.25 a pound) to create 30 pounds of a blend that sells for $15 a pound. How many pounds of each should the florist use?

55. CONFETTI   How many pounds of $14.50-per-pound small flake confetti should be mixed with $24.50-per-pound mylar confetti stars to obtain one ton of a confetti mix that would be worth $20 per pound?   (*Hint:* How many pounds equal one ton?)

56. BATH SALTS   The owner of a kitchen and bath store wants to combine $7.92-per-pound eucalyptus bath salts with 88¢-per-pound Epsom salt to create 40 pounds of a $2.64-per-pound bath salt mix. How many pounds of each should be used?

57. ANTIFREEZE   How many pints of a 10% antifreeze solution and how many pints of a 40% antifreeze solution must be mixed to obtain 24 pints of a 30% solution?

58. MIXING SOLUTIONS   How many ounces of the two alcohol solutions must be mixed to obtain 100 ounces of a 12.2% solution?

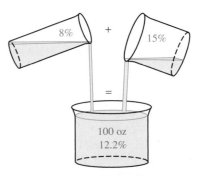

59. DERMATOLOGY   Tests of an antibacterial face wash cream showed that a mixture containing 0.3% Triclosan (active ingredient) gave the best results. How many grams of cream from each tube shown in the illustration should be used to make an equal-size tube of the 0.3% cream?

Contents: 185 g
Daily Face Wash
0.2%
Triclosan

Contents: 185 g
Daily Face Wash
0.7%
Triclosan

60. SALADS   A chef wants to make 1 gallon (128 ounces) of a 50% vinegar-to-oil salad dressing. He only has pure vinegar and a mild 4% vinegar-to-oil salad dressing on hand. How many ounces of each should he mix to make the desired dressing?

WRITING

61. Write a problem that can be solved by solving the system:
$$\begin{cases} x + y = 36 \\ \$1.29x + \$2.29y = \$72.44 \end{cases}$$

62. Write a problem to fit the information given in the table.

	Ounces · Strength =		Amount of insecticide
Weak	x	0.02	0.02x
Strong	y	0.10	0.10y
Mixture	80	0.07	0.07(80)

63. What is a *break point*? Give an example.

64. A woman paid $219 for two blouses and four pairs of pants. If we let x = the cost of a blouse and y = the cost of a pair of pants, an equation modeling the purchase is $2x + 4y = 219$. Explain why there is not enough information to determine the cost of a blouse or the cost of a pair of pants.

65. To solve mixture problems, do you prefer the one-variable or two-variable solution strategy? Explain why.

66. Write a system of two equations in two variables to attempt to solve the following problem. Then explain why the problem has no solution.

>   *How many gallons of a 20% salt solution and how many gallons of a 30% salt solution should be mixed to obtain 10 gallons of a 50% salt solution?*

REVIEW

Fill in the blanks.

67. A _____ number is any number that can be written as a fraction with an integer numerator and a nonzero integer denominator.

68. The _____ of the term $-8c$ is -8.

69. An equation that is true for all values of its variable is called an _____.

70. The _____ of a three-dimensional geometric solid is the amount of space it encloses.

71. If a triangle has exactly two sides with equal measures, it is called an _____ triangle.

72. The _____ of $\frac{5}{9}$ is $\frac{9}{5}$.

CHALLENGE PROBLEMS

73. MANAGING AN APARTMENT   The manager of an apartment complex is also a tenant. He pays only three-fourths of the monthly rent that each of the remaining 5 tenants pays. Each month, a total of $5,520 in rent is paid by the 6 occupants. How much rent does the manager pay?

74. DIGITS PROBLEM   The sum of the digits of a two-digit number is 10. If we interchange the digits, then the new number formed is 54 less than the original. Find the original number.

SECTION 3.4
Solving Systems of Equations in Three Variables

Objectives

1 Determine whether an ordered triple is a solution of a system.

2 Solve systems of three linear equations in three variables.

3 Solve systems of equations with missing variable terms.

4 Identify inconsistent systems and dependent equations.

In previous sections, we solved systems of linear equations in two variables. We will now extend this discussion to consider systems of linear equations in *three* variables.

Determine Whether an Ordered Triple Is a Solution of a System.

The equation $x - 5y + 7z = 10$, where each variable is raised to the first power, is an example of a linear equation in three variables. In general, we have the following definition.

Standard (General) Form	A **linear equation in three variables** is an equation that can be written in the form $$Ax + By + Cz = D$$ where A, B, C, and D are real numbers and A, B, and C are not all 0.

A solution of a linear equation in three variables is an **ordered triple** of numbers of the form (x, y, z) whose coordinates satisfy the equation. For example, $(2, 0, 1)$ is a solution of $x + y + z = 3$ because a true statement results when we substitute 2 for x, 0 for y, and 1 for z: $2 + 0 + 1 = 3$.

A **solution of a system of three linear equations** in three variables is an ordered triple that satisfies each equation of the system.

EXAMPLE 1   Determine whether $(-4, 2, 5)$ is a solution of the system:

$$\begin{cases} 2x + 3y + 4z = 18 \\ 3x + 4y + z = 1 \\ x + y + 3z = 13 \end{cases}$$

Strategy   We will substitute the x-, y-, and z-coordinates of $(-4, 2, 5)$ for the corresponding variables in each equation of the system.

Why   If each equation is satisfied by the x-, y-, and z-coordinates, the ordered triple is a solution of the system.

Solution   We substitute -4 for x, 2 for y, and 5 for z in each equation.

The first equation	The second equation	The third equation
$2x + 3y + 4z = 18$	$3x + 4y + z = 1$	$x + y + 3z = 13$
$2(-4) + 3(2) + 4(5) \overset{?}{=} 18$	$3(-4) + 4(2) + 5 \overset{?}{=} 1$	$-4 + 2 + 3(5) \overset{?}{=} 13$
$-8 + 6 + 20 \overset{?}{=} 18$	$-12 + 8 + 5 \overset{?}{=} 1$	$-4 + 2 + 15 \overset{?}{=} 13$
$18 = 18$ True	$1 = 1$ True	$13 = 13$ True

Since $(-4, 2, 5)$ satisfies each equation, it is a solution of the system.

Self Check 1   Determine whether $(6, -3, 1)$ is a solution of the system:

$$\begin{cases} x - y + z = 10 \\ x + 4y - z = -7 \\ 3x - y + 4z = 24 \end{cases}$$

Now Try   **Problem 11**

The graph of an equation of the form $Ax + By + Cz = D$ is a flat surface called a **plane.** A system of three linear equations with three variables is consistent or inconsistent, depending on how the three planes corresponding to the three equations intersect. The following illustration shows some of the possibilities. A system of three linear equations in three variables can have exactly one solution, no solution, or infinitely many solutions.

Consistent system   Consistent system   Inconsistent systems

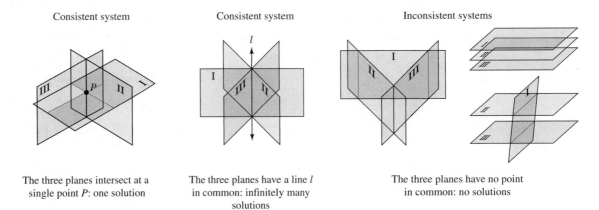

The three planes intersect at a single point P: one solution

The three planes have a line l in common: infinitely many solutions

The three planes have no point in common: no solutions

(a)   (b)   (c)

Solve Systems of Three Linear Equations in Three Variables.

To **solve a system of three linear equations** in three variables means to find all of the solutions of the system. Solving such a system by graphing is not practical because we would need a coordinate system with three axes.

The substitution method is useful to solve systems of three equations where one or more equations have only two variables. However, the best way to solve systems of three linear equations in three variables is usually the elimination method.

Solving a System of Three Linear Equations by Elimination	1. Write each equation in standard form $Ax + By + Cz = D$ and clear any decimals or fractions.
	2. Pick any two equations and eliminate a variable.
	3. Pick a different pair of equations and eliminate the same variable as in step 1.
	4. Solve the resulting pair of two equations in two variables.
	5. To find the value of the third variable, substitute the values of the two variables found in step 4 into any equation containing all three variables and solve the equation.
	6. Check the proposed solution in all three of the original equations. Write the solution as an ordered triple.

EXAMPLE 2   Solve the system: $\begin{cases} 2x + y + 4z = 12 \\ x + 2y + 2z = 9 \\ 3x - 3y - 2z = 1 \end{cases}$

Strategy   Since the coefficients of the z-terms are opposites in the second and third equations, we will add the left and right sides of those equations to eliminate z. Then we will choose another pair of equations and eliminate z again.

Why   The result will be a system of two equations in x and y that we can solve by elimination.

Solution

Step 1: We can skip step 1 because each equation is written in standard form and there are no fractions or decimals to clear. We will number each equation and move to step 2.

(1) $\quad \begin{cases} 2x + y + 4z = 12 \\ \textbf{(2)} \quad x + 2y + 2z = 9 \\ \textbf{(3)} \quad 3x - 3y - 2z = 1 \end{cases}$

Step 2: If we pick equations 2 and 3 and add them, the variable z is eliminated.

(2) $\quad x + 2y + 2z = \ \ 9$
(3) $\quad \underline{3x - 3y - 2z = \ \ 1}$
(4) $\quad 4x - \ y \qquad \ = 10$ This equation does not contain z.

Step 3: We now pick a different pair of equations (equations 1 and 3) and eliminate z again. If each side of equation 3 is multiplied by 2, and the resulting equation is added to equation 1, z is eliminated.

(1) $\quad 2x + \ y + 4z = 12$
$\qquad \underline{6x - 6y - 4z = \ \ 2}$ This is 2(3x − 3y − 2z) = 2(1).
(5) $\quad 8x - 5y \qquad \ = 14$ This equation does not contain z.

Step 4: Equations 4 and 5 form a system of two equations in x and y.

$$\textbf{(4)} \quad \begin{cases} 4x - y = 10 \\ \textbf{(5)} \quad 8x - 5y = 14 \end{cases}$$

To solve this system, we multiply equation 4 by -5 and add the resulting equation to equation 5 to eliminate y.

$$-20x + 5y = -50 \qquad \text{This is } -5(4x - y) = -5(10).$$

$$\textbf{(5)} \quad \underline{8x - 5y = 14}$$

$$-12x = -36$$

$$x = 3 \qquad \text{Divide both sides by } -12. \text{ This is the x-value of the solution.}$$

To find y, we substitute 3 for x in any equation containing x and y (such as equation 5) and solve for y:

$$\textbf{(5)} \quad 8x - 5y = 14$$

$$8(\textbf{3}) - 5y = 14 \qquad \text{Substitute 3 for x.}$$

$$24 - 5y = 14 \qquad \text{Simplify.}$$

$$-5y = -10 \qquad \text{Subtract 24 from both sides.}$$

$$y = 2 \qquad \text{Divide both sides by } -5. \text{ This is the y-value of the solution.}$$

Step 5: To find z, we substitute 3 for x and 2 for y in any equation containing x, y, and z (such as equation 1) and solve for z:

$$\textbf{(1)} \quad 2x + y + 4z = 12$$

$$2(\textbf{3}) + \textbf{2} + 4z = 12 \qquad \text{Substitute 3 for x and 2 for y.}$$

$$8 + 4z = 12 \qquad \text{Simplify.}$$

$$4z = 4 \qquad \text{Subtract 8 from both sides.}$$

$$z = 1 \qquad \text{Divide both sides by 4. This is the z-value of the solution.}$$

Step 6: To verify that the solution is (3, 2, 1), we substitute 3 for x, 2 for y, and 1 for z in the three equations of the original system. The solution set is written as $\{(3, 2, 1)\}$. Since this system has a solution, it is a consistent system.

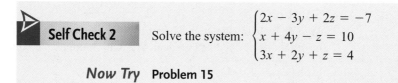

Self Check 2   Solve the system: $\begin{cases} 2x - 3y + 2z = -7 \\ x + 4y - z = 10 \\ 3x + 2y + z = 4 \end{cases}$

Now Try   **Problem 15**

---

③   **Solve Systems of Equations with Missing Variable Terms.**

When one or more of the equations of a system is missing a variable term, the elimination of a variable that is normally performed in step 2 of the solution process can be skipped.

EXAMPLE 3   Solve the system: $\begin{cases} 3x = 6 - 2y + z \\ -y - 2z = -8 - x \\ x = 1 - 2z \end{cases}$

Strategy   Since the third equation does not contain the variable y, we will work with the first and second equations to obtain another equation that does not contain y.

Why Then we can use the elimination method to solve the resulting system of two equations in x and z.

Solution

Step 1: We use the addition property of equality to write each equation in the standard form $Ax + By + Cz = D$ and number each equation.

(1) $\quad 3x + 2y - z = 6 \qquad$ Add 2y and subtract z from both sides of $3x = 6 - 2y + z$.

(2) $\quad x - y - 2z = -8 \qquad$ Add x to both sides of $-y - 2z = -8 - x$.

(3) $\quad x + 2z = 1 \qquad$ Add 2z to both sides of $x = 1 - 2z$.

Step 2: Since equation 3 does not have a y-term, we can skip to step 3, where we will find another equation that does not contain a y-term.

Step 3: If each side of equation 2 is multiplied by 2 and the resulting equation is added to equation 1, y is eliminated.

$$
\begin{array}{llrl}
(1) & 3x + 2y - & z = & 6 \\
& \underline{2x - 2y - 4z = -16} & & \text{This is } 2(x - y - 2z) = 2(-8). \\
(4) & 5x \qquad\quad -5z = & -10 &
\end{array}
$$

Step 4: Equations 3 and 4 form a system of two equations in x and z:

(3) $\quad x + 2z = 1$

(4) $\quad 5x - 5z = -10$

To solve this system, we multiply equation 3 by -5 and add the resulting equation to equation 4 to eliminate x:

$$
\begin{array}{lrl}
& -5x - 10z = -5 & \text{This is } -5(x + 2z) = -5(1). \\
(4) & \underline{5x - 5z = -10} & \\
& \qquad -15z = -15 & \\
& \qquad\quad\; z = 1 & \text{Divide both sides by } -15. \text{ This is the z-value of the solution.}
\end{array}
$$

To find x, we substitute 1 for z in equation 3.

$$
\begin{array}{lll}
(3) & x + 2z = 1 & \\
& x + 2(1) = 1 & \text{Substitute 1 for z.} \\
& x + 2 = 1 & \text{Multiply.} \\
& x = -1 & \text{Subtract 2 from both sides.}
\end{array}
$$

Step 5: To find y, we substitute -1 for x and 1 for z in equation 1:

$$
\begin{array}{lll}
(1) & 3x + 2y - z = 6 & \\
& 3(-1) + 2y - 1 = 6 & \text{Substitute } -1 \text{ for x and 1 for z.} \\
& -3 + 2y - 1 = 6 & \text{Multiply.} \\
& 2y = 10 & \text{Simplify and add 4 to both sides.} \\
& y = 5 & \text{Divide both sides by 2.}
\end{array}
$$

The solution of the system is $(-1, 5, 1)$ and the solution set is $\{(-1, 5, 1)\}$.

Step 6: Check the proposed solution in all three of the original equations.

Success Tip

We don't have to find the values of the variables in alphabetical order. In step 2, choose the variable that is the easiest to eliminate. In this example, the value of z is found first.

STUDY SET
3.4

VOCABULARY

Fill in the blanks.

1. $\begin{cases} 2x + y - 3z = 0 \\ 3x - y + 4z = 5 \\ 4x + 2y - 6z = 0 \end{cases}$ is called a _____ of three linear equations in three variables. Each equation is written in _____ $Ax + By + Cz = D$ form.

2. If the first two equations of the system in Exercise 1 are added, the variable y is _____.

3. Solutions of a system of three equations in three variables, x, y, and z, are written in the form (x, y, z) and are called ordered _____.

4. The graph of the equation $2x + 3y + 4z = 5$ is a flat surface called a _____.

5. When three planes coincide, the equations of the system are _____, and there are infinitely many solutions.

6. When three planes intersect in a line, the system will have _____ many solutions.

CONCEPTS

7. For each graph of a system of three equations, determine whether the solution set contains one solution, infinitely many solutions, or no solution.

a. 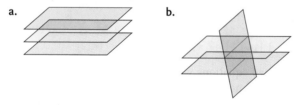   b.

8. Consider the system: $\begin{array}{l}(1)\\(2)\\(3)\end{array} \begin{cases} -2x + y + 4z = 3 \\ x - y + 2z = 1 \\ x + y - 3z = 2 \end{cases}$

   a. What is the result if equation 1 and equation 2 are added?

   b. What is the result if equation 2 and equation 3 are added?

   c. What variable was eliminated in the steps performed in parts (a) and (b)?

NOTATION

9. For the following system, clear the equations of any fractions or decimals and write each equation in $Ax + By + Cz = D$ form.

$$\begin{cases} x + y = 3 - 4z & \longrightarrow \\ 0.7x - 0.2y + 0.8z = 1.5 & \longrightarrow \\ \dfrac{x}{2} + \dfrac{y}{3} - \dfrac{z}{6} = \dfrac{2}{3} & \longrightarrow \end{cases} \begin{cases} \\ \\ \\ \end{cases}$$

10. What is the purpose of the numbers shown in red in front of the equations below?

$\begin{array}{l}(1)\\(2)\\(3)\end{array} \begin{cases} x + y - z = 6 \\ 2x - y + z = 3 \\ 5x + 3y - z = -2 \end{cases}$

GUIDED PRACTICE

Determine whether the ordered triple is a solution of the system. See Example 1.

11. $(2, 1, 1)$, $\begin{cases} x - y + z = 2 \\ 2x + y - z = 4 \\ 2x - 3y + z = 2 \end{cases}$

12. $(-3, 2, -1)$, $\begin{cases} 3x + y - z = -6 \\ 2x + 2y + 3z = -1 \\ x + y + 2z = 1 \end{cases}$

13. $(6, -7, -5)$, $\begin{cases} 3x - 2y - z = 37 \\ x - 3y = 27 \\ 2x + 7y + 2z = -48 \end{cases}$

14. $(-4, 0, 9)$, $\begin{cases} x + 2y - 3z = -31 \\ 2x + 6z = 46 \\ 3x - y = -12 \end{cases}$

Solve each system. See Example 2.

15. $\begin{cases} x + y + z = 4 \\ 2x + y - z = 1 \\ 2x - 3y + z = 1 \end{cases}$ 16. $\begin{cases} x + y + z = 4 \\ x - y + z = 2 \\ x - y - 2z = -1 \end{cases}$

17. $\begin{cases} 3x + 2y - 5z = 3 \\ 4x - 2y - 3z = -10 \\ 5x - 2y - 2z = -11 \end{cases}$ 18. $\begin{cases} 5x + 4y + 2z = -2 \\ 3x + 4y - 3z = -27 \\ 2x - 4y - 7z = -23 \end{cases}$

19. $\begin{cases} 2x + 6y + 3z = 9 \\ 5x - 3y - 5z = 3 \\ 4x + 3y + 2z = 15 \end{cases}$ 20. $\begin{cases} 4x - 3y + 5z = 23 \\ 2x - 5y - 3z = 13 \\ -4x - 6y + 7z = 7 \end{cases}$

21. $\begin{cases} 4x - 5y - 8z = -52 \\ 2x - 3y - 4z = -26 \\ 3x + 7y + 8z = 31 \end{cases}$

22. $\begin{cases} 2x + 6y + 3z = -20 \\ 5x - 3y - 5z = 47 \\ 4x + 3y + 2z = 4 \end{cases}$

Solve each system. See Example 3.

23. $\begin{cases} 3x + 3z = 6 - 4y \\ 7x - 5z = 46 + 2y \\ 4x = 31 - z \end{cases}$

24. $\begin{cases} 5x + 6z = 4y - 21 \\ 9x + 2y = 3z - 47 \\ 3x + y = -19 \end{cases}$

25. $\begin{cases} 2x + z = -2 + y \\ 8x - 3y = -2 \\ 6x - 2y + 3z = -4 \end{cases}$

26. $\begin{cases} 3y + z = -1 \\ -x + 2z = -9 + 6y \\ 9y + 3z = -9 + 2x \end{cases}$

Solve each system using elimination. See Example 4.

27. $\begin{cases} x + y + 3z = 35 \\ -x - 3y = 20 \\ 2y + z = -35 \end{cases}$

28. $\begin{cases} x + 2y + 3z = 11 \\ 5x - y = 13 \\ 2x - 3z = -11 \end{cases}$

29. $\begin{cases} 3x + 2y - z = 7 \\ 6x - 3y = -2 \\ 3y - 2z = 8 \end{cases}$

30. $\begin{cases} 2x + y = 4 \\ -x - 2y + 8z = 7 \\ -y + 4z = 5 \end{cases}$

Solve each system using substitution. See Example 5.

31. $\begin{cases} r + s - 3t = 21 \\ r + 4s = 9 \\ 5s + t = -4 \end{cases}$

32. $\begin{cases} r - s + 6t = 12 \\ r + 6s = -28 \\ 7s + t = -26 \end{cases}$

33. $\begin{cases} x - 8z = -30 \\ 3x + y - 4z = 5 \\ y + 7z = 30 \end{cases}$

34. $\begin{cases} x + 6z = -36 \\ 5x + 3y - 2z = -20 \\ y + 4z = -20 \end{cases}$

Solve each system. If a system is inconsistent or if the equations are dependent, state this. See Examples 6 and 7.

35. $\begin{cases} 7a + 9b - 2c = -5 \\ 5a + 14b - c = -11 \\ 2a - 5b - c = 3 \end{cases}$

36. $\begin{cases} 3x + 4y + z = 10 \\ x - 2y + z = -3 \\ 2x + y + z = 5 \end{cases}$

37. $\begin{cases} 7x - y - z = 10 \\ x - 3y + z = 2 \\ x + 2y - z = 1 \end{cases}$

38. $\begin{cases} 2a - b + c = 6 \\ -5a - 2b - 4c = -30 \\ a + b + c = 8 \end{cases}$

Solve each system, if possible. If a system is inconsistent or if the equations are dependent, state this.

39. $\begin{cases} 2a + 3b - 2c = 18 \\ 5a - 6b + c = 21 \\ 4b - 2c - 6 = 0 \end{cases}$

40. $\begin{cases} r - s + t = 4 \\ r + 2s - t = -1 \\ r + s - 3t = -2 \end{cases}$

41. $\begin{cases} 2x + 2y - z = 2 \\ x + 3z - 24 = 0 \\ y = 7 - 4z \end{cases}$

42. $\begin{cases} r - 3t = -11 \\ r + s + t = 13 \\ s - 4t = -12 \end{cases}$

43. $\begin{cases} b + 2c = 7 - a \\ a + c = 2(4 - b) \\ 2a + b + c = 9 \end{cases}$

44. $\begin{cases} 0.02a = 0.02 - 0.03b - 0.01c \\ 4a + 6b + 2c - 5 = 0 \\ a + c = 3 + 2b \end{cases}$

45. $\begin{cases} 2x + y - z = 1 \\ x + 2y + 2z = 2 \\ 4x + 5y + 3z = 3 \end{cases}$

46. $\begin{cases} 2x + 2y + 3z = 10 \\ 3x + y - z = 0 \\ x + y + 2z = 6 \end{cases}$

47. $\begin{cases} 0.4x + 0.3z = 0.4 \\ 2y - 6z = -1 \\ 4(2x + y) = 9 - 3z \end{cases}$

48. $\begin{cases} a + b + c = 180 \\ \dfrac{a}{4} + \dfrac{b}{2} + \dfrac{c}{3} = 60 \\ 2b + 3c - 330 = 0 \end{cases}$

49. $\begin{cases} r + s + 4t = 3 \\ 3r + 7t = 0 \\ 3s + 5t = 0 \end{cases}$

50. $\begin{cases} x - y = 3 \\ 2x - y + z = 1 \\ x + z = -2 \end{cases}$

51. $\begin{cases} 0.5a + 0.3b = 2.2 \\ 1.2c - 8.5b = -24.4 \\ 3.3c + 1.3a = 29 \end{cases}$

52. $\begin{cases} 4a - 3b = 1 \\ 6a - 8c = 1 \\ 2b - 4c = 0 \end{cases}$

53. $\begin{cases} 2x + 3y = 6 - 4z \\ 2x = 3y + 4z - 4 \\ 4x + 6y + 8z = 12 \end{cases}$

54. $\begin{cases} -x + 5y - 7z = 0 \\ 4x + y - z = 0 \\ x + y - 4z = 0 \end{cases}$

55. $\begin{cases} a + b = 2 + c \\ a = 3 + b - c \\ -a + b + c - 4 = 0 \end{cases}$

56. $\begin{cases} 0.1x - 0.3y + 0.4z = 0.2 \\ 2x + y + 2z = 3 \\ 4x - 5y + 10z = 7 \end{cases}$

57. $\begin{cases} x + \dfrac{1}{3}y + z = 13 \\ \dfrac{1}{2}x - y + \dfrac{1}{3}z = -2 \\ x + \dfrac{1}{2}y - \dfrac{1}{3}z = 2 \end{cases}$

58. $\begin{cases} x - \dfrac{1}{5}y - z = 9 \\ \dfrac{1}{4}x + \dfrac{1}{5}y - \dfrac{1}{2}z = 5 \\ 2x + y + \dfrac{1}{6}z = 12 \end{cases}$

APPLICATIONS

59. GRAPHS OF SYSTEMS Explain how each of the following pictures could be thought of as an example of the graph of a system of three equations. Then describe the solution, if there is any.

a.   **b.**

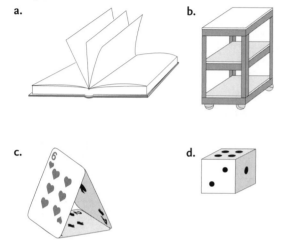

c.   **d.**

60. ZOOLOGY An X-ray of a mouse revealed a cancerous tumor located at the intersection of the coronal, sagittal, and transverse planes. From this description, would you expect the tumor to be at the base of the tail, on the back, in the stomach, on the tip of the right ear, or in the mouth of the mouse?

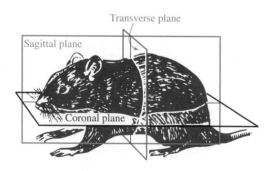

61. NBA RECORDS The three highest one-game point totals by one player in a National Basketball Association game are shown below.

Pts	Player, team	Date
x	Wilt Chamberlain, Philadelphia	3/2/1962
y	Kobe Bryant, Los Angeles	1/22/2006
z	Wilt Chamberlain, Philadelphia	12/8/1961

Solve the following system to find x, y, and z.

$\begin{cases} x + y + z = 259 \\ x - y = 19 \\ x - z = 22 \end{cases}$

62. BICYCLE FRAMES The angle measures of the triangular part of the bicycle frame shown can be found by solving the following system. Find x, y, and z.

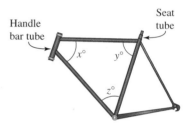

$\begin{cases} x + y + z = 180 \\ x + y = 120 \\ y + z = 135 \end{cases}$

WRITING

63. Explain how a system of three equations in three variables can be reduced to a system of two equations in two variables.

64. What makes a system of three equations with three variables inconsistent?

65. What does the graph of a linear equation in three variables such as $2x - 3y + 9z = 10$ look like?

66. What situation discussed in this section looks like two walls of a room and the floor meeting in a corner?

REVIEW

Graph each of the basic functions.

67. $f(x) = |x|$ **68.** $g(x) = x^2$

69. $h(x) = x^3$ **70.** $S(x) = x$

CHALLENGE PROBLEMS

Solve each system.

71. $\begin{cases} w + x + y + z = 3 \\ w - x + y + z = 1 \\ w + x - y + z = 1 \\ w + x + y - z = 3 \end{cases}$

72. $\begin{cases} \dfrac{1}{x} + \dfrac{1}{y} + \dfrac{1}{z} = 3 \\ \dfrac{2}{x} + \dfrac{1}{y} - \dfrac{1}{z} = 0 \\ \dfrac{1}{x} - \dfrac{2}{y} + \dfrac{4}{z} = 21 \end{cases}$

SECTION 3.5
Problem Solving Using Systems of Three Equations

Objectives   ❶ Assign variables to three unknowns.

❷ Use systems to solve curve-fitting problems.

Problems that involve three unknown quantities can be solved using a strategy similar to that for solving problems involving two unknowns. To solve such problems, we will write three equations in three variables to model the situation and then we will use the methods of Section 3.4 to solve the system formed by the three equations.

 Assign Variables to Three Unknowns.

> **EXAMPLE 1**   *Tool Manufacturing.* A company makes three types of hammers, which are marketed as "good," "better," and "best." The cost of manufacturing each type of hammer is $4, $6, and $7, respectively, and the hammers sell for $6, $9, and $12. Each day, the cost of manufacturing 100 hammers is $520, and the daily revenue from their sale is $810. How many hammers of each type are manufactured?

Analyze the Problem   We need to find how many of each type of hammer are manufactured daily. Since there are three unknowns, we must write three equations to find them.

Form Three Equations   Let $x =$ the number of good hammers, $y =$ the number of better hammers, and $z =$ the number of best hammers. We know that

The cost of manufacturing
- the good hammers is $4x$ ($4 times x hammers).
- the better hammers is $6y$ ($6 times y hammers).
- the best hammers is $7z$ ($7 times z hammers).

The revenue received by selling
- the good hammers is $6x$ ($6 times x hammers).
- the better hammers is $9y$ ($9 times y hammers).
- the best hammers is $12z$ ($12 times z hammers).

We can use the facts of the problem to write three equations.

The number of good hammers	plus	the number of better hammers	plus	the number of best hammers	is	the total number of hammers.
x	$+$	y	$+$	z	$=$	100

The cost of good hammers	plus	the cost of better hammers	plus	the cost of best hammers	is	the total cost.
$4x$	$+$	$6y$	$+$	$7z$	$=$	520

The revenue from good hammers	plus	the revenue from better hammers	plus	the revenue from best hammers	is	the total revenue.
$6x$	$+$	$9y$	$+$	$12z$	$=$	810

Check the Result   The sum of $103 + 92 + 63$ is 258. Furthermore, 103 is 11 more than 92, and 92 is 29 more than 63. The results check.

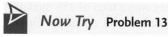 **Now Try   Problem 13**

2 **Use Systems to Solve Curve-Fitting Problems.**

The process of determining an equation whose graph contains given points is called **curve fitting**.

EXAMPLE 3   The equation of a parabola opening upward or downward is of the form $y = ax^2 + bx + c$. Find the equation of the parabola graphed on the left by determining the values of a, b, and c.

Strategy   We will substitute the x- and y-coordinates of three points that lie on the graph into the equation $y = ax^2 + bx + c$. This will produce a system of three equations in three variables that we can solve to find a, b, and c.

Why   Once we know a, b, and c, we can write the equation.

Solution   Since the parabola passes through the points $(-1, 5)$, $(1, 1)$, and $(2, 2)$, each pair of coordinates must satisfy the equation $y = ax^2 + bx + c$. If we substitute each pair into $y = ax^2 + bx + c$, we will get a system of three equations in three variables.

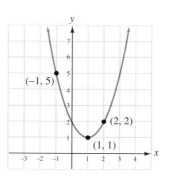

Success Tip
If a point lies on the graph of an equation, it is a solution of the equation, and the coordinates of the point satisfy the equation.

Substitute $(-1, 5)$	*Substitute* $(1, 1)$	*Substitute* $(2, 2)$
$y = ax^2 + bx + c$	$y = ax^2 + bx + c$	$y = ax^2 + bx + c$
$5 = a(-1)^2 + b(-1) + c$	$1 = a(1)^2 + b(1) + c$	$2 = a(2)^2 + b(2) + c$
$5 = a - b + c$	$1 = a + b + c$	$2 = 4a + 2b + c$
This is equation 1.	This is equation 2.	This is equation 3.

The three equations above give the system, which we can solve to find a, b, and c.

$$\begin{array}{ll} (1) \\ (2) \\ (3) \end{array} \begin{cases} a - b + c = 5 \\ a + b + c = 1 \\ 4a + 2b + c = 2 \end{cases}$$

If we add equations 1 and 2, we obtain

$$\begin{array}{l} a - b + c = 5 \\ \underline{a + b + c = 1} \end{array}$$
$$(4) \quad 2a + 2c = 6$$

If we multiply equation 1 by 2 and add the result to equation 3, we get

$$2a - 2b + 2c = 10$$
$$(3) \quad \underline{4a + 2b + c = 2}$$
$$(5) \quad 6a + 3c = 12$$

We can then divide both sides of equation 4 by 2 to get equation 6 and divide both sides of equation 5 by 3 to get equation 7. We now have the system

$$\begin{array}{ll} (6) \\ (7) \end{array} \begin{cases} a + c = 3 \\ 2a + c = 4 \end{cases}$$

To eliminate c, we multiply equation 6 by -1 and add the result to equation 7. We get

$$-a - c = -3 \qquad \text{This is } -1(a + c) = -1(3).$$
$$\underline{2a + c = 4}$$
$$a = 1$$

To find c, we can substitute 1 for a in equation 6 and find that $c = 2$. To find b, we can substitute 1 for a and 2 for c in equation 2 and find that $b = -2$.

After we substitute these values of a, b, and c into the equation $y = ax^2 + bx + c$, we have the equation of the parabola.

$$y = ax^2 + bx + c$$
$$y = 1x^2 - 2x + 2$$
$$y = x^2 - 2x + 2$$

▷ **Now Try**   **Problem 25**

STUDY SET
3.5

VOCABULARY

Fill in the blanks.

1. If a point lies on the graph of an equation, it is a solution of the equation, and the coordinates of the point _____ the equation.

2. The process of determining an equation whose graph contains given points is called curve _____.

CONCEPTS

Write a system of three equations in three variables that models the situation. Do not solve the system.

3. DESSERTS   A bakery makes three kinds of pies: chocolate cream, which sells for $5; apple, which sells for $6; and cherry, which sells for $7. The cost to make the pies is $2, $3, and $4, respectively. Let $x =$ the number of chocolate cream pies made daily, $y =$ the number of apple pies made daily, and $z =$ the number of cherry pies made daily.
   - Each day, the bakery makes 50 pies.
   - Each day, the revenue from the sale of the pies is $295.
   - Each day, the cost to make the pies is $145.

4. FAST FOODS   Let $x =$ the number of calories in a Big Mac hamburger, $y =$ the number of calories in a small order of French fries, and $z =$ the number of calories in a medium Coca-Cola.
   - The total number of calories in a Big Mac hamburger, a small order of French fries, and a medium Coke is 1,000.
   - The number of calories in a Big Mac is 260 more than in a small order of French fries.
   - The number of calories in a small order of French fries is 40 more than in a medium Coke. (Source: McDonald's USA)

5. What equation results when the coordinates of the point $(2, -3)$ are substituted into $y = ax^2 + bx + c$?

6. The equation $y = 5x^2 - 6x + 1$ is written in the form $y = ax^2 + bx + c$. What are a, b, and c?

APPLICATIONS

7. MAKING STATUES   An artist makes three types of ceramic statues (large, medium, and small) at a monthly cost of $650 for 180 statues. The manufacturing costs for the three types are $5, $4, and $3. If the statues sell for $20, $12, and $9, respectively, how many of each type should be made to produce $2,100 in monthly revenue?

8. PUPPETS   A toy company makes a total of 500 puppets in three sizes during a production run. The small puppets cost $5 to make and sell for $8 each, the standard size puppets cost $10 to make and sell for $16 each, and the super-size puppets cost $15 to make and sell for $25. The total cost to make the puppets is $4,750 and the revenue from their sale is $7,700. How many small, standard, and super-size puppets are made during a production run?

9. NUTRITION   A dietitian is to design a meal that will provide a patient with exactly 14 grams (g) of fat, 9 g of carbohydrates, and 9 g of protein. She is to use a combination of the three foods listed in the table. If one ounce of each of the foods has the nutrient content shown in the table, how many ounces of each food should be used?

Food	Fat	Carbohydrates	Protein
A	2 g	1 g	2 g
B	3 g	2 g	1 g
C	1 g	1 g	2 g

(g stands for gram)

10. NUTRITIONAL PLANNING   One ounce of each of three foods has the vitamin and mineral content shown in the table. How many ounces of each must be used to provide exactly 22 milligrams (mg) of niacin, 12 mg of zinc, and 20 mg of vitamin C?

Food	Niacin	Zinc	Vitamin C
A	1 mg	1 mg	2 mg
B	2 mg	1 mg	1 mg
C	2 mg	1 mg	2 mg

(mg stands for milligram)

11. *from Campus to Careers*
Fashion Designer

A clothing manufacturer makes coats, shirts, and slacks. The time required for cutting, sewing, and packaging each item is shown in the table. How many of each should be made to use all available labor hours?

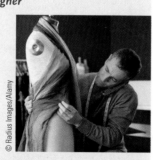
© Radius Images/Alamy

	Coats	Shirts	Slacks	Time available
Cutting	20 min	15 min	10 min	115 hr
Sewing	60 min	30 min	24 min	280 hr
Packaging	5 min	12 min	6 min	65 hr

12. SCULPTING   A wood sculptor carves three types of statues with a chainsaw. The number of hours required for carving, sanding, and painting a totem pole, a bear, and a deer are shown in the table. How many of each should be produced to use all available labor hours?

	Totem pole	Bear	Deer	Time available
Carving	2 hr	2 hr	1 hr	14 hr
Sanding	1 hr	2 hr	2 hr	15 hr
Painting	3 hr	2 hr	2 hr	21 hr

13. NFL RECORDS   Jerry Rice, who played the majority of his career with the San Francisco 49ers and the Oakland Raiders, holds the all-time record for touchdown (TD) passes caught. Here are some interesting facts about this feat.

- He caught 30 more TD passes from Steve Young than he did from Joe Montana.
- He caught 39 more TD passes from Joe Montana than he did from Rich Gannon.
- He caught a total of 156 TD passes from Young, Montana, and Gannon.

Determine the number of touchdown passes Rice has caught from Young, from Montana, and from Gannon.

14. HOT DOGS   In 12 minutes, the top three finishers in the 2007 Nathan's Hot Dog Eating Contest consumed a total of 178 hot dogs. The winner, Joey Chestnut, ate 3 more hot dogs than the runner-up, Takeru Kobayashi. Pat Bertoletti finished a distant third, 14 hot dogs behind Kobayashi. How many hot dogs did each person eat?

15. EARTH'S ATMOSPHERE   Use the information in the circle graph to determine what percent of Earth's atmosphere is nitrogen, is oxygen, and is other gases.

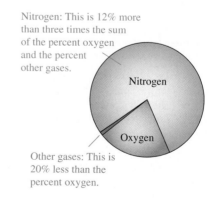

Nitrogen: This is 12% more than three times the sum of the percent oxygen and the percent other gases.

Other gases: This is 20% less than the percent oxygen.

16. DECEASED CELEBRITIES   Between October 2005 and October 2006, the estates of Kurt Cobain, Elvis Presley, and Charles M. Schultz (Snoopy cartoonist) earned a total of $127 million. Together, the Presley and Schultz estates earned $27 million more than the Cobain estate. If the Schultz estate had earned $15 million more, it would equal the value of the Cobain estate. Use this information to label to vertical axis of the graph below. (Source: Forbes.com)

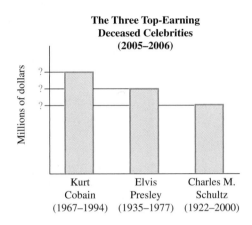

The Three Top-Earning Deceased Celebrities (2005–2006)

17. TRIANGLES   The sum of the measures of the angles of any triangle is 180°. In $\triangle ABC$, $\angle A$ measures 100° less than the sum of the measures of $\angle B$ and $\angle C$, and the measure of $\angle C$ is 40° less than twice the measure of $\angle B$. Find the measure of each angle of the triangle.

18. QUADRILATERALS   A quadrilateral is a four-sided polygon. The sum of the measures of the angles of any quadrilateral is 360°. In the illustration below, the measures of $\angle A$ and $\angle B$ are the same. The measure of $\angle C$ is 20° greater than the measure of $\angle A$, and the measure of $\angle D$ is 60° less than $\angle B$. Find the measure of $\angle A$, $\angle B$, $\angle C$, and $\angle D$.

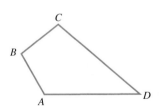

19. TV HISTORY   *X-Files, Will & Grace,* and *Seinfeld* are three of the most popular television shows of all time. The total number of episodes of these three shows is 575. There are 21 more episodes of *X-Files* than *Seinfeld,* and the difference between the number of episodes of *Will & Grace* and *Seinfeld* is 14. Find the number of episodes of each show.

20. TRAFFIC LIGHTS   At a traffic light, one cycle through green-yellow-red lasts for 80 seconds. The green light is on eight times longer than the yellow light, and the red light is on eleven times longer than the yellow light. For how long is each colored light on during one cycle?

21. POTPOURRI   The owner of a home decorating shop wants to mix dried rose petals selling for $6 per pound, dried lavender selling for $5 per pound, and buckwheat hulls selling for $4 per pound to get 10 pounds of a mixture that would sell for $5.50 per pound. She wants to use twice as many pounds of rose petals as lavender. How many pounds of each should she use?

22. MIXING NUTS   The owner of a candy store wants to mix some peanuts worth $3 per pound, some cashews worth $9 per pound, and some Brazil nuts worth $9 per pound to get 50 pounds of a mixture that will sell for $6 per pound. She uses 15 fewer pounds of cashews than peanuts. How many pounds of each did she use?

23. PIGGY BANKS   When a child breaks open her piggy bank, she finds a total of 64 coins, consisting of nickels, dimes, and quarters. The total value of the coins is $6. If the nickels were dimes, and the dimes were nickels, the value of the coins would be $5. How many nickels, dimes, and quarters were in the piggy bank?

24. THEATER SEATING   The illustration shows the cash receipts and the ticket prices from two sold-out Sunday performances of a play. Find the number of seats in each of the three sections of the 800-seat theater.

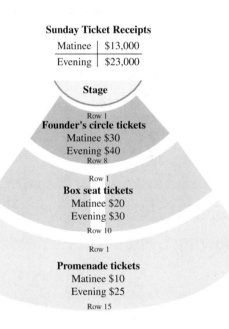

Sunday Ticket Receipts	
Matinee	$13,000
Evening	$23,000

25. ASTRONOMY   Comets have elliptical orbits, but the orbits of some comets are so large that they are indistinguishable from parabolas. Find an equation of the form $y = ax^2 + bx + c$ for the parabola that closely describes the orbit of the comet shown in the illustration.

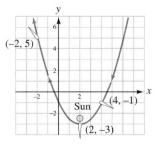

26. CURVE FITTING Find an equation of the form $y = ax^2 + bx + c$ for the parabola shown in the illustration.

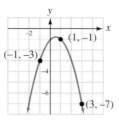

27. WALKWAYS A circular sidewalk is to be constructed in a city park. The walk is to pass by three particular areas of the park, as shown in the illustration. If an equation of a circle is of the form $x^2 + y^2 + Cx + Dy + E = 0$, find an equation that describes the path of the sidewalk by determining C, D, and E.

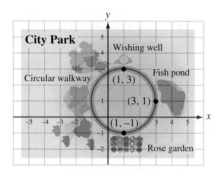

28. CURVE FITTING The equation of a circle is of the form $x^2 + y^2 + Cx + Dy + E = 0$. Find an equation of the circle shown in the illustration by determining C, D, and E.

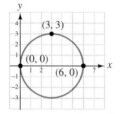

WRITING

29. Explain why the following problem does not give enough information to answer the question: The sum of three integers is 48. If the first integer is doubled, the sum is 60. Find the integers.

30. Write an application problem that can be solved using a system of three equations in three variables.

REVIEW

Determine whether each equation defines y to be a function of x. If it does not, find two ordered pairs where more than one value of y corresponds to a single value of x.

31. $y = \dfrac{1}{x}$

32. $y^4 = x$

33. $xy = 9$

34. $y = |x|$

35. $x + 1 = |y|$

36. $y = \dfrac{1}{x^2}$

37. $y^2 = x$

38. $x = |y|$

CHALLENGE PROBLEMS

39. DIGITS PROBLEM The sum of the digits of a three-digit number is 8. Twice the hundreds digit plus the tens digit is equal to the ones digit. If the digits of the number are reversed, the new number is 82 more than twice the original number. What is the three-digit number?

40. PURCHASING PETS A pet store owner spent $100 to buy 100 animals. He bought at least one iguana, one guinea pig, and one mouse, but no other kinds of animals. If an iguana cost $10.00, a guinea pig cost $3.00, and a mouse cost $0.50, how many of each did he buy?

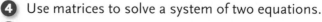

SECTION 3.6
Solving Systems of Equations Using Matrices

Objectives

1 Define a matrix and determine its order.

2 Write the augmented matrix for a system.

3 Perform elementary row operations on matrices.

4 Use matrices to solve a system of two equations.

5 Use matrices to solve a system of three equations.

6 Use matrices to identify inconsistent systems and dependent equations.

In this section, we will discuss another way to solve systems of linear equations. This technique uses a mathematical tool called a *matrix* in a series of steps that are based on the elimination (addition) method.

Define a Matrx and Determine Its Order.

Another way to solve systems of equations involves rectangular arrays of numbers called *matrices* (plural of matrix).

Matrices	A **matrix** is any rectangular array of numbers arranged in rows and columns, written within brackets.

The Language of Algebra
An *array* is an orderly arrange-ment. For example, a jewelry store might display an impressive *array* of gemstones.

Some examples of matrices are

$$A = \begin{bmatrix} 1 & -3 & 8 \\ 2 & 5 & -1 \end{bmatrix} \begin{matrix} \leftarrow \text{Row 1} \\ \leftarrow \text{Row 2} \end{matrix} \qquad B = \begin{bmatrix} 1 & 4 & -2 & -4 \\ 6 & -2 & 6 & 1 \\ 3 & 8 & -3 & 12 \end{bmatrix} \begin{matrix} \leftarrow \text{Row 1} \\ \leftarrow \text{Row 2} \\ \leftarrow \text{Row 3} \end{matrix}$$

Column Column Column   Column Column Column Column
1   2   3   1   2   3   4

Each number in a matrix is called an **element** or an **entry** of the matrix. A matrix with m rows and n columns has **order** $m \times n$, which is read as "m by n." Because matrix A has two rows and three columns, its order is 2×3. The order of matrix B is 3×4 because it has three rows and four columns.

2 **Write the Augmented Matrix for a System.**

To show how to use matrices to solve systems of linear equations, we consider the system

$$\begin{cases} x - y = 4 \\ 2x + y = 5 \end{cases}$$

Caution
The equations of a system must be written in standard form before the corresponding augmented matrix can be written.

which can be represented by the following matrix, called an **augmented matrix:**

$$\begin{bmatrix} 1 & -1 & \vdots & 4 \\ 2 & 1 & \vdots & 5 \end{bmatrix}$$

Each row of the augmented matrix represents one equation of the system. The first two columns of the augmented matrix are determined by the coefficients of x and y in the equations of the system. The last column is determined by the constants in the equations.

$$\begin{bmatrix} 1 & -1 & \vdots & 4 \\ 2 & 1 & \vdots & 5 \end{bmatrix}$$ This row represents the equation $x - y = 4$.
                          This row represents the equation $2x + y = 5$.

Coefficients   Coefficients   Constants
of x   of y

EXAMPLE 1   Represent each system using an augmented matrix:

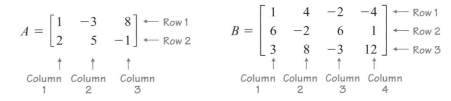

   a. $\begin{cases} 3x + y = 11 \\ x - 8y = 0 \end{cases}$ **b.** $\begin{cases} 2a + b - 3c = -3 \\ 9a + 4c = 2 \\ a - b - 6c = -7 \end{cases}$

Strategy   We will write the coefficients of the variables and the constants from each equation in rows to form a matrix. The coefficients are written to the left of a vertical dashed line and constants to the right.

Why In an augmented matrix, each row represents one equation of the system.

Solution Since the equations of each system are written in standard form, we can easily write the corresponding augmented matrices.

a. $\begin{cases} 3x + y = 11 \\ x - 8y = 0 \end{cases}$ $\leftrightarrow$ $\begin{bmatrix} 3 & 1 & | & 11 \\ 1 & -8 & | & 0 \end{bmatrix}$

b. $\begin{cases} 2a + b - 3c = -3 \\ 9a + 4c = 2 \\ a - b - 6c = -7 \end{cases}$ $\begin{matrix} \leftrightarrow \\ \leftrightarrow \\ \leftrightarrow \end{matrix}$ $\begin{bmatrix} 2 & 1 & -3 & | & -3 \\ 9 & 0 & 4 & | & 2 \\ 1 & -1 & -6 & | & -7 \end{bmatrix}$ In the second row, 0 is entered as the coefficient of the missing b-term.

Self Check 1 Represent each system using an augmented matrix:

a. $\begin{cases} 2x - 4y = 9 \\ 5x - y = -2 \end{cases}$ **b.** $\begin{cases} a + b - c = -4 \\ -2b + 7c = 0 \\ 10a + 8b - 4c = 5 \end{cases}$

Now Try Problem 13

③ Perform Elementary Row Operations on Matrices.

The Language of Algebra
Two matrices are *equivalent* if they represent systems that have the same solution set.

To solve a 2×2 system of equations using matrices, we transform the augmented matrix into an equivalent matrix that has 1's down its main diagonal and a 0 below the 1 in the first column. A matrix written in this form is said to be in **row echelon form.** We can easily determine the solution of the associated system of equations when an augmented matrix is written in this form.

$\begin{bmatrix} 1 & a & | & b \\ 0 & 1 & | & c \end{bmatrix}$ a, b, and c represent real numbers.

Main diagonal

To write an augmented matrix in row echelon form, we use three operations called *elementary row operations.*

Elementary Row Operations

Type 1: Any two rows of a matrix can be interchanged.
Type 2: Any row of a matrix can be multiplied by a nonzero constant.
Type 3: Any row of a matrix can be changed by adding a nonzero constant multiple of another row to it.

None of these row operations affect the solution of a given system of equations. The changes to the augmented matrix produce equivalent matrices that correspond to systems with the same solution.

- A type 1 row operation corresponds to interchanging two equations of the system.
- A type 2 row operation corresponds to multiplying both sides of an equation by a nonzero constant.
- A type 3 row operation corresponds to adding a nonzero multiple of one equation to another.

EXAMPLE 2 Perform the following elementary row operations.

$$A = \begin{bmatrix} 2 & 4 & \vdots & -3 \\ 1 & -8 & \vdots & 0 \end{bmatrix} \qquad B = \begin{bmatrix} 1 & -1 & \vdots & 2 \\ 4 & -8 & \vdots & 0 \end{bmatrix} \qquad C = \begin{bmatrix} 2 & 1 & -8 & \vdots & 4 \\ 0 & 1 & 4 & \vdots & -2 \\ 0 & 0 & -6 & \vdots & 24 \end{bmatrix}$$

a. Type 1: Interchange rows 1 and 2 of matrix A.

b. Type 2: Multiply row 3 of matrix C by $-\frac{1}{6}$.

c. Type 3: To the numbers in row 2 of matrix B, add the results of multiplying each number in row 1 by -4.

Strategy We will perform elementary row operations on each matrix as if we were performing those operations on the equations of a system.

Why The rows of an augmented matrix correspond to the equations of a system.

Solution

a. Interchanging rows 1 and 2 of matrix $A = \begin{bmatrix} 2 & 4 & \vdots & -3 \\ 1 & -8 & \vdots & 0 \end{bmatrix}$ gives $\begin{bmatrix} 1 & -8 & \vdots & 0 \\ 2 & 4 & \vdots & -3 \end{bmatrix}$.

We can represent the instruction to interchange rows 1 and 2 with the symbol $R_1 \leftrightarrow R_2$.

b. We multiply each number in row 3 of matrix $C = \begin{bmatrix} 2 & 1 & -8 & \vdots & 4 \\ 0 & 1 & 4 & \vdots & -2 \\ 0 & 0 & -6 & \vdots & 24 \end{bmatrix}$ by $-\frac{1}{6}$. Note

that rows 1 and 2 remain unchanged.

$$\begin{bmatrix} 2 & 1 & -8 & \vdots & 4 \\ 0 & 1 & 4 & \vdots & -2 \\ 0 & 0 & 1 & \vdots & -4 \end{bmatrix}$$

We can represent the instruction to multiply the third row by $-\frac{1}{6}$ with the symbol $-\frac{1}{6}R_3$.

c. We multiply each number from the first row of matrix $B = \begin{bmatrix} 1 & -1 & \vdots & 2 \\ 4 & -8 & \vdots & 0 \end{bmatrix}$ by -4 to get

$$-4 \qquad 4 \qquad -8 \qquad \text{This is } -4R_1.$$

We then add these numbers to the entries in row 2 of matrix B. (Note that row 1 remains unchanged.)

$$\begin{bmatrix} 1 & -1 & \vdots & 2 \\ 4 + (-4) & -8 + 4 & \vdots & 0 + (-8) \end{bmatrix}$$

This procedure is represented by $-4R_1 + R_2$, which means "Multiply row 1 by -4 and add the result to row 2."

After simplifying the bottom row, we have the matrix $\begin{bmatrix} 1 & -1 & \vdots & 2 \\ 0 & -4 & \vdots & -8 \end{bmatrix}$.

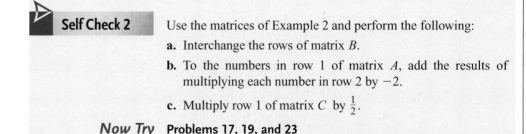

Self Check 2 Use the matrices of Example 2 and perform the following:

a. Interchange the rows of matrix B.

b. To the numbers in row 1 of matrix A, add the results of multiplying each number in row 2 by -2.

c. Multiply row 1 of matrix C by $\frac{1}{2}$.

Now Try **Problems 17, 19, and 23**

Use Matrices to Solve a System of Two Equations.

We can solve a system of two linear equations using a series of elementary row operations.

EXAMPLE 3   Use matrices to solve the system: $\begin{cases} 2x + y = 5 \\ x - y = 4 \end{cases}$

Strategy   We will represent the system with an augmented matrix and use a series of elementary row operations to produce an equivalent matrix in row echelon form.

Why   When the resulting row echelon form matrix is written as a system of two equations, we will know the value of one variable, and the value of the other can be found using substitution.

Solution   We can represent the system with the following augmented matrix:

$$\begin{bmatrix} 2 & 1 & \vdots & 5 \\ 1 & -1 & \vdots & 4 \end{bmatrix}$$

First, we want to get a 1 in the top row of the first column where the shaded 2 is. This can be done by applying a type 1 row operation and interchanging rows 1 and 2.

$$\begin{bmatrix} 1 & -1 & \vdots & 4 \\ 2 & 1 & \vdots & 5 \end{bmatrix} \quad R_1 \leftrightarrow R_2$$

To get a 0 in the first column where the shaded 2 is, we use a type 3 row operation and multiply each entry in row 1 by -2 to get

$$-2 \qquad 2 \qquad -8$$

and add these numbers to the entries in row 2.

$$\begin{bmatrix} 1 & -1 & \vdots & 4 \\ 2 + (-2) & 1 + 2 & \vdots & 5 + (-8) \end{bmatrix} \quad \text{This is } -2R_1 + R_2.$$

After simplifying the bottom row, we have

$$\begin{bmatrix} 1 & -1 & \vdots & 4 \\ 0 & 3 & \vdots & -3 \end{bmatrix}$$

To get a 1 in the bottom row of the second column where the shaded 3 is, we use a type 2 row operation and multiply row 2 by $\frac{1}{3}$.

$$\begin{bmatrix} 1 & -1 & \vdots & 4 \\ 0 & 1 & \vdots & -1 \end{bmatrix} \quad \frac{1}{3}R_2$$

This augmented matrix represents the system of equations

$$\begin{cases} 1x - 1y = 4 \\ 0x + 1y = -1 \end{cases}$$

Writing the equations without the coefficients of 1 and -1 and dropping the $0x$ term, we have

(1) $\begin{cases} x - y = 4 \\ y = -1 \end{cases}$
(2)

From equation 2, we see that $y = -1$. We can back substitute -1 for y in equation 1 to find x.

$$x - y = 4$$

$$x - (-1) = 4 \quad \text{Substitute } -1 \text{ for } y.$$

$$x + 1 = 4 \quad \text{Simplify: } -(-1) = 1.$$

$$x = 3 \quad \text{To solve for x, subtract 1 from both sides.}$$

The solution of the system is $(3, -1)$ and the solution set is $\{(3, -1)\}$. Verify that this ordered pair satisfies both equations in the original system.

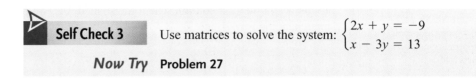

Self Check 3 Use matrices to solve the system: $\begin{cases} 2x + y = -9 \\ x - 3y = 13 \end{cases}$

Now Try **Problem 27**

When a system of linear equations has one solution, we can use the following steps to solve it.

Solving Systems of Linear Equations Using Matrices

1. Write an augmented matrix for the system.
2. Use elementary row operations to transform the augmented matrix into a matrix in row echelon form with 1's down its main diagonal and 0's under the 1's.
3. When step 2 is complete, write the resulting system and use *back substitution* to find the solution.
4. Check the proposed solution in the equations of the original system.

5 **Use Matrices to Solve a System of Three Equations.**

To show how to use matrices to solve systems of three linear equations containing three variables, we consider the following system that can be represented by the augmented matrix to its right.

$$\begin{cases} 3x + y + 5z = 8 \\ 2x + 3y - z = 6 \\ x + 2y + 2z = 10 \end{cases} \qquad \begin{bmatrix} 3 & 1 & 5 & | & 8 \\ 2 & 3 & -1 & | & 6 \\ 1 & 2 & 2 & | & 10 \end{bmatrix}$$

To solve the 3×3 system of equations, we transform the augmented matrix into a matrix with 1's down its main diagonal and 0's below its main diagonal.

$$\begin{bmatrix} 1 & a & b & | & c \\ 0 & 1 & d & | & e \\ 0 & 0 & 1 & | & f \end{bmatrix} \qquad a, b, c, \ldots, f \text{ represent real numbers.}$$

Main diagonal

EXAMPLE 4 Use matrices to solve the system:

$$\begin{cases} 3x + y + 5z = 8 \\ 2x + 3y - z = 6 \\ x + 2y + 2z = 10 \end{cases}$$

Strategy We will represent the system with an augmented matrix and use a series of elementary row operations to produce an equivalent matrix in row echelon form.

Why When the resulting matrix in row echelon form is written as a system of three equations, the value of one variable is known, and the values of the other two can be found using substitution.

Solution   This system can be represented by the augmented matrix

$$\left[\begin{array}{ccc|c} 3 & 1 & 5 & 8 \\ 2 & 3 & -1 & 6 \\ 1 & 2 & 2 & 10 \end{array}\right]$$

Success Tip

Follow this order in getting 1's and 0's in the proper positions of the augmented matrix.

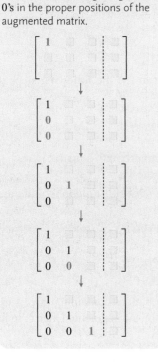

To get a 1 in the first column where the shaded 3 is, we perform a type 1 row operation by interchanging rows 1 and 3.

$$\left[\begin{array}{ccc|c} 1 & 2 & 2 & 10 \\ 2 & 3 & -1 & 6 \\ 3 & 1 & 5 & 8 \end{array}\right] \quad R_1 \leftrightarrow R_3$$

To get a 0 where the shaded 2 is, we perform a type 3 row operation by multiplying each entry in row 1 by -2 to get

$$-2 \qquad -4 \qquad -4 \qquad -20$$

and add these numbers to the entries in row 2.

$$\left[\begin{array}{ccc|c} 1 & 2 & 2 & 10 \\ 2+(-2) & 3+(-4) & -1+(-4) & 6+(-20) \\ 3 & 1 & 5 & 8 \end{array}\right] \quad \text{This is } -2R_1 + R_2.$$

After simplifying the second row, we have

$$\left[\begin{array}{ccc|c} 1 & 2 & 2 & 10 \\ 0 & -1 & -5 & -14 \\ 3 & 1 & 5 & 8 \end{array}\right]$$

To get a 0 where the shaded 3 is, we perform another type 3 row operation by multiplying the entries in row 1 by -3 and adding the results to row 3.

$$\left[\begin{array}{ccc|c} 1 & 2 & 2 & 10 \\ 0 & -1 & -5 & -14 \\ 0 & -5 & -1 & -22 \end{array}\right] \quad -3R_1 + R_3$$

To get a 1 where the shaded -1 is, we perform a type 2 row operation by multiplying row 2 by -1.

$$\left[\begin{array}{ccc|c} 1 & 2 & 2 & 10 \\ 0 & 1 & 5 & 14 \\ 0 & -5 & -1 & -22 \end{array}\right] \quad -1R_2$$

To get a 0 where the shaded -5 is, we perform a type 3 row operation by multiplying the entries in row 2 by 5 and adding the results to row 3.

$$\left[\begin{array}{ccc|c} 1 & 2 & 2 & 10 \\ 0 & 1 & 5 & 14 \\ 0 & 0 & 24 & 48 \end{array}\right] \quad 5R_2 + R_3$$

To get a 1 where the shaded 24 is, we perform a type 2 row operation by multiplying row 3 by $\frac{1}{24}$.

$$\left[\begin{array}{ccc|c} 1 & 2 & 2 & 10 \\ 0 & 1 & 5 & 14 \\ 0 & 0 & 1 & 2 \end{array}\right] \quad \frac{1}{24}R_3$$

The final augmented matrix represents the system

$$\begin{cases} 1x + 2y + 2z = 10 \\ 0x + 1y + 5z = 14 \\ 0x + 0y + 1z = 2 \end{cases} \text{ which can be written as } \begin{cases} x + 2y + 2z = 10 \quad \textbf{(1)} \\ y + 5z = 14 \quad \textbf{(2)} \\ z = 2 \quad \textbf{(3)} \end{cases}$$

From equation 3, we can see that $z = 2$. To find y, we back substitute 2 for z in equation 2 and solve for y:

$$y + 5z = 14 \qquad \text{This is equation 2.}$$
$$y + 5(2) = 14 \qquad \text{Substitute 2 for z.}$$
$$y + 10 = 14$$
$$y = 4 \qquad \text{To solve for y, subtract 10 from both sides.}$$

Thus, $y = 4$. To find x, we back substitute 2 for z and 4 for y in equation 1 and solve for x:

$$x + 2y + 2z = 10 \qquad \text{This is equation 1.}$$
$$x + 2(4) + 2(2) = 10 \qquad \text{Substitute 2 for z and 4 for y.}$$
$$x + 8 + 4 = 10$$
$$x + 12 = 10$$
$$x = -2 \qquad \text{To solve for x, subtract 12 from both sides.}$$

Thus, $x = -2$. The solution of the given system is $(-2, 4, 2)$ and the solution set is $\{(-2, 4, 2)\}$. Verify that this ordered triple satisfies each equation of the original system.

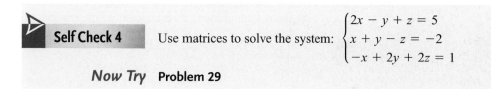

Self Check 4   Use matrices to solve the system: $\begin{cases} 2x - y + z = 5 \\ x + y - z = -2 \\ -x + 2y + 2z = 1 \end{cases}$

Now Try **Problem 29**

 Use Matrices to Identify Inconsistent Systems and Dependent Equations.

In the next example, we will see how to recognize inconsistent systems and systems of dependent equations when matrices are used to solve them.

EXAMPLE 5   If possible, use matrices to solve the system.

a. $\begin{cases} x + y = -1 \\ -3x - 3y = -5 \end{cases}$ **b.** $\begin{cases} 2x - y = 4 \\ -6x + 3y = -12 \end{cases}$

Strategy   We will represent the system with an augmented matrix and use a series of elementary row operations to produce an equivalent matrix in row echelon form.

Why   When the resulting matrix in row echelon form is written as a system of two equations, we can determine whether the system is consistent or inconsistent and whether the equations are dependent or independent.

Solution

a. The system $\begin{cases} x + y = -1 \\ -3x - 3y = -5 \end{cases}$ can be represented by the augmented matrix

$$\begin{bmatrix} 1 & 1 & \vdots & -1 \\ -3 & -3 & \vdots & -5 \end{bmatrix}$$

Since the matrix has a 1 in the top row of the first column, we proceed to get a 0 where the shaded -3 is by multiplying row 1 by 3 and adding the results to row 2.

$$\begin{bmatrix} 1 & 1 & \vdots & -1 \\ 0 & 0 & \vdots & -8 \end{bmatrix} \quad 3R_1 + R_2$$

This augmented matrix represents the system

$$\begin{cases} x + y = -1 \\ 0 + 0 = -8 \end{cases}$$

The equation in the second row of this system simplifies to $0 = -8$. This false statement indicates that the system is inconsistent and has no solution. The solution set is $\varnothing$.

b. The system $\begin{cases} 2x - y = 4 \\ -6x + 3y = -12 \end{cases}$ can be represented by the augmented matrix

$$\begin{bmatrix} 2 & -1 & \vdots & 4 \\ -6 & 3 & \vdots & -12 \end{bmatrix}$$

To get a 1 where the shaded 2 is, we perform a type 2 row operation by multiplying row 1 by $\frac{1}{2}$.

$$\begin{bmatrix} 1 & -\dfrac{1}{2} & \vdots & 2 \\ -6 & 3 & \vdots & -12 \end{bmatrix} \quad \tfrac{1}{2}R_1$$

To get a 0 where the shaded -6 is, we perform a type 3 row operation by multiplying the entries in row 1 by 6 and adding the results to the entries in row 2.

$$\begin{bmatrix} 1 & -\dfrac{1}{2} & \vdots & 2 \\ 0 & 0 & \vdots & 0 \end{bmatrix} \quad 6R_1 + R_2$$

This augmented matrix represents the system

$$\begin{cases} x - \dfrac{1}{2}y = 2 \\ 0 + 0 = 0 \end{cases}$$

The equation in the second row of the system simplifies to $0 = 0$. This true statement indicates that the equations are dependent and that the system has infinitely many solutions. The solution set is $\{(x, y) \mid 2x - y = 4\}$.

Self Check 5   If possible, use matrices to solve the system.

a. $\begin{cases} 4x - 8y = 9 \\ x - 2y = -5 \end{cases}$

b. $\begin{cases} x - 3y = 6 \\ -4x + 12y = -24 \end{cases}$

Now Try **Problems 33 and 35**

STUDY SET
3.6

VOCABULARY

Fill in the blanks.

1. A _____ is a rectangular array of numbers written within brackets.

2. Each number in a matrix is called an _____ or entry of the matrix.

3. If the order of a matrix is 3×4, it has 3 _____ and 4 _____ . We read 3×4 as "3 ___ 4."

4. Elementary _____ operations are used to produce equivalent matrices that lead to the solution of a system.

5. A matrix that represents the equations of a system is called an _____ matrix.

6. The matrix $\begin{bmatrix} 1 & 3 & \vdots & -2 \\ 0 & 1 & \vdots & 4 \end{bmatrix}$, with 1's down its main _____ and a 0 below the 1 in the first column, is in row echelon form.

CONCEPTS

7. For each matrix, determine the number of rows and the number of columns.

   a. $\begin{bmatrix} 4 & 6 & \vdots & -1 \\ 1 & 9 & \vdots & -3 \end{bmatrix}$ **b.** $\begin{bmatrix} 1 & -2 & 3 & \vdots & 1 \\ 0 & 1 & 6 & \vdots & 4 \\ 0 & 0 & 1 & \vdots & \frac{1}{3} \end{bmatrix}$

8. Fill in the blanks to complete each elementary row operation:

   a. Type 1: Any two rows of a matrix can be _____.

   b. Type 2: Any row of a matrix can be _____ by a nonzero constant.

   c. Type 3: Any row of a matrix can be changed by _____ a nonzero constant multiple of another row to it.

9. Matrices were used to solve a system. The final augmented matrix is shown. Fill in the blanks.

   a. $\begin{bmatrix} 1 & 1 & \vdots & 10 \\ 0 & 1 & \vdots & 6 \end{bmatrix}$ represents $\begin{cases} \quad + \quad = 10 \\ \quad = 6 \end{cases}$

   Therefore, $y =$ ▢ . Using back substitution, we find that $x =$ ▢ .

   b. $\begin{bmatrix} 1 & -2 & 1 & \vdots & -16 \\ 0 & 1 & 2 & \vdots & 8 \\ 0 & 0 & 1 & \vdots & 4 \end{bmatrix}$ represents $\begin{cases} \quad - 2y + \quad = -16 \\ y + \quad = 8 \\ \quad = 4 \end{cases}$

   Therefore, $z =$ ▢ . Using back substitution, we find that $y =$ ▢ and $x =$ ▢ .

10. **a.** Which matrix shown below indicates that its associated system of equations has no solution?

    b. Which matrix indicates that the equations of its associated system are dependent?

    i. $\begin{bmatrix} 1 & 2 & \vdots & -4 \\ 0 & 0 & \vdots & 0 \end{bmatrix}$ **ii.** $\begin{bmatrix} 1 & 3 & \vdots & 6 \\ 0 & 1 & \vdots & 0 \end{bmatrix}$ **iii.** $\begin{bmatrix} 1 & 2 & \vdots & -4 \\ 0 & 0 & \vdots & 2 \end{bmatrix}$

NOTATION

11. Explain what each symbolism means.

    a. $R_1 \leftrightarrow R_2$

    b. $\frac{1}{2} R_1$

    c. $6R_2 + R_3$

12. Complete the solution.

    Solve using matrices: $\begin{cases} 4x - y = 14 \\ x + y = 6 \end{cases}$

    $\begin{bmatrix} 4 & \quad & \vdots & 14 \\ 1 & 1 & \vdots & 6 \end{bmatrix}$

    $\begin{bmatrix} \quad & 1 & \vdots & 6 \\ 4 & -1 & \vdots & 14 \end{bmatrix} \quad R_1 \leftrightarrow R_2$

    $\begin{bmatrix} 1 & 1 & \vdots & 6 \\ 0 & \quad & \vdots & -10 \end{bmatrix} \quad -4R_1 + R_2$

    $\begin{bmatrix} 1 & 1 & \vdots & 6 \\ 0 & 1 & \vdots & 2 \end{bmatrix} \quad -\frac{1}{5} R_2$

    This matrix represents the system

    $\begin{cases} x + y = 6 \\ \quad = 2 \end{cases}$

    The solution is (▢ , 2).

GUIDED PRACTICE

Represent each system using an augmented matrix. See Example 1.

13. $\begin{cases} x + 2y = 6 \\ 3x - y = -10 \end{cases}$

14. $\begin{cases} x + y + z = 4 \\ 2x + y - z = 1 \\ 2x - 3y = 1 \end{cases}$

For each augmented matrix, give the system of equations that it represents.

15. $\begin{bmatrix} 1 & 6 & | & 7 \\ 0 & 1 & | & 4 \end{bmatrix}$

16. $\begin{bmatrix} 1 & -2 & 9 & | & 1 \\ 0 & 1 & 4 & | & 0 \\ 0 & 0 & 1 & | & -7 \end{bmatrix}$

Perform each of the following elementary row operations on the augmented matrix $\begin{bmatrix} -3 & 1 & | & -6 \\ 1 & -4 & | & 4 \end{bmatrix}$**. See Example 2.**

17. $R_1 \leftrightarrow R_2$

18. $5R_2$

19. $-\dfrac{1}{3}R_1$

20. $3R_2 + R_1$

Perform each of the following elementary row operations on the augmented matrix $\begin{bmatrix} 3 & 6 & -9 & | & 0 \\ 1 & 5 & -2 & | & 1 \\ -2 & 2 & -2 & | & 5 \end{bmatrix}$**. See Example 2.**

21. $R_2 \leftrightarrow R_3$

22. $-\dfrac{1}{2}R_3$

23. $-R_1 + R_2$

24. $2R_2 + R_3$

Use matrices to solve each system of equations. See Example 3.

25. $\begin{cases} x + y = 2 \\ x - y = 0 \end{cases}$

26. $\begin{cases} x + y = 3 \\ x - y = -1 \end{cases}$

27. $\begin{cases} 2x + y = 1 \\ x + 2y = -4 \end{cases}$

28. $\begin{cases} 5x - 4y = 10 \\ x - 7y = 2 \end{cases}$

Use matrices to solve each system of equations. See Example 4.

29. $\begin{cases} x + y + z = 6 \\ x + 2y + z = 8 \\ x + y + 2z = 7 \end{cases}$

30. $\begin{cases} x + y + z = 6 \\ x + 2y + z = 8 \\ x + y + 2z = 9 \end{cases}$

31. $\begin{cases} 3x + y - 3z = 5 \\ x - 2y + 4z = 10 \\ x + y + z = 13 \end{cases}$

32. $\begin{cases} 2x + y - 3z = -1 \\ 3x - 2y - z = -5 \\ x - 3y - 2z = -12 \end{cases}$

Use matrices to solve each system of equations. If the equations of a system are dependent or if a system is inconsistent, state this. See Example 5.

33. $\begin{cases} x - 3y = 9 \\ -2x + 6y = 18 \end{cases}$

34. $\begin{cases} -6x + 12y = 10 \\ 2x - 4y = 8 \end{cases}$

35. $\begin{cases} -4x - 4y = -12 \\ x + y = 3 \end{cases}$

36. $\begin{cases} 5x - 15y = 10 \\ x - 3y = 2 \end{cases}$

TRY IT YOURSELF

Use matrices to solve each system of equations. If the equations of a system are dependent or if a system is inconsistent, state this.

37. $\begin{cases} 2x + 3y - z = -8 \\ x - y - z = -2 \\ -4x + 3y + z = 6 \end{cases}$

38. $\begin{cases} 2a + b + 3c = 3 \\ -2a - b + c = 5 \\ 4a - 2b + 2c = 2 \end{cases}$

39. $\begin{cases} 2x - y = -1 \\ x - 2y = 1 \end{cases}$

40. $\begin{cases} 2x - y = 0 \\ x + y = 3 \end{cases}$

41. $\begin{cases} 3x + 4y = -12 \\ 9x - 2y = 6 \end{cases}$

42. $\begin{cases} 2x - 3y = 16 \\ -4x + y = -22 \end{cases}$

43. $\begin{cases} 2x + y - z = 1 \\ x + 2y + 2z = 2 \\ 4x + 5y + 3z = 3 \end{cases}$

44. $\begin{cases} x - y = 1 \\ 2x - z = 0 \\ 2y - z = -2 \end{cases}$

45. $\begin{cases} 8x - 2y = 4 \\ 4x - y = 2 \end{cases}$

46. $\begin{cases} 9x - 3y = 6 \\ 3x - y = 8 \end{cases}$

47. $\begin{cases} 2x + y - 2z = 6 \\ 4x - y + z = -1 \\ 6x - 2y + 3z = -5 \end{cases}$

48. $\begin{cases} 2x - 3y + 3z = 14 \\ 3x + 3y - z = 2 \\ -2x + 6y + 5z = 9 \end{cases}$

49. $\begin{cases} 6x + y - z = -2 \\ x + 2y + z = 5 \\ 5y - z = 2 \end{cases}$ **50.** $\begin{cases} 2x + 3y - 2z = 18 \\ 5x - 6y + z = 21 \\ 4y - 2z = 6 \end{cases}$

51. $\begin{cases} 5x + 3y = 4 \\ 3y - 4z = 4 \\ x + z = 1 \end{cases}$ **52.** $\begin{cases} y + 2z = -2 \\ x + y = 1 \\ 2x - z = 0 \end{cases}$

APPLICATIONS

53. DIGITAL PHOTOGRAPHY
A digital camera stores the
black and white photograph
shown on the right as a
512×512 matrix. Each
element of the matrix
corresponds to a small dot of
grey scale shading, called a
pixel, in the picture. How
many elements does a
512×512 matrix have?

54. DIGITAL IMAGING   A scanner stores a black and white
photograph as a matrix that has a total of 307,200 elements. If
the matrix has 480 rows, how many columns does it have?

*Write a system of equations to solve each problem. Use matrices
to solve the system.*

55. COMPLEMENTARY ANGLES   One angle measures 46°
more than the measure of its complement. Find the measure of
each angle.   (*Hint:* The sum of the measures of complementary angles is 90°.)

56. SUPPLEMENTARY ANGLES   One angle measures 14° more
than the measure of its supplement. Find the measure of each
angle.   (*Hint:* The sum of the measures of supplementary
angles is 180°.)

57. TRIANGLES   In the illustration, $\angle B$ measures 25° more than
the measure of $\angle A$, and the
measure of $\angle C$ is 5° less than
twice the measure of $\angle A$. Find
the measure of each angle of the
triangle.

58. TRIANGLES   In the illustration, $\angle A$ measures 10° less than
the measure of $\angle B$, and the
measure of $\angle B$ is 10° less than
the measure of $\angle C$. Find the
measure of each angle of the
triangle.

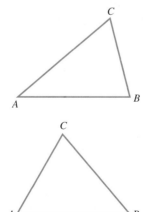

59. PHYSICAL THERAPY   After
an elbow injury, a volleyball
player has restricted movement
of her arm. Her range of motion
(the measure of $\angle 1$) is 28° less
than the measure of $\angle 2$. Find
the measure of each angle.

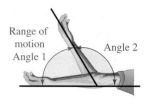

60. ICE SKATING   Three circles are traced out by a figure skater
during her performance. If the centers of the circles are the
given distances apart, find the radius of each circle.

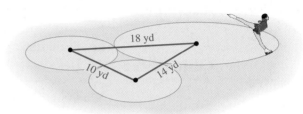

WRITING

61. For the system $\begin{cases} 2x - 3y = 5 \\ 4x + 8 = y \end{cases}$, explain what is wrong with
writing its corresponding augmented matrix as $\begin{bmatrix} 2 & -3 & \vdots & 5 \\ 4 & 8 & \vdots & 1 \end{bmatrix}$.
How should it be written?

62. Explain what is meant by the phrase *back substitution.* Give an
example of how it was used in this section.

63. Explain how a type 3 row operation is similar to the elimination
(addition) method of solving a system of equations.

64. If the system represented by the following augmented matrix
has no solution, what do you know about k? Explain your
answer.

$$\begin{bmatrix} 1 & 1 & 0 & \vdots & 1 \\ 0 & 0 & 1 & \vdots & 2 \\ 0 & 0 & 0 & \vdots & k \end{bmatrix}$$

REVIEW

65. What is the formula used to find the slope of a line, given two
points on the line?

66. What is the form of the equation of a horizontal line? Of a
vertical line?

67. What is the point–slope form of the equation of a line?

68. What is the slope–intercept form of the equation of a line?

CHALLENGE PROBLEMS

Use matrices to solve the system.

69. $\begin{cases} x^2 + y^2 + z^2 = 14 \\ 2x^2 + 3y^2 - 2z^2 = -7 \\ x^2 - 5y^2 + z^2 = 8 \end{cases}$

70. $\begin{cases} w + x + y + z = 0 \\ w - 2x + y - 3z = -3 \\ 2w + 3x + y - 2z = -1 \\ 2w - 2x - 2y + z = -12 \end{cases}$

SECTION 3.7
Solving Systems of Equations Using Determinants

Objectives

1 Evaluate 2 × 2 and 3 × 3 determinants.

2 Use Cramer's rule to solve systems of two equations.

3 Use Cramer's rule to solve systems of three equations.

In this section, we will discuss another method for solving systems of linear equations. With this method, called *Cramer's rule,* we work with combinations of the coefficients and the constants of the equations written as *determinants.*

 Evaluate 2 × 2 and 3 × 3 Determinants.

An idea related to the concept of matrix is the **determinant.** A determinant is a number that is associated with a **square matrix,** a matrix that has the same number of rows and columns. For any square matrix A, the symbol $|A|$ represents the determinant of A. To write a determinant, we put the elements of a square matrix between two vertical lines.

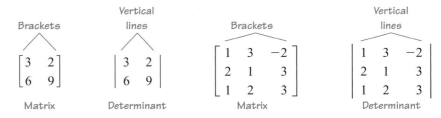

Like matrices, determinants are classified according to the number of rows and columns they contain. The determinant above, on the left, is a 2 × 2 determinant. The other is a 3 × 3 determinant.

The determinant of a 2 × 2 matrix is the number that is equal to the product of the numbers on the main diagonal minus the product of the numbers on the other diagonal.

Main diagonal   Other diagonal

Value of a 2 × 2 Determinant

If a, b, c, and d are numbers, the **determinant** of the matrix $\begin{bmatrix} a & b \\ c & d \end{bmatrix}$ is

$$\begin{vmatrix} a & b \\ c & d \end{vmatrix} = ad - bc.$$

EXAMPLE 1 Evaluate each determinant: **a.** $\begin{vmatrix} 3 & 2 \\ 6 & 9 \end{vmatrix}$ **b.** $\begin{vmatrix} -20 & 1 \\ -8 & 4 \end{vmatrix}$

Strategy We will find the product of the numbers on the main diagonal and the product of the numbers along the other diagonal and subtract the results.

Why The value of a determinant of the form $\begin{vmatrix} a & b \\ c & d \end{vmatrix}$ is $ad - bc$.

Solution To evaluate the determinant, we proceed as follows:

a. This is always minus.

$\begin{vmatrix} 3 & 2 \\ 6 & 9 \end{vmatrix} = 3(9) - 2(6) = 27 - 12 = 15$

b. $\begin{vmatrix} -20 & 1 \\ -8 & 4 \end{vmatrix} = -20(4) - 1(-8) = -80 - (-8) = -80 + 8 = -72$

▷ **Self Check 1** Evaluate: $\begin{vmatrix} 4 & -3 \\ 2 & 1 \end{vmatrix}$

Now Try **Problems 15 and 17**

A 3×3 determinant is evaluated by **expanding by minors.** The following definition shows how we can evaluate a 3×3 determinant by expanding by minors along the first row.

Value of a 3 × 3 Determinant

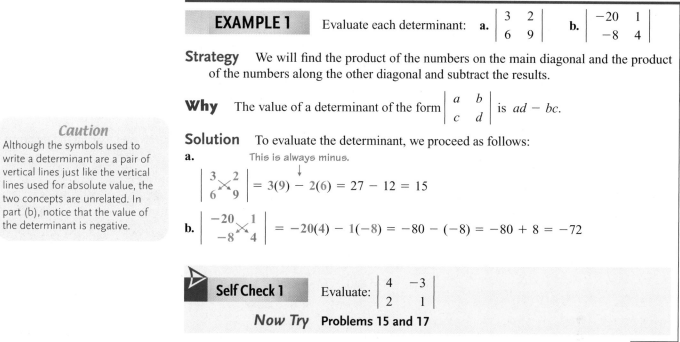

$$\begin{vmatrix} a_1 & b_1 & c_1 \\ a_2 & b_2 & c_2 \\ a_3 & b_3 & c_3 \end{vmatrix} = a_1 \begin{vmatrix} b_2 & c_2 \\ b_3 & c_3 \end{vmatrix} - b_1 \begin{vmatrix} a_2 & c_2 \\ a_3 & c_3 \end{vmatrix} + c_1 \begin{vmatrix} a_2 & b_2 \\ a_3 & b_3 \end{vmatrix}$$

To find the minor of a_1, we cross out the elements of the determinant that are in the same row and column as a_1:

$\begin{vmatrix} \cancel{a_1} & \cancel{b_1} & \cancel{c_1} \\ \cancel{a_2} & b_2 & c_2 \\ \cancel{a_3} & b_3 & c_3 \end{vmatrix}$ The minor of a_1 is $\begin{vmatrix} b_2 & c_2 \\ b_3 & c_3 \end{vmatrix}$.

To find the minor of b_1, we cross out the elements of the determinant that are in the same row and column as b_1:

$\begin{vmatrix} \cancel{a_1} & \cancel{b_1} & \cancel{c_1} \\ a_2 & \cancel{b_2} & c_2 \\ a_3 & \cancel{b_3} & c_3 \end{vmatrix}$ The minor of b_1 is $\begin{vmatrix} a_2 & c_2 \\ a_3 & c_3 \end{vmatrix}$.

To find the minor of c_1, we cross out the elements of the determinant that are in the same row and column as c_1:

$\begin{vmatrix} \cancel{a_1} & \cancel{b_1} & \cancel{c_1} \\ a_2 & b_2 & \cancel{c_2} \\ a_3 & b_3 & \cancel{c_3} \end{vmatrix}$ The minor of c_1 is $\begin{vmatrix} a_2 & b_2 \\ a_3 & b_3 \end{vmatrix}$.

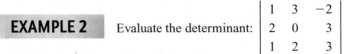

EXAMPLE 2 Evaluate the determinant: $\begin{vmatrix} 1 & 3 & -2 \\ 2 & 0 & 3 \\ 1 & 2 & 3 \end{vmatrix}$

Strategy We will expand the determinant along the first row using the numbers in the first row and their corresponding minors.

Why We can then evaluate the resulting 2×2 determinants and simplify.

Solution To evaluate the determinant, we can use the first row and expand the determinant by minors:

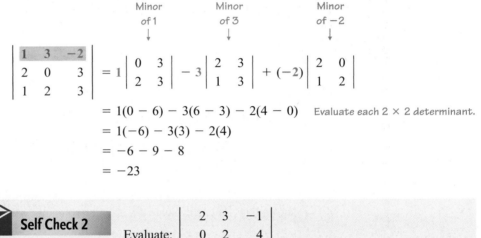

$$\begin{vmatrix} 1 & 3 & -2 \\ 2 & 0 & 3 \\ 1 & 2 & 3 \end{vmatrix} = 1\begin{vmatrix} 0 & 3 \\ 2 & 3 \end{vmatrix} - 3\begin{vmatrix} 2 & 3 \\ 1 & 3 \end{vmatrix} + (-2)\begin{vmatrix} 2 & 0 \\ 1 & 2 \end{vmatrix}$$

$$= 1(0 - 6) - 3(6 - 3) - 2(4 - 0) \quad \text{Evaluate each } 2 \times 2 \text{ determinant.}$$

$$= 1(-6) - 3(3) - 2(4)$$

$$= -6 - 9 - 8$$

$$= -23$$

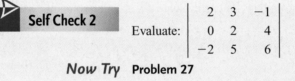

Self Check 2 Evaluate: $\begin{vmatrix} 2 & 3 & -1 \\ 0 & 2 & 4 \\ -2 & 5 & 6 \end{vmatrix}$

Now Try Problem 27

We can evaluate a 3×3 determinant by expanding it by the minors of any row or column. We will get the same value. To determine the signs between the terms of the expansion of a 3×3 determinant, we use the following array of signs.

Array of Signs for a 3 × 3 Determinant

$$\begin{array}{ccc} + & - & + \\ - & + & - \\ + & - & + \end{array}$$

This array of signs is often called the checkerboard pattern.

To remember the sign pattern, note that there is a $+$ sign in the upper left position and that the signs alternate for all of the positions that follow.

EXAMPLE 3 Evaluate the determinant $\begin{vmatrix} 1 & 3 & -2 \\ 2 & 0 & 3 \\ 1 & 2 & 3 \end{vmatrix}$ by expanding by the minors of the middle column. (This is the determinant of Example 2.)

Strategy We will expand the determinant using the numbers in the middle column and their corresponding minors. We will use the sign pattern $- + -$ between the terms of the expansion.

Why We can then evaluate the resulting 2×2 determinants and simplify.

Solution To evaluate the determinant, we proceed as follows:

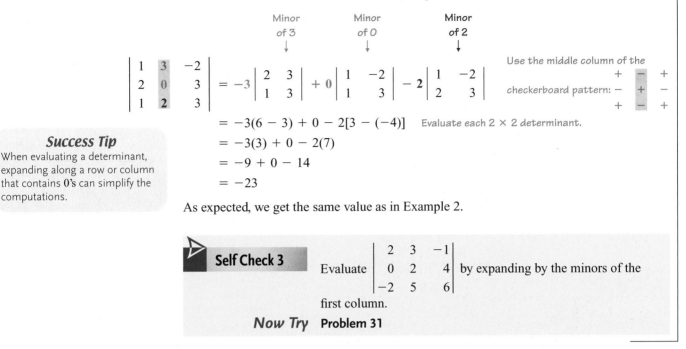

$$= -3(6 - 3) + 0 - 2[3 - (-4)] \quad \text{Evaluate each 2 × 2 determinant.}$$
$$= -3(3) + 0 - 2(7)$$
$$= -9 + 0 - 14$$
$$= -23$$

As expected, we get the same value as in Example 2.

Success Tip
When evaluating a determinant, expanding along a row or column that contains 0's can simplify the computations.

Self Check 3 Evaluate $\begin{vmatrix} 2 & 3 & -1 \\ 0 & 2 & 4 \\ -2 & 5 & 6 \end{vmatrix}$ by expanding by the minors of the first column.

Now Try **Problem 31**

Using Your Calculator

Evaluating Determinants

It is possible to use a graphing calculator to evaluate determinants. For example, to evaluate the determinant in Example 3, we first enter the matrix by pressing the ⬚MATRIX⬚ key, selecting EDIT, and pressing the ⬚ENTER⬚ key. Next, we enter the dimensions and the elements of the matrix to get figure (a). We then press ⬚2nd⬚ ⬚QUIT⬚ to clear the screen, press ⬚MATRIX⬚, select MATH, and press 1 to get figure (b). We then press ⬚MATRIX⬚, select NAMES, press 1, and press ⬚)⬚ and ⬚ENTER⬚ to get the value of the determinant. Figure (c) shows that the value of the determinant is −23.

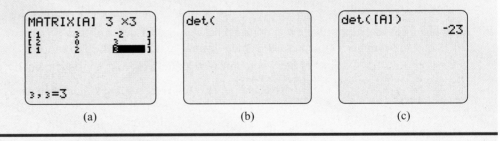

(a)   (b)   (c)

---

2 ### Use Cramer's Rule to Solve Systems of Two Equations.

The method of using determinants to solve systems of linear equations is called **Cramer's rule,** named after the 18th-century Swiss mathematician Gabriel Cramer. To develop Cramer's rule, we consider the system

$$\begin{cases} ax + by = e \\ cx + dy = f \end{cases}$$

where x and y are variables and a, b, c, d, e, and f are constants.

If we multiply both sides of the first equation by d and multiply both sides of the second equation by $-b$, we can add the equations and eliminate y:

$$
\begin{array}{ll}
adx + bdy = ed & \text{This is } d(ax + by) = d(e). \\
\underline{-bcx - bdy = -bf} & \text{This is } -b(cx + dy) = -b(f). \\
adx - bcx \qquad = ed - bf &
\end{array}
$$

To solve for x, we use the distributive property to write $adx - bcx$ as $(ad - bc)x$ on the left side and divide each side by $ad - bc$:

$$(ad - bc)x = ed - bf$$

$$x = \frac{ed - bf}{ad - bc} \qquad \text{where } ad - bc \neq 0$$

We can find y in a similar way. After eliminating the variable x, we get

$$y = \frac{af - ec}{ad - bc} \qquad \text{where } ad - bc \neq 0$$

Note that the denominator for both x and y is

$$\begin{vmatrix} a & b \\ c & d \end{vmatrix} = ad - bc$$

The numerators can be expressed as determinants also:

$$x = \frac{ed - bf}{ad - bc} = \frac{\begin{vmatrix} e & b \\ f & d \end{vmatrix}}{\begin{vmatrix} a & b \\ c & d \end{vmatrix}} \qquad \text{and} \qquad y = \frac{af - ec}{ad - bc} = \frac{\begin{vmatrix} a & e \\ c & f \end{vmatrix}}{\begin{vmatrix} a & b \\ c & d \end{vmatrix}}$$

Cramer's Rule for Two Equations in Two Variables

The solution of the system $\begin{cases} ax + by = e \\ cx + dy = f \end{cases}$ is given by

$$x = \frac{D_x}{D} = \frac{\begin{vmatrix} e & b \\ f & d \end{vmatrix}}{\begin{vmatrix} a & b \\ c & d \end{vmatrix}} \qquad \text{and} \qquad y = \frac{D_y}{D} = \frac{\begin{vmatrix} a & e \\ c & f \end{vmatrix}}{\begin{vmatrix} a & b \\ c & d \end{vmatrix}}$$

If every determinant is 0, the system is consistent, but the equations are dependent.

If $D = 0$ and D_x or D_y is nonzero, the system is inconsistent. If $D \neq 0$, the system is consistent, and the equations are independent.

The following observations are helpful when memorizing the three determinants of Cramer's rule.

- The denominator determinant, D, is formed by using the coefficients a, b, c, and d of the variables in the equations.

$$D = \begin{vmatrix} a & b \\ c & d \end{vmatrix}$$

x-term ⌐   ⌐ y-term
coefficients   coefficients

- The numerator determinants, D_x and D_y, are the same as the denominator determinant, D, except that the column of coefficients of the variable for which we are solving is replaced with the column of constants e and f.

$$D_x = \begin{vmatrix} e & b \\ f & d \end{vmatrix} \qquad\qquad D_y = \begin{vmatrix} a & e \\ c & f \end{vmatrix}$$

↑
Replace the x-term coefficients
with the constants.   ↑
Replace the y-term coefficients
with the constants.

EXAMPLE 4 Use Cramer's rule to solve the system: $\begin{cases} 4x - 3y = 6 \\ -2x + 5y = 4 \end{cases}$

Strategy We will evaluate three determinants, D, D_x, and D_y.

Why The x-value of the solution of the system is the quotient of D_x and D and the y-value of the solution is the quotient of two determinants, D_y and D.

Solution The denominator determinant D is made up of the coefficients of x and y:

$$D = \begin{vmatrix} 4 & -3 \\ -2 & 5 \end{vmatrix}$$

Success Tip

We can now solve systems of linear equations in five ways:
- graphing
- substitution
- elimination (addition)
- matrices
- Cramer's rule

To solve for x, we form the numerator determinant D_x from D by replacing its first column (the coefficients of x) with the column of constants (**6** and **4**).

To solve for y, we form the numerator determinant D_y from D by replacing the second column (the coefficients of y) with the column of constants (**6** and **4**).

To find the values of x and y, we evaluate each determinant:

$$x = \frac{D_x}{D} = \frac{\begin{vmatrix} 6 & -3 \\ 4 & 5 \end{vmatrix}}{\begin{vmatrix} 4 & -3 \\ -2 & 5 \end{vmatrix}} = \frac{6(5) - (-3)(4)}{4(5) - (-3)(-2)} = \frac{30 + 12}{20 - 6} = \frac{42}{14} = 3$$

$$y = \frac{D_y}{D} = \frac{\begin{vmatrix} 4 & 6 \\ -2 & 4 \end{vmatrix}}{\begin{vmatrix} 4 & -3 \\ -2 & 5 \end{vmatrix}} = \frac{4(4) - 6(-2)}{14} = \frac{16 + 12}{14} = \frac{28}{14} = 2$$

The solution of the system is $(3, 2)$. Verify that it satisfies both equations.

▷ **Self Check 4** Use Cramer's rule to solve the system: $\begin{cases} 2x - 3y = -16 \\ 3x + 5y = 14 \end{cases}$

Now Try Problem 39

In the next example, we will see how to recognize inconsistent systems when Cramer's rule is used to solve them.

EXAMPLE 5 Use Cramer's rule to solve $\begin{cases} 7x = 8 - 4y \\ 2y = 3 - \frac{7}{2}x \end{cases}$, if possible.

Strategy We will evaluate three determinants, D, D_x, and D_y.

Why   The *x*-value of the solution of the system is the quotient of D_x and D, and the *y*-value of the solution is the quotient of two determinants, D_y and D.

Solution   Before we can form the required determinants, the equations of the system must be written in standard form: $Ax + By = C$.

$$\begin{cases} 7x = 8 - 4y \\ 2y = 3 - \frac{7}{2}x \end{cases} \quad \xrightarrow[\text{and add 7x to both sides.}]{\substack{\text{Add 4y to both sides.} \\ \\ \text{Multiply both sides by 2}}} \quad \begin{cases} 7x + 4y = 8 \\ 7x + 4y = 6 \end{cases}$$

Success Tip

If any two rows or any two columns of a determinant are identical, the value of the determinant is **0**.

When we attempt to use Cramer's rule to solve this system for *x*, we obtain

$$x = \frac{D_x}{D} = \frac{\begin{vmatrix} 8 & 4 \\ 6 & 4 \end{vmatrix}}{\begin{vmatrix} 7 & 4 \\ 7 & 4 \end{vmatrix}} = \frac{32 - 24}{28 - 28} = \frac{8}{0}, \text{ which is undefined.}$$

Since the denominator determinant D is 0 and the numerator determinant D_x is not 0, the system is inconsistent. It has no solution and the solution set is $\varnothing$.

We can see directly from the system that it is inconsistent. For any values of *x* and *y*, it is impossible that 7 times *x* plus 4 times *y* could be both 8 and 6.

Self Check 5   Use Cramer's rule to solve the system $\begin{cases} 3x = 8 - 4y \\ y = \frac{5}{2} - \frac{3}{4}x \end{cases}$, if possible.

Now Try   Problem 47

③   Use Cramer's Rule to Solve Systems of Three Equations.

Cramer's rule can be extended to solve systems of three linear equations with three variables.

Cramer's Rule for Three Equations in Three Variables

The solution of the system $\begin{cases} ax + by + cz = j \\ dx + ey + fz = k \\ gx + hy + iz = l \end{cases}$ is given by

$$x = \frac{D_x}{D}, \qquad y = \frac{D_y}{D}, \qquad \text{and} \qquad z = \frac{D_z}{D}, \qquad \text{where}$$

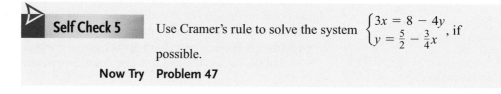

$$D = \begin{vmatrix} a & b & c \\ d & e & f \\ g & h & i \end{vmatrix} \quad \substack{\text{Only the coefficients} \\ \text{of the variables.}} \qquad D_x = \begin{vmatrix} j & b & c \\ k & e & f \\ l & h & i \end{vmatrix} \quad \substack{\text{Replace the x-term} \\ \text{coefficients with the} \\ \text{constants.}}$$

$$D_y = \begin{vmatrix} a & j & c \\ d & k & f \\ g & l & i \end{vmatrix} \quad \substack{\text{Replace the y-term} \\ \text{coefficients with the} \\ \text{constants.}} \qquad D_z = \begin{vmatrix} a & b & j \\ d & e & k \\ g & h & l \end{vmatrix} \quad \substack{\text{Replace the z-term} \\ \text{coefficients with the} \\ \text{constants.}}$$

If every determinant is 0, the system is consistent, but the equations are dependent.

If $D = 0$ and D_x or D_y or D_z is nonzero, the system is inconsistent. If $D \neq 0$, the system is consistent, and the equations are independent.

EXAMPLE 6   Use Cramer's rule to solve the system: $\begin{cases} 2x + y + 4z = 12 \\ x + 2y + 2z = 9 \\ 3x - 3y - 2z = 1 \end{cases}$

Strategy   We will evaluate four determinants, D, D_x, D_y, and D_z.

Why   The x-value of the solution of the system is the quotient of D_x and D, the y-value of the solution is the quotient of two determinants, D_y and D, and the z-value of the solution is the quotient of two determinants, D_z and D.

Solution   The denominator determinant D is the determinant formed by the coefficients of the variables. The numerator determinants, D_x, D_y, and D_z, are formed by replacing the coefficients of the variable being solved for by the column of constants. We form the quotients for x, y, and z and evaluate each determinant by expanding by minors about the first row:

$$x = \frac{D_x}{D} = \frac{\begin{vmatrix} 12 & 1 & 4 \\ 9 & 2 & 2 \\ 1 & -3 & -2 \end{vmatrix}}{\begin{vmatrix} 2 & 1 & 4 \\ 1 & 2 & 2 \\ 3 & -3 & -2 \end{vmatrix}} = \frac{12\begin{vmatrix} 2 & 2 \\ -3 & -2 \end{vmatrix} - 1\begin{vmatrix} 9 & 2 \\ 1 & -2 \end{vmatrix} + 4\begin{vmatrix} 9 & 2 \\ 1 & -3 \end{vmatrix}}{2\begin{vmatrix} 2 & 2 \\ -3 & -2 \end{vmatrix} - 1\begin{vmatrix} 1 & 2 \\ 3 & -2 \end{vmatrix} + 4\begin{vmatrix} 1 & 2 \\ 3 & -3 \end{vmatrix}} = \frac{12(2) - 1(-20) + 4(-29)}{2(2) - 1(-8) + 4(-9)} = \frac{-72}{-24} = 3$$

$$y = \frac{D_y}{D} = \frac{\begin{vmatrix} 2 & 12 & 4 \\ 1 & 9 & 2 \\ 3 & 1 & -2 \end{vmatrix}}{\begin{vmatrix} 2 & 1 & 4 \\ 1 & 2 & 2 \\ 3 & -3 & -2 \end{vmatrix}} = \frac{2\begin{vmatrix} 9 & 2 \\ 1 & -2 \end{vmatrix} - 12\begin{vmatrix} 1 & 2 \\ 3 & -2 \end{vmatrix} + 4\begin{vmatrix} 1 & 9 \\ 3 & 1 \end{vmatrix}}{-24} = \frac{2(-20) - 12(-8) + 4(-26)}{-24} = \frac{-48}{-24} = 2$$

$$z = \frac{D_z}{D} = \frac{\begin{vmatrix} 2 & 1 & 12 \\ 1 & 2 & 9 \\ 3 & -3 & 1 \end{vmatrix}}{\begin{vmatrix} 2 & 1 & 4 \\ 1 & 2 & 2 \\ 3 & -3 & -2 \end{vmatrix}} = \frac{2\begin{vmatrix} 2 & 9 \\ -3 & 1 \end{vmatrix} - 1\begin{vmatrix} 1 & 9 \\ 3 & 1 \end{vmatrix} + 12\begin{vmatrix} 1 & 2 \\ 3 & -3 \end{vmatrix}}{-24} = \frac{2(29) - 1(-26) + 12(-9)}{-24} = \frac{-24}{-24} = 1$$

The solution of this system is $(3, 2, 1)$. Verify that it satisfies the three original equations.

Self Check 6   Use Cramer's rule to solve the system: $\begin{cases} x + y + 2z = 6 \\ 2x - y + z = 9 \\ x + y - 2z = -6 \end{cases}$

Now Try   **Problem 51**

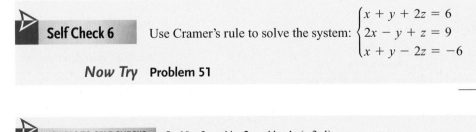
ANSWERS TO SELF CHECKS   **1.** 10   **2.** -44 **3.** -44 **4.** $(-2, 4)$
5. No solution, $\varnothing$; inconsistent system   **6.** $(2, -2, 3)$

STUDY SET
3.7

VOCABULARY

Fill in the blanks.

1. $\begin{vmatrix} 4 & 9 \\ -6 & 1 \end{vmatrix}$ is a 2 × 2 _____. The numbers 4 and 1 lie along its main _____.

2. A determinant is number that is associated with a _____ matrix. A 3 × 3 determinant has 3 _____ and 3 _____.

3. The _____ of b_1 in $\begin{vmatrix} a_1 & b_1 & c_1 \\ a_2 & b_2 & c_2 \\ a_3 & b_3 & c_3 \end{vmatrix}$ is $\begin{vmatrix} a_2 & c_2 \\ a_3 & c_3 \end{vmatrix}$.

4. _____ rule uses determinants to solve systems of linear equations.

CONCEPTS

Fill in the blanks.

5. $\begin{vmatrix} a & b \\ c & d \end{vmatrix} = \boxed{} - \boxed{}$

6. To find the minor of 5, we cross out the elements of the determinant that are in the same row and column as ___.

$$\begin{vmatrix} 3 & 5 & 1 \\ 6 & -2 & 2 \\ 8 & -1 & 4 \end{vmatrix}$$

7. In evaluating the determinant below, about what row or column was it expanded?

$$\begin{vmatrix} 5 & 1 & -1 \\ 8 & 7 & 4 \\ 9 & 7 & 6 \end{vmatrix} = -1\begin{vmatrix} 8 & 7 \\ 9 & 7 \end{vmatrix} - 4\begin{vmatrix} 5 & 1 \\ 9 & 7 \end{vmatrix} + 6\begin{vmatrix} 5 & 1 \\ 8 & 7 \end{vmatrix}$$

8. What is the denominator determinant D for the system $\begin{cases} 3x + 4y = 7 \\ 2x - 3y = 5 \end{cases}$?

9. What is the denominator determinant D for the system $\begin{cases} x + 2y = -8 \\ 3x + y - z = -2 \\ 8x + 4y - z = 6 \end{cases}$?

10. For the system $\begin{cases} 3x + 2y = 1 \\ 4x - y = 3 \end{cases}$, $D_x = -7$, $D_y = 5$, and $D = -11$. Find the solution of the system.

11. For the system $\begin{cases} 2x + 3y - z = -8 \\ x - y - z = -2 \\ -4x + 3y + z = 6 \end{cases}$, $D_x = -28$, $D_y = -14$, $D_z = 14$, and $D = 14$. Find the solution.

12. Fill in the blank. If the denominator determinant D for a system of equations is 0, the equations of the system are dependent or the system is _____.

NOTATION

Complete the evaluation of each determinant.

13. $\begin{vmatrix} 5 & -2 \\ -2 & 6 \end{vmatrix} = 5() - (-2)(-2)$

    $\qquad = \boxed{} - 4$

    $\qquad = 26$

14. $\begin{vmatrix} 2 & 1 & 3 \\ 3 & 4 & 2 \\ 1 & 5 & 3 \end{vmatrix}$

    $= 2\begin{vmatrix} 4 & \\ 5 & 3 \end{vmatrix} - \boxed{} 1\begin{vmatrix} 3 & 2 \\ & 3 \end{vmatrix} + 3\begin{vmatrix} 3 & 4 \\ & 1 \end{vmatrix}$

    $= 2(- 10) - 1(9 -) + 3(15 -)$

    $= 2(2) - 1() + \boxed{}(11)$

    $= 4 - 7 + \boxed{}$

    $= 30$

GUIDED PRACTICE

Evaluate each determinant. See Example 1.

15. $\begin{vmatrix} 2 & 3 \\ 2 & 5 \end{vmatrix}$

16. $\begin{vmatrix} 3 & 2 \\ 2 & 4 \end{vmatrix}$

17. $\begin{vmatrix} -9 & 7 \\ 4 & -2 \end{vmatrix}$

18. $\begin{vmatrix} -1 & 2 \\ 3 & -4 \end{vmatrix}$

19. $\begin{vmatrix} 5 & 20 \\ 10 & 6 \end{vmatrix}$

20. $\begin{vmatrix} 10 & 15 \\ 15 & 5 \end{vmatrix}$

21. $\begin{vmatrix} -6 & -2 \\ 15 & 4 \end{vmatrix}$

22. $\begin{vmatrix} 3 & -2 \\ 12 & -8 \end{vmatrix}$

23. $\begin{vmatrix} -9 & -1 \\ -10 & -5 \end{vmatrix}$

24. $\begin{vmatrix} -7 & -7 \\ -6 & -4 \end{vmatrix}$

25. $\begin{vmatrix} 8 & 8 \\ -9 & -9 \end{vmatrix}$

26. $\begin{vmatrix} 20 & -3 \\ 20 & -3 \end{vmatrix}$

Evaluate each determinant. See Examples 2 and 3.

27. $\begin{vmatrix} 3 & 2 & 1 \\ 4 & 1 & 2 \\ 5 & 3 & 1 \end{vmatrix}$

28. $\begin{vmatrix} 6 & 2 & 3 \\ 1 & 5 & 4 \\ 2 & 3 & 5 \end{vmatrix}$

29. $\begin{vmatrix} 1 & -2 & 3 \\ -2 & 1 & 1 \\ -3 & -2 & 1 \end{vmatrix}$

30. $\begin{vmatrix} 1 & 1 & 2 \\ 2 & 1 & -2 \\ 3 & 1 & 3 \end{vmatrix}$

31. $\begin{vmatrix} -2 & 5 & 1 \\ 0 & 3 & 4 \\ -1 & 2 & 6 \end{vmatrix}$ **32.** $\begin{vmatrix} 4 & -1 & 2 \\ 6 & -1 & 0 \\ 1 & -3 & 4 \end{vmatrix}$

33. $\begin{vmatrix} 1 & -4 & 1 \\ 3 & 0 & -2 \\ 3 & 1 & -2 \end{vmatrix}$ **34.** $\begin{vmatrix} 8 & -3 & 1 \\ 1 & 0 & 2 \\ 3 & -9 & 4 \end{vmatrix}$

35. $\begin{vmatrix} 1 & 2 & 1 \\ -3 & 7 & 3 \\ -4 & 3 & -5 \end{vmatrix}$ **36.** $\begin{vmatrix} 1 & 4 & 7 \\ 2 & 5 & 8 \\ 3 & 6 & 9 \end{vmatrix}$

37. $\begin{vmatrix} 1 & 2 & 0 \\ 0 & 1 & 2 \\ 0 & 0 & 1 \end{vmatrix}$ **38.** $\begin{vmatrix} 1 & 0 & 1 \\ 0 & 1 & 0 \\ 1 & 1 & 1 \end{vmatrix}$

Use Cramer's rule to solve each system of equations. See Example 4.

39. $\begin{cases} x + y = 6 \\ x - y = 2 \end{cases}$ **40.** $\begin{cases} x - y = 4 \\ 2x + y = 5 \end{cases}$

41. $\begin{cases} x + 2y = -21 \\ x - 2y = 11 \end{cases}$ **42.** $\begin{cases} 5x + 2y = 11 \\ 7x + 6y = 9 \end{cases}$

43. $\begin{cases} 3x - 4y = 9 \\ x + 2y = 8 \end{cases}$ **44.** $\begin{cases} 2x + 2y = -1 \\ 3x + 4y = 0 \end{cases}$

45. $\begin{cases} 2x + 3y = 31 \\ 3x + 2y = 39 \end{cases}$ **46.** $\begin{cases} 5x + 3y = 72 \\ 3x + 5y = 56 \end{cases}$

Use Cramer's rule to solve each system of equations, if possible. If a system is inconsistent or if the equations are dependent, so indicate. See Example 5.

47. $\begin{cases} 3x + 2y = 11 \\ 6x + 4y = 11 \end{cases}$ **48.** $\begin{cases} 5x - 4y = 20 \\ 10x - 8y = 30 \end{cases}$

49. $\begin{cases} \dfrac{5}{6}x = 2 - y \\ 10x + 12y = 24 \end{cases}$ **50.** $\begin{cases} 16x - 8y = 32 \\ x - 2 = \dfrac{y}{2} \end{cases}$

Use Cramer's rule to solve each system of equations. See Example 6.

51. $\begin{cases} x + y + z = 4 \\ x + y - z = 0 \\ x - y + z = 2 \end{cases}$ **52.** $\begin{cases} x + y + z = 4 \\ x - y + z = 2 \\ x - y - z = 0 \end{cases}$

53. $\begin{cases} 3x + 2y - z = -8 \\ 2x - y + 7z = 10 \\ 2x + 2y - 3z = -10 \end{cases}$ **54.** $\begin{cases} x + 2y + 2z = 10 \\ 2x + y + 2z = 9 \\ 2x + 2y + z = 1 \end{cases}$

TRY IT YOURSELF

Use Cramer's rule to solve each system of equations, if possible. If a system is inconsistent or if the equations are dependent, so indicate.

55. $\begin{cases} 2x + y + z = 5 \\ x - 2y + 3z = 10 \\ x + y - 4z = -3 \end{cases}$ **56.** $\begin{cases} x + y + 2z = 7 \\ x + 2y + z = 8 \\ 2x + y + z = 9 \end{cases}$

57. $\begin{cases} y = \dfrac{-2x + 1}{3} \\ 3x - 2y = 8 \end{cases}$ **58.** $\begin{cases} 2x + 3y = -1 \\ x = \dfrac{y - 9}{4} \end{cases}$

59. $\begin{cases} 4x - 3y = 1 \\ 6x - 8z = 1 \\ 2y - 4z = 0 \end{cases}$ **60.** $\begin{cases} 4x + 3z = 4 \\ 2y - 6z = -1 \\ 8x + 4y + 3z = 9 \end{cases}$

61. $\begin{cases} 2x + y - z - 1 = 0 \\ x + 2y + 2z - 2 = 0 \\ 4x + 5y + 3z - 3 = 0 \end{cases}$ **62.** $\begin{cases} 2x - y + 4z + 2 = 0 \\ 5x + 8y + 7z = -8 \\ x + 3y + z + 3 = 0 \end{cases}$

63. $\begin{cases} 3x - 16 = 5y \\ -3x + 5y - 33 = 0 \end{cases}$ **64.** $\begin{cases} 2x + 5y - 13 = 0 \\ -2x + 13 = 5y \end{cases}$

65. $\begin{cases} x + y = 1 \\ \dfrac{1}{2}y + z = \dfrac{5}{2} \\ x - z = -3 \end{cases}$ **66.** $\begin{cases} \dfrac{1}{2}x + y + z + \dfrac{3}{2} = 0 \\ x + \dfrac{1}{2}y + z - \dfrac{1}{2} = 0 \\ x + y + \dfrac{1}{2}z + \dfrac{1}{2} = 0 \end{cases}$

67. $\begin{cases} 2x + 3y = 0 \\ 4x - 6y = -4 \end{cases}$ **68.** $\begin{cases} 4x - 3y = -1 \\ 8x + 3y = 4 \end{cases}$

69. $\begin{cases} 2x + 3y + 4z = 6 \\ 2x - 3y - 4z = -4 \\ 4x + 6y + 8z = 12 \end{cases}$ **70.** $\begin{cases} x - 3y + 4z - 2 = 0 \\ 2x + y + 2z - 3 = 0 \\ 4x - 5y + 10z - 7 = 0 \end{cases}$

Use a calculator with matrix capabilities. Evaluate each determinant. See Using Your Calculator: Evaluating Determinants.

71. $\begin{vmatrix} 25 & -36 & 44 \\ -11 & 21 & 54 \\ 37 & -31 & 19 \end{vmatrix}$ **72.** $\begin{vmatrix} 13 & -27 & 62 \\ -38 & 27 & -52 \\ 10 & -300 & 42 \end{vmatrix}$

73. $\begin{vmatrix} -280 & 191 & -356 \\ -211 & -102 & -422 \\ 400 & -213 & -333 \end{vmatrix}$ **74.** $\begin{vmatrix} 4.1 & 2.2 & -3.3 \\ 2.7 & -5.9 & 6.8 \\ 2.3 & 5.3 & 0.6 \end{vmatrix}$

REVIEW EXERCISES

29. Determine whether $(2, -1, 1)$ is a solution of the system:
$$\begin{cases} x - y + z = 4 \\ x + 2y - z = -1 \\ x + y - 3z = -1 \end{cases}$$

30. A system of three linear equations in three variables is graphed on the right. Does the system have a solution? If so, how many solutions does it have?

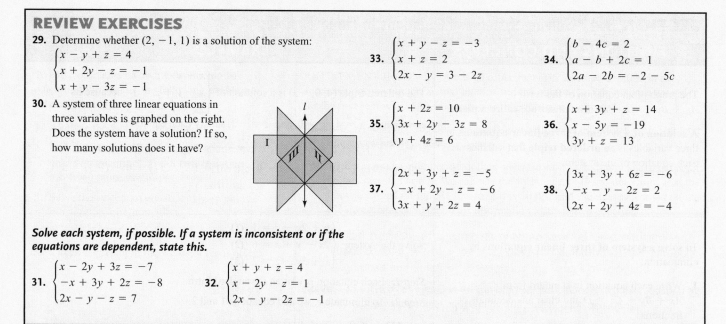

33. $\begin{cases} x + y - z = -3 \\ x + z = 2 \\ 2x - y = 3 - 2z \end{cases}$

34. $\begin{cases} b - 4c = 2 \\ a - b + 2c = 1 \\ 2a - 2b = -2 - 5c \end{cases}$

35. $\begin{cases} x + 2z = 10 \\ 3x + 2y - 3z = 8 \\ y + 4z = 6 \end{cases}$

36. $\begin{cases} x + 3y + z = 14 \\ x - 5y = -19 \\ 3y + z = 13 \end{cases}$

37. $\begin{cases} 2x + 3y + z = -5 \\ -x + 2y - z = -6 \\ 3x + y + 2z = 4 \end{cases}$

38. $\begin{cases} 3x + 3y + 6z = -6 \\ -x - y - 2z = 2 \\ 2x + 2y + 4z = -4 \end{cases}$

Solve each system, if possible. If a system is inconsistent or if the equations are dependent, state this.

31. $\begin{cases} x - 2y + 3z = -7 \\ -x + 3y + 2z = -8 \\ 2x - y - z = 7 \end{cases}$

32. $\begin{cases} x + y + z = 4 \\ x - 2y - z = 1 \\ 2x - y - 2z = -1 \end{cases}$

SECTION 3.5   Problem Solving Using Systems of Three Equations

DEFINITIONS AND CONCEPTS	EXAMPLES

DEFINITIONS AND CONCEPTS

Problems that involve **three unknown quantities** can be solved using a strategy similar to that for solving problems involving two unknowns.

EXAMPLES

BATTERIES   A hardware store sells three types of batteries: AA size for $1 each, C size for $1.50 each, and D size for $2.00 each. One Saturday, the store sold 25 batteries for a total of $34. If the number of C batteries that were sold was four less than the number of AA batteries that were sold, how many of each size battery were sold?

Analyze   To find the three unknowns we will write a system of three equations in three variables.

Form   Let A = the number of AA batteries sold, C = the number of C batteries sold, and D = the number of D batteries sold. The given information leads to three equations:

$$\begin{cases} A + C + D = 25 \\ 1A + 1.50C + 2D = 34 \\ C = A - 4 \end{cases}$$
The total number of batteries sold was 25.
The total value of the batteries sold was $34.
The number of C batteries sold was 4 less than AA batteries sold.

If we multiply the second equation by 10 to clear the decimal and write the third equation in standard form, we have the system:

(1) $\begin{cases} A + C + D = 25 \\ $ **(2)** $ 10A + 15C + 20D = 340 \\ $ **(3)** $ -A + C = -4 \end{cases}$

Solve   Since equation 3 does not contain a D-term, we will find another equation that does not contain a D-term. If each side of equation 1 is multiplied by -20 and the resulting equation is added to the equation 2, D is eliminated, and we obtain

$$-20A - 20C - 20D = -500 \qquad \text{This is } -20(A + C + D) = -20(25).$$
(2) $\underline{\quad 10A + 15C + 20D = \quad 340 \quad}$
(4) $-10A - 5C \qquad\qquad = -160$

SECTION 3.5 Problem Solving Using Systems of Three Equations–*continued*

DEFINITIONS AND CONCEPTS	EXAMPLES
	Equations 3 and 4 form a system of two equations in *A* and *C* that can be solved in the usual manner. (The remaining work is left to the reader.)

$$\begin{cases} -10A - 5C = -160 \\ -A + C = -4 \end{cases}$$

State   There were 12 AA batteries, 8 C batteries, and 5 D batteries sold.

Check   Verify that these results are correct by checking them in the words of the problem.

REVIEW EXERCISES

39. TEDDY BEARS   A toy company produces three sizes of teddy bears. Each day, the total cost to produce the bears is $850, the total time needed to stuff them is 480 minutes, and the total time needed to sew them is 1,260 minutes. Use the information in the table to determine how many of each type of teddy bear are produced daily.

Size of teddy bear	Production cost	Stuffing time	Sewing time
Small	$3	2 min	6 min
Medium	$5	3 min	8 min
Large	$10	5 min	12 min

40. VETERINARY MEDICINE   The daily requirements of a balanced diet for an animal are shown in the nutritional pyramid. The number of grams per cup of nutrients in three food mixes are shown in the table. How many cups of each mix should be used to meet the daily requirements for protein, carbohydrates, and essential fatty acids in the animal's diet?

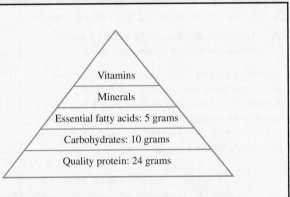

	Grams per cup		
	Protein	Carbohydrates	Fatty acids
Mix A	5	2	1
Mix B	6	3	2
Mix C	8	3	1

41. FINANCIAL PLANNING   A financial planner invested $22,000 in three accounts, paying 5%, 6%, and 7% annual interest. She invested $2,000 more at 6% than at 5%. If the total interest earned in one year was $1,370, how much was invested at each rate?

42. BALLISTICS   The path of a thrown object is a parabola with an equation of $y = ax^2 + bx + c$. The parabola passes through the points (0, 0), (8, 12), and (12, 15). Find *a*, *b*, and *c*.

SECTION 3.6 Solving Systems of Equations Using Matrices

DEFINITIONS AND CONCEPTS	EXAMPLES
A **matrix** is a rectangular array of numbers. Each number in a matrix is called an **element** or an **entry** of the matrix. A matrix with *m* rows and *n* columns has **order** $m \times n$.	A 2 × 3 matrix: $\begin{bmatrix} 2 & -7 & 5 \\ -3 & 4 & 1 \end{bmatrix}$ A 3 × 3 matrix: $\begin{bmatrix} 5 & -3 & 12 \\ 4 & 7 & -5 \\ 1 & -4 & 2 \end{bmatrix}$

CHAPTER 3
TEST

1. Fill in the blanks.

  a. $\begin{cases} 2x - 7y = 1 \\ 4x - y = -8 \end{cases}$ is called a _____ of linear equations.

  b. The matrix $\begin{bmatrix} -10 & 3 \\ 4 & 9 \end{bmatrix}$ has 2 _____ and 2 _____.

  c. Solutions of a system of three equations in three variables, x, y, and z, are written in the form (x, y, z) and are called ordered _____.

  d. The graph of the equation $2x + 3y + 4z = 5$ is a flat surface called a _____.

  e. A _____ is a rectangular array of numbers written within brackets.

2. Solve the system by graphing: $\begin{cases} 2x + y = 5 \\ y = 2x - 3 \end{cases}$

3. Determine whether $\left(-\frac{1}{2}, -\frac{2}{3}\right)$ is a solution of the system:
$\begin{cases} 10x - 12y = 3 \\ 18x - 15y = 1 \end{cases}$

4. POLITICS   Explain the importance of the point of intersection of the graphs shown below.

Governor Arnold Schwarzenegger's Job Performance

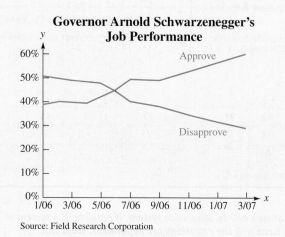

Source: Field Research Corporation

5. Use the vocabulary of this chapter to describe each system of two linear equations in two variables graphed below. Does the system have a solution (or solutions)?

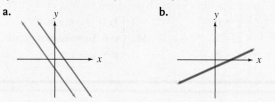

6. Use the graphs in the illustration to solve $3(x - 2) - 2(-2 + x) = 1$.

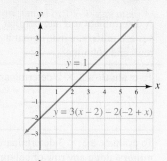

7. Use substitution to solve the system: $\begin{cases} 2x - 4y = 14 \\ x + 2y = 7 \end{cases}$

8. Use elimination (addition) to solve the system:
$\begin{cases} 2c + 3d = -5 \\ 3c - 2d = 12 \end{cases}$

Solve each system by any method, if possible. If a system is inconsistent or if the equations are dependent, state this.

9. $\begin{cases} 3(x + y) = x - 3 \\ -y = \dfrac{2x + 3}{3} \end{cases}$

10. $\begin{cases} 0.6x + 0.5y = 1.2 \\ x - \dfrac{4}{9}y + \dfrac{5}{9} = 0 \end{cases}$

Write a system of two equations in two variables to solve each problem.

11. TRAFFIC SIGNS   In the sign, find x and y, if y is 15 more than x.

12. ANTIFREEZE   How much of a 40% antifreeze solution must a mechanic mix with an 80% antifreeze solution if 20 gallons of a 50% antifreeze solution are needed?

13. BREAK POINTS   A metal stamping plant is considering purchasing a new piece of equipment. The machine they currently use has a setup cost of $1,775 and a cost of $5.75 per impression. The new machine has a setup cost of $3,975 and a cost of $4.15 per impression. Find the break point.

14. Determine whether $\left(-1, -\frac{1}{2}, 5\right)$ is a solution of:
$\begin{cases} x - 2y + z = 5 \\ 2x + 4y = -4 \\ -6y + 4z = 22 \end{cases}$

15. Solve the system: $\begin{cases} x + y + z = 4 \\ x + y - z = 6 \\ 2x - 3y + z = -1 \end{cases}$

16. Solve the system: $\begin{cases} z - 2y = 1 \\ x + y + z = 1 \\ x + 5y = 4 \end{cases}$

17. MOVIE TICKETS   The receipts for one showing of a movie were \$410 for an audience of 100 people. The ticket prices are given in the table. If twice as many children's tickets as general admission tickets were purchased, how many of each type of ticket were sold?

Ticket prices	
Children	\$3
General admission	\$6
Seniors	\$5

18. Let $A = \begin{bmatrix} 1 & 7 & \vdots & -3 \\ 3 & -1 & \vdots & 13 \end{bmatrix}$. Write the matrix obtained when the elementary row operations $-3R_1 + R_2$ are performed on matrix A.

Use matrices to solve each system, if possible. If a system is inconsistent or if the equations are dependent, state this.

19. $\begin{cases} x + y = 4 \\ 2x - y = 2 \end{cases}$

20. $\begin{cases} x - 3y + 2z = 1 \\ x - 2y + 3z = 5 \\ 2x - 6y + 4z = 3 \end{cases}$

Evaluate each determinant.

21. $\begin{vmatrix} 2 & -3 \\ -4 & 5 \end{vmatrix}$

22. $\begin{vmatrix} 1 & 2 & 0 \\ 2 & 0 & 3 \\ 1 & -2 & 2 \end{vmatrix}$

23. Use Cramer's rule to solve the system:
$$\begin{cases} x - y = -6 \\ 3x + y = -6 \end{cases}$$

24. Solve the following system for z only, using Cramer's rule.
$$\begin{cases} x + y + z = 4 \\ x + y - z = 6 \\ 2x - 3y + z = -1 \end{cases}$$

GROUP PROJECT

METHODS OF SOLUTION

Overview: In this activity, you will explore the advantages and disadvantages of several methods for solving a system of linear equations.

Instructions: Form groups of 5 students. Have each member of your group solve the system

$$\begin{cases} x - y = 4 \\ 2x + y = 5 \end{cases}$$

in a different way. The methods to use are graphing, substitution, elimination, matrices, and Cramer's rule. Have each person briefly explain his or her method of solution to the group. After everyone has presented a solution, discuss the advantages and drawbacks of each method. Then rank the five methods, from most desirable to least desirable.

CHAPTERS 1–3
Cumulative Review

1. Complete the illustration by labeling the rational numbers, irrational numbers, integers, and whole numbers. [Section 1.2]

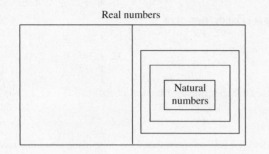

Real numbers

Natural numbers

2. Insert either a $<$ or a $>$ symbol to make a true statement. [Section 1.2]

   a. -5.96 -5.95 **b.** $-(-1)$ $-|-18|$

Evaluate each expression for $a = -3$ and $b = -5$.

3. $-|b| - ab^2$

[Section 1.3]

4. $\dfrac{14 + 2[2a - (b - a)]}{-b - 2}$

[Section 1.3]

Simplify each expression.

5. $40\left(\dfrac{3}{8}m - \dfrac{1}{4}\right) + 40\left(\dfrac{4}{5}\right)$ [Section 1.4]

6. $3[2(a + 2)] - 5[3(a - 5)] + 5a$ [Section 1.4]

Solve each equation, if possible. If an equation is an identity or a contradiction, state this.

7. $\dfrac{3}{4}x + 1.5 = -19.5$ [Section 1.5]

8. $1 + 3[-2 + 6(4 - 2x)] = -(x + 3)$ [Section 1.5]

9. $\dfrac{x + 7}{3} = \dfrac{x - 2}{5} - \dfrac{x}{15} + \dfrac{7}{3}$ [Section 1.5]

10. $3p - 6 = 4(p - 2) + 2 - p$ [Section 1.5]

Solve each equation for the specified variable.

11. $\lambda = Ax + AB$ for B [Section 1.6]

12. $v = \dfrac{d_1 - d_2}{t}$ for d_2 [Section 1.6]

13. ANTIQUE SHOWS A traveling antique show will be on the road for 17 weeks, visiting three cities. They will be in Los Angeles for 2 weeks longer than they will be in Las Vegas. Their stay in Dallas will be 1 week less than twice that in Las Vegas. How many weeks will they be in each city? [Section 1.7]

14. U.S. FEDERAL BUDGET The proposed federal spending for the fiscal year 2008 was $2,902 billion. The illustration shows how a typical dollar of the budget was to be spent. Determine the total amount to be spent on Social Security, Medicare, and Medicaid. Round to the nearest billion dollars. [Section 1.8]

Social Security:	Defense:	Other entitlements:	Medicare:	Interest:	Medicaid:	Other spending:
21¢	21¢	14¢	13¢	9¢	7¢	15¢

Source: Office of Management and Budget

15. COMMUTING Use the following facts to determine a commuter's average speed when she drives to work.
   - If she drives her car, it takes a quarter of an hour to get to work.
   - If she rides the bus, it takes half an hour to get to work.
   - When she drives, her average speed is 10 miles per hour faster than that of the bus. [Section 1.8]

16. DRIED FRUITS Dried apple slices cost $4.60 per pound, and dried banana chips sell for $3.40 per pound. How many pounds of each should be used to create a 10-pound mixture that sells for $4 per pound? [Section 1.8]

Graph each equation.

17. $3x = 4y - 11$
[Section 2.2]

18. $y = -4$
[Section 2.2]

19. Find the slope of the line, shown on the right.
[Section 2.3]

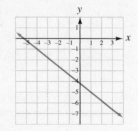

20. Determine whether the line passing through $(-3, -1)$ and $(-3, 4)$ and the line passing through $(-4, -2)$ and $(5, -2)$ are parallel, perpendicular, or neither. [Section 2.3]

21. Find an equation of the line that passes through $(4, 5)$ and is parallel to the graph of $y = -3x$. Write the equation in slope–intercept form. [Section 2.4]

22. COLLECTIBLES A collector buys the Hummel figurine shown in the illustration anticipating that it will be worth $650 in 20 years. Assuming straight-line appreciation, write an equation that gives the value v of the figurine x years after it is purchased. [Section 2.4]

Price: $300.00

© Christa Knijff

If $f(x) = -x^2 - \frac{x}{2}$, find each of the following.

23. $f(10)$ [Section 2.5]

24. $f(r)$ [Section 2.5]

25. We can think of a function as a machine. (See the illustration.) Write a function that turns the given input into the given output. [Section 2.5]

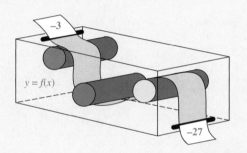

−3

$y = f(x)$

−27

26. Does the table define y as a function of x? [Section 2.5]

x	y
-2	5
-1	2
0	2
5	5

27. Use the graph of function f to find each of the following. [Section 2.6]
   a. $f(-1)$
   b. The value of x for which $f(x) = 3$

28. Determine whether the graph on the right is the graph of a function. Explain why or why not. [Section 2.6]

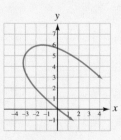

29. Graph $f(x) = (x + 4)^2$. Give the domain and range of the function. [Section 2.6]

30. Solve the system by graphing: $\begin{cases} 3x - y = -3 \\ y = -2x - 7 \end{cases}$ [Section 3.1]

31. a. What is an inconsistent system of two linear equations in two variables? [Section 3.1]

   b. What are dependent linear equations in two variables? [Section 3.1]

32. Solve the system: $\begin{cases} y = \frac{-2x + 1}{3} \\ 3x - 2y = 8 \end{cases}$ [Section 3.2]

33. INVENTORIES The table shows an end-of-the-year inventory report for a warehouse that supplies electronics stores. If the warehouse stocks two models of cordless telephones, one valued at $67 and the other at $100, how many of each model of phone did the warehouse have at the time of the inventory? [Section 3.3]

Item	Number	Merchandise value
Cordless phones	360	$29,400

34. Solve the system: $\begin{cases} -x + 3y + 2z = 5 \\ 3x + 2y + z = -1 \\ 2x - y + 3z = 4 \end{cases}$ [Section 3.4]

35. INVESTING A woman wants to earn $2,200 the first year that she invests $30,000 in three certificates of deposit. (See the table.) She wants to invest five times as much in the 36-month CD as in the 12-month CD. How much should she invest in each CD? [Section 3.5]

Type of CD	Annual rate of return
12-month	6%
24-month	7%
36-month	8%

36. Use matrices to solve the system: $\begin{cases} x - y = 5 \\ 2x - 5y = 1 \end{cases}$ [Section 3.6]

37. Evaluate each determinant. [Section 3.6]

   a. $\begin{vmatrix} 6 & 2 \\ -2 & 6 \end{vmatrix}$

   b. $\begin{vmatrix} 2 & 1 & -3 \\ -2 & 2 & 4 \\ 1 & -2 & 2 \end{vmatrix}$

38. Use Cramer's rule to solve the system: $\begin{cases} x + 2y = 6 \\ x - y = 4 \end{cases}$

[Section 3.7]

CHAPTER 4

Inequalities

© Andrew Brookes/Corbis

4.1 Solving Linear Inequalities in One Variable

4.2 Solving Compound Inequalities

4.3 Solving Absolute Value Equations and Inequalities

4.4 Linear Inequalities in Two Variables

4.5 Systems of Linear Inequalities

CHAPTER SUMMARY AND REVIEW

CHAPTER TEST

Group Project

CUMULATIVE REVIEW

from *Campus to Careers*
Heating, Ventilation, and Air Conditioning Technician

HVAC technicians make sure that we are warm in the winter and cool in the summer. They install, maintain, and repair heating and cooling systems in residential, commercial, and industrial buildings. HVAC technicians constantly work with numbers as they take measurements, read blueprints, and prepare work estimates. The installation instructions and diagrams they follow require strong mathematical skills in algebra and geometry.

HVAC technicians adjust heating /cooling system controls to recommended temperature settings. In **Problem 75** of **Study Set 4.2**, you will use interval notation to express the temperature range for a room by interpreting the settings on a thermostat.

JOB TITLE:
Heating, Ventilation, and Air Conditioning Technician

EDUCATION:
Technical school training or completion of an apprenticeship.

JOB OUTLOOK:
Employment opportunities are projected to increase by 18% to 26% through the year 2014.

ANNUAL EARNINGS:
Starting salaries range from $25,000 to $35,000.

FOR MORE INFORMATION:
www.bls.gov/oco/home.htm

319

variable. For example, the set of real numbers greater than -5 is written in set-builder notation as

$$\{x \mid x > -5\}$$

The set of all real numbers x   such that   x is greater than -5

The interval shown in the following figure is the graph of the real numbers less than or equal to 7. It contains the numbers that satisfy the inequality $x \leq 7$. The right **bracket** at 7 indicates that 7 is included in the interval. To express this interval in interval notation, we write $(-\infty, 7]$, where $-\infty$ (read as **negative infinity**) indicates that the interval extends indefinitely to the left. The bracket is used to show that 7 is included in the interval. To describe the interval using set-builder notation, we write $\{x \mid x \leq 7\}$.

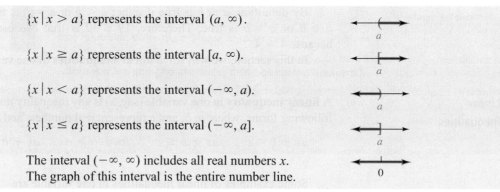

If an interval extends forever in one direction, as in the previous examples, it is called an **unbounded interval.** The following chart illustrates the various types of unbounded intervals and shows how they are described using set-builder notation and a graph.

Unbounded Intervals

$\{x \mid x > a\}$ represents the interval (a, ∞).

$\{x \mid x \geq a\}$ represents the interval $[a, \infty)$.

$\{x \mid x < a\}$ represents the interval $(-\infty, a)$.

$\{x \mid x \leq a\}$ represents the interval $(-\infty, a]$.

The interval $(-\infty, \infty)$ includes all real numbers x.
The graph of this interval is the entire number line.

When graphing intervals, an open circle can be used to show that a point is not included in a graph, and a solid circle can be used to show that a point is included. For example,

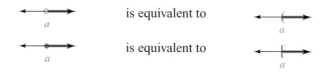

is equivalent to

is equivalent to

We will use parentheses and brackets when graphing intervals because they are consistent with interval notation.

EXAMPLE 1   Represent each set using a graph, interval notation, and set-builder notation.   **a.** The set of real numbers greater than or equal to 8   **b.** The set of real numbers less than -3

Strategy   To graph the interval on a number line, we will determine whether the endpoint is in the graph. Then we will determine whether the real numbers to the right or the left of the endpoint should be shaded.

Why We draw the graph first because the corresponding interval notation and set-builder notation follow directly from it.

Solution

a. The set of real numbers *greater than or equal to* 8 is graphed below. Because equality with 8 is allowed, a bracket is used to include the endpoint 8. The numbers to the right of 8 are shaded because they are greater than 8. The interval is written as $[8, \infty)$ and the set-builder notation is written as $\{x \mid x \geq 8\}$.

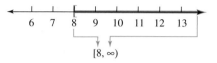

b. The set of real numbers *less than* -3 is graphed below. Since the elements of the set must be strictly less than -3, we exclude -3 using a parenthesis and shade the real numbers to its left. The interval notation is written as $(-\infty, -3)$ and the set-builder notation is written as $\{x \mid x < -3\}$.

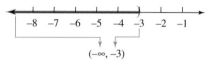

 Self Check 1   Represent each set using a graph, interval notation, and set-builder notation.
   a. The set of real numbers greater than 1
   b. The set of real numbers less than or equal to -2

Now Try   **Problems 17 and 19**

---

3   **Solve Linear Inequalities Using Properties of Inequality.**

To **solve a linear inequality** means to find the values of its variable that make the inequality true. The set of all solutions of an inequality is called its **solution set.** We will use the following properties to solve inequalities in one variable.

Addition and Subtraction Properties of Inequality	Adding the same number to, or subtracting the same number from, both sides of an inequality does not change the solutions.

Adding the same number to, or subtracting the same number from, both sides of an inequality does not change the solutions.
   For any real numbers a, b, and c,

   If $a < b$, then $a + c < b + c$.

   If $a < b$, then $a - c < b - c$.

Similar statements can be made for the symbols $\leq$, $>$, or $\geq$.

As with equations, there are properties for multiplying and dividing both sides of an inequality by the same number. To develop the **multiplication property of inequality,** we consider the true statement $2 < 5$. If both sides are multiplied by a positive number, such as 3, another true inequality results.

$$2 < 5$$
$$3 \cdot 2 < 3 \cdot 5 \qquad \text{Multiply both sides by 3.}$$
$$6 < 15 \qquad \text{This is a true inequality.}$$

However, if we multiply both sides of $2 < 5$ by a negative number, such as -3, the direction of the inequality symbol must be reversed to produce another true inequality.

$$2 < 5$$
$$-3 \cdot 2 > -3 \cdot 5 \qquad \text{Multiply both sides by the negative number } -3 \text{ and change } < \text{ to } >.$$
$$-6 > -15 \qquad \text{This is a true inequality.}$$

Dividing both sides of a true inequality by the same positive number leads to another true inequality. However, dividing both sides of a true inequality by the same negative number requires that the direction of the inequality symbol be reversed to produce another true inequality.

$$6 > -4 \qquad \text{This is a true inequality.}$$
$$\frac{6}{-2} < \frac{-4}{-2} \qquad \text{Divide both sides by the negative number } -2 \text{ and change } > \text{ to } <.$$
$$-3 < 2 \qquad \text{This is a true inequality.}$$

These examples illustrate the multiplication and division properties of inequality.

Multiplication and Division Properties of Inequality	Multiplying or dividing both sides of an inequality by the same positive number does not change the solutions. For any real numbers a, b, and c, where c is **positive,** $\qquad$ If $a < b$, $\quad$ then $\quad ac < bc \quad$ and $\quad \dfrac{a}{c} < \dfrac{b}{c}$. If we multiply or divide both sides of an inequality by a negative number, the direction of the inequality symbol must be reversed for the inequalities to have the same solutions. For any real numbers a, b, and c, where c is **negative,** $\qquad$ If $a < b$, $\quad$ then $\quad ac > bc \quad$ and $\quad \dfrac{a}{c} > \dfrac{b}{c}$. Similar statements can be made for the symbols $\leq$, $>$, or $\geq$.

After applying one of the properties of inequality, the resulting inequality is equivalent to the original one. Like equivalent equations, **equivalent inequalities** have the same solution set.

EXAMPLE 2 Solve: $6x - 27 < 9$. Graph the solution set and write it using interval notation.

Strategy We will use properties of inequality to isolate the variable on one side.

Why Once we have obtained an equivalent inequality of the form $x < a$, with the variable isolated on one side, the solution set is obvious.

The Language of Algebra
Because $<$ requires one number to be strictly less than another number and $>$ requires one number to be strictly greater than another number, $<$ and $>$ are called *strict inequalities*.

Solution We use the same steps to solve inequalities as we used to solve equations.

$$6x - 27 < 9 \qquad \text{This is the inequality to solve.}$$
$$6x - 27 + 27 < 9 + 27 \qquad \text{To isolate the variable term } 6x, \text{ undo the subtraction of 27 on the left side by adding 27 to both sides.}$$
$$6x < 36 \qquad \text{Simplify each side.}$$

$$\frac{6x}{6} < \frac{36}{6}$$ To isolate x on the left side, undo the multiplication by 6 by dividing both sides by 6.

$$x < 6$$ Simplify each side.

The graph of the solution set is shown on the right. It can be written in interval notation as $(-\infty, 6)$ and in set-builder notation as $\{x \mid x < 6\}$.

5   6   7

Since the solution set contains infinitely many real numbers, we cannot check all of them to see whether they satisfy the original inequality. However, as an informal check, we can pick one number in the graph, near the endpoint, such as 5, and see whether it satisfies the inequality. We can also pick one number not in the graph, but near the endpoint, such as 7, and see whether it fails to satisfy the inequality.

Check a value in the graph: $x = 5$	*Check a value not in the graph*: $x = 7$
$6x - 27 < 9$	$6x - 27 < 9$
$6(5) - 27 \overset{?}{<} 9$ Substitute 5 for x.	$6(7) - 27 \overset{?}{<} 9$ Substitute 7 for x.
$30 - 27 \overset{?}{<} 9$	$42 - 27 \overset{?}{<} 9$
$3 < 9$ True	$15 < 9$ False

Since 5 satisfies $6x - 27 < 9$ and 7 does not, the solution set appears to be correct.

Self Check 2   Solve: $8x + 4 < -44$. Graph the solution set and write it using interval notation.

Now Try   **Problem 25**

EXAMPLE 3   Solve: $2x - 9 - 10x \le 3 + 4x + 12$. Graph the solution set and write it using interval notation.

Strategy   We will combine the like terms on each side of the inequality and use properties of inequality to isolate the variable on one side.

Why   Once we have obtained an equivalent inequality, with the variable isolated on one side, the solution set is obvious.

Solution

$2x - 9 - 10x \le 3 + 4x + 12$	This is the inequality to solve.
$-8x - 9 \le 4x + 15$	Combine like terms: $2x - 10x = -8x$ and $3 + 12 = 15$.
$-8x - 9 - 4x \le 4x + 15 - 4x$	To eliminate 4x on the right side, subtract 4x from both sides.
$-12x - 9 \le 15$	Simplify each side.
$-12x - 9 + 9 \le 15 + 9$	To isolate the variable term −12x, undo the subtraction of 9 on the left side by adding 9 to both sides.
$-12x \le 24$	Simplify each side.
$\dfrac{-12x}{-12} \ge \dfrac{24}{-12}$	To isolate x, undo the multiplication by −12 by dividing both sides by −12. Because of the division by a negative number, reverse the $\le$ symbol.
$x \ge -2$	Simplify each side.

The graph of the solution set is shown on the right. It can be written in interval notation as $[-2, \infty)$ and in set-builder notation as $\{x \mid x \ge -2\}$.

−3   −2   −1

> **Self Check 3**   Solve: $x + 4 - 5x \le 1 + 2x - 15$. Graph the solution set and write it using interval notation.
>
> **Now Try**   **Problem 31**

EXAMPLE 4   Solve: $-7 > \dfrac{16}{15}t + 1$. Graph the solution set and write it using interval notation.

Strategy   We will use properties of inequality to isolate the variable on one side.

Why   Once we have obtained an equivalent inequality, with the variable isolated on one side, the solution set is obvious.

Solution

$$-7 > \frac{16}{15}t + 1 \qquad \text{This is the inequality to solve.}$$

$$-8 > \frac{16}{15}t \qquad \text{Subtract 1 from both sides.}$$

$$\frac{15}{16}(-8) > \frac{15}{16}\left(\frac{16}{15}t\right) \qquad \begin{array}{l}\text{To undo the multiplication by } \frac{16}{15} \text{ and isolate } t, \\ \text{multiply both sides by the reciprocal, which is } \frac{15}{16}.\end{array}$$

$$-\frac{15}{2} > t \qquad \text{Simplify each side.}$$

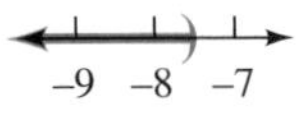

> **Caution**
> When solving inequalities, the variable can end up on the right side. This causes some students trouble when graphing the solution set. It is helpful to write such an inequality in an equivalent form with the variable on the left and then draw the graph.

We can write an equivalent inequality with the variable on the left side.

$$t < -\frac{15}{2} \qquad \text{If } -\frac{15}{2} \text{ is greater than } t, \text{ then } t \text{ must be less than } -\frac{15}{2}.$$

The graph of the solution set is shown on the right. It can be written in interval notation as $\left(-\infty, -\frac{15}{2}\right)$ and in set-builder notation as $\left\{t \mid t < -\frac{15}{2}\right\}$.

-9 -8 -7
    -15/2

> **Self Check 4**   Solve: $7 < \dfrac{10}{9}s + 2$. Graph the solution set and write it using interval notation.
>
> **Now Try**   **Problem 33**

EXAMPLE 5   Solve: $\dfrac{2}{3}(x + 2) > \dfrac{4}{5}(x - 3)$. Graph the solution set and write it using interval notation.

Strategy   We will clear the inequality of fractions by multiplying both sides by the LCD of $\frac{2}{3}$ and $\frac{4}{5}$.

Why   It's easier to solve an inequality that involves only integers.

Solution   The LCD of $\frac{2}{3}$ and $\frac{4}{5}$ is 15.

$$\frac{2}{3}(x + 2) > \frac{4}{5}(x - 3)$$ *This is the inequality to solve.*

$$15 \cdot \frac{2}{3}(x + 2) > 15 \cdot \frac{4}{5}(x - 3)$$ *Multiply both sides by 15.*

$$10(x + 2) > 12(x - 3)$$ *Simplify: $15 \cdot \frac{2}{3} = 10$ and $15 \cdot \frac{4}{5} = 12$.*

$$10x + 20 > 12x - 36$$ *Distribute the multiplication by 10 and 12.*

$$-2x + 20 > -36$$ *To eliminate 12x on the right side, subtract 12x from both sides.*

$$-2x > -56$$ *To isolate the variable term $-2x$, undo the addition of 20 by subtracting 20 from both sides.*

$$\frac{-2x}{-2} < \frac{-56}{-2}$$ *To isolate x, undo the multiplication by -2 by dividing both sides by -2 and reverse the $>$ symbol.*

$$x < 28$$ *Simplify each side.*

The graph of the solution set is shown on the right. It can be written in interval notation as $(-\infty, 28)$ and in set-builder notation as $\{x \mid x < 28\}$.

27 28 29

> **Self Check 5**   Solve: $\frac{3}{2}(x + 2) \le \frac{3}{5}(x - 3)$. Graph the solution set and write it using interval notation.
>
> **Now Try**   **Problem 37**

When solving equations, we have seen that some are true for all real numbers while others have no solution. Similar situations can occur when solving inequalities.

EXAMPLE 6   Solve each inequality. Graph the solution set and write it using interval notation.

a. $\dfrac{3a - 4}{-5} > \dfrac{3a + 15}{-5}$ **b.** $1 - 2a \ge 2(1 - a)$

Strategy   We will use properties of inequality to isolate the variable on one side.

Why   Once we have obtained an equivalent inequality, with the variable isolated on one side, the solution set is obvious.

Solution

a.

$$\frac{3a - 4}{-5} > \frac{3a + 15}{-5}$$ *This is the inequality to solve.*

$$-5\left(\frac{3a - 4}{-5}\right) < -5\left(\frac{3a + 15}{-5}\right)$$ *To clear the inequality of fractions, multiply both sides by -5. Since we are multiplying both sides by a negative number, reverse the direction of the inequality.*

$$3a - 4 < 3a + 15$$ *Simplify each side.*

$$3a - 4 - 3a < 3a + 15 - 3a$$ *Subtract 3a from both sides.*

$$-4 < 15$$ *This is a true statement.*

The terms involving the variable a drop out. The resulting true statement indicates that the original inequality is true for all values of a. Therefore, the solution set is the set of real numbers, denoted $(-\infty, \infty)$ or $\mathbb{R}$, and its graph is as shown.

−1   0   1

b.

$1 - 2a \geq 2(1 - a)$	This is the inequality to solve.
$1 - 2a \geq 2 - 2a$	Distribute the multiplication by 2.
$1 - 2a + 2a \geq 2 - 2a + 2a$	Add 2a to both sides.
$1 \geq 2$	This is a false statement.

The terms involving the variable *a* drop out. The resulting false statement indicates that the original inequality is false for all values of *a*. Therefore, the inequality has no solution. The solution set has no elements and is denoted $\varnothing$.

Self Check 6   Solve each inequality. Graph the solution set and write it using interval notation.   **a.** $-8n + 10 \geq 1 - 2(4n - 2)$
b. $\frac{4d - 5}{-10} > \frac{2(2d - 3)}{-10}$

Now Try   **Problems 41 and 43**

Using Your Calculator

Solving Linear Inequalities in One Variable

There are several ways to solve linear inequalities graphically. For example, to solve $3(2x - 9) < 9$ we can subtract 9 from both sides and solve the equivalent inequality $3(2x - 9) - 9 < 0$. Using standard window settings of $[-10, 10]$ for *x* and $[-10, 10]$ for *y*, we graph $y = 3(2x - 9) - 9$ and then use TRACE. Moving the cursor closer and closer to the *x*-axis, as shown in figure (a), we see that the graph is below the *x*-axis for *x*-values in the interval $(-\infty, 6)$. This interval is the solution, because in this interval, $3(2x - 9) - 9 < 0$.

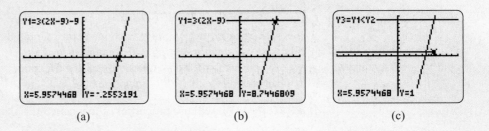

(a)   (b)   (c)

Another way to solve $3(2x - 9) < 9$ is to graph $y = 3(2x - 9)$ and $y = 9$. We can then trace to see that the graph of $y = 3(2x - 9)$ is below the graph of $y = 9$ for *x*-values in the interval $(-\infty, 6)$. See figure (b). This interval is the solution, because in this interval, $3(2x - 9) < 9$.

A third approach is to enter and then graph

$Y_1 = 3(2x - 9)$

$Y_2 = 9$

$Y_3 = Y_1 < Y_2$ To do this, use the VARS key. Consult your owner's manual for the specific directions.

The graphs of $y = 3(2x - 9)$, $y = 9$, and a horizontal line 1 unit above the *x*-axis will be displayed, as shown in figure (c). In the TRACE mode, we then move the cursor to the rightmost endpoint of the horizontal line to determine that the interval $(-\infty, 6)$ is the solution of $3(2x - 9) < 9$.

Use Linear Inequalities to Solve Problems.

To solve problems involving inequalities, we will use the familiar five-step problem-solving strategy. Here are some key words and phrases that translate to inequalities.

The statement	*Translates to*	*The statement*	*Translates to*
a does not exceed *b*.	$a \leq b$	*a* will exceed *b*.	$a > b$
a is at most *b*.	$a \leq b$	*a* is at least *b*.	$a \geq b$
a is no more than *b*.	$a \leq b$	*a* is not less than *b*.	$a \geq b$

EXAMPLE 7 *Political Contributions.* Some volunteers are making long-distance telephone calls to ask for contributions for their candidate. The calls are billed at the rate of 25¢ for the first three minutes and 7¢ for each additional minute or part thereof. If the campaign chairperson has ordered that the cost of each call is not to exceed $1.00, for how many minutes can a volunteer talk to a prospective donor on the phone?

Analyze the Problem   We are given the rate at which a call is billed. Since the cost of a call is not to exceed $1.00, the cost must be *less than or equal to* $1.00. This phrase indicates that we should write an inequality to find how long a volunteer can talk to a prospective donor.

Form an Inequality   We will let x = the total number of minutes that a call can last. Then the cost of a call will be 25¢ for the first 3 minutes plus 7¢ times the number of additional minutes, where the number of *additional* minutes is $x - 3$ (the total number of minutes minus the first 3 minutes). With this information, we can form an inequality.

The cost of the first 3 minutes	plus	the cost of the additional minutes	is not to exceed	$1.00.
0.25	+	0.07(x − 3)	≤	1

Solve the Inequality   To simplify the computations, we first clear the inequality of decimals.

$$0.25 + 0.07(x - 3) \leq 1$$
$$25 + 7(x - 3) \leq 100 \qquad \text{To eliminate the decimals, multiply both sides by 100.}$$
$$25 + 7x - 21 \leq 100 \qquad \text{Distribute the multiplication by 7.}$$
$$7x + 4 \leq 100 \qquad \text{Combine like terms.}$$
$$7x \leq 96 \qquad \text{Subtract 4 from both sides.}$$
$$x \leq 13.71428571 \ldots \qquad \text{Divide both sides by 7.}$$

State the Conclusion   Since the phone company doesn't bill for part of a minute, the longest time a call can last is 13 minutes. If a call lasts for $x = 13.71428571 \ldots$ minutes, it will be charged as a 14-minute call, and the cost will be $0.25 + $0.07(11) = $1.02.

Check the Result   If the call lasts 13 minutes, the cost will be $0.25 + $0.07(10) = $0.95. This is less than $1.00. The result checks.

 Now Try **Problem 85**

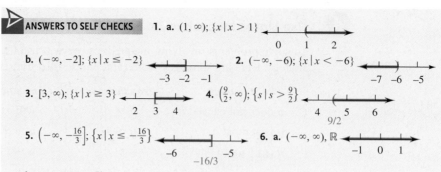

ANSWERS TO SELF CHECKS

1. a. $(1, \infty)$; $\{x \mid x > 1\}$

b. $(-\infty, -2]$; $\{x \mid x \leq -2\}$ **2.** $(-\infty, -6)$; $\{x \mid x < -6\}$

3. $[3, \infty)$; $\{x \mid x \geq 3\}$ **4.** $\left(\frac{9}{2}, \infty\right)$; $\left\{s \mid s > \frac{9}{2}\right\}$

5. $\left(-\infty, -\frac{16}{3}\right]$; $\left\{x \mid x \leq -\frac{16}{3}\right\}$ **6. a.** $(-\infty, \infty)$, $\mathbb{R}$

b. No solution, $\varnothing$

STUDY SET
4.1

VOCABULARY

Fill in the blanks.

1. $<$, $>$, $\leq$, and $\geq$ are _____ symbols.

2. $3x + 2 \geq 7$ is an example of a _____ inequality in one variable.

3. The graph of a set of real numbers that is a portion of a number line is called an _____.

4. In $(-\infty, 5)$, the right _____ is used to show that 5 is not included in the interval. In $[12, \infty)$, the left _____ is used to show that 12 is included in the interval.

5. We read the set-_____ notation $\{x \mid x < 1\}$ as "the set of all real numbers x _____ _____ x is less than 1."

6. To _____ an inequality means to find all values of the variable that make the inequality true.

CONCEPTS

7. Which of the following are inequalities?

$$6 - x = 8 \qquad 5 + a \qquad 7t - 5 > 4 \qquad \frac{x}{2} \leq -1$$

8. Perform each step listed below on the inequality $4 > -2$ and give the resulting true inequality.

    a. Add 2 to both sides.   **b.** Subtract 4 from both sides.

    c. Multiply both sides by 4.   **d.** Divide both sides by -2.

9. Determine whether each number is a solution of $3x + 6 \leq 6$.

    a. 0   **b.** $\dfrac{2}{3}$

    c. -10 **d.** 1.5

10. The solution set of a linear inequality in x is graphed on the right. Determine whether a true or false statement results when

    a. -4 is substituted for x.

    b. -3 is substituted for x.

    c. 0 is substituted for x.

11. a. Suppose that when solving a linear inequality, the variable drops out, and the result is $6 \leq 10$. Write the solution set in interval notation and graph it.

    b. Suppose that when solving a linear inequality, the variable drops out, and the result is $7 < -1$. What symbol is used to represent the solution set?

12. Insert the correct symbol, $\geq$ or $\leq$, in each blank.

    a. As many as 16 people were seriously injured:

        The number of people seriously injured _____ 16.

    b. There were no fewer than 8 references to carpools in the speech:

        The number of carpool references _____ 8.

NOTATION

Complete the solution to solve the inequality.

13. $-5x - 1 \geq -11$

$$-5x \geq \boxed{}$$

$$\frac{-5x}{-5} \; \boxed{} \; \frac{-10}{\boxed{}}$$

$$x \leq \boxed{}$$

The solution set is $\left(\boxed{}, 2\right]$. Using set-builder notation, it is $\{x \mid \boxed{}\}$.

14. Match each interval with its graph.

    a. $(-\infty, -1]$ **i.**

    b. $(-\infty, 1)$ **ii.**

    c. $[-1, \infty)$ **iii.**

15. Fill in the blank: If $-10 > x$ then x _____ -10.

16. Fill in the blank: ∞ is a symbol representing positive _____.

GUIDED PRACTICE

Represent each set using a graph, interval notation, and set-builder notation. See Example 1.

17. The set of all real numbers less than 14

18. The set of all real numbers greater than 6

19. The set of all real numbers greater than or equal to -2

20. The set of all real numbers less than or equal to -7

Solve each inequality. Graph the solution set and write it using interval notation. See Example 2.

21. $x + 4 < 5$ **22.** $x - 5 > 2$

23. $3x > -9$ **24.** $4x < -36$

25. $2x - 7 \geq -29$ **26.** $6x + 8 \leq -16$

27. $9a + 11 \leq 29$ **28.** $3b - 26 \geq 4$

Solve each inequality. Graph the solution set and write it using interval notation. See Example 3.

29. $2x + 4 + 6x > 2 - 3x + 2$

30. $5x + 6 + 2x \geq 2 - x + 4$

31. $t + 1 - 3t \geq t - 20$

32. $a + 4 - 10a > a - 16$

Solve each inequality. Graph the solution set and write it using interval notation. See Example 4.

33. $4 \leq \dfrac{9}{10}x + 1$ **34.** $2 \leq \dfrac{9}{4}x + 8$

35. $-3 > \dfrac{7}{8}x - 1$ **36.** $-10 > \dfrac{11}{2}x - 6$

Solve each inequality. Graph the solution set and write it using interval notation. See Example 5.

37. $\dfrac{3}{4}(x - 3) < \dfrac{1}{3}(x - 4)$ **38.** $\dfrac{5}{16}(x + 1) \geq \dfrac{1}{4}(x - 3)$

39. $\dfrac{2}{5}(3 - 2n) \geq \dfrac{3}{8}(2 - 3n)$ **40.** $\dfrac{1}{5}(1 - n) < \dfrac{1}{3}(2 - n)$

Solve each inequality. Graph the solution set and write it using interval notation, if possible. See Example 6.

41. $2(5x - 6) > 4x - 15 + 6x$ **42.** $3(4x - 2) > 14x - 7 - 2x$

43. $\dfrac{5x + 2}{-4} > \dfrac{5x + 1}{-4}$ **44.** $\dfrac{7 - n}{-6} > \dfrac{1 - n}{-6}$

TRY IT YOURSELF

Solve each inequality. Graph the solution set and write it using interval notation.

45. $-5t + 3 \leq 5$ **46.** $-9t + 6 \geq 16$

47. $\dfrac{2}{5} > \dfrac{4}{5}x$

48. $\dfrac{5}{9} < \dfrac{11}{9}x$

69. $\dfrac{1}{2}y + 2 \geq \dfrac{1}{3}y - 4$

70. $\dfrac{1}{4}x - \dfrac{1}{3} \leq x + 2$

49. $-0.6x \leq -36$

50. $-0.2x > -8$

71. $-11(2 - b) < 4(2b + 2)$

72. $-9(h - 3) + 2h \leq 8(4 - h)$

51. $7 < \dfrac{5}{3}a - 3$

52. $5 > \dfrac{7}{2}a - 9$

73. $\dfrac{2}{3}x + \dfrac{3}{2}(x - 5) \leq x$

74. $\dfrac{5}{9}(x + 3) - \dfrac{4}{3}(x - 3) \geq x - 1$

53. $-7y + 5 > -5y - 1$

54. $-2s - 105 \leq -7s - 205$

75. $5[3t - (t - 4)] - 11 \leq -12(t - 6) - (-t)$

55. $\dfrac{6 - d}{-2} \leq -6$

56. $\dfrac{9 - 3b}{-8} < 3$

76. $2 - 2[3h - (7 - h)] > 6[-(19 + h) - (1 - h)]$

Use a graphing calculator to solve each inequality. Write the solution set using interval notation. See Using Your Calculator: Solving Linear Inequalities in One Variable.

57. $0.4x + 0.4 \leq 0.1x + 0.85$

58. $0.05 + 0.8x \leq 0.5x - 0.7$

77. $2x + 3 < 5$

78. $3x - 2 > 4$

79. $5x + 2 \geq 4x - 2$

80. $3x - 4 \leq 2x + 4$

59. $3(z - 2) \leq 2(z + 7)$

60. $5(3 + z) > -3(z + 3)$

APPLICATIONS

81. REAL ESTATE   Refer to the graph below. For which regions of the country was the following inequality true in the year 2007?

Median sales price < U.S. median price

61. $\dfrac{3b + 7}{3} \leq \dfrac{2b - 9}{2}$

62. $-\dfrac{5x}{4} > \dfrac{3 - 5x}{4}$

63. $\dfrac{x - 7}{2} - \dfrac{x - 1}{5} \geq -\dfrac{x}{4}$

64. $\dfrac{3a + 1}{3} - \dfrac{4 - 3a}{5} \geq -\dfrac{1}{15}$

2007 Median Price of Existing Single-Family Homes

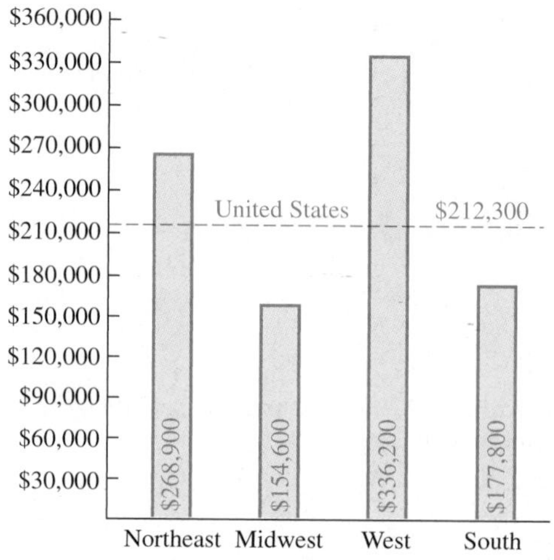

United States $212,300

Northeast	$268,900
Midwest	$154,600
West	$336,200
South	$177,800

Source: National Association of Realtors (First Quarter)

65. $\dfrac{1}{2}x + 6 \geq 4 + 2x$

66. $\dfrac{1}{3}x + 1 < 4 + 5x$

67. $5(2n + 2) - n > 3n - 3(1 - 2n)$

68. $-1 + 4(y - 1) + 2y \leq \dfrac{1}{2}(12y - 30) + 15$

82. PUBLIC EDUCATION Refer to the illustration. For which years is the following inequality true?

$$\frac{\text{Enrollment}}{\text{in grade 4}} \geq \frac{\text{Enrollment}}{\text{in grade 1}}$$

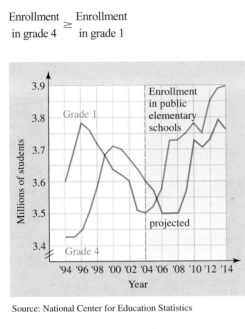

Source: National Center for Education Statistics

83. COMPUTER PROGRAMMING Flowcharts like the one shown are used by programmers to show the step-by-step instructions of a computer program. Work through the steps of the flow chart using $a = 9$, $b = -12$, and $c = 4$ and tell what statement the computer would print.

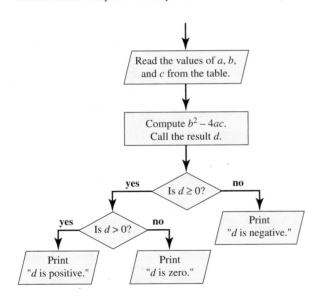

84. WIKIPEDIA The web-based encyclopedia called Wikipedia was launched in January 2001. The size of the English-language edition can be modeled by the equation $a = 0.56t + 0.40$, where a is the number of articles in millions and t is the number of years since 2005. If the current trends continue, when will the number of articles exceed 6 million? (Source: Wikipedia article: *Size of Wikipedia*)

The Wikipedia logo and trademark are used by permission from the Wikipedia Foundation.

85. FUND-RAISING A school PTA wants to rent a dunking tank for its annual school fund-raising carnival. The cost is $85.00 for the first three hours and then $19.50 for each additional hour or part thereof. How long can the tank be rented if up to $185 is budgeted for this expense?

86. AVERAGING GRADES A student has scores of 70, 77, and 85 on three government exams. What score does she need on a fourth exam to give her an average of 80 or better?

87. WORK SCHEDULES A student works two part-time jobs. He earns $8 an hour for working at the college library and $15 an hour for construction work. To save time for study, he limits his work to 25 hours a week. If he enjoys the work at the library more, how many hours can he work at the library and still earn at least $300 a week?

88. SCHEDULING EQUIPMENT An excavating company charges $300 an hour for the use of a backhoe and $500 an hour for the use of a bulldozer. (Part of an hour counts as a full hour.) The company employs one operator for 40 hours per week to operate the machinery. If the company wants to bring in at least $18,500 each week from equipment rental, how many hours per week can it schedule the operator to use a backhoe?

89. VIDEO GAME SYSTEMS A student who can afford to spend up to $1,000 sees the ad shown in the illustration. If she decides to buy the video game system, find the greatest number of video games that she can also purchase. (Disregard sales tax.)

VIDEO GAME SYSTEM
only **$449⁹⁹**

YOUR FAVORITE GAMES
only **$45⁹⁹ each**

90. INVESTMENTS   If a woman has invested $10,000 at 8% annual interest, how much more must she invest at 9% so that her annual income will exceed $1,250?

WRITING

91. How are the methods for solving linear equations and linear inequalities similar? How are they different?

92. Explain how the symbol ∞ is used in this section. Is ∞ a real number?

93. Explain what is wrong with the following statement:

When solving inequalities involving negative numbers, the direction of the inequality symbol must be reversed.

94. In each case, determine what is wrong with the interval notation.
   a. $(\infty, -3)$
   b. $[-\infty, -3)$

95. GEOMETRY   The **triangle inequality** states an important relationship between the sides of any triangle:

The sum of the lengths of two sides of a triangle $>$ the length of the third side.

Use the triangle inequality to explain why the dimensions of the shuffleboard court shown in the illustration must be mislabeled.

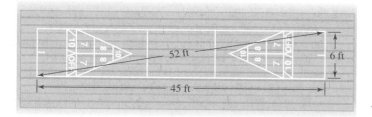

96. Explain how to use the following graph to solve $2x + 1 < 3$.

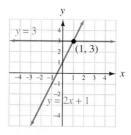

REVIEW

Use the graph of the function to find $f(-1)$, $f(0)$, and $f(2)$.

97.   **98.**

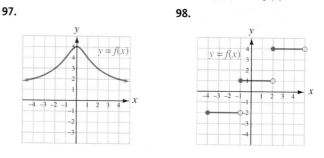

CHALLENGE PROBLEMS

99. In the illustration, which of the following are true?
   i. $b > 0$ **ii.** $a - b < 0$ **iii.** $ab > 0$

100. Consider the following "solution" of the inequality $\frac{1}{3} > \frac{1}{x}$, where it appears that the solution set is the interval $(3, \infty)$.

$$\frac{1}{3} > \frac{1}{x}$$

$$3x\left(\frac{1}{3}\right) > 3x\left(\frac{1}{x}\right)$$

$$x > 3$$

   a. Show that $x = -1$ makes the original inequality true.
   b. If $x = -1$ makes the original inequality true, there must be an error in the solution. Where is it?

101. MEDICAL PLANS   A college provides its employees with a choice of the two medical plans shown in the following table. For what size hospital bills is Plan 2 better for the employee than Plan 1? (*Hint:* The cost to the employee includes both the deductible payment and the employee's coinsurance payment.)

Plan 1	Plan 2
Employee pays $100	Employee pays $200
Plan pays 70% of the rest	Plan pays 80% of the rest

102. MEDICAL PLANS   To save costs, the college in Exercise 101 raised the employee deductible, as shown in the following table. For what size hospital bills is Plan 2 better for the employee than Plan 1? (*Hint:* The cost to the employee includes both the deductible payment and the employee's coinsurance payment.)

Plan 1	Plan 2
Employee pays $200	Employee pays $400
Plan pays 70% of the rest	Plan pays 80% of the rest

SECTION 4.2
Solving Compound Inequalities

Objectives

 1 Find the intersection and the union of two sets.

2 Solve compound inequalities containing the word *and.*

3 Solve double linear inequalities.

4 Solve compound inequalities containing the word *or.*

A label on a first-aid cream warns the user about the temperature at which the medication should be stored. A careful reading reveals that the storage instructions consist of two parts:

The storage temperature should be at least 59°F

and

the storage temperature should be at most 77°F

> DIRECTIONS: Clean the affected area thoroughly. Apply a small amount of this product (an amount equal to the surface area of the tip of a finger) on the area 1 to 3 times daily. Do not use in eyes. **Store at 59° to 77°F.** Do not use longer than 1 week. Keep this and all drugs out of the reach of children.

When the word *and* or the word *or* is used to connect pairs of inequalities, we call the statement a *compound inequality.* To solve compound inequalities, we need to know how to find the *intersection* and *union* of two sets.

1 **Find the Intersection and the Union of Two Sets.**

Just as operations such as addition and multiplication are performed on real numbers, operations can also be performed on sets. The operation of intersection of two sets produces a new third set that consists of all of the elements that the two given sets have in common.

The Intersection of Two Sets	The **intersection of set *A* and set *B*,** written $A \cap B$, is the set of all elements that are common to set *A* and set *B*.

The operation of union of two sets produces a third set that is a combination of all of the elements of the two given sets.

The Union of Two Sets	The **union of set *A* and set *B*,** written $A \cup B$, is the set of elements that belong to set *A* or set *B* or both.

Venn diagrams can be used to illustrate the intersection and union of sets. The area shown in purple in figure (a) represents $A \cap B$ and the area shown in both shades of red in figure (b) represents $A \cup B$.

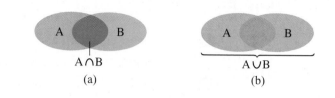

(a)   (b)

EXAMPLE 1   Let $A = \{0, 1, 2, 3, 4, 5, 6\}$ and $B = \{-4, -2, 0, 2, 4\}$.
a. Find $A \cap B$.   **b.** Find $A \cup B$.

Strategy   In part (a), we will find the elements that sets A and B have in common, and in part (b), we will find the elements that are in one set or the other.

Why   The symbol $\cap$ means intersection, and the symbol $\cup$ means union.

Solution
a. Since the numbers 0, 2, and 4 are common to both sets A and B, we have

$A \cap B = \{0, 2, 4\}$

b. Since the numbers in either or both sets are $-4, -2, 0, 1, 2, 3, 4, 5$, and 6, we have

$A \cup B = \{-4, -2, 0, 1, 2, 3, 4, 5, 6\}$

 Self Check 1   Let $C = \{8, 9, 10, 11\}$ and $D = \{3, 6, 9, 12, 15\}$.
a. Find $C \cap D$.   **b.** Find $C \cup D$.

Now Try   **Problems 17 and 21**

---

②   **Solve Compound Inequalities Containing the Word *And*.**

When two inequalities are joined with the word *and*, we call the statement a **compound inequality.** Some examples are

$x \geq -3$ and $x \leq 6$

$\dfrac{x}{2} + 1 > 0$ and $2x - 3 < 5$

$x + 3 \leq 2x - 1$ and $3x - 2 < 5x - 4$

The solution set of a compound inequality containing the word *and* includes all numbers that make both of the inequalities true. That is, it is the intersection of their solution sets. We can find the solution set of the compound inequality $x \geq -3$ and $x \leq 6$, for example, by graphing the solution sets of each inequality on the same number line and looking for the numbers common to both graphs.

In the following figure, the graph of the solution set of $x \geq -3$ is shown in red, and the graph of the solution set of $x \leq 6$ is shown in blue.

The figure below shows the graph of the solution of the compound inequality $x \geq -3$ and $x \leq 6$. The purple shaded interval, where the red and blue graphs intersect, represents the numbers that are common to the graphs of $x \geq -3$ and $x \leq 6$.

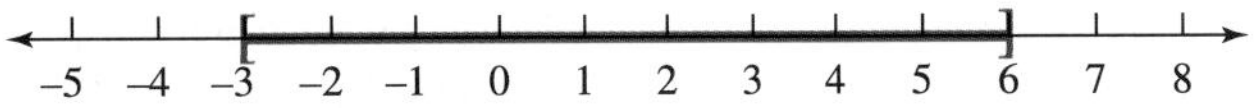

The solution set of $x \geq -3$ and $x \leq 6$ is the **bounded interval** $[-3, 6]$, where the brackets indicate that the endpoints, -3 and 6, are included. It represents all real numbers between -3 and 6, including -3 and 6. Intervals such as this, which contain both endpoints, are called **closed intervals.**

Since the solution set of $x \geq -3$ and $x \leq 6$ is the intersection of the solution sets of the two inequalities, we can write

$$[-3, \infty) \cap (-\infty, 6] = [-3, 6]$$

The solution set of the compound inequality $x \geq -3$ and $x \leq 6$ can be expressed in several ways:

1. As a graph:

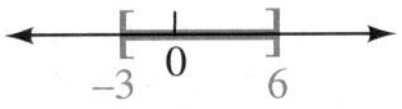

2. In words: all real numbers between -3 and 6, including -3 and 6

3. In interval notation: $[-3, 6]$

4. Using set-builder notation: $\{x \mid x \geq -3 \text{ and } x \leq 6\}$

> **Success Tip**
>
> Check with your instructor to see his or her preferred way (or ways) to present solutions of compound inequalities.

EXAMPLE 2   Solve $\dfrac{x}{2} + 1 > 0$ and $2x - 3 < 5$. Graph the solution set and write it using interval notation and set-builder notation.

Strategy   We will solve each inequality separately. Then we will graph the two solution sets on the same number line and determine their intersection.

Why   The solution set of a compound inequality containing the word *and* is the intersection of the solution sets of the two inequalities.

Solution   In each case, we can use properties of inequality to isolate the variable on one side of the inequality.

$$\dfrac{x}{2} + 1 > 0 \qquad \text{and} \qquad 2x - 3 < 5 \qquad \text{\small This is the compound inequality to solve.}$$

$$\dfrac{x}{2} > -1 \qquad\qquad\qquad 2x < 8$$

$$x > -2 \qquad\qquad\qquad\quad x < 4$$

Next, we graph the solutions of each inequality on the same number line and determine their intersection.

> **Notation**
>
> When graphing on a number line, $(-2, 4)$ represents an *interval*. When graphing on a rectangular coordinate system, $(-2, 4)$ is an *ordered pair* that gives the coordinates of a point.

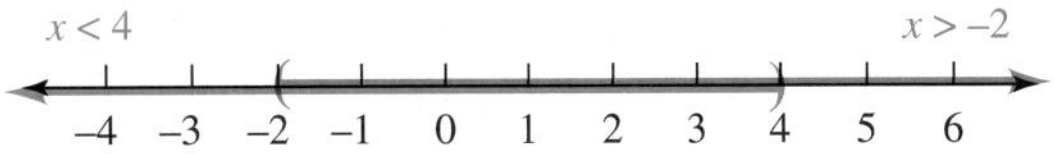

We see that the intersection of the graphs is the set of all real numbers between -2 and 4. The solution set of the compound inequality is the interval $(-2, 4)$, whose graph is shown below. This bounded interval, which does not include either endpoint, is called an **open interval.** Written using set-builder notation, the solution set is $\{x \mid x > -2 \text{ and } x < 4\}$.

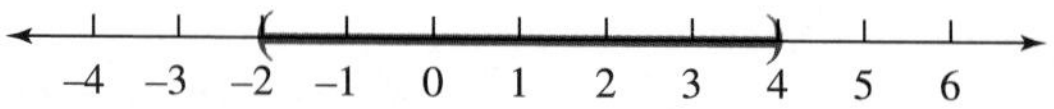

$\xleftarrow{\quad\quad} \overset{(\quad\quad\quad\quad\quad)}{\underset{-4\ -3\ -2\ -1\ \ 0\ \ 1\ \ 2\ \ 3\ \ 4\ \ 5\ \ 6}{}} \xrightarrow{\quad\quad}$

> ▷ **Self Check 2**   Solve $3x > -18$ and $\frac{x}{5} - 1 \le 1$. Graph the solution set and write it using interval notation.
>
> **_Now Try_**   **Problem 27**

The solution of the compound inequality in the Self Check of Example 2 is the interval $(-6, 10]$. A bounded interval such as this, which includes only one endpoint, is called a **half-open interval.** The following chart shows the various types of bounded intervals, along with the inequalities and interval notation that describe them.

Intervals	**Open intervals**	The interval (a, b) includes all real numbers x such that $a < x < b$.	
	Half-open intervals	The interval $[a, b)$ includes all real numbers x such that $a \le x < b$.	
		The interval $(a, b]$ includes all real numbers x such that $a < x \le b$.	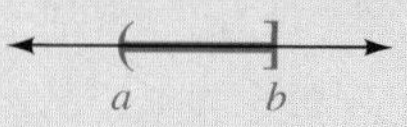
	Closed intervals	The interval $[a, b]$ includes all real numbers x such that $a \le x \le b$.	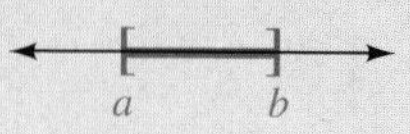

EXAMPLE 3   Solve $x + 3 \le 2x - 1$ and $3x - 2 < 5x - 4$. Graph the solution set and write it using interval notation and set-builder notation.

Strategy   We will solve each inequality separately. Then we will graph the two solution sets on the same number line and determine their intersection.

Why   The solution set of a compound inequality containing the word *and* is the intersection of the solution sets of the two inequalities.

Solution   In each case, we can use properties of inequality to isolate the variable on one side.

$$x + 3 \le 2x - 1 \quad \text{and} \quad 3x - 2 < 5x - 4 \qquad \text{\small This is the compound inequality to solve.}$$
$$4 \le x \qquad\qquad\qquad\quad 2 < 2x$$
$$x \ge 4 \qquad\qquad\qquad\quad 1 < x$$
$$\qquad\qquad\qquad\qquad\qquad\qquad x > 1$$

The graph of $x \ge 4$ is shown below in red and the graph of $x > 1$ is shown below in blue.

$$\overset{x > 1 \qquad\qquad\qquad\qquad x \ge 4}{\xleftarrow{\quad} \underset{0\ \ 1\ \ 2\ \ 3\ \ 4\ \ 5\ \ 6\ \ 7}{(\quad\quad\quad[\quad\quad\quad)} \xrightarrow{\quad}}$$

Only those values of x where $x \ge 4$ and $x > 1$ are in the solution set of the compound inequality. Since all numbers greater than or equal to 4 are also greater than 1, the solutions are the numbers x where $x \geq 4$. The solution set is the interval $[4, \infty)$, whose graph is shown below. Written using set-builder notation, the solution set is $\{x \mid x \geq 4\}$.

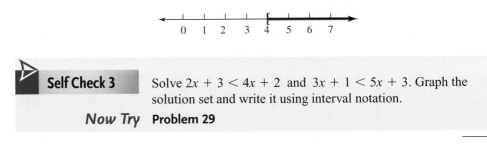

Self Check 3   Solve $2x + 3 < 4x + 2$ and $3x + 1 < 5x + 3$. Graph the solution set and write it using interval notation.

Now Try   **Problem 29**

EXAMPLE 4   Solve $x - 1 > -3$ and $2x < -8$, if possible.

Strategy   We will solve each inequality separately. Then we will graph the two solution sets on the same number line and determine their intersection, if any.

Why   The solution set of a compound inequality containing the word *and* is the intersection of the solution sets of the two inequalities.

Solution   In each case, we can use properties of inequality to isolate the variable on one side.

$$x - 1 > -3 \quad \text{and} \quad 2x < -8 \quad \text{This is the compound inequality to solve.}$$
$$x > -2 \qquad \mid \qquad x < -4$$

Notation

The graphs of two linear inequalities can intersect at a single point, as shown below. The interval notation used to describe this point of intersection is $[3, 3]$.

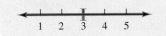

The graphs of the solution sets shown below do not intersect. Since there are no numbers that make both parts of the original compound inequality true, $x - 1 > -3$ and $2x < -8$ has no solution.

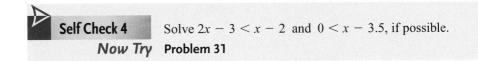

The solution set of the compound inequality is the empty set, which can be written as $\varnothing$.

Self Check 4   Solve $2x - 3 < x - 2$ and $0 < x - 3.5$, if possible.

Now Try   **Problem 31**

---

3   **Solve Double Linear Inequalities.**

Inequalities that contain exactly two inequality symbols are called **double inequalities.** An example is

$$-3 \leq 2x + 5 < 7 \quad \text{Read as "-3 is less than or equal to $2x + 5$ and $2x + 5$ is less than 7."}$$

Any double linear inequality can be written as a compound inequality containing the word *and.* In general, the following is true.

Double Linear Inequalities

The compound inequality $c < x < d$ is equivalent to $c < x$ and $x < d$.

EXAMPLE 5 Solve $-3 \leq 2x + 5 < 7$. Graph the solution set and write it using interval notation and set-builder notation.

Strategy We will solve the double inequality by applying properties of inequality to *all three of its parts* to isolate x in the middle.

Why This double inequality $-3 \leq 2x + 5 < 7$ means that $-3 \leq 2x + 5$ and $2x + 5 < 7$. We can solve it more easily by leaving it in its original form.

Solution

$$-3 \leq 2x + 5 < 7 \qquad \text{This is the double inequality to solve.}$$

$$-3 - 5 \leq 2x + 5 - 5 < 7 - 5 \qquad \text{To undo the addition of 5, subtract 5 from all three parts.}$$

$$-8 \leq 2x < 2 \qquad \text{Perform the subtractions.}$$

$$\frac{-8}{2} \leq \frac{2x}{2} < \frac{2}{2} \qquad \text{To isolate } x, \text{ undo the multiplication by 2 by dividing all three parts by 2.}$$

$$-4 \leq x < 1 \qquad \text{Perform the divisions.}$$

The solution set of the double linear inequality is the half-open interval $[-4, 1)$, whose graph is shown below. Written using set-builder notation, the solution set is $\{x \mid -4 \leq x < 1\}$.

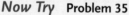

Self Check 5 Solve $-5 \leq 3x - 8 \leq 7$. Graph the solution set and write it using interval notation.

Now Try **Problem 35**

Caution When multiplying or dividing all three parts of a double inequality by a negative number, don't forget to reverse the direction of both inequalities. As an example, we solve $-15 < -5x \leq 25$.

$$-15 < -5x \leq 25$$

$$\frac{-15}{-5} > \frac{-5x}{-5} \geq \frac{25}{-5} \qquad \text{Divide all three parts by } -5 \text{ to isolate } x \text{ in the middle. Reverse both inequality signs.}$$

$$3 > x \geq -5 \qquad \text{Perform the divisions.}$$

$$-5 \leq x < 3 \qquad \text{Write an equivalent double inequality with the smaller number, } -5, \text{ on the left.}$$

 Solve Compound Inequalities Containing the Word *Or.*

A warning on the water temperature gauge of a commercial dishwasher cautions the operator to shut down the unit if

The water temperature goes below 140°

or

The water temperature goes above 160°

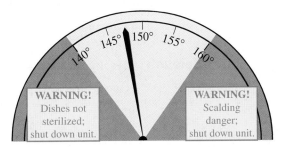

When two inequalities are joined with the word *or*, we also call the statement a compound inequality. Some examples are

$$x < 140 \quad \text{or} \quad x > 160$$

$$x \leq -3 \quad \text{or} \quad x \geq 2$$

$$\frac{x}{3} > \frac{2}{3} \quad \text{or} \quad -(x - 2) > 3$$

> **Caution**
>
> It is incorrect to write the statement $x \leq -3$ or $x \geq 2$ as the double inequality $2 \leq x \leq -3$, because that would imply that $2 \leq -3$, which is false.

The solution set of a compound inequality containing the word *or* includes all numbers that make one or the other or both inequalities true. That is, it is the union of their solution sets. We can find the solution set of $x \leq -3$ or $x \geq 2$, for example, by drawing the graphs of each inequality on the same number line.

In the following figure, the graph of the solution set of $x \leq -3$ is shown in red, and the graph of the solution set of $x \geq 2$ is shown in blue.

The figure below shows the graph of the solution set of $x \leq -3$ or $x \geq 2$. This graph is a union of the graph of $x \leq -3$ with the graph of $x \geq 2$.

For the compound inequality $x \leq -3$ or $x \geq 2$, we can write the solution set as the union of two intervals:

$$(-\infty, -3] \cup [2, \infty)$$

We can express the solution set of the compound inequality $x \leq -3$ or $x \geq 2$ in several ways:

1. As a graph:

2. In words: all real numbers less than or equal to -3 *or* greater than or equal to 2

3. As the union of two intervals: $(-\infty, -3] \cup [2, \infty)$

4. Using set-builder notation: $\{x \mid x \leq -3 \text{ or } x \geq 2\}$

EXAMPLE 6   Solve $\frac{x}{3} > \frac{2}{3}$ or $-(x - 2) > 3$. Graph the solution set and write it using interval notation and set-builder notation.

Strategy   We will solve each inequality separately. Then we will graph the two solution sets on the same number line to show their union.

Why   The solution set of a compound inequality containing the word *or* is the union of the solution sets of the two inequalities.

Solution   To solve each inequality, we proceed as follows:

$$\frac{x}{3} > \frac{2}{3} \quad \text{or} \quad -(x - 2) > 3 \qquad \text{This is the compound inequality to solve.}$$

$$x > 2 \qquad \qquad -x + 2 > 3$$

$$-x > 1$$

$$x < -1$$

Next, we graph the solutions of each inequality on the same number line and determine their union.

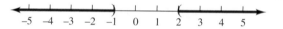

$$x < -1 \qquad\qquad\qquad x > 2$$
$$-5 \quad -4 \quad -3 \quad -2 \quad -1 \quad 0 \quad 1 \quad 2 \quad 3 \quad 4 \quad 5$$

The union of the two solution sets consists of all real numbers less than -1 or greater than 2. The solution set of the compound inequality is the union of two intervals: $(-\infty, -1) \cup (2, \infty)$. Its graph appears below. Written using set-builder notation, the solution set is $\{x \mid x > 2 \text{ or } x < -1\}$.

$$-5 \quad -4 \quad -3 \quad -2 \quad -1 \quad 0 \quad 1 \quad 2 \quad 3 \quad 4 \quad 5$$

Self Check 6   Solve $\frac{x}{2} > 2$ or $-3(x - 2) > 0$. Graph the solution set and write it using interval notation.

Now Try   Problem 41

EXAMPLE 7   Solve $x + 3 \geq -3$ or $-x > 0$. Graph the solution set and write it using interval notation and set-builder notation.

Strategy   We will solve each inequality separately. Then we will graph the two solution sets on the same number line to show their union.

Why   The solution set of a compound inequality containing the word *or* is the union of the solution sets of the two inequalities.

Solution   To solve each inequality, we proceed as follows:

$$x + 3 \geq -3 \quad \text{or} \quad -x > 0 \qquad \text{This is the compound inequality to solve.}$$

$$x \geq -6 \qquad\qquad x < 0$$

We graph the solution set of each inequality on the same number line and determine their union.

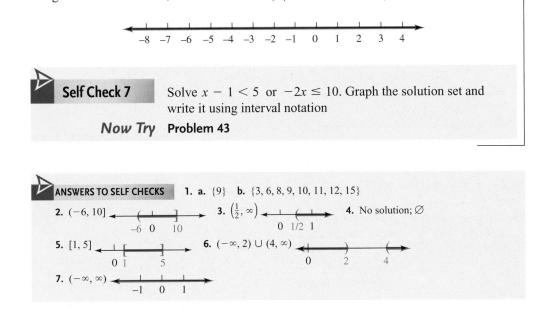

Since the entire number line is shaded, all real numbers satisfy the original compound inequality and the solution set is denoted as $(-\infty, \infty)$ or $\mathbb{R}$. Its graph is shown below. Written using set-builder notation, the solution set is $\{x \mid x \text{ is a real number}\}$.

Self Check 7   Solve $x - 1 < 5$ or $-2x \le 10$. Graph the solution set and write it using interval notation

Now Try   **Problem 43**

ANSWERS TO SELF CHECKS   **1. a.** $\{9\}$ **b.** $\{3, 6, 8, 9, 10, 11, 12, 15\}$
2. $(-6, 10]$ **3.** $\left(\frac{1}{2}, \infty\right)$ **4.** No solution; $\varnothing$
5. $[1, 5]$ **6.** $(-\infty, 2) \cup (4, \infty)$
7. $(-\infty, \infty)$

STUDY SET
4.2

VOCABULARY

Fill in the blanks.

1. The _____ of two sets is the set of elements that are common to both sets and the _____ of two sets is the set of elements that are in one set, or the other, or both.

2. $x \ge 3$ and $x < 4$ is a _____ inequality.

3. $-6 < x + 1 \le 1$ is a _____ linear inequality.

4. $(2, 8)$ is an example of an open _____, $[-4, 0]$ is an example of a _____ interval, and $(0, 9]$ is an example of a half-_____ interval.

CONCEPTS

Fill in the blanks.

5. **a.** The solution set of a compound inequality containing the word *and* includes all numbers that make _____ inequalities true.

   b. The solution set of a compound inequality containing the word *or* includes all numbers that make _____, or the other, or _____ inequalities true.

6. The double inequality $4 < 3x + 5 \le 15$ is equivalent to $4 < 3x + 5$ _____ $3x + 5 \le 15$.

7. **a.** When solving a compound inequality containing the word *and,* the solution set is the _____ of the solution sets of the inequalities.

   b. When solving a compound inequality containing the word *or,* the solution set is the _____ of the solution sets of the inequalities.

8. When multiplying or dividing all three parts of a double inequality by a negative number, the direction of both inequality symbols must be _____.

9. In each case, determine whether -3 is a solution of the compound inequality.

   a. $\dfrac{x}{3} + 1 \ge 0$ and $2x - 3 < -10$

   b. $2x \le 0$ or $-3x < -5$

10. In each case, determine whether -3 is a solution of the double linear inequality.

    a. $-1 < -3x + 4 < 12$

    b. $-1 < -3x + 4 < 14$

11. Use interval notation, if possible, to describe the intersection of each pair of graphs.

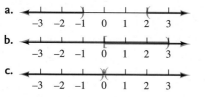

12. Use interval notation to describe the union of each pair of graphs.

NOTATION

13. Fill in the blanks: We read ∪ as _____ and ∩ as _____.

14. Match each interval with its corresponding graph.

   a. [2, 3)   i.

   b. (2, 3)   ii.

   c. [2, 3]   iii.

15. What set is represented by the interval notation $(-\infty, \infty)$? Graph it.

16. a. Graph: $(-\infty, 2) \cup [3, \infty)$
    b. Graph: $(-\infty, 3) \cap [-2, \infty)$

GUIDED PRACTICE

Let $A = \{0, 1, 2, 3, 4, 5, 6\}$, $B = \{4, 6, 8, 10\}$, $C = \{-3, -1, 0, 1, 2\}$, and $D = \{-3, 1, 2, 5, 8\}$. Find each set. See Example 1.

17. $A \cap B$ 18. $A \cap D$

19. $C \cap D$ 20. $B \cap C$

21. $B \cup C$ 22. $A \cup C$

23. $A \cup D$ 24. $C \cup D$

Solve each compound inequality, if possible. Graph the solution set (if one exists) and write it using interval notation. See Examples 2–4.

25. $x > -2$ and $x \le 5$
26. $x \le -4$ and $x \ge -7$
27. $2x - 1 > 3$ and $x + 8 \le 11$
28. $5x - 3 \ge 2$ and $6 \ge 4x - 3$
29. $6x + 1 < 5x - 3$ and $\frac{x}{2} + 9 \le 6$

30. $\frac{2}{3}x + 1 > -9$ and $\frac{3}{4}x - 1 > -10$

31. $x + 2 < -\frac{1}{3}x$ and $-6x < 9x$

32. $\frac{3}{2}x + \frac{1}{5} < 5$ and $2x + 1 > 9$

Solve each double inequality. Graph the solution set and write it using interval notation. **See Example 5.**

33. $4 \le x + 3 \le 7$ 34. $-5.3 \le x - 2.3 \le -1.3$

35. $0.9 < 2x - 0.7 < 1.5$ 36. $7 < 3x - 2 < 25$

Solve each compound inequality. Graph the solution set and write it using interval notation. **See Examples 6 and 7.**

37. $x \le -2$ or $x > 6$
38. $x \ge -1$ or $x \le -3$
39. $x - 3 < -4$ or $-x + 2 < 0$
40. $4x < -12$ or $\frac{x}{2} > 4$
41. $3x + 2 < 8$ or $2x - 3 > 11$
42. $3x + 4 < -2$ or $3x + 4 > 10$
43. $2x > x + 3$ or $\frac{x}{8} + 1 < \frac{13}{8}$

44. $2(x + 2) < x - 11$ or $-\frac{x}{5} < 20$

TRY IT YOURSELF

Solve each compound inequality, if possible. Graph the solution set (if one exists) and write it using interval notation.

45. $-4(x + 2) \ge 12$ or $3x + 8 < 11$
46. $4.5x - 1 < -10$ or $6 - 2x \ge 12$
47. $2.2x < -19.8$ and $-4x < 40$
48. $\frac{1}{2}x \le 2$ and $0.75x \ge -6$
49. $-2 < -b + 3 < 5$
50. $2 < -t - 2 < 9$
51. $4.5x - 2 > 2.5$ or $\frac{1}{2}x \le 1$
52. $0 < x$ or $3x - 5 > 4x - 7$
53. $5(x - 2) \ge 0$ and $-3x < 9$
54. $x - 1 \le 2(x + 2)$ and $x \le 2x - 5$
55. $-x < -2x$ and $3x > 2x$
56. $-\frac{x}{4} > -2.5$ and $9x > 2(4x + 5)$
57. $-6 < -3(x - 4) \le 24$
58. $-4 \le -2(x + 8) < 8$
59. $2x + 1 \ge 5$ and $-3(x + 1) \ge -9$

60. $2(-2) \le 3x - 1$ and $3x - 1 \le -1 - 3$

61. $\dfrac{4.5x - 12}{2} < x$ or $-15.3 > -3(x - 1.4)$

62. $y + 0.52 < 1.05y$ or $9.8 - 15y > -15.7$

63. $\dfrac{x}{0.7} + 5 > 4$ and $-4.8 \le \dfrac{3x}{-0.125}$

64. $5(x + 1) \le 4(x + 3)$ and $x + 12 < -3$

65. $-24 < \dfrac{3}{2}x - 6 \le -15$

66. $-4 > \dfrac{2}{3}x - 2 > -6$

67. $\dfrac{x}{3} - \dfrac{x}{4} > \dfrac{1}{6}$ or $\dfrac{x}{2} + \dfrac{2}{3} \le \dfrac{3}{4}$

68. $\dfrac{a}{2} + \dfrac{7}{4} > 5$ or $\dfrac{3}{8} + \dfrac{a}{3} \le \dfrac{5}{12}$

69. $0 \le \dfrac{4 - x}{3} \le 2$

70. $-2 \le \dfrac{5 - 3x}{2} \le 2$

71. $x \le 6 - \dfrac{1}{2}x$ and $\dfrac{1}{2}x + 1 \ge 3$

72. $3\left(x + \dfrac{2}{3}\right) \le -7$ and $2(x + 2) \ge -2$

APPLICATIONS

73. BABY FURNITURE   Refer to the illustration. A company manufactures various sizes of play yard cribs having perimeters between 128 and 192 inches, inclusive.

a. Complete the double inequality that describes the range of the perimeters of the play yard shown.

$$\boxed{} \le 4s \le \boxed{}$$

b. Solve the double inequality to find the range of the side lengths of the play yard.

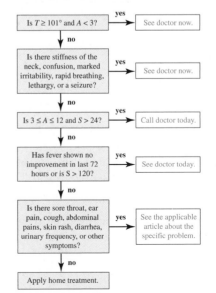

74. TRUCKING   The distance that a truck can travel in 8 hours, at a constant rate of r mph, is given by $8r$. A trucker wants to travel at least 350 miles, and company regulations don't allow him to exceed 450 miles in one 8-hour shift.

a. Complete the double inequality that describes the mileage range of the truck.

$$\boxed{} \le 8r \le \boxed{}$$

b. Solve the double inequality to find the range of the average rate (speed) of the truck for the 8-hour trip.

75. *from Campus to Careers*
Heating, Ventilation, and Air Conditioning Technician

Suppose that you are an HVAC technician and that you are repairing a thermostat in a commercial building. During business hours, as shown in figure (a), the *Temp range* control is set at 5. This means that the heater comes on when the room temperature gets 5 degrees below the *Temp setting* and the air conditioner comes on when the room temperature gets 5 degrees above the *Temp setting*.

© Andrew Brookes/Corbis

a. Use interval notation to describe the temperature range when neither the heater nor the air conditioner will come on during business hours.

b. After business hours, the Temp range setting is changed to save energy. See figure (b). Use interval notation to describe the after business hours temperature range when neither the heater nor the air conditioner will be come on.

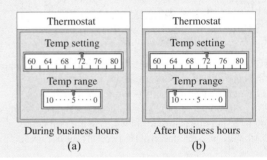

During business hours   After business hours
(a)   (b)

76. TREATING FEVERS   Use the flow chart to determine what action should be taken for a 13-month-old child who has had a 99.8° temperature for 3 days and is not suffering any other symptoms. T represents the child's temperature, A the child's age in months, and S the number of hours the child has experienced the symptoms.

Is $T \ge 101°$ and $A < 3$? —**yes**→ See doctor now.
↓ **no**

Is there stiffness of the neck, confusion, marked irritability, rapid breathing, lethargy, or a seizure? —**yes**→ See doctor now.
↓ **no**

Is $3 \le A \le 12$ and $S > 24$? —**yes**→ Call doctor today.
↓ **no**

Has fever shown no improvement in last 72 hours or is $S > 120$? —**yes**→ See doctor today.
↓ **no**

Is there sore throat, ear pain, cough, abdominal pains, skin rash, diarrhea, urinary frequency, or other symptoms? —**yes**→ See the applicable article about the specific problem.
↓ **no**

Apply home treatment.

Based on information from *Take Care of Yourself* (Addison-Wesley, 1993)

77. U.S. HEALTH CARE   Refer to the following graph. Let P represent the percent of children covered by private insurance, M the percent covered by Medicare/Medicaid, and N the percent not covered. For what years are the following true?

a. $P \geq 65$ and $M \geq 20$

b. $P > 63$ or $M \geq 25$

c. $M \geq 25$ and $N \leq 10$

d. $M \geq 26$ or $N < 11$

U.S. Health Care Coverage for Persons Under 18 Years of Age (in percent)

Private insurance ☐   Medicaid ☐

Not covered ☐

2001	66.7	21.2	11.0
2002	63.5	24.8	10.9
2003	63.0	26.0	9.8
2004	63.2	26.4	9.2

Source: U.S. Department of Health and Human Services

78. POLLS   For each response to the poll question shown below, the *margin of error* is $+/-$ (read as "plus or minus") 3.4%. This means that for the statistical methods used to do the polling, the actual response could be as much as 3.4 points more or 3.4 points less than shown. Use interval notation to describe the possible interval (in percent) for each response.

882 high school students were asked, "which one of these environmental problems do you think is the most serious?"

Air pollution	33.4%
Not enough energy	22.4%
Water pollution	18.9%
Toxic dumps	14.7%
Climate change	10.6%

Source: Zogby International

79. STREET INTERSECTIONS   Refer to the illustration in the next column.

a. Shade the area that represents the intersection of the two streets shown in the illustration.

b. Shade the area that represents the union of the two streets.

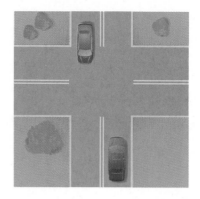

80. TRAFFIC SIGNS   The pair of signs shown below are a real-life example of which concept discussed in this section?

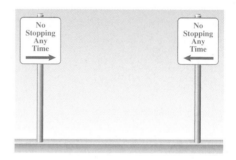

WRITING

81. Explain how to find the union and how to find the intersection of $(-\infty, 5)$ and $(-2, \infty)$ graphically.

82. Explain why the double inequality

$$2 < x < 8$$

can be written in the equivalent form

$$2 < x \text{ and } x < 8$$

83. Explain the meaning of the notation (2, 3) for each type of graph.

a.

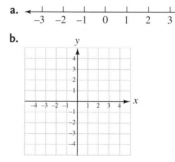

b.

84. The meaning of the word *or* in a compound inequality differs from our everyday use of the word. Explain the difference.

85. Describe each set in words.

   a. $(-3, 3)$

   b. $[7, 12]$

   c. $(-\infty, 5] \cup (6, \infty)$

86. What is incorrect about the double inequality
$3 < -3x + 4 < -3$?

REVIEW

For Exercises 87 and 88, refer to the following table that lists the number of earthquakes of magnitude 6.0 or higher in the United States for the years 1994–2006.

'94	'95	'96	'97	'98	'99	'00	'01	'02	'03	'04	'05	'06
5	6	6	6	3	8	10	6	6	9	2	5	6

Source: U.S. Geological Survey

87. The collection of earthquake data in the table has an interesting characteristic. Find its mean, median, and mode.

88. a. What was the percent increase in the number of earthquakes of magnitude 6.0 or higher from 1999 to 2000?

   b. What was the percent decrease in the number of earthquakes of magnitude 6.0 or higher from 1997 to 1998?

CHALLENGE PROBLEMS

Solve each compound inequality. Graph the solution set and write it in interval notation.

89. $-5 < \dfrac{x + 2}{-2} < 0$ or $2x + 10 \geq 30$

90. $-2 \leq \dfrac{x - 4}{3} \leq 0$ and $\dfrac{x - 5}{2} \geq -3$

SECTION 4.3
Solving Absolute Value Equations and Inequalities

Objectives

1 Solve equations of the form $|X| = k$.

2 Solve equations with two absolute values.

3 Solve inequalities of the form $|X| < k$.

4 Solve inequalities of the form $|X| > k$.

Many quantities studied in mathematics, science, and engineering are expressed as positive numbers. To guarantee that a quantity is positive, we often use absolute value. In this section, we will consider equations and inequalities involving the absolute value of an algebraic expression. Some examples are

$$|3x - 2| = 5, \qquad |2x - 3| < 9, \qquad \text{and} \qquad \left|\dfrac{3 - x}{5}\right| \geq 6$$

To solve these *absolute value equations* and *inequalities,* we write and then solve equivalent compound equations and inequalities.

1 **Solve Equations of the Form $|X| = k$.**

Recall that the absolute value of a real number is its distance from 0 on a number line. To solve the **absolute value equation** $|x| = 5$, we must find all real numbers x whose distance from 0 on the number line is 5. There are two such numbers: 5 and -5. It follows that the solutions of $|x| = 5$ are 5 and -5 and the solution set is $\{5, -5\}$.

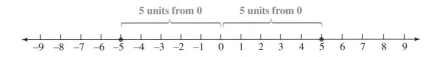

The results from this example suggest the following approach for solving absolute value equations.

Solving Absolute Value Equations

For any positive number k and any algebraic expression X:
To solve $|X| = k$, solve the equivalent compound equation

$$X = k \qquad \text{or} \qquad X = -k$$

The statement $X = k$ or $X = -k$ is called a **compound equation** because it consists of two equations joined with the word *or*.

EXAMPLE 1   Solve: **a.** $|x| = 8$ **b.** $|s| = 0.003$

Strategy   To solve each of these absolute value equations, we will write and solve an equivalent compound equation.

Why   We can use this approach because an equation of the form $|x| = k$, where k is positive, is equivalent to $x = k$ or $x = -k$.

The Language of Algebra
When we say that the absolute value equation and a compound equation are *equivalent*, we mean that they have the same solution(s).

Solution
a. The absolute value equation $|x| = 8$ is equivalent to the compound equation

$$x = 8 \qquad \text{or} \qquad x = -8$$

Therefore, the solutions of $|x| = 8$ are 8 and -8 and the solution set is $\{8, -8\}$.

b. The absolute value equation $|s| = 0.003$ is equivalent to the compound equation

$$s = 0.003 \qquad \text{or} \qquad s = -0.003$$

Therefore, the solutions of $|s| = 0.003$ are 0.003 and -0.003 and the solution set is $\{0.003, -0.003\}$.

Self Check 1   Solve: **a.** $|y| = 24$ **b.** $|x| = \frac{1}{2}$
Now Try   Problems 17 and 19

The equation-solving procedure discussed in Example 1 can often be used when the expression within absolute value bars is more complicated than a single variable.

EXAMPLE 2   Solve: **a.** $|3x - 2| = 5$ **b.** $|10 - x| = -40$

Strategy   To solve the first equation, we will write and then solve an equivalent compound equation. We will solve the second equation by inspection.

Why   Both equations are of the form $|X| = k$. However, the standard method for solving absolute value equations cannot be applied to $|10 - x| = -40$ because k is negative.

Solution
a. The absolute value equation $|3x - 2| = 5$ is equivalent to the compound equation

$$3x - 2 = 5 \qquad \text{or} \qquad 3x - 2 = -5$$

Now we solve each equation for x:

$$3x - 2 = 5 \quad \text{or} \quad 3x - 2 = -5$$
$$3x = 7 \qquad\qquad 3x = -3$$
$$x = \frac{7}{3} \qquad\qquad x = -1$$

The results must be checked separately to see whether each of them produces a true statement. We substitute $\frac{7}{3}$ for x and then -1 for x in the original equation.

Check: *For $x = \dfrac{7}{3}$*   *For $x = -1$*

$$|3x - 2| = 5 \qquad\qquad\qquad |3x - 2| = 5$$
$$\left|3\left(\frac{7}{3}\right) - 2\right| \overset{?}{=} 5 \qquad\qquad |3(-1) - 2| \overset{?}{=} 5$$
$$|7 - 2| \overset{?}{=} 5 \qquad\qquad\qquad |-3 - 2| \overset{?}{=} 5$$
$$|5| \overset{?}{=} 5 \qquad\qquad\qquad\quad |-5| \overset{?}{=} 5$$
$$5 = 5 \quad \text{True} \qquad\qquad\qquad 5 = 5 \quad \text{True}$$

The resulting true statements indicate that the equation has two solutions: $\frac{7}{3}$ and -1.

b. Since an absolute value can never be negative, there are no real numbers x that make $|10 - x| = -40$ true. The equation has no solution and the solution set is $\varnothing$.

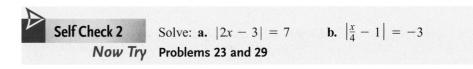

Self Check 2   Solve: **a.** $|2x - 3| = 7$ **b.** $\left|\frac{x}{4} - 1\right| = -3$

Now Try   **Problems 23 and 29**

Caution   When solving absolute value equations (or inequalities), isolate the absolute value expression on one side *before* writing the equivalent compound statement.

EXAMPLE 3   Solve: $\left|\frac{2}{3}x + 3\right| + 4 = 10$

Strategy   We will first isolate $\left|\frac{2}{3}x + 3\right|$ on the left side of the equation and then write and solve an equivalent compound equation.

Why   After isolating the absolute value expression on the left, the resulting equation will have the desired form $|X| = k$.

Solution

$$\left|\frac{2}{3}x + 3\right| + 4 = 10 \quad \text{\small This is the equation to solve.}$$

$$\left|\frac{2}{3}x + 3\right| = 6 \quad \text{\small To isolate the absolute value expression, subtract 4}$$
$$\text{\small from both sides. The equation is in the form } |X| = k.$$

With the absolute value now isolated, we can solve $\left|\frac{2}{3}x + 3\right| = 6$ by writing and solving an equivalent compound equation:

$$\frac{2}{3}x + 3 = 6 \quad \text{or} \quad \frac{2}{3}x + 3 = -6$$

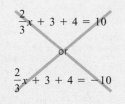

Now we solve each equation for x:

$$\frac{2}{3}x + 3 = 6 \quad \text{or} \quad \frac{2}{3}x + 3 = -6$$

$$\frac{2}{3}x = 3 \qquad\qquad \frac{2}{3}x = -9$$

$$2x = 9 \qquad\qquad 2x = -27$$

$$x = \frac{9}{2} \qquad\qquad x = -\frac{27}{2}$$

Verify that both solutions, $\frac{9}{2}$ and $-\frac{27}{2}$, check by substituting them into the original equation.

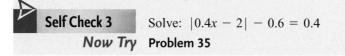

Self Check 3 Solve: $|0.4x - 2| - 0.6 = 0.4$

Now Try **Problem 35**

EXAMPLE 4 Solve: $3\left|\frac{1}{2}x - 5\right| - 4 = -4$

Strategy We will first isolate $\left|\frac{1}{2}x - 5\right|$ on the left side of the equation and then write and solve an equivalent compound equation.

Why After isolating the absolute value expression, the resulting equation will have the desired form $|X| = k$.

Solution

> **Success Tip**
>
> To solve most absolute value equations, we must consider two cases. However, if an absolute value is equal to 0, we need only consider one: the case when the expression within the absolute value bars is equal to 0.

$$3\left|\frac{1}{2}x - 5\right| - 4 = -4 \quad \text{This is the equation to solve.}$$

$$3\left|\frac{1}{2}x - 5\right| = 0 \quad \text{Add 4 to both sides.}$$

$$\left|\frac{1}{2}x - 5\right| = 0 \quad \begin{array}{l}\text{To isolate the absolute value expression, divide both sides by 3.}\\ \text{The equation is in the form } |X| = k.\end{array}$$

Since 0 is the only number whose absolute value is 0, the expression $\frac{1}{2}x - 5$ must be 0, and we have

$$\frac{1}{2}x - 5 = 0$$

$$\frac{1}{2}x = 5 \quad \text{Add 5 to both sides.}$$

$$x = 10 \quad \text{To solve for x, multiply both sides by 2.}$$

The solution is 10. Verify that it satisfies the original equation.

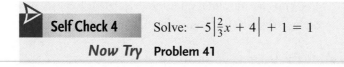

Self Check 4 Solve: $-5\left|\frac{2}{3}x + 4\right| + 1 = 1$

Now Try **Problem 41**

In Section 2.6 we discussed absolute value functions and their graphs. If we are given an output of an absolute value function, we can work in reverse to find the corresponding input(s).

EXAMPLE 5   Let $f(x) = |x + 4|$. For what value(s) of x is $f(x) = 20$?

Strategy   We will substitute 20 for $f(x)$ and solve for x.

Why   In the equation, there are two unknowns, x and $f(x)$. If we replace $f(x)$ with 20, we can solve the resulting absolute value equation for x.

Solution

$$f(x) = |x + 4|$$
$$20 = |x + 4| \text{Substitute 20 for f(x).}$$

To solve $20 = |x + 4|$, we write and then solve an equivalent compound equation:

$$20 = x + 4   \text{or}   -20 = x + 4$$

Now we solve each equation for x:

$$
\begin{array}{ccc}
20 = x + 4 & \text{or} & -20 = x + 4 \\
16 = x & & -24 = x
\end{array}
$$

To check, substitute 16 and then -24 for x in $f(x) = |x + 4|$, and verify that $f(x) = 20$ in each case.

Self Check 5   For what value(s) of x is $f(x) = 11$?

Now Try   **Problem 77**

❷ Solve Equations with Two Absolute Values.

Equations can contain two absolute value expressions. To develop a strategy to solve them, consider the following examples.

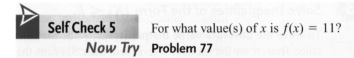

These four possible cases are really just two cases: *For two expressions to have the same absolute value, they must either be equal or be opposites of each other.* This observation suggests the following approach for solving equations having two absolute value expressions.

Solving Equations with Two Absolute Values

For any algebraic expressions X and Y:

To solve $|X| = |Y|$, solve the compound equation $X = Y$ or $X = -Y$.

EXAMPLE 6   Solve: $|5x + 3| = |3x + 25|$

Strategy   To solve this equation, we will write and then solve an equivalent compound equation.

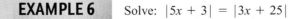

The solution set is the interval [2.885, 2.915]. This means that the distance between the two holes should be between 2.885 and 2.915 inches, inclusive. If the distance is less than 2.885 inches or more than 2.915 inches, the part should be rejected.

 **Now Try   Problem 103**

EXAMPLE 9   Solve: $|4x - 5| < -2$

Strategy   We will solve this inequality by inspection.

Why   The inequality $|4x - 5| < -2$ is of the form $|X| < k$. However, the standard method for solving absolute value inequalities cannot be used because k is negative.

Solution   Since $|4x - 5|$ is always greater than or equal to 0 for any real number x, this absolute value inequality has no solution. The solution set is $\varnothing$.

Self Check 9   Solve: $|6x + 24| < -51$
Now Try   Problem 65

---

4   **Solve Inequalities of the Form $|X| > k$.**

To solve the absolute value inequality $|x| > 5$, we must find all real numbers x whose distance from 0 on the number line is greater than 5. From the following graph, we see that there are many such numbers. For example, $-5.001, -6, -7.5$, and $-8\frac{3}{8}$, as well as $5.001, 6.2, 7, 8$, and $9\frac{1}{2}$ all meet this requirement. We conclude that the solution set is all numbers less than -5 or greater than 5, which can be written as the union of two intervals: $(-\infty, -5) \cup (5, \infty)$.

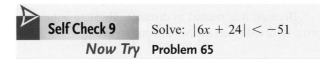

Since x is less than -5 or greater than 5, it follows that $|x| > 5$ is equivalent to $x < -5$ or $x > 5$. This observation suggests the following approach for solving absolute value inequalities of the form $|X| > k$ and $|X| \geq k$.

Solving $|X| > k$ and $|X| \geq k$	For any positive number k and any algebraic expression X:
	To solve $|X| > k$, solve the equivalent compound inequality $X < -k$ or $X > k$.
	To solve $|X| \geq k$, solve the equivalent compound inequality $X \leq -k$ or $X \geq k$.

EXAMPLE 10   Solve $\left|\dfrac{3 - x}{5}\right| \geq 6$ and graph the solution set.

Strategy   To solve this absolute value inequality, we will write and solve an equivalent compound inequality

Why   We can use this approach because the equation is of the form $|X| \geq k$, and k is positive.

Solution   The absolute value inequality $\left|\frac{3-x}{5}\right| \geq 6$ is equivalent to the compound inequality

$$\frac{3-x}{5} \leq -6 \quad \text{or} \quad \frac{3-x}{5} \geq 6$$

Now we solve each inequality for x:

$\dfrac{3-x}{5} \leq -6$	or	$\dfrac{3-x}{5} \geq 6$	
$3 - x \leq -30$		$3 - x \geq 30$	To clear the fraction, multiply both sides by 5.
$-x \leq -33$		$-x \geq 27$	To isolate the variable term $-x$, subtract 3 from both sides.
$x \geq 33$		$x \leq -27$	To isolate x, divide both sides by −1 and reverse the direction of the inequality symbol.

The solution set is the union of two intervals: $(-\infty, -27] \cup [33, \infty)$. Its graph appears on the right.

Self Check 10   Solve $\left|\frac{2-x}{4}\right| \geq 1$ and graph the solution set.

Now Try   **Problems 69 and 71**

EXAMPLE 11   Solve $\left|\frac{2}{3}x - 2\right| - 3 > 6$ and graph the solution set.

Strategy   We will first isolate $\left|\frac{2}{3}x - 2\right|$ on the left side of the inequality and write and solve an equivalent compound inequality.

Why   After isolating the absolute value expression, the resulting inequality will have the desired form $|X| > k$.

Solution   We add 3 to both sides to isolate the absolute value on the left side.

$$\left|\frac{2}{3}x - 2\right| - 3 > 6 \quad \text{This is the inequality to solve.}$$

$$\left|\frac{2}{3}x - 2\right| > 9 \quad \text{Add 3 to both sides to isolate the absolute value.}$$

To solve this absolute value inequality, we write and solve an equivalent compound inequality:

$\dfrac{2}{3}x - 2 < -9$	or	$\dfrac{2}{3}x - 2 > 9$	
$\dfrac{2}{3}x < -7$		$\dfrac{2}{3}x > 11$	Add 2 to both sides.
$2x < -21$		$2x > 33$	Multiply both sides by 3.
$x < -\dfrac{21}{2}$		$x > \dfrac{33}{2}$	To isolate x, divide both sides by 2.

The solution set is the union of two intervals: $\left(-\infty, -\frac{21}{2}\right) \cup \left(\frac{33}{2}, \infty\right)$. Its graph appears on the right.

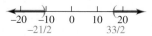

> **Self Check 11** Solve $\left|\frac{3}{4}x + 2\right| - 1 > 3$ and graph the solution set.
>
> **Now Try** **Problem 73**

EXAMPLE 12 Solve $\left|\frac{x}{8} - 1\right| \geq -4$ and graph the solution set.

Strategy We will solve this inequality by inspection.

Why The inequality $\left|\frac{x}{8} - 1\right| \geq -4$ is of the form $|x| \geq k$. However, the standard method for solving absolute value inequalities cannot be used because k is negative.

Solution Since $\left|\frac{x}{8} - 1\right|$ is always greater than or equal to 0 for any real number x, this absolute value inequality is true for all real numbers. The solution set is the interval $(-\infty, \infty)$ or $\mathbb{R}$. Its graph appears on the right.

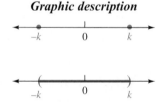

> **Self Check 12** Solve $|-x - 9| > -0.5$ and graph the solution set.
>
> **Now Try** **Problem 75**

The following summary shows how we can interpret absolute value in three ways. Assume $k > 0$.

Geometric description	*Graphic description*	*Algebraic description*
1. $\|x\| = k$ means that x is k units from 0 on the number line.		$\|x\| = k$ is equivalent to $x = k$ or $x = -k$.
2. $\|x\| < k$ means that x is less than k units from 0 on the number line.		$\|x\| < k$ is equivalent to $-k < x < k$.
3. $\|x\| > k$ means that x is more than k units from 0 on the number line.		$\|x\| > k$ is equivalent to $x > k$ or $x < -k$.

Using Your Calculator | **Solving Absolute Value Equations and Inequalities**

We can solve absolute value equations and inequalities with a graphing calculator. For example, to solve $|2x - 3| = 9$, we graph the equations $y = |2x - 3|$ and $y = 9$ on the same coordinate system, as shown in the figure. The equation $|2x - 3| = 9$ will be true for all x-coordinates of points that lie on *both* graphs. Using the TRACE or the INTERSECT feature, we can see that the graphs intersect at the points $(-3, 9)$ and $(6, 9)$. Thus, the solutions of the absolute value equation are -3 and 6.

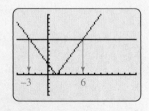

The inequality $|2x - 3| < 9$ will be true for all x-coordinates of points that lie on the graph of $y = |2x - 3|$ and *below* the graph of $y = 9$. We see that these values of x are between -3 and 6. Thus, the solution set is the interval $(-3, 6)$.

The inequality $|2x - 3| > 9$ will be true for all x-coordinates of points that lie on the graph of $y = |2x - 3|$ and *above* the graph of $y = 9$. We see that these values of x are less than -3 or greater than 6. Thus, the solution set is the union of two intervals: $(-\infty, -3) \cup (6, \infty)$.

ANSWERS TO SELF CHECKS   **1. a.** 24, -24 **b.** $\frac{1}{2}, -\frac{1}{2}$ **2. a.** 5, -2 **b.** No solution, $\varnothing$ **3.** 7.5, 2.5

4. -6 **5.** 7, -15 **6.** $-1, -6$ **7.** $\left(-2, \frac{2}{3}\right)$ **9.** No solution, $\varnothing$

10. $(-\infty, -2] \cup [6, \infty)$

11. $(-\infty, -8) \cup \left(\frac{8}{3}, \infty\right)$ **12.** $(-\infty, \infty)$

STUDY SET
4.3

VOCABULARY

Fill in the blanks.

1. The _____ _____ of a number is its distance from 0 on a number line.

2. $|2x - 1| = 10$ is an absolute value _____ and $|2x - 1| > 10$ is an absolute value _____.

3. To _____ the absolute value in $|3 - x| - 4 = 5$, we add 4 to both sides.

4. When we say that the absolute value equation and a compound equation are equivalent, we mean that they have the same _____.

5. When two equations are joined by the word *or,* such as $x + 1 = 5$ or $x + 1 = -5$, we call the statement a _____ equation.

6. $f(x) = |6x - 2|$ is called an absolute value _____.

CONCEPTS

Fill in the blanks.

7. To solve these absolute value equations and inequalities, we write and solve equivalent _____ equations and inequalities.

8. For two expressions to have the same absolute value, they must either be equal or _____ of each other.

9. Consider the following real numbers:
$-3, -2.01, -2, -1.99, -1, 0, 1, 1.99, 2, 2.01, 3$
a. Which of them make $|x| = 2$ true?
b. Which of them make $|x| < 2$ true?
c. Which of them make $|x| > 2$ true?

10. Determine whether -3 is a solution of the given equation or inequality.
a. $|x - 1| = 4$ **b.** $|x - 1| > 4$
c. $|x - 1| \le 4$ **d.** $|5 - x| = |x + 12|$

11. For each absolute value equation, write an equivalent compound equation.
a. $|x - 7| = 8$ is equivalent to

$$x - 7 = \boxed{} \quad \text{or} \quad x - 7 = \boxed{}$$

b. $|x + 10| = |x - 3|$ is equivalent to

$$x + 10 = \boxed{} \quad \text{or} \quad x + 10 = \boxed{}$$

12. For each absolute value inequality, write an equivalent compound inequality.
a. $|x + 5| < 1$ is equivalent to

$$\boxed{} < x + 5 < \boxed{}$$

b. $|x - 6| \ge 3$ is equivalent to

$$x - 6 \le \boxed{} \quad \text{or} \quad x - 6 \ge \boxed{}$$

13. For each absolute value equation or inequality, write an equivalent compound equation or inequality.

    a. $|x| = 8$ **b.** $|x| \geq 8$

    c. $|x| \leq 8$ **d.** $|5x - 1| = |x + 3|$

14. Perform the necessary steps to isolate the absolute value expression on one side of the equation. *Do not solve.*

    a. $|3x + 2| - 7 = -5$

    b. $6 + |5x - 19| \leq 40$

NOTATION

15. Match each equation or inequality with its graph.

    a. $|x| = 1$ **i.** ![number line with open interval between -1 and 1, marks at -1 0 1]

    b. $|x| > 1$ **ii.** ![number line with closed points at -1 and 1, marks at -1 0 1]

    c. $|x| < 1$ **iii.** ![number line with rays going outward from -1 and 1, marks at -1 0 1]

16. Describe the set graphed below using interval notation.

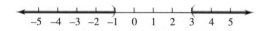

        $-5 \quad -4 \quad -3 \quad -2 \quad -1 \quad 0 \quad 1 \quad 2 \quad 3 \quad 4 \quad 5$

GUIDED PRACTICE

Solve each equation. See Example 1.

17. $|x| = 23$ **18.** $|x| = 90$

19. $|x| = \dfrac{3}{4}$ **20.** $|x| = 6.95$

Solve each equation. See Example 2.

21. $|x - 5| = 8$ **22.** $|x - 7| = 4$

23. $|3x + 2| = 16$ **24.** $|5x - 3| = 22$

25. $\left|\dfrac{x}{5}\right| = 10$ **26.** $\left|\dfrac{x}{7}\right| = 2$

27. $|2x + 3.6| = 9.8$ **28.** $|4x - 24.8| = 32.4$

29. $|50.4 - 3x| = -1$ **30.** $\left|75 - \dfrac{1}{3}x\right| = -1$

31. $\left|\dfrac{7}{2}x + 3\right| = -5$ **32.** $|x - 2.1| = -16.3$

Solve each equation. See Example 3.

33. $|x - 3| - 19 = 3$ **34.** $|x - 10| + 30 = 50$

35. $|3x - 7| + 8 = 22$ **36.** $|6x - 3| + 7 = 28$

37. $|3 - 4x| + 1 = 6$ **38.** $|8 - 5x| - 8 = 10$

39. $\left|\dfrac{7}{8}x + 5\right| - 2 = 7$ **40.** $\left|\dfrac{3}{4}x + 4\right| - 5 = 11$

Solve each equation. See Example 4.

41. $\left|\dfrac{1}{5}x + 2\right| - 8 = -8$ **42.** $\left|\dfrac{1}{9}x + 4\right| + 25 = 25$

43. $2|3x + 24| = 0$ **44.** $8\left|\dfrac{2x}{3} + 10\right| = 0$

45. $-5|2x - 9| + 14 = 14$ **46.** $-10|16x + 4| - 3 = -3$

47. $6 - 3|10x + 5| = 6$ **48.** $15 - |12x + 12| = 15$

Solve each equation. See Example 6.

49. $|5x - 12| = |4x - 16|$ **50.** $|4x - 7| = |3x - 21|$

51. $|10x| = |x - 18|$ **52.** $|6x| = |x + 45|$

53. $|2 - x| = |3x + 2|$ **54.** $|4x + 3| = |9 - 2x|$

55. $|5x - 7| = |4(x + 1)|$ **56.** $|2x + 1| = |3(x + 1)|$

Solve each inequality. Graph the solution set and write it using interval notation. See Examples 7 and 9.

57. $|x| < 4$ **58.** $|x| < 9$

59. $|x + 9| \leq 12$ **60.** $|x - 8| \leq 12$

61. $\left|\dfrac{x}{4} + 3\right| \leq 3$ **62.** $\left|\dfrac{x}{10} - 1\right| \leq 1$

63. $|3x - 2| < 10$ **64.** $|4 - 3x| \leq 13$

65. $|5x - 12| < -5$ **66.** $|3x + 2| \leq -3$

67. $|3.4x| + 19.7 \leq 19.6$ **68.** $|1.9x| - 3.1 < -3.2$

Solve each inequality. Graph the solution set and write it using interval notation. See Examples 10–12.

69. $|x| > 3$

70. $|x| > 7$

71. $|x - 12| > 24$

72. $|x + 5| \geq 7$

73. $|5x - 1| - 2 \geq 0$

74. $|6x - 3| - 5 \geq 0$

75. $|4x + 3| \geq -5$

76. $|7x + 2| \geq -8$

See Examples 5, 7, and 10.

77. Let $f(x) = |x + 3|$. For what value(s) of x is $f(x) = 3$?

78. Let $g(x) = |2 - x|$. For what value(s) of x is $g(x) = 2$?

79. Let $f(x) = |2(x - 1) + 4|$. For what value(s) of x is $f(x) < 4$?

80. Let $h(x) = \left|\frac{x}{5} - \frac{1}{2}\right|$. For what value(s) of x is $h(x) > \frac{9}{10}$?

TRY IT YOURSELF

Solve each equation and inequality. For the inequalities, graph the solution set and write it using interval notation.

81. $|3x + 2| + 1 > 15$

82. $|2x - 5| - 5 > 20$

83. $6\left|\frac{x - 2}{3}\right| \leq 24$

84. $8\left|\frac{x - 2}{3}\right| > 32$

85. $-7 = 2 - |0.3x - 3|$

86. $-1 = 1 - |0.1x + 8|$

87. $|2 - 3x| \geq -8$

88. $|-1 - 2x| > 5$

89. $|7x + 12| = |x - 6|$

90. $|8 - x| = |x + 2|$

91. $3|2 - 3x| + 2 \leq 2$

92. $|15x - 45| + 7 \leq 7$

93. $-14 = |x - 3|$

94. $-75 = |x + 4|$

95. $\frac{6}{5} = \left|\frac{3x}{5} + \frac{x}{2}\right|$

96. $\frac{11}{12} = \left|\frac{x}{3} - \frac{3x}{4}\right|$

97. $-|2x - 3| < -7$

98. $-|3x + 1| < -8$

99. $|0.5x + 1| < -23$

100. $15 \geq 7 - |1.4x + 9|$

APPLICATIONS

101. TEMPERATURE RANGES   The temperatures on a sunny summer day satisfied the inequality $|t - 78°| \leq 8°$, where t is a temperature in degrees Fahrenheit. Solve this inequality and express the range of temperatures as a double inequality.

102. OPERATING TEMPERATURES   A car CD player has an operating temperature of $|t - 40°| < 80°$, where t is a temperature in degrees Fahrenheit. Solve the inequality and express this range of temperatures as an interval.

103. AUTO MECHANICS   On most cars, the bottoms of the front wheels are closer together than the tops, creating a *camber angle*. This lessens road shock to the steering system. (See the illustration.) The specifications for a certain car state that the camber angle c of its wheels should be $0.6° \pm 0.5°$.

   a. Express the range with an inequality containing absolute value symbols.

   b. Solve the inequality and express this range of camber angles as an interval.

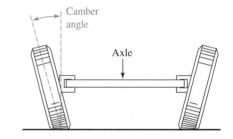

104. STEEL PRODUCTION   A sheet of steel is to be 0.250 inch thick with a tolerance of 0.025 inch.

   a. Express this specification with an inequality containing absolute value symbols, using x to represent the thickness of a sheet of steel.

   b. Solve the inequality and express the range of thickness as an interval.

105. ERROR ANALYSIS   In a lab, students measured the percent of copper p in a sample of copper sulfate. The students know that copper sulfate is actually 25.46% copper by mass. They are to compare their results to the actual value and find the amount of *experimental error.* Which measurements shown in the illustration satisfy the absolute value inequality $|p - 25.46| \leq 1.00$?

Lab 4	Section A
Title:	
"Percent copper (Cu) in copper sulfate ($CuSO_4 \cdot 5H_2O$)"	
Results	
	% Copper
Trial #1:	22.91%
Trial #2:	26.45%
Trial #3:	26.49%
Trial #4:	24.76%

106. ERROR ANALYSIS See Exercise 105. Which measurements satisfy the absolute value inequality $|p - 25.46| > 1.00$?

WRITING

107. Explain the error.

Solve: $|x| + 2 = 6$

~~$x + 2 = 6$~~ or ~~$x + 2 = -6$~~
~~$x = 4$~~ | ~~$x = -8$~~

108. Explain why the equation $|x - 4| = -5$ has no solution.

109. Explain the differences between the solution sets of $|x| < 8$ and $|x| > 8$.

110. Explain how to use the graph in the illustration to solve the following.

a. $|x - 2| = 3$
b. $|x - 2| \leq 3$
c. $|x - 2| \geq 3$

REVIEW

111. RAILROAD CROSSINGS The warning sign in the illustration is to be painted on the street in front of a railroad crossing. If y is 30° more than twice x, find x and y.

112. GEOMETRY Refer to the illustration. What is the value of $2x + 2y$?

CHALLENGE PROBLEMS

113. a. For what values of k does $|x| + k = 0$ have exactly two solutions?

b. For what values of k does $|x| + k = 0$ have exactly one solution?

114. Solve: $2^{|2x - 3|} = 64$

SECTION 4.4
Linear Inequalities in Two Variables

Objectives

1. Graph linear inequalities.
2. Graph inequalities with a boundary through the origin.
3. Graph inequalities having horizontal and vertical boundary lines.
4. Solve applied problems involving linear inequalities in two variables.

In the first three sections of this chapter, we worked with linear inequalities in one variable. Some examples are

$$x \geq -7, \qquad 10 < 2a - 4, \qquad \text{and} \qquad 5(3 + z) > -3(z + 3)$$

These inequalities have infinitely many solutions. When their solutions are graphed on a number line, we obtain an interval.

In this section, we will discuss **linear inequalities in two variables.** Some examples are

$$y > 3x + 2, \qquad 2x - 3y \leq 6, \qquad \text{and} \qquad y < 2x$$

Linear Inequalities in Two Variables	A **linear inequality** in x and y is any inequality that can be written in the form
	$Ax + By < C$ or $Ax + By > C$ or $Ax + By \leq C$ or $Ax + By \geq C$
	where A, B, and C are real numbers and A and B are not both 0.

The solutions of linear inequalities in two variables are ordered pairs whose coordinates satisfy the inequality. For example, $(-2, 4)$ is a solution of $y > 3x + 2$ because a true statement results when we substitute -2 for x and 4 for y:

$$y > 3x + 2$$
$$4 \overset{?}{>} 3(-2) + 2$$
$$4 \overset{?}{>} -6 + 2$$
$$4 > -4 \qquad \text{True}$$

We can graph the solutions of linear inequalities in two variables on a rectangular coordinate system.

 Graph Linear Inequalities.

The **graph of a linear inequality** in x and y is the graph of all ordered pairs (x, y) whose coordinates **satisfy** the inequality.

EXAMPLE 1   Graph: $y > 3x + 2$

Strategy   We will graph the *equation* $y = 3x + 2$ to establish a boundary line between two regions of the coordinate plane. Then we will determine which region contains points whose coordinates satisfy $y > 3x + 2$.

Why   To *graph a linear inequality* in two variables means to draw a "picture" of the ordered pairs (x, y) that make the inequality true.

Solution   The graph of a linear inequality in two variables such as $y > 3x + 2$ is a region of the coordinate plane on one side of a boundary line. To find the boundary line, we graph $y = 3x + 2$. Since the symbol $>$ does *not* include an equal symbol, the points on the graph of $y = 3x + 2$ are not part of the graph of $y > 3x + 2$. Therefore, the boundary line should be dashed, as shown in part (a) of the illustration on the next page. Note that the boundary line divides the coordinate plane into two **half-planes.**

To find which half-plane is the graph of $y > 3x + 2$, we can substitute the coordinates of any point in either half-plane. We will choose the origin as the test point because its coordinates, $(0, 0)$, make the computations easy. We substitute 0 for x and 0 for y into the inequality and simplify.

> **Caution**
> Remember to substitute the coordinates of the test point into the given inequality, not the equation for the boundary.

> **The Language of Algebra**
> The boundary line is also called an *edge* of the half-plane.

Check the test point (0, 0):

$$y > 3x + 2 \qquad \text{This is the original inequality.}$$
$$0 \overset{?}{>} 3(0) + 2 \qquad \text{Substitute 0 for y and 0 for x.}$$
$$0 > 2 \qquad \text{False}$$

Since the coordinates of the origin don't satisfy the given inequality, the origin is not part of its graph. This result suggests that the coordinates of every point in the half-plane *on the other side of the boundary* do satisfy $y > 3x + 2$. To show this, we shade that region. See figure (b) on the next page.

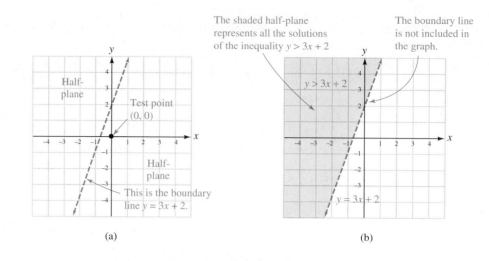

(a)   (b)

Success Tip

All the points in the region below the boundary line have coordinates that satisfy $y < 3x + 2$.

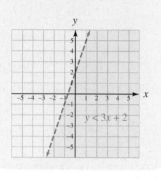

Since the solution set contains infinitely many ordered pairs, we cannot check all of them to see whether they satisfy $y > 3x + 2$. However, as an informal check, we can pick one point in the graph that is near the boundary, such as $(-1, 1)$, and see whether it satisfies the inequality. We can also pick one point not in the graph but near the boundary, such as $(1, 3)$, and see whether it fails to satisfy the inequality.

Check a point in the graph: $(-1, 1)$

$y > 3x + 2$

$1 \overset{?}{>} 3(-1) + 2$ Substitute for x and y.

$1 \overset{?}{>} -3 + 2$

$1 > -1$ True

Check a point not in the graph: $(1, 3)$

$y > 3x + 2$

$3 \overset{?}{>} 3(1) + 2$ Substitute for x and y.

$3 \overset{?}{>} 3 + 2$

$3 > 5$ False

Since $(-1, 1)$ satisfies $y > 3x + 2$ and $(1, 3)$ does not, the graph appears to be correct.

Self Check 1   Graph: $y > 2x - 4$

Now Try   **Problem 11**

EXAMPLE 2   Graph: $2x - 3y \le 6$

Strategy   We will graph the *equation* $2x - 3y = 6$ to establish a boundary line between two regions of the coordinate plane. Then we will determine which region contains points whose coordinates satisfy the inequality $2x - 3y \le 6$.

Why   To *graph a linear inequality* in two variables means to draw a "picture" of the ordered pairs (x, y) that make the inequality true.

Solution   The inequality $2x - 3y \le 6$ means that

$2x - 3y = 6$

    or

$2x - 3y < 6$

We begin by graphing $2x - 3y = 6$ to find the boundary line that separates the two half-planes. This time, we draw the solid line shown in figure (a), because equality is

Success Tip
Draw a solid boundary line if the inequality has ≤ or ≥. Draw a dashed line if the inequality has < or >.

permitted by the symbol ≤. To decide which half-plane to shade, we check to see whether the coordinates of the origin satisfy the inequality.

Check the test point (0, 0):

$$2x - 3y \leq 6 \quad \text{This is the original inequality.}$$
$$2(0) - 3(0) \overset{?}{\leq} 6 \quad \text{Substitute 0 for x and 0 for y.}$$
$$0 \leq 6 \quad \text{True}$$

The coordinates of the origin satisfy the inequality. In fact, the coordinates of every point on the same side of the boundary line as the origin satisfy the inequality. We then shade that half-plane to complete the graph of $2x - 3y \leq 6$, shown in figure (b).

The shaded half-plane and the solid boundary represent all the solutions of the inequality $2x - 3y \leq 6$.

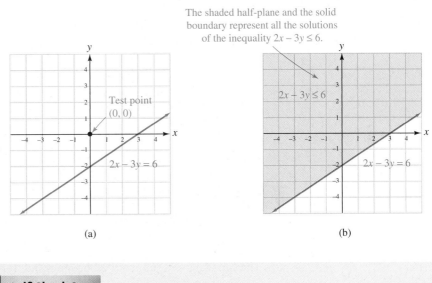

(a)   (b)

> ▷ **Self Check 2**   Graph: $3x - 2y \geq 12$
>
> *Now Try*   **Problem 15**

2   **Graph Inequalities with a Boundary through the Origin.**

EXAMPLE 3   Graph: $y < 2x$

Strategy   We will graph the equation $y = 2x$ to establish the boundary line. Then we will determine which region contains points whose coordinates satisfy the inequality $y < 2x$.

Why   To *graph a linear inequality* in two variables means to draw a "picture" of the ordered pairs (x, y) that make the inequality true.

Solution   To graph $y = 2x$, we use the fact that the equation is in slope–intercept form and that $m = 2 = \frac{2}{1}$ and $b = 0$. Since the symbol < does not include an equal symbol, the points on the graph of $y < 2x$ are not on the graph of $y = 2x$. We draw the boundary line as a dashed line to show this, as in figure (a) on the next page.

To decide which half-plane is the graph of $y < 2x$, we check to see whether the coordinates of some fixed point satisfy the inequality. We cannot use the origin as a test point, because the boundary line passes through the origin. However, we can choose a different point—say, (2, 0).

Success Tip
The origin (0, 0) is a smart choice for a test point because computations involving 0 are usually easy. If the origin is on the boundary, choose a test point not on the boundary that has one coordinate that is 0, such as (0, 1) or (2, 0).

Check the test point **(2, 0):**

$$y < 2x$$

$$0 \overset{?}{<} 2(2) \qquad \text{Substitute 2 for x and 0 for y.}$$

$$0 < 4 \qquad \text{True}$$

Since $0 < 4$ is a true statement, the point $(2, 0)$ satisfies the inequality and is in the graph of $y < 2x$. We then shade the half-plane containing $(2, 0)$, as shown in figure (b).

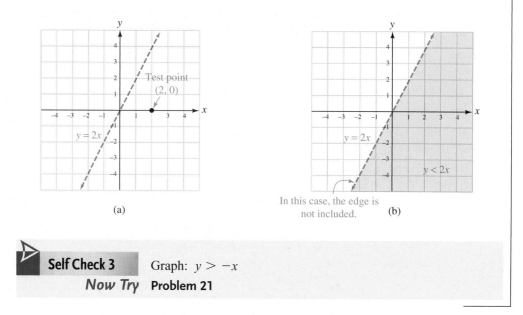

(a)

In this case, the edge is
not included.   (b)

Self Check 3   Graph: $y > -x$
      Now Try   **Problem 21**

The following is a summary of the procedure for graphing linear inequalities.

**Graphing Linear
Inequalities in
Two Variables**

1. Graph the boundary line of the region. If the inequality allows equality (the symbol is either $\leq$ or $\geq$), draw the boundary line as a solid line. If equality is not allowed ($<$ or $>$), draw the boundary line as a dashed line.

2. Pick a test point that is on one side of the boundary line. (Use the origin if possible.) Replace x and y in the original inequality with the coordinates of that point. If the inequality is satisfied, shade the side that contains the point. If the inequality is not satisfied, shade the other side of the boundary.

3 **Graph Inequalities Having Horizontal and Vertical Boundary Lines.**

Recall that the graph of $x = a$ is a vertical line with x-intercept at $(a, 0)$, and the graph of $y = b$ is a horizontal line with y-intercept at $(0, b)$.

EXAMPLE 4   Graph: $x \geq -1$

Strategy   We will graph the equation $x = -1$ to establish the boundary line. Then we will determine which region contains points whose coordinates satisfy the inequality $x \geq -1$.

Why   To *graph a linear inequality* in two variables means to draw a "picture" of the ordered pairs (x, y) that make the inequality true.

Solution   The graph of the boundary $x = -1$ is a vertical line passing through $(-1, 0)$. We draw the boundary as a solid line to show that it is part of the solution. See figure (a).

In this case, we need not pick a test point. The inequality $x \geq -1$ is satisfied by points with an x-coordinate greater than or equal to -1. Points satisfying this condition lie to the right of the boundary. We shade that half-plane, as shown in figure (b), to complete the graph of $x \geq -1$.

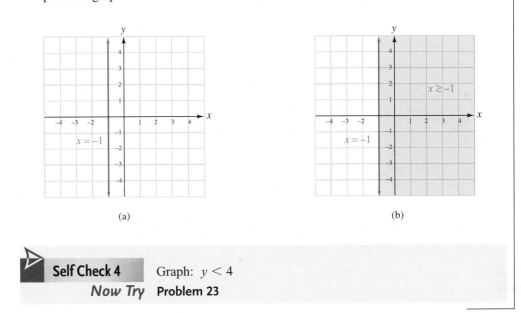

(a)   (b)

 Self Check 4   Graph: $y < 4$
Now Try   **Problem 23**

④   **Solve Applied Problems Involving Linear Inequalities in Two Variables.**

When solving applied problems, phrases such as *at least, at most,* and *should not exceed* indicate that an inequality should be used. In the next example, we solve a problem by writing a linear inequality in two variables to model a situation mathematically.

EXAMPLE 5   ***Social Security.***   Retirees, ages 62–65, can earn as much as $12,960 and still receive their full Social Security benefits. If their annual earnings exceed $12,960, their benefits are reduced. A 64-year-old retired woman receiving Social Security works two part-time jobs: one at the library, paying $540 per week, and another at a pet store, paying $405 per week. Write an inequality representing the number of weeks the woman can work at each job during the year without losing any of her benefits.

Analyze the Problem   We need to find the various combinations of weeks that she can work at the library and at the pet store so that her annual income is less than or equal to $12,960.

Form an Inequality   If we let $x =$ the number of weeks she works at the library, she will earn $540x$ annually working there. If we let $y =$ the number of weeks she works at the pet store, she will earn $405y$ annually working there. Combining the income from these jobs, the total is not to exceed $12,960.

The weekly rate on the library job	·	the weeks worked on the library job	plus	the weekly rate on the pet store job	·	the weeks worked on the pet store job	should not exceed	$12,960 .
$540	·	x	+	$405	·	y	≤	$12,960

Solve the Inequality The graph of $540x + 405y \leq$ 12,960 is shown in the figure. Since she cannot work a negative number of weeks, the graph has no meaning when x or y is negative, so only the first quadrant is used.

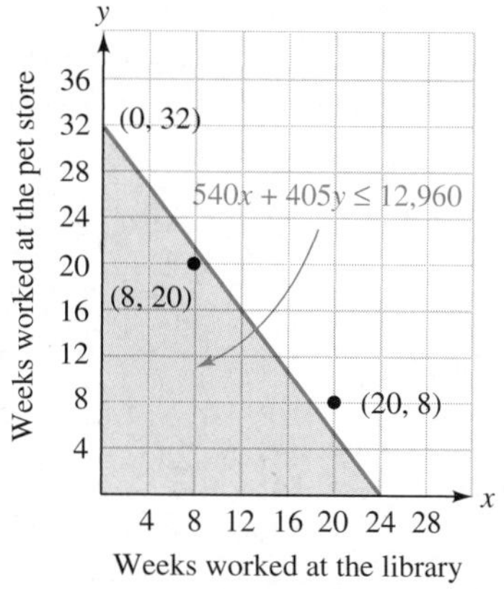

State the Conclusion Any point in the shaded region represents a way that she can schedule her work weeks and earn $12,960 or less annually. For example, the point (8, 20), which is in the graph, indicates that she can work 8 weeks at the library and 20 weeks at the pet store.

Check the Result As an informal check, we will consider the earnings represented by a point that lies in the graph and a point that does not. For (8, 20), a point in the graph, she will earn

$540(8) + \$405(20) = \$4,320 + \$8,100$ This represents 8 weeks at the library and 20 weeks at the pet store.

$= \$12,420$ The amount does not exceed $12,960.

For (20, 8), a point not in the graph, she will earn

$540(20) + \$405(8) = \$10,800 + \$3,240$ This represents 20 weeks at the library and 8 weeks at the pet store.

$= \$14,040$ The amount exceeds $12,960.

From these results, it appears that the graph is correct.

Now Try Problem 49

Using Your Calculator

Graphing Inequalities

Some graphing calculators (such as the TI-84) have a graph style icon in the $y =$ editor. Some of the different graph styles are

\	line	A straight line or curved graph is shown.	\$Y_1 =$
◥	above	Shading covers the area above a graph.	◥$Y_1 =$
◣	below	Shading covers the area below a graph.	◣$Y_1 =$

We can change the icon in front of Y_1 by placing the cursor on it and pressing the $\boxed{\text{ENTER}}$ key.

To graph $2x - 3y \leq 6$ of Example 2, we first write it in an equivalent form, with y isolated on the left side.

$$2x - 3y \leq 6$$

$$-3y \leq -2x + 6 \qquad \text{Subtract } 2x \text{ from both sides.}$$

$$y \geq \frac{2}{3}x - 2 \qquad \begin{array}{l} \text{Divide both sides by } -3. \text{ Change the direction} \\ \text{of the inequality symbol.} \end{array}$$

We then change the graph style icon to above (◥), because the inequality $y \geq \frac{2}{3}x - 2$ contains a $\geq$ symbol. Using window settings of $[-10, 10]$ for x and $[-10, 10]$ for y, we enter the boundary equation $y = \frac{2}{3}x - 2$. See figure (a). Finally, we press the $\boxed{\text{GRAPH}}$ key to get figure (b).

To graph $y < 2x$ from Example 3, we change the graph style icon to below (◣), because the inequality contains a $<$ symbol. Using window settings of $[-10, 10]$ for x and $[-10, 10]$ for y, we enter the boundary equation $y = 2x$ and press the $\boxed{\text{GRAPH}}$ key to get figure (c).

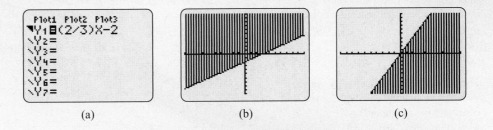

    (a)   (b)   (c)

If your calculator does not have a graph style icon, you can graph linear inequalities with a SHADE feature. Graphing calculators do not distinguish between solid and dashed lines to show whether the edge of a region is included in the graph.

ANSWERS TO SELF CHECKS

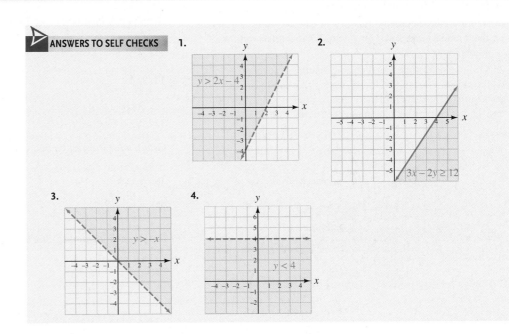

STUDY SET
4.4

VOCABULARY

Fill in the blanks.

1. $4x - 2y \geq -8$ is an example of a _____ inequality in _____ variables.

2. Graphs of linear inequalities are half-_____.

3. The boundary line of a half-plane is called an _____.

4. The graph of a linear inequality in x and y is the graph of all ordered pairs (x, y) whose coordinates _____ the inequality.

CONCEPTS

5. Decide whether each ordered pair is a solution of $3x - 2y \geq 5$.
   a. $(3, 1)$ b. $(0, 3)$
   c. $(-1, -4)$ d. $\left(1, \dfrac{1}{2}\right)$

6. The graph of a linear inequality in two variables is shown. Tell whether each point satisfies the inequality.
   a. $(-1, 4)$
   b. $(3, -2)$
   c. $(0, 0)$
   d. $(-3, -3)$

7. a. To graph the inequality $y > 3x - 1$, we begin by graphing the boundary line $y = 3x - 1$. What is the slope m of the line? What is its y-intercept?

   b. To graph the inequality $2x + 3y \leq -6$, we begin by graphing the boundary line $2x + 3y = -6$. What are its x- and y-intercepts?

8. ZOOS   To determine the allowable number of juvenile chimpanzees x and adult chimpanzees y that can live in an enclosure, a zookeeper refers to the illustration. Can 7 juvenile and 4 adult chimps be kept in the enclosure?

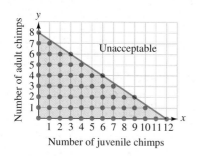

NOTATION

9. Tell whether the graph of each inequality includes the boundary line. In each case, would the boundary be a solid or a dashed line?
   a. $y < 3x - 1$ b. $2x + 3y \geq -6$
   c. $y \leq -10$ d. $x > 1$

10. Match each inequality or compound inequality with the correct graph of its solution set.
    a. $2x + 4 \geq 8$ b. $x \leq 2$ or $x \geq 4$
    c. $x \leq 4$ and $x \geq 2$ d. $2x + 4y \geq 8$

    i.
    ii.
    iii.
    iv.

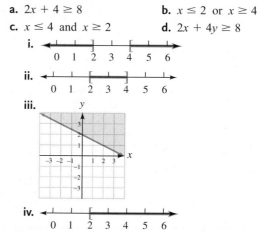

GUIDED PRACTICE

Graph each inequality. See Examples 1 and 2.

11. $y > x + 1$ 12. $y < 2x - 1$
13. $y \geq -\dfrac{3}{2}x + 1$ 14. $y < \dfrac{x}{3} - 1$
15. $2x + y \leq 6$ 16. $x - 2y \geq 4$
17. $3x + 5y > -9$ 18. $5x + 3y < 10$

Graph each inequality. See Example 3.

19. $y \geq x$ 20. $y \leq 2x$
21. $y < -\dfrac{x}{2}$ 22. $y > \dfrac{x}{3}$

Graph each inequality. See Example 4.

23. $x < 4$ 24. $y \geq -2$
25. $y < 0$ 26. $x \geq 0$

TRY IT YOURSELF

Graph each inequality.

27. $3x \geq -y + 3$ 28. $2x \leq -3y - 12$
29. $3x + y > 2 + x$ 30. $3x - y > 6 + y$

31. $y \geq \dfrac{8}{3}$

32. $x < -3.75$

33. $y + 4x \geq 0$

34. $y \geq -\dfrac{5}{4}x$

35. $\dfrac{x}{2} + \dfrac{y}{2} \leq 2$

36. $\dfrac{x}{3} - \dfrac{y}{2} \geq 1$

37. $y - 4.5 < 0$

38. $x - 3 \geq 0$

39. $x < -\dfrac{1}{2}y$

40. $x < \dfrac{1}{6}y$

41. $0.3x + 0.4y \geq -1.2$

42. $0.8x - 0.3y < 2.4$

Use a graphing calculator to graph each inequality. See Using Your Calculator: Graphing Inequalities.

43. $y < 0.27x - 1$

44. $y > -3.5x + 2.7$

45. $y \geq -2.37x + 1.5$

46. $y \leq 3.37x - 1.7$

APPLICATIONS

47. GEOGRAPHY   A region of the continental United States is shaded in the following map.

   a. What is the boundary that separates the shaded and unshaded regions?

   b. In words, describe the shaded area with respect to the boundary.

48. THE KOREAN WAR   After World War II, the 38th parallel of north latitude was established as the boundary between North Korea and South Korea. In the illustration, shade the region of the Korean Peninsula south of the 38th parallel.

49. RESTAURANT SEATING   As part of a remodeling project, a restaurant owner will be installing new booths that seat 4 persons, and new tables that seat 6 persons. The overall seating must conform to the sign shown below. Write an inequality that describes the possible combinations of the number of booths (x) and the number of tables (y) that the owner can install. Graph the inequality and give three ordered-pair solutions.

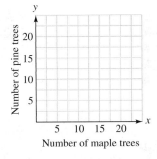

MAXIMUM OCCUPANCY
NOT TO EXCEED
120
By order of Clake County Fire Marshal

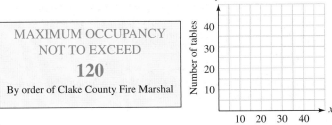

50. GARDENING   During an Arbor Day sale, a garden store sold more than $2,000 worth of 6-foot maple and 5-foot pine trees. A 6-foot maple sold for $100 and a 5-foot pine sold for $125. Write an inequality that describes the possible combinations of the number of the maple trees (x) and the number of pine trees (y) that could have been sold during the sale. Graph the inequality and give three ordered-pair solutions.

51. SPORTING GOODS   A sporting goods manufacturer allocates at least 1,200 units of time per day to make fishing rods and reels. It takes 10 units of time to make a rod and 15 units of time to make a reel. Write an inequality that describes the possible combinations of the number of units of time devoted to make rods (x) and the number of units of time devoted to make reels (y). Graph the inequality and give three ordered-pair solutions.

52. HOUSEKEEPING   One housekeeper charges $12 per hour, and another charges $9 per hour. Sarah, who can afford no more than $54 per month to clean her house, decides to hire the housekeepers. Write an inequality that describes the possible combinations of the number of hours that she can hire the first housekeeper (x) and the number of hours that she can hire second housekeeper (y). Graph the inequality and give three ordered-pair solutions.

WRITING

53. Explain how to decide whether the boundary of the graph of a linear inequality should be drawn as a solid or a dashed line.

54. Explain how to decide which side of the boundary of the graph of a linear inequality should be shaded.

55. The boundary for the graph of a linear inequality is shown in the illustration. Why can't the origin be used as a test point to decide which side to shade?

56. Explain the difference between the graph of the solution set of $x + 2 > 6$, an inequality in *one variable,* and the graph of $x + 2y > 6$, an inequality in *two variables.*

REVIEW

Decide whether the ordered pair $(-4, 3)$ *is a solution of the system of linear equations.*

57. $\begin{cases} 4x - y = -19 \\ 3x + 2y = -6 \end{cases}$

58. $\begin{cases} y = 2x + 11 \\ \frac{x}{2} + y = 0 \end{cases}$

Solve each system of equations.

59. $\begin{cases} x = \frac{2}{3}y \\ y = 4x + 5 \end{cases}$

60. $\begin{cases} x - \frac{y}{2} = -2 \\ 0.01x + 0.02y = 0.03 \end{cases}$

CHALLENGE PROBLEMS

61. Can an inequality in two variables be an identity, one that is satisfied by all pairs (x, y)? Illustrate.

62. Can an inequality in two variables have no solutions? Illustrate.

Find the equation of the boundary line. Then give the inequality whose graph is shown.

63. a.   **b.**

64. a.   **b.**

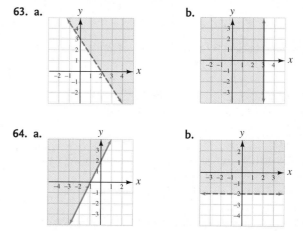

SECTION 4.5
Systems of Linear Inequalities

Objectives

1 Solve systems of linear inequalities.

2 Graph compound inequalities.

3 Solve problems involving systems of linear inequalities.

We have discussed how to solve systems of linear equations by the graphing method. For example, to solve

$$\begin{cases} y = -x + 1 \\ 2x - y = 2 \end{cases}$$

we graph both equations on the same rectangular coordinate system and then find the coordinates of the point of intersection of the straight lines. We will now discuss how to solve **systems of linear inequalities** graphically, such as

$$\begin{cases} y \le -x + 1 \\ 2x - y > 2 \end{cases}$$

1 **Solve Systems of Linear Inequalities.**

When we graph the solution of a linear inequality in x and y, the result is a half-plane. To solve a system of linear inequalities, we graph each of the inequalities on one set of coordinate axes and find the intersection of the shaded half-planes.

EXAMPLE 1   Graph the solution set of $\begin{cases} y \leq -x + 1 \\ 2x - y > 2 \end{cases}$.

Strategy   We will graph the solutions of $y \leq -x + 1$ in one color and the solutions of $2x - y > 2$ in another color on the same rectangular coordinate system.

Why   We can then see where the graphs of the two inequalities intersect. This is the solution set of the system.

Solution   To graph $y \leq -x + 1$, we graph the boundary $y = -x + 1$, as shown in figure (a). Since the edge is to be included, we draw it as a solid line. To determine which half-plane to shade, we use the origin as a test point. Because the coordinates of the origin satisfy $y \leq -x + 1$, we shade (in red) the half-plane containing the origin.

In figure (b), we superimpose the graph of $2x - y > 2$ on the graph of $y \leq -x + 1$ so that we can determine the points that the graphs have in common. To graph $2x - y > 2$, we graph the boundary $2x - y = 2$ as a dashed line. Since the test point $(0, 0)$ does not satisfy $2x - y > 2$, we then shade (in blue) the half-plane that does not contain $(0, 0)$.

The area that is shaded twice represents the solutions of the given system. Any point in the doubly shaded region in purple (including the purple portion of one of the boundaries) has coordinates that satisfy both inequalities.

> **The Language of Algebra**
> To solve a system of linear inequalities, we *superimpose* the graphs of the inequalities. That is, we place one graph over the other. Perhaps you have seen a video camcorder that can *superimpose* the date and time over the picture being recorded.

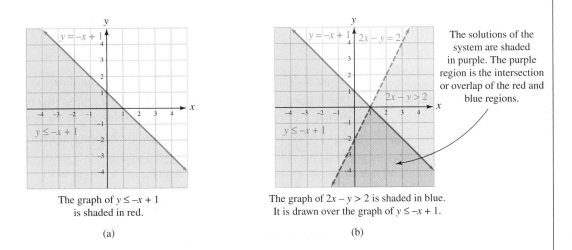

The graph of $y \leq -x + 1$
is shaded in red.

(a)

The graph of $2x - y > 2$ is shaded in blue.
It is drawn over the graph of $y \leq -x + 1$.

(b)

The solutions of the system are shaded in purple. The purple region is the intersection or overlap of the red and blue regions.

Since there are infinitely many solutions, we cannot check each of them. However, as an informal check, we can select several points near the boundaries that lie in the doubly shaded region and show that their coordinates satisfy both inequalities of the system. For example, to check $(0, -3)$, we have:

$$y \leq -x + 1 \quad \text{The first inequality.}$$
$$-3 \overset{?}{\leq} -(0) + 1 \quad \text{Substitute.}$$
$$-3 \leq 1 \quad \text{True}$$

$$2x - y > 2 \quad \text{The second inequality.}$$
$$2(0) - (-3) \overset{?}{>} 2 \quad \text{Substitute.}$$
$$3 > 2 \quad \text{True}$$

Self Check 1   Graph the solution set of $\begin{cases} x + y \geq 1 \\ 2x - y < 2 \end{cases}$.

Now Try   **Problem 9**

EXAMPLE 3   Graph $2 < x \leq 5$ on a rectangular coordinate system.

Strategy   We will write the double inequality $2 < x \leq 5$ as an equivalent system of inequalities. Then we will graph the system.

Why   The compound inequality $2 < x \leq 5$ is equivalent to $2 < x$ and $x \leq 5$.

Solution   The compound inequality $2 < x \leq 5$ is equivalent to the following system of two linear inequalities:

$$\begin{cases} 2 < x \\ x \leq 5 \end{cases}$$

The graph of $2 < x$, shown in the figure, is the half-plane to the right of the vertical line $x = 2$. The graph of $x \leq 5$, shown in the figure, includes the line $x = 5$ and the half-plane to its left. The graph of $2 < x \leq 5$ will contain all points in the plane that satisfy the inequalities $2 < x$ and $x \leq 5$ simultaneously. These points are in the purple-shaded region of the figure.

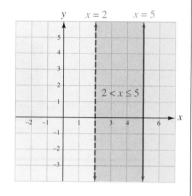

> **Self Check 3**   Graph $-2 \leq y < 3$ on the rectangular coordinate plane.
> **Now Try**   **Problem 17**

EXAMPLE 4   Graph $y \geq -1$ or $y < -3$ on a rectangular coordinate system.

Strategy   We will graph the inequalities $y \geq -1$ and $y < -3$ on the same rectangular coordinate system and determine their union.

Why   When solving a compound inequality containing the word *or*, the solution set is the union of the solution sets of the two inequalities.

Solution   Since the operation of union combines the graphs, we can draw them in the same color. The graph of $y \geq -1$, shown in red, includes the line $y = -1$ and the half plane above it. The graph of $y < -3$, also shown in the figure in red, is the half-plane below the line $y = -3$. The graph of $y \geq -1$ or $y < -3$ contains all the points in the plane that satisfy one or the other inequalities.

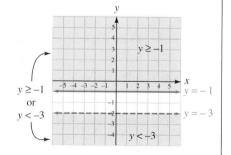

> **Self Check 4**   Graph $x \leq -2$ or $x > 3$ on a rectangular coordinate system.
> **Now Try**   **Problem 19**

3 Solve Problems Involving Systems of Linear Inequalities.

© Chuck Franklin/Alamy

EXAMPLE 5   *Landscaping.*   A homeowner has a budget of $300 to $600 for trees and bushes to landscape his yard. After shopping, he finds that good trees cost $150 and mature bushes cost $75. What combinations of trees and bushes can he afford to buy?

Analyze the Problem   We must find the number of trees and bushes that the home-owner can afford. This suggests we should use two variables. We know that he is willing to spend *at least* $300 and *at most* $600 for trees and bushes. These phrases suggest that we should write two inequalities that model the situation.

Form Two Inequalities   If x = the number of trees purchased, $150x$ will be the cost of the trees. If y = the number of bushes purchased, $75y$ will be the cost of the bushes. The homeowner wants the sum of these costs to be from $300 to $600. Using this information, we can form the following system of linear inequalities.

The cost of a tree	times	the number of trees purchased	plus	the cost of a bush	times	the number of bushes purchased	should be at least	$300.
$150	·	x	+	$75	·	y	$\geq$	$300

The cost of a tree	times	the number of trees purchased	plus	the cost of a bush	times	the number of bushes purchased	should be at most	$600.
$150	·	x	+	$75	·	y	$\leq$	$600

Solve the System   We graph the system

$$\begin{cases} 150x + 75y \geq 300 \\ 150x + 75y \leq 600 \end{cases}$$

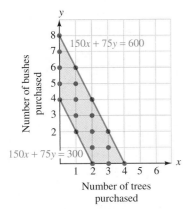

Number of bushes purchased

150x + 75y = 600

150x + 75y = 300

Number of trees purchased

as shown on the left. The coordinates of each point highlighted in the graph give a possible combination of trees (x) and bushes (y) that can be purchased.

State the Conclusion   The possible combinations of trees and bushes that can be purchased are given by

(0, 4), (0, 5), (0, 6), (0, 7), (0, 8)
(1, 2), (1, 3), (1, 4), (1, 5), (1, 6)
(2, 0), (2, 1), (2, 2), (2, 3), (2, 4)
(3, 0), (3, 1), (3, 2), (4, 0)

The ordered pair (1, 6), for example, indicates that the homeowner can afford 1 tree and 6 bushes.

Only these points can be used, because the homeowner cannot buy a portion of a tree or a bush.

Check the Result   Check some of the ordered pairs to verify that they satisfy both inequalities.

Because the homeowner cannot buy a negative number of trees or bushes, we graph the system for $x \geq 0$ and $y \geq 0$.

Now Try   Problem 49

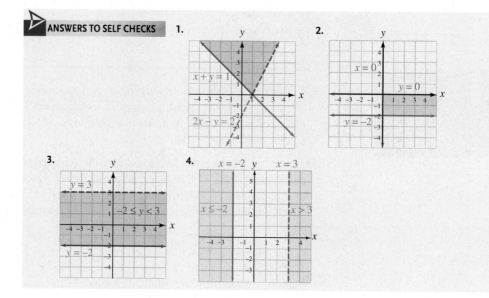

ANSWERS TO SELF CHECKS

1.

2.

3.

4.

STUDY SET
4.5

VOCABULARY

Fill in the blanks.

1. $\begin{cases} x + y \le 2 \\ x - 3y > 10 \end{cases}$ is a system of linear _____ in two variables.

2. To solve a system of inequalities by graphing, we graph each inequality. The solution is the region where the graphs overlap or _____.

3. To determine which half-plane to shade when graphing a linear inequality, we see whether the coordinates of a test _____ satisfy the inequality.

4. When we graph a system of two linear inequalities, any point in the doubly shaded region has coordinates that _____ both inequalities.

CONCEPTS

5. Determine whether each point satisfies the system of linear inequalities $\begin{cases} x + y \le 2 \\ x - 3y > 10 \end{cases}$.

   a. $(2, -3)$ **b.** $(12, -1)$

   c. $(0, -3)$ **d.** $(-0.5, -5)$

6. **a.** Determine whether $(-3, 10)$ satisfies the compound inequality $-5 < x \le 8$ in the rectangular coordinate system.

   b. Determine whether $(-3, 3)$ satisfies the compound inequality $y \le 0$ or $y > 4$ in the rectangular coordinate system.

7. In the illustration, the solution of one linear inequality is shaded in red, and the solution of a second is shaded in blue. Determine whether a true or false statement results if the coordinates of the given point are substituted into the given inequality.

   a. A, inequality 1   **b.** A, inequality 2

   c. B, inequality 1   **d.** B, inequality 2

   e. C, inequality 1   **f.** C, inequality 2

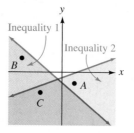

8. Match each equation, inequality, or system with the graph of its solution shown on the next page.

   a. $2x + y = 2$ **b.** $2x + y \ge 2$

   c. $\begin{cases} 2x + y = 2 \\ 2x - y = 2 \end{cases}$ **d.** $\begin{cases} 2x + y \ge 2 \\ 2x - y \le 2 \end{cases}$

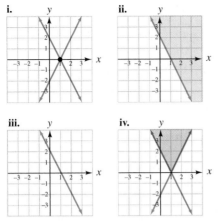

i.   ii.   iii.   iv.

GUIDED PRACTICE

Graph the solution set of each system of inequalities. See Example 1.

9. $\begin{cases} y < 3x + 2 \\ 2x + y < 3 \end{cases}$

10. $\begin{cases} y \leq x - 2 \\ 2x - y \leq -1 \end{cases}$

11. $\begin{cases} 3x + y \leq 1 \\ -x + 2y \geq 6 \end{cases}$

12. $\begin{cases} 3x + 2y > 6 \\ x + 3y \leq 2 \end{cases}$

Graph the solution set of each system of inequalities. See Example 2.

13. $\begin{cases} y - x \leq 2 \\ y > -2 \\ x < 2 \end{cases}$

14. $\begin{cases} x - y < 4 \\ y \leq 0 \\ x \geq 0 \end{cases}$

15. $\begin{cases} 2x + 3y \leq 6 \\ 3x + y \leq 1 \\ x \leq 0 \end{cases}$

16. $\begin{cases} 2x + y \leq 2 \\ y \geq x \\ x \geq 0 \end{cases}$

Graph each inequality on a rectangular coordinate system. See Examples 3 and 4.

17. $-2 \leq x < 0$

18. $-3 < y \leq -1$

19. $y < -2$ or $y > 3$

20. $x \leq 1$ or $x > 5$

TRY IT YOURSELF

Graph the solution set of each system of inequalities on a rectangular coordinate system.

21. $\begin{cases} 2x < 3y \\ 2x + 3y \geq 12 \end{cases}$

22. $\begin{cases} x > 2 - y \\ x - y < -2 \end{cases}$

23. $\begin{cases} x > 0 \\ y > 0 \end{cases}$

24. $\begin{cases} x \leq 0 \\ y < 0 \end{cases}$

25. $\begin{cases} y \geq x \\ y \leq \frac{1}{3}x + 1 \\ x > -3 \end{cases}$

26. $\begin{cases} x - y \leq 6 \\ x + 2y \leq 6 \\ x \geq 0 \end{cases}$

27. $5 > y \geq 2$

28. $0 \geq x > -4$

29. $\begin{cases} x + y < 2 \\ x + y \leq 1 \end{cases}$

30. $\begin{cases} x + 2y < 3 \\ 2x + 4y < 8 \end{cases}$

31. $\begin{cases} y < -\frac{3}{2}x - 3 \\ 3x + 2y \geq 2 \end{cases}$

32. $\begin{cases} y > \frac{1}{4}x + 3 \\ x - 4y > 4 \end{cases}$

33. $\begin{cases} 3y - 5x < 0 \\ 5x - 3y \geq -12 \end{cases}$

34. $\begin{cases} x + y < 2 \\ x \leq 1 - y \end{cases}$

35. $-x \leq 1$ or $x \geq 2$

36. $y < 1$ or $y \leq -5$

37. $\begin{cases} x < 1 \\ x > -1 \\ x - y + 4 \geq 0 \\ y - x > -4 \end{cases}$

38. $\begin{cases} x \geq 0 \\ y \geq 0 \\ 9x + 3y \leq 18 \\ 3x + 6y \leq 18 \end{cases}$

39. $\begin{cases} 2x - 3y \leq 3 \\ 3y \leq 2x - 3 \end{cases}$

40. $\begin{cases} x \leq 3 \\ x \geq 3 \end{cases}$

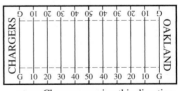 *Use a graphing calculator to solve each system. See Using Your Calculator: Solving Systems of Inequalities.*

41. $\begin{cases} y < 3x + 2 \\ y < -2x + 3 \end{cases}$

42. $\begin{cases} y > -x + 2 \\ y < -x + 4 \end{cases}$

43. $\begin{cases} 2x + y \geq 6 \\ y \leq 2(2x - 3) \end{cases}$

44. $\begin{cases} 3x + y < -2 \\ y > 3(1 - x) \end{cases}$

APPLICATIONS

45. **FOOTBALL**   In 2006, the San Diego Chargers scored either a touchdown or a field goal 95.2% of the time when their offense was in the *red zone*. This was the best record in the NFL! Refer to the illustration below. If x represents the yard line the football is on, the Chargers' red zone is the area on the left end of the field that can be described by the following system:

$$\begin{cases} x > 0 \\ x \leq 20 \end{cases}$$

Shade the red zone.

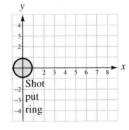

CHARGERS   OAKLAND

G 10 20 30 40 50 40 30 20 10 G

← Chargers moving this direction

46. **TRACK AND FIELD**   In the shot put, the solid metal ball must land in a marked sector for it to be a fair throw. In the illustration, graph the system of inequalities that describes the region in which a shot must land.

$$\begin{cases} y \leq \frac{3}{8}x \\ y \geq -\frac{3}{8}x \\ x \geq 1 \end{cases}$$

Shot put ring

47. NO-FLY ZONES   After the first Gulf War, U.S. and Allied forces enforced northern and southern "no-fly" zones over Iraq. Iraqi aircraft was prohibited from flying in this air space. If y represents the north latitude parallel measurement, the no-fly zones can be described by

$$y \geq 36 \quad \text{or} \quad y \leq 33$$

On the map, shade the regions of Iraq over which there was a no-fly zone.

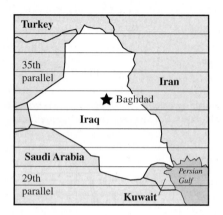

48. CARDIOVASCULAR FITNESS   The graph in the illustration shows the range of pulse rates that persons ages 20–90 should maintain during aerobic exercise to get the most benefit from the training. The shaded region "Effective Training Heart Rate Zone" can be described by a system of linear inequalities. Determine what inequality symbol should be inserted in each blank.

$$\begin{cases} x \;\underline{\hspace{1em}}\; 20 \\ x \;\underline{\hspace{1em}}\; 90 \\ y \;\underline{\hspace{1em}}\; -0.87x + 191 \\ y \;\underline{\hspace{1em}}\; -0.72x + 158 \end{cases}$$

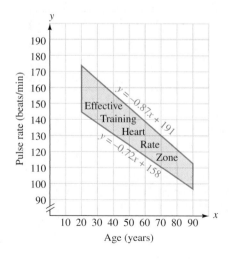

49. COMPACT DISCS   A music store has compact discs on sale for either $10 or $15. A customer wants to spend at least $30 but no more than $60 on CDs at the store. Use the illustration to graph a system of inequalities that will show the possible combinations of the number of $10 CDs ($x$) and the number of $15 CDs ($y$) that the customer can buy. Give two possible solutions.

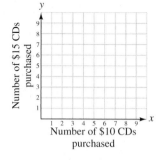

50. BOAT SALES   Dry Boat Works wholesales aluminum boats for $800 and fiberglass boats for $600. Northland Marina wants to order at least $2,400 worth but no more than $4,800 worth of boats from them. Use the illustration to graph a system of inequalities that will show the possible combinations of the number of aluminum boats (x) and the number of fiberglass boats (y) that can be ordered. Give two possible solutions.

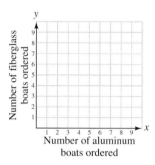

51. FURNITURE SALES   A distributor wholesales desk chairs for $150 and side chairs for $100. Best Furniture wants to order no more than $900 worth of chairs, including more side chairs than desk chairs, from that distributor. Use the illustration to graph a system of inequalities that will show the possible combinations of the number of desk chairs (x) and the number of side chairs (y) that can be ordered. Give two possible solutions.

52. FURNACE EQUIPMENT   J. Bolden Heating Company wants to order no more than $2,000 worth of electronic air cleaners and humidifiers from a wholesaler that charges $500 for air cleaners and $200 for humidifiers. If Bolden wants more humidifiers than air cleaners, use the illustration to graph a system of inequalities that will show the possible combinations of the number of air cleaners (x) and the number of humidifiers (y) that can be ordered. Give two possible solutions.

WRITING

53. Explain how to solve a system of two linear inequalities graphically.

54. Explain how a system of two linear inequalities might have no solution.

55. A student graphed the following system as shown. Explain how to informally check the result.

$$\begin{cases} -x + 3y > 0 \\ x + 3y < 3 \end{cases}$$

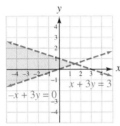

56. Describe the result when $-3 \le x < 4$ is graphed on a number line. Describe the result when $-3 \le x < 4$ is graphed on the rectangular coordinate plane.

REVIEW

Use the given conditions to determine in which quadrant of a rectangular coordinate system each point (x, y) is located.

57. $x > 0$ and $y < 0$

58. $x < 0$ and $y < 0$

59. $x < 0$ and $y > 0$

60. $x > 0$ and $y > 0$

CHALLENGE PROBLEMS

61. Write a compound inequality whose graph is shown.

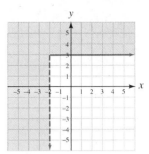

62. Write a system of linear inequalities in two variables whose graph is shown.

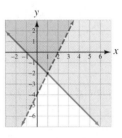

63. The solution of a system of inequalities in two variables is *bounded* if it is possible to draw a circle around the solution.

a. Can the solution of two linear inequalities be bounded?

b. Can the solution of three linear inequalities be bounded?

64. The solution of $\begin{cases} y \ge |x| \\ y \le k \end{cases}$ has an area of 25. Find k.

CHAPTER 4
Summary & Review

SECTION 4.1 Solving Linear Inequalities in One Variable

DEFINITIONS AND CONCEPTS	EXAMPLES
Inequalities are statements that contain one or more **inequality symbols.**	Each of the following inequalities is true: $5 \neq 7$ 5 is not equal to 7. $-1 < 9$ −1 is less than 9. $8 > 5$ 8 is greater than 5. $3 \leq 6$ 3 is less than or equal to 6. $4 \geq 4$ 4 is greater than or equal to 4.
A **linear inequality in one variable** is any inequality that can be written in the form $ax + b < c$, where a, b, and c represent real numbers and $a \neq 0$. The inequality symbols $>$, $\leq$, and $\geq$ can also be used.	Linear inequalities in one variable: $\quad 2x + 3 < 6 \qquad 8x > 2x - 7 \qquad 5(x + 9) \leq 6(x - 6)$
The **solution of a linear inequality** is a number that satisfies the inequality.	The number 5 is a solution of $12 < 3x + 11$ because 5 satisfies the inequality. $\quad 12 < 3x + 11$ $\quad 12 \overset{?}{<} 3(5) + 11 \qquad$ Substitute 5 for x. $\quad 12 \overset{?}{<} 15 + 11$ $\quad 12 < 26 \qquad\qquad$ True

To **solve a linear inequality** in one variable we use **properties of inequality** to find the values of its variable that make the inequality true.
 	
Adding the same number to, or subtracting the same number from, both sides of an inequality does not change the solutions.
 	
Multiplying or dividing both sides of an inequality by the same positive number does not change the solutions.
 	
The set of all solutions of an inequality is called its **solution set.**
 	
If we multiply or divide both sides of an inequality by a negative number, the direction of the **inequality symbol must be reversed** for the inequalities to have the same solutions.
	Solve: $3x - 12 \geq 12$
	
	$\quad 3x - 12 + 12 \geq 12 + 12 \qquad$ Add 12 to both sides.
	$\qquad\qquad\quad 3x \geq 24$
	$\qquad\qquad\quad \dfrac{3x}{3} \geq \dfrac{24}{3} \qquad$ Divide both sides by 3.
	$\qquad\qquad\quad x \geq 8$
	
	The solution set can be expressed in three ways:
	
	Graph Interval notation Set-builder notation
	                     $[8, \infty)$ $\{x \mid x \geq 8\}$
	7 8 9
	
	Solve: $-5x + 7 > 22$
	
	$\quad -5x + 7 - 7 > 22 - 7 \qquad$ Subtract 7 from both sides.
	$\qquad\qquad -5x > 15$
	$\qquad\qquad \dfrac{-5x}{-5} < \dfrac{15}{-5} \qquad$ Divide both sides by −5 and reverse the direction
	$\qquad\qquad\qquad\qquad\qquad$ of the inequality symbol.
	$\qquad\qquad x < -3$
	
	The solution set is:
	
	Graph Interval notation Set-builder notation
	                     $(-\infty, -3)$ $\{x \mid x < -3\}$
	−4 −3 −2

REVIEW EXERCISES

1. Determine whether -2 is a solution of each inequality.

   a. $3x - 6 < x - 10$ **b.** $\dfrac{x}{2} - 3 \geq 4(x + 1)$

2. Represent the set of real numbers greater than or equal to -5 with a graph, using interval notation, and using set-builder notation.

Solve each inequality. Graph the solution set and write it using interval notation and set-builder notation.

3. $5(x - 2) \leq 5$

4. $0.3x - 0.4 \geq 1.2 - 0.1x$

5. $-16 < -\dfrac{4}{5}t$

6. $\dfrac{7}{4}(x + 3) < \dfrac{3}{8}(x - 3)$

7. $7 - [6t - 5(t - 3)] > 2(t - 3) - 3(t + 1)$

8. $\dfrac{2b + 7}{2} \leq \dfrac{3b - 1}{3}$

9. INVESTMENTS   A woman has invested $10,000 at 6% annual interest. How much more must she invest at 7% so that her annual income is at least $2,000?

10. ICE SKATING   For the free-skating portion of a competition, an ice skater received scores of 5.3, 4.8, 4.7, 4.9, and 5.1 from the first five judges. What score must she receive from the sixth and final judge to average better than 5.0 for her performance?

11. LAWYERS   A lawyer earns $200 an hour for telephone consultations and $300 an hour for office consultations with clients. To save time for court appearances, she limits her consulting to 15 hours a week. What is the greatest number of hours that she can spend on the phone and still earn at least $4,000 in consulting fees a week?

12. Explain how to use the graphs of $y = 1$ and $y = x - 3$ to solve $x - 3 \leq 1$.

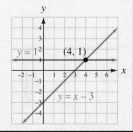

SECTION 4.2   Solving Compound Inequalities

DEFINITIONS AND CONCEPTS	EXAMPLES
The **intersection** of two sets A and B, written $A \cap B$, is the set of all elements that are common to set A and set B.	Let $A = \{-2, 0, 3, 5\}$ and $B = \{-3, 0, 5, 7\}$. $A \cap B = \{0, 5\}$ The intersection contains the elements that the sets have in common.
The **union** of two sets A and B, written $A \cup B$, is the set of all elements that are in set A, set B, or both.	$A \cup B = \{-3, -2, 0, 3, 5, 7\}$ The union contains the elements that are in one or the other set, or both.
When the word *and* or the word *or* is used to connect pairs of inequalities, we call the statement a **compound inequality.** The solution set of a **compound inequality containing the word *and*** includes all numbers that make both of the inequalities true. That is, it is the intersection of their solution sets.	Solve: $2x - 1 \leq 5$ and $5x + 1 > 4$ We solve each inequality separately. Then we graph the two solution sets on the same number line and determine their intersection. $\begin{array}{ll} 2x - 1 \leq 5 & \text{and} \quad 5x + 1 > 4 \\ 2x \leq 6 & \qquad\quad 5x > 3 \\ x \leq 3 & \qquad\quad x > \dfrac{3}{5} \end{array}$ The purple-shaded interval is where the red and blue graphs overlap. Thus, the solution set is: Interval notation: $\left(\dfrac{3}{5}, 3\right]$

SECTION 4.2   Solving Compound Inequalities—*continued*

DEFINITIONS AND CONCEPTS	EXAMPLES
Inequalities that contain exactly two inequality symbols are called **double inequalities.** Any double linear inequality can be written as a compound inequality containing the word *and.* For example: $c < x < d$ is equivalent to $c < x$ and $x < d$	Solve: $-7 \le 3x - 1 < 5$ We apply properties of inequality to *all three of its parts* to isolate x in the middle. $$-7 \le 3x - 1 < 5$$ $$-7 + 1 \le 3x - 1 + 1 < 5 + 1 \quad \text{Add 1 to all three parts.}$$ $$-6 \le 3x < 6$$ $$\frac{-6}{3} \le \frac{3x}{3} < \frac{6}{3} \quad \text{Divide each part by 3.}$$ $$-2 \le x < 2$$ The solution set is: Interval notation: $[-2, 2)$
The solution set of a **compound inequality containing the word** *or* includes all numbers that make one or the other, or both, inequalities true. That is, it is the union of their solution sets.	Solve: $2x - 1 > 5$ or $-(5x - 7) \ge 2$ We solve each inequality separately. Then we graph the two solution sets on the same number line to show their union. $2x - 1 > 5 \quad$ or $\quad -(5x - 7) \ge 2$ $2x > 6 \qquad\qquad\quad -5x + 7 \ge 2$ $x > 3 \qquad\qquad\qquad -5x \ge -5$ $\qquad\qquad\qquad\qquad\quad x \le 1$ The solution set is: Interval notation: $(-\infty, 1] \cup (3, \infty)$ This is the union of two intervals.

REVIEW EXERCISES

Let $A = \{-6, -3, 0, 3, 6\}$ and $B = \{-5, -3, 3, 8\}$.

13. Find $A \cap B$.   **14.** Find $A \cup B$.

Determine whether -4 *is a solution of the compound inequality.*

15. $x < 0$ and $x > -5$

16. $x + 3 < -3x - 1$ and $4x - 3 > 3x$

Graph each set.

17. $(-3, 3) \cup [1, 6]$ **18.** $(-\infty, 2] \cap [1, 4)$

Solve each compound inequality. Graph the solution set and write it using interval notation.

19. $-2x > 8$ and $x + 4 \ge -6$

20. $5(x + 2) \le 4(x + 1)$ and $11 + x < 0$

21. $\dfrac{2}{5}x - 2 < -\dfrac{4}{5}$ and $\dfrac{x}{-3} < -1$

22. $4\left(x - \dfrac{1}{4}\right) \le 3x - 1$ and $x \ge 0$

Solve each double inequality. Graph the solution set and write it using interval notation.

23. $3 < 3x + 4 < 10$ **24.** $-2 \le \dfrac{5 - x}{2} \le 2$

Determine whether -4 *is a solution of the compound inequality.*

25. $x < 1.6$ or $x > -3.9$

26. $x + 1 < 2x - 1$ or $4x - 3 > 3x$

Solve each compound inequality. Graph the solution set and write it using interval notation.

27. $x + 1 < -4$ or $x - 4 > 0$ **28.** $\dfrac{x}{2} + 3 > -2$ or $4 - x > 4$

29. RUGS   A manufacturer makes a line of decorator rugs that are 4 feet wide and of varying lengths x (in feet). The floor area covered by the rugs ranges from 17 ft^2 to 25 ft^2. Write and then solve a double linear inequality to find the range of the lengths of the rugs.

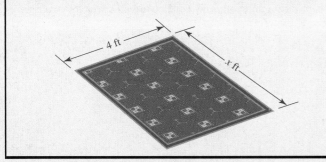

30. Match each word in Column I with *two* associated items in Column II.

Column I	*Column II*
a. or	**i.** $\cap$
	ii. $\cup$
b. and	**iii.** intersection
	iv. union

SECTION 4.3   Solving Absolute Value Equations and Inequalities

DEFINITIONS AND CONCEPTS	EXAMPLES
To **solve absolute value equations** of the form $\lvert X \rvert = k$, where $k > 0$, solve the equivalent **compound equation** $\qquad X = k \qquad$ or $\qquad X = -k$ If k is negative, then $\lvert X \rvert = k$ has no solution.	Solve: $\lvert 2x + 1 \rvert = 7$ This absolute value equation is equivalent to the following compound equation, which we can solve: $\quad 2x + 1 = 7 \quad$ or $\quad 2x + 1 = -7$ $\qquad\quad 2x = 6 \qquad\qquad\quad 2x = -8$ $\qquad\qquad x = 3 \qquad\qquad\qquad x = -4$ This equation has two solutions: 3 and -4. The solution set is $\{-4, 3\}$. Solve: $\lvert 4x - 5 \rvert = -3$ Since an absolute value can never be negative, there are no real numbers x that make $\lvert 4x - 5 \rvert = -3$ true. The equation has no solution and the solution set is $\varnothing$.
To **solve absolute value equations** of the form $\lvert X \rvert = \lvert Y \rvert$, solve the compound equation $\qquad X = Y \qquad$ or $\qquad X = -Y$	Solve: $\lvert 3x - 2 \rvert = \lvert 2x + 4 \rvert$ This equation is equivalent to the following compound equation, which we can solve: $\quad 3x - 2 = 2x + 4 \quad$ or $\quad 3x - 2 = -(2x + 4)$ $\qquad x - 2 = 4 \qquad\qquad\quad 3x - 2 = -2x - 4$ $\qquad\qquad x = 6 \qquad\qquad\qquad 5x - 2 = -4$ $\qquad\qquad\qquad\qquad\qquad\qquad\quad 5x = -2$ $\qquad\qquad\qquad\qquad\qquad\qquad\qquad x = -\dfrac{2}{5}$ This equation has two solutions: 6 and $-\dfrac{2}{5}$. The solution set is $\left\{-\dfrac{2}{5}, 6\right\}$.

SECTION 4.3   Solving Absolute Value Equations and Inequalities—*continued*

DEFINITIONS AND CONCEPTS	EXAMPLES
To **solve absolute value inequalities** of the form $\lvert X \rvert < k$, where $k > 0$, solve the equivalent double inequality $-k < X < k$. Use a similar approach to solve $\lvert X \rvert \le k$.	Solve: $\lvert 4x - 3 \rvert < 9$ This inequality is equivalent to the following double inequality which we can solve: $$-9 < 4x - 3 < 9$$ $$-6 < 4x < 12 \qquad \text{Add 3 to all three parts.}$$ $$-\frac{3}{2} < x < 3 \qquad \text{Divide each part by 4 and simplify.}$$ The solution set is: Interval notation: $\left(-\frac{3}{2}, 3\right)$
To **solve absolute value inequalities** of the form $\lvert X \rvert \ge k$, where $k > 0$, solve the equivalent compound inequality $X \le -k$ or $X \ge k$. Use a similar approach to solve $X > k$.	Solve: $\lvert 3x + 1 \rvert \ge 7$ This inequality is equivalent to the following compound inequality, which we can solve: $3x + 1 \le -7 \qquad \text{or} \qquad 3x + 1 \ge 7$ $3x \le -8 \qquad\qquad\qquad 3x \ge 6$ $x \le -\dfrac{8}{3} \qquad\qquad\qquad x \ge 2$ Interval notation: $\left(-\infty, -\frac{8}{3}\right] \cup [2, \infty)$ This is the union of two intervals.

REVIEW EXERCISES

Solve each absolute value equation.

31. $\lvert 4x \rvert = 8$

32. $2\lvert 3x + 1 \rvert - 1 = 19$

33. $\left\lvert \dfrac{3}{2}x - 4 \right\rvert - 10 = -1$

34. $\left\lvert \dfrac{2 - x}{3} \right\rvert = -4$

35. $\lvert -4(2x - 6) \rvert = 0$

36. $\left\lvert \dfrac{3}{8} + \dfrac{x}{3} \right\rvert = \dfrac{5}{12}$

37. $\lvert 3x + 2 \rvert = \lvert 2x - 3 \rvert$

38. $\left\lvert \dfrac{2(1 - x) + 1}{2} \right\rvert = \left\lvert \dfrac{3x - 2}{3} \right\rvert$

Solve each absolute value inequality. Graph the solution set and write it using interval notation.

39. $\lvert x \rvert \le 3$

40. $\lvert 2x + 7 \rvert < 3$

41. $2\lvert 5 - 3x \rvert \le 28$

42. $\left\lvert \dfrac{2}{3}x + 14 \right\rvert + 6 < 6$

43. $\lvert x \rvert > 1$

44. $\left\lvert \dfrac{1 - 5x}{3} \right\rvert \ge 7$

45. $\lvert 3x - 8 \rvert - 4 > 0$

46. $\left\lvert \dfrac{3}{2}x - 14 \right\rvert \ge 0$

47. Explain why $\lvert 0.04x - 8.8 \rvert < -2$ has no solution.

48. Explain why the solution set of $\left\lvert \dfrac{3x}{50} + \dfrac{1}{45} \right\rvert \ge -\dfrac{4}{5}$ is the set of all real numbers.

49. PRODUCE   Before packing, freshly picked tomatoes are weighed on the scale shown. Tomatoes having a weight w (in ounces) that falls within the highlighted range are sold to grocery stores.

   a. Complete the following absolute value inequality that expresses the acceptable weight range:

   $|w - \quad| \leq$

   b. Solve the inequality from part (a) and express the acceptable weight range using interval notation.

50. Let $f(x) = \frac{1}{3}|6x| - 1$. For what value(s) of x is $f(x) = 5$?

SECTION 4.4   Linear Inequalities In Two Variables

DEFINITIONS AND CONCEPTS

The graph of a **linear inequality in x and y** is the graph of all ordered pairs (x, y) whose coordinates satisfy the inequality.

To **graph a linear inequality in x and y,** graph the **boundary line.** Draw a solid boundary line if the inequality has $\leq$ or $\geq$. Draw a dashed line if the inequality has $<$ or $>$.

Then use a **test point** to decide which side of the boundary should be shaded. If the inequality is satisfied, shade the side that contains the test point. If the inequality is not satisfied, shade the other side of the boundary.

EXAMPLES

Graph: $y < 3x + 2$

To find the boundary line, we graph $y = 3x + 2$. Since the inequality symbol $<$ does not include an $=$ symbol, points on the boundary line are not included in the graph, and we draw a dashed boundary line.

To find which half-plane is the graph of $y < 3x + 2$, we choose the origin $(0, 0)$ as the test point and see whether its coordinates satisfy the inequality.

$y < 3x + 2$

$0 < 3(0) + 2$ Substitute 0 for x and 0 for y.

$0 < 2$ True

Since the coordinates of the origin satisfy the inequality, the origin is in the graph. In fact, the coordinates of every point on the same side of the boundary line as the origin satisfy the inequality. We then shade that half-plane to complete the graph.

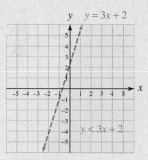

REVIEW EXERCISES

51. Determine whether $(-1, -4)$ is a solution of the linear inequality $6x - 4y \geq 15$.

52. Does the graph of $6x - 4y \geq 15$ include the boundary line?

Graph each inequality in the rectangular coordinate system.

53. $2x + 3y > 6$

54. $y \leq 4 - x$

55. $y < \frac{1}{2}x$

56. $x \geq -\frac{3}{2}$

57. CONCERT TICKETS
Tickets to a concert cost $6 for reserved seats and $4 for general admission. If receipts must be at least $10,200 to meet expenses, find an inequality that shows the possible combinations of the number of reserved seats (x) and the number of general admission tickets (y) that the box office can sell. Then graph the inequality for nonnegative values of x and y and give three ordered pairs that satisfy the inequality.

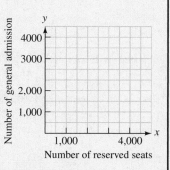

58. Find the equation of the boundary line. Then give the inequality whose graph is shown.

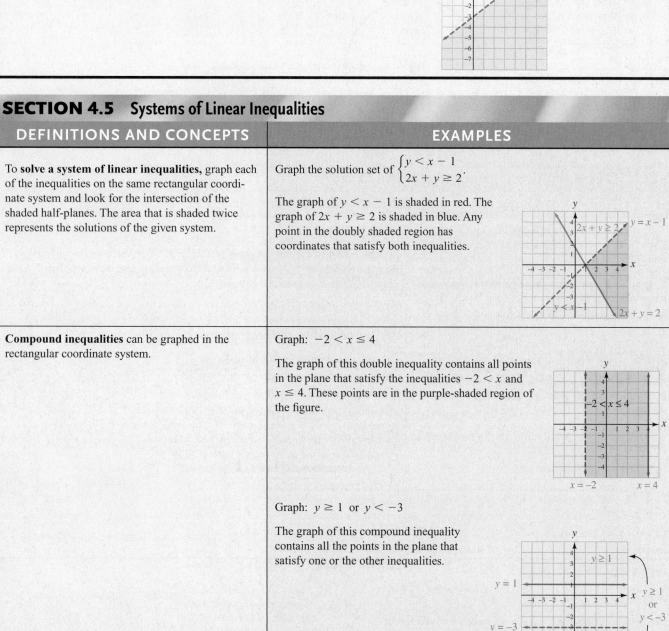

SECTION 4.5 Systems of Linear Inequalities

DEFINITIONS AND CONCEPTS	EXAMPLES
To **solve a system of linear inequalities,** graph each of the inequalities on the same rectangular coordinate system and look for the intersection of the shaded half-planes. The area that is shaded twice represents the solutions of the given system.	Graph the solution set of $\begin{cases} y < x - 1 \\ 2x + y \geq 2 \end{cases}$. The graph of $y < x - 1$ is shaded in red. The graph of $2x + y \geq 2$ is shaded in blue. Any point in the doubly shaded region has coordinates that satisfy both inequalities.
Compound inequalities can be graphed in the rectangular coordinate system.	Graph: $-2 < x \leq 4$ The graph of this double inequality contains all points in the plane that satisfy the inequalities $-2 < x$ and $x \leq 4$. These points are in the purple-shaded region of the figure. Graph: $y \geq 1$ or $y < -3$ The graph of this compound inequality contains all the points in the plane that satisfy one or the other inequalities.

REVIEW EXERCISES

59. Determine whether $(1, -2)$ is a solution of the system of linear
inequalities $\begin{cases} y \le -x + 1 \\ 2x - y > 2 \end{cases}$.

60. In the illustration, the
solution of one linear
inequality is shaded in red,
and the solution of a second
is shaded in blue. Determine
whether a true or false
statement results if the
coordinates of the given
point are substituted into the
given inequality.

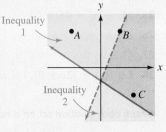

  a. *A*, inequality 1   **b.** *A*, inequality 2
  c. *B*, inequality 1   **d.** *B*, inequality 2
  e. *C*, inequality 1   **f.** *C*, inequality 2

Graph the solution set of each system of inequalities.

61. $\begin{cases} y \ge x + 1 \\ 3x + 2y < 6 \end{cases}$ **62.** $\begin{cases} x - y < 3 \\ y \le 0 \\ x \ge 0 \end{cases}$

Graph each compound inequality in the rectangular coordinate system.

63. $-2 < x < 4$ **64.** $y \le -2$ or $y > 1$

65. INVENTORY   A men's clothing store carries shirts that sell for
$20 and some that sell for $30. The store manager wants to stock at
least $300 but no more than $600 in shirts at the store. Graph a
system of inequalities that will show the possible combinations of
the number of $20 shirts ($x$) and the number of $30 shirts ($y$) that
the store can stock. Give two possible solutions.

66. PETROLEUM EXPLORATION   Organic matter converts to oil
and gas within a specific range of temperature and depth called the
petroleum window. The petroleum window shown can be described
by a system of linear inequalities, where x is the temperature in °C
of the soil at a depth of y meters. Determine what inequality
symbol should be inserted in each blank.

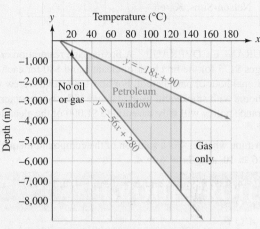

Based on data from *The Blue Planet* (Wiley, 1995)

$\begin{cases} x & 35 \\ x & 130 \\ y & -56x + 280 \\ y & -18x + 90 \end{cases}$

CHAPTER 4
Test

1. Fill in the blanks.

  a. $<$, $>$, $\le$, and $\ge$ are _____ symbols.

  b. ∞ is a symbol representing _____.

  c. $x + 1 > 2$ or $2x - 3 \le 8$ is a _____ inequality.

  d. We read $\cup$ as _____ and $\cap$ as _____.

  e. $\begin{cases} x + y > 10 \\ 3x - 2y \ge 4 \end{cases}$ is a _____ of linear inequalities in _____
    variables.

2. Determine whether -2 is a solution of the inequality:

  $3(x - 2) \le 2(x + 7)$

Solve each inequality. Graph the solution set and write it using interval notation and set-builder notation.

3. $\dfrac{2}{3}t - 1 > 7$

4. $-2(2x + 3) \ge 14$

5. $\dfrac{x}{4} - \dfrac{1}{3} > \dfrac{5}{6} + \dfrac{x}{3}$

6. $4 - 4[3t - 2(3 - t)] \le -15t - (5t - 28)$

Solve each compound inequality. Graph the solution set and write it using interval notation.

37. $3x + 2 < 8$ or $2x - 3 > 11$ [Section 4.2]

38. $5x - 3 \geq 2$ and $6 \geq 4x - 3$ [Section 4.2]

Solve each equation.

39. $2|4x - 3| + 1 = 19$ [Section 4.3]

40. $|2x - 1| = |3x + 4|$ [Section 4.3]

Solve each inequality. Graph the solution set and write it using interval notation.

41. $|3x - 2| \leq 4$ [Section 4.3]

42. $|2x + 3| - 1 > 4$ [Section 4.3]

Graph the solution set.

43. $2x - 3y \leq 12$ [Section 4.4]

44. $\begin{cases} y < x + 2 \\ 3x + y \leq 6 \end{cases}$ [Section 4.5]

CHAPTER 5

Exponents, Polynomials, and Polynomial Functions

© Leslie Harris/Photolibrary

5.1 Exponents

5.2 Scientific Notation

5.3 Polynomials and Polynomial Functions

5.4 Multiplying Polynomials

5.5 The Greatest Common Factor and Factoring by Grouping

5.6 Factoring Trinomials

5.7 The Difference of Two Squares; the Sum and Difference of Two Cubes

5.8 Summary of Factoring Techniques

5.9 Solving Equations by Factoring

CHAPTER SUMMARY AND REVIEW

CHAPTER TEST

Group Project

from **Campus to Careers**
Landscape Architect

Whether it's a community park, a college campus, or simply someone's backyard, landscape architects are skilled at creating outdoor areas that are both functional and beautiful. They use algebra and geometry to prepare working drawings, design scale models, and estimate costs. Throughout the planning and construction phases, they make computations to find everything from drainage slopes and sunlight angles to walkway elevations.

In **Problem 93** of **Study Set 5.9,** you will determine the dimensions of a concrete walkway around a fountain.

JOB TITLE:
Landscape Architect

EDUCATION:
A bachelor's degree in landscape architecture and some experience is required.

JOB OUTLOOK:
Excellent; it is expected to increase 18% to 26% through 2014.

ANNUAL EARNINGS:
The median salary in 2007 was $67,862.

FOR MORE INFORMATION:
www.bls.gov/oco/

Study Skills Workshop
Attending Class Regularly

It's not uncommon for students' enthusiasm to lessen toward the middle of the term. Sometimes, their effort and attendance begin to slip. Realize that missing even one class can have a great effect on your grade. Being tardy takes its toll as well. If you are just a few minutes late, or miss an entire class, you risk getting behind. So, keep the following tips in mind.

ARRIVE ON TIME, OR A LITTLE EARLY: When you arrive, get out your note-taking materials and homework. Identify any questions or comments that you plan to ask your instructor once the class starts.

IF YOU MUST MISS A CLASS: Get a set of notes, the homework assignments, and any handouts that the instructor may have provided for the day(s) that you missed.

STUDY THE MATERIAL YOU MISSED: Take advantage of the help that comes with this textbook, such as the CD and the videotapes. Watch the explanations of the material from the section(s) that you missed.

Now Try This

1. Plan ahead! List five possible situations that could cause you to be late to class or miss a class. (Some examples are: parking/traffic delays, lack of a babysitter, over-sleeping, and job responsibilities.) What can you do ahead of time so that these situations won't cause you to be tardy or absent?

2. Watch one section on the CD or the videotape series that accompanies this book. Take notes as you watch the explanations.

SECTION 5.1
Exponents

Objectives

1. Identify bases and exponents.
2. Use the product and power rules for exponents.
3. Use the zero and the negative integer exponent rules.
4. Use the quotient rule for exponents.
5. Simplify quotients raised to negative powers.

We have evaluated exponential expressions having natural-number exponents. In this section, we will extend the definition of exponent to include zero exponents, as in 3^0, and negative-integer exponents, as in 3^{-2}. We will also develop several rules that simplify work with exponents.

1 Identify Bases and Exponents.

The *exponential expression* x^n is called a *power of x*, and we read it as "*x* to the *n*th power." In this expression, *x* is called the *base,* and *n* is called the *exponent.*

Base $\longrightarrow x^n \longleftarrow$ Exponent

Exponents provide a way to write products of *repeated factors* in compact form.

Natural-Number Exponents	A natural-number exponent indicates how many times its base is to be used as a factor. For any number x and any natural number n,

$$\overbrace{x^n = x \cdot x \cdot x \cdot \cdots \cdot x}^{n \text{ factors of } x}$$

EXAMPLE 1   Identify the base and the exponent in each expression:

$$\textbf{a. } (5x)^3 \quad \textbf{b. } 5x^3 \quad \textbf{c. } -a^4 \quad \textbf{d. } \left(\frac{2b^8}{9c}\right)^4 \quad \textbf{e. } (x - 7)^2$$

Strategy   To identify the base and exponent, we will look for the form  .

Why   The exponent is the small raised number to the right of the base.

Solution

> **Notation**
> An exponent of 1 means the base is to be used as a factor 1 time. For example, $x^1 = x$.

a. When an exponent is written outside parentheses, the expression within the parentheses is the base. For $(5x)^3$, $5x$ is the base and 3 is the exponent: $(5x)^3 = (5x)(5x)(5x)$.

$$\underset{\underset{\text{Base}}{\uparrow}}{(5x)^3} \longleftarrow \text{Exponent}$$

b. $5x^3$ means $5 \cdot x^3$. Thus, x is the base and 3 is the exponent: $5x^3 = 5 \cdot x \cdot x \cdot x$.

c. $-a^4$ means $-1 \cdot a^4$. Thus, a is the base and 4 is the exponent: $-a^4 = -1(a \cdot a \cdot a \cdot a)$.

d. Because of the parentheses, $\left(\frac{2b^8}{9c}\right)$ is the base and 4 is the exponent:

$$\left(\frac{2b^8}{9c}\right)^4 = \left(\frac{2b^8}{9c}\right)\left(\frac{2b^8}{9c}\right)\left(\frac{2b^8}{9c}\right)\left(\frac{2b^8}{9c}\right)$$

e. Because of the parentheses, $x - 7$ is the base and 2 is the exponent:

$$(x - 7)^2 = (x - 7)(x - 7)$$

Self Check 1   Identify the base and the exponent in each expression:

$$\textbf{a. } (kt)^4 \qquad \qquad \textbf{b. } \pi r^2 \qquad \qquad \textbf{c. } -h^8$$

$$\textbf{d. } \left(\frac{3n}{2m^5}\right)^5 \qquad \qquad \textbf{e. } (y + 1)^3$$

Now Try   **Problems 15, 17, and 23**

---

②   Use the Product and Power Rules for Exponents.

Several rules for exponents come directly from the definition of exponent. To develop the first rule, we consider $x^5 \cdot x^3$, the product of two exponential expressions having the same base. Since x^5 means that x is to be used as a factor five times, and since x^3 means that x is to be used as a factor three times, $x^5 \cdot x^3$ means that x will be used as a factor eight times.

5 factors of x   3 factors of x   8 factors of x

$$x^5 x^3 = \overbrace{x \cdot x \cdot x \cdot x \cdot x}^{} \cdot \overbrace{x \cdot x \cdot x}^{} = \overbrace{x \cdot x \cdot x \cdot x \cdot x \cdot x \cdot x \cdot x}^{} = x^8$$

In general,

m factors of x   n factors of x   m + n factors of x

$$x^m x^n = \overbrace{x \cdot x \cdot x \cdot \ \cdots \ \cdot x}^{} \cdot \overbrace{x \cdot x \cdot x \cdot \ \cdots \ \cdot x}^{} = \overbrace{x \cdot x \cdot x \cdot x \cdot \ \cdots \ \cdot x}^{} = x^{m+n}$$

This result is called the *product rule for exponents.*

Product Rule for Exponents	To multiply exponential expressions with the same base, keep the common base and add the exponents.
	
	   For any real number x and any natural numbers m and n,
	
	   $$x^m \cdot x^n = x^{m+n}$$

EXAMPLE 2   Simplify each expression:

a. $x^{11}x^5$ **b.** $y^5 y^4 y$ **c.** $a^2 b^3 a^3 b^2$ **d.** $-8x^4(x^3)$

Strategy   Since there are products of the form $x^m \cdot x^n$ in each of these expressions, we will use the product rule for exponents and keep the common base and add the exponents to simplify them.

Why   We use the product rule to multiply exponential expressions with the same base.

Solution

The Language of Algebra
Here, the instruction *Simplify* means to write an equivalent expression where each base occurs only once.

a. $x^{11}x^5 = x^{11+5}$ Keep the common base x. Add the exponents.

$\qquad = x^{16}$

b. $y^5 y^4 y = (y^5 y^4)y$

$\qquad = y^9 y^1$

$\qquad = y^{10}$

c. $a^2 b^3 a^3 b^2 = a^2 a^3 b^3 b^2$ Reorder the factors.

$\qquad = a^5 b^5$

d. $-8x^4(x^3) = -8(x^4 x^3)$ Regroup the factors.

$\qquad = -8x^7$

Self Check 2   Simplify each expression:

a. $2^3 2^5$ **b.** $k \cdot k^4$

c. $a^2 b^3 a^3 b^4$ **d.** $-8a^4(a^2 b)$

Now Try   Problems 27, 29, and 33

Caution   Here are examples of two common errors associated with the product rule:

$$3^2 \cdot 3^4 \neq 9^6 \qquad\qquad\qquad 2^3 \cdot 5^2 \neq 10^5$$

Do not multiply the common bases. Keep   The power rule does not apply
the common base and add exponents to get 3^6.   because the bases are not the same.

To develop another rule for exponents, we consider $(x^4)^3$, which means x^4 cubed.

The Language of Algebra
An exponential expression raised to a power, such as $(x^4)^3$, is called a power of a power.

$$(x^4)^3 = x^4 \cdot x^4 \cdot x^4 = \overbrace{x \cdot x \cdot x \cdot x}^{x^4} \cdot \overbrace{x \cdot x \cdot x \cdot x}^{x^4} \cdot \overbrace{x \cdot x \cdot x \cdot x}^{x^4} = x^{12}$$

In general, we have

$$(x^m)^n = \overbrace{x^m \cdot x^m \cdot x^m \cdot \,\cdots\, \cdot x^m}^{n \text{ factors of } x^m} = \overbrace{x \cdot x \cdot x \cdot x \cdot x \cdot \,\cdots\, \cdot x}^{mn \text{ factors of } x} = x^{mn}$$

This result is called the *power rule for exponents.*

Power Rule for Exponents

To raise an exponential expression to a power, keep the base and multiply the exponents. For any real number x and any natural numbers m and n,

$$(x^m)^n = x^{m \cdot n} = x^{mn}$$

EXAMPLE 3 Simplify each expression:
  a. $(3^2)^3$ **b.** $(x^{11})^5$ **c.** $(x^2 x^3)^6$ **d.** $(x^2)^4 (x^3)^2$

Strategy Since there are powers of the form $(x^m)^n$ in each of these expressions, we will use the power rule for exponents and keep the base and multiply the exponents to simplify them.

Why We use the power rule to raise an exponential expression to a power.

The Language of Algebra
Here, the instruction *Simplify* means to write an equivalent expression such that:
• No powers are raised to powers
• No parentheses appear
• Each base occurs only once

Solution
a. $(3^2)^3 = 3^{2 \cdot 3}$ Keep the base and multiply the exponents.

$\quad\quad = 3^6$

$\quad\quad = 729$

b. $(x^{11})^5 = x^{11 \cdot 5}$

$\quad\quad = x^{55}$

c. $(x^2 x^3)^6 = (x^5)^6$ Within the parentheses, keep the common base and add the exponents.

$\quad\quad = x^{30}$ Keep the base and multiply the exponents.

d. $(x^2)^4 (x^3)^2 = x^8 x^6$

$\quad\quad = x^{14}$

▷ **Self Check 3** Simplify each expression:
  a. $(a^5)^8$ **b.** $(6^3)^5$ **c.** $(a^4 a^3)^3$ **d.** $(a^3)^3 (a^2)^3$

Now Try Problems 35, 39, and 41

To develop a third rule for exponents, we consider $(3x)^2$, which means $3x$ squared.

$$(3x)^2 = (3x)(3x) = 3 \cdot 3 \cdot x \cdot x = 3^2 x^2 = 9x^2$$

In general, we have

$$(xy)^n = \overbrace{(xy)(xy)(xy) \cdot \,\cdots\, \cdot (xy)}^{n \text{ factors of } xy} = \overbrace{xxx \cdot \,\cdots\, \cdot x}^{n \text{ factors of } x} \cdot \overbrace{yyy \cdot \,\cdots\, \cdot y}^{n \text{ factors of } y} = x^n y^n$$

To develop a fourth rule for exponents, we consider $\left(\frac{x}{3}\right)^3$, which means $\frac{x}{3}$ cubed.

$$\left(\frac{x}{3}\right)^3 = \frac{x}{3} \cdot \frac{x}{3} \cdot \frac{x}{3} = \frac{x \cdot x \cdot x}{3 \cdot 3 \cdot 3} = \frac{x^3}{3^3} = \frac{x^3}{27}$$

In general, we have

$$\left(\frac{x}{y}\right)^n = \overbrace{\left(\frac{x}{y}\right)\left(\frac{x}{y}\right)\left(\frac{x}{y}\right) \cdot \ \cdots \ \cdot \left(\frac{x}{y}\right)}^{n \text{ factors of } \frac{x}{y}} \qquad \text{where } y \neq 0$$

$$= \frac{\overbrace{xxx \cdot \ \cdots \ \cdot x}^{n \text{ factors of } x}}{\underbrace{yyy \cdot \ \cdots \ \cdot y}_{n \text{ factors of } y}} \qquad \text{Multiply the numerators and multiply the denominators.}$$

$$= \frac{x^n}{y^n}$$

> **Caution**
> In this section, as we work with many types of quotients, we will assume that no denominators are 0.

The previous results are called the *power of a product* and the *power of a quotient rules.*

> **Powers of a Product and a Quotient**
>
> To raise a product to a power, raise each factor of the product to that power. To raise a quotient to a power, raise the numerator and denominator to that power.
> For any real numbers x and y, and any natural number n,
>
> $$(xy)^n = x^n y^n \qquad \text{and} \qquad \left(\frac{x}{y}\right)^n = \frac{x^n}{y^n}, \qquad \text{where } y \neq 0$$

EXAMPLE 4   Simplify each expression:

  a. $(x^2 y)^3$ **b.** $(2y^4)^5$ **c.** $\left(\dfrac{x}{y^2}\right)^4$ **d.** $\left(\dfrac{6x^3}{5y^4}\right)^2$

Strategy   Since these expressions have the form $(xy)^n$ and $\left(\frac{x}{y}\right)^n$, we will use the power of a product rule and the power of a quotient rule to simplify them.

Why   We use the power of a product rule to raise a product to a power and the power of a quotient rule is used to raise a quotient to a power.

> **The Language of Algebra**
> Here, the instruction *Simplify* means to write an equivalent expression such that:
> • No powers are raised to powers
> • No parentheses appear
> • Each base occurs only once

Solution

a. $(x^2 y)^3 = (x^2)^3 y^3$ Raise each factor of the product x^2y to the 3rd power.

$\qquad\qquad\ \ = x^6 y^3$

b. $(2y^4)^5 = (2)^5(y^4)^5$

$\qquad\qquad\ = 32y^{20}$

c. $\left(\dfrac{x}{y^2}\right)^4 = \dfrac{x^4}{(y^2)^4}$ Raise the numerator and denominator to the 4th power.

$\qquad\qquad = \dfrac{x^4}{y^8}$

d. $\left(\dfrac{6x^3}{5y^4}\right)^2 = \dfrac{6^2(x^3)^2}{5^2(y^4)^2}$

$\qquad\qquad\quad = \dfrac{36x^6}{25y^8}$

▷ **Self Check 4**   Simplify each expression:

  a. $(a^4 b^5)^2$ **b.** $\left(\dfrac{-6a^5}{b^7}\right)^3$ **c.** $(-2d^5)^4$

Now Try   Problems 43 and 49

Use the Zero and Negative Integer Exponent Rules.

To develop the definition of a zero exponent, we consider the expression $x^0 \cdot x^n$, where x is not 0. By the product rule,

$$x^0 \cdot x^n = x^{0+n} = x^n = 1x^n$$

For the product rule to hold true for 0 exponents, $x^0 \cdot x^n$ must equal $\mathbf{1}x^n$. Comparing factors (as shown with the colored arrows), it follows that $x^0 = 1$. This result suggests the following definition.

Zero Exponents	A nonzero base raised to the 0 power is 1.
	For any nonzero base x,
	$$x^0 = 1$$

The Language of Algebra
Because 0^0 is undefined, it is called an *indeterminate form*.

For example, if no variables are equal to zero, then

$$3^0 = 1 \qquad (-7)^0 = 1 \qquad (3ax^3)^0 = 1 \qquad \left(\frac{1}{2}x^5y^7z^9\right)^0 = 1$$

EXAMPLE 5   Simplify each expression:
  a. $(5x)^0$ **b.** $5x^0$ **c.** $-(5cd)^0$ **d.** $5a^0b$

Strategy   Since there are factors of the form x^0 in each of these expressions, we will use the zero exponent rule to simplify them.

Why   Any nonzero base raised to the 0 power is 1.

The Language of Algebra
Here, the instruction *Simplify* means to write an equivalent expression such that no zero exponent appears.

Solution
a. $(5x)^0 = 1$ The base is 5x and the exponent is 0.
b. $5x^0 = 5 \cdot x^0 = 5 \cdot 1 = 5$ The base is x and the exponent is 0.
c. $-(5cd)^0 = -1$ The base is 5cd and the exponent is 0.
d. $5a^0b = 5 \cdot a^0 \cdot b = 5 \cdot 1 \cdot b = 5b$ The base is a and the exponent is 0.

Self Check 5   Simplify each expression: **a.** $2xy^0$ **b.** $-(xy)^0$
Now Try   Problems 51 and 55

To develop the definition of a negative integer exponent, we consider the expression $x^{-n} \cdot x^n$, where $x \neq 0$. By the product rule,

$$x^{-n} \cdot x^n = x^{-n+n} = x^0 = 1$$

Since the product is 1, x^{-n} and x^n must be reciprocals. It is also true that $\frac{1}{x^n}$ and x^n are reciprocals and their product is 1.

Comparing factors (as shown with the colored arrows), it follows that x^{-n} must equal $\frac{1}{x^n}$. This result suggests the following definition.

Negative Exponents

For any nonzero real number x and any integer n,

$$x^{-n} = \frac{1}{x^n}$$

In words, x^{-n} is the reciprocal of x^n.

From the definition, we see that another way to write x^{-n} is to write its reciprocal and change the sign of the exponent. For example,

Caution

A negative exponent does not indicate a negative number. It indicates a reciprocal.

$3^{-2} = \frac{1}{3^2}$ Write the reciprocal of 3^{-2}, which is $\frac{1}{3^{-2}}$. Then change the sign of the exponent.

$= \frac{1}{9}$

EXAMPLE 6 Simplify each expression. Write answers using positive exponents.
a. 4^{-3} **b.** $(-2)^{-5}$ **c.** $7m^{-8}$ **d.** $-n^{-4}$

Strategy Since there are factors of the form x^{-n} in each of these expressions, we will use the negative integer exponent rule to write equivalent expressions with positive exponents.

Why This rule enables us to rid these expressions of negative exponents by writing the reciprocal of the base and changing the sign of the exponent.

Solution

a. $4^{-3} = \frac{1}{4^3}$ Write the reciprocal of 4^{-3} and change the sign of the exponent.

$= \frac{1}{64}$ Evaluate: $4^3 = 64$.

b. $(-2)^{-5} = \frac{1}{(-2)^5}$

$= -\frac{1}{32}$

Caution

Don't confuse negative numbers with negative exponents. For example, the expressions -2 and 2^{-1} are not the same.

$2^{-1} = \frac{1}{2^1} = \frac{1}{2}$

c. $7m^{-8} = 7 \cdot m^{-8}$ Since there are no parentheses, the base is m.

$= 7 \cdot \frac{1}{m^8}$ Write the reciprocal of m^{-8} and change the sign of the exponent.

$= \frac{7}{m^8}$

d. $-n^{-4} = -1 \cdot n^{-4}$

$= -1 \cdot \frac{1}{n^4}$

$= -\frac{1}{n^4}$

Self Check 6 Simplify each expression. Write answers using positive exponents.
a. 8^{-2} **b.** $(-3)^{-3}$ **c.** $12h^{-9}$ **d.** $-c^{-1}$

Now Try Problems 59 and 63

The rules for exponents involving products and powers are also true for negative exponents.

EXAMPLE 7   Simplify each expression. Write answers using positive exponents.
  a. $x^{-5}x^3$ **b.** $(x^{-3})^{-2}$

Strategy   We will use the product rule and power rule for exponents to simplify these expressions.

Why   The first expression has the form $x^m \cdot x^n$ and the second has the form $(x^m)^n$.

Solution

a. $x^{-5}x^3 = x^{-5+3}$ Use the product rule: Keep the common base x and add the exponents.

   $= x^{-2}$

   $= \dfrac{1}{x^2}$

b. $(x^{-3})^{-2} = x^{(-3)(-2)}$ Use the power rule: Keep the base x and multiply the exponents.

   $= x^6$

Self Check 7   Simplify each expression. Write answers using positive exponents only.   **a.** $a^{-7}a^3$ **b.** $(a^{-5})^{-3}$

Now Try   **Problems 67 and 69**

Negative exponents can appear in the numerator and/or the denominator of a fraction. To develop rules to apply to such situations, consider the following example.

$$\frac{x^{-4}}{y^{-3}} = \frac{\dfrac{1}{x^4}}{\dfrac{1}{y^3}} = \frac{1}{x^4} \cdot \frac{y^3}{1} = \frac{y^3}{x^4}$$

We can obtain this result in a simpler way. Beginning with $\dfrac{x^{-4}}{y^{-3}}$, move x^{-4} to the denominator and change the sign of its exponent. Then move y^{-3} to the numerator and change the sign of its exponent.

This example illustrates the following rules.

Changing from Negative to Positive Exponents

A factor can be moved from the denominator to the numerator or from the numerator to the denominator of a fraction if the sign of its exponent is changed.
     For any nonzero real numbers x and y, and any integers m and n,

$$\frac{1}{x^{-n}} = x^n \qquad \text{and} \qquad \frac{x^{-m}}{y^{-n}} = \frac{y^n}{x^m}$$

These rules streamline the process when simplifying fractions involving negative exponents.

EXAMPLE 8   Simplify each expression. Write answers using positive exponents.

$$\textbf{a. } \frac{1}{c^{-10}} \qquad \textbf{b. } \frac{2^{-3}}{3^{-4}} \qquad \textbf{c. } -\frac{s^{-2}}{5t^{-9}}$$

Strategy   Since these expressions have the form $\frac{1}{x^{-n}}$ or $\frac{x^{-m}}{y^{-n}}$, we will use the rule for changing exponents from negative to positive to write equivalent expressions with positive exponents only.

Why   This rule enables us to rid these expressions of negative exponents by moving factors with negative exponents to the other side of the fraction bar and changing the sign of the exponent to positive.

Solution

a. $\dfrac{1}{c^{-10}} = c^{10}$ Move c^{-10} to the numerator and change the sign of the exponent.

b. $\dfrac{2^{-3}}{3^{-4}} = \dfrac{3^4}{2^3}$ Move 2^{-3} to the denominator and change the sign of the exponent. Move 3^{-4} to the numerator and change the sign of the exponent.

$= \dfrac{81}{8}$ Evaluate 3^4 and 2^3.

c. $-\dfrac{s^{-2}}{5t^{-9}} = -\dfrac{t^9}{5s^2}$ Move s^{-2} to the denominator and change the sign of the exponent. Since $5t^{-9}$ has no parentheses, t is the base. Move t^{-9} to the numerator and change the sign of the exponent.

> **Caution**
> This rule does not allow us to move terms that have negative exponents. For example,
> $$\frac{3^{-2} + 8}{5} \neq \frac{8}{3^2 \cdot 5}$$

Self Check 8   Simplify each expression. Write answers using positive exponents.

$$\textbf{a. } \frac{1}{t^{-9}} \qquad \textbf{b. } \frac{5^{-2}}{4^{-3}} \qquad \textbf{c. } -\frac{h^{-6}}{8r^{-7}}$$

Now Try   **Problems 71 and 77**

---

④   Use the Quotient Rule for Exponents.

To develop a rule for dividing exponential expressions, we proceed as follows:

$$\frac{x^m}{x^n} = x^m\left(\frac{1}{x^n}\right) = x^m x^{-n} = x^{m+(-n)} = x^{m-n}$$

This result is called the *quotient rule for exponents.*

Quotient Rule for Exponents	To divide exponential expressions with the same base, keep the common base and subtract the exponents. For any nonzero number x and any integers m and n, $$\frac{x^m}{x^n} = x^{m-n}$$

EXAMPLE 9 Simplify each expression. Write answers using positive exponents.

a. $\dfrac{a^5}{a^3}$ **b.** $\dfrac{2x^{-5}}{x^{11}}$

Strategy Since these expressions have the form $\frac{x^m}{x^n}$, we will use the quotient rule for exponents to simplify them.

Why We use the quotient rule to divide exponential expressions with the same base.

Solution

a. $\dfrac{a^5}{a^3} = a^{5-3}$ Keep the common base a. Subtract the exponents.

$= a^2$

b. $\dfrac{2x^{-5}}{x^{11}} = 2x^{-5-11}$

$= 2x^{-16}$

$= \dfrac{2}{x^{16}}$

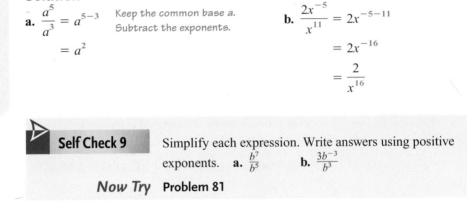

Self Check 9 Simplify each expression. Write answers using positive exponents. **a.** $\dfrac{b^7}{b^5}$ **b.** $\dfrac{3b^{-3}}{b^3}$

Now Try Problem 81

> **Success Tip**
>
> We can also simplify $\frac{2x^{-5}}{x^{11}}$ by moving x^{-5} to the denominator and changing the sign of the exponent.
>
> $\dfrac{2x^{-5}}{x^{11}} = \dfrac{2}{x^{11}x^5} = \dfrac{2}{x^{16}}$

EXAMPLE 10 Simplify each expression. Write answers using positive exponents.

a. $\dfrac{x^4x^3}{x^{-5}}$ **b.** $\dfrac{(x^2)^3}{(x^3)^2}$ **c.** $\dfrac{x^2y^3}{7xy^4}$ **d.** $\left(\dfrac{2a^{-2}b^3}{3a^5b^4}\right)^3$

Strategy To simplify these expressions, we must use more than one rule for exponents.

Why The expressions involve products, powers, and quotients of exponential expressions with the same base as well as negative exponents.

Solution

a. $\dfrac{x^4x^3}{x^{-5}} = \dfrac{x^7}{x^{-5}}$

$= x^{7-(-5)}$

$= x^{12}$

b. $\dfrac{(x^2)^3}{(x^3)^2} = \dfrac{x^6}{x^6}$

$= x^{6-6}$

$= x^0$

$= 1$

c. $\dfrac{x^2y^3}{7xy^4} = \dfrac{x^{2-1}y^{3-4}}{7}$

$= \dfrac{xy^{-1}}{7}$

$= \dfrac{x}{7y}$

d. $\left(\dfrac{2a^{-2}b^3}{3a^5b^4}\right)^3 = \left(\dfrac{2a^{-2-5}b^{3-4}}{3}\right)^3$

$= \left(\dfrac{2a^{-7}b^{-1}}{3}\right)^3$

$= \left(\dfrac{2}{3a^7b}\right)^3$

$= \dfrac{8}{27a^{21}b^3}$

> **Success Tip**
>
> When more than one rule for exponents is involved in a simplification, more than one approach can often be used. In Example 10a, we obtain the same result with this alternate approach:
>
> $\dfrac{x^4x^3}{x^{-5}} = x^4x^3x^5$
>
> $= x^{4+3+5}$
>
> $= x^{12}$

Simplify each expression. Write answers using positive
exponents.   **a.** $\dfrac{(a^{-2})^3}{(a^2)^{-3}}$ **b.** $\left(\dfrac{a^{-2}b^5}{5b^8}\right)^{-3}$

Now Try   Problem 85

5   **Simplify Quotients Raised to Negative Powers.**

To develop the final rule for exponents, we consider the following simplification:

The exponent is the opposite of -4.
$$\left(\frac{2}{3}\right)^{-4} = \frac{1}{\left(\frac{2}{3}\right)^4} = \frac{1}{\frac{2^4}{3^4}} = 1 \div \frac{2^4}{3^4} = 1 \cdot \frac{3^4}{2^4} = \frac{3^4}{2^4} = \left(\frac{3}{2}\right)^4$$

The base is the reciprocal of $\frac{2}{3}$.

This process can be streamlined using the following rule.

| **Negative Exponents and Reciprocals** | A fraction raised to a power is equal to the reciprocal of the fraction raised to the opposite power. |

For any nonzero real numbers x and y, and any integer n,
$$\left(\frac{x}{y}\right)^{-n} = \left(\frac{y}{x}\right)^n$$

EXAMPLE 11   Simplify each expression. Write answers using positive exponents.

   a. $\left(\dfrac{2}{3}\right)^{-4}$ **b.** $\left(\dfrac{y^2}{x^3}\right)^{-3}$ **c.** $\left(\dfrac{a^{-2}b^3}{a^2a^3b^4}\right)^{-3}$ **d.** $\left(\dfrac{2x^2}{3y^{-3}}\right)^{-4}$

Strategy   To simplify these expressions, we must use more than one rule for exponents.

Why   The expressions involve fractions to negative powers as well as products, powers, and quotients of exponential expressions with the same base.

Solution

The Language of Algebra
Here, the instruction *Simplify* means to write an equivalent expression such that:
- No powers are raised to powers
- No parentheses appear
- Each base occurs only once
- No zero or negative exponents appear

a. $\left(\dfrac{2}{3}\right)^{-4} = \left(\dfrac{3}{2}\right)^4$ Write the reciprocal of $\frac{2}{3}$ and change the exponent to 4.

$= \dfrac{3^4}{2^4}$

$= \dfrac{81}{16}$

b. $\left(\dfrac{y^2}{x^3}\right)^{-3} = \left(\dfrac{x^3}{y^2}\right)^3$

$= \dfrac{x^9}{y^6}$

c. $\left(\dfrac{a^{-2}b^3}{a^2a^3b^4}\right)^{-3} = \left(\dfrac{a^2a^3b^4}{a^{-2}b^3}\right)^3$ Write the reciprocal of the fraction and change the exponent to 3.

$= \left(\dfrac{a^5b^4}{a^{-2}b^3}\right)^3$

$= (a^{5-(-2)}b^{4-3})^3$

$= (a^7b)^3$

$= a^{21}b^3$

d. $\left(\dfrac{2x^2}{3y^{-3}}\right)^{-4} = \left(\dfrac{3y^{-3}}{2x^2}\right)^4$

$= \dfrac{3^4y^{-12}}{2^4x^8}$

$= \dfrac{81}{16x^8y^{12}}$

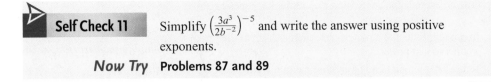

We summarize the rules for exponents as follows.

Summary of Exponent Rules

If m and n represent integers and there are no divisions by 0, then

Product rule
$$x^m \cdot x^n = x^{m+n}$$

Power rule
$$(x^m)^n = x^{mn}$$

Power of a product
$$(xy)^n = x^n y^n$$

Quotient rule
$$\frac{x^m}{x^n} = x^{m-n}$$

Power of a quotient
$$\left(\frac{x}{y}\right)^n = \frac{x^n}{y^n}$$

Exponents of 0 and 1
$$x^0 = 1 \text{ and } x^1 = x$$

Negative exponent
$$x^{-n} = \frac{1}{x^n}$$

Negative exponents appearing in fractions
$$\frac{1}{x^{-n}} = x^n \qquad \frac{x^{-m}}{y^{-n}} = \frac{y^n}{x^m} \qquad \left(\frac{x}{y}\right)^{-n} = \left(\frac{y}{x}\right)^n$$

ANSWERS TO SELF CHECKS 1. a. $kt; 4$ b. $r; 2$ c. $h; 8$ d. $\frac{3n}{2m^5}; 5$ e. $y+1; 3$ 2. a. $2^8 = 256$
b. k^5 c. $a^5 b^7$ d. $-8a^6 b$ 3. a. a^{40} b. 6^{15} c. a^{21} d. a^{15} 4. a. $a^8 b^{10}$ b. $-\frac{216a^{15}}{b^{21}}$ c. $16d^{20}$
5. a. $2x$ b. -1 6. a. $\frac{1}{64}$ b. $-\frac{1}{27}$ c. $\frac{12}{h^9}$ d. $-\frac{1}{c}$ 7. a. $\frac{1}{a^4}$ b. a^{15} 8. a. t^9 b. $\frac{64}{25}$ c. $-\frac{r^7}{8h^6}$
9. a. b^2 b. $\frac{3}{b^6}$ 10. a. 1 b. $125a^6 b^9$ 11. $\frac{32}{243a^{15}b^{10}}$

STUDY SET 5.1

VOCABULARY

Fill in the blanks.

1. Expressions such as x^4, 10^3, and $(5t)^2$ are called _____ expressions.
2. In the exponential expression x^n, the _____ is x and n is called the _____.
3. The expression x^4 represents a repeated multiplication where x is to be written as a _____ four times.
4. $3^4 \cdot 3^8$ is a _____ of exponential expressions with the same base, and $\frac{x^4}{x^2}$ is a _____ of exponential expressions with the same base.
5. $(h^3)^7$ is a _____ of an exponential expression.
6. In the expression 5^{-1}, the exponent is a _____ integer.

CONCEPTS

Complete the rules for exponents. Assume that there are no divisions by 0.

7. a. $x^m x^n =$ ___ b. $(x^m)^n =$ ___
   c. $(xy)^n =$ ___ d. $\left(\frac{x}{y}\right)^n =$ ___
   e. $x^0 =$ ___ f. $x^{-n} =$ ___
   g. $\frac{x^m}{x^n} =$ ___ h. $\left(\frac{x}{y}\right)^{-n} = \left(\frac{\ }{\ }\right)$ ___
   i. $\frac{x^{-m}}{y^{-n}} =$ ___

8. a. To multiply exponential expressions with the same base, keep the common base and _____ the exponents.
   b. To divide exponential expressions with the same base, keep the common base and _____ the exponents.

9. To raise an exponential expression to a power, keep the base and _____ the exponents.

10. a. To raise a product to a power, raise each _____ of the product to that power.

  b. To raise a quotient to a power, raise the numerator and the _____ to that power.

11. a. Any nonzero base raised to the 0 power is ___.

  b. Another way to write x^{-n} is to write its _____ and change the sign of the exponent.

12. a. A factor can be moved from the denominator to the numerator or from the numerator to the denominator of a fraction if the _____ of its exponent is changed.

  b. A fraction raised to a power is equal to the reciprocal of the fraction raised to the _____ power.

NOTATION

Complete each simplification.

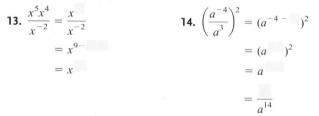

13. $\dfrac{x^5 x^4}{x^{-2}} = \dfrac{x}{x^{-2}}$

  $= x^{9-}$

  $= x$

14. $\left(\dfrac{a^{-4}}{a^3}\right)^2 = (a^{-4-})^2$

  $= (a^{})^2$

  $= a$

  $= \dfrac{}{a^{14}}$

GUIDED PRACTICE

Identify the base and the exponent. See Example 1.

15. $(6x)^3$

16. -7^2

17. $-x^5$

18. $(-t)^4$

19. $2b^6$

20. $12a^2$

21. $\left(\dfrac{n}{4}\right)^3$

22. $\left(\dfrac{3x}{y}\right)^0$

23. $(m-8)^6$

24. $(t+7)^8$

25. 10^0

26. -9^0

Simplify each expression. See Example 2.

27. $x^2 x^3$

28. $y^3 y^4$

29. $y^3 y^7 y^2$

30. $x^2 x^3 x^5$

31. $-6t^8(t^{15})$

32. $-9n^{20}(n^{14})$

33. $aba^3 b^4$

34. $x^2 y^3 x^3 y^2$

Simplify each expression. See Example 3.

35. $(2^3)^2$

36. $(4^2)^3$

37. $(x^4)^7$

38. $(y^7)^5$

39. $(r^8 r^3)^5$

40. $(w^5 w^2)^4$

41. $(g^4)^5 (g^2)^6$

42. $(s^5)^2 (s^3)^8$

Simplify each expression. See Example 4.

43. $(x^5 y)^4$

44. $(a^6 b)^5$

45. $(4m^7)^3$

46. $(9t^8)^2$

47. $\left(\dfrac{m^{10}}{n}\right)^8$

48. $\left(\dfrac{c}{d^9}\right)^4$

49. $\left(\dfrac{3a^{16}}{7n^{11}}\right)^2$

50. $\left(\dfrac{5t^8}{3m^3}\right)^3$

Simplify each expression. See Example 5.

51. 8^0

52. 1^0

53. $(-6t)^0$

54. $(-9s)^0$

55. $60h^0$

56. $75b^0$

57. $-3s^0 t$

58. $-9mn^0$

Simplify each expression. Write answers using positive exponents. See Example 6.

59. 5^{-2}

60. 3^{-4}

61. $(-3)^{-3}$

62. $(-4)^{-3}$

63. $8x^{-9}$

64. $12y^{-5}$

65. $-h^{-1}$

66. $-w^{-2}$

Simplify each expression. Write answers using positive exponents. See Example 7.

67. $m^{-4} \cdot m^{-6}$

68. $n^{-9} \cdot n^{-2}$

69. $(s^2)^{-3}$

70. $(-t^2)^{-5}$

Simplify each expression. Write answers using positive exponents. See Example 8.

71. $\dfrac{1}{a^{-5}}$

72. $\dfrac{3}{b^{-5}}$

73. $\dfrac{1}{7^{-2}}$

74. $\dfrac{1}{4^{-3}}$

75. $\dfrac{2^{-4}}{1^{-10}}$

76. $\dfrac{3^{-4}}{1^{-9}}$

77. $-\dfrac{t^{-6}}{12p^{-8}}$

78. $-\dfrac{x^{-9}}{20k^{-7}}$

Simplify each expression. Write answers using positive exponents. See Example 9.

79. $\dfrac{m^{15}}{m^3}$

80. $\dfrac{n^{18}}{n^6}$

81. $\dfrac{33y^{-2}}{y^{10}}$

82. $\dfrac{24k^{-4}}{k^8}$

Simplify each expression. Write answers using positive exponents. See Example 10.

83. $\dfrac{t^4 t^9}{t^{-1}}$

84. $\dfrac{r^6 r^2}{r^{-7}}$

85. $\dfrac{m^3 n^2}{6mn^{11}}$

86. $\dfrac{a^{11} b^7}{100a^{15} b}$

Simplify each expression. Write answers using positive exponents. See Example 11.

87. $\left(\dfrac{2}{3}\right)^{-2}$

88. $\left(\dfrac{4}{5}\right)^{-3}$

89. $\left(\dfrac{g^{20}}{t^{30}}\right)^{-4}$

90. $\left(\dfrac{y^8}{z^6}\right)^{-2}$

TRY IT YOURSELF

Simplify each expression. Write answers using positive exponents.

91. $(2a^2a^3b^0)^4$

92. $(3bb^2b^3c^0)^4$

93. $\left(\dfrac{4a^{-2}b}{3ab^{-3}}\right)^3$

94. $\left(\dfrac{2ab^{-3}}{3a^{-2}b^2}\right)^2$

95. $-\dfrac{8t^{-3}\cdot t^{-11}}{t^{-14}}$

96. $-\dfrac{4x^{-9}\cdot x^{-3}}{x^{-12}}$

97. $(-x^8)^2y^4x^3x^0$

98. $(-x^2)^5y^7y^3x^{-2}y^0$

99. $\left(\dfrac{x^{-5}}{x^2}\right)^{-4}\left(\dfrac{x^7}{x^{-8}}\right)^3$

100. $\left(\dfrac{a^{-3}}{a^6}\right)^{-3}\left(\dfrac{a^6}{a^{-2}}\right)^6$

101. $5^2r^{-5}(r^6)^3$

102. $8^2d^{-8}(d^9)^2$

103. $m^{-4}\cdot m^2\cdot m^{-8}$

104. $n^{-9}\cdot n^5\cdot n^{-7}$

105. $\left(\dfrac{4a^2b^3z^{-4}}{3a^{-2}b^{-1}z^3}\right)^{-3}$

106. $\left(\dfrac{-3pqr^{-4}}{2p^2q^{-3}r^2}\right)^{-2}$

107. $\dfrac{(4c^{-3}d)^0}{25(d+3)^0}$

108. $\dfrac{b^0}{2(a+b)^0}$

109. $\dfrac{(3x^2)^{-2}}{x^3x^{-4}x^0}$

110. $\dfrac{y^{-3}y^{-4}y^0}{(2y^{-2})^3}$

111. $\left(\dfrac{3(d^{-1})^{-5}}{8(d^{-4})^{-2}}\right)^{-2}$

112. $\left(\dfrac{(c^{-2})^{-4}}{15(c^{-3})^{-7}}\right)^{-1}$

113. $\dfrac{(-3cd^2)^3(c^{-1}d^{-3})^3}{(c^3d)^5}$

114. $\dfrac{(c^3t^{-4})^2(2c^4t^4)^{-2}}{(c^2t^5)^{-3}}$

Use a calculator to verify that each statement is true by showing that the values on either side of the equation are equal.

115. $(3.68)^0 = 1$

116. $(2.1)^4(2.1)^3 = (2.1)^7$

117. $\left(\dfrac{5.4}{2.7}\right)^{-4} = \left(\dfrac{2.7}{5.4}\right)^4$

118. $(7.23)^{-3} = \dfrac{1}{(7.23)^3}$

APPLICATIONS

119. LICENSE PLATES   The number of different Ohio license plates of the form three letters followed by four digits, as shown in the illustration, is $26\cdot 26\cdot 26\cdot 10\cdot 10\cdot 10\cdot 10$. Write this expression using exponents. Then evaluate it.

120. ASTRONOMY   See the illustration. The distance d, in miles, of the nth planet from the sun is given by the formula

$$d = 9{,}275{,}200[3(2^{n-2}) + 4]$$

Find the distance of Earth and the distance of Mercury from the sun.

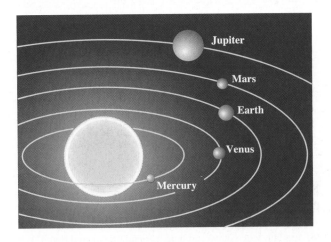

121. MICROSCOPES   The illustration shows the relative sizes of some chemical and biological structures, expressed as fractions of a meter (m). Beginning at the bottom of the scale, express each fraction shown in the illustration as a negative integer power of 10.

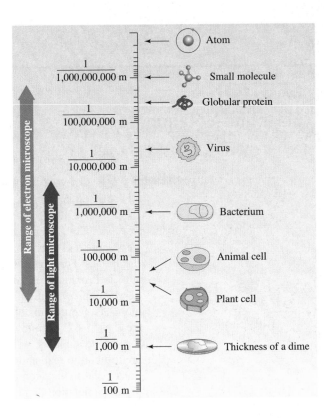

122. PHYSICS Albert Einstein's work in the area of special relativity resulted in the observation that the total energy E of a body is equal to its total mass m times the square of the speed of light c. This relationship is given by the famous equation $E = mc^2$. Identify the base and exponent on the right side.

123. GEOMETRY A cube is shown on the right.

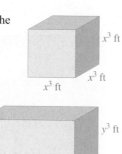

x^3 ft

x^3 ft

x^3 ft

   a. Find the area of its base.

   b. Find its volume.

124. GEOMETRY A rectangular solid is shown on the right.

y^3 ft

y^2 ft

y^4 ft

   a. Find the area of its base.

   b. Find its volume.

WRITING

125. Explain how you would help a friend understand that 3^{-2} is not equal to -9.

126. Explain how you would help a friend understand that 4^0 is not equal to 0.

127. Explain how an exponential expression with a negative exponent can be expressed as an equivalent expression with a positive exponent. Give an example.

128. Explain the error in the following solution.

Write $-8ab^{-3}$ using positive exponents only.

$$-8ab^{-3} = \frac{a}{8b^3}$$

REVIEW

Solve each inequality. Graph the solution set and write it using interval notation.

129. $-9x + 5 \geq 15$

130. $\frac{1}{4}p - \frac{1}{3} \leq p + 2$

CHALLENGE PROBLEMS

Evaluate each expression.

131. $(2^{-1} + 3^{-1} - 4^{-1})^{-1}$

132. $(3^{-1} + 4^{-1})^{-2}$

Simplify each expression. Assume there are no divisions by 0.

133. $\dfrac{8^{5a}(8^{6a})^5}{8^{-2a} \cdot 8^a \cdot 8^{4a}}$

134. $\left(\dfrac{(y^{5x})^2(y^{4x})^4}{(y^{2x} \cdot y^x)^{-3}} \right)^{-2}$

SECTION 5.2
Scientific Notation

Objectives

1 Write numbers in scientific notation.

2 Convert from scientific notation to standard notation.

3 Perform computations with scientific notation.

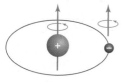

Hydrogen atom

Very large and very small numbers occur in science and other disciplines. For example, the star nearest Earth (excluding the sun) is Proxima Centauri, about 24,793,000,000,000 (read as "24 trillion, 793 billion") miles away. The mass of a hydrogen atom is approximately 0.00000000000000000000001673 (read as "1,673 octillionths") of a gram.

These numbers, written in **standard** or **decimal notation,** are difficult to read and cumbersome to work with in computations because they contain many zeros. In this section, we will discuss a notation that enables us to express such numbers in a more manageable form.

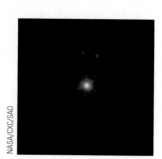

NASA/CXC/SAO

Write Numbers in Scientific Notation.

Scientific notation provides a compact way of writing very large or very small numbers.

Scientific Notation	A positive number is written in **scientific notation** when it is written in the form $N \times 10^n$, where $1 \leq N < 10$ and n is an integer.

Some examples of numbers written in scientific notation are

$$3.67 \times 10^6 \qquad 2.2 \times 10^{-4} \qquad 9.875 \times 10^{22}$$

Every positive number written in scientific notation is the product of a decimal number that is at least 1, but less than 10, and a power of 10.

> **Notation**
> A raised dot · is sometimes used when writing scientific notation.
> $$3.67 \times 10^6 = 3.67 \cdot 10^6$$

A decimal that is at
least 1, but less than 10   An integer exponent

$$\boxed{}. \boxed{} \times 10^{\boxed{}}$$

To write numbers in scientific notation, you need to be familiar with **powers of 10,** like those listed in the table below.

Power of 10	10^{-3}	10^{-2}	10^{-1}	10^0	10^1	10^2	10^3
Value	$\frac{1}{1,000} = 0.001$	$\frac{1}{100} = 0.01$	$\frac{1}{10} = 0.1$	1	10	100	1,000

EXAMPLE 1   Write each number in scientific notation:
a. 24,793,000,000,000   **b.** 0.0000000000000000000001673

Strategy   We will write each number as the product of a number between 1 and 10 and an integer power of 10.

Why   Numbers written in scientific notation have the form $N \times 10^n$ where $1 \leq N < 10$ and n is an integer.

Solution

a. The number 2.4793 is between 1 and 10. To get 24,793,000,000,000, the decimal point in 2.4793 must be moved 13 places to the *right*.

2.4,793,000,000,000.   Start with a decimal point (shown in red) to the right of the first
⌣⌣⌣⌣⌣⌣⌣⌣⌣↗   nonzero digit, which is 2.
13 places

We can move the red decimal point 13 places to the right by multiplying 2.4793 by 10^{13}.

24,793,000,000,000 = 2.4793×10^{13} This is the distance (in miles) from Proxima
Centauri to Earth.

b. The number 1.673 is between 1 and 10. To get 0.0000000000000000000001673, the decimal point in 1.673 must be moved 24 places to the *left*.

0.0000000000000000000000001.673   Start with a decimal point (shown in red) to
↖⌣⌣⌣⌣⌣⌣⌣⌣⌣⌣⌣⌣⌣⌣   the right of the first nonzero digit, which is 1.
24 places

We can move the red decimal point 24 places to the left by multiplying 1.673 by 10^{-24}.

$$0.000000000000000000000001673 = 1.673 \times 10^{-24}$$ This is the mass (in grams) of a hydrogen atom.

Self Check 1

Write each italicized number in scientific notation.
a. In 2006, the country earning the most money from tourism was the United States, *$85,700,000,000*.
b. DNA molecules contain and transmit the information that allows cells to reproduce. They are only *0.000000002* meter wide.

Now Try Problems 13 and 15

When a number is written in scientific notation, the first factor must be least 1, but less than 10.

EXAMPLE 2 Write each number in scientific notation:
  a. 47.2×10^3 **b.** 0.063×10^{-2}

Strategy To write 47.2×10^3 in scientific notation, we will write the first factor, 47.2, in scientific notation and then multiply the powers of 10. We will answer part (b) in a similar way.

Why This approach is necessary because neither 47.2×10^3 nor 0.063×10^{-2} are written in scientific notation—the first factors (47.2 and 0.063) are not between 1 and 10.

Solution

> **Notation**
>
> When writing numbers in scientific notation, keep the negative exponents. Don't apply the negative exponent rule.
>
> 6.3×10^{-4} $\cancel{6.3 \times \dfrac{1}{10^4}}$

a. $47.2 \times 10^3 = (4.72 \times 10^1) \times 10^3$ Write 47.2 in scientific notation.

$= 4.72 \times (10^1 \times 10^3)$ Group the powers of 10 together.

$= 4.72 \times 10^4$ Use the product rule for exponents: $10^1 \times 10^3 = 10^{1+3} = 10^4$.

b. $0.063 \times 10^{-2} = (6.3 \times 10^{-2}) \times 10^{-2}$ Write 0.063 in scientific notation.

$= 6.3 \times (10^{-2} \times 10^{-2})$

$= 6.3 \times 10^{-4}$

Self Check 2

Write each number in scientific notation:
  a. 17.3×10^2 **b.** 0.0045×10^{-3}

Now Try Problems 21 and 25

---

2 **Convert from Scientific Notation to Standard Notation.**

Each of the following numbers is written in scientific and standard notation. In each case, the exponent gives the number of places that the decimal point moves, and the sign of the exponent indicates the direction that it moves:

$$5.32 \times 10^4 = 5.3\,2\,0\,0.$$ $$6.45 \times 10^7 = 6.4\,5\,0\,0\,0\,0\,0.$$

       4 places to the right   7 places to the right

Success Tip

Since $10^0 = 1$, scientific notation involving 10^0 is easily simplified. For example,

$$4.8 \times 10^0 = 4.8 \times 1 = 4.8$$

$$2.37 \times 10^{-4} = 0.0\,0\,0\,2.3\,7$$
4 places to the left

$$9.234 \times 10^{-2} = 0.0\,9.2\,3\,4$$
2 places to the left

$$4.8 \times 10^0 = 4.8$$
No movement of the decimal point

These results suggest the following steps for changing a number written in scientific notation to standard notation.

Converting from Scientific to Standard Notation

1. If the exponent is positive, move the decimal point the same number of places to the right as the exponent.

2. If the exponent is negative, move the decimal point the same number of places to the left as the absolute value of the exponent.

EXAMPLE 3 Convert to standard notation: **a.** 8.706×10^5 **b.** 1.1×10^{-3}

Strategy In each case, we need to identify the exponent on the power of 10 and consider its sign.

Why The exponent gives the number of decimal places that we should move the decimal point. The sign of the exponent indicates whether it should be moved to the right or the left.

Solution

a. Since the exponent in 10^5 is 5, the red decimal point moves 5 places to the right. (Multiplication by 10^5, which is 100,000, moves the decimal point 5 places to the right.)

$$8.706 \times 10^5 = 8.7\,0\,6\,0\,0. = 870,600$$
5 places to the right

b. Since the exponent in 10^{-3} is -3, the red decimal point moves 3 places to the left. (Multiplication by 10^{-3}, which is 0.001, moves the decimal point 3 places to the left.)

$$1.1 \times 10^{-3} = 0.0\,0\,1\,.\,1 = 0.0011$$
3 places to the left

Self Check 3 Convert each number written in scientific notation to standard notation. **a.** In 2006, the world's forest areas were estimated to occupy 9.5×10^9 acres. **b.** The average distance between molecules of air in a room is 3.937×10^{-7} inch.

Now Try Problems 33 and 39

The results from the previous examples suggest the following forms to use when converting numbers from standard to scientific notation.

For real numbers between 0 and 1: $\square \times 10^{\text{negative integer}}$

For real numbers at least 1, but less than 10: $\square \times 10^0$

For real numbers greater than or equal to 10: $\square \times 10^{\text{positive integer}}$

Perform Computations with Scientific Notation.

Scientific notation is useful when multiplying and dividing very large or very small numbers.

EXAMPLE 4   ***Astronomy.***   The galaxy in which we live is called the *Milky Way*. It has a diameter of approximately 100,000 light years. (A light year is the distance light travels in a vacuum in one year: 9.46×10^{15} meters.) Find the diameter of the Milky Way in meters.

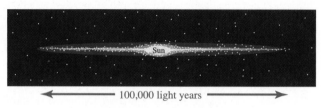

100,000 light years

A cross-sectional representation of the Milky Way Galaxy

Strategy   To find the diameter of the Milky Way, we will convert the number of light years, 100,000, to scientific notation and multiply it by the number of meters per light year, 9.46×10^{15}.

Why   When the numbers to be multiplied (100,000 and 9.46×10^{15}) are written in scientific notation, we can use the product rule for exponents to simplify the computation.

Solution   After writing 100,000 in scientific notation as 1.0×10^5, we perform the arithmetic on the decimals and the exponential expressions separately.

$\mathbf{100{,}000(9.46 \times 10^{15})}$

$= (\mathbf{1.0 \times 10^5})(9.46 \times 10^{15})$ Multiply the number of light years by the number of meters per light year.

$= (1.0 \cdot 9.46) \times (10^5 \cdot 10^{15})$ Use the commutative and associative properties of multiplication to group the decimal factors together and the powers of 10 together.

$= 9.46 \times 10^{5+15}$ Perform the multiplication: $1.0 \cdot 9.46 = 9.46$. For the powers of 10, keep the base and add the exponents.

$= 9.46 \times 10^{20}$

The Milky Way Galaxy is about 9.46×10^{20} meters in diameter.

▷ **Self Check 4**   A light year is 5.88×10^{12} miles. Find the diameter of the Milky Way in miles.

     Now Try   **Problems 47 and 49**

EXAMPLE 5   ***World Oil Reserves/Demand.***   According to estimates in the *Oil and Gas Journal,* there were 1,290,000,000,000 (read as "1 trillion, 290 billion") barrels of crude oil reserves in the ground at the start of 2006. At that time, world demand was 30,800,000,000 (read as "30 billion, 800 million") barrels per year according to the Energy Information Administration. If annual demand as of 2006 remains the same and if no new oil discoveries are made, when will the world's oil supply run out?

Strategy   To find the number of years of crude oil that remains, we will convert the number of barrels in reserves to scientific notation and divide it by the annual number of barrels of demand, also written is scientific notation.

Why   When the numbers to be divided are written in scientific notation, we can use the quotient rule for exponents to simplify the computation.

Solution   First, we write 1,290,000,000,000 in scientific notation as 1.29×10^{12} and 30,800,000,000 as 3.08×10^{10}. Then we perform the arithmetic on the decimals and the exponential expressions separately.

$$\frac{1,290,000,000,000}{30,800,000,000} = \frac{1.29 \times 10^{12}}{3.08 \times 10^{10}} \qquad \text{Divide the number of barrels in reserve by the number of barrels used each year.}$$

$$= \frac{1.29}{3.08} \times \frac{10^{12}}{10^{10}} \qquad \text{Separate the factors to divide the decimals and divide the powers of 10.}$$

$$\approx 0.42 \times 10^{12-10} \qquad \text{Perform the division: } \frac{1.29}{3.08} \approx 0.42. \text{ For the powers of 10, keep the base and subtract the exponents.}$$

$$\approx 0.42 \times 10^{2}$$

$$\approx 42 \qquad \text{Write } 0.42 \times 10^{2} \text{ in standard notation by moving the decimal point 2 places to the right.}$$

According to industry estimates, as of 2006, there were 42 years of crude oil reserves left if the demand remained constant. Under those conditions, the world's crude oil supply will run out in the year 2048.

▷ *Now Try*   **Problems 55 and 59**

EXAMPLE 6   Use scientific notation to evaluate: $\dfrac{(0.00000064)(24,000,000,000)}{(400,000,000)(0.0000000012)}$

Strategy   After writing each number in scientific notation, we will perform the arithmetic on the decimals and the exponential expressions separately.

Why   When the numbers to be multiplied and divided are written in scientific notation, we can use the product and quotient rules for exponents to simplify the computation.

Solution

$$\frac{(0.00000064)(24,000,000,000)}{(400,000,000)(0.0000000012)} = \frac{(6.4 \times 10^{-7})(2.4 \times 10^{10})}{(4.0 \times 10^{8})(1.2 \times 10^{-9})} \qquad \text{Convert each number to scientific notation.}$$

$$= \frac{(6.4)(2.4)}{(4)(1.2)} \times \frac{10^{-7}10^{10}}{10^{8}10^{-9}} \qquad \text{Separate the decimal factors and the power of 10 factors.}$$

$$= \frac{15.36}{4.8} \times \frac{10^{3}}{10^{-1}} \qquad \text{Simplify: } 10^{-7}10^{10} = 10^{-7+10} = 10^{3} \text{ and } 10^{8}10^{-9} = 10^{8+(-9)} = 10^{-1}.$$

$$= 3.2 \times 10^{3-(-1)} \qquad \text{Use the quotient rule for exponents.}$$

$$= 3.2 \times 10^{4}$$

The result is 3.2×10^{4}. In standard notation, this is 32,000.

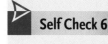

Self Check 6   Use scientific notation to evaluate: $\dfrac{(320)(25,000)}{0.00004}$

Now Try   **Problem 69**

Using Your Calculator

Using Scientific Notation

Scientific and graphing calculators often give answers in scientific notation. For example, if we use a calculator to find 301.2^8, the display will read

> 6.77391496^{19} *On a scientific calculator*

> 301.2 ∧ 8
>   6.773914961E19   *On a graphing calculator*

In either case, the answer is given in scientific notation and means

$6.77391496 \times 10^{19}$

Numbers can be entered into a calculator in scientific notation. For example, to enter 24,000,000,000 (which is 2.4×10^{10} in scientific notation), we enter these numbers and press these keys:

2.4 $\boxed{\text{EXP}}$ 10   *On some scientific calculators*

2.4 $\boxed{\text{EE}}$ 10   *On a graphing calculator and on most scientific calculators*

To use a scientific calculator to evaluate

$$\frac{(24,000,000,000)(0.00000006495)}{0.00000004824}$$

we enter each number in scientific notation, because each number has too many digits to be entered directly. In scientific notation, the three numbers are

$$2.4 \times 10^{10} \qquad 6.495 \times 10^{-8} \qquad 4.824 \times 10^{-8}$$

Using a scientific calculator, we enter these numbers and press these keys:

2.4 $\boxed{\text{EE}}$ 10 $\boxed{\times}$ 6.495 $\boxed{\text{EE}}$ 8 $\boxed{+/-}$ $\boxed{\div}$ 4.824 $\boxed{\text{EE}}$ 8 $\boxed{+/-}$ $\boxed{=}$

The display will read $\boxed{3.231343284\ ^{10}}$. In standard notation, the answer is 32,313,432,840. The steps are similar on a graphing calculator.

ANSWERS TO SELF CHECKS   **1. a.** 8.57×10^{10} **b.** 2.0×10^{-9} **2. a.** 1.73×10^{3} **b.** 4.5×10^{-6}
3. a. 9,500,000,000 **b.** 0.0000003937 **4.** 5.88×10^{17} mi **6.** $2.0 \times 10^{11} = 200,000,000,000$

**STUDY SET
5.2**

VOCABULARY

Fill in the blanks.

1. 7.4×10^6 is written in _____ notation and 7,400,000 is written in _____ notation.
2. 10^{-3}, 10^0, 10^1, and 10^4 are _____ of 10.

CONCEPTS

Fill in the blanks.

3. A positive number is written in scientific notation when it is written in the form $N \times$ _____, where $1 \le N < 10$ and *n* is an _____.
4. Insert > or <: 5.3×10^2 ____ 5.3×10^{-2}
5. To change 6.31×10^{-4} to standard notation, we move the decimal point four places to the _____.
6. To change 9.7×10^3 to standard notation, we move the decimal point three places to the _____.

NOTATION

7. **a.** Explain why the number 60.22×10^{22} is not written in scientific notation.
   b. Explain why the number 0.6022×10^{24} is not written in scientific notation.
8. Determine what type of exponent must be used when writing each of the three categories of real numbers in scientific notation.
   a. For real numbers between 0 and 1:
   ■ $\times 10$
   b. For numbers at least 1, but less than 10:
   ■ $\times 10$
   c. For real numbers greater than or equal to 10:
   ■ $\times 10$

GUIDED PRACTICE

Write each number in scientific notation. See Example 1.

9. 3,900
10. 1,700
11. 0.0078
12. 0.068
13. 173,000,000,000,000
14. 89,800,000,000
15. 0.0000096
16. 0.000000046
17. 0.00000000203
18. 0.0000000000301
19. 50,160,000,000,000,000
20. 220,000,000,000,000,000

Write each number in scientific notation. See Example 2.

21. 23.65×10^6
22. 75.6×10^5
23. 90.09×10^{-11}
24. 20.08×10^{-13}
25. 0.0317×10^{-2}
26. 0.0012×10^{-3}
27. 0.0527×10^5
28. 0.0298×10^3
29. 323×10^5
30. 689×10^9
31. $6,000 \times 10^{-7}$
32. 765×10^{-5}

Write each number in standard notation. See Example 3.

33. 2.7×10^4
34. 7.2×10^3
35. 3.23×10^{-3}
36. 6.48×10^{-2}
37. 7.96×10^8
38. 9.67×10^9
39. 3.5×10^{-7}
40. 4.12×10^{-10}
41. 5.23×10^0
42. 8.67×10^0
43. 8.0×10^{13}
44. 4.0×10^{14}

Multiply. Give all answers in scientific notation. See Example 4.

45. $(1.3 \times 10^4)(2.0 \times 10^5)$
46. $(3.0 \times 10^8)(2.2 \times 10^3)$
47. $(7.9 \times 10^5)(2.3 \times 10^6)$
48. $(6.1 \times 10^8)(3.9 \times 10^5)$
49. $(9.1 \times 10^{-5})(5.5 \times 10^{12})$
50. $(8.4 \times 10^{-13})(4.8 \times 10^9)$
51. $(9.0 \times 10^{-1})(8.0 \times 10^{-6})$
52. $(8.1 \times 10^{-4})(2.4 \times 10^{-15})$

Divide. Give all answers in scientific notation. See Example 5.

53. $\dfrac{8.6 \times 10^{15}}{2.0 \times 10^6}$
54. $\dfrac{9.6 \times 10^{20}}{3.0 \times 10^{10}}$
55. $\dfrac{2.193 \times 10^{32}}{4.3 \times 10^{20}}$
56. $\dfrac{1.107 \times 10^{16}}{4.1 \times 10^2}$
57. $\dfrac{2.686 \times 10^{10}}{7.9 \times 10^{-7}}$
58. $\dfrac{4.216 \times 10^{31}}{6.8 \times 10^{-14}}$
59. $\dfrac{4.2 \times 10^{-12}}{8.4 \times 10^{-5}}$
60. $\dfrac{1.21 \times 10^{-17}}{1.1 \times 10^{-2}}$

Write each number in scientific notation and perform the operations. Give all answers in scientific notation and in standard notation. See Example 6.

61. $\dfrac{4,500,000,000,000}{0.0002}$

62. $\dfrac{6{,}150{,}000{,}000}{0.003}$

63. $\dfrac{0.00000128}{0.0004}$

64. $\dfrac{0.000000000117}{0.00039}$

65. $(89{,}000{,}000{,}000)(4{,}500{,}000{,}000)$

66. $(0.000000061)(3{,}500{,}000{,}000)$

67. $\dfrac{(640{,}000)(2{,}700{,}000)}{120{,}000}$

68. $\dfrac{(220{,}000)(0.000009)}{0.00033}$

69. $\dfrac{(15{,}000{,}000)(7{,}000{,}000{,}000)}{25{,}000{,}000}$

70. $\dfrac{(4{,}900{,}000)(2{,}700)}{63{,}000}$

71. $\dfrac{(0.0000000039)(0.00095)}{(0.0195)(4{,}000)}$

72. $\dfrac{(0.00024)(96{,}000{,}000)}{(640{,}000{,}000)(0.025)}$

APPLICATIONS

73. FIVE-CARD POKER   The odds against being dealt the hand shown in the illustration are about 2.6×10^6 to 1. Express the odds using standard notation.

74. ENERGY   See the illustration. Express each of the following using scientific notation. (1 quadrillion is 10^{15}.)

   a. U.S. energy consumption

   b. U.S. energy production

   c. The difference in 2006 consumption and production

2006 U.S. Energy Consumption and Production
(petroleum, natural gas, coal, hydroelectric, nuclear, geothermal, solar, wind)

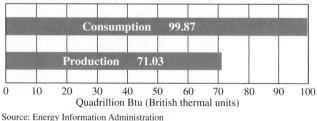

Source: Energy Information Administration

75. ATOMS   A hydrogen atom is so small that a single drop of water contains more than a million million billion hydrogen atoms. Express this number in scientific notation.

76. ASTRONOMY   The American Physical Society recently honored first-year graduate student Gwen Bell for coming up with what it considers the most accurate estimate of the mass of the Milky Way. In pounds, her estimate is a 3 with 42 zeros after it. Express this number in scientific notation.

77. NATIONAL DEBT   As of July 2007, the U.S. national debt was approximately \$8,969,400,000,000. The estimated population of the United States at that time was approximately 302,000,000. Express each person's share of the debt in scientific and standard notation.

78. WARP SPEED   In the series *Star Trek,* the *U.S.S. Enterprise* traveled at warp speeds. To convert a warp speed, W, to an equivalent velocity in miles per second, v, we can use the equation

$$v = W^3 c$$

where c is the speed of light, 1.86×10^5 miles per second. Find the velocity of a spacecraft traveling at warp 2.

79. ATOMS   A simple model of a helium atom is shown. If a proton has a mass of 1.7×10^{-24} grams, and if the mass of an electron is only about $\frac{1}{2{,}000}$ that of a proton, find the mass of an electron.

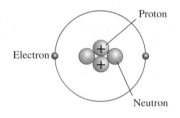

80. OCEANS   The mass of the Earth's oceans is only about $\frac{1}{4{,}400}$ that of the Earth. If the mass of the Earth is 6.578×10^{21} tons, find the mass of the oceans.

81. LIGHT YEAR   Light travels about 300,000,000 meters per second. A *light year* is the distance that light can travel in one year. Estimate the number of meters in one light year.

82. AQUARIUMS   Express the volume of the fish tank in scientific notation.

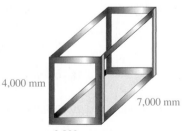

4,000 mm

7,000 mm

3,000 mm

83. THE BIG DIPPER   One star in the Big Dipper is named Merak. It is approximately 4.65×10^{14} miles from Earth.

a. If light travels about 1.86×10^5 miles/sec, how many seconds does it take light emitted from Merak to reach the Earth?   (*Hint:* Use the formula $t = \frac{d}{r}$.)

Merak

b. Convert your result from part (a) to years.

84. BIOLOGY   A paramecium is a single-celled organism that propels itself with hair-like projections called *cilia*. Use the scale in the illustration below to estimate the length of the paramecium. Express the result in scientific and in standard notation.

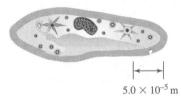

5.0×10^{-5} m

85. COMETS   On March 23, 1997, Comet Hale-Bopp made its closest approach to Earth, coming within 1.3 *astronomical units*. One astronomical unit (AU) is the distance from the Earth to the sun—about 9.3×10^7 miles. Express this distance in miles, using scientific notation.

86. DIAMONDS   The approximate number of atoms of carbon in a $\frac{1}{2}$-carat diamond is given by the following expression. Find the number of carbon atoms in scientific and in standard notation.

$$\frac{6.0 \times 10^{23}}{1.2 \times 10^2}$$

WRITING

87. Explain how to change a number from standard notation to scientific notation.

88. Explain how to change a number from scientific notation to standard notation.

89. Explain why 9.99×10^n represents a number less than 1 but greater than 0 if n is a negative integer.

90. Explain the advantages of writing very large and very small numbers in scientific notation.

91. a. To multiply a number by 10^3, we move the decimal point. Which way, and how far?

   b. To multiply a number by 10^{-3}, we move the decimal point. Which way, and how far?

92. Explain why 437.9×10^{23} is not written in scientific notation.

REVIEW

Solve each compound inequality. Graph the solution set and write it using interval notation.

93. $4x \geq -x + 5$ and $6 \geq 4x - 3$

94. $15 > 2x - 7 > 9$

95. $3x + 2 < 8$ or $2x - 3 > 11$

96. $-4(x + 2) \geq 12$ or $3x + 8 < 11$

CHALLENGE PROBLEMS

97. What is the reciprocal of the opposite of 2.5×10^{-24}? Write the result in scientific notation.

98. Solve: $(1.1 \times 10^{-16})x - (1.2 \times 10^{10}) = (6.5 \times 10^{10})$ and write the solution in scientific notation.

SECTION 5.3
Polynomials and Polynomial Functions

Objectives

1 Define and classify polynomials.

2 Evaluate polynomial functions.

3 Graph polynomial functions.

4 Simplify polynomials by combining like terms.

5 Add polynomials.

6 Subtract polynomials.

In arithmetic, we add, subtract, multiply, divide, and find powers of real numbers. In algebra, we perform these operations on algebraic expressions called *polynomials*.

1 **Define and Classify Polynomials.**

Recall from Chapter 1 that a **term** is a product or quotient of numbers and/or variables. A single number or variable is also a term. Examples of terms are:

$$4, \quad y, \quad 6r, \quad -w^3, \quad 3.7x^5, \quad \text{and} \quad -15ab^2$$

If a term contains only a number, such as 4, it is called a **constant term,** or simply a **constant.**

The **numerical coefficient,** or simply the **coefficient,** is the numerical factor of a term. For example, the coefficient of $6r$ is 6 and the coefficient of $-15ab^2$ is -15. The coefficient of a constant term is that constant.

Polynomials	A **polynomial** is a single term or the sum of terms in which all variables have whole-number exponents. No variable appears in a denominator.

Here are some examples of polynomials:

$$5x + 3, \quad 4n^2 - 6n - 8, \quad p^3 + 3p^2q + 3pq^2 + q^3, \quad \text{and} \quad -\frac{5}{2}rs^2t^4$$

The Language of Algebra
The prefix *poly* means many. A *poly*gon is a many-sided figure and *poly*unsaturated fats are molecules having many strong chemical bonds.

The polynomial $5x + 3$ is the sum of two terms, $5x$ and 3, and we say it is a **polynomial in one variable, x.** Since $4n^2 - 6n - 8$ can be written as the sum $4n^2 + (-6n) + (-8)$, it has three terms, $4n^2$, $-6n$, and -8. It is written in **descending powers of n,** because the exponents on n decrease from left to right. When a polynomial is written in descending powers, the first term, in this case $4n^2$, is called the **leading term.** The coefficient of the leading term, in this case 4, is called the **leading coefficient.**

A polynomial can have more than one variable. For example, $p^3 + 3p^2q + 3pq^2 + q^3$ is a **polynomial in two variables,** p and q. It has four terms and is written in descending powers of p and **ascending powers** of q. Its leading term is p^3 and its leading coefficient is 1.

The polynomial $-\frac{5}{2}rs^2t^4$ is a **polynomial in three variables,** r, s, and t. It has one term.

Caution The following expressions are not polynomials:

$$\frac{2x}{x^2 + 1}, \quad y^{1/2} - 8, \quad \text{and} \quad c^{-3} + 2c + 24$$

The first expression is a quotient with a variable in the denominator. The last two contain variables with exponents that are not whole numbers.

The Language of Algebra
The prefix *mono* means one; Jay Leno begins the *Tonight Show* with a monologue. The prefix *bi* means two, as in bilingual or bifocals. The prefix *tri* means three, as in triangle or the *Star Wars Trilogy*.

Polynomials can be classified according to their number of terms. A polynomial with one term is called a **monomial,** a polynomial with two terms is called a **binomial,** and a polynomial with three terms is called a **trinomial.** Polynomials with four or more terms have no special names.

Polynomials		
Monomials	*Binomials*	*Trinomials*
$2x^3$	$2x + 5$	$2x^2 + 4x + 3$
a^2b	$-\dfrac{17}{2}x^4 - \dfrac{3}{5}x$	$3mn^3 - m^2n^3 + 7n$
$3x^3y^5z^2$	$3.2x^{13}y^5z^3 + 4.7x^3yz$	$-12x^5y^2 + 13x^4y^3 - 7x^3y^3$

Polynomials and their terms can be classified according to the exponents on their variables.

Degree of a Term of a Polynomial	The **degree of a term** of a polynomial in one variable is the value of the exponent on the variable. If a polynomial is in more than one variable, the **degree of a term** is the sum of the exponents on the variables in that term. The **degree of a nonzero constant** is 0. The constant 0 has no defined degree.

Here are some examples:

$15x^4$ has degree **4**.

$-2.9y^2$ has degree **2**.

$\frac{7}{8}m^8 n^6$ has degree **14** because $8 + 6 = 14$.

6 has degree **0** since it can be written as $6x^0$.

We determine the *degree of a polynomial* by considering the degrees of each of its terms.

Degree of a Polynomial	The **degree of a polynomial** is the same as the highest degree of any term of the polynomial.

EXAMPLE 1   Use the vocabulary of this section to describe each polynomial:

a. $x^4 - 2x^2 + 4$ **b.** $\frac{1}{5}y^3 - y$

c. $-2.7m^{12} - 4.5m^{10}n^4 + 9.1m^8 n^6 + mn^9$

Strategy   First, we will identify the variable(s) in each polynomial and determine whether it is written in ascending or descending powers. Then we will count its number of terms and determine the degree of each term.

Why   The number of terms determines the type of polynomial. The highest degree of any term of the polynomial determines its degree.

Solution

a. $x^4 - 2x^2 + 4$ is a polynomial in one variable that is written in descending powers of x. Because it has three terms, x^4, $-2x^2$, and 4, it is a trinomial. Because the highest degree of any of its terms is 4, it is of degree 4. We can list the characteristics of the polynomial in a table.

Term	Coefficient	Degree
x^4	1	4
$-2x^2$	-2	2
4	4	0

Degree of the polynomial: **4**
Type of polynomial: Trinomial

b. $\frac{1}{5}y^3 - y$ is a polynomial in one variable. It is written in descending powers of y. Since it has two terms, $\frac{1}{5}y^3$ and $-y$, it is a binomial. The highest degree of any of its terms is 3, so it is of degree 3.

Term	Coefficient	Degree
$\frac{1}{5}y^3$	$\frac{1}{5}$	3
$-y$	-1	1

Degree of the polynomial: **3**
Type of polynomial: Binomial

The Language of Algebra
The word *degree* is also used in other disciplines for classification. For example, doctors speak of second- and third-*degree* burns.

c. $-2.7m^{12} - 4.5m^{10}n^4 + 9.1m^8n^6 + mn^9$ is a polynomial in two variables, m and n. It is written in descending powers of m and ascending powers of n. It has four terms, and therefore, has no special name. The highest degree of any term is 14, so it is of degree 14.

Term	Coefficient	Degree
$-2.7m^{12}$	-2.7	12
$-4.5m^{10}n^4$	-4.5	14
$9.1m^8n^6$	9.1	14
mn^9	1	10

Degree of the polynomial: **14**
Type of polynomial: No special name

Self Check 1 Use the vocabulary of this section to describe each polynomial:
a. $10t^2 + 6t - 8$ **b.** $-5.6p^6q + p^5q^3$

Now Try **Problems 17 and 19**

❷ Evaluate Polynomial Functions.

We have seen that linear functions are defined by equations of the form $f(x) = mx + b$. Some examples of linear functions are

$$f(x) = 3x + 1, \qquad g(x) = -\frac{1}{2}x - 1, \qquad \text{and} \qquad h(x) = 5x$$

In each case, the right side of the equation is a polynomial. For this reason, linear functions are members of a larger class of functions known as *polynomial functions.*

Polynomial Functions

A **polynomial function** is a function whose equation is defined by a polynomial in one variable.

Another example of a polynomial function is $f(x) = x^2 + 6x - 8$. This is a second-degree polynomial function, called a **quadratic function.** Quadratic functions are of the form $f(x) = ax^2 + bx + c$, where $a \neq 0$.

An example of a third-degree polynomial function is $f(x) = x^3 - 3x^2 - 9x + 2$. Third-degree polynomial functions, also called **cubic functions,** are of the form $f(x) = ax^3 + bx^2 + cx + d$, where $a \neq 0$.

Polynomial functions can be used to model many real-life situations. If we are given a polynomial function model, we can learn more about the situation by evaluating the function at specific values.

EXAMPLE 2 *Rocketry.* If a toy rocket is shot straight up with an initial velocity of 128 feet per second, its height, in feet, t seconds after being launched is given by the function

$$h(t) = -16t^2 + 128t$$

Find the height of the rocket:

a. 2 seconds after being launched.

b. 7.9 seconds after being launched.

Strategy   We will find $h(2)$ and $h(7.9)$.

Why   The notation $h(2)$ represents the height of the rocket 2 seconds after being launched and $h(7.9)$ represents the height of the rocket 7.9 seconds after being launched.

Solution

a. To find the height of the rocket 2 seconds after being launched, we find $h(2)$ as follows:

$$h(t) = -16t^2 + 128t \quad \text{This is the given function.}$$
$$h(2) = -16(2)^2 + 128(2) \quad \text{Substitute 2 for each } t. \text{ (The input is 2.)}$$
$$= -16(4) + 256 \quad \text{Evaluate the right side.}$$
$$= -64 + 256$$
$$= 192 \quad \text{The output is 192.}$$

We have found that $h(2) = 192$. Thus, 2 seconds after it is launched, the height of the rocket is 192 feet.

b. To find the height of the rocket 7.9 seconds after it is launched, we find $h(7.9)$ as follows:

$$h(t) = -16t^2 + 128t \quad \text{This is the given function.}$$
$$h(7.9) = -16(7.9)^2 + 128(7.9) \quad \text{Substitute 7.9 for each } t. \text{ (The input is 7.9.)}$$
$$= -16(62.41) + 1{,}011.2 \quad \text{Evaluate the right side.}$$
$$= -998.56 + 1{,}011.2$$
$$= 12.64 \quad \text{The output is 12.64.}$$

At 7.9 seconds, the height of the rocket is 12.64 feet. It has almost fallen back to Earth.

Self Check 2   Find the height of the rocket 4 seconds after it is launched.

Now Try   **Problems 23 and 83**

---

3   **Graph Polynomial Functions.**

The graphs of three basic polynomial functions are shown below. The domain and range of the functions are expressed in interval notation.

> **The Language of Algebra**
> $f(x) = x$ is called the *identity function* because it assigns each real number to itself. Note that the graph passes through $(-2, -2)$, $(0, 0)$, $(1, 1)$, and so on.

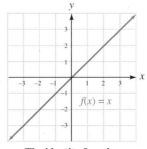

The identity function
The domain is $(-\infty, \infty)$.
The range is $(-\infty, \infty)$.

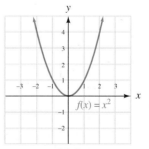

The squaring function
The domain is $(-\infty, \infty)$.
The range is $[0, \infty)$.

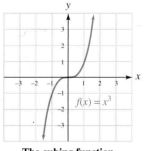

The cubing function
The domain is $(-\infty, \infty)$.
The range is $(-\infty, \infty)$.

When graphing a linear function, we need to plot only two points, because the graph is a straight line. The graphs of polynomial functions of degree greater than 1 are smooth, continuous curves. To graph them, we must plot more than two points.

> **EXAMPLE 3**   Graph $f(x) = x^3 - 3x^2 - 9x + 2$ and find its domain and range.
>
> **Strategy**   We will graph the function by creating a table of function values and plotting the corresponding ordered pairs.
>
> **Why**   After drawing a smooth curve though the plotted points, we will have the graph.
>
> **Solution**   To graph this cubic function, we begin by evaluating it for $x = -3$.
>
> $$f(x) = x^3 - 3x^2 - 9x + 2$$
> $$f(-3) = (-3)^3 - 3(-3)^2 - 9(-3) + 2 \quad \text{Substitute } -3 \text{ for each } x.$$
> $$= -27 - 3(9) - 9(-3) + 2$$
> $$= -27 - 27 + 27 + 2$$
> $$= -25$$
>
> In the following table, we enter the ordered pair $(-3, -25)$. We continue the function evaluation process for $x = -2, -1, 0, 1, 2, 3, 4,$ and 5, and list the results in the table. After plotting the ordered pairs, we draw a smooth curve through the points to get the graph of function f.

Success Tip

When constructing a table of function values, select positive and negative integer values for x, as well as 0.

Success Tip

The graphs of many polynomial functions of degree 3 and higher have "peaks" and "valleys" as shown here.

$$f(x) = x^3 - 3x^2 - 9x + 2$$

x	$f(x)$	
-3	-25	$(-3, -25)$
-2	0	$(-2, 0)$
-1	7	$(-1, 7)$
0	2	$(0, 2)$
1	-9	$(1, -9)$
2	-20	$(2, -20)$
3	-25	$(3, -25)$
4	-18	$(4, -18)$
5	7	$(5, 7)$

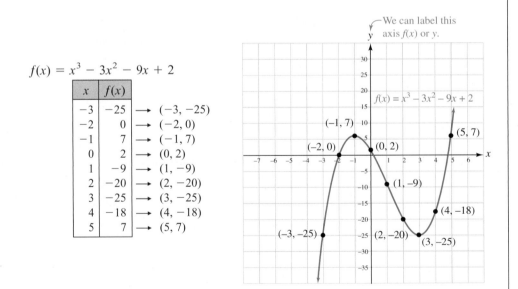

To find the domain of $f(x) = x^3 - 3x^2 - 9x + 2$, we project its graph onto the x-axis as shown in figure (a) on the next page. Because the graph extends indefinitely to the left and right, the projection includes all real numbers. Therefore, the domain of the function is the set of real numbers, which can written in interval notation as $(-\infty, \infty)$.

To determine the range of the same polynomial function, we project the graph onto the y-axis, as shown in figure (b) on the next page. Because the graph of the function extends indefinitely upward and downward, the projection includes all real numbers. Therefore the range of the function is the set of real numbers, written $(-\infty, \infty)$.

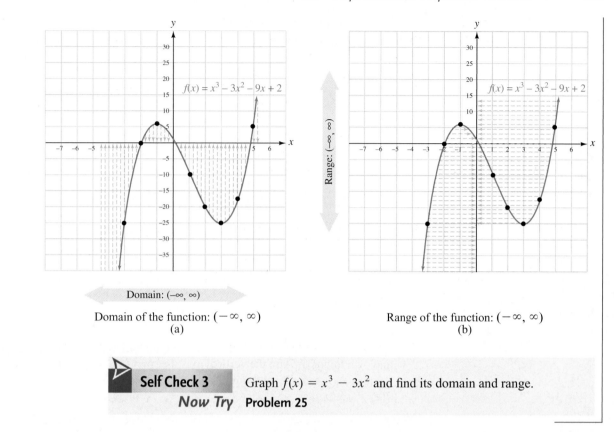

Domain: $(-\infty, \infty)$

Domain of the function: $(-\infty, \infty)$
(a)

Range of the function: $(-\infty, \infty)$
(b)

Self Check 3   Graph $f(x) = x^3 - 3x^2$ and find its domain and range.

Now Try   **Problem 25**

Using Your Calculator   ***Graphing Polynomial Functions***

We can graph polynomial functions with a graphing calculator. For example, to graph $f(x) = x^3 - 3x^2 - 9x + 2$ from Example 3, we enter the right side of the function notation as shown in figure (a). Using window settings of $[-8, 8]$ for x and $[-50, 50]$ for y, we get the graph shown in figure (b).

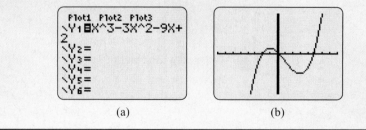

(a)   (b)

EXAMPLE 4   ***Labor Statistics.***   The number of manufacturing jobs in the United States, in millions, is approximated by the polynomial function

$$J(x) = 0.00118x^4 - 0.03867x^3 + 0.37551x^2 - 1.16766x + 17.80937$$

where x is the number of years after 1990. Use the graph of the function in figure (a) on the next page to answer the following questions.

a. Find $J(5)$. Explain what the result means.

b. Find the value of x for which $J(x) = 15.5$. Explain what the result means.

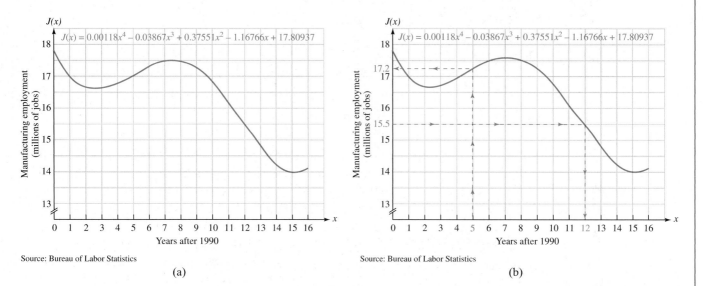

(a)   (b)

Strategy   For part (a), we will begin at 5 on the horizontal axis of the graph and scan up and over to find $J(5)$ on the vertical axis. For part (b), we will begin at 15.5 on the vertical axis and scan over and down to find the value of x on the horizontal axis for which $J(x) = 15.5$.

Why   The horizontal axis is associated with values of x (inputs) and the vertical axis is associated with the function values (outputs).

Solution

a. Refer to figure (b). To find $J(5)$, we use the dashed red lines to determine that $J(5) \approx 17.2$. This means 5 years after 1990, or in 1995, there were approximately 17.2 million manufacturing jobs in the United States.

b. Refer again to figure (b). To find the input value x that is assigned the output value 15.5, we used the dashed blue lines to determine that $J(12) \approx 15.5$. This means 12 years after 1990, or in 2002, there were approximately 15.5 million manufacturing jobs in the United States.

 Now Try   Problem 91

---

④   **Simplify Polynomials by Combining Like Terms.**

Recall that **like terms** have the same variables with the same exponents:

Like terms	*Unlike terms*	
$-7x$ and $15x$	$-7x$ and $15a$	Different variables.
$4y^3$ and $16y^3$	$4y^3$ and $16y^2$	Different exponents on the same variable.
$\dfrac{1}{2}xy^2$ and $-\dfrac{1}{3}xy^2$	$\dfrac{1}{2}xy^2$ and $-\dfrac{1}{3}x^2y$	Different exponents on different variables.

Also recall that to **combine like terms,** we combine their coefficients and keep the same variables with the same exponents. For example,

$$4y + 5y = (4 + 5)y \qquad\qquad 8x^2 - x^2 = (8 - 1)x^2$$
$$= 9y \qquad\qquad\qquad\qquad = 7x^2$$

Polynomials with like terms can be simplified by combining like terms.

EXAMPLE 5   Simplify each polynomial by combining like terms.

    a. $4x^4 + 81x^4$ **b.** $17x^2y^2 + 2x^2y - 6x^2y^2$

    c. $r - 3r^2 - 4r^2 + 8r^2$ **d.** $\dfrac{3}{5}ab + \dfrac{4}{3}a - 7 + \dfrac{1}{2}ab - \dfrac{1}{6}a + 4$

Strategy   We will use the distributive property in reverse to add (or subtract) the coefficients of the like terms. We will keep the same variables raised to the same powers.

Why   To *combine like terms* means to add or subtract the like terms in an expression.

Solution

a. $4x^4 + 81x^4 = 85x^4$ Think: $(4 + 81)x^4 = 85x^4$.

b. The first and third terms are like terms.

$$17x^2y^2 + 2x^2y - 6x^2y^2 = 11x^2y^2 + 2x^2y \quad \text{Think: } (17 - 6)x^2y^2 = 11x^2y^2.$$

> **Caution**
> When combining like terms, the exponents on the variables *stay the same*. Don't incorrectly add the exponents.

c. The last three terms are like terms.

$$r - 3r^2 - 4r^2 + 8r^2 = r + r^2 \quad \text{Think: } (-3 - 4 + 8)r^2 = 1r^2 = r^2.$$
$$= r^2 + r \quad \text{Write the result in descending powers of } r.$$

d. The first and fourth terms are like terms, the second and fifth terms are like terms, and the third and sixth terms are like terms.

$$\dfrac{3}{5}ab + \dfrac{4}{3}a - 7 + \dfrac{1}{2}ab - \dfrac{1}{6}a + 4$$

$$= \left(\dfrac{3}{5} + \dfrac{1}{2}\right)ab + \left(\dfrac{4}{3} - \dfrac{1}{6}\right)a - 7 + 4 \qquad \text{Combine like terms.}$$

$$= \left(\dfrac{6}{10} + \dfrac{5}{10}\right)ab + \left(\dfrac{8}{6} - \dfrac{1}{6}\right)a - 7 + 4 \qquad \begin{array}{l}\text{To add and subtract the fractions,}\\\text{build equivalent fractions: } \frac{3}{5}\cdot\frac{2}{2} = \frac{6}{10},\\ \frac{1}{2}\cdot\frac{5}{5} = \frac{5}{10}, \text{ and } \frac{4}{3}\cdot\frac{2}{2} = \frac{8}{6}.\end{array}$$

$$= \dfrac{11}{10}ab + \dfrac{7}{6}a - 3 \qquad \text{Do the additions and the subtraction.}$$

Caution   Do not try to clear this expression of fractions by multiplying it by the LCD 30. That strategy works only when we multiply *both sides of an equation* by the LCD.

$$\cancel{30}\left(\dfrac{3}{5}ab + \dfrac{4}{3}a - 7 + \dfrac{1}{2}ab - \dfrac{1}{6}a + 4\right)$$

▷ **Self Check 5**   Simplify each polynomial by combining like terms.

    a. $6m^4 + 3m^4$ **b.** $17s^3t + 3s^2t - 6s^3t$

    c. $x - 19x^2 + 22x^2 - x^2$

    d. $\dfrac{7}{8}rs + \dfrac{7}{9}r + 1 - \dfrac{3}{4}rs + \dfrac{4}{3}r - 2$

Now Try   Problems 31 and 37

5 **Add Polynomials.**

When adding polynomials horizontally, each polynomial is usually enclosed within parentheses. For example, $(3x^2 - 2x + 4) + (2x^2 + 4x - 3)$ is the sum of two trinomials.

Adding Polynomials	To add polynomials, drop the parentheses and combine their like terms.

EXAMPLE 6 Add: **a.** $(3x^2 - 2x + 4) + (2x^2 + 4x - 3)$

b. $(-5x^3y^2 - 4x^2y^3) + (2x^3y^2 + x^3y + 5x^2y^3)$

c. $\left(\dfrac{1}{4}m^4 + \dfrac{1}{2}m^3\right) + \left(\dfrac{3}{4}m^4 - \dfrac{7}{3}m^3\right)$

Strategy We will drop the parentheses and combine like terms.

Why To add polynomials means to combine their like terms.

Solution

a. $(3x^2 - 2x + 4) + (2x^2 + 4x - 3)$

$= 3x^2 - 2x + 4 + 2x^2 + 4x - 3$ Drop the parentheses.

$= 5x^2 + 2x + 1$ Combine like terms.

> **Notation**
> When performing operations on polynomials, it is standard practice to write the terms of a result in descending powers of one variable.

b. $(-5x^3y^2 - 4x^2y^3) + (2x^3y^2 + x^3y + 5x^2y^3)$ This is the sum of a binomial and a trinomial.

$= -5x^3y^2 - 4x^2y^3 + 2x^3y^2 + x^3y + 5x^2y^3$ Drop the parentheses.

$= -3x^3y^2 + x^3y + x^2y^3$ Combine like terms.

c. $\left(\dfrac{1}{4}m^4 + \dfrac{1}{2}m^3\right) + \left(\dfrac{3}{4}m^4 - \dfrac{7}{3}m^3\right)$

$= \dfrac{1}{4}m^4 + \dfrac{1}{2}m^3 + \dfrac{3}{4}m^4 - \dfrac{7}{3}m^3$ Drop the parentheses.

$= \dfrac{1}{4}m^4 + \dfrac{3}{6}m^3 + \dfrac{3}{4}m^4 - \dfrac{14}{6}m^3$ To add and subtract the fractions, build equivalent fractions $\frac{1}{2} \cdot \frac{3}{3} = \frac{3}{6}$ and $\frac{7}{3} \cdot \frac{2}{2} = \frac{14}{6}$.

$= \dfrac{4}{4}m^4 - \dfrac{11}{6}m^3$ Combine like terms.

$= m^4 - \dfrac{11}{6}m^3$ Simplify: $\frac{4}{4}m^4 = 1m^4 = m^4$.

Self Check 6 Add: **a.** $(2a^2 - 3a + 5) + (5a^2 + 4a - 2)$

b. $(-6a^2b^3 - 5a^3b^2) + (3a^2b^3 + 2a^3b^2 + ab^2)$

c. $\left(\frac{5}{8}t^5 - \frac{1}{6}t^4\right) + \left(\frac{3}{8}t^5 + \frac{7}{4}t^4\right)$

Now Try **Problems 39, 41, and 43**

The first two additions in Example 6 can be done by aligning the terms vertically and combining like terms column by column.

$$
\begin{array}{r}
3x^2 - 2x + 4 \\
+\ 2x^2 + 4x - 3 \\
\hline
5x^2 + 2x + 1
\end{array}
\qquad
\begin{array}{r}
-5x^3y^2 \qquad\quad -\ 4x^2y^3 \\
+\ 2x^3y^2 + x^3y + 5x^2y^3 \\
\hline
-3x^3y^2 + x^3y + x^2y^3
\end{array}
$$

 Subtract Polynomials.

Recall from Chapter 1 that we can use the distributive property to find the opposite of several terms enclosed within parentheses. For example, we consider $-(2x^3 - 3x^2)$.

$$-(2x^3 - 3x^2) = -1(2x^3 - 3x^2) \quad \text{Replace the } - \text{ symbol in front of the parentheses with } -1.$$
$$= -2x^3 + 3x^2 \quad \text{Use the distributive property to remove parentheses.}$$

This result illustrates that we can remove a $-$ sign preceding parentheses by dropping the $-$ sign and the parentheses and *changing the sign of every term within the parentheses*. This observation suggests a way to subtract polynomials.

Subtracting Polynomials	To subtract two polynomials, change the signs of the terms of the polynomial being subtracted, drop the parentheses, and combine like terms.

EXAMPLE 7 Subtract: **a.** $(8x^3 + 2x^2) - (2x^3 - 3x^2)$
  b. $(3.1rt^2 + 4.3r^2t^2) - (8.7rt^2 - 4.3r^2t^2 + 5.9r^3t^2)$

Strategy In each case, we will change the signs of the terms of the polynomial being subtracted, drop the parentheses, and combine like terms.

Why We can remove a $-$ sign preceding parentheses by dropping the $-$ sign and the parentheses and changing the sign of every term within the parentheses.

Solution

a. $(8x^3 + 2x^2) - (2x^3 - 3x^2)$ This is the difference of two binomials.

$\quad = 8x^3 + 2x^2 - 2x^3 + 3x^2$ Change the sign of each term of $2x^3 - 3x^2$ and drop the parentheses.

$\quad = 6x^3 + 5x^2$ Combine like terms.

b. $(3.1rt^2 + 4.3r^2t^2) - (8.7rt^2 - 4.3r^2t^2 + 5.9r^3t^2)$ This is the difference of a binomial and a trinomial.

$\quad = 3.1rt^2 + 4.3r^2t^2 - 8.7rt^2 + 4.3r^2t^2 - 5.9r^3t^2$ Change the signs of the terms of the polynomial being subtracted.

$\quad = -5.6rt^2 + 8.6r^2t^2 - 5.9r^3t^2$ Combine like terms.

$\quad = -5.9r^3t^2 + 8.6r^2t^2 - 5.6rt^2$ Write the terms of the result in descending powers of r.

Self Check 7 Subtract: **a.** $(9m^4 + 16m^2) - (12m^4 - 18m^2)$
  b. $(6.4a^2b^3 - 2.7a^2b^2) - (-2.5a^2b^3 + 8.1a^2b^2)$

Now Try **Problems 47 and 51**

Just as real numbers have opposites, so do polynomials. To find the opposite of a polynomial, multiply each of its terms by -1. This changes the sign of each term of the polynomial.

A polynomial		*Its opposite*
$2x^2 - 4x + 5$	$\xrightarrow{\text{Multiply by } -1}$	$-(2x^2 - 4x + 5)$ or $-2x^2 + 4x - 5$

To subtract polynomials in vertical form, we add the opposite of the polynomial that is being subtracted.

$$\begin{array}{r} 8x^3 + 2x^2 \\ - \underline{(2x^3 - 3x^2)} \end{array} \xrightarrow[\text{and add}]{\text{Change signs}} \begin{array}{r} 8x^3 + 2x^2 \\ + \underline{-2x^3 + 3x^2} \\ 6x^3 + 5x^2 \end{array} \quad \text{This is the opposite of } 2x^3 - 3x^2.$$

EXAMPLE 8   Subtract $16x^4y - 9x^3y$ from the sum of $x^4y + 10x^3y$ and $7x^4y - 8x^3y$.

Strategy   First, we will translate the words of the problem into mathematical symbols. Then we will perform the indicated operations.

Why   The words of the problem contain the key phrases *subtract from* and *sum*.

Solution   Since $16x^4y - 9x^3y$, is to be subtracted from the sum, the order must be reversed when we translate to mathematical symbols.

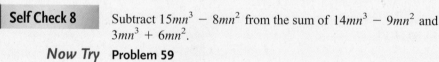

Subtract $16x^4y - 9x^3y$ from   the sum of $x^4y + 10x^3y$ and $7x^4y - 8x^3y$.

$$[(x^4y + 10x^3y) + (7x^4y - 8x^3y)] - (16x^4y - 9x^3y) \quad \text{Use brackets [ ] to enclose the sum. Don't forget the parentheses.}$$

Next, we remove the grouping symbols to obtain

$$= x^4y + 10x^3y + 7x^4y - 8x^3y - 16x^4y + 9x^3y \quad \text{Change the sign of each term within } (16x^4y - 9x^3y) \text{ and drop the parentheses.}$$

$$= -8x^4y + 11x^3y \quad \text{Combine like terms.}$$

> **Self Check 8**   Subtract $15mn^3 - 8mn^2$ from the sum of $14mn^3 - 9mn^2$ and $3mn^3 + 6mn^2$.
>
> **Now Try**   Problem 59

ANSWERS TO SELF CHECKS   **1. a.** A trinomial in one variable of degree 2 written in descending powers of t; terms: $10t^2$, $6t$, -8; coefficients: $10, 6, -8$; degree: $2, 1, 0$ **b.** A binomial in two variables of degree 8 written in descending powers of p and ascending powers of q; terms: $-5.6p^6q$, p^5q^3; coefficients: $-5.6, 1$; degree: $7, 8$ **2.** 256 ft   **3.**   D: $(-\infty, \infty)$; R: $(-\infty, \infty)$ **5. a.** $9m^4$ **b.** $11s^3t + 3s^2t$ **c.** $2x^2 + x$ **d.** $\frac{1}{8}rs + \frac{19}{9}r - 1$ **6. a.** $7a^2 + a + 3$ **b.** $-3a^2b^3 - 3a^3b^2 + ab^2$ **c.** $t^5 + \frac{19}{12}t^4$ **7. a.** $-3m^4 + 34m^2$ **b.** $8.9a^2b^3 - 10.8a^2b^2$ **8.** $2mn^3 + 5mn^2$

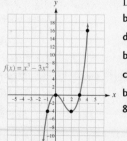

STUDY SET
5.3

VOCABULARY

Fill in the blanks.

1. A _____ is the sum of one or more algebraic terms whose variables have whole-number exponents.

2. $x^3 - 8x^2 - x + 9$ is a polynomial in _____ variable, and is written in _____ powers of x. The polynomial $m^3 + 10m^2n - n^2$ is in _____ variables and is written in _____ powers of n.

3. For the polynomial $7x^2 - 5x - 12$, the _____ term is $7x^2$, and the leading _____ is 7. The _____ term is -12.

4. A _____ is a polynomial with one term. A _____ is a polynomial with two terms. A _____ is a polynomial with three terms.

5. The _____ of the term $6x^5$ is 5 because x appears as a factor 5 times: $6 \cdot x \cdot x \cdot x \cdot x \cdot x$.

6. A second-degree polynomial function is also called a _____ function. A third-degree polynomial function is also called a _____ function.

7. Terms having the same variables with the same exponents are called _____ terms.

8. The _____ of $y^2 + 4y - 2$ is $-y^2 - 4y + 2$.

CONCEPTS

9. Determine whether each expression is a polynomial.

   a. $\dfrac{3}{x^2} + \dfrac{4}{x} + 2$ **b.** $16a^2 - 25b^2$

   c. $y^{-2} - 5y^{-1}$ **d.** $t^4 + t^3 + t^2 + t + 1$

10. Classify each polynomial as a monomial, a binomial, a trinomial, or none of these.

   a. $cdy^2 + 4cd - 3c$ **b.** $3 - 5x^2$

   c. $-\dfrac{15}{16}z^{18}$ **d.** $\dfrac{3}{5}x^4 - \dfrac{2}{5}x^3 + \dfrac{3}{5}x - 1$

11. Fill in the blank so that the term has degree 6.

   a. $-15x$ **b.** $\dfrac{9}{8}ab$

12. Decide whether the terms are like or unlike terms. If they are like terms, combine them.

   a. $12x,\ 5x$ **b.** $9u^2,\ 10u^2$

   c. $6x^2y^3,\ 6x^2y^2$ **d.** $-27x^6y^4z,\ 8x^6y^4z^2$

13. Use the graph of function f to find each of the following
   a. $f(1)$
   b. $f(-3)$
   c. The values of x for which $f(x) = 0$.
   d. The domain and range of f.

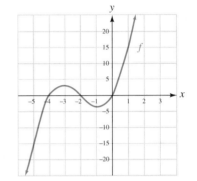

14. Use the graph of function f to find each of the following
   a. $f(-1)$
   b. $f(0)$
   c. The values of x for which $f(x) = 3$.
   d. The domain and range of f.

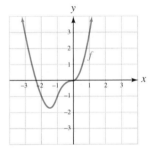

NOTATION

Complete the evaluation.

15. If $h(t) = -t^3 - t^2 + 2t + 1$, find $h(3)$.

   $$h(\quad) = -(\quad)^3 - (\quad)^2 + 2(3) + 1$$
   $$= \boxed{} - 9 + 6 + 1$$
   $$= \boxed{}$$

16. **a.** Write $3x - 2x^4 + 7 - 5x^2$ with the exponents on x in descending order.

   b. Write $x^3y^2 + x^2y^3 - 2x^3y + x^7y^6 - 3x^6$ with the exponents on y in ascending order.

GUIDED PRACTICE

Complete each table. See Example 1.

17. $25x^2 - x + 4$

Term	Coefficient	Degree

Degree of the polynomial:
Type of polynomial:

18. $6a^5 + a^3 - 9a^2 + 7a$

Term	Coefficient	Degree

Degree of the polynomial:
Type of polynomial:

19. $5a^7b^4 - 33a^2b$

Term	Coefficient	Degree

Degree of the polynomial:
Type of polynomial:

20. $\dfrac{3}{4}u^4 + \dfrac{5}{8}u^2v^2 - \dfrac{1}{2}v^4$

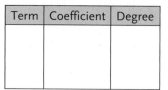

Term	Coefficient	Degree

Degree of the polynomial:
Type of polynomial:

Find each function value. See Example 2.

21. Let $f(x) = 5x^4 - 2x^2 + 9x + 5$.
   a. $f(-1)$ **b.** $f(2)$

22. Let $g(t) = -t^4 + 2t^3 - 5t + 8$.
   a. $g(-2)$ **b.** $g(3)$

23. Let $h(t) = \dfrac{1}{4}t^2 - \dfrac{5}{8}t$.

   a. $h(-4)$ **b.** $h(8)$

24. Let $s(x) = 2.5x^3 - 3.6x^2 + 1.1x$.
   a. $s(-10)$ **b.** $s(10)$

Complete each table of values. Then graph each polynomial function. Write the domain and range in interval notation. See Example 3.

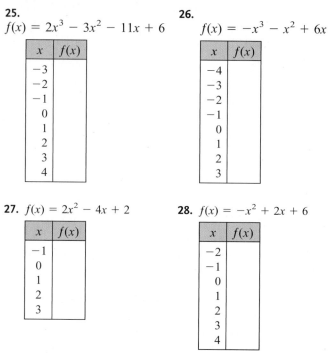

25.
$f(x) = 2x^3 - 3x^2 - 11x + 6$

x	f(x)
-3	
-2	
-1	
0	
1	
2	
3	
4	

26.
$f(x) = -x^3 - x^2 + 6x$

x	f(x)
-4	
-3	
-2	
-1	
0	
1	
2	
3	

27. $f(x) = 2x^2 - 4x + 2$

x	f(x)
-1	
0	
1	
2	
3	

28. $f(x) = -x^2 + 2x + 6$

x	f(x)
-2	
-1	
0	
1	
2	
3	
4	

Use a graphing calculator to graph each polynomial function. Use window settings of $[-4, 6]$ for x and $[-5, 5]$ for y.

29. $f(x) = 2.75x^2 - 4.7x + 1.5$
30. $f(x) = 0.37x^3 - 1.4x + 1.5$

Simplify each polynomial. See Example 5.

31. $15x^2 + 4x - 5x + 5x^2 + 9$
32. $8m^3 - 5m^2 - 5 + 3m^2 - 7m^3$
33. $-7y^2 - 11y^3 - 4 + y^3 - y^2$
34. $2y^3 - 3y^4 - 5 + 4y^4 - 3y^3$
35. $1.4ab^2 - 4.8ab - 0.2a + 5.4ab + 3.7ab^2$
36. $8.7c^4d + 5.4cd - 7.9cd + 1.8 - 6.3c^4d$
37. $\dfrac{9}{4}rst^2 - \dfrac{5}{3}rst - \dfrac{1}{2}rst^2 + \dfrac{5}{6}rst$
38. $\dfrac{1}{9}m^4np - \dfrac{5}{2}mnp - \dfrac{2}{3}m^4np + \dfrac{3}{4}mnp$

Add. See Example 6.

39. $(3x^2 + 2x + 1) + (2x^2 - 7x + 5)$
40. $(2a^2 + 4a - 7) + (3a^2 - a - 2)$
41. $(9p^2q^2 + p - q) + (-p^2q^2 - p - q + 8)$
42. $(6a^2x^3 - 2ax^2 + 3a^3) + (-4a^2x^3 - 2a^3)$
43. $\left(\dfrac{1}{5}h^6 + \dfrac{3}{4}h^2\right) + \left(\dfrac{2}{3}h^6 - \dfrac{1}{12}h^2\right)$
44. $\left(\dfrac{9}{16}n^8 - \dfrac{5}{6}n^4\right) + \left(\dfrac{1}{4}n^8 + \dfrac{7}{3}n^4\right)$

45.
$$6a^4 + 9a^2 + a$$
$$+ 2a^4 - 13a^2 + a$$

46.
$$13b^3 - 11b^2 + 7b$$
$$+ 2b^3 + 9b^2 + 8b$$

Subtract. See Example 7.

47. $(6x^3 + 3x - 2) - (2x^3 + 3x^2 + 5)$

48. $(x^2 - 3x + 8) - (3x^2 + x + 3)$

49. $(2m^2n^2 + 2m - n) - (-2m^2n^2 - 2m + n)$

50. $(2x^2y^3 + 6xy + 5y^2) - (-4x^2y^3 - 7xy + 2y^2 + y)$

51. $(7.1y^3 + 4.9y^2 + 0.1y) - (-8.4y^3 - 0.1y)$

52. $(-8.9p^3 - 2.4p) - (2.1p^3 + 0.8p^2 - p)$

53. $\left(\dfrac{5}{2}w^3 + \dfrac{1}{4}w^2 + \dfrac{3}{5}\right) - \left(\dfrac{1}{3}w^3 + \dfrac{1}{2}w^2 - \dfrac{1}{5}\right)$

54. $\left(\dfrac{3}{8}t^9 - \dfrac{1}{10}t^3 - \dfrac{13}{4}t\right) - \left(\dfrac{1}{8}t^9 - \dfrac{1}{2}t^3 - \dfrac{5}{4}t\right)$

55.
$$3x^2 - 4x + 17$$
$$- (2x^2 + 4x - 5)$$

56.
$$7y^2 - 4y + 3$$
$$- (3y^2 + 10y - 5)$$

57.
$$4x^3 + 6a$$
$$- (4x^3 - 2x^2 - a)$$

58.
$$-2a^3 - 7b$$
$$- (-2a^3 + 3a^2 - 6b)$$

Perform the indicated operations. See Example 8.

59. Subtract $3x^2y^3 + 4xy^2 - 3x^2$ from the sum of $-2x^2y^3 - xy^2 + 7x^2$ and $5x^2y^3 + 3xy^2 - x^2$.

60. Subtract $8m^3n^3 + 2m^2n - n^2$ from the sum of $m^2n + mn^2 + 2n^2$ and $2m^3n^3 - mn^2 + 9n^2$.

61. Find the sum when the difference of $2x^2 - 4x + 3$ and $8x^2 + 5x - 3$ is added to $-2x^2 + 7x - 4$.

62. Find the sum when the difference of $7x^3 - 4x$ and $x^2 + 2$ is added to $x^2 + 3x + 5$.

TRY IT YOURSELF

Perform the indicated operations.

63. $(a^2 + 2a) - (4a^2 - 2a - 1)$

64. $(8w^2 - 7w) - (-10w^2 + 6w + 4)$

65. $(2.8b^2 + 1.2bc + 4.2c^2) + (5.1b^2 - 7.6bc - 3.9c^2)$

66. $(4.4t^2 - 2.9bt + 1.5b^2) + (5.6t^2 + 1.2bt - 3.3b^2)$

67. $(3x^2 + 4x - 3) + (2x^2 - 3x - 1) - (x^2 + x + 7)$

68. $(-2x^2 + 6x + 5) - (-4x^2 - 7x + 2) - (4x^2 + 10x + 5)$

69.
$$3x^3 - 2x^2 + 4x - 3$$
$$-2x^3 + 3x^2 + 3x - 2$$
$$+ 5x^3 - 7x^2 + 7x - 12$$

70.
$$7a^3 + 3a + 7$$
$$-2a^3 + 4a^2 - 13$$
$$+ 3a^3 - 3a^2 + 4a + 5$$

71. Find the difference when $(ay^3 - 2ay^2 + 2a)$ is subtracted from the sum of $(3ay^3 + ay^2)$ and $(-ay^3 + 6ay - 3a)$.

72. Find the difference when $(-3mn^3 - 4mn + 7m)$ is subtracted from the sum of $(2mn^2 + 3mn - 7m)$ and $(-4mn^3 - 2mn - 3m)$.

73. $\left(\dfrac{1}{3}y^6 - \dfrac{1}{6}y^4 - \dfrac{4}{3}y^2\right) + \left(-\dfrac{1}{6}y^6 - \dfrac{1}{2}y^4 + \dfrac{5}{6}y^2\right)$

74. $\left(-\dfrac{2}{5}d^5 + \dfrac{2}{5}d^3 - \dfrac{2}{5}d\right) - \left(\dfrac{1}{3}d^5 + \dfrac{1}{3}d^3 - \dfrac{1}{3}d\right)$

75. $(0.2xy^7 + 0.8xy^5) - (0.5xy^7 - 0.6xy^5 + 0.2xy)$

76. $(5.8g^3t + 0.9g^2t) - (1.3g^3t - 7.6g^2t + 9.9gt)$

77.
$$-5y^3 + 4y^2 - 11y + 3$$
$$- (-2y^3 - 14y^2 + 17y - 32)$$

78.
$$17x^4 - 3x^2 - 65x - 12$$
$$- (23x^4 + 14x^2 + 3x - 23)$$

79. $(1 - 2x - x^2 + 4x^3) + (x^3 - 5x^2 + x + 8)$

80. $(3 - 2z - z^2 + 7z^3) + (3z^3 - z^2 + z + 1)$

81. Subtract $(k^3 - 5k)$ from $(6k^4 + 2k^2 - 16)$.

82. Subtract $(-m^4 + 9m^2)$ from $(m^3 + 2m - 4)$.

APPLICATIONS

83. JUGGLING   During a performance, a juggler tosses one ball straight upward while continuing to juggle three others. The height $f(t)$, in feet, of the ball is given by the polynomial function $f(t) = -16t^2 + 32t + 4$, where t is the time in seconds since the ball was thrown. Find the height of the ball 1 second after it is tossed upward.

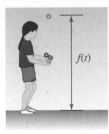

84. STOPPING DISTANCES   The number of feet that a car travels before stopping depends on the driver's reaction time and the braking distance. For one driver, the stopping distance $d(v)$, in feet, is given by the polynomial function $d(v) = 0.04v^2 + 0.9v$, where v is the velocity of the car in mph. Find the stopping distance at 60 mph.

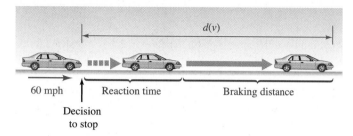

85. STORAGE TANKS   The volume $V(r)$ of the gasoline storage tank, in cubic feet, is given by the polynomial function $V(r) = 4.2r^3 + 37.7r^2$, where r is the radius in feet of the cylindrical part of the tank. What is the capacity of the tank if its radius is 4 feet?

86. ROLLER COASTERS   The polynomial function $f(x) = 0.001x^3 - 0.12x^2 + 3.6x + 10$ models the path of a portion of the track of a roller coaster. Find the height of the track for $x = 0$, 20, 40, and 60.

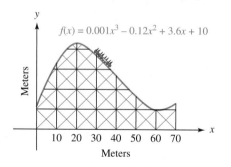

87. PACKAGING   To make boxes, a manufacturer cuts equal-sized squares from each corner of the 10 in. × 12 in. piece of cardboard shown below and then folds up the sides. The polynomial function $f(x) = 4x^3 - 44x^2 + 120x$ gives the volume (in cubic inches) of the resulting box when a square with sides x inches long is cut from each corner. Find the volume of a box if 3-inch squares are cut out.

Fold on dashed lines.

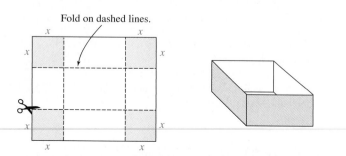

88. RAIN GUTTERS   A rectangular sheet of metal will be used to make a rain gutter by bending up its sides, as shown. If the ends are covered, the capacity $f(x)$ of the gutter is a polynomial function of x: $f(x) = -240x^2 + 1,440x$. Find the capacity of the gutter if x is 3 inches.

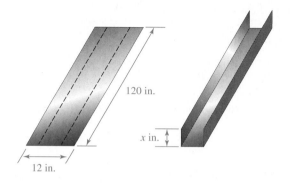

89. GEOMETRY   The formulas for the volume of a cylinder and a cone are given. If the figures below have the same height h and the same base radius r, find a formula for their combined volume.

Volume of a cylinder: Volume of a cone:

$$V = \pi r^2 h \qquad\qquad\qquad V = \frac{1}{3}\pi r^2 h$$

90. CUSTOMER SERVICE   A software service hotline has found that on Mondays, the polynomial function $C(t) = -0.0625t^4 + t^3 - 6t^2 + 16t$ approximates the number of callers to the hotline at any one time. Here, t represents the time, in hours, since the hotline opened at 8:00 A.M. How many service technicians should be on duty on Mondays at noon if the company doesn't want any callers to the hotline waiting to be helped by a technician?

91. TRANSPORTATION ENGINEERING   The polynomial function $A(x) = -0.000000000002x^3 + 0.00000008x^2 - 0.0006x + 2.45$ approximates the number of accidents per mile in one year on a 4-lane interstate, where x is the average daily traffic in number of vehicles. Use the graph of the function on the next page to answer the following questions.

a. Find $A(20,000)$. Explain what the result means.

b. Find the value of x for which $A(x) = 2$. Explain what the result means.

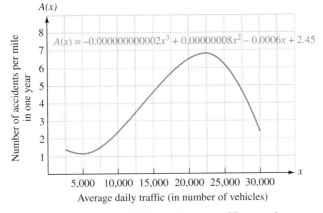

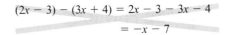

Source: Highway Safety Manual, Colorado Department of Transportation

92. BUSINESS EXPENSES   A company purchased two vehicles for its sales force to use. The following functions give the respective values of the vehicles after x years.

Toyota Camry LE: $T(x) = -2{,}500x + 21{,}075$

Ford Explorer Sport: $F(x) = -2{,}900x + 25{,}845$

   a. Find one polynomial function V that will give the combined value of both cars after x years.

   b. Use your answer in part (a) to find the combined value of the two cars after 3 years.

WRITING

93. Explain why the terms $5x^2y$ and $5xy^2$ are not like terms.

94. Explain why the range of the polynomial function graphed below is not $(-\infty, \infty)$.

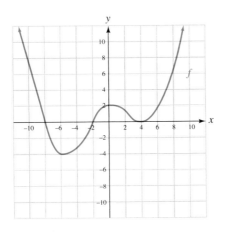

95. Explain the error in the following solution.

   Subtract $2x - 3$ from $3x + 4$.

   $$(2x - 3) - (3x + 4) = 2x - 3 - 3x - 4$$
   $$= -x - 7$$

96. Explain why $f(x) = \dfrac{1}{x + 1}$ is not a polynomial function.

97. Use the word *descending* in a sentence in which the context is not mathematical. Do the same for the word *ascending*.

98. Look up the meaning of the prefix *poly* in a dictionary. Why do you think the name *polynomial* was given to expressions such as $x^3 - x^2 + 2x + 15$?

REVIEW

Solve each inequality. Graph the solution set and write it using interval notation.

99. $|x| \leq 5$

100. $|x| > 7$

101. $|x - 4| < 5$

102. $|2x + 1| \geq 7$

CHALLENGE PROBLEMS

103. What polynomial should be subtracted from $5x^3y - 5xy + 2x$ to obtain the polynomial $8x^3y - 7xy + 11x$?

104. Find two trinomials such that their sum is a binomial and their difference is a monomial.

3 **Multiply Two Binomials.**

The distributive property can also be used to multiply binomials. For example, to multiply $3x + 2$ and $4x + 9$, we think of $3x + 2$ as a single quantity and distribute it over each term of $4x + 9$.

$$(3x + 2)(4x + 9) = (3x + 2)4x + (3x + 2)9$$

$$= (3x + 2)4x + (3x + 2)9 \qquad \text{The distributive property must be used two more times.}$$

$$= (3x)4x + (2)4x + (3x)9 + (2)9 \qquad \text{Distribute the multiplication by } 4x \text{ and by } 9.$$

$$= 12x^2 + 8x + 27x + 18 \qquad \text{Multiply the monomials.}$$

$$= 12x^2 + 35x + 18 \qquad \text{Combine like terms.}$$

In the third line of the solution, each term of $3x + 2$ has been multiplied by each term of $4x + 9$. This example suggests the following rule.

Multiplying Binomials	To multiply two binomials, multiply each term of one binomial by each term of the other binomial, and then combine like terms.

A shortcut method, called the **FOIL method,** can be used to multiply each term of one binomial by each term of the other. FOIL is an acronym for **F**irst terms, **O**uter terms, **I**nner terms, **L**ast terms. To use the FOIL method to multiply $3x + 2$ by $4x + 9$, we

1. Multiply the **F**irst terms $3x$ and $4x$ to obtain $12x^2$,
2. Multiply the **O**uter terms $3x$ and 9 to obtain $27x$,
3. Multiply the **I**nner terms 2 and $4x$ to obtain $8x$, and
4. Multiply the **L**ast terms 2 and 9 to obtain 18.

Then we simplify the resulting polynomial, if possible.

$$(3x + 2)(4x + 9) = 3x(4x) + 3x(9) + 2(4x) + 2(9)$$

$$= 12x^2 + 27x + 8x + 18 \qquad \text{Multiply the monomials.}$$

$$= 12x^2 + 35x + 18 \qquad \text{Combine like terms.}$$

EXAMPLE 3 Multiply: **a.** $(2x - 3)(3x + 2)$ **b.** $(7y^2 - 4)(2y^2 - 1)$

c. $\left(6a - \dfrac{1}{3}\right)\left(3a + \dfrac{5}{6}\right)$ **d.** $(4x^3y + 5)(2x^2 - 3y)$

Strategy We will use the FOIL method to multiply the binomials.

Why In each case we are to find the product of two binomials, and the FOIL method is a shortcut for multiplying two binomials.

Solution

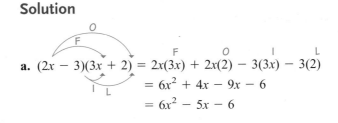

a. $(2x - 3)(3x + 2) = 2x(3x) + 2x(2) - 3(3x) - 3(2)$

$\qquad\qquad\qquad\quad = 6x^2 + 4x - 9x - 6$ Multiply the monomials.

$\qquad\qquad\qquad\quad = 6x^2 - 5x - 6$ Combine like terms.

The Language of Algebra
The acronym FOIL helps us remember the order to follow when multiplying two binomials.

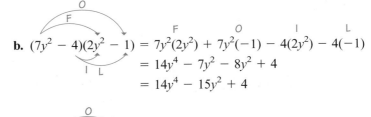

b. $(7y^2 - 4)(2y^2 - 1) = 7y^2(2y^2) + 7y^2(-1) - 4(2y^2) - 4(-1)$

$\qquad\qquad\qquad\qquad\quad = 14y^4 - 7y^2 - 8y^2 + 4$

$\qquad\qquad\qquad\qquad\quad = 14y^4 - 15y^2 + 4$

c. $\left(6a - \dfrac{1}{3}\right)\left(3a + \dfrac{5}{6}\right) = 6a(3a) + 6a\left(\dfrac{5}{6}\right) - \dfrac{1}{3}(3a) - \dfrac{1}{3}\left(\dfrac{5}{6}\right)$

$\qquad\qquad\qquad\qquad\qquad\; = 18a^2 + 5a - a - \dfrac{5}{18}$ Multiply the monomials.

$\qquad\qquad\qquad\qquad\qquad\; = 18a^2 + 4a - \dfrac{5}{18}$ Combine like terms.

d. $(4x^3y + 5)(2x^2 - 3y) = 4x^3y(2x^2) + 4x^3y(-3y) + 5(2x^2) + 5(-3y)$

$\qquad\qquad\qquad\qquad\qquad\;\; = 8x^5y - 12x^3y^2 + 10x^2 - 15y$

Since the terms of $8x^5y - 12x^3y^2 + 10x^2 - 15y$ are unlike, we cannot simplify this result.

▷ **Self Check 3**

Multiply: **a.** $(3a + 4)(2a - 9)$

b. $(4c^4 + 5)(3c^4 + 7)$

c. $\left(10w - \dfrac{1}{5}\right)\left(5w + \dfrac{3}{10}\right)$

d. $(5a^2b - 3b)(8a^2 - b)$

Now Try   Problems 27, 31, and 33

---

4   **Multiply Any Two Polynomials.**

To develop a general rule for multiplying any two polynomials, we will find the product of $x + 4$ and $x^2 + 2x + 1$. In the solution, the distributive property is used four times.

$(x + 4)(x^2 + 2x + 1) = (x + 4)x^2 + (x + 4)2x + (x + 4)1$ Distribute.

$\qquad\qquad\qquad\qquad\quad = (x + 4)x^2 + (x + 4)2x + (x + 4)1$

$\qquad\qquad\qquad\qquad\quad = (x)x^2 + (4)x^2 + (x)2x + (4)2x + (x)1 + (4)1$ Distribute.

$$= x^3 + 4x^2 + 2x^2 + 8x + x + 4 \qquad \text{Multiply the monomials.}$$
$$= x^3 + 6x^2 + 9x + 4 \qquad \text{Combine like terms.}$$

In the third line of the solution, each term of $x^2 + 2x + 1$ has been multiplied by each term of $x + 4$. This result suggests the following rule.

Multiplying Polynomials	To multiply two polynomials, multiply each term of one polynomial by each term of the other polynomial, and then combine like terms.

EXAMPLE 4

Multiply: $(2a + b)(3a^2 - 4ab - b^2)$

Strategy We will multiply each term of the trinomial, $3a^2 - 4ab - b^2$, by each term of the binomial, $2a + b$.

Why To multiply two polynomials, we must multiply each term of one polynomial by each term of the other polynomial.

Solution

> **Success Tip**
> The FOIL method cannot be applied here, only to products of two binomials.

$$(2a + b)(3a^2 - 4ab - b^2)$$

$$= 2a(3a^2) + 2a(-4ab) + 2a(-b^2) + b(3a^2) + b(-4ab) + b(-b^2)$$
$$= 6a^3 - 8a^2b - 2ab^2 + 3a^2b - 4ab^2 - b^3 \qquad \text{Multiply the monomials.}$$
$$= 6a^3 - 5a^2b - 6ab^2 - b^3 \qquad \text{Combine like terms.}$$

▷ **Self Check 4** Multiply: $(3y^2 + y)(4y^2 - 2y + 5)$
Now Try Problem 35

It is often convenient to multiply polynomials using a vertical form.

EXAMPLE 5

Use vertical form to multiply:

a. $(3x + 2)(4x + 9)$ **b.** $(3a^2 - 4ab + b^2)(2a - b)$
c. $(-2y^3 - 6y^2 + 1)(5y^2 - 10y - 2)$

Strategy We will write one polynomial underneath the other and multiply each term of the upper polynomial by each term of the lower polynomial. Then we will combine like terms column-by-column.

Why *Vertical form* means to use an approach similar to that used in arithmetic to multiply two whole numbers.

> **Success Tip**
> Multiplying two polynomials in vertical form is much like multiplying two numbers in arithmetic.
>
> $$\begin{array}{r} 32 \\ \times\ 49 \\ \hline 288 \\ +\ 128 \\ \hline 1568 \end{array}$$

Solution

a. Multiply:

$$
\begin{array}{r}
3x + 2 \\
4x + 9 \\
\hline
27x + 18 \\
12x^2 + 8x \\
\hline
12x^2 + 35x + 18
\end{array}
$$

← This is the result of $9(3x + 2)$.

← This is the result of $4x(3x + 2)$.

In each column, combine like terms.

b. Multiply:

$$
\begin{array}{r}
3a^2 - 4ab + b^2 \\
2a - b \\
\hline
-3a^2b + 4ab^2 - b^3 \\
6a^3 - 8a^2b + 2ab^2 \\
\hline
6a^3 - 11a^2b + 6ab^2 - b^3
\end{array}
$$

← This row is $-b(3a^2 - 4ab + b^2)$.

← This row is $2a(3a^2 - 4ab + b^2)$.

In each column, combine like terms.

c. With this method, it is often necessary to leave a space for a missing term to vertically align like terms. We begin by multiplying $-2y^3 - 6y^2 + 1$ by -2; then we multiply $-2y^3 - 6y^2 + 1$ by $-10y$; and finally we multiply $-2y^3 - 6y^2 + 1$ by $5y^2$. Then we combine like terms, column by column.

$$
\begin{array}{r}
-2y^3 - 6y^2 + 1 \\
5y^2 - 10y - 2 \\
\hline
4y^3 + 12y^2 \qquad - 2 \\
20y^4 + 60y^3 \qquad - 10y \\
-10y^5 - 30y^4 \qquad + 5y^2 \\
\hline
-10y^5 - 10y^4 + 64y^3 + 17y^2 - 10y - 2
\end{array}
$$

There is no y-term; leave a space.

There is no y^2-term; leave a space.

There is no y^3-term; leave a space.

In each column, combine like terms.

 Self Check 5   Multiply: **a.** $\dfrac{3x^2 + 2x - 5}{2x + 1}$ **b.** $\dfrac{2a^2 + 6a - 1}{3a^2 + 9a - 5}$

Now Try   **Problem 41**

⑤   **Multiply Three Polynomials.**

When finding the product of three polynomials, we begin by multiplying *any* two of them, and then we multiply that result by the third polynomial.

EXAMPLE 6   Multiply: $5cd(c + 6d)(3c - 8d)$

Strategy   We will find the product of $c + 6d$ and $3c - 8d$ and then multiply that result by $5cd$.

Why   It is wise to perform the most difficult multiplication first. (In this case, this is the product of two binomials). Save the simpler multiplication by $5cd$ for last.

Solution

$$5cd(c + 6d)(3c - 8d) = 5cd(3c^2 - 8cd + 18cd - 48d^2)$$

Use the FOIL method to find $(c + 6d)(3c - 8d)$.

$$= 5cd(3c^2 + 10cd - 48d^2)$$

Combine like terms: $-8cd + 18cd = 10cd$.

$$= 15c^3d + 50c^2d^2 - 240cd^3$$

Distribute the multiplication by $5cd$.

 Self Check 6 Multiply: $-2r(r - 2s)(5r - 4s)$

Now Try **Problem 43**

6 **Find Special Products.**

To find the square of a binomial, we can use the FOIL method. For example, to find $(x + y)^2$ and $(x - y)^2$, we proceed as follows.

$$\begin{aligned}(x + y)^2 &= (x + y)(x + y) \\ &= x^2 + xy + xy + y^2 \\ &= x^2 + 2xy + y^2\end{aligned} \qquad \begin{aligned}(x - y)^2 &= (x - y)(x - y) \\ &= x^2 - xy - xy + y^2 \\ &= x^2 - 2xy + y^2\end{aligned}$$

In each case, we see that the square of the binomial is the square of its first term, twice the product of its two terms, and the square of its last term.

The figure shows how $(x + y)^2$ can be found geometrically.

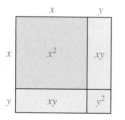

The area of the largest square is the product of its length and width: $(x + y)(x + y) = (x + y)^2$.

The area of the largest square is also the sum of its four parts: $x^2 + xy + xy + y^2 = x^2 + 2xy + y^2$.

Thus, $(x + y)^2 = x^2 + 2xy + y^2$.

Another common binomial product is the product of the sum and difference of the same two terms. An example of such a product is $(x + y)(x - y)$. To find this product, we use the FOIL method.

$$\begin{aligned}(x + y)(x - y) &= x^2 - xy + xy - y^2 \\ &= x^2 - y^2\end{aligned}$$

Combine like terms: $-xy + xy = 0$.

We see that the product of the sum and the difference of the same two terms is the square of the first term minus the square of the second term.

The results from these three examples suggest the following **special-product rules.**

Special-Product Rules		
	$(A + B)^2 = A^2 + 2AB + B^2$	The square of a sum.
	$(A - B)^2 = A^2 - 2AB + B^2$	The square of a difference.
	$(A + B)(A - B) = A^2 - B^2$	The product of the sum and difference of two terms.

Caution   Remember that the square of a binomial is a *trinomial*. A common error when squaring a binomial is to forget the middle term of the product. For example,

$$(x + y)^2 \neq x^2 + y^2 \qquad \text{and} \qquad (x - y)^2 \neq x^2 - y^2$$

Missing 2xy   Missing −2xy —— Should be + symbol

Also remember that the product $(x + y)(x - y)$ is the binomial $x^2 - y^2$. And since $(x + y)(x - y) = (x - y)(x + y)$ by the commutative property of multiplication,

$$(x - y)(x + y) = x^2 - y^2$$

EXAMPLE 7   Multiply: **a.** $(5c + 3d)^2$

b. $\left(\dfrac{1}{2}a^4 - b^2\right)^2$ **c.** $(0.2m^3 + 2.5n)(0.2m^3 - 2.5n)$

Strategy   To find the product of each pair of binomials, we will use a special-product rule.

Why   This approach is faster than using the FOIL method.

Solution

a. To find $(5c + 3d)^2$ using the square of a sum rule, we begin by noting that the first term of the binomial is $5c$ and the last term is $3d$.

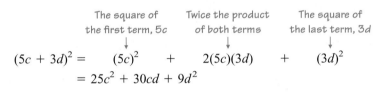

The square of   Twice the product   The square of
the first term, 5c   of both terms   the last term, 3d

$$(5c + 3d)^2 = \quad (5c)^2 \quad + \quad 2(5c)(3d) \quad + \quad (3d)^2$$

$$= 25c^2 + 30cd + 9d^2$$

b. To find $\left(\dfrac{1}{2}a^4 - b^2\right)^2$ using the square of a difference rule, we begin by noting that the first term of the binomial is $\dfrac{1}{2}a^4$ and the last term is $-b^2$.

The square of   Twice the product   The square of
the first term, ½a⁴   of both terms   the last term, −b²

$$\left(\frac{1}{2}a^4 - b^2\right)^2 = \quad \left(\frac{1}{2}a^4\right)^2 \quad + \quad 2\left(\frac{1}{2}a^4\right)(-b^2) \quad + \quad (-b^2)^2$$

$$= \frac{1}{4}a^8 - a^4b^2 + b^4$$

c. $(0.2m^3 + 2.5n)(0.2m^3 - 2.5n)$ is the product of the sum and the difference of the same two terms: $0.2m^3$ and $2.5n$. Using a special-product rule, we proceed as follows.

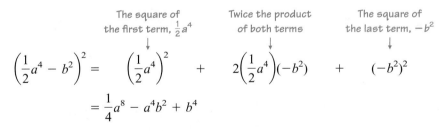

The square of the   The square of the
first term, 0.2m³   second term, 2.5n

$$(0.2m^3 + 2.5n)(0.2m^3 - 2.5n) = \quad (0.2m^3)^2 \quad - \quad (2.5n)^2$$

$$= 0.04m^6 - 6.25n^2$$

Self Check 7   Multiply: **a.** $(8r + 2s)^2$
b. $\left(\dfrac{1}{3}a^3 - b^6\right)^2$
c. $(0.4x + 1.2y^4)(0.4x - 1.2y^4)$

Now Try   Problems 47, 51, and 55

The special-product rules can be helpful in finding more complicated powers and products.

EXAMPLE 8   Multiply: **a.** $[(5x + y) + 4]^2$

b. $[12 + (c - d)][12 - (c - d)]$ **c.** $(a + b)^3$

Strategy   For each problem, we will use a special-product rule to simplify the calculations.

Why   Using a special-product rule often produces the result more quickly than multiplying each term of one polynomial by each term of the other.

Solution

a. It is helpful to think of $(5x + y)$ as the first term and 4 as the second term of the square of a sum. We can then match the given expression to a special-product rule.

$$(\quad A \quad + B)^2 = \quad A^2 \quad + 2 \quad A \quad B + B^2$$
$$[(5x + y) + 4]^2 = (5x + y)^2 + 2(5x + y)(4) + 4^2$$
$$= 25x^2 + 10xy + y^2 + 8(5x + y) + 16 \quad \text{Square } (5x + y). \text{ Simplify.}$$
$$= 25x^2 + 10xy + y^2 + 40x + 8y + 16$$

b. It is helpful to think of 12 as the first term and $(c - d)$ as the second term of the sum and difference of the same two terms. We can then match the given expression to a special-product rule.

$$(A + \quad B \quad)(A - \quad B \quad) = A^2 - \quad B^2$$
$$[12 + (c - d)][12 - (c - d)] = 12^2 - (c - d)^2$$
$$= 144 - (c^2 - 2cd + d^2) \quad \text{Square 12 and } (c - d).$$
$$= 144 - c^2 + 2cd - d^2$$

c. Since $(a + b)^3$ can be written as $(a + b)(a + b)^2$, we can use a special-product rule to quickly find $(a + b)^2$.

$$(a + b)^3 = (a + b)(a + b)^2$$
$$= (a + b)(a^2 + 2ab + b^2) \qquad \text{Square } (a + b).$$
$$= a^3 + 2a^2b + ab^2 + a^2b + 2ab^2 + b^3 \qquad \text{Multiply each term of}$$
$$\qquad\qquad a^2 + 2ab + b^2 \text{ by each term of}$$
$$\qquad\qquad a + b.$$
$$= a^3 + 3a^2b + 3ab^2 + b^3 \qquad \text{Combine like terms.}$$

▷ **Self Check 8**   Multiply: **a.** $[(s + 4t) + 2]^2$
b. $[9 + (m - n)][9 - (m - n)]$
c. $(4x - y)^3$

Now Try   Problems 59, 63, and 65

Polynomial multiplication is also encountered when evaluating functions.

EXAMPLE 9   If $f(x) = x^2 + 9x - 5$, find $f(a + 4)$.

Strategy   To find $f(a + 4)$, we will substitute $a + 4$ for each x in $f(x) = x^2 + 9x - 5$ and simplify the right side.

Why Whatever expression appears within the parentheses in $f(\ )$ is to be substituted for each x in $f(x) = x^2 + 9x - 5$.

Solution

$$f(x) = x^2 + 9x - 5$$
$$f(a + 4) = (a + 4)^2 + 9(a + 4) - 5 \qquad \text{Use a special-product rule to find } (a + 4)^2.$$
$$= a^2 + 8a + 16 + 9a + 36 - 5$$
$$= a^2 + 17a + 47$$

Self Check 9 If $f(x) = x^2 - 6x + 1$, find $f(a - 8)$.

Now Try **Problem 69**

---

7 **Use Multiplication to Simplify Expressions.**

The procedures discussed in this section are helpful when we simplify algebraic expressions that involve the multiplication of polynomials.

EXAMPLE 10 Simplify: $(5x - 4)^2 - (x - 7)(x + 1)$

Strategy We will use a special-product rule to find $(5x - 4)^2$ and the FOIL method to find $(x - 7)(x + 1)$. Then we will combine like terms.

Why To simplify expressions, we follow the order of operations rule. We find powers and perform the multiplication before performing the subtraction.

Solution

$$(5x - 4)^2 - (x - 7)(x + 1) = 25x^2 - 40x + 16 - (x^2 - 6x - 7) \quad \text{Don't forget to write parentheses around } x^2 - 6x - 7.$$
$$= 25x^2 - 40x + 16 - x^2 + 6x + 7$$
$$= 24x^2 - 34x + 23 \qquad \text{Combine like terms.}$$

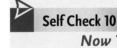

Self Check 10 Simplify: $(y - 7)(y + 7) - (4y + 3)^2$

Now Try **Problem 71**

---

8 **Solve Problems Using Multiplication of Polynomials.**

11 in.

8.5 in.

EXAMPLE 11 **Scrapbooks.** A 8.5-inch by 11-inch scrapbook page has a border of uniform width surrounding a rectangular-shaped birth announcement. Write a polynomial that represents the area of the birth announcement.

Strategy We will write polynomials that represent the length and the width of the rectangular-shaped announcement and find their product.

Why The area of a rectangle is the product of its length and width.

11 in.

8.5 in.

Solution Let x represent the width of the border, in inches. Then the length of the birth announcement is $11 - x - x$ or $(11 - 2x)$ inches and the width is $8.5 - x - x$ or $(8.5 - 2x)$ inches. The area of the announcement is the product of its length and width:

$A = lw$ 　　　　This is the formula for the area of a rectangle.

$= (11 - 2x)(8.5 - 2x)$ 　　　Substitute for l and w.

$= 11(8.5) - 11(2x) - 2x(8.5) - 2x(-2x)$ 　　Use the FOIL method.

$= 93.5 - 22x - 17x + 4x^2$ 　　Multiply the monomials.

$= 93.5 - 39x + 4x^2$ 　　Combine like terms.

The polynomial $93.5 - 39x + 4x^2$ or $4x^2 - 39x + 93.5$ represents the area of the birth announcement in square inches.

> **Now Try** Problem 111

ANSWERS TO SELF CHECKS 1. a. $8y^5$ b. $4m^{12}n^7$ c. $-90a^4b^6c^4$ 2. a. $2a^8 - 10a^5 + 6a^3$
b. $-32c^6d^4 - 64c^5d^3 + 8cd^7$ c. $6m^{10}n^{10}p^7 - 54m^3n^9p^4$ 3. a. $6a^2 - 19a - 36$
b. $12c^8 + 43c^4 + 35$ c. $50w^2 + 2w - \frac{3}{50}$ d. $40a^4b - 5a^2b^2 - 24a^2b + 3b^2$
4. $12y^4 - 2y^3 + 13y^2 + 5y$ 5. a. $6x^3 + 7x^2 - 8x - 5$ b. $6a^4 + 36a^3 + 41a^2 - 39a + 5$
6. $-10r^3 + 28r^2s - 16rs^2$ 7. a. $64r^2 + 32rs + 4s^2$ b. $\frac{1}{9}a^6 - \frac{2}{3}a^3b^6 + b^{12}$ c. $0.16x^2 - 1.44y^8$
8. a. $s^2 + 8st + 16t^2 + 4s + 16t + 4$ b. $81 - m^2 + 2mn - n^2$ c. $64x^3 - 48x^2y + 12xy^2 - y^3$
9. $a^2 - 22a + 113$ 10. $-15y^2 - 24y - 58$

STUDY SET
5.4

VOCABULARY

Fill in the blanks.

1. The expression $(2x^3)(3x^4)$ is the product of two _____ and the expression $(x + 4)(x - 5)$ is the product of two _____.

2. $(x + 4)^2$ is the _____ of a sum and $(m - 9)^2$ is the square of a _____. The expression $(b + 1)(b - 1)$ is the product of the _____ and difference of two terms.

3. FOIL is an acronym for _____ terms, _____ terms, _____ terms, and _____ terms.

4. Since $x^2 + 16x + 64$ is the square of $x + 8$, it is called a _____-square trinomial.

CONCEPTS

Fill in the blanks.

5. a. To multiply a monomial by a monomial, we multiply the numerical _____ and then multiply the variable factors.

   b. To multiply a polynomial by a monomial, we multiply each _____ of the polynomial by the monomial.

   c. To multiply a polynomial by a polynomial, we multiply each _____ of one polynomial by each term of the other polynomial.

6. a. The square of a binomial is the _____ of its first term, _____ the product of its two terms, plus the _____ of its last term.

$$(x + y)^2 = (x + y)(x + y) = \quad$$

$$(x - y)^2 = (x - y)(x - y) = \quad$$

7. The product of the sum and difference of the same two terms is the _____ of the first term minus the _____ of the second term.

$$(x + y)(x - y) = \quad$$

8. Consider $(2x + 4)(4x - 3)$. The first terms are _____ and _____ , the outer terms are _____ and _____ , the inner terms are _____ and _____ , and the last terms are _____ and _____ .

9. Perform the indicated operation.
   a. $(4b - 1) + (2b - 1)$
   b. $(4b - 1) - (2b - 1)$
   c. $(4b - 1)(2b - 1)$

10. To find $-2a^2b(a^2 + b^2)(a^2 - b^2)$, which two polynomials would you multiply first? Explain why.

GUIDED PRACTICE

Multiply. See Example 1.

11. $(2a^2)(3a^5)$

12. $(3x^2)(3x^9)$

13. $(-3ab^2)(5ab)$

14. $(-2m^2n)(4mn^3)$

15. $\left(\frac{1}{6}g^4h^5\right)(36g^7h^{11})$

16. $\left(\frac{1}{16}r^9s^{10}\right)(32r^2s^{10})$

17. $(2x^2y^3z^5)(4xy^5z)(-5y^6z^6)$

18. $(4a^2bc^6)(-5a^3b^2c^2)(6a^4c^4)$

Multiply. See Example 2.

19. $10x^5(3x^5 + 2x^2)$

20. $5a^4(6a^7 + 2a^2)$

21. $-2d^4(3d^3 - 3d^2 + 2d)$

22. $-3a^8(4a^4 + 3a^3 - 4a^2)$

23. $7rs^3t(r^2 + s^2 - t^2)$

24. $3x^2yz(x^2 - 2y + 3z^2)$

25. $(7x^6y^3z - 4x^2yz^4)4xy^6z^3$

26. $(8abc^{20} - 6a^2bc^2)5ab^8c^9$

Multiply. See Example 3.

27. $(7t - 2)(2t + 3)$

28. $(5p + 3)(3p - 4)$

29. $(3y^3 - 4)(2y^3 - 1)$

30. $(2m^5 - 7)(3m^5 - 1)$

31. $\left(4b + \frac{1}{2}\right)\left(8b - \frac{3}{4}\right)$

32. $\left(6x + \frac{1}{3}\right)\left(6x - \frac{1}{6}\right)$

33. $(9b^3c - c)(3b^2 - bc)$

34. $(h^5k - k)(4h^3 - hk)$

Multiply. See Example 4.

35. $(3y + 1)(2y^2 + 3y + 2)$

36. $(a + 2)(3a^2 + 4a - 2)$

37. $(x - y)(x^2 + xy + y^2)$

38. $(x + y)(x^2 - xy + y^2)$

Use vertical form to multiply the polynomials. See Example 5.

39. $\begin{array}{r} 3a^2 + 4a - 2 \\ \underline{2a + 3} \end{array}$

40. $\begin{array}{r} 5y^2 - y - 2 \\ \underline{2y - 1} \end{array}$

41. $\begin{array}{r} 2x^2 - 10x + 14 \\ \underline{x^2 + 2x - 3} \end{array}$

42. $\begin{array}{r} 4x^2 - 12x + 4 \\ \underline{x^2 - 2x + 7} \end{array}$

Multiply. See Example 6.

43. $6p^2(3p - 4)(p + 3)$

44. $4a(2a + 3)(3a - 2)$

45. $4my(3m - y)(2m - y)$

46. $-3hz(2h - z)(3h - z)$

Use a special product formula to find each product. See Example 7.

47. $(2a + b)^2$

48. $(a - 2b)^2$

49. $(5r^2 + 6)^2$

50. $(6p^2 - 3)^2$

51. $\left(\frac{1}{4}b - 2\right)^2$

52. $\left(\frac{2}{3}y - 7\right)^2$

53. $(9ab^2 - 4)^2$

54. $(2yz^2 + 5)^2$

55. $(5y + 2.4)(5y - 2.4)$

56. $(9t + 0.7)(9t - 0.7)$

57. $\left(x^2y - \frac{6}{5}\right)\left(x^2y + \frac{6}{5}\right)$

58. $\left(a^4b - \frac{1}{2}c\right)\left(a^4b + \frac{1}{2}c\right)$

Multiply. See Example 8.

59. $[(6a + b) + 4]^2$

60. $[(3m + n) + 10]^2$

61. $[7 - (x - y)]^2$

62. $[12 - (r - s)]^2$

63. $[5 + (2n + p)][5 - (2n + p)]$

64. $[4 + (w + 3z)][4 - (w + 3z)]$

65. $(3x - 2)^3$

66. $(p - 2q)^3$

Evaluate each function. See Example 9

67. If $f(x) = x^2 - 8x + 2$, find $f(b + 1)$.

68. If $f(x) = x^2 - 6x + 10$, find $f(n + 2)$.

69. If $f(x) = x^2 + 4x - 9$, find $f(a - 6)$.

70. If $f(x) = x^2 + 12x - 8$, find $f(c - 3)$.

Simplify each expression. See Example 10.

71. $(7x - 1)^2 - (x - 4)(x + 3)$

72. $(9x - 2)^2 - (x + 9)(x - 7)$

73. $(3x - 4)^2 - (2x + 3)^2$

74. $(3y + 1)^2 + (2y - 4)^2$

Use a calculator to help find each product.

75. $(3.21x - 7.85)(2.87x + 4.59)$

76. $(7.44y + 56.7)(-2.1y - 67.3)$

77. $(-17.3y + 4.35)^2$

78. $(-0.31x + 29.3)(-0.31x - 29.3)$

TRY IT YOURSELF

Perform the indicated operations.

79. $(2a - b)(4a^2 + 2ab + b^2)$
80. $(x - 3y)(x^2 + 3xy + 9y^2)$
81. $(2b + 3)(5b - 1) - (b + 2)^2$
82. $(x + 3)^2 - (2x - 1)(4x + 2)$
83. $(11m^2 + 3n^3)(5m + 2n^2)$
84. $(50m^4 - 3n^4)(2m + 2n^3)$
85. $(-5s^2tu)(-3s^4t^4u)$
86. $(-2a^3b^2c^5)(-3ab^4c^2)$
87. $(a + b + c)(2a - b - 2c)$
88. $(x + 2y + 3z)^2$
89. $(a + b)(a - b)(a - 3b)$
90. $(x - y)(x + 2y)(x - 2y)$
91. $(4k - 13)^2$
92. $(5k - 6)^2$
93. $\left(\dfrac{1}{2}x - 4\right)\left(\dfrac{1}{2}x + 16\right)$
94. $\left(9h^2 - \dfrac{2}{3}\right)\left(3h^2 + \dfrac{2}{3}\right)$
95. $(0.4t - 3)(0.5t - 3)$
96. $(0.7d - 2)(0.1d + 3)$
97. $(y^3 + 2)(y^3 - 2)$
98. $(y^4 + 3)(y^4 - 3)$
99. $(3a^3 + 4c)^2$
100. $(2y^5 + 5z)^2$
101. $\dfrac{1}{2}m^2n(m^2 + n^3)(-12mn)$

102. $\dfrac{1}{3}a^2b^3(3a^3 + b^4)(-6b)$

103. $(5s - t^3)^3$
104. $(3a - 2b)^3$
105. $[(3d + f) + 2]^2$
106. $[8 - (a - 5b)]^2$

APPLICATIONS

107. GEOMETRY   Write a polynomial that represents the area of the figure.

a.

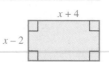

b.

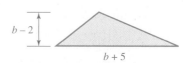

108. GEOMETRY   Write a polynomial that represents the area of the figure.

a.

b.

109. THE YELLOW PAGES   Refer to the illustration.

  a. Describe the area occupied by the ads for movers by using a product of two binomials.

  b. Describe the area occupied by the ad for Budget Moving Co. by using a product. Then perform the multiplication.

  c. Describe the area occupied by the ad for Snyder Movers by using a product. Then perform the multiplication.

  d. Explain why your answer to part (a) is equal to the sum of your answers to parts (b) and (c). What special product does this exercise illustrate?

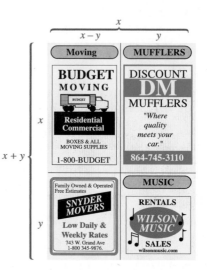

110. HELICOPTER PADS   To determine the amount of fluorescent paint needed to paint the circular ring on the landing pad design shown in the illustration on the next page, painters must find its area. The area of the ring is given by the expression $\pi(R + r)(R - r)$.

  a. Find the product $\pi(R + r)(R - r)$.

  b. If $R = 25$ feet and $r = 20$ feet, find the area to be painted. Round to the nearest tenth.

  c. If a quart of fluorescent paint covers 65 ft^2, how many quarts will be needed to paint the ring?

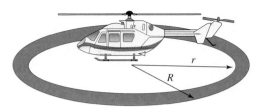

111. QUILTS   A 36-inch by 46-inch baby crib quilt has a border of uniform width x inches surrounding a rectangular-shaped center panel. Write a polynomial that represents the area of the center panel.

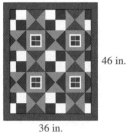

46 in.

36 in.

112. GIFT BOXES   The corners of a 12-in. by 12-in. piece of cardboard are creased, folded inward, and glued to make a gift box. (See the illustration.) Write a polynomial that gives the volume of the resulting box.

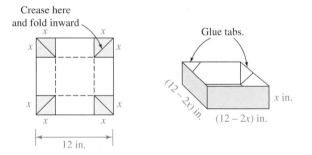

Crease here
and fold inward

x x

x x

Glue tabs.

x x

x x

12 in.

$(12 - 2x)$ in.

x in.

$(12 - 2x)$ in.

WRITING
113. Explain how to use the FOIL method.
114. Explain how you would multiply two trinomials.
115. On a test, when asked to find $(x - y)^2$, a student answered $x^2 - y^2$. What error did the student make?
116. Describe each expression in words:

$$(x + y)^2 \qquad (x - y)^2 \qquad (x + y)(x - y)$$

REVIEW

Graph the solution set of each inequality or system of inequalities on a rectangular coordinate system.

117. $2x + y \le 2$ **118.** $x \ge 2$

119. $\begin{cases} y - 2 < 3x \\ y + 2x < 3 \end{cases}$ **120.** $\begin{cases} y < 0 \\ x < 0 \end{cases}$

CHALLENGE PROBLEMS

Multiply. Write all answers without negative exponents.

121. $ab^{-2}c^{-3}(a^{-4}bc^3 + a^{-3}b^4c^3)$

122. $(5x^{-4} - 4y^2)(5x^2 - 4y^{-4})$

Multiply. Assume n is a natural number.

123. $a^{2n}(a^n + a^{2n})$
124. $(a^{3n} - b^{3n})(a^{3n} + b^{3n})$
125. If $f(x) = x^2 - 4x - 7$, find $f(a + h) - f(a)$.
126. If $f(x) = x^3 + x$, find $f(a + h) - f(a)$.

SECTION 5.5
The Greatest Common Factor and Factoring by Grouping

Objectives

1. Find the greatest common factor of a list of terms.
2. Factor out the greatest common factor.
3. Factor by grouping.
4. Use factoring to solve formulas for a specified variable.

In Section 5.4, we discussed ways of multiplying polynomials. In this section, we will discuss the reverse process—*factoring* polynomials. When factoring a polynomial, the first step is to determine whether its terms have any common factors.

1 **Find the Greatest Common Factor of a List of Terms.**

To determine whether two or more integers have common factors, it is helpful to write them as products of prime numbers. For example, the prime factorizations of 42 and 90 are given below.

$$42 = 2 \cdot 3 \cdot 7 \qquad 90 = 2 \cdot 3 \cdot 3 \cdot 5$$

The highlighting shows that 42 and 90 have one factor of 2 and one factor of 3 in common. To find their *greatest common factor (GCF)*, we multiply the common factors: $2 \cdot 3 = 6$. Thus, the GCF of 42 and 90 is 6.

The Greatest Common Factor (GCF)	The **greatest common factor (GCF)** of a list of integers is the largest common factor of those integers.

Recall from arithmetic that the factors of a number divide the number exactly, leaving no remainder. Therefore, the greatest common factor of two or more integers is the largest natural number that divides each of the integers exactly.

To find the greatest common factor of a list of terms, we can use the following approach.

Strategy for Finding the GCF	1. Write each coefficient as a product of prime factors. 2. Identify the numerical and variable factors common to each term. 3. Multiply the common numerical and variable factors identified in Step 2 to obtain the GCF. If there are no common factors, the GCF is 1.

EXAMPLE 1 Find the GCF of $6a^2b^3c$, $9a^3b^2c$, and $18a^4c^3$.

Strategy We will prime factor each coefficient of each term in the list. Then we will identify the numerical and variable factors common to each term and find their product.

Why The product of the common factors is the GCF of the terms in the list.

Solution We begin by factoring each term.

$$6a^2b^3c = 2 \cdot 3 \cdot a \cdot a \cdot b \cdot b \cdot b \cdot c \qquad \text{This can be written as } 2 \cdot 3 \cdot a^2 \cdot b^3 \cdot c.$$
$$9a^3b^2c = 3 \cdot 3 \cdot a \cdot a \cdot a \cdot b \cdot b \cdot c \qquad \text{This can be written as } 3^2 \cdot a^3 \cdot b^2 \cdot c.$$
$$18a^4c^3 = 2 \cdot 3 \cdot 3 \cdot a \cdot a \cdot a \cdot a \cdot c \cdot c \cdot c \qquad \text{This can be written as } 2 \cdot 3^2 \cdot a^4 \cdot c^3.$$

Since each term has one factor of 3, two factors of a, and one factor of c in common, the GCF is

$$3 \cdot a \cdot a \cdot c = 3a^2c$$

> **Success Tip**
> The exponent on any variable in a GCF is the smallest exponent that appears on that variable in all of the terms under consideration.

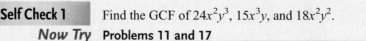

Self Check 1 Find the GCF of $24x^2y^3$, $15x^3y$, and $18x^2y^2$.

Now Try Problems 11 and 17

The concept of greatest common factor is used to factor polynomials.

 Factor Out the Greatest Common Factor.

We have seen that the distributive property provides a method for multiplying a polynomial by a monomial. For example, to multiply $3x^2 + 4$ by $2x^3$, we proceed as follows:

$$\overset{\text{Multiplication}}{\longrightarrow}$$

$$2x^3(3x^2 + 4) = 2x^3 \cdot 3x^2 + 2x^3 \cdot 4$$
$$= 6x^5 + 8x^3$$

In this section, we reverse the previous steps and determine what factors were multiplied to obtain $6x^5 + 8x^3$. We call that process *factoring the polynomial.*

$$\overset{\text{Factoring}}{\longrightarrow}$$

$$6x^5 + 8x^3 = 2x^3 \cdot 3x^2 + 2x^3 \cdot 4 \quad \text{Write } 6x^5 \text{ and } 8x^3 \text{ as the product of the GCF, } 2x^3\text{, and one}$$
$$\text{other factor.}$$

$$= 2x^3(3x^2 + 4) \quad \text{Write an expression so that the multiplication by } 2x^3$$
$$\text{distributes over the terms } 3x^2 \text{ and 4.}$$

Since $2x^3$ is the GCF of the terms of $6x^5 + 8x^3$, this method is called **factoring out the greatest common factor.** When we factor a polynomial, we write a sum of terms as a product of factors.

$$\underbrace{6x^5 + 8x^3}_{\text{Sum of terms}} = \underbrace{2x^3(3x^2 + 4)}_{\text{Product of factors}}$$

EXAMPLE 2   Factor: **a.** $16y^2 + 24y$ **b.** $25a^3b - 15ab^3$
c. $3xy^2z^3 + 6xyz^3 + 3xz^2$

Strategy   We will determine the GCF of the terms of the polynomial. Then we will write each term of the polynomial as the product of the GCF and one other factor.

Why   We can then use the distributive property to factor out the GCF.

Solution

a. Since the GCF of $16y^2$ and $24y$ is $8y$, we write $16y^2$ and $24y$ as the product of $8y$ and one other factor and proceed as follows:

$$16y^2 + 24y = 8y \cdot 2y + 8y \cdot 3$$

$$= 8y(2y + 3) \qquad \text{Factor out the GCF, } 8y.$$

To check, we multiply: $8y(2y + 3) = 8y \cdot 2y + 8y \cdot 3 = 16y^2 + 24y$. Since we obtain the original polynomial, $16y^2 + 24y$, the factorization is correct.

Caution   Remember to factor out the greatest common factor, not just a common factor. If we factored out $4y$ in the previous example, we would get

$$16y^2 + 24y = 4y(4y + 6)$$

However, the terms in red within parentheses have a common factor of 2, indicating that the factoring is not complete.

b. We begin by factoring each term of $25a^3b - 15ab^3$:

$$25a^3b = 5 \cdot 5 \cdot a \cdot a \cdot a \cdot b$$
$$-15ab^3 = -1 \cdot 5 \cdot 3 \cdot a \cdot b \cdot b \cdot b$$

> **Success Tip**
>
> Always verify a factorization by performing the indicated multiplication. The result should be the original polynomial.

Since each term has one factor of 5, one factor of a, and one factor of b in common, and there are no other common factors, $5ab$ is the GCF of the two terms. We write $25a^3b$ and $15ab^3$ as the product of the GCF, $5ab$, and one other factor and proceed as follows:

$$25a^3b - 15ab^3 = 5ab \cdot 5a^2 - 5ab \cdot 3b^2$$
$$= 5ab(5a^2 - 3b^2) \qquad \text{Factor out the GCF, } 5ab.$$

To check, we multiply: $5ab(5a^2 - 3b^2) = 25a^3b - 15ab^3$.

c. We begin by factoring each term of $3xy^2z^3 + 6xyz^3 + 3xz^2$:

$$3xy^2z^3 = 3 \cdot x \cdot y \cdot y \cdot z \cdot z \cdot z$$
$$6xyz^3 = 2 \cdot 3 \cdot x \cdot y \cdot z \cdot z \cdot z$$
$$3xz^2 = 3 \cdot x \cdot z \cdot z$$

Since each term has one factor of 3, one factor of x, and two factors of z in common, and because there are no other common factors, $3xz^2$ is the GCF of the three terms. We write each term as the product of the GCF, $3xz^2$, and one other factor and proceed as follows:

$$3xy^2z^3 + 6xyz^3 + 3xz^2 = 3xz^2 \cdot y^2z + 3xz^2 \cdot 2yz + 3xz^2 \cdot 1$$
$$= 3xz^2(y^2z + 2yz + 1)$$

Check the factorization using multiplication.

Self Check 2   Factor: **a.** $30t^2 + 20t$
b. $9x^4y^2 - 12x^3y^3$
c. $4a^4b^2 + 6a^3b^2 + 2a^2b$

Now Try   **Problems 23, 29, and 33**

A polynomial that cannot be factored is called a **prime polynomial** or an **irreducible polynomial.**

EXAMPLE 3   Factor $9x + 16$, if possible.

Strategy   First, we will determine the GCF of the terms of the polynomial.

Why   If the terms have no common factors (other than 1), this polynomial does not factor.

Solution   We factor each term of $9x + 16$:

$$9x = 3 \cdot 3 \cdot x \qquad 16 = 2 \cdot 2 \cdot 2 \cdot 2$$

Since there are no common factors other than 1, this polynomial cannot be factored. It is a prime polynomial.

Self Check 3   Factor $8a + 27$, if possible.
Now Try   **Problem 31**

EXAMPLE 4   Factor -1 from each polynomial: **a.** $-n^3 + 2n^2 - 8$ **b.** $a - b$

Strategy   We will write each term of the polynomial as the product of -1 and one other factor.

Why   We can then use the distributive property to factor out the -1.

Solution
a. First, we write each term of the polynomial as the product of -1 and another factor. Then we factor out the common factor, -1.

$$-n^3 + 2n^2 - 8 = (-1)n^3 + (-1)(-2n^2) + (-1)8$$
$$= -1(n^3 - 2n^2 + 8) \qquad \text{Factor out } -1.$$
$$= -(n^3 - 2n^2 + 8) \qquad \text{The 1 need not be written.}$$

b. $a - b = -1(-a) + (-1)(b)$
$$= -1(-a + b) \qquad \text{Factor out } -1.$$
$$= -(b - a) \qquad \text{The 1 need not be written. Within the parentheses, write the binomial with the } b\text{-term first.}$$

> **Success Tip**
> To factor out -1, simply change the sign of each term of $-n^3 + 2n^2 - 8$ and write a $-$ sign in front of the parentheses.

Self Check 4   Factor -1 from each polynomial:
a. $-b^4 - 3b^2 + 2$ **b.** $5 - x^2$

Now Try   Problems 37 and 39

EXAMPLE 5   Factor the opposite of the GCF from $-6u^2v^3 + 8u^3v^2$.

Strategy   We will determine the GCF of the terms of the polynomial. Then we will write each term as the product of the opposite of the GCF and one other factor.

Why   We can use the distributive property to factor out the opposite of the GCF.

Solution   Because the GCF of the two terms is $2u^2v^2$, the opposite of the GCF is $-2u^2v^2$. To factor out $-2u^2v^2$, we proceed as follows:

$$-6u^2v^3 + 8u^3v^2 = -2u^2v^2 \cdot 3v - (-2u^2v^2)4u$$
$$= -2u^2v^2(3v - 4u) \qquad \text{Note that the leading coefficient of the polynomial within the parentheses is positive.}$$

> **Success Tip**
> It is standard practice to factor in such a way that the leading coefficient of the polynomial within the parentheses is positive.

Self Check 5   Factor out the opposite of the GCF from $-8a^2b^2 - 12ab^3$.
Now Try   Problem 47

A common factor can have more than one term.

EXAMPLE 6   Factor: **a.** $x(x + 1) + y(x + 1)$
b. $a(x - y + z) - b(x - y + z) + 3(x - y + z)$

Strategy   We will identify the terms of the expression and find their GCF.

Why We can then use the distributive property to factor out the GCF.

Solution

a. The expression has two terms: $\underbrace{x(x + 1)}_{\text{The first term}} + \underbrace{y(x + 1)}_{\text{The second term}}$

The GCF of the terms is the binomial $x + 1$, which can be factored out.

$x(x + 1) + y(x + 1) = (x + 1)(x + y)$ Factor out the GCF, $x + 1$.

b. We can factor out the GCF of the three terms, which is $(x - y + z)$.

$$a(x - y + z) - b(x - y + z) + 3(x - y + z)$$
$$= (x - y + z)(a - b + 3)$$

Self Check 6   Factor: **a.** $c(y^2 + 1) + d(y^2 + 1)$
b. $x(a + b - c) - y(a + b - c)$

Now Try   Problems 51 and 55

3   **Factor by Grouping.**

Suppose that we wish to factor

$$ac + ad + bc + bd$$

Although there is no factor common to all four terms, there is a common factor of a in the first two terms and a common factor of b in the last two terms. We can factor out these common factors to get

$$\boxed{ac + ad} + \boxed{bc + bd} = a(c + d) + b(c + d)$$

We can now factor out the common factor of $c + d$ on the right side:

$$= (c + d)(a + b)$$

The grouping in this type of problem is not always unique. For example, if we rearrange the terms of the polynomial as follows

$$ac + bc + ad + bd$$

and factor c from the first two terms and d from the last two terms, we obtain

$$\boxed{ac + bc} + \boxed{ad + bd} = c(a + b) + d(a + b)$$
$$= (a + b)(c + d)$$ This is equivalent to $(c + d)(a + b)$.

The method used in the previous examples is called **factoring by grouping.**

Factoring by Grouping	1. Group the terms of the polynomial so that each group has a common factor.
	2. Factor out the common factor from each group.
	3. Factor out the resulting common factor. If there is no common factor, regroup the terms of the polynomial and repeat steps 2 and 3.

EXAMPLE 7   Factor by grouping: $2m - 2n + mn - n^2$

Strategy   We will follow the steps for factoring by grouping.

Why   Since the four terms of the polynomial do not have a common factor (other than 1), we will attempt to factor it by grouping.

Solution

Step 1: If we group the terms as shown, the first two terms have a common factor of 2 and the last two terms have a common factor of n.

Step 2: When we factor out the common factor from each group, a common factor of $m - n$ appears.

$$2m - 2n + mn - n^2 = 2(m - n) + n(m - n) \quad \text{Factor out 2 from } 2m - 2n \text{ and } n \text{ from } mn - n^2.$$

Step 3: $= (m - n)(2 + n)$ \quad Factor out the common binomial factor, $m - n$.

We can check the factorization by multiplying:

$$(m - n)(2 + n) = 2m + mn - 2n - n^2$$
$$= 2m - 2n + mn - n^2 \quad \text{Rearrange the terms to get the original polynomial.}$$

Self Check 7   Factor by grouping: $7r - 7s + rs - s^2$

Now Try   **Problems 59 and 63**

By the multiplication property of 1, we know that 1 is a factor of every term. We can use this fact to factor certain polynomials by grouping.

EXAMPLE 8   Factor: **a.** $y^3 + 3y^2 + y + 3$ **b.** $x^2 - bx - x + b$

Strategy   We will follow the steps for factoring by grouping.

Why   Since the four terms of the polynomials do not have a common factor (other than 1), we will attempt to factor them by grouping.

Solution

a. The first two terms, y^3 and $3y^2$, have a common factor of y^2. The only common factor of the last two terms, y and 3, is 1.

$$y^3 + 3y^2 + y + 3 = y^2(y + 3) + 1(y + 3) \quad \text{Factor out } y^2 \text{ from } y^3 + 3y^2.$$
$$\text{Factor out 1 from } y + 3.$$
$$= (y + 3)(y^2 + 1) \quad \text{Factor out the common binomial factor, } y + 3.$$

b. Since x is a common factor of the first two terms, we can factor it out and proceed as follows.

$$x^2 - bx - x + b = x(x - b) - x + b \quad \text{Factor out } x \text{ from } x^2 - bx.$$

When factoring four terms by grouping, if the coefficient of the third term is negative, we often factor out a negative coefficient from the last two terms. If we factor -1 from

$-x + b$, a common binomial factor $x - b$ appears within the second set of parentheses, which we can factor out.

$$x^2 - bx - x + b = x(x - b) - 1(x - b)$$

To factor the first two terms, we factor out x. To factor the last two terms, we factor out -1.

$$= (x - b)(x - 1)$$

Factor out the common factor, $x - b$.

> **Self Check 8** Factor: **a.** $x^5 + 17x^4 + x + 17$
> **b.** $a^2 - ab - a + b$
>
> **Now Try** Problems 67 and 69

EXAMPLE 9 Factor: $5x^3 - 8 + 10x^2 - 4x$

Strategy We will follow the steps for factoring by grouping.

Why Since the four terms of the polynomial do not have a common factor (other than 1), we will attempt to factor it by grouping.

Solution Since the first two terms do not have a common factor, we cannot factor the polynomial in its current form.

$$\underbrace{5x^3 - 8}_{\substack{\text{No common factor} \\ \text{(other than 1)}}} + \underbrace{10x^2 - 4x}_{\text{GCF} = 2x}$$

However, we can use the commutative property of addition to reorder the terms and try to factor by grouping again.

$$5x^3 - 8 + 10x^2 - 4x = \underbrace{5x^3 + 10x^2}\; \underbrace{- 8 - 4x}$$

Reorder the middle terms and regroup.

$$= 5x^2(x + 2) - 4(2 + x)$$

Factor $5x^2$ from $5x^3 + 10x^2$ and -4 from $-8 - 4x$.

$$= 5x^2(x + 2) - 4(x + 2)$$

Reorder the terms of $(2 + x)$.

$$= (x + 2)(5x^2 - 4)$$

Factor out the GCF, $x + 2$.

> **Self Check 9** Factor: $y^3 - 6 + 3y^2 - 2y$
>
> **Now Try** Problem 71

Success Tip
An equivalent factorization, $(5x^2 - 4)(x + 2)$, results if the terms are arranged as $5x^3 - 4x + 10x^2 - 8$ or as $5x^3 - 4x - 8 + 10x^2$.

To **factor a polynomial completely,** it is often necessary to factor more than once. When factoring a polynomial, *always look for a common factor first.*

EXAMPLE 10 Factor: $3x^3y - 4x^2y^2 - 6x^2y + 8xy^2$

Strategy Since all four terms have a common factor of xy, we factor it out first. Then we will try to factor the resulting polynomial by grouping.

Why Factoring out the GCF first makes factoring by any method easier.

Solution   We begin by factoring out the common factor of xy.

$$3x^3y - 4x^2y^2 - 6x^2y + 8xy^2 = xy(3x^2 - 4xy - 6x + 8y)$$

We can now factor $3x^2 - 4xy - 6x + 8y$ by grouping:

$$3x^3y - 4x^2y^2 - 6x^2y + 8xy^2$$

$$= xy(3x^2 - 4xy - 6x + 8y)$$

$$= xy[x(3x - 4y) - 2(3x - 4y)] \qquad \text{Factor } x \text{ from } 3x^2 - 4xy \text{ and } -2 \text{ from } -6x + 8y.$$

$$= xy[(3x - 4y)(x - 2)] \qquad\qquad \text{Factor out } 3x - 4y.$$

$$= xy(3x - 4y)(x - 2) \qquad\qquad\; \text{The brackets [ ] are not needed.}$$

Because xy, $3x - 4y$, and $x - 2$ are prime, no further factoring can be done; The factorization is complete.

▷ **Self Check 10**   Factor: $3a^3b + 3a^2b - 2a^2b^2 - 2ab^2$

    Now Try   **Problem 75**

4   **Use Factoring to Solve Formulas for a Specified Variable.**

Factoring is often required to solve a formula for one of its variables.

EXAMPLE 11   *Electronics.*   The formula $r_1r_2 = rr_2 + rr_1$ is used in electronics to relate the combined resistance, r, of two resistors wired in parallel. The variable r_1 represents the resistance of the first resistor, and the variable r_2 represents the resistance of the second. Solve for r_2.

Strategy   To isolate r_2 on one side of the equation, we will get all the terms involving r_2 on the left side and all the terms not involving r_2 on the right side.

Why   To *solve a formula for a specified variable* means to isolate that variable on one side of the equation, with all other variables and constants on the opposite side.

Solution

$$\overbrace{r_1r_2 = rr_2}^{} + rr_1 \qquad\qquad \text{We want to isolate this variable on one side of the equation.}$$

$$r_1r_2 - rr_2 = rr_1 \qquad\qquad \text{To eliminate } rr_2 \text{ on the right side, subtract } rr_2 \text{ from both sides.}$$

$$r_2(r_1 - r) = rr_1 \qquad\qquad \text{On the left side, factor out } r_2 \text{ from } r_1r_2 - rr_2.$$

$$\frac{\overset{1}{r_2(\cancel{r_1 - r})}}{\underset{1}{\cancel{r_1 - r}}} = \frac{rr_1}{r_1 - r} \qquad\qquad \text{To isolate } r_2 \text{ on the left side, divide both sides by } r_1 - r.$$

$$r_2 = \frac{rr_1}{r_1 - r}$$

▷ **Self Check 11**   Solve $A = p + prt$ for p.

    Now Try   **Problems 79 and 83**

▷ **ANSWERS TO SELF CHECK**   **1.** $3x^2y$ **2. a.** $10t(3t + 2)$ **b.** $3x^3y^2(3x - 4y)$ **c.** $2a^2b(2a^2b + 3ab + 1)$
3. A prime polynomial   **4. a.** $-(b^4 + 3b^2 - 2)$ **b.** $-(x^2 - 5)$ **5.** $-4ab^2(2a + 3b)$
6. a. $(y^2 + 1)(c + d)$ **b.** $(a + b - c)(x - y)$ **7.** $(r - s)(7 + s)$ **8. a.** $(x + 17)(x^4 + 1)$
b. $(a - b)(a - 1)$ **9.** $(y + 3)(y^2 - 2)$ **10.** $ab(3a - 2b)(a + 1)$ **11.** $p = \dfrac{A}{1 + rt}$

STUDY SET
5.5

VOCABULARY

Fill in the blanks.

1. When we write $2x + 4$ as $2(x + 2)$, we say that we have
   _____ $2x + 4$.
2. When we factor a polynomial, we write a sum of terms as a
   _____ of factors.
3. The abbreviation GCF stands for _____ _____ _____.
4. If a polynomial cannot be factored, it is called a _____
   polynomial or an irreducible polynomial.
5. The terms $t(t - 5)$ and $9(t - 5)$ have the common _____
   factor $t - 5$.
6. To factor $ab + 6a + 2b + 12$ by _____, we begin by
   writing $a(b + 6) + 2(b + 6)$.

CONCEPTS

7. The prime factorizations of three terms are shown. Find their
   GCF.

   $$12x^2y^3 = 2 \cdot 2 \cdot 3 \cdot x \cdot x \cdot y \cdot y \cdot y$$
   $$18xy^4 = 2 \cdot 3 \cdot 3 \cdot x \cdot y \cdot y \cdot y \cdot y$$
   $$126x^3y^2 = 2 \cdot 3 \cdot 3 \cdot 7 \cdot x \cdot x \cdot x \cdot y \cdot y$$

8. Fill in the blanks to complete each factorization.
   a. $8x^3 + 6x^2 + 2x = 2x(4x^2 + 3x + \quad)$
   b. $-9x + 6 = -(9x \quad 6)$
9. Consider the polynomial $5t - 25 + st - 5s$.
   a. How many terms does the polynomial have?
   b. Is there a common factor of all the terms, other than 1?
   c. What is the GCF of the first two terms and what is the GCF
   of the last two terms?
10. Check to determine whether each factorization is correct.
    a. $9a^4 + 15a^2 = 3a^2(3a^2 + 5)$
    b. $3x^3 + 2x^2 + 6x + 4 = (3x + 1)(x^2 + 4)$

GUIDED PRACTICE

Find the GCF of each list of terms. See Example 1.

11. $14x$, $24x^2$
12. $20a^2$, $35a$

13. $45a^5b$, $30a^4$
14. $36x^3y^2$, $27xy$

15. $16y^4$, $40y^2$, $24y^3$
16. $16m^4$, $40m^6$, $28m^3$

17. $24r^3s^5$, $36r^8s^4$, $48r^4s^4$
18. $28m^6n^6$, $56m^4n^5$, $42m^7n^6$

19. $18x^4y^3z^9$, $54xy^2z^6$, $36xy^5z^8$
20. $56x^2y^6z^7$, $24x^3y^7z^9$, $40x^2y^5z^3$

21. $10(c - d)$, $c(c - d)$
22. $16(x + y)$, $7x(x + y)$

Factor, if possible. See Examples 2 and 3.

23. $24x + 16$
24. $24y - 30$

25. $3y^3 + 5y^2 - 9y$
26. $2x^3 - 6x^2 + 11x$

27. $45a^2 - 9a$
28. $21x^2 + 7x$

29. $15x^2y - 10x^2y^2$
30. $13b^2c^3 - 26b^2c$

31. $14r + 15$
32. $100m + 33$

33. $27a^9b^4c^5 - 12a^7b^4c^2 + 30a^6b^4c^3$

34. $25t^{10}u^2v^5 - 10t^7u^2v^4 - 55t^6u^2v^4$

Factor -1 from each polynomial. See Example 4.

35. $-a - b$
36. $-2x - y$

37. $-5xy + y - 4$
38. $-7m - 12n + 16$

39. $x - 2y$
40. $r - 3s$

41. $p^2 + p - 17$
42. $2x^3 - x^2 + 1$

*Factor each polynomial by factoring out the opposite of the GCF.
See Example 5.*

43. $-8a - 16$
44. $-6b - 30$

45. $-6x^2 - 3xy$

46. $-15y^3 - 25y^2$

47. $-18a^2b + 12ab^2$

48. $-21t^5 - 28t^3$

49. $-8a^4c^8 + 28a^3c^8 - 20a^2c^9$ **50.** $-30x^{10}y + 24x^9y^2 - 60x^8y^2$

Factor. See Example 6.

51. $4(x + y) + t(x + y)$
52. $5(a - b) - t(a - b)$
53. $(a - b)r - (a - b)s + (a - b)t$
54. $(x + y)u + (x + y)v - (x + y)z$
55. $3(m + n + p) + x(m + n + p)$
56. $x(x - y - z) + y(x - y - z)$
57. $3x(x + 7)^2 - 2(x + 7)^2$
58. $15x^2(x - y)^3 - 11(x - y)^3$

Factor by grouping. See Example 7.

59. $ax + 8x + ay + 8y$ **60.** $cd + 2c + 5d + 10$

61. $3c - c^2 + 3d - cd$ **62.** $x^2 - xy + 4x - 4y$

63. $a^2 + ab - 4a - 4b$ **64.** $7u + uv - 7v - v^2$

65. $t^3 - 3t^2 - 7t + 21$ **66.** $b^3 - 4b^2 - 3b + 12$

Factor by grouping. See Example 8.

67. $x^2 + yx + x + y$ **68.** $cd + d^2 + c + d$

69. $1 - n - m + mn$ **70.** $1 - c - 3t + 3ct$

Rearrange the terms and factor by grouping. See Example 9.

71. $y^3 - 12 + 3y - 4y^2$ **72.** $h^3 - 8 + h - 8h^2$

73. $st + rv + sv + rt$ **74.** $2tx + 3dy + ty + 6dx$

Factor by grouping. Factor out the GCF first. See Example 10.

75. $28a^3b^3c + 14a^3c - 4b^3c - 2c$
76. $12x^3z + 12xy^2z - 8x^2yz - 8y^3z$
77. $mpx + mqx + npx + nqx$
78. $abd - abe + acd - ace$

Solve for the specified variable or expression. See Example 11.

79. $d_1d_2 = fd_2 + fd_1$ for f **80.** $2g = ch + dh$ for h

81. $b^2x^2 + a^2y^2 = a^2b^2$ for a^2 **82.** $d_1d_2 = fd_2 + fd_1$ for d_1

83. $rx - ty = by$ for y **84.** $Ar - c^2d = dm^2$ for d

85. $S(1 - r) = a - lr$ for r **86.** $Sn = (n - 2)180$ for n

Factor.

87. $-63u^3 + 28u^2$ **88.** $-56x^4 - 72x^3$

89. $a^3b^2 - 3 + a^3 - 3b^2$ **90.** $2ax^2 - 4 + a - 8x^2$

91. $45x^{10}y^3 - 63x^7y^7 + 81x^{10}y^{10}$ **92.** $48u^6v^6 - 16u^4v^4 - 3u^6v^3$

93. $ar^2t - br^2t + as^2t - bs^2t$ **94.** $2a^2x^2 - 4x^2 + 10a^2 - 20$

95. $16a^3b^3 - 27c^3d^3$ **96.** $2x^2 + x - 9y^2$

97. $\dfrac{3}{5}ax^4 + \dfrac{1}{5}bx^2 - \dfrac{4}{5}ax^3$ **98.** $\dfrac{3}{2}t^2y^4 - \dfrac{1}{2}ty^4 - \dfrac{5}{2}ry^3$

99. $a(2x + y) - b(2x + y) + c(2x + y)$
100. $b(a - b + c) - c(a - b + c)$
101. $45y^{12} + 30y^{10} + 25y^8 - 5y^6$
102. $2a^4 + 4a^8 + 6a^{12} + 8a^{16}$
103. $a^2x + bx - a^2 - b$
104. $x^2y - ax - xy + a$
105. $ab + b^2 + bc + ac + bc + c^2$
106. $x^2 + xy + xz + xy + y^2 + zy$

107. GEOMETRIC FORMULAS

   a. Write an expression that gives the area of the part of the figure that is shaded red.

   b. Do the same for the part of the figure that is shaded blue.

   c. Add the results from parts (a) and (b) and then factor that expression. What important formula from geometry do you obtain?

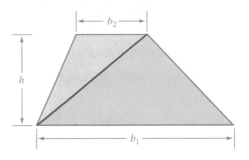

b. The trinomial $9a^4 - 30a^2b^2 + 25b^4$ is a perfect-square trinomial because

- The first term $9a^4$ is the square of $3a^2$: $(3a^2)^2 = 9a^4$.
- The last term $25b^4$ is the square of $-5b^2$: $(-5b^2)^2 = 25b^4$.
- The middle term $-30a^2b^2$ is twice the product of $3a^2$ and $-5b^2$: $2(3a^2)(-5b^2) = -30a^2b^2$.

We can match the trinomial to a special-product rule to find the factorization.

$$9a^4 - 30a^2b^2 + 25b^4 = (3a^2)^2 + 2 \cdot 3a^2 \cdot (-5b^2) + (-5b^2)^2 = (3a^2 - 5b^2)^2$$

Thus, $9a^4 - 30a^2b^2 + 25b^4 = (3a^2 - 5b^2)^2$. Check by multiplying.

Self Check 1   Factor: **a.** $m^2 + 12m + 36$
                             b. $49b^4 - 28b^2c^2 + 4c^4$

***Now Try*   Problems 19 and 23**

2 **Factor Trinomials of the Form $x^2 + bx + c$.**

To begin the discussion of trinomial factoring, we consider trinomials of the form $x^2 + bx + c$, such as

$$x^2 + 10x + 24, \qquad x^2 + 14x + 24, \qquad t^2 - 7t + 12, \qquad \text{and} \qquad y^2 + 2xy - 15x^2$$

In each case, the **leading coefficient**—the coefficient of the squared variable—is 1.

To develop a method for factoring such trinomials, we will find the product of $x + 6$ and $x + 4$ and make some observations about the result.

$$(x + 6)(x + 4) = x \cdot x + 4x + 6x + 6 \cdot 4 \qquad \text{\textit{Use the FOIL method to multiply.}}$$
$$= x^2 + 10x + 24$$

First term ⌐ Middle term ⌐ Last term

The result is a trinomial, where

- The first term, x^2, is the product of x and x.
- The last term, 24, is the product of 6 and 4.
- The coefficient of the middle term, 10, is the sum of 6 and 4.

These observations suggest a strategy to use to factor trinomials with a leading coefficient of 1.

Factoring Trinomials Whose Leading Coefficient is 1	To factor a trinomial of the form $x^2 + bx + c$, find two numbers whose product is c and whose sum is b.

1. If c is positive, the numbers have the same sign.
2. If c is negative, the numbers have different signs.

Then write the trinomial as a product of two binomials. You can check by multiplying.

$$x^2 + bx + c = \left(x \,\boxed{}\right)\left(x \,\boxed{}\right)$$

The product of these numbers must
be c and their sum must be b.

EXAMPLE 2   Factor: $x^2 + 11x + 24$

Strategy   We will assume that $x^2 + 11x + 24$ is the product of two binomials. We must find the terms of the binomials.

Why   Since the terms of $x^2 + 11x + 24$ do not have a common factor (other than 1), the only option is to try to factor it as the product of two binomials.

Solution   We represent the binomials using two sets of parentheses. Since the first term of the trinomial is x^2, we enter x and x as the first terms of the binomial factors.

$$x^2 + 11x + 24 = \left(x \; \boxed{}\right)\left(x \; \boxed{}\right)$$ Because $x \cdot x$ will give x^2.

The second terms of the binomials must be two integers whose product is 24 and whose sum is 11. Since the integers must have a positive product and a positive sum, we consider only pairs of positive integer factors of 24. All such possible integer pairs are listed in the table.

Positive factors of 24	Sum of the positive factors of 24	
$1 \cdot 24 = 24$	$1 + 24 = 25$	
$2 \cdot 12 = 24$	$2 + 12 = 14$	
$3 \cdot 8 = 24$	$3 + 8 = 11$	⟵ This is the one to choose.
$6 \cdot 4 = 24$	$6 + 4 = 10$	

The third row of the table contains the correct pair of integers 3 and 8, whose product is 24 and whose sum is 11. To complete the factorization, we enter 3 and 8 as the second terms of the binomial factors.

$$x^2 + 11x + 24 = (x + 3)(x + 8)$$

We can check this factorization by multiplying:

$$(x + 3)(x + 8) = x^2 + 8x + 3x + 24$$ Use the FOIL method.
$$= x^2 + 11x + 24$$ This is the original trinomial.

By the commutative property of multiplication, the order of binomial factors in a factorization does not matter. Thus, we can also write $x^2 + 11x + 24 = (x + 8)(x + 3)$.

> **The Language of Algebra**
> Make sure you understand the following vocabulary: Many trinomials factor as the product of two binomials.
>
>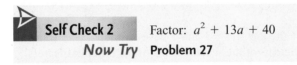

> ⊳ **Self Check 2**   Factor: $a^2 + 13a + 40$
>
> *Now Try*   **Problem 27**

EXAMPLE 3   Factor: $-8t^2 + t^4 + 12$

Strategy   We will write the terms of the trinomial in descending powers of t.

Why   It is easier to factor a trinomial if its terms are written in descending powers of the variable.

> *Caution*
> Always write a trinomial in descending powers of one variable before factoring.

Solution   We can use the commutative property of addition to reorder the terms:

$$-8t^2 + t^4 + 12 = t^4 - 8t^2 + 12$$

Since the first term of the trinomial is t^4, the first term of each binomial factor must be t^2.

$$t^4 - 8t^2 + 12 = \left(t^2 \;\boxed{}\;\right)\!\left(t^2 \;\boxed{}\;\right)$$ Because $t^2 \cdot t^2$ will give t^4.

The second terms of the binomials must be two integers whose product is 12 and whose sum is -8. Since the integers must have a positive product and a negative sum, we only consider pairs of negative integer factors of 12. The possible pairs are listed in the table.

Factors of 12	Sum of factors
$-1(-12) = 12$	$-1 + (-12) = -13$
$-2(-6) = 12$	$-2 + (-6) = -8$
$-3(-4) = 12$	$-3 + (-4) = -7$

The second row of the table contains the correct pair of integers -2 and -6, whose product is 12 and whose sum is -8. To complete the factorization, we enter -2 and -6 as the second terms of the binomial factors.

$$t^4 - 8t^2 + 12 = (t^2 - 2)(t^2 - 6)$$ Check using multiplication.

> **Self Check 3**   Factor: $-9m^2 + m^4 + 18$
>
> **Now Try**   Problem 33

EXAMPLE 4   Factor: $2x^2y^2 + 4xy^3 - 30y^4$

Strategy   We will factor out the GCF of $2y^2$ and factor the resulting trinomial.

Why   The first step in factoring any polynomial is to factor out the GCF. Factoring out the GCF first makes factoring easier.

Solution   Each term of the trinomial has a common factor of $2y^2$, which we can factor out.

$$2x^2y^2 + 4xy^3 - 30y^4 = 2y^2(x^2 + 2xy - 15y^2)$$

Next, we factor $x^2 + 2xy - 15y^2$, which is a trinomial in two variables, x and y. Since its first term is x^2, the first term of each binomial factor must be x. Since its third term contains y^2, the last term of each binomial factor must contain y. We need to determine the coefficient of each y-term.

$$2y^2(x^2 + 2xy - 15y^2) = 2y^2\!\left(x \;\boxed{}\; y\right)\!\left(x \;\boxed{}\; y\right)$$ Because $x \cdot x$ will give x^2 and $y \cdot y$ will give y^2.

> **Caution**
> When factoring a polynomial, always factor out the GCF first. For multistep factorizations, remember to write the GCF in the final factored form:
>
> $$2y^2(x - 3y)(x + 5y)$$
> ↑
> GCF

Since the integers -3 and 5 have a product is -15 and a sum of 2, they are entered as the coefficients of the second terms of the binomial factors.

$$2y^2(x^2 + 2xy - 15y^2) = 2y^2(x - 3y)(x + 5y)$$

We can check by multiplying.

> **Self Check 4**   Factor: $3a^2b^2 + 6ab^3 - 105b^4$
>
> **Now Try**   Problem 41

 3   **Factor Trinomials of the Form $ax^2 + bx + c$.**

There are more combinations of factors to consider when factoring trinomials with leading coefficients other than 1.

EXAMPLE 5   Factor: $5x^2 + 7x + 2$

Strategy   We will assume that this trinomial is the product of two binomials. To find their terms, we will make educated guesses and then check them using multiplication.

Why   Since the terms of the trinomial do not have a common factor (other than 1), the only option is to try to factor it as the product of two binomials.

Solution   We represent the binomial factors using two sets of parentheses.

Since the first term of the trinomial $5x^2 + 7x + 2$ is $5x^2$, the first terms of the binomial factors must be $5x$ and x.

The second terms of the binomials must be two integers whose product is 2.

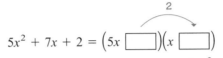

Because the coefficients of the terms of $5x^2 + 7x + 2$ are positive, we only consider pairs of positive integer factors of 2. Since there is only one such pair, $1 \cdot 2$, we can use 1 and 2 as the second terms of the binomials, or we can reverse the order and enter 2 and 1.

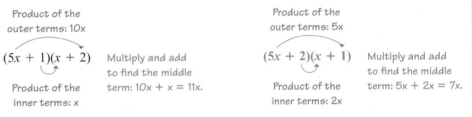

The second possibility shown in blue is correct, because it gives a middle term of $7x$. Thus,

$$5x^2 + 7x + 2 = (5x + 2)(x + 1)$$

We can verify this result by multiplication:

> ***Check:*** $(5x + 2)(x + 1) = 5x^2 + 5x + 2x + 2$
> $= 5x^2 + 7x + 2$ This is the original trinomial.

Self Check 5   Factor: $5w^2 + 11w + 2$

Now Try   Problem 45

EXAMPLE 6   Factor: $3p^2 - 4p - 4$

Strategy   We will assume that this trinomial is the product of two binomials. To find their terms, we will make educated guesses and then check them using multiplication.

Why Since the terms of the trinomial do not have a common factor (other than 1), the only option is to try to factor it as the product of two binomials.

Solution To factor the trinomial, we note that the first terms of the binomial factors must be $3p$ and p to give the first term of $3p^2$.

$$3p^2 - 4p - 4 = \left(3p \,\boxed{}\right)\left(p \,\boxed{}\right) \quad \text{Because } 3p \cdot p \text{ will give } 3p^2.$$

The second terms of the binomials must be two integers whose product is -4. There are three such pairs: $1(-4)$, $-1(4)$, and $-2(2)$. When these pairs are entered, and then reversed, as second terms of the binomials, there are six possibilities to consider.

For 1 and −4:

$(3p + 1)(p - 4)$ $(3p - 4)(p + 1)$

$-12p + p = -11p$ $3p + (-4p) = -p$

For −1 and 4:

$(3p - 1)(p + 4)$ $(3p + 4)(p - 1)$

$12p + (-p) = 11p$ $-3p + 4p = p$

For −2 and 2:

$(3p - 2)(p + 2)$ $(3p + 2)(p - 2)$

$6p + (-2p) = 4p$ $-6p + 2p = -4p$

Of these possibilities, only the one in blue gives the required middle term of $-4p$. Thus,

$$3p^2 - 4p - 4 = (3p + 2)(p - 2)$$

> **Notation**
> By the commutative property of multiplication, the factors of a trinomial can be written in either order. Thus, we could also write:
>
> $3p^2 - 4p - 4 = (p - 2)(3p + 2)$

▷ **Self Check 6**   Factor: $2q^2 - 17q - 9$

   Now Try   **Problem 49**

EXAMPLE 7   Factor $9t^2 - 15t + 7$, if possible.

Strategy We will assume that the trinomial is the product of two binomials and find the terms of the binomials.

Why Since the terms of the trinomial do not have a common factor (other than 1), the only option is to try to factor it as the product of two binomials.

Solution To try to factor the trinomial, we note that the first terms of the binomial factors could be $9t$ and t to give the first term of $9t^2$.

$$9t^2 - 15t + 7 = \left(9t \,\boxed{}\right)\left(t \,\boxed{}\right) \quad \text{Because } 9t \cdot t \text{ will give } 9t^2.$$

The second terms of the binomials must be two integers whose product is 7. Since the last term of $9t^2 - 15t + 7$ is positive and the coefficient of the middle term is negative, we only consider negative integer factors of the last term. There is only one such pair, $-1(-7)$. When this pair is entered, and then reversed, as second terms of the binomials, there are two possibilities to consider.

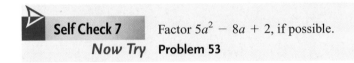

$$(9t - 1)(t - 7) \quad -63t + (-t) = -64t \qquad (9t - 7)(t - 1) \quad -9t + (-7t) = -16t$$

Since neither of the possible factorizations gives the correct middle term of $-15t$, the trinomial $9t^2 - 15t + 7$ does not factor. It is a prime trinomial.

Self Check 7   Factor $5a^2 - 8a + 2$, if possible.
Now Try **Problem 53**

Because guesswork is often necessary, it is difficult to give specific rules for factoring trinomials with leading coefficients other than 1. However, the following hints are helpful when using the **trial-and-check method.**

Factoring Trinomials with Leading Coefficients Other Than 1

To factor trinomials with leading coefficients other than 1:

1. Factor out any GCF (including -1 if that is necessary to make a positive in a trinomial of the form $ax^2 + bx + c$).
2. Write the trinomial as a product of two binomials. The coefficients of the first terms of each binomial factor must be factors of a, and the last terms must be factors of c.

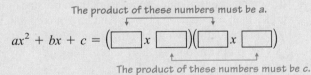

The product of these numbers must be *a*.

$$ax^2 + bx + c = (\boxed{}x \boxed{})(\boxed{}x \boxed{})$$

The product of these numbers must be *c*.

3. If c is positive, the signs within the binomial factors match the sign of b. If c is negative, the signs within the binomial factors are opposites.
4. Try combinations of first terms and second terms until you find the one that gives the correct middle term. If no combination works, the trinomial is prime.
5. Check by multiplying.

EXAMPLE 8   Factor: $-15x^2 + 25xy + 60y^2$

Strategy   We will factor out the opposite of the GCF first. Then we will factor the resulting trinomial.

Why   It is easier to factor trinomials that have a positive leading coefficient.

Solution   Factor out -5 from each term.

$$-15x^2 + 25xy + 60y^2 = -5(3x^2 - 5xy - 12y^2) \quad \text{The opposite of the GCF is } -5.$$

To factor $3x^2 - 5xy - 12y^2$, we examine its terms.

- Since the first term is $3x^2$, the first terms of the binomial factors must be $3x$ and x.
- Since the sign of the first term of the trinomial is positive and the sign of the last term is negative, the signs within the binomial factors will be opposites.
- Since the last term of the trinomial contains y^2, the second terms of the binomial factors must contain y.

$$-5(3x^2 - 5xy - 12y^2) = -5\left(3x\boxed{}y\right)\left(x\boxed{}y\right)$$

The product of the last terms must be $-12y^2$, and the sum of the product of the outer terms and the product of the inner terms must be $-5xy$.

$$-5(3x^2 - 5xy - 12y^2) = -5\left(3x\boxed{}y\right)\left(x\boxed{}y\right)$$

Outer + Inner = $-5xy$

Since $1(-12)$, $2(-6)$, $3(-4)$, $-1(12)$, $-2(6)$, and $-3(4)$ all give a product of -12, there are 12 combinations to consider.

$(3x + 1y)(x - 12y)$	$(3x - 12y)(x + 1y)$ ← $3x - 12y$ has a common factor 3.
$(3x + 2y)(x - 6y)$	$(3x - 6y)(x + 2y)$ ← $3x - 6y$ has a common factor 3.
$(3x + 3y)(x - 4y)$	$(3x - 4y)(x + 3y)$ ← $3x + 3y$ has a common factor 3.
$(3x - 1y)(x + 12y)$	$(3x + 12y)(x - 1y)$ ← $3x + 12y$ has a common factor 3.
$(3x - 2y)(x + 6y)$	$(3x + 6y)(x - 2y)$ ← $3x + 6y$ has a common factor 3.
$(3x - 3y)(x + 4y)$	$(3x + 4y)(x - 3y)$ ← This is the one to choose.

The combinations in blue cannot work, because one of the factors has a common factor. This implies that $3x^2 - 5xy - 12y^2$ would have a common factor, which it doesn't.

After mentally trying the remaining combinations, we find that only $(3x + 4y)(x - 3y)$ gives the proper middle term of $-5xy$. Thus,

$$-15x^2 + 25xy + 60y^2 = -5(3x^2 - 5xy - 12y^2)$$
$$= -5(3x + 4y)(x - 3y)$$

▷ **Self Check 8**   Factor: $-6x^2 - 15xy - 6y^2$

 Now Try   **Problem 61**

EXAMPLE 9   Factor: $6y^3 + 13x^2y^3 + 6x^4y^3$

Strategy   We write the expression in descending powers of x.

Why   It is easier to factor a trinomial if its terms are written in descending powers of one variable.

Solution   We write the expression in descending powers of x and factor out the greatest common factor, y^3.

$$6y^3 + 13x^2y^3 + 6x^4y^3 = 6x^4y^3 + 13x^2y^3 + 6y^3$$
$$= y^3(6x^4 + 13x^2 + 6)$$

To factor $6x^4 + 13x^2 + 6$, we examine its terms.

- Since the first term is $6x^4$, the first terms of the binomial factors must be either $2x^2$ and $3x^2$ or x^2 and $6x^2$.

$$6x^4 + 13x^2 + 6 = \left(2x^2 \;\boxed{}\right)\left(3x^2 \;\boxed{}\right) \text{ or } \left(x^2 \;\boxed{}\right)\left(6x^2 \;\boxed{}\right)$$

- Since the signs of the middle term and the last term of the trinomial are positive, the signs within each binomial factor will be positive.

- Since the product of the last terms of the binomial factors must be 6, we must find two numbers whose product is 6 that will lead to a middle term of $13x^2$.

After trying some combinations, we find the one that works.

$$6x^4y^3 + 13x^2y^3 + 6y^3 = y^3(6x^4 + 13x^2 + 6)$$
$$= y^3(2x^2 + 3)(3x^2 + 2)$$

▷ **Self Check 9**   Factor: $4b + 11a^2b + 6a^4b$

Now Try **Problem 65**

4 **Use Substitution to Factor Trinomials.**

For more complicated expressions, especially those involving a quantity within parentheses, a substitution sometimes helps to simplify the factoring process.

EXAMPLE 10   Factor: $(x + y)^2 + 7(x + y) + 12$

Strategy   We will use a substitution where we will replace each expression $x + y$ with the variable z and factor the resulting trinomial.

Why   The resulting trinomial will be easier to factor because it will be in only one variable, z.

Solution   If we use the substitution $z = x + y$, we obtain

$$(x + y)^2 + 7(x + y) + 12 = z^2 + 7z + 12 \quad \text{Replace x + y with z.}$$
$$= (z + 4)(z + 3) \quad \text{Factor the trinomial.}$$

To find the factorization of $(x + y)^2 + 7(x + y) + 12$, we substitute $x + y$ for each z in the expression $(z + 4)(z + 3)$.

$$(z + 4)(z + 3) = (x + y + 4)(x + y + 3)$$

Thus, $(x + y)^2 + 7(x + y) + 12 = (x + y + 4)(x + y + 3)$

▷ **Self Check 10**   Factor: $(a + b)^2 - 3(a + b) - 10$

Now Try **Problem 73**

Notation
In Example 11b, the middle term, $13xy$, may be expressed as $15xy - 2xy$ or as $-2xy + 15xy$ when using factoring by grouping. The resulting factorizations will be equivalent.

$$10x^2 + 13xy - 3y^2 = 10x^2 + 15xy - 2xy - 3y^2 \quad \text{Express } 13xy \text{ as } 15xy - 2xy.$$
$$= 5x(2x + 3y) - y(2x + 3y) \quad \begin{array}{l}\text{Factor out } 5x \text{ from } 10x^2 + 15xy.\\ \text{Factor out } -y \text{ from } -2xy - 3y^2.\end{array}$$
$$= (2x + 3y)(5x - y) \quad \text{Factor out the GCF, } 2x + 3y.$$

Thus, $10x^2 + 13xy - 3y^2 = (2x + 3y)(5x - y)$. Check by multiplying.

Self Check 11   Factor by grouping: **a.** $m^2 + 13m + 42$
b. $15a^2 + 17ab - 4b^2$

Now Try   Problems 29, 45, 49, and 57

EXAMPLE 12   Factor by grouping: $12x^5 - 17x^4 + 6x^3$

Strategy   We will factor out the GCF of x^3 and factor the resulting trinomial using the grouping method.

Why   The first step in factoring any polynomial is to factor out the GCF.

Solution   The GCF of the three terms of the trinomial is x^3.

$$12x^5 - 17x^4 + 6x^3 = x^3(12x^2 - 17x + 6)$$

To factor $12x^2 - 17x + 6$, we must find two integers whose product is $12(6) = 72$ and whose sum is -17. Two such numbers are -8 and -9. They serve as the coefficients of $-8x$ and $-9x$, the two terms that we use to represent the middle term, $-17x$, of the trinomial.

Key number $= 72$ $b = -17$

Factors	Sum
$-8(-9) = 72$	$-8 + (-9) = -17$

$$12x^2 - 17x + 6 = 12x^2 - 8x - 9x + 6 \quad \begin{array}{l}\text{Express } -17x \text{ as } -8x - 9x.\\ (-9x - 8x \text{ could also be used.})\end{array}$$
$$= 4x(3x - 2) - 3(3x - 2) \quad \text{Factor out } 4x \text{ and factor out } -3.$$
$$= (3x - 2)(4x - 3) \quad \text{Factor out } 3x - 2.$$

The complete factorization of the original trinomial is

$$12x^5 - 17x^4 + 6x^3 = x^3(3x - 2)(4x - 3) \quad \text{Don't forget to write the GCF, } x^3.$$

Check the factorization by multiplying.

Self Check 12   Factor by grouping: $21a^4 - 13a^3 + 2a^2$
Now Try   Problem 65

ANSWERS TO SELF CHECKS   **1. a.** $(m + 6)^2$ **b.** $(7b^2 - 2c^2)^2$ **2.** $(a + 5)(a + 8)$
3. $(m^2 - 3)(m^2 - 6)$ **4.** $3b^2(a - 5b)(a + 7b)$ **5.** $(5w + 1)(w + 2)$ **6.** $(2q + 1)(q - 9)$
7. A prime polynomial   **8.** $-3(x + 2y)(2x + y)$ **9.** $b(2a^2 + 1)(3a^2 + 4)$ **10.** $(a + b + 2)(a + b - 5)$
11. a. $(m + 7)(m + 6)$ **b.** $(3a + 4b)(5a - b)$ **12.** $a^2(7a - 2)(3a - 1)$

STUDY SET
5.6

VOCABULARY

Fill in the blanks.

1. Since $y^2 + 2y + 1 = (y + 1)^2$, we call $y^2 + 2y + 1$ a _____-square trinomial.

2. The statement $x^2 - x - 12 = (x - 4)(x + 3)$ shows that the trinomial $x^2 - x - 12$ factors into the product of two _____.

3. The _____ coefficient of the trinomial $x^2 - 3x + 2$ is 1, the _____ of the middle term is -3, and the last term is ___ . The trinomial is written in _____ powers of x.

4. A trinomial is factored _____ when no factor can be factored further. A _____ polynomial cannot be factored by using only integers.

CONCEPTS

5. Consider $3x^2 - x + 16$. What is the sign of the
  a. First term?
  b. Middle term?
  c. Last term?

6. Find two integers whose
  a. Product is 10 and whose sum is 7.
  b. Product is 20 and whose sum is -9.
  c. Product is -6 and whose sum is 1.
  d. Product is -32 and whose sum is -4.

7. Complete the table and the sentence below it.

Factors of 8	Sum of the factors of 8
$1(8) = 8$	
$2(4) = 8$	
$-1(-8) = 8$	
$-2(-4) = 8$	

The numbers -1 and -8 are two integers whose _____ is 8 and whose _____ is -9.

8. Fill in the blanks. When factoring a trinomial, we write it in _____ powers of the variable. Then we factor out any _____ (including -1 if that is necessary to make the leading coefficient _____).

9. Use a check to determine whether $(3t - 1)(5t - 6)$ is the correct factorization of $15t^2 - 19t + 6$.

10. Complete the key number table and the sentence below.

Key number $= 12$

Negative factors of 12	Sum of factors of 12
$-1(-12) = 12$	
	$-2 + (-6) = -8$
$-3(-4) = 12$	

The numbers -3 and -4 are two integers whose _____ is 12 and whose _____ is -7.

Complete each factorization.

11. $x^2 - 6x + 8 = (x - 4)(x \quad 2)$
12. $x^2 - 3x - 18 = (x - 6)(x \quad 3)$
13. $2a^2 + 9a + 4 = (2a \quad 1)(a + 4)$
14. $6p^2 - 5p - 4 = (3p \quad 4)(2p + 1)$

NOTATION

15. The trinomial $4m^2 - 4m + 1$ is written in the form $ax^2 + bx + c$. Identify a, b, and c.

16. Use the substitution $x = a + b$ to rewrite the trinomial $6(a + b)^2 - 17(a + b) - 3$.

GUIDED PRACTICE

Use a special-product rule to factor each perfect-square trinomial. See Example 1.

17. $a^2 + 18a + 81$ **18.** $b^2 + 16b + 64$

19. $4y^2 + 28y + 49$ **20.** $25x^2 + 60x + 36$

21. $y^4 - 10y^2 + 25$ **22.** $a^4 - 14a^2 + 49$

23. $9b^4 - 12b^2c^2 + 4c^4$ **24.** $4a^4 - 12a^2b^2 + 9b^4$

Factor each trinomial. See Examples 2 and 3 or Example 11.

25. $x^2 + 5x + 6$ **26.** $y^2 + 8y + 15$

27. $t^2 + 14t + 48$ **28.** $m^2 + 13m + 36$

29. $x^2 - 16x + 55$ **30.** $c^2 - 24c + 44$

31. $y^2 - 17y + 72$ **32.** $t^2 - 11t + 28$

33. $-13y^2 + 30 + y^4$ **34.** $-13b^2 + 42 + b^4$

35. $50 + 15x^2 + x^4$ **36.** $27 + 12d^2 + d^4$

Factor each trinomial. Factor out the GCF first. See Example 4 or Example 11.

37. $3x^2 + 12x - 63$ **38.** $2y^2 + 4y - 48$

39. $15x^2 + 45x + 30$ **40.** $9y^2 + 90y + 81$

41. $2x^2y - 12xy^2 - 14y^3$ **42.** $5a^2b + 20ab^2 - 25b^3$

43. $3s^2t^2 + 18st^3 - 48t^4$ **44.** $4h^2k^2 - 12hk^3 - 160k^4$

Factor each trinomial, if possible. See Examples 5–7 or Example 11.

45. $5x^2 + 13x + 6$ **46.** $5x^2 + 18x + 9$

47. $7a^2 + 12a + 5$ **48.** $7a^2 + 36a + 5$

49. $3r^2 + 13r - 10$ **50.** $3r^2 - r - 10$

51. $11y^2 + 32y - 3$ **52.** $2y^2 - 9y - 18$

53. $2t^2 + 9t - 6$ **54.** $2t^2 - t - 9$

55. $13a^2 + 2t + 1$ **56.** $7z^2 + 8z + 2$

57. $8a^2 - 18ay + 7y^2$ **58.** $6n^2 - 17ny + 5y^2$

59. $7g^2 - 12gh - 4h^2$ **60.** $9q^2 + 25qt - 6t^2$

Factor each trinomial. Factor out the opposite of the GCF first. See Example 8 or Example 12.

61. $-18p^2 + 14pq + 4q^2$ **62.** $-6m^2 + 3mn + 3n^2$

63. $-30x^2 + 25xy + 20y^2$ **64.** $-72y^2 + 12yz + 40z^2$

Factor each trinomial. Factor out the GCF first. See Example 9 or Example 12.

65. $9a^2b^4 + 15a^2b^2 + 4a^2$ **66.** $9d^2h^4 + 37d^2h^2 + 4d^2$

67. $8b^4c^8 + 14b^2c^8 - 15c^8$ **68.** $8t^5u^4 - 30t^5u^2 - 27t^5$

69. $10b^6 - 19b^4 + 6b^2$ **70.** $12x^6 - 4x^4 - 21x^2$

71. $9m^7 + 6m^5 - 8m^3$ **72.** $15s^7 + 19s^5 - 10s^3$

Use a substitution to help factor each expression. See Example 10.

73. $(x + a)^2 + 2(x + a) + 1$

74. $(a + b)^2 - 2(a + b) + 1$

75. $(a + b)^2 - 2(a + b) - 24$

76. $(x - y)^2 + 3(x - y) - 10$

77. $14(q - r)^2 - 17(q - r) - 6$

78. $8(h + s)^2 + 34(h + s) + 35$

79. $16(s + t)^2 - 6(s + t) - 27$

80. $15(w + z)^2 - 31(w + z) + 10$

TRY IT YOURSELF

Factor each expression, if possible. Factor out any GCF first (including −1 if the leading coefficient is negative).

81. $32 - a^2 + 4a$ **82.** $15 - x^2 - 2x$

83. $6z^2 + 17z + 12$ **84.** $10x^2 + 19x + 6$

85. $25y^2 - 10y + 1$ **86.** $49m^2 - 14m + 1$

87. $64h^6 + 24h^5 - 4h^4$ **88.** $27x^2yz + 90xyz - 72yz$

89. $5x^2 + 4x + 1$ **90.** $3 + 4a^2 + 20a$

91. $b^4x^2 - 12b^2x^2 + 35x^2$ **92.** $c^3x^4 + 11c^3x^2 - 42c^3$

93. $-3a^2 + ab + 2b^2$ **94.** $-2x^2 + 3xy + 5y^2$

95. $56a^2 + 42a - 70$ **96.** $150b^2 + 40b - 20$

97. $6(t + w)^2 + 11(t + w) - 10$

98. $4(x - y)^2 + 13(x - y) + 3$

99. $12y^6 + 23y^3 + 10$ **100.** $5m^8 + 29m^4 - 42$

101. $6a^2(m + n) + 13a(m + n) - 15(m + n)$

102. $15n^2(q - r) - 17n(q - r) - 18(q - r)$

103. $20a^2 - 60ab + 45b^2$ **104.** $-12x^2 - 27 + 36x$

105. $-3a^2x^2 + 15a^2x - 18a^2$ **106.** $-2bc^2y^2 - 16bc^2y + 40bc^2$

107. $25m^8 - 60m^4n + 36n^2$ **108.** $49s^6 + 84s^3n^2 + 36n^4$

APPLICATIONS

109. CHECKERS   The area of a square checkerboard is represented by the polynomial $25x^2 - 40x + 16$. Use factoring to find an expression that represents the length of a side.

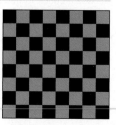

110. ICE   The surface area of a cubical block of ice is represented by the polynomial $6x^2 + 36x + 54$. Use factoring to find an expression that represents the length of an edge of the block.   (*Hint:* The surface area of a cube is 6 times the surface area of one face.)

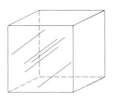

WRITING

111. Explain the error.

> Factor:
>
> $$2x^2 - 4x - 6 = (2x + 2)(x - 3)$$

112. How do you know when a polynomial has been factored completely?

113. How was substitution used in this section?

114. How does one determine whether a trinomial is a perfect-square trinomial?

REVIEW

115. If $f(x) = |2x - 1|$, find $f(-2)$.

116. If $g(x) = 2x^2 - 1$, find $g(-2)$.

117. Let $f(x) = -2x + 5$. For what value of x does $f(x) = -7$?

118. Let $f(x) = \frac{3}{2}x - 2$. For what value of x does $f(x) = \frac{2}{3}$?

CHALLENGE PROBLEMS

119. What are the only integer values of b for which $9m^2 + bm - 1$ can be factored?

120. Find the missing factors:

$$17y^2 + 1{,}496y - 11{,}305 = ?(y - 7)?$$

Factor. Assume that n is a natural number.

121. $x^{2n} + 2x^n + 1$

122. $2a^{6n} - 3a^{3n} - 2$

123. $x^{4n} + 2x^{2n}y^{2n} + y^{4n}$

124. $6x^{2n} + 7x^n - 3$

SECTION 5.7
The Difference of Two Squares; the Sum and Difference of Two Cubes

Objectives

1 Factor the difference of two squares.

2 Factor the sum and difference of two cubes.

We will now discuss some special rules of factoring. These rules are applied to polynomials that can be written as the difference of two squares or as the sum or difference of two cubes. To use these factoring methods, we must first be able to recognize such polynomials. We begin with a discussion that will help you recognize polynomials with terms that are *perfect squares.*

1   **Factor the Difference of Two Squares.**

Recall the special-product rule for multiplying the sum and difference of the same two terms:

$$(A + B)(A - B) = A^2 - B^2$$

The binomial $A^2 - B^2$ is called a **difference of two squares,** because A^2 is the square of A and B^2 is the square of B. If we reverse this rule, we obtain a method for factoring a difference of two squares.

The Language of Algebra
The expression $A^2 - B^2$ is a *difference of two squares,* whereas $(A - B)^2$ is the *square of a difference*. They are not equivalent because $(A - B)^2 \neq A^2 - B^2$.

Factoring ⟶

$$A^2 - B^2 = (A + B)(A - B)$$

This pattern is easy to remember if we think of a difference of two squares as the square of a **F**irst quantity minus the square of a **L**ast quantity.

Factoring a Difference of Two Squares	To factor the square of a First quantity minus the square of a Last quantity, multiply the First plus the Last by the First minus the Last. $$F^2 - L^2 = (F + L)(F - L)$$

To factor the difference of two squares, it is helpful to know the first twenty **perfect-square integers.** The number 400, for example, is a perfect square, because $400 = 20^2$.

$1 = 1^2$	$25 = 5^2$	$81 = 9^2$	$169 = 13^2$	$289 = 17^2$
$4 = 2^2$	$36 = 6^2$	$100 = 10^2$	$196 = 14^2$	$324 = 18^2$
$9 = 3^2$	$49 = 7^2$	$121 = 11^2$	$225 = 15^2$	$361 = 19^2$
$16 = 4^2$	$64 = 8^2$	$144 = 12^2$	$256 = 16^2$	$400 = 20^2$

EXAMPLE 1   Factor: $49x^2 - 16$

Strategy   The terms of this binomial do not have a common factor (other than 1). The only option is to attempt to factor it as a difference of two squares.

Why   If a binomial is a difference of two squares, we can factor it using a special product.

Solution $49x^2 - 16$ is the difference of two squares because it can be written as $(7x)^2 - (4)^2$. We can match it to the rule for factoring a difference of two squares to find the factorization.

$$\begin{array}{ccccccc} F^2 & - & L^2 & = & (F & + & L)(F & - & L) \\ \downarrow & & \downarrow & & \downarrow & & \downarrow & & \downarrow \\ (7x)^2 & - & (4)^2 & = & (7x & + & 4)(7x & - & 4) \end{array}$$

We can verify this result using the FOIL method.

$$(7x + 4)(7x - 4) = 49x^2 - 28x + 28x - 16$$
$$= 49x^2 - 16 \qquad \text{This is the original binomial.}$$

Self Check 1   Factor: $81p^2 - 25$
Now Try   Problems 11 and 17

Expressions such as a^4 and x^6y^8 are also perfect squares, because they can be written as the square of another quantity:

$$a^4 = (a^2)^2 \qquad \text{and} \qquad x^6y^8 = (x^3y^4)^2$$

EXAMPLE 2   Factor: $64a^4 - 25b^2$

Strategy   The terms of this binomial do not have a common factor (other than 1). The only option is to attempt to factor it as a difference of two squares.

Why   If a binomial is a difference of two squares, we can factor it using a special-product rule.

Solution   We can write $64a^4 - 25b^2$ in the form $(8a^2)^2 - (5b)^2$ and use the rule for factoring the difference of two squares.

$$\begin{array}{ccccccc} F^2 & - & L^2 & = & (F & + & L)(F & - & L) \\ \downarrow & & \downarrow & & \downarrow & & \downarrow & & \downarrow \\ (8a^2)^2 & - & (5b)^2 & = & (8a^2 & + & 5b)(8a^2 & - & 5b) \end{array}$$

Self Check 2   Factor: $36r^4 - s^2$
Now Try   **Problems 21 and 25**

EXAMPLE 3   Factor: $x^4 - 1$

Strategy   The terms of $x^4 - 1$ do not have a common factor (other than 1). To factor this binomial, we will write it in a form that shows it is a difference of two squares.

Why   We can then use a special-product rule to factor it.

Solution   Because the binomial is the difference of the squares of x^2 and 1, it factors into the sum of x^2 and 1 and the difference of x^2 and 1.

$$\begin{aligned} x^4 - 1 &= (x^2)^2 - (1)^2 \\ &= (x^2 + 1)(x^2 - 1) \end{aligned}$$

The factor $x^2 + 1$ is the sum of two quantities and is prime. However, the factor $x^2 - 1$ is the difference of two squares and can be factored as $(x + 1)(x - 1)$. Thus,

$$\begin{aligned} x^4 - 1 &= (x^2 + 1)(x^2 - 1) \\ &= (x^2 + 1)(x + 1)(x - 1) \end{aligned}$$

Self Check 3   Factor: $a^4 - 81$
Now Try   **Problem 29**

EXAMPLE 4   Factor: $(x + y)^4 - z^4$

Strategy   We will use a substitution to factor this difference of two squares.

Why   For more complicated expressions, especially those involving a quantity within parentheses, a substitution often helps simplify the factoring process.

Solution   If we use the substitution $a = x + y$, we obtain

$$\begin{aligned} (x + y)^4 - z^4 &= a^4 - z^4 && \text{Replace } x + y \text{ with } a. \\ &= (a^2 + z^2)(a^2 - z^2) && \text{Factor the difference of two squares.} \\ &= (a^2 + z^2)(a + z)(a - z) && \text{Factor } a^2 - z^2. \end{aligned}$$

To find the factorization of $(x + y)^4 - z^4$, we substitute $x + y$ for each a in the expression $(a^2 + z^2)(a + z)(a - z)$.

$$(a^2 + z^2)(a + z)(a - z) = [(x + y)^2 + z^2](x + y + z)(x + y - z)$$

Caution
When factoring a polynomial, be sure to factor it completely. Always check to see whether any of the factors of your result can be factored further.

Thus, $(x + y)^4 - z^4 = [(x + y)^2 + z^2](x + y + z)(x + y - z)$.

If we square the binomial within the brackets, we have

$$(x + y)^4 - z^4 = [x^2 + 2xy + y^2 + z^2](x + y + z)(x + y - z)$$

▷ **Self Check 4**   Factor: $(a - b)^4 - c^4$
Now Try   **Problem 35**

When possible, we always factor out a common factor before factoring the difference of two squares. The factoring process is easier when all common factors are factored out first.

EXAMPLE 5   Factor: $2x^4y - 32y$

Strategy   We will factor out the GCF of $2y$ and factor the resulting difference of two squares.

Why   The first step in factoring any polynomial is to factor out the GCF.

Solution

$$
\begin{aligned}
2x^4y - 32y &= 2y(x^4 - 16) &&\text{Factor out the GCF, which is 2y.}\\
&= 2y(x^2 + 4)(x^2 - 4) &&\text{Factor } x^4 - 16.\\
&= 2y(x^2 + 4)(x + 2)(x - 2) &&\text{Factor } x^2 - 4.
\end{aligned}
$$

▷ **Self Check 5**   Factor: $3a^4 - 3$
Now Try   **Problems 37 and 43**

EXAMPLE 6   Factor: **a.** $x^2 - y^2 + x - y$ **b.** $x^2 + 6x + 9 - z^2$

Strategy   The terms of each expression do not have a common factor (other than 1) and traditional factoring by grouping will not work. Instead, in part (a), we will group only the first two terms of the polynomial and in part (b) we will group the first three terms.

Why   Hopefully, those steps will produce equivalent expressions that can be factored.

Solution   We group the first two terms, factor them as a difference of two squares, and look for a common factor.

a. $x^2 - y^2 + x - y = (x + y)(x - y) + (x - y)$ Factor $x^2 - y^2$. The terms of the resulting expression have a common binomial factor, $x - y$, that can be factored out.

$$= (x - y)(x + y + 1)$$ Factor out the GCF, $x - y$.

b. We group the first three terms and factor that trinomial to get:

$$x^2 + 6x + 9 - z^2 = (x + 3)(x + 3) - z^2$$ $x^2 + 6x + 9$ is a perfect-square trinomial.

$$= (x + 3)^2 - z^2$$ The expression that results is a difference of two squares.

$$= (x + 3 + z)(x + 3 - z)$$ Factor the difference of two squares.

Self Check 6   Factor: **a.** $a^2 - b^2 + a + b$
b. $a^2 + 4a + 4 - b^2$

Now Try   **Problems 45 and 49**

2 **Factor the Sum and Difference of Two Cubes.**

The number 64 is called a perfect cube, because $4^3 = 64$. To factor the sum or difference of two cubes, it is helpful to know the first ten **perfect-cube integers:**

$1 = 1^3$	$27 = 3^3$	$125 = 5^3$	$343 = 7^3$	$729 = 9^3$
$8 = 2^3$	$64 = 4^3$	$216 = 6^3$	$512 = 8^3$	$1{,}000 = 10^3$

Expressions such as b^6 and $x^9 y^{12}$ are also perfect cubes, because they can be written as the cube of another quantity:

$$b^6 = (b^2)^3 \quad \text{and} \quad x^9 y^{12} = (x^3 y^4)^3$$

To find rules for factoring the sum of two cubes and the difference of two cubes, we need to find the products shown below. Note that each term of the trinomial is multiplied by each term of the binomial.

$$(x + y)(x^2 - xy + y^2) = x^3 - x^2y + xy^2 + x^2y - xy^2 + y^3$$
$$= x^3 + y^3 \quad \textit{Combine like terms: } -x^2y + x^2y = 0 \textit{ and } xy^2 - xy^2 = 0.$$

$$(x - y)(x^2 + xy + y^2) = x^3 + x^2y + xy^2 - x^2y - xy^2 - y^3$$
$$= x^3 - y^3 \quad \textit{Combine like terms.}$$

We have found that

$$x^3 + y^3 = (x + y)(x^2 - xy + y^2) \quad \text{and} \quad x^3 - y^3 = (x - y)(x^2 + xy + y^2)$$

The Language of Algebra
The expression $x^3 + y^3$ is a *sum of two cubes,* whereas $(x + y)^3$ is the *cube of a sum.* If you expand $(x + y)^3$, you will see that they are not equivalent.

The binomial $x^3 + y^3$ is called the **sum of two cubes,** because x^3 represents the cube of x, y^3 represents the cube of y, and $x^3 + y^3$ represents the sum of these cubes. Similarly, $x^3 - y^3$ is called the **difference of two cubes.**

These results justify the rules for factoring the **sum and difference of two cubes.** They are easier to remember if we think of a sum (or a difference) of two cubes as the cube of a **First** quantity plus (or minus) the cube of the **Last** quantity.

Factoring the Sum and Difference of Two Cubes	To factor the cube of a First quantity plus the cube of a Last quantity, multiply the First plus the Last by the First squared, minus the First times the Last, plus the Last squared.

$$F^3 + L^3 = (F + L)(F^2 - FL + L^2)$$

To factor the cube of a First quantity minus the cube of a Last quantity, multiply the First minus the Last by the First squared, plus the First times the Last, plus the Last squared.

$$F^3 - L^3 = (F - L)(F^2 + FL + L^2)$$

EXAMPLE 7   Factor: $a^3 + 8$

Strategy   We will write the binomial in a form that shows it is the sum of two cubes.

Why   We can then use the rule for factoring the sum of two cubes.

Solution   Since $a^3 + 8$ can be written as $a^3 + 2^3$, it is a sum of two cubes, which factors as follows:

$$F^3 + L^3 = (F + L)(F^2 - FL + L^2)$$
$$a^3 + 2^3 = (a + 2)(a^2 - a2 + 2^2)$$
$$= (a + 2)(a^2 - 2a + 4) \qquad a^2 - 2a + 4 \text{ does not factor.}$$

Therefore, $a^3 + 8 = (a + 2)(a^2 - 2a + 4)$. We can check by multiplying.

$$(a + 2)(a^2 - 2a + 4) = a^3 - 2a^2 + 4a + 2a^2 - 4a + 8$$
$$= a^3 + 8 \qquad \text{This is the original binomial.}$$

> **Caution**
>
> In Example 7, a common error is to try to factor $a^2 - 2a + 4$. It is not a perfect-square trinomial, because the middle term needs to be $-4a$. Furthermore, it cannot be factored by the methods of Section 5.6. It is prime.

▷ **Self Check 7**   Factor: $p^3 + 27$
   Now Try   **Problem 53**

You should memorize the rules for factoring the sum and the difference of two cubes. Note that each has the form (a binomial)(a trinomial) and that there is a relationship between the signs that appear in these forms.

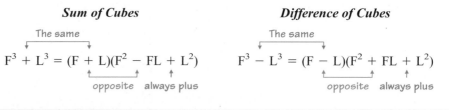

Sum of Cubes

The same

$$F^3 + L^3 = (F + L)(F^2 - FL + L^2)$$

opposite   always plus

Difference of Cubes

The same

$$F^3 - L^3 = (F - L)(F^2 + FL + L^2)$$

opposite   always plus

EXAMPLE 8   Factor: $27a^3 - 64b^6$

Strategy   We will write the binomial in a form that shows it is the difference of two cubes.

Why   We can then use the rule for factoring the difference of two cubes.

Solution   Since $27a^3 - 64b^6$ can be written as $(3a)^3 - (4b^2)^3$, it is a difference of two cubes, which factors as follows:

$$F^3 - L^3 = (F - L)(F^2 + FL + L^2)$$
$$(3a)^3 - (4b^2)^3 = (3a - 4b^2)[(3a)^2 + (3a)(4b^2) + (4b^2)^2]$$
$$= (3a - 4b^2)(9a^2 + 12ab^2 + 16b^4)$$

Thus, $27a^3 - 64b^6 = (3a - 4b^2)(9a^2 + 12ab^2 + 16b^4)$.

▷ **Self Check 8**   Factor: $8c^6 - 125d^3$
   Now Try   **Problem 57**

EXAMPLE 9   Factor: $a^3 - (c + d)^3$

Strategy   To factor this expression, we will use the rule for factoring the difference of two cubes.

Why   The terms a^3 and $(c + d)^3$ are perfect cubes.

Solution

$$\underset{\downarrow}{F^3} - \underset{\downarrow}{L^3} = (\underset{\downarrow}{F} - \underset{\downarrow}{L})(\underset{\downarrow}{F^2} + \underset{\downarrow}{F}\ \underset{\downarrow}{L} + \underset{\downarrow}{L^2})$$

$$a^3 - (c + d)^3 = [a - (c + d)][a^2 + a(c + d) + (c + d)^2]$$

Now we simplify the expressions inside both sets of brackets.

$$a^3 - (c + d)^3 = (a - c - d)(a^2 + ac + ad + c^2 + 2cd + d^2)$$

> ### Success Tip
> We could also use the substitution $x = c + d$, factor $a^3 - x^3$ as $(a - x)(a^2 + ax + x^2)$, and then replace each x with $c + d$.

▷ **Self Check 9**   Factor: $(p + q)^3 - r^3$

Now Try **Problem 61**

EXAMPLE 10   Factor: $x^6 - 64$

Strategy   This binomial is both the difference of two squares and the difference of two cubes. We will write it in a form that shows it is a difference of two squares to begin the factoring process.

Why   It is easier to factor it as the difference of two squares first.

Solution
$$x^6 - 64 = (x^3)^2 - 8^2$$
$$= (x^3 + 8)(x^3 - 8)$$

Each of these factors can be factored further. One is the sum of two cubes and the other is the difference of two cubes.

$$x^6 - 64 = (x + 2)(x^2 - 2x + 4)(x - 2)(x^2 + 2x + 4)$$

▷ **Self Check 10**   Factor: $1 - x^6$

Now Try **Problem 65**

EXAMPLE 11   Factor: $2a^5 + 250a^2$

Strategy   We will factor out the GCF of $2a^2$ and factor the resulting sum of two cubes.

Why   The first step in factoring any polynomial is to factor out the GCF.

Solution   We first factor out the common factor of $2a^2$ to obtain

$$2a^5 + 250a^2 = 2a^2(a^3 + 125)$$

Then we factor $a^3 + 125$ as the sum of two cubes to obtain

$$2a^5 + 250a^2 = 2a^2(a + 5)(a^2 - 5a + 25)$$

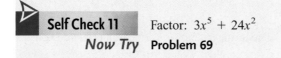

Self Check 11 Factor: $3x^5 + 24x^2$

Now Try Problem 69

ANSWERS TO SELF CHECKS **1.** $(9p + 5)(9p - 5)$ **2.** $(6r^2 + s)(6r^2 - s)$ **3.** $(a^2 + 9)(a + 3)(a - 3)$
4. $[(a - b)^2 + c^2](a - b + c)(a - b - c)$ **5.** $3(a^2 + 1)(a + 1)(a - 1)$ **6. a.** $(a + b)(a - b + 1)$
b. $(a + 2 + b)(a + 2 - b)$ **7.** $(p + 3)(p^2 - 3p + 9)$ **8.** $(2c^2 - 5d)(4c^4 + 10c^2d + 25d^2)$
9. $(p + q - r)(p^2 + 2pq + q^2 + pr + qr + r^2)$ **10.** $(1 + x)(1 - x + x^2)(1 - x)(1 + x + x^2)$
11. $3x^2(x + 2)(x^2 - 2x + 4)$

STUDY SET
5.7

VOCABULARY

Fill in the blanks.

1. When the polynomial $4x^2 - 25$ is written as $(2x)^2 - (5)^2$, we see that it is the difference of two _____.

2. When the polynomial $8x^3 + 125$ is written as $(2x)^3 + (5)^3$, we see that it is the sum of two _____.

CONCEPTS

3. a. Write the first ten perfect-integer squares.

b. Write the first ten perfect-integer cubes.

4. a. Use multiplication to verify that the sum of two squares $x^2 + 25$ does not factor as $(x + 5)(x + 5)$.

b. Use multiplication to verify that the difference of two squares $x^2 - 25$ factors as $(x + 5)(x - 5)$.

5. Complete each factorization.
a. $F^2 - L^2 = (F + L)()$
b. $F^3 + L^3 = (F + L)()$
c. $F^3 - L^3 = (F - L)()$

6. Factor each binomial.
a. $5p^2 + 20$
b. $5p^2 - 20$
c. $5p^3 + 20$
d. $5p^3 + 40$

NOTATION

7. Give an example of each.
a. a difference of two squares
b. a square of a difference
c. a sum of two squares
d. a sum of two cubes
e. a cube of a sum

8. Fill in the blanks.
a. $36y^2 - 49m^4 = ()^2 - ()^2$
b. $125h^3 - 27k^6 = ()^3 - ()^3$

GUIDED PRACTICE

Factor, if possible. **See Example 1.**

9. $x^2 - 16$

10. $y^2 - 49$

11. $9y^2 - 64$

12. $16x^2 - 81$

13. $144 - c^2$

14. $25 - t^2$

15. $100m^2 - 1$

16. $144x^2 - 1$

17. $81a^2 - 49b^2$

18. $64r^2 - 121s^2$

19. $x^2 + 25$

20. $a^2 + 36$

Factor each difference of two squares. **See Example 2.**

21. $9r^4 - 121s^2$

22. $81a^4 - 16b^2$

23. $16t^2 - 25w^4$

24. $9r^2 - 25s^4$

25. $100r^2s^4 - t^4$

26. $400x^2z^4 - a^4$

27. $36x^4y^2 - 49z^6$

28. $4a^2b^4 - 9d^6$

Factor completely. See Example 3.

29. $x^4 - y^4$

30. $16n^4 - 1$

31. $16a^4 - 81b^4$

32. $81m^4 - 256n^4$

Factor. See Example 4.

33. $(x + y)^2 - z^2$

34. $a^2 - (b - c)^2$

35. $(r - s)^2 - t^4$

36. $(m + n)^2 - p^4$

Factor each expression. Factor out any GCF first. See Example 5.

37. $2x^2 - 288$

38. $8x^2 - 72$

39. $3x^3 - 243x$

40. $2x^3 - 32x$

41. $5ab^4 - 5a$

42. $3ac^4 - 243a$

43. $64b - 4b^5$

44. $1,250n - 2n^5$

Factor by first grouping the appropriate terms. See Example 6.

45. $c^2 - d^2 + c + d$

46. $s^2 - t^2 + s - t$

47. $a^2 - b^2 + 2a - 2b$

48. $m^2 - n^2 + 3m + 3n$

49. $x^2 + 12x + 36 - y^2$

50. $x^2 - 6x + 9 - 4y^2$

51. $x^2 - 2x + 1 - 9z^2$

52. $x^2 + 10x + 25 - 16z^2$

Factor each sum of cubes. See Example 7.

53. $a^3 + 125$

54. $b^3 + 64$

55. $8r^3 + s^3$

56. $27t^3 + u^3$

Factor each difference of cubes. See Example 8.

57. $64t^6 - 27v^3$

58. $125m^3 - x^6$

59. $x^3 - 216y^6$

60. $8c^6 - 343w^3$

Factor. See Example 9.

61. $(a - b)^3 + 27$

62. $(b - c)^3 - 1,000$

63. $64 - (a + b)^3$

64. $1 - (x + y)^3$

Factor each expression completely. Factor a difference of two squares first. See Example 10.

65. $x^6 - 1$

66. $x^6 - y^6$

67. $x^{12} - y^6$

68. $a^{12} - 64$

Factor each sum or difference of cubes. Factor out the GCF first. See Example 11.

69. $5x^3 + 625$

70. $2x^3 - 128$

71. $4x^5 - 256x^2$

72. $2x^6 + 54x^3$

TRY IT YOURSELF

Factor each expression, if possible.

73. $64a^3 - 125b^6$

74. $8x^6 - 27y^3$

75. $288b^2 - 2b^6$

76. $98x - 2x^5$

77. $x^2 - y^2 + 8x + 8y$

78. $5m - 5n + m^2 - n^2$

79. $x^9 + y^9$

80. $x^6 + y^6$

81. $144a^2t^2 - 169b^6$

82. $25x^6 - 81y^2z^2$

83. $100a^2 + 9b^2$

84. $25s^4 + 16t^2$

85. $81c^4d^4 - 16t^4$

86. $256x^4 - 81y^4$

87. $128u^2v^3 - 2t^3u^2$

88. $56rs^2t^3 + 7rs^2v^6$

89. $y^2 - (2x - t)^2$

90. $(15 - r)^2 - s^2$

91. $x^2 + 20x + 100 - 9z^2$

92. $49a^2 - b^2 - 14b - 49$

93. $(c - d)^3 + 216$

94. $1 - (x + y)^3$

95. $\dfrac{1}{36} - y^4$

96. $\dfrac{4}{81} - m^4$

97. $m^6 - 64$

98. $y^6 - 1$

99. $(a + b)x^3 + 27(a + b)$

100. $(c - d)r^3 - (c - d)s^3$

101. $x^9 - y^{12}z^{15}$

102. $r^{12} + s^{18}t^{24}$

APPLICATIONS

103. CANDY   To find the amount of chocolate used in the outer coating of the malted-milk ball shown, we can find the volume V of the chocolate shell using the formula

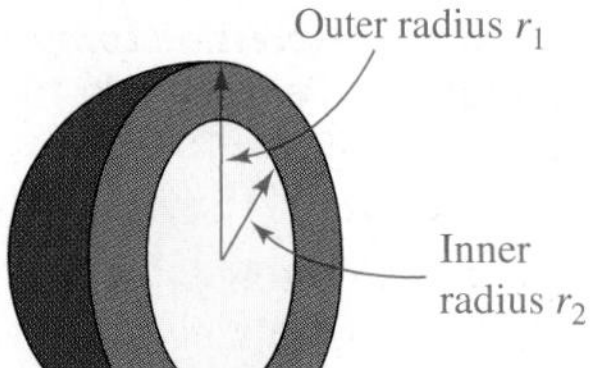

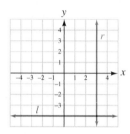

Outer radius r_1

Inner radius r_2

$$V = \frac{4}{3}\pi r_1{}^3 - \frac{4}{3}\pi r_2{}^3.$$

Factor the expression on the right side of the formula.

104. MOVIE STUNTS   The function that gives the distance a stuntwoman is above the ground t seconds after she falls over the side of a 144-foot tall building is $h(t) = 144 - 16t^2$. Factor the right side.

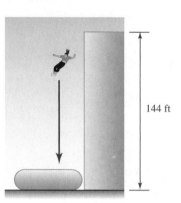

144 ft

WRITING

105. Explain how the patterns used to factor the sum and difference of two cubes are similar and how they differ.

106. Explain why the factorization is not complete.

Factor: $1 - t^8 = (1 + t^4)(1 - t^4)$

107. Explain the error.

Factor: $4g^2 - 16 = (2g + 4)(2g - 4)$

108. When asked to factor $81t^2 - 16$, one student answered $(9t - 4)(9t + 4)$, and another answered $(9t + 4)(9t - 4)$. Explain why both students are correct.

REVIEW

Graph the line with the given characteristics.

109. Passing through $(-2, -1)$; slope $-\dfrac{2}{3}$

110. y-intercept $(0, -4)$; slope 3

111. Write the equation of line l shown below.

112. Write the equation of line r shown below.

CHALLENGE PROBLEMS

Factor. Assume all variables represent natural numbers.

113. $4x^{2n} - 9y^{2n}$

114. $25 - x^{6n}$

115. $a^{3b} - c^{3b}$

116. $27x^{3n} + y^{3n}$

117. Factor: $x^{32} - y^{32}$

118. Find the error in this proof that $2 = 1$.

$$x = y$$
$$x^2 = xy$$
$$x^2 - y^2 = xy - y^2$$
$$(x + y)(x - y) = y(x - y)$$
$$\frac{(x + y)(x - y)}{(x - y)} = \frac{y(x - y)}{(x - y)}$$
$$x + y = y$$
$$y + y = y$$
$$2y = y$$
$$\frac{2y}{y} = \frac{y}{y}$$
$$2 = 1$$

SECTION 5.8
Summary of Factoring Techniques

Objective   ❶ Factor random polynomials.

The factoring methods discussed so far will be used in the remaining chapters to simplify expressions and solve equations. In such cases, we must determine the factoring method—it will not be specified. This section will give you practice in selecting the appropriate factoring method to use given a randomly chosen polynomial.

 Factor Random Polynomials.

The following strategy is helpful when factoring polynomials.

Steps for Factoring a Polynomial

1. Is there a common factor? If so, factor out the GCF, or the opposite of the GCF, so that the leading coefficient is positive.
2. How many terms does the polynomial have?

    If it has *two terms,* look for the following problem types:

    a. The difference of two squares

    b. The sum of two cubes

    c. The difference of two cubes

    If it has *three terms,* look for the following problem types:

    a. A perfect-square trinomial

    b. If the trinomial is not a perfect square, use the trial-and-check method or the grouping method

    If it has *four* or more terms, try to factor by grouping.
3. Can any factors be factored further? If so, factor them completely.
4. Does the factorization check? Check by multiplying.

For more complicated expressions, a substitution sometimes helps to simplify the factoring process.

EXAMPLE 1   Factor: $12x^2y^2z^3 - 2xy^2z^3 - 4y^2z^3$

Strategy   We will answer the four questions listed in the *Steps for Factoring a Polynomial.*

Why   The answers will help us determine which factoring techniques to use.

The Language of Algebra
Remember that the instruction to factor means to factor completely. A polynomial is factored completely when no factor can be factored further.

Solution   *Is there a common factor?* Yes. Factor out the greatest common factor $2y^2z^3$.

$$12x^2y^2z^3 - 2xy^2z^3 - 4y^2z^3 = 2y^2z^3(6x^2 - x - 2)$$

How many terms does it have? The polynomial within the parentheses has three terms. We can factor $6x^2 - x - 2$, using the trial-and-check method or the key number method, to get

$$12x^2y^2z^3 - 2xy^2z^3 - 4y^2z^3 = 2y^2z^3(6x^2 - x - 2)$$
$$= 2y^2z^3(3x - 2)(2x + 1)$$

└─ Don't forget to write the
GCF from the first step.

Is it factored completely? Yes. Since each of the individual factors is prime, the factorization is complete.

Does it check? To check, we multiply.

$$2y^2z^3(3x - 2)(2x + 1) = 2y^2z^3(6x^2 - x - 2)$$ Multiply the binomials first.
$$= 12x^2y^2z^3 - 2xy^2z^3 - 4y^2z^3$$ Distribute the multiplication by $2y^2z^3$.

Since we obtain the original polynomial, the factorization is correct.

▷ **Self Check 1**   Factor: $30a^2b^3c - 27ab^3c + 6b^3c$
Now Try   **Problem 13**

EXAMPLE 2   Factor: $48a^4c^3 - 3b^4c^3$

Strategy   We will answer the four questions listed in the *Steps for Factoring a Polynomial.*

Why   The answers will help us determine which factoring techniques to use.

Solution   ***Is there a common factor?*** Yes. Factor out the greatest common factor $3c^3$.

$$48a^4c^3 - 3b^4c^3 = 3c^3(16a^4 - b^4)$$

How many terms does it have? The polynomial within the parentheses, $16a^4 - b^4$, has two terms. It is the difference of two squares and factors as $(4a^2 + b^2)(4a^2 - b^2)$.

$$48a^4c^3 - 3b^4c^3 = 3c^3(16a^4 - b^4)$$
$$= 3c^3(4a^2 + b^2)(4a^2 - b^2)$$

Is it factored completely? No. The binomial $4a^2 + b^2$ is the sum of two squares and is prime. However, $4a^2 - b^2$ is the difference of two squares and factors as $(2a + b)(2a - b)$.

$$48a^4c^3 - 3b^4c^3 = 3c^3(16a^4 - b^4)$$
$$= 3c^3(4a^2 + b^2)(4a^2 - b^2)$$
$$= 3c^3(4a^2 + b^2)(2a + b)(2a - b)$$

Since each of the individual factors is prime, the factorization is now complete.

Does it check? Multiply to verify that this factorization is correct.

▷ **Self Check 2**   Factor: $3p^4r^3 - 3q^4r^3$
Now Try   **Problem 19**

> ### EXAMPLE 3   Factor: $x^5y + x^2y^4 - x^3y^3 - y^6$
>
> **Strategy**   We will answer the four questions listed in the *Steps for Factoring a Polynomial.*
>
> **Why**   The answers will help us determine which factoring techniques to use.
>
> **Solution**   *Is there a common factor?* Yes. Factor out the greatest common factor of y.
>
> $$x^5y + x^2y^4 - x^3y^3 - y^6 = y(x^5 + x^2y^3 - x^3y^2 - y^5)$$
>
> *How many terms does it have?* The polynomial $x^5 + x^2y^3 - x^3y^2 - y^5$ has four terms. We try factoring by grouping to obtain
>
> $$x^5y + x^2y^4 - x^3y^3 - y^6$$
> $$= y(\; x^5 + x^2y^3 \;\; - x^3y^2 - y^5 \;) \qquad \text{Factor out } y.$$
> $$= y[x^2(x^3 + y^3) - y^2(x^3 + y^3)] \qquad \text{Factor by grouping.}$$
> $$= y(x^3 + y^3)(x^2 - y^2) \qquad \text{Factor out } x^3 + y^3.$$
>
> *Is it factored completely?* No. We can factor $x^3 + y^3$ (the sum of two cubes) and $x^2 - y^2$ (the difference of two squares) to obtain
>
> $$x^5y + x^2y^4 - x^3y^3 - y^6 = y(x + y)(x^2 - xy + y^2)(x + y)(x - y)$$
>
> Because each of the individual factors is prime, the factorization is complete.
>
> *Does it check?* Multiply to verify that this factorization is correct.
>
> ⊳ | **Self Check 3**   Factor: $9a^5 - 9a^3b^2 + 9a^2b^3 - 9b^5$
> *Now Try*   **Problem 31**

> ### EXAMPLE 4   Factor: $x^3 + 5x^2 + 6x + x^2y + 5xy + 6y$
>
> **Strategy**   We will answer the four questions listed in the *Steps for Factoring a Polynomial.*
>
> **Why**   The answers will help us determine which factoring techniques to use.
>
> **Solution**   *Is there a common factor?* No. There is no common factor (other than 1).
>
> *How many terms does it have?* Since there are more than three terms, we try factoring by grouping. We can factor x from the first three terms and y from the last three terms.
>
> $$x^3 + 5x^2 + 6x \;\; + \;\; x^2y + 5xy + 6y$$
> $$= x(x^2 + 5x + 6) + y(x^2 + 5x + 6)$$
> $$= (x^2 + 5x + 6)(x + y) \qquad \text{Factor out } x^2 + 5x + 6.$$
>
> *Is it factored completely?* No. We can factor the trinomial $x^2 + 5x + 6$ to obtain
>
> $$x^3 + 5x^2 + 6x + x^2y + 5xy + 6y = (x + 3)(x + 2)(x + y)$$
>
> Since each of the individual factors is prime, the factorization is now complete.
>
> *Does it check?* Multiply to verify that the factorization is correct.

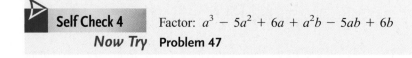

Self Check 4   Factor: $a^3 - 5a^2 + 6a + a^2b - 5ab + 6b$

Now Try   **Problem 47**

EXAMPLE 5   Factor: $4x^4 + 4x^3 + x^2 + 2x + 1$

Strategy   We will answer the four questions listed in the *Steps for Factoring a Polynomial*.

Why   The answers will help us determine which factoring techniques to use.

Solution   *Is there a common factor?* There is no common factor (other than 1).

How many terms does it have? Since there are more than three terms, we try factoring by grouping. We can factor x^2 from the first three terms, and group the last two terms together.

$$\underbrace{4x^4 + 4x^3 + x^2} + \underbrace{2x + 1} = x^2(4x^2 + 4x + 1) + (2x + 1)$$

Is it factored completely? No. We recognize $4x^2 + 4x + 1$ as a perfect-square trinomial, because $4x^2 = (2x)^2$, $1 = (1)^2$, and $4x = 2 \cdot 2x \cdot 1$. Therefore, it factors as $(2x + 1)(2x + 1)$.

$$
\begin{aligned}
4x^4 + 4x^3 + x^2 + 2x + 1 &= x^2(4x^2 + 4x + 1) + (2x + 1) \\
&= x^2(2x + 1)(2x + 1) + (2x + 1)
\end{aligned}
$$

Finally, we factor out the common factor $2x + 1$.

$$
\begin{aligned}
4x^4 + 4x^3 + x^2 + 2x + 1 &= x^2(4x^2 + 4x + 1) + (2x + 1) \\
&= x^2(2x + 1)(2x + 1) + (2x + 1) \\
&= (2x + 1)[x^2(2x + 1) + 1] \\
&= (2x + 1)(2x^3 + x^2 + 1)
\end{aligned}
$$

Within the brackets, distribute the multiplication by x^2.

Since each of the individual factors is prime, the factorization is complete.

Does it check? Multiply to verify that this factorization is correct.

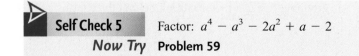

Self Check 5   Factor: $a^4 - a^3 - 2a^2 + a - 2$

Now Try   **Problem 59**

ANSWERS TO SELF CHECKS   **1.** $3b^3c(5a - 2)(2a - 1)$ **2.** $3r^3(p^2 + q^2)(p + q)(p - q)$
3. $9(a + b)(a^2 - ab + b^2)(a + b)(a - b)$ **4.** $(a - 2)(a - 3)(a + b)$ **5.** $(a - 2)(a^3 + a^2 + 1)$

STUDY SET
5.8

VOCABULARY

Fill in the blanks.

1. Each factor of a completely factored expression will be _____.
2. When we factor a polynomial, we write a sum of terms as an equivalent product of _____.
3. $x^3 + y^3$ is called a sum of two _____ and $x^3 - y^3$ is called a difference of two _____.
4. $x^2 - y^2$ is called a _____ of two squares.

CONCEPTS

Fill in the blanks.

5. In any factoring problem, always factor out any _____ factors first.
6. If a polynomial has two terms, check to see whether the problem type is the _____ of two squares, the sum of two _____, or the _____ of two cubes.
7. If a polynomial has three terms, try to factor it as a _____.
8. If a polynomial has four or more terms, try factoring it by _____.
9. Explain how to verify that $y^2z^3(x + 6)(x + 1)$ is the factored form of $x^2y^2z^3 + 7xy^2z^3 + 6y^2z^3$.

10. Why is the polynomial $x^2 + 6$ classified as prime?

NOTATION

Complete each factorization.

11. $18a^3b + 3a^2b^2 - 6ab^3 = \quad (6a^2 + ab - 2b^2)$

$$= 3ab(3a + \quad)(\quad - b)$$

12. $2x^4 - 1{,}250 = 2(\qquad)$

$$= 2(\qquad)(x^2 - 25)$$

$$= 2(x^2 + 25)(x + 5)(\qquad)$$

TRY IT YOURSELF

Factor each expression completely. If an expression is prime, so indicate.

13. $4a^2bc + 4abc - 120bc$
14. $8x^3y^4 - 27y$
15. $-3x^2y - 6xy^2 + 12xy$
16. $xy - ty + xs^2 - ts^2$
17. $y^3(y^2 - 1) - 27(y^2 - 1)$
18. $b^2c + b^2 + bcd + bd$
19. $36x^4 - 36$

20. $27x^9 - y^3$
21. $16c^2g^2 + h^4$
22. $12x^2 + 14x - 6$
23. $-14x + 8 + 6x^2$
24. $-13m^2 + m^4 + 36$
25. $4x^2y^2 + 4xy^2 + y^2$
26. $x^3 + a^6y^3$
27. $4x^2y^2z^2 - 26x^2y^2z^3$
28. $-2x^3 + 54$
29. $9a^6 - 48a^4 + 64a^2$
30. $-4m^7 + 36m^4 - 81m$
31. $6a^5 - 6a^3b^2 - 6a^2b^3 + 6b^5$
32. $2(x + y)^2 + (x + y) - 3$
33. $(x - y)^3 + 125$
34. $625x^4 - 256y^4$
35. $2(a - b)^2 + 5(a - b) + 3$
36. $5x + 4y + 25x^2 - 16y^2$
37. $6x^2 - 63 - 13x$
38. $a^4b^2 - 20a^2b^2 + 64b^2$
39. $-17x^2 + 16 + x^4$
40. $x^2 + 6x + 9 - y^2$
41. $x^2 + 10x + 25 - y^8$
42. $4x^2 + 4x + 1 - 4y^2$
43. $9x^2 - 6x + 1 - 25y^2$
44. $x^2 - y^2 - 2y - 1$
45. $a^2 - b^2 + 4b - 4$
46. $60q^2r^2s^4 + 78qr^2s^4 - 18r^2s^4$
47. $ax^2 - 2axy + ay^2 - x^2 + 2xy - y^2$
48. $32x^{10} + 48x^9 + 18x^8$
49. $\dfrac{81}{16}x^4 - y^{40}$
50. $\dfrac{d^2x^2}{2} - \dfrac{f^2x^2}{2} - \dfrac{c^2d^2}{2} + \dfrac{c^2f^2}{2}$
51. $16m^{16} - 16$
52. $8(4 - a^2) - x^3(4 - a^2)$
53. $9y^5 + 6y^4 + y^3 + 3y + 1$
54. $25m^4 - 10m^3 + m^2 + 5m - 1$
55. $x^3 - xy^2 - 4x^2 + 4y^2$
56. $m^2n - 9n + 9m^2 - 81$
57. $c^3 - 4a^2c + 4abc - b^2c$
58. $4t^3 - s^2t - 6stz - 9tz^2$
59. $9x^4 + 6x^3 + x^2 + 3x + 1$
60. $z^3(y^2 - 4) + 8(y^2 - 4)$
61. $(2x - 1)^2 + 4(2x - 1) + 4$

62. $(3z + 2)^2 - 12(3z + 2) + 36$

63. $a^2 + b^2 + 25$

64. $4x^3y^3 + 256$

APPLICATIONS

65. GRAPHIC DESIGN   The logo for a bookkeeping firm consists of small gray squares (each with side length x inches) positioned within a large square (side length 2 inches). Write a polynomial that represents the number of square inches of white space in the logo and then factor it.

Sharon West Smith & Associates
BOOKKEEPERS

66. GRAPHIC DESIGN   The logo for a marketing firm consists of small green squares (each with side length x inches) positioned within a large square (side length 1 inch). Write a polynomial that represents the number of square inches of white space in the logo and then factor it.

Jones-Kennedy
INCORPORATED

WRITING

67. What is your strategy for factoring a polynomial?

68. For the factorization below, explain why the polynomial is not factored completely.

$$48a^4c^3 - 3b^4c^3 = 3c^3(16a^4 - b^4)$$

REVIEW

Evaluate each determinant.

69. $\begin{vmatrix} -6 & -2 \\ 15 & 4 \end{vmatrix}$

70. $\begin{vmatrix} 3 & -2 \\ 12 & -8 \end{vmatrix}$

71. $\begin{vmatrix} -1 & 2 & 1 \\ 2 & 1 & -3 \\ 1 & 1 & 1 \end{vmatrix}$

72. $\begin{vmatrix} 1 & 0 & 1 \\ 0 & 1 & 0 \\ 1 & 1 & 1 \end{vmatrix}$

CHALLENGE PROBLEMS

73. Factor: $x^4 + x^2 + 1$ (*Hint:* Add and subtract x^2.)

74. Factor: $x^4 + 7x^2 + 16$ (*Hint:* Add and subtract x^2.)

Factor. Assume that n is a natural number.

75. $2a^{2n} + 2a^n - 24$

76. $ma^{2n} - mb^{2n}$

77. $54a^{3n} + 16b^{3n}$

78. $-a^{2n} - 2a^nb^n - b^{2n}$

79. $12m^{4n} + 10m^{2n} + 2$

80. $nx^{4n} + 2nx^{2n}y^{2n} + ny^{4n}$

SECTION 5.9
Solving Equations by Factoring

Objectives

① Solve quadratic equations using the zero-factor property.

② Solve higher-degree polynomial equations by factoring.

③ Use quadratic equations to solve problems.

We have previously solved linear equations in one variable such as $3x - 1 = 5$ and $10y + 9 = y + 4$. These equations are also called **polynomial equations** because they involve two polynomials that are set equal to each other. Some other examples of polynomial equations are

$$3x^2 = -6x, \qquad 6x^3 - x^2 = 2x, \qquad \text{and} \qquad x^4 + 4 - 5x^2 = 0$$

In this section, we will discuss a method for solving polynomial equations like these.

① **Solve Quadratic Equations Using the Zero-Factor Property.**

A second-degree polynomial equation in one variable is called a *quadratic equation.*

Quadratic Equations	A **quadratic equation** is an equation that can be written in the **standard form**
	$$ax^2 + bx + c = 0$$
	where a, b, and c represent real numbers and $a \neq 0$.

Here are some examples of quadratic equations. Only the last one is written in standard form.

$$2x^2 + 6x = 15 \qquad y^2 = -16y \qquad 4m^2 - 7m + 2 = 0$$

A *solution of a quadratic equation* is a value of the variable that makes the equation true. To *solve a quadratic equation* means to find all of its solutions. Many quadratic equations can be solved by factoring and using the zero-factor property.

The Zero-Factor Property	When the product of two real numbers is 0, at least one of them is 0.
	If $ab = 0$, then $a = 0$ or $b = 0$.
	This property also applies to three or more factors.

EXAMPLE 1   Solve: $x^2 + 5x + 6 = 0$

Strategy   To solve this equation, we will factor the trinomial on the left side and use the zero-factor property.

Why   To use the zero-factor property, we need one side of the equation to be factored completely and the other side to be 0.

Solution

$$x^2 + 5x + 6 = 0 \quad \text{This is the equation to solve.}$$
$$(x + 3)(x + 2) = 0 \quad \text{Factor the left side.}$$

Since the product of $x + 3$ and $x + 2$ is 0, at least one of the factors must be 0. Thus, we can set each factor equal to 0 and solve each resulting linear equation for x:

$$x + 3 = 0 \qquad \text{or} \qquad x + 2 = 0$$
$$x = -3 \qquad\qquad x = -2$$

To check these solutions, we substitute -3 and -2 for x in the original equation and verify that each number satisfies the equation.

Check:

$$\begin{array}{ccc} x^2 + 5x + 6 = 0 & \text{or} & x^2 + 5x + 6 = 0 \\ (-3)^2 + 5(-3) + 6 \stackrel{?}{=} 0 & & (-2)^2 + 5(-2) + 6 \stackrel{?}{=} 0 \\ 9 - 15 + 6 \stackrel{?}{=} 0 & & 4 - 10 + 6 \stackrel{?}{=} 0 \\ 0 = 0 \quad \text{True} & & 0 = 0 \quad \text{True} \end{array}$$

Both -3 and -2 are solutions, because they satisfy the equation. The solution set is $\{-3, -2\}$.

> **Success Tip**
> When you see the word *solve* in this example, you probably think of steps from Chapter 1 such as combining like terms, distributing, or doing something to both sides. However, to solve this quadratic equation, we begin by factoring $x^2 + 5x + 6$.

Self Check 1   Solve: $x^2 - 4x - 45 = 0$

Now Try **Problems 15 and 19**

The previous example suggests the following strategy to solve quadratic equations by factoring.

The Factoring Method for Solving a Quadratic Equation	1. Write the equation in standard form: $ax^2 + bx + c = 0$.
	2. Factor the polynomial.
	3. Use the zero-factor property to set each factor equal to zero.
	4. Solve each resulting linear equation.
	5. Check the results in the original equation.

EXAMPLE 2   Solve: **a.** $3x^2 = 6x$ **b.** $9x^2 = 25$

Strategy   To solve each equation, we will get 0 on the right side, factor the resulting binomial on the left side, and use the zero-factor property.

Why   To use the zero-factor property, we need one side of the equation to be factored completely and the other side to be 0.

Solution

a. We begin by writing the equation in $ax^2 + bx + c = 0$ form.

$$3x^2 = 6x \quad \text{This is the equation to solve.}$$

$$3x^2 - 6x = 0 \quad \text{To get 0 on the right side, subtract } 6x \text{ from both sides.}$$

To solve the equation, we factor the left side, set each factor equal to 0, and solve each resulting equation for x.

$$3x^2 - 6x = 0$$

$$3x(x - 2) = 0 \quad \text{Factor out the GCF, } 3x.$$

$$3x = 0 \quad \text{or} \quad x - 2 = 0 \quad \text{By the zero-factor property, at least one of the factors must be equal to zero.}$$

$$x = 0 \qquad\qquad x = 2 \quad \text{Solve each linear equation.}$$

The solutions are 0 and 2. Check each one in the original equation.

b.

$$9x^2 = 25 \quad \text{This is the equation to solve.}$$

$$9x^2 - 25 = 0 \quad \text{To get 0 on the right side, subtract 25 from both sides.}$$

$$(3x + 5)(3x - 5) = 0 \quad \text{Factor the difference of two squares.}$$

$$3x + 5 = 0 \quad \text{or} \quad 3x - 5 = 0 \quad \text{Set each factor equal to 0.}$$

$$3x = -5 \qquad\qquad 3x = 5 \quad \text{Solve each linear equation.}$$

$$x = -\frac{5}{3} \qquad\qquad x = \frac{5}{3}$$

The solutions are $-\frac{5}{3}$ and $\frac{5}{3}$. Check each one in the original equation.

Self Check 2   Solve: **a.** $4p^2 = 12p$ **b.** $16a^2 = 49$

Now Try **Problems 25 and 29**

EXAMPLE 3   Solve: $x = \dfrac{6}{5} - \dfrac{6}{5}x^2$

Strategy   We will multiply both sides of the equation by 5 to clear it of fractions and use factoring to solve the resulting quadratic equation.

Caution

A creative, but incorrect, approach is to divide both sides of $3x^2 = 6x$ by $3x$

$$\frac{\cancel{3x^2}}{\cancel{3x}} = \frac{\cancel{6x}}{\cancel{3x}}$$

You will obtain $x = 2$. However, you will lose the second solution, 0.

Notation

$3x^2 - 6x = 0$ is a quadratic equation in standard $ax^2 + bx + c = 0$ form, where $a = 3$, $b = -6$, and $c = 0$. Although $9x^2 - 25 = 0$ is missing an x-term, it is also a quadratic equation in standard form, where $a = 9$, $b = 0$, and $c = -25$.

Why   It is easier to factor a polynomial that contains integers than one that contains fractions.

Solution   We clear the equation of fractions and proceed as follows:

$$x = \frac{6}{5} - \frac{6}{5}x^2$$ This is the equation to solve.

$$5(x) = 5\left(\frac{6}{5} - \frac{6}{5}x^2\right)$$ Multiply both sides by 5.

$$5x = 5 \cdot \frac{6}{5} - 5 \cdot \frac{6}{5}x^2$$ Distribute the multiplication by 5.

$$5x = 6 - 6x^2$$ Simplify.

To use factoring to solve this quadratic equation, one side of the equation must be 0. Since it is easier to factor a second-degree polynomial if the coefficient of the squared term is positive, we add $6x^2$ to both sides and subtract 6 from both sides to obtain

$$6x^2 + 5x - 6 = 0$$

$$(3x - 2)(2x + 3) = 0$$ Factor the trinomial.

$$3x - 2 = 0 \quad \text{or} \quad 2x + 3 = 0$$ Set each factor equal to 0 and solve for x.

$$3x = 2 \qquad\qquad 2x = -3$$ Solve each linear equation.

$$x = \frac{2}{3} \qquad\qquad x = -\frac{3}{2}$$

The solutions are $\frac{2}{3}$ and $-\frac{3}{2}$. Check them in the original equation.

> **The Language of Algebra**
>
> Quadratic equations involve the square of a variable, not the fourth power as "quad" might suggest. Why the inconsistency? A closer look at the origin of the word quadratic reveals that it comes from the Latin word quadratus, meaning square.

Self Check 3   Solve: $x = \frac{6}{7}x^2 - \frac{3}{7}$

Now Try   **Problem 31**

Caution   To solve a quadratic equation by factoring, set the quadratic polynomial equal to 0 before factoring and using the zero-factor property. Don't make the following error:

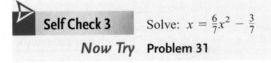

$$6x^2 + 5x = 6$$

$$x(6x + 5) = 6$$ If the product of two numbers is 6, neither number need be 6. For example, $2 \cdot 3 = 6$.

$$x = 6 \quad \text{or} \quad 6x + 5 = 6$$

$$x = \frac{1}{6}$$

Neither solution checks.

Many equations that don't appear to be quadratic can be written in standard form and then solved by factoring.

EXAMPLE 4   Solve: $(3x + 1)^2 = 6(3x - 1) + 3$

Strategy   We will square the binomial on the left side and distribute and combine like terms on the right side.

Why   We want to write an equivalent equation in standard form and use factoring to solve it.

Solution

$$(3x + 1)^2 = 6(3x - 1) + 3$$ This is the equation to solve.

$$9x^2 + 6x + 1 = 18x - 6 + 3$$ Square $3x + 1$ and distribute the multiplication by 6.

$$9x^2 + 6x + 1 = 18x - 3$$ Combine like terms: $-6 + 3 = 3$.

$$9x^2 - 12x + 4 = 0$$ To get 0 on the right side, subtract 18x from and add 3 to both sides.

$$(3x - 2)(3x - 2) = 0$$ Factor the perfect-square trinomial.

$$3x - 2 = 0 \quad \text{or} \quad 3x - 2 = 0$$ Set each factor equal to 0.

$$x = \frac{2}{3} \qquad\qquad x = \frac{2}{3}$$ Solve each equation.

We see that the two solutions are the same. We call $\frac{2}{3}$ a *repeated solution*. Check by substituting it into the original equation.

> **Self Check 4**   Solve: $(2x + 1)^2 = -8(2x + 1) - 16$
>
> **Now Try**   **Problem 39**

Using Your Calculator

Solving Quadratic Equations

To solve a quadratic equation such as $x^2 + 4x - 5 = 0$ with a graphing calculator, we can use window settings of $[-10, 10]$ for x and $[-10, 10]$ for y and graph the quadratic function $y = x^2 + 4x - 5$, as shown in figure (a). We can then trace to find the x-coordinates of the x-intercepts of the parabola. See figures (b) and (c). For better results, we can zoom in. Since these are the numbers x that make $y = 0$, they are the solutions of the equation.

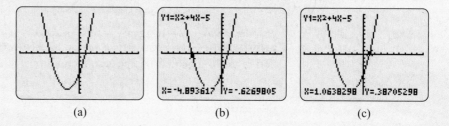

(a)   (b)   (c)

We can also find the x-intercepts of the graph of $y = x^2 + 4x - 5$ by using the ZERO feature found on most graphing calculators (Press $\boxed{\text{2nd}}$, CALC, 2). Figures (d) and (e) show how this feature locates the x-intercept and displays its coordinates.

From the displays, we can conclude that -5 and 1 are solutions of $x^2 + 4x - 5 = 0$.

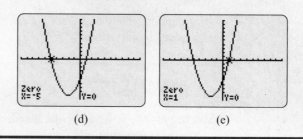

(d)   (e)

EXAMPLE 5   Let $f(x) = x^2 - 71$. For what value(s) of x is $f(x) = 10$?

Strategy   We will substitute 10 for $f(x)$ and solve for x.

Why   In the equation, there are two unknowns, x and $f(x)$. If we replace $f(x)$ with 10, we can use equation solving techniques to find x.

Solution

$$f(x) = x^2 - 71$$
$$10 = x^2 - 71 \quad \text{Substitute 10 for } f(x).$$

To solve this quadratic equation, we first write it in standard form. Then we factor the difference of two squares on the right side, set each factor equal to 0, and solve each resulting equation.

$$0 = x^2 - 81 \qquad\qquad \text{To get 0 on the left side, subtract 10 from both sides.}$$
$$0 = (x + 9)(x - 9) \qquad \text{Factor the difference of two squares.}$$
$$x + 9 = 0 \quad \text{or} \quad x - 9 = 0 \qquad \text{Set each factor equal to 0.}$$
$$x = -9 \quad\ \mid\quad\ x = 9 \qquad \text{Solve each equation.}$$

If x is -9 or 9, then $f(x) = 10$. Check these results by substituting -9 and 9 for x in $f(x) = x^2 - 71$.

 Self Check 5   Let $g(x) = x^2 - 21$. For what value(s) of x is $g(x) = 100$?

Now Try   **Problem 43**

---

2   **Solve Higher-Degree Polynomial Equations by Factoring.**

Some equations involving polynomials with degrees higher than 2 can be solved by factoring. In such cases, we use the extension of the zero-factor property: When the product of two *or more* real numbers is 0, at least one of them is 0.

EXAMPLE 6   Solve: $6x^3 - x^2 = 2x$

Strategy   This equation is not quadratic, because it contains a term involving x^3. To solve it, we will get 0 on the right side, factor the polynomial on the left side, and use the zero-factor property.

Why   To use the zero-factor property, we need one side of the equation to be factored completely and the other side to be 0.

Solution

The Language of Algebra
Since the highest degree of any term in $6x^3 - x^2 = 2x$ is 3, it is called a *third-degree* polynomial equation. Note that it has three solutions.

$$6x^3 - x^2 = 2x \quad \text{This is the equation to solve.}$$
$$6x^3 - x^2 - 2x = 0 \quad \text{To get 0 on the right side, subtract 2x from both sides.}$$

Next, we factor x from each term of the polynomial on the left side and proceed as follows:

$$6x^3 - x^2 - 2x = 0$$
$$x(6x^2 - x - 2) = 0 \quad \text{Factor out the GCF, } x.$$
$$x(3x - 2)(2x + 1) = 0 \quad \text{Factor } 6x^2 - x - 2.$$

$$x = 0 \quad \text{or} \quad 3x - 2 = 0 \quad \text{or} \quad 2x + 1 = 0$$

Set each of the three factors equal to 0.

$$x = \frac{2}{3} \qquad\qquad x = -\frac{1}{2}$$ *Solve each equation.*

The solutions are 0, $\frac{2}{3}$, and $-\frac{1}{2}$. Check each one in the original equation.

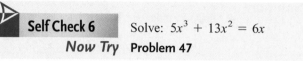

Self Check 6   Solve: $5x^3 + 13x^2 = 6x$

Now Try   **Problem 47**

EXAMPLE 7   Solve: **a.** $-x^4 + 5x^2 - 4 = 0$ **b.** $x^3 - x^2 - 9x + 9 = 0$

Strategy   To solve each equation, we will factor the polynomial on the left side and use the zero-factor property.

Why   To use the zero-factor property, we need one side of the equation to be factored completely and the other side to be 0.

Solution

a. To make the leading coefficient positive, we multiply both sides of the equation by -1. Then we factor the resulting trinomial and set the factors equal to 0.

$$-x^4 + 5x^2 - 4 = 0 \qquad \text{This is the equation to solve.}$$
$$-1(-x^4 + 5x^2 - 4) = -1(0) \qquad \text{Multiply both sides by } -1.$$
$$x^4 - 5x^2 + 4 = 0 \qquad \text{This is a fourth-degree polynomial equation.}$$
$$(x^2 - 1)(x^2 - 4) = 0 \qquad \text{Factor the trinomial.}$$
$$(x + 1)(x - 1)(x + 2)(x - 2) = 0 \qquad \text{Factor each difference of two squares.}$$
$$x + 1 = 0 \quad \text{or} \quad x - 1 = 0 \quad \text{or} \quad x + 2 = 0 \quad \text{or} \quad x - 2 = 0$$
$$x = -1 \quad | \quad x = 1 \quad | \quad x = -2 \quad | \quad x = 2$$

The solutions are -1, 1, -2, and 2. Check each one in the original equation.

b. We can using grouping to factor the polynomial on the left side.

$$x^3 - x^2 - 9x + 9 = 0 \qquad \text{This is the equation to solve.}$$
$$x^2(x - 1) - 9(x - 1) = 0 \qquad \text{Factor } x^2 \text{ from } x^3 - x^2 \text{ and } -9 \text{ from } -9x + 9.$$
$$(x - 1)(x^2 - 9) = 0 \qquad \text{Factor out the GCF, } x - 1.$$
$$(x - 1)(x + 3)(x - 3) = 0 \qquad \text{Factor the difference of two squares.}$$
$$x - 1 = 0 \quad \text{or} \quad x + 3 = 0 \quad \text{or} \quad x - 3 = 0 \qquad \text{Set each factor equal to 0.}$$
$$x = 1 \quad | \quad x = -3 \quad | \quad x = 3 \qquad \text{Solve each equation.}$$

The solutions are 1, -3, and 3. Check each one in the original equation.

Self Check 7   Solve: **a.** $-a^4 - 36 + 13a^2 = 0$
b. $x^3 + x^2 - 100x - 100 = 0$

Now Try   **Problems 51 and 53**

Using Your Calculator   *Solving Equations*

To solve $x^4 - 5x^2 + 4 = 0$ with a graphing calculator, we can use window settings of $[-6, 6]$ for x and $[-5, 10]$ for y and graph the polynomial function $y = x^4 - 5x^2 + 4$ as shown in the figure. We can then read the values of x that make $y = 0$. They are $x = -2, -1, 1$, and 2. If the x-coordinates of the x-intercepts were not obvious, we could approximate their values by using TRACE and ZOOM or by using the ZERO feature.

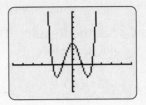

3 **Use Quadratic Equations to Solve Problems.**

EXAMPLE 8   ***Stained Glass.*** The length of the base of a stained glass window is 3 times its height, and its area is 96 square feet. Find the length of its base and its height.

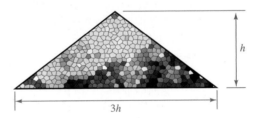

Analyze the Problem   The formula that gives the area of a triangle is $A = \frac{1}{2}bh$, where b is the length of the base and h the height.

Form an Equation   We can let $h =$ the height of the window, in feet. Then $3h =$ the length of the base, in feet. To form an equation in terms of h, we can substitute $3h$ for b and 96 for A in the formula for the area of a triangle.

$$A = \frac{1}{2}bh$$

$$96 = \frac{1}{2}(3h)h$$

$$96 = \frac{3}{2}h^2 \qquad \text{Simplify the right side.}$$

Solve the Equation   To solve this quadratic equation, we will get 0 on the left side, factor the resulting binomial on the right side and use the zero-factor property.

$$96 = \frac{3}{2}h^2$$

$$192 = 3h^2 \qquad \text{To clear the equation of the fraction, multiply both sides by 2.}$$

$$64 = h^2 \qquad \text{Divide both sides by 3.}$$

$$0 = h^2 - 64 \qquad \text{To get 0 on the left side, subtract 64 from both sides.}$$

$$0 = (h + 8)(h - 8) \qquad \text{Factor the difference of two squares.}$$

$$h + 8 = 0 \quad \text{or} \quad h - 8 = 0$$

$$h = -8 \qquad \qquad h = 8$$

State the Conclusion   The solutions of the equation are -8 and 8. Since h represents the height of the window, and the height cannot be negative, we discard -8. Thus, the height of the stained glass window is 8 feet, and the length of its base is $3(8)$, or 24 feet.

Check the Result   The area of a triangle with a base of 24 feet and a height of 8 feet is 96 square feet:

$$A = \frac{1}{2}bh = \frac{1}{2}(24)(8) = 12(8) = 96$$

The result checks.

 Now Try   **Problem 89**

EXAMPLE 9   *Recreation.*   A rectangular-shaped spa, 5 feet wide and 6 feet long, is surrounded by decking of uniform width, as shown in the illustration. If the total area of the deck is 60 ft^2, how wide is the decking?

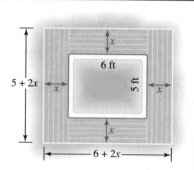

Analyze the Problem   Since the dimensions of the rectangular spa are 5 feet by 6 feet, its surface area is $6 \cdot 5 = 30$ ft^2. The decking has an area of 60 ft^2.

Form an Equation   Let $x = $ the width of the decking, in feet. Then the width of the outer rectangle (the pool and the decking) is $x + 5 + x$ or $(5 + 2x)$ feet, and the length of the outer rectangle is $x + 6 + x$ or $(6 + 2x)$ feet. The area of the outer rectangle is the product of its length and width: $(6 + 2x)(5 + 2x)$ ft^2. We can now form an equation.

The area of the outer rectangle	minus	the area of the spa	equals	the area of the decking.
$(6 + 2x)(5 + 2x)$	$-$	30	$=$	60

Solve the Equation

$$(6 + 2x)(5 + 2x) - 30 = 60$$

$$30 + 12x + 10x + 4x^2 - 30 = 60 \qquad \text{Multiply the binomials on the left side.}$$

$$4x^2 + 22x = 60 \qquad \text{Simplify the left side.}$$

$$4x^2 + 22x - 60 = 0 \qquad \text{Subtract 60 from both sides to get 0 on the right side.}$$

$$2(2x^2 + 11x - 30) = 0 \qquad \text{Factor out the GCF, 2.}$$

$$2(2x + 15)(x - 2) = 0 \qquad \text{Factor the trinomial.}$$

Success Tip

An alternate way to solve the equation $4x^2 + 22x - 60 = 0$ is to divide both sides by the GCF of the terms on its left side. When we divide both sides by the GCF, which is 2, we obtain

$$\frac{4x^2}{2} + \frac{22x}{2} - \frac{60}{2} = \frac{0}{2}$$

$$2x^2 + 11x - 30 = 0$$

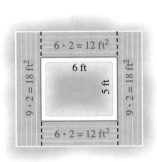

$$2x + 15 = 0 \qquad \text{or} \qquad x - 2 = 0$$

Since 2 cannot equal 0, discard that possibility. Set each of the remaining factors that contain a variable equal to 0.

$$2x = -15 \qquad\qquad x = 2$$ Solve each equation.

$$\cancel{x = -\frac{15}{2}}$$

State the Conclusion   The solutions of the equation are $-\frac{15}{2}$ and 2. Since x represents the width of the decking, and the width cannot be negative, we discard $-\frac{15}{2}$. Thus, the width of the decking is 2 feet.

Check the Result   The illustration shows that if the decking is 2 feet wide, the total area of the decking is $18 + 12 + 18 + 12 = 60$ ft^2. The result checks.

  Now Try   **Problem 93**

EXAMPLE 10   *Ballistics.*   If the initial velocity of an object launched from the ground straight up into the air is 176 feet per second, when will the object strike the ground?

Analyze the Problem   The height of an object launched straight up into the air from the ground with an initial velocity of v feet per second is given by the formula

$$h = -16t^2 + vt$$

The height h is in feet, and t represents the number of seconds since the object was released. When the object hits the ground, its height will be 0.

Form an Equation   In the formula, we set h equal to 0 and set v equal to 176.

$$h = -16t^2 + vt$$
$$0 = -16t^2 + 176t$$

Solve the Equation   To solve this quadratic equation, we will use the factoring method.

$$0 = -16t^2 + 176t$$
$$0 = -16t(t - 11)$$ Factor out $-16t$.
$$-16t = 0 \quad \text{or} \quad t - 11 = 0$$ Set each factor equal to 0.
$$t = 0 \qquad\qquad t = 11$$ Solve each equation.

State the Conclusion   When t is 0, the object's height above the ground is 0 feet, because it has not been released. When t is 11, the height is again 0 feet, and the object has returned to the ground. The solution is 11 seconds.

Check the Result   If we substitute 11 for t in the formula $h = -16t^2 + 176t$, we have

$$h = -16(11)^2 + 176(11) = -1{,}936 + 1{,}936 = 0$$

The result checks.

Success Tip
When solving equations of this type, it is helpful to determine if the coefficient of the t-term has 16 as a factor. If it does, factor out -16 from each term.

$$0 = -16t^2 + 176t$$
$$\uparrow$$
This is $16 \cdot 11$.

  Now Try   **Problem 97**

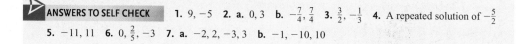

STUDY SET
5.9

VOCABULARY

Fill in the blanks.

1. A _____ equation is any equation that can be written in the form $ax^2 + bx + c = 0$, where $a \neq 0$.

2. To _____ a quadratic equation means to find all the values of the variable that make the equation true.

3. To write the quadratic equation $x^2 - 3x = 15$ in _____ form, we subtract 15 from both sides.

4. $2x^2 - 4x = 0$, $3x^3 - x^2 - 6x = 0$, and $x^4 - 5x^2 + 4 = 0$ are examples of _____ equations.

CONCEPTS

5. Determine whether each equation is a quadratic equation.
   a. $w^2 + 7w + 12 = 0$ **b.** $6t + 11 = 0$
   c. $x(x + 3) = -2$ **d.** $k^3 - 4k^2 + k - 15 = 0$

6. Write each equation in standard form.
   a. $x^2 - x = 6$
   b. $x(x + 9) = -1$

7. **a.** If the product of two numbers is 0, what must be true about at least one of the numbers?
   b. Fill in the blanks: By the _____-factor property, if $ab = 0$, then $a = $ or $b = $.

8. Use the zero-factor property to solve each equation.
   a. $(x - 3)(x + 5) = 0$
   b. $6x(x - 1)(2x + 5) = 0$

9. Use a check to determine whether 4 and -5 are solutions of $a^2 - 9a + 20 = 0$.

10. What *first step* should be performed to solve each quadratic equation?
    a. $x^2 + 24 = -11x$
    b. $x^2 + x + \dfrac{1}{4} = 0$
    c. $-2x^2 + 7x + 4 = 0$
    d. $m(m + 3) = 2$

11. **a.** Write an expression that represents the width of the outer rectangle.
    b. Write an expression that represents the length of the outer rectangle.

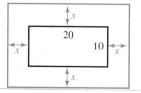

12. **a.** Use the graph to solve $x^2 - 2x - 3 = 0$.

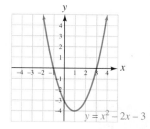

b. Use the graph to solve $x^3 - 4x^2 + 4x = 0$

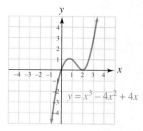

NOTATION

Complete each solution to solve each equation.

13. Solve:
$$y^2 - 2y - 8 = 0$$
$$(y - 4)(\quad) = 0$$
$$\quad = 0 \quad \text{or} \quad y + 2 = 0$$
$$y = 4 \quad | \quad y = \quad$$

14. Solve:
$$2x^2 - 3x - 1 = 1$$
$$2x^2 - 3x - 2 = \quad$$
$$(\quad)(x - 2) = 0$$
$$2x + 1 = \quad \text{or} \quad \quad = 0$$
$$2x = -1 \quad | \quad x = 2$$
$$x = \quad |$$

GUIDED PRACTICE

Solve each equation. **See Example 1.**

15. $z^2 + 8z + 15 = 0$ 16. $w^2 + 7w + 12 = 0$

17. $x^2 + 6x + 8 = 0$ **18.** $x^2 + 9x + 20 = 0$

19. $2x^2 - 3x + 1 = 0$ **20.** $2x^2 - x - 15 = 0$

21. $3m^2 + 10m + 3 = 0$ **22.** $3r^2 + 7r + 2 = 0$

Solve each equation. See Example 2.

23. $x^2 + x = 0$ **24.** $x^2 + 5x = 0$

25. $4x^2 = 8x$ **26.** $3x^2 = 9x$

27. $y^2 - 16 = 0$ **28.** $y^2 - 25 = 0$

29. $16y^2 = 9$ **30.** $81y^2 = 25$

Solve each equation by first clearing it of fractions. See Example 3.

31. $\dfrac{3a^2}{2} = \dfrac{1}{2} - a$ **32.** $x^2 + 1 = \dfrac{5}{2}x$

33. $x^2 - \dfrac{2}{5} = -\dfrac{9}{5}x$ **34.** $\dfrac{4}{5}x^2 + x = \dfrac{6}{5}$

35. $\dfrac{8}{3}a^2 = 1 - \dfrac{10}{3}a$ **36.** $\dfrac{5}{6}z^2 = 1 - \dfrac{13}{6}z$

37. $\dfrac{3}{16}m^2 - \dfrac{27}{16} = 0$ **38.** $\dfrac{2}{7}n^2 - \dfrac{128}{7} = 0$

Solve each equation. See Example 4.

39. $(x + 7)^2 = -2(x + 7) - 1$

40. $2(7x + 18) - 1 = (x + 6)^2$

41. $(m + 4)(2m + 3) - 22 = 10m$

42. $(d - 2)(d + 1) - d = 1$

See Example 5.

43. Let $f(x) = x^2 - 3x + 3$. For what value(s) of x is $f(x) = 1$?

44. Let $f(x) = 6x^2 + 5x + 2$. For what value(s) of x is $f(x) = 6$?

45. Let $f(x) = x^3 - 6x^2 + 8x + 2$. For what value(s) of x is $f(x) = 2$?

46. Let $f(x) = x^3 - 2x^2 - 8x + 10$. For what value(s) of x is $f(x) = 10$?

Solve each equation. See Example 6.

47. $x^3 - 4x^2 = 21x$ **48.** $x^3 + 8x^2 = 9x$

49. $y^3 - 49y = 0$ **50.** $2z^3 - 200z = 0$

Solve each equation. See Example 7.

51. $-z^4 + 37z^2 - 36 = 0$ **52.** $-y^4 + 10y^2 - 9 = 0$

53. $x^3 - 3x^2 - 4x + 12 = 0$ **54.** $x^3 + 4x^2 - 25x - 100 = 0$

Use a graphing calculator to find the solutions of each equation, if one exists. If an answer is not exact, give the answer to the nearest hundredth.

55. $2x^2 - 7x + 4 = 0$ **56.** $x^2 - 4x + 7 = 0$

57. $-3x^3 - 2x^2 + 5 = 0$ **58.** $-2x^3 - 3x - 5 = 0$

TRY IT YOURSELF

Solve each equation.

59. $b(6b - 7) = 10$ **60.** $2y(4y + 3) = 9$

61. $x^3 + x^2 = 0$ **62.** $2x^4 + 8x^3 = 0$

63. $\dfrac{x^2}{5} - \dfrac{4}{5} = -\dfrac{3}{5}x$ **64.** $\dfrac{x^2}{9} = \dfrac{8}{9}x - \dfrac{7}{9}$

65. $a^3 - 2a^2 - 16a + 32 = 0$ **66.** $b^3 - 5b^2 - 9b + 45 = 0$

67. $6y^2 = 25y$ **68.** $15x^2 = 7x$

69. $a^3 - 18a^2 = -81a$ **70.** $y^3 + 22y^2 = -121y$

71. $7t^2 - 2t - 5 = 0$ **72.** $15x^2 + 22x - 5 = 0$

73. $3x^3 + 3x^2 = 12(x + 1)$ **74.** $9(y + 4) = y^3 + 4y^2$

75. $64t^2 - 81 = 0$ **76.** $36t^2 - 1 = 0$

77. $\dfrac{x^2(6x + 37)}{35} = x$ **78.** $x^2 = -\dfrac{4x^3(3x + 5)}{3}$

79. $0 = d^2 - 5d - 66$ **80.** $0 = t^2 - 2t - 63$

81. $n(3n - 4) = (n - 6)^2 + 11n - 1$

82. $s(2s + 7) = (s + 1)^2 + 71 - s$

83. $(x - 5.5)(x + 3) = 0$ **84.** $(2y + 11)(y - 4) = 0$

85. $-x^4 + 34x^2 - 225 = 0$ **86.** $-x^4 + 26x^2 - 25 = 0$

APPLICATIONS

87. INTEGER PROBLEM   The product of two positive consecutive even integers is 288. Find the integers.   (*Hint:* Let x = the smaller even integer and $x + 2$ = the larger even integer.)

88. INTEGER PROBLEM   The product of two positive consecutive odd integers is 143. Find the integers.   (*Hint:* Let x = the smaller odd integer and $x + 2$ = the larger odd integer.)

Reading the Textbook

Reading an algebra textbook is different from reading a newspaper or a novel. Here are two ways that you should be reading this textbook.

SKIMMING FOR AN OVERVIEW: This is a quick way to look at material just *before* it is covered in class. It helps you become familiar with the new vocabulary and notation that will be used by your instructor in the lecture. It lays a foundation.

READING FOR UNDERSTANDING: This in-depth type of reading is done more slowly, with a pencil and paper at hand. Don't skip anything—every word counts! You should do just as much writing as you do reading. Highlight the important points and work each example. If you become confused, stop and reread the material until you understand it.

Now Try This

Choose a section from this chapter and . . .
1. Quickly skim it. Write down any terms in boldface type and the titles of any properties, definitions, or strategies that are given in the colored boxes.
2. Work each *Self Check* problem. Your solutions should look like those in the *Examples*. Be sure to include author notes (the sentences in red to the right of each step of a solution).

SECTION 6.1

Rational Functions and Simplifying Rational Expressions

Objectives

 1 Define rational expressions.

2 Evaluate rational functions.

3 Graph rational functions.

4 Find the domain of a rational function.

5 Simplify rational expressions.

6 Simplify rational expressions that have factors that are opposites.

Linear and polynomial functions can be used to model many real-world situations. In this section, we introduce another family of functions known as *rational functions*. Rational functions get their name from the fact that their defining equation contains a *ratio* (fraction) of two polynomials.

1 **Define Rational Expressions.**

Fractions that are the quotient of two integers are *rational numbers*. Fractions that are the quotient of two polynomials are called *rational expressions*.

| **Rational Expressions** | A **rational expression** is an expression of the form $\frac{A}{B}$, where A and B are polynomials and B does not equal 0. |

Some examples of rational expressions are

$$\frac{3x}{x-7}, \qquad \frac{8yz^4}{6y^2z^2}, \qquad \frac{5m+n}{8m+16}, \qquad \text{and} \qquad \frac{6a^2-13a+6}{3a^2+a-2}$$

 Evaluate Rational Functions.

Rational expressions in one variable often define functions. For example, if the cost of subscribing to an online research network is $6 per month plus $1.50 per hour of access time, the average (mean) hourly cost of the service is the total monthly cost, divided by the number of hours of access time used that month:

$$\frac{C}{n} = \frac{1.50n+6}{n} \qquad \text{\small C is the total monthly cost, and n is the number}$$
$$\text{\small of hours the service is used that month.}$$

The right side of this equation is a rational expression: the quotient of the binomial $1.50n+6$ and the monomial n.

The rational function that gives the average hourly cost of using the network for n hours per month can be written

$$f(n) = \frac{1.50n+6}{n}$$

We are assuming that at least one access call will be made each month, so the function is defined for $n > 0$.

Caution

Since division by 0 is undefined, the value of a polynomial in the denominator of a rational expression cannot be 0. For example, x cannot be 7 in the rational expression $\frac{3x}{x-7}$, because the value of the denominator would be 0. In $\frac{5m+n}{8m+16}$, m cannot be -2, because the value of the denominator would be 0.

| **Rational Functions** | A **rational function** is a function whose equation is defined by a rational expression in one variable, where the value of the polynomial in the denominator is never zero. |

EXAMPLE 1   Use the function $f(n) = \dfrac{1.50n+6}{n}$ to find the average hourly cost when the network described earlier is used for:

a. 1 hour   **b.** 9 hours

Strategy   We will find $f(1)$ and $f(9)$.

Why   The notation $f(1)$ represents the average hourly cost for 1 hour of use and $f(9)$ represents average hourly cost for 9 hours of use.

Solution

a. To find the average hourly cost for 1 hour of time, we find $f(1)$:

$$f(1) = \frac{1.50(1)+6}{1} = 7.5 \qquad \text{\small Input 1 for n and simplify.}$$

The average hourly cost for 1 hour of access time is $7.50.

b. To find the average hourly cost for 9 hours of time, we find $f(9)$:

$$f(9) = \frac{1.50(9)+6}{9} = 2.166666666\ldots \qquad \text{\small Input 9 for n and simplify.}$$

The average hourly cost for 9 hours of access time is approximately $2.17.

Self Check 1   Find the average hourly cost when the network is used for:
a. 3 hours   **b.** 100 hours

Now Try   **Problem 91**

③   **Graph Rational Functions.**

To graph the rational function $f(n) = \dfrac{1.50n + 6}{n}$, we substitute values for n (the inputs) in the function, compute the corresponding values of $f(n)$ (the outputs), and express the results as ordered pairs. From the evaluations in Example 1 and its Self Check, we know four such ordered pairs are: $(1, 7.50)$, $(3, 3.50)$, $(9, 2.17)$, and $(100, 1.56)$. Those pairs and others are listed in the table below. We then plot the points and draw a smooth curve through them to get the graph.

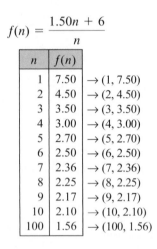

$$f(n) = \frac{1.50n + 6}{n}$$

n	$f(n)$	
1	7.50	$\rightarrow (1, 7.50)$
2	4.50	$\rightarrow (2, 4.50)$
3	3.50	$\rightarrow (3, 3.50)$
4	3.00	$\rightarrow (4, 3.00)$
5	2.70	$\rightarrow (5, 2.70)$
6	2.50	$\rightarrow (6, 2.50)$
7	2.36	$\rightarrow (7, 2.36)$
8	2.25	$\rightarrow (8, 2.25)$
9	2.17	$\rightarrow (9, 2.17)$
10	2.10	$\rightarrow (10, 2.10)$
100	1.56	$\rightarrow (100, 1.56)$

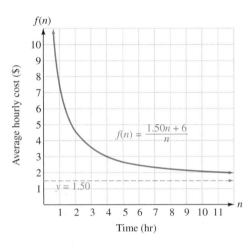

As the access time increases, the graph approaches the line y = 1.50, which indicates that the average hourly cost approaches $1.50 as the hours of use increase.

From the graph, we can see that the average hourly cost decreases as the number of hours of access time increases. Since the cost of each extra hour of access time is $1.50, the average hourly cost can approach $1.50 but never drop below it. Thus, the graph of the function approaches the line $y = 1.5$ as n increases. When a graph approaches a line, we call the line an **asymptote**. The line $y = 1.5$ is a **horizontal asymptote** of the graph.

As n gets smaller and approaches 0, the graph approaches the y-axis. The y-axis is a **vertical asymptote** of the graph.

④   **Find the Domain of a Rational Function.**

Since division by 0 is undefined, any values that make the denominator 0 in a rational function must be excluded from the domain of the function.

EXAMPLE 2   Find the domain of: $f(x) = \dfrac{3x + 2}{x^2 + x - 6}$

Strategy   We will set $x^2 + x - 6$ equal to 0 and solve for x.

Why   We don't need to examine the numerator of the rational expression; it can be any value, including 0. The domain of the function includes all real numbers, except those that make the *denominator equal to 0*.

The Language of Algebra

Another way that Example 2 could be phrased is: *State the restrictions on the variable.* For $\frac{3x+2}{x^2+x-6}$, we can write $x \ne -3$ and $x \ne 2$.

Solution

$$x^2 + x - 6 = 0 \qquad \text{Set the denominator equal to 0.}$$

$$(x + 3)(x - 2) = 0 \qquad \text{Factor the trinomial.}$$

$$x + 3 = 0 \quad \text{or} \quad x - 2 = 0 \qquad \text{Set each factor equal to 0.}$$

$$x = -3 \qquad \qquad x = 2 \qquad \text{Solve each linear equation.}$$

Thus, the domain of the function is the set of all real numbers except -3 and 2. Using set-builder notation we can describe the domain as $\{x \mid x \text{ is a real number and } x \ne -3, x \ne 2\}$. In interval notation, the domain is $(-\infty, -3) \cup (-3, 2) \cup (2, \infty)$.

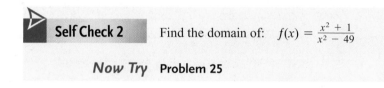

Self Check 2   Find the domain of: $f(x) = \dfrac{x^2 + 1}{x^2 - 49}$

Now Try   **Problem 25**

Using Your Calculator

Finding the Domain and Range of a Rational Function

We can find the domain and range of the function in Example 2 by looking at its graph. If we use window settings of $[-10, 10]$ for x and $[-10, 10]$ for y and graph the function

$$f(x) = \frac{3x + 2}{x^2 + x - 6}$$

we will obtain the graph shown on the next page.

From the figure, we can see that

- As x approaches -3 from the left, the values of y decrease, and the graph approaches the vertical line $x = -3$. As x approaches -3 from the right, the values of y increase, and the graph approaches the vertical line $x = -3$.

From the figure, we can also see that

- As x approaches 2 from the left, the values of y decrease, and the graph approaches the vertical line $x = 2$. As x approaches 2 from the right, the values of y increase, and the graph approaches the vertical line $x = 2$.

The lines $x = -3$ and $x = 2$ are vertical asymptotes. These asymptotes seem to appear in the figure. This is because graphing calculators draw graphs by connecting dots whose x-coordinates are close together. Often when two such points straddle a vertical asymptote and their y-coordinates are far apart, the calculator draws a line between them, producing what appears to be a vertical asymptote. If instead of connected mode, you set your calculator to dot mode by pressing $\boxed{\text{MODE}}$, pressing ▼ five times, pressing ▶ once, and then pressing $\boxed{\text{ENTER}}$, the vertical lines will not appear.

From figure (a), we can also see that

- As x increases to the right of 2, the values of y decrease and approach the line $y = 0$.
- As x decreases to the left of -3, the values of y increase and approach the line $y = 0$.

The line $y = 0$ (the x-axis) is a horizontal asymptote. Graphing calculators do not draw lines that appear to be horizontal asymptotes.

From the graph, we can see that every real number x, except -3 and 2, gives a value of y. This observation confirms that the domain of the function is $(-\infty, -3) \cup (-3, 2) \cup (2, \infty)$. We can also see that y can be any value. Thus, the range is $(-\infty, \infty)$.

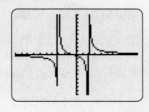

---

5 ## Simplify Rational Expressions.

When working with rational expressions, we will use some familiar rules from arithmetic.

Properties of Fractions	If a, b, c, d, and k represent real numbers, and if there are no divisions by 0, then

1. $\dfrac{a}{b} = \dfrac{c}{d}$ if and only if $ad = bc$ 2. $\dfrac{a}{1} = a$ and $\dfrac{a}{a} = 1$

3. $\dfrac{ak}{bk} = \dfrac{a}{b} \cdot \dfrac{k}{k} = \dfrac{a}{b}$ 4. $-\dfrac{a}{b} = \dfrac{-a}{b} = \dfrac{a}{-b}$

Property 3 is true because any number times 1 is that number.

$$\frac{ak}{bk} = \frac{a}{b} \cdot \frac{k}{k} = \frac{a}{b} \cdot 1 = \frac{a}{b} \quad \text{where} \quad b \neq 0 \quad \text{and} \quad k \neq 0$$

The Language of Algebra
Property 3 is known as the *fundamental property of fractions*. Stated in another way, it enables us to divide out factors that are common to the numerator and denominator of a fraction.

To streamline this process, we can replace $\frac{k}{k}$ in $\frac{ak}{bk}$ with the equivalent fraction $\frac{1}{1}$.

$$\frac{ak}{bk} = \frac{a\overset{1}{\cancel{k}}}{b\underset{1}{\cancel{k}}} = \frac{a}{b} \qquad \tfrac{k}{k} = \tfrac{1}{1} = 1$$

We say that we have simplified $\frac{ak}{bk}$ by *removing a factor equal to 1*.

To **simplify a rational expression** means to write it so that the numerator and denominator have no common factors other than 1.

Simplifying Rational Expressions	1. Factor the numerator and denominator completely to determine their common factors.
	2. Remove factors equal to 1 by replacing each pair of factors common to the numerator and denominator with the equivalent fraction $\frac{1}{1}$.
	3. Multiply the remaining factors in the numerator and in the denominator.

EXAMPLE 3   Simplify: $\dfrac{8yz^4}{6y^2z^2}$

Strategy   We will begin by writing the numerator and denominator in factored form. Then we will remove any factors common to the numerator and denominator.

Why   The rational expression is simplified when the numerator and denominator have no common factors other than 1.

Solution

$$\dfrac{8yz^4}{6y^2z^2} = \dfrac{2 \cdot 2 \cdot 2 \cdot y \cdot z \cdot z \cdot z \cdot z}{2 \cdot 3 \cdot y \cdot y \cdot z \cdot z}$$ Factor $8yz^4$ and $6y^2z^2$ completely.

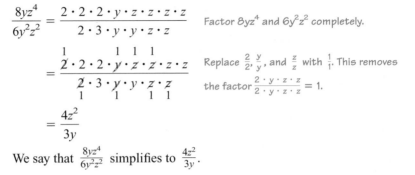

$$= \dfrac{\overset{1}{2} \cdot 2 \cdot 2 \cdot \overset{1}{y} \cdot \overset{1}{z} \cdot \overset{1}{z} \cdot z \cdot z}{\underset{1}{2} \cdot 3 \cdot \underset{1}{y} \cdot y \cdot \underset{1}{z} \cdot \underset{1}{z}}$$ Replace $\frac{2}{2}, \frac{y}{y}$, and $\frac{z}{z}$ with $\frac{1}{1}$. This removes the factor $\frac{2 \cdot y \cdot z \cdot z}{2 \cdot y \cdot z \cdot z} = 1$.

$$= \dfrac{4z^2}{3y}$$

We say that $\dfrac{8yz^4}{6y^2z^2}$ simplifies to $\dfrac{4z^2}{3y}$.

The Language of Algebra
When a rational expression is simplified, the result is an *equivalent expression*.

An alternate approach is to use rules for exponents to simplify the rational expression.

$$\dfrac{8yz^4}{6y^2z^2} = \dfrac{\overset{1}{2} \cdot 2 \cdot 2 \cdot y^{1-2}z^{4-2}}{\underset{1}{2} \cdot 3} = \dfrac{4y^{-1}z^2}{3} = \dfrac{4z^2}{3y}$$ To divide exponential expressions with the same base, keep the base and subtract the exponents.

Self Check 3   Simplify: $\dfrac{12a^4b^2}{20ab^4}$

Now Try   **Problem 33**

To simplify rational expressions, we often make use of the factoring methods discussed in Chapter 5.

EXAMPLE 4   Simplify: **a.** $\dfrac{6x^3}{3x^4 - 9x^3}$ **b.** $\dfrac{x^2 - 16}{2x^2 + 8x}$

Strategy   We will begin by factoring the numerator and denominator. Then we will remove any factors common to the numerator and denominator.

Why   We need to make sure that the numerator and denominator have no common factors other than 1. If that is the case, then the rational expression is simplified.

Solution

a. $\dfrac{6x^3}{3x^4 - 9x^3} = \dfrac{2 \cdot 3 \cdot x^3}{3x^3(x - 3)}$ Factor the numerator.
In the denominator, factor out the GCF, $3x^3$.

$$= \dfrac{2 \cdot \overset{1}{3} \cdot \overset{1}{x^3}}{\underset{1}{3} \cdot \underset{1}{x^3} \cdot (x - 3)}$$ Remove the factors common to the numerator and denominator.

$$= \dfrac{2}{x - 3}$$

b. In the numerator, we factor the difference of two squares. In the denominator, we factor out the GCF, $2x$.

$$\frac{x^2 - 16}{2x^2 + 8x} = \frac{\overset{1}{\cancel{(x + 4)}}(x - 4)}{2x\underset{1}{\cancel{(x + 4)}}}$$ Remove the binomial factor $x + 4$ that is common to the numerator and denominator.

$$= \frac{x - 4}{2x}$$ This rational expression does not simplify further.

> ▷ **Self Check 4**   Simplify: **a.** $\dfrac{28x^4}{7x^5 - 14x^4}$ **b.** $\dfrac{x^2 - 9}{5x^2 - 15x}$
>
> **Now Try**   Problems 35 and 39

When simplifying rational expressions, we can only remove factors common to the entire numerator and denominator. It is incorrect to remove terms common to the numerator and denominator.

$$\frac{\overset{1}{\cancel{x}} - 4}{\underset{1}{\cancel{2x}}} \qquad\qquad \frac{a^2 - 3a + \overset{1}{\cancel{2}}}{a + \underset{1}{\cancel{2}}} \qquad\qquad \frac{\overset{1}{\cancel{y^2}} - 36}{\underset{1}{\cancel{y^2}} - y - 7}$$

x is a term of $x - 4$.   2 is a term of $a^2 - 3a + 2$ and a term of $a + 2$. y^2 is a term of $y^2 - 36$ and a term of $y^2 - y - 7$.

Using Your Calculator

Checking an Algebraic Simplification

After simplifying an expression, we can use a scientific calculator to check the answer. One way to check whether $\frac{x^2 - 16}{2x^2 + 8x} = \frac{x - 4}{2x}$ is correct in Example 4b is to evaluate $\frac{x^2 - 16}{2x^2 + 8x}$ and $\frac{x - 4}{2x}$ for a value of x (say, **2**). The expressions should give identical results.

For $\dfrac{x^2 - 16}{2x^2 + 8x}$:

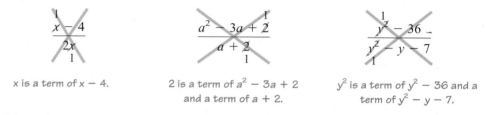

For $\dfrac{x - 4}{2x}$:

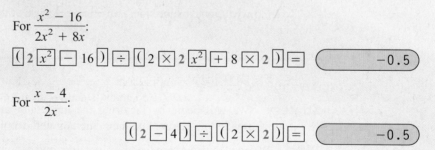

The results of the evaluations are indeed the same. Evaluate the expressions for several other values of x. If the results differ for any given value, the original expression was not simplified correctly.

We can also use a graphing calculator to show that the simplification in Example 4b is correct. We enter the functions $f(x) = \frac{x^2 - 16}{2x^2 + 8x}$ and $g(x) = \frac{x - 4}{2x}$ as Y_1 and Y_2, respectively. See figure (a) on the next page. Then select the TABLE feature. Reading across the table, the values of Y_1 and Y_2 should be the same for each value of x as shown in figure (b). Note for $x = -4$ the Y_1 value says error while the Y_2 value is 1. This happens as a result of removing the common factor $x + 4$ in the simplification.

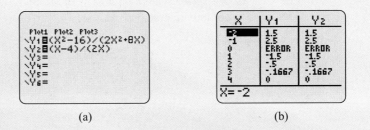

(a)   (b)

A third method to informally check the simplification is to compare the graphs of $f(x) = \frac{x^2 - 16}{2x^2 + 8x}$, shown in figure (c), and $g(x) = \frac{x - 4}{2x}$, shown in figure (d). Since the graphs appear to be the same, we can conclude that the simplification is probably correct.

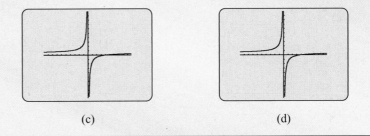

(c)   (d)

EXAMPLE 5   Simplify: **a.** $\dfrac{x^2 - 10x + 25}{8x - 40}$ **b.** $\dfrac{6a^2 - 13a + 6}{3a^2 + a - 2}$

c. $\dfrac{y^3 - 8}{y^3 - 2y^2 + 3y - 6}$

Strategy   We will begin by factoring the numerator and denominator completely. Then we will remove any factors common to the numerator and denominator.

Why   We need to make sure that the numerator and denominator have no common factors other than 1. If that is the case, then the rational expression is simplified.

Solution

a. We factor the perfect-square trinomial in the numerator. In the denominator, we factor out the GCF, 8. Then we remove the common factor, $x - 5$.

$$\frac{x^2 - 10x + 25}{8x - 40} = \frac{\overset{1}{\cancel{(x - 5)}}(x - 5)}{8\underset{1}{\cancel{(x - 5)}}} \qquad \text{Remove a factor equal to 1: } \tfrac{x - 5}{x - 5} = 1.$$

$$= \frac{x - 5}{8}$$

b. We factor the trinomials in the numerator and the denominator and then remove the common factor, $3a - 2$.

$$\frac{6a^2 - 13a + 6}{3a^2 + a - 2} = \frac{\overset{1}{\cancel{(3a - 2)}}(2a - 3)}{\underset{1}{\cancel{(3a - 2)}}(a + 1)}$$ Remove a factor equal to 1: $\frac{3a - 2}{3a - 2} = 1$.

$$= \frac{2a - 3}{a + 1}$$ This expression does not simplify further.

c. $\dfrac{y^3 - 8}{y^3 - 2y^2 + 3y - 6} = \dfrac{(y - 2)(y^2 + 2y + 4)}{y^2(y - 2) + 3(y - 2)}$ In the numerator, factor the sum of two cubes. In the denominator, begin the process of factoring by grouping.

$$= \frac{(y - 2)(y^2 + 2y + 4)}{(y - 2)(y^2 + 3)}$$ In the denominator, complete the factoring by grouping.

$$= \frac{\overset{1}{\cancel{(y - 2)}}(y^2 + 2y + 4)}{\underset{1}{\cancel{(y - 2)}}(y^2 + 3)}$$ Remove the factor common to the numerator and denominator: $\frac{y - 2}{y - 2} = 1$.

$$= \frac{y^2 + 2y + 4}{y^2 + 3}$$ This expression does not simplify further

Self Check 5   Simplify: **a.** $\dfrac{x^2 - 6x + 9}{6x - 18}$ **b.** $\dfrac{2b^2 + 7b - 15}{2b^2 + 13b + 15}$

c. $\dfrac{a^3 - 1}{a^3 - a^2 + 6a - 6}$

Now Try   **Problems 43, 47, and 51**

Sometimes we will encounter rational expressions that are already in simplified form. For example, to attempt to simplify

$$\frac{x^2 + xa + 2x + 2a}{x^2 + x - 6}$$

we factor the numerator and denominator:

$$\frac{x^2 + xa + 2x + 2a}{x^2 + x - 6} = \frac{x(x + a) + 2(x + a)}{(x - 2)(x + 3)}$$ In the numerator, begin the process of factoring by grouping. In the denominator, factor the trinomial.

$$= \frac{(x + a)(x + 2)}{(x - 2)(x + 3)}$$ In the numerator, complete the factoring by grouping.

Since there are no common factors in the numerator and denominator, the rational expression is in *lowest terms*. It cannot be simplified.

6   **Simplify Rational Expressions That Have Factors That Are Opposites.**

If the terms of two polynomials are the same, except that they are opposite in sign, the polynomials are *opposites*. For example, $b - a$ and $a - b$ are opposites.

To simplify $\dfrac{b-a}{a-b}$, the quotient of opposites, we factor -1 from the numerator and remove any factors common to both the numerator and the denominator:

$$\frac{b-a}{a-b} = \frac{-a+b}{a-b} \qquad \text{Rewrite the numerator.}$$

$$= \frac{\overset{1}{-(\cancel{a-b})}}{\underset{1}{(\cancel{a-b})}} \qquad \begin{array}{l}\text{Factor out } -1 \text{ from each term in the numerator} \\ \text{and remove the common factor } a-b.\end{array}$$

$$= \frac{-1}{1}$$

$$= -1$$

In general, we have the following principle.

The Quotient of Opposites	The quotient of any nonzero polynomial and its opposite is -1.

Success Tip

When a difference is reversed, the original binomial and the resulting binomial are opposites. Here are some pairs of opposites:

$b-a$ and $a-b$

$y-6$ and $6-y$

x^2-4 and $4-x^2$

EXAMPLE 6   Simplify: $\dfrac{3x^2 - 10xy - 8y^2}{4y^2 - xy}$

Strategy   We will begin by factoring the numerator and denominator. Then we look for common factors, or factors that are opposites, and remove them.

Why   We need to make sure that the numerator and denominator have no common factors other than 1. If that is the case, then the rational expression is simplified.

Solution   We factor the numerator and denominator. Because $x-4y$ and $4y-x$ are opposites, their quotient is -1.

$$\frac{3x^2 - 10xy - 8y^2}{4y^2 - xy} = \frac{(3x+2y)\overset{-1}{(\cancel{x-4y})}}{y\underset{1}{(\cancel{4y-x})}} \qquad \begin{array}{l}\text{Since } x-4y \text{ and } 4y-x \text{ are opposites,} \\ \text{simplify by replacing } \frac{x-4y}{4y-x} \text{ with the equivalent} \\ \text{fraction } \frac{-1}{1} = -1.\end{array}$$

$$= \frac{-(3x+2y)}{y}$$

This result can also be written as $-\dfrac{3x+2y}{y}$ or $\dfrac{-3x-2y}{y}$.

Caution

A $-$ symbol preceding a fraction may be applied to the numerator or to the denominator, but not to both. For example,

$$-\frac{3x+2y}{y} \neq \frac{-3x-2y}{-y}$$

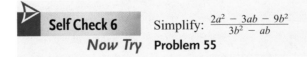

Self Check 6   Simplify: $\dfrac{2a^2 - 3ab - 9b^2}{3b^2 - ab}$

Now Try   Problem 55

ANSWERS TO SELF CHECKS   **1. a.** $3.50 **b.** $1.56

2. The domain is the set of all real numbers except -7 and 7: $(-\infty, -7) \cup (-7, 7) \cup (7, \infty)$.

3. $\dfrac{3a^3}{5b^2}$ **4. a.** $\dfrac{4}{x-2}$ **b.** $\dfrac{x+3}{5x}$ **5. a.** $\dfrac{x-3}{6}$ **b.** $\dfrac{2b-3}{2b+3}$ **c.** $\dfrac{a^2+a+1}{a^2+6}$ **6.** $-\dfrac{2a+3b}{b}$ or $\dfrac{-2a-3b}{b}$

STUDY SET
6.1

VOCABULARY

Fill in the blanks.

1. A quotient of two polynomials, such as $\frac{x^2 + x}{x^2 - 3x}$, is called a _____ expression.

2. In the rational expression $\frac{(x + 2)(3x - 1)}{(x + 2)(4x + 2)}$, $x + 2$ is a common _____ of the numerator and the denominator.

3. To _____ a rational expression, we remove factors common to the numerator and denominator.

4. Because of the division by 0, the expression $\frac{8}{0}$ is _____.

5. The binomials $x - 15$ and $15 - x$ are called _____, because their terms are the same, except that they are opposite in sign.

6. The graph of the function shown in Problem 7 below approaches the positive x-axis. When a graph approaches a line, we call the line an _____.

CONCEPTS

7. The graph of rational function f for $x > 0$ is shown in the illustration. Find each of the following.

    a. $f(1)$ b. $f(2)$

    c. The value(s) of x for which $f(x) = 4$

    d. The domain and range of f

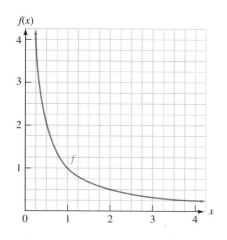

8. Fill in the blanks to show that $\frac{x - y}{y - x} = -1$ by factoring out -1 from each term in the numerator.

$$\frac{x - y}{y - x} = \frac{-y + \boxed{}}{y - x} = \frac{\overset{1}{(y - x)}}{\underset{1}{(y - x)}} = \boxed{}$$

9. Simplify each expression.

    a. $\dfrac{3 \cdot 5 \cdot x \cdot y \cdot y}{5 \cdot 7 \cdot x \cdot x \cdot x \cdot y}$

    b. $\dfrac{(x + 8)(x - 3)}{(x + 2)(x + 8)}$

    c. $\dfrac{a^3(a - 9)}{(9 - a)(9 + a)}$

10. Simplify each rational expression, if possible.

    a. $\dfrac{x + 8}{x}$ b. $\dfrac{3a^2 + 23}{a^2}$

11. MANUFACTURING   Each graph shows the average cost to manufacture a certain item for a given number of units produced. Which graph is best described as the graph of a

    a. linear function?   b. quadratic function?

    c. rational function?   d. polynomial function?

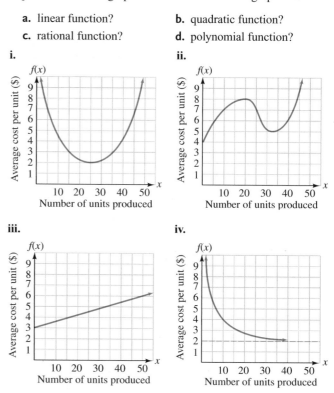

12. Refer to the graphs in Problem 11. Complete the description of the graph by filling in each blank with the word *decreases* or *increases*.

    a. Graph **i.** decreases then steadily _____ .

    b. Graph **ii.** increases, then _____ , and then steadily

    _____ .

    c. Graph **iii.** steadily _____ .

    d. Graph **iv.** steadily _____ approaching a cost of $2.00 per unit.

NOTATION

13. A student checks his answers with those in the back of his textbook. Determine whether they are equivalent.

Answer	Book's answer	Equivalent?
$\dfrac{-3}{x+3}$	$-\dfrac{3}{x+3}$	
$\dfrac{-x+4}{6x+1}$	$\dfrac{-(x-4)}{6x+1}$	
$\dfrac{x+7}{(x-4)(x+2)}$	$\dfrac{x+7}{(x+2)(x-4)}$	
$\dfrac{x-4}{x+4}$	$\dfrac{4-x}{x+4}$	
$\dfrac{a-3b}{2b-a}$	$\dfrac{3b-a}{a-2b}$	

14. a. In $\dfrac{(x+5)\overset{1}{\cancel{(x-5)}}}{x\underset{1}{\cancel{(x-5)}}}$, what do the slashes show?

    b. In $\dfrac{(x-3)\overset{-1}{\cancel{(x-7)}}}{(x+3)\underset{1}{\cancel{(7-x)}}}$, what do the slashes show?

GUIDED PRACTICE

Complete the table of values for each rational function (round to the nearest hundredth when appropriate). Then graph the function. Each function is defined for $x > 0$. Label the horizontal asymptote. See Example 1 and Objective 3.

15. $f(x) = \dfrac{6}{x}$

x	f(x)
1	
2	
4	
6	
8	
10	
12	

16. $f(x) = \dfrac{12}{x}$

x	f(x)
1	
4	
8	
12	
16	
20	
24	

17. $f(x) = \dfrac{x+2}{x}$

x	f(x)
1	
2	
4	
6	
8	
10	
12	

18. $f(x) = \dfrac{2x+4}{x}$

x	f(x)
1	
4	
8	
12	
16	
20	
24	

Find the domain of each rational function. Express your answer in words and using interval notation. See Example 2.

19. $f(x) = \dfrac{2}{x}$

20. $f(x) = \dfrac{8}{x-1}$

21. $f(x) = \dfrac{2x}{x+2}$

22. $f(x) = \dfrac{2x+1}{x^2-2x}$

23. $f(x) = \dfrac{3x-1}{x-x^2}$

24. $f(x) = \dfrac{x^2+36}{x^2-36}$

25. $f(x) = \dfrac{x^2+3x+2}{x^2-x-56}$

26. $f(x) = \dfrac{2x^2-3x-2}{x^2+2x-24}$

Simplify each rational expression. See Example 3.

27. $\dfrac{12a^3}{18a}$

28. $\dfrac{25b^4}{55b}$

29. $\dfrac{15a^2}{25a^8}$

30. $\dfrac{12x}{16x^7}$

31. $\dfrac{27st}{36st^2}$

32. $\dfrac{49xy^2}{21xy}$

33. $\dfrac{24x^3y^{10}}{18x^4y^3}$

34. $\dfrac{15a^5b^4}{21a^8b^3}$

Simplify each rational expression. See Example 4.

35. $\dfrac{4x^2}{2x^3 - 12x^2}$

36. $\dfrac{-15y^2}{5y^3 + 15y^2}$

37. $\dfrac{24n^4}{16n^4 + 24n^3}$

38. $\dfrac{18m^4}{36m^4 - 9m^3}$

39. $\dfrac{2x + 18}{x^2 - 81}$

40. $\dfrac{6x - 12}{x^2 - 4}$

41. $\dfrac{4a^2 - 25}{20a - 50}$

42. $\dfrac{9b^2 - 16}{21b + 28}$

Simplify each rational expression. See Example 5.

43. $\dfrac{5x^2 - 10x}{x^2 - 4x + 4}$

44. $\dfrac{x^2 + 6x + 9}{2x^2 + 6x}$

45. $\dfrac{x^2 + 2x + 1}{x^2 + 4x + 3}$

46. $\dfrac{y^2 - 4y + 4}{y^2 - 8y + 12}$

47. $\dfrac{3d^2 + 13d + 4}{3d^2 + 7d + 2}$

48. $\dfrac{10r^2 + 17r + 3}{2r^2 + 17r + 21}$

49. $\dfrac{2h^2 + 9h - 5}{4h^2 - 4h + 1}$

50. $\dfrac{6x^2 + x - 2}{8x^2 + 2x - 3}$

51. $\dfrac{t^3 + 27}{t^3 + 3t^2 + 4t + 12}$

52. $\dfrac{m^3 + 64}{m^3 + 4m^2 + 3m + 12}$

53. $\dfrac{s^3 + s^2 - 6s - 6}{s^3 + 1}$

54. $\dfrac{d^3 + 5d^2 - 5d - 25}{d^3 + 125}$

Simplify each rational expression. See Example 6.

55. $\dfrac{3m^2 - 2mn - n^2}{mn - m^2}$

56. $\dfrac{5s^2 - 4st - t^2}{st - s^2}$

57. $\dfrac{b^2 - a^2}{a - b}$

58. $\dfrac{d^2 - 16c^2}{4c - d}$

59. $\dfrac{4 - x^2}{x^2 - x - 2}$

60. $\dfrac{x^2 - 2x - 15}{25 - x^2}$

61. $\dfrac{20x^3 - 20x^4}{x^2 - 2x + 1}$

62. $\dfrac{16m^5 - 2m^6}{m^2 - 16m + 64}$

Use a graphing calculator to graph each rational function. From the graph, determine its domain and range. Answer using interval notation. See Using Your Calculator: Finding the Domain and Range of a Rational Function.

63. $f(x) = \dfrac{x}{x - 2}$

64. $f(x) = \dfrac{x + 2}{x}$

65. $f(x) = \dfrac{x + 1}{x^2 - 4}$

66. $f(x) = \dfrac{x - 2}{x^2 - 3x - 4}$

TRY IT YOURSELF

Simplify each expression. If an expression cannot be simplified, write "Does not simplify."

67. $\dfrac{x^2 + x - 30}{3x^2 - 3x - 60}$

68. $\dfrac{4x^2 + 24x + 32}{16x^2 + 8x - 48}$

69. $\dfrac{a^2 - 4}{a^3 - 8}$

70. $\dfrac{x^3 - 27}{3x^2 - 8x - 3}$

71. $\dfrac{m^3 - mn^2}{mn^2 + m^2n - 2m^3}$

72. $\dfrac{a^3 - ab^2}{ab^2 - 4a^2b + 3a^3}$

73. $\dfrac{sx + 4s - 3x - 12}{sx + 4s + 6x + 24}$

74. $\dfrac{ax + by + ay + bx}{a + b}$

75. $\dfrac{2x^2 - 3x - 9}{2x^2 + 3x - 9}$

76. $\dfrac{6x^2 - 7x - 5}{2x^2 + 5x + 2}$

77. $\dfrac{3x + 6y}{x + 2y}$

78. $\dfrac{y - xy}{xy - x}$

79. $\dfrac{x^4 + 3x^3 + 9x^2}{x^3 - 27}$

80. $\dfrac{x^3 + 8}{x^4 - 2x^3 + 4x^2}$

81. $\dfrac{2x^2 + 2x - 12}{x^3 + 3x^2 - 4x - 12}$

82. $\dfrac{3x^2 - 3y^2}{x^2 + 2y + 2x + yx}$

83. $\dfrac{4x^2 + 8x + 3}{6 + x - 2x^2}$

84. $\dfrac{6x^2 + 13x + 6}{6 - 5x - 6x^2}$

85. $\dfrac{x^2 - 6x + 9}{81 - x^4}$

86. $\dfrac{y^2 - 2y + 1}{1 - y^4}$

87. $\dfrac{16p^3q^2}{24pq^8}$

88. $\dfrac{30a^3b^{15}}{18a^9b^{10}}$

89. $\dfrac{t^3 - 5t^2 + 6t}{9t - t^3}$

90. $\dfrac{a^4 - 27a}{36a - 4a^3}$

APPLICATIONS

91. ENVIRONMENTAL CLEANUP   Suppose the cost (in dollars) of removing $p\%$ of the pollution in a river is given by the rational function

$$f(p) = \frac{50{,}000p}{100 - p} \quad \text{where } 0 \le p < 100$$

Find the cost of removing each percent of pollution.
a. 50%   **b.** 80%

92. DIRECTORY COSTS   The average (mean) cost for a service club to publish a directory of its members is given by the rational function

$$f(x) = \frac{1.25x + 700}{x}$$

where x is the number of directories printed. Find the average cost per directory if
a. 500 directories are printed.
b. 2,000 directories are printed.

93. UTILITY COSTS   An electric company charges $7.50 per month plus 9¢ for each kilowatt hour (kwh) of electricity used.
a. Find a linear function that gives the total cost of n kwh of electricity.
b. Find a rational function that gives the average cost per kwh when using n kwh.
c. Find the average cost per kwh when 775 kwh are used.

94. SCHEDULING WORK CREWS   The rational function

$$f(t) = \frac{t^2 + 2t}{2t + 2}$$

gives the number of days it would take two construction crews, working together, to frame a house that crew 1 (working alone) could complete in t days and crew 2 (working alone) could complete in $t + 2$ days.
a. If crew 1 could frame a certain house in 15 days, how long would it take both crews working together?
b. If crew 2 could frame a certain house in 20 days, how long would it take both crews working together?

95. FILLING A POOL   The rational function

$$f(t) = \frac{t^2 + 3t}{2t + 3}$$

gives the number of hours it would take two pipes, working together, to fill a pool that the larger pipe (working alone) could fill in t hours and the smaller pipe (working alone) could fill in $t + 3$ hours.
a. If the smaller pipe could fill a pool in 7 hours, how long would it take both pipes to fill the pool?
b. If the larger pipe could fill a pool in 8 hours, how long would it take both pipes to fill the pool?

96. RETENTION STUDY   After learning a list of words, two subjects were tested over a 28-day period to see what percent of the list they remembered. In both cases, their percent recall could be modeled by rational functions, as shown in the illustration.

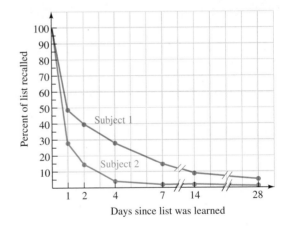

a. Use the graphs to complete the table.

Days since learning	0	1	2	4	7	14	28
% recall—subject 1							
% recall—subject 2							

b. After 28 days, which subject had the better recall?

WRITING

97. A student simplified $\frac{6x^2 - 7x - 5}{2x^2 + 5x + 2}$ and obtained $\frac{3x - 5}{x - 2}$. As a check, she graphed $Y_1 = \frac{6x^2 - 7x - 5}{2x^2 + 5x + 2}$ and $Y_2 = \frac{3x - 5}{x - 2}$. What conclusion can be drawn from the graphs? Explain your answer.

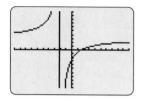

98. Simplify: $\frac{6x^2 + x - 2}{8x^2 + 2x - 3}$. Then explain how the table of values for $Y_1 = \frac{6x^2 + x - 2}{8x^2 + 2x - 3}$ and $Y_2 = \frac{3x + 2}{4x + 3}$ shown in the illustration can be used to check your result.

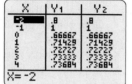

REVIEW

Perform each operation.

99. $(a^2 - 4a - 3)(a - 2)$

100. $(3c^2 + 5c) + (7 - c^2 - 5c)$

101. $-3mn^2(m^3 - 7mn - 2m^2)$

102. $(4u^2 + z^2 - 3u^2z^2) - (u^3 + 3z^2 - 3u^2z^2)$

CHALLENGE PROBLEMS

Simplify each expression.

103. $\frac{x^{32} - 1}{x^{16} - 1}$

104. $\frac{20m^2(m^2 - 1) - 47m(1 - m^2) + 24(m^2 - 1)}{4m^2 - m - 3}$

105. $\frac{a^6 - 64}{(a^2 + 2a + 4)(a^2 - 2a + 4)}$

106. $\frac{(p + q)^3 + 64}{(p + q)^2 - 16}$

SECTION 6.2
Multiplying and Dividing Rational Expressions

Objectives

1 Multiply rational expressions.

2 Find powers of rational expressions.

3 Divide rational expressions.

4 Perform mixed operations.

In this section, we review the rules for multiplying and dividing arithmetic fractions—fractions whose numerators and denominators are integers. Then we use these rules, in combination with the simplification skills learned in Section 6.1, to multiply and divide rational expressions.

1 **Multiply Rational Expressions.**

Recall that to multiply fractions, we multiply the numerators and multiply the denominators. For example,

$$\frac{3}{5} \cdot \frac{2}{7} = \frac{3 \cdot 2}{5 \cdot 7} \qquad \frac{4}{7} \cdot \frac{5}{8} = \frac{4 \cdot 5}{7 \cdot 8}$$

$$= \frac{6}{35} \qquad\qquad = \frac{\overset{1}{\cancel{2}} \cdot \overset{1}{\cancel{2}} \cdot 5}{7 \cdot \underset{1}{\cancel{2}} \cdot \underset{1}{\cancel{2}} \cdot 2}$$

Factor 4 as $2 \cdot 2$. Factor 8 as $2 \cdot 2 \cdot 2$. Then simplify.

$$= \frac{5}{14}$$

We use the same procedure to multiply rational expressions.

Multiplying Rational Expressions	To multiply rational expressions, multiply their numerators and their denominators. Then, if possible, factor and simplify. For any two rational expressions, $\frac{A}{B}$ and $\frac{C}{D}$, $$\frac{A}{B} \cdot \frac{C}{D} = \frac{AC}{BD}$$

EXAMPLE 1   Multiply: $\dfrac{25a^3}{11b} \cdot \dfrac{b}{5a}$

Strategy   To find the product, we will use the rule for multiplying rational expressions. In the process, we must be prepared to factor the numerators and denominators so that any common factors can be removed.

Why   We want to give the result in simplified form.

Solution

$$\frac{25a^3}{11b} \cdot \frac{b}{5a} = \frac{25a^3 \cdot b}{11b \cdot 5a} \qquad \begin{array}{l}\text{Multiply the numerators.}\\ \text{Multiply the denominators.}\end{array}$$

It is obvious that the numerator and denominator of $\frac{25a^3 \cdot b}{11b \cdot 5a}$ have several common factors, such as 5, a, and b. These common factors become more apparent when we factor the numerator and denominator completely.

$$\frac{25a^3 \cdot b}{11b \cdot 5a} = \frac{5 \cdot 5 \cdot a \cdot a \cdot a \cdot b}{11 \cdot b \cdot 5 \cdot a} \qquad \text{Factor } 25a^3.$$

$$= \frac{\overset{1}{\cancel{5}} \cdot 5 \cdot \cancel{a} \cdot a \cdot a \cdot \cancel{b}}{11 \cdot \cancel{b} \cdot \cancel{5} \cdot \cancel{a}} \qquad \begin{array}{l}\text{Simplify by replacing } \frac{5}{5}, \frac{a}{a}, \text{ and } \frac{b}{b} \text{ with the equivalent fraction}\\ \frac{1}{1}. \text{ This removes the factor } \frac{5 \cdot a \cdot b}{5 \cdot a \cdot b} = 1.\end{array}$$

$$= \frac{5a^2}{11} \qquad \begin{array}{l}\text{Multiply the remaining factors in the numerator.}\\ \text{Multiply the remaining factors in the denominator.}\end{array}$$

> **Success Tip**
> We could also use rules for exponents to simplify the product:
> $$\frac{25a^3 \cdot b}{11b \cdot 5a} = \frac{\overset{1}{\cancel{5}} \cdot 5 \cdot a^{3-1} \cdot b^{1-1}}{11 \cdot \underset{1}{\cancel{5}}}$$
> $$= \frac{5a^2 b^0}{11}$$
> $$= \frac{5a^2}{11}$$

Self Check 1   Multiply: $\dfrac{x^7}{16y} \cdot \dfrac{24y}{17x^3}$

Now Try   **Problem 17**

EXAMPLE 2   Multiply: **a.** $\dfrac{9x^2 - 6x + 9}{20x} \cdot \dfrac{5x^2}{6x - 18}$

   b. $\dfrac{x^2 - x - 6}{x^2 - 4} \cdot \dfrac{x^2 + x - 6}{x^2 - 9}$

Strategy   To find the product, we will use the rule for multiplying rational expressions. In the process, we must be prepared to factor the numerators and denominators so that any common factors can be removed.

Why   We want to give the result in simplified form.

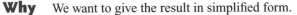

Solution

a. $\dfrac{x^2 - 6x + 9}{20x} \cdot \dfrac{5x^2}{6x - 18} = \dfrac{(x^2 - 6x + 9)5x^2}{20x(6x - 18)}$ Multiply the numerators.
     Multiply the denominators.

$= \dfrac{(x - 3)(x - 3)5xx}{4 \cdot 5 \cdot x \cdot 6(x - 3)}$ Factor the numerator.
     Factor the denominator.

$= \dfrac{\overset{1}{\cancel{(x - 3)}}(x - 3)\overset{11}{\cancel{5}}xx}{\underset{1}{\cancel{4}} \cdot \underset{1}{\cancel{5}} \cdot x \cdot 6\underset{1}{\cancel{(x - 3)}}}$ Simplify by removing common factors of the numerator and denominator.

$= \dfrac{x(x - 3)}{24}$ Multiply the remaining monomial factors in the numerator.
     Multiply the remaining factors in the denominator.

b. $\dfrac{x^2 - x - 6}{x^2 - 4} \cdot \dfrac{x^2 + x - 6}{x^2 - 9} = \dfrac{(x^2 - x - 6)(x^2 + x - 6)}{(x^2 - 4)(x^2 - 9)}$ Multiply the numerators.
     Multiply the denominators.

$= \dfrac{(x - 3)(x + 2)(x + 3)(x - 2)}{(x + 2)(x - 2)(x + 3)(x - 3)}$ Factor the polynomials.

$= \dfrac{\overset{1}{\cancel{(x - 3)}}\overset{1}{\cancel{(x + 2)}}\overset{1}{\cancel{(x + 3)}}\overset{1}{\cancel{(x - 2)}}}{\underset{1}{\cancel{(x + 2)}}\underset{1}{\cancel{(x - 2)}}\underset{1}{\cancel{(x + 3)}}\underset{1}{\cancel{(x - 3)}}}$ Simplify by removing common factors of the numerator and denominator.

$= 1$

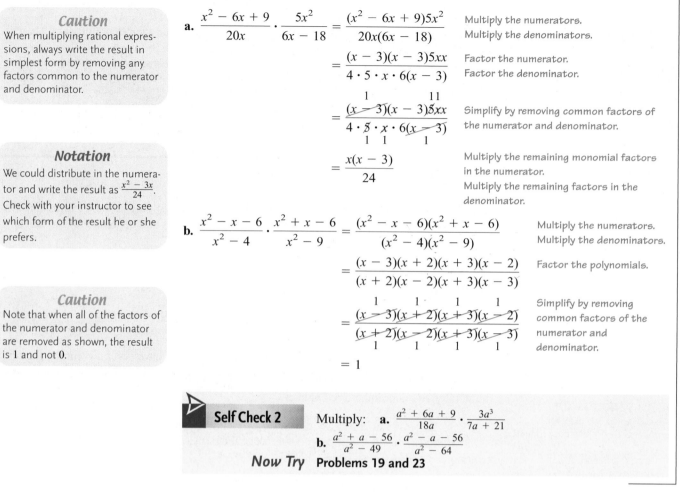

> **Self Check 2**   Multiply: **a.** $\dfrac{a^2 + 6a + 9}{18a} \cdot \dfrac{3a^3}{7a + 21}$
> **b.** $\dfrac{a^2 + a - 56}{a^2 - 49} \cdot \dfrac{a^2 - a - 56}{a^2 - 64}$
>
> ***Now Try***   Problems 19 and 23

Using Your Calculator

Checking an Algebraic Simplification

We can check the simplification in Example 2(a) by graphing the functions $f(x) = \left(\dfrac{x^2 - 6x + 9}{20x}\right)\left(\dfrac{5x^2}{6x - 18}\right)$, shown in figure (a), and $g(x) = \dfrac{x(x - 3)}{24}$, shown in figure (b), and observing that the graphs are the same, except that 0 and 3 are not included in the domain of the function f.

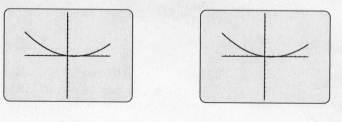

           (a)   (b)

We can use the split-screen G-T (graph, table) mode to check the result of a multiplication. To set the split-screen feature on a graphing calculator, press $\boxed{\text{MODE}}$, press ▼ seven times, press ► twice, then press $\boxed{\text{ENTER}}$. If we enter $Y_3 = Y_1 - Y_2$, use the cursor to highlight the $=$ sign as shown in figure (c), and then press $\boxed{\text{GRAPH}}$, we get the display shown in figure (d). The zeros under the Y_3 column indicate that the value of $\left(\frac{x^2 - 6x + 9}{20x}\right)\left(\frac{5x^2}{6x - 18}\right)$ and the value of $\frac{x(x - 3)}{24}$ are the same for different values of x. (The error message is given because when $x = 0$ and $x = 3$, $\left(\frac{x^2 - 6x + 9}{20x}\right)\left(\frac{5x^2}{6x - 18}\right)$ is undefined.)

The graph of $Y_3 = Y_1 - Y_2$ is difficult to see because it lies on the x-axis. The graph indicates that for all x-values (except those that make the rational expressions undefined), $Y_3 = 0$, or more specifically, $\left(\frac{x^2 - 6x + 9}{20x}\right)\left(\frac{5x^2}{6x - 18}\right) = \frac{x(x - 3)}{24}$.

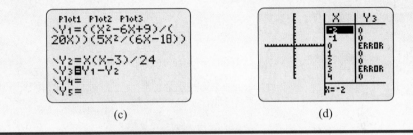

(c)   (d)

EXAMPLE 3   Multiply: $\dfrac{6x^2 + 5xy - 4y^2}{2x^2 + 5xy + 3y^2} \cdot \dfrac{8x^2 + 6xy - 9y^2}{12x^2 + 7xy - 12y^2}$

Strategy   To find the product, we will use the rule for multiplying rational expressions. In the process, we must be prepared to factor the numerators and denominators so that any common factors can be removed.

Why   We want to give the result in simplified form.

Solution

$$\frac{6x^2 + 5xy - 4y^2}{2x^2 + 5xy + 3y^2} \cdot \frac{8x^2 + 6xy - 9y^2}{12x^2 + 7xy - 12y^2}$$

$$= \frac{(6x^2 + 5xy - 4y^2)(8x^2 + 6xy - 9y^2)}{(2x^2 + 5xy + 3y^2)(12x^2 + 7xy - 12y^2)} \qquad \text{Multiply the numerators.}$$
$$\qquad\qquad\qquad\qquad\qquad\qquad\qquad\qquad \text{Multiply the denominators.}$$

$$= \frac{(3x + 4y)(2x - y)(4x - 3y)(2x + 3y)}{(2x + 3y)(x + y)(3x + 4y)(4x - 3y)} \qquad \text{Factor the trinomials.}$$

$$= \frac{\overset{1}{\cancel{(3x + 4y)}}(2x - y)\overset{1}{\cancel{(4x - 3y)}}\overset{1}{\cancel{(2x + 3y)}}}{\underset{1}{\cancel{(2x + 3y)}}(x + y)\underset{1}{\cancel{(3x + 4y)}}\underset{1}{\cancel{(4x - 3y)}}} \qquad \begin{array}{l}\text{Simplify by removing common factors}\\ \text{of the numerator and denominator.}\end{array}$$

$$= \frac{2x - y}{x + y} \qquad \begin{array}{l}\text{Multiply the remaining factors in the}\\ \text{numerator. Multiply the remaining}\\ \text{factors in the denominator.}\end{array}$$

Self Check 3   Multiply: $\dfrac{2a^2 + 5ab - 12b^2}{2a^2 + 11ab + 12b^2} \cdot \dfrac{2a^2 - 3ab - 9b^2}{2a^2 - ab - 3b^2}$

Now Try   **Problems 29 and 31**

EXAMPLE 4   Multiply: $(2x - x^2) \cdot \dfrac{x}{5x^3 - 10x^2 + 20x - 40}$

Strategy   We will write $2x - x^2$ as a rational expression with denominator 1. (Remember, any number divided by 1 remains unchanged.) Then we will use the rule for multiplying rational expressions.

Why   Writing $2x - x^2$ as $\dfrac{2x - x^2}{1}$ is helpful during the multiplication process when we multiply numerators and multiply denominators.

Solution

$(2x - x^2) \cdot \dfrac{x}{5x^3 - 10x^2 + 20x - 40}$

$= \dfrac{2x - x^2}{1} \cdot \dfrac{x}{5x^3 - 10x^2 + 20x - 40}$ Write $2x - x^2$ as $\frac{2x - x^2}{1}$.

$= \dfrac{(2x - x^2)x}{1(5x^3 - 10x + 20x - 40)}$ Multiply the numerators. Multiply the denominators.

$= \dfrac{x(2 - x)x}{1 \cdot 5(x^3 - 2x^2 + 4x - 8)}$ Factor out x in the numerator. Factor out 5 in the denominator.

$= \dfrac{x(2 - x)x}{1 \cdot 5[x^2(x - 2) + 4(x - 2)]}$ In the denominator, begin factoring by grouping.

$= \dfrac{x(2 - x)x}{1 \cdot 5(x - 2)(x^2 + 4)}$ In the denominator, complete the factoring by grouping. The brackets [ ] are no longer needed.

$= \dfrac{x\overset{-1}{\cancel{(2 - x)}}x}{1 \cdot 5\underset{1}{\cancel{(x - 2)}}(x^2 + 4)}$ Simplify. Recall that the quotient of any nonzero quantity and its opposite is -1: $\frac{2 - x}{x - 2} = -1$.

$= \dfrac{-x^2}{5(x^2 + 4)}$ Multiply the remaining factors in the numerator. Multiply the remaining factors in the denominator.

Since the $-$ sign can be written in front of the fraction, this result can be expressed as

$-\dfrac{x^2}{5(x^2 + 4)}$

Self Check 4   Multiply: $\dfrac{x}{8x^3 - 32x^2 + 8x - 32}(4x - x^2)$

Now Try   Problem 39

2   **Find Powers of Rational Expressions.**

EXAMPLE 5   Find: $\left(\dfrac{x^2 + x - 1}{2x + 3}\right)^2$

Strategy   We will find the product $\left(\dfrac{x^2 + x - 1}{2x + 3}\right)\left(\dfrac{x^2 + x - 1}{2x + 3}\right)$ using the rule for multiplying rational expressions.

Why   The exponent 2 means the base, $\dfrac{x^2 + x - 1}{2x + 3}$, should be written as a factor two times.

Solution

$$\left(\frac{x^2 + x - 1}{2x + 3}\right)^2 = \left(\frac{x^2 + x - 1}{2x + 3}\right)\left(\frac{x^2 + x - 1}{2x + 3}\right)$$

$$= \frac{(x^2 + x - 1)(x^2 + x - 1)}{(2x + 3)(2x + 3)} \qquad \text{Multiply the numerators.}$$
Multiply the denominators.

$$= \frac{x^4 + 2x^3 - x^2 - 2x + 1}{4x^2 + 12x + 9}$$

 Self Check 5 Find: $\left(\frac{x + 5}{x^2 - 6x}\right)^2$

Now Try **Problem 43**

3 **Divide Rational Expressions.**

Recall that one number is called the **reciprocal** of another if their product is 1. To find the reciprocal of a fraction, we invert its numerator and denominator. We have seen that to divide fractions, we multiply the first fraction by the reciprocal of the second fraction.

$$\frac{3}{5} \div \frac{8}{9} = \frac{3}{5} \cdot \frac{9}{8} \qquad\qquad \frac{4}{7} \div \frac{2}{21} = \frac{4}{7} \cdot \frac{21}{2}$$

$$= \frac{3 \cdot 9}{5 \cdot 8} \qquad\qquad = \frac{4 \cdot 21}{7 \cdot 2}$$

$$= \frac{27}{40} \qquad\qquad = \frac{\overset{1}{\cancel{2}} \cdot 2 \cdot 3 \cdot \overset{1}{\cancel{7}}}{\underset{1}{\cancel{7}} \cdot \underset{1}{\cancel{2}}} \quad \begin{array}{l}\text{Factor 4 as } 2 \cdot 2. \text{ Factor 21} \\ \text{as } 3 \cdot 7. \text{ Then simplify.}\end{array}$$

$$= 6$$

We use the same procedure to divide rational expressions.

Dividing Rational Expressions

To divide two rational expressions, multiply the first by the reciprocal of the second. Then, if possible, factor and simplify.

For any two rational expressions, $\frac{A}{B}$ and $\frac{C}{D}$, where $\frac{C}{D} \neq 0$,

$$\frac{A}{B} \div \frac{C}{D} = \frac{A}{B} \cdot \frac{D}{C} = \frac{AD}{BC}$$

EXAMPLE 6 Divide: $\dfrac{6}{y^3 z^2} \div \dfrac{20}{yz^3}$

Strategy We will use the rule for dividing rational expressions. After multiplying by the reciprocal, we will use rules for exponents to simplify the result.

Why We want to give the result in simplified form.

Solution

$$\frac{6}{y^3z^2} \div \frac{20}{yz^3} = \frac{6}{y^3z^2} \cdot \frac{yz^3}{20}$$

Multiply the first rational expression by the reciprocal of the second.

$$= \frac{6yz^3}{20y^3z^2}$$

Multiply the numerators.
Multiply the denominators.

$$= \frac{2 \cdot 3 \cdot y^{1-3}z^{3-2}}{2 \cdot 10}$$

Factor 6 and 20. To divide exponential expressions with the same base, keep the base and subtract the exponents.

$$= \frac{\overset{1}{\cancel{2}} \cdot 3 \cdot y^{-2}z^1}{\underset{1}{\cancel{2}} \cdot 10}$$

Remove the common factor of 2. Simplify the exponents.

$$= \frac{3z}{10y^2}$$

Write the result without the negative exponent.

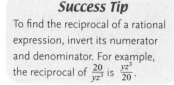 **Self Check 6**   Divide: $\frac{8a^4}{t^5} \div \frac{28a^3}{t^2}$

Now Try   **Problem 51**

EXAMPLE 7   Divide: $\dfrac{x^3 + 8}{4x + 4} \div \dfrac{x^2 - 2x + 4}{2x^2 - 2}$

Strategy   To find the quotient, we will use the rule for dividing rational expressions. After multiplying by the reciprocal, we will factor each polynomial that is not prime and remove any common factors of the numerator and denominator.

Why   We want to give the result in simplified form.

Solution

$$\frac{x^3 + 8}{4x + 4} \div \frac{x^2 - 2x + 4}{2x^2 - 2}$$

$$= \frac{x^3 + 8}{4x + 4} \cdot \frac{2x^2 - 2}{x^2 - 2x + 4}$$

Multiply the first rational expression by the reciprocal of the second.

$$= \frac{(x^3 + 8)(2x^2 - 2)}{(4x + 4)(x^2 - 2x + 4)}$$

Multiply the numerators.
Multiply the denominators.

$$= \frac{(x + 2)(\overset{1}{\cancel{x^2 - 2x + 4}})2(\overset{1}{\cancel{x + 1}})(x - 1)}{2 \cdot \underset{1}{\cancel{2}}(\underset{1}{\cancel{x + 1}})(\underset{1}{\cancel{x^2 - 2x + 4}})}$$

Factor completely. $x^2 - 2x + 4$ is prime. Then simplify.

$$= \frac{(x + 2)(x - 1)}{2}$$

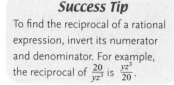 **Self Check 7**   Divide: $\dfrac{x^3 - 8}{9x - 9} \div \dfrac{x^2 + 2x + 4}{3x^2 - 3x}$

Now Try   **Problem 55**

EXAMPLE 8   Divide: $\dfrac{b^3 - 4b}{x - 1} \div (b - 2)$

Strategy   We will begin by writing $b - 2$ as a rational expression by inserting a denominator 1. Then we will use the rule for dividing rational expressions.

Why   Writing $b - 2$ over 1 is helpful when we invert its numerator and denominator to find its reciprocal.

Solution

$$\dfrac{b^3 - 4b}{x - 1} \div (b - 2) = \dfrac{b^3 - 4b}{x - 1} \div \dfrac{b - 2}{1}$$ Write $b - 2$ as a fraction with a denominator of 1.

$$= \dfrac{b^3 - 4b}{x - 1} \cdot \dfrac{1}{b - 2}$$ Multiply the first rational expression by the reciprocal of the second.

$$= \dfrac{b^3 - 4b}{(x - 1)(b - 2)}$$ Multiply the numerators. Multiply the denominators.

$$= \dfrac{b(b + 2)\overset{1}{\cancel{(b - 2)}}}{(x - 1)\underset{1}{\cancel{(b - 2)}}}$$ Factor $b^3 - 4b$ and then simplify.

$$= \dfrac{b(b + 2)}{x - 1}$$ Multiply the remaining factors in the numerator. Multiply the remaining factors in the denominator.

 Self Check 8   Divide: $\dfrac{m^4 - 9m^2}{a^2 - 3a} \div (m^2 + 3m)$

Now Try   **Problem 63**

4   **Perform Mixed Operations.**

EXAMPLE 9   Simplify: $\dfrac{x^2 + 2x - 3}{6x^2 + 5x + 1} \div \dfrac{2x^2 - 2}{2x^2 - 5x - 3} \cdot \dfrac{6x^2 + 4x - 2}{x^2 - 2x - 3}$

Strategy   We will consider the division first by multiplying the first rational expression by the reciprocal of the second. Then we will find the product of the three rational expressions.

Why   By the rules for the order of operations, we must perform division and multiplication in order from left to right.

Solution   Since multiplications and divisions are done in order from left to right, we begin by focusing on the division. We introduce grouping symbols to emphasize this. To divide the expressions within the parentheses, we invert $\dfrac{2x^2 - 2}{2x^2 - 5x - 3}$ and multiply.

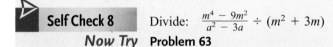

$$\left(\dfrac{x^2 + 2x - 3}{6x^2 + 5x + 1} \div \dfrac{2x^2 - 2}{2x^2 - 5x - 3} \right) \dfrac{6x^2 + 4x - 2}{x^2 - 2x - 3} = \left(\dfrac{x^2 + 2x - 3}{6x^2 + 5x + 1} \cdot \dfrac{2x^2 - 5x - 3}{2x^2 - 2} \right) \dfrac{6x^2 + 4x - 2}{x^2 - 2x - 3}$$

Next, we multiply the three rational expressions and simplify the result.

$$= \frac{(x^2 + 2x - 3)(2x^2 - 5x - 3)(6x^2 + 4x - 2)}{(6x^2 + 5x + 1)(2x^2 - 2)(x^2 - 2x - 3)}$$

$$= \frac{(x + 3)\cancel{(x - 1)}\cancel{(2x + 1)}\cancel{(x - 3)}2\cancel{(3x - 1)}\cancel{(x + 1)}}{(3x + 1)\cancel{(2x + 1)}2(x + 1)\cancel{(x - 1)}\cancel{(x - 3)}\cancel{(x + 1)}} \qquad \begin{array}{l}\text{Factor each polynomial} \\ \text{completely and simplify.}\end{array}$$

$$= \frac{(x + 3)(3x - 1)}{(3x + 1)(x + 1)}$$

Self Check 9   Simplify: $\dfrac{x^2 - 25}{4x^2 + 12x + 9} \div \dfrac{x^2 - 5x}{3x - 1} \cdot \dfrac{2x + 3}{3x^2 + 14x - 5}$

Now Try   **Problem 67**

ANSWERS TO SELF CHECKS   **1.** $\frac{3x^4}{34}$ **2. a.** $\frac{a^2(a + 3)}{42}$ **b.** 1   **3.** $\frac{a - 3b}{a + b}$ **4.** $-\frac{x^2}{8(x^2 + 1)}$

5. $\frac{x^2 + 10x + 25}{x^4 - 12x^3 + 36x^2}$ **6.** $\frac{2a}{7t^3}$ **7.** $\frac{x(x - 2)}{3}$ **8.** $\frac{m(m - 3)}{a(a - 3)}$ **9.** $\frac{1}{x(2x + 3)}$

STUDY SET
6.2

VOCABULARY

Fill in the blanks.

1. $\frac{a^2 - 9}{a^2 - 49} \cdot \frac{a - 7}{a + 3}$ is the product of two _____ expressions.

2. The _____ of $\frac{a + 3}{a + 7}$ is $\frac{a + 7}{a + 3}$.

3. To find the reciprocal of a rational expression, we _____ its numerator and denominator.

4. To simplify a rational expression, remove any factors _____ to the numerator and denominator.

CONCEPTS

Fill in the blanks.

5. To multiply rational expressions, multiply their _____ and multiply their _____. In symbols,

$$\frac{A}{B} \cdot \frac{C}{D} = \underline{\quad\quad}$$

6. To divide two rational expressions, multiply the first by the _____ of the second. In symbols,

$$\frac{A}{B} \div \frac{C}{D} = \frac{A}{B} \cdot \underline{\quad\quad} = \underline{\quad\quad}$$

NOTATION

Complete each solution.

7. $\dfrac{x^2 + 3x}{5x - 25} \cdot \dfrac{x - 5}{x + 3} = \dfrac{(x^2 + 3x)}{(5x - 25)}$

$$= \frac{(x + 3)(x - 5)}{(x - 5)(x + 3)}$$

$$= \frac{\quad}{5}$$

8. $\dfrac{x^2 - x - 6}{4x^2 + 16x} \div \dfrac{x - 3}{x + 4} = \dfrac{x^2 - x - 6}{4x^2 + 16x} \cdot \underline{\quad\quad}$

$$= \frac{(x^2 - x - 6)\,\underline{\quad}}{(4x^2 + 16x)\,\underline{\quad}}$$

$$= \frac{(x + 2)(x + 4)}{(x - 3)\,\underline{\quad}}$$

$$= \frac{x + 2}{\underline{\quad}}$$

9. A student checks her answers with those in the back of her textbook. Determine whether they are equivalent.

Student's answer	Book's answer	Equivalent?
$\dfrac{-x^{10}}{y^2}$	$-\dfrac{x^{10}}{y^2}$	
$\dfrac{x-3}{x+3}$	$\dfrac{3-x}{x+3}$	
$\dfrac{a+b}{(2-x)(c+d)}$	$-\dfrac{a+b}{(x-2)(c+d)}$	

10. a. Write $5x^2 + 35x$ as a fraction.

b. What is the reciprocal of $5x^2 + 35x$?

GUIDED PRACTICE

Multiply, and then simplify, if possible. **See Objective 1.**

11. $\dfrac{3}{4} \cdot \dfrac{11}{3}$

12. $\dfrac{13}{6} \cdot \dfrac{6}{21}$

13. $\dfrac{15}{24} \cdot \dfrac{16}{25}$

14. $\dfrac{49}{36} \cdot \dfrac{18}{35}$

Multiply, and then simplify, if possible. **See Example 1.**

15. $\dfrac{3a}{10} \cdot \dfrac{2}{15a^4}$

16. $\dfrac{4p}{21} \cdot \dfrac{7}{12p^6}$

17. $\dfrac{12x^6}{7y^4} \cdot \dfrac{y}{8x^2}$

18. $\dfrac{b^6}{27a^2} \cdot \dfrac{18a^4}{5b^9}$

Multiply, and then simplify, if possible. **See Example 2.**

19. $\dfrac{y^2 + 6y + 9}{15y} \cdot \dfrac{3y^2}{2y + 6}$

20. $\dfrac{3p^2}{6p + 24} \cdot \dfrac{p^2 - 16}{6p}$

21. $\dfrac{x^2 + x - 6}{5x} \cdot \dfrac{5x - 10}{x + 3}$

22. $\dfrac{z^2 + 4z - 5}{25z - 25} \cdot \dfrac{5z}{z + 5}$

23. $\dfrac{x^2 + 2x + 1}{9x^3} \cdot \dfrac{2x^2 - 2x}{2x^2 - 2}$

24. $\dfrac{a^4 + 6a^3}{a^2 - 16} \cdot \dfrac{3a - 12}{3a + 18}$

25. $\dfrac{t^2 + t - 6}{t^2 - 6t + 9} \cdot \dfrac{t^2 - 9}{t^2 - 4}$

26. $\dfrac{s^2 - 5s + 6}{s^2 - 10s + 16} \cdot \dfrac{s^2 - 6s - 16}{s^2 + 2s}$

Multiply, and then simplify, if possible. **See Example 3.**

27. $\dfrac{2x^2 - x - 3}{x^2 - 1} \cdot \dfrac{x^2 + x - 2}{2x^2 + x - 6}$

28. $\dfrac{2p^2 - 5p - 3}{p^2 - 9} \cdot \dfrac{2p^2 + 5p - 3}{2p^2 + 5p + 2}$

29. $\dfrac{3t^2 - t - 2}{6t^2 - 5t - 6} \cdot \dfrac{4t^2 - 9}{2t^2 + 5t + 3}$

30. $\dfrac{9x^2 + 3x - 20}{3x^2 - 7x + 4} \cdot \dfrac{3x^2 - 5x + 2}{9x^2 + 18x + 5}$

31. $\dfrac{x^2 + 4xy + 4y^2}{2x^2 + 4xy} \cdot \dfrac{3x - 6y}{x^2 - 4y^2}$

32. $\dfrac{x^2 - y^2}{xy} \cdot \dfrac{x^2}{x^2 + 2xy + y^2}$

33. $\dfrac{3a^2 + 7ab + 2b^2}{a^2 + 2ab} \cdot \dfrac{a^2 - ab}{3a^2 + ab}$

34. $\dfrac{a^2 + 3ab + 2b^2}{a^2 - 3ab - 4b^2} \cdot \dfrac{a^2 - 4ab}{ab^2 + 2b^3}$

Multiply, and then simplify, if possible. **See Example 4.**

35. $15x\left(\dfrac{x+1}{15x}\right)$

36. $30t\left(\dfrac{t-7}{30t}\right)$

37. $12y\left(\dfrac{y+8}{6y}\right)$

38. $16x\left(\dfrac{3x+8}{4x}\right)$

39. $(6a - a^2) \cdot \dfrac{a^3}{2a^3 - 12a^2 + 6a - 36}$

40. $(10n - n^2) \cdot \dfrac{n^6}{n^4 - 10n^3 - 2n^2 + 20n}$

41. $(x^2 + x - 2cx - 2c) \cdot \dfrac{x^2 + 3x + 2}{4c^2 - x^2}$

42. $(2ax - 10x + a - 5) \cdot \dfrac{x}{2x^2 + x}$

Find each power. **See Example 5.**

43. $\left(\dfrac{x-3}{x^2+4}\right)^2$

44. $\left(\dfrac{2t^2+t}{t-1}\right)^2$

45. $\left(\dfrac{2m^2 - m - 3}{x^2 - 1}\right)^2$

46. $\left(\dfrac{k^4 + 3k}{x^2 - x + 1}\right)^2$

Divide, and then simplify, if possible. **See Objective 3.**

47. $\dfrac{6}{11} \div \dfrac{36}{55}$

48. $\dfrac{17}{12} \div \dfrac{34}{3}$

49. $\dfrac{12}{5} \div \dfrac{24}{45}$

50. $\dfrac{18}{7} \div \dfrac{54}{35}$

Divide, and then simplify, if possible. **See Example 6.**

51. $\dfrac{22x^3}{y^2} \div \dfrac{33x^9}{y^7}$

52. $\dfrac{24a^6}{b} \div \dfrac{64a^9}{b^2}$

53. $\dfrac{pq^2}{50} \div \dfrac{p^{10}q^2}{15}$

54. $\dfrac{s^3t^3}{12} \div \dfrac{s^3t^{11}}{144}$

Divide, and then simplify, if possible. **See Example 7.**

55. $\dfrac{x^{12}}{x^3 - 8} \div \dfrac{x^2}{x^2 - 2x}$

56. $\dfrac{x^9}{x^3 + 125} \div \dfrac{x^4}{x^2 + 5x}$

57. $\dfrac{x^2 - 16}{x^2 - 25} \div \dfrac{5x + 20}{10x^2 - 50x}$

58. $\dfrac{a^2 - 9}{a^2 - 49} \div \dfrac{9a^2 + 27a}{3a + 21}$

59. $\dfrac{3n^2 + 5n - 2}{12n^2 - 13n + 3} \div \dfrac{n^2 + 3n + 2}{4n^2 + 5n - 6}$

60. $\dfrac{8y^2 - 14y - 15}{6y^2 - 11y - 10} \div \dfrac{4y^2 - 9y - 9}{3y^2 - 7y - 6}$

61. $\dfrac{5cd + d^2}{6d^2} \div \dfrac{125c^3 + d^3}{6c + 6d}$

62. $\dfrac{6m - 8n}{9m^3} \div \dfrac{27m^3 - 64n^3}{9m + 9n}$

Divide, and then simplify, if possible. **See Example 8.**

63. $\dfrac{y^3 - 9y}{y + 2} \div (y - 3)$

64. $\dfrac{x - 2}{x} \div (x^2 - 4)$

65. $(x + 1) \div \dfrac{x^2 + 2x + 1}{2}$

66. $(y + 4) \div \dfrac{y^2 + 8y + 16}{ab}$

Perform each operation and simplify, if possible. **See Example 9.**

67. $\dfrac{6a^2 - 7a - 3}{a^2 - 1} \div \dfrac{4a^2 - 12a + 9}{a^2 - 1} \cdot \dfrac{2a^2 - a - 3}{3a^2 - 2a - 1}$

68. $\dfrac{x^2 - x - 12}{x^2 + x - 2} \div \dfrac{x^2 - 6x + 8}{x^2 - 3x - 10} \cdot \dfrac{x^2 - 3x + 2}{x^2 - 2x - 15}$

69. $\dfrac{2x^2 - 2x - 4}{x^2 + 2x - 8} \cdot \dfrac{3x^2 + 15x}{x + 1} \div \dfrac{4x^2 - 100}{x^2 - x - 20}$

70. $\dfrac{4a^2 - 10a + 6}{a^4 - 3a^3} \div \dfrac{3 - 2a}{2a^3} \cdot \dfrac{a - 3}{2a - 2}$

TRY IT YOURSELF

Perform the operations and simplify.

71. $\dfrac{x^2 - 6x + 9}{4 - x^2} \div \dfrac{x^2 - 9}{x^2 - 8x + 12}$

72. $\dfrac{x^3 + 1}{4} \div \dfrac{x + 1}{2}$

73. $\dfrac{2x^2 - 2x - 12}{x^2 - 4} \cdot \dfrac{x^2 - x - 2}{x^3 - 9x}$

74. $\dfrac{x^2 + 2x - 35}{12x^3} \cdot \dfrac{x^2 + 4x - 21}{x^2 - 3x}$

75. $\dfrac{p^3 - q^3}{p^2 - q^2} \cdot \dfrac{q^2 + pq}{p^3 + p^2q + pq^2}$

76. $\dfrac{x^2 - 4}{2b - bx} \div \dfrac{x^2 + 4x + 4}{2b + bx}$

77. $\dfrac{10r^2s}{6rs^2} \cdot \dfrac{3r^3}{2rs}$

78. $\dfrac{3a^3b}{25cd^3} \cdot \dfrac{5cd^2}{6ab}$

79. $10(h - 9)\dfrac{h - 3}{9 - h}$

80. $r(r - 25)\dfrac{r + 4}{r - 25}$

81. $\dfrac{2x^2 + 5xy + 3y^2}{3x^2 - 5xy + 2y^2} \div \dfrac{2x^2 + xy - 3y^2}{3x^2 - 5xy + 2y^2}$

82. $\dfrac{2p^2 - 5pq - 3q^2}{p^2 - 9q^2} \div \dfrac{2p^2 + 5pq + 2q^2}{2p^2 + 5pq - 3q^2}$

83. $(4x^2 - 9) \div \dfrac{2x^2 + 5x + 3}{x + 2} \cdot \dfrac{1}{2x - 3}$

84. $(4x + 12) \div \dfrac{2x - 6}{x^2} \cdot \dfrac{x - 3}{2}$

85. $\dfrac{x^3 - 3x^2 - 25x + 75}{x^3 - 27} \cdot \dfrac{2x^3 + 6x^2 + 18x}{x^2 + 10x + 25}$

86. $\dfrac{x^2 + 3x + xy + 3y}{x^2 - 9} \cdot \dfrac{3 - x}{x^3 + 3x^2}$

APPLICATIONS

87. PHYSICS EXPERIMENTS   The following table contains data from a physics experiment. Complete the table.

Trial	Rate (m/sec)	Time (sec)	Distance (m)
1	$\dfrac{k_1^2 + 3k_1 + 2}{k_1 - 3}$	$\dfrac{k_1^2 - 3k_1}{k_1 + 1}$	
2	$\dfrac{k_2^2 + 6k_2 + 5}{k_2 + 1}$		$k_2^2 + 11k_2 + 30$

88. GEOMETRY   Find a simplified rational expression that represents the volume of the rectangular solid shown here.

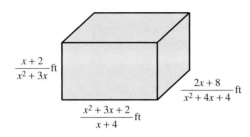

$\dfrac{x+2}{x^2+3x}$ ft

$\dfrac{2x+8}{x^2+4x+4}$ ft

$\dfrac{x^2+3x+2}{x+4}$ ft

WRITING

89. Explain how to multiply two rational expressions.

90. Write some comments to the student who wrote the following solution, explaining the error.

$$\frac{x^2+x-2}{x^2-4}\cdot\frac{x-2}{x-1}=\frac{\cancel{(x+2)}\cancel{(x-1)}\cancel{(x-2)}}{\cancel{(x+2)}\cancel{(x-2)}\cancel{(x-1)}}$$
$$=0$$

91. The graph of $Y_3 = Y_1 - Y_2$, where

$$Y_1 = \frac{2x^2-5x-3}{x^2-9}\cdot\frac{2x^2+5x-3}{2x^2+5x+2}$$

$$Y_2 = \frac{2x-1}{x+2}$$

is shown. Explain how the graph and table can be used to verify that

$$\frac{2x^2-5x-3}{x^2-9}\cdot\frac{2x^2+5x-3}{2x^2+5x+2}=\frac{2x-1}{x+2}$$

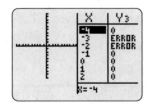

92. A student obtained an answer of $\frac{x+3}{x+7}$ after performing $\frac{x^2-9}{x^2-49}\div\frac{x+3}{x+7}$. As a check, he graphed $Y_3 = Y_1 - Y_2$, where

$$Y_1 = \left(\frac{x^2-9}{x^2-49}\right)\div\left(\frac{x+3}{x+7}\right)$$

$$Y_2 = \frac{x+3}{x+7}$$

The graph is shown. Explain what conclusion can be drawn from the graph and the table.

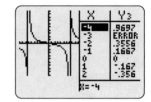

REVIEW

Complete the rules for exponents. Assume that there are no divisions by 0.

93. $x^m x^n =$

94. $(x^m)^n =$

95. $(xy)^n =$

96. $\left(\dfrac{x}{y}\right)^n =$

97. $x^0 =$

98. $x^{-n} =$

99. $\dfrac{x^m}{x^n} =$

100. $\left(\dfrac{x}{y}\right)^{-n} = \left(\ \ \right)$

101. $\dfrac{x^{-m}}{y^{-n}} = $ ——

102. $x^1 =$

CHALLENGE PROBLEMS

Insert either a multiplication symbol · or a division symbol ÷ in each blank to make a true statement.

103. $\dfrac{x^2}{y}\ \ \dfrac{x}{y^2}\ \ \dfrac{x^2}{y^2}=\dfrac{x^3}{y}$

104. $\dfrac{x^2}{y}\ \ \dfrac{x}{y^2}\ \ \dfrac{x^2}{y^2}=\dfrac{y^3}{x}$

SECTION 6.3
Adding and Subtracting Rational Expressions

Objectives

① Add and subtract rational expressions with like denominators.

② Add and subtract rational expressions with unlike denominators.

③ Find the least common denominator.

④ Perform mixed operations.

The methods used to add and subtract rational expressions are based on the rules for adding and subtracting arithmetic fractions. In this section, we will add and subtract rational expressions with *like* and *unlike* denominators.

 Add and Subtract Rational Expressions with Like Denominators.

To add or subtract fractions with a common denominator, we add or subtract their numerators and write the sum or difference over the common denominator. For example,

$$\frac{3}{7} + \frac{2}{7} = \frac{3+2}{7} \qquad\qquad \frac{3}{7} - \frac{2}{7} = \frac{3-2}{7}$$

$$= \frac{5}{7} \qquad\qquad\qquad\qquad = \frac{1}{7}$$

We use the same procedure to add and subtract rational expressions with like denominators.

| **Adding and Subtracting Rational Expressions That Have the Same Denominator** | To add (or subtract) rational expressions that have same denominator, add (or subtract) their numerators and write the sum (or difference) over the common denominator. Then, if possible, factor and simplify. |

If $\frac{A}{D}$ and $\frac{B}{D}$ are rational expressions,

$$\frac{A}{D} + \frac{B}{D} = \frac{A+B}{D} \qquad \text{and} \qquad \frac{A}{D} - \frac{B}{D} = \frac{A-B}{D}$$

EXAMPLE 1   Perform the operations: **a.** $\dfrac{4}{3x} + \dfrac{7}{3x}$ **b.** $\dfrac{a^2}{a^2-1} - \dfrac{a}{a^2-1}$

Strategy   In part (a), we will add the numerators and write the sum over the common denominator. In part (b), we will subtract the numerators and write the difference over the common denominator. Then, if possible, we will factor and simplify.

Why   These are the rules for adding and subtracting rational expressions that have the *same* denominator.

Solution

a. $\dfrac{4}{3x} + \dfrac{7}{3x} = \dfrac{4+7}{3x}$ Add the numerators. Write the sum over the common denominator, 3x.

$= \dfrac{11}{3x}$ The result does not simplify.

b. $\dfrac{a^2}{a^2-1} - \dfrac{a}{a^2-1} = \dfrac{a^2-a}{a^2-1}$ Subtract the numerators. Write the difference over the common denominator, $a^2 - 1$.

We note that the polynomials in the numerator and the denominator of the result factor.

$$= \frac{a(a-1)}{(a+1)(a-1)}$$

$$= \frac{\overset{1}{a(\cancel{a-1})}}{(a+1)(\cancel{a-1})} \quad \text{Simplify.}$$

$$= \frac{a}{a+1}$$

Caution

When adding or subtracting rational expressions, always write the result in simplest form, by removing any factors common to the numerator and denominator.

Self Check 1   Perform the operations:

a. $\dfrac{17}{22t} + \dfrac{13}{22t}$ **b.** $\dfrac{a^2}{a^2-2a} - \dfrac{4}{a^2-2a}$

Now Try   Problems 17 and 23

Using Your Calculator   **Checking Algebra**

We can check the subtraction in Example 1(b) by replacing each a with x and graphing the rational functions $f(x) = \dfrac{x^2}{x^2-1} - \dfrac{x}{x^2-1}$, shown in figure (a), and $g(x) = \dfrac{x}{x+1}$, shown in figure (b), and observing that the graphs are the same. Note that -1 and 1 are not in the domain of the first function and that -1 is not in the domain of the second function.

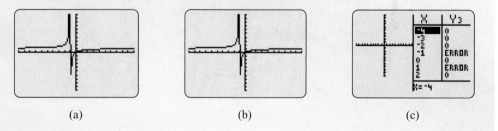

(a)   (b)   (c)

Figure (c) shows the display when the G-T mode is used to check the simplification. Here, $Y_3 = Y_1 - Y_2$, where $Y_1 = \dfrac{x^2}{x^2-1} - \dfrac{x}{x^2-1}$ and $Y_2 = \dfrac{x}{x+1}$.

---

2   **Add and Subtract Rational Expressions with Unlike Denominators.**

Recall that writing a fraction as an equivalent fraction with a larger denominator is called *building the fraction*. For example, to write $\frac{3}{5}$ as an equivalent fraction with a denominator of 35, we multiply it by 1 in the form of $\frac{7}{7}$. When a number is multiplied by 1, its value does not change.

$$\frac{3}{5} = \frac{3}{5} \cdot \frac{7}{7} = \frac{21}{35}$$

To add and subtract rational expressions with different denominators, we write them as equivalent expressions having a common denominator. To do so, we build rational expressions.

Building Rational Expressions	To build a rational expression, multiply it by 1 in the form of $\frac{c}{c}$, where c is any nonzero number or expression.

The following steps summarize how to add or subtract rational expressions with different denominators.

Adding and Subtracting Rational Expressions with Unlike Denominators	1. Find the LCD.
	2. Write each rational expression as an equivalent expression with the LCD as the denominator. To do so, build each rational expression using a form of 1 that involves any factor(s) needed to obtain the LCD.
	3. Add or subtract the numerators and write the sum or difference over the LCD.
	4. Simplify the resulting rational expression, if possible.

EXAMPLE 2   Add: $\dfrac{3}{x} + \dfrac{4}{y}$

Strategy   The LCD for the rational expressions is xy. We will multiply each one by the appropriate form of 1 to build it into an equivalent rational expression with a denominator of xy.

Why   Since the denominators are different, we cannot add these rational expressions in their present form.

Solution

$$\frac{3}{x} + \frac{4}{y} = \frac{3}{x} \cdot \frac{y}{y} + \frac{4}{y} \cdot \frac{x}{x}$$ Build the rational expressions so that each has a denominator of xy.

$$= \frac{3y}{xy} + \frac{4x}{xy}$$ Multiply the numerators. Multiply the denominators.

$$= \frac{3y + 4x}{xy}$$ Add the numerators. Write the sum over the common denominator, xy.

Self Check 2   Add: $\frac{5}{a} + \frac{7}{b}$

Now Try   **Problem 25**

EXAMPLE 3   Subtract: $\dfrac{4x}{x + 2} - \dfrac{7x}{x - 2}$

Strategy   The LCD for the rational expressions is $(x + 2)(x - 2)$. We will multiply each one by the appropriate form of 1 to build it into an equivalent rational expression with a denominator of $(x + 2)(x - 2)$.

Why Since the denominators are different, we cannot subtract these rational expressions in their present form.

Solution

Success Tip

We use the distributive property to multiply the numerators of $\frac{4x}{x+2}$ and $\frac{x-2}{x-2}$. We don't multiply out the denominators.

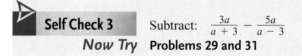

$$\frac{4x}{x+2} \cdot \frac{x-2}{x-2}$$

$$= \frac{4x^2 - 8x}{(x+2)(x-2)}$$

$$\frac{4x}{x+2} - \frac{7x}{x-2}$$

$$= \frac{4x}{x+2} \cdot \frac{x-2}{x-2} - \frac{7x}{x-2} \cdot \frac{x+2}{x+2} \qquad \text{Build each rational expression.}$$

$$= \frac{4x^2 - 8x}{(x+2)(x-2)} - \frac{7x^2 + 14x}{(x+2)(x-2)} \qquad \begin{array}{l}\text{Multiply the numerators.}\\\text{Multiply the denominators.}\end{array}$$

This numerator is written within parentheses to make sure that we subtract both of its terms.

$$= \frac{(4x^2 - 8x) - (7x^2 + 14x)}{(x+2)(x-2)} \qquad \begin{array}{l}\text{Subtract the numerators. Write the difference}\\\text{over the common denominator.}\end{array}$$

$$= \frac{4x^2 - 8x - 7x^2 - 14x}{(x+2)(x-2)} \qquad \begin{array}{l}\text{To subtract the polynomials in the numerator,}\\\text{add the first and the opposite of the second.}\end{array}$$

$$= \frac{-3x^2 - 22x}{(x+2)(x-2)} \qquad \text{Combine like terms in the numerator.}$$

Notation

The numerator of the result may be written in two forms:

Not factored | Factored
$\dfrac{-3x^2 - 22x}{(x+2)(x-2)}$ | $\dfrac{-x(3x+22)}{(x+2)(x-2)}$

If the common factor of $-x$ is factored out of the terms in the numerator, this result can be written in two other equivalent forms.

$$\frac{-3x^2 - 22x}{(x+2)(x-2)} = \frac{-x(3x+22)}{(x+2)(x-2)} = -\frac{x(3x+22)}{(x+2)(x-2)}$$

Self Check 3 Subtract: $\frac{3a}{a+3} - \frac{5a}{a-3}$

Now Try **Problems 29 and 31**

We can use the following fact to add or subtract rational expressions whose denominators are opposites.

Multiplying by -1

When a polynomial is multiplied by -1, the result is its opposite.

EXAMPLE 4 Add: $\frac{x}{x-y} + \frac{y}{y-x}$

Strategy Since the denominators are opposites, either one can serve as the LCD. If we choose $x - y$, we can multiply $\frac{y}{y-x}$ by $\frac{-1}{-1}$ to build it into an equivalent rational expression with the denominator $x - y$.

Why When $y - x$ is multiplied by -1, the subtraction is reversed, and the result is $x - y$.

Solution

$$\frac{x}{x-y} + \frac{y}{y-x} = \frac{x}{x-y} + \frac{y}{y-x} \cdot \frac{-1}{-1}$$

Build $\frac{y}{y-x}$ so that it has a denominator of $x - y$.

$$= \frac{x}{x-y} + \frac{-y}{-y+x}$$

Multiply the numerators.
Multiply the denominators.

$$= \frac{x}{x-y} + \frac{-y}{x-y}$$

Write the second denominator, $-y + x$, as $x - y$. The rational expressions now have a common denominator.

$$= \frac{x-y}{x-y}$$

Add the numerators. Write the result over the common denominator, $x - y$.

$$= 1$$

Simplify.

> **Self Check 4**   Add: $\dfrac{2a}{a-b} + \dfrac{b}{b-a}$
>
> **Now Try**   Problem 33

EXAMPLE 5   Subtract: $3 - \dfrac{7}{x-15}$

Strategy   We will begin by writing 3 as $\frac{3}{1}$.

Why   Then we can multiply $\frac{3}{1}$ by the appropriate form of 1 to build it into an equivalent rational expression with a denominator of $x - 15$.

Solution

$$3 - \frac{7}{x-15} = \frac{3}{1} - \frac{7}{x-15}$$

$3 = \frac{3}{1}$

$$= \frac{3}{1} \cdot \frac{x-15}{x-15} - \frac{7}{x-15}$$

Build $\frac{3}{1}$ to a rational expression with a denominator of $x - 15$.

$$= \frac{3x-45}{x-15} - \frac{7}{x-15}$$

Distribute the multiplication by 3.

$$= \frac{3x-45-7}{x-15}$$

Subtract the numerators. Write the difference over the common denominator, $x - 15$.

$$= \frac{3x-52}{x-2}$$

Combine like terms in the numerator. The result does not simplify.

> **Self Check 5**   Subtract: $6 - \dfrac{5y}{6-y}$
>
> **Now Try**   Problem 37

---

3   **Find the Least Common Denominator.**

When adding or subtracting rational expressions with unlike denominators, it is easiest if we write the rational expressions in terms of the smallest common denominator possible, called the *least* (or lowest) *common denominator (LCD)*. To find the least common denominator of several rational expressions, we follow these steps.

Finding the LCD	1. Factor each denominator completely.
	2. The LCD is a product that uses each different factor obtained in step 1 the greatest number of times it appears in any one factorization.

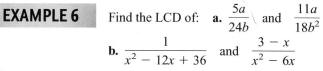

EXAMPLE 6 Find the LCD of: **a.** $\dfrac{5a}{24b}$ and $\dfrac{11a}{18b^2}$

b. $\dfrac{1}{x^2 - 12x + 36}$ and $\dfrac{3 - x}{x^2 - 6x}$

Strategy We begin by factoring completely the denominator of each rational expression.

Why Since the LCD must contain the factors of each denominator, we need to write each denominator in factored form.

Solution

a. We write each denominator as the product of prime numbers and variables.

$$24b = 2 \cdot 2 \cdot 2 \cdot 3 \cdot b = 2^3 \cdot 3 \cdot b$$
$$18b^2 = 2 \cdot 3 \cdot 3 \cdot b \cdot b = 2 \cdot 3^2 \cdot b^2$$

To find the LCD, we form a product using each of these factors the greatest number of times it appears in any one factorization.

┌─ The greatest number of times the factor 2 appears is three times.
  ┌─ The greatest number of times the factor 3 appears is twice.
    ┌─ The greatest number of times the factor *b* appears is twice.

$$\text{LCD} = 2 \cdot 2 \cdot 2 \cdot 3 \cdot 3 \cdot b \cdot b = 72b^2$$

b. We factor each denominator completely:

$$x^2 - 12x + 36 = (x - 6)(x - 6) = (x - 6)^2$$
$$x^2 - 6x = x(x - 6)$$

To find the LCD, we form a product using the highest power of each of the factors:

┌─ The greatest number of times the factor *x* appears is once.
  ┌─ The greatest number of times the factor *x* − 6 appears is twice.

$$\text{LCD} = x(x-6)^2$$

Self Check 6 Find the LCD of: **a.** $\dfrac{5x}{28z^3}$ and $\dfrac{1}{21z}$

b. $\dfrac{a - 1}{a^2 - 25}$ and $\dfrac{3 - a^2}{a^2 + 7a + 10}$

Now Try **Problems 41 and 45**

EXAMPLE 7 Add: $\dfrac{5a}{24b} + \dfrac{11a}{18b^2}$

Strategy In Example 6, we saw that the LCD of these rational expressions is $72b^2$. We will multiply each one by the appropriate form of 1 to build it into an equivalent rational expression with a denominator of $72b^2$.

Success Tip

Note that the highest power of each factor is used to form the LCD:

$$24b = 2^{\textcircled{3}} \cdot 3 \cdot b$$
$$18b^2 = 2 \cdot 3^{\textcircled{2}} \cdot b^{\textcircled{2}}$$
$$\text{LCD} = 2^3 \cdot 3^2 \cdot b^2 = 72b^2$$

Why   Since the denominators are different, we cannot add these rational expressions in their present form.

Solution

$$\frac{5a}{24b} + \frac{11a}{18b^2} = \frac{5a}{24b} \cdot \frac{3b}{3b} + \frac{11a}{18b^2} \cdot \frac{4}{4}$$ Build the rational expressions so that each has a denominator of $72b^2$.

$$= \frac{15ab}{72b^2} + \frac{44a}{72b^2}$$ Multiply the numerators.
Multiply the denominators.

$$= \frac{15ab + 44a}{72b^2}$$ Add the numerators. Write the sum over the common denominator. The result does not simplify.

Self Check 7   Add: $\frac{5x}{28z^3} + \frac{1}{21z}$

Now Try   **Problems 49 and 51**

EXAMPLE 8   Subtract: $\dfrac{x + 1}{x^2 - 2x + 1} - \dfrac{x - 4}{x^2 - 1}$

Strategy   We will factor each denominator, find the LCD, and build the rational expressions so each one has the LCD as its denominator.

Why   Since the denominators are different, we cannot subtract these rational expressions in their present form.

Solution   We factor each denominator to find the LCD:

$$\left. \begin{array}{l} x^2 - 2x + 1 = (x - 1)(x - 1) = (x - 1)^2 \\ x^2 - 1 = (x + 1)(x - 1) \end{array} \right\}$$

The greatest number of times $x - 1$ appears is twice.
The greatest number of times $x + 1$ appears is once.

The LCD is $(x - 1)^2(x + 1)$ or $(x - 1)(x - 1)(x + 1)$.

We now write each rational expression with its denominator in factored form. Then we multiply each numerator and denominator by the missing factor, so that each rational expression has a denominator of $(x - 1)(x - 1)(x + 1)$.

$$\frac{x + 1}{x^2 - 2x + 1} - \frac{x - 4}{x^2 - 1}$$

Success Tip

To build each rational expression, we use the FOIL method to multiply the numerators. Note that we don't multiply out the denominators. For example, to build the second rational expression, we have:

$$\frac{x - 4}{(x + 1)(x - 1)} \cdot \frac{x - 1}{x - 1}$$

The result is: $\frac{x^2 - 5x + 4}{(x + 1)(x - 1)(x - 1)}$

$$= \frac{x + 1}{(x - 1)(x - 1)} - \frac{x - 4}{(x + 1)(x - 1)}$$ Write each denominator in factored form.

$$= \frac{x + 1}{(x - 1)(x - 1)} \cdot \frac{x + 1}{x + 1} - \frac{x - 4}{(x + 1)(x - 1)} \cdot \frac{x - 1}{x - 1}$$ Build each rational expression.

$$= \frac{x^2 + 2x + 1}{(x - 1)(x - 1)(x + 1)} - \frac{x^2 - 5x + 4}{(x - 1)(x - 1)(x + 1)}$$ Multiply the numerators using the FOIL method. Multiply the denominators.

$$= \frac{(x^2 + 2x + 1) - (x^2 - 5x + 4)}{(x - 1)(x - 1)(x + 1)}$$ Subtract the numerators. Write the difference over the common denominator.

$$= \frac{x^2 + 2x + 1 - x^2 + 5x - 4}{(x - 1)(x - 1)(x + 1)}$$ In the numerator, subtract the trinomials.

$$= \frac{7x - 3}{(x - 1)(x - 1)(x + 1)}$$ Combine like terms. The result does not simplify.

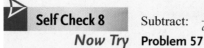

Self Check 8   Subtract: $\dfrac{a+2}{a^2-4a+4} - \dfrac{a-3}{a^2-4}$

Now Try **Problem 57**

4 **Perform Mixed Operations.**

EXAMPLE 9   Perform the operations: $\dfrac{2x}{x^2-4} - \dfrac{1}{x^2-3x+2} + \dfrac{x+1}{x^2+x-2}$

Strategy   We will factor each denominator, find the LCD, and build the rational expressions so each one has the LCD as its denominator.

Why   Since the denominators are different, we cannot add or subtract these rational expressions in their present form.

Solution   We factor each denominator to find the LCD and note that the greatest number of times each factor appears is once.

$$\left.\begin{array}{l} x^2 - 4 = (x-2)(x+2) \\ x^2 - 3x + 2 = (x-2)(x-1) \\ x^2 + x - 2 = (x-1)(x+2) \end{array}\right\} \quad \text{LCD} = (x-2)(x+2)(x-1)$$

We then write each rational expression as an equivalent rational expression with the LCD as its denominator and do the subtraction and addition.

> **Caution**
> Always write the result in simplest form by removing any factors common to the numerator and denominator.

$$\dfrac{2x}{x^2-4} - \dfrac{1}{x^2-3x+2} + \dfrac{x+1}{x^2+x-2}$$

$$= \dfrac{2x}{(x-2)(x+2)} - \dfrac{1}{(x-2)(x-1)} + \dfrac{x+1}{(x-1)(x+2)} \qquad \text{Factor the denominators.}$$

$$= \dfrac{2x}{(x-2)(x+2)} \cdot \dfrac{x-1}{x-1} - \dfrac{1}{(x-2)(x-1)} \cdot \dfrac{x+2}{x+2} + \dfrac{x+1}{(x-1)(x+2)} \cdot \dfrac{x-2}{x-2}$$

$$= \dfrac{2x(x-1) - 1(x+2) + (x+1)(x-2)}{(x+2)(x-2)(x-1)} \qquad \begin{array}{l}\text{Write the sum and difference over the}\\ \text{common denominator.}\end{array}$$

$$= \dfrac{2x^2 - 2x - x - 2 + x^2 - x - 2}{(x+2)(x-2)(x-1)}$$

$$= \dfrac{3x^2 - 4x - 4}{(x+2)(x-2)(x-1)} \qquad \text{Combine like terms.}$$

$$= \dfrac{(3x+2)\overset{1}{\cancel{(x-2)}}}{(x+2)\underset{1}{\cancel{(x-2)}}(x-1)} \qquad \text{Factor the trinomial and simplify.}$$

$$= \dfrac{3x+2}{(x+2)(x-1)}$$

Self Check 9   Perform the operations: $\dfrac{5a}{a^2-25} - \dfrac{7}{a-5} + \dfrac{2}{a+5}$

Now Try **Problem 63**

STUDY SET
6.3

VOCABULARY

Fill in the blanks.

1. The rational expressions $\frac{7}{6n}$ and $\frac{n+1}{6n}$ have a common
_____ of $6n$.

2. The least _____ _____ of $\frac{x-8}{x+6}$ and $\frac{6-5x}{x}$ is
$x(x+6)$.

3. To _____ a rational expression, we multiply it by a form of 1.
For example, $\frac{2}{n^2} \cdot \frac{8}{8} = \frac{16}{8n^2}$.

4. The polynomials $x-y$ and $y-x$ are _____ because their
terms are the same but opposite in sign.

CONCEPTS

Fill in the blanks.

5. To add or subtract rational expressions that have the same
denominator, add or subtract the _____, and write the
sum or difference over the common _____.
     In symbols, if $\frac{A}{D}$ and $\frac{B}{D}$ are rational expressions,

$$\frac{A}{D} + \frac{B}{D} = \frac{}{D} \qquad \text{and} \qquad \frac{A}{D} - \frac{B}{D} = \frac{}{D}$$

6. When a number is multiplied by ___ , its value does not change.

7. To find the least common denominator of several rational
expressions, _____ each denominator completely. The LCD is
a product that uses each different factor the _____ number of
times it appears in any one factorization.

8. $\frac{x^2+3x}{x-1} - \frac{2x-1}{x-1} = \frac{x^2+3x-()}{x-1}$

9. Consider the following two procedures.

i. $\dfrac{x^2-2x}{x^2+4x-12} = \dfrac{\overset{1}{\cancel{x(x-2)}}}{(x+6)\underset{1}{\cancel{(x-2)}}} = \dfrac{x}{x+6}$

ii. $\dfrac{x}{x+6} = \dfrac{x}{x+6} \cdot \dfrac{x-2}{x-2} = \dfrac{x^2-2x}{(x+6)(x-2)}$

a. In which of these procedures are we *building* a rational
expression?

b. For what type of problem is this procedure often necessary?

c. What name is used to describe the other procedure?

10. The LCD for $\frac{2x+1}{x^2+5x+6}$ and $\frac{3x}{x^2-4}$ is

$$\text{LCD} = (x+2)(x+3)(x-2)$$

If we want to subtract these rational expressions, what form of
1 should be used:
a. to build $\frac{2x+1}{x^2+5x+6}$?
b. to build $\frac{3x}{x^2-4}$?

11. Consider the following factorizations.

$$2 \cdot 3 \cdot 3 \cdot (x-2)$$
$$3(x-2)(x+1)$$

a. What is the greatest number of times the factor 3 appears in
any one factorization?

b. What is the greatest number of times the factor $x-2$
appears in any one factorization?

12. The factorizations of the denominators of two rational
expressions follow. Find the LCD.

$$2 \cdot 3 \cdot a \cdot a \cdot a$$
$$2 \cdot 3 \cdot 3 \cdot a \cdot a$$

13. Factor each denominator completely.

   a. $\dfrac{17}{40x^2}$

   b. $\dfrac{x + 25}{2x^2 - 6x}$

   c. $\dfrac{n^2 + 3n - 4}{n^2 - 64}$

14. By what must $y - 4$ be multiplied to obtain $4 - y$?

NOTATION

Complete each solution.

15. $\dfrac{6x - 1}{3x - 1} + \dfrac{3x - 2}{3x - 1} = \dfrac{6x - 1 + \boxed{}}{3x - 1}$

$= \dfrac{9x - \boxed{}}{3x - 1}$

$= \dfrac{3(\boxed{})}{3x - 1}$

$= \boxed{}$

16. $\dfrac{8}{3v} - \dfrac{1}{4v^2} = \dfrac{8}{3v} \cdot \dfrac{\boxed{}}{\boxed{}} - \dfrac{1}{4v^2} \cdot \dfrac{\boxed{}}{\boxed{}}$

$= \dfrac{\boxed{}}{12v^2} - \dfrac{3}{\boxed{}}$

$= \dfrac{32v - 3}{\boxed{}}$

GUIDED PRACTICE

Add or subtract, and then simplify, if possible. **See Example 1.**

17. $\dfrac{8}{3x} + \dfrac{5}{3x}$ **18.** $\dfrac{3}{4y} + \dfrac{8}{4y}$

19. $\dfrac{t}{4r} + \dfrac{t}{4r}$ **20.** $\dfrac{16x}{3z^2} - \dfrac{x}{3z^2}$

21. $\dfrac{4y}{y - 4} - \dfrac{16}{y - 4}$ **22.** $\dfrac{3x}{2x + 2} + \dfrac{x + 4}{2x + 2}$

23. $\dfrac{3x}{x^2 - 9} - \dfrac{9}{x^2 - 9}$ **24.** $\dfrac{9x}{x^2 - 1} - \dfrac{9}{x^2 - 1}$

Add or subtract, and then simplify, if possible. **See Example 2.**

25. $\dfrac{15}{p} + \dfrac{2}{q}$ **26.** $\dfrac{2}{a} + \dfrac{19}{b}$

27. $\dfrac{7}{2b} - \dfrac{11}{3a}$ **28.** $\dfrac{5}{3n} - \dfrac{7}{4m}$

Add or subtract, and then simplify, if possible. **See Example 3.**

29. $\dfrac{3}{x + 2} + \dfrac{5}{x - 4}$

30. $\dfrac{6}{a + 4} - \dfrac{2}{a + 3}$

31. $\dfrac{6x}{x + 3} - \dfrac{4x}{x - 3}$

32. $\dfrac{t}{t + 2} + \dfrac{8}{t - 2}$

Add or subtract, and then simplify, if possible. **See Example 4.**

33. $\dfrac{5x}{x - 3} + \dfrac{4x}{3 - x}$ **34.** $\dfrac{8x}{x - 4} - \dfrac{10x}{4 - x}$

35. $\dfrac{9m}{m - n} - \dfrac{2}{n - m}$ **36.** $\dfrac{3s}{s - x} + \dfrac{1}{x - s}$

Add or subtract, and then simplify, if possible. **See Example 5.**

37. $4 + \dfrac{1}{x - 2}$ **38.** $2 - \dfrac{1}{x + 1}$

39. $x + \dfrac{4x}{7x - 3}$ **40.** $x - \dfrac{3x}{3x - 2}$

The denominators of several fractions are given. Find the LCD. **See Example 6.**

41. $12xy, \; 18x^2y$

42. $15ab^2, \; 27a^2b$

43. $x^2 + 3x, \; x^2 - 9$

44. $3y^2 - 6y, \; 3y(y - 4)$

45. $x^3 + 27, \; x^2 + 6x + 9$

46. $x^3 - 8, \; x^2 - 4x + 4$

47. $2x^2 + 5x + 3, \; 4x^2 + 12x + 9, \; x^2 + 2x + 1$

48. $2x^2 + 5x + 3, \; 4x^2 + 12x + 9, \; 4x + 6$

Perform the operations and simplify the result when possible. **See Example 7.**

49. $\dfrac{11}{5m} - \dfrac{5}{6m}$ **50.** $\dfrac{5}{9s} - \dfrac{1}{4s}$

51. $\dfrac{3}{4ab^2} - \dfrac{5}{2a^2b}$ **52.** $\dfrac{1}{5xy^3} - \dfrac{2}{15x^2y}$

Perform the operations and simplify the result when possible. **See Example 8.**

53. $\dfrac{1}{x + 3} + \dfrac{2}{x^2 + 4x + 3}$

54. $\dfrac{4}{y^2 + 8y + 12} + \dfrac{1}{y + 6}$

55. $\dfrac{m}{m^2 + 9m + 20} - \dfrac{4}{m^2 + 7m + 12}$

56. $\dfrac{t}{t^2 + 5t + 6} - \dfrac{2}{t^2 + 3t + 2}$

57. $\dfrac{x}{x^2 + 5x + 6} + \dfrac{x}{x^2 - 4}$

58. $\dfrac{2a}{a^2 - 2a - 8} + \dfrac{3}{a^2 - 5a + 4}$

59. $\dfrac{x+2}{6x-42} - \dfrac{x-3}{5x-35}$

60. $\dfrac{x-1}{4x-24} - \dfrac{3x-2}{5x-30}$

Perform the operations and simplify the result when possible.
See Example 9.

61. $\dfrac{5x}{x+1} + \dfrac{3}{x+1} - \dfrac{2x}{x+1}$

62. $\dfrac{4}{a+4} - \dfrac{2a}{a+4} + \dfrac{3a}{a+4}$

63. $\dfrac{8}{x^2-9} + \dfrac{2}{x-3} - \dfrac{6}{x}$

64. $\dfrac{x}{x^2-4} - \dfrac{x}{x+2} + \dfrac{2}{x}$

65. $\dfrac{3x}{2x-1} + \dfrac{x+1}{3x+2} - \dfrac{2x}{6x^3+x^2-2x}$

66. $\dfrac{2}{x-2} + \dfrac{3}{x+2} - \dfrac{x-1}{x^2-4}$

67. $\dfrac{1}{x+y} - \dfrac{1}{x-y} - \dfrac{2y}{y^2-x^2}$

68. $\dfrac{a}{a-b} + \dfrac{b}{a+b} + \dfrac{a^2+b^2}{b^2-a^2}$

TRY IT YOURSELF

Perform the operations and simplify the result when possible.

69. $\dfrac{s+7}{s+3} - \dfrac{s-3}{s+7}$

70. $\dfrac{t+5}{t-5} - \dfrac{t-5}{t+5}$

71. $\dfrac{x-y}{2} + \dfrac{x+y}{3}$

72. $\dfrac{a+b}{3} + \dfrac{a-b}{7}$

73. $\dfrac{3x^2+3x}{x^2-5x+6} - \dfrac{3x^2-3x+12}{x^2-5x+6}$

74. $\dfrac{2m^2-7}{m^4-9} + \dfrac{4-m^2}{m^4-9}$

75. $\dfrac{a^2+ab}{a^3-b^3} - \dfrac{b^2}{b^3-a^3}$

76. $\dfrac{y^2-3xy}{x^3-y^3} - \dfrac{x^2+4xy}{y^3-x^3}$

77. $2x+3 + \dfrac{1}{x+1}$

78. $x+1 + \dfrac{1}{x-1}$

79. $\dfrac{4}{x^2-2x-3} - \dfrac{x}{3x^2-7x-6}$

80. $\dfrac{x+3}{2x^2-5x+2} - \dfrac{3x-1}{x^2-x-2}$

81. $\dfrac{3}{x+1} - \dfrac{2}{x-1} + \dfrac{x+3}{x^2-1}$

82. $\dfrac{7n^2}{m-n} + \dfrac{3m}{n-m} - \dfrac{3m^2-n}{m^2-2mn+n^2}$

83. $\dfrac{8}{9y^2} + \dfrac{1}{6y^4}$

84. $\dfrac{5}{6a^3} + \dfrac{7}{8a^2}$

Perform the operations and simplify the result when possible.
Be careful to apply the correct method, because these problems
involve addition, subtraction, multiplication, and division of
rational expressions.

85. $\dfrac{6}{b^2-9} \cdot \dfrac{b+3}{2b+4}$

86. $\dfrac{3a^2-22a+7}{a-a^2} \cdot \dfrac{8a^2-8a}{a^2+a-56}$

87. $\dfrac{4a}{a-5} + a$

88. $\dfrac{10z}{z+4} + z$

89. $\dfrac{2a+1}{3a-2} - \dfrac{a-4}{2-3a}$

90. $\dfrac{2x+1}{x^4-81} + \dfrac{2-x}{x^4-81}$

91. $\dfrac{x^2+x}{3x-15} \div \dfrac{(x+1)^2}{6x-30}$

92. $\dfrac{z^2-9}{z^2+4z+3} \div \dfrac{z^2-3z}{(z+1)^2}$

93. $\dfrac{m}{m^2+5m+6} - \dfrac{2}{m^2+3m+2}$

94. $\dfrac{1}{m+1} + \dfrac{1}{m-1} + \dfrac{2}{m^2-1}$

95. $\dfrac{27p^4}{35q} \div \dfrac{9p}{21q}$

96. $\dfrac{12t}{25s^5} \div \dfrac{10t}{15s^2}$

97. $\dfrac{6}{5d^2-5d} - \dfrac{3}{5d-5}$

98. $\dfrac{9}{2r^2-2r} - \dfrac{5}{2r-2}$

99. $\dfrac{s^3t}{4s^2-9t^2} \cdot \dfrac{4s^2-12st+9t^2}{s^3t^2}$

100. $\dfrac{25x^2-40xy+16y^2}{x^2y^4} \cdot \dfrac{xy^4}{25x^2-16y^2}$

APPLICATIONS

101. DRAFTING   Among the tools used in drafting are the 45°–45°–90° and the 30°–60°–90° triangles shown. Find the perimeter of each triangle. Express each result as a single rational expression.

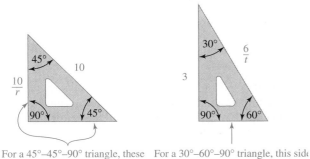

For a 45°–45°–90° triangle, these two sides are the same length.

For a 30°–60°–90° triangle, this side is half as long as the hypotenuse.

102. THE AMAZON   The Amazon River flows in an easterly direction to the Atlantic Ocean. In Brazil, when the river is at low stage, the rate of flow is about 5 mph. Suppose that a river guide can canoe in still water at a rate of r mph.

a. Complete the table to find rational expressions that represent the time it would take the guide to canoe 3 miles downriver and to canoe 3 miles upriver on the Amazon.

	Rate (mph)	Time (hr)	Distance (ml)
Downriver	$r + 5$		3
Upriver	$r - 5$		3

b. Find the difference in the times for the trips upriver and downriver. Express the result as a single rational expression.

WRITING

103. Explain how to find the least common denominator of a set of rational expressions.

104. Add the rational expressions by expressing them in terms of a common denominator $24b^3$. (*Note:* This is not the LCD.)

$$\frac{r}{4b^2} + \frac{s}{6b}$$

An extra step had to be performed because the lowest common denominator was not used. What was the step?

105. Write some comments to the student who wrote the following solution, explaining his misunderstanding.

$$\text{Multiply: } \frac{1}{x} \cdot \frac{3}{2} = \frac{1 \cdot 2}{x \cdot 2} \cdot \frac{3 \cdot x}{2 \cdot x}$$

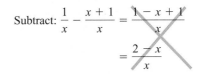

$$= \frac{2}{2x} \cdot \frac{3x}{2x}$$

$$= \frac{6x}{2x}$$

106. Write some comments to the student who wrote the following solution, pointing out where she made an error.

$$\text{Subtract: } \frac{1}{x} - \frac{x + 1}{x} = \frac{1 - x + 1}{x}$$

$$= \frac{2 - x}{x}$$

REVIEW

Solve each equation.

107. $a(a - 6) = -9$

108. $x^2 - \frac{1}{2}(x + 1) = 0$

109. $y^3 + y^2 = 0$

110. $5x^2 = 6 - 13x$

CHALLENGE PROBLEMS

111. Find two rational expressions, each with denominator $x^2 + 5x + 6$, such that their sum is $\frac{1}{x + 2}$.

112. Add: $x^{-1} + x^{-2} + x^{-3} + x^{-4} + x^{-5}$

Perform the operations and simplify the result when possible.

113. $\left(\frac{3}{x - 3} - \frac{1}{x}\right) \div \frac{12x + 18}{x^3 - 9x}$

114. $\frac{13x + 39}{4x^2 + 24x + 36} \div \left(\frac{7}{3x + 9} - \frac{5}{4x + 12}\right)$

115. $\left(\frac{3x}{x + 1} - \frac{6}{x^2 - 1} + \frac{4}{x - 1}\right)\left(\frac{x^3 - 1}{9x^2 - 4}\right)$

116. $\left(\frac{3}{a + 2b} - \frac{2b}{a^2 + 2ab}\right) \div \left(\frac{3b}{a^2 + 2ab} + \frac{5}{a}\right)$

SECTION 6.4
Simplifying Complex Fractions

Objectives

1 Simplify complex fractions using division.

2 Simplify complex fractions using the LCD.

A rational expression whose numerator and/or denominator contain rational expressions is called a **complex rational expression** or, more simply, a **complex fraction.** The expression above the main fraction bar of a complex fraction is the numerator, and the expression below the main fraction bar is the denominator. Two examples are:

$$\dfrac{\dfrac{3a}{b}}{\dfrac{6ac}{b^2}} \quad \begin{matrix}\leftarrow \\ \leftarrow \\ \leftarrow\end{matrix} \quad \begin{matrix}\text{Numerator} \\ \text{Main fraction bar} \\ \text{Denominator}\end{matrix} \quad \begin{matrix}\rightarrow \\ \rightarrow \\ \rightarrow\end{matrix} \quad \dfrac{\dfrac{1}{x}+\dfrac{1}{y}}{\dfrac{1}{x}-\dfrac{1}{y}}$$

In this section, we will discuss two methods for simplifying complex fractions. To **simplify a complex fraction** means to write it in the form $\dfrac{A}{B}$, where A and B are polynomials that have no common factors.

1 **Simplify Complex Fractions Using Division.**

One method for simplifying complex fractions uses the fact that the main fraction bar indicates division.

Simplifying Complex Fractions	***Method 1: Using Division***
	1. Add or subtract in the numerator and/or denominator so that the numerator is a single fraction and the denominator is a single fraction.
	2. Perform the indicated division by multiplying the numerator of the complex fraction by the reciprocal of the denominator.
	3. Simplify the result, if possible.

EXAMPLE 1 Use Method 1 to simplify: $\dfrac{\dfrac{3a}{b}}{\dfrac{6ac}{b^2}}$

Strategy We will perform the division indicated by the main fraction bar using the procedure for dividing rational expressions from Section 6.2.

Why We can skip the first step of Method 1 and immediately divide because the numerator and the denominator of the complex fraction are already single fractions.

Solution

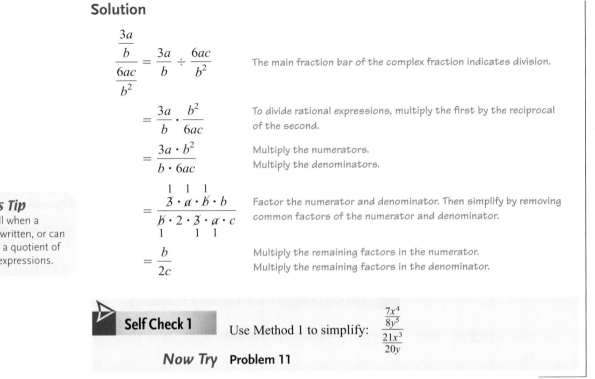

$$\dfrac{\dfrac{3a}{b}}{\dfrac{6ac}{b^2}} = \dfrac{3a}{b} \div \dfrac{6ac}{b^2} \qquad \text{The main fraction bar of the complex fraction indicates division.}$$

$$= \dfrac{3a}{b} \cdot \dfrac{b^2}{6ac} \qquad \text{To divide rational expressions, multiply the first by the reciprocal of the second.}$$

$$= \dfrac{3a \cdot b^2}{b \cdot 6ac} \qquad \begin{array}{l}\text{Multiply the numerators.}\\ \text{Multiply the denominators.}\end{array}$$

$$= \dfrac{\overset{1}{\cancel{3}} \cdot \overset{1}{\cancel{a}} \cdot \overset{1}{\cancel{b}} \cdot b}{\underset{1}{\cancel{b}} \cdot 2 \cdot \underset{1}{\cancel{3}} \cdot \underset{1}{\cancel{a}} \cdot c} \qquad \text{Factor the numerator and denominator. Then simplify by removing common factors of the numerator and denominator.}$$

$$= \dfrac{b}{2c} \qquad \begin{array}{l}\text{Multiply the remaining factors in the numerator.}\\ \text{Multiply the remaining factors in the denominator.}\end{array}$$

> ### Success Tip
> Method 1 works well when a complex fraction is written, or can be easily written, as a quotient of two single rational expressions.

 Self Check 1   Use Method 1 to simplify: $\dfrac{\dfrac{7x^4}{8y^5}}{\dfrac{21x^3}{20y}}$

Now Try   **Problem 11**

---

2 ## Simplify Complex Fractions Using the LCD.

A second method for simplifying complex fractions uses the concepts of LCD and multiplication by a form of 1. The multiplication by 1 produces a simpler, equivalent expression, which will not contain fractions in its numerator or denominator.

Simplifying Complex Fractions	**Method 2: Multiplying by the LCD**
	1. Find the LCD of all fractions within the complex fraction.
	2. Multiply the complex fraction by 1 in the form $\dfrac{\text{LCD}}{\text{LCD}}$.
	3. Perform the operations in the numerator and denominator. No fractional expressions should remain within the complex fraction.
	4. Simplify the result, if possible.

EXAMPLE 2   Use Method 2 to simplify: $\dfrac{\dfrac{3a}{b}}{\dfrac{6ac}{b^2}}$

Strategy   We will use Method 2 to rework Example 1. We can eliminate the fractions in the numerator and denominator of the complex fraction by multiplying it by 1, written in the form $\dfrac{b^2}{b^2}$.

Why   We use $\dfrac{b^2}{b^2}$ because b^2 is the LCD of $\dfrac{3a}{b}$ and $\dfrac{6ac}{b^2}$.

We can also check the simplification in Example 3 by graphing the functions $f(x) = \dfrac{\frac{2}{x} + 5}{3 + x}$ shown in figure (a) and $g(x) = \dfrac{2 + 5x}{x^2 + 3x}$ shown in figure (b) and observing that the graphs are the same. Each graph has window settings of $[-10, 10]$ for x and $[-10, 10]$ for y.

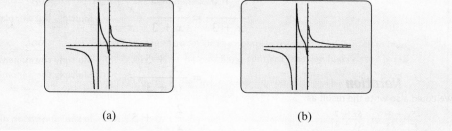

(a)   (b)

EXAMPLE 4

Simplify: $\dfrac{\dfrac{2}{x} + \dfrac{3}{y}}{\dfrac{3}{x} - \dfrac{4}{y}}$

Strategy   We will simplify the complex fraction using both Method 1 and Method 2.

Why   We want to show that using either method, the result is the same.

Solution

Method 1

To write the numerator and denominator of the complex fraction as single fractions, we add the rational expressions in the numerator and subtract the rational expressions in the denominator.

$$\dfrac{\dfrac{2}{x} + \dfrac{3}{y}}{\dfrac{3}{x} - \dfrac{4}{y}} = \dfrac{\dfrac{2}{x} \cdot \dfrac{y}{y} + \dfrac{3}{y} \cdot \dfrac{x}{x}}{\dfrac{3}{x} \cdot \dfrac{y}{y} - \dfrac{4}{y} \cdot \dfrac{x}{x}}$$

← The LCD for the numerator is xy. Build each rational expression so that each has a denominator of xy.

← The LCD for the denominator is xy. Build each rational expression so that each has a denominator of xy.

$$= \dfrac{\dfrac{2y}{xy} + \dfrac{3x}{xy}}{\dfrac{3y}{xy} - \dfrac{4x}{xy}}$$

Multiply the numerators.
Multiply the denominators.

$$= \dfrac{\dfrac{2y + 3x}{xy}}{\dfrac{3y - 4x}{xy}}$$

Add the rational expressions in the numerator and subtract the rational expressions in the denominator.

$$= \dfrac{2y + 3x}{xy} \div \dfrac{3y - 4x}{xy}$$

Write the division indicated by the main fraction bar using a $\div$ symbol.

$$= \dfrac{2y + 3x}{xy} \cdot \dfrac{xy}{3y - 4x}$$

Multiply by the reciprocal of $\frac{3y - 4x}{xy}$.

$$= \frac{(2y + 3x) \cdot \overset{1}{\cancel{x}} \cdot \overset{1}{\cancel{y}}}{\underset{1}{\cancel{x}} \cdot \underset{1}{\cancel{y}} \cdot (3y - 4x)}$$ Multiply the rational expressions and simplify the result.

$$= \frac{2y + 3x}{3y - 4x}$$

Method 2

The LCD of the fractions appearing in the complex fraction is xy. We multiply the complex fraction by 1 in the form of $\frac{\text{LCD}}{\text{LCD}}$.

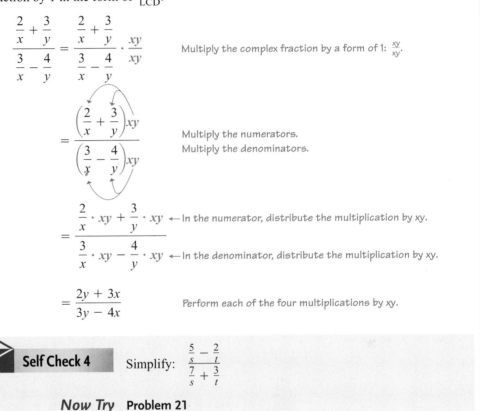

$$\frac{\dfrac{2}{x} + \dfrac{3}{y}}{\dfrac{3}{x} - \dfrac{4}{y}} = \frac{\dfrac{2}{x} + \dfrac{3}{y}}{\dfrac{3}{x} - \dfrac{4}{y}} \cdot \frac{xy}{xy}$$ Multiply the complex fraction by a form of 1: $\frac{xy}{xy}$.

$$= \frac{\left(\dfrac{2}{x} + \dfrac{3}{y}\right)xy}{\left(\dfrac{3}{x} - \dfrac{4}{y}\right)xy}$$ Multiply the numerators.
Multiply the denominators.

$$= \frac{\dfrac{2}{x} \cdot xy + \dfrac{3}{y} \cdot xy}{\dfrac{3}{x} \cdot xy - \dfrac{4}{y} \cdot xy}$$ ← In the numerator, distribute the multiplication by xy.
← In the denominator, distribute the multiplication by xy.

$$= \frac{2y + 3x}{3y - 4x}$$ Perform each of the four multiplications by xy.

> **Self Check 4**   Simplify: $\dfrac{\dfrac{5}{s} - \dfrac{2}{t}}{\dfrac{7}{s} + \dfrac{3}{t}}$
>
> *Now Try*   **Problem 21**

If negative exponents occur, we write an equivalent expression involving positive exponents and simplify using Method 1 or Method 2.

EXAMPLE 5   Simplify: $\dfrac{x^{-1} + y^{-1}}{x^{-2} - y^{-2}}$

Strategy   We will use the rule for exponents $a^{-m} = \dfrac{1}{a^m}$ to write each term of the numerator and denominator without negative exponents.

Why   We can then simplify the resulting complex fraction.

Solution   We write the complex fraction without using negative exponents and use Method 2 to simplify.

$$\frac{x^{-1} + y^{-1}}{x^{-2} - y^{-2}} = \frac{\dfrac{1}{x} + \dfrac{1}{y}}{\dfrac{1}{x^2} - \dfrac{1}{y^2}}$$

Use the rule for negative exponents to write x^{-1}, y^{-1}, x^{-2}, and y^{-2} using positive exponents. See the Caution box.

$$= \frac{\dfrac{1}{x} + \dfrac{1}{y}}{\dfrac{1}{x^2} - \dfrac{1}{y^2}} \cdot \frac{x^2 y^2}{x^2 y^2}$$

The LCD of all rational expressions in the complex fraction is $x^2 y^2$. Multiply the complex fraction by $\dfrac{LCD}{LCD}$.

$$= \frac{\left(\dfrac{1}{x} + \dfrac{1}{y}\right) x^2 y^2}{\left(\dfrac{1}{x^2} - \dfrac{1}{y^2}\right) x^2 y^2}$$

Multiply the numerators.
Multiply the denominators.

$$= \frac{\dfrac{1}{x} \cdot x^2 y^2 + \dfrac{1}{y} \cdot x^2 y^2}{\dfrac{1}{x^2} \cdot x^2 y^2 - \dfrac{1}{y^2} \cdot x^2 y^2}$$

← In the numerator, distribute the multiplication by $x^2 y^2$.

← In the denominator, distribute the multiplication by $x^2 y^2$.

$$= \frac{xy^2 + yx^2}{y^2 - x^2}$$

Perform each of the four multiplications by $x^2 y^2$.

$$= \frac{xy \overset{1}{\cancel{(y + x)}}}{\cancel{(y + x)}(y - x)}$$

Factor the numerator and denominator.

$$= \frac{xy}{y - x}$$

Simplify the rational expression by removing the factor $y + x$ that is common to the numerator and denominator.

Self Check 5   Simplify: $\dfrac{x^{-2} + y^{-2}}{x^{-1} - y^{-1}}$

Now Try   Problem 29

EXAMPLE 6   Simplify: $\dfrac{\dfrac{1}{a^2 - 3a + 2}}{\dfrac{3}{a - 2} - \dfrac{2}{a - 1}}$

Strategy   We will factor $a^2 - 3a + 2$ and determine the LCD for all the fractions appearing in the complex fraction. Then we will use Method 2 to simplify.

Why   Method 2 works well when the complex fraction has sums and/or differences in the numerator or denominator.

Solution   To find the LCD for all the fractions appearing in the complex fraction, we must factor $a^2 - 3a + 2$.

$$\frac{\dfrac{1}{a^2 - 3a + 2}}{\dfrac{3}{a-2} - \dfrac{2}{a-1}} = \frac{\dfrac{1}{(a-2)(a-1)}}{\dfrac{3}{a-2} - \dfrac{2}{a-1}} \qquad \text{Factor the trinomial } a^2 - 3a + 2.$$

The LCD of the fractions in the numerator and denominator of the complex fraction is $(a - 2)(a - 1)$. We multiply the numerator and the denominator by the LCD.

$$= \frac{\dfrac{1}{(a-2)(a-1)}}{\dfrac{3}{a-2} - \dfrac{2}{a-1}} \cdot \frac{(a-2)(a-1)}{(a-2)(a-1)}$$

$$= \frac{\left[\dfrac{1}{(a-2)(a-1)}\right](a-2)(a-1)}{\left(\dfrac{3}{a-2} - \dfrac{2}{a-1}\right)(a-2)(a-1)} \qquad \begin{array}{l}\text{Multiply the numerators.}\\ \text{Multiply the denominators.}\end{array}$$

$$= \frac{\dfrac{(a-2)(a-1)}{(a-2)(a-1)}}{\dfrac{3(a-2)(a-1)}{a-2} - \dfrac{2(a-2)(a-1)}{a-1}} \qquad \begin{array}{l}\text{Perform the mutiplication in the numerator.}\\ \text{In the denominator, distribute the LCD,}\\ (a-2)(a-1).\end{array}$$

$$= \frac{1}{3(a-1) - 2(a-2)} \qquad \begin{array}{l}\text{Simplify each of the three rational expressions}\\ \text{highlighted in blue.}\end{array}$$

$$= \frac{1}{3a - 3 - 2a + 4} \qquad \begin{array}{l}\text{In the denominator, use the distributive}\\ \text{property.}\end{array}$$

$$= \frac{1}{a + 1} \qquad \text{In the denominator, combine like terms.}$$

The Language of Algebra

After multiplying a complex fraction by $\frac{\text{LCD}}{\text{LCD}}$ and performing the multiplications, the numerator and denominator of the complex fraction will be *cleared of fractions.*

> **Self Check 6** Simplify: $\dfrac{\dfrac{b}{b+4} + \dfrac{2}{b+3}}{\dfrac{b}{b^2 + 7b + 12}}$

> **Now Try** Problem 33

> **ANSWERS TO SELF CHECKS** 1. $\dfrac{5x}{6y^4}$ 2. $\dfrac{5x}{6y^4}$ 3. $\dfrac{3 + 2m}{m^2 + 2m}$ 4. $\dfrac{5t - 2s}{7t + 3s}$ 5. $\dfrac{y^2 + x^2}{xy^2 - x^2 y}$ or $\dfrac{x^2 + y^2}{xy(y - x)}$
> 6. $\dfrac{b^2 + 5b + 8}{b}$

STUDY SET
6.4

VOCABULARY

Fill in the blanks.

1. $\dfrac{\dfrac{x}{y} + \dfrac{1}{x}}{\dfrac{1}{y} + \dfrac{2}{x}}$ and $\dfrac{\dfrac{5a^2}{b}}{\dfrac{b}{2a^3}}$ are examples of complex _____ expressions, or more simply, _____ fractions.

2. To _____ a complex fraction means to express it in the form $\frac{A}{B}$, where A and B are polynomials with no common factors.

CONCEPTS

3. To simplify the following complex fraction, it is multiplied by what form of 1?

$$\frac{\dfrac{4}{t^2} + \dfrac{b}{t}}{\dfrac{3b}{t}} = \frac{\dfrac{4}{t^2} + \dfrac{b}{t}}{\dfrac{3b}{t}} \cdot \frac{t^2}{t^2}$$

4. Determine the LCD of the rational expressions appearing in each complex fraction.

a. $\dfrac{1 + \dfrac{4}{c}}{\dfrac{2}{c} + c}$

b. $\dfrac{\dfrac{6}{m^2} + \dfrac{1}{2m}}{\dfrac{m^2 - 1}{4}}$

c. $\dfrac{\dfrac{p}{p+2} + \dfrac{12}{p+3}}{\dfrac{p-1}{p^2 + 5p + 6}}$

d. $\dfrac{2 + \dfrac{3}{x+1}}{\dfrac{1}{x} + x + x^2}$

NOTATION

Complete each solution to simplify the rational expression.

5. $\dfrac{\dfrac{5m^2}{6}}{\dfrac{25m}{3}} = \dfrac{5m^2}{6} \boxed{} \dfrac{25m}{3}$ Use Method 1

$$= \frac{5m^2}{6} \cdot \frac{\boxed{}}{}$$

$$= \frac{5 \cdot \boxed{} \cdot m \cdot \boxed{}}{2 \cdot \boxed{} \cdot 5 \cdot 5 \cdot \boxed{}}$$

$$= \frac{m}{\boxed{}}$$

6. $\dfrac{\dfrac{2}{a} - \dfrac{1}{b}}{\dfrac{5}{a} + \dfrac{3}{b}} = \dfrac{\dfrac{2}{a} - \dfrac{1}{b}}{\dfrac{5}{a} + \dfrac{3}{b}} \cdot \dfrac{\boxed{}}{\boxed{}}$ Use Method 2

$$= \frac{\left(\dfrac{2}{a} - \dfrac{1}{b} \right) \boxed{}}{\left(\dfrac{5}{a} + \dfrac{3}{b} \right) \boxed{}}$$

$$= \frac{\dfrac{2}{a} \cdot \boxed{} - \dfrac{1}{b} \cdot \boxed{}}{\dfrac{5}{a} \cdot \boxed{} + \dfrac{3}{b} \cdot \boxed{}}$$

$$= \frac{2b - \boxed{}}{\boxed{} + 3a}$$

7. a. Fill in the blank: The expression $\dfrac{\dfrac{a}{b}}{\dfrac{c}{d}}$ is equivalent to $\dfrac{a}{b} \boxed{} \dfrac{c}{d}$.

   b. What is the numerator and what is the denominator of the following complex fraction?

$$\frac{6 - k - \dfrac{5}{k}}{k^2 - 9}$$

8. A student checks her answers with those in the back of her textbook. Determine whether they are equivalent.

Student's answer	Book's answer	Equivalent?
$\dfrac{3 + 2t}{t^2 + 2t}$	$\dfrac{2t + 3}{t(t + 2)}$	
$\dfrac{5 - 3x^2}{x + x^2}$	$-\dfrac{3x^2 - 5}{x^2 + x}$	
$\dfrac{3xy(y + x)}{(2y - x)(2y + 3x)}$	$\dfrac{3xy^2 + 3x^2y}{(2y + x)(2y - 3x)}$	

GUIDED PRACTICE

Simplify each complex fraction. **See Examples 1 and 2.**

9. $\dfrac{\dfrac{a^6}{2}}{\dfrac{3a}{4}}$

10. $\dfrac{\dfrac{3b^7}{4}}{\dfrac{b^9}{2}}$

11. $\dfrac{\dfrac{20x}{y}}{\dfrac{36x}{y^2}}$

12. $\dfrac{\dfrac{5t^4}{9x^2}}{\dfrac{2t}{18x}}$

13. $\dfrac{\dfrac{18x^5}{35y^2}}{\dfrac{2x^8}{21y^5}}$

14. $\dfrac{\dfrac{32m}{45n^{10}}}{\dfrac{8m^7}{3n^{11}}}$

15. $\dfrac{\dfrac{16c}{77d^4}}{\dfrac{28c^7}{55d^0}}$

16. $\dfrac{\dfrac{8m^{10}}{27n^0}}{\dfrac{32m}{63n^9}}$

Simplify each complex fraction. See Example 3.

17. $\dfrac{\dfrac{3}{a} - 2}{a - 3}$

18. $\dfrac{\dfrac{5}{t} - 5}{t - 5}$

19. $\dfrac{4p - \dfrac{4}{p}}{12 - \dfrac{4}{p}}$

20. $\dfrac{3s - \dfrac{3}{s}}{6 + \dfrac{3}{s}}$

Simplify each complex fraction. See Example 4.

21. $\dfrac{\dfrac{y}{x} - \dfrac{x}{y}}{\dfrac{1}{x} + \dfrac{1}{y}}$

22. $\dfrac{\dfrac{y}{x} - \dfrac{x}{y}}{\dfrac{1}{y} - \dfrac{1}{x}}$

23. $\dfrac{\dfrac{1}{a} - \dfrac{1}{b}}{\dfrac{a}{b} - \dfrac{b}{a}}$

24. $\dfrac{\dfrac{1}{a} + \dfrac{1}{b}}{\dfrac{a}{b} - \dfrac{b}{a}}$

25. $\dfrac{\dfrac{2}{a^2} + \dfrac{1}{a}}{\dfrac{2}{a} + \dfrac{1}{a^2}}$

26. $\dfrac{\dfrac{3}{y^2} - \dfrac{4}{y}}{\dfrac{1}{y} + \dfrac{15}{y^2}}$

27. $\dfrac{\dfrac{3}{b^2} - \dfrac{4}{b} + 1}{1 - \dfrac{1}{b} - \dfrac{6}{b^2}}$

28. $\dfrac{\dfrac{18}{c^2} + \dfrac{11}{c} + 1}{1 - \dfrac{3}{c} - \dfrac{10}{c^2}}$

Simplify each complex fraction. See Example 5.

29. $\dfrac{x^{-2} - y^{-2}}{x^{-1} - y^{-1}}$

30. $\dfrac{x^{-1} + y^{-1}}{x^{-1} - y^{-1}}$

31. $\dfrac{a - b^{-2}}{b - a^{-2}}$

32. $\dfrac{m^{-1} + 2n}{m^{-1} - n}$

Simplify each complex fraction. See Example 6.

33. $\dfrac{\dfrac{3}{z - 3} + \dfrac{2}{z - 2}}{\dfrac{5z}{z^2 - 5z + 6}}$

34. $\dfrac{\dfrac{h}{h^2 + 3h + 2}}{\dfrac{4}{h + 2} - \dfrac{4}{h + 1}}$

35. $\dfrac{\dfrac{2}{x + 3} - \dfrac{1}{x - 3}}{\dfrac{3}{x^2 - 9}}$

36. $\dfrac{2 + \dfrac{1}{x^2 - 1}}{1 + \dfrac{1}{x - 1}}$

TRY IT YOURSELF

Simplify each complex fraction.

37. $\dfrac{1 + \dfrac{x}{y}}{1 - \dfrac{x}{y}}$

38. $\dfrac{\dfrac{x}{y} + 1}{1 - \dfrac{x}{y}}$

39. $\dfrac{\dfrac{x^2 + 5x + 6}{3xy}}{\dfrac{9 - x^2}{6xy}}$

40. $\dfrac{\dfrac{x - y}{xy}}{\dfrac{y - x}{x}}$

41. $\dfrac{1 + \dfrac{6}{x} + \dfrac{8}{x^2}}{1 + \dfrac{1}{x} - \dfrac{12}{x^2}}$

42. $\dfrac{1 - x - \dfrac{2}{x}}{\dfrac{6}{x^2} + \dfrac{1}{x} - 1}$

43. $\dfrac{\dfrac{ac - ad - c + d}{a^3 - 1}}{\dfrac{c^2 - 2cd + d^2}{a^2 + a + 1}}$

44. $\dfrac{\dfrac{2x - tx + 2y - ty}{x^2 + 2xy + y^2}}{\dfrac{t^3 - 8}{15x + 15y}}$

45. $\dfrac{\dfrac{1}{a + 1} + 1}{\dfrac{3}{a - 1} + 1}$

46. $\dfrac{2 + \dfrac{4}{y - 7}}{\dfrac{4}{y - 7}}$

47. $\dfrac{5ab^2}{\dfrac{ab}{25}}$

48. $\dfrac{\dfrac{6a^2b}{4t}}{3a^2b^2}$

49. $\dfrac{a - 4 + \dfrac{1}{a}}{-\dfrac{1}{a} - a + 4}$

50. $\dfrac{a + 1 + \dfrac{1}{a^2}}{\dfrac{1}{a^2} + a - 1}$

EXAMPLE 1 Simplify: **a.** $\dfrac{21x^5}{7x^2}$ **b.** $\dfrac{10r^6s}{6rs^3}$ Write answers using positive exponents.

Strategy First, we will simplify the rational expression by factoring and removing any factors that are common to the numerator and the denominator. Then we will simplify the rational expression using the rules for exponents.

Why We want to show that the result is the same using either method.

Solution

By simplifying fractions

a. $\dfrac{21x^5}{7x^2} = \dfrac{3 \cdot \overset{1}{\cancel{7}} \cdot \overset{1}{\cancel{x}} \cdot \overset{1}{\cancel{x}} \cdot x \cdot x \cdot x}{\underset{1}{\cancel{7}} \cdot \underset{1}{\cancel{x}} \cdot \underset{1}{\cancel{x}}}$

$= 3x^3$

b. $\dfrac{10r^6s}{6rs^3} = \dfrac{\overset{1}{\cancel{2}} \cdot 5 \cdot \overset{1}{\cancel{r}} \cdot r \cdot r \cdot r \cdot r \cdot r \cdot \overset{1}{\cancel{s}}}{\underset{1}{\cancel{2}} \cdot 3 \cdot \underset{1}{\cancel{r}} \cdot \underset{1}{\cancel{s}} \cdot s \cdot s}$

$= \dfrac{5r^5}{3s^2}$

Using the rules for exponents

$\dfrac{21x^5}{7x^2} = 3x^{5-2}$ Divide the coefficients. Keep the common base x and subtract exponents.

$= 3x^3$

$\dfrac{10r^6s}{6rs^3} = \dfrac{5}{3}r^{6-1}s^{1-3}$ Simplify $\frac{10}{6}$. Keep each base and subtract exponents.

$= \dfrac{5}{3}r^5s^{-2}$

$= \dfrac{5r^5}{3s^2}$ Move s^{-2} to the denominator and change the sign of the exponent.

 Self Check 1 Simplify: **a.** $\dfrac{30y^4}{5y^2}$ **b.** $\dfrac{8c^2d^6}{32c^5d^2}$

Now Try Problems 15 and 17

2 **Divide a Polynomial by a Monomial.**

Recall that to add two fractions with the same denominator, we add their numerators and keep the common denominator.

$$\frac{a}{d} + \frac{b}{d} = \frac{a+b}{d}$$

We can use this rule in reverse to divide polynomials by monomials.

Dividing a Polynomial by a Monomial

To divide a polynomial by a monomial, divide each term of the polynomial by the monomial. Let A, B, and D represent monomials, where D is not 0,

$$\frac{A+B}{D} = \frac{A}{D} + \frac{B}{D}$$

| EXAMPLE 2 | Divide: **a.** $\dfrac{9x^2 + 6x}{3x}$ **b.** $\dfrac{12a^4b^3 - 18a^3b^2 + 2a^2}{6a^2b^2}$ |

Strategy We will divide each term of the polynomial in the numerator by the monomial in the denominator.

Why A fraction bar indicates division of the numerator by the denominator.

Solution

a. Here, we have a binomial divided by a monomial.

The Language of Algebra
The names of the parts of a division statement are

Dividend
↓
$\dfrac{9x^2 + 6x}{3x} = 3x + 2$
↑ ↑
Divisor   Quotient

$$\dfrac{9x^2 + 6x}{3x} = \dfrac{9x^2}{3x} + \dfrac{6x}{3x} \qquad \text{Divide each term of the numerator, } 9x^2 + 6x, \text{ by the denominator, } 3x.$$

$$= 3x^{2-1} + 2x^{1-1} \qquad \text{Perform each monomial division. Divide the coefficients. Keep each base and subtract the exponents.}$$

$$= 3x + 2 \qquad \text{Recall that } x^0 = 1.$$

Check: We multiply the divisor, $3x$, and the quotient, $3x + 2$. The result should be the dividend, $9x^2 + 6x$.

$$3x(3x + 2) = 9x^2 + 6x \qquad \text{The answer checks.}$$

b. Here we have a trinomial divided by a monomial.

$$\dfrac{12a^4b^3 - 18a^3b^2 + 2a^2}{6a^2b^2} = \dfrac{12a^4b^3}{6a^2b^2} - \dfrac{18a^3b^2}{6a^2b^2} + \dfrac{2a^2}{6a^2b^2} \qquad \begin{array}{l}\text{Divide each term of the} \\ \text{numerator by the} \\ \text{denominator, } 6a^2b^2.\end{array}$$

$$= 2a^{4-2}b^{3-2} - 3a^{3-2}b^{2-2} + \dfrac{a^{2-2}}{3b^2} \qquad \begin{array}{l}\text{Perform each monomial} \\ \text{division.} \\ \text{Simplify: } \frac{2}{6} = \frac{1}{3}.\end{array}$$

$$= 2a^2b - 3a + \dfrac{1}{3b^2}$$

Success Tip
The sum, difference, and product of two polynomials are always polynomials. However, as seen in Example 2(b), the quotient of two polynomials is not always a polynomial.

Since the variables in a polynomial must have whole-number exponents, $2a^2b - 3a + \frac{1}{3b^2}$ is not a polynomial because the last term can be written as $\frac{1}{3}b^{-2}$.

Check: $6a^2b^2\left(2a^2b - 3a + \dfrac{1}{3b^2}\right) = 12a^4b^3 - 18a^3b^2 + 2a^2 \qquad \text{The answer checks.}$

| ▷ **Self Check 2** | Divide: **a.** $\dfrac{50h^3 + 15h^2}{5h^2}$
 b. $\dfrac{22s^5t^2 - s^4t^3 + 44s^2t}{11s^2t^2}$ |

Now Try Problems 21 and 25

---

3 **Divide a Polynomial by a Polynomial.**

To divide a polynomial by a polynomial (other than a monomial), we use a method similar to long division in arithmetic. To use long division to divide $x^2 + 7x + 12$ (the *dividend*) by $x + 4$ (the *divisor*), we proceed as follows:

Step 1: $\overset{\displaystyle x}{x + 4 \overline{\smash{\big)}x^2 + 7x + 12}}$ $\qquad$ Divide the first term of the dividend by the first term of the divisor: $\frac{x^2}{x} = x$. Write the result, x, above the long division symbol.

Success Tip

Notice that this method is much like that used for division of whole numbers.

$$
\begin{array}{r}
13 \\
12\overline{)156} \\
-12\downarrow \\
\hline
36 \\
-36 \\
\hline
0
\end{array}
$$

Hundreds ┘│↑
Tens ┘│
Ones ┘

Success Tip

The long division method aligns like terms vertically.

$$
\begin{array}{r}
x + 3 \\
x + 4\overline{)x^2 + 7x + 12} \\
-(x^2 + 4x) \\
\hline
3x + 12 \\
-(3x + 12) \\
\hline
0
\end{array}
$$

x^2-terms ┘│↑
x-terms ┘│
constants ┘

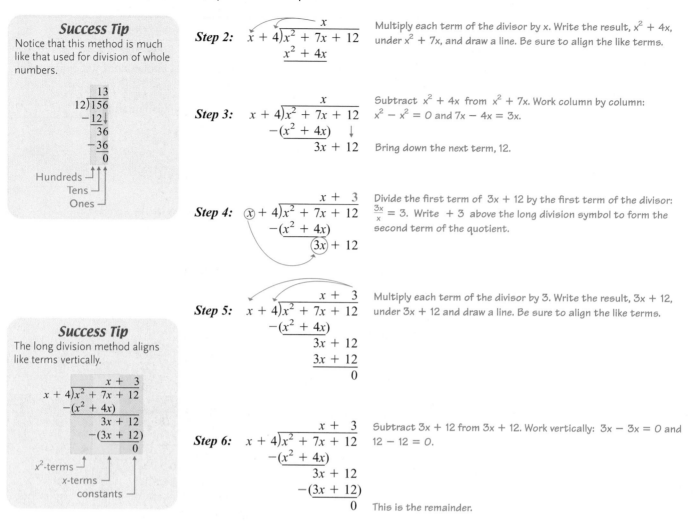

Step 2:

$$
\begin{array}{r}
x \\
x + 4\overline{)x^2 + 7x + 12} \\
x^2 + 4x
\end{array}
$$

Multiply each term of the divisor by x. Write the result, $x^2 + 4x$, under $x^2 + 7x$, and draw a line. Be sure to align the like terms.

Step 3:

$$
\begin{array}{r}
x \\
x + 4\overline{)x^2 + 7x + 12} \\
-(x^2 + 4x) \downarrow \\
\hline
3x + 12
\end{array}
$$

Subtract $x^2 + 4x$ from $x^2 + 7x$. Work column by column: $x^2 - x^2 = 0$ and $7x - 4x = 3x$.

Bring down the next term, 12.

Step 4:

$$
\begin{array}{r}
x + 3 \\
\textcircled{x} + 4\overline{)x^2 + 7x + 12} \\
-(x^2 + 4x) \\
\hline
\textcircled{3x} + 12
\end{array}
$$

Divide the first term of $3x + 12$ by the first term of the divisor: $\frac{3x}{x} = 3$. Write $+3$ above the long division symbol to form the second term of the quotient.

Step 5:

$$
\begin{array}{r}
x + 3 \\
x + 4\overline{)x^2 + 7x + 12} \\
-(x^2 + 4x) \\
\hline
3x + 12 \\
3x + 12 \\
\hline
0
\end{array}
$$

Multiply each term of the divisor by 3. Write the result, $3x + 12$, under $3x + 12$ and draw a line. Be sure to align the like terms.

Step 6:

$$
\begin{array}{r}
x + 3 \\
x + 4\overline{)x^2 + 7x + 12} \\
-(x^2 + 4x) \\
\hline
3x + 12 \\
-(3x + 12) \\
\hline
0
\end{array}
$$

Subtract $3x + 12$ from $3x + 12$. Work vertically: $3x - 3x = 0$ and $12 - 12 = 0$.

This is the remainder.

Step 7:   The division process stops when the result of the subtraction is a constant or a polynomial with degree less than the degree of the divisor. Here, the quotient is $x + 3$ and the remainder is 0.

We can check the answer using the fact that for any division:

Divisor · quotient + remainder = dividend

$$
\underbrace{\text{Divisor} \cdot \text{quotient}}_{} + \text{remainder} = \underbrace{\text{dividend}}_{}
$$

Check: $(x + 4)(x + 3) + \quad 0 \quad = x^2 + 7x + 12$ The answer checks.

EXAMPLE 3   Divide: $\dfrac{2a^3 + 13a^2 + 11a - 6}{2a + 3}$

Strategy   We will use the long division method. The dividend is $2a^3 + 13a^2 + 11a - 6$ and the divisor is $2a + 3$.

Why   Since the divisor has more than one term, we must use the long division method to divide the polynomials.

Success Tip

The long division process is a series of four steps that are repeated: divide, multiply, subtract, and bring down.

Solution

$$\begin{array}{r} a^2 \\ 2a + 3 \overline{)\, 2a^3 + 13a^2 + 11a - 6\,} \end{array}$$

Divide the first term of the dividend by the first term of the divisor: $\frac{2a^3}{2a} = a^2$. Write the result, a^2 above the long division symbol.

$$\begin{array}{r} a^2 \\ 2a + 3 \overline{)\, 2a^3 + 13a^2 + 11a - 6\,} \\ -(2a^3 + 3a^2) \\ \hline 10a^2 + 11a \end{array}$$

Multiply each term in the divisor by a^2 to get $2a^3 + 3a^2$. Subtract $2a^3 + 3a^2$ from $2a^3 + 13a^2$ and bring down the 11a.

$$\begin{array}{r} a^2 + 5a \\ 2a + 3 \overline{)\, 2a^3 + 13a^2 + 11a - 6\,} \\ -(2a^3 + 3a^2) \\ \hline 10a^2 + 11a \end{array}$$

Divide the first term of $10a^2 + 11a$ by the first term of the divisor: $\frac{10a^2}{2a} = 5a$. Write $+5a$ above the long division symbol to form the second term of the quotient.

$$\begin{array}{r} a^2 + 5a \\ 2a + 3 \overline{)\, 2a^3 + 13a^2 + 11a - 6\,} \\ -(2a^3 + 3a^2) \\ \hline 10a^2 + 11a \\ -(10a^2 + 15a) \\ \hline -4a - 6 \end{array}$$

Multiply each term in the divisor by $5a$ to get $10a^2 + 15a$. Subtract $10a^2 + 15a$ from $10a^2 + 11a$ and bring down the -6.

$$\begin{array}{r} a^2 + 5a - 2 \\ 2a + 3 \overline{)\, 2a^3 + 13a^2 + 11a - 6\,} \\ -(2a^3 + 3a^2) \\ \hline 10a^2 + 11a \\ -(10a^2 + 15a) \\ \hline -4a - 6 \end{array}$$

Divide the first term of $-4a - 6$ by the first term of the divisor: $\frac{-4a}{2a} = -2$. Write -2 above the long division symbol to form the third term of the quotient.

$$\begin{array}{r} a^2 + 5a - 2 \\ 2a + 3 \overline{)\, 2a^3 + 13a^2 + 11a - 6\,} \\ 2a^3 + 3a^2 \\ \hline 10a^2 + 11a \\ -(10a^2 + 15a) \\ \hline -4a - 6 \\ -(-4a - 6) \\ \hline 0 \end{array}$$

Multiply each term in the divisor by -2 to get $-4a - 6$. Subtract $-4a - 6$ from $-4a - 6$ to get 0.

This is the remainder.

Since the remainder is 0, the quotient is $a^2 + 3a - 2$. We can check the quotient by verifying that

$$\overbrace{\text{Divisor} \cdot \text{quotient}} + \text{remainder} = \overbrace{\text{dividend}}$$

Check: $(2a + 3)(a^2 + 5a - 2) + 0 = 2a^3 + 13a^2 + 11a - 6$ The quotient checks.

Self Check 3   Divide: $\dfrac{4x^3 + 11x^2 + 2x - 3}{4x + 3}$

Now Try   **Problems 27 and 33**

The long division method used in algebra can have a remainder just as long division in arithmetic does.

EXAMPLE 4   Divide: $\dfrac{3x^3 + 2x^2 - 3x - 28}{x - 2}$

Strategy   We will use the long division method. The dividend is $3x^3 + 2x^2 - 3x - 28$ and the divisor is $x - 2$.

Why   Since the divisor has more than one term, we must use the long division method to divide the polynomials.

Solution

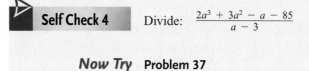

$$\begin{array}{r} 3x^2 + 8x + 13 \\ x - 2 \overline{)\,3x^3 + 2x^2 - 3x - 28} \\ -(3x^3 - 6x^2) \\ \hline 8x^2 - 3x \\ -(8x^2 - 16x) \\ \hline 13x - 28 \\ -(13x - 26) \\ \hline -2 \end{array}$$

The first division: $\frac{3x^3}{x} = 3x^2$.

The second division: $\frac{8x^2}{x} = 8x$.

The third division: $\frac{13x}{x} = 13$.

Subtract: $-28 - (-26) = -2$. The remainder is -2.

This division gives a quotient of $3x^2 + 8x + 13$ and a remainder of -2. It is common to form a fraction with the remainder as the numerator and the divisor as the denominator and to write the result as

$$3x^2 + 8x + 13 + \frac{-2}{x - 2} \quad \text{or} \quad 3x^2 + 8x + 13 - \frac{2}{x - 2}$$

To check, we verify that

$$(x - 2)\left(3x^2 + 8x + 13 + \frac{-2}{x - 2}\right) = 3x^3 + 2x^2 - 3x - 28.$$

> **Success Tip**
> The long division method for polynomials continues until the degree of the remainder is less than the degree of the divisor. Here, the remainder, -2, has degree 0. The divisor, $x - 2$, has degree 1. Therefore, the division process ends.

▶ **Self Check 4**   Divide: $\dfrac{2a^3 + 3a^2 - a - 85}{a - 3}$

Now Try   **Problem 37**

The division method works best when the terms of the divisor and the dividend are written in descending powers of the variable. If the powers in the dividend or divisor are not in descending order, we use the commutative property of addition to write them that way.

EXAMPLE 5   Divide: $(-10x + 12x^3 + x^2 - 3) \div (1 + 3x)$

Strategy   We will write the dividend and divisor in descending powers of x and use the long division method.

Why   It is easier to align like terms in columns when the powers of the variable in the dividend and divisor are written in descending order.

Solution   After writing the terms of the dividend and divisor in descending powers of x, we proceed as follows.

$$
\begin{array}{r}
4x^2 - x - 3 \\
(3x) + 1\overline{)12x^3 + x^2 - 10x - 3} \\
-\underline{(12x^3 + 4x^2)} \downarrow \\
-3x^2 - 10x \\
-\underline{(-3x^2 - x)} \downarrow \\
-9x - 3 \\
-\underline{(-9x - 3)} \\
0
\end{array}
$$

The first division: $\frac{12x^3}{3x} = 4x^2$.

The second division: $\frac{-3x^2}{3x} = -x$.

The third division: $\frac{-9x}{3x} = -3$.

Thus,

$$\frac{-10x + 12x^3 + x^2 - 3}{1 + 3x} = 4x^2 - x - 3$$

Self Check 5   Divide: $2 + 3a\overline{)-4a + 15a^2 + 18a^3 - 4}$

Now Try   **Problems 39 and 43**

4   **Divide Polynomials with Missing Terms.**

If a power of the variable is missing in the dividend, we will insert placeholder terms or leave space. This keeps like terms in the same column, which is necessary when performing the subtraction in vertical form.

EXAMPLE 6   Divide $64x^3 + 1$ by $4x + 1$.

Strategy   The dividend, $64x^3 + 1$, does not have an x^2-term or an x-term. We will insert a $0x^2$ term and a $0x$ term as placeholders and use the long division method.

Why   We insert placeholder terms so that like terms will be aligned in the same column when we subtract.

Solution   After inserting placeholder terms of $0x^2$ and $0x$ in the dividend, we proceed as follows.

$$
\begin{array}{r}
16x^2 - 4x + 1 \\
(4x) + 1\overline{)64x^3 + 0x^2 + 0x + 1} \\
-\underline{(64x^3 + 16x^2)} \downarrow \\
-16x^2 + 0x \\
-\underline{(-16x^2 - 4x)} \downarrow \\
4x + 1 \\
-\underline{(4x + 1)} \\
0
\end{array}
$$

The first division: $\frac{64x^3}{4x} = 16x^2$.

The second division: $\frac{-16x^2}{4x} = -4x$.

The third division: $\frac{4x}{4x} = 1$.

The Language of Algebra
Notice how the instruction to divide by $4x + 1$ translates:

Divide $64x^3 + 1$ by $4x + 1$.

$4x + 1\overline{)64x^3 + 1}$

Thus,

$$\frac{64x^3 + 1}{4x + 1} = 16x^2 - 4x + 1$$

Self Check 6 Divide $125a^3 - 1$ by $5a - 1$.

Now Try Problem 47

EXAMPLE 7 Divide: $\dfrac{-17x^2 + 5x + x^4 + 2}{x^2 - 1 + 4x}$

Strategy We will write the terms of the divisor and the dividend in descending powers of x. Then we will insert any needed placeholder terms and use the long division method.

Why We write the terms of the divisor and the dividend in descending powers of x (and insert necessary placeholder terms) so that like terms will be aligned in the same column when we subtract.

Solution After writing the divisor and the dividend in descending powers of x and inserting $0x^3$ for the missing term in the dividend, we proceed as follows:

$$
\begin{array}{r}
x^2 - 4x \\
(x^2) + 4x - 1\,\overline{)\,x^4 + 0x^3 - 17x^2 + 5x + 2} \\
-(x^4 + 4x^3 - x^2) \\
\hline
-4x^3 - 16x^2 + 5x \\
-(-4x^3 - 16x^2 + 4x) \\
\hline
x + 2
\end{array}
$$

The first division: $\dfrac{x^4}{x^2} = x^2$.

The second division: $\dfrac{-4x^3}{x^2} = -4x$.

This is the remainder.

The degree of the remainder, $x + 2$, is 1 and the degree of the divisor, $x^2 + 4x - 1$, is 2. Since the degree of the remainder is less than the divisor, the division process stops. Thus,

$$
\frac{-17x^2 + 5x + x^4 + 2}{x^2 - 1 + 4x} = x^2 - 4x + \frac{x + 2}{x^2 + 4x - 1}
$$

Self Check 7 Divide: $\dfrac{2a^2 + 3a^3 + a^4 - 6 + a}{a^2 + 1 - 2a}$

Now Try Problem 51

ANSWERS TO SELF CHECKS 1. a. $6y^2$ b. $\dfrac{d^4}{4c^3}$ 2. a. $10h + 3$ b. $2s^3 - \dfrac{s^2t}{11} + \dfrac{4}{t}$ 3. $x^2 + 2x - 1$
4. $2a^2 + 9a + 26 + \dfrac{-7}{a-3}$ or $2a^2 + 9a + 26 - \dfrac{7}{a-3}$ 5. $6a^2 + a - 2$ 6. $25a^2 + 5a + 1$
7. $a^2 + 5a + 11 + \dfrac{18a - 17}{a^2 - 2a + 1}$

STUDY SET
6.5

VOCABULARY

Fill in the blanks.

1. The expression $\frac{18x^7}{9x^4}$ is a monomial divided by a _____.

The expression $\frac{6x^3 - 4x^2 + 8x - 2}{2x^4}$ is a _____ divided by a monomial. The expression $\frac{x^2 - 8x + 12}{x - 6}$ is a trinomial divided by a _____.

2. The powers of x in $2x^4 + 3x^3 + 4x^2 - 7x - 8$ are written in _____ order.

3.

$$
\begin{array}{r}
x - 2 \\
x - 6 \overline{)\, x^2 - 8x - 4} \\
-(x^2 - 6x) \\
\hline
-2x - 4 \\
-(-2x + 12) \\
\hline
-16
\end{array}
$$

4. Since $5x^2 + 6$ is missing an x-term, we insert a _____ $0x$ term in a division and write the polynomial as $5x^2 + 0x + 6$.

CONCEPTS

Fill in the blanks.

5. a. To divide a polynomial by a monomial, divide each _____ of the polynomial by the monomial.

b. $\dfrac{18x + 9}{9} = \dfrac{18x}{} + \dfrac{9}{}$

c. $\dfrac{30x^2 + 12x - 24}{6} = \dfrac{30x^2}{} + \dfrac{12x}{} - \dfrac{24}{}$

6. Divisor · _____ + remainder = dividend

7. Suppose that after dividing $2x^3 + 5x^2 - 11x + 4$ by $2x - 1$, you obtain $x^2 + 3x - 4$. Show how multiplication can be used to check the result.

8. Consider the first step of the division process for

$$2x^2 - 1 \overline{)\, 4x^4 + 0x^3 + 0x^2 + 0x - 1}$$

How many times does $2x^2$ divide $4x^4$?

NOTATION

Complete each solution.

9.

$$
\begin{array}{r}
x + 7 \\
x + 4 \overline{)\, x^2 + 11x + 28} \\
-(+ 4x) \\
\hline
+ 28 \\
-(7x +) \\
\hline
0
\end{array}
$$

10.

$$
\begin{array}{r}
2x - 1 \\
3x + 4 \overline{)\, 6x^2 + 5x - 4} \\
-(6x^2 +) \\
\hline
- 4 \\
-(-3x -) \\
\end{array}
$$

11. If a polynomial is divided by $3a - 2$ and the quotient is $3a^2 + 5$ with a remainder of 6, how do we write the result?

12. A polynomial is divided by $3a - 2$. The quotient is $3a^2 + 5$ with a remainder of -6. Write the answer to the division in two ways.

13. List three ways we can use symbols to write $x^2 - x - 12$ divided by $x - 4$.

14. Is the following statement true or false? Justify your answer.

$$2x^3 - 9 = 2x^3 + 0x^2 + 0x - 9$$

GUIDED PRACTICE

Simplify. Write answers using positive exponents. See Example 1.

15. $\dfrac{4x^2y^3}{8x^5y^2}$

16. $\dfrac{25x^4y^7}{5xy^9}$

17. $\dfrac{33a^2b^2}{44a^4b^2}$

18. $\dfrac{63a^4}{81a^6b^3}$

Perform each division. See Example 2.

19. $\dfrac{4x^4 + 6x}{2}$

20. $\dfrac{11a^3 - 99a^2}{11}$

21. $\dfrac{4x^2 - x^3}{6x}$

22. $\dfrac{5y^4 + 45y^3}{15y^2}$

23. $\dfrac{54a^3y^2 - 18a^4y^3}{27a^2y^2}$

24. $\dfrac{12x^2y^3 + x^3y^2}{6xy}$

25. $\dfrac{24x^6y^7 - 12x^5y^{12} + 36xy}{-48x^2y^3}$

26. $\dfrac{9x^4y^3 + 18x^2y - 27xy^4}{-9x^3y^3}$

Perform each division. See Objective 3 and Example 3.

27. $\dfrac{x^2 + 5x + 6}{x + 3}$

28. $\dfrac{x^2 + 10x + 21}{x + 7}$

29. $\dfrac{x^2 - 10x + 21}{x - 7}$

30. $\dfrac{x^2 - 5x + 6}{x - 3}$

31. $\dfrac{16x^2 - 16x - 5}{4x + 1}$

32. $\dfrac{6x^2 - x - 12}{2x - 3}$

33. $\dfrac{6x^3 - x^2 - 6x - 9}{2x - 3}$

34. $\dfrac{16x^3 + 16x^2 - 9x - 5}{4x + 5}$

Perform each division. See Example 4.

35. $\dfrac{t^3 + 8t^2 + 13t + 9}{t + 6}$

36. $\dfrac{s^3 + 10s^2 + 17s + 12}{s + 8}$

37. $\dfrac{6x^3 + 11x^2 - 19x - 2}{3x - 2}$

38. $\dfrac{6x^3 + 11x^2 - 9x - 20}{2x + 3}$

Perform each division. See Example 5.

39. $(2a + 1 + a^2) \div (a + 1)$

40. $(a - 15 + 6a^2) \div (2a - 3)$

41. $(6y - 4 + 10y^2) \div (5y - 2)$

42. $(-10x + x^2 + 16) \div (x - 2)$

43. $\dfrac{3x^2 + 9x^3 + 4x + 4}{2 + 3x}$

44. $\dfrac{3 + 5x + 6x^3 + 11x^2}{3 + 2x}$

45. $\dfrac{13x + 16x^4 + 3x^2 + 3}{3 + 4x}$

46. $\dfrac{4x^3 - 12x^2 + 17x - 12}{2x - 3}$

Perform each division. See Example 6.

47. Divide $8a^3 + 1$ by $2a + 1$.

48. Divide $27a^3 - 8$ by $3a - 2$.

49. Divide $15a^3 - 29a^2 + 16$ by $3a - 4$.

50. Divide $15c^3 - 19c^2 + 4$ by $5c + 2$.

Perform each division. See Example 7.

51. Divide $7x^2 - x + x^4 + 5x^3 - 12$ by $x^2 - 3 + 2x$.

52. Divide $x^4 + 2 + 4x^2 + 3x + 2x^3$ by $x^2 + 2 + x$.

53. $3x^2 - 7x + 4 \overline{)7x - 1 + 6x^3 - 5x^2}$

54. $3m^2 - m + 4 \overline{)5m - 11 + 9m^3 - 6m^2 + 5m}$

TRY IT YOURSELF

Perform each division.

55. $y - 2 \overline{)-24y + 24 + 6y^2}$

56. $a - 3 \overline{)54 - 21a + a^2}$

57. $\dfrac{4a^3 + a^2 - 3a + 7}{a + 1}$

58. $\dfrac{3x^3 - 2x^2 + x - 6}{x - 1}$

59. $(x^6 - x^4 + 2x^2 - 8) \div (x^2 - 2)$

60. $(x^3 + 3x + 5x^2 + 6 + x^4) \div (x^2 + 3)$

61. $\dfrac{5a^5 - 10a}{25a^3}$

62. $\dfrac{24b^7 - 32b^2}{16b^5}$

63. Divide $2s^2 + 13s + 5$ by $2s + 3$.

64. Divide $4s^2 + 6s + 1$ by $2s - 1$.

65. $\dfrac{40m^{17}n^{20}}{35m^{15}n^{30}}$

66. $\dfrac{34s^{30}t^{15}}{14s^{40}t^{12}}$

67. Divide $m^3 - 4m^2 + 2m - 1$ by $m^2 + 1$.

68. Divide $6m^3 + 2m^2 + m + 4$ by $2m^2 - 3$.

69. $(y^3 - 64) \div (y - 4)$

70. $(8w^3 + 1) \div (2w + 1)$

71. $\dfrac{a^8 + a^6 - 4a^4 + 5a^2 - 3}{a^4 + 2a^2 - 3}$

72. $\dfrac{2x^4 + 3x^3 + 3x^2 - 5x - 3}{2x^2 - x - 1}$

73. $\dfrac{40x^3z^2 - 8x^2z - 4z}{4xz}$

74. $\dfrac{22a^2b^2 - 18a^2b - 52a}{2ab}$

75. $\dfrac{x^5 + 3x + 2}{x^3 + 1 + 2x}$

76. $\dfrac{9a^4 + 6a^3 + 55a^2 + 18a + 81}{3a^2 + a + 9}$

77. Divide $11x^2 - 4x + 8x^4 - 6x^3 + 3$ by $3 + 4x^2 - x$.

78. Divide $4x - 2 + x^4 - x^2 - x^3$ by $x - 1 + x^2$.

79. $\dfrac{15x^2 + 9x - 3}{27}$

80. $\dfrac{8x^2 + 12x + 9}{6}$

81. $x^2 + 3 \overline{)x^6 + 2x^4 - 6x^2 - 9}$

82. $m^2 - 2 \overline{)m^4 - 3m^2 + 10}$

APPLICATIONS

83. ADVERTISING Find the length of one of the longer sides of the rectangular-shaped billboard if its area is represented by $x^3 - 4x^2 + x + 6$.

84. MASONRY The trowel shown is in the shape of an isosceles triangle. Find the height if its area is represented by $6 + 18t + t^2 + 3t^3$.

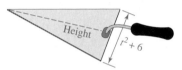

85. WINTER TRAVEL Complete the following table, which lists the rate (mph), time traveled (hr), and distance traveled (mi) by an Alaskan trail guide using two different means of travel.

	r	$\cdot$	t	$=$	d
Dog sled			$4x + 7$		$12x^2 + 13x - 14$
Snowshoes	$3x + 4$				$3x^2 + 19x + 20$

86. PRICING Complete the table for two items sold at a produce store.

	Price per lb	$\cdot$	Number of lb	$=$	Total Value
Cashews	$x^2 + 2x + 4$				$x^4 + 4x^2 + 16$
Sunflower seeds			$x^2 + 6$		$x^4 - x^2 - 42$

WRITING

87. Explain how to check to determine if
$$(3x^2 - 15) \div (x + 3) = 3x - 9 + \frac{12}{x + 3}.$$

88. Explain the error in the following long division. Use the word *degree* in your answer.

$$\begin{array}{r} 9x + 3 \\ x + \\ 3x + 1 \\ \hline 3x + 1\overline{)3x^2 + 10x + 3} \\ -(3x^2 + x) \\ \hline 9x + 3 \end{array}$$

REVIEW

Simplify each expression.

89. $2(x^2 + 4x - 1) + 3(2x^2 - 2x + 2)$

90. $3(2a^2 - 3a + 2) - 4(2a^2 + 4a - 7)$

91. $-2(3y^3 - 2y + 7) - (y^2 + 2y - 4) + 4(y^3 + 2y - 1)$

92. $3(4y^3 + 3y - 2) + 2(3y^2 - y + 3) - 5(2y^3 - y^2 - 2)$

CHALLENGE PROBLEMS

Perform each division.

93. $\left(3c^2 - \dfrac{7}{4}c - 3\right) \div (4c + 3)$

94. Divide $x^5 - 1$ by $x - 1$.

95. $\dfrac{c^4 - c^2d^2 + 10c^2 - 6d^2 + 23}{c^2 + 6}$

96. $\dfrac{0.03a^2 + 0.17a + 0.1}{0.03a + 0.02}$

(*Hint:* Think of a way to simplify the division.)

97. $x - 2\overline{)9.8x^2 - 3.2x - 69.3}$

98. $2.5x - 3.7\overline{)-22.25x^2 - 38.9x - 16.65}$

SECTION 6.6
Synthetic Division

Objectives

1 Perform synthetic division.

2 Use the remainder theorem to evaluate polynomials.

3 Use the factor theorem to factor polynomials.

We have discussed how to divide polynomials by polynomials using a long division process. We will now discuss a shortcut method, called **synthetic division,** that we can use to divide a polynomial by a binomial of the form $x - k$.

1 Perform Synthetic Division.

To see how synthetic division works, we consider the division of $4x^3 - 5x^2 - 11x + 20$ by $x - 2$.

$$
\begin{array}{r}
4x^2 + \ 3x - 5 \\
x - 2{\overline{\smash{\big)}\,4x^3 - 5x^2 - 11x + 20}} \\
\underline{-(4x^3 - 8x^2)} \\
3x^2 - 11x \\
\underline{-(3x^2 - \ 6x)} \\
- 5x + 20 \\
\underline{-(- 5x + 10)} \\
10 \ \text{(remainder)}
\end{array}
$$

$$
\begin{array}{r}
4 \ \ 3 - \ 5 \\
1 - 2{\overline{\smash{\big)}\,4 - 5 - 11 \ \ \ 20}} \\
\underline{-(4 - 8)} \\
3 - 11 \\
\underline{-(3 - \ 6)} \\
- 5 \ \ 20 \\
\underline{-(- 5 \ \ 10)} \\
10 \ \text{(remainder)}
\end{array}
$$

On the left is the long division, and on the right is the same division with the variables and their exponents removed and the coefficients of the quotient moved to the left. We can remember the various powers of x without actually writing them, because the exponents of the terms in the divisor, dividend, and quotient were written in descending order.

We can further shorten the version on the right. The numbers printed in color need not be written, because they are duplicates of the numbers above them. If we remember to perform subtraction at the proper times, the minus symbols and the parentheses can also be dropped. Thus, we can write the division in the following form:

$$
\begin{array}{r}
4 \ \ \ 3 - \ 5 \\
1 - 2{\overline{\smash{\big)}\,4 - 5 - 11 \ \ \ 20}} \\
\underline{- \ 8} \\
3 \\
\underline{- \ 6} \\
- \ 5 \\
\underline{} \\
10 \\
\underline{10}
\end{array}
$$

We can shorten the process further by compressing the work vertically and eliminating the 1 (the coefficient of x in the divisor):

$$
\begin{array}{r}
4 \ \ \ \ 3 \ \ \ -5 \\
-2{\overline{\smash{\big)}\,4 \ \ -5 \ \ -11 \ \ \ 20}} \\
\underline{-8 \ \ \ -6 \ \ \ 10} \\
3 \ \ \ -5 \ \ \ 10
\end{array}
$$

If we write the 4 in the quotient on the bottom line, the bottom line gives the coefficients of the quotient and the remainder. If we eliminate the top line, the division appears as follows:

$$\begin{array}{r|rrrr} -2 & 4 & -5 & -11 & 20 \\ & & -8 & -6 & 10 \\ \hline & 4 & 3 & -5 & 10 \end{array}$$

The bottom line is obtained by subtracting the middle line from the top line. If we replace the -2 in the divisor by 2, the division process will reverse the signs of every entry in the middle line, and then the bottom line can be obtained by addition. This gives the final form of the synthetic division.

$$\begin{array}{r|rrrr} 2 & 4 & -5 & -11 & 20 \\ & & 8 & 6 & -10 \\ \hline & 4 & 3 & -5 & 10 \end{array}$$

These are the coefficients of the dividend.

These are the coefficients of the quotient and the remainder.

$$4x^2 + 3x - 5 + \dfrac{10}{x - 2}$$ Read the result from the bottom row.

Thus,

$$\frac{4x^3 - 5x^2 - 11x + 20}{x - 2} = 4x^2 + 3x - 5 + \frac{10}{x - 2}$$

The Language of Algebra
Synthetic division is used to divide a polynomial by a binomial of the form $x - k$. We call k the **synthetic divisor.** In this example, we are dividing by $x - 2$, so k is 2.

EXAMPLE 1   Divide: $(6x^2 - 29x - 5) \div (x - 5)$

Strategy   We will use synthetic division to divide.

Why   When a polynomial is divided by a binomial of the form $x - k$, synthetic division produces the quotient (and possible remainder) with less effort than long division.

Solution   We write the coefficients in the dividend and the 5 in the divisor in the following form:

Since we are dividing the polynomial by $x - 5$, the synthetic divisor is 5. →

$$\begin{array}{r|rrr} 5 & 6 & -29 & -5 \end{array}$$ ←This represents the dividend $6x^2 - 29x - 5$.

Then we follow these steps:

$$\begin{array}{r|rrr} 5 & 6 & -29 & -5 \\ & \downarrow & & \\ \hline & 6 & & \end{array}$$ Begin by bringing down the 6.

$$\begin{array}{r|rrr} 5 & 6 & -29 & -5 \\ & & 30 & \\ \hline & 6 & & \end{array}$$ Multiply 5 by 6 to get 30.

$$\begin{array}{r|rrr} 5 & 6 & -29 & -5 \\ & & 30 & \\ \hline & 6 & 1 & \end{array}$$ Add -29 and 30 to get 1.

$$\begin{array}{r|rrr} 5 & 6 & -29 & -5 \\ & & 30 & 5 \\ \hline & 6 & 1 & \end{array}$$ Multiply 1 by 5 to get 5.

$$\begin{array}{r|rrr} 5 & 6 & -29 & -5 \\ & & 30 & 5 \\ \hline & 6 & 1 & 0 \end{array}$$ Add -5 and 5 to get 0.

The remainder is 0.

Success Tip
In this process, numbers below the line are multiplied by the synthetic divisor and that product is carried above the line to the next column. Numbers above the horizontal line are added.

The numbers 6 and 1 represent the quotient $6x + 1$, and 0 is the remainder. Thus,

$$\frac{6x^2 - 29x - 5}{x - 5} = 6x + 1$$

Self Check 1   Use synthetic division to divide: $5x^2 - 4x - 33$ by $x - 3$

Now Try   **Problem 15**

EXAMPLE 2   Divide: $\dfrac{x^3 + x^2 - 1}{x - 3}$

Strategy   We will use synthetic division to divide.

Why   When a polynomial is divided by a binomial of the form $x - k$, synthetic division produces the quotient (and possible remainder) with less effort than long division.

Solution   We begin by writing

$$\underline{3|}\quad 1\quad 1\quad 0\quad -1\qquad \textit{Write 0 for the coefficient of x, the missing term.}$$

and complete the division as follows.

Multiply, then add.   Multiply, then add.   Multiply, then add.

The numbers 1, 4, and 12 represent the quotient $x^2 + 4x + 12$ and 35 is the remainder. Thus,

$$\frac{x^3 + x^2 - 1}{x - 3} = x^2 + 4x + 12 + \frac{35}{x - 3}$$

Self Check 2   Use synthetic division to divide: $\dfrac{x^3 + 3x - 62}{x - 4}$

Now Try   **Problems 21 and 25**

EXAMPLE 3   Divide $5a^2 + 6a^3 + 2 - 4a$ by $a + 2$.

Strategy   Here, the dividend and the divisor are polynomials in the variable a. Since $a + 2$ can be written as $a - (-2)$, the divisor can be written in the form $a - k$ and we can use synthetic division.

Why   We will use synthetic division because it produces the quotient (and possible remainder) with less effort than long division.

Solution   First, we write the dividend with the powers of x in descending order.

$$6a^3 + 5a^2 - 4a + 2$$

Then we write the divisor in $a - k$ form: $a - (-2)$. Thus, $k = -2$. Using synthetic division, we begin by writing

This represents division by $a + 2$. →

$$-2 \underline{|\quad 6 \quad\quad 5 \quad -4 \quad\quad 2}$$

and complete the division.

$$\begin{array}{r}
-2 \underline{|\quad 6 \quad\quad 5 \quad -4 \quad\quad 2} \\
\underline{-12 \quad 14 \quad -20} \\
6 \quad -7 \quad 10 \quad -18
\end{array}$$

The remainder is negative.

Notation
Because the remainder is negative, we can also write the result as
$$6a^2 - 7a + 10 - \frac{18}{a + 2}$$

Thus,

$$\frac{5a^2 + 6a^3 + 2 - 4a}{a + 2} = 6a^2 - 7a + 10 + \frac{-18}{a + 2}$$

 Self Check 3   Use synthetic division to divide $2a - 4a^2 + 3a^3 - 3$ by $a + 1$.

Now Try Problems 29 and 33

2 **Use the Remainder Theorem to Evaluate Polynomials.**

Synthetic division is important because of the **remainder theorem**.

Remainder Theorem

If a polynomial $P(x)$ is divided by $x - k$, the remainder is $P(k)$.

It follows from the remainder theorem that we can evaluate polynomials using synthetic division. We illustrate this in the following example.

EXAMPLE 4   Let $P(x) = 2x^3 - 3x^2 - 2x + 1$. Find each value: **a.** $P(3)$
b. the remainder when $P(x)$ is divided by $x - 3$

Strategy   To find $P(3)$, we will substitute 3 for x in $P(x)$ and simplify. To find the remainder when $P(x)$ is divided by $x - 3$, we will use synthetic division.

Why   After finding the remainder in two ways, we will see that the method using synthetic division is easier.

Solution   To find $P(3)$ we evaluate the function for $x = 3$.

Notation
Naming the function with the letter P, instead of f, stresses that we are working with a polynomial function.

a. $P(x) = 2x^3 - 3x^2 - 2x + 1$

$\quad P(3) = 2(3)^3 - 3(3)^2 - 2(3) + 1$ Substitute 3 for x.

$\quad\quad\quad = 2(27) - 3(9) - 6 + 1$

$\quad\quad\quad = 54 - 27 - 6 + 1$

$\quad\quad\quad = 22$

Thus, $P(3) = 22$.

Success Tip
It is often easier to find $P(k)$ by using synthetic division than by substituting k for x in $P(x)$. This is especially true if k is a decimal.

b. We use synthetic division to find the remainder when $2x^3 - 3x^2 - 2x + 1$ is divided by $x - 3$.

$$\begin{array}{r|rrrr} 3 & 2 & -3 & -2 & 1 \\ & & 6 & 9 & 21 \\ \hline & 2 & 3 & 7 & 22 \end{array}$$

Thus, the remainder is 22.

The same results in parts (a) and (b) show that instead of substituting 3 for x in $P(x) = 2x^3 - 3x^2 - 2x + 1$, we can divide the polynomial $2x^3 - 3x^2 - 2x + 1$ by $x - 3$ to find $P(3)$.

Self Check 4 Let $P(x) = 5x^3 - 3x^2 + x + 6$. Find each value:
a. $P(1)$ **b.** use synthetic division to find the remainder when $P(x)$ is divided by $x - 1$

Now Try **Problem 41**

3 Use the Factor Theorem to Factor Polynomials.

If two quantities are multiplied, each is called a *factor* of the product. Thus, $x - 2$ is a factor of $6x - 12$, because $6(x - 2) = 6x - 12$. A theorem, called the *factor theorem,* tells us how to find one factor of a polynomial if the remainder of a certain division is 0.

Factor Theorem

If $P(x)$ is a polynomial in x, then

$$P(k) = 0 \text{ if and only if } x - k \text{ is a factor of } P(x)$$

If $P(x)$ is a polynomial in x and if $P(k) = 0$, k is called a **zero of the polynomial function**.

EXAMPLE 5 Let $P(x) = 3x^3 - 5x^2 + 3x - 10$. Show that **a.** $P(2) = 0$ **b.** $x - 2$ is a factor of $P(x)$

Strategy We will substitute 2 for x in $P(x)$ to verify that $P(2) = 0$. We will then use synthetic division to divide $P(x)$ by $x - 2$.

Why Since $P(x) = 0$, the division has a remainder of 0. This means that the divisor and the quotient are factors of the dividend.

Solution

The Language of Algebra
The phrase *if and only if* in the factor theorem means:

If $P(2) = 0$, then $x - 2$ is a factor of $P(x)$
and
If $x - 2$ is a factor of $P(x)$, then $P(2) = 0$.

a. Use the remainder theorem to evaluate $P(2)$ by dividing $P(x) = 3x^3 - 5x^2 + 3x - 10$ by $x - 2$.

$$\begin{array}{r|rrrr} 2 & 3 & -5 & 3 & -10 \\ & & 6 & 2 & 10 \\ \hline & 3 & 1 & 5 & 0 \end{array}$$

The remainder in this division is 0. By the remainder theorem, the remainder is $P(2)$. Thus, $P(2) = 0$, and 2 is a zero of the polynomial.

b. Because the remainder is 0, the numbers 3, 1, and 5 in the synthetic division in part (a) represent the quotient $3x^2 + x + 5$. Thus,

$$\underbrace{(x - 2)}_{\text{Divisor}} \cdot \underbrace{(3x^2 + x + 5)}_{\text{quotient}} + \underbrace{0}_{\text{remainder}} = \underbrace{3x^3 - 5x^2 + 3x - 10}_{\text{the dividend, } P(x)}$$

or

$$(x - 2)(3x^2 + x + 5) = 3x^3 - 5x^2 + 3x - 10$$

Thus, $x - 2$ is a factor of $3x^3 - 5x^2 + 3x - 10$.

Self Check 5   Let $P(x) = x^3 - 4x^2 + x + 6$. Show that $x + 1$ is a factor of $P(x)$ using synthetic division.

Now Try   **Problem 65**

The result in Example 5 is true, because the remainder, $P(2)$, is 0. If the remainder had not been 0, then $x - 2$ would not have been a factor of $P(x)$.

Using Your Calculator

Approximating Zeros of Polynomials

We can use a graphing calculator to approximate the real zeros of a polynomial function. For example, to find the real zeros of $f(x) = 2x^3 - 6x^2 + 7x - 21$, we graph the function as in the figure.

It is clear from the display that the function f has a zero at $x = 3$.

$$f(3) = 2(3)^3 - 6(3)^2 + 7(3) - 21 \qquad \text{Substitute 3 for x.}$$
$$= 2(27) - 6(9) + 21 - 21$$
$$= 0$$

From the factor theorem, we know that $x - 3$ is a factor of the polynomial. To find the other factor, we can synthetically divide by 3.

$$\begin{array}{r|rrrr} 3 & 2 & -6 & 7 & -21 \\ & & 6 & 0 & 21 \\ \hline & 2 & 0 & 7 & 0 \end{array}$$

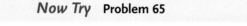

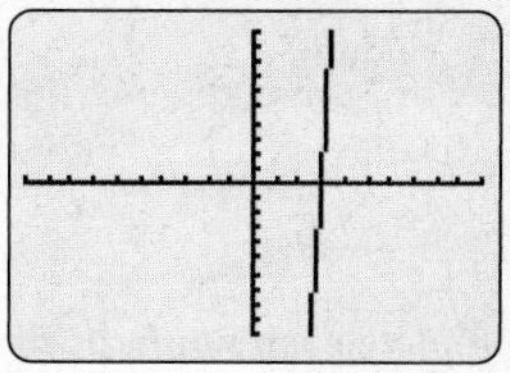

Thus, $f(x) = (x - 3)(2x^2 + 7)$. Since $2x^2 + 7$ cannot be factored over the real numbers, we can conclude that 3 is the only real zero of the polynomial function.

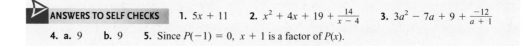

ANSWERS TO SELF CHECKS   **1.** $5x + 11$ **2.** $x^2 + 4x + 19 + \frac{14}{x - 4}$ **3.** $3a^2 - 7a + 9 + \frac{-12}{a + 1}$
4. a. 9   **b.** 9   **5.** Since $P(-1) = 0$, $x + 1$ is a factor of $P(x)$.

STUDY SET
6.6

VOCABULARY

Fill in the blanks.

1. The method of dividing $x^2 + 2x - 9$ by $x - 4$ shown below is called _____ division.

$$
\begin{array}{r|rrr}
4 & 1 & 2 & -9 \\
  &   & 4 & 24 \\
\hline
  & 1 & 6 & 15
\end{array}
$$

2. Synthetic division is used to divide a polynomial by a _____ of the form $x - k$.
3. In Exercise 1, the synthetic _____ is 4.
4. By the _____ theorem, if a polynomial $P(x)$ is divided by $x - k$, the remainder is $P(k)$.
5. The factor _____ tells us how to find one factor of a polynomial if the remainder of a certain division is 0.
6. If $P(x)$ is a polynomial and if $P(k) = 0$, then k is called a _____ of the polynomial.

CONCEPTS

7. **a.** What division is represented below?

   b. What is the answer?

$$
\begin{array}{r|rrrr}
-2 & 5 & 0 & 1 & -3 \\
   &   & -10 & 20 & -42 \\
\hline
   & 5 & -10 & 21 & -45
\end{array}
$$

Fill in the blanks.

8. In the synthetic division process, numbers below the line are _____ by the synthetic divisor and that product is carried above the line to the next column. Numbers above the horizontal line are _____.
9. Rather than substituting 8 for x in $P(x) = 6x^3 - x^2 - 17x + 9$, we can divide the polynomial _____ by _____ to find $P(8)$.
10. For $P(x) = x^3 - 4x^2 + x + 6$, suppose we know that $P(3) = 0$. Then _____ is a factor of $x^3 - 4x^2 + x + 6$.

NOTATION

Complete each synthetic division.

11. Divide $6x^3 + x^2 - 23x + 2$ by $x - 2$.

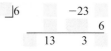

12. Divide $2x^3 - 4x^2 - 25x + 15$ by $x + 3$.

$$
\begin{array}{r|rrrr}
\rule{0.5cm}{0pt} & 2 & -4 & \rule{0.5cm}{0pt} & 15 \\
   &   &   & 30 & \\
\hline
   &   &   &   & 0
\end{array}
$$

GUIDED PRACTICE

Use synthetic division to perform each division. See Example 1.

13. $(2x^2 + x - 3) \div (x - 1)$
14. $(4x^2 - 5x - 6) \div (x - 2)$
15. $(5x^2 - 27x + 10) \div (x - 5)$
16. $(6x^2 - 29x + 20) \div (x - 4)$
17. $(3x^2 - 13x + 12) \div (x - 3)$
18. $(2x^2 - 23x + 63) \div (x - 7)$
19. $(5x^2 - 24x - 36) \div (x - 6)$
20. $(3x^2 - 14x - 24) \div (x - 6)$

Use synthetic division to perform each division. See Example 2.

21. $\dfrac{a^3 - 3a^2 + 4}{a - 2}$

22. $\dfrac{a^3 - 2a^2 - 9}{a - 3}$

23. $\dfrac{3a^3 - 47a - 4}{a - 4}$

24. $\dfrac{2a^3 - 7a + 5}{a - 1}$

25. $\dfrac{3b^3 - 31b + 13}{b - 3}$

26. $\dfrac{4c^3 - 107c + 37}{c - 5}$

27. $\dfrac{4t^3 - t - 18}{t - 2}$

28. $\dfrac{m^3 + 2m + 5}{m - 2}$

Use synthetic division to perform each division. See Example 3.

29. Divide $x - 4x^2 + x^3 + 6$ by $x + 1$.
30. Divide $4x^2 - 10x + 12 + x^3$ by $x + 6$.
31. Divide $20x^2 - 36x - 42 + 3x^3$ by $x + 8$.
32. Divide $3x - 6x^2 + 5x^3 + 10$ by $x + 1$.
33. Divide $8 - 3x + 7x^2 + 2x^3$ by $x + 5$.
34. Divide $1 - 4x + 7x^2 + 3x^3$ by $x + 3$.
35. Divide $27 + x^3 - 17x + 8x^2$ by $x + 10$.
36. Divide $1 + x^3 - 23x + 5x^2$ by $x + 8$.

Use a calculator and synthetic division to perform each division. See Examples 1–3.

37. $\dfrac{7.2x^2 - 2.1x + 0.5}{x - 0.2}$

38. $\dfrac{2.7x^2 + x - 5.2}{x + 1.7}$

39. $\dfrac{9x^3 - 25}{x + 57}$

40. $\dfrac{0.5x^3 + x}{x - 2.3}$

Let $P(x) = 2x^3 - 4x^2 + 2x - 1$. Evaluate $P(x)$ by substituting the given value of x into the polynomial and simplifying. Then evaluate the polynomial by using the remainder theorem and synthetic division. See Example 4.

41. $P(1)$ **42.** $P(2)$

43. $P(-2)$ **44.** $P(-1)$

45. $P(3)$ **46.** $P(-4)$

47. $P(0)$ **48.** $P(4)$

Let $Q(x) = x^4 - 3x^3 + 2x^2 + x - 3$. Evaluate $Q(x)$ by substituting the given value of x into the polynomial and simplifying. Then evaluate the polynomial by using the remainder theorem and synthetic division. See Example 4.

49. $Q(-1)$ **50.** $Q(1)$

51. $Q(2)$ **52.** $Q(-2)$

53. $Q(3)$ **54.** $Q(0)$

55. $Q(-3)$ **56.** $Q(-4)$

Use the remainder theorem and synthetic division to find each function value. See Example 4.

57. $P(x) = x^3 - 4x^2 + x - 2$; find $P(2)$

58. $P(x) = x^3 - 3x^2 + x + 1$; find $P(1)$

59. $P(x) = 2x^3 + x + 2$; find $P(3)$

60. $P(x) = x^3 + x^2 + 1$; find $P(-2)$

61. $P(x) = x^4 - 2x^3 + x^2 - 3x + 2$; find $P(-2)$

62. $P(x) = x^5 + 3x^4 - x^2 + 1$; find $P(-1)$

63. $P(x) = 3x^5 + 1$; find $P\left(-\dfrac{1}{2}\right)$

64. $P(x) = 5x^7 - 7x^4 + x^2 + 1$; find $P(2)$

Use the factor theorem and determine whether the first expression is a factor of $P(x)$. See Example 5.

65. $x - 3$; $P(x) = x^3 - 3x^2 + 5x - 15$

66. $x + 1$; $P(x) = x^3 + 2x^2 - 2x - 3$
(*Hint:* Write $x + 1$ as $x - (-1)$.)

67. $x + 2$; $P(x) = 3x^2 - 7x + 4$
(*Hint:* Write as $x - (-2)$.)

68. x; $P(x) = 7x^3 - 5x^2 - 8x$
(*Hint:* $x = x - 0$.)

TRY IT YOURSELF

Use synthetic division to perform each division.

69. $\dfrac{5x^2 + 4 + 6x^3}{x + 1}$

70. $\dfrac{-4 + 3x^2 - x}{x - 4}$

71. $(x^2 - 5x + 14) \div (x + 2)$ **72.** $(x^2 + 13x + 42) \div (x + 6)$

73. Divide $a^5 - 1$ by $a - 1$.

74. Divide $b^4 - 81$ by $b - 3$.

75. $\dfrac{-6c^5 + 14c^4 + 38c^3 + 4c^2 + 25c - 36}{c - 4}$

76. $\dfrac{-5x^5 + 4x^4 + 30x^3 + 2x^2 + 20x + 3}{x - 3}$

77. $\dfrac{9a^3 + 3a^2 - 21a - 7}{a + \dfrac{1}{3}}$ **78.** $\dfrac{8t^3 - 4t^2 + 2t - 1}{t - \dfrac{1}{2}}$

79. $\dfrac{4x^4 + 12x^3 - x^2 - x + 12}{x + 3}$

80. $\dfrac{x^4 - 9x^3 + x^2 - 7x - 20}{x - 9}$

81. $\dfrac{3x^3 - 25x^2 + 10x - 16}{x - 8}$

82. $\dfrac{2x^3 + 3x^2 - 8x + 3}{x + 3}$

83. $(2x^3 - 50 - 16x^2 - 35x) \div (x - 10)$

84. $(m^3 - m^2 - m - 1) \div (m - 1)$

85. $(4x^3 - 1 + 5x^2) \div (x + 2)$

86. $(t^3 + t^2 + t + 2) \div (t + 1)$

87. Divide $8a^3 - 10a^2 - 32a - 15$ by $a + \dfrac{3}{4}$.

88. Divide $4a^3 - 2a^2 - 18a - 9$ by $a + \dfrac{3}{2}$.

WRITING

89. When dividing a polynomial by a binomial of the form $x - k$, synthetic division is considered to be faster than long division. Explain why.

90. Let $P(x) = x^3 - 6x^2 - 9x + 4$. You now know two ways to find $P(6)$. What are they? Which method do you prefer?

91. Explain the factor theorem.

92. This section includes a feature entitled *Using Your Calculator: Approximating Zeros of Polynomials*. What is a *zero* of a polynomial?

REVIEW

Evaluate each expression for $x = -3$, $y = -5$, and $z = 0$.

93. $x^2 z(y^3 - z)$

94. $|y^3 - z|$

95. $\dfrac{x - y^2}{2y - 1 + x}$

96. $\dfrac{2y + 1}{x} - x$

CHALLENGE PROBLEMS

Suppose that $P(x) = x^{100} - x^{99} + x^{98} - x^{97} + \cdots + x^2 - x + 1$.

97. Find the remainder when $P(x)$ is divided by $x - 1$.

98. Find the remainder when $P(x)$ is divided by $x + 1$.

99. Find 2^6 by using synthetic division to evaluate the polynomial $P(x) = x^6$ at $x = 2$.

100. Find $(-3)^5$ by using synthetic division to evaluate the polynomial $P(x) = x^5$ at $x = -3$.

SECTION 6.7
Solving Rational Equations

Objectives

1. Solve rational equations.
2. Solve rational equations with extraneous solutions.
3. Solve formulas for a specified variable.

In Chapter 1, we solved equations such as $\frac{1}{6}x + \frac{5}{2} = \frac{1}{3}$ by multiplying both sides by the LCD. With this approach, the equation that results is equivalent to the original equation, but easier to solve because it is cleared of fractions.

In this section, we will extend the fraction-clearing strategy to solve another type of equation, called a *rational equation*.

Solve Rational Equations.

If an equation contains one or more rational expressions, it is called a **rational equation.** Rational equations often have a variable in a denominator. Some examples are:

$$\frac{3}{5} + \frac{7}{x + 2} = 2, \qquad \frac{x + 3}{x - 3} = \frac{2}{x^2 - 4}, \qquad \text{and} \qquad \frac{-x^2 + 10}{x^2 - 1} + \frac{3x}{x - 1} = \frac{2x}{x + 1}$$

To solve a rational equation, we find all the values of the variable that make the equation true. Any value of the variable that makes a denominator in a rational equation equal to 0 cannot be a solution of the equation. Such a number must be rejected, because division by 0 is undefined.

EXAMPLE 1   Solve: $\dfrac{3}{5} + \dfrac{7}{x + 2} = 2$

Strategy   This equation contains a rational expression that has a variable in the denominator. We begin by asking, "What value(s) of x make that denominator 0?"

Why   If a number makes the denominator of a rational expression 0, that number cannot be a solution of the equation because division by 0 is undefined.

Solution   We note that x cannot be -2, because this would produce a 0 in the denominator of $\dfrac{7}{x + 2}$.

Since the denominators of the rational expressions in the equation are 5 and $x + 2$, we multiply both sides by the LCD, $5(x + 2)$, to clear the equation of fractions.

Success Tip

To *simplify the expression*
$\frac{3}{5} + \frac{7}{x+2}$, we build each fraction
to have the LCD $5(x + 2)$, add the
numerators, and write the sum
over the LCD.

To *solve the equation*
$\frac{3}{5} + \frac{7}{x+2} = 2$, we multiply both
sides by the LCD $5(x + 2)$ to
eliminate the denominators.

$$\frac{3}{5} + \frac{7}{x + 2} = 2 \qquad \text{This is the equation to solve.}$$

$$5(x + 2)\left(\frac{3}{5} + \frac{7}{x + 2}\right) = 5(x + 2)(2) \qquad \begin{array}{l}\text{Write each side of the equation} \\ \text{within parentheses and then} \\ \text{multiply both sides by the LCD.}\end{array}$$

$$5(x + 2)\left(\frac{3}{5}\right) + 5(x + 2)\left(\frac{7}{x + 2}\right) = 5(x + 2)(2) \qquad \begin{array}{l}\text{On the left side, distribute the} \\ \text{multiplication by } 5(x + 2).\end{array}$$

$$\overset{1}{\cancel{5}}(x + 2)\left(\frac{3}{\cancel{5}}\right) + 5\overset{1}{\cancel{(x + 2)}}\left(\frac{7}{\cancel{(x + 2)}}\right) = 5(x + 2)(2) \qquad \begin{array}{l}\text{On the left side, simplify: } \frac{5}{5} = 1 \text{ and} \\ \frac{x + 2}{x + 2} = 1.\end{array}$$

$$3(x + 2) + 5(7) = 10(x + 2) \qquad \text{Simplify each side.}$$

The resulting equation does not contain any fractions. We now solve this linear equation for x.

$$3x + 6 + 35 = 10x + 20 \qquad \text{Use the distributive property and simplify.}$$

$$3x + 41 = 10x + 20 \qquad \text{Combine like terms.}$$

$$-7x = -21 \qquad \text{Subtract 10x and 41 from both sides.}$$

$$x = 3 \qquad \text{Divide both sides by } -7.$$

The solution is 3 and the solution set is $\{3\}$. To check, we substitute 3 for x in the original equation and simplify:

Check: $\dfrac{3}{5} + \dfrac{7}{x + 2} = 2$

$$\frac{3}{5} + \frac{7}{3 + 2} \overset{?}{=} 2$$

$$\frac{3}{5} + \frac{7}{5} \overset{?}{=} 2$$

$$2 = 2 \qquad \text{True}$$

Self Check 1   Solve: $\dfrac{2}{5} + \dfrac{8}{x - 4} = 2$

Now Try **Problem 13**

Using Your Calculator

Solving Rational Equations Graphically

To use a graphing calculator to solve $\frac{3}{5} + \frac{7}{x + 2} = 2$, we graph the functions $f(x) = \frac{3}{5} + \frac{7}{x + 2}$ and $g(x) = 2$. If we trace and move the cursor closer to the intersection point of the two graphs, we will get the approximate value of x shown in figure (a) on the next page. If we zoom twice and trace again, we get the results shown in figure (b). As we saw in Example 1, the exact solution is 3.

An alternate way of finding the point of intersection of the two graphs is to use the INTERSECT feature. In figure (c), the display shows that the graphs intersect at the point (3, 2). This implies that the solution of the rational equation is 3.

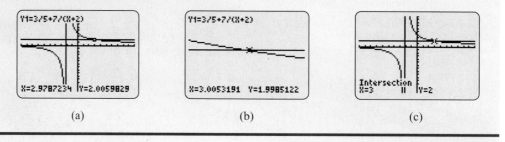

(a)   (b)   (c)

EXAMPLE 2   Solve: $\dfrac{-x^2 + 10}{x^2 - 1} + \dfrac{3x}{x - 1} = \dfrac{2x}{x + 1}$

Strategy   We will begin by factoring the first denominator.

Why   To determine any restrictions on the variable and to find the LCD, we need to write $x^2 - 1$ in factored form.

Solution   Since $x^2 - 1$ factors as $(x + 1)(x - 1)$, we can write the given equation as:

$$\frac{-x^2 + 10}{(x + 1)(x - 1)} + \frac{3x}{x - 1} = \frac{2x}{x + 1} \qquad \text{Factor the denominator } x^2 - 1.$$

We see that -1 and 1 cannot be solutions of the equation because they make rational expressions in the equation undefined.

We can clear the equation of fractions by multiplying both sides by $(x + 1)(x - 1)$, which is the LCD of the three rational expressions.

> **Caution**
> After multiplying both sides by the LCD and simplifying, the equation should not contain any fractions. If it does, check for an algebraic error, or perhaps your LCD is incorrect.

$$(x + 1)(x - 1)\left[\frac{-x^2 + 10}{(x + 1)(x - 1)} + \frac{3x}{x - 1}\right] = (x + 1)(x - 1)\left(\frac{2x}{x + 1}\right) \quad \text{Multiply both sides by the LCD.}$$

$$(x + 1)(x - 1)\left[\frac{-x^2 + 10}{(x + 1)(x - 1)}\right] + (x + 1)(x - 1)\left(\frac{3x}{x - 1}\right) = (x + 1)(x - 1)\left(\frac{2x}{x + 1}\right) \quad \begin{array}{l}\text{On the left side, distribute}\\ (x + 1)(x - 1).\end{array}$$

$$\overset{1}{(x + 1)}\overset{1}{(x - 1)}\left[\frac{-x^2 + 10}{(x + 1)(x - 1)}\right] + (x + 1)\overset{1}{(x - 1)}\left(\frac{3x}{x - 1}\right) = \overset{1}{(x + 1)}(x - 1)\left(\frac{2x}{x + 1}\right) \quad \begin{array}{l}\text{Remove common factors of the}\\ \text{numerator and denominator.}\end{array}$$

$$-x^2 + 10 + 3x(x + 1) = 2x(x - 1) \qquad \begin{array}{l}\text{Simplify. The resulting equation}\\ \text{does not contain any fractions.}\end{array}$$

$$-x^2 + 10 + 3x^2 + 3x = 2x^2 - 2x \qquad \text{Use the distributive property.}$$

$$2x^2 + 10 + 3x = 2x^2 - 2x \qquad \text{Combine like terms on each side.}$$

$$10 + 3x = -2x \qquad \text{Subtract } 2x^2 \text{ from both sides.}$$

$$10 + 5x = 0 \qquad \text{Add } 2x \text{ to both sides.}$$

$$5x = -10 \qquad \text{Subtract 10 from both sides.}$$

$$x = -2 \qquad \text{Divide both sides by 5.}$$

The solution is -2. Verify that it satisfies the original equation.

Self Check 2   Solve: $\dfrac{2}{x-3} = \dfrac{-x}{x^2-9} + \dfrac{4}{x+3}$

Now Try   Problem 17

We can summarize the procedure used to solve rational equations.

Solving Rational Equations

1. Factor all denominators.
2. Determine which numbers cannot be solutions of the equation.
3. Multiply both sides of the equation by the LCD of all rational expressions in the equation.
4. Use the distributive property to remove parentheses, remove any factors equal to 1, and write the result in simplified form.
5. Solve the resulting equation.
6. Check all possible solutions in the original equation.

Using Your Calculator

Checking Apparent Solutions

We can use a scientific calculator to check the solution -2 found in Example 2 by evaluating

$$\dfrac{-x^2+10}{x^2-1} + \dfrac{3x}{x-1} \quad \text{and} \quad \dfrac{2x}{x+1}$$

In each case, the result is 4. Since the results are the same, -2 is a solution of the equation. We can also check by using a graphing calculator. One way of doing this is to enter

$$Y_1 = \dfrac{-x^2+10}{x^2-1} + \dfrac{3x}{x-1} \quad \text{and} \quad Y_2 = \dfrac{2x}{x+1}$$

and compare the values of the expressions when $x = -2$ in the table mode. See the figure. We know that -2 is a solution of

$$\dfrac{-x^2+10}{x^2-1} + \dfrac{3x}{x-1} = \dfrac{2x}{x+1}$$

because the value of Y1 and Y2 are the same (namely, 4) for $x = -2$.

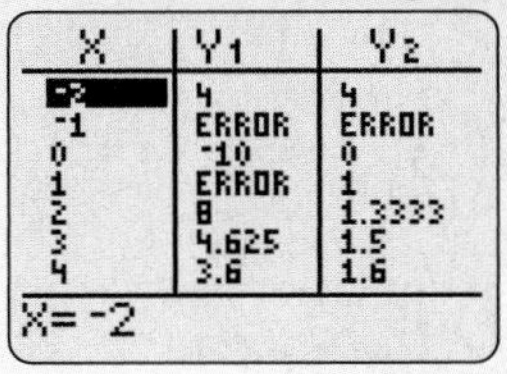

X	Y₁	Y₂
-2	4	4
-1	ERROR	ERROR
0	-10	0
1	ERROR	1
2	8	1.3333
3	4.625	1.5
4	3.6	1.6

X= -2

EXAMPLE 3   Solve: $\dfrac{a}{2} = \dfrac{a-6}{3a-9} - \dfrac{1}{3}$

Strategy   We will begin by factoring the second denominator.

Why   To determine any restrictions on the variable and to find the LCD, we need to write $3a - 9$ in factored form.

Solution   Since the binomial $3a - 9$ factors as $3(a - 3)$, we can write the given equation as:

$$\frac{a}{2} = \frac{a - 6}{3(a - 3)} - \frac{1}{3} \qquad \textit{Factor the denominator } 3a - 9.$$

Caution
When solving rational equations, each term on both sides must be multiplied by the LCD.

We see that 3 cannot be a solution of the equation, because it makes one of the rational expressions in the equation undefined.

We can clear the equation of fractions by multiplying both sides by $2 \cdot 3 \cdot (a - 3)$, which is the LCD of the three rational expressions.

$$2 \cdot 3 \cdot (a - 3)\left(\frac{a}{2}\right) = 2 \cdot 3 \cdot (a - 3)\left[\frac{a - 6}{3(a - 3)} - \frac{1}{3}\right] \qquad \textit{Multiply both sides by the LCD, } 2 \cdot 3 \cdot (a - 3).$$

$$2 \cdot 3 \cdot (a - 3)\left(\frac{a}{2}\right) = 2 \cdot 3 \cdot (a - 3)\left[\frac{a - 6}{3(a - 3)}\right] - 2 \cdot 3 \cdot (a - 3)\left(\frac{1}{3}\right) \qquad \textit{On the right side, distribute } 2 \cdot 3 \cdot (a - 3).$$

$$\overset{1}{\cancel{2}} \cdot 3(a - 3)\left(\frac{a}{\cancel{2}}\right) = 2 \cdot \cancel{3}\cancel{(a - 3)}\left[\frac{a - 6}{\cancel{3}\cancel{(a - 3)}}\right] - 2 \cdot \cancel{3}(a - 3)\left(\frac{1}{\cancel{3}}\right) \qquad \textit{Remove common factors of the numerator and denominator.}$$

$$3a(a - 3) = 2(a - 6) - 2(a - 3)$$

$$3a^2 - 9a = 2a - 12 - 2a + 6 \qquad \textit{Use the distributive property.}$$

$$3a^2 - 9a = -6 \qquad \textit{Combine like terms.}$$

To use factoring to solve the resulting quadratic equation, we must write it in standard form $ax^2 + bx + c = 0$.

$$3a^2 - 9a + 6 = 0 \qquad \textit{To get 0 on the right side, add 6 to both sides.}$$

$$a^2 - 3a + 2 = 0 \qquad \textit{Divide both sides by 3.}$$

$$(a - 1)(a - 2) = 0 \qquad \textit{Factor the trinomial.}$$

$$a - 1 = 0 \quad \text{or} \quad a - 2 = 0 \qquad \textit{Set each factor equal to 0.}$$

$$a = 1 \qquad\qquad a = 2$$

Verify that 1 and 2 both satisfy the original equation.

Self Check 3   Solve: $\frac{b}{5} = \frac{b - 14}{2b - 16} - \frac{1}{2}$

Now Try   **Problem 25**

Recall that the quotient of a polynomial and its opposite is -1. For example, $\frac{y - 1}{1 - y} = -1$. We can use this fact when solving rational equations whose denominators contain factors that are opposites.

EXAMPLE 4   Solve: $\dfrac{1}{6y - 6} + \dfrac{1}{1 - y} = \dfrac{1}{6}$

Strategy   We will begin by factoring the first denominator.

Why   To determine any restrictions on the variable and to find the LCD, we need to write $6y - 6$ in factored form.

Solution   Since the binomial $6y - 6$ factors as $6(y - 1)$, we can write the given equation as:

$$\frac{1}{6(y - 1)} + \frac{1}{1 - y} = \frac{1}{6}$$

We see that 1 cannot be a solution of the equation because it makes two of the rational expressions in the equation undefined.

We note that $y - 1$ and $1 - y$ are opposites. We can clear the equation of fractions by multiplying both sides by $6(y - 1)$.

$$6(y - 1)\left[\frac{1}{6(y - 1)} + \frac{1}{1 - y}\right] = 6(y - 1)\left(\frac{1}{6}\right) \qquad \text{Multiply both sides by the LCD, } 6(y-1).$$

$$6(y - 1)\left[\frac{1}{6(y - 1)}\right] + 6(y - 1)\left(\frac{1}{1 - y}\right) = 6(y - 1)\left(\frac{1}{6}\right) \qquad \begin{array}{l}\text{On the left side,}\\ \text{distribute } 6(y-1).\end{array}$$

$$\overset{1}{\cancel{6}}\overset{1}{\cancel{(y - 1)}}\left[\frac{1}{\cancel{6}\cancel{(y - 1)}}\right]_{1\ 1} + 6\overset{-1}{\cancel{(y - 1)}}\left(\frac{1}{\underset{1}{\cancel{1 - y}}}\right) = \overset{1}{\cancel{6}}(y - 1)\left(\frac{1}{\underset{1}{\cancel{6}}}\right) \qquad \text{Simplify: } \tfrac{y-1}{1-y} = -1.$$

$$1 - 6 = y - 1$$

$$-5 = y - 1 \qquad \text{Combine like terms.}$$

$$-4 = y \qquad \text{Add 1 to both sides.}$$

The solution is -4. Verify that it satisfies the original equation.

 | **Self Check 4**   Solve: $\dfrac{1}{2h - 8} + \dfrac{11}{4 - h} = \dfrac{3}{2}$
Now Try   **Problem 33**

---

2   **Solve Rational Equations with Extraneous Solutions.**

When we multiply both sides of an equation by a quantity that contains a variable, we can get false solutions, called *extraneous solutions*. This happens when we multiply both sides of an equation by 0 and get a solution that gives a 0 in the denominator of a rational expression. Extraneous solutions must be discarded.

The Language of Algebra
Extraneous means not a vital part. Mathematicians speak of *extraneous* solutions. Rock groups don't want any *extraneous* sounds (like humming or feedback) coming from their amplifiers. Artists erase any *extraneous* marks on their sketches.

EXAMPLE 5   Solve: $3 - \dfrac{1 - 2t}{t + 2} = \dfrac{t - 3}{t + 2}$

Strategy   We will clear the equation of fractions by multiplying both sides by the LCD, $t + 2$.

Why   Equations that contain only integers are usually easier to solve than equations that contain fractions.

Solution   We note that t cannot be -2, because this would give a 0 in a denominator.

$$3 - \frac{1 - 2t}{t + 2} = \frac{t - 3}{t + 2}$$

$$(t + 2)\left(3 - \frac{1 - 2t}{t + 2}\right) = (t + 2)\left(\frac{t - 3}{t + 2}\right) \qquad \text{Multiply both sides by the LCD.}$$

$$(t + 2)(3) - (t + 2)\left(\frac{1 - 2t}{t + 2}\right) = (t + 2)\left(\frac{t - 3}{t + 2}\right)$$

On the left side, distribute the multiplication by $t + 2$.

$$(t + 2)(3) - (\cancel{t + 2})\left(\frac{1 - 2t}{\cancel{t + 2}}\right) = (\cancel{t + 2})\left(\frac{t - 3}{\cancel{t + 2}}\right)$$

Remove common factors of the numerator and denominator.

$$3t + 6 - (1 - 2t) = t - 3$$ Simplify.

$$3t + 6 - 1 + 2t = t - 3$$

$$5t + 5 = t - 3$$ Combine like terms.

$$4t = -8$$

$$t = -2$$

Since t cannot be -2, it is an extraneous solution and must be discarded. This equation has no solution.

 Self Check 5   Solve: $2 - \frac{2a}{a - 1} = \frac{3a - 5}{a - 1}$

Now Try **Problem 37**

3   Solve Formulas for a Specified Variable.

Many formulas involve rational expressions. We can use the fraction-clearing method of this section to solve such formulas for a specified variable.

EXAMPLE 6   *Physics.*   The *law of gravitation,* formulated by Sir Isaac Newton in 1684, states that if two masses, m_1 and m_2, are separated by a distance of r, the force F exerted by one mass on the other is

$$F = \frac{Gm_1m_2}{r^2}$$

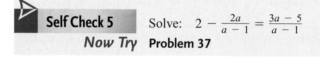

where G is the gravitational constant. Solve for m_2.

Strategy   To solve for m_2, we will treat it as if it were the only variable in the equation. To isolate this variable, we will use the same strategy that we used in previous examples to solve rational equations in one variable.

Why   We can solve a formula as if it were an equation in one variable because all the other variables are treated as if they were numbers (constants).

Solution

$$F = \frac{Gm_1m_2}{r^2}$$

$$r^2(F) = \overset{1}{\cancel{r^2}}\left(\frac{Gm_1m_2}{\cancel{r^2}}\right)$$ Multiply both sides by the LCD, r^2. Simplify: $\frac{r^2}{r^2} = 1$.

$$\frac{r^2F}{Gm_1} = \frac{Gm_1m_2}{Gm_1}$$ To isolate m_2, divide both sides by Gm_1.

$$\frac{r^2 F}{Gm_1} = m_2$$

Simplify the right side by removing the factors G and m_1, which are common to the numerator and denominator.

$$m_2 = \frac{r^2 F}{Gm_1}$$

Reverse the sides of the equation so that m_2 is on the left.

Self Check 6   Solve the law of gravitation formula for r^2.

Now Try   **Problem 41**

EXAMPLE 7   *Electronics.*   In electronic circuits, resistors oppose the flow of an electric current. The total resistance R of a parallel combination of two resistors as shown is given by

$$\frac{1}{R} = \frac{1}{R_1} + \frac{1}{R_2}$$

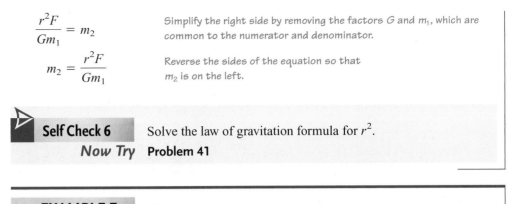

Resistor 1

Current →   ← Total resistance?

Resistor 2

where R_1 is the resistance of the first resistor and R_2 is the resistance of the second resistor. Solve for R.

Strategy   To solve for R, we will treat it as if it were the only variable in the equation. To isolate this variable, we will use the same strategy that we used in previous examples to solve rational equations in one variable; we will clear the equation of fractions.

Why   We can solve a formula as if it were an equation in one variable because all the other variables are treated as if they were numbers (constants).

Solution   We begin by clearing the equation of fractions by multiplying both sides by the LCD, which is RR_1R_2.

$$\frac{1}{R} = \frac{1}{R_1} + \frac{1}{R_2}$$

$$RR_1R_2\left(\frac{1}{R}\right) = RR_1R_2\left(\frac{1}{R_1} + \frac{1}{R_2}\right)$$ Multiply both sides by the LCD.

$$RR_1R_2\left(\frac{1}{R}\right) = RR_1R_2\left(\frac{1}{R_1}\right) + RR_1R_2\left(\frac{1}{R_2}\right)$$ On the right side, distribute RR_1R_2.

$$\overset{1}{\cancel{R}}R_1R_2\left(\frac{1}{\cancel{R}}\right) = R\overset{1}{\cancel{R_1}}R_2\left(\frac{1}{\cancel{R_1}}\right) + RR_1\overset{1}{\cancel{R_2}}\left(\frac{1}{\cancel{R_2}}\right)$$ Remove common factors of the numerator and denominator.

$$R_1R_2 = RR_2 + RR_1$$ Simplify.

$$R_1R_2 = R(R_2 + R_1)$$ Factor out R on the right side.

$$\frac{R_1R_2}{R_2 + R_1} = R$$ To isolate R, divide both sides by $R_2 + R_1$.

$$R = \frac{R_1R_2}{R_2 + R_1}$$ Reverse the sides of the equation to write R on the left side.

Self Check 7   Solve $\frac{1}{x} - \frac{1}{y} = \frac{1}{z}$ for z.

Now Try   **Problem 49**

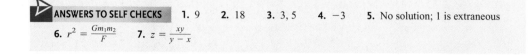
STUDY SET
6.7

VOCABULARY

Fill in the blanks.

1. Equations that contain one or more rational expressions, such as $\dfrac{x}{x + 2} = 4 + \dfrac{10}{x + 1}$, are called _____ equations.

2. When solving a rational equation, if we obtain a number that does not satisfy the original equation, the number is called an _____ solution.

CONCEPTS

3. Is 2 a solution of the following equations?

a. $\dfrac{x + 2}{x + 3} + \dfrac{1}{x^2 + 2x - 3} = 1$

b. $\dfrac{x + 2}{x - 2} + \dfrac{1}{x^2 - 4} = 1$

4. Consider the rational equation $\dfrac{x}{x - 3} = \dfrac{1}{x} + \dfrac{2}{x - 3}$.

   a. What values of x make a denominator 0?

   b. What values of x make a rational expression undefined?

   c. What numbers can't be solutions of the equation?

5. To clear the following equation of fractions, by what should both sides be multiplied?

$$\frac{4}{10} + y = \frac{4y - 50}{5y - 25}$$

6. Perform each multiplication.

   a. $4x\left(\dfrac{3}{4x}\right)$ **b.** $(x + 6)(x - 2)\left(\dfrac{3}{x - 2}\right)$

   c. $8(x + 4)\left(\dfrac{7x}{2(x + 4)}\right)$ **d.** $6(m - 5)\left(\dfrac{7}{5 - m}\right)$

NOTATION

Complete each solution.

7.
$$\frac{10}{3y} - \frac{7}{30} = \frac{9}{2y}$$

$$\left(\frac{10}{3y} - \frac{7}{30}\right) = 30y\left(\boxed{}\right)$$

$$\left(\frac{10}{3y}\right) - \boxed{}\left(\frac{7}{30}\right) = \boxed{}\left(\frac{9}{2y}\right)$$

$$100 - \boxed{} = 135$$

$$-7y = \boxed{}$$

$$y = -5$$

8.
$$\frac{2}{u - 1} + \frac{1}{u} = \frac{1}{u^2 - u}$$

$$\frac{2}{u - 1} + \frac{1}{u} = \frac{1}{u(\boxed{})}$$

$$\left(\frac{2}{u - 1} + \frac{1}{u}\right) = \boxed{}\left[\frac{1}{u(u - 1)}\right]$$

$$\left(\frac{2}{u - 1}\right) + \boxed{}\left(\frac{1}{u}\right) = \boxed{}\left[\frac{1}{u(u - 1)}\right]$$

$$\boxed{} + u - 1 = 1$$

$$\boxed{} = 2$$

$$u = \boxed{}$$

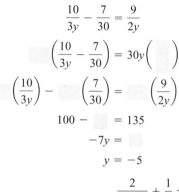

GUIDED PRACTICE

Solve each equation. See Example 1.

9. $\dfrac{1}{4} + \dfrac{9}{x} = 1$ **10.** $\dfrac{1}{3} - \dfrac{10}{x} = -3$

11. $\dfrac{1}{a} = \dfrac{1}{3} - \dfrac{2}{3a}$ **12.** $\dfrac{1}{b} = \dfrac{1}{8} - \dfrac{3}{8b}$

13. $\dfrac{18}{y + 1} + \dfrac{2}{5} = 4$ **14.** $\dfrac{2}{3} + \dfrac{10}{a + 2} = 4$

15. $\dfrac{1}{2} + \dfrac{x}{x - 1} = 3$ **16.** $\dfrac{2}{3} + \dfrac{a}{a - 2} = 5$

Solve each equation. **See Example 2.**

17. $\dfrac{4}{t+3} + \dfrac{8}{t^2-9} = \dfrac{2}{t-3}$

18. $\dfrac{5}{x-1} = \dfrac{1}{x^2-1} + \dfrac{1}{x-1}$

19. $\dfrac{4}{x^2-4} - \dfrac{5}{x-2} = \dfrac{1}{x+2}$

20. $\dfrac{1}{m+3} - \dfrac{m}{m^2-9} = \dfrac{-2}{m-3}$

21. $\dfrac{2}{x-2} + \dfrac{10}{x+5} = \dfrac{2x}{x^2+3x-10}$

22. $\dfrac{2}{a+4} + \dfrac{2a-1}{a^2+2a-8} = \dfrac{1}{a-2}$

23. $\dfrac{1}{n+2} - \dfrac{2}{n-3} = \dfrac{-2n}{n^2-n-6}$

24. $\dfrac{2x}{x^2+9x+20} - \dfrac{3}{x+4} = \dfrac{2}{x+5}$

Solve each equation. **See Example 3.**

25. $\dfrac{2}{5x-5} + \dfrac{x-2}{15} = \dfrac{4}{5x-5}$

26. $\dfrac{3}{2x+4} = \dfrac{x-2}{2} + \dfrac{x-5}{2x+4}$

27. $\dfrac{p-1}{2} + 1 = \dfrac{3}{p}$

28. $\dfrac{b+1}{2} - \dfrac{3}{2} = \dfrac{4}{b}$

29. $\dfrac{16}{t+3} + \dfrac{7}{t-2} = 3$

30. $\dfrac{17}{s-4} - \dfrac{10}{s+2} = 2$

31. $\dfrac{5}{x-2} = 2 - \dfrac{6}{x+2}$

32. $\dfrac{-10}{t+3} = 1 - \dfrac{11}{t-3}$

Solve each equation. **See Example 4.**

33. $\dfrac{1}{3x-18} + \dfrac{5}{6-x} = \dfrac{1}{3}$

34. $\dfrac{1}{2x-16} + \dfrac{14}{8-x} = \dfrac{3}{2}$

35. $\dfrac{7}{3x-9} + \dfrac{1}{3-x} = \dfrac{4}{9}$

36. $\dfrac{1}{2d-4} - \dfrac{1}{2-d} = \dfrac{1}{4}$

Solve each equation. If a solution is extraneous, so indicate. **See Example 5.**

37. $4 - \dfrac{3x}{x-9} = \dfrac{5x-72}{x-9}$

38. $2 - \dfrac{2x}{x-10} = \dfrac{4x-60}{x-10}$

39. $\dfrac{6}{x+3} + \dfrac{48}{x^2-2x-15} - \dfrac{7}{x-5} = 0$

40. $\dfrac{3}{x-4} + \dfrac{2}{x+5} + \dfrac{18}{x^2+x-20} = 0$

Solve each formula for the specified variable. **See Examples 6 and 7.**

41. $Q = \dfrac{A-I}{L}$ for A (from banking)

42. $z = \dfrac{x-\bar{x}}{s}$ for x (from statistics)

43. $I = \dfrac{E}{R_L + r}$ for r (from physics)

44. $P = \dfrac{R-C}{n}$ for C (from business)

45. $\mu_R = \dfrac{n_1(n_1+n_2+1)}{2}$ for n_2 (from statistics)

46. $\dfrac{P_1V_1}{T_1} = \dfrac{P_2V_2}{T_2}$ for T_2 (from chemistry)

47. $P = \dfrac{Q_1}{Q_2-Q_1}$ for Q_1 (from refrigeration/heating)

48. $S = \dfrac{a-\ell r}{1-r}$ for r (from mathematics)

49. $\dfrac{1}{R} = \dfrac{1}{R_1} + \dfrac{1}{R_2} + \dfrac{1}{R_3}$ for R (from electronics)

50. $\dfrac{x}{a} + \dfrac{y}{b} = 1$ for a (from mathematics)

51. $\dfrac{E}{e} = \dfrac{R+r}{r}$ for r (from engineering)

52. $P + \dfrac{a}{V^2} = \dfrac{RT}{V-b}$ for b (from physics)

TRY IT YOURSELF

Solve each equation. If a solution is extraneous, so indicate.

53. $\dfrac{x+2}{x+3} - 1 = \dfrac{-1}{x^2+2x-3}$

54. $\dfrac{m+6}{3m-12} + \dfrac{5}{4-m} = \dfrac{2}{3}$

55. $\dfrac{3}{y} + \dfrac{7}{2y} = 13$

56. $\dfrac{2}{x} + \dfrac{1}{2} = \dfrac{7}{2x}$

57. $\dfrac{3}{r} + \dfrac{12}{r^2-4r} = \dfrac{-7}{r-4}$

58. $\dfrac{4t^2+36}{t^2-9} - \dfrac{4t}{t+3} = \dfrac{-12}{t-3}$

59. $\dfrac{x+4}{2x+14} - \dfrac{x}{2x+6} = \dfrac{3}{16}$

60. $\dfrac{30}{y-2} + \dfrac{24}{y-5} = 13$

61. $\dfrac{3}{m} = 2 - \dfrac{m}{m-2}$

62. $\dfrac{n}{2} = 1 + \dfrac{12}{n}$

63. $\dfrac{x+2}{2x-6} + \dfrac{3}{3-x} = \dfrac{x}{2}$

64. $\dfrac{3}{4x-8} = \dfrac{1}{36} - \dfrac{2}{6-3x}$

65. $\dfrac{2}{x} + \dfrac{1}{2} = \dfrac{9}{4x} - \dfrac{1}{2x}$

66. $\dfrac{7}{5x} - \dfrac{1}{2} = \dfrac{5}{6x} + \dfrac{1}{3}$

67. $\dfrac{3 - 5y}{2 + y} = \dfrac{-5y - 3}{y - 2}$

68. $\dfrac{a - 3}{a + 1} = \dfrac{a - 6}{a + 5}$

69. $\dfrac{21}{x^2 - 4} - \dfrac{14}{x + 2} = \dfrac{3}{2 - x}$

70. $\dfrac{-5}{c + 2} = \dfrac{3}{2 - c} + \dfrac{2c}{c^2 - 4}$

71. $\dfrac{x - 4}{x - 3} - \dfrac{x - 2}{3 - x} = x - 3$

72. $\dfrac{5}{x + 4} + \dfrac{1}{x + 4} = x - 1$

73. $\dfrac{a + 2}{a + 1} = \dfrac{a - 4}{a - 3}$

74. $\dfrac{z + 2}{z + 8} = \dfrac{z - 3}{z - 2}$

75. $\dfrac{5}{y - 1} + \dfrac{3}{y - 3} = \dfrac{8}{y - 2}$

76. $\dfrac{3 + 2a}{a^2 + 6 + 5a} + \dfrac{2 - 5a}{a^2 - 4} = \dfrac{2 - 3a}{a^2 - 6 + a}$

77. $\dfrac{3}{s - 2} + \dfrac{s - 14}{2s^2 - 3s - 2} - \dfrac{4}{2s + 1} = 0$

78. $\dfrac{1}{y^2 - 2y - 3} + \dfrac{1}{y^2 - 4y + 3} - \dfrac{1}{y^2 - 1} = 0$

79. $\dfrac{x}{x + 2} = 1 - \dfrac{3x + 2}{x^2 + 4x + 4}$

80. $\dfrac{a - 1}{a + 3} - \dfrac{1 - 2a}{3 - a} = \dfrac{2 - a}{a - 3}$

81. $\dfrac{5}{2z^2 + z - 3} - \dfrac{2}{2z + 3} = \dfrac{z + 1}{z - 1} - 1$

82. $\dfrac{x}{x - 5} + \dfrac{5}{x} = \dfrac{11}{6}$

83. $\dfrac{5}{3x + 12} - \dfrac{1}{9} = \dfrac{x - 1}{3x}$

84. $\dfrac{1}{y + 5} = \dfrac{1}{3y + 6} - \dfrac{y + 2}{y^2 + 7y + 10}$

For each expression in part (a), perform the indicated operations and then simplify, if possible. Solve each equation in part (b) and check the result.

85. a. $\dfrac{11}{12} - \dfrac{3}{2x} + \dfrac{4}{x}$ **b.** $\dfrac{11}{12} - \dfrac{3}{2x} = \dfrac{4}{x}$

86. a. $\dfrac{1}{6x} - \dfrac{2}{x - 6}$ **b.** $\dfrac{1}{6x} = \dfrac{2}{x - 6}$

87. a. $\dfrac{m}{m - 2} - \dfrac{1}{m - 3}$

  b. $\dfrac{m}{m - 2} - \dfrac{1}{m - 3} = 1$

88. a. $\dfrac{a^2 + 1}{a^2 - a} - \dfrac{a}{a - 1}$

  b. $\dfrac{a^2 + 1}{a^2 - a} - \dfrac{a}{a - 1} = \dfrac{1}{a}$

APPLICATIONS

89. PHOTOGRAPHY The illustration shows the relationship between distances when taking a photograph. The design of a camera lens uses the equation

$$\frac{1}{f} = \frac{1}{s_1} + \frac{1}{s_2}$$

which relates the focal length f of a lens to the image distance s_1 and the object distance s_2.

a. Solve the formula for f.

b. Find the focal length of the lens in the illustration. (*Hint:* Convert feet to inches.)

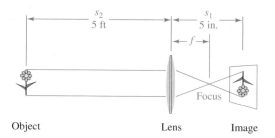

90. OPTICS See the illustration. The focal length, f, of a lens is given by the lensmaker's formula,

$$\frac{1}{f} = 0.6\left(\frac{1}{r_1} + \frac{1}{r_2}\right)$$

where f is the focal length of the lens and r_1 and r_2 are the radii of the two circular surfaces.

a. Solve the formula for f.

b. Find the focal length of the lens in the illustration.

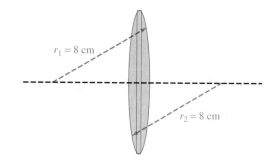

91. ACCOUNTING As a piece of equipment gets older, its value usually lessens. One way to calculate *depreciation* is to use the formula

$$V = C - \left(\frac{C - S}{L}\right)N$$

where V denotes the value of the equipment at the end of year N, L is its useful lifetime (in years), C is its cost new, and S is its salvage value at the end of its useful life.

a. Solve the formula for L.

b. Determine what an accountant considered the useful life-time of a forklift that cost $25,000 new, was worth $13,000 after 4 years, and has a salvage value of $1,000.

92. ENGINEERING The equation

$$a = \frac{9.8m_2 - f}{m_2 + m_1}$$

models the system shown, where a is the acceleration of the suspended block, m_1 and m_2 are the masses of the blocks, and f is the friction force. Solve for m_2.

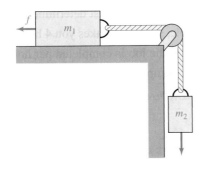

WRITING

93. Why is it necessary to check the solutions of a rational equation?

94. Explain what it means to *clear* a rational equation of fractions. Give an example.

95. Would you use the same approach to answer the following problems? Explain why or why not.

$$\text{Simplify: } \frac{x^2 - 10}{x^2 - 1} - \frac{3x}{x - 1} - \frac{2x}{x + 1}$$

$$\text{Solve: } \frac{x^2 - 10}{x^2 - 1} - \frac{3x}{x - 1} = -\frac{2x}{x + 1}$$

96. Explain how to solve the rational equation graphically:

$$\frac{3x}{x - 2} + \frac{1}{5} = 2$$

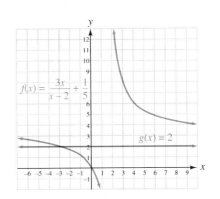

REVIEW

Write each italicized number in scientific notation.

97. OIL The total cost of the Alaskan pipeline, running 800 miles from Prudhoe Bay to Valdez, was $*9,000,000,000*.

98. NATURAL GAS The TransCanada Pipeline transported a record *2,352,000,000,000* cubic feet of gas in 1995.

99. RADIOACTIVITY The least stable radioactive isotope is lithium 5, which decays in *0.00000000000000000000044* second.

100. BALANCES The finest balances in the world are made in Germany. They can weigh objects to an accuracy of 35×10^{-11} ounce.

CHALLENGE PROBLEMS

101. Solve: $\left(\dfrac{1}{2}\right)^{-1} = \dfrac{5b^{-1}}{2} + 2b(b + 1)^{-1}$

102. Write a rational equation that has an extraneous solution of 3.

103. Let $f(x) = \dfrac{x^3 - 3x^2 + 12}{x}$. For what values of x is $f(x) = 4$?

104. Let $f(x) = \dfrac{x^3 + 2x^2 - 32}{x}$. For what values of x is $f(x) = 16$?

105. Let $f(x) = \dfrac{2x^3 + x^2}{98x + 49}$. For what values of x is $f(x) = 1$?

106. Let $f(x) = \dfrac{x^3 + 4x^2}{25x + 100}$. For what values of x is $f(x) = 1$?

$$540 - 45c + 540 + 45c = 8(144 - c^2)$$

On the left side, distribute. On the right side, use the FOIL method.

$$1{,}080 = 1{,}152 - 8c^2$$

Combine like terms and multiply. This is a quadratic equation.

$$8c^2 - 72 = 0$$

To get 0 on the right side, add $8c^2$ and subtract 1,152 from both sides.

$$c^2 - 9 = 0$$

Divide both sides by 8.

$$(c + 3)(c - 3) = 0$$

Factor the difference of two squares.

$$c + 3 = 0 \quad \text{or} \quad c - 3 = 0$$

Set each factor equal to 0.

$$\cancel{c = -3} \quad | \quad c = 3$$

State the Conclusion   Since the current cannot be negative, the solution -3 must be discarded. The current in the Rock River is 3 mph.

Check the Result   The downstream trip is at $12 + 3 = 15$ mph for $\frac{9}{12 + 3} = \frac{3}{5}$ hr. Thus, the distance traveled is $15 \cdot \frac{3}{5} = 9$ miles. The upstream trip is at $12 - 3 = 9$ mph for $\frac{9}{12 - 3} = 1$ hr. Thus, the distance traveled is $9 \cdot 1 = 9$ miles. Since both distances are 9 miles, the result checks.

Now Try   **Problem 39**

STUDY SET
6.8

VOCABULARY

Fill in the blanks.

1. In this section, we call problems that involve:
- people or machines completing jobs, shared-_____ problems.
- moving vehicles, uniform _____ problems.

2. When a boat travels _____, the speed of the boat is increased by the current. When a boat travels _____, the speed of the boat is decreased by the current.

CONCEPTS

3. Fill in the blank: If a job can be completed in x hours, then the rate of work can be expressed as $\frac{1}{\quad}$ of the job is completed per hour.

4. a. It takes a night security officer 35 minutes to check each of the doors in an office building to make sure they are locked. What is the officer's rate of work?

   b. It takes a high school mathematics teacher 4 hours to make out the semester report cards. What part of the job does she complete in x hours?

5. Complete the table.

	Rate	· Time	= Work completed
1st crew	$\frac{1}{15}$	x	
2nd crew	$\frac{1}{8}$	x	

6. Solve $d = rt$ for t.

7. Complete the table.

	r	· t	= d
Running	x		12
Bicycling	$x + 15$		12

8. A boat can cruise at 30 mph in still water.
   a. What is its cruising speed upstream against a current of 4 mph?
   b. What is its cruising speed downstream with a current of 4 mph?

NOTATION

9. Write $\frac{41}{9}$ hours using a mixed number.

10. Fill in the blanks: In the formula $W = rt$, the variable W stands for the _____ completed, r is the _____, and t is the _____.

APPLICATIONS

11. ROOFING   A homeowner estimates that it will take him 7 days to roof his house. A professional roofer estimates that he could roof the house in 4 days. How long will it take if the homeowner helps the roofer?

12. DECORATING   One crew can put up holiday decorations in a department store in 12 hours. A second crew can put up the decorations in 15 hours. How long will it take if both crews work together to decorate the store?

13. HOUSEPAINTING   The illustration shows two bids to paint a house.

   a. To get the job done quicker, the homeowner hired both the painters who submitted bids. How long will it take them to paint the house working together?

   b. What will the homeowner have to pay each painter?

Santos Painting
Residential Bid:
3 days
@ $220 a day
Total: $660

Mays ⬆
House ⬆ Painting
Bid:
$200 per day
5 days work
Total: $1,000

14. GROUNDSKEEPING   It takes a groundskeeper 45 minutes to prepare a Little League baseball field for a game. It takes his assistant 55 minutes to prepare the same field. How long will it take if they work together to prepare the field?

15. FARMING   In 10 minutes, a conveyor belt can move 1,000 bushels of corn into the storage bin shown. A smaller belt can move 1,000 bushels to the storage bin in 14 minutes. If both belts are used, how long will it take to move 1,000 bushels to the storage bin?

16. BOTTLING   At a packaging plant, the older of two machines can fill 5,000 bottles of shampoo in 6 hours. A newer machine can fill 5,000 bottles in 4 hours. If both machines are used, how long will it take to fill 5,000 bottles of shampoo?

17. THRILL RIDES   At the end of an amusement park ride, a boat lands in a pool, splashing out a lot of water. Three inlet pipes, each working alone, can fill the pool in 10 seconds, 15 seconds, and 20 seconds, respectively. How long would it take to fill the pool if all three inlet pipes are used?

© Phil Degginger/Alamy

18. SMOKE DAMAGE   Three ventilation fans, each working alone, can clear the smoke out of a room in 12 hours, 16 hours, and 24 hours, respectively. How long would it take to clear out the smoke in the room if all three fans are used?

19. FILLING PONDS   One pipe can fill a pond in 3 weeks, and a second pipe can fill it in 5 weeks. However, evaporation and seepage can empty the pond in 10 weeks. If both pipes are used, how long will it take to fill the pond?

20. HOUSECLEANING   Sally can clean the house in 6 hours, her father can clean the house in 4 hours, and her younger brother, Dennis, can completely mess up the house in 8 hours. If Sally and her father clean and Dennis plays, how long will it take to clean the house?

21. FINE DINING   It takes a waiter 5 minutes less time than a busboy to fold the napkins used for the dinner seating in an upscale restaurant. Working together, they can fold the napkins in 6 minutes. How long would it take each person working alone to fold the napkins?

22. FIRE DRILL   If the east and west exit doors of a banquet hall are open, the occupants can clear out in 2 minutes. It takes 3 minutes longer to clear the hall if just the east door is open as it does if just the west door is open. How long does it take to clear the hall if just the west door is open?

23. FUND-RAISING LETTERS   Working together, two secretaries can stuff the envelopes for a political fund-raising letter in 4 hours. Working alone, it takes the slower worker 6 hours longer to do the job than the faster worker. How long does it take each to do the job alone?

24. SURVEYS   It takes one team 9 days less than another to survey 1,000 people. If the teams work together, it takes them 20 days to complete such a survey. How long will it take each to do the survey alone?

25. PLUMBING   An experienced plumber can install the plumbing in a new apartment twice as fast as his apprentice. Working together, they can complete the plumbing job in 4 days. How long would it take each, working alone, to complete the plumbing?

26. NEWSLETTERS   An elementary school teacher can assemble and staple the weekly newsletter three times faster than her student aide. Working together, they can assemble and staple the letters in 12 minutes. How long would it take each, working alone, to complete the job?

27. DETAILING A CAR   It takes a man 3 hours to wash and wax the family car. If his teenage son helps him, it only takes 1 hour. How long would it take the son, working alone, to wash and wax the car?

28. CLEANUP CREWS   It takes one crew 4 hours to clean an auditorium after an event. If a second crew helps, it only takes 1.5 hours. How long would it take the second crew, working alone, to clean the auditorium?

29. OYSTERS   According to the *Guinness Book of World Records,* the record for opening oysters is 100 in 140 seconds by Mike Racz in Invercargill, New Zealand, on July 16, 1990. If it would take a novice $8\frac{1}{2}$ minutes to perform the same task, how long would it take them working together to open 100 oysters? (*Hint:* Work in terms of seconds.)

30. END ZONES   One groundskeeper can paint the end zone of a football field in 2 hours. Another can paint it in 1 hour 20 minutes. How many minutes will it take them working together to paint the end zone?

31. TRUCK DELIVERIES   A trucker drove 120 miles to make a delivery and returned home on the same route. Because of foggy conditions, his average speed on the return trip was 10 mph less than his average speed going. If the return trip took 1 hour longer, how fast did he drive in each direction?

32. MOVING HOUSES   A house mover towed a historic Victorian home 45 miles to locate it on a new site. On his return, without the heavy house in tow, his average speed was 30 mph faster and the trip was 2 hours shorter. How fast did he drive in each direction?

33. TRAIN TRAVEL   A train traveled 120 miles from Freeport to Chicago and returned the same distance in a total time of 5 hours. If the train traveled 20 mph slower on the return trip, how fast did the train travel in each direction?

34. BOXING   For his morning workout, a boxer bicycles for 8 miles and then jogs back to camp along the same route. If he bicycles 6 mph faster than he jogs, and the entire workout lasts 2 hours, how fast does he jog?

35. RATES OF SPEED   Two trains made the same 315-mile run. Since one train traveled 10 mph faster than the other, it arrived 2 hours earlier. Find the speed of each train.

36. DELIVERIES   A FedEx delivery van traveled from Rockford to Chicago in 3 hours less time than it took a second van to travel from Rockford to St. Louis. If the vans traveled at the same average speed, use the information in the map to help determine how long the first driver was on the road.

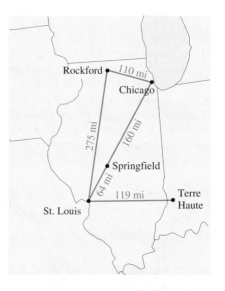

37. COMPARING TRAVEL   A plane can fly 600 miles in the same time as it takes a car to go 240 miles. If the car travels 90 mph slower than the plane, find the speed of the plane.

38. COMPARING TRAVEL   A bicyclist can travel 40 miles in the same time that a motorcyclist can travel 60 miles. If the bicyclist travels 12 mph slower than the motorcyclist, find the speed of the motorcyclist.

Self Check 4 S
y

Now Try P

---

5 **Solve Problems Involvi**

To introduce direct variatio

$$C = \pi D$$

where C is the circumferenc
of a circle, we determine an

$$C_1 = \pi(2D) = 2\pi D =$$

Thus, doubling the diamete
diameter, we will triple the

In the formula, $C = \pi L$
directly proportional. This
this example, the constan
proportionality.

Direct Variation

The words "y varies directly
some nonzero constant k. Th
of proportionality.

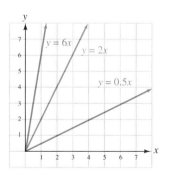

Since the formula for
always a line with a y-interc
margin for three positive val

One example of direct v
distance a spring will stretch

If d represents a distanc
be expressed as

$$d = kf$$ The direct varia

where k is the constant of va
when a weight of 6 pounds i

$$d = kf$$
$$10 = k(6)$$ Substitute 1
$$\frac{5}{3} = k$$

To find the force required to
35 inches, we can solve the
$d = 35$ and $k = \frac{5}{3}$.

---

39. BOATING It takes 6 hours for a boater to travel 16 miles upstream and back. If the speed of the boat in still water is 6 mph, what is the speed of the current?

40. RIVER TOURS A wave runner trip begins by going 60 miles upstream against a current. There, the driver turns around and returns with the current. If the still-water speed of the wave runner is set at 25 mph and the entire trip takes 5 hours, what is the speed of the current?

© Joe Fox/Alamy

41. BOATING A man can drive a motorboat 45 miles down the Colorado River in the same amount of time that he can drive 27 miles upstream. Find the speed of the current if the speed of the boat is 12 mph in still water.

42. CROP DUSTING A helicopter spraying fertilizer over a field can fly 0.5 mile downwind in the same time as it can fly 0.4 mile upwind. Find the speed of the wind if the helicopter travels 45 mph in still air when dusting crops.

WRITING

43. In Example 1, one crew could drywall a house in 4 days, and another crew could drywall the same house in 5 days. We were asked to find how long it would take them to drywall the house working together. Explain why each of the following approaches is incorrect.

The time it would take to drywall the house
- is the *sum* of the lengths of time it takes each crew to drywall the house: 4 days + 5 days = 9 day.
- is the *difference* in lengths of time it takes each crew to drywall the house: 5 days − 4 days = 1 day.
- is the *average* of the lengths of time it takes each crew to drywall the house: $\frac{4\ \text{days}\ +\ 5\ \text{days}}{2} = \frac{9}{2}$ days $= 4\frac{1}{2}$ days.

---

44. Write a shared-work problem that can be modeled by the equation

$$\frac{x}{3} + \frac{x}{4} = 1$$

REVIEW

Simplify each expression. Write answers using positive exponents.

45. $\left(\dfrac{m^{10}}{n}\right)^{8}$ **46.** $\left(\dfrac{g^{20}}{t^{30}}\right)^{-4}$

47. $-w^{-2}$ **48.** $-3s^0 t$

49. $-\dfrac{4x^{-9} \cdot x^{-3}}{x^{-12}}$ **50.** $\dfrac{y^{-3}y^{-4}y^{0}}{(2y^{-2})^3}$

51. $(-x^2)^5 y^7 y^3 x^{-2} y^0$ **52.** $5^2 r^{-5}(r^6)^3$

CHALLENGE PROBLEMS

53. FIREPLACES A mason and his assistant work together for 6 hours on a brick fireplace before the mason has to leave the job. The assistant finishes the job alone in 10 hours. If the mason can construct a fireplace in 18 hours working alone, how long does it take his assistant working alone to construct a fireplace?

54. EXTENDED VACATION Use the facts in the e-mail message to determine how long the student had originally planned to stay in Europe. (*Hint:* Unit cost · number = total cost.)

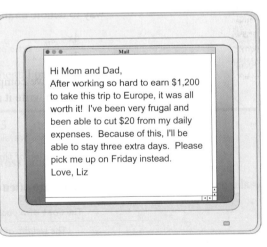

Mail

Hi Mom and Dad,
After working so hard to earn $1,200 to take this trip to Europe, it was all worth it! I've been very frugal and been able to cut $20 from my daily expenses. Because of this, I'll be able to stay three extra days. Please pick me up on Friday instead.
Love, Liz

Success Tip

Similar triangles do not have to be positioned the same. When they are placed differently, be careful to match their corresponding letters correctly. For example, in the illustration below,

ΔRST is similar to ΔMNO

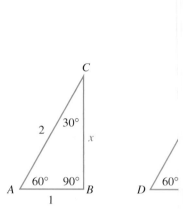

The following triangles are

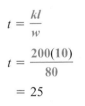

The properties of similar tr
triangles indirectly. For exampl
ground.

EXAMPLE 4   Heigh
time as
height

Strategy   We will use the fact

Why   Three of the entries of
unknown height of the tree,

Solution   Refer to the figure,
shadow and the yardstick a
they are similar, and the mea
let h = the height of the tre
lowing proportion: h is to 3

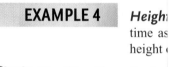

$$\text{Height of the tree} \rightarrow \dfrac{h}{3} =$$
$$\text{Height of the yardstick} \rightarrow$$

$2.5h = 3(29)$ Find each cro

$2.5h = 87$ Multiply.

$h = 34.8$ To isolate h, d

The tree is about 35 feet tall.

Strategy   We will use the combined variation model $t = \dfrac{kl}{w}$, where t represents the time in days, l represents the length of road built in miles, and w represents the number of workers.

Why   The words *the time it takes to build a highway varies directly as the length of the road, and inversely with the number of workers* indicate that this type of model should be used.

Solution   The relationship between these variables can be expressed by the equation

$$t = \dfrac{kl}{w} \quad \text{This is a combined variation model.}$$

We substitute 4 for t, 100 for w, and 2 for l to find k:

$$4 = \dfrac{k(2)}{100}$$

$400 = 2k$ Multiply both sides by 100.

$200 = k$ Divide both sides by 2 to solve for k.

We now substitute 80 for w, 10 for l, and 200 for k in the equation $t = \dfrac{kl}{w}$ and simplify:

$$t = \dfrac{kl}{w}$$

$$t = \dfrac{200(10)}{80}$$

$$= 25$$

It will take 25 weeks for 80 workers to build 10 miles of highway.

Self Check 8   How long will it take 60 workers to build 6 miles of highway?

Now Try   Problem 91

ANSWERS TO SELF CHECKS   **1.** $\frac{3}{2}$ **2.** $\frac{2}{3}$, 1   **3.** $68   **4.** 24 ft   **5.** 735 British pounds
6. 16 foot-candles   **8.** 20 weeks

STUDY SET 6.9

VOCABULARY

Fill in the blanks.

1. A _____ is the quotient of two numbers or two quantities with the same units.

2. An equation that states that two ratios are equal, such as $\frac{1}{2} = \frac{4}{8}$, is called a _____.

3. In $\frac{50}{3} = \frac{x}{9}$, the terms 50 and 9 are called the _____ and the terms 3 and x are called the _____ of the proportion. In a proportion, the product of the _____ is equal to the product of the _____.

4. The _____ products for the proportion $\frac{10}{3} = \frac{5}{x}$ are 10x and 15.

5. If two angles of one triangle have the same measure as two angles of a second triangle, the triangles are _____.

6. The equation $y = kx$ defines _____ variation: As x increases,
y _____.

7. The equation $y = \frac{k}{x}$ defines _____ variation: As x increases, y _____.

8. The equation $y = kxz$ defines _____ variation, and $y = \frac{kz}{x}$ defines _____ variation.

CONCEPTS

Determine whether direct or inverse variation applies and sketch a possible graph for the situation.

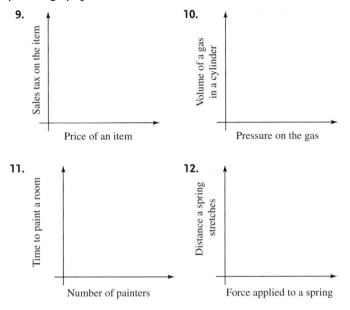

9. Sales tax on the item / Price of an item

10. Volume of a gas in a cylinder / Pressure on the gas

11. Time to paint a room / Number of painters

12. Distance a spring stretches / Force applied to a spring

NOTATION

Complete each solution.

13. Solve: $\dfrac{7}{6} = \dfrac{x+3}{12}$

$\quad\quad (12) = \quad (x+3)$

$\quad\quad 84 = 6x + \quad$

$\quad\quad \quad = 6x$

$\quad\quad \quad = x$

14. Solve: $\dfrac{18}{2x+1} = \dfrac{3}{14}$

$\quad\quad (14) = (\quad)3$

$\quad\quad 252 = \quad + 3$

$\quad\quad 249 = \quad$

$\quad\quad \quad = x$

GUIDED PRACTICE

Solve each proportion. See Example 1.

15. $\dfrac{x}{5} = \dfrac{15}{25}$

16. $\dfrac{4}{y} = \dfrac{6}{27}$

17. $\dfrac{r-2}{3} = \dfrac{r}{5}$

18. $\dfrac{x+1}{x-1} = \dfrac{6}{4}$

19. $\dfrac{5}{5z+3} = \dfrac{3}{2z+6}$

20. $\dfrac{9t+6}{t} = \dfrac{7}{3}$

21. $\dfrac{x-2}{x} = \dfrac{x+1}{x+2}$

22. $\dfrac{a}{a+1} = \dfrac{a+2}{a}$

Solve each proportion. See Example 2.

23. $\dfrac{2}{3x} = \dfrac{6x}{36}$

24. $\dfrac{y}{4} = \dfrac{4}{y}$

25. $\dfrac{2}{c} = \dfrac{c-3}{2}$

26. $\dfrac{2}{x+6} = \dfrac{-2x}{5}$

27. $\dfrac{1}{x+3} = \dfrac{-2x}{x+5}$

28. $\dfrac{x-1}{x+1} = \dfrac{2}{3x}$

29. $\dfrac{2b}{b+5} = \dfrac{-b}{3b+8}$

30. $\dfrac{-3c}{c-2} = \dfrac{c}{c+2}$

Express each verbal model in symbols. See Objectives 5 and 6.

31. A varies directly as the square of p.

32. t varies directly as s.

33. z varies inversely as the cube of t.

34. v varies inversely as the square of r.

Express each verbal model in symbols. See Objectives 7 and 8.

35. C varies jointly as x, y, and z.

36. d varies jointly as r and t.

37. P varies directly as the square of a and inversely as the cube of j.

38. M varies inversely as the cube of n and jointly as x and the square of z.

Express each variation model in words. In each equation, k is the constant of variation. See Objectives 5 and 6.

39. $r = kt$

40. $A = kr^3$

41. $b = \dfrac{k}{h}$

42. $d = \dfrac{k}{W^4}$

Express each variation model in words. In each equation, k is the constant of variation. See Objectives 7 and 8.

43. $U = krs^2t$

44. $L = kmn$

45. $P = \dfrac{km}{n}$

46. $R = \dfrac{kL}{d^2}$

TRY IT YOURSELF

Solve each proportion.

47. $\dfrac{b+4}{5} = \dfrac{3b-6}{3}$

48. $\dfrac{2y+6}{3} = \dfrac{4y-16}{5}$

49. $\dfrac{5}{b+3} = \dfrac{b}{2}$

50. $\dfrac{p+2}{p+5} = \dfrac{p-3}{p-2}$

51. $\dfrac{9z+6}{z^2+3z} = \dfrac{7}{z+3}$

52. $\dfrac{3}{n^2+3n} = \dfrac{2}{n^2+4n+3}$

53. $\dfrac{h^2}{5} = \dfrac{h}{2h-9}$

54. $\dfrac{b^2}{5} = \dfrac{b}{6b-13}$

55. $\dfrac{x}{x+2} = \dfrac{6}{x+2}$

56. $\dfrac{a}{a-3} = \dfrac{5}{a-3}$

57. $\dfrac{t^2-1}{5} = \dfrac{1-t^2}{2t}$

58. $\dfrac{n^2}{6} = \dfrac{n}{n-1}$

59. $\dfrac{2.5x+1}{2} = \dfrac{4.5}{12}$

60. $\dfrac{2}{5} = \dfrac{1.5x-2}{0.25}$

61. $\dfrac{t}{10} = \dfrac{10}{t}$

62. $\dfrac{-6}{m} = \dfrac{m}{-6}$

APPLICATIONS

Use a proportion to solve each problem.

63. CAFFEINE   Many convenience stores sell super-size 44-ounce soft drinks in refillable cups. For each of the products listed in the table, find the amount of caffeine contained in one of the large cups. Round to the nearest milligram.

Soft drink, 12 oz	Caffeine (mg)
Mountain Dew	55
Pepsi	38
Coca-Cola Classic	34

64. TELEPHONES   As of 2007, Luxembourg, in Europe, had 1,500 mobile cellular telephones per 1,000 people—the highest rate of any country in the world. If Luxembourg's population is about 480,200, how many mobile cellular telephones does the country have?

65. WALLPAPERING   Read the instructions on the label of wallpaper adhesive. Estimate the amount of adhesive needed to paper 500 square feet of kitchen walls if a heavy wallpaper will be used.

> COVERAGE: One-half gallon will hang approximately 4 single rolls (140 sq ft), depending on the weight of the wall covering and the condition of the wall.

66. RECOMMENDED DOSAGES   The recommended child's dose of the sedative hydroxine is 0.006 gram per kilogram of body mass. Find the dosage for a 30-kg child in grams and in milligrams.

67. ERGONOMICS   The science of ergonomics coordinates the design of working conditions with the requirements of the worker. The illustration gives guidelines for the dimensions (in inches) of a computer workstation to be used by a person whose height is 69 inches. Find a set of workstation dimensions for a person 5 feet 11 inches tall. Round to the nearest tenth.

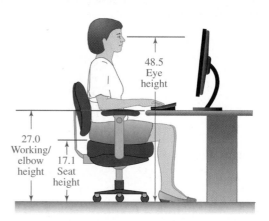

68. SHOPPING   A recipe for guacamole dip calls for 5 avocados. If they are advertised at 3 for $1.98, what will 5 avocados cost?

69. DRAWING   See the illustration. To make an enlargement of the sailboat, an artist drew a grid over the smaller picture and transferred the contents of each small box to its corresponding larger box on another sheet of paper. If the smaller picture is 3 in. × 5 in. and if the width of the enlargement is 7.5 in., what is the length of the enlargement?

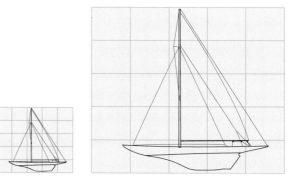

70. DRAFTING   In a scale drawing, a 280-foot antenna tower is drawn $7\frac{1}{2}$ inches high. The building next to it is drawn $2\frac{1}{4}$ inches high. How tall is the actual building?

Use similar triangles to solve each problem.

71. FLAGPOLES A man places a mirror on the ground and sees the reflection of the top of a flagpole, as in the illustration. The two triangles in the illustration are similar. Find the height *h* of the flagpole.

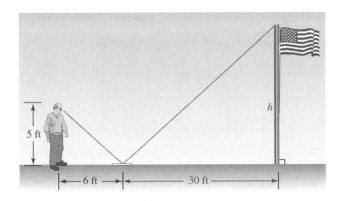

72. WASHINGTON, D.C. The Washington Monument casts a shadow of $166\frac{1}{2}$ feet at the same time as a 5-foot-tall tourist casts a shadow of $1\frac{1}{2}$ feet. Find the height of the monument.

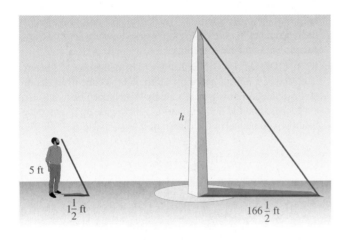

73. WIDTH OF A RIVER Use the dimensions in the illustration to find *w*, the width of the river. The two triangles in the illustration are similar.

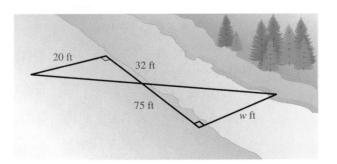

74. FLIGHT PATHS An airplane ascends 150 feet as it flies a horizontal distance of 1,000 feet. How much altitude will it gain as it flies a horizontal distance of 1 mile? (*Hint:* 5,280 feet = 1 mile.)

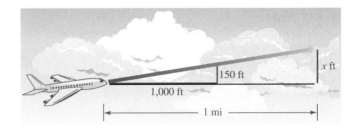

75. GRAPHIC ARTS The compass in the illustration is used to draw circles with different radii (plural for radius). For the setting shown, what radius will the resulting circle have?

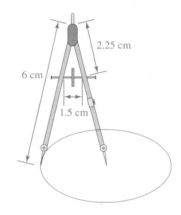

76. SKI RUNS A ski course with $\frac{1}{2}$ mile of horizontal run falls 100 feet in every 300 feet of run. Find the height of the hill.

77. *Campus to Careers*
Webmaster

The language of variation is often used to describe various aspects of the Internet and websites. Determine whether each statement, generally speaking, is true or false.

    a. The dollar amount of sales that an Internet website receives is inversely proportional to the amount of Internet traffic that visits the website.

    b. The download time of an Internet website varies directly with the bandwidth being used.

    c. Search engines like Google place a value on a website that is directly proportional to the number of sites that link to it.

Solve each problem by writing a variation model.

78. GRAVITY   The force of gravity acting on an object varies directly as the mass of the object. The force on a mass of 5 kilograms is 49 newtons. What is the force acting on a mass of 12 kilograms?

79. FREE FALL   An object in free fall travels a distance s that is directly proportional to the square of the time t. If an object falls 1,024 feet in 8 seconds, how far will it fall in 10 seconds?

80. FINDING DISTANCE   The distance that a car can go varies directly as the number of gallons of gasoline it consumes. If a car can go 288 miles on 12 gallons of gasoline, how far can it go on a full tank of 18 gallons?

81. FARMING   The number of days that a given number of bushels of corn will last when feeding cattle varies inversely as the number of animals. If x bushels will feed 25 cows for 10 days, how long will the feed last for 10 cows?

82. ORGAN PIPES   The frequency of vibration of air in an organ pipe is inversely proportional to the length of the pipe. If a pipe 2 feet long vibrates 256 times per second, how many times per second will a 6-foot pipe vibrate?

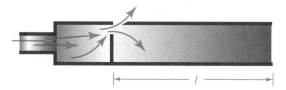

83. GAS PRESSURE   Under constant temperature, the volume occupied by a gas varies inversely to the pressure applied. If the gas occupies a volume of 20 cubic inches under a pressure of 6 pounds per square inch, find the volume when the gas is subjected to a pressure of 10 pounds per square inch.

84. REAL ESTATE   The following table shows the listing price for three homes in the same general locality. Write the variation model (direct or inverse) that describes the relationship between the listing price and the number of square feet of a house in this area.

Number of square feet	Listing price
1,720	$180,600
1,205	$126,525
1,080	$113,400

85. TRUCKING COSTS   The costs of a trucking company vary jointly as the number of trucks in service and the number of hours they are used. When 4 trucks are used for 6 hours each, the costs are $1,800. Find the costs of using 10 trucks, each for 12 hours.

86. OIL STORAGE   The number of gallons of oil that can be stored in a cylindrical tank varies jointly as the height of the tank and the square of the radius of its base. The constant of proportionality is 23.5. Find the number of gallons that can be stored in the cylindrical tank shown.

20 ft

15 ft

87. ELECTRONICS   The voltage (in volts) measured across a resistor is directly proportional to the current (in amperes) flowing through the resistor. The constant of variation is the **resistance** (in ohms). If 6 volts is measured across a resistor carrying a current of 2 amperes, find the resistance.

88. ELECTRONICS   The power (in watts) lost in a resistor (in the form of heat) varies directly as the square of the current (in amperes) passing through it. The constant of proportionality is the resistance (in ohms). What power is lost in a 5-ohm resistor carrying a 3-ampere current?

89. STRUCTURAL ENGINEERING   The deflection of a beam is inversely proportional to its width and the cube of its depth. If the deflection of a 4-inch wide by 4-inch deep beam is 1.1 inches, find the deflection of a 2-inch wide by 8-inch deep beam positioned as in the illustration.

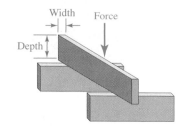

Width   Force

Depth

90. STRUCTURAL ENGINEERING   Find the deflection of the beam in Exercise 89 when the beam is positioned as in the illustration.

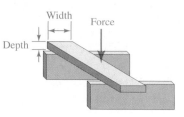

Width   Force

Depth

91. TENSION IN A STRING   When playing with a Skip It toy, a child swings a weighted ball on the end of a string in a circular motion around one leg while jumping over the revolving string with the other leg. See the illustration. The tension T in the string is directly proportional to the square of the speed s of the ball and inversely proportional to the radius r of the circle. If the tension in the string is 6 pounds when the speed of the ball is 6 feet per second and the radius is 3 feet, find the tension when the speed is 8 feet per second and the radius is 2.5 feet.

92. GAS PRESSURE   The pressure of a certain amount of gas is directly proportional to the temperature (measured on the Kelvin scale) and inversely proportional to the volume. A sample of gas at a pressure of 1 atmosphere occupies a volume of 1 cubic meter at a temperature of 273 Kelvin. When heated, the gas expands to twice its volume, but the pressure remains constant. To what temperature is it heated?

WRITING

93. Distinguish between a *ratio* and a *proportion*.

94. Give examples of two quantities from everyday life that vary directly and two quantities that vary inversely.

REVIEW

Perform the indicated operations.

95. $\left(\dfrac{5}{2}w^3 + \dfrac{1}{4}w^2 + \dfrac{3}{5}\right) - \left(\dfrac{1}{3}w^3 + \dfrac{1}{2}w^2 - \dfrac{1}{5}\right)$

96. $(6a^2x^3 - 2ax^2 + 3a^3) + (-4a^2x^3 - 2a^3)$

97. $(3y + 1)(2y^2 + 3y + 2)$

98. $(5k - 6m^2)^2$

CHALLENGE PROBLEMS

99. As the cost of a purchase that is less than \$5 increases, the amount of change received from a five-dollar bill decreases. Is this inverse variation? Explain.

100. You've probably heard of Murphy's first law:

If anything can go wrong, it will.

Another of Murphy's laws is:

The chances of a piece of bread falling with the grapejelly side down varies directly with the cost of the carpet.

Write one of your own witty sayings using the phrase *varies directly*.

CHAPTER 6
Summary & Review

SECTION 6.1   Rational Functions and Simplifying Rational Expressions

DEFINITIONS AND CONCEPTS	EXAMPLES
A **rational expression** is an expression of the form $\frac{A}{B}$, where A and B are polynomials and B does not equal 0.	Rational expressions: $$\frac{3x^2}{xy}, \quad \frac{5b - 15}{b^2 - 25}, \quad \text{and} \quad \frac{a + 2}{a^2 - 3a - 4}$$
A **rational function** is a function whose equation is defined by a rational expression in one variable.	Rational functions: $f(x) = \dfrac{6x}{x - 2}$ and $f(n) = \dfrac{n + 3}{n^3 + 2n - 9}$
Since division by 0 is undefined, any values that make the denominator 0 in a rational function must be excluded from the **domain** of the function.	Find the domain of the rational function: $f(x) = \dfrac{x + 3}{x^2 - 4}$ $x^2 - 4 = 0$ *Set the denominator equal to 0.* $(x + 2)(x - 2) = 0$ *Factor the difference of two squares.* $x + 2 = 0$ or $x - 2 = 0$ *Set each factor equal to 0.* $x = -2$ $x = 2$ *Solve each equation.* The domain of the function is the set of all real numbers except -2 and 2. In interval notation, the domain is $(-\infty, -2) \cup (-2, 2) \cup (2, \infty)$.
To simplify a rational expression: **1.** Factor the numerator and denominator completely. **2.** Remove factors equal to 1 by replacing each pair of factors common to the numerator and denominator with the equivalent fraction $\frac{1}{1}$. **3.** Multiply the remaining factors in the numerator and in the denominator.	Simplify: $\dfrac{x^2 - 4}{2x + 4} = \dfrac{\overset{1}{\cancel{(x + 2)}}(x - 2)}{2\underset{1}{\cancel{(x + 2)}}} = \dfrac{x - 2}{2}$ $\dfrac{2a^3 - 5a^2 - 12a}{2a^3 - 11a^2 + 12a} = \dfrac{\cancel{a}(2a + 3)\overset{1}{\cancel{(a - 4)}}}{\underset{1}{\cancel{a}}(2a - 3)\underset{1}{\cancel{(a - 4)}}} = \dfrac{2a + 3}{2a - 3}$
The quotient of any nonzero expression and its **opposite** is -1.	$\dfrac{7x - 6}{6 - 7x} = -1$ *Because $7x - 6$ and $6 - 7x$ are opposites.* Simplify: $\dfrac{10 - 2b}{b^2 - 5b} = \dfrac{2(5 - b)}{b(b - 5)} = \dfrac{2\overset{-1}{\cancel{(5 - b)}}}{b\underset{1}{\cancel{(b - 5)}}} = -\dfrac{2}{b}$

REVIEW EXERCISES

1. Complete the table of values for the rational function $f(x) = \frac{4}{x}$ where $x > 0$. Round to the nearest hundredth when appropriate. Then graph the function. Label the horizontal asymptote.

x	$f(x)$
$\frac{1}{2}$	
1	
2	
3	
4	
5	
6	
7	
8	

4. Use a graphing calculator to graph the rational function $f(x) = \frac{3x + 2}{x}$. From the graph, determine the equations of the horizontal and vertical asymptotes and the domain and range.

Simplify each rational expression, if possible.

5. $\dfrac{48x^2y}{76xy^8}$

6. $\dfrac{x^2 - 49}{x^2 + 14x + 49}$

7. $\dfrac{x^2 - 2x + 4}{2x^5 + 16x^2}$

8. $\dfrac{x^2 + 6x + 36}{x^7 - 216x^4}$

2. Use the graph of function f to find each of the following:

   a. $f(12)$

   b. The value(s) of x for which $f(x) = 6$

   c. The domain and range of f

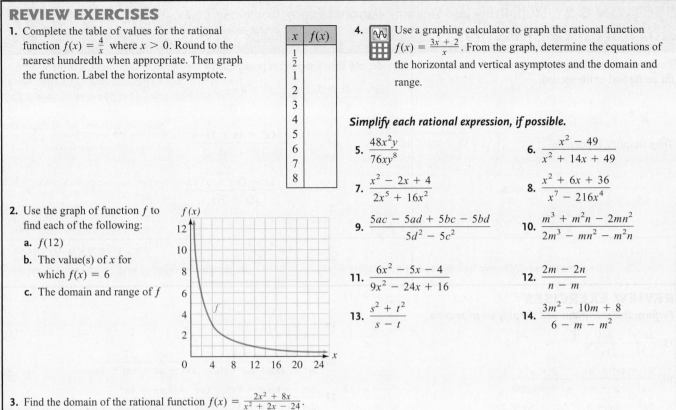

9. $\dfrac{5ac - 5ad + 5bc - 5bd}{5d^2 - 5c^2}$

10. $\dfrac{m^3 + m^2n - 2mn^2}{2m^3 - mn^2 - m^2n}$

11. $\dfrac{6x^2 - 5x - 4}{9x^2 - 24x + 16}$

12. $\dfrac{2m - 2n}{n - m}$

13. $\dfrac{s^2 + t^2}{s - t}$

14. $\dfrac{3m^2 - 10m + 8}{6 - m - m^2}$

3. Find the domain of the rational function $f(x) = \frac{2x^2 + 8x}{x^2 + 2x - 24}$. Express your answer in words and using interval notation.

SECTION 6.2   Multiplying and Dividing Rational Expressions

DEFINITIONS AND CONCEPTS	EXAMPLES
To **multiply rational expressions,** multiply the numerators and multiply the denominators. $$\frac{A}{B} \cdot \frac{C}{D} = \frac{AC}{BD}$$ Then simplify, if possible.	Multiply, and then simplify, if possible. $$\frac{8z^2}{y^3} \cdot \frac{y}{4z} = \frac{8z^2 \cdot y}{y^3 \cdot 4z} = \frac{2 \cdot \overset{1}{\cancel{4}} \cdot \overset{1}{\cancel{z}} \cdot z \cdot \overset{1}{\cancel{y}}}{\underset{1}{\cancel{y}} \cdot y \cdot y \cdot \underset{1}{\cancel{4}} \cdot \underset{1}{\cancel{z}}} = \frac{2z}{y^2}$$ $\dfrac{x^2 - 4}{x + 3} \cdot \dfrac{3x + 9}{x + 2} = \dfrac{(x^2 - 4)(3x + 9)}{(x + 3)(x + 2)}$ Multiply the numerators. Multiply the denominators. $= \dfrac{(x + 2)(x - 2) \cdot 3 \cdot (x + 3)}{(x + 3)(x + 2)}$ Factor completely and then simplify. $= 3(x - 2)$ Multiply the remaining factors in the numerator. Multiply the remaining factors in the denominator.

SECTION 6.4   Simplifying Complex Fractions—*continued*

DEFINITIONS AND CONCEPTS	EXAMPLES

Two methods are used to simplify **complex fractions.**

Method 1: Write the numerator and denominator as single fractions. Then divide the fractions and simplify.

This method works well when a complex fraction is written, or can be easily written, as a quotient of two single rational expressions.

Simplify:

$$\dfrac{\dfrac{4x^2}{y^3}}{\dfrac{14x}{y}} = \dfrac{4x^2}{y^3} \div \dfrac{14x}{y}$$

The main fraction bar of the complex fraction indicates division.

$$= \dfrac{4x^2}{y^3} \cdot \dfrac{y}{14x}$$

To divide rational expressions, multiply the first by the reciprocal of the second.

$$= \dfrac{4x^2 \cdot y}{y^3 \cdot 14x}$$

Multiply the numerators.
Multiply the denominators.

$$= \dfrac{\overset{1}{\cancel{2}} \cdot 2 \cdot \overset{1}{\cancel{x}} \cdot x \cdot \overset{1}{\cancel{y}}}{\underset{1}{\cancel{y}} \cdot y \cdot y \cdot \underset{1}{\cancel{2}} \cdot 7 \cdot \underset{1}{\cancel{x}}}$$

Factor the numerator and denominator. Then simplify by removing common factors of the numerator and denominator.

$$= \dfrac{2x}{7y^2}$$

Multiply the remaining factors in the numerator.
Multiply the remaining factors in the denominator.

Method 2: Determine the LCD of all the rational expressions in the complex fraction and multiply the complex fraction by 1, written in the form $\dfrac{LCD}{LCD}$.

This method works well when the complex fraction has sums and/or differences in the numerator or denominator.

Simplify:

$$\dfrac{\dfrac{1}{x} - y}{\dfrac{5}{2x}} = \dfrac{\dfrac{1}{x} - y}{\dfrac{5}{2x}} \cdot \dfrac{2x}{2x}$$

The LCD of all the rational expressions in the complex fraction is $2x$. Multiply the complex fraction by 1 in the form $\dfrac{2x}{2x}$.

$$= \dfrac{\left(\dfrac{1}{x} - y\right)2x}{\left(\dfrac{5}{2x}\right)2x}$$

Multiply the numerators.
Multiply the denominators.

$$= \dfrac{\dfrac{1}{x} \cdot 2x - y \cdot 2x}{5}$$

In the numerator, distribute the multiplication by $2x$.
In the denominator, perform the multiplication by $2x$.

$$= \dfrac{2 - 2xy}{5}$$

In the numerator, perform each multiplication by $2x$.

REVIEW EXERCISES

Simplify each complex fraction.

39. $\dfrac{\dfrac{4a^3b^2}{9c}}{\dfrac{14a^3b}{9c^4}}$

40. $\dfrac{\dfrac{p^2 - 9}{6pt}}{\dfrac{p^2 + 5p + 6}{3pt}}$

43. $\dfrac{(x - y)^{-2}}{x^{-2} - y^{-2}}$

44. $\dfrac{1 + \dfrac{1}{b + d}}{\dfrac{1}{b + d} - 1}$

41. $\dfrac{\dfrac{1}{a} + \dfrac{2}{b}}{\dfrac{2}{a} - \dfrac{1}{b}}$

42. $\dfrac{1 - \dfrac{1}{x} - \dfrac{2}{x^2}}{1 + \dfrac{4}{x} + \dfrac{3}{x^2}}$

45. $\dfrac{\dfrac{2b}{b - 1} - \dfrac{3}{b}}{\dfrac{1}{b - 1} + \dfrac{2}{b}}$

46. $\dfrac{\dfrac{8}{r + 3}}{\dfrac{4}{r - 2} - \dfrac{2}{r^2 + r - 6}}$

SECTION 6.5   Dividing Polynomials

DEFINITIONS AND CONCEPTS	EXAMPLES

To divide monomials, use the method for simplifying fractions or use the rules for exponents.

Divide the monomials:

$$\frac{8p^2q}{20pq^3} = \frac{2 \cdot \overset{1}{\cancel{4}} \cdot \overset{1}{\cancel{p}} \cdot p \cdot \overset{1}{\cancel{q}}}{\underset{1}{\cancel{4}} \cdot 5 \cdot \underset{1}{\cancel{p}} \cdot \underset{1}{\cancel{q}} \cdot q \cdot q} \quad \text{or} \quad \frac{8p^2q}{20pq^3} = \frac{2}{5}p^{2-1}q^{1-3}$$

Keep each base and subtract the exponents.

$$= \frac{2p}{5q^2} \qquad\qquad = \frac{2p^1q^{-2}}{5}$$

$$= \frac{2p}{5q^2}$$

Move q^{-2} to the denominator and change the sign of the exponent.

To divide a polynomial by a monomial, divide each term of the numerator by the denominator.

Divide: $\dfrac{9c^9d^4 - 12c^3d^7}{27cd^5} = \dfrac{9c^9d^4}{27cd^5} - \dfrac{12c^3d^7}{27cd^5}$

$$= \frac{c^8}{3d} - \frac{4c^2d^2}{9} \qquad \text{Do each monomial division.}$$

Long division can be used to **divide a polynomial by a polynomial** (other than a monomial). The long division method is a series of four steps that are repeated: Divide, multiply, subtract, and bring down the next term.

When the division has a remainder, write the answer in the form Quotient $+ \frac{remainder}{divisor}$.

Divide $6x^3 - x^2 + 6x + 5$ by $2x + 1$.

$$
\begin{array}{r}
3x^2 - 2x + 4 \\
2x + 1 \overline{\smash{)}\ 6x^3 - x^2 + 6x + 5} \\
-(6x^3 + 3x^2) \\
\hline
-4x^2 + 6x \\
-(-4x^2 - 2x) \\
\hline
8x + 5 \\
-(8x + 4) \\
\hline
1
\end{array}
$$

The first division: $\frac{6x^3}{2x} = 3x^2$.

The second division: $\frac{-4x^2}{2x} = -2x$.

The third division: $\frac{8x}{2x} = 4$.

The remainder is 1.

Thus $\dfrac{6x^3 - x^2 + 6x + 5}{2x + 1} = 3x^2 - 2x + 4 + \dfrac{1}{2x + 1}$.

The long division method works best when the terms of the divisor and the dividend are written in **descending powers of the variable.**

When the dividend has **missing terms,** insert such terms with a coefficient of 0, or leave a blank space.

To divide $\dfrac{5x + x^3 + 3 + 3x^2}{x + 1}$, write: $x + 1 \overline{\smash{)}\ x^3 + 3x^2 + 5x + 3}$

To divide $\dfrac{x^2 - 9}{x - 3}$, write: $x - 3 \overline{\smash{)}\ x^2 + 0x - 9}$

REVIEW EXERCISES

Perform each division. Write answers using positive exponents.

47. $\dfrac{25h^4k^7}{55hk^9}$

48. $(5x^3y^3z^{10}) \div (10x^3y^6z^{20})$

Perform each division.

49. $\dfrac{36a + 32}{6}$

50. $\dfrac{30x^3y^2 - 15x^2y - 10xy^2}{-10xy}$

51. $b + 5 \overline{\smash{)}\ b^2 + 9b + 20}$

52. $\dfrac{-33v - 8v^2 + 3v^3 - 10}{1 + 3v}$

53. $x + 2 \overline{\smash{)}\ x^3 + 8}$

54. Divide $(8m^2 - 18m - 9)$ by $4m + 1$.

55. $(3a^3 - 2a^2 - 8) \div (a^2 + 5)$

56. $\dfrac{m^8 + m^6 - 4m^4 + 5m^2 - 1}{m^4 + 2m^2 - 3}$

SECTION 6.6   Synthetic Division

DEFINITIONS AND CONCEPTS	EXAMPLES
Synthetic division is used to divide a polynomial by a binomial of the form $x - k$.	Use synthetic division to divide: $\dfrac{5x^3 - x^2 + 4x - 3}{x + 1}$ We write the divisor in $x - k$ form as $x - (-1)$ to determine that $k = -1$ and proceed as follows: This represents → $\underline{-1}$ 5 -1 4 -3 division by x + 1. $\qquad\qquad\qquad\quad -5\quad 6\quad -10$ $\qquad\qquad\overline{\quad 5\ \ -6\ \ \ 10\ \ -13}$ ← This is the remainder. Thus, $\frac{5x^3 - x^2 + 4x - 3}{x + 1} = 5x^2 - 6x + 10 + \frac{-13}{x + 1}$.
Remainder theorem: If a polynomial $P(x)$ is divided by $x - k$, the remainder is $P(k)$. It follows from the remainder theorem that a polynomial can be evaluated using synthetic division.	Since the remainder in the previous division is -13, the remainder theorem guarantees that $P(-1) = -13$. We can check this by evaluating the function at $x = -1$: $P(x) = 5x^3 - x^2 + 4x - 3$ $P(-1) = 5(-1)^3 - (-1)^2 + 4(-1) - 3$ $\qquad\ = -5 - 1 - 4 - 3$ $\qquad\ = -13$
Factor theorem: If $P(x)$ is divided by $x - k$, then $P(k) = 0$, if and only if $x - k$ is a factor of $P(x)$.	Determine whether $x - 3$ is a factor of $2x^3 - 8x^2 + 9x - 9$. We can use synthetic division to find the remainder when the polynomial is divided by $x - 3$. If the remainder is 0, $x - 3$ is a factor. If the remainder is not 0, $x - 3$ is not a factor. $\underline{3}$ 2 -8 9 -9 $\qquad\qquad 6\ \ -6\quad 9$ $\overline{\ \ 2\ \ -2\quad 3\quad\ \ 0}$ Since the remainder is 0, $x - 3$ is a factor of $2x^3 - 8x^2 + 9x - 9$.

REVIEW EXERCISES

Use synthetic division to perform each division.

57. $(x^2 - 13x + 42) \div (x - 6)$

58. Divide $m^3 - 6m^2 + 11m - 6$ by $m - 3$.

59. $\dfrac{-3n^5 + 10n^4 + 7n^3 + 2n^2 + 9n - 4}{n - 4}$

60. $\dfrac{4x^3 + 5x^2 - 1}{x + 2}$

61. Divide $3a - a^2 + 3a^3 + 3a^4 + 10$ by $a + 1$.

62. $\dfrac{x^4 + 1}{x - 3}$

63. Let $P(x) = x^4 - 2x^3 + x^2 - 3x + 12$. Use the remainder theorem and synthetic division to find $P(-2)$.

64. Let $P(x) = x^3 - 13x^2 - 27$. Use the remainder theorem and synthetic division to find $P(5)$.

Use the factor theorem to determine whether the first expression is a factor of P(x).

65. $x - 5$; $P(x) = x^3 - 3x^2 - 8x - 10$

66. $x + 5$; $P(x) = x^3 + 4x^2 - 5x + 5$
   Hint: Write $x + 5$ as $x - (-5)$.

SECTION 6.7   Solving Rational Equations

DEFINITIONS AND CONCEPTS	EXAMPLES

If an equation contains one or more rational expressions, it is called a **rational equation.**

To **solve a rational equation:**

1. Factor all denominators.

2. Determine which numbers cannot be solutions of the equation.

3. Multiply both sides of the equation by the LCD of all rational expressions in the equation.

4. Use the distributive property to remove parentheses, remove any factors equal to 1, and write the result in simplified form.

5. Solve the resulting equation.

6. Check all possible solutions in the original equation.

Rational equations: $\dfrac{4}{5} = \dfrac{x}{x-1}$ and $\dfrac{2a}{a^2 + 9a + 20} = \dfrac{2}{a+5} + \dfrac{3}{a+4}$

Solve: $\dfrac{3}{2} + \dfrac{1}{a-4} = \dfrac{5}{2a-8}$

If we factor the last denominator, the equation can be written as:

$$\frac{3}{2} + \frac{1}{a-4} = \frac{5}{2(a-4)}$$

We see that 4 cannot be a solution of the equation, because it makes at least one of the rational expressions in the equation undefined.

We can clear the equation of fractions by multiplying both sides by $2(a-4)$, which is the LCD of the three rational expressions.

$$2(a-4)\left(\frac{3}{2} + \frac{1}{a-4}\right) = 2(a-4)\left[\frac{5}{2(a-4)}\right] \qquad \text{Multiply both sides by the LCD.}$$

$$2(a-4)\left(\frac{3}{2}\right) + 2(a-4)\left(\frac{1}{a-4}\right) = 2(a-4)\left[\frac{5}{2(a-4)}\right] \qquad \text{Distribute.}$$

$$\overset{1}{2}(a-4)\left(\frac{3}{2}\right) + 2(a-4)\left(\frac{1}{a-4}\right) = \overset{1}{2}(a-4)\left[\frac{5}{2(a-4)}\right] \qquad \begin{array}{l}\text{Remove common}\\\text{factors.}\end{array}$$

$$(a-4)3 + 2 = 5 \qquad \text{Simplify.}$$

$$3a - 12 + 2 = 5 \qquad \text{Distribute.}$$

$$3a - 10 = 5 \qquad \text{Combine like terms.}$$

$$3a = 15$$

$$a = 5$$

The solution is 5. Verify that it satisfies the original equation.

Multiplying both sides of an equation by a quantity that contains a variable can lead to **extraneous solutions**.

All possible solutions of a rational equation must be checked.

Solve: $\dfrac{x+6}{x-1} = 2 + \dfrac{7}{x-1}$

$$(x-1)\left(\frac{x+6}{x-1}\right) = (x-1)\left(2 + \frac{7}{x-1}\right) \qquad \begin{array}{l}\text{Multiply both sides by the}\\\text{LCD, } x-1.\end{array}$$

$$(x-1)\left(\frac{x+6}{x-1}\right) = (x-1)(2) + (x-1)\left(\frac{7}{x-1}\right) \qquad \text{Distribute.}$$

$$(x-1)\left(\frac{x+6}{x-1}\right) = (x-1)(2) + (x-1)\left(\frac{7}{x-1}\right) \qquad \text{Remove common factors.}$$

$$x + 6 = (x-1)(2) + 7 \qquad \text{Simplify.}$$

$$x + 6 = 2x - 2 + 7 \qquad \text{Distribute.}$$

$$x + 6 = 2x + 5 \qquad \text{Combine like terms.}$$

$$1 = x \qquad \text{Solve for } x.$$

Since 1 makes the denominator of a rational expression 0, it does not check. It is an extraneous solution. The equation has no solution.

REVIEW EXERCISES

Solve each equation, if possible.

67. $\dfrac{4}{x} - \dfrac{1}{10} = \dfrac{7}{2x}$

68. $\dfrac{11}{t} = \dfrac{6}{t - 7}$

69. $\dfrac{3}{y} - \dfrac{2}{y + 1} = \dfrac{1}{2}$

70. $\dfrac{2}{3x + 15} - \dfrac{1}{18} = \dfrac{1}{3x + 12}$

71. $\dfrac{3}{x + 2} = \dfrac{1}{2 - x} + \dfrac{2}{x^2 - 4}$

72. $\dfrac{x + 3}{x - 5} + \dfrac{2x^2 + 6}{x^2 - 7x + 10} = \dfrac{3x}{x - 2}$

73. $\dfrac{5a}{a - 3} - 7 = \dfrac{15}{a - 3}$

74. a. Simplify: $\dfrac{10}{x^2 - 4x} - \dfrac{4}{x} + \dfrac{5}{x - 4}$

   b. Solve: $\dfrac{10}{x^2 - 4x} - \dfrac{4}{x} = \dfrac{5}{x - 4}$

Solve each formula for the indicated variable or expression.

75. $H = \dfrac{2ab}{a + b}$ for b

76. $\dfrac{x^2}{a^2} - \dfrac{y^2}{b^2} = 1$ for y^2

77. $\dfrac{1}{R} = \dfrac{1}{R_1} + \dfrac{1}{R_2}$ for R

78. $k = \dfrac{ma}{F}$ for F

SECTION 6.8   Problem Solving Using Rational Equations

DEFINITIONS AND CONCEPTS

Rate of Work: If a job can be completed in x units of time, the rate of work can be expressed as $\frac{1}{x}$ of the job is completed per unit of time.

Shared-work problems:

Work completed = rate of work · time worked

EXAMPLES

PRINTERS   Working alone, a 300-A model printer can print a company's payroll checks in 30 minutes. A 500-X model can print the same checks in 20 minutes. How long will it take if the printers work together to print the checks?

Analyze the problem   Let $x =$ the number of minutes it will take the printers, working together, to print the checks. Enter the data in a table.

	Rate ·	Time	= Work Completed
Model 300-A	$\dfrac{1}{30}$	x	$\dfrac{x}{30}$
Model 500-X	$\dfrac{1}{20}$	x	$\dfrac{x}{20}$

Form an equation   The part of the job done by the 300-A model plus the part of the job done by the 500-X model equals 1 job completed.

$$\dfrac{x}{30} + \dfrac{x}{20} = 1$$

Solve the equation

$$60\left(\dfrac{x}{30} + \dfrac{x}{20}\right) = 60(1) \qquad \text{Multiply both sides by the LCD, 60.}$$

$$60\left(\dfrac{x}{30}\right) + 60\left(\dfrac{x}{20}\right) = 60(1) \qquad \text{Distribute the multiplication by 60.}$$

$$2x + 3x = 60 \qquad \text{Perform each multiplication by 60.}$$

$$5x = 60 \qquad \text{Combine like terms.}$$

$$x = \dfrac{60}{5} = 12$$

State the conclusion   Working together, it will take the printers 12 minutes to print the checks.

SECTION 6.8   Problem Solving Using Rational Equations—*continued*

DEFINITIONS AND CONCEPTS	EXAMPLES
	Check the result In 12 minutes, the 300-A model will do $\frac{12}{30} = \frac{2}{5}$ of the job and the 500-X model will do $\frac{12}{20} = \frac{3}{5}$ of the job. Together they will do $\frac{2}{5} + \frac{3}{5} = 1$ whole job. The result checks.
Uniform motion problems: $$\text{Time} = \frac{\text{distance}}{\text{rate}}$$ When a boat travels *downstream,* the speed of the boat is increased by the current. When a boat travels *upstream,* the speed of the boat is decreased by the current.	See pages 599-601 for examples.

REVIEW EXERCISES

79. a. If a painter can complete a job in 10 hours, what is the painter's rate of work?

   b. If the painter works for x hours, how much of the job is completed?

80. DRAINING A TANK   If one outlet pipe can drain a tank in 24 hours and another pipe can drain the tank in 36 hours, how long will it take for both pipes to drain the tank?

81. ELECTRICIANS   It takes an apprentice 5 days more than an experienced electrician to wire a 1,500-square foot house. Working together, they can wire such a house in 6 days. How long would it take each person working alone to wire the house?

82. SINKS   A faucet can fill a garage sink in 2 minutes. It takes 3 minutes for the drain to empty the sink when it is full. How long will it take to fill the sink if the drain is open and the faucet is on?

83. ADVERTISING   A small plane pulling a banner can fly at a rate of 75 mph in calm air. Flying down the coast, with a tailwind, the plane flew 40 miles in the same time that it took to fly 35 miles up the coast, into a headwind. Find the rate of the wind.

84. TRIP LENGTH   Heavy traffic reduced a driver's usual speed by 10 mph, which lengthened her 200-mile trip by 1 hour. Find the driver's usual speed.

SECTION 6.9   Proportion and Variation

DEFINITIONS AND CONCEPTS	EXAMPLES
A **proportion** is a statement that two ratios are equal. In the proportion $\frac{a}{b} = \frac{c}{d}$, a and d are the **extremes** and b and c are the **means.** In any proportion, the product of the extremes is equal to the product of the means. (The **cross products** are equal.)	Proportion: $\frac{4}{9} = \frac{28}{63}$ Extremes: 4 and 63 Means: 9 and 28 Proportion: $\frac{4}{9} \diagup\!\!\!\!\diagdown \frac{28}{63}$ Cross product: $4 \cdot 63 = 252$ Cross product: $9 \cdot 28 = 252$

CHAPTER 6
Test

1. Fill in the blanks.

  a. A quotient of two polynomials, such as $\dfrac{x-8}{x^2-2x-3}$, is called a _____ expression.

  b. The _____ of $\dfrac{x+1}{x-7}$ is $\dfrac{x-7}{x+1}$.

  c. $\dfrac{\frac{x}{4}+\frac{1}{x}}{\frac{1}{8}+\frac{1}{x}}$ is an example of a _____ fraction.

  d. When solving a rational equation, if we obtain a number that does not satisfy the original equation, the number is called an _____ solution.

  e. An equation that states that two ratios are equal, such as $\frac{1}{2}=\frac{3}{6}$, is called a _____.

2. Explain the error that was made in the solution shown below.

$$\text{Simplify: } \frac{2(x+2)+3(x-3)}{x+2} = \frac{2\cancel{(x+2)}^{1}+3(x-3)}{\cancel{x+2}_{1}}$$

$$= \frac{2+3x-9}{1}$$

$$= 3x-7$$

Simplify each rational expression.

3. $\dfrac{12x^2y^3z^2}{18x^3y^4z^2}$

4. $\dfrac{2x+4}{x^2-4}$

5. $\dfrac{3y-6z}{2z-y}$

6. $\dfrac{2x^2+7xy+3y^2}{4xy+12y^2}$

7. Graph the rational function $f(x)=\dfrac{2}{x}$ for $x>0$. Label the horizontal asymptote.

8. Find the domain of the rational function $f(x)=\dfrac{x^2+6x+5}{x-x^2}$. Express it using interval notation.

Perform the operations and simplify when necessary. Write all answers using positive exponents only.

9. $\dfrac{x^2}{x^3z^2y^2}\cdot\dfrac{x^2z^4}{y^2z}$

10. $\dfrac{a^2+5a+6}{a^2-4}\cdot\dfrac{a^2-5a+6}{a^2-9}$

11. $\dfrac{xu+2u+3x+6}{u^2-9}\cdot\dfrac{13u-39}{x^2+3x+2}$

12. $\dfrac{x^3+y^3}{16x^2}\div\dfrac{x^2-xy+y^2}{8x^2+8xy}$

13. $\dfrac{a^2+7a+12}{a+3}\div\dfrac{16-a^2}{a-4}$

14. $\dfrac{(2x-3)^3}{x^2-2x+1}\div\dfrac{3x^2+7x+2}{3x^2-2x-1}\cdot\dfrac{x^2+x-2}{2x^7-3x^6}$

15. $\dfrac{-3t+4}{t^2+t-20}+\dfrac{6+5t}{t^2+t-20}$

16. $\dfrac{3wx}{wx-5}+\dfrac{wx+10}{5-wx}$

17. $8b-5+\dfrac{5b+4}{3b+1}$

18. $\dfrac{a+3}{a^2-a-2}-\dfrac{a-4}{a^2-2a-3}$

Simplify.

19. $\dfrac{\dfrac{2u^2w^3}{v^2}}{\dfrac{4uw^4}{uv}}$

20. $\dfrac{\dfrac{4}{3k}+\dfrac{k}{k+1}}{\dfrac{k}{k+1}-\dfrac{3}{k}}$

21. Divide: $\dfrac{18x^2y^3-12x^3y^2+9xy}{-3xy^4}$

22. Divide: $(y^3-48)\div(y+2)$

23. Let $P(x)=4x^3+3x^2+2x-7$. Use synthetic division to find $P(2)$.

24. Use the factor theorem to decide whether $x+3$ is a factor of $P(x)=x^4+3x^3-16x^2-27x+63$.

Solve each equation.

25. $\dfrac{34}{x^2}+\dfrac{13}{20x}=\dfrac{3}{2x}$

26. $\dfrac{u-2}{u-3}+3=u+\dfrac{u-4}{3-u}$

27. $\dfrac{3}{x-2}=\dfrac{x+3}{2x}$

28. $\dfrac{4}{m^2-9}+\dfrac{5}{m^2-m-12}=\dfrac{7}{m^2-7m+12}$

Solve each formula for the indicated variable or expression.

29. $\dfrac{1}{r}=\dfrac{1}{r_1}+\dfrac{1}{r_2}$ for r_2

30. $\dfrac{x^2}{a^2}+\dfrac{y^2}{b^2}=1$ for a^2

31. ROOFING One crew can finish a 2,800-square-foot roof in 12 hours, and another crew can do the job in 10 hours. If they work together, can they finish before a predicted rain in 5 hours? If not, how long will they have to work in the rain?

32. HOSPITALS It takes a technician 45 minutes longer than his supervisor to clean an outpatient surgery room. Working together, they can clean the room in 30 minutes. How long would it take each person working alone to clean the room?

33. TOURING THE COUNTRYSIDE A man bicycles 5 mph faster than he can walk. He bicycles a distance of 24 miles and then hikes back along the same route. If the entire trip takes 11 hours, how fast does he walk?

34. RIVER CRUISES A paddleboat can make an 8-mile trip up the Mississippi River and return in a total of 3 hours. If the boat travels 6 mph in still water, find the speed of the current.

35. SHADOWS Refer to the illustration. Find the height of the tree.

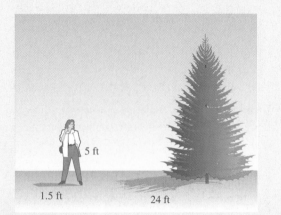

36. ANNIVERSARY GIFTS A florist sells a dozen long-stemmed red roses for $57.99. In honor of their 16th wedding anniversary, a man wants to buy 16 roses for his wife. What will the roses cost?

37. SOUND Sound intensity (loudness) varies inversely as the square of the distance from the source. If a rock band has a sound intensity of 100 decibels 30 feet away from the amplifier, find the sound intensity 60 feet away from the amplifier.

38. Draw a possible graph showing that the weekly earnings of a person *varies directly* with the number of hours worked during the week. Label the axes.

GROUP PROJECT

USING PROPORTIONS WHEN COOKING

Overview: In this activity, you will gain experience with proportions.

Instructions: Each student should bring to class the recipe for a favorite food written on the front of a 3 × 5 card. Form groups of 2 or 3 students. Working together, the group should write proportions to determine the amount of each ingredient needed to make the recipe for the exact number of people in the class. For instance, suppose there are 32 students in your class. If a recipe that serves 8 calls for 3 cups of flour, the amount of flour needed for the recipe to serve 32 is given by the proportion

$$\text{Cups of flour} \rightarrow \frac{3}{8} = \frac{x}{32} \leftarrow \text{Cups of flour}$$
$$\text{People served} \rightarrow \qquad \leftarrow \text{People served}$$

Write the new ingredient list on the back of each recipe card. Since the divisions involved might not be exact, be prepared to round when making the calculations. After all of the students' recipes have been adjusted, exchange recipe cards with someone in your class.

CONTINUED FRACTIONS

Overview: In this activity, as you gain experience simplifying complex fractions, you will make an interesting discovery about continued fractions.

Instructions: Form groups of 2 or 3 students. Working as a group, simplify each expression in the following list. Note that the third, fourth, fifth, and all the subsequent fractions in the list have a complex fraction in their denominator. These expressions are called *continued fractions*.

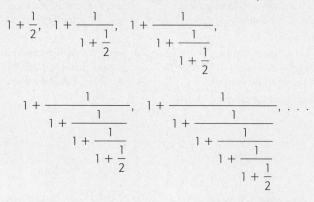

Each of these expressions can be simplified by using the value of the expression preceding it. For example, to simplify the second expression in the list, replace $1 + \frac{1}{2}$ with $\frac{3}{2}$. Show that the expressions in the list simplify to $\frac{3}{2}, \frac{5}{3}, \frac{8}{5}, \frac{13}{8}, \frac{21}{13}, \dots$. Then write the next 3 continued fractions in the list. From what you have learned, predict the answers if each of them were simplified.

CUMULATIVE REVIEW
Chapters 1–6

1. Evaluate: $12 - 6[(130 - 4^3) - 2]$ [Section 1.3]

2. Simplify: $9(a^3 + 3a) - 5(3a - a^3) - 8(-a - a^3)$
[Section 1.4]

3. Solve: $\frac{3x - 4}{6} - \frac{x - 2}{2} = \frac{-2x - 3}{3}$ [Section 1.5]

4. Solve $l = a + (n - 1)d$ for n. [Section 1.6]

5. Find the area of a circle with diameter 25 inches. Round to one decimal place. [Section 1.6]

6. TRUCK SALES   See the following graph. A truck dealership is going to order 150 new Toyota Tundra trucks. According to the survey, exactly how many silver trucks should be purchased to meet the expected customer demand? [Section 1.7]

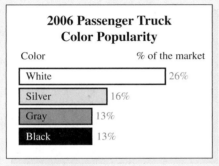

2006 Passenger Truck Color Popularity

Color	% of the market
White	26%
Silver	16%
Gray	13%
Black	13%

Source: Infoplease.com

7. LIFE EXPECTANCY   Determine the predicted rate of change in the life expectancy of females during the years 2000–2050, as shown in the graph. [Section 2.3]

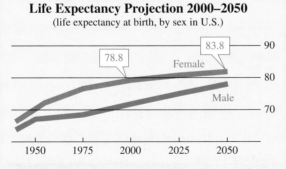

Life Expectancy Projection 2000–2050
(life expectancy at birth, by sex in U.S.)

Based on data from the Social Security Administration, Office of Chief Actuary

Find an equation of the line with the given properties. Write the equation in slope–intercept form.

8. Passing through $(7, 5)$ and perpendicular to the line whose equation is $y = \frac{1}{7}x - 4$ [Section 2.4]

9. Passing through $(-4, 5)$ and $(2, -6)$ [Section 2.4]

10. The graph of a line is shown on the graphing calculator screen below. The axes are scaled in units of 1.

a. Give the x- and y-intercepts of the line. [Section 2.2]

b. What is the slope of the line? [Section 2.3]

c. The line doesn't lie in one quadrant. Which quadrant is that? [Section 2.1]

d. What is the equation of the line? [Section 2.4]

e. Does the line pass through $(15, -42)$? [Section 2.2]

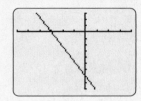

11. If $g(x) = -3x^3 + x - 4$, find $g(-2)$. [Section 2.5]

12. Determine whether the equation $x = |y|$ defines y to be a function of x. If it does not, find two ordered pairs where more than one value of y corresponds to a single value of x. [Section 2.5]

13. Graph $f(x) = |x + 4|$ and give its domain and range in interval notation. [Section 2.6]

14. Solve the system $\begin{cases} x = y + 3 \\ \frac{1}{4}x - \frac{1}{6}y = \frac{1}{3} \end{cases}$ by graphing. [Section 3.1]

15. ENGINEERING   The tensions T_1 and T_2 (in pounds) in each of the ropes shown in the illustration can be found by solving the system

$$\begin{cases} 0.6T_1 - 0.8T_2 = 0 \\ 0.8T_1 + 0.6T_2 = 100 \end{cases}$$

Find T_1 and T_2. [Section 3.2]

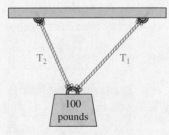

16. JEWELRY   A jeweler wants to make angel pins from an alloy that has a 40% silver content. The jeweler has on hand two alloys, one with 50% silver content and one with a 25% silver content. How many ounces of each alloy should be used to make 20 ounces of the 40% silver alloy? Use two variables to solve the problem. [Section 3.3]

17. Solve: $\begin{cases} x + 2y + 3z = 11 \\ 5x - y = 13 \\ 2x - 3z = -11 \end{cases}$ [Section 3.4]

18. Evaluate: $\begin{vmatrix} 3 & -2 \\ -2 & 4 \end{vmatrix}$ [Section 3.7]

Solve. Write the solution set in interval notation and then graph it.

19. $\frac{1}{2}x + 6 \geq 4 + 2x$ [Section 4.1]

20. $5(x + 2) \leq 4(x + 1)$ and $11 + x < 0$ [Section 4.2]

21. $-4(x + 2) \geq 12$ or $3x + 8 < 11$ [Section 4.2]

22. $\left| \frac{3a}{5} - 2 \right| + 1 \geq \frac{6}{5}$ [Section 4.3]

23. Graph: $y < 4 - x$ [Section 4.4]

24. Without graphing, determine if $(4, -2)$ is a solution of the system

$\begin{cases} 3x + 2y > 6 \\ x + y \leq 2 \end{cases}$ [Section 4.5]

Simplify each expression. Write answers using positive exponents only.

25. $(3bb^2b^3c^0)^4$ [Section 5.1]

26. $9^2d^{-8}(d^9)^2$ [Section 5.1]

27. $-\dfrac{x^{-9}}{20k^{-7}}$ [Section 5.1]

28. $\left(\dfrac{2x^{-2}y^3}{x^2x^3y^4} \right)^{-3}$ [Section 5.1]

29. MONEY   Express the *dollar value* of each type of U.S. coin and currency shown in the illustration, using a power of 10. For example, one hundred dollars can be expressed as $\$10^2$. [Section 5.2]

30. Write each number in scientific notation and perform the indicated operation: $\dfrac{6,150,000,000}{0.003}$. [Section 5.2]

31. SUNDIALS   Refer to the illustration. The graph shows the correction that must be made to a sundial reading to obtain accurate clock time. The difference is caused by the Earth's orbit and tilted axis. [Section 5.3]

   a. Is this the graph of a function?

   b. During the year, what is the maximum number of minutes the sundial reading gets ahead of a clock?

   c. During the year, what is the maximum number of minutes the sundial reading falls behind a clock?

   d. How many times during a year is the sundial reading exactly the same as a clock?

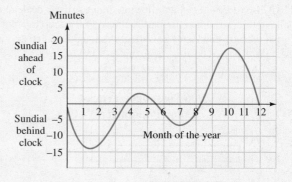

32. Graph: $f(x) = 2x^2 - 3$ and give its domain and range in interval notation. [Section 5.3]

33. Find the degree of $17x^3y^4 + x^2y + 3$. [Section 5.3]

34. Simplify the polynomial: $\frac{9}{4}c^2 - \frac{5}{3}c - \frac{1}{2}c^2 + \frac{5}{6}c$ [Section 5.3]

Perform the indicated operations.

35. $(x^3 + 3x^2 - 2x + 7) + (x^3 - 2x^2 + 2x + 5)$ [Section 5.3]

36. $(2m^2n^2 + 2m - n) - (-2m^2n^2 - 2m + n)$ [Section 5.3]

37. $\left(\dfrac{1}{16}r^9s^{10} \right)(32r^2s^{10})$ [Section 5.4]

38. $-3a^8(4a^4 + 3a^3 - 4a^2)$ [Section 5.4]

39. $(2x^3 - 1)^2$ [Section 5.4]

40. $(a + b + c)(2a - b - 2c)$ [Section 5.4]

Factor each expression.

41. $3r^2s^3 - 6rs^4$ [Section 5.5]

42. $5(x - y) - a(x - y)$ [Section 5.5]

43. $xu + yv + xv + yu$ [Section 5.5]

44. $3 - 10x^2 + 8x^4$ [Section 5.6]

45. $(x - y)^2 + 3(x - y) - 10$ [Section 5.6]

46. $81x^4 - 16y^4$ [Section 5.7]

47. $8x^3 - 27y^6$ [Section 5.7]

48. $x^2 + 10x + 25 - 16z^8$ [Section 5.8]

49. Solve: $b^2x^2 + a^2y^2 = a^2b^2$ for b^2. [Section 5.5]

50. Solve: $6x^2 + 7 = -23x$. [Section 5.9]

51. Solve: $x^3 - 4x = 0$. [Section 5.9]

52. CAMPING   The rectangular-shaped cooking surface of a small camping stove is 108 in.^{2}. If its length is 3 inches longer than its width, what are its dimensions? [Section 5.9]

53. Simplify: $\dfrac{2x^2y + xy - 6y}{3x^2y + 5xy - 2y}$ [Section 6.1]

54. Find the domain of the function $f(x) = \dfrac{2x + 1}{x^2 - 2x}$. Express your answer in words and using interval notation. [Section 6.1]

Perform the indicated operations and then simplify, if possible.

55. $(10n - n^2) \cdot \dfrac{n^6}{n^4 - 10n^3 - 2n^2 + 20n}$ [Section 6.2]

56. $\dfrac{p^3 - q^3}{q^2 - p^2} \div \dfrac{p^3 + p^2q + pq^2}{q^2 + pq}$ [Section 6.2]

57. $\dfrac{2}{x + y} + \dfrac{3}{x - y} - \dfrac{x - 3y}{x^2 - y^2}$ [Section 6.3]

58. $\dfrac{\dfrac{18}{c^2} + \dfrac{11}{c} + 1}{1 - \dfrac{3}{c} - \dfrac{10}{c^2}}$ [Section 6.4]

Perform the division.

59. $\dfrac{5y^4 + 45y^3}{15y^2}$ [Section 6.5]

60. $\dfrac{16x^3 + 16x^2 - 9x - 5}{4x + 5}$ [Section 6.5]

61. Solve: $\dfrac{3}{x - 2} + \dfrac{x^2}{x^2 + x - 6} = \dfrac{x + 4}{x + 3}$ [Section 6.7]

62. Solve: $\dfrac{5x - 3}{x + 2} = \dfrac{5x + 3}{x - 2}$ [Section 6.9]

63. ECONOMICS   The controversial Phillips curve shown below depicts the trade-off between unemployment and inflation as seen by one school of economists. If unemployment drops to very low levels, what does the theoretical model predict will happen to the inflation rate? [Section 6.9]

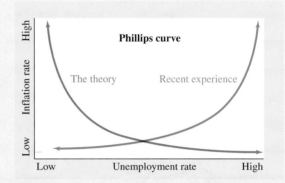

64. ECONOMICS   See Exercise 63. The graph shows that economic factors have not followed the Phillips model in recent years. As unemployment has dropped to very low levels, what has happened to the inflation rate? [Section 6.9]

65. COOKING   A recipe for brownies calls for 4 eggs and $1\frac{1}{2}$ cups of flour. If the recipe makes 15 brownies, how many cups of flour will be needed to make 130 brownies? [Section 6.9]

66. FARMING   The number of days a given number of bushels of corn will last when feeding chickens varies inversely with the number of animals. If a certain number of bushels will feed 300 chickens for 4 days, how long will the feed last for 1,200 chickens? [Section 6.9]

CHAPTER 7

Radical Expressions and Equations

© Tetra Images/Alamy

7.1 Radical Expressions and Radical Functions

7.2 Rational Exponents

7.3 Simplifying and Combining Radical Expressions

7.4 Multiplying and Dividing Radical Expressions

7.5 Solving Radical Equations

7.6 Geometric Applications of Radicals

7.7 Complex Numbers

CHAPTER SUMMARY AND REVIEW

CHAPTER TEST

Group Project

from *Campus to Careers*
General Contractor

The growing popularity of remodeling has created a boom for general contractors. If it's an additional bedroom you need or a makeover of a dated kitchen or bathroom, they can provide design and construction expertise, as well as knowledge of local building code requirements. From the planning stages of a project through its completion, general contractors use mathematics every step of the way.

In **Problem 135** of **Study Set 7.2,** you will use concepts from this chapter to examine the movement of construction materials through a tight hallway.

JOB TITLE:
General Contractor

EDUCATION:
Courses in mathematics, science, drafting, business math, and English are important. Certificate programs are also available.

JOB OUTLOOK:
In general, employment is expected to increase between 9% to 17% through the year 2014.

ANNUAL EARNINGS:
Mean annual salary $76,699

FOR MORE INFORMATION:
http://www.careers.stateuniversity.com

Study groups give students an opportunity to ask their classmates questions, share ideas, compare lecture notes, and review for tests. If something like this interests you, here are some suggestions for forming an effective study group.

GROUP SIZE: A study group should be small—from 3 to 6 people is best.

TIME AND PLACE: You should meet on a regular basis in a place where you can spread out and talk without disturbing others.

GROUND RULES: The study group will be more effective if, early on, you agree on some rules to follow.

Now Try This

Would you like to begin a study group? If so, you need to answer the following questions.

Who will be in your group? Where will your group meet? How often will it meet? For how long will each session last? Will you have a group leader? If so, what will be the leader responsibilities? What will you try to accomplish each session? How will the members prepare for each meeting? Will you follow a set agenda each session? How will the members share contact information? When will you discuss ways to improve the study sessions?

SECTION 7.1
Radical Expressions and Radical Functions

Objectives

1. Find square roots.
2. Find square roots of expressions containing variables.
3. Graph the square root function.
4. Find cube roots.
5. Graph the cube root function.
6. Find nth roots.

In this section, we will reverse the squaring process and learn how to find *square roots* of numbers. Then we will generalize the concept of root and consider cube roots, fourth roots, and so on. We will also discuss a new family of functions, called *radical functions*.

1 Find Square Roots.

When we raise a number to the second power, we are squaring it, or finding its **square.**

- The square of 5 is 25 because $5^2 = 25$.
- The square of -5 is 25, because $(-5)^2 = 25$.

We can reverse the squaring process to find **square roots** of numbers. For example, to find the square roots of 25, we ask ourselves "What number, when squared, is equal to 25?" There are two possible answers.

- 5 is a square root of 25, because $5^2 = 25$.
- -5 is a square root of 25, because $(-5)^2 = 25$.

In general, we have the following definition.

Square Root of a

The number b is a **square root** of the number a if $b^2 = a$.

Every positive number has two square roots, one positive and one negative. For example, the two square roots of 9 are 3 and -3, and the two square roots of 144 are 12 and -12. The number 0 is the only real number with exactly one square root. In fact, it is its own square root, because $0^2 = 0$.

The Language of Algebra

We can read $\sqrt{9}$ as "the square root of 9" or as "radical 9."

A **radical symbol** $\sqrt{}$ represents the **positive** or **principal square root** of a number. Since 3 is the positive square root of 9, we can write

$$\sqrt{9} = 3$$

The symbol $-\sqrt{}$ represents the **negative square root** of a number. It is the opposite of the principal square root. Since -12 is the negative square root of 144, we can write

$$-\sqrt{144} = -12 \qquad \text{Read as "the negative square root of 144 is } -12 \text{" or}$$
$$\text{"the opposite of the square root of 144 is } -12 \text{."}$$

Square Root Notation

If a is a positive real number,

1. $\sqrt{a}$ represents the **positive** or **principal square root** of a. It is the positive number we square to get a.

2. $-\sqrt{a}$ represents the **negative square root** of a. It is the opposite of the principal square root of a: $-\sqrt{a} = -1 \cdot \sqrt{a}$.

3. The principal square root of 0 is 0: $\sqrt{0} = 0$.

The number or variable expression under a radical symbol is called the **radicand.** Together, the radical symbol and radicand are called a **radical.** An algebraic expression containing a radical is called a **radical expression.**

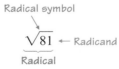

To evaluate square root radical expressions, it is helpful to memorize the whole numbers that are perfect squares.

$1 = 1^2$	$25 = 5^2$	$81 = 9^2$	$169 = 13^2$	$289 = 17^2$
$4 = 2^2$	$36 = 6^2$	$100 = 10^2$	$196 = 14^2$	$324 = 18^2$
$9 = 3^2$	$49 = 7^2$	$121 = 11^2$	$225 = 15^2$	$361 = 19^2$
$16 = 4^2$	$64 = 8^2$	$144 = 12^2$	$256 = 16^2$	$400 = 20^2$

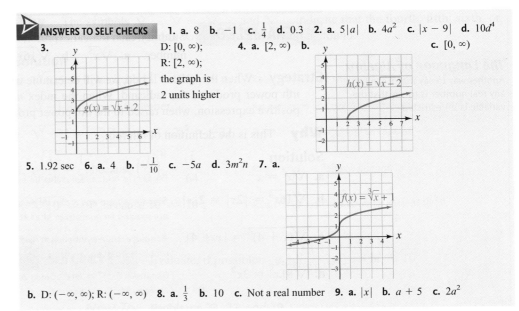

ANSWERS TO SELF CHECKS   **1. a.** 8   **b.** -1 **c.** $\frac{1}{4}$ **d.** 0.3   **2. a.** $5|a|$ **b.** $4a^2$ **c.** $|x - 9|$ **d.** $10d^4$

3.   D: $[0, \infty)$;   **4. a.** $[2, \infty)$ **b.**   **c.** $[0, \infty)$
R: $[2, \infty)$;
the graph is
2 units higher

$g(x) = \sqrt{x} + 2$ $h(x) = \sqrt{x} - 2$

5. 1.92 sec   **6. a.** 4   **b.** $-\frac{1}{10}$ **c.** $-5a$ **d.** $3m^2n$ **7. a.**

$f(x) = \sqrt[3]{x} + 1$

b. D: $(-\infty, \infty)$; R: $(-\infty, \infty)$ **8. a.** $\frac{1}{3}$ **b.** 10   **c.** Not a real number   **9. a.** $|x|$ **b.** $a + 5$ **c.** $2a^2$

STUDY SET
7.1

VOCABULARY

Fill in the blanks.

1. $5x^2$ is the _____ root of $25x^4$ because $(5x^2)^2 = 25x^4$. The _____ root of 216 is 6 because $6^3 = 216$.

2. The symbol $\sqrt{}$ is called a _____ symbol or a _____ root symbol.

3. In the expression $\sqrt[3]{27x^6}$, the _____ is 3 and $27x^6$ is the _____.

4. When we write $\sqrt{b^4} = b^2$, we say that we have _____ the radical expression.

5. When n is an odd number, $\sqrt[n]{x}$ represents an _____ root. When n is an _____ number, $\sqrt[n]{x}$ represents an even root.

6. $f(x) = \sqrt{x}$ and $g(x) = \sqrt[3]{x}$ are _____ functions.

CONCEPTS

Fill in the blanks.

7. b is a square root of a if $b^2 = $ ▓ .

8. $\sqrt{0} = $ ▓ and $\sqrt[3]{0} = $ ▓ .

9. The number 25 has _____ square roots. The principal square root of 25 is ▓ .

10. $\sqrt{-4}$ is not a real number, because no real number _____ equals -4.

11. $\sqrt[3]{x} = y$ if $y^3 = $ ▓ .

12. $\sqrt{x^2} = $ ▓ and $\sqrt[3]{x^3} = $ ▓

13. The graph of $g(x) = \sqrt{x} + 3$ is the graph of $f(x) = \sqrt{x}$ translated ▓ units _____.

14. The graph of $g(x) = \sqrt{x + 5}$ is the graph of $f(x) = \sqrt{x}$ translated ▓ units to the _____.

15. The graph of a square root function f is shown. Find each of the following.

a. $f(11)$ **b.** $f(2)$

c. The value(s) of x for which $f(x) = 2$

d. The domain and range of f

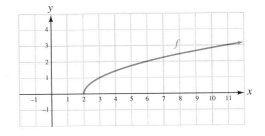

16. The graph of a cube root function f is shown on the next page. Find each of the following.

a. $f(-8)$ **b.** $f(0)$

c. The value(s) of x for which $f(x) = -2$

d. The domain and range of f

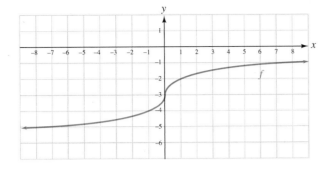

NOTATION

Translate each sentence into mathematical symbols.

17. The square root of x squared is the absolute value of x.

18. The cube root of x cubed is x.

19. f of x equals the square root of the quantity x minus five.

20. The fifth root of negative thirty-two is negative two.

GUIDED PRACTICE

Evaluate each square root, if possible, without using a calculator. **See Objective 1 and Example 1.**

21. $\sqrt{100}$ **22.** $\sqrt{49}$

23. $-\sqrt{64}$ **24.** $-\sqrt{1}$

25. $\sqrt{\dfrac{1}{9}}$ **26.** $\sqrt{\dfrac{4}{25}}$

27. $\sqrt{0.25}$ **28.** $\sqrt{0.16}$

29. $\sqrt{-81}$ **30.** $-\sqrt{-49}$

31. $\sqrt{121}$ **32.** $\sqrt{144}$

Use a calculator to find each square root. Give each answer to four decimal places. **See Objective 1.**

33. $\sqrt{12}$ **34.** $\sqrt{340}$

35. $\sqrt{679.25}$ **36.** $\sqrt{0.0063}$

Simplify each expression. Assume that all variables are unrestricted and use absolute value symbols when necessary. **See Example 2.**

37. $\sqrt{4x^2}$ **38.** $\sqrt{64t^2}$

39. $\sqrt{81h^4}$ **40.** $\sqrt{36y^4}$

41. $\sqrt{36s^6}$ **42.** $\sqrt{9y^6}$

43. $\sqrt{144m^8}$ **44.** $\sqrt{4n^8}$

45. $\sqrt{y^2 - 2y + 1}$ **46.** $\sqrt{b^2 - 14b + 49}$

47. $\sqrt{a^4 + 6a^2 + 9}$ **48.** $\sqrt{x^4 + 10x^2 + 25}$

49. Let $f(x) = \sqrt{x - 4}$. Find each function value. See Example 3.
     a. $f(8)$ **b.** $f(29)$

50. Let $f(x) = \sqrt{x^2 + 1}$. Find each function value. Use a calculator to approximate each answer to four decimal places. See Example 3.
     a. $f(4)$ **b.** $f(2.35)$

51. Let $g(x) = \sqrt[3]{x - 4}$. Find each function value. See Example 7.
     a. $g(12)$ **b.** $g(-23)$

52. Let $g(x) = \sqrt[3]{x^2 + 1}$. Find each function value. Use a calculator to approximate each answer to four decimal places. See Using Your Calculator: Finding Roots.
     a. $g(6)$ **b.** $g(21.57)$

Complete each table and graph the function. Find the domain and range. **See Example 3.**

53. $f(x) = -\sqrt{x}$

x	y
0	
1	
4	
9	
16	

54. $f(x) = \sqrt{x} + 2$

x	y
0	
1	
4	
9	
16	

Find the domain of each function, graph it, and find its range. **See Example 4.**

55. $f(x) = \sqrt{x + 4}$ **56.** $f(x) = \sqrt{x - 1}$

Simplify each cube root. **See Example 6.**

57. $\sqrt[3]{1}$ **58.** $\sqrt[3]{8}$

59. $\sqrt[3]{-125}$ **60.** $\sqrt[3]{-27}$

61. $\sqrt[3]{\dfrac{8}{27}}$ **62.** $\sqrt[3]{\dfrac{125}{64}}$

63. $\sqrt[3]{64}$ **64.** $\sqrt[3]{1,000}$

65. $\sqrt[3]{-216a^3}$ **66.** $\sqrt[3]{-512x^3}$

67. $\sqrt[3]{-1,000p^6q^3}$ **68.** $\sqrt[3]{-343a^6b^3}$

Complete each table and graph the function. Give the domain and range. **See Example 7.**

69. $f(x) = \sqrt[3]{x} - 3$

x	y
-8	
-1	
0	
1	
8	

70. $f(x) = -\sqrt[3]{x}$

x	y
-8	
-1	
0	
1	
8	

Graph each function. Give the domain and range. **See Example 7.**

71. $f(x) = \sqrt[3]{x} - 3$

72. $f(x) = \sqrt[3]{x} + 3$

Evaluate each radical expression, if possible, without using a calculator. **See Example 8.**

73. $\sqrt[4]{81}$

74. $\sqrt[6]{64}$

75. $-\sqrt[5]{243}$

76. $-\sqrt[4]{625}$

77. $\sqrt[6]{-256}$

78. $\sqrt[6]{-729}$

79. $\sqrt[7]{-\dfrac{1}{128}}$

80. $\sqrt[5]{-\dfrac{243}{32}}$

Simplify each radical expression. Assume all variables are unrestricted. **See Example 9.**

81. $\sqrt[5]{32a^5}$

82. $\sqrt[5]{-32x^5}$

83. $\sqrt[4]{81a^4}$

84. $\sqrt[8]{t^8}$

85. $\sqrt[6]{k^{12}}$

86. $\sqrt[6]{64b^6}$

87. $\sqrt[4]{(m+4)^8}$

88. $\sqrt[4]{(x-7)^8}$

TRY IT YOURSELF

Simplify each radical expression, if possible. Assume all variables are unrestricted.

89. $\sqrt[3]{64s^9t^6}$

90. $\sqrt[3]{1,000a^6b^6}$

91. $-\sqrt{49b^8}$

92. $-\sqrt{144t^4}$

93. $-\sqrt[5]{-\dfrac{1}{32}}$

94. $-\sqrt[5]{-243}$

95. $\sqrt[3]{-125m^6}$

96. $\sqrt[3]{-216z^9} \quad -6z^3$

97. $\sqrt[6]{64a^6b^6}$

98. $\sqrt{169p^4q^2}$

99. $\sqrt[7]{(x+2)^7}$

100. $\sqrt[6]{(x+4)^6}$

101. $\sqrt[4]{-81}$

102. $\sqrt[6]{-1}$

103. $\sqrt{n^2 + 12n + 36}$

104. $\sqrt{s^2 - 20s + 100}$

APPLICATIONS

Use a calculator to solve each problem. Round answers to the nearest tenth.

105. EMBROIDERY   The radius r of a circle is given by the formula

$$r = \sqrt{\dfrac{A}{\pi}}$$

where A is its area. Find the diameter of the embroidery hoop if there are 38.5 in.^{2} of stretched fabric on which to embroider.

106. PENDULUMS   Find the period of a pendulum with length 1 foot. *See Example 5.*

107. SHOELACES   The formula $S = 2\left[H + L + (p-1)\sqrt{H^2 + V^2}\right]$ can be used to calculate the correct shoelace length for the criss-cross lacing pattern shown in the illustration, where p represents the number of *pairs* of eyelets. Find the correct shoelace length if H (horizontal distance) = 50 millimeters, L (length of end) = 250 millimeters, and V (vertical distance) = 20 millimeters. (*Source:* Ian's Shoelace Site at www.fieggen.com)

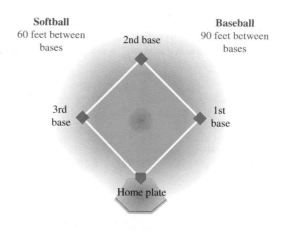

108. BASEBALL   The length of a diagonal of a square is given by the function $d(s) = \sqrt{2s^2}$, where s is the length of a side of the square. Find the distance from home plate to second base on a softball diamond and on a baseball diamond. The illustration gives the dimensions of each type of infield.

Softball
60 feet between bases

Baseball
90 feet between bases

2nd base

3rd base

1st base

Home plate

109. PULSE RATES   The approximate pulse rate (in beats per minute) of an adult who is t inches tall is given by the function

$$p(t) = \dfrac{590}{\sqrt{t}}$$

The *Guinness Book of World Records 2008* lists Leonid Stadnyk of Ukraine as the tallest living man, at 8 feet, 5.5 inches. Find his approximate pulse rate as predicted by the function.

110. THE GRAND CANYON   The time t (in seconds) that it takes for an object to fall a distance of s feet is given by the formula

$$t = \frac{\sqrt{s}}{4}$$

In some places, the Grand Canyon is one mile (5,280 feet) deep. How long would it take a stone dropped over the edge of the canyon to hit bottom?

111. BIOLOGY   Scientists will place five rats inside a clear plastic hemisphere and control the environment to study the rats' behavior. The function

$$d(V) = \sqrt[3]{12\left(\frac{V}{\pi}\right)}$$

gives the diameter of a hemisphere with volume V. Use the function to determine the diameter of the base of the hemisphere, if each rat requires 125 cubic feet of living space.

112. AQUARIUMS   The function

$$s(g) = \sqrt[3]{\frac{g}{7.5}}$$

determines how long (in feet) an edge of a cube-shaped tank must be if it is to hold g gallons of water. What dimensions should a cube-shaped aquarium have if it is to hold 1,250 gallons of water?

113. COLLECTIBLES   The *effective rate of interest* r earned by an investment is given by the formula

$$r = \sqrt[n]{\frac{A}{P}} - 1$$

where P is the initial investment that grows to value A after n years. Determine the effective rate of interest earned by a collector on a Lladró porcelain figurine purchased for $800 and sold for $950 five years later.

114. LAW ENFORCEMENT   The graphs of the two radical functions shown in the illustration can be used to estimate the speed (in mph) of a car involved in an accident. Suppose a police accident report listed skid marks to be 220 feet long but failed to give the road conditions. Estimate the possible speeds the car was traveling prior to the brakes being applied.

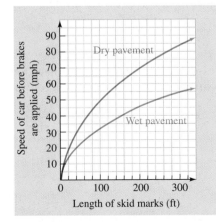

WRITING

115. If x is any real number, that is, if x is unrestricted, then $\sqrt{x^2} = x$ is not correct. Explain.

116. Explain why $\sqrt{36}$ is just 6, and not -6.

117. Explain what is wrong with the graph in the illustration if it is supposed to be the graph of $f(x) = \sqrt{x}$.

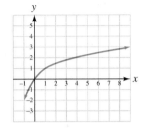

118. Explain how to estimate the domain and range of the radical function shown below.

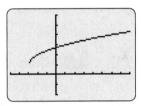

REVIEW

Perform the operations and simplify when possible.

119. $\dfrac{x^2 - x - 6}{x^2 - 2x - 3} \cdot \dfrac{x^2 - 1}{x^2 + x - 2}$

120. $\dfrac{x^2 - 3xy - 4y^2}{x^2 + cx - 2yx - 2cy} \div \dfrac{x^2 - 2xy - 3y^2}{x^2 + cx - 4yx - 4cy}$

121. $\dfrac{3}{m + 1} + \dfrac{3m}{m - 1}$

122. $\dfrac{2x + 3}{3x - 1} - \dfrac{x - 4}{2x + 1}$

CHALLENGE PROBLEMS

123. Graph $f(x) = -\sqrt{x - 2} + 3$ and find the domain and range.

124. Simplify $\sqrt{9a^{16} + 12a^8 b^{25} + 4b^{50}}$ and assume that $a > 0$ and $b > 0$.

SECTION 7.2
Rational Exponents

Objectives

1. Simplify expressions of the form $a^{1/n}$.
2. Simplify expressions of the form $a^{m/n}$.
3. Convert between radicals and rational exponents.
4. Simplify expressions with negative rational exponents.
5. Use rules for exponents to simplify expressions.
6. Simplify radical expressions.

In this section, we will extend the definition of exponent to include rational (fractional) exponents. We will see how expressions such as $9^{1/2}$, $\left(\frac{1}{16}\right)^{3/4}$, and $(-32x^5)^{-2/5}$ can be simplified by writing them in an equivalent radical form using two new rules for exponents.

1   Simplify Expressions of the Form $a^{1/n}$.

It is possible to raise numbers to fractional powers. To give meaning to rational exponents, we first consider $\sqrt{7}$. Because $\sqrt{7}$ is the positive number whose square is 7, we have

The Language of Algebra
Rational exponents are also called *fractional exponents.*

$$\left(\sqrt{7}\right)^2 = 7$$

We now consider the notation $7^{1/2}$. If rational exponents are to follow the same rules as integer exponents, the square of $7^{1/2}$ must be 7, because

$$
\begin{aligned}
(7^{1/2})^2 &= 7^{1/2 \cdot 2} \quad &\text{Keep the base and multiply the exponents.}\\
&= 7^1 \quad &\text{Do the multiplication: } \tfrac{1}{2}\cdot 2 = 1.\\
&= 7
\end{aligned}
$$

Since the square of $7^{1/2}$ and the square of $\sqrt{7}$ are both equal to 7, we define $7^{1/2}$ to be $\sqrt{7}$. Similarly,

$$7^{1/3} = \sqrt[3]{7}, \qquad 7^{1/4} = \sqrt[4]{7}, \qquad \text{and} \qquad 7^{1/5} = \sqrt[5]{7}$$

In general, we have the following definition.

The Definition of $x^{1/n}$

A **rational exponent** of $\frac{1}{n}$ indicates the nth root of its base.
If n represents a positive integer greater than 1 and $\sqrt[n]{x}$ represents a real number,

$$x^{1/n} = \sqrt[n]{x}$$

We can use this definition to simplify exponential expressions that have rational exponents with a numerator of 1. For example, to simplify $8^{1/3}$, we write it as an equivalent expression in radical form and proceed as follows:

Index

$$8^{1/3} = \sqrt[3]{8} = 2 \quad \text{The base of the exponential expression, 8, is the radicand of the radical}$$

Radicand

expression. The denominator of the fractional exponent, 3, is the index of the radical.

Thus, $8^{1/3} = 2$.

EXAMPLE 1   Evaluate: **a.** $9^{1/2}$ **b.** $(-64)^{1/3}$ **c.** $16^{1/4}$ **d.** $-\left(\dfrac{1}{32}\right)^{1/5}$

Strategy   First, we will identify the base and the exponent of the exponential expression. Then we will write the expression in an equivalent radical form using the rule for rational exponents $x^{1/n} = \sqrt[n]{x}$.

Why   We can then use the methods from Section 7.1 to evaluate the resulting square root, cube root, fourth root, and fifth root.

Solution

a. $9^{1/2} = \sqrt{9} = 3$ *Because the denominator of the exponent is 2,
find the square root of the base, 9.*

b. $(-64)^{1/3} = \sqrt[3]{-64} = -4$ *Because the denominator of the exponent is 3,
find the cube root of the base, −64.*

c. $16^{1/4} = \sqrt[4]{16} = 2$ *Because the denominator of the exponent is 4,
find the fourth root of the base, 16.*

d. $-\left(\dfrac{1}{32}\right)^{1/5} = -\sqrt[5]{\dfrac{1}{32}} = -\dfrac{1}{2}$ *Because the denominator of the exponent is 5,
find the fifth root of the base, $\frac{1}{32}$.*

Self Check 1   Evaluate: **a.** $16^{1/2}$ **b.** $\left(-\dfrac{27}{8}\right)^{1/3}$

c. $-(81)^{1/4}$ **d.** $1^{1/5}$

Now Try   **Problems 17, 21, 23, and 25**

As with radicals, when n is an *odd natural number* in the expression $x^{1/n}$, where $n > 1$, there is exactly one real nth root, and we don't need to use absolute value symbols.

When n is an *even natural number,* there are two nth roots. Since we want the expression $x^{1/n}$ to represent the positive nth root, we must often use absolute value symbols to guarantee that the simplified result is positive. Thus, if n is even,

$$(x^n)^{1/n} = |x|$$

When n is even and x is negative, the expression $x^{1/n}$ is not a real number.

EXAMPLE 2   Simplify each expression, if possible. Assume that the variables can be any real number.   **a.** $(-27x^3)^{1/3}$ **b.** $(256a^8)^{1/8}$
c. $[(y + 4)^2]^{1/2}$ **d.** $(25b^4)^{1/2}$ **e.** $(-256x^4)^{1/4}$

Strategy   We will write each exponential expression in an equivalent radical form using the rule for rational exponents $x^{1/n} = \sqrt[n]{x}$.

Why   We can then use the methods of Section 7.1 to simplify the resulting radical expression.

Solution

a. $(-27x^3)^{1/3} = \sqrt[3]{-27x^3} = -3x$ *Because $(-3x)^3 = -27x^3$. Since n is odd, no absolute value
symbols are needed.*

b. $(256a^8)^{1/8} = \sqrt[8]{256a^8} = 2|a|$ *Because $(2|a|)^8 = 256a^8$. Since n is even and a can be any
real number, 2a can be negative. Thus, absolute value
symbols are needed.*

c. $[(y + 4)^2]^{1/2} = \sqrt{(y + 4)^2} = |y + 4|$ Because $|y + 4|^2 = (y + 4)^2$. Since n is even and y can be any real number, $y + 4$ can be negative. Thus, absolute value symbols are needed.

d. $(25b^4)^{1/2} = \sqrt{25b^4} = 5b^2$ Because $(5b^2)^2 = 25b^4$. Since $b^2 \geq 0$, no absolute value symbols are needed.

e. $(-256x^4)^{1/4} = \sqrt[4]{-256x^4}$, which is not a real number   Because no real number raised to the 4th power is -256.

Self Check 2   Simplify each expression, if possible. Assume that the variables can be any real number.
    a. $(-8n^3)^{1/3}$ **b.** $(625a^4)^{1/4}$
    c. $(b^4)^{1/2}$ **d.** $(-64b^{12})^{1/6}$

Now Try   Problems 27, 35, 37, and 39

If we were told that the variables represent positive real numbers in parts (b) and (c) of Example 2, the absolute value symbols in the answers would not be needed.

$(256a^8)^{1/8} = 2a$ If a represents a positive real number, then $2a$ is positive.

$[(y + 4)^2]^{1/2} = y + 4$ If y represents a positive real number, then $y + 4$ is positive.

We summarize the cases as follows.

Summary of the Definitions of $x^{1/n}$

If n is a natural number greater than 1 and x is a real number,

   If $x > 0$, then $x^{1/n}$ is the real number such that $(x^{1/n})^n = x$.
   If $x = 0$, then $x^{1/n} = 0$.

   If $x < 0$ $\begin{cases} \text{and } n \text{ is odd, then } x^{1/n} \text{ is the negative number such that } (x^{1/n})^n = x. \\ \text{and } n \text{ is even, then } x^{1/n} \text{ is not a real number.} \end{cases}$

---

2   **Simplify Expressions of the Form $a^{m/n}$.**

We can extend the definition of $x^{1/n}$ to include fractional exponents with numerators other than 1. For example, since $8^{2/3}$ can be written as $(8^{1/3})^2$, we have

$8^{2/3} = (8^{1/3})^2$

     $= (\sqrt[3]{8})^2$ Write $8^{1/3}$ in radical form.

     $= 2^2$ Find the cube root first: $\sqrt[3]{8} = 2$.

     $= 4$ Then find the power.

Thus, we can simplify $8^{2/3}$ by finding the second power of the cube root of 8.

The numerator of the rational exponent is the power.

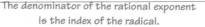

$8^{2/3} = \left(\sqrt[3]{8}\right)^2$ The base of the exponential expression is the radicand.

The denominator of the rational exponent is the index of the radical.

We can also simplify $8^{2/3}$ by taking the cube root of 8 squared.

$$8^{2/3} = (8^2)^{1/3}$$

$$= 64^{1/3} \qquad \text{Find the power first: } 8^2 = 64.$$

$$= \sqrt[3]{64} \qquad \text{Write } 64^{1/3} \text{ in radical form.}$$

$$= 4 \qquad \text{Now find the cube root.}$$

In general, we have the following definition.

The Definition of $x^{m/n}$

If m and n represent positive integers ($n \neq 1$) and $\sqrt[n]{x}$ represents a real number,

$$x^{m/n} = \left(\sqrt[n]{x}\right)^m \qquad \text{and} \qquad x^{m/n} = \sqrt[n]{x^m}$$

Because of the previous definition, we can interpret $x^{m/n}$ in two ways:

1. $x^{m/n}$ means the nth root of the mth power of x.

2. $x^{m/n}$ means the mth power of the nth root of x.

We can use this definition to evaluate exponential expressions that have rational exponents with a numerator that is not 1. To avoid large numbers, we usually find the root of the base first and then calculate the power using the rule $x^{m/n} = \left(\sqrt[n]{x}\right)^m$.

EXAMPLE 3   Evaluate: **a.** $32^{2/5}$ **b.** $81^{3/4}$ **c.** $(-64)^{2/3}$ **d.** $-\left(\dfrac{1}{25}\right)^{3/2}$

Strategy   First, we will identify the base and the exponent of the exponential expression. Then we will write the expression in an equivalent radical form using the rule for rational exponents $x^{m/n} = \left(\sqrt[n]{x}\right)^m$.

Why   We know how to evaluate square roots, cube roots, fourth roots, and fifth roots.

Solution

a. To evaluate $32^{2/5}$, we write it in an equivalent radical form. The denominator of the rational exponent is the same as the index of the corresponding radical. The numerator of the rational exponent indicates the power to which the radical base is raised.

$$32^{2/5} = \left(\sqrt[5]{32}\right)^2 = (2)^2 = 4 \qquad \text{Because the exponent is 2/5, find the fifth root of the base, 32, to get 2. Then find the second power of 2.}$$

b. $81^{3/4} = \left(\sqrt[4]{81}\right)^3 = (3)^3 = 27 \qquad \text{Because the exponent is 3/4, find the fourth root of the base, 81, to get 3. Then find the third power of 3.}$

c. For $(-64)^{2/3}$, the base is -64.

$$(-64)^{2/3} = \left(\sqrt[3]{-64}\right)^2 = (-4)^2 = 16 \qquad \text{Because the exponent is 2/3, find the cube root of the base, } -64, \text{ to get } -4. \text{ Then find the second power of } -4.$$

Caution

We can also evaluate $x^{m/n}$ using $\sqrt[n]{x^m}$, however the resulting radicand is often extremely large. For example,

$$81^{3/4} = \sqrt[4]{81^3}$$

$$= \sqrt[4]{531{,}441}$$

$$= 27$$

d. For $-\left(\dfrac{1}{25}\right)^{3/2}$, the base is $\dfrac{1}{25}$, not $-\dfrac{1}{25}$.

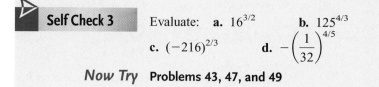

$$-\left(\frac{1}{25}\right)^{3/2} = -\left(\sqrt[2]{\frac{1}{25}}\right)^3 = -\left(\frac{1}{5}\right)^3 = -\frac{1}{125}$$

Because the exponent is 3/2, find the square root of the base, $\frac{1}{25}$, to get $\frac{1}{5}$. Then find the third power of $\frac{1}{5}$.

> ▷ **Self Check 3**   Evaluate: **a.** $16^{3/2}$ **b.** $125^{4/3}$
>
>   **c.** $(-216)^{2/3}$ **d.** $-\left(\dfrac{1}{32}\right)^{4/5}$
>
>   *Now Try*   **Problems 43, 47, and 49**

EXAMPLE 4   Simplify each expression. All variables represent positive real numbers.   **a.** $(36m^4)^{3/2}$ **b.** $(-8x^3)^{4/3}$ **c.** $-(x^5y^5)^{2/5}$

Strategy   We will write each exponential expression in an equivalent radical form using the rule for rational exponents $x^{m/n} = \left(\sqrt[n]{x}\right)^m$.

Why   We can then use the methods of Section 7.1 to simplify the resulting radical expression.

Solution

Power
  Root

a. $(36m^4)^{3/2} = \left(\sqrt[2]{36m^4}\right)^3 = (6m^2)^3 = 216m^6$ Because the exponent is 3/2, find the square root of the base, $36m^4$, to get $6m^2$. Then find the third power of $6m^2$.

Power
  Root

b. $(-8x^3)^{4/3} = \left(\sqrt[3]{-8x^3}\right)^4 = (-2x)^4 = 16x^4$ Because the exponent is 4/3, find the cube root of the base, $-8x^3$, to get $-2x$. Then find the fourth power of $-2x$.

Power
  Root

c. $-(x^5y^5)^{2/5} = -\left(\sqrt[5]{x^5y^5}\right)^2 = -(xy)^2 = -x^2y^2$ Because the exponent is 2/5, find the fifth root of the base, x^5y^5, to get xy. Then find the second power of xy.

> ▷ **Self Check 4**   Simplify each expression. All variables represent positive real numbers.   **a.** $(4c^4)^{3/2}$ **b.** $(-27m^3n^3)^{2/3}$
>
>   **c.** $-(32a^{10})^{3/5}$
>
>   *Now Try*   **Problems 51 and 53**

| **Using Your Calculator** | ***Rational Exponents*** |

We can evaluate expressions containing rational exponents using the exponential key $\boxed{y^x}$ or $\boxed{x^y}$ on a scientific calculator. For example, to evaluate $10^{2/3}$, we enter

$$10 \;\boxed{y^x}\;\boxed{(}\;2\;\boxed{\div}\;3\;\boxed{)}\;\boxed{=} \qquad \boxed{4.641588834}$$

Note that parentheses were used when entering the power. Without them, the calculator would interpret the entry as $10^2 \div 3$.

To evaluate the exponential expression using a graphing calculator, we use the $\boxed{\wedge}$ key, which raises a base to a power. Again, we use parentheses when entering the power.

$$10 \;\boxed{\wedge}\;\boxed{(}\;2\;\boxed{\div}\;3\;\boxed{)}\;\boxed{\text{ENTER}}$$

```
10^(2/3)
          4.641588834
```

To the nearest hundredth, $10^{2/3} \approx 4.64$.

---

3   **Convert Between Radicals and Rational Exponents.**

We can use the rules for rational exponents to convert expressions from radical form to exponential form, and vice versa.

EXAMPLE 5   Write $\sqrt{5xyz}$ as an exponential expression with a rational exponent.

Strategy   We will use the first rule for rational exponents in reverse: $\sqrt[n]{x} = x^{1/n}$.

Why   We are given a radical expression and we want to write an equivalent exponential expression.

Solution   The radicand is $5xyz$, so the base of the exponential expression is $5xyz$. The index of the radical is an understood 2, so the denominator of the fractional exponent is 2.

$$\sqrt{5xyz} = (5xyz)^{1/2} \qquad \text{Recall: } \sqrt[2]{5xyz} = \sqrt{5xyz}.$$

   Self Check 5   Write $\sqrt[6]{7ab}$ as an exponential expression with a rational exponent.

Now Try   Problem 59

Rational exponents appear in formulas used in many disciplines, such as science and engineering.

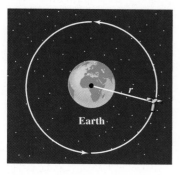

EXAMPLE 6 *Satellites.* The formula

$$r = \left(\frac{GMP^2}{4\pi^2}\right)^{1/3}$$

gives the orbital radius (in meters) of a satellite circling Earth, where G and M are constants and P is the time in seconds for the satellite to make one complete revolution. Write the formula using a radical.

Strategy We will use the first rule for rational exponents: $x^{1/n} = \sqrt[n]{x}$.

Why We are given an exponential expression involving a rational exponent with a numerator of 1 and we want to write an equivalent radical expression.

Solution The fractional exponent $\frac{1}{3}$, with a numerator of 1 and a denominator of 3, indicates that we are to find the cube root of the base of the exponential expression. So we have

$$r = \sqrt[3]{\frac{GMP^2}{4\pi^2}}$$

 Now Try **Problem 63**

4 **Simplify Expressions with Negative Rational Exponents.**

To be consistent with the definition of negative integer exponents, we define $x^{-m/n}$ as follows.

Definition of $x^{-m/n}$

If m and n are positive integers, $\frac{m}{n}$ is in simplified form, and $x^{1/n}$ is a real number, then

$$x^{-m/n} = \frac{1}{x^{m/n}} \quad \text{and} \quad \frac{1}{x^{-m/n}} = x^{m/n} \quad (x \neq 0)$$

From the definition, we see that another way to write $x^{-m/n}$ is to write its reciprocal and change the sign of the exponent.

EXAMPLE 7 Simplify each expression. Assume that x can represent any nonzero real number. **a.** $64^{-1/2}$ **b.** $(-16)^{-5/4}$ **c.** $-625^{-3/4}$ **d.** $(-32x^5)^{-2/5}$ **e.** $\dfrac{1}{25^{-3/2}}$

Strategy We will use one of the rules $x^{-m/n} = \dfrac{1}{x^{m/n}}$ or $\dfrac{1}{x^{-m/n}} = x^{m/n}$ to write the reciprocal of each exponential expression and change the exponent's sign to positive.

Why If we can produce an equivalent expression having a positive rational exponent, we can use the methods of this section to simplify it.

Solution

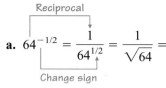

a. $64^{-1/2} = \dfrac{1}{64^{1/2}} = \dfrac{1}{\sqrt{64}} = \dfrac{1}{8}$ Because the exponent is negative, write the reciprocal of $64^{-1/2}$, and change the sign of the exponent.

b. $(-16)^{-5/4}$ is not a real number because $(-16)^{1/4}$ is not a real number.

c. In $-625^{-3/4}$, the base is 625.

$$-625^{-3/4} = -\dfrac{1}{625^{3/4}} = -\dfrac{1}{\left(\sqrt[4]{625}\right)^3} = -\dfrac{1}{(5)^3} = -\dfrac{1}{125}$$

d. $(-32x^5)^{-2/5} = \dfrac{1}{(-32x^5)^{2/5}} = \dfrac{1}{\left(\sqrt[5]{-32x^5}\right)^2} = \dfrac{1}{(-2x)^2} = \dfrac{1}{4x^2}$

e. $\dfrac{1}{25^{-3/2}} = 25^{3/2} = (\sqrt{25})^3 = (5)^3 = 125$ Because the exponent is negative, write the reciprocal of $\frac{1}{25^{-3/2}}$, and change the sign of the exponent.

> **Self Check 7**   Simplify. Assume that a can represent any nonzero real number.
> **a.** $9^{-1/2}$ **b.** $(36)^{-3/2}$
> **c.** $(-27a^3)^{-2/3}$ **d.** $-\dfrac{1}{81^{-3/4}}$
> ***Now Try***   **Problems 67, 73, and 79**

5   ## Use the Rules for Exponents to Simplify Expressions.

We can use the rules for exponents to simplify many expressions with fractional exponents. If all variables represent positive real numbers, absolute value symbols are not needed.

EXAMPLE 8   Simplify each expression. All variables represent positive real numbers. Write all answers using positive exponents only.

a. $5^{2/7}5^{3/7}$ **b.** $(11^{2/7})^3$ **c.** $(a^{2/3}b^{1/2})^6$ **d.** $\dfrac{a^{8/3}a^{1/3}}{a^2}$

Strategy   We will use the product, power, and quotient rules for exponents to simplify each expression.

Why   The familiar rules for exponents discussed in Chapter 5 are valid for rational exponents.

Solution

a. $5^{2/7}5^{3/7} = 5^{2/7+3/7}$ Use the rule $x^m x^n = x^{m+n}$.

$\qquad\qquad\ = 5^{5/7}$ Add: $\frac{2}{7} + \frac{3}{7} = \frac{5}{7}$.

b. $(11^{2/7})^3 = 11^{(2/7)(3)}$ Use the rule $(x^m)^n = x^{mn}$.

$\qquad\qquad\ = 11^{6/7}$ Multiply: $\frac{2}{7}(3) = \frac{6}{7}$.

c. $(a^{2/3}b^{1/2})^6 = (a^{2/3})^6(b^{1/2})^6$ Use the rule $(xy)^n = x^n y^n$.

$\qquad\qquad\quad = a^{12/3}b^{6/2}$ Use the rule $(x^m)^n = x^{mn}$ twice.

$\qquad\qquad\quad = a^4 b^3$ Simplify the exponents.

d. $\dfrac{a^{8/3}a^{1/3}}{a^2} = a^{8/3+1/3-2}$ Use the rules $x^m x^n = x^{m+n}$ and $\dfrac{x^m}{x^n} = x^{m-n}$.

$= a^{8/3+1/3-6/3}$ To establish an LCD, write -2 as $-\dfrac{6}{3}$.

$= a^{3/3}$ Simplify: $\dfrac{8}{3} + \dfrac{1}{3} - \dfrac{6}{3} = \dfrac{3}{3}$.

$= a$ Simplify: $\dfrac{3}{3} = 1$.

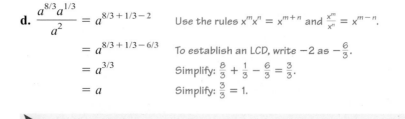

Self Check 8   Simplify. All variables represent positive real numbers.
     a. $2^{1/5}2^{2/5}$ **b.** $(12^{1/3})^4$
     c. $(x^{1/3}y^{3/2})^6$ **d.** $\dfrac{x^{5/3}x^{2/3}}{x^{1/3}}$

Now Try   Problems 83, 87, and 91

EXAMPLE 9   Perform each multiplication and simplify when possible. Assume all variables represent positive real numbers. Write all answers using positive exponents only.
     a. $a^{4/5}(a^{1/5} + a^{3/5})$ **b.** $x^{1/2}(x^{-1/2} - x^{1/2})$

Strategy   We will use the distributive property and multiply each term within the parentheses by the term outside the parentheses.

Why   The first expression has the form $a(b + c)$ and the second has the form $a(b - c)$.

Solution

a. $a^{4/5}(a^{1/5} + a^{3/5}) = a^{4/5}a^{1/5} + a^{4/5}a^{3/5}$ Use the distributive property.

$= a^{4/5+1/5} + a^{4/5+3/5}$ Use the rule $x^m x^n = x^{m+n}$.

$= a^{5/5} + a^{7/5}$ Add the exponents.

$= a + a^{7/5}$ We cannot add these terms because they are not like terms.

b. $x^{1/2}(x^{-1/2} - x^{1/2}) = x^{1/2}x^{-1/2} - x^{1/2}x^{1/2}$ Use the distributive property.

$= x^{1/2+(-1/2)} - x^{1/2+1/2}$ Use the rule $x^m x^n = x^{m+n}$.

$= x^0 - x^1$ Add the exponents.

$= 1 - x$ Simplify: $x^0 = 1$.

Self Check 9   Simplify: $t^{5/8}(t^{3/8} - t^{-5/8})$. Assume t represents a positive real number.

Now Try   Problem 97

---

6   **Simplify Radical Expressions.**

We can simplify many radical expressions by using the following steps.

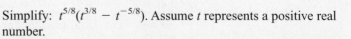

Using Rational Exponents to Simplify Radicals	1. Change the radical expression into an exponential expression.
	2. Simplify the rational exponents.
	3. Change the exponential expression back into a radical.

EXAMPLE 10   Simplify: **a.** $\sqrt[4]{3^2}$ **b.** $\sqrt[8]{x^6}$ **c.** $\sqrt[9]{27x^6y^3}$ **d.** $\sqrt[5]{\sqrt[3]{t}}$

Strategy   We will write each radical expression as an equivalent exponential expression and use rules for exponents to simplify it. Then we will change that result back into a radical.

Why   When the given expression is written in an equivalent exponential form, we can use rules for exponents and our arithmetic skills with fractions to simplify the exponents.

Solution

a. $\sqrt[4]{3^2} = (3^2)^{1/4}$ *Change the radical to an exponential expression.*

$\phantom{\sqrt[4]{3^2}} = 3^{2/4}$ *Use the rule $(x^m)^n = x^{mn}$.*

$\phantom{\sqrt[4]{3^2}} = 3^{1/2}$ *Simplify the fractional exponent: $\frac{2}{4} = \frac{1}{2}$.*

$\phantom{\sqrt[4]{3^2}} = \sqrt{3}$ *Change back to radical form.*

b. $\sqrt[8]{x^6} = (x^6)^{1/8}$ *Change the radical to an exponential expression.*

$\phantom{\sqrt[8]{x^6}} = x^{6/8}$ *Use the rule $(x^m)^n = x^{mn}$.*

$\phantom{\sqrt[8]{x^6}} = x^{3/4}$ *Simplify the fractional exponent: $\frac{6}{8} = \frac{3}{4}$.*

$\phantom{\sqrt[8]{x^6}} = (x^3)^{1/4}$ *Write $\frac{3}{4}$ as $3\left(\frac{1}{4}\right)$.*

$\phantom{\sqrt[8]{x^6}} = \sqrt[4]{x^3}$ *Change back to radical form.*

c. $\sqrt[9]{27x^6y^3} = (3^3x^6y^3)^{1/9}$ *Write 27 as 3^3 and change the radical to an exponential expression.*

$\phantom{\sqrt[9]{27x^6y^3}} = 3^{3/9}x^{6/9}y^{3/9}$ *Raise each factor to the $\frac{1}{9}$ power by multiplying the fractional exponents.*

$\phantom{\sqrt[9]{27x^6y^3}} = 3^{1/3}x^{2/3}y^{1/3}$ *Simplify each fractional exponent.*

$\phantom{\sqrt[9]{27x^6y^3}} = (3x^2y)^{1/3}$ *Use the rule $(xy)^n = x^ny^n$.*

$\phantom{\sqrt[9]{27x^6y^3}} = \sqrt[3]{3x^2y}$ *Change back to radical form.*

d. $\sqrt[5]{\sqrt[3]{t}} = \sqrt[5]{t^{1/3}}$ *Change the radical $\sqrt[3]{t}$ to exponential notation.*

$\phantom{\sqrt[5]{\sqrt[3]{t}}} = (t^{1/3})^{1/5}$ *Change the radical $\sqrt[5]{t^{1/3}}$ to exponential notation.*

$\phantom{\sqrt[5]{\sqrt[3]{t}}} = t^{1/15}$ *Use the rule $(x^m)^n = x^{mn}$. Multiply: $\frac{1}{3} \cdot \frac{1}{5} = \frac{1}{15}$.*

$\phantom{\sqrt[5]{\sqrt[3]{t}}} = \sqrt[15]{t}$ *Change back to radical form.*

▷ **Self Check 10**   Simplify: **a.** $\sqrt[6]{3^3}$ **b.** $\sqrt[4]{49x^2y^2}$

       c. $\sqrt[3]{\sqrt[4]{m}}$

    Now Try   **Problems 99, 105, and 107**

STUDY SET
7.2

VOCABULARY

Fill in the blanks.

1. The expressions $4^{1/2}$ and $(-8)^{-2/3}$ have _____ exponents.

2. In the exponential expression $27^{4/3}$, 27 is the _____ and 4/3 is the _____.

3. In the radical expression $\sqrt[4]{16x^8}$, 4 is the _____, and $16x^8$ is the _____.

4. $32^{4/5}$ means the fourth _____ of the fifth _____ of 32.

CONCEPTS

5. Complete the table by writing the given expression in the alternate form.

Radical form	Exponential form
$\sqrt[5]{25}$	
	$(-27)^{2/3}$
$\left(\sqrt[4]{16}\right)^{-3}$	
	$81^{3/2}$
$-\sqrt{\dfrac{9}{64}}$	

6. In your own words, explain the three rules for rational exponents illustrated in the diagrams below.

a. $(-32)^{1/5} = \sqrt[5]{-32}$

b. $125^{4/3} = (\sqrt[3]{125})^4$

c. $8^{-1/3} = \dfrac{1}{8^{1/3}}$

7. Graph each number on the number line.

$$\left\{ 8^{2/3}, (-125)^{1/3}, -16^{-1/4}, 4^{3/2}, -\left(\frac{9}{100}\right)^{-1/2} \right\}$$

8. a. Evaluate $25^{3/2}$ by writing in the form $(25^{1/2})^3$.

 b. Evaluate $25^{3/2}$ by writing in the form $(25^3)^{1/2}$.

 c. Which way was easier?

Complete each rule for exponents.

9. $x^{1/n} = $

10. $x^{m/n} = $ $= \sqrt[n]{x^m}$

11. $x^{-m/n} = $

12. $\dfrac{1}{x^{-m/n}} = $

NOTATION

Complete each solution.

13. Simplify:

$$(100a^4)^{3/2} = \left(\sqrt{}\right)^3$$
$$= \left(\right)^3$$
$$= 1{,}000a^6$$

14. Simplify:

$$(m^{1/3}n^{1/2})^6 = \left(\right)^6 (n^{1/2})^6$$
$$= m^{}n^{6/2}$$
$$= m^2n^3$$

GUIDED PRACTICE

Evaluate each expression. See Example 1.

15. $125^{1/3}$

16. $8^{1/3}$

17. $81^{1/4}$

18. $625^{1/4}$

19. $32^{1/5}$

20. $0^{1/5}$

21. $(-216)^{1/3}$

22. $(-1{,}000)^{1/3}$

23. $-16^{1/4}$

24. $-125^{1/3}$

25. $\left(\dfrac{1}{4}\right)^{1/2}$

26. $\left(\dfrac{1}{16}\right)^{1/2}$

Simplify each expression, if possible. Assume that the variables can be any real number, and use absolute value symbols when necessary. See Example 2.

27. $(4x^4)^{1/2}$

28. $(25a^8)^{1/2}$

29. $(x^2)^{1/2}$

30. $(x^3)^{1/3}$

31. $(m^4)^{1/2}$

32. $(a^4)^{1/4}$

33. $(-64p^8)^{1/2}$

34. $(-16q^4)^{1/2}$

35. $(-27n^9)^{1/3}$

36. $(-64t^9)^{1/3}$

37. $(16x^4)^{1/4}$

38. $(-x^4)^{1/4}$

39. $(-64x^8)^{1/8}$

40. $(243x^{10})^{1/5}$

41. $[(x + 1)^6]^{1/6}$

42. $[(x + 5)^8]^{1/8}$

Evaluate each expression. See Example 3.

43. $36^{3/2}$

44. $27^{2/3}$

45. $16^{3/4}$

46. $-100^{3/2}$

47. $\left(-\dfrac{1}{216}\right)^{2/3}$

48. $\left(\dfrac{4}{9}\right)^{3/2}$

49. $-4^{5/2}$

50. $(-125)^{4/3}$

Simplify each expression. All variables represent positive real numbers. See Example 4.

51. $(25x^4)^{3/2}$

52. $(27a^3b^3)^{2/3}$

53. $(-8x^6y^3)^{2/3}$

54. $(-32x^{10}y^5)^{4/5}$

55. $(81x^4y^8)^{3/4}$

56. $\left(\dfrac{1}{16}x^8y^4\right)^{3/4}$

57. $-\left(\dfrac{x^5}{32}\right)^{4/5}$

58. $-\left(\dfrac{27}{64y^6}\right)^{2/3}$

Change each radical to an exponential expression. See Example 5.

59. $\sqrt[5]{8abc}$

60. $\sqrt[7]{7p^2q}$

61. $\sqrt[3]{a^2 - b^2}$

62. $\sqrt{x^2 + y^2}$

Change each exponential expression to a radical. See Example 6.

63. $(6x^3y)^{1/4}$

64. $(7a^2b^2)^{1/5}$

65. $(2s^2 - t^2)^{1/2}$

66. $(x^3 + y^3)^{1/3}$

Simplify each expression. All variables represent positive real numbers. See Example 7.

67. $4^{-1/2}$

68. $49^{-1/2}$

69. $125^{-1/3}$

70. $8^{-1/3}$

71. $16^{-3/2}$

72. $(16)^{-5/4}$

73. $-(1{,}000y^3)^{-2/3}$

74. $-(81c^4)^{-3/2}$

75. $\left(-\dfrac{27}{8}\right)^{-4/3}$

76. $\left(\dfrac{25}{49}\right)^{-3/2}$

77. $\left(\dfrac{16}{81y^4}\right)^{-3/4}$

78. $\left(-\dfrac{8x^3}{27}\right)^{-1/3}$

79. $\dfrac{1}{32^{-1/5}}$

80. $\dfrac{1}{64^{-1/6}}$

81. $\dfrac{1}{9^{-5/2}}$

82. $\dfrac{1}{16^{-5/2}}$

Simplify each expression. Write the answers without negative exponents. All variables represent positive real numbers. See Example 8.

83. $9^{3/7}9^{2/7}$

84. $4^{2/5}4^{2/5}$

85. $6^{-2/3}6^{-4/3}$

86. $5^{1/3}5^{-5/3}$

87. $(m^{2/3}m^{1/3})^6$

88. $(b^{3/5}b^{2/5})^8$

89. $(a^{1/2}b^{1/3})^{3/2}$

90. $(mn^{-2/3})^{-3/5}$

91. $\dfrac{3^{4/3}3^{1/3}}{3^{2/3}}$

92. $\dfrac{2^{5/6}2^{1/3}}{2^{1/2}}$

93. $\dfrac{a^{3/4}a^{3/4}}{a^{1/2}}$

94. $\dfrac{b^{4/5}b^{4/5}}{b^{3/5}}$

Perform the multiplications. All variables represent positive real numbers. See Example 9.

95. $y^{1/3}(y^{2/3} + y^{5/3})$

96. $y^{2/5}(y^{-2/5} + y^{3/5})$

97. $x^{3/5}(x^{7/5} - x^{-3/5} + 1)$

98. $x^{4/3}(x^{2/3} + 3x^{5/3} - 4)$

Use rational exponents to simplify each radical. All variables represent positive real numbers. See Example 10.

99. $\sqrt[4]{5^2}$

100. $\sqrt[6]{7^3}$

101. $\sqrt[9]{11^3}$

102. $\sqrt[12]{13^4}$

103. $\sqrt[6]{p^3}$

104. $\sqrt[8]{q^2}$

105. $\sqrt[10]{x^2y^2}$

106. $\sqrt[6]{x^2y^2}$

107. $\sqrt[9]{\sqrt{c}}$

108. $\sqrt[4]{\sqrt{x}}$

109. $\sqrt[5]{\sqrt[3]{7m}}$

110. $\sqrt[3]{\sqrt[4]{21x}}$

Use a calculator to evaluate each expression. Round to the nearest hundredth. See Using Your Calculator: Rational Exponents.

111. $15^{1/3}$

112. $(50.5)^{1/4}$

113. $(1.045)^{2/5}$

114. $(-1{,}000)^{3/5}$

TRY IT YOURSELF

Simplify each expression. All variables represent positive real numbers.

115. $(25y^2)^{1/2}$

116. $(-27x^3)^{1/3}$

117. $-\left(\dfrac{a^4}{81}\right)^{3/4}$

118. $-\left(\dfrac{b^8}{625}\right)^{3/4}$

119. $\dfrac{p^{8/5}p^{7/5}}{p^2}$

120. $\dfrac{c^{2/3}c^{2/3}}{c^{1/3}}$

121. $(-27x^6)^{-1/3}$

122. $(16a^4)^{-1/2}$

123. $n^{1/5}(n^{2/5} - n^{-1/5})$

124. $t^{4/3}(t^{5/3} + t^{-4/3})$

125. $\dfrac{1}{9^{-5/2}}$

126. $\dfrac{1}{16^{-5/2}}$

127. $\sqrt[4]{25b^2}$

128. $\sqrt[9]{8x^6}$

129. $-(8a^3b^6)^{-2/3}$

130. $-(25s^4t^6)^{-3/2}$

APPLICATIONS

131. BALLISTIC PENDULUMS   The formula

$$v = \frac{m + M}{m}(2gh)^{1/2}$$

gives the velocity (in ft/sec) of a bullet with weight m fired into a block with weight M, that raises the height of the block h feet after the collision. The letter g represents a constant, 32. Find the velocity of the bullet to the nearest ft/sec.

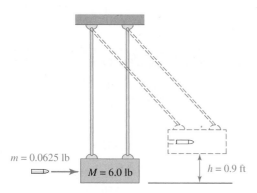

$m = 0.0625$ lb

$M = 6.0$ lb

$h = 0.9$ ft

132. GEOGRAPHY   The formula

$$A = [s(s - a)(s - b)(s - c)]^{1/2}$$

gives the area of a triangle with sides of length a, b, and c, where s is one-half of the perimeter. Estimate the area of Virginia (to the nearest square mile) using the data given in the illustration.

Virginia

370 mi

220 mi

★
Richmond

430 mi

133. RELATIVITY   One concept of relativity theory is that an object moving past an observer at a speed near the speed of light appears to have a larger mass because of its motion. If the mass of the object is m_0 when the object is at rest relative to the observer, its mass m will be given by the formula

$$m = m_0\left(1 - \frac{v^2}{c^2}\right)^{-1/2}$$

when it is moving with speed v (in miles per second) past the observer. The variable c is the speed of light, 186,000 mi/sec. If a proton with a rest mass of 1 unit is accelerated by a nuclear accelerator to a speed of 160,000 mi/sec, what mass will the technicians observe it to have? Round to the nearest hundredth.

134. LOGGING   The width w and height h of the strongest rectangular beam that can be cut from a cylindrical log of radius a are given by

$$w = \frac{2a}{3}(3^{1/2}) \quad \text{and} \quad h = a\left(\frac{8}{3}\right)^{1/2}$$

Find the width, height, and cross-sectional area of the strongest beam that can be cut from a log with *diameter* 4 feet. Round to the nearest hundredth.

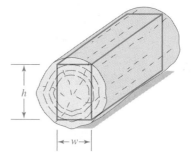

h

w

135. *from Campus to Careers*
                General Contractor

The length L of the longest board that can be carried horizontally around the right-angle corner of two intersecting hallways is given by the formula

$$L = (a^{2/3} + b^{2/3})^{3/2}$$

where a and b represent the widths of the hallways. Find the longest shelf that a carpenter can carry around the corner if $a = 40$ in. and $b = 64$ in. Give your result in inches and in feet. In each case, round to the nearest tenth.

© Tetra Images/Alamy

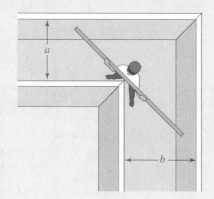

a

b

136. CUBICLES The area of the base of a cube is given by the function $A(V) = V^{2/3}$, where V is the volume of the cube. In a preschool room, 18 children's cubicles like the one shown are placed on the floor around the room. Estimate how much floor space is lost to the cubicles. Give your answer in square inches and in square feet.

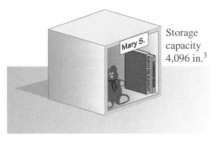

Storage capacity 4,096 in.³

WRITING

137. What is a rational exponent? Give some examples.

138. Explain how the root key on a scientific calculator can be used in combination with other keys to evaluate the expression $16^{3/4}$.

REVIEW

139. COMMUTING TIME The time it takes a car to travel a certain distance varies inversely with its rate of speed. If a certain trip takes 3 hours at 50 miles per hour, how long will the trip take at 60 miles per hour?

140. BANKRUPTCY After filing for bankruptcy, a company was able to pay its creditors only 15 cents on the dollar. If the company owed a lumberyard $9,712, how much could the lumberyard expect to be paid?

CHALLENGE PROBLEMS

141. The fraction $\frac{2}{4}$ is equal to $\frac{1}{2}$. Is $16^{2/4}$ equal to $16^{1/2}$? Explain.

142. Explain how would you evaluate an expression with a mixed-number exponent? For example, what is $8^{1\frac{1}{3}}$? What is $25^{2\frac{1}{2}}$?

SECTION 7.3
Simplifying and Combining Radical Expressions

Objectives

❶ Use the product rule to simplify radical expressions.
❷ Use prime factorization to simplify radical expressions.
❸ Use the quotient rule to simplify radical expressions.
❹ Add and subtract radical expressions.

In algebra, it is often helpful to replace an expression with a simpler equivalent expression. This is certainly true when working with radicals. In most cases, radical expressions should be written in simplified form. We use two rules for radicals to do this.

❶ **Use the Product Rule to Simplify Radical Expressions.**

To introduce the product rule for radicals, we will find $\sqrt{4 \cdot 25}$ and $\sqrt{4}\sqrt{25}$, and compare the results.

Square root of a product
$$\sqrt{4 \cdot 25} = \sqrt{100}$$
$$= 10$$

Product of square roots
$$\sqrt{4}\sqrt{25} = 2 \cdot 5$$
$$= 10$$

In each case, the answer is 10. Thus, $\sqrt{4 \cdot 25} = \sqrt{4}\sqrt{25}$.

Similarly, we will find $\sqrt[3]{8 \cdot 27}$ and $\sqrt[3]{8}\sqrt[3]{27}$ and compare the results.

Cube root of a product	**Product of cube roots**
$\sqrt[3]{8 \cdot 27} = \sqrt[3]{216}$	$\sqrt[3]{8}\sqrt[3]{27} = 2 \cdot 3$
$= 6$	$= 6$

In each case, the answer is 6. Thus, $\sqrt[3]{8 \cdot 27} = \sqrt[3]{8}\sqrt[3]{27}$. These results illustrate the *product rule for radicals*.

The Product Rule for Radicals

The *n*th root of the product of two numbers is equal to the product of their *n*th roots. If $\sqrt[n]{a}$ and $\sqrt[n]{b}$ are real numbers,

$$\sqrt[n]{ab} = \sqrt[n]{a}\sqrt[n]{b}$$

Caution   The product rule for radicals applies to the *n*th root of a product. There is no such property for sums or differences. For example,

$$\sqrt{9 + 4} \ne \sqrt{9} + \sqrt{4} \qquad\qquad \sqrt{9 - 4} \ne \sqrt{9} - \sqrt{4}$$
$$\sqrt{13} \ne 3 + 2 \qquad\qquad\qquad \sqrt{5} \ne 3 - 2$$
$$\sqrt{13} \ne 5 \qquad\qquad\qquad\qquad \sqrt{5} \ne 1$$

Thus, $\sqrt{a + b} \ne \sqrt{a} + \sqrt{b}$ and $\sqrt{a - b} \ne \sqrt{a} - \sqrt{b}$.

The product rule for radicals can be used to simplify radical expressions. When a radical expression is written in **simplified form,** each of the following is true.

Simplified Form of a Radical Expression

1. Each factor in the radicand is to a power that is less than the index of the radical.
2. The radicand contains no fractions or negative numbers.
3. No radicals appear in the denominator of a fraction.

To simplify radical expressions, we must often factor the radicand using two natural-number factors. To simplify square-root, cube-root, and fourth-root radicals, it is helpful to memorize the following lists.

Perfect squares: **1, 4, 9, 16, 25, 36, 49, 64, 81, 100, 121, 144, 169, 196, 225,** . . .

Perfect cubes: **1, 8, 27, 64, 125, 216, 343, 512, 729, 1,000,** . . .

Perfect-fourth powers: **1, 16, 81, 256, 625,** . . .

EXAMPLE 1   Simplify: **a.** $\sqrt{12}$ **b.** $\sqrt{98}$ **c.** $\sqrt[3]{54}$ **d.** $-\sqrt[4]{48}$

Strategy   We will factor each radicand into two factors, one of which is a perfect square, perfect cube, or perfect-fourth power, depending on the index of the radical. Then we can use the product rule for radicals to simplify the expression.

Why   Factoring the radicand in this way leads to a square root, cube root, or fourth root of a perfect square, perfect cube, or perfect-fourth power that we can easily simplify.

Solution

a. To simplify $\sqrt{12}$, we first factor 12 so that one factor is the largest perfect square that divides 12. Since 4 is the largest perfect-square factor of 12, we write 12 as $4 \cdot 3$, use the product rule for radicals, and simplify.

$$\sqrt{12} = \sqrt{4 \cdot 3} \qquad \text{Write 12 as } 12 = 4 \cdot 3.$$

Write the perfect-square factor first.

$$= \sqrt{4}\sqrt{3} \qquad \text{The square root of a product is equal to the product of the square roots.}$$

$$= 2\sqrt{3} \qquad \text{Evaluate } \sqrt{4}. \text{ Read as "2 times the square root of 3" or as "2 radical 3."}$$

We say that $2\sqrt{3}$ is the simplified form of $\sqrt{12}$.

b. The largest perfect-square factor of 98 is 49. Thus,

$$\sqrt{98} = \sqrt{49 \cdot 2} \qquad \text{Write 98 in factored form: } 98 = 49 \cdot 2.$$

$$= \sqrt{49}\sqrt{2} \qquad \text{The square root of a product is equal to the product}$$
$$\text{of the square roots: } \sqrt{49 \cdot 2} = \sqrt{49}\sqrt{2}.$$

$$= 7\sqrt{2} \qquad \text{Evaluate } \sqrt{49}.$$

c. Since the largest perfect-cube factor of 54 is 27, we have

$$\sqrt[3]{54} = \sqrt[3]{27 \cdot 2} \qquad \text{Write 54 as } 27 \cdot 2.$$

$$= \sqrt[3]{27}\sqrt[3]{2} \qquad \text{The cube root of a product is equal to the product}$$
$$\text{of the cube roots: } \sqrt[3]{27 \cdot 2} = \sqrt[3]{27}\sqrt[3]{2}.$$

$$= 3\sqrt[3]{2} \qquad \text{Evaluate } \sqrt[3]{27}.$$

d. The largest perfect-fourth power factor of 48 is 16. Thus,

$$-\sqrt[4]{48} = -\sqrt[4]{16 \cdot 3} \qquad \text{Write 48 as } 16 \cdot 3.$$

$$= -\sqrt[4]{16}\sqrt[4]{3} \qquad \text{The fourth root of a product is equal to the product}$$
$$\text{of the fourth roots: } \sqrt[4]{16 \cdot 3} = \sqrt[4]{16} \cdot \sqrt[4]{3}.$$

$$= -2\sqrt[4]{3} \qquad \text{Evaluate } \sqrt[4]{16}.$$

> **Success Tip**
> In Example 1, a radical of a product is written as a product of radicals:
> $$\sqrt[n]{ab} = \sqrt[n]{a}\sqrt[n]{b}$$

Self Check 1   Simplify: **a.** $\sqrt{20}$ **b.** $\sqrt[3]{24}$
c. $\sqrt[5]{128}$

Now Try   **Problems 13, 17, and 19**

Variable expressions can also be perfect squares, perfect cubes, perfect-fourth powers, and so on. For example,

$$\text{Perfect squares: } x^2, x^4, x^6, x^8, x^{10}, \ldots$$
$$\text{Perfect cubes: } x^3, x^6, x^9, x^{12}, x^{15}, \ldots$$
$$\text{Perfect-fourth powers: } x^4, x^8, x^{12}, x^{16}, x^{20}, \ldots$$

EXAMPLE 2   Simplify: **a.** $\sqrt{m^9}$ **b.** $\sqrt{128a^5}$ **c.** $\sqrt[3]{-24x^5}$ **d.** $\sqrt[5]{a^9b^5}$
All variables represent positive real numbers.

Strategy   We will factor each radicand into two factors, one of which is a perfect nth power.

Why   We can then apply the rule *the nth root of a product is the product of the nth roots* to simplify the radical expression.

Solution

a. The largest perfect-square factor of m^9 is m^8.

$$\sqrt{m^9} = \sqrt{m^8 \cdot m} \qquad \text{Write } m^9 \text{ in factored form as } m^8 \cdot m.$$

$$= \sqrt{m^8}\sqrt{m} \qquad \text{Use the product rule for radicals.}$$

$$= m^4\sqrt{m} \qquad \text{Simplify } \sqrt{m^8}.$$

b. Since the largest perfect-square factor of 128 is 64 and the largest perfect-square factor of a^5 is a^4, the largest perfect-square factor of $128a^5$ is $64a^4$. We write $128a^5$ as $64a^4 \cdot 2a$ and proceed as follows:

$$\sqrt{128a^5} = \sqrt{64a^4 \cdot 2a} \qquad \text{Write } 128a^5 \text{ in factored form as } 64a^4 \cdot 2a.$$

$$= \sqrt{64a^4}\sqrt{2a} \qquad \text{Use the product rule for radicals.}$$

$$= 8a^2\sqrt{2a} \qquad \text{Simplify } \sqrt{64a^4}.$$

c. We write $-24x^5$ as $-8x^3 \cdot 3x^2$ and proceed as follows:

$$\sqrt[3]{-24x^5} = \sqrt[3]{-8x^3 \cdot 3x^2} \qquad \begin{array}{l} 8x^3 \text{ is the largest perfect-cube factor of } 24x^5. \text{ Since} \\ \text{the radicand is negative, we factor it using } -8x^3. \end{array}$$

$$= \sqrt[3]{-8x^3}\sqrt[3]{3x^2} \qquad \text{Use the product rule for radicals.}$$

$$= -2x\sqrt[3]{3x^2} \qquad \text{Simplify } \sqrt[3]{-8x^3}.$$

d. The largest perfect-fifth power factor of a^9 is a^5, and b^5 is a perfect-fifth power.

> **The Language of Algebra**
> Perfect-fifth powers of a are
> $$a^5, a^{10}, a^{15}, a^{20}, a^{25}, \ldots$$

$$\sqrt[5]{a^9b^5} = \sqrt[5]{a^5b^5 \cdot a^4} \qquad a^5b^5 \text{ is the largest perfect-fifth power factor of } a^9b^5.$$

$$= \sqrt[5]{a^5b^5}\sqrt[5]{a^4} \qquad \text{Use the product rule for radicals.}$$

$$= ab\sqrt[5]{a^4} \qquad \text{Simplify } \sqrt[5]{a^5b^5}.$$

> **Self Check 2**   Simplify. All variables represent positive real numbers.
> **a.** $\sqrt{98b^3}$ **b.** $\sqrt[3]{-54y^5}$
> **c.** $\sqrt[4]{t^8u^{15}}$
>
> **Now Try**   Problems 21, 29, and 31

②   Use Prime Factorization to Simplify Radical Expressions.

When simplifying radical expressions, prime factorization can be helpful in determining how to factor the radicand.

EXAMPLE 3   Simplify. All variables represent positive real numbers.
a. $\sqrt{150}$ **b.** $\sqrt[3]{297b^4}$ **c.** $\sqrt[4]{224s^8t^7}$

Strategy   In each case, the way to factor the radicand is not obvious. Another approach is to prime-factor the coefficient of the radicand and look for groups of like factors.

Why   Identifying groups of like factors of the radicand leads to a factorization of the radicand that can be easily simplified.

Solution

a. $\sqrt{150} = \sqrt{2 \cdot 3 \cdot 5 \cdot 5}$ Write 150 in prime-factored form.

$\phantom{\sqrt{150}} = \sqrt{2 \cdot 3}\sqrt{5 \cdot 5}$ Group the pair of like factors together and use the product rule for radicals.

$\phantom{\sqrt{150}} = \sqrt{2 \cdot 3}\sqrt{5^2}$ Write $5 \cdot 5$ as 5^2.

$\phantom{\sqrt{150}} = \sqrt{6 \cdot 5}$ Evaluate $\sqrt{5^2}$.

$\phantom{\sqrt{150}} = 5\sqrt{6}$ Write the factor 5 first.

b. $\sqrt[3]{297b^4} = \sqrt[3]{3 \cdot 3 \cdot 3 \cdot 11 \cdot b^3 \cdot b}$ Write 297 in prime-factored form. The largest perfect-cube factor of b^4 is b^3.

$\phantom{\sqrt[3]{297b^4}} = \sqrt[3]{3 \cdot 3 \cdot 3 \cdot b^3}\sqrt[3]{11b}$ Group the three like factors of 3 together and use the product rule for radicals.

$\phantom{\sqrt[3]{297b^4}} = \sqrt[3]{3^3 b^3}\sqrt[3]{11b}$ Write $3 \cdot 3 \cdot 3$ as 3^3.

$\phantom{\sqrt[3]{297b^4}} = 3b\sqrt[3]{11b}$ Simplify $\sqrt[3]{3^3 b^3}$.

c. $\sqrt[4]{224s^8t^7} = \sqrt[4]{2 \cdot 2 \cdot 2 \cdot 2 \cdot 2 \cdot 7 \cdot s^8 \cdot t^4 \cdot t^3}$ Write 224 in prime-factored form. The largest perfect-fourth power factor of t^7 is t^4.

$\phantom{\sqrt[4]{224s^8t^7}} = \sqrt[4]{2 \cdot 2 \cdot 2 \cdot 2 \cdot s^8 \cdot t^4}\sqrt[4]{2 \cdot 7 \cdot t^3}$ Group the four like factors of 2 together and use the product rule for radicals.

$\phantom{\sqrt[4]{224s^8t^7}} = \sqrt[4]{2^4 s^8 t^4}\sqrt[4]{2 \cdot 7 \cdot t^3}$ Write $2 \cdot 2 \cdot 2 \cdot 2$ as 2^4.

$\phantom{\sqrt[4]{224s^8t^7}} = 2s^2 t\sqrt[4]{14t^3}$ Simplify $\sqrt[4]{2^4 s^8 t^4}$.

> **Self Check 3**   Simplify: **a.** $\sqrt{275}$ **b.** $\sqrt[3]{189c^4d^3}$
>
> ***Now Try***   **Problems 33 and 39**

3   **Use the Quotient Rule to Simplify Radical Expressions.**

To introduce the quotient rule for radicals, we will find $\sqrt{\dfrac{100}{4}}$ and $\dfrac{\sqrt{100}}{\sqrt{4}}$ and compare the results.

Square root of a quotient

$$\sqrt{\frac{100}{4}} = \sqrt{25}$$
$$= 5$$

Quotient of square roots

$$\frac{\sqrt{100}}{\sqrt{4}} = \frac{10}{2}$$
$$= 5$$

Since the answer is 5 in each case, $\sqrt{\dfrac{100}{4}} = \dfrac{\sqrt{100}}{\sqrt{4}}$.

Similarly, we will find $\sqrt[3]{\dfrac{64}{8}}$ and $\dfrac{\sqrt[3]{64}}{\sqrt[3]{8}}$, and compare the results.

Cube root of a quotient

$$\sqrt[3]{\frac{64}{8}} = \sqrt[3]{8}$$
$$= 2$$

Quotient of cube roots

$$\frac{\sqrt[3]{64}}{\sqrt[3]{8}} = \frac{4}{2}$$
$$= 2$$

Since the answer is 2 in each case, $\sqrt[3]{\dfrac{64}{8}} = \dfrac{\sqrt[3]{64}}{\sqrt[3]{8}}$. These results illustrate the *quotient rule for radicals*.

| **The Quotient Rule for Radicals** | The *n*th root of the quotient of two numbers is equal to the quotient of their *n*th roots. If $\sqrt[n]{a}$ and $\sqrt[n]{b}$ are real numbers, then $$\sqrt[n]{\dfrac{a}{b}} = \dfrac{\sqrt[n]{a}}{\sqrt[n]{b}} \qquad (b \neq 0)$$ |

EXAMPLE 4   Simplify each expression: **a.** $\sqrt{\dfrac{7}{64}}$ **b.** $\sqrt{\dfrac{15}{49x^2}}$ **c.** $\sqrt[3]{\dfrac{10x^2}{27y^6}}$ All variables represent positive real numbers.

Strategy   In each case, the radical is not in simplified form because the radicand contains a fraction. To write each of these expressions in simplified form, we will use the quotient rule for radicals.

Why   Writing these expressions in $\dfrac{\sqrt[n]{a}}{\sqrt[n]{b}}$ form leads to square roots of perfect squares and cube roots of perfect cubes that we can easily simplify.

Solution

a. We can use the quotient rule for radicals to simplify each expression.

$$\sqrt{\dfrac{7}{64}} = \dfrac{\sqrt{7}}{\sqrt{64}} \qquad \text{The square root of a quotient is equal to the quotient of the square roots.}$$

$$= \dfrac{\sqrt{7}}{8} \qquad \text{Evaluate } \sqrt{64}.$$

b. $\sqrt{\dfrac{15}{49x^2}} = \dfrac{\sqrt{15}}{\sqrt{49x^2}} \qquad \text{The square root of a quotient is equal to the quotient of the square roots.}$

$$= \dfrac{\sqrt{15}}{7x} \qquad \text{Simplify the denominator: } \sqrt{49x^2} = 7x.$$

c. $\sqrt[3]{\dfrac{10x^2}{27y^6}} = \dfrac{\sqrt[3]{10x^2}}{\sqrt[3]{27y^6}} \qquad \text{The cube root of a quotient is equal to the quotient of the cube roots.}$

$$= \dfrac{\sqrt[3]{10x^2}}{3y^2} \qquad \text{Simplify the denominator.}$$

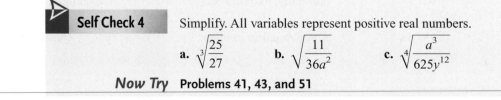

Self Check 4   Simplify. All variables represent positive real numbers.

a. $\sqrt[3]{\dfrac{25}{27}}$ **b.** $\sqrt{\dfrac{11}{36a^2}}$ **c.** $\sqrt[4]{\dfrac{a^3}{625y^{12}}}$

Now Try   Problems 41, 43, and 51

Sucess Tip

In Example 4, a radical of a quotient is written as a quotient of radicals:

$$\sqrt[n]{\dfrac{a}{b}} = \dfrac{\sqrt[n]{a}}{\sqrt[n]{b}}$$

EXAMPLE 5 Simplify each expression. All variables represent positive real numbers. **a.** $\dfrac{\sqrt{45xy^2}}{\sqrt{5x}}$ **b.** $\dfrac{\sqrt[3]{-432x^5}}{\sqrt[3]{8x}}$

Strategy We will use the quotient rule for radicals in reverse: $\dfrac{\sqrt[n]{a}}{\sqrt[n]{b}} = \sqrt[n]{\dfrac{a}{b}}$.

Why When the radicands are written under a single radical symbol, the result is a rational expression. Our hope is that the rational expression can be simplified.

Solution

a. We can write the quotient of the square roots as the square root of a quotient.

$$\frac{\sqrt{45xy^2}}{\sqrt{5x}} = \sqrt{\frac{45xy^2}{5x}} \quad \text{Use the quotient rule for radicals. Note that the resulting radicand is a rational expression.}$$

$$= \sqrt{9y^2} \quad \text{Simplify the radicand: } \frac{45xy^2}{5x} = \frac{\overset{1}{\cancel{5}} \cdot 9 \cdot \overset{1}{\cancel{x}} \cdot y^2}{\underset{1}{\cancel{5}} \cdot \underset{1}{\cancel{x}}} = 9y^2.$$

$$= 3y \quad \text{Simplify the radical.}$$

b. We can write the quotient of the cube roots as the cube root of a quotient.

$$\frac{\sqrt[3]{-432x^5}}{\sqrt[3]{8x}} = \sqrt[3]{-\frac{432x^5}{8x}} \quad \text{Use the quotient rule for radicals. Note that the resulting radicand is a rational expression.}$$

$$= \sqrt[3]{-54x^4} \quad \text{Simplify the radicand: } -\frac{432x^5}{8x} = -54x^4.$$

$$= \sqrt[3]{-27x^3 \cdot 2x} \quad 27x^3 \text{ is the largest perfect cube that divides } 54x^4.$$

$$= \sqrt[3]{-27x^3}\sqrt[3]{2x} \quad \text{Use the product rule for radicals.}$$

$$= -3x\sqrt[3]{2x} \quad \text{Simplify: } \sqrt[3]{-27x^3} = -3x.$$

 Self Check 5 Simplify each expression. All variables represent positive real numbers. **a.** $\dfrac{\sqrt{50ab^2}}{\sqrt{2a}}$ **b.** $\dfrac{\sqrt[3]{-2{,}000x^5v^3}}{\sqrt[3]{2x}}$

Now Try **Problems 55 and 59**

Success Tip

In Example 5, a quotient of radicals is written as a radical of a quotient.

$$\frac{\sqrt[n]{a}}{\sqrt[n]{b}} = \sqrt[n]{\frac{a}{b}}$$

---

④ Add and Subtract Radical Expressions.

Radical expressions with the same index and the same radicand are called **like** or **similar radicals.** For example, $3\sqrt{2}$ and $2\sqrt{2}$ are like radicals. However,

- $3\sqrt{5}$ and $4\sqrt{2}$ are not like radicals, because the radicands are different.
- $3\sqrt[4]{5}$ and $2\sqrt[3]{5}$ are not like radicals, because the indices are different.

For an expression with two or more radical terms, we should attempt to combine like radicals, if possible. For example, to simplify the expression $3\sqrt{2} + 2\sqrt{2}$, we use the distributive property to factor out $\sqrt{2}$ and simplify.

Success Tip

Combining like radicals is similar to combining like terms.

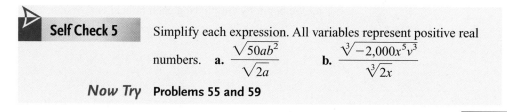

$$3\sqrt{2} + 2\sqrt{2} = 5\sqrt{2}$$

$$3x + 2x = 5x$$

$$3\sqrt{2} + 2\sqrt{2} = (3 + 2)\sqrt{2}$$
$$= 5\sqrt{2}$$

Radicals with the same index but different radicands can often be written as like radicals. For example, to simplify the expression $\sqrt{75} - \sqrt{27}$, we simplify both radicals first and then combine the like radicals.

$$\sqrt{75} - \sqrt{27} = \sqrt{25 \cdot 3} - \sqrt{9 \cdot 3} \qquad \text{Write 75 and 27 in factored form.}$$
$$= \sqrt{25}\sqrt{3} - \sqrt{9}\sqrt{3} \qquad \text{Use the product rule for radicals.}$$
$$= 5\sqrt{3} - 3\sqrt{3} \qquad \text{Evaluate } \sqrt{25} \text{ and } \sqrt{9}.$$
$$= (5 - 2)\sqrt{3} \qquad \text{Factor out } \sqrt{3}.$$
$$= 3\sqrt{3}$$

As the previous examples suggest, we can add or subtract radicals as follows.

Adding and Subtracting Radicals

To add or subtract radicals, simplify each radical, if possible, and combine like radicals.

EXAMPLE 6 Simplify: **a.** $2\sqrt{12} - 3\sqrt{48}$ **b.** $\sqrt[3]{16} + \sqrt[3]{54} - \sqrt[3]{24}$

Strategy Since the radicals in each part are unlike radicals, we cannot add or subtract them in their current form. However, we will simplify the radicals and hope that like radicals result.

Why Like radicals can be combined.

Solution

a. We begin by simplifying each radical expression:

$$2\sqrt{12} - 3\sqrt{48} = 2\sqrt{4 \cdot 3} - 3\sqrt{16 \cdot 3}$$
$$= 2\sqrt{4}\sqrt{3} - 3\sqrt{16}\sqrt{3}$$
$$= 2(2)\sqrt{3} - 3(4)\sqrt{3}$$
$$= 4\sqrt{3} - 12\sqrt{3} \qquad \text{Both expressions have the same index and radicand.}$$
$$= (4 - 12)\sqrt{3} \qquad \text{Combine like radicals.}$$
$$= -8\sqrt{3}$$

b. We begin by simplifying each radical expression:

$$\sqrt[3]{16} + \sqrt[3]{54} - \sqrt[3]{24} = \sqrt[3]{8 \cdot 2} + \sqrt[3]{27 \cdot 2} - \sqrt[3]{8 \cdot 3}$$
$$= \sqrt[3]{8}\sqrt[3]{2} + \sqrt[3]{27}\sqrt[3]{2} - \sqrt[3]{8}\sqrt[3]{3}$$
$$= 2\sqrt[3]{2} + 3\sqrt[3]{2} - 2\sqrt[3]{3}$$

Now we combine the two radical expressions that have the same index and radicand.

$$\sqrt[3]{16} + \sqrt[3]{54} - \sqrt[3]{24} = (2 + 3)\sqrt[3]{2} - 2\sqrt[3]{3} \qquad \text{Combine like radicals.}$$
$$= 5\sqrt[3]{2} - 2\sqrt[3]{3}$$

Caution Even though the expressions $5\sqrt[3]{2}$ and $2\sqrt[3]{3}$ have the same index, we cannot combine them, because their radicands are different. Neither can we combine radical expressions having the same radicand but a different index. For example, the expression $\sqrt[3]{2} + \sqrt[4]{2}$ cannot be simplified.

Self Check 6   Simplify: **a.** $3\sqrt{75} - 2\sqrt{12} + 2\sqrt{48}$

b. $\sqrt[3]{24} - \sqrt[3]{16} + \sqrt[3]{54}$

Now Try **Problems 67 and 71**

EXAMPLE 7   Simplify: $\sqrt[3]{16x^4} + \sqrt[3]{54x^4} - \sqrt[3]{-128x^4}$

Strategy   Since the radicals are unlike radicals, we cannot add or subtract them in their current form. However, we will simplify the radicals and hope that like radicals result.

Why   Like radicals can be combined.

Solution   We begin by simplifying each radical expression.

$$\sqrt[3]{16x^4} + \sqrt[3]{54x^4} - \sqrt[3]{-128x^4}$$

$$= \sqrt[3]{8x^3 \cdot 2x} + \sqrt[3]{27x^3 \cdot 2x} - \sqrt[3]{-64x^3 \cdot 2x}$$

$$= \sqrt[3]{8x^3}\sqrt[3]{2x} + \sqrt[3]{27x^3}\sqrt[3]{2x} - \sqrt[3]{-64x^3}\sqrt[3]{2x}$$

$$= 2x\sqrt[3]{2x} + 3x\sqrt[3]{2x} + 4x\sqrt[3]{2x} \quad \text{All three radicals have the same index and radicand.}$$

$$= (2x + 3x + 4x)\sqrt[3]{2x} \quad\quad\quad \text{Combine like radicals.}$$

$$= 9x\sqrt[3]{2x} \quad\quad\quad\quad\quad\quad\quad \text{Within the parentheses, combine like terms.}$$

Self Check 7   Simplify: $\sqrt{32x^3} + \sqrt{50x^3} - \sqrt{18x^3}$

Now Try **Problems 81 and 83**

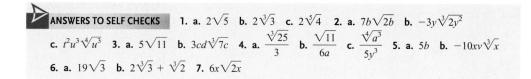

ANSWERS TO SELF CHECKS   **1. a.** $2\sqrt{5}$ **b.** $2\sqrt[3]{3}$ **c.** $2\sqrt[5]{4}$ **2. a.** $7b\sqrt{2b}$ **b.** $-3y\sqrt[3]{2y^2}$

c. $t^2u^3\sqrt[4]{u^3}$ **3. a.** $5\sqrt{11}$ **b.** $3cd\sqrt[3]{7c}$ **4. a.** $\dfrac{\sqrt[3]{25}}{3}$ **b.** $\dfrac{\sqrt{11}}{6a}$ **c.** $\dfrac{\sqrt[4]{a^3}}{5y^3}$ **5. a.** $5b$ **b.** $-10xv\sqrt[3]{x}$

6. a. $19\sqrt{3}$ **b.** $2\sqrt[3]{3} + \sqrt[3]{2}$ **7.** $6x\sqrt{2x}$

STUDY SET
7.3

VOCABULARY

Fill in the blanks.

1. Radical expressions such as $\sqrt[3]{4}$ and $6\sqrt[3]{4}$ with the same index and the same radicand are called _____ radicals.

2. Numbers such as 1, 4, 9, 16, 25, and 36 are called perfect _____. Numbers such as 1, 8, 27, 64, and 125 are called perfect _____. Numbers such as 1, 16, 81, 256, and 625 are called perfect-fourth _____.

3. The largest perfect-square _____ of 27 is 9. The largest _____-cube factor of 16 is 8.

4. To _____ $\sqrt{24}$ means to write it as $2\sqrt{6}$.

CONCEPTS

Fill in the blanks.

5. The product rule for radicals: $\sqrt[n]{ab} =$ _____ . In words, the nth root of the _____ of two numbers is equal to the product of their nth _____.

6. The quotient rule for radicals: $\sqrt[n]{\dfrac{a}{b}} =$ _____ . In words, the nth root of the _____ of two numbers is equal to the quotient of their nth _____.

7. Consider the expressions

$$\sqrt{4\cdot 5} \quad \text{and} \quad \sqrt{4}\sqrt{5}$$

Which expression is

a. the square root of a product?

b. the product of square roots?

c. How are these two expressions related?

8. Consider the expressions

$$\frac{\sqrt[3]{a}}{\sqrt[3]{x^2}} \quad \text{and} \quad \sqrt[3]{\frac{a}{x^2}}$$

Which expression is

a. the cube root of a quotient?

b. the quotient of cube roots?

c. How are these two expressions related?

9. a. Write two radical expressions that have the same radicand but a different index. Can the expressions be added?

b. Write two radical expressions that have the same index but a different radicand. Can the expressions be added?

10. Fill in the blanks.

a. $5\sqrt{6} + 3\sqrt{6} = ($ ___ $+$ ___ $)\sqrt{6} =$ ___ $\sqrt{6}$

b. $9\sqrt[3]{n} - 2\sqrt[3]{n} = ($ ___ $-$ ___ $)\sqrt[3]{n} = 7$ ___

NOTATION

Complete each solution.

11. Simplify:

$$\sqrt[3]{32k^4} = \sqrt[3]{} \cdot 4k$$
$$= \sqrt[3]{}\,\sqrt[3]{4k}$$
$$= 2k\sqrt[3]{}$$

12. Simplify:

$$\frac{\sqrt{80s^2t^4}}{\sqrt{5s^2}} = \sqrt{\frac{80s^2t^4}{}}$$
$$= \sqrt{}$$
$$= 4t^2$$

GUIDED PRACTICE

Simplify each expression. See Example 1.

13. $\sqrt{50}$

14. $\sqrt{28}$

15. $\sqrt{45}$

16. $\sqrt{54}$

17. $\sqrt[3]{32}$

18. $\sqrt[3]{40}$

19. $\sqrt[4]{48}$

20. $\sqrt[4]{32}$

Simplify each radical expression. All variables represent positive real numbers. See Example 2.

21. $\sqrt{75a^2}$

22. $\sqrt{50x^2}$

23. $\sqrt{32b}$

24. $\sqrt{80c}$

25. $\sqrt{128a^3b^5}$

26. $\sqrt{75b^8c}$

27. $\sqrt{300xy}$

28. $\sqrt{200x^2y}$

29. $\sqrt[3]{-54x^6}$

30. $\sqrt[3]{-81a^3}$

31. $\sqrt[4]{32x^{12}y^4}$

32. $\sqrt[5]{64x^{10}y^5}$

Simplify each radical expression. All variables represent positive real numbers. See Example 3.

33. $\sqrt{242}$

34. $\sqrt{363}$

35. $\sqrt{112a^3}$

36. $\sqrt{147a^5}$

37. $-\sqrt[5]{96a^4}$

38. $-\sqrt[7]{256t^6}$

39. $\sqrt[3]{405x^{12}y^4}$

40. $\sqrt[3]{280a^5b^6}$

Simplify each radical expression. All variables represent positive real numbers. See Example 4.

41. $\sqrt{\dfrac{11}{9}}$

42. $\sqrt{\dfrac{3}{4}}$

43. $\sqrt[3]{\dfrac{7}{64}}$

44. $\sqrt[3]{\dfrac{4}{125}}$

45. $\sqrt[4]{\dfrac{3}{625}}$

46. $\sqrt[5]{\dfrac{2}{243}}$

47. $\sqrt[5]{\dfrac{3x^{10}}{32}}$

48. $\sqrt[6]{\dfrac{5y^{12}}{64}}$

49. $\sqrt{\dfrac{z^2}{16x^2}}$

50. $\sqrt{\dfrac{b^4}{64a^8}}$

51. $\sqrt[4]{\dfrac{5x}{16z^4}}$

52. $\sqrt[3]{\dfrac{11a^2}{125b^6}}$

Simplify each expression. All variables represent positive real numbers. See Example 5.

53. $\dfrac{\sqrt{500}}{\sqrt{5}}$

54. $\dfrac{\sqrt{128}}{\sqrt{2}}$

55. $\dfrac{\sqrt{98x^3}}{\sqrt{2x}}$

56. $\dfrac{\sqrt{75y^5}}{\sqrt{3y}}$

57. $\dfrac{\sqrt[3]{48x^7}}{\sqrt[3]{6x}}$

58. $\dfrac{\sqrt[3]{64y^8}}{\sqrt[3]{8y^2}}$

59. $\dfrac{\sqrt[3]{189a^5}}{\sqrt[3]{7a}}$

60. $\dfrac{\sqrt[3]{243x^8}}{\sqrt[3]{9x}}$

Simplify by combining like radicals. All variables represent positive real numbers. See Objective 4 and Example 6.

61. $5\sqrt{7} + 3\sqrt{7}$

62. $11\sqrt{3} + 2\sqrt{3}$

63. $20\sqrt[3]{4} - 15\sqrt[3]{4}$

64. $30\sqrt[3]{6} - 10\sqrt[3]{6}$

65. $\sqrt{8} + \sqrt{2}$

66. $\sqrt{45} + \sqrt{20}$

67. $\sqrt{98} - \sqrt{50} - \sqrt{72}$

68. $\sqrt{20} + \sqrt{125} - \sqrt{80}$

69. $\sqrt[3]{32} - \sqrt[3]{108}$

70. $\sqrt[3]{80} - \sqrt[3]{10{,}000}$

71. $2\sqrt[3]{125} - 5\sqrt[3]{64}$

72. $3\sqrt[3]{27} + 12\sqrt[3]{216}$

73. $14\sqrt[4]{32} - 15\sqrt[4]{2}$

74. $23\sqrt[4]{3} + \sqrt[4]{48}$

75. $\sqrt{80} + \sqrt{45} - \sqrt{27}$

76. $\sqrt{63} + \sqrt{72} - \sqrt{28}$

Simplify by combining like radicals. All variables represent positive real numbers. See Example 7.

77. $4\sqrt{2x} + 6\sqrt{2x}$

78. $6\sqrt[3]{5y} + 3\sqrt[3]{5y}$

79. $8\sqrt[5]{7a^2} - 7\sqrt[5]{7a^2}$

80. $10\sqrt[6]{12xy} - \sqrt[6]{12xy}$

81. $\sqrt{18t} + \sqrt{300t} - \sqrt{243t}$

82. $\sqrt{80m} - \sqrt{128m} + \sqrt{288m}$

83. $2\sqrt[3]{16} - \sqrt[3]{54} - 3\sqrt[3]{128}$

84. $\sqrt[3]{250} - 4\sqrt[3]{5} + \sqrt[3]{16}$

85. $2\sqrt[3]{64a} + 2\sqrt[3]{8a}$

86. $3\sqrt[4]{x^4y} - 2\sqrt[4]{x^4y}$

87. $\sqrt[4]{64} + 5\sqrt[4]{4} - \sqrt[4]{324}$

88. $\sqrt[4]{48} - \sqrt[4]{243} - \sqrt[4]{768}$

TRY IT YOURSELF

Simplify each expression. All variables represent positive real numbers.

89. $\sqrt[6]{m^{11}}$

90. $\sqrt[6]{n^{13}}$

91. $\sqrt{8y^7} + \sqrt{32y^7} - \sqrt{2y^7}$

92. $\sqrt{y^5} - \sqrt{9y^5} - \sqrt{25y^5}$

93. $\sqrt{\dfrac{125n^5}{64n}}$

94. $\sqrt{\dfrac{72q^7}{25q^3}}$

95. $\sqrt[5]{x^6y^2} + \sqrt[5]{32x^6y^2} + \sqrt[5]{x^6y^2}$

96. $\sqrt[3]{xy^4} + \sqrt[3]{8xy^4} - \sqrt[3]{27xy^4}$

97. $\sqrt[4]{208m^4n}$

98. $\sqrt[4]{128p^8q^3}$

99. $\sqrt[3]{\dfrac{a^7}{64a}}$

100. $\sqrt[3]{\dfrac{b^3c^8}{125c^5}}$

101. $\sqrt[5]{64t^{11}}$

102. $\sqrt[5]{243r^{22}}$

103. $\sqrt[3]{24x} + \sqrt[3]{3x}$

104. $\sqrt[3]{16y} + \sqrt[3]{128y}$

APPLICATIONS

First give the exact answer, expressed as a simplified radical expression. Then give an approximation, rounded to the nearest tenth.

105. UMBRELLAS The surface area of a cone is given by the formula $S = \pi r \sqrt{r^2 + h^2}$, where r is the radius of the base and h is its height. Use this formula to find the number of square feet of waterproof cloth used to make the umbrella shown.

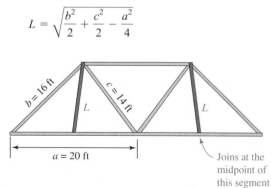

106. STRUCTURAL ENGINEERING Engineers have determined that two additional supports need to be added to strengthen the truss shown. Find the length L of each new support using the formula

$$L = \sqrt{\frac{b^2}{2} + \frac{c^2}{2} - \frac{a^2}{4}}$$

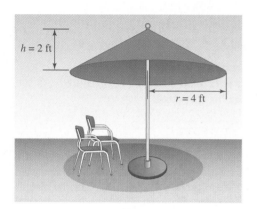

107. BLOW DRYERS The current I (in amps), the power P (in watts), and the resistance R (in ohms) are related by the formula $I = \sqrt{\frac{P}{R}}$. What current is needed for a 1,200-watt hair dryer if the resistance is 16 ohms?

108. COMMUNICATIONS SATELLITES Engineers have determined that a spherical communications satellite needs to have a capacity of 565.2 cubic feet to house all of its operating systems. The volume V of a sphere is related to its radius r by the formula $r = \sqrt[3]{\frac{3V}{4\pi}}$. What radius must the satellite have to meet the engineer's specification? Use 3.14 for π.

109. DUCTWORK The following pattern is laid out on a sheet of galvanized tin. Then it is cut out with snips and bent to make an air conditioning duct connection. Find the total length of the cut that must be made with the tin snips. (All measurements are in inches.)

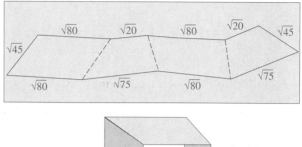

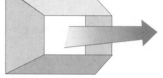

110. OUTDOOR COOKING The diameter of a circle is given by the function $d(A) = 2\sqrt{\frac{A}{\pi}}$, where A is the area of the circle. Find the difference between the diameters of the barbecue grills.

WRITING

111. Explain why each expression is not in simplified form.

a. $\sqrt[3]{9x^4}$ b. $\sqrt{\frac{24m}{25}}$ c. $\dfrac{\sqrt[4]{c^3}}{\sqrt[4]{16}}$

112. How are the procedures used to simplify $3x + 4x$ and $3\sqrt{x} + 4\sqrt{x}$ similar?

113. Explain the mistake in the student's solution shown below.
Simplify: $\sqrt[3]{54}$

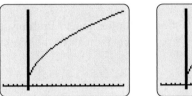

$$\sqrt[3]{54} = \sqrt[3]{27 + 27}$$
$$= \sqrt[3]{27} + \sqrt[3]{27}$$
$$= 3 + 3$$
$$= 6$$

114. Explain how the graphs of $Y_1 = 3\sqrt{24x} + \sqrt{54x}$ (on the left) and $Y_1 = 9\sqrt{6x}$ (on the right) can be used to verify the simplification $3\sqrt{24x} + \sqrt{54x} = 9\sqrt{6x}$. In each graph, settings of $[-5, 20]$ for x and $[-5, 100]$ for y were used.

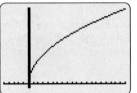

REVIEW

Perform each operation.

115. $3x^2y^3(-5x^3y^{-4})$

116. $(2x^2 - 9x - 5) \cdot \dfrac{x}{2x^2 + x}$

117. $2p - 5\overline{)6p^2 - 7p - 25}$

118. $\dfrac{xy}{\dfrac{1}{x} - \dfrac{1}{y}}$

CHALLENGE PROBLEMS

119. Can you find any numbers a and b such that
$$\sqrt{a + b} = \sqrt{a} + \sqrt{b}?$$

120. Find the sum:
$$\sqrt{3} + \sqrt{3^2} + \sqrt{3^3} + \sqrt{3^4} + \sqrt{3^5}$$

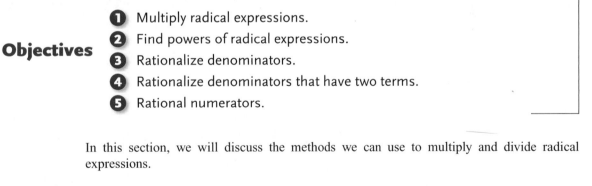

SECTION 7.4
Multiplying and Dividing Radical Expressions

Objectives

① Multiply radical expressions.
② Find powers of radical expressions.
③ Rationalize denominators.
④ Rationalize denominators that have two terms.
⑤ Rational numerators.

In this section, we will discuss the methods we can use to multiply and divide radical expressions.

① **Multiply Radical Expressions.**

We have used the *product rule for radicals* to write radical expressions in simplified form. We can also use this rule to multiply radical expressions that have the same index.

The Product Rule for Radicals	The product of the nth roots of two nonnegative numbers is equal to the nth root of the product of those numbers. If $\sqrt[n]{a}$ and $\sqrt[n]{b}$ are real numbers, $$\sqrt[n]{a} \cdot \sqrt[n]{b} = \sqrt[n]{a \cdot b}$$

EXAMPLE 1   Multiply and then simplify, if possible:
  a. $\sqrt{5}\sqrt{10}$ **b.** $3\sqrt{6}\left(2\sqrt{3}\right)$ **c.** $-2\sqrt[3]{7x} \cdot 6\sqrt[3]{49x^2}$

Strategy   In each expression, we will use the product rule for radicals to multiply factors of the form $\sqrt[n]{a}$ and $\sqrt[n]{b}$.

Why   The product rule for radicals is used to multiply radicals that have the same index.

Solution

a. $\sqrt{5}\sqrt{10} = \sqrt{5 \cdot 10}$ Use the product rule for radicals.

$\qquad\qquad = \sqrt{50}$ Multiply under the radical. Note that $\sqrt{50}$ can be simplified.

$\qquad\qquad = \sqrt{25 \cdot 2}$ Prepare to simplify: factor 50.

$\qquad\qquad = 5\sqrt{2}$ Simplify: $\sqrt{25 \cdot 2} = \sqrt{25}\sqrt{2} = 5\sqrt{2}$.

b. We use the commutative and associative properties of multiplication to multiply the integer factors and the radicals separately. Then we simplify any radicals in the product, if possible.

$3\sqrt{6}\left(2\sqrt{3}\right) = 3(2)\sqrt{6}\sqrt{3}$ Multiply the integer factors, 3 and 2, and multiply the radicals.

$\qquad\qquad = 6\sqrt{18}$ Use the product rule for radicals.

$\qquad\qquad = 6\sqrt{9}\sqrt{2}$ Simplify: $\sqrt{18} = \sqrt{9 \cdot 2} = \sqrt{9}\sqrt{2}$.

$\qquad\qquad = 6(3)\sqrt{2}$ Evaluate: $\sqrt{9} = 3$.

$\qquad\qquad = 18\sqrt{2}$ Multiply.

c. $-2\sqrt[3]{7x} \cdot 6\sqrt[3]{49x^2} = -2(6)\sqrt[3]{7x}\sqrt[3]{49x^2}$ Write the integer factors together and the radicals together.

$\qquad\qquad = -12\sqrt[3]{7x \cdot 49x^2}$ Multiply the integer factors, -2 and 6, and multiply the radicals.

$\qquad\qquad = -12\sqrt[3]{7x \cdot 7^2x^2}$ Write 49 as 7^2.

$\qquad\qquad = -12\sqrt[3]{7^3x^3}$ Prepare to simplify: write $7x \cdot 7^2x^2$ as 7^3x^3.

$\qquad\qquad = -12(7x)$ Simplify: $\sqrt[3]{7^3x^3} = 7x$.

$\qquad\qquad = -84x$ Multiply.

Self Check 1   Multiply. All variables represent positive real numbers.
  a. $\sqrt{7}\sqrt{14}$
  b. $-2\sqrt[3]{2}\left(5\sqrt[3]{12}\right)$ **c.** $\sqrt[4]{4x^3} \cdot 9\sqrt[4]{8x^2}$

Now Try   **Problems 11, 19, and 21**

Recall that to multiply a polynomial by a monomial, we use the distributive property. We use the same technique to multiply a radical expression that has two or more terms by a radical expression that has only one term.

EXAMPLE 2   Multiply and then simplify, if possible: $3\sqrt{3}\left(4\sqrt{8} - 5\sqrt{10}\right)$

Strategy   We will use the distributive property and multiply each term within the parentheses by the term outside the parentheses.

Why The given expression has the form $a(b - c)$.

Solution

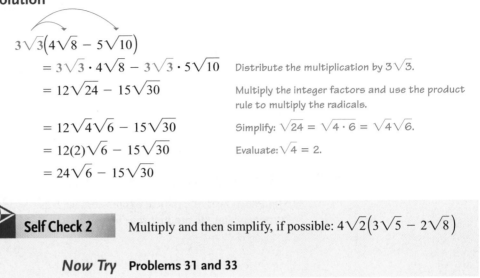

$$3\sqrt{3}\left(4\sqrt{8} - 5\sqrt{10}\right)$$
$$= 3\sqrt{3} \cdot 4\sqrt{8} - 3\sqrt{3} \cdot 5\sqrt{10} \qquad \text{Distribute the multiplication by } 3\sqrt{3}.$$
$$= 12\sqrt{24} - 15\sqrt{30} \qquad \text{Multiply the integer factors and use the product rule to multiply the radicals.}$$
$$= 12\sqrt{4}\sqrt{6} - 15\sqrt{30} \qquad \text{Simplify: } \sqrt{24} = \sqrt{4 \cdot 6} = \sqrt{4}\sqrt{6}.$$
$$= 12(2)\sqrt{6} - 15\sqrt{30} \qquad \text{Evaluate: } \sqrt{4} = 2.$$
$$= 24\sqrt{6} - 15\sqrt{30}$$

▷ **Self Check 2** Multiply and then simplify, if possible: $4\sqrt{2}\left(3\sqrt{5} - 2\sqrt{8}\right)$

Now Try **Problems 31 and 33**

Recall that to multiply two binomials, we multiply each term of one binomial by each term of the other binomial and simplify. We multiply two radical expressions, each having two terms, in the same way.

EXAMPLE 3 Multiply and then simplify, if possible:
  a. $\left(\sqrt{7} + \sqrt{2}\right)\left(\sqrt{7} - 3\sqrt{2}\right)$
  b. $\left(\sqrt[3]{x^2} - 4\sqrt[3]{5}\right)\left(\sqrt[3]{x} + \sqrt[3]{2}\right)$

Strategy As with binomials, we will multiply each term within the first set of parentheses by each term within the second set of parentheses.

Why This is an application of the FOIL method for multiplying binomials.

Solution

a. $\left(\sqrt{7} + \sqrt{2}\right)\left(\sqrt{7} - 3\sqrt{2}\right)$

$$\qquad\quad\; F \qquad\quad O \qquad\quad I \qquad\quad L$$
$$= \sqrt{7}\sqrt{7} - 3\sqrt{7}\sqrt{2} + \sqrt{2}\sqrt{7} - 3\sqrt{2}\sqrt{2} \qquad \text{Use the FOIL method.}$$
$$= 7 - 3\sqrt{14} + \sqrt{14} - 3(2) \qquad \text{Perform each multiplication.}$$
$$= 7 - 2\sqrt{14} - 6 \qquad \text{Combine like radicals.}$$
$$= 1 - 2\sqrt{14} \qquad \text{Combine like terms.}$$

b. $\left(\sqrt[3]{x^2} - 4\sqrt[3]{5}\right)\left(\sqrt[3]{x} + \sqrt[3]{2}\right)$

$$= \sqrt[3]{x^2}\sqrt[3]{x} + \sqrt[3]{x^2}\sqrt[3]{2} - 4\sqrt[3]{5}\sqrt[3]{x} - 4\sqrt[3]{5}\sqrt[3]{2} \qquad \text{Use the FOIL method.}$$
$$= \sqrt[3]{x^3} + \sqrt[3]{2x^2} - 4\sqrt[3]{5x} - 4\sqrt[3]{10} \qquad \text{Perform each multiplication.}$$
$$= x + \sqrt[3]{2x^2} - 4\sqrt[3]{5x} - 4\sqrt[3]{10} \qquad \text{Simplify the first term.}$$

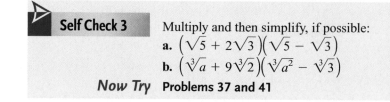

 Find Powers of Radical Expressions.

To find the power of a radical expression, such as $\left(\sqrt{5}\right)^2$ or $\left(\sqrt[3]{2}\right)^3$, we can use the definition of exponent and the product rule for radicals.

$$\left(\sqrt{5}\right)^2 = \sqrt{5}\sqrt{5} \qquad\qquad \left(\sqrt[3]{2}\right)^3 = \sqrt[3]{2} \cdot \sqrt[3]{2} \cdot \sqrt[3]{2}$$
$$= \sqrt{25} \qquad\qquad\qquad\qquad = \sqrt[3]{8}$$
$$= 5 \qquad\qquad\qquad\qquad\quad = 2$$

These results illustrate the following property of radicals.

The *n*th Power of the *n*th Root	If $\sqrt[n]{a}$ is a real number, $$\left(\sqrt[n]{a}\right)^n = a$$

EXAMPLE 4   Find: **a.** $\left(\sqrt{5}\right)^2$ **b.** $\left(2\sqrt[3]{7x^2}\right)^3$ **c.** $\left(\sqrt{m+1}+2\right)^2$

Strategy   In part (a), we will use the definition of square root. In part (b), we will use a power rule for exponents. In part (c), we will use the FOIL method.

Why   Part (a) is the square of a square root, part (b) has the form $(xy)^n$, and part (c) has the form $(x + y)^2$.

Solution

a. $\left(\sqrt{5}\right)^2 = 5$ Because the square of the square root of 5 is 5.

b. We can use the power of a product rule for exponents to find $\left(2\sqrt[3]{7x^2}\right)^3$.

$$\left(2\sqrt[3]{7x^2}\right)^3 = 2^3\left(\sqrt[3]{7x^2}\right)^3 \quad \text{Raise each factor of } 2\sqrt[3]{7x^2} \text{ to the 3rd power.}$$
$$= 8(7x^2) \qquad\quad \text{Evaluate: } 2^3 = 8. \text{ Use } \left(\sqrt[n]{a}\right)^n = a.$$
$$= 56x^2$$

c. We can use the FOIL method to find the product.

$$\left(\sqrt{m+1}+2\right)^2 = \left(\sqrt{m+1}+2\right)\left(\sqrt{m+1}+2\right)$$
$$= \left(\sqrt{m+1}\right)^2 + 2\sqrt{m+1} + 2\sqrt{m+1} + 2\cdot 2$$
$$= m + 1 + 2\sqrt{m+1} + 2\sqrt{m+1} + 4 \quad \text{Use } \left(\sqrt[n]{a}\right)^n = a.$$
$$= m + 4\sqrt{m+1} + 5 \qquad\qquad\qquad \text{Combine like terms.}$$

Success Tip

Since $\left(\sqrt{m+1}+2\right)^2$ has the form $(x + y)^2$, we could also use a special-product rule to find this square of a sum quickly.

Self Check 4   Find: **a.** $\left(\sqrt{11}\right)^2$ **b.** $\left(3\sqrt[3]{4y}\right)^3$
c. $\left(\sqrt{x-8}-5\right)^2$

Now Try   Problems 43, 49, and 51

3   **Rationalize Denominators.**

We have seen that when a radical expression is written in simplified form, each of the following statements is true.

1. Each factor in the radicand is to a power that is less than the index of the radical.
2. The radicand contains no fractions or negative numbers.
3. No radicals appear in the denominator of a fraction.

The Language of Algebra

Since $\sqrt{3}$ is an irrational number, $\dfrac{\sqrt{5}}{\sqrt{3}}$ has an irrational denominator. Since 3 is a rational number, $\dfrac{\sqrt{15}}{3}$ has a rational denominator.

We now consider radical expressions that do not satisfy requirements 2 or 3. We will introduce an algebraic technique, called *rationalizing the denominator,* that is used to write such expressions in an equivalent simplified form.

To divide radical expressions, we **rationalize the denominator** of a fraction to replace the denominator with a rational number. For example, to divide $\sqrt{5}$ by $\sqrt{3}$, we write the division as the fraction

$$\frac{\sqrt{5}}{\sqrt{3}}$$ This radical expression is not in simplified form, because a radical appears in the denominator.

We want to find a fraction equivalent to $\dfrac{\sqrt{5}}{\sqrt{3}}$ that does not have a radical in its denominator. If we multiply $\dfrac{\sqrt{5}}{\sqrt{3}}$ by $\dfrac{\sqrt{3}}{\sqrt{3}}$, the denominator becomes $\sqrt{3}\cdot\sqrt{3}=3$, a rational number.

Success Tip

As an informal check, we can use a calculator to evaluate each expression.

$$\frac{\sqrt{5}}{\sqrt{3}} \approx 1.290994449$$

$$\frac{\sqrt{15}}{3} \approx 1.290994449$$

$$\frac{\sqrt{5}}{\sqrt{3}} = \frac{\sqrt{5}}{\sqrt{3}}\cdot\frac{\sqrt{3}}{\sqrt{3}}$$ To build an equivalent fraction, multiply by $\dfrac{\sqrt{3}}{\sqrt{3}}=1$.

$$= \frac{\sqrt{15}}{3}$$ Multiply the numerators: $\sqrt{5}\cdot\sqrt{3}=\sqrt{15}$.
Multiply the denominators: $\sqrt{3}\cdot\sqrt{3}=\left(\sqrt{3}\right)^2=3$.

Thus, $\dfrac{\sqrt{5}}{\sqrt{3}} = \dfrac{\sqrt{15}}{3}$. These equivalent fractions represent the same number, but have different forms. Since there is no radical in the denominator, and $\sqrt{15}$ is in simplest form, the expression $\dfrac{\sqrt{15}}{3}$ is in simplified form.

EXAMPLE 5   Rationalize the denominator: **a.** $\sqrt{\dfrac{20}{7}}$ **b.** $\dfrac{4}{\sqrt[3]{2}}$

Strategy   We look at each denominator and ask, "By what must we multiply it to obtain a rational number?" Then we will multiply each expression by a carefully chosen form of 1.

Why   We want to produce an equivalent expression that does not have a radical in its denominator.

Solution

a. This radical expression is not in simplified form, because the radicand contains a fraction. We begin by writing the square root of the quotient as the quotient of two square roots:

$$\sqrt{\frac{20}{7}} = \frac{\sqrt{20}}{\sqrt{7}}$$ Use the division property of radicals: $\sqrt[n]{\frac{a}{b}} = \frac{\sqrt[n]{a}}{\sqrt[n]{b}}$.

> **Caution**
>
> Do not attempt to remove a common factor of 7 from the numerator and denominator of $\frac{2\sqrt{35}}{7}$. The numerator, $2\sqrt{35}$, does not have a factor of 7.
>
> $$\frac{2\sqrt{35}}{7} = \frac{2 \cdot \sqrt{5 \cdot 7}}{7}$$

To rationalize the denominator, we proceed as follows:

$$\frac{\sqrt{20}}{\sqrt{7}} = \frac{\sqrt{20}}{\sqrt{7}} \cdot \frac{\sqrt{7}}{\sqrt{7}}$$ To build an equivalent fraction, multiply by $\frac{\sqrt{7}}{\sqrt{7}} = 1$.

$$= \frac{\sqrt{140}}{7}$$ Multiply the numerators.
Multiply the denominators: $\sqrt{7} \cdot \sqrt{7} = \left(\sqrt{7}\right)^2 = 7$.

$$= \frac{2\sqrt{35}}{7}$$ Simplify: $\sqrt{140} = \sqrt{4 \cdot 35} = \sqrt{4}\sqrt{35} = 2\sqrt{35}$.

b. This expression is not in simplified form because a radical appears in the denominator of a fraction. Here, we must rationalize a denominator that is a cube root. We multiply the numerator and the denominator by a number that will give a perfect cube under the radical. Since $2 \cdot 4 = 8$ is a perfect cube, $\sqrt[3]{4}$ is such a number.

> **Caution**
>
> Multiplying $\frac{4}{\sqrt[3]{2}}$ by $\frac{\sqrt[3]{2}}{\sqrt[3]{2}}$ does not rationalize the denominator.
>
> $$\frac{4}{\cancel{\sqrt[3]{2}}} \cdot \frac{\sqrt[3]{2}}{\sqrt[3]{2}} = \frac{4\sqrt[3]{2}}{\sqrt[3]{4}}$$
> ↑
>
> Since 4 is not a perfect cube, this radical does not simplify.

$$\frac{4}{\sqrt[3]{2}} = \frac{4}{\sqrt[3]{2}} \cdot \frac{\sqrt[3]{4}}{\sqrt[3]{4}}$$ To build an equivalent fraction, multiply by $\frac{\sqrt[3]{4}}{\sqrt[3]{4}} = 1$.

$$= \frac{4\sqrt[3]{4}}{\sqrt[3]{8}}$$ Multiply the numerators. Multiply the denominators.
← This radicand is now a perfect cube.

$$= \frac{4\sqrt[3]{4}}{2}$$ Evaluate the denominator: $\sqrt[3]{8} = 2$.

$$= 2\sqrt[3]{4}$$ Simplify the fraction: $\frac{4\sqrt[3]{4}}{2} = \frac{\overset{2}{\cancel{4}} \cdot 2\sqrt[3]{4}}{\underset{1}{\cancel{2}}} = 2\sqrt[3]{4}$.

> **Self Check 5** Rationalize the denominator: **a.** $\sqrt{\frac{8}{5}}$ **b.** $\frac{5}{\sqrt[3]{3}}$
>
> **Now Try** Problems 57, 59, and 63

EXAMPLE 6 Rationalize the denominator: $\dfrac{\sqrt{5xy^2}}{\sqrt{xy^3}}$

Strategy We will begin by using the quotient rule for radicals in reverse $\dfrac{\sqrt[n]{a}}{\sqrt[n]{b}} = \sqrt[n]{\dfrac{a}{b}}$.

Why When the radicands are written under a single radical symbol, the result is a rational expression. Our hope is that the rational expression can be simplified, which could possibly make rationalizing the denominator easier.

Solution   There are two methods we can use to rationalize the denominator. In each method, we simplify the expression first.

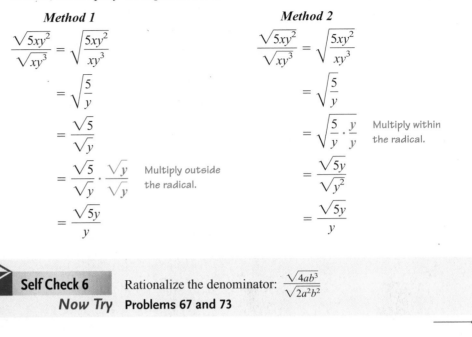

Method 1

$$\frac{\sqrt{5xy^2}}{\sqrt{xy^3}} = \sqrt{\frac{5xy^2}{xy^3}}$$

$$= \sqrt{\frac{5}{y}}$$

$$= \frac{\sqrt{5}}{\sqrt{y}}$$

$$= \frac{\sqrt{5}}{\sqrt{y}} \cdot \frac{\sqrt{y}}{\sqrt{y}} \quad \text{Multiply outside the radical.}$$

$$= \frac{\sqrt{5y}}{y}$$

Method 2

$$\frac{\sqrt{5xy^2}}{\sqrt{xy^3}} = \sqrt{\frac{5xy^2}{xy^3}}$$

$$= \sqrt{\frac{5}{y}}$$

$$= \sqrt{\frac{5}{y} \cdot \frac{y}{y}} \quad \text{Multiply within the radical.}$$

$$= \frac{\sqrt{5y}}{\sqrt{y^2}}$$

$$= \frac{\sqrt{5y}}{y}$$

Self Check 6   Rationalize the denominator: $\dfrac{\sqrt{4ab^3}}{\sqrt{2a^2b^2}}$

Now Try   **Problems 67 and 73**

EXAMPLE 7   Rationalize the denominator: $\dfrac{11}{\sqrt{20q^5}}$

Strategy   We will simplify the radical expression in the denominator before rationalizing the denominator.

Why   We could begin by multiplying $\dfrac{11}{\sqrt{20q^5}}$ by $\dfrac{\sqrt{20q^5}}{\sqrt{20q^5}}$. However, to work with smaller numbers and simpler radical expressions, it is easier if we simplify $\sqrt{20q^5}$ first, and then rationalize the denominator.

Solution

$$\frac{11}{\sqrt{20q^5}} = \frac{11}{\sqrt{4q^4 \cdot 5q}} \qquad \text{To prepare to simplify } \sqrt{20q^5}, \text{ factor } 20q^5 \text{ as } 4q^4 \cdot 5q.$$

$$= \frac{11}{2q^2\sqrt{5q}} \qquad \text{Simplify: } \sqrt{4q^4 \cdot 5q} = \sqrt{4q^4}\sqrt{5q} = 2q^2\sqrt{5q}.$$

$$= \frac{11}{2q^2\sqrt{5q}} \cdot \frac{\sqrt{5q}}{\sqrt{5q}} \qquad \text{To rationalize the denominator, multiply by } \frac{\sqrt{5q}}{\sqrt{5q}} = 1.$$

$$= \frac{11\sqrt{5q}}{2q^2(5q)} \qquad \begin{array}{l}\text{Multiply the numerators.}\\ \text{Multiply the denominators: } \sqrt{5q} \cdot \sqrt{5q} = \left(\sqrt{5q}\right)^2 = 5q.\end{array}$$

$$= \frac{11\sqrt{5q}}{10q^3} \qquad \text{Multiply in the denominator.}$$

Self Check 7   Rationalize the denominator: $\dfrac{7}{\sqrt{18c^3}}$

Now Try   **Problems 75 and 81**

EXAMPLE 8   Rationalize each denominator: **a.** $\dfrac{5}{\sqrt[3]{6n^2}}$ **b.** $\dfrac{\sqrt[4]{2}}{\sqrt[4]{9a}}$

Strategy   In part (a), we will examine the radicand in the denominator and ask, "By what must we multiply it to obtain a perfect cube?" In part (b), we will examine the radicand in the denominator and ask, "By what must we multiply it to obtain a perfect-fourth power?"

Why   The answers to those questions will determine what form of 1 we use to rationalize each denominator.

Solution

a. To rationalize the denominator $\sqrt[3]{6n^2}$, we need the radicand to be a perfect cube. Since $6n^2 = 6 \cdot n \cdot n$, the radicand needs two more factors of 6 and one more factor of n. It follows that we should multiply the given expression by $\dfrac{\sqrt[3]{36n}}{\sqrt[3]{36n}}$.

$$\frac{5}{\sqrt[3]{6n^2}} = \frac{5}{\sqrt[3]{6n^2}} \cdot \frac{\sqrt[3]{36n}}{\sqrt[3]{36n}} \qquad \text{Multiply by a form of 1 to rationalize the denominator.}$$

$$= \frac{5\sqrt[3]{36n}}{\sqrt[3]{216n^3}} \longleftarrow \qquad \text{Multiply the numerators. Multiply the denominators.}$$
$$\text{This radicand is now a perfect cube.}$$

$$= \frac{5\sqrt[3]{36n}}{6n} \qquad \text{Simplify the denominator: } \sqrt[3]{216n^3} = 6n.$$

b. To rationalize the denominator $\sqrt[4]{9a}$, we need the radicand to be a perfect-fourth power. Since $9a = 3 \cdot 3 \cdot a$, the radicand needs two more factors of 3 and three more factors of a. It follows that we should multiply the given expression by $\dfrac{\sqrt[4]{9a^3}}{\sqrt[4]{9a^3}}$.

$$\frac{\sqrt[4]{2}}{\sqrt[4]{9a}} = \frac{\sqrt[4]{2}}{\sqrt[4]{9a}} \cdot \frac{\sqrt[4]{9a^3}}{\sqrt[4]{9a^3}} \qquad \text{Multiply by a form of 1 to rationalize the denominator.}$$

$$= \frac{\sqrt[4]{18a^3}}{\sqrt[4]{81a^4}} \longleftarrow \qquad \text{Multiply the numerators. Multiply the denominators.}$$
$$\text{This radicand is now a perfect-fourth power.}$$

$$= \frac{\sqrt[4]{18a^3}}{3a} \qquad \text{Simplify the denominator: } \sqrt[4]{81a^4} = 3a.$$

Self Check 8   Rationalize each denominator:

a. $\dfrac{27}{\sqrt[3]{100a}}$ **b.** $\dfrac{\sqrt[4]{3}}{\sqrt[4]{4y^2}}$

Now Try   **Problems 83 and 87**

4   **Rationalize Denominators That Have Two Terms.**

So far, we have rationalized denominators that have only one term. We will now discuss a method to rationalize denominators that have two terms.

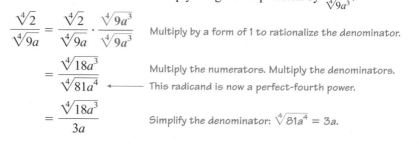

One-termed denominators	*Two-termed denominators*
$\dfrac{\sqrt{5}}{\sqrt{3}}, \ \dfrac{11}{\sqrt{20q^5}}, \ \dfrac{4}{\sqrt[3]{2}}$	$\dfrac{1}{\sqrt{2}+1}, \ \dfrac{\sqrt{x}+\sqrt{2}}{\sqrt{x}-\sqrt{2}}$

To rationalize the denominator of $\dfrac{1}{\sqrt{2} + 1}$, for example, we multiply the numerator and denominator by $\sqrt{2} - 1$, because the product $\left(\sqrt{2} + 1\right)\left(\sqrt{2} - 1\right)$ contains no radicals.

$$\left(\sqrt{2} + 1\right)\left(\sqrt{2} - 1\right) = \left(\sqrt{2}\right)^2 - (1)^2 \quad \text{\small Use a special-product rule.}$$
$$= 2 - 1$$
$$= 1$$

Radical expressions that involve the sum and difference of the same two terms, such as $\sqrt{2} + 1$ and $\sqrt{2} - 1$, are called **conjugates.**

EXAMPLE 9   Rationalize the denominator: **a.** $\dfrac{1}{\sqrt{2} + 1}$ **b.** $\dfrac{\sqrt{x} + \sqrt{2}}{\sqrt{x} - \sqrt{2}}$

Strategy   In each part, we will rationalize the denominator by multiplying the numerator and the denominator by the conjugate of the denominator.

Why   Multiplying each denominator by its conjugate will produce a new denominator that does not contain radicals.

Solution

a. To find a fraction equivalent to $\dfrac{1}{\sqrt{2} + 1}$ that does not have a radical in its denominator, we multiply $\dfrac{1}{\sqrt{2} + 1}$ by a form of 1 that uses the conjugate of $\sqrt{2} + 1$.

$$\frac{1}{\sqrt{2} + 1} = \frac{1}{\sqrt{2} + 1} \cdot \frac{\sqrt{2} - 1}{\sqrt{2} - 1}$$

$$= \frac{\sqrt{2} - 1}{\left(\sqrt{2}\right)^2 - (1)^2} \qquad \begin{array}{l}\text{\small Multiply the numerators.}\\ \text{\small Multiply the denominators using a}\\ \text{\small special-product rule.}\end{array}$$

$$= \frac{\sqrt{2} - 1}{2 - 1}$$

$$= \frac{\sqrt{2} - 1}{1}$$

$$= \sqrt{2} - 1$$

b. We multiply the numerator and denominator by $\sqrt{x} + \sqrt{2}$, which is the conjugate of $\sqrt{x} - \sqrt{2}$, and simplify.

$$\frac{\sqrt{x} + \sqrt{2}}{\sqrt{x} - \sqrt{2}} = \frac{\sqrt{x} + \sqrt{2}}{\sqrt{x} - \sqrt{2}} \cdot \frac{\sqrt{x} + \sqrt{2}}{\sqrt{x} + \sqrt{2}}$$

$$= \frac{x + \sqrt{2x} + \sqrt{2x} + 2}{\left(\sqrt{x}\right)^2 - \left(\sqrt{2}\right)^2} \qquad \begin{array}{l}\text{\small Multiply the numerators.}\\ \text{\small Multiply the denominators.}\end{array}$$

$$= \frac{x + \sqrt{2x} + \sqrt{2x} + 2}{x - 2}$$

$$= \frac{x + 2\sqrt{2x} + 2}{x - 2} \qquad \text{\small In the numerator, combine like radicals.}$$

Self Check 9   Rationalize the denominator: $\dfrac{\sqrt{x} - \sqrt{2}}{\sqrt{x} + \sqrt{2}}$

Now Try   **Problems 91 and 99**

Rationalize Numerators.

In calculus, we sometimes have to rationalize a numerator by multiplying the numerator and denominator of the fraction by the conjugate of the numerator.

EXAMPLE 10   Rationalize the numerator: $\dfrac{\sqrt{x} - 3}{\sqrt{x}}$

Strategy   To rationalize the numerator, we will multiply the numerator and the denominator by the conjugate of the numerator.

Why   After rationalizing the numerator, we can simplify the expression. Although the result will not be in simplified form, this nonsimplified form is often desirable in calculus.

Solution   We multiply the numerator and denominator by $\sqrt{x} + 3$, which is the conjugate of the numerator.

$$\frac{\sqrt{x} - 3}{\sqrt{x}} = \frac{\sqrt{x} - 3}{\sqrt{x}} \cdot \frac{\sqrt{x} + 3}{\sqrt{x} + 3} \qquad \text{Multiply by a form of 1 to rationalize the numerator.}$$

$$= \frac{\left(\sqrt{x}\right)^2 - (3)^2}{x + 3\sqrt{x}} \qquad \begin{array}{l}\text{Multiply the numerators using a special-product rule.}\\ \text{Multiply the denominators.}\end{array}$$

$$= \frac{x - 9}{x + 3\sqrt{x}}$$

Self Check 10   Rationalize the numerator: $\dfrac{\sqrt{x} + 3}{\sqrt{x}}$

Now Try   **Problem 105**

ANSWERS TO SELF CHECKS   **1. a.** $7\sqrt{2}$ **b.** $-20\sqrt[3]{3}$ **c.** $18x\sqrt[4]{2x}$ **2.** $12\sqrt{10} - 32$
3. a. $-1 + \sqrt{15}$ **b.** $a - \sqrt[3]{3a} + 9\sqrt[3]{2a^2} - 9\sqrt[3]{6}$ **4. a.** 11 **b.** $108y$ **c.** $x - 10\sqrt{x - 8} + 17$
5. a. $\frac{2\sqrt{10}}{5}$ **b.** $\frac{5\sqrt[3]{9}}{3}$ **6.** $\frac{\sqrt{2ab}}{a}$ **7.** $\frac{7\sqrt{2c}}{6c^2}$ **8. a.** $\frac{27\sqrt[3]{10a^2}}{10a}$ **b.** $\frac{\sqrt[4]{12y^2}}{2y}$ **9.** $\frac{x - 2\sqrt{2x} + 2}{x - 2}$ **10.** $\frac{x - 9}{x - 3\sqrt{x}}$

STUDY SET
7.4

VOCABULARY

Fill in the blanks.

1. To multiply $\left(\sqrt{3} + \sqrt{2}\right)\left(\sqrt{3} - 2\sqrt{2}\right)$, we can use the _____ method.

2. To multiply $2\sqrt{5}\left(3\sqrt{8} + \sqrt{3}\right)$, use the _____ property.

3. The denominator of the fraction $\dfrac{4}{\sqrt{5}}$ is an _____ number.

4. The _____ of $\sqrt{x} + 1$ is $\sqrt{x} - 1$.

5. To obtain a _____-cube radicand in the denominator of $\dfrac{\sqrt[3]{7}}{\sqrt[3]{5n}}$, we multiply the fraction by $\dfrac{\sqrt[3]{25n^2}}{\sqrt[3]{25n^2}}$.

6. To _____ the denominator of $\dfrac{4}{\sqrt{5}}$, we multiply the fraction by $\dfrac{\sqrt{5}}{\sqrt{5}}$.

CONCEPTS

7. Perform each operation, if possible.

   a. $4\sqrt{6} + 2\sqrt{6}$ **b.** $4\sqrt{6}\left(2\sqrt{6}\right)$

   c. $3\sqrt{2} - 2\sqrt{3}$ **d.** $3\sqrt{2}\left(-2\sqrt{3}\right)$

8. Perform each operation, if possible.

   a. $5 + 6\sqrt[3]{6}$ **b.** $5\left(6\sqrt[3]{6}\right)$

   c. $\dfrac{30\sqrt[3]{15}}{5}$ **d.** $\dfrac{\sqrt[3]{15}}{5}$

NOTATION

Fill in the blanks.

9. Multiply:

$$
\begin{aligned}
5\sqrt{8} \cdot 7\sqrt{6} &= 5(7)\sqrt{8} \\
&= 35\sqrt{} \\
&= 35\sqrt{\cdot 3} \\
&= 35()\sqrt{3} \\
&= 140\sqrt{3}
\end{aligned}
$$

10. Rationalize the denominator:

$$
\begin{aligned}
\frac{9}{\sqrt[3]{4a^2}} &= \frac{9}{\sqrt[3]{4a^2}} \cdot \frac{\sqrt[3]{2a}}{} \\
&= \frac{9\sqrt[3]{2a}}{\sqrt[3]{}} \\
&= \frac{9\sqrt[3]{2a}}{}
\end{aligned}
$$

GUIDED PRACTICE

Multiply and simplify, if possible. All variables represent positive real numbers. See Example 1.

11. $\sqrt{3}\sqrt{15}$ **12.** $\sqrt{5}\sqrt{15}$

13. $2\sqrt{3}\sqrt{6}$ **14.** $-3\sqrt{11}\sqrt{33}$

15. $\left(3\sqrt[3]{9}\right)\left(2\sqrt[3]{3}\right)$ **16.** $\left(2\sqrt[3]{16}\right)\left(-\sqrt[3]{4}\right)$

17. $\sqrt[3]{2} \cdot \sqrt[3]{12}$ **18.** $\sqrt[3]{3} \cdot \sqrt[3]{18}$

19. $6\sqrt{ab^3}\left(8\sqrt{ab}\right)$ **20.** $3\sqrt{8x}\left(2\sqrt{2x^3y}\right)$

21. $\sqrt[4]{5a^3}\sqrt[4]{125a^2}$ **22.** $\sqrt[4]{2r^3}\sqrt[4]{8r^2}$

23. $-4\sqrt[3]{5r^2s}\left(5\sqrt[3]{2r}\right)$ **24.** $-\sqrt[3]{3xy^2}\left(-\sqrt[3]{9x^3}\right)$

25. $\sqrt{x(x+3)}\sqrt{x^3(x+3)}$ **26.** $\sqrt{y^2(x+y)}\sqrt{(x+y)^3}$

Multiply and simplify, if possible. All variables represent positive real numbers. See Example 2.

27. $3\sqrt{5}\left(4 - \sqrt{5}\right)$ **28.** $2\sqrt{7}\left(3 - \sqrt{7}\right)$

29. $\sqrt{2}\left(4\sqrt{6} + 2\sqrt{7}\right)$ **30.** $-\sqrt{3}\left(\sqrt{7} - \sqrt{15}\right)$

31. $-2\sqrt{5x}\left(4\sqrt{2x} - 3\sqrt{3}\right)$ **32.** $3\sqrt{7t}\left(2\sqrt{7t} + 3\sqrt{3t^2}\right)$

33. $\sqrt[3]{2}\left(4\sqrt[3]{4} + \sqrt[3]{12}\right)$ **34.** $\sqrt[3]{3}\left(2\sqrt[3]{9} + \sqrt[3]{18}\right)$

Multiply and simplify, if possible. All variables represent positive real numbers. See Example 3.

35. $\left(\sqrt{2} + 1\right)\left(\sqrt{2} - 3\right)$
36. $\left(2\sqrt{3} + 1\right)\left(\sqrt{3} - 1\right)$
37. $\left(\sqrt{3x} - \sqrt{2y}\right)\left(\sqrt{3x} + \sqrt{2y}\right)$
38. $\left(\sqrt{3m} + \sqrt{2n}\right)\left(\sqrt{3m} - \sqrt{2n}\right)$
39. $\left(2\sqrt[3]{4} - 3\sqrt[3]{2}\right)\left(3\sqrt[3]{4} + 2\sqrt[3]{10}\right)$

40. $\left(4\sqrt[3]{9} - 3\sqrt[3]{3}\right)\left(4\sqrt[3]{3} + 2\sqrt[3]{6}\right)$

41. $\left(\sqrt[3]{5z} + \sqrt[3]{3}\right)\left(\sqrt[3]{5z} + 2\sqrt[3]{3}\right)$
42. $\left(\sqrt[3]{3p} - 2\sqrt[3]{2}\right)\left(\sqrt[3]{3p} + \sqrt[3]{2}\right)$

Square or cube each quantity and simplify the result, if possible. See Example 4.

43. $\left(\sqrt{7}\right)^2$ **44.** $\left(\sqrt{11}\right)^2$
45. $\left(\sqrt[3]{12}\right)^3$ **46.** $\left(\sqrt[3]{9}\right)^3$
47. $\left(3\sqrt{2}\right)^2$ **48.** $\left(2\sqrt{5}\right)^2$
49. $\left(-2\sqrt[3]{2x^2}\right)^3$ **50.** $\left(-3\sqrt[3]{10y^3}\right)^3$
51. $\left(3\sqrt{2r} - 2\right)^2$
52. $\left(2\sqrt{3t} + 5\right)^2$
53. $\left(\sqrt{3x} + \sqrt{3}\right)^2$
54. $\left(\sqrt{5x} - \sqrt{3}\right)^2$

Rationalize each denominator. See Example 5.

55. $\sqrt{\dfrac{2}{7}}$ **56.** $\sqrt{\dfrac{5}{3}}$

57. $\sqrt{\dfrac{8}{3}}$ **58.** $\sqrt{\dfrac{8}{7}}$

59. $\dfrac{4}{\sqrt{6}}$

60. $\dfrac{8}{\sqrt{10}}$

61. $\dfrac{1}{\sqrt[3]{2}}$

62. $\dfrac{2}{\sqrt[3]{6}}$

63. $\dfrac{3}{\sqrt[3]{9}}$

64. $\dfrac{2}{\sqrt[3]{a}}$

65. $\dfrac{1}{\sqrt[4]{4}}$

66. $\dfrac{1}{\sqrt[5]{2}}$

Rationalize each denominator. All variables represent positive real numbers. See Example 6.

67. $\dfrac{\sqrt{10y^2}}{\sqrt{2y^3}}$

68. $\dfrac{\sqrt{15b^2}}{\sqrt{5b^3}}$

69. $\dfrac{\sqrt{48x^2}}{\sqrt{8x^2y}}$

70. $\dfrac{\sqrt{9xy}}{\sqrt{3x^2y}}$

71. $\dfrac{\sqrt[3]{12t^3}}{\sqrt[3]{54t^2}}$

72. $\dfrac{\sqrt[3]{15m^4}}{\sqrt[3]{12m^3}}$

73. $\dfrac{\sqrt[3]{4a^6}}{\sqrt[3]{2a^5b}}$

74. $\dfrac{\sqrt[3]{9x^5y^4}}{\sqrt[3]{3x^5y^5}}$

Rationalize each denominator. All variables represent positive real numbers. See Example 7.

75. $\dfrac{23}{\sqrt{50p^5}}$

76. $\dfrac{11}{\sqrt{75s^5}}$

77. $\dfrac{7}{\sqrt{24b^3}}$

78. $\dfrac{13}{\sqrt{32n^3}}$

79. $\sqrt[3]{\dfrac{5}{16}}$

80. $\sqrt[3]{\dfrac{2}{81}}$

81. $\sqrt[3]{\dfrac{4}{81}}$

82. $\sqrt[3]{\dfrac{7}{16}}$

Rationalize each denominator. All variables represent positive real numbers. See Example 8.

83. $\dfrac{19}{\sqrt[3]{5c^2}}$

84. $\dfrac{1}{\sqrt[3]{4m^2}}$

85. $\dfrac{\sqrt[3]{3}}{\sqrt[3]{2r}}$

86. $\dfrac{\sqrt[3]{7}}{\sqrt[3]{100s}}$

87. $\dfrac{\sqrt[4]{2}}{\sqrt[4]{3t^2}}$

88. $\dfrac{\sqrt[4]{3}}{\sqrt[4]{5b^3}}$

89. $\dfrac{25}{\sqrt[4]{8a}}$

90. $\dfrac{4}{\sqrt[4]{9t}}$

Rationalize each denominator. All variables represent positive real numbers. See Example 9.

91. $\dfrac{\sqrt{2}}{\sqrt{5}+3}$

92. $\dfrac{\sqrt{3}}{\sqrt{3}-2}$

93. $\dfrac{2}{\sqrt{x}+1}$

94. $\dfrac{3}{\sqrt{x}-2}$

95. $\dfrac{\sqrt{7}-\sqrt{2}}{\sqrt{2}+\sqrt{7}}$

96. $\dfrac{\sqrt{3}+\sqrt{2}}{\sqrt{3}-\sqrt{2}}$

97. $\dfrac{3\sqrt{2}-5\sqrt{3}}{2\sqrt{3}-3\sqrt{2}}$

98. $\dfrac{3\sqrt{6}+5\sqrt{5}}{2\sqrt{5}-3\sqrt{6}}$

99. $\dfrac{\sqrt{x}-\sqrt{y}}{\sqrt{x}+\sqrt{y}}$

100. $\dfrac{\sqrt{x}+\sqrt{y}}{\sqrt{x}-\sqrt{y}}$

101. $\dfrac{2z-1}{\sqrt{2z}-1}$
(*Hint:* Do not perform the multiplication of the numerators.)

102. $\dfrac{3t-1}{\sqrt{3t}+1}$
(*Hint:* Do not perform the multiplication of the numerators.)

Rationalize each numerator. All variables represent positive real numbers. See Example 10.

103. $\dfrac{\sqrt{x}+3}{x}$

104. $\dfrac{2+\sqrt{x}}{5x}$

105. $\dfrac{\sqrt{x}+\sqrt{y}}{\sqrt{x}}$

106. $\dfrac{\sqrt{x}-\sqrt{y}}{\sqrt{x}+\sqrt{y}}$

TRY IT YOURSELF

The following problems involve addition, subtraction, and multiplication of radical expressions, as well as rationalizing the denominator. Perform the operations and simplify, if possible. All variables represent positive real numbers.

107. $\sqrt{x}\left(\sqrt{14x}+\sqrt{2}\right)$

108. $2\sqrt[3]{16}-3\sqrt[3]{128}-\sqrt[3]{54}$

109. $\left(10\sqrt[3]{2x}\right)^3$

110. $\dfrac{\sqrt{3}}{\sqrt{98x^2}}$

111. $\left(3p+\sqrt{5}\right)^2$

112. $\sqrt{288t}+\sqrt{80t}-\sqrt{128t}$

113. $\sqrt{\dfrac{72m^8}{25m^3}}$

114. $\left(\sqrt{14x}+\sqrt{3}\right)\left(\sqrt{14x}-\sqrt{3}\right)$

115. $\sqrt[4]{3n^2}\sqrt[4]{27n^3}$

116. $\dfrac{\sqrt{y}-2}{\sqrt{y}+3}$

117. $\dfrac{\sqrt[3]{x}}{\sqrt[3]{9}}$

118. $\sqrt[5]{\dfrac{2}{243}}$

where k_1 and k_2 indicate the stiffness of the springs and m is the mass of the block. Rationalize the right side and restate the formula.

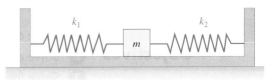

APPLICATIONS

119. STATISTICS   An example of a normal distribution curve, or *bell-shaped* curve, is shown. A fraction that is part of the equation that models this curve is $\dfrac{1}{\sigma\sqrt{2\pi}}$, where σ is a letter from the Greek alphabet. Rationalize the denominator of the fraction.

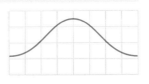

120. ANALYTIC GEOMETRY   The length of the perpendicular segment drawn from $(-2, 2)$ to the line with equation $2x - 4y = 4$ is given by

$$L = \frac{|2(-2) + (-4)(2) + (-4)|}{\sqrt{(2)^2 + (-4)^2}}$$

Find L. Express the result in simplified radical form. Then give an approximation to the nearest tenth.

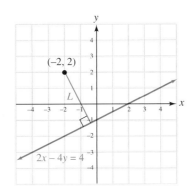

121. TRIGONOMETRY   In trigonometry, we must often find the ratio of the lengths of two sides of right triangles. Use the information in the illustration to find the ratio

$$\frac{\text{length of side } AC}{\text{length of side } AB}$$

Write the result in simplified radical form.

122. ENGINEERING   Refer to the illustration in the next column that shows a block connected to two walls by springs. A measure of how fast the block will oscillate when the spring system is set in motion is given by the formula

$$\omega = \sqrt{\frac{k_1 + k_2}{m}}$$

WRITING

123. Consider $\dfrac{\sqrt{3}}{\sqrt{7}} = \dfrac{\sqrt{3}}{\sqrt{7}} \cdot \dfrac{\sqrt{7}}{\sqrt{7}}$. Explain why the expressions on the left side and the right side of the equation are equal.

124. To rationalize the denominator of $\dfrac{\sqrt[4]{12}}{\sqrt[4]{3}}$, why wouldn't we multiply the numerator and denominator by $\dfrac{\sqrt[4]{3}}{\sqrt[4]{3}}$?

125. Explain why $\dfrac{\sqrt[3]{12}}{\sqrt[3]{5}}$ is not in simplified form.

126. Explain why $\sqrt{\dfrac{3a}{11k}}$ is not in simplified form.

127. Explain why $\sqrt{m} \cdot \sqrt{m} = m$ but $\sqrt[3]{m} \cdot \sqrt[3]{m} \neq m$. Assume that m represents a positive number.

128. Explain why the product of $\sqrt{m} + 3$ and $\sqrt{m} - 3$ does not contain a radical.

REVIEW

Solve each equation.

129. $\dfrac{8}{b-2} + \dfrac{3}{2-b} = -\dfrac{1}{b}$

130. $\dfrac{2}{x-2} + \dfrac{1}{x+1} = \dfrac{1}{(x+1)(x-2)}$

CHALLENGE PROBLEMS

131. Multiply: $\sqrt{2} \cdot \sqrt[3]{2}$.   (*Hint:* Keep in mind two things. The indices (plural for *index*) must be the same to use the product rule for radicals, and radical expressions can be written using rational exponents.)

132. Show that $\dfrac{\sqrt[3]{a^2} + \sqrt[3]{a}\sqrt[3]{b} + \sqrt[3]{b^2}}{\sqrt[3]{a^2} + \sqrt[3]{a}\sqrt[3]{b} + \sqrt[3]{b^2}}$ can be used to rationalize the denominator of $\dfrac{1}{\sqrt[3]{a} - \sqrt[3]{b}}$.

37. $(m^3 + 26)^{1/3} = m + 2$ **38.** $(x^3 + 56)^{1/3} = x + 2$

39. $(5r + 14)^{1/3} = 4$ **40.** $(2b + 29)^{1/3} = 3$

71. $L_A = L_B\sqrt{1 - \dfrac{v^2}{c^2}}$ for v^2 **72.** $R_1 = \sqrt{\dfrac{A}{\pi} - R_2{}^2}$ for A

See Example 6.

41. Let $f(x) = \sqrt[4]{3x + 1}$. For what value(s) of x is $f(x) = 4$?

42. Let $f(x) = \sqrt{2x^2 - 7x}$. For what value(s) of x is $f(x) = 2$?

43. Let $f(x) = \sqrt[3]{3x - 6}$. For what value(s) of x is $f(x) = -3$?

44. Let $f(x) = \sqrt[5]{4x - 4}$. For what value(s) of x is $f(x) = -2$?

Solve each equation. See Example 7.

45. $\sqrt{3x + 12} = \sqrt{5x - 12}$ **46.** $\sqrt{m + 4} = \sqrt{2m - 5}$

47. $2\sqrt{4x + 1} = \sqrt{x + 4}$ **48.** $\sqrt{6 - 2x} = 4\sqrt{x - 3}$

49. $\sqrt{6t + 9} = 3\sqrt{t}$ **50.** $\sqrt{12x + 24} = 6\sqrt{x}$

51. $(34x + 26)^{1/3} = 4(x - 1)^{1/3}$ **52.** $(a^2 + 2a)^{1/3} = 2(a - 1)^{1/3}$

Solve each equation. Write all proposed solutions. Cross out those that are extraneous. See Example 8.

53. $\sqrt{x - 5} + \sqrt{x} = 5$ **54.** $\sqrt{x - 7} + \sqrt{x} = 7$

55. $\sqrt{z + 3} - \sqrt{z} = 1$ **56.** $\sqrt{x + 12} + \sqrt{x} = 6$

57. $\sqrt{x + 5} + \sqrt{x - 3} = 4$ **58.** $\sqrt{b + 7} - \sqrt{b - 5} = 2$

59. $\sqrt{r + 16} + \sqrt{r + 9} = 7$ **60.** $\sqrt{x + 8} - \sqrt{x - 4} = 2$

61. $3 = \sqrt{y + 4} - \sqrt{y + 7}$ **62.** $3 = \sqrt{u - 3} - \sqrt{u}$

63. $2 = \sqrt{2u + 7} - \sqrt{u}$ **64.** $1 = \sqrt{4s + 5} - \sqrt{2s + 2}$

Solve each equation for the specified variable or expression. See Example 9.

65. $v = \sqrt{2gh}$ for h **66.** $d = 1.4\sqrt{h}$ for h

67. $T = 2\pi\sqrt{\dfrac{l}{32}}$ for l **68.** $d = \sqrt[3]{\dfrac{12V}{\pi}}$ for V

69. $r = \sqrt[3]{\dfrac{A}{P}} - 1$ for A **70.** $r = \sqrt[3]{\dfrac{A}{P}} - 1$ for P

TRY IT YOURSELF

Solve each equation. Write all proposed solutions. Cross out those that are extraneous.

73. $2\sqrt{x} = \sqrt{5x - 16}$ **74.** $3\sqrt{x} = \sqrt{3x + 54}$

75. $n = (n^3 + n^2 - 1)^{1/3}$ **76.** $(m^4 + m^2 - 25)^{1/4} = m$

77. $\sqrt{y + 2} + y = 4$ **78.** $\sqrt{22y + 86} - y = 9$

79. $\sqrt[3]{x + 8} = -2$ **80.** $\sqrt[3]{x + 4} = -1$

81. $1 = \sqrt{x + 5} - \sqrt{x}$ **82.** $2 = \sqrt{x + 8} - \sqrt{x}$

83. $x = \dfrac{\sqrt{12x - 5}}{2}$ **84.** $x = \dfrac{\sqrt{16x - 12}}{2}$

85. $(n^2 + 6n + 3)^{1/2} = (n^2 - 6n - 3)^{1/2}$

86. $(m^2 - 12m - 3)^{1/2} = (m^2 + 12m + 3)^{1/2}$

87. $\sqrt{x - 5} - \sqrt{x + 3} = 4$ **88.** $\sqrt{x + 8} - \sqrt{x - 4} = -2$

89. $\sqrt[4]{10y + 6} = 2\sqrt[4]{y}$ **90.** $\sqrt[4]{21a + 39} = 3\sqrt[4]{a - 1}$

91. $\sqrt{-5x + 24} = 6 - x$ **92.** $-s - 3 = 2\sqrt{5 - s}$

93. $\sqrt{2x + 5} = 1$ **94.** $\sqrt{3x + 10} = 1$

95. $\sqrt{6x + 2} - \sqrt{5x + 3} = 0$ **96.** $\sqrt{5x + 2} - \sqrt{x + 10} = 0$

APPLICATIONS

97. HIGHWAY DESIGN   A curved road will accommodate traffic traveling s mph if the radius of the curve is r feet, according to the formula $s = 3\sqrt{r}$. If engineers expect 40-mph traffic, what radius should they specify? Give the result to the nearest foot.

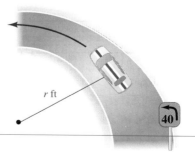

r ft

1 **Use the Pythagorean Theorem to**

If we know the lengths of two legs of a r
(the side opposite the 90° angle) by usir

| The Pythagorean Theorem | If a and b are the lengths of the legs of
c is the length of the hypotenuse,

$$a^2 + b^2 = c^2$$ |

In words, the Pythagorean theorem is e:

In any right triangle, the square of
two legs.

Suppose the right triangle shown in
length of the hypotenuse, we use the Py

$$a^2 + b^2 = c^2$$
$$3^2 + 4^2 = c^2 \quad \text{Substitute 3 for a a}$$
$$9 + 16 = c^2$$
$$25 = c^2$$

To find c, we ask "What number,
numbers: the positive square root of 25
the length of the hypotenuse, and it can
root of 25.

$$\sqrt{25} = c \quad \text{Recall that a radical syn}$$
$$\text{the positive, or principal}$$

$$5 = c$$

The length of the hypotenuse is 5 units.

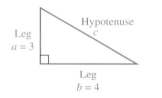

Leg
$a = 3$
Hypotenuse
c
Leg
$b = 4$

The Language of Algebra

A *theorem* is a mathematical
statement that can be proved. The
Pythagorean theorem is named
after Pythagoras, a Greek
mathematician who lived about
2,500 years ago. He is thought to
have been the first to prove the
theorem.

EXAMPLE 1   *Firefighting.*
clear a rectan
tration on the next page. Crews ar
3,000-yard range. Can crews at po

Strategy   We will use the Pythago
A and B.

Why   If this distance is less than 3,0
3,000 yards, they cannot communi

Solution   The line segments connec
the distance c from point A to poir
ing 2,400 for a and 1,000 for b anc

---

98. FORESTRY   The taller a lookout tower, the farther an
observer can see. That distance d (called the *horizon distance,*
measured in miles) is related to the height h of the observer
(measured in feet) by the formula $d = 1.22\sqrt{h}$. How tall
must a lookout tower be to see the edge of the forest, 25 miles
away? (Round to the nearest foot.)

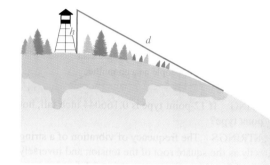

99. WIND POWER   The power generated by a windmill is
related to the velocity of the wind by the formula

$$v = \sqrt[3]{\frac{P}{0.02}}$$

where P is the power (in watts) and v is the velocity of the
wind (in mph). Find how much power the windmill is
generating when the wind is 29 mph.

100. DIAMONDS   The *effective rate of interest* r earned by an
investment is given by the formula

$$r = \sqrt[n]{\frac{A}{P}} - 1$$

where P is the initial investment that grows to value A after n
years. If a diamond buyer got \$4,000 for a 1.73-carat diamond
that he had purchased 4 years earlier, and earned an annual
rate of return of 6.5% on the investment, what did he
originally pay for the diamond?

101. THEATER PRODUCTIONS   The ropes, pulleys, and sand-
bags shown in the illustration are part of a mechanical system
used to raise and lower scenery for a stage play. For the scenery
to be in the proper position, the following formula must apply:

$$w_2 = \sqrt{w_1{}^2 + w_3{}^2}$$

If $w_2 = 12.5$ lb and $w_3 = 7.5$ lb, find w_1.

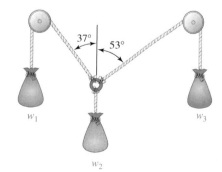

102. CARPENTRY   During construction, carpenters often brace
walls as shown in the illustration, where the length L of the
brace is given by the formula

$$L = \sqrt{f^2 + h^2}$$

If a carpenter nails a 10-ft brace to the wall 6 feet above the
floor, how far from the base of the wall should he nail the
brace to the floor?

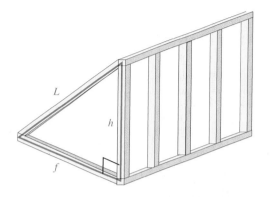

103. SUPPLY AND DEMAND   The number of wrenches that
will be produced at a given price can be predicted by the
formula $s = \sqrt{5x}$, where s is the supply (in thousands)
and x is the price (in dollars). The demand d for wrenches
can be predicted by the formula $d = \sqrt{100 - 3x^2}$. Find the
equilibrium price—that is, find the price at which supply will
equal demand.

104. SUPPLY AND DEMAND   The number of mirrors that will
be produced at a given price can be predicted by the formula
$s = \sqrt{23x}$, where s is the supply (in thousands) and x is the
price (in dollars). The demand d for mirrors can be predicted
by the formula $d = \sqrt{312 - 2x^2}$. Find the equilibrium
price—that is, find the price at which supply will equal
demand.

WRITING

105. What is wrong with the work shown below?

Solve: $\sqrt{x + 1} - 3 = 8$

$$\cancel{\sqrt{x + 1} = 11}$$
$$\left(\sqrt{x + 1}\right)^2 = 11$$
$$\cancel{x + 1 = 11}$$
$$x = 10$$

106. The first step of a student's solution is shown below. What is a
better way to begin the solution?

Solve: $\sqrt{x} + \sqrt{x + 22} = 12$

$$\left(\sqrt{x} + \sqrt{x + 22}\right)^2 = 12^2$$

107. Explain why it is immediately apparent that $\sqrt{8x - 7} = -2$ has no solution.

108. Explain the error in the following work.

$$\text{Solve:}\quad \sqrt{2y + 1} = \sqrt{y + 7} + 3$$
$$\left(\sqrt{2y + 1}\right)^2 = \left(\sqrt{y + 7} + 3\right)^2$$
$$2y + 1 = y + 7 + 9$$

109. To solve $\sqrt{2x + 7} = \sqrt{x}$ we need only square both sides once. To solve $\sqrt{2x + 7} = \sqrt{x} + 2$ we have to square both sides twice. Why does the second equation require more work?

110. Explain how $\sqrt{2x - 1} = x$ can be solved graphically.

111. Explain how the table can be used to solve $\sqrt{4x - 3} - 2 = \sqrt{2x - 5}$ if $Y_1 = \sqrt{4x - 3} - 2$ and $Y_2 = \sqrt{2x - 5}$.

X	Y₁	Y₂
2	.23607	ERROR
3	1	1
4	1.6056	1.7321
5	2.1231	2.2361
6	2.5826	2.6458
7	3	3
8	3.3852	3.3166

X=2

112. Explain how to use the graph of $f(x) = \sqrt[3]{x - 0.5} - 1$, shown in the illustration, to approximate the solution of $\sqrt[3]{x - 0.5} = 1$.

REVIEW

113. LIGHTING   The intensity of light from a lightbulb varies inversely as the square of the distance from the bulb. If you are 5 feet away from a bulb and the intensity is 40 foot-candles, what will the intensity be if you move 20 feet away from the bulb?

114. COMMIT
tration? A
what happ

115. TYPESE
tall is 30-

116. GUITAR
varies dir
as the len
under a te
Find k, th

CHALLEN

**Solve each equ
that are extra**

117. $\sqrt[3]{2x} = $

118. $\sqrt[4]{x} = \sqrt{}$

119. $\sqrt{x + 2}$

120. $\sqrt{8 - x}$

Effectiveness of

SECTION 7.6
Geometric Applications of Radicals

Objectives

   Use the Pythagorean theorem

❷   Solve problems involving 45°–

❸   Solve problems involving 30°–

❹   Use the distance formula to so

We will now consider applications of square
between two points on a rectangular coordina
root. We begin by considering an important

CHAPTER 8

Quadratic Equations, Functions, and Inequalities

© Robert E Daemmrich/Getty Images

8.1 The Square Root Property and Completing the Square

8.2 The Quadratic Formula

8.3 The Discriminant and Equations That Can Be Written in Quadratic Form

8.4 Quadratic Functions and Their Graphs

8.5 Quadratic and Other Nonlinear Inequalities

CHAPTER SUMMARY AND REVIEW

CHAPTER TEST

Group Project

CUMULATIVE REVIEW

from *Campus to Careers*
Police Patrol Officer

The responsibilities of a police patrol officer are extremely broad. Quite often, he or she must make a split-second decision while under enormous pressure. One internet vocational website cautions anyone considering such a career to "pay attention in your mathematics and science classes. Those classes help sharpen your ability to think things through and solve problems—an important part of police work."

Police patrol officers often use yellow "DO NOT CROSS" barricade tape to keep the public from entering a crime scene. In **problem 89** of **Study Set 8.4,** you will determine the maximum rectangular area that can be sealed off using a 300-foot roll of barricade tape.

JOB TITLE:
Police Patrol Officer

EDUCATION:
A basic high school education is required, however, an associate's or bachelor's degree is recommended.

JOB OUTLOOK:
In general, employment is expected to increase between 9% to 17% through the year 2014.

ANNUAL EARNINGS:
Mean annual base salary $48,120 with an opportunity for overtime pay

FOR MORE INFORMATION:
www.bls.gov/oco/ocos160.htm

Study Skills Workshop
Organizing Your Notebook

If you're like most students, your algebra notebook could probably use some attention at this stage of the course. You will definitely appreciate a well-organized notebook when it comes time to study for the final exam. Here are some suggestions to put it in tip-top shape.

ORGANIZE YOUR NOTEBOOK INTO SECTIONS: Create a separate section in the notebook for each chapter (or unit of study) that your class has covered this term.

ORGANIZE THE PAPERS WITHIN EACH SECTION: One recommended order is to begin each section with your class notes, followed by your completed homework assignments, then any study sheets or handouts, and, finally, all graded quizzes and tests.

Now Try This

1. Organize your algebra notebook using the guidelines given above.
2. Write a Table of Contents to place at the beginning of your notebook. List each chapter (or unit of study) and include the dates over which the material was covered.
3. Compare your completed notebook with those of other students in your class. Have you overlooked any important items that would be useful when studying for the final exam?

SECTION 8.1
The Square Root Property and Completing the Square

Objectives

1. Use the square root property to solve quadratic equations.
2. Solve quadratic equations by completing the square.
3. Use quadratic equations to solve application problems.

Recall that a *quadratic equation* is an equation of the form $ax^2 + bx + c = 0$, where a, b, and c are real numbers and $a \neq 0$. We have solved quadratic equations using factoring and the zero-factor property.

EXAMPLE 1   Solve: $6x^2 - 7x - 3 = 0$

Strategy   We will factor the trinomial on the left side of the equation and use the zero-factor property to solve for x.

Why   To use the zero-factor property, we need one side of the equation to be factored completely and the other side to be 0.

Solution

$$6x^2 - 7x - 3 = 0$$

$$(2x - 3)(3x + 1) = 0 \qquad \text{Factor the trinomial.}$$

$$2x - 3 = 0 \quad \text{or} \quad 3x + 1 = 0 \qquad \text{Set each factor equal to 0.}$$

$$x = \frac{3}{2} \qquad \qquad x = -\frac{1}{3} \qquad \text{Solve each linear equation.}$$

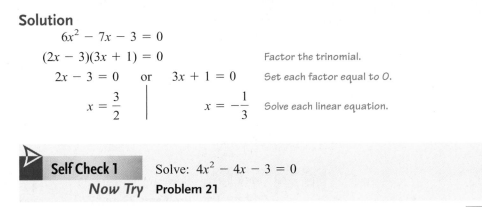

Self Check 1 Solve: $4x^2 - 4x - 3 = 0$

Now Try **Problem 21**

Many expressions do not factor as easily as $6x^2 - 7x - 3$. For example, it would be difficult to solve $2x^2 + 4x + 1 = 0$ by factoring, because $2x^2 + 4x + 1$ cannot be factored by using only integers. We will now develop a more general method that enables us to solve any quadratic equation. It is based on the *square root property*.

 Use the Square Root Property to Solve Quadratic Equations.

To develop general methods for solving quadratic equations, we first consider the equation $x^2 = c$. If $c \geq 0$, we can find the real solutions of $x^2 = c$ as follows:

$$x^2 = c$$

$$x^2 - c = 0 \qquad \text{Subtract } c \text{ from both sides.}$$

$$x^2 - \left(\sqrt{c}\right)^2 = 0 \qquad \text{Replace } c \text{ with } \left(\sqrt{c}\right)^2, \text{ since } c = \left(\sqrt{c}\right)^2.$$

$$\left(x + \sqrt{c}\right)\left(x - \sqrt{c}\right) = 0 \qquad \text{Factor the difference of two squares.}$$

$$x + \sqrt{c} = 0 \quad \text{or} \quad x - \sqrt{c} = 0 \qquad \text{Set each factor equal to 0.}$$

$$x = -\sqrt{c} \qquad \qquad x = \sqrt{c} \qquad \text{Solve each linear equation.}$$

The solutions of $x^2 = c$ are $\sqrt{c}$ and $-\sqrt{c}$.

The Square Root Property

For any nonnegative real number c, if $x^2 = c$, then

$$x = \sqrt{c} \quad \text{or} \quad x = -\sqrt{c}$$

EXAMPLE 2 Solve: $x^2 - 12 = 0$

Strategy We will add 12 to both sides of the equation and use the square root property to solve for x.

Why After adding 12 to both sides, the resulting equivalent equation will have the desired form $x^2 = c$.

Notation

We can use **double-sign notation** $\pm$ to write the solutions in compact form as $\pm 2\sqrt{3}$. Read $\pm$ as "positive or negative."

Solution

$$x^2 - 12 = 0 \qquad \text{This is the equation to solve.}$$

$$x^2 = 12 \qquad \text{To isolate } x^2 \text{ on the left side, add 12 to both sides.}$$

$$x = \sqrt{12} \quad \text{or} \quad x = -\sqrt{12} \qquad \text{Use the square root property.}$$

$$x = 2\sqrt{3} \qquad \qquad x = -2\sqrt{3} \qquad \text{Simplify: } \sqrt{12} = \sqrt{4}\sqrt{3} = 2\sqrt{3}.$$

Check:

$$x^2 - 12 = 0 \qquad\qquad x^2 - 12 = 0$$
$$\left(2\sqrt{3}\right)^2 - 12 \stackrel{?}{=} 0 \qquad\qquad \left(-2\sqrt{3}\right)^2 - 12 \stackrel{?}{=} 0$$
$$12 - 12 \stackrel{?}{=} 0 \qquad\qquad 12 - 12 \stackrel{?}{=} 0$$
$$0 = 0 \quad \text{True} \qquad\qquad 0 = 0 \quad \text{True}$$

The exact solutions are $2\sqrt{3}$ and $-2\sqrt{3}$ and the solution set is $\{2\sqrt{3}, -2\sqrt{3}\}$. We can use a calculator to approximate the solutions. To the nearest hundredth, they are ± 3.46.

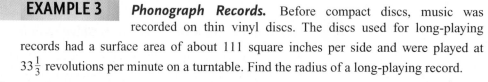

Self Check 2   Solve: $x^2 - 18 = 0$

Now Try   **Problems 23 and 27**

EXAMPLE 3   ***Phonograph Records.*** Before compact discs, music was recorded on thin vinyl discs. The discs used for long-playing records had a surface area of about 111 square inches per side and were played at $33\frac{1}{3}$ revolutions per minute on a turntable. Find the radius of a long-playing record.

Record

CD $\longleftarrow r \longrightarrow$

Strategy   The area A of a circle with radius r is given by the formula $A = \pi r^2$. We will find the radius of a record by substituting 111 for A and dividing both sides by π. Then we will use the square root property to solve for r.

Why   After substituting 111 for A and dividing both sides by π, the resulting equivalent equation will have the desired form $r^2 = c$.

Solution

$$A = \pi r^2 \qquad\qquad \text{This is the formula for the area of a circle.}$$

$$111 = \pi r^2 \qquad\qquad \text{Substitute 111 for } A.$$

$$\frac{111}{\pi} = r^2 \qquad\qquad \text{To undo the multiplication by } \pi, \text{ divide both sides by } \pi.$$

$$r = \sqrt{\frac{111}{\pi}} \quad \text{or} \quad r = -\sqrt{\frac{111}{\pi}} \qquad \begin{array}{l}\text{Use the square root property. Since the radius of the}\\ \text{record cannot be negative, discard the second solution.}\end{array}$$

The radius of a long-playing record is $\sqrt{\frac{111}{\pi}}$ inches—to the nearest tenth, 5.9 inches.

Now Try   **Problem 99**

EXAMPLE 4   Use the square root property to solve $(x - 1)^2 = 16$.

Strategy   Instead of a variable squared on the left side, we have a quantity squared. We still use the square root property to solve the equation.

Why   We want to eliminate the square on the binomial, so that we can eventually isolate the variable on one side of the equation.

Solution

$$(x - 1)^2 = 16 \qquad \text{This is the equation to solve.}$$

$$x - 1 = \pm \sqrt{16} \qquad \text{Use the square root property.}$$

$$x - 1 = \pm 4 \qquad \text{Simplify: } \sqrt{16} = 4.$$

$$x = 1 \pm 4 \qquad \text{To isolate } x, \text{ add 1 to both sides.}$$

$$x = 1 + 4 \quad \text{or} \quad x = 1 - 4 \qquad \text{To find one solution, use } +. \text{ To find the other, use } -.$$

$$x = 5 \quad | \quad x = -3 \qquad \text{Add (subtract).}$$

Verify that 5 and -3 satisfy the original equation.

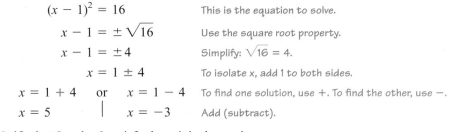

Self Check 4   Use the square root property to solve $(x + 2)^2 = 9$.

Now Try   **Problems 31 and 35**

Some quadratic equations have solutions that are not real numbers.

EXAMPLE 5   Solve: $4x^2 + 25 = 0$

Strategy   We will subtract 25 from both sides of the equation and divide both sides by 4. Then we will use the square root property to solve for x.

Why   After subtracting 25 from both sides and dividing both sides by 4, the resulting equivalent equation will have the desired form $x^2 = c$.

Solution

$$4x^2 + 25 = 0 \qquad \text{This is the equation to solve.}$$

$$x^2 = -\frac{25}{4} \qquad \text{To isolate } x^2, \text{ subtract 25 from both sides and divide both sides by 4.}$$

$$x = \pm \sqrt{-\frac{25}{4}} \qquad \text{Use the square root property.}$$

Since

$$\sqrt{-\frac{25}{4}} = \sqrt{-1 \cdot \frac{25}{4}} = \sqrt{-1}\frac{\sqrt{25}}{\sqrt{4}} = \frac{5}{2}i$$

we have

$$x = \pm \frac{5}{2}i$$

Since the solutions are $\frac{5}{2}i$ and $-\frac{5}{2}i$, the solution set is $\left\{\frac{5}{2}i, -\frac{5}{2}i\right\}$.

Check:

$$4x^2 + 25 = 0$$

$$4\left(\frac{5}{2}i\right)^2 + 25 \stackrel{?}{=} 0$$

$$4\left(\frac{25}{4}\right)i^2 + 25 \stackrel{?}{=} 0$$

$$25(-1) + 25 \stackrel{?}{=} 0$$

$$0 = 0 \quad \text{True}$$

$$4x^2 + 25 = 0$$

$$4\left(-\frac{5}{2}i\right)^2 + 25 \stackrel{?}{=} 0$$

$$4\left(\frac{25}{4}\right)i^2 + 25 \stackrel{?}{=} 0$$

$$25(-1) + 25 \stackrel{?}{=} 0$$

$$0 = 0 \quad \text{True}$$

The Language of Algebra

The $\pm$ symbol is often seen in political polls. A candidate with 48% ($\pm 4\%$) support could be between $48 + 4 = 52\%$ and $48 - 4 = 44\%$.

Self Check 5   Solve: $16x^2 + 49 = 0$

Now Try   **Problem 41**

2   **Solve Quadratic Equations by Completing the Square.**

When the polynomial in a quadratic equation doesn't factor easily, we can solve the equation by *completing the square*. This method is based on the following special products:

$$x^2 + 2bx + b^2 = (x + b)^2 \qquad \text{and} \qquad x^2 - 2bx + b^2 = (x - b)^2$$

In each of these perfect-square trinomials, the third term is the square of one-half of the coefficient of x.

The Language of Algebra
Recall that trinomials that are the square of a binomial are called *perfect-square trinomials*.

- In $x^2 + 2bx + b^2$, the coefficient of x is $2b$. If we find $\frac{1}{2} \cdot 2b$, which is b, and square it, we get the third term, b^2.

- In $x^2 - 2bx + b^2$, the coefficient of x is $-2b$. If we find $\frac{1}{2}(-2b)$, which is $-b$, and square it, we get the third term: $(-b)^2 = b^2$.

We can use these observations to change certain binomials into perfect-square trinomials. For example, to change $x^2 + 12x$ into a perfect-square trinomial, we find one-half of the coefficient of x, square the result, and add the square to $x^2 + 12x$.

$$x^2 + 12x + \boxed{}$$

Find one-half of the   Add the square
coefficient of x.   to the binomial.

$$\frac{1}{2} \cdot 12 = 6 \qquad 6^2 = 36$$

Square the result.

We obtain the perfect-square trinomial $x^2 + 12x + 36$, which factors as $(x + 6)^2$. By adding 36 to $x^2 + 12x$, we say that we have *completed the square on* $x^2 + 12x$.

Completing the Square

To complete the square on $x^2 + bx$, add the square of one-half of the coefficient of x:

$$x^2 + bx + \left(\frac{1}{2}b\right)^2$$

EXAMPLE 6   Complete the square and factor the resulting perfect-square trinomial: **a.** $x^2 + 10x$ **b.** $x^2 - 11x$

Strategy   We will add the square of one-half of the coefficient of x to the given binomial.

Why   Adding such a term will change the binomial into a perfect-square trinomial that will factor.

Caution
Realize that when we complete the square on a binomial, we are not writing an equivalent trinomial expression. Thus, it would be incorrect to use an $=$ symbol between the two.

~~$x^2 + 10x = x^2 + 10x + 25$~~

Solution

a. To make $x^2 + 10x$ a perfect-square trinomial, we find one-half of 10, square it, and add the result to $x^2 + 10x$.

$$x^2 + 10x + 25 \qquad \text{$\frac{1}{2} \cdot 10 = 5$ and $5^2 = 25$. Add 25 to the binomial.}$$

This trinomial factors as $(x + 5)^2$.

b. To make $x^2 - 11x$ a perfect-square trinomial, we find one-half of -11, square it, and add the result to $x^2 - 11x$.

$$x^2 - 11x + \frac{121}{4} \qquad \tfrac{1}{2}(-11) = -\tfrac{11}{2} \text{ and } \left(-\tfrac{11}{2}\right)^2 = \tfrac{121}{4}. \text{ Add } \tfrac{121}{4} \text{ to the binomial.}$$

This trinomial factors as $\left(x - \frac{11}{2}\right)^2$.

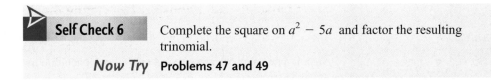

Self Check 6   Complete the square on $a^2 - 5a$ and factor the resulting trinomial.

Now Try   **Problems 47 and 49**

To solve an equation of the form $ax^2 + bx + c = 0$ by completing the square, we use the following steps.

Completing the Square to Solve a Quadratic Equation in x

1. If the coefficient of x^2 is 1, go to step 2. If it is not, make it 1 by dividing both sides of the equation by the coefficient of x^2.
2. Get all variable terms on one side of the equation and constants on the other side.
3. Complete the square by finding one-half of the coefficient of x, squaring the result, and adding the square to both sides of the equation.
4. Factor the perfect-square trinomial as the square of a binomial.
5. Solve the resulting equation using the square root property.
6. Check your answers in the original equation.

EXAMPLE 7   Solve $x^2 + 8x + 7 = 0$ by completing the square.

Strategy   We will begin by subtracting 7 from both sides of the equation. Then we will proceed to complete the square to solve for x.

Why   We subtract 7 from both sides to isolate the variable terms, x^2 and $8x$, on the left side of the equation and the constant term on the right side.

Solution

Step 1: In this example, the coefficient of x^2 is 1.

The Language of Algebra
In $x^2 + 8x + 7 = 0$, x^2 and $8x$ are called *variable terms* and 7 is called the *constant term.*

Step 2: To prepare to complete the square, we subtract 7 from both sides.

$$\begin{aligned} x^2 + 8x + 7 &= 0 \qquad \text{This is the equation to solve.} \\ x^2 + 8x &= -7 \end{aligned}$$

Step 3: The coefficient of x is 8, one-half of 8 is 4, and $4^2 = 16$. To complete the square, we add 16 to both sides.

$$x^2 + 8x + 16 = 16 - 7$$
$$(1) \quad x^2 + 8x + 16 = 9 \qquad \text{Simplify: } 16 - 7 = 9.$$

Step 4: Since the left side of Equation 1 is a perfect-square trinomial, we can factor it to get $(x + 4)^2$.

$$x^2 + 8x + 16 = 9$$
$$(2) \qquad (x + 4)^2 = 9$$

Notation
It is standard practice to write the subtraction of the 4 in front of the ± symbol.

$$x = -4 \pm 3$$

not

~~$x = \pm 3 - 4$~~

Step 5: We can solve Equation 2 by using the square root property.

$$x + 4 = \pm\sqrt{9}$$

$$x + 4 = \pm 3 \qquad \text{Simplify: } \sqrt{9} = 3.$$

$$x = -4 \pm 3 \qquad \text{To isolate } x \text{, subtract 4 from both sides.}$$

This result represents two solutions. To find the first solution we add 3, and to find the second solution, we subtract 3.

$$x = -4 + 3 \quad \text{or} \quad x = -4 - 3 \qquad \text{± represents + or −.}$$

$$x = -1 \qquad\qquad x = -7 \qquad \text{Add (subtract).}$$

Step 6: The solutions are -1 and -7. Check each one in the original equation.

Success Tip
We could also solve $x^2 + 8x + 7 = 0$ using factoring:

$$(x + 1)(x + 7) = 0$$
$$x = -1 \text{ or } x = -7$$

▷ **Self Check 7** Solve $x^2 + 12x + 11 = 0$ by completing the square.
Now Try Problem 51

If the coefficient of the squared variable (called the **leading coefficient**) of a quadratic equation is not 1, we must make it 1 before we can complete the square.

EXAMPLE 8 Solve $6x^2 + 5x - 6 = 0$ by completing the square.

Strategy We will begin by dividing both sides of the equation by 6.

Why This will create a leading coefficient that is 1 so that we can proceed to complete the square to solve the equation.

Solution
Step 1: To make the coefficient of x^2 equal to 1, we divide both sides of the equation by 6.

$$6x^2 + 5x - 6 = 0 \qquad \text{This is the equation to solve.}$$

$$\frac{6x^2}{6} + \frac{5}{6}x - \frac{6}{6} = \frac{0}{6} \qquad \text{Divide both sides by 6, term-by-term.}$$

$$x^2 + \frac{5}{6}x - 1 = 0 \qquad \text{Simplify.}$$

Success Tip
When solving a quadratic equation using the factoring method, one side of the equation must be 0. When we complete the square, we are not concerned with that requirement.

Step 2: To have the constant term on one side of the equation and the variable terms on the other, add 1 to both sides.

$$x^2 + \frac{5}{6}x = 1$$

Step 3: The coefficient of x is $\frac{5}{6}$, one-half of $\frac{5}{6}$ is $\frac{5}{12}$, and $\left(\frac{5}{12}\right)^2 = \frac{25}{144}$. To complete the square, we add $\frac{25}{144}$ to both sides.

$$x^2 + \frac{5}{6}x + \frac{25}{144} = 1 + \frac{25}{144}$$

$$(3) \quad x^2 + \frac{5}{6}x + \frac{25}{144} = \frac{169}{144} \qquad \text{Simplify the right side: } 1 + \frac{25}{144} = \frac{144}{144} + \frac{25}{144} = \frac{169}{144}.$$

Step 4: Factor the left side of Equation 3.

(4) $\left(x + \dfrac{5}{12}\right)^2 = \dfrac{169}{144}$ $x^2 + \frac{5}{6}x + \frac{25}{144}$ is a perfect-square trinomial.

Step 5: We can solve Equation 4 by using the square root property.

$$x + \dfrac{5}{12} = \pm\sqrt{\dfrac{169}{144}}$$

$$x + \dfrac{5}{12} = \pm\dfrac{13}{12}$$ Simplify: $\pm\sqrt{\frac{169}{144}} = \pm\frac{13}{12}$.

$$x = -\dfrac{5}{12} \pm \dfrac{13}{12}$$ To isolate x, subtract $\frac{5}{12}$ from both sides.

$$x = -\dfrac{5}{12} + \dfrac{13}{12} \quad \text{or} \quad x = -\dfrac{5}{12} - \dfrac{13}{12}$$

$$x = \dfrac{8}{12} \qquad\qquad x = -\dfrac{18}{12}$$ Add (subtract) the fractions.

$$x = \dfrac{2}{3} \qquad\qquad x = -\dfrac{3}{2}$$ Simplify each fraction.

Step 6: Check each solution in the original equation.

▷ **Self Check 8**   Solve: $3x^2 + 2x - 8 = 0$

Now Try   Problem 59

EXAMPLE 9   Solve: $2x^2 + 4x + 1 = 0$

Strategy   We will follow the steps for solving a quadratic equation by completing the square.

Why   Since the trinomial $2x^2 + 4x + 1$ cannot be factored using only integers, solving the equation by completing the square is our only option at this time.

Solution

$$2x^2 + 4x + 1 = 0$$ This is the equation to solve.

$$x^2 + 2x + \dfrac{1}{2} = 0$$ Divide both sides by 2 to make the coefficient of x^2 equal to 1.

$$x^2 + 2x \qquad = -\dfrac{1}{2}$$ Subtract $\frac{1}{2}$ from both sides.

$$x^2 + 2x + 1 = -\dfrac{1}{2} + 1$$ Square one-half of the coefficient of x and add it to both sides.

$$(x + 1)^2 = \dfrac{1}{2}$$ Factor and combine like terms.

$$x + 1 = \pm\sqrt{\dfrac{1}{2}}$$ Use the square root property.

$$x = -1 \pm \sqrt{\dfrac{1}{2}}$$ To isolate x, subtract 1 from both sides.

To write $\sqrt{\frac{1}{2}}$ in simplified radical form, we write it as a quotient of square roots and then rationalize the denominator.

$$x = -1 + \frac{\sqrt{2}}{2} \quad \text{or} \quad x = -1 - \frac{\sqrt{2}}{2} \qquad \sqrt{\tfrac{1}{2}} = \frac{\sqrt{1}}{\sqrt{2}} = \frac{1 \cdot \sqrt{2}}{\sqrt{2}\sqrt{2}} = \frac{\sqrt{2}}{2}.$$

We can express each solution in an alternate form if we write -1 as a fraction with a denominator of 2.

$$x = -\frac{2}{2} + \frac{\sqrt{2}}{2} \quad \text{or} \quad x = -\frac{2}{2} - \frac{\sqrt{2}}{2} \qquad \text{Write } -1 \text{ as } -\tfrac{2}{2}.$$

$$x = \frac{-2 + \sqrt{2}}{2} \qquad\qquad x = \frac{-2 - \sqrt{2}}{2} \qquad \text{Add (subtract) the numerators and keep the common denominator 2.}$$

> **Caution**
> Recall that to simplify a fraction, we remove common factors of the numerator and denominator. Since -2 is a term of the numerator of $\frac{-2 + \sqrt{2}}{2}$, no further simplification of this expression can be made.

The exact solutions are $\frac{-2 + \sqrt{2}}{2}$ and $\frac{-2 - \sqrt{2}}{2}$, or simply, $\frac{-2 \pm \sqrt{2}}{2}$. We can use a calculator to approximate them. To the nearest hundredth, they are -0.29 and -1.71.

> **Self Check 9**   Solve: $3x^2 + 6x + 1 = 0$
> **Now Try**   Problem 63

Using Your Calculator

Checking Solutions of Quadratic Equations

We can use a graphing calculator to check the solutions of the quadratic equation $2x^2 + 4x + 1 = 0$ found in Example 9. After entering $Y_1 = 2x^2 + 4x + 1$, we call up the home screen by pressing $\boxed{\text{2nd}}$ $\boxed{\text{QUIT}}$. Then we press $\boxed{\text{VARS}}$, arrow to Y-VARS, press $\boxed{\text{ENTER}}$, and enter 1 to get the display shown in figure (a). We evaluate $2x^2 + 4x + 1$ for $x = \frac{-2 + \sqrt{2}}{2}$ by entering the solution using function notation, as shown in figure (b). When $\boxed{\text{Enter}}$ is pressed, the result of 0 is confirmation that $x = \frac{-2 + \sqrt{2}}{2}$ is a solution of the equation.

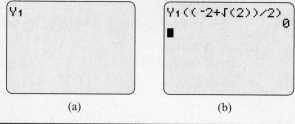

(a)   (b)

In the next example, the solutions of the equation are two complex numbers that contain i.

EXAMPLE 10   Solve: $3x^2 + 2x + 2 = 0$

Strategy   We will follow the steps for solving a quadratic equation by completing the square.

Why   Since the trinomial $3x^2 + 2x + 2$ cannot be factored using only integers, solving the equation by completing the square is our only option at this time.

Solution

$3x^2 + 2x + 2 = 0$	This is the equation to solve.
$x^2 + \dfrac{2}{3}x + \dfrac{2}{3} = \dfrac{0}{3}$	Divide both sides by 3 to make the coefficient of x^2 equal to 1.
$x^2 + \dfrac{2}{3}x \quad\quad = -\dfrac{2}{3}$	Subtract $\dfrac{2}{3}$ from both sides.
$x^2 + \dfrac{2}{3}x + \dfrac{1}{9} = \dfrac{1}{9} - \dfrac{2}{3}$	$\dfrac{1}{2} \cdot \dfrac{2}{3} = \dfrac{1}{3}$ and $\left(\dfrac{1}{3}\right)^2 = \dfrac{1}{9}$. Add $\dfrac{1}{9}$ to both sides.
$\left(x + \dfrac{1}{3}\right)^2 = -\dfrac{5}{9}$	Factor the left side and combine terms: $\dfrac{1}{9} - \dfrac{2}{3} = \dfrac{1}{9} - \dfrac{6}{9} = -\dfrac{5}{9}$.
$x + \dfrac{1}{3} = \pm\sqrt{-\dfrac{5}{9}}$	Use the square root property.
$x = -\dfrac{1}{3} \pm \sqrt{-\dfrac{5}{9}}$	To isolate x, subtract $\dfrac{1}{3}$ from both sides.

Since

$$\sqrt{-\frac{5}{9}} = \sqrt{-1 \cdot \frac{5}{9}} = \sqrt{-1}\frac{\sqrt{5}}{\sqrt{9}} = \frac{\sqrt{5}}{3}i$$

we have

$$x = -\frac{1}{3} \pm \frac{\sqrt{5}}{3}i$$

The solutions are $-\dfrac{1}{3} + \dfrac{\sqrt{5}}{3}i$ and $-\dfrac{1}{3} - \dfrac{\sqrt{5}}{3}i$.

Self Check 10 Solve: $x^2 + 4x + 6 = 0$

Now Try Problem 67

③ **Use Quadratic Equations to Solve Application Problems.**

EXAMPLE 11 *Graduation Announcements.* To create the announcement shown, a graphic artist must follow two design requirements:

- A border of uniform width should surround the text.
- Equal areas should be devoted to the text and to the border.

To meet these requirements, how wide should the border be?

Analyze the Problem The text occupies $4 \cdot 3 = 12$ in.² of space. The border must also have an area of 12 in.².

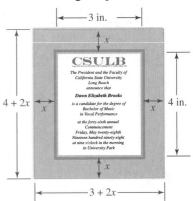

Form an Equation   If we let x = the width of the border in inches, the length of the announcement is $(4 + 2x)$ inches and the width is $(3 + 2x)$ inches. We can now form an equation.

The area of the announcement	minus	the area of the text	equals	the area of the border.
$(4 + 2x)(3 + 2x)$	$-$	12	$=$	12

Solve the Equation

$$(4 + 2x)(3 + 2x) - 12 = 12$$

$$12 + 8x + 6x + 4x^2 - 12 = 12 \qquad \text{On the left side, use the FOIL method.}$$

$$4x^2 + 14x = 12 \qquad \text{Combine like terms.}$$

$$2x^2 + 7x - 6 = 0 \qquad \text{Subtract 12 from both sides. Then divide both sides of } 4x^2 + 14x - 12 = 0 \text{ by 2.}$$

Since the trinomial on the left side does not factor, we will solve the equation by completing the square.

$$x^2 + \frac{7}{2}x - 3 = 0 \qquad \text{Divide both sides by 2 so that the coefficient of } x^2 \text{ is 1.}$$

$$x^2 + \frac{7}{2}x \quad\;\; = 3 \qquad \text{Add 3 to both sides.}$$

$$x^2 + \frac{7}{2}x + \frac{49}{16} = 3 + \frac{49}{16} \qquad \text{One-half of } \frac{7}{2} \text{ is } \frac{7}{4}. \text{ Square } \frac{7}{4}, \text{ which is } \frac{49}{16}, \text{ and add it to both sides.}$$

$$\left(x + \frac{7}{4}\right)^2 = \frac{97}{16} \qquad \text{On the left side, factor the trinomial. On the right side, } 3 = \frac{3 \cdot 16}{1 \cdot 16} = \frac{48}{16} \text{ and } \frac{48}{16} + \frac{49}{16} = \frac{97}{16}.$$

$$x + \frac{7}{4} = \pm\sqrt{\frac{97}{16}} \qquad \text{Use the square root property.}$$

$$x = -\frac{7}{4} \pm \frac{\sqrt{97}}{4} \qquad \text{Subtract } \frac{7}{4} \text{ from both sides and simplify: } \sqrt{\frac{97}{16}} = \frac{\sqrt{97}}{\sqrt{16}} = \frac{\sqrt{97}}{4}.$$

$$x = \frac{-7 + \sqrt{97}}{4} \quad \text{or} \quad x = \frac{-7 - \sqrt{97}}{4} \qquad \text{Write each expression as a single fraction.}$$

State the Conclusion   The width of the border should be $\dfrac{-7 + \sqrt{97}}{4} \approx 0.71$ inch. $\left(\text{We discard the solution } \dfrac{-7 - \sqrt{97}}{4}, \text{ since it is negative.}\right)$

Check the Result   If the border is 0.71 inch wide, the announcement has an area of about $5.42 \cdot 4.42 \approx 23.96$ in.^{2}. If we subtract the area of the text from the area of the announcement, we get $23.96 - 12 = 11.96$ in.^{2}. This represents the area of the border, which was to be 12 in.^{2}. The answer seems reasonable.

 Now Try   **Problem 101**

STUDY SET
8.1

VOCABULARY

Fill in the blanks.

1. An equation of the form $ax^2 + bx + c = 0$, where $a \neq 0$, is called a _____ equation.

2. $x^2 + 6x + 9$ is called a _____-square trinomial because it factors as $(x + 3)^2$.

3. When we add 16 to $x^2 + 8x$, we say that we have completed the _____ on $x^2 + 8x$.

4. The _____ coefficient of $5x^2 - 2x + 7$ is 5 and the _____ term is 7.

CONCEPTS

Fill in the blanks.

5. For any nonnegative number c, if $x^2 = c$, then $x = \boxed{}$ or $x = \boxed{}$.

6. To complete the square on $x^2 + 10x$, add the square of _____ of the coefficient of x.

7. Find one-half of the given number and square the result.
   a. 12   **b.** -5

8. Fill in the blanks to complete the square. Then factor the resulting perfect-square trinomial.
   a. $x^2 + 8x + \boxed{} = (x + \boxed{})^2$
   b. $x^2 - 9x + \boxed{} = \left(x - \boxed{}\right)^2$

9. What is the first step to solve each equation by completing the square? **Do not solve.**
   a. $x^2 + 9x + 7 = 0$
   b. $4x^2 + 5x - 16 = 0$

10. Determine whether $-2 + \sqrt{2}$ is a solution of $x^2 + 4x + 2 = 0$.

11. Determine whether each statement is true or false.
   a. Any quadratic equation can be solved by the factoring method.
   b. Any quadratic equation can be solved by completing the square.

12. **a.** Write an expression that represents the width of the larger rectangle shown in red.
    b. Write an expression that represents the length of the larger rectangle shown in red.

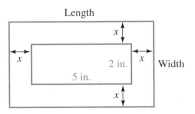

NOTATION

13. We read $8 \pm \sqrt{3}$ as "eight _____ ___ _____ the square root of 3."

14. When solving a quadratic equation, a student obtains $x = \dfrac{-5 \pm \sqrt{7}}{3}$.
   a. How many solutions are represented by this notation? List them.
   b. Approximate the solutions to the nearest hundredth.

GUIDED PRACTICE

Use factoring to solve each equation. See Example 1.

15. $6x^2 + 12x = 0$ 16. $5x^2 + 11x = 0$
17. $y^2 - 25 = 0$ 18. $y^2 - 16 = 0$
19. $r^2 + 6r + 8 = 0$ 20. $x^2 + 9x + 20 = 0$
21. $2z^2 = -2 + 5z$ 22. $3x^2 = 8 - 10x$

Use the square root property to solve each equation. See Example 2.

23. $t^2 - 81 = 0$ 24. $w^2 - 49 = 0$
25. $x^2 = 36$ 26. $x^2 = 144$
27. $z^2 - 50 = 0$ 28. $u^2 - 24 = 0$
29. $3x^2 - 16 = 0$ 30. $5x^2 - 49 = 0$

Use the square root property to solve each equation. See Example 4.

31. $(x + 5)^2 = 9$ 32. $(x - 1)^2 = 4$
33. $(t + 4)^2 = 16$ 34. $(s - 7)^2 = 9$
35. $(x + 5)^2 - 3 = 0$ 36. $(x + 3)^2 - 7 = 0$
37. $(a - 2)^2 = 8$ 38. $(c - 2)^2 = 12$

Use the square root property to solve each equation. See Example 5.

39. $p^2 = -16$ 40. $q^2 = -25$
41. $4m^2 + 81 = 0$ 42. $9n^2 + 121 = 0$
43. $(x - 3)^2 = -5$ 44. $(x + 2)^2 = -3$
45. $(y - 1)^2 + 4 = 0$ 46. $(t + 4)^2 + 49 = 0$

Complete the square and factor the resulting perfect-square trinomial. See Example 6.

47. $x^2 + 24x$
48. $y^2 - 18y$
49. $a^2 - 7a$
50. $b^2 + 11b$

Use completing the square to solve each equation. See Example 7.

51. $x^2 + 2x - 8 = 0$ 52. $x^2 + 6x + 5 = 0$
53. $k^2 - 8k + 12 = 0$ 54. $p^2 - 4p + 3 = 0$
55. $g^2 + 5g - 6 = 0$ 56. $s^2 + 5s - 14 = 0$
57. $x^2 - 3x - 4 = 0$ 58. $x^2 - 7x + 12 = 0$

Use completing the square to solve each equation. See Example 8.

59. $2x^2 - x - 1 = 0$ **60.** $2x^2 - 5x + 2 = 0$

61. $12t^2 - 5t - 3 = 0$ **62.** $5m^2 + 13m - 6 = 0$

Use completing the square to solve each equation. See Example 9.

63. $3x^2 - 12x + 1 = 0$ **64.** $6x^2 - 12x + 1 = 0$

65. $2x^2 + 5x - 2 = 0$ **66.** $2x^2 - 8x + 5 = 0$

Use completing the square to solve each equation. See Example 10.

67. $p^2 + 2p + 2 = 0$ **68.** $x^2 - 6x + 10 = 0$

69. $y^2 + 8y + 18 = 0$ **70.** $n^2 + 10n + 28 = 0$

TRY IT YOURSELF

Solve each equation.

71. $(3x - 1)^2 = 25$ **72.** $(5x - 2)^2 = 64$

73. $3x^2 - 6x = 1$ **74.** $2x^2 - 6x = -3$

75. $x^2 + 8x + 6 = 0$ **76.** $x^2 + 6x + 4 = 0$

77. $6x^2 + 72 = 0$ **78.** $5x^2 + 40 = 0$

79. $x^2 - 2x = 17$ **80.** $x^2 + 10x = 7$

81. $m^2 - 7m + 3 = 0$ **82.** $m^2 - 5m + 3 = 0$

83. $7h^2 = 35$ **84.** $9n^2 = 99$

85. $\dfrac{7x + 1}{5} = -x^2$ **86.** $\dfrac{3}{8}x^2 = \dfrac{1}{8} - x$

87. $t^2 + t + 3 = 0$ **88.** $b^2 - b + 5 = 0$

89. $(8x + 5)^2 = 24$ **90.** $(3y - 2)^2 = 18$

91. $r^2 - 6r - 27 = 0$ **92.** $s^2 - 6s - 40 = 0$

93. $4p^2 + 2p + 3 = 0$ **94.** $3m^2 - 2m + 3 = 0$

APPLICATIONS

95. MOVIE STUNTS   According to the *Guinness Book of World Records,* stuntman Dan Koko fell a distance of 312 feet into an airbag after jumping from the Vegas World Hotel and Casino. The distance d in feet traveled by a free-falling object in t seconds is given by the formula $d = 16t^2$. To the nearest tenth of a second, how long did the fall last?

96. ACCIDENTS   The height h (in feet) of an object that is dropped from a height of s feet is given by the formula $h = s - 16t^2$, where t is the time the object has been falling. A 5-foot-tall woman on a sidewalk looks directly overhead and sees a window washer drop a bottle from four stories up. How long does she have to get out of the way? Round to the nearest tenth. (A story is 12 feet.)

97. GEOGRAPHY   The surface area S of a sphere is given by the formula $S = 4\pi r^2$, where r is the radius of the sphere. An almanac lists the surface area of the Earth as 196,938,800 square miles. Assuming the Earth to be spherical, what is its radius to the nearest mile?

98. FLAGS   In 1912, an order by President Taft fixed the width and length of the U.S. flag in the ratio 1 to 1.9. If 100 square feet of cloth are to be used to make a U.S. flag, estimate its dimensions to the nearest $\frac{1}{4}$ foot.

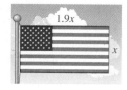

99. AUTOMOBILE ENGINES   As the piston shown moves upward, it pushes a cylinder of a gasoline/air mixture that is ignited by the spark plug. The formula that gives the volume of a cylinder is $V = \pi r^2 h$, where r is the radius and h the height. Find the radius of the piston (to the nearest hundredth of an inch) if it displaces 47.75 cubic inches of gasoline/air mixture as it moves from its lowest to its highest point.

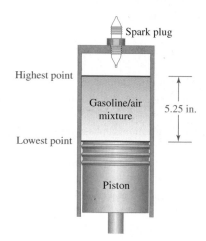

100. INVESTMENTS   If P dollars are deposited in an account that pays an annual rate of interest r, then in n years, the amount of money A in the account is given by the formula $A = P(1 + r)^n$. A savings account was opened on January 3, 2006, with a deposit of $10,000 and closed on January 2, 2008, with an ending balance of $11,772.25. Find the rate of interest.

101. PICTURE FRAMING The matting around the picture has a uniform width. How wide is the matting if its area equals the area of the picture? Round to the nearest hundredth of an inch.

102. SWIMMING POOLS In the advertisement shown, how wide will the free concrete decking be if a uniform width is constructed around the perimeter of the pool? Round to the nearest hundredth of a yard. (*Hint:* Note the difference in units. Convert the dimensions of the pool to yards.)

103. DIMENSIONS OF A RECTANGLE A rectangle is 4 feet longer than it is wide, and its area is 20 square feet. Find its dimensions to the nearest tenth of a foot.

104. DIMENSIONS OF A TRIANGLE The height of a triangle is 4 meters longer than twice its base. Find the base and height if the area of the triangle is 10 square meters. Round to the nearest hundredth of a meter.

WRITING

105. Give an example of a perfect-square trinomial. Why do you think the word "perfect" is used to describe it?

106. Explain why completing the square on $x^2 + 5x$ is more difficult than completing the square on $x^2 + 4x$.

107. Explain the error in the work shown below.

a. $\dfrac{4 \pm \sqrt{3}}{8} = \dfrac{\overset{1}{\cancel{4}} \pm \sqrt{3}}{\underset{1}{\cancel{4} \cdot 2}} = \dfrac{1 \pm \sqrt{3}}{2}$

b. $\dfrac{1 \pm \sqrt{5}}{5} = \dfrac{1 \pm \overset{1}{\cancel{\sqrt{5}}}}{\underset{1}{\cancel{5}}} = \dfrac{1 \pm 1}{1}$

108. Explain the steps involved in expressing $8 \pm \dfrac{\sqrt{15}}{2}$ as a single fraction with denominator 2.

REVIEW

Simplify each expression. All variables represent positive real numbers.

109. $\sqrt[3]{40a^3b^6}$ **110.** $\sqrt[8]{x^{24}}$

111. $\sqrt[4]{\dfrac{16}{625}}$ **112.** $\sqrt{175a^2b^3}$

CHALLENGE PROBLEMS

113. What number must be added to $x^2 + \sqrt{3}x$ to make a perfect-square trinomial?

114. Solve $x^2 + \sqrt{3}x - \frac{1}{4} = 0$ by completing the square.

Solve for the specified variable. Assume that all variables represent positive numbers. Express all radicals in simplified form.

115. $E = mc^2$ for c **116.** $A = \pi r^2$ for r

SECTION 8.2
The Quadratic Formula

Objectives

1 Derive the quadratic formula.

2 Solve quadratic equations using the quadratic formula.

3 Write equivalent equations to make quadratic formula computations easier.

4 Use the quadratic formula to solve application problems.

We can solve quadratic equations by completing the square, but the work is often tedious. In this section, we will develop a formula, called the *quadratic formula*, that enables us solve quadratic equations with less effort.

1   **Derive the Quadratic Formula.**

To develop a formula that will produce the solutions of a quadratic equation, we start with the **general quadratic equation** $ax^2 + bx + c = 0$ with $a > 0$, and solve it for x by completing the square.

$$ax^2 + bx + c = 0$$

$$\frac{ax^2}{a} + \frac{bx}{a} + \frac{c}{a} = \frac{0}{a} \qquad \text{Divide both sides by } a \text{ so that the coefficient of } x^2 \text{ is 1.}$$

$$x^2 + \frac{b}{a}x + \frac{c}{a} = 0 \qquad \text{Simplify: } \frac{ax^2}{a} = x^2. \text{ Write } \frac{bx}{a} \text{ as } \frac{b}{a}x.$$

$$x^2 + \frac{b}{a}x \qquad = -\frac{c}{a} \qquad \text{Subtract } \frac{c}{a} \text{ from both sides so that only the terms involving } x \text{ are on the left side of the equation.}$$

We can complete the square on $x^2 + \frac{b}{a}x$ by adding the square of one-half of the coefficient of x. Since the coefficient of x is $\frac{b}{a}$, we have $\frac{1}{2} \cdot \frac{b}{a} = \frac{b}{2a}$ and $\left(\frac{b}{2a}\right)^2 = \frac{b^2}{4a^2}$.

$$x^2 + \frac{b}{a}x + \frac{b^2}{4a^2} = -\frac{c}{a} + \frac{b^2}{4a^2} \qquad \text{To complete the square, add } \frac{b^2}{4a^2} \text{ to both sides.}$$

$$x^2 + \frac{b}{a}x + \frac{b^2}{4a^2} = -\frac{4ac}{4aa} + \frac{b^2}{4a^2} \qquad \text{Multiply } -\frac{c}{a} \text{ by } \frac{4a}{4a}. \text{ Now the fractions on the right side have the common denominator } 4a^2.$$

$$\left(x + \frac{b}{2a}\right)^2 = \frac{b^2 - 4ac}{4a^2} \qquad \text{On the left side, factor. On the right side, add the fractions.}$$

$$x + \frac{b}{2a} = \pm\sqrt{\frac{b^2 - 4ac}{4a^2}} \qquad \text{Use the square root property.}$$

$$x + \frac{b}{2a} = \pm\frac{\sqrt{b^2 - 4ac}}{\sqrt{4a^2}} \qquad \text{The square root of a quotient is the quotient of square roots.}$$

$$x + \frac{b}{2a} = \pm\frac{\sqrt{b^2 - 4ac}}{2a} \qquad \text{Since } a > 0, \ \sqrt{4a^2} = 2a.$$

$$x = -\frac{b}{2a} \pm \frac{\sqrt{b^2 - 4ac}}{2a} \qquad \text{To isolate } x, \text{ subtract } \frac{b}{2a} \text{ from both sides.}$$

$$x = \frac{-b \pm \sqrt{b^2 - 4ac}}{2a} \qquad \text{Combine the fractions.}$$

This result is called the *quadratic formula*. To develop this formula, we assumed that a was positive. If a is negative, similar steps are used, and we obtain the same result.

> **The Language of Algebra**
> To *derive* means to obtain by reasoning. To *derive* the quadratic formula means to solve $ax^2 + bx + c = 0$ for x, using the series of steps shown here, to obtain
> $$x = \frac{-b \pm \sqrt{b^2 - 4ac}}{2a}$$

Quadratic Formula	The solutions of $ax^2 + bx + c = 0$, with $a \neq 0$, are $$x = \frac{-b \pm \sqrt{b^2 - 4ac}}{2a}$$

2   **Solve Quadratic Equations Using the Quadratic Formula.**

In the next example, we will use the quadratic formula to solve a quadratic equation.

| **EXAMPLE 1** | Solve $2x^2 - 5x - 3 = 0$ by using the quadratic formula. |

Strategy   We will begin by comparing $2x^2 - 5x - 3 = 0$ to the standard form $ax^2 + bx + c = 0$.

Why   To use the quadratic formula, we need to identify the values of a, b, and c.

Solution

$$2x^2 - 5x - 3 = 0 \quad \text{This is the equation to solve.}$$
$$\uparrow \qquad \uparrow \qquad \uparrow$$
$$ax^2 + bx + c = 0$$

We see that $a = 2$, $b = -5$, and $c = -3$. To find the solutions of the equation, we substitute these values into the quadratic formula and evaluate the right side.

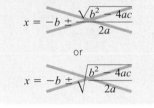

Caution
When writing the quadratic formula, be careful to draw the fraction bar so that it includes the entire numerator. Do not write

$$x = -b \pm \frac{\sqrt{b^2 - 4ac}}{2a}$$

or

$$x = -b \pm \sqrt{\frac{b^2 - 4ac}{2a}}$$

$$x = \frac{-b \pm \sqrt{b^2 - 4ac}}{2a} \qquad \text{This is the quadratic formula.}$$

$$x = \frac{-(-5) \pm \sqrt{(-5)^2 - 4(2)(-3)}}{2(2)} \qquad \text{Substitute 2 for } a, \ -5 \text{ for } b, \text{ and } -3 \text{ for } c.$$

$$x = \frac{5 \pm \sqrt{25 - (-24)}}{4} \qquad \text{Simplify: } -(-5) = 5. \text{ Evaluate the power and multiply within the radical. Multiply in the denominator.}$$

$$x = \frac{5 \pm \sqrt{49}}{4} \qquad \text{Simplify within the radical.}$$

$$x = \frac{5 \pm 7}{4} \qquad \text{Simplify: } \sqrt{49} = 7.$$

To find the first solution, we evaluate the expression using the $+$ symbol. To find the second solution, we evaluate the expression using the $-$ symbol.

$$x = \frac{5 + 7}{4} \qquad \text{or} \qquad x = \frac{5 - 7}{4}$$

$$x = \frac{12}{4} \qquad\qquad\qquad x = \frac{-2}{4}$$

$$x = 3 \qquad\qquad\qquad x = -\frac{1}{2}$$

The solutions are 3 and $-\frac{1}{2}$ and the solution set is $\left\{3, -\frac{1}{2}\right\}$. Check each solution in the original equation.

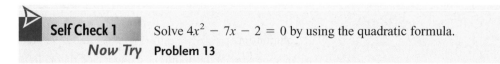

| **Self Check 1** | Solve $4x^2 - 7x - 2 = 0$ by using the quadratic formula. |
| *Now Try* **Problem 13** |

To solve a quadratic equation in x using the quadratic formula, we follow these steps.

Solving a Quadratic Equation in x Using the Quadratic Formula	1. Write the equation in standard form: $ax^2 + bx + c = 0$. 2. Identify a, b, and c. 3. Substitute the values for a, b, and c in the quadratic formula $$x = \frac{-b \pm \sqrt{b^2 - 4ac}}{2a}$$ and evaluate the right side to obtain the solutions.

EXAMPLE 2   Solve: $2x^2 = -4x - 1$

Strategy   We will write the equation in standard form $ax^2 + bx + c = 0$. Then we will identify the values of a, b, and c, and substitute these values into the quadratic formula.

Why   The quadratic equation must be in standard form to identify the values of a, b, and c.

Solution   To write the equation in standard form, we need to have all nonzero terms on the left side and 0 on the right side.

$$2x^2 = -4x - 1 \qquad \text{This is the equation to solve.}$$

$$2x^2 + 4x + 1 = 0 \qquad \text{To get 0 on the right side, add } 4x \text{ and 1 to both sides.}$$

In this equation, $a = 2$, $b = 4$, and $c = 1$.

$$x = \frac{-b \pm \sqrt{b^2 - 4ac}}{2a} \qquad \text{This is the quadratic formula.}$$

$$x = \frac{-4 \pm \sqrt{4^2 - 4(2)(1)}}{2(2)} \qquad \text{Substitute 2 for } a, \text{ 4 for } b, \text{ and 1 for } c.$$

$$x = \frac{-4 \pm \sqrt{16 - 8}}{4} \qquad \text{Evaluate the expression within the radical. Multiply in the denominator.}$$

$$x = \frac{-4 \pm \sqrt{8}}{4}$$

$$x = \frac{-4 \pm 2\sqrt{2}}{4} \qquad \text{Simplify: } \sqrt{8} = \sqrt{4 \cdot 2} = 2\sqrt{2}.$$

> **Success Tip**
> Perhaps you noticed that Example 1 could be solved by factoring. However, that is not the case for Example 2. These observations illustrate that the quadratic formula can be used to solve any quadratic equation.

We can write the solutions in simpler form by factoring out 2 from the terms in the numerator and removing the common factor of 2 in the numerator and denominator.

$$x = \frac{-4 \pm 2\sqrt{2}}{4} = \frac{2\left(-2 \pm \sqrt{2}\right)}{4} = \frac{\overset{1}{\cancel{2}}\left(-2 \pm \sqrt{2}\right)}{\underset{1}{\cancel{2} \cdot 2}} = \frac{-2 \pm \sqrt{2}}{2}$$

> **Notation**
> The solutions can also be written as
> $$-\frac{2}{2} \pm \frac{\sqrt{2}}{2} = -1 \pm \frac{\sqrt{2}}{2}$$

The solutions are $\frac{-2 + \sqrt{2}}{2}$ and $\frac{-2 - \sqrt{2}}{2}$ and the solution set is $\left\{ \frac{-2 + \sqrt{2}}{2}, \frac{-2 - \sqrt{2}}{2} \right\}$. We can approximate the solutions using a calculator. To two decimal places, they are -0.29 and -1.71.

Self Check 2   Solve $3x^2 = 2x + 3$. Approximate the solutions to two decimal places.

Now Try   **Problem 25**

The solutions to the next example are imaginary numbers.

EXAMPLE 3   Solve: $x^2 + x = -1$

Strategy   We will write the equation in standard form $ax^2 + bx + c = 0$. Then we will identify the values of a, b, and c, and substitute these values into the quadratic formula.

Why   The quadratic equation must be in standard form to identify the values of a, b, and c.

Solution   To write $x^2 + x = -1$ in standard form, we add 1 to both sides, to get

$x^2 + x + 1 = 0$ *This is the equation to solve.*

In this equation, $a = 1$, $b = 1$, and $c = 1$:

$$x = \frac{-b \pm \sqrt{b^2 - 4ac}}{2a}$$

$$x = \frac{-1 \pm \sqrt{1^2 - 4(1)(1)}}{2(1)}$$ *Substitute 1 for a, 1 for b, and 1 for c.*

$$x = \frac{-1 \pm \sqrt{1 - 4}}{2}$$ *Evaluate the expression within the radical.*

$$x = \frac{-1 \pm \sqrt{-3}}{2}$$

$$x = \frac{-1 \pm i\sqrt{3}}{2}$$ $\sqrt{-3} = \sqrt{-1 \cdot 3} = \sqrt{-1}\sqrt{3} = i\sqrt{3}.$

> **Notation**
> The solutions are written in complex number form $a + bi$. They could also be written as
> $$\frac{-1 \pm i\sqrt{3}}{2}$$

The   solutions   are $-\frac{1}{2} + \frac{\sqrt{3}}{2}i$ and $-\frac{1}{2} - \frac{\sqrt{3}}{2}i$ and   the   solution   set   is $\left\{ -\frac{1}{2} + \frac{\sqrt{3}}{2}i, -\frac{1}{2} - \frac{\sqrt{3}}{2}i \right\}$

Self Check 3   Solve: $a^2 + 3a + 5 = 0$

Now Try   Problem 29

---

3   **Write Equivalent Equations to Make Quadratic Formula Computations Easier.**

When solving a quadratic equation by the quadratic formula, we can often simplify the computations by solving a simpler, but equivalent equation.

EXAMPLE 4   For each equation below, write an equivalent equation so that the quadratic formula computations will be simpler.

a. $-2x^2 + 4x - 1 = 0$ **b.** $x^2 + \frac{4}{5}x - \frac{1}{3} = 0$

c. $20x^2 - 60x - 40 = 0$ **d.** $0.03x^2 - 0.04x - 0.01 = 0$

Strategy   We will multiply both sides of each equation by a carefully chosen number.

Why   In each case, the objective is to find an equivalent equation whose values of a, b, and c are easier to work with than those of the given equation.

Solution

a. It is often easier to solve a quadratic equation using the quadratic formula if a is positive. If we multiply (or divide) both sides of $-2x^2 + 4x - 1 = 0$ by -1, we obtain an equivalent equation with $a > 0$.

$$-2x^2 + 4x - 1 = 0 \qquad \text{Here, } a = -2.$$
$$-1(-2x^2 + 4x - 1) = -1(0)$$
$$2x^2 - 4x + 1 = 0 \qquad \text{Now } a = 2.$$

b. For $x^2 + \frac{4}{5}x - \frac{1}{3} = 0$, two coefficients are fractions: $b = \frac{4}{5}$ and $c = -\frac{1}{3}$. We can multiply both sides of the equation by their least common denominator, 15, to obtain an equivalent equation having coefficients that are integers.

$$x^2 + \frac{4}{5}x - \frac{1}{3} = 0 \qquad \text{Here, } a = 1, b = \frac{4}{5}, \text{ and } c = -\frac{1}{3}.$$
$$15\left(x^2 + \frac{4}{5}x - \frac{1}{3}\right) = 15(0)$$
$$15x^2 + 12x - 5 = 0 \qquad \text{Now } a = 15, b = 12, \text{ and } c = -5.$$

c. For $20x^2 - 60x - 40 = 0$, the coefficients 20, -60, and -40 have a common factor of 20. If we divide both sides of the equation by their GCF, we obtain an equivalent equation having smaller coefficients.

$$20x^2 - 60x - 40 = 0 \qquad \text{Here, } a = 20, b = -60, \text{ and } c = -40.$$
$$\frac{20x^2}{20} - \frac{60x}{20} - \frac{40}{20} = \frac{0}{20}$$
$$x^2 - 3x - 2 = 0 \qquad \text{Now } a = 1, b = -3, \text{ and } c = -2.$$

d. For $0.03x^2 - 0.04x - 0.01 = 0$, all three coefficients are decimals. We can multiply both sides of the equation by 100 to obtain an equivalent equation having coefficients that are integers.

$$0.03x^2 - 0.04x - 0.01 = 0 \qquad \text{Here, } a = 0.03, b = -0.04, \text{ and } c = -0.01.$$
$$100(0.03x^2 - 0.04x - 0.01) = 100(0)$$
$$3x^2 - 4x - 1 = 0 \qquad \text{Now } a = 3, b = -4, \text{ and } c = -1.$$

Self Check 4   For each equation, write an equivalent equation so that the quadratic formula computations will be simpler.

a. $-6x^2 + 7x - 9 = 0$

b. $\frac{1}{3}x^2 - \frac{2}{3}x - \frac{5}{6} = 0$

c. $44x^2 + 66x - 99 = 0$

d. $0.08x^2 - 0.07x - 0.02 = 0$

Now Try **Problems 37 and 39**

④   **Use the Quadratic Formula to Solve Application Problems.**

A variety of real-world applications can be modeled by quadratic equations. However, such equations are often difficult or even impossible to solve using the factoring method. In those cases, we can use the quadratic formula to solve the equation.

> **EXAMPLE 5**   ***Shortcuts.***   Instead of using the hallways, students are wearing a path through a planted quad area to walk 195 feet directly from the classrooms to the cafeteria. If the length of the hallway from the office to the cafeteria is 105 feet longer than the hallway from the office to the classrooms, how much walking are the students saving by taking the shortcut?

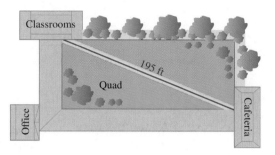

Analyze the Problem   The two hallways and the shortcut form a right triangle with a hypotenuse 195 feet long. We will use the Pythagorean theorem to solve this problem.

Form an Equation   If we let x = the length (in feet) of the hallway from the classrooms to the office, then the length of the hallway from the office to the cafeteria is $(x + 105)$ feet. Substituting these lengths into the Pythagorean theorem, we have

$$a^2 + b^2 = c^2 \qquad \text{This is the Pythagorean theorem.}$$

$$x^2 + (x + 105)^2 = 195^2 \qquad \text{Substitute } x \text{ for } a, \ (x + 105) \text{ for } b, \text{ and } 195 \text{ for } c.$$

$$x^2 + x^2 + 105x + 105x + 11{,}025 = 38{,}025 \qquad \text{Find } (x + 105)^2.$$

$$2x^2 + 210x + 11{,}025 = 38{,}025 \qquad \text{Combine like terms.}$$

$$2x^2 + 210x - 27{,}000 = 0 \qquad \text{To get 0 on the right side, subtract } 38{,}025 \text{ from both sides.}$$

$$x^2 + 105x - 13{,}500 = 0 \qquad \text{The coefficients have a common factor of } 2\text{: divide both sides by } 2.$$

Solve the Equation   To solve $x^2 + 105x - 13{,}500 = 0$, we will use the quadratic formula with $a = 1$, $b = 105$, and $c = -13{,}500$.

$$x = \frac{-b \pm \sqrt{b^2 - 4ac}}{2a}$$

$$x = \frac{-105 \pm \sqrt{105^2 - 4(1)(-13{,}500)}}{2(1)}$$

$$x = \frac{-105 \pm \sqrt{65{,}025}}{2} \qquad \text{Simplify: } 105^2 - 4(1)(-13{,}500) = 11{,}025 + 54{,}000 = 65{,}025.$$

$$x = \frac{-105 \pm 255}{2} \qquad \text{Use a calculator: } \sqrt{65{,}025} = 255.$$

$$x = \frac{150}{2} \quad \text{or} \quad x = \frac{-360}{2}$$

$$x = 75 \qquad\qquad x = -180 \qquad \text{Since the length of the hallway can't be negative, discard the solution } -180.$$

Classrooms

195 ft

x ft

Office

$(x + 105)$ ft

Cafeteria

State the Conclusion   The length of the hallway from the classrooms to the office is 75 feet. The length of the hallway from the office to the cafeteria is $75 + 105 = 180$ feet. Instead of using the hallways, a distance of $75 + 180 = 255$ feet, the students are taking the 195-foot shortcut to the cafeteria, a savings of $(255 - 195)$, or 60 feet.

Check the Result   The length of the 180-foot hallway is 105 feet longer than the length of the 75-foot hallway. The sum of the squares of the lengths of the hallways is $75^2 + 180^2 = 38,025$. This equals the square of the length of the 195-foot shortcut: $195^2 = 38,025$. The result checks.

 Now Try   **Problem 73**

EXAMPLE 6   ***Mass Transit.***   A bus company has 4,000 passengers daily, each currently paying a 75¢ fare. For each 15¢ fare increase, the company estimates that it will lose 50 passengers. If the company needs to bring in $6,570 per day to stay in business, what fare must be charged to produce this amount of revenue?

Analyze the Problem   To understand how a fare increase affects the number of passengers, let's consider what happens if there are two fare increases. We organize the data in a table. The fares are expressed in terms of dollars.

Number of increases	New fare	Number of passengers
One $0.15 increase	$0.75 + $0.15(1) = $0.90	4,000 − 50(1) = 3,950
Two $0.15 increases	$0.75 + $0.15(2) = $1.05	4,000 − 50(2) = 3,900

In general, the new fare will be the old fare ($0.75) plus the number of fare increases times $0.15. The number of passengers who will pay the new fare is 4,000 minus 50 times the number of $0.15 fare increases.

Form an Equation   If we let $x =$ the number of $0.15 fare increases necessary to bring in $6,570 daily, then $(0.75 + 0.15x)$ is the fare that must be charged. The number of passengers who will pay this fare is $4,000 - 50x$. We can now form an equation.

The bus fare	times	the number of passengers who will pay that fare	equals	$6,570.
$(0.75 + 0.15x)$	·	$(4,000 - 50x)$	=	6,570

Solve the Equation

$$(0.75 + 0.15x)(4,000 - 50x) = 6,570$$

$$3,000 - 37.5x + 600x - 7.5x^2 = 6,570 \qquad \text{Multiply the binomials.}$$

$$-7.5x^2 + 562.5x + 3,000 = 6,570 \qquad \text{Combine like terms: } -37.5x + 600x = 562.5x.$$

$$-7.5x^2 + 562.5x - 3,570 = 0 \qquad \text{To get 0 on the right side, subtract 6,570 from both sides.}$$

$$7.5x^2 - 562.5x + 3,570 = 0 \qquad \text{Multiply both sides by } -1 \text{ so that the value of } a, 7.5, \text{ is positive.}$$

To solve this equation, we will use the quadratic formula.

$$x = \frac{-b \pm \sqrt{b^2 - 4ac}}{2a}$$

$$x = \frac{-(-562.5) \pm \sqrt{(-562.5)^2 - 4(7.5)(3,570)}}{2(7.5)}$$ Substitute 7.5 for a, -562.5 for b, and 3,570 for c.

$$x = \frac{562.5 \pm \sqrt{209,306.25}}{15}$$ Simplify: $(-562.5)^2 - 4(7.5)(3,570) =$ 316,406.25 − 107.100 = 209,306.25.

$$x = \frac{562.5 \pm 457.5}{15}$$ Use a calculator: $\sqrt{209,306.25} = 457.5$.

$$x = \frac{1,020}{15} \quad \text{or} \quad x = \frac{105}{15}$$

$$x = 68 \qquad\qquad x = 7$$

State the Conclusion   If there are 7 fifteen-cent increases in the fare, the new fare will be $0.75 + $0.15(7) = $1.80. If there are 68 fifteen-cent increases in the fare, the new fare will be $0.75 + $0.15(68) = $10.95. Although this fare would bring in the necessary revenue, a $10.95 bus fare is unreasonable, so we discard it.

Check the Result   A fare of $1.80 will be paid by [4,000 − 50(7)] = 3,650 bus riders. The amount of revenue brought in would be $1.80(3,650) = $6,570. The result checks.

Now Try   **Problem 83**

EXAMPLE 7   *Lawyers.*   The number of lawyers in the United States is approximated by the function $N(x) = 45x^2 + 21,000x + 560,000$, where $N(x)$ is the number of lawyers and x is the number of years after 1980. In what year does this model indicate that the United States had one million lawyers? (Based on data from the American Bar Association)

Strategy   We will substitute 1,000,000 for $N(x)$ in the equation and solve for x.

Why   The value of x will give the number of years after 1980 that the United States had 1,000,000 lawyers.

Solution

$$N(x) = 45x^2 + 21,000x + 560,000$$

$$1,000,000 = 45x^2 + 21,000x + 560,000$$ Replace $N(x)$ with 1,000,000.

$$0 = 45x^2 + 21,000x - 440,000$$ To get 0 on the left side, subtract 1,000,000 from both sides.

We can simplify the computations by dividing both sides of the equation by 5, which is the greatest common factor of 45, 21,000, and 440,000.

$$9x^2 + 4,200x - 88,000 = 0$$ Divide both sides by 5.

We solve this equation using the quadratic formula.

$$x = \frac{-b \pm \sqrt{b^2 - 4ac}}{2a}$$

$$x = \frac{-4,200 \pm \sqrt{(4,200)^2 - 4(9)(-88,000)}}{2(9)}$$ Substitute 9 for a, 4,200 for b, and $-88,000$ for c.

$$x = \frac{-4,200 \pm \sqrt{20,808,000}}{18}$$ Evaluate the expression within the radical.

$$x \approx \frac{362}{18} \quad \text{or} \quad x \approx \frac{-8{,}762}{18} \qquad \text{Use a calculator.}$$

$$x \approx 20.1 \qquad \cancel{x \approx -486.8} \qquad \text{Since the model is defined only for positive values of } x, \text{ we discard the second solution.}$$

In 20.1 years after 1980, or in early 2000, the model predicts that United States had approximately 1,000,000 lawyers.

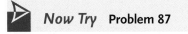 **Now Try** Problem 87

ANSWERS TO SELF CHECKS   **1.** $2, -\frac{1}{4}$ **2.** $\frac{1 \pm \sqrt{10}}{3}; -0.72, 1.39$ **3.** $-\frac{3}{2} \pm \frac{\sqrt{11}}{2}i$

4. a. $6x^2 - 7x + 9 = 0$ **b.** $2x^2 - 4x - 5 = 0$ **c.** $4x^2 + 6x - 9 = 0$ **d.** $8x^2 - 7x - 2 = 0$

STUDY SET
8.2

VOCABULARY

Fill in the blanks.

1. A _____ equation in one variable is any equation that can be written in the standard form $ax^2 + bx + c = 0$.

2. $x = \dfrac{-b \pm \sqrt{b^2 - 4ac}}{2a}$ is called the _____ formula.

CONCEPTS

3. Write each equation in standard form.

   a. $x^2 + 2x = -5$ **b.** $3x^2 = -2x + 1$

4. For each quadratic equation, find the values of a, b, and c.

   a. $x^2 + 5x + 6 = 0$ **b.** $8x^2 - x = 10$

5. Determine whether each statement is true or false.

   a. Any quadratic equation can be solved by using the quadratic formula.

   b. Any quadratic equation can be solved by completing the square.

6. What is wrong with the beginning of the solution shown below?

   Solve: $x^2 - 3x = 2$

   $a = 1 \quad b = -3 \quad c = 2$

Evaluate each expression.

7. $\dfrac{-2 \pm \sqrt{2^2 - 4(1)(-8)}}{2(1)}$

8. $\dfrac{-(-1) \pm \sqrt{(-1)^2 - 4(2)(-4)}}{2(2)}$

9. A student used the quadratic formula to solve a quadratic equation and obtained $x = \frac{-2 \pm \sqrt{3}}{2}$.

   a. How many solutions does the equation have? What are they exactly?

   b. Graph the solutions on a number line.

10. Simplify each of the following.

   a. $\dfrac{3 \pm 6\sqrt{2}}{3}$ **b.** $\dfrac{-12 \pm 4\sqrt{7}}{8}$

NOTATION

11. On a quiz, students were asked to write the quadratic formula. What is wrong with each answer shown below?

   a. $x = -b \pm \dfrac{\sqrt{b^2 - 4ac}}{2a}$

   b. $x = \dfrac{-b\sqrt{b^2 - 4ac}}{2a}$

12. When reading $\frac{-b \pm \sqrt{b^2 - 4ac}}{2a}$, we say, "The _____ of b, plus or _____ the square _____ of b _____ minus $\quad$ times a times c, all _____ $2a$."

GUIDED PRACTICE

Use the quadratic formula to solve each equation. **See Example 1.**

13. $x^2 - 3x + 2 = 0$ **14.** $x^2 + 3x + 2 = 0$

15. $x^2 + 12x = -36$

16. $x^2 - 18x + 81 = 0$

17. $2x^2 + x - 3 = 0$

18. $6x^2 - x - 1 = 0$

19. $12t^2 - 5t - 2 = 0$

20. $12z^2 + 5z - 3 = 0$

Solve each equation. Approximate the solutions to two decimal places. See Example 2.

21. $x^2 = x + 7$

22. $t^2 = t + 4$

23. $5x^2 + 5x = -1$

24. $2x^2 + 7x = -1$

25. $3y^2 + 1 = -6y$

26. $4w^2 + 1 = -6w$

27. $4m^2 = 4m + 19$

28. $3y^2 = 12y - 4$

Solve each equation. See Example 3.

29. $2x^2 + x + 1 = 0$

30. $2x^2 + 3x + 5 = 0$

31. $3x^2 - 2x + 1 = 0$

32. $3x^2 - 2x + 5 = 0$

33. $x^2 - 2x + 2 = 0$

34. $x^2 - 4x + 8 = 0$

35. $4a^2 + 4a + 5 = 0$

36. $4b^2 + 4b + 17 = 0$

For each equation, write an equivalent quadratic equation that will be easier to solve. Do not solve the equation. See Example 4.

37. a. $-5x^2 + 9x - 2 = 0$
  b. $1.6t^2 + 2.4t - 0.9 = 0$

38. a. $\dfrac{1}{8}x^2 + \dfrac{1}{2}x - \dfrac{3}{4} = 0$
  b. $33y^2 + 99y - 66 = 0$

39. a. $45x^2 + 30x - 15 = 0$
  b. $\dfrac{1}{3}m^2 - \dfrac{1}{2}m - \dfrac{1}{3} = 0$

40. a. $0.6t^2 - 0.1t - 0.2 = 0$
  b. $-a^2 - 15a + 12 = 0$

TRY IT YOURSELF

Solve each equation.

41. $x^2 - \dfrac{14}{15}x = \dfrac{8}{15}$

42. $x^2 = -\dfrac{5}{4}x + \dfrac{3}{2}$

43. $3x^2 - 4x = -2$

44. $2x^2 + 3x = -3$

45. $-16y^2 - 8y + 3 = 0$

46. $-16x^2 - 16x - 3 = 0$

47. $2x^2 - 3x - 1 = 0$

48. $3x^2 - 9x - 2 = 0$

49. $-x^2 + 10x = 18$

50. $-x^2 - 6x - 2 = 0$

51. $x^2 - 6x = 391$

52. $x^2 - 27x - 280 = 0$

53. $x^2 + 5x - 5 = 0$

54. $x^2 - 3x - 27 = 0$

55. $9h^2 - 6h + 7 = 0$

56. $5x^2 = 2x - 1$

57. $50x^2 + 30x - 10 = 0$

58. $120b^2 + 120b - 40 = 0$

59. $0.6x^2 + 0.03 - 0.4x = 0$

60. $2x^2 + 0.1x = 0.04$

61. $\dfrac{1}{8}x^2 - \dfrac{1}{2}x + 1 = 0$

62. $\dfrac{1}{2}x^2 + 3x + \dfrac{13}{2} = 0$

63. $\dfrac{a^2}{10} - \dfrac{3a}{5} + \dfrac{7}{5} = 0$

64. $\dfrac{c^2}{4} + c + \dfrac{11}{4} = 0$

65. $\dfrac{x^2}{2} + \dfrac{5}{2}x = -1$

66. $\dfrac{x^2}{8} - \dfrac{x}{4} = \dfrac{1}{2}$

67. $900x^2 - 8,100x = 1,800$

68. $14x^2 - 21x = 49$

Solve each equation. Approximate the solutions to two decimal places.

69. $\dfrac{1}{4}x^2 - \dfrac{1}{6}x - \dfrac{1}{6} = 0$

70. $81x^2 + 12x - 80 = 0$

71. $0.7x^2 - 3.5x - 25 = 0$

72. $-4.5x^2 + 0.2x + 3.75 = 0$

APPLICATIONS

73. CROSSWALKS   Refer to the illustration on the next page. Instead of using the Main Street and First Avenue crosswalks to get from Nordstroms to Best Buy, a shopper uses the diagonal crosswalk to walk 97 feet directly from one corner to the other. If the length of the Main Street crosswalk is 7 feet longer than the First Avenue crosswalk, how much walking does the shopper save by using the diagonal crosswalk?

The Discriminant	For a quadratic equation of the form $ax^2 + bx + c = 0$ with real-number coefficients and $a \neq 0$, the expression $b^2 - 4ac$ is called the **discriminant** and can be used to determine the number and type of the solutions of the equation.

Discriminant: $b^2 - 4ac$	Number and type of solutions
Positive	Two different real numbers
0	One repeated solution, a rational number
Negative	Two different imaginary numbers that are complex conjugates

Discriminant: $b^2 - 4ac$	Number and type of solutions
A perfect square	Two different rational numbers
Positive and not a perfect square	Two different irrational numbers

---

EXAMPLE 1   Determine the number and type of solutions for each equation:
a. $x^2 + x + 1 = 0$ **b.** $3x^2 + 5x + 2 = 0$

Strategy   We will identify the values of a, b, and c in each equation. Then we will use those values to compute $b^2 - 4ac$, the discriminant.

Why   Once we know whether the discriminant is positive, 0, or negative, and whether it is a perfect square, we can determine the number and type of the solutions of the equation.

Solution

a. For $x^2 + x + 1 = 0$, the discriminant is:

$$b^2 - 4ac = 1^2 - 4(1)(1) \quad \text{Substitute: } a = 1, b = 1, \text{ and } c = 1.$$
$$= -3 \quad \text{The result is a negative number.}$$

Since $b^2 - 4ac < 0$, the solutions of $x^2 + x + 1 = 0$ are two different imaginary numbers that are complex conjugates.

b. For $3x^2 + 5x + 2 = 0$, the discriminant is:

$$b^2 - 4ac = 5^2 - 4(3)(2) \quad \text{Substitute: } a = 3, b = 5, \text{ and } c = 2.$$
$$= 25 - 24$$
$$= 1 \quad \text{The result is a positive number.}$$

Since $b^2 - 4ac > 0$ and $b^2 - 4ac$ is a perfect square, the solutions of $3x^2 + 5x + 2 = 0$ are two different rational numbers.

> **Success Tip**
> The discriminant can also be used to determine factorability. The trinomial $ax^2 + bx + c$ with integer coefficients and $a \neq 0$ will factor as the product of two binomials with integer coefficients if the value of $b^2 - 4ac$ is a perfect square. If $b^2 - 4ac = 0$, the factors will be the same.

Self Check 1   Determine the number and type of solutions for
a. $x^2 + x - 1 = 0$
b. $3x^2 + 4x + 2 = 0$

Now Try   Problems 11, 13, and 15

---

2   **Solve Equations That Are Quadratic in Form.**

We have discussed four methods that are used to solve quadratic equations. The table on the next page shows some advantages and disadvantages of each method.

Method	Advantages	Disadvantages	Examples
Factoring and the zero-factor property	It can be very fast. When each factor is set equal to 0, the resulting equations are usually easy to solve.	Some polynomials may be difficult to factor and others impossible.	$x^2 - 2x - 24 = 0$ $4a^2 - a = 0$
Square root property	It is the fastest way to solve equations of the form $ax^2 = n$ ($n =$ a number) or $(ax + b)^2 = n$.	It only applies to equations that are in these forms.	$x^2 = 27$ $(2y + 3)^2 = 25$
Completing the square*	It can be used to solve any quadratic equation. It works well with equations of the form $x^2 + bx = n$, where b is even.	It involves more steps than the other methods. The algebra can be cumbersome if the leading coefficient is not 1.	$x^2 + 4x + 1 = 0$ $t^2 - 14t - 9 = 0$
Quadratic formula	It can be used to solve any quadratic equation.	It involves several computations where sign errors can be made. Often the result must be simplified.	$x^2 + 3x - 33 = 0$ $4s^2 - 10s + 5 = 0$

*The quadratic formula is just a condensed version of completing the square and is usually easier to use. However, you need to know how to complete the square because it is used in more advanced mathematics courses.

To determine the most efficient method for a given equation, we can use the following strategy.

Strategy for Solving Quadratic Equations

1. See whether the equation is in a form such that the **square root method** is easily applied.
2. See whether the equation is in a form such that the **completing the square method** is easily applied.
3. If neither Step 1 nor Step 2 is reasonable, write the equation in $ax^2 + bx + c = 0$ form.
4. See whether the equation can be solved using the **factoring method.**
5. If you can't factor, solve the equation by the **quadratic formula.**

Many nonquadratic equations can be written in quadratic form ($ax^2 + bx + c = 0$) and solved using the techniques discussed in previous sections. For example, a careful inspection of the equation $x^4 - 5x^2 + 4 = 0$ leads to the following observations:

The leading term, x^4, is the square of the expression x^2 in the middle term: $x^4 = (x^2)^2$.

$$x^4 - 5x^2 + 4 = 0$$

The last term is a constant.

Equations that contain an expression, the same expression squared, and a constant term are said to be *quadratic in form.* One method used to solve such equations is to make a substitution.

EXAMPLE 2   Solve: $x^4 - 3x^2 - 4 = 0$

Strategy   Since the leading term, x^4, is the square of the expression x^2 in the middle term, we will substitute y for x^2.

Why   Our hope is that such a substitution will produce an equation that we can solve using one of the methods previously discussed.

Solution   If we write x^4 as $(x^2)^2$, the equation takes the form

$$(x^2)^2 - 3x^2 - 4 = 0$$

and it is said to be *quadratic in* x^2. We can solve this equation by letting $y = x^2$.

$$y^2 - 3y - 4 = 0 \quad \text{Replace each } x^2 \text{ with } y.$$

We can solve this quadratic equation by factoring.

$$(y - 4)(y + 1) = 0 \qquad \text{Factor } y^2 - 3y - 4.$$
$$y - 4 = 0 \quad \text{or} \quad y + 1 = 0 \qquad \text{Set each factor equal to 0.}$$
$$y = 4 \qquad \qquad \quad y = -1$$

Notation
The choice of the letter y for the substitution is arbitrary. We could just as well let $b = x^2$.

These are not the solutions for x. To find x, we reverse the substitution by replacing each y with x^2 and proceed as follows:

$$x^2 = 4 \qquad \text{or} \qquad x^2 = -1 \qquad \text{Substitute } x^2 \text{ for } y.$$
$$x = \pm\sqrt{4} \qquad \qquad x = \pm\sqrt{-1} \qquad \text{Use the square root property.}$$
$$x = \pm 2 \qquad \qquad \quad x = \pm i$$

Caution
If you are solving an equation in x, you can't answer with values of y. Remember to reverse any substitutions, and solve for the variable in the original equation.

This equation has four solutions: 2, -2, i, and $-i$. Check each one in the original equation.

> **Self Check 2**   Solve: $x^4 - 5x^2 - 36 = 0$
> **Now Try**   **Problem 23**

EXAMPLE 3   Solve: $x - 7\sqrt{x} + 12 = 0$

Strategy   Since the leading term, x, is the square of the expression $\sqrt{x}$ in the middle term, we will substitute y for $\sqrt{x}$.

Why   Our hope is that such a substitution will produce an equation that we can solve using one of the methods previously discussed.

Solution   We examine the leading term and the middle term.

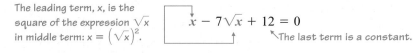

The leading term, x, is the square of the expression $\sqrt{x}$ in middle term: $x = \left(\sqrt{x}\right)^2$. $x - 7\sqrt{x} + 12 = 0$ The last term is a constant.

If we write x as $\left(\sqrt{x}\right)^2$, the equation takes the form

$$\left(\sqrt{x}\right)^2 - 7\sqrt{x} + 12 = 0$$

and it is said to be *quadratic in* $\sqrt{x}$. We can solve this equation by letting $y = \sqrt{x}$ and factoring.

$$y^2 - 7y + 12 = 0 \qquad \text{Replace each } \sqrt{x} \text{ with } y.$$
$$(y - 3)(y - 4) = 0 \qquad \text{Factor } y^2 - 7y + 12.$$
$$y - 3 = 0 \quad \text{or} \quad y - 4 = 0 \qquad \text{Set each factor equal to 0.}$$
$$y = 3 \qquad \qquad \quad y = 4$$

The Language of Algebra
Equations such as
$x - 7\sqrt{x} + 12 = 0$ that are quadratic in form are also said to be *reducible* to a quadratic.

To find x, we reverse the substitution and replace each y with $\sqrt{x}$. Then we solve the resulting radical equations by squaring both sides.

$$\sqrt{x} = 3 \quad \text{or} \quad \sqrt{x} = 4$$
$$x = 9 \quad \mid \quad x = 16$$

The solutions are 9 and 16. Check each solution in the original equation.

Self Check 3   Solve: $x + \sqrt{x} - 6 = 0$
Now Try   Problem 27

EXAMPLE 4   Solve: $2m^{2/3} - 2 = 3m^{1/3}$

Strategy   We will write the equation in descending powers of m and look for a possible substitution to make.

Why   Our hope is that a substitution will produce an equation that we can solve using one of the methods previously discussed.

Solution   After writing the equation in descending powers of m, we see that

$$2m^{2/3} - 3m^{1/3} - 2 = 0$$

is *quadratic in* $m^{1/3}$, because $m^{2/3} = (m^{1/3})^2$. We will use the substitution $y = m^{1/3}$ to write this equation in quadratic form.

$$2m^{2/3} - 3m^{1/3} - 2 = 0$$
$$2(m^{1/3})^2 - 3m^{1/3} - 2 = 0 \qquad \text{Write } m^{2/3} \text{ as } (m^{1/3})^2.$$
$$2y^2 - 3y - 2 = 0 \qquad \text{Replace each } m^{1/3} \text{ with } y.$$
$$(2y + 1)(y - 2) = 0 \qquad \text{Factor } 2y^2 - 3y - 2.$$
$$2y + 1 = 0 \quad \text{or} \quad y - 2 = 0 \quad \text{Set each factor equal to 0.}$$
$$y = -\frac{1}{2} \quad \mid \quad y = 2$$

To find m, we reverse the substitution and replace each y with $m^{1/3}$. Then we solve the resulting equations by cubing both sides.

$$m^{1/3} = -\frac{1}{2} \quad \text{or} \quad m^{1/3} = 2$$
$$(m^{1/3})^3 = \left(-\frac{1}{2}\right)^3 \quad \mid \quad (m^{1/3})^3 = (2)^3 \qquad \text{Recall that } m^{1/3} = \sqrt[3]{m}. \text{ To solve for } m, \text{ cube both sides.}$$
$$m = -\frac{1}{8} \quad \mid \quad m = 8$$

The solutions are $-\frac{1}{8}$ and 8. Check each solution in the original equation.

Self Check 4   Solve: $a^{2/3} = -3a^{1/3} + 10$
Now Try   Problem 31

EXAMPLE 5 Solve: $(4t + 2)^2 - 30(4t + 2) + 224 = 0$

Strategy Since the leading term, $(4t + 2)^2$, is the square of the expression $4t + 2$ in the middle term, we will substitute y for $4t + 2$.

Why Our hope is that such a substitution will produce an equation that we can solve using one of the methods previously discussed.

Solution This equation is *quadratic in* $4t + 2$. If we make the substitution $y = 4t + 2$, we have

$$y^2 - 30y + 224 = 0$$

which can be solved by using the quadratic formula.

$$y = \frac{-b \pm \sqrt{b^2 - 4ac}}{2a}$$

$$y = \frac{-(-30) \pm \sqrt{(-30)^2 - 4(1)(224)}}{2(1)}$$ Substitute 1 for a, -30 for b, and 224 for c.

$$y = \frac{30 \pm \sqrt{900 - 896}}{2}$$ Simplify within the radical.

$$y = \frac{30 \pm 2}{2}$$ $\sqrt{900 - 896} = \sqrt{4} = 2.$

$$y = 16 \quad \text{or} \quad y = 14$$

To find t, we reverse the substitution and replace y with $4t + 2$. Then we solve for t.

$$
\begin{array}{c|c}
4t + 2 = 16 \quad \text{or} & 4t + 2 = 14 \\
4t = 14 & 4t = 12 \\
t = 3.5 & t = 3
\end{array}
$$

Verify that 3.5 and 3 satisfy the original equation.

Self Check 5 Solve: $(n + 3)^2 - 6(n + 3) = -8$

Now Try **Problem 35**

EXAMPLE 6 Solve: $15a^{-2} - 8a^{-1} + 1 = 0$

Strategy We will write the equation with positive exponents and look for a possible substitution to make.

Why Our hope is that a substitution will produce an equation that we can solve using one of the methods previously discussed.

Solution When we write the terms $15a^{-2}$ and $-8a^{-1}$ using positive exponents, we see that this equation is *quadratic in* $\frac{1}{a}$.

$$\frac{15}{a^2} - \frac{8}{a} + 1 = 0$$ Think of this equation as $15 \cdot \left(\frac{1}{a}\right)^2 - 8 \cdot \frac{1}{a} + 1 = 0.$

Success Tip
We could also solve this equation by multiplying both sides by the LCD, a^2.

If we let $y = \frac{1}{a}$, the resulting quadratic equation can be solved by factoring.

$$15y^2 - 8y + 1 = 0 \quad \text{Substitute } y^2 \text{ for } \tfrac{1}{a^2} \text{ and } y \text{ for } \tfrac{1}{a}.$$

$$(5y - 1)(3y - 1) = 0 \quad \text{Factor } 15y^2 - 8y + 1 = 0.$$

$$5y - 1 = 0 \quad \text{or} \quad 3y - 1 = 0$$

$$y = \frac{1}{5} \qquad \qquad y = \frac{1}{3}$$

To find a, we reverse the substitution and replace each y with $\frac{1}{a}$. Then we proceed as follows:

$$\frac{1}{a} = \frac{1}{5} \quad \text{or} \quad \frac{1}{a} = \frac{1}{3}$$

$$5 = a \qquad \qquad 3 = a \quad \text{Solve the proportions.}$$

The solutions are 5 and 3. Check each solution in the original equation.

Self Check 6   Solve: $28c^{-2} - 3c^{-1} - 1 = 0$

Now Try   Problem 41

3   Solve Problems Involving Quadratic Equations.

EXAMPLE 7   ***Household Appliances.***   The illustration shows a temperature control on a washing machine. When the *warm* setting is selected, both the hot and cold water pipes open to fill the tub in 2 minutes 15 seconds. When the *cold* setting is chosen, the tub fills 45 seconds faster than when the *hot* setting is used. How long does it take to fill the washing machine with hot water?

Electronic Temperature Control

Water Temp

Analyze the Problem   It is helpful to organize the facts of this shared-work problem in a table.

Form an Equation   Let $x =$ the number of seconds it takes to fill the tub with hot water. Since the cold water inlet fills the tub in 45 seconds less time, $x - 45 =$ the number of seconds it takes to fill the tub with cold water. The hot and cold water inlets will be open for the same time: 2 minutes 15 seconds, or 135 seconds.

To determine the work completed by each inlet, multiply the rate by the time.

	Rate $\cdot$ Time $=$ Work completed		
Hot water	$\dfrac{1}{x}$	135	$\dfrac{135}{x}$
Cold water	$\dfrac{1}{x - 45}$	135	$\dfrac{135}{x - 45}$

Enter this information first.   Multiply to get each of these entries: $W = rt$.

Success Tip

An alternate way to form an equation is to note that what the hot water inlet can do in 1 second plus what the cold water inlet can do in 1 second equals what they can do together in 1 second:

$$\frac{1}{x} + \frac{1}{x - 45} = \frac{1}{135}$$

In shared-work problems, 1 represents one whole job completed. So we have,

The fraction of tub filled with hot water	plus	the fraction of the tub filled with cold water	equals	1 tub filled.
$\dfrac{135}{x}$	$+$	$\dfrac{135}{x - 45}$	$=$	1

Solve the Equation

$$\frac{135}{x} + \frac{135}{x - 45} = 1$$

$$x(x - 45)\left(\frac{135}{x} + \frac{135}{x - 45}\right) = x(x - 45)(1)$$

Multiply both sides by the LCD $x(x - 45)$ to clear the rational equation of fractions.

$$x(x - 45)\frac{135}{x} + x(x - 45)\frac{135}{x - 45} = x(x - 45)(1)$$

Distribute the multiplication by $x(x - 45)$.

$$135(x - 45) + 135x = x(x - 45)$$

Simplify each side.

$$135x - 6{,}075 + 135x = x^2 - 45x$$

Distribute the multiplication by 135 and by x.

$$270x - 6{,}075 = x^2 - 45x$$

Combine like terms.

$$0 = x^2 - 315x + 6{,}075$$

To get 0 on the left side, subtract 270x from both sides, and add 6,075 to both sides.

To solve this equation, we will use the quadratic formula, with $a = 1$, $b = -315$, and $c = 6{,}075$.

$$x = \frac{-b \pm \sqrt{b^2 - 4ac}}{2a}$$

$$x = \frac{-(-315) \pm \sqrt{(-315)^2 - 4(1)(6{,}075)}}{2(1)}$$

Substitute 1 for a, −315 for b, and 6,075 for c.

$$x = \frac{315 \pm \sqrt{99{,}225 - 24{,}300}}{2}$$

Simplify within the radical.

$$x = \frac{315 \pm \sqrt{74{,}925}}{2}$$

$$x \approx \frac{589}{2} \quad \text{or} \quad x \approx \frac{41}{2}$$

$$x \approx 294 \quad \Big| \quad x \approx 21$$

State the Conclusion We can discard the solution of 21 seconds, because this would imply that the cold water inlet fills the tub in a negative number of seconds $(21 - 45 = -24)$. Therefore, the hot water inlet fills the washing machine tub in about 294 seconds, which is 4 minutes 54 seconds.

Check the Result Use estimation to check the result.

> ▷ **Now Try** Problem 83

STUDY SET
8.3

VOCABULARY

Fill in the blanks.

1. For the quadratic equation $ax^2 + bx + c = 0$, the _____ is $b^2 - 4ac$.

2. We can solve $x - 2\sqrt{x} - 8 = 0$ by making the _____ $y = \sqrt{x}$.

CONCEPTS

Consider the quadratic equation $ax^2 - bx + c = 0$, where a, b, and c represent rational numbers, and fill in the blanks.

3. If $b^2 - 4ac < 0$, the solutions of the equation are two different imaginary numbers that are complex _____.

4. If $b^2 - 4ac = $ ___, the equation has one repeated rational-number solution.

5. If $b^2 - 4ac$ is a perfect square, the solutions of the equation are two different _____ numbers.

6. If $b^2 - 4ac$ is positive and not a perfect square, the solutions of the equation are two different _____ numbers.

7. For each equation, determine the substitution that should be made to write the equation in quadratic form.

   a. $x^4 - 12x^2 + 27 = 0$ Let $y = $

   b. $x - 13\sqrt{x} + 40 = 0$ Let $y = $

   c. $x^{2/3} + 2x^{1/3} - 3 = 0$ Let $y = $

   d. $x^{-2} - x^{-1} - 30 = 0$ Let $y = $

   e. $(x + 1)^2 - (x + 1) - 6 = 0$ Let $y = $

8. Fill in the blanks.

   a. $x^4 = \left(\right)^2$ **b.** $x = \left(\right)^2$

   c. $x^{2/3} = \left(\right)^2$ **d.** $\dfrac{1}{x^2} = \left(\right)^2$

NOTATION

Complete the solution.

9. To find the type of solutions for the equation $x^2 + 5x + 6 = 0$, we compute the discriminant.

$$b^2 - = ^2 - 4(1)\left(\right)$$
$$= 25 - $$
$$= 1$$

Since a, b, and c are rational numbers and the value of the discriminant is a perfect square, the solutions are two different _____ numbers.

10. Fill in the blanks to write each equation in quadratic form.

   a. $x^4 - 2x^2 - 15 = 0 \rightarrow \left(\right)^2 - 2 - 15 = 0$

   b. $x - 2\sqrt{x} + 3 = 0 \rightarrow \left(\right)^2 - 2 + 3 = 0$

   c. $8m^{2/3} - 10m^{1/3} - 3 = 0 \rightarrow 8\left(\right)^2 - 10 - 3 = 0$

GUIDED PRACTICE

Use the discriminant to determine the number and type of solutions for each equation. Do not solve. See Example 1.

11. $4x^2 - 4x + 1 = 0$ **12.** $6x^2 - 5x - 6 = 0$

13. $5x^2 + x + 2 = 0$ **14.** $3x^2 + 10x - 2 = 0$

15. $2x^2 = 4x - 1$ **16.** $9x^2 = 12x - 4$

17. $x(2x - 3) = 20$ **18.** $x(x - 3) = -10$

19. $3x^2 - 10 = 0$ **20.** $5x^2 - 24 = 0$

21. $x^2 - \dfrac{14}{15}x = \dfrac{8}{15}$ **22.** $x^2 = -\dfrac{5}{4}x + \dfrac{3}{2}$

Solve each equation. See Example 2.

23. $x^4 - 17x^2 + 16 = 0$ **24.** $x^4 - 10x^2 + 9 = 0$

25. $x^4 + 5x^2 - 36 = 0$ **26.** $x^4 - 15x^2 - 16 = 0$

Solve each equation. See Example 3.

27. $x - 13\sqrt{x} + 40 = 0$ **28.** $x - 9\sqrt{x} + 18 = 0$

29. $2x + \sqrt{x} - 3 = 0$ **30.** $2x - \sqrt{x} - 1 = 0$

Solve each equation. See Example 4.

31. $a^{2/3} - 2a^{1/3} = 3$ **32.** $r^{2/3} + 4r^{1/3} = 5$

33. $x^{2/3} + 2x^{1/3} - 8 = 0$ **34.** $x^{2/3} - 7x^{1/3} + 12 = 0$

Solve each equation. See Example 5.

35. $(c + 1)^2 - 4(c + 1) + 3 = 0$

36. $(a - 5)^2 - 4(a - 5) - 21 = 0$

37. $2(2x + 1)^2 - 7(2x + 1) + 6 = 0$

38. $3(2 - x)^2 + 10(2 - x) - 8 = 0$

Solve each equation. See Example 6.

39. $m^{-2} + m^{-1} - 6 = 0$ **40.** $t^{-2} + t^{-1} - 42 = 0$

41. $8x^{-2} - 10x^{-1} - 3 = 0$ **42.** $2x^{-2} - 5x^{-1} - 3 = 0$

Solve each equation. See Example 7.

43. $1 - \dfrac{5}{x} = \dfrac{10}{x^2}$ **44.** $1 - \dfrac{3}{x} = \dfrac{5}{x^2}$

45. $\dfrac{1}{2} + \dfrac{1}{b} = \dfrac{1}{b - 7}$ **46.** $\dfrac{1}{4} - \dfrac{1}{n} = \dfrac{1}{n + 3}$

TRY IT YOURSELF

Solve each equation.

47. $2x - \sqrt{x} = 3$ **48.** $3x + 4\sqrt{x} = 4$

49. $x^{-2} + 2x^{-1} - 3 = 0$ **50.** $x^{-2} + 2x^{-1} - 8 = 0$

51. $x^4 + 19x^2 + 18 = 0$ **52.** $t^4 + 4t^2 - 5 = 0$

53. $(k - 7)^2 + 6(k - 7) + 10 = 0$
54. $(d + 9)^2 - 4(d + 9) + 8 = 0$

55. $\dfrac{2}{x - 1} + \dfrac{1}{x + 1} = 3$ **56.** $\dfrac{3}{x - 2} - \dfrac{1}{x + 2} = 5$

57. $x - 6x^{1/2} = -8$ **58.** $x - 5x^{1/2} + 4 = 0$

59. $(y^2 - 9)^2 + 2(y^2 - 9) - 99 = 0$
60. $(a^2 - 4)^2 - 4(a^2 - 4) - 32 = 0$

61. $x^{-4} - 2x^{-2} + 1 = 0$ **62.** $4x^{-4} + 1 = 5x^{-2}$

63. $t^4 + 3t^2 = 28$ **64.** $3h^4 + h^2 - 2 = 0$

65. $2x^{2/5} - 5x^{1/5} = -3$ **66.** $2x^{2/5} + 3x^{1/5} = -1$

67. $9\left(\dfrac{3m + 2}{m}\right)^2 - 30\left(\dfrac{3m + 2}{m}\right) + 25 = 0$

68. $4\left(\dfrac{c - 7}{c}\right)^2 - 12\left(\dfrac{c - 7}{c}\right) + 9 = 0$

69. $\dfrac{3}{a - 1} = 1 - \dfrac{2}{a}$ **70.** $1 + \dfrac{4}{x} = \dfrac{3}{x^2}$

71. $\left(8 - \sqrt{a}\right)^2 + 6\left(8 - \sqrt{a}\right) - 7 = 0$
72. $\left(10 - \sqrt{t}\right)^2 - 4\left(10 - \sqrt{t}\right) - 45 = 0$

73. $x + \dfrac{2}{x - 2} = 0$ **74.** $x + \dfrac{x + 5}{x - 3} = 0$

75. $3x + 5\sqrt{x} + 2 = 0$ **76.** $3x - 4\sqrt{x} + 1 = 0$

77. $x^4 - 6x^2 + 5 = 0$ **78.** $2x^4 - 26x^2 + 24 = 0$

79. $8(t + 1)^{-2} - 30(t + 1)^{-1} + 7 = 0$
80. $2(s - 2)^{-2} + 3(s - 2)^{-1} - 5 = 0$

81. $\dfrac{1}{x + 2} + \dfrac{24}{x + 3} = 13$ **82.** $\dfrac{3}{x} + \dfrac{4}{x + 1} = 2$

APPLICATIONS

83. CROWD CONTROL   After a performance at a county fair, security guards have found that the grandstand area can be emptied in 6 minutes if both the east and west exits are opened. If just the east exit is used, it takes 4 minutes longer to clear the grandstand than it does if just the west exit is opened. How long does it take to clear the grandstand if everyone must file through the west exit? Round to the nearest tenth of a minute.

84. PAPER ROUTES   When a father, in a car, and his son, on a bicycle, work together to distribute the morning newspaper, it takes them 35 minutes to complete the route. Working alone, it takes the son 25 minutes longer than the father. To the nearest minute, how long does it take the son to cover the route on his bicycle?

85. ASSEMBLY LINES   A newly manufactured product traveled 300 feet on a high-speed conveyor belt at a rate of r feet per second. It could have traveled the 300 feet in 3 seconds less time if the speed of the conveyor belt was increased by 5 feet per second. Find r.

86. BICYCLING   Tina bicycles 160 miles at the rate of r mph. The same trip would have taken 2 hours longer if she had decreased her speed by 4 mph. Find r.

87. ARCHITECTURE   A **golden rectangle** is one of the most visually appealing of all geometric forms. The Parthenon, built by the Greeks in the 5th century B.C., fits into a golden rectangle if its ruined triangular pediment is included. See the illustration. In a golden rectangle, the length l and width w must satisfy the equation

$$\frac{l}{w} = \frac{w}{l - w}$$

If a rectangular billboard is to have a width of 20 feet, what should its length be so that it is a golden rectangle? Round to the nearest tenth.

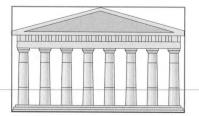

88. ENTRY DOOR DESIGNS   An architect needs to determine the height h of the window shown in the illustration. The radius r, the width w, and the height h of the circular-shaped window are related by the formula

$$r = \frac{4h^2 + w^2}{8h}$$

If w is to be 34 inches and r is to be 18 inches, find h to the nearest tenth of an inch.

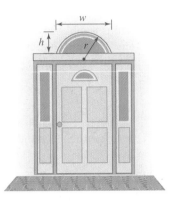

WRITING

89. Describe how to predict what type of solutions the equation $3x^2 - 4x + 5 = 0$ will have.

90. What error is made in the following solution?

Solve: $x^4 - 12x^2 + 27 = 0$

Let $y = x^2$

$$y^2 - 12y + 27 = 0$$
$$(y - 9)(y - 3) = 0$$

$y - 9 = 0$ or $y - 3 = 0$
$y = 9$ | $y = 3$

The solutions of $x^4 - 12x^2 + 27 = 0$ are 9 and 3.

REVIEW

91. Write an equation of the vertical line that passes through $(3, 4)$.

92. Find an equation of the line that passes through $(-1, -6)$ and $(-2, -1)$. Write the equation in slope–intercept form.

93. Write an equation of the line with slope $\frac{2}{3}$ that passes through the origin.

94. Find an equation of the line that passes through $(2, -3)$ and is perpendicular to the line whose equation is $y = \frac{x}{5} + 6$. Write the equation in slope–intercept form.

CHALLENGE PROBLEMS

95. Solve: $x^6 + 17x^3 + 16 = 0$

96. Find the real-number solutions of $x^4 - 3x^2 - 2 = 0$. Rationalize the denominators of the solutions.

SECTION 8.4
Quadratic Functions and Their Graphs

Objectives

1 Graph functions of the form $f(x) = ax^2$.

2 Graph functions of the form $f(x) = ax^2 + k$.

3 Graph functions of the form $f(x) = a(x - h)^2$.

4 Graph functions of the form $f(x) = a(x - h)^2 + k$.

5 Graph functions of the form $f(x) = ax^2 + bx + c$ by completing the square.

6 Find the vertex using $-\frac{b}{2a}$.

7 Determine minimum and maximum values.

8 Solve quadratic equations graphically.

In this section, we will discuss methods for graphing *quadratic functions*.

Quadratic Functions	A **quadratic function** is a second-degree polynomial function that can be written in the form $$f(x) = ax^2 + bx + c$$ where a, b, and c are real numbers and $a \neq 0$.

Quadratic functions are often written in another form, called **standard form,**

$$f(x) = a(x - h)^2 + k$$

where a, h, and k are real numbers and $a \neq 0$. This form is useful because a, h, and k give us important information about the graph of the function. To develop a strategy for graphing quadratic functions written in standard form, we will begin by considering the simplest case, $f(x) = ax^2$.

> **Notation**
> Since $y = f(x)$, quadratic functions can also be written as $y = ax^2 + bx + c$ and $y = a(x - h)^2 + k$.

1   **Graph Functions of the Form $f(x) = ax^2$.**

One way to graph quadratic functions is to plot points.

EXAMPLE 1   Graph: **a.** $f(x) = x^2$ **b.** $g(x) = 3x^2$ **c.** $s(x) = \frac{1}{3}x^2$

Strategy   We can make a table of values for each function, plot each point, and connect them with a smooth curve.

Why   At this time, this method is our only option.

Solution   After graphing each curve, we see that the graph of $g(x) = 3x^2$ is narrower than the graph of $f(x) = x^2$, and the graph of $s(x) = \frac{1}{3}x^2$ is wider than the graph of $f(x) = x^2$. For $f(x) = ax^2$, the smaller the value of $|a|$, the wider the graph.

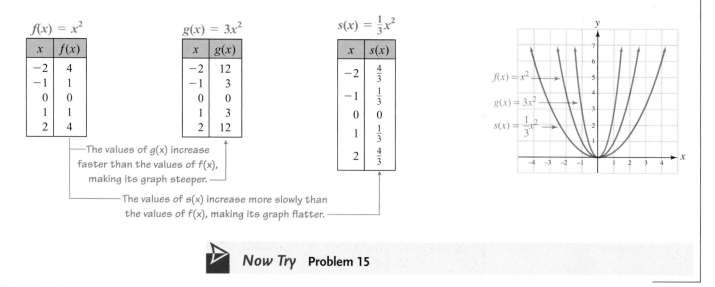

$f(x) = x^2$

x	$f(x)$
-2	4
-1	1
0	0
1	1
2	4

$g(x) = 3x^2$

x	$g(x)$
-2	12
-1	3
0	0
1	3
2	12

$s(x) = \frac{1}{3}x^2$

x	$s(x)$
-2	$\frac{4}{3}$
-1	$\frac{1}{3}$
0	0
1	$\frac{1}{3}$
2	$\frac{4}{3}$

The values of $g(x)$ increase faster than the values of $f(x)$, making its graph steeper.

The values of $s(x)$ increase more slowly than the values of $f(x)$, making its graph flatter.

Now Try   **Problem 15**

EXAMPLE 2 Graph: $f(x) = -3x^2$

Strategy We make a table of values for the function, plot each point, and connect them with a smooth curve.

Why At this time, this method is our only option.

Solution After graphing the curve, we see that it opens downward and has the same shape as the graph of $g(x) = 3x^2$ that was graphed in Example 1.

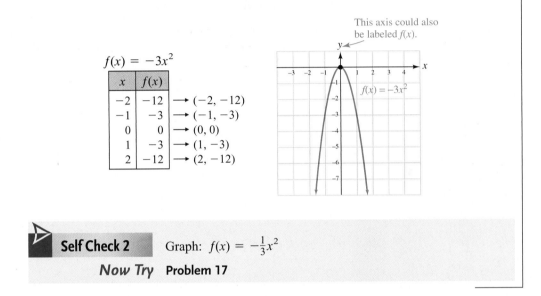

$f(x) = -3x^2$

x	$f(x)$	
-2	-12	$\rightarrow (-2, -12)$
-1	-3	$\rightarrow (-1, -3)$
0	0	$\rightarrow (0, 0)$
1	-3	$\rightarrow (1, -3)$
2	-12	$\rightarrow (2, -12)$

Self Check 2 Graph: $f(x) = -\frac{1}{3}x^2$

Now Try **Problem 17**

The graphs of functions of the form $f(x) = ax^2$ are **parabolas.** The lowest point on a parabola that opens upward, or the highest point on a parabola that opens downward, is called the **vertex** of the parabola. The vertical line, called an **axis of symmetry,** that passes through the vertex divides the parabola into two congruent halves. If we fold the paper along the axis of symmetry, the two sides of the parabola will match.

The Language of Algebra
An axis of symmetry divides a parabola into two matching sides. The sides are said to be *mirror images* of each other.

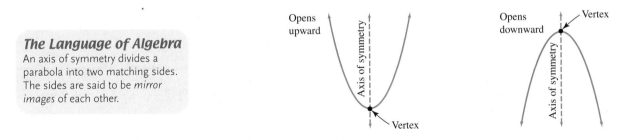

The results from Examples 1 and 2 confirm the following facts.

The Graph of
$f(x) = ax^2$

The graph of $f(x) = ax^2$ is a parabola opening upward when $a > 0$ and downward when $a < 0$, with vertex at the point $(0, 0)$ and axis of symmetry the line $x = 0$.

2   **Graph Functions of the Form $f(x) = ax^2 + k$.**

EXAMPLE 3   Graph: **a.** $f(x) = 2x^2$ **b.** $g(x) = 2x^2 + 3$
c. $s(x) = 2x^2 - 3$

Strategy   We make a table of values for each function, plot each point, and connect them with a smooth curve.

Why   At this time, this method is our only option.

Solution   After graphing the curves, we see that the graph of $g(x) = 2x^2 + 3$ is identical to the graph of $f(x) = 2x^2$, except that it has been translated 3 units upward. The graph of $s(x) = 2x^2 - 3$ is identical to the graph of $f(x) = 2x^2$, except that it has been translated 3 units downward. In each case, the axis of symmetry is the line $x = 0$.

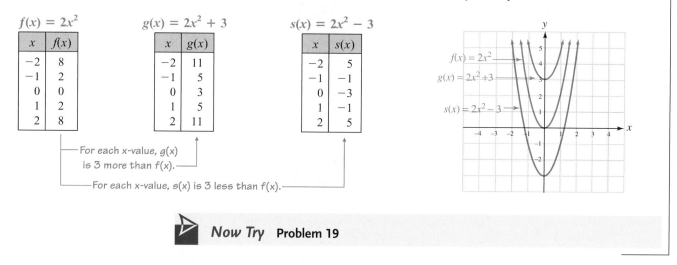

$f(x) = 2x^2$

x	$f(x)$
-2	8
-1	2
0	0
1	2
2	8

$g(x) = 2x^2 + 3$

x	$g(x)$
-2	11
-1	5
0	3
1	5
2	11

$s(x) = 2x^2 - 3$

x	$s(x)$
-2	5
-1	-1
0	-3
1	-1
2	5

For each x-value, g(x) is 3 more than f(x).

For each x-value, s(x) is 3 less than f(x).

Now Try   **Problem 19**

The results of Example 3 confirm the following facts.

The Graph of $f(x) = ax^2 + k$	The graph of $f(x) = ax^2 + k$ is a parabola having the same shape as $f(x) = ax^2$ but translated k units upward if k is positive and $|k|$ units downward if k is negative. The vertex is at the point $(0, k)$, and the axis of symmetry is the line $x = 0$.

3   **Graph Functions of the Form $f(x) = a(x - h)^2$.**

EXAMPLE 4   Graph: **a.** $f(x) = 2x^2$ **b.** $g(x) = 2(x - 3)^2$
c. $s(x) = 2(x + 3)^2$

Strategy   We make a table of values for each function, plot each point, and connect them with a smooth curve.

Why   At this time, this method is our only option.

Solution   We note that the graph of $g(x) = 2(x - 3)^2$ on the next page is identical to the graph of $f(x) = 2x^2$, except that it has been translated 3 units to the right. The graph of $s(x) = 2(x + 3)^2$ is identical to the graph of $f(x) = 2x^2$, except that it has been translated 3 units to the left.

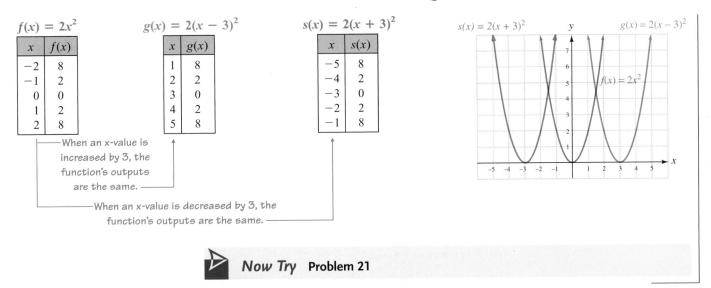

$f(x) = 2x^2$

x	$f(x)$
-2	8
-1	2
0	0
1	2
2	8

$g(x) = 2(x - 3)^2$

x	$g(x)$
1	8
2	2
3	0
4	2
5	8

$s(x) = 2(x + 3)^2$

x	$s(x)$
-5	8
-4	2
-3	0
-2	2
-1	8

When an x-value is increased by 3, the function's outputs are the same.

When an x-value is decreased by 3, the function's outputs are the same.

 Now Try   **Problem 21**

The results of Example 4 confirm the following facts.

The Graph of $f(x) = a(x - h)^2$

The graph of $f(x) = a(x - h)^2$ is a parabola having the same shape as $f(x) = ax^2$ but translated h units to the right if h is positive and $|h|$ units to the left if h is negative. The vertex is at the point $(h, 0)$, and the axis of symmetry is the line $x = h$.

4 **Graph Functions of the Form $f(x) = a(x - h)^2 + k$.**

The results of Examples 1–4 suggest a general strategy for graphing quadratic functions that are written in the form $f(x) = a(x - h)^2 + k$.

Graphing a Quadratic Function in Standard Form

The graph of the quadratic function

$$f(x) = a(x - h)^2 + k \quad \text{where } a \neq 0$$

is a parabola with vertex at (h, k). The axis of symmetry is the line $x = h$. The parabola opens upward when $a > 0$ and downward when $a < 0$.

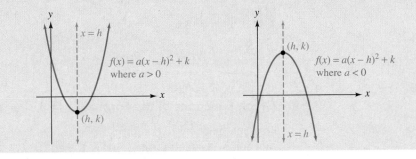

EXAMPLE 5   Graph: $f(x) = 2(x - 3)^2 - 4$. Label the vertex and draw the axis of symmetry.

Strategy   We will determine whether the graph opens upward or downward and find its vertex and axis of symmetry. Then we will plot some points and complete the graph.

Why   This method will be more efficient than plotting many points.

Solution   The graph of $f(x) = 2(x - 3)^2 - 4$ is identical to the graph of $g(x) = 2(x - 3)^2$, except that it has been translated 4 units downward. The graph of $g(x) = 2(x - 3)^2$ is identical to the graph of $s(x) = 2x^2$, except that it has been translated 3 units to the right.

We can learn more about the graph of $f(x) = 2(x - 3)^2 - 4$ by determining a, h, and k.

$$\left. \begin{array}{c} f(x) = 2(x - 3)^2 - 4 \\ \uparrow \qquad \uparrow \qquad \uparrow \\ f(x) = a(x - h)^2 + k \end{array} \right\} a = 2, h = 3, \text{ and } k = -4$$

Upward/downward: Since $a = 2$ and $2 > 0$, the parabola opens upward.

Vertex: The vertex of the parabola is $(h, k) = (3, -4)$, as shown below.

Axis of symmetry: Since $h = 3$, the axis of symmetry is the line $x = 3$, as shown below.

Plotting points: We can construct a table of values to determine several points on the parabola. Since the x-coordinate of the vertex is 3, we choose the x-values of 4 and 5, find $f(4)$ and $f(5)$, and record the results in a table. Then we plot $(4, -2)$ and $(5, 4)$, and use symmetry to locate two other points on the parabola: $(2, -2)$ and $(1, 4)$. Finally, we draw a smooth curve through the points to get the graph.

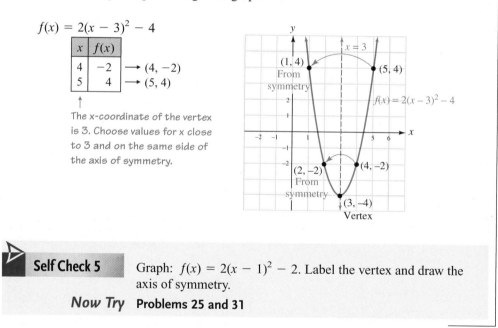

$f(x) = 2(x - 3)^2 - 4$

x	$f(x)$
4	-2	$\longrightarrow (4, -2)$
5	4	$\longrightarrow (5, 4)$

The x-coordinate of the vertex is 3. Choose values for x close to 3 and on the same side of the axis of symmetry.

 Self Check 5   Graph: $f(x) = 2(x - 1)^2 - 2$. Label the vertex and draw the axis of symmetry.

Now Try   Problems 25 and 31

5   **Graph Functions of the Form $f(x) = ax^2 + bx + c$ by Completing the Square.**

To graph functions of the form $f(x) = ax^2 + bx + c$, we can complete the square to write the function in standard form $f(x) = a(x - h)^2 + k$.

EXAMPLE 6   Determine the vertex and the axis of symmetry of the graph of $f(x) = x^2 + 8x + 21$. Will the graph open upward or downward?

Strategy   To find the vertex and the axis of symmetry, we will complete the square on x and write the equation of the function in standard form.

Why Once the equation is written in standard form, we can determine the values of a, h, and k. The coordinates of the vertex will be (h, k) and the equation of the axis of symmetry will by $x = h$. The graph will open upward if $a > 0$ or downward if $a < 0$.

Solution To determine the vertex and the axis of symmetry of the graph, we complete the square on the right side so we can write the function in $f(x) = a(x - h)^2 + k$ form.

$$f(x) = x^2 + 8x + 21$$
$$f(x) = (x^2 + 8x \quad) + 21 \qquad \text{Prepare to complete the square on x by writing parentheses around } x^2 + 8x.$$

To complete the square on $x^2 + 8x$, we note that one-half of the coefficient of x is $\frac{1}{2} \cdot 8 = 4$, and $4^2 = 16$. If we add 16 to $x^2 + 8x$, we obtain a perfect-square trinomial within the parentheses. Since this step adds 16 to the right side, we must also subtract 16 from the right side so that it remains in an equivalent form.

Add 16 to the right side.   Subtract 16 from the right side to counteract the addition of 16.

$$f(x) = (x^2 + 8x + 16) + 21 - 16$$
$$f(x) = (x + 4)^2 + 5 \qquad \text{Factor } x^2 + 8x + 16 \text{ and combine like terms.}$$

The function is now written in standard form, and we can determine a, h, and k.

The standard form requires a minus symbol here.

$$f(x) = \left[(x - (-4))\right]^2 + 5 \qquad \text{Write } x + 4 \text{ as } x - (-4) \text{ to determine h.}$$
$$a = 1 \qquad h = -4 \qquad k = 5$$

The vertex is $(h, k) = (-4, 5)$ and the axis of symmetry is the line $x = -4$. Since $a = 1$ and $1 > 0$, the parabola opens upward.

Self Check 6   Determine the vertex and the axis of symmetry of the graph of $f(x) = x^2 + 4x + 10$. Will the graph open upward or downward?

Now Try **Problem 39**

EXAMPLE 7   Graph: $f(x) = 2x^2 - 4x - 1$

Strategy We will complete the square on x and write the equation of the function in standard form, $f(x) = a(x - h)^2 + k$.

Why When the equation is in standard form, we can identify the values of a, h, and k from the equation. This information will help us sketch the graph.

Solution Recall that to complete the square on $2x^2 - 4x$, the coefficient of x^2 must be equal to 1. Therefore, we factor 2 from $2x^2 - 4x$.

$$f(x) = 2x^2 - 4x - 1$$
$$f(x) = 2(x^2 - 2x \quad) - 1$$

To complete the square on $x^2 - 2x$, we note that one-half of the coefficient of x is $\frac{1}{2}(-2) = -1$, and $(-1)^2 = 1$. If we add 1 to $x^2 - 2x$, we obtain a perfect-square trinomial within the parentheses. Since this step adds 2 to the right side, we must also subtract 2 from the right side so that it remains in an equivalent form.

By the distributive property, when 1 is added to the expression within the parentheses, $2 \cdot 1 = 2$ is added to the right side.

Subtract 2 to counteract the addition of 2.

$$f(x) = 2(x^2 - 2x + 1) - 1 - 2$$

$$f(x) = 2(x - 1)^2 - 3 \qquad \text{Factor } x^2 - 2x + 1 \text{ and combine like terms.}$$

We see that $a = 2$, $h = 1$, and $k = -3$. Thus, the vertex is at the point $(1, -3)$, and the axis of symmetry is $x = 1$. Since $a = 2$ and $2 > 0$, the parabola opens upward. We plot the vertex and axis of symmetry as shown below.

Finally, we construct a table of values, plot the points, use symmetry to plot the corresponding points, and then draw the graph.

$$f(x) = 2x^2 - 4x - 1$$

or

$$f(x) = 2(x - 1)^2 - 3$$

x	$f(x)$	
2	-1	$\rightarrow (2, -1)$
3	5	$\rightarrow (3, 5)$

↑ The x-coordinate of the vertex is 1. Choose values for x close to 1 and on the same side of the axis of symmetry.

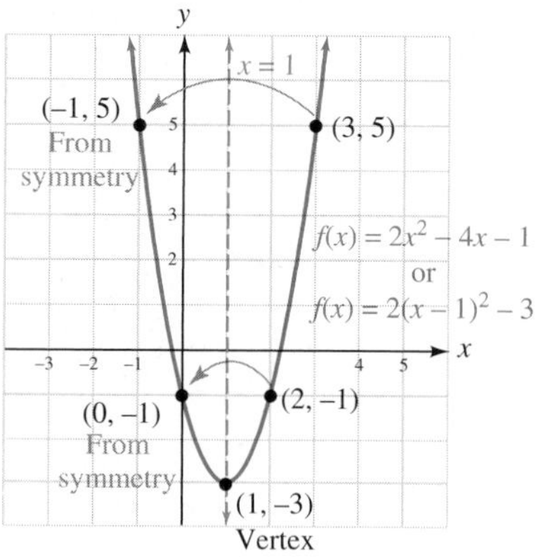

Success Tip
To find additional points on the graph, select values of x that are close to the x-coordinate of the vertex.

Self Check 7 Graph: $f(x) = 3x^2 - 12x + 8$.

Now Try Problem 45

6 Find the Vertex Using $-\frac{b}{2a}$.

Because of symmetry, if a parabola has two x-intercepts, the x-coordinate of the vertex is exactly midway between them. We can use this fact to derive a formula to find the vertex of a parabola.

In general, if a parabola has two x-intercepts, they can be found by solving $0 = ax^2 + bx + c$ for x. We can use the quadratic formula to find the solutions. They are

$$x = \frac{-b - \sqrt{b^2 - 4ac}}{2a} \quad \text{and} \quad x = \frac{-b + \sqrt{b^2 - 4ac}}{2a}$$

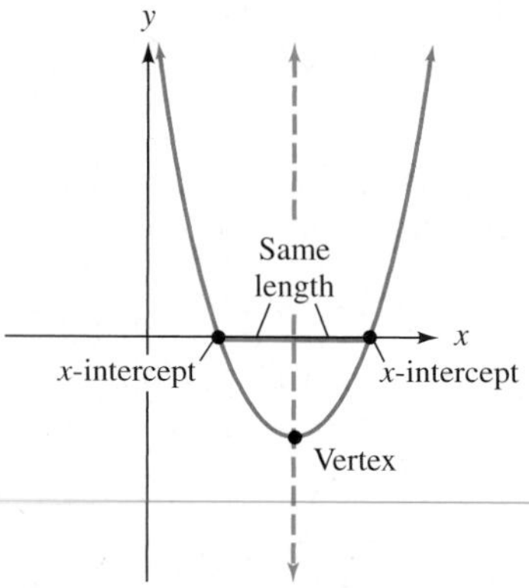

Thus, the parabola's x-intercepts are $\left(\frac{-b - \sqrt{b^2 - 4ac}}{2a}, 0\right)$ and $\left(\frac{-b + \sqrt{b^2 - 4ac}}{2a}, 0\right)$.

Since the x-value of the vertex of a parabola is halfway between the two x-intercepts, we can find this value by finding the average, or $\frac{1}{2}$ of the sum of the x-coordinates of the x-intercepts.

$$x = \frac{1}{2}\left(\frac{-b - \sqrt{b^2 - 4ac}}{2a} + \frac{-b + \sqrt{b^2 - 4ac}}{2a}\right)$$

$$x = \frac{1}{2}\left(\frac{-b - \sqrt{b^2 - 4ac} + (-b) + \sqrt{b^2 - 4ac}}{2a}\right)$$ Add the numerators and keep the common denominator.

$$x = \frac{1}{2}\left(\frac{-2b}{2a}\right)$$ Combine like terms: $-b + (-b) = -2b$ and $-\sqrt{b^2 - 4ac} + \sqrt{b^2 - 4ac} = 0$.

$$x = -\frac{b}{2a}$$ Remove the common factor of 2 in the numerator and denominator and simplify.

This result is true even if the graph has no *x*-intercepts.

Formula for the Vertex of a Parabola

The vertex of the graph of the quadratic function $f(x) = ax^2 + bx + c$ is

$$\left(-\frac{b}{2a},\, f\left(-\frac{b}{2a}\right)\right)$$

and the axis of symmetry of the parabola is the line $x = -\frac{b}{2a}$.

EXAMPLE 8   Find the vertex of the graph of $f(x) = 2x^2 - 4x - 1$.

Strategy   We will determine the values of *a* and *b* and substitute into the formula for the vertex of a parabola.

Why   It is easier to find the coordinates of the vertex using the formula than it is to complete the square on $2x^2 - 4x - 1$.

Solution   The function is written in $f(x) = ax^2 + bx + c$ form, where $a = 2$ and $b = -4$. To find the vertex of its graph, we compute

Success Tip
We can find the vertex of the graph of a quadratic function by completing the square or by using the formula.

$$-\frac{b}{2a} = -\frac{-4}{2(2)}$$

$$= -\frac{-4}{4}$$

$$= 1 \qquad \text{This is the x-coordinate of the vertex.}$$

$$f\left(-\frac{b}{2a}\right) = f(1)$$

$$= 2(1)^2 - 4(1) - 1$$

$$= -3 \qquad \text{This is the y-coordinate of the vertex.}$$

The vertex is the point $(1, -3)$. This agrees with the result we obtained in Example 7 by completing the square.

Self Check 8   Find the vertex of the graph of $f(x) = 3x^2 - 12x + 8$.

Now Try   **Problem 55**

Using Your Calculator   *Finding the Vertex*

We can use a graphing calculator to graph the function $f(x) = 2x^2 + 6x - 3$ and find the coordinates of the vertex and the axis of symmetry of the parabola. If we enter the function, we will obtain the graph shown in figure (a).

We then trace to move the cursor to the lowest point on the graph, as shown in figure (b). By zooming in, we can see that the vertex is the point $(-1.5, -7.5)$, or $\left(-\frac{3}{2}, -\frac{15}{2}\right)$, and that the line $x = -\frac{3}{2}$ is the axis of symmetry.

Some calculators have an fmin or fmax feature that can also be used to find the vertex.

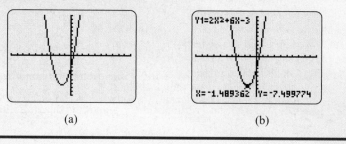

(a)   (b)

We can determine much about the graph of $f(x) = ax^2 + bx + c$ from the coefficients a, b, and c. This information is summarized as follows:

Graphing a Quadratic Function
$f(x) = ax^2 + bx + c$

- Determine whether the parabola opens upward or downward by finding the value of a.
- The x-coordinate of the vertex of the parabola is $x = -\frac{b}{2a}$.
- To find the y-coordinate of the vertex, substitute $-\frac{b}{2a}$ for x and find $f\left(-\frac{b}{2a}\right)$.
- The axis of symmetry is the vertical line passing through the vertex.
- The y-intercept is determined by the value of $f(x)$ when $x = 0$: the y-intercept is $(0, c)$.
- The x-intercepts (if any) are determined by the values of x that make $f(x) = 0$. To find them, solve the quadratic equation $ax^2 + bx + c = 0$.

EXAMPLE 9   Graph: $f(x) = -2x^2 - 8x - 8$

Strategy   We will follow the steps for graphing a quadratic function.

Why   This is the most efficient way to graph a general quadratic function.

Solution

Step 1: *Determine whether the parabola opens upward or downward.* The function is in the form $f(x) = ax^2 + bx + c$, with $a = -2$, $b = -8$, and $c = -8$. Since $a < 0$, the parabola opens downward.

Step 2: *Find the vertex and draw the axis of symmetry.* To find the coordinates of the vertex, we compute

$$x = -\frac{b}{2a}$$

$$x = -\frac{-8}{2(-2)} \quad \substack{\text{Substitute } -2 \text{ for } a \\ \text{and } -8 \text{ for } b.}$$

$$= -2 \quad \substack{\text{This is the } x\text{-coordinate} \\ \text{of the vertex.}}$$

$$f\left(-\frac{b}{2a}\right) = f(-2)$$

$$= -2(-2)^2 - 8(-2) - 8$$

$$= -8 + 16 - 8$$

$$= 0 \quad \substack{\text{This is the } y\text{-coordinate} \\ \text{of the vertex.}}$$

Success Tip

An easy way to remember the vertex formula is to note that $x = \frac{-b}{2a}$ is part of the quadratic formula:

$$x = \frac{-b \pm \sqrt{b^2 - 4ac}}{2a}$$

The vertex of the parabola is the point $(-2, 0)$. This point is in blue on the graph. The axis of symmetry is the line $x = -2$.

Step 3: *Find the x- and y-intercepts.* Since $c = -8$, the y-intercept of the parabola is $(0, -8)$. The point $(-4, -8)$, two units to the left of the axis of symmetry, must also be on the graph. We plot both points in black on the graph.

To find the x-intercepts, we set $f(x)$ equal to 0 and solve the resulting quadratic equation.

$$f(x) = -2x^2 - 8x - 8$$

$0 = -2x^2 - 8x - 8$ Set f(x) = 0.

$0 = x^2 + 4x + 4$ Divide both sides by −2.

$0 = (x + 2)(x + 2)$ Factor the trinomial.

$x + 2 = 0$ or $x + 2 = 0$ Set each factor equal to 0.

$x = -2$ $\mid$ $x = -2$

Since the solutions are the same, the graph has only one x-intercept: $(-2, 0)$. This point is the vertex of the parabola and has already been plotted.

Step 4: *Plot another point.* Finally, we find another point on the parabola. If $x = -3$, then $f(-3) = -2$. We plot $(-3, -2)$ and use symmetry to determine that $(-1, -2)$ is also on the graph. Both points are in green.

Step 5: Draw a smooth curve through the points, as shown.

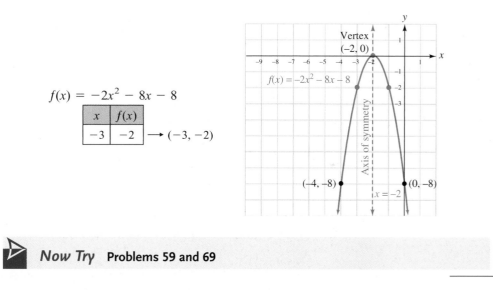

$$f(x) = -2x^2 - 8x - 8$$

x	$f(x)$
-3	-2	$\longrightarrow (-3, -2)$

 Now Try Problems 59 and 69

7 **Determine Minimum and Maximum Values.**

It is often useful to know the smallest or largest possible value a quantity can assume. For example, companies try to minimize their costs and maximize their profits. If the quantity is expressed by a quadratic function, the y-coordinate of the vertex of the graph of the function gives its minimum or maximum value.

EXAMPLE 10 *Minimizing Costs.* A glassworks that makes lead crystal vases has daily production costs given by the function $C(x) = 0.2x^2 - 10x + 650$, where x is the number of vases made each day. How many vases should be produced to minimize the per-day costs? What will the costs be?

Strategy We will find the vertex of the graph of the quadratic function.

Why The x-coordinate of the vertex indicates the number of vases to make to keep costs at a minimum, and the y-coordinate indicates the minimum cost.

Solution   The graph of $C(x) = 0.2x^2 - 10x + 650$ is a parabola opening upward. The vertex is the lowest point on the graph. To find the vertex, we compute

$$-\frac{b}{2a} = -\frac{-10}{2(0.2)} \qquad b = -10 \text{ and } a = 0.2.$$

$$= -\frac{-10}{0.4}$$

$$= 25$$

$$f\left(-\frac{b}{2a}\right) = f(25)$$

$$= 0.2(25)^2 - 10(25) + 650$$

$$= 525$$

The vertex is (25, 525), and it indicates that the costs are a minimum of $525 when 25 vases are made daily.

To solve this problem with a graphing calculator, we graph the function $C(x) = 0.2x^2 - 10x + 650$. By using TRACE and ZOOM, we can locate the vertex of the graph. The coordinates of the vertex indicate that the minimum cost is $525 when the number of vases produced is 25.

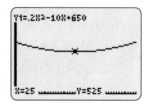

 Now Try   **Problem 85**

8   **Solve Quadratic Equations Graphically.**

When solving quadratic equations graphically, we must consider three possibilities. If the graph of the associated quadratic function has two x-intercepts, the quadratic equation has two real-number solutions. Figure (a) shows an example of this. If the graph has one x-intercept, as shown in figure (b), the equation has one repeated real-number solution. Finally, if the graph does not have an x-intercept, as shown in figure (c), the equation does not have any real-number solutions.

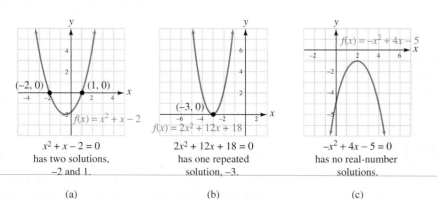

(a)	(b)	(c)
$x^2 + x - 2 = 0$ has two solutions, -2 and 1.	$2x^2 + 12x + 18 = 0$ has one repeated solution, -3.	$-x^2 + 4x - 5 = 0$ has no real-number solutions.

Using Your Calculator | *Solving Quadratic Equations Graphically*

We can use a graphing calculator to find approximate solutions of quadratic equations. For example, the solutions of $0.7x^2 + 2x - 3.5 = 0$ are the numbers x that will make $y = 0$ in the quadratic function $f(x) = 0.7x^2 + 2x - 3.5$. To approximate these numbers, we graph the quadratic function and read the x-intercepts from the graph using the ZERO feature. (The ZERO feature can be found by pressing $\boxed{2^{nd}}$, CALC, and then 2.) In the figure, we see that the x-coordinate of the left-most x-intercept of the graph is given as -4.082025. This means that an approximate solution of the equation is -4.08. To find the positive x-intercept, we use similar steps.

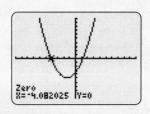

▷ **ANSWERS TO SELF CHECKS**

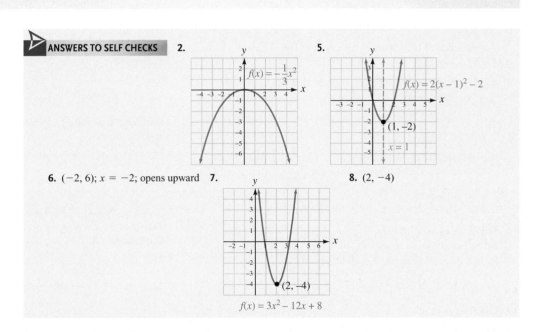

2.

5.

$f(x) = -\frac{1}{3}x^2$

$f(x) = 2(x - 1)^2 - 2$

$(1, -2)$

$x = 1$

6. $(-2, 6)$; $x = -2$; opens upward   7.

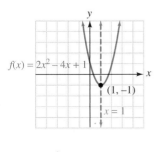

8. $(2, -4)$

$(2, -4)$

$f(x) = 3x^2 - 12x + 8$

STUDY SET
8.4

VOCABULARY

Refer to the graph. Fill in the blanks.

1. $f(x) = 2x^2 - 4x + 1$ is called a _____ function. Its graph is a cup-shaped figure called a _____.

2. The lowest point on the graph is $(1, -1)$. This is called the _____ of the parabola.

3. The vertical line $x = 1$ divides the parabola into two halves. This line is called the _____ ___ _____.

$f(x) = 2x^2 - 4x + 1$

$(1, -1)$

$x = 1$

4. $f(x) = a(x - h)^2 + k$ is called the _____ form of the equation of a quadratic function.

CONCEPTS

5. Refer to the graph.
   a. What are the x-intercepts of the graph?
   b. What is the y-intercept of the graph?
   c. What is the vertex?
   d. What is the axis of symmetry?
   e. What is the domain and the range of the function?

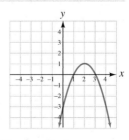

6. The vertex of a parabola is at $(1, -3)$, its y-intercept is $(0, -2)$, and it passes through the point $(3, 1)$, as shown in the illustration. Use the axis of symmetry shown in blue to help determine two other points on the parabola.

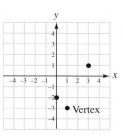

7. Draw the graph of a quadratic function using the given facts about its graph.

- Opens upward
- y-intercept: $(0, -3)$
- Vertex: $(-1, -4)$
- x-intercepts: $(-3, 0)$, $(1, 0)$

x	$f(x)$
2	5

8. For $f(x) = -x^2 + 6x - 7$, the value of $-\frac{b}{2a}$ is 3. Find the y-coordinate of the vertex of the graph of this function.

9. Fill in the blanks.

a. To complete the square on the right side of $f(x) = 2x^2 + 12x + 11$, what should be factored from the first two terms?

$$f(x) = \boxed{}(x^2 + 6x) + 11$$

b. To complete the square on $x^2 + 6x$ shown below, what should be added within the parentheses and what should be subtracted outside the parentheses?

$$f(x) = 2(x^2 + 6x + \boxed{}) + 11 - \boxed{}$$

10. Fill in the blanks. To complete the square on $x^2 + 4x$ shown below, what should be added within the parentheses and what should be added outside the parentheses?

$$f(x) = -5(x^2 + 4x + \boxed{}) + 7 + \boxed{}$$

11. Use the graph of $f(x) = \frac{1}{10}x^2 - \frac{1}{5}x - \frac{3}{2}$, shown below, to estimate the solutions of the equation $\frac{1}{10}x^2 - \frac{1}{5}x - \frac{3}{2} = 0$.

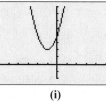

12. Three quadratic equations are to be solved graphically. The graphs of their associated quadratic functions are shown here. Determine which graph indicates that the equation has

a. two real solutions.

b. one repeated real solution.

c. no real solutions.

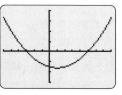

(i)

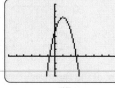

(ii)

(iii)

13. The function $f(x) = 2(x + 1)^2 + 6$ is written in the form $f(x) = a(x - h)^2 + k$. Is $h = -1$ or is $h = 1$? Explain.

14. Consider the function $f(x) = 2x^2 + 4x - 8$.

a. What are a, b, and c?

b. Find $-\frac{b}{2a}$.

GUIDED PRACTICE

Graph each group of functions on the same coordinate system. See Example 1.

15. $f(x) = x^2$, $g(x) = 2x^2$, $s(x) = \frac{1}{2}x^2$

16. $f(x) = x^2$, $g(x) = 4x^2$, $s(x) = \frac{1}{4}x^2$

Graph each pair of functions on the same coordinate system. See Example 2.

17. $f(x) = 2x^2$, $g(x) = -2x^2$ **18.** $f(x) = \frac{1}{2}x^2$, $g(x) = -\frac{1}{2}x^2$

Graph each group of functions on the same coordinate system. See Example 3.

19. $f(x) = 4x^2$, $g(x) = 4x^2 + 3$, $s(x) = 4x^2 - 2$

20. $f(x) = \frac{1}{3}x^2$, $g(x) = \frac{1}{3}x^2 + 4$, $s(x) = \frac{1}{3}x^2 - 3$

Graph each group of functions on the same coordinate system and describe how the graphs are similar and how they are different. See Example 4.

21. $f(x) = 3x^2$, $g(x) = 3(x + 2)^2$, $s(x) = 3(x - 3)^2$

22. $f(x) = \frac{1}{2}x^2$, $g(x) = \frac{1}{2}(x + 3)^2$, $s(x) = \frac{1}{2}(x - 2)^2$

Find the vertex and the axis of symmetry of the graph of each function. Do not graph the equation, but determine whether the graph will open upward or downward. See Example 5.

23. $f(x) = (x - 1)^2 + 2$

24. $f(x) = 2(x - 2)^2 - 1$

25. $f(x) = -2(x + 3)^2 - 4$

26. $f(x) = -3(x + 1)^2 + 3$

27. $f(x) = -0.5(x - 7.5)^2 + 8.5$

28. $f(x) = -\frac{3}{2}\left(x + \frac{1}{4}\right)^2 + \frac{7}{8}$

29. $f(x) = 2x^2 - 4$

30. $f(x) = 3x^2 - 3$

Determine the vertex and the axis of symmetry of the graph of each function. Then plot several points and complete the graph. See Example 5.

31. $f(x) = (x - 3)^2 + 2$ **32.** $f(x) = (x + 1)^2 - 2$

33. $f(x) = -(x - 2)^2$ **34.** $f(x) = -(x + 2)^2$

35. $f(x) = -2(x + 3)^2 + 4$ **36.** $f(x) = -2(x - 2)^2 - 4$

37. $f(x) = \dfrac{1}{2}(x + 1)^2 - 3$ **38.** $f(x) = \dfrac{1}{3}(x - 1)^2 + 2$

Determine the vertex and the axis of symmetry of the graph of each function. Will the graph open upward or downward? **See Example 6.**

39. $f(x) = x^2 + 4x + 5$

40. $f(x) = x^2 - 4x - 1$

41. $f(x) = -x^2 + 6x - 15$

42. $f(x) = -x^2 - 6x + 3$

Complete the square to write each function in $f(x) = a(x - h)^2 + k$ form. Determine the vertex and the axis of symmetry of the graph of the function. Then plot several points and complete the graph. **See Examples 6 and 7.**

43. $f(x) = x^2 + 2x - 3$

44. $f(x) = x^2 + 6x + 5$

45. $f(x) = 4x^2 + 24x + 37$

46. $f(x) = 3x^2 - 12x + 10$

47. $f(x) = x^2 + x - 6$

48. $f(x) = x^2 - x - 6$

49. $f(x) = -4x^2 + 16x - 10$

50. $f(x) = -2x^2 + 4x + 3$

51. $f(x) = 2x^2 + 8x + 6$

52. $f(x) = 3x^2 - 12x + 9$

53. $f(x) = -x^2 - 8x - 17$

54. $f(x) = -x^2 + 6x - 8$

Use the vertex formula to find the vertex of the graph of each function. **See Example 8.**

55. $f(x) = x^2 + 2x - 5$ **56.** $f(x) = -x^2 + 4x - 5$

57. $f(x) = 2x^2 - 3x + 4$ **58.** $f(x) = 2x^2 - 7x - 4$

Find the x- and y-intercepts of the graph of the quadratic function. **See Example 9.**

59. $f(x) = x^2 - 2x - 35$ **60.** $f(x) = -x^2 - 10x - 21$

61. $f(x) = -2x^2 + 4x$ **62.** $f(x) = 3x^2 + 6x - 9$

Determine the coordinates of the vertex of the graph of each function using the vertex formula. Then determine the x- and y-intercepts of the graph. Finally, plot several points and complete the graph. **See Example 9.**

63. $f(x) = x^2 + 4x + 4$

64. $f(x) = x^2 - 6x + 9$

65. $f(x) = -x^2 + 2x - 1$

66. $f(x) = -x^2 - 2x - 1$

67. $f(x) = x^2 - 2x$

68. $f(x) = x^2 + x$

69. $f(x) = 2x^2 - 8x + 6$

70. $f(x) = 3x^2 - 12x + 12$

71. $f(x) = -6x^2 - 12x - 8$

72. $f(x) = -2x^2 + 8x - 10$

73. $f(x) = 4x^2 - 12x + 9$

74. $f(x) = 4x^2 + 4x - 3$

Use a graphing calculator to find the coordinates of the vertex of the graph of each quadratic function. Round to the nearest hundredth. **See Using Your Calculator: Finding the Vertex.**

75. $f(x) = 2x^2 - x + 1$ **76.** $f(x) = x^2 + 5x - 6$

77. $f(x) = -x^2 + x + 7$ **78.** $f(x) = 2x^2 - 3x + 2$

Use a graphing calculator to solve each equation. If an answer is not exact, round to the nearest hundredth. **See Using Your Calculator: Solving Quadratic Equations Graphically.**

79. $x^2 + x - 6 = 0$ **80.** $2x^2 - 5x - 3 = 0$

81. $0.5x^2 - 0.7x - 3 = 0$ **82.** $2x^2 - 0.5x - 2 = 0$

APPLICATIONS

83. CROSSWORD PUZZLES
Darken the appropriate squares to the right of the dashed red line so that the puzzle has symmetry with respect to that line.

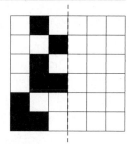

axis of symmetry

84. GRAPHIC ARTS   Draw an axis of symmetry over the letter shown here.

M

85. OPERATING COSTS The cost C in dollars of operating a certain concrete-cutting machine is related to the number of minutes n the machine is run by the function

$$C(n) = 2.2n^2 - 66n + 655$$

For what number of minutes is the cost of running the machine a minimum? What is the minimum cost?

86. WATER USAGE The height (in feet) of the water level in a reservoir over a 1-year period is modeled by the function

$$H(t) = 3.3(t - 9)^2 + 14$$

where $t = 1$ represents January, $t = 2$ represents February, and so on. How low did the water level get that year, and when did it reach the low mark?

87. FIREWORKS A fireworks shell is shot straight up with an initial velocity of 120 feet per second. Its height s after t seconds is given by the equation $s = 120t - 16t^2$. If the shell is designed to explode when it reaches its maximum height, how long after being fired, and at what height, will the fireworks appear in the sky?

88. BALLISTICS From the top of the building in the illustration, a ball is thrown straight up with an initial velocity of 32 feet per second. The equation

$$s = -16t^2 + 32t + 48$$

gives the height s of the ball t seconds after it is thrown. Find the maximum height reached by the ball and the time it takes for the ball to hit the ground.

89. *from Campus to Careers*
Police Patrol Officer

Suppose you are a police patrol officer and you have a 300-foot-long roll of yellow "DO NOT CROSS" barricade tape to seal off an automobile accident, as shown in the illustration. What dimensions should you use to seal off the maximum rectangular area around the collision? What is the maximum area?

90. RANCHING See the illustration. A farmer wants to fence in three sides of a rectangular field with 1,000 feet of fencing. The other side of the rectangle will be a river. If the enclosed area is to be maximum, find the dimensions of the field.

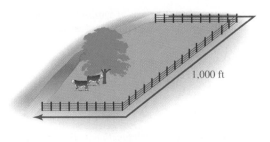

91. MILITARY HISTORY The function

$$N(x) = -0.0534x^2 + 0.337x + 0.97$$

gives the number of active-duty military personnel in the United States Army (in millions) for the years 1965–1972, where $x = 0$ corresponds to 1965, $x = 1$ corresponds to 1966, and so on. For this period, when was the army's personnel strength level at its highest, and what was it? Historically, can you explain why?

92. SCHOOL ENROLLMENT After peaking in 1970, school enrollment in the United States fell during the 1970s and 1980s. The total annual enrollment (in millions) in U.S. elementary and secondary schools for the years 1975–1996 is given by the model

$$E(x) = 0.058x^2 - 1.162x + 50.604$$

where $x = 0$ corresponds to 1975, $x = 1$ corresponds to 1976, and so on. For this period, when was enrollment the lowest? What was the enrollment?

93. MAXIMIZING REVENUE The revenue R received for selling x stereos is given by the formula

$$R = -\frac{x^2}{5} + 80x - 1,000$$

How many stereos must be sold to obtain the maximum revenue? Find the maximum revenue.

94. MAXIMIZING REVENUE When priced at $30 each, a toy has annual sales of 4,000 units. The manufacturer estimates that each $1 increase in price will decrease sales by 100 units. Find the unit price that will maximize total revenue. (*Hint:* Total revenue = price · the number of units sold.)

WRITING

95. Use the example of a stream of water from a drinking fountain to explain the concepts of the vertex and the axis of symmetry of a parabola. Draw a picture.

96. What are some quantities that are good to maximize? What are some quantities that are good to minimize?

97. A mirror is held against the y-axis of the graph of a quadratic function. What fact about parabolas does this illustrate?

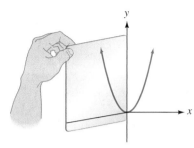

98. The vertex of a quadratic function $f(x) = ax^2 + bx + c$ is given by the formula $\left(-\frac{b}{2a},\ f\left(-\frac{b}{2a}\right)\right)$. Explain what is meant by the notation $f\left(-\frac{b}{2a}\right)$.

99. A table of values for $f(x) = 2x^2 - 4x + 3$ is shown. Explain why it appears that the vertex of the graph of f is the point $(1, 1)$.

100. The illustration shows the graph of the quadratic function $f(x) = -4x^2 + 12x$ with domain $[0, 3]$. Explain how the value of $f(x)$ changes as the value of x increases from 0 to 3.

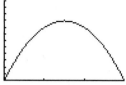

REVIEW

Simplify each expression. Assume all variables represent positive numbers.

101. $\dfrac{\sqrt{3}}{\sqrt{50}}$

102. $\dfrac{3}{\sqrt[3]{9}}$

103. $3\left(\sqrt{5b} - \sqrt{3}\right)^2$

104. $-2\sqrt{5b}\left(4\sqrt{2b} - 3\sqrt{3}\right)$

CHALLENGE PROBLEMS

105. Find a number between 0 and 1 such that the difference of the number and its square is a maximum.

106. Determine a quadratic function whose graph has x-intercepts of $(2, 0)$ and $(-4, 0)$.

SECTION 8.5
Quadratic and Other Nonlinear Inequalities

Objectives

1 Solve quadratic inequalities.

2 Solve rational inequalities.

3 Graph nonlinear inequalities in two variables.

We have previously solved *linear* inequalities in one variable such as $2x + 3 > 8$ and $6x - 7 < 4x - 9$. To find their solution sets, we used properties of inequalities to isolate the variable on one side of the inequality.

In this section, we will solve *quadratic* inequalities in one variable such as $x^2 + x - 6 < 0$ and $x^2 + 4x \geq 5$. We will use an interval testing method on the number line to determine their solution sets.

1 **Solve Quadratic Inequalities.**

Recall that a quadratic equation can be written in the form $ax^2 + bx + c = 0$. If we replace the $=$ symbol with an inequality symbol, we have a quadratic inequality.

Quadratic Inequalities

A **quadratic inequality** can be written in one of the standard forms

$$ax^2 + bx + c < 0 \qquad ax^2 + bx + c > 0 \qquad ax^2 + bx + c \leq 0 \qquad ax^2 + bx + c \geq 0$$

where a, b, and c are real numbers and $a \neq 0$.

To solve a quadratic inequality in one variable, we will use the following steps to find the values of the variable that make the inequality true.

Solving Quadratic Inequalities

1. Write the inequality in standard form and solve its related quadratic equation.
2. Locate the solutions (called **critical numbers**) of the related quadratic equation on a number line.
3. Test each interval on the number line created in step 2 by choosing a test value from the interval and determining whether it satisfies the inequality. The solution set includes the interval(s) whose test value makes the inequality true.
4. Determine whether the endpoints of the intervals are included in the solution set.

EXAMPLE 1   Solve: $x^2 + x - 6 < 0$

Strategy   We will solve the related quadratic equation $x^2 + x - 6 = 0$ by factoring to determine the critical numbers. These critical numbers will separate the number line into intervals.

Why   We can test each interval to see whether numbers in the interval are in the solution set of the inequality.

Solution   The expression $x^2 + x - 6$ can be positive, negative, or 0, depending on what value is substituted for x. Solutions of the inequality are x-values that make $x^2 + x - 6$ less than 0. To find them, we will follow the steps for solving quadratic inequalities.

Step 1:   *Solve the related quadratic equation.* For the quadratic inequality $x^2 + x - 6 < 0$, the related quadratic equation is $x^2 + x - 6 = 0$.

$$x^2 + x - 6 = 0$$
$$(x + 3)(x - 2) = 0 \qquad \text{Factor the trinomial.}$$
$$x + 3 = 0 \quad \text{or} \quad x - 2 = 0 \qquad \text{Set each factor equal to 0.}$$
$$x = -3 \qquad\qquad x = 2 \qquad \text{Solve each equation.}$$

The solutions of $x^2 + x - 6 = 0$ are -3 and 2. These solutions are the critical numbers.

Step 2:   *Locate the critical numbers on a number line.* When we highlight -3 and 2 on a number line, they separate the number line into three intervals:

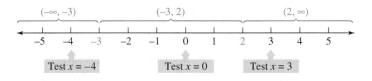

Step 3:   *Test each interval.* To determine whether the numbers in $(-\infty, -3)$ are solutions of the inequality, we choose a number from that interval, substitute it for x, and see whether it satisfies $x^2 + x - 6 < 0$. *If one number in that interval satisfies the inequality, all numbers in that interval will satisfy the inequality.*

If we choose -4 from $(-\infty, -3)$, we have:

$$x^2 + x - 6 < 0 \qquad \text{This is the original inequality.}$$
$$(-4)^2 + (-4) - 6 \overset{?}{<} 0 \qquad \text{Substitute } -4 \text{ for } x.$$
$$16 + (-4) - 6 \overset{?}{<} 0$$
$$6 < 0 \qquad \text{False}$$

Since -4 does not satisfy the inequality, none of the numbers in $(-\infty, -3)$ are solutions. To test the second interval, $(-3, 2)$, we choose $x = 0$.

$x^2 + x - 6 < 0$ This is the original inequality.

$0^2 + 0 - 6 \overset{?}{<} 0$ Substitute 0 for x.

$-6 < 0$ True

Since 0 satisfies the inequality, all of the numbers in $(-3, 2)$ are solutions. To test the third interval, $(2, \infty)$, we choose $x = 3$.

$x^2 + x - 6 < 0$ This is the original inequality.

$3^2 + 3 - 6 \overset{?}{<} 0$ Substitute 3 for x.

$9 + 3 - 6 \overset{?}{<} 0$

$6 < 0$ False

Since 3 does not satisfy the inequality, none of the numbers in $(2, \infty)$ are solutions.

Success Tip

If a quadratic inequality contains $\leq$ or $\geq$, the endpoints of the intervals are included in the solution set. If the inequality contains $<$ or $>$, they are not.

Step 4: *Are the endpoints included?* From the interval testing, we see that only numbers from $(-3, 2)$ satisfy $x^2 + x - 6 < 0$. The endpoints -3 and 2 are not included in the solution set because they do not satisfy the inequality. (Recall that -3 and 2 make $x^2 + x - 6$ equal to 0.) The solution set is the interval $(-3, 2)$ as graphed on the right.

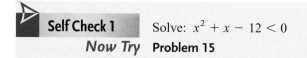

Self Check 1   Solve: $x^2 + x - 12 < 0$

Now Try   Problem 15

EXAMPLE 2   Solve: $x^2 + 4x \geq 5$

Strategy   This inequality is not in standard form because it does not have 0 on the right side. We will write it in standard form and solve its related quadratic equation to find any critical numbers. These critical numbers will separate the number line into intervals.

Why   We can then test each interval to see whether numbers in the interval are in the solution set of the inequality.

Solution   To get 0 on the right side, we subtract 5 from both sides.

$x^2 + 4x \geq 5$ This is the inequality to solve.

$x^2 + 4x - 5 \geq 0$ Write the inequality in the equivalent form $ax^2 + bx + c \geq 0$.

We can solve the related quadratic equation $x^2 + 4x - 5 = 0$ by factoring.

$x^2 + 4x - 5 = 0$

$(x + 5)(x - 1) = 0$ Factor the trinomial.

$x + 5 = 0$ or $x - 1 = 0$ Set each factor equal to 0.

$x = -5$ | $x = 1$

The critical numbers -5 and 1 separate the number line into three intervals. We pick a test value from each interval to see whether it satisfies $x^2 + 4x - 5 \geq 0$.

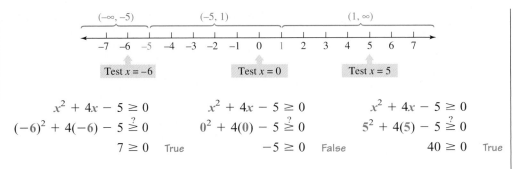

The numbers in the intervals $(-\infty, -5)$ and $(1, \infty)$ satisfy the inequality. Since the endpoints -5 and 1 also satisfy $x^2 + 4x - 5 \geq 0$, they are included in the solution set. (Recall that -5 and 1 make $x^2 + 4x - 5$ equal to 0.) Thus, the solution set is the union of two intervals: $(-\infty, -5] \cup [1, \infty)$. The graph of the solution set is shown on the right.

▷ **Self Check 2** Solve: $x^2 + 3x \geq 40$

Now Try Problem 19

2 **Solve Rational Inequalities.**

Rational inequalities in one variable such as $\frac{9}{x} < 8$ and $\frac{x^2 + x - 2}{x - 4} \geq 0$ can also be solved using the interval testing method.

Solving Rational Inequalities

1. Write the inequality in standard form with a single quotient on the left side and 0 on the right side. Then solve its related rational equation.

2. Set the denominator equal to zero and solve that equation.

3. Locate the solutions (called *critical numbers*) found in steps 1 and 2 on a number line.

4. Test each interval on the number line created in step 3 by choosing a test value from the interval and determining whether it satisfies the inequality. The solution set includes the interval(s) whose test value makes the inequality true.

5. Determine whether the endpoints of the intervals are included in the solution set. Exclude any values that make the denominator 0.

EXAMPLE 3 Solve: $\frac{9}{x} < 8$

Strategy This rational inequality is not in standard form because it does not have 0 on the right side. We will write it in standard form and solve its related rational equation to find any critical numbers. These critical numbers will separate the number line into intervals.

Why We can test each interval to see whether numbers in the interval are in the solution set of the inequality.

Solution To get 0 on the right side, we subtract 8 from both sides. We then find a common denominator to write the left side as a single quotient.

$$\frac{9}{x} < 8 \qquad \text{This is the inequality to solve.}$$

$$\frac{9}{x} - 8 < 0 \qquad \text{Subtract 8 from both sides.}$$

$$\frac{9}{x} - 8 \cdot \frac{x}{x} < 0 \qquad \text{To write the left side as a single quotient, build 8 to a fraction with denominator } x.$$

$$\frac{9}{x} - \frac{8x}{x} < 0$$

$$\frac{9 - 8x}{x} < 0 \qquad \text{Subtract the numerators and keep the common denominator, } x.$$

Now we solve the related rational equation.

$$\frac{9 - 8x}{x} = 0$$

$$9 - 8x = 0 \qquad \text{If } x \neq 0, \text{ we can clear the equation of the fraction by multiplying both sides by } x.$$

$$-8x = -9 \qquad \text{Subtract 9 from both sides.}$$

$$x = \frac{9}{8} \qquad \text{This is a critical number.}$$

If we set the denominator of $\frac{9 - 8x}{x}$ equal to 0, we obtain a second critical number, $x = 0$. When graphed, the critical numbers 0 and $\frac{9}{8}$ separate the number line into three intervals. We pick a test value from each interval to see whether it satisfies $\frac{9 - 8x}{x} < 0$.

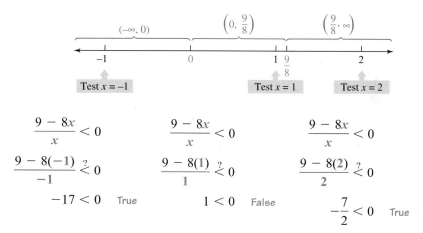

$$\frac{9 - 8x}{x} < 0 \qquad\qquad \frac{9 - 8x}{x} < 0 \qquad\qquad \frac{9 - 8x}{x} < 0$$

$$\frac{9 - 8(-1)}{-1} \overset{?}{<} 0 \qquad \frac{9 - 8(1)}{1} \overset{?}{<} 0 \qquad \frac{9 - 8(2)}{2} \overset{?}{<} 0$$

$$-17 < 0 \quad \text{True} \qquad\quad 1 < 0 \quad \text{False} \qquad\quad -\frac{7}{2} < 0 \quad \text{True}$$

The numbers in the intervals $(-\infty, 0)$ and $\left(\frac{9}{8}, \infty\right)$ satisfy the inequality. We do not include the endpoint 0 in the solution set, because it makes the denominator of the original inequality 0. Neither do we include $\frac{9}{8}$, because it does not satisfy $\frac{9 - 8x}{x} < 0$. (Recall that $\frac{9}{8}$ makes $\frac{9 - 8x}{x}$ equal to 0.) Thus, the solution set is the union of two intervals: $(-\infty, 0) \cup \left(\frac{9}{8}, \infty\right)$. Its graph is shown on the right.

Self Check 3   Solve: $\frac{3}{x} < 5$

Now Try   **Problem 23**

EXAMPLE 4	Solve: $\dfrac{x^2 + x - 2}{x - 4} \geq 0$

Strategy   This inequality is in standard form. We will solve its related rational equation to find any critical numbers. These critical numbers will separate the number line into intervals.

Why   We can test each interval to see whether numbers in the interval are in the solution set of the inequality.

Solution   To solve the related rational equation, we proceed as follows:

$$\frac{x^2 + x - 2}{x - 4} = 0$$

$$x^2 + x - 2 = 0 \qquad \text{If } x \neq 4, \text{ we can clear the equation of the fraction by multiplying both sides by } x - 4.$$

$$(x + 2)(x - 1) = 0 \qquad \text{Factor the trinomial.}$$

$$x + 2 = 0 \quad \text{or} \quad x - 1 = 0 \qquad \text{Set each factor equal to 0.}$$

$$x = -2 \qquad \qquad x = 1 \qquad \text{These are critical numbers.}$$

If we set the denominator of $\dfrac{x^2 + x - 2}{x - 4}$ equal to 0, we see that $x = 4$ is also a critical number. When graphed, the critical numbers, -2, 1, and 4, separate the number line into four intervals. We pick a test value from each interval to see whether it satisfies $\dfrac{x^2 + x - 2}{x - 4} \geq 0$.

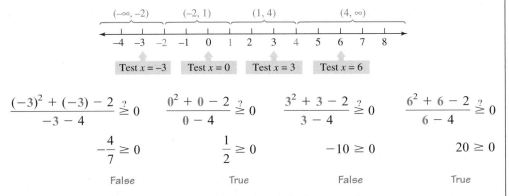

$$\frac{(-3)^2 + (-3) - 2}{-3 - 4} \overset{?}{\geq} 0 \qquad \frac{0^2 + 0 - 2}{0 - 4} \overset{?}{\geq} 0 \qquad \frac{3^2 + 3 - 2}{3 - 4} \overset{?}{\geq} 0 \qquad \frac{6^2 + 6 - 2}{6 - 4} \overset{?}{\geq} 0$$

$$-\frac{4}{7} \geq 0 \qquad\qquad \frac{1}{2} \geq 0 \qquad\qquad -10 \geq 0 \qquad\qquad 20 \geq 0$$

    False   True   False   True

The numbers in the intervals $(-2, 1)$ and $(4, \infty)$ satisfy the inequality. We include the endpoints -2 and 1 in the solution set because they satisfy the inequality. We do not include 4 because it makes the denominator of the inequality 0. Thus, the solution set is the union of two intervals $[-2, 1] \cup (4, \infty)$, as graphed on the right.

▷ **Self Check 4**	Solve: $\dfrac{x + 2}{x^2 - 2x - 3} \geq 0$
Now Try	**Problem 27**

EXAMPLE 5	Solve: $\dfrac{3}{x - 1} < \dfrac{2}{x}$

Strategy   We will subtract $\dfrac{2}{x}$ from both sides to get 0 on the right side and solve the resulting related rational equation to find any critical numbers. These critical numbers will separate the number line into intervals.

Why We can test each interval to see whether numbers in the interval are in the solution set of the inequality.

Solution

$$\frac{3}{x-1} < \frac{2}{x} \qquad \text{This is the inequality to solve.}$$

$$\frac{3}{x-1} - \frac{2}{x} < 0 \qquad \text{Subtract } \tfrac{2}{x} \text{ from both sides.}$$

$$\frac{3}{x-1} \cdot \frac{x}{x} - \frac{2}{x} \cdot \frac{x-1}{x-1} < 0 \qquad \text{To get a single quotient on the left side, build each rational expression to have the common denominator } x(x-1).$$

$$\frac{3x - 2x + 2}{x(x-1)} < 0 \qquad \text{Subtract the numerators and keep the common denominator.}$$

$$\frac{x+2}{x(x-1)} < 0 \qquad \text{Combine like terms.}$$

The only solution of the related rational equation $\frac{x+2}{x(x-1)} = 0$ is -2. Thus, -2 is a critical number. When we set the denominator equal to 0 and solve $x(x-1) = 0$, we find two more critical numbers, 0 and 1. These three critical numbers create four intervals to test.

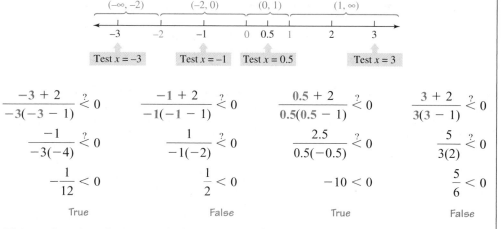

$$\frac{-3+2}{-3(-3-1)} \overset{?}{<} 0 \qquad \frac{-1+2}{-1(-1-1)} \overset{?}{<} 0 \qquad \frac{0.5+2}{0.5(0.5-1)} \overset{?}{<} 0 \qquad \frac{3+2}{3(3-1)} \overset{?}{<} 0$$

$$\frac{-1}{-3(-4)} \overset{?}{<} 0 \qquad \frac{1}{-1(-2)} \overset{?}{<} 0 \qquad \frac{2.5}{0.5(-0.5)} \overset{?}{<} 0 \qquad \frac{5}{3(2)} \overset{?}{<} 0$$

$$-\frac{1}{12} < 0 \qquad \frac{1}{2} < 0 \qquad -10 < 0 \qquad \frac{5}{6} < 0$$

| True | False | True | False |

The numbers 0 and 1 are not included in the solution set because they make the denominator 0, and the number -2 is not included because it does not satisfy the inequality. The solution set is the union of two intervals $(-\infty, -2) \cup (0, 1)$, as graphed on the right.

Self Check 5 Solve: $\frac{2}{x+1} > \frac{1}{x}$

Now Try **Problem 31**

Using Your Calculator ## *Solving Inequalities Graphically*

We can solve $x^2 + 4x \geq 5$ (Example 2) graphically by writing the inequality as $x^2 + 4x - 5 \geq 0$ and graphing the quadratic function $f(x) = x^2 + 4x - 5$, as shown in figure (a). The solution set of the inequality will be those values of x for which the graph lies on or above the x-axis. We can trace to determine that this is the union of two intervals: $(-\infty, -5] \cup [1, \infty)$.

To solve $\frac{3}{x-1} < \frac{2}{x}$ (Example 5) graphically, we first write the inequality in the form $\frac{x+2}{x(x-1)} < 0$ and then graph the rational function $f(x) = \frac{x+2}{x(x-1)}$, as shown in figure (b). The solution of the inequality will be those values of x for which the graph lies below the axis.

We can trace to see that the graph is below the x-axis when x is less than -2. Since we cannot see the graph in the interval $0 < x < 1$, we redraw the graph using window settings of $[-1, 2]$ for x and $[-25, 10]$ for y, as shown in figure (c).

Now we see that the graph is below the x-axis in the interval $(0, 1)$. Thus, the solution set of the inequality is the union of the two intervals: $(-\infty, -2) \cup (0, 1)$.

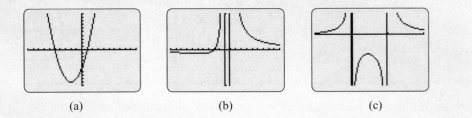

(a)   (b)   (c)

③ Graph Nonlinear Inequalities in Two Variables.

We have previously graphed linear inequalities in two variables such as $y > 3x + 2$ and $2x - 3y \le 6$ using the following steps.

Graphing Inequalities in Two Variables	1. Graph the related equation to find the boundary line of the region. If the inequality allows equality (the symbol is either $\le$ or $\ge$), draw the boundary as a solid line. If equality is not allowed ($<$ or $>$), draw the boundary as a dashed line.
	2. Pick a test point that is on one side of the boundary line. (Use the origin if possible.) Replace x and y in the original inequality with the coordinates of that point. If the inequality is satisfied, shade the side that contains that point. If the inequality is not satisfied, shade the other side of the boundary.

We use the same procedure to graph *nonlinear* inequalities in two variables.

EXAMPLE 6   Graph: $y < -x^2 + 4$

Strategy   We will graph the related equation $y = -x^2 + 4$ to establish a boundary parabola. Then we will determine which side of the boundary parabola represents the solution set of the inequality.

Why   To *graph a nonlinear inequality* in two variables means to draw a "picture" of the ordered pairs (x, y) that make the inequality true.

Solution   The graph of the boundary $y = -x^2 + 4$ is a parabola opening downward, with vertex at $(0, 4)$ and axis of symmetry $x = 0$ (the y-axis). Since the inequality contains an $<$ symbol and equality is not allowed, we draw the parabola using a dashed curve.

To determine which region to shade, we pick the test point $(0, 0)$ and substitute its coordinates into the inequality. We shade the region containing $(0, 0)$ because its coordinates satisfy $y < -x^2 + 4$.

Graph the boundary

$$y = -x^2 + 4$$

Compare to $y = a(x - h)^2 + k$

$a = -1$: Opens downward

$h = 0$ and $k = 4$: Vertex $(0, 4)$

Axis of symmetry $x = 0$

x	y
1	3
2	0

Shading: Use the test point (0, 0)

$$y < -x^2 + 4$$

$$0 \overset{?}{<} -0^2 + 4$$

$$0 < 4 \qquad \text{True}$$

Since $0 < 4$ is true, $(0, 0)$ is a solution of $y < -x^2 + 4$.

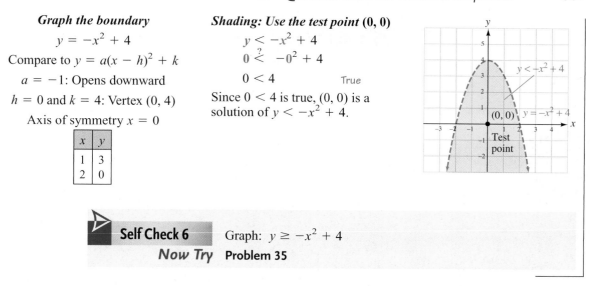

▷ **Self Check 6**   Graph: $y \geq -x^2 + 4$

Now Try   **Problem 35**

EXAMPLE 7   Graph: $x \leq |y|$

Strategy   We will graph the related equation $x = |y|$ to establish a boundary. Then we will determine which side of the boundary represents the solution set of the inequality.

Why   To *graph a nonlinear inequality* in two variables means to draw a "picture" of the ordered pairs (x, y) that make the inequality true.

Solution   To graph the boundary, $x = |y|$, we construct a table of solutions, as shown in figure (a). In figure (b), the boundary is graphed using a solid line because the inequality contains a $\leq$ symbol and equality is permitted. Since the origin is on the graph, we cannot use it as a test point. However, any other point, such as $(1, 0)$, will do. We substitute 1 for x and 0 for y into the inequality to get

$$x \leq |y|$$

$$1 \overset{?}{\leq} |0|$$

$$1 \leq 0 \qquad \text{False}$$

Since $1 \leq 0$ is a false statement, the point $(1, 0)$ does not satisfy the inequality and is not part of the graph. Thus, the graph of $x \leq |y|$ is to the left of the boundary.

The complete graph is shown in figure (c).

$x = |y|$

x	y
0	0
1	1
1	-1
2	2
2	-2

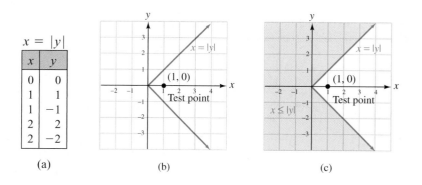

(a)   (b)   (c)

▷ **Self Check 7**   Graph: $x \geq -|y|$
      Now Try   **Problem 39**

▷ **ANSWERS TO SELF CHECKS**

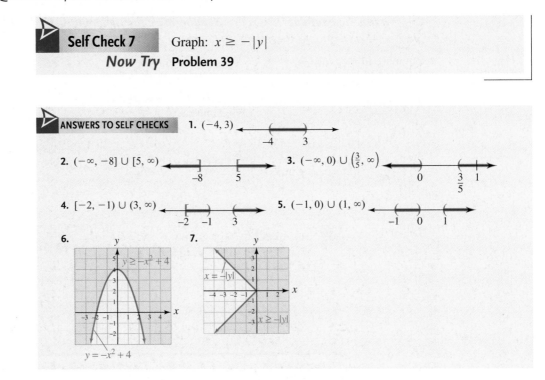

1. $(-4, 3)$

2. $(-\infty, -8] \cup [5, \infty)$

3. $(-\infty, 0) \cup \left(\frac{3}{5}, \infty\right)$

4. $[-2, -1) \cup (3, \infty)$

5. $(-1, 0) \cup (1, \infty)$

6.

7.

STUDY SET
8.5

VOCABULARY

Fill in the blanks.

1. $x^2 + 3x - 18 < 0$ is an example of a _____ inequality in one variable.

2. $\frac{x - 1}{x^2 - x - 20} \leq 0$ is an example of a _____ inequality in one variable.

3. $y \leq x^2 - 4x + 3$ is an example of a nonlinear inequality in _____ variables.

4. The set of real numbers greater than 3 can be represented using the _____ notation $(3, \infty)$.

CONCEPTS

5. The critical numbers of a quadratic inequality are highlighted in red on the number line shown below. Use interval notation to represent each interval that must be tested to solve the inequality.

6. Graph each of the following solution sets.

  a. $(-2, 4)$ **b.** $(-\infty, -2) \cup (3, 5]$

7. The graph of the solution set of a rational inequality in one variable is shown. Determine whether each of the following numbers is a solution of the inequality.

  a. -10 **b.** -5
  c. 0 **d.** 4

8. What are the critical numbers for each inequality?

  a. $x^2 - 2x - 48 \geq 0$ **b.** $\dfrac{x - 3}{x(x + 4)} > 0$

9. a. The results after interval testing for a quadratic inequality containing a $>$ symbol are shown below. (The critical numbers are highlighted in red.) What is the solution set?

  b. The results after interval testing for a quadratic inequality containing a $\leq$ symbol are shown below. (The critical numbers are highlighted in red.) What is the solution set?

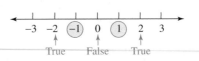

10. Fill in the blank to complete this important fact about the interval testing method discussed in this section: *If one number in an interval satisfies the inequality, _____ numbers in that interval will satisfy the inequality.*

11. a. When graphing the solution of $y \leq x^2 + 2x + 1$, should the boundary be solid or dashed?

  b. Does the test point $(0, 0)$ satisfy the inequality?

12. a. Estimate the solution of $x^2 - x - 6 > 0$ using the graph of $y = x^2 - x - 6$ shown in figure (a) below.

  b. Estimate the solution of $\frac{x-3}{x} \leq 0$ using the graph of $y = \frac{x-3}{x}$ shown in figure (b) below.

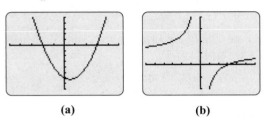

(a)   (b)

NOTATION

13. Write the quadratic inequality $x^2 - 6x \geq 7$ in standard form.

14. The solution set of a rational inequality consists of the intervals $(-1, 4]$ and $(7, \infty)$. When writing the solution set, what symbol is used between the two intervals?

GUIDED PRACTICE

Solve each inequality. Write the solution set in interval notation and graph it. See Example 1.

15. $x^2 - 5x + 4 < 0$

16. $x^2 + 2x - 8 < 0$

17. $x^2 - 8x + 15 > 0$

18. $x^2 - 3x - 4 > 0$

Solve each inequality. Write the solution set in interval notation and graph it. See Example 2.

19. $x^2 - x \geq 42$

20. $x^2 - x \geq 72$

21. $x^2 + x \leq 12$

22. $x^2 - 8x \leq -15$

Solve each inequality. Write the solution set in interval notation and graph it. See Example 3.

23. $\frac{1}{x} < 2$

24. $\frac{1}{x} < 3$

25. $\frac{5}{x} \geq -3$

26. $\frac{4}{x} \geq 8$

Solve each inequality. Write the solution set in interval notation and graph it. See Example 4.

27. $\frac{x^2 - x - 12}{x - 1} < 0$

28. $\frac{x^2 + x - 6}{x - 4} \geq 0$

29. $\frac{6x^2 - 5x + 1}{2x + 1} \geq 0$

30. $\frac{6x^2 + 11x + 3}{3x - 1} < 0$

Solve each inequality. Write the solution set in interval notation and graph it. See Example 5.

31. $\frac{3}{x - 2} < \frac{4}{x}$

32. $\frac{-6}{x + 1} \geq \frac{1}{x}$

33. $\frac{7}{x - 3} \geq \frac{2}{x + 4}$

34. $\frac{-5}{x - 4} < \frac{3}{x + 1}$

Graph each inequality. See Example 6.

35. $y < x^2 + 1$

36. $y > x^2 - 3$

37. $y \leq x^2 + 5x + 6$

38. $y \geq x^2 + 5x + 4$

Graph each inequality. See Example 7.

39. $y < |x + 4|$

40. $y \leq |x - 3|$

41. $y \geq -|x| + 2$

42. $y > |x| - 2$

Use a graphing calculator to solve each inequality. Write the solution set in interval notation. See Using Your Calculator: Solving Inequalities Graphically.

43. $x^2 - 2x - 3 < 0$

44. $x^2 + x - 6 > 0$

45. $\frac{x + 3}{x - 2} > 0$

46. $\frac{3}{x} < 2$

TRY IT YOURSELF

Solve each inequality. Write the solution set in interval notation and graph it.

47. $\frac{x}{x + 4} \leq \frac{1}{x + 1}$

48. $\frac{x}{x + 9} \geq \frac{1}{x + 1}$

49. $x^2 \geq 9$

50. $x^2 \geq 16$

51. $x^2 + 6x \geq -9$

52. $x^2 + 8x < -16$

53. $\frac{x^2 + x - 2}{x - 3} > 0$

54. $\frac{x - 2}{x^2 - 1} > 0$

55. $2x^2 - 50 < 0$

56. $3x^2 - 243 < 0$

57. $\dfrac{2x - 3}{3x + 1} < 0$

58. $\dfrac{x - 5}{x + 1} < 0$

59. $x^2 - 6x + 9 < 0$

60. $x^2 + 4x + 4 > 0$

61. $\dfrac{5}{x + 1} > \dfrac{3}{x - 4}$

62. $\dfrac{3}{x - 2} \le -\dfrac{2}{x + 3}$

APPLICATIONS

63. BRIDGES   If an x-axis is superimposed over the roadway of the Golden Gate Bridge, with the origin at the center of the bridge, the length L in feet of a vertical support cable can be approximated by the formula

$$L = \dfrac{1}{9,000}x^2 + 5$$

For the Golden Gate Bridge, $-2,100 < x < 2,100$. For what intervals along the x-axis are the vertical cables more than 95 feet long?

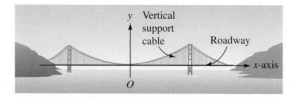

64. MALLS   The number of people n in a mall is modeled by the formula

$$n = -100x^2 + 1,200x$$

where x is the number of hours since the mall opened. If the mall opened at 9 A.M., when were there 2,000 or more people in it?

WRITING

65. How are critical numbers used when solving a quadratic inequality in one variable?

66. Explain how to graph $y \ge x^2$.

67. The graph of $f(x) = x^2 - 3x + 4$ is shown below. Explain why the quadratic inequality $x^2 - 3x + 4 < 0$ has no solution.

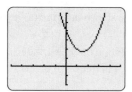

68. Describe the following solution set of a rational inequality in words: $(-\infty, 4] \cup (6, 7)$.

REVIEW

Translate each statement into an equation.

69. x varies directly with y.

70. y varies inversely with t.

71. t varies jointly with x and y.

72. d varies directly with t and inversely with u^2.

CHALLENGE PROBLEMS

73. a. Solve: $x^2 - x - 12 > 0$

   b. Find a rational inequality in one variable that has the same solution set as the quadratic inequality in part (a).

74. a. Solve: $\frac{1}{x} < 1$

   b. Now incorrectly "solve" $\frac{1}{x} < 1$ by multiplying both sides by x to clear it of the fraction. What part of the solution set is not obtained with this incorrect approach?

CHAPTER 8
Summary & Review

SECTION 8.1   The Square Root Property and Completing the Square

DEFINITIONS AND CONCEPTS	EXAMPLES

We can use the **square root property** to solve equations of the form $x^2 = c$, where $c > 0$. The two solutions are

$$x = \sqrt{c} \quad \text{or} \quad x = -\sqrt{c}$$

We can write $x = \sqrt{c}$ or $x = -\sqrt{c}$ in more compact form using **double-sign notation:**

$$x = \pm \sqrt{c}$$

Solve: $x^2 = 24$

$\qquad x = \sqrt{24}$ or $x = -\sqrt{24}$ — Use the square root property.

$\qquad x = \pm \sqrt{24}$ — Use double-sign notation.

$\qquad x = \pm 2\sqrt{6}$ — Simplify $\sqrt{24}$.

The solutions are $2\sqrt{6}$ and $-2\sqrt{6}$.

Solve: $(x - 3)^2 = -81$

$\qquad x - 3 = \pm \sqrt{-81}$ — Use the square root property and double-sign notation.

$\qquad x = 3 \pm \sqrt{-81}$ — To isolate x, add 3 to both sides.

$\qquad x = 3 \pm 9i$ — Simplify the radical expression.

The solutions are $3 + 9i$ and $3 - 9i$.

To **complete the square** on $x^2 + bx$, add the square of one-half of the coefficient of x.

$$x^2 + bx + \left(\frac{1}{2}b\right)^2$$

Complete the square on $x^2 + 8x$ and factor the resulting perfect-square trinomial.

$\qquad x^2 + 8x + 16$ — The coefficient of x is 8. To complete the square: $\frac{1}{2} \cdot 8 = 4$ and $4^2 = 16$. Add 16 to the binomial.

Now we factor: $x^2 + 8x + 16 = (x + 4)^2$

To **solve a quadratic equation in x by completing the square:**

1. If necessary, divide both sides of the equation by the coefficient of x^2 to make its coefficient 1.

2. Get all variable terms on one side of the equation and all constants on the other side.

3. Complete the square.

4. Factor the perfect-square trinomial.

5. Solve the resulting equation by using the square root property.

6. Check your answers in the original equation.

Solve: $3x^2 - 12x + 6 = 0$

$\qquad \dfrac{3x^2}{3} - \dfrac{12x}{3} + \dfrac{6}{3} = \dfrac{0}{3}$ — To make the leading coefficient 1, divide both sides by 3.

$\qquad x^2 - 4x + 2 = 0$ — Do the divisions.

$\qquad x^2 - 4x = -2$ — Subtract 2 from both sides so that the constant term, -2, is on the right side.

$\qquad x^2 - 4x + 4 = -2 + 4$ — The coefficient of x is -4. To complete the square: $\frac{1}{2}(-4) = -2$ and $(-2)^2 = 4$. Add 4 to both sides.

$\qquad (x - 2)^2 = 2$ — Factor the perfect-square trinomial on the left side. Add on the right side.

$\qquad x - 2 = \pm \sqrt{2}$ — Use the square root property.

$\qquad x = 2 \pm \sqrt{2}$ — To isolate x, add 2 to both sides.

The solutions are $2 + \sqrt{2}$ and $2 - \sqrt{2}$.

REVIEW EXERCISES

Solve each equation by factoring.

1. $x^2 + 9x + 20 = 0$
2. $6x^2 + 17x + 5 = 0$

Solve each equation using the square root property.

3. $x^2 = 28$
4. $(t + 2)^2 = 36$
5. $a^2 + 25 = 0$
6. $5x^2 - 49 = 0$

7. Solve $A = \pi r^2$ for r. Assume all variables represent positive numbers. Express the result in simplified radical form.

8. Complete the square on $x^2 - x$ and then factor the resulting perfect-square trinomial.

Solve each equation by completing the square.

9. $x^2 + 6x + 8 = 0$
10. $2x^2 - 6x + 3 = 0$
11. $6a^2 - 12a = -1$
12. $x^2 - 2x = -13$

13. Explain why completing the square on $x^2 + 7x$ is more difficult than completing the square on $x^2 + 6x$.

14. Explain the error: $\dfrac{2 \pm \sqrt{7}}{2} = \dfrac{\overset{1}{\cancel{2}} \pm \sqrt{7}}{\underset{1}{\cancel{2}}}$

15. a. Write an expression that represents the width of the larger rectangle shown in red.

b. Write an expression that represents the length of the larger rectangle shown in red.

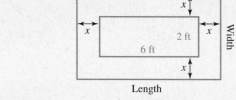

16. HAPPY NEW YEAR   As part of a New Year's Eve celebration, a huge ball is to be dropped from the top of a 605-foot-tall building at the proper moment so that it strikes the ground at exactly 12:00 midnight. The distance d in feet traveled by a free-falling object in t seconds is given by the formula $d = 16t^2$. To the nearest second, when should the ball be dropped from the building?

SECTION 8.2   The Quadratic Formula

DEFINITIONS AND CONCEPTS	EXAMPLES

To **solve a quadratic equation in x using the quadratic formula:**

1. Write the equation in standard form: $ax^2 + bx + c = 0$.

2. Identify a, b, and c.

3. Substitute the values for a, b, and c in the quadratic formula

$$x = \dfrac{-b \pm \sqrt{b^2 - 4ac}}{2a}$$

and evaluate the right side to obtain the solutions.

Solve: $3x^2 - 2x - 2 = 0$

Here, $a = 3$, $b = -2$, and $c = -2$.

$x = \dfrac{-b \pm \sqrt{b^2 - 4ac}}{2a}$ This is the quadratic formula.

$x = \dfrac{-(-2) \pm \sqrt{(-2)^2 - 4(3)(-2)}}{2(3)}$ Substitute 3 for a, -2 for b, and -2 for c.

$x = \dfrac{2 \pm \sqrt{4 - (-24)}}{6}$ Evaluate the power and multiply within the radical. Multiply in the denominator.

$x = \dfrac{2 \pm \sqrt{28}}{6}$ Add the opposite: $4 - (-24) = 4 + 24 = 28$.

$x = \dfrac{2 \pm 2\sqrt{7}}{6}$ Simplify the radical: $\sqrt{28} = \sqrt{4}\sqrt{7} = 2\sqrt{7}$.

$x = \dfrac{\overset{1}{\cancel{2}}(1 \pm \sqrt{7})}{\underset{1}{2 \cdot 3}}$ Factor out the GCF, 2, from the two terms in the numerator. In the denominator, factor 6 as $2 \cdot 3$. Remove the common factor, 2.

$x = \dfrac{1 \pm \sqrt{7}}{3}$

The exact solutions are $\dfrac{1 + \sqrt{7}}{3}$ and $\dfrac{1 - \sqrt{7}}{3}$. We can use a calculator to approximate them. To two decimal places, they are 1.22 and -0.55.

When solving a quadratic equation using the quadratic formula, we can often **simplify the computations** by solving an equivalent equation that does not involve fractions or decimals, and whose leading coefficient is positive.	**Before solving . . .** **do this . . .** **to get this**

Before solving . . .	do this . . .	to get this
$-3x^2 + 5x - 1 = 0$	Multiply both sides by -1	$3x^2 - 5x + 1 = 0$
$x^2 + \dfrac{7}{8}x - \dfrac{1}{2} = 0$	Multiply both sides by 8	$8x^2 + 7x - 4 = 0$
$60x^2 - 40x + 90 = 0$	Divide both sides by 10	$6x^2 - 4x + 9 = 0$
$0.05x^2 + 0.16x + 0.71 = 0$	Multiply both sides by 100	$5x^2 + 16x + 71 = 0$

REVIEW EXERCISES

Solve each equation using the quadratic formula.

17. $2x^2 + 13x = 7$

18. $-x^2 + 10x - 18 = 0$

19. $x^2 - 10x = 0$

20. $3y^2 = 26y - 2$

21. $\frac{1}{3}p^2 + \frac{1}{2}p + \frac{1}{2} = 0$

22. $3{,}000t^2 - 4{,}000t = -2{,}000$

23. $0.5x^2 + 0.3x - 0.1 = 0$

24. $x^2 - 3x - 27 = 0$

25. TUTORING   A private tutoring company charges $20 for a 1-hour session. Currently, 300 students are tutored each week. Since the company is losing money, the owner has decided to increase the price. For each 50¢ increase, she estimates that 5 fewer students will participate. If the company needs to bring in $6,240 per week to stay in business, what price must be charged for a 1-hour tutoring session to produce this amount of revenue?

26. POSTERS   The specifications for a poster of Cesar Chavez call for a 615-square-inch photograph to be surrounded by a green border. The borders on the top and bottom of the poster are to be twice as wide as those on the sides. Find the width of each border.

35 in.

◄— 23 in. —►

© Hulton-Deutsch Collection/ CORBIS

27. ACROBATS   To begin his routine on a trapeze, an acrobat is catapulted upward as shown in the illustration. His distance d (in feet) from the arena floor during this maneuver is given by the formula $d = -16t^2 + 40t + 5$, where t is the time in seconds since being launched. If the trapeze bar is 25 feet in the air, at what two times will he be able to grab it? Round to the nearest tenth.

28. TRIANGLES   The length of the longer leg of a right triangle exceeds the length of the shorter leg by 23 inches and the length of the hypotenuse is 65 inches. Find the length of each leg of the triangle.

SECTION 8.3   The Discriminant and Equations That Can Be Written in Quadratic Form

DEFINITIONS AND CONCEPTS	EXAMPLES
The **discriminant** predicts the type of solutions of $ax^2 + bx + c = 0$, where a, b, and c are real numbers and $a \neq 0$: **1.** If $b^2 - 4ac > 0$, there are two different real-number solutions. If $b^2 - 4ac$ is a perfect square, there are two different rational-number solutions. If $b^2 - 4ac$ is not a perfect square, there are two different irrational-number solutions. **2.** If $b^2 - 4ac = 0$, there is one repeated solution, a rational number. **3.** If $b^2 - 4ac < 0$, there are two different imaginary-number solutions that are complex conjugates.	In the quadratic equation $2x^2 - 5x - 3 = 0$, we have $a = 2$, $b = -5$, and $c = -3$. So the value of the discriminant is $$b^2 - 4ac = (-5)^2 - 4(2)(-3) = 25 + 24 = 49$$ Since the value of the discriminant is positive and a perfect square, the equation $2x^2 - 5x - 3 = 0$ has two different rational-number solutions.

SECTION 8.3   The Discriminant and Equations That Can Be Written in Quadratic Form—*continued*

DEFINITIONS AND CONCEPTS	EXAMPLES
Equations that contain an expression, the same expression squared, and a constant term are said to be **quadratic in form.** One method used to solve such equations is to make a **substitution.**	Solve: $x^{2/3} - 6x^{1/3} + 5 = 0$ The equation can be written in quadratic form: $(x^{1/3})^2 - 6x^{1/3} + 5 = 0$ We substitute y for $x^{1/3}$ and use factoring to solve the resulting quadratic equation $y^2 - 6y + 5 = 0$. $(y - 1)(y - 5) = 0$ Let $y = x^{1/3}$. $y = 1$ or $y = 5$ Now we reverse the substitution $y = x^{1/3}$ and solve for x. $x^{1/3} = 1$ or $x^{1/3} = 5$ $(x^{1/3})^3 = (1)^3$ $(x^{1/3})^3 = (5)^3$ $x = 1$ $x = 125$ The solutions are 1 and 125. Check both in the original equation.

REVIEW EXERCISES

Use the discriminant to determine the number and type of solutions for each equation.

29. $3x^2 + 4x - 3 = 0$

30. $4x^2 - 5x + 7 = 0$

31. $3x^2 - 4x + \dfrac{4}{3} = 0$

32. $m(2m - 3) = 20$

Solve each equation.

33. $x - 13\sqrt{x} + 12 = 0$

34. $a^{2/3} + a^{1/3} - 6 = 0$

35. $3x^4 + x^2 - 2 = 0$

36. $\dfrac{6}{x + 2} + \dfrac{6}{x + 1} = 5$

37. $(x - 7)^2 + 6(x - 7) + 10 = 0$

38. $m^{-4} - 2m^{-2} + 1 = 0$

39. $4\left(\dfrac{x + 1}{x}\right)^2 + 12\left(\dfrac{x + 1}{x}\right) + 9 = 0$

40. $2m^{2/5} - 5m^{1/5} + 2 = 0$

41. WEEKLY CHORES   Working together, two sisters can do the yard work at their house in 45 minutes. When the older girl does it all herself, she can complete the job in 20 minutes less time than it takes the younger girl working alone. How long does it take the older girl to do the yard work?

42. ROAD TRIPS   A woman drives her automobile 150 miles at a rate of r mph. She could have gone the same distance in 2 hours less time if she had increased her speed by 20 mph. Find r.

SECTION 8.4   Quadratic Functions and Their Graphs

DEFINITIONS AND CONCEPTS	EXAMPLES
A **quadratic function** is a second-degree polynomial function of the form $f(x) = ax^2 + bx + c$	Quadratic functions: $f(x) = 2x^2 - 3x + 5$, $g(x) = -x^2 + 4x$, and $s(x) = \dfrac{1}{4}x^2 - 10$

SECTION 8.4   Quadratic Functions and Their Graphs—*continued*

DEFINITIONS AND CONCEPTS	EXAMPLES

The graph of the quadratic function $f(x) = a(x - h)^2 + k$ where $a \neq 0$ is a **parabola** with **vertex** at (h, k). The **axis of symmetry** is the line $x = h$. The parabola opens upward when $a > 0$ and downward when $a < 0$.

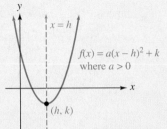

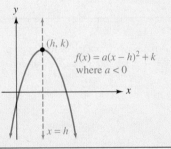

Graph: $f(x) = 2(x + 1)^2 - 8$

$$f(x) = 2[x - (-1)]^2 - 8$$
$$f(x) = a(x - \quad h)^2 + k$$

We see that $a = 2$, $h = -1$, and $k = -8$. The graph is a parabola with vertex $(h, k) = (-1, -8)$ and axis of symmetry $x = -1$. Since a is positive, the parabola opens upward.

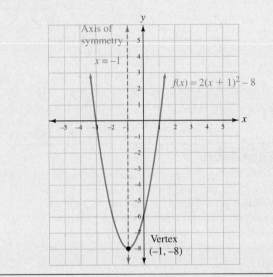

The **vertex** of the graph of $f(x) = ax^2 + bx + c$ is

$$\left(-\frac{b}{2a}, f\left(-\frac{b}{2a} \right) \right)$$

and the axis of symmetry is the line

$$x = -\frac{b}{2a}$$

The y-coordinate of the vertex of the graph of a quadratic function gives the **minimum or maximum value** of the function.

Graph: $f(x) = -x^2 + 3x + 4$

Here, $a = -1$, $b = 3$, and $c = 4$.

- Since $a < 0$, the graph opens downward.

- The x-coordinate of the vertex of the graph is

$$-\frac{b}{2a} = -\frac{3}{2(-1)} = \frac{3}{2}$$

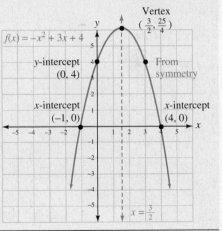

SECTION 8.4 Quadratic Functions and Their Graphs—*continued*

DEFINITIONS AND CONCEPTS	EXAMPLES
The **y-intercept** is determined by the value of $f(x)$ when $x = 0$: the y-intercept is $(0, c)$. To find the **x-intercepts**, let $f(x) = 0$ and solve $ax^2 + bx + c = 0$.	To find the y-coordinate of the vertex, we substitute $\frac{3}{2}$ for x in the function. $$f(x) = -x^2 + 3x + 4 = -\left(\frac{3}{2}\right)^2 + 3\left(\frac{3}{2}\right) + 4 = \frac{25}{4}$$ The vertex of the parabola is the point $\left(\frac{3}{2}, \frac{25}{4}\right)$. • The y-intercept is the value of the function when $x = 0$. Thus, the y-intercept is $(0, 4)$. • To find the x-intercepts, we solve: $\qquad -x^2 + 3x + 4 = 0$ $\qquad x^2 - 3x - 4 = 0$ Multiply both sides by -1. $\qquad (x + 1)(x - 4) = 0$ Factor. $\qquad x = -1$ or $x = 4$ The x-intercepts are $(-1, 0)$ and $(4, 0)$.

REVIEW EXERCISES

43. HOSPITALS   The annual number of in-patient admissions to U.S. community hospitals for the years 1980–2004 can be modeled by the quadratic function $A(x) = 0.03x^2 - 0.82x + 36.31$, where $A(x)$ is the number of admissions in millions and x is the number of years after 1980. Use the function to estimate the number of in-patient admissions for the year 1992. Round to the nearest tenth of one million.

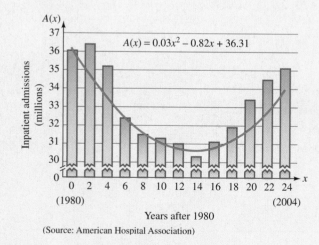

Years after 1980
(Source: American Hospital Association)

44. Fill in the blanks. The graph of the quadratic function $f(x) = a(x - h)^2 + k$ is a parabola with vertex at (,). The axis of symmetry is the line $x = h$. The parabola opens upward when $a > 0$ and downward when $a < 0$.

Graph each pair of functions on the same coordinate system.

45. $f(x) = 2x^2$, $g(x) = 2x^2 - 3$

46. $f(x) = -\frac{1}{4}x^2$, $g(x) = -\frac{1}{4}(x + 2)^2$

47. Find the vertex and the axis of symmetry of the graph of $f(x) = -2(x - 1)^2 + 4$. Then plot several points and complete the graph.

48. Complete the square to write $f(x) = 4x^2 + 16x + 9$ in the form $f(x) = a(x - h)^2 + k$. Determine the vertex and the axis of symmetry of the graph. Then plot several points and complete the graph.

49. Find the vertex of the graph of $f(x) = -2x^2 + 4x - 8$ using the vertex formula.

50. First determine the coordinates of the vertex and the axis of symmetry of the graph of $f(x) = x^2 + x - 2$ using the vertex formula. Then determine the x- and y-intercepts of the graph. Finally, plot several points and complete the graph.

51. FARMING   The number of farms in the United States for the years 1870–1970 is approximated by

$$N(x) = -1{,}526x^2 + 155{,}652x + 2{,}500{,}200$$

where $x = 0$ represents 1870, $x = 1$ represents 1871, and so on. For this period, when was the number of U.S. farms a maximum? How many farms were there?

52. Estimate the solutions of $-3x^2 - 5x + 2 = 0$ from the graph of $f(x) = -3x^2 - 5x + 2$, shown here.

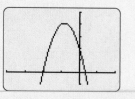

SECTION 8.5   Quadratic and Other Nonlinear Inequalities

DEFINITIONS AND CONCEPTS	EXAMPLES
To solve a quadratic inequality, get 0 on the right side and solve the related quadratic equation. Then locate the **critical numbers** on a number line, test each interval, and check the endpoints.	To solve $x^2 - x - 6 \geq 0$, we solve the related quadratic equation $x^2 - x - 6 = 0$. $$x^2 - x - 6 = 0$$ $$(x - 3)(x + 2) = 0 \quad \text{Factor.}$$ $$x = 3 \quad \text{or} \quad x = -2$$ These are the critical numbers that divide the number line into three intervals. $(-\infty, -2)$ $\qquad$ $(-2, 3)$ $\qquad$ $(3, \infty)$ $-4 \quad -2 \quad\quad 0 \quad\quad 3 \quad 5$ Test $x = -4$ $\quad$ Test $x = 0$ $\quad$ Test $x = 5$ True $\qquad\qquad$ False $\qquad\qquad$ True After testing each interval and noting that 3 and -2 satisfy the inequality, we see that the solution set is $(-\infty, -2] \cup [3, \infty)$.
To solve a rational inequality, get 0 on the right side and solve the related rational equation. Then locate the **critical numbers** (including any values that make the denominator 0) on a number line, test each interval, and check the endpoints.	To solve $\frac{x+1}{x-4} < 0$, we solve the related rational equation $\frac{x+1}{x-4} = 0$ to obtain the solution $x = -1$, which is a critical number. Another critical number is $x = 4$, the value that makes the denominator 0. These critical numbers divide the number line into three intervals. $(-\infty, -1)$ $\qquad$ $(-1, 4)$ $\qquad$ $(4, \infty)$ $-3 \quad -1 \quad 0 \quad\quad\quad 4 \quad\quad 6$ Test $x = -3$ $\quad$ Test $x = 0$ $\qquad\quad$ Test $x = 6$ False $\qquad\quad$ True $\qquad\qquad\quad$ False After testing each interval and noting that -1 and 4 do not satisfy the inequality, we see that the solution set is the interval $(-1, 4)$.
To graph a nonlinear inequality in two variables, first graph the boundary. Then use a test point to determine which side of the boundary to shade.	This is the graph of $y \leq x^2 + 5x + 4$. Since the inequality contains the symbol $\leq$, and equality is allowed, we draw the parabola determined by $y = x^2 + 5x + 4$ using a solid line. We shade the region containing the test point $(0, 0)$ because its coordinates satisfy $y \leq x^2 + 5x + 4$.

REVIEW EXERCISES

Solve each inequality. Write the solution set in interval notation and graph it.

53. $x^2 + 2x - 35 > 0$

54. $x^2 \le 81$

55. $\dfrac{3}{x} \le 5$

56. $\dfrac{2x^2 - x - 28}{x - 1} > 0$

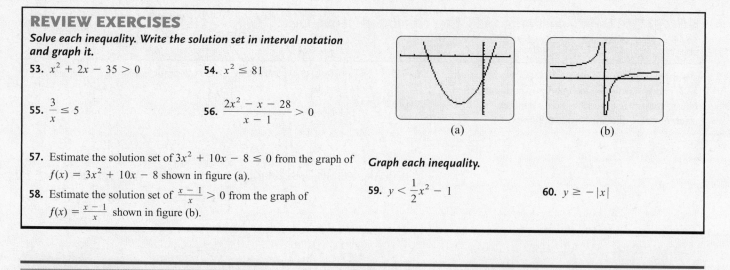

(a)   (b)

57. Estimate the solution set of $3x^2 + 10x - 8 \le 0$ from the graph of $f(x) = 3x^2 + 10x - 8$ shown in figure (a).

58. Estimate the solution set of $\frac{x-1}{x} > 0$ from the graph of $f(x) = \frac{x-1}{x}$ shown in figure (b).

Graph each inequality.

59. $y < \dfrac{1}{2}x^2 - 1$

60. $y \ge -|x|$

CHAPTER 8
Test

1. Fill in the blanks.

   a. An equation of the form $ax^2 + bx + c = 0$, where $a \ne 0$, is called a _____ equation.

   b. When we add 81 to $x^2 + 18x$, we say that we have _____ the _____ on $x^2 + 18x$.

   c. The lowest point on a parabola that opens upward, or the highest point on a parabola that opens downward, is called the _____ of the parabola.

   d. $\dfrac{x - 5}{x^2 - x - 56} > 0$ is an example of a _____ inequality in one variable.

   e. $y \le x^2 - 4x + 3$ is an example of a _____ inequality in two variables.

2. Solve $x^2 - 63 = 0$ using the square root property. Approximate the solutions to the nearest hundredth.

Solve each equation using the square root property.

3. $(a + 7)^2 = 50$

4. $m^2 + 4 = 0$

5. Add a number to make $x^2 + 11x$ a perfect-square trinomial. Then factor the result

6. Solve $4x^2 - 16x + 15 = 0$ by completing the square.

Use the quadratic formula to solve each equation.

7. $4x^2 + 4x - 1 = 0$

8. $\dfrac{1}{8}t^2 - \dfrac{1}{4}t = \dfrac{1}{2}$

9. $-t^2 + 4t - 13 = 0$

10. $0.01x^2 = -0.08x - 0.15$

Solve each equation by any method.

11. $2y - 3\sqrt{y} + 1 = 0$

12. $3 = m^{-2} - 2m^{-1}$

13. $x^4 - x^2 - 12 = 0$

14. $4\left(\dfrac{x + 2}{3x}\right)^2 - 4\left(\dfrac{x + 2}{3x}\right) - 3 = 0$

15. $\dfrac{1}{n + 2} = \dfrac{1}{3} - \dfrac{1}{n}$

16. $5a^{2/3} + 11a^{1/3} = -2$

17. Solve $E = mc^2$ for c. Assume that all variables represent positive numbers. Express any radical in simplified form.

18. Use the discriminant to determine the number and type of solutions for each equation.

   a. $3x^2 + 5x + 17 = 0$

   b. $9m^2 - 12m = -4$

19. TABLECLOTHS   In 1990, Sportex of Highland, Illinois, made what was at the time the world's longest tablecloth. Find the dimensions of the rectangular tablecloth if it covered an area of 6,759 square feet and its length was 8 feet more than 332 times its width.

20. COOKING   Working together, a chef and his assistant can make a pastry dessert in 25 minutes. When the chef makes it himself, it takes him 8 minutes less time than it takes his assistant working alone. How long does it take the chef to make the dessert?

21. DRAWING   An artist uses four equal-sized right triangles to block out a perspective drawing of an old hotel. See the illustration on the next page. For each triangle, the leg on the horizontal line is 14 inches longer than the leg on the center line. The length of each hypotenuse is 26 inches. On the centerline of the drawing, what is the length of the segment extending from the ground to the top of the building?

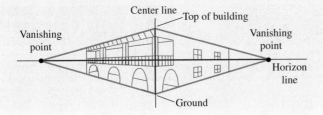

Center line — Top of building
Vanishing point · Vanishing point · Horizon line · Ground

22. ANTHROPOLOGY Anthropologists refer to the shape of the human jaw as a *parabolic dental arcade*. Which function is the best mathematical model of the parabola shown in the illustration?

i. $f(x) = -\dfrac{3}{8}(x-4)^2 + 6$

ii. $f(x) = -\dfrac{3}{8}(x-6)^2 + 4$

iii. $f(x) = -\dfrac{3}{8}x^2 + 6$

iv. $f(x) = \dfrac{3}{8}x^2 + 6$

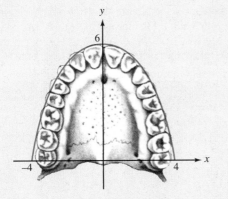

23. Find the vertex and the axis of symmetry of the graph of $f(x) = -3(x-1)^2 + 2$. Then plot several points and complete the graph.

24. Complete the square to write the function $f(x) = 5x^2 + 10x - 1$ in the form $f(x) = a(x-h)^2 + k$. Determine the vertex and the axis of symmetry of the graph. Then plot several points and complete the graph.

25. First determine the coordinates of the vertex and the axis of symmetry of the graph of $f(x) = 2x^2 + x - 1$ using the vertex formula. Then determine the x- and y-intercepts of the graph. Finally, plot several points and complete the graph.

26. DISTRESS SIGNALS A flare is fired directly upward into the air from a boat that is experiencing engine problems. The height of the flare (in feet) above the water, t seconds after being fired, is given by the formula $h = -16t^2 + 112t + 15$. If the flare is designed to explode when it reaches its highest point, at what height will this occur?

Solve each inequality. Write the solution set in interval notation and then graph it.

27. $x^2 - 2x > 8$

28. $\dfrac{x-2}{x+3} \le 0$

29. WATER USAGE The average amount of water used per month by a single-family residential customer in Tucson, Arizona, for the year 2004 is modeled by the function $W(m) = -235m^2 + 2{,}095m + 6{,}540$, where $W(m)$ is the number of gallons and m is the number of months *after March*. Use the function to approximate the average number of gallons of water used in July, which is typically Tucson's warmest month. (Based on data from the City of Tucson Water Department)

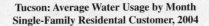
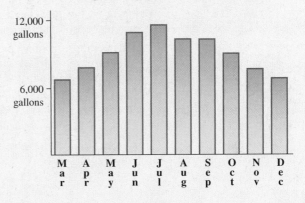

Tucson: Average Water Usage by Month
Single-Family Residental Customer, 2004

30. Graph: $y \le -x^2 + 3$

31. The graph of a quadratic function of the form $f(x) = ax^2 + bx + c$ is shown. Estimate the solutions of the corresponding quadratic equation $ax^2 + bx + c = 0$.

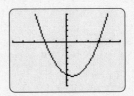

32. See Exercise 31. Estimate the solution of the quadratic inequality $ax^2 + bx + c \le 0$.

GROUP PROJECT

PICTURE FRAMING

Overview: When framing pictures, mats are often used to enhance the images and give them a sense of depth. In this activity, you will use the quadratic formula to design the matting for several pictures.

Instructions: Form groups of 3 students. Each person in your group is to bring a picture to class. You can use a picture from a magazine or newspaper, a picture postcard, or a photograph that is no larger than 5 in. × 7 in. You will also need a pair of scissors, a ruler, glue, and three pieces of construction paper (12 in. × 18 in.).

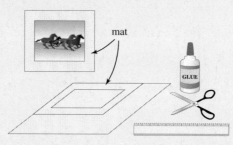

Select one of the pictures and find its area. A mat of uniform width is to be placed around the picture. The area of the mat should equal the area of the picture. To determine the proper width of the matting, follow the steps of Example 11 in Section 8.1. However, use the quadratic formula, instead of completing the square, to solve the equation. Once you have determined the proper width, cut out the mat from the construction paper and glue it to the picture.

Then, choose another picture and find its area. Determine the uniform width that a matting should have so that its area is double that of the picture. Cut out the proper-size matting from the construction paper and glue it to the second picture.

Finally, find the area of the third picture and determine the uniform width that a matting should have so that its area is one-half that of the picture. Cut out the proper-size matting from the construction paper and glue it to the third picture.

Is one size matting more visually appealing than another? Discuss this among the members of your group.

CUMULATIVE REVIEW
Chapters 1–8

1. Solve: $3(x + 2) - 2 = -(5 + x) + x$ [Section 1.5]

2. **PHARMACISTS** How many liters of a 1% glucose solution should a pharmacist mix with 2 liters of a 5% glucose solution to obtain a 2% glucose solution? [Section 1.8]

Find an equation of the line with the given properties. Write the equation in slope–intercept form.

3. Slope 3, passing through $(-2, -4)$ [Section 2.4]

4. Parallel to the graph of $2x + 3y = 6$ and passing through $(0, -2)$ [Section 2.4]

5. **SHORTAGE OF NURSES** Use the data in the graph to find the projected rates of change in the supply and demand for registered nurses (RNs) in the United States for the years 2010–2020. [Section 2.3]

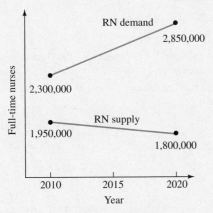

(Source: American Hospital Association)

6. **TIDES** The illustration shows the graph of a function f, which gives the height of the tide for a 24-hour period in Seattle, Washington. (Note that military time is used on the x-axis: 3 A.M. = 3, noon = 12, 9 P.M. = 21, and so on.) [Section 2.6]

   a. Find the domain of the function.

   b. Find $f(6)$.

   c. What information does $f(12)$ give?

   d. Estimate the values of x for which $f(x) = 0$.

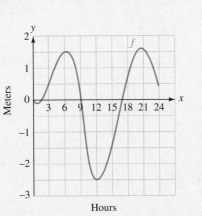

7. Solve the system by graphing:

$$\begin{cases} y = -\dfrac{5}{2}x + \dfrac{1}{2} \\ 2x - \dfrac{3}{2}y = 5 \end{cases}$$ [Section 3.1]

8. Use substitution to solve the system:

$$\begin{cases} x - y = -5 \\ 3x - 2y = -7 \end{cases}$$ [Section 3.2]

9. Solve the system:

$$\begin{cases} x - y + z = 4 \\ x + 2y - z = -1 \\ x + y - 3z = -2 \end{cases}$$ [Section 3.3]

10. Evaluate the determinant: $\begin{vmatrix} -6 & -2 \\ 15 & 4 \end{vmatrix}$ [Section 3.5]

Solve each inequality. Write the solution set in interval notation and then graph it.

11. $5(-2x + 2) > 20 - x$ [Section 4.1]

12. $5x - 3 \geq 2$ and $6 \geq 4x - 3$ [Section 4.2]

13. $|2x - 5| \geq 25$ [Section 4.3]

14. Graph the solution set of the system:

$$\begin{cases} 3x + 2y > 6 \\ x + 3y \leq 2 \end{cases}$$ [Section 4.5]

15. Simplify: $\left(\dfrac{2a^2 b^3 c^{-4}}{5a^{-2} b^{-1} c^3}\right)^{-3}$ [Section 5.1]

16. Write each number in scientific notation and perform the operations. Give the answer in scientific notation and in standard notation:

$$\dfrac{(1,280,000,000)(2,700,000)}{240,000}$$ [Section 5.2]

Perform the indicated operations.

17. $(-8.9t^3 - 2.4t) - (2.1t^3 + 0.8t^2 - t)$ [Section 5.3]

18. $(2a - b)(4a^2 + 2ab + b^2)$ [Section 5.4]

Factor each expression.

19. $x^2 + 4y - xy - 4x$ [Section 5.5]

20. $30a^4 - 4a^3 - 16a^2$ [Section 5.6]

21. $49s^6 - 84s^3 n^2 + 36n^4$ [Section 5.6]

22. $x^2 + 10x + 25 - y^8$ [Section 5.7]

23. $x^4 - 16y^4$ [Section 5.7]

24. $8x^6 + 125y^3$ [Section 5.7]

Solve each equation.

25. $(m + 4)(2m + 3) - 22 = 10m$ [Section 5.9]

26. $6a^3 - 2a = a^2$ [Section 5.9]

Graph each function and give its domain and range.

27. $f(x) = x^3 + x^2 - 6x$ [Section 5.3]

28. $f(x) = \dfrac{4}{x}$ for $x > 0$ [Section 6.1]

29. Simplify: $\dfrac{6x^2 - 7x - 5}{2x^2 + 5x + 2}$ [Section 6.1]

30. Divide: $\dfrac{x^3 + y^3}{x^3 - y^3} \div \dfrac{x^2 - xy + y^2}{x^2 + xy + y^2}$ [Section 6.2]

31. Perform the operations: $\dfrac{1}{x + y} - \dfrac{1}{x - y} + \dfrac{2y}{x^2 - y^2}$

[Section 6.3]

32. Simplify: $\dfrac{\dfrac{1}{r^2 + 4r + 4}}{\dfrac{r}{r + 2} + \dfrac{r}{r + 2}}$ [Section 6.4]

33. Divide: $\dfrac{24x^6y^7 - 12x^5y^{12} + 36xy}{48x^2y^3}$

[Section 6.5]

34. Divide: $3a - 4 \overline{)15a^3 - 29a^2 + 16}$
[Section 6.5]

35. Solve: $\dfrac{x - 4}{x - 3} + \dfrac{x - 2}{x - 3} = x - 3$
[Section 6.7]

36. Solve for R: $\dfrac{1}{R} = \dfrac{1}{R_1} + \dfrac{1}{R_2} + \dfrac{1}{R_3}$

[Section 6.7]

37. SINKS   A sink has two faucets, one for cold water and one for hot water. It can be filled by the cold-water faucet in 30 seconds and by the hot-water faucet in 45 seconds. How long will it take to fill the sink if both faucets are opened? [Section 6.7]

38. SNOW REMOVAL   A state highway department uses a 7-to-2 sand-to-salt mix in the winter months for spreading across roadways covered with snow and ice. If they have 6 tons of salt in storage, how many tons of sand should be added to obtain the proper mix? [Section 6.8]

39. DELIVERIES   The costs of a delivery company vary jointly with the number of trucks in service and the number of hours they are used. When 8 trucks are used for 12 hours each, the costs are $3,600. Find the costs of using 20 trucks, each for 12 hours. [Section 6.8]

40. Graph the function $f(x) = \sqrt{x - 2}$ and give its domain and range. [Section 7.1]

Simplify each expression.

41. $\sqrt[3]{-27x^3}$ [Section 7.1]

42. $\sqrt{48t^3}$ [Section 7.1]

43. $64^{-2/3}$ [Section 7.2]

44. $\dfrac{x^{5/3}x^{1/2}}{x^{3/4}}$ [Section 7.2]

45. $-3\sqrt[4]{32} - 2\sqrt[4]{162} + 5\sqrt[4]{48}$ [Section 7.3]

46. $3\sqrt{2}\left(2\sqrt{3} - 4\sqrt{12}\right)$ [Section 7.4]

47. $\dfrac{\sqrt{x} + 2}{\sqrt{x} - 1}$ [Section 7.4]

48. $\dfrac{5}{\sqrt[3]{x}}$ [Section 7.4]

Solve each equation.

49. $5\sqrt{x + 2} = x + 8$ [Section 7.5]

50. $\sqrt{x} + \sqrt{x + 2} = 2$ [Section 7.5]

51. Find the length of the hypotenuse of the right triangle in figure (a). [Section 7.6]

52. Find the length of the hypotenuse of the right triangle in figure (b). [Section 7.6]

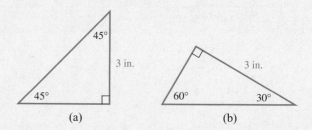

(a)   (b)

53. Find the distance between $(-2, 6)$ and $(4, 14)$. [Section 7.6]

54. Simplify: i^{43} [Section 7.7]

Perform the indicated operations. Write each result in $a + bi$ form.

55. $\left(-7 + \sqrt{-81}\right) - \left(-2 - \sqrt{-64}\right)$ [Section 7.7]

56. $\dfrac{5}{3 - i}$ [Section 7.7]

57. $(2 + i)^2$ [Section 7.7]

58. $\dfrac{-4}{6i^7}$ [Section 7.7]

Solve each equation.

59. $x^2 = 28$ [Section 8.1]

60. $(x - 19)^2 = -5$ [Section 8.1]

61. Use the method of completing the square to solve $2x^2 - 6x + 3 = 0$. [Section 8.1]

62. Use the quadratic formula to solve $a^2 - \frac{2}{5}a = -\frac{1}{5}$.
[Section 8.2]

63. COMMUNITY GARDENS   Residents of a community can work their own 16-ft × 24-ft plot of city-owned land if they agree to the following conditions:

- The area of the garden cannot exceed 180 square feet.
- A path of uniform width must be maintained around the garden.

Find the dimensions of the largest possible garden.
[Section 8.2]

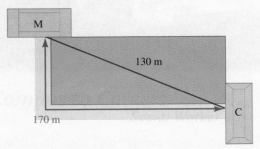

24 ft

16 ft

64. SIDEWALKS   A 170-meter-long sidewalk from the mathematics building M to the student center C is shown in red in the illustration. However, students prefer to walk directly from M to C, across a lawn. How long are the two segments of the existing sidewalk? [Section 8.2]

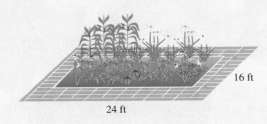

M

130 m

170 m

C

Solve each equation.

65. $t^{2/3} - t^{1/3} = 6$ [Section 8.3]

66. $x^{-4} - 2x^{-2} + 1 = 0$ [Section 8.3]

67. First determine the vertex and the axis of symmetry of the graph of $f(x) = -x^2 - 4x$ using the vertex formula. Then determine the x- and y-intercepts of the graph. Finally, plot several points and complete the graph. [Section 8.4]

Solve each inequality. Write the solution set in interval notation and then graph it.

68. $x^2 - 81 < 0$ [Section 8.5]

69. $\dfrac{1}{x+1} \geq \dfrac{x}{x+4}$ [Section 8.5]

70. a. The graph of $f(x) = 16x^2 + 24x + 9$ is shown below. Estimate the solution(s) of $16x^2 + 24x + 9 = 0$.
[Section 8.5]

b. Use the graph to determine the solution of $16x^2 + 24x + 9 < 0$.

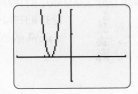

To be in the domain of the composite function $f \circ g$, a number x has to be in the domain of g and the output of g must be in the domain of f. Thus, the domain of $f \circ g$ consists of those numbers x that are in the domain of g, and for which $g(x)$ is in the domain of f.

Composite Functions	The **composite function $f \circ g$** is defined by $$(f \circ g)(x) = f(g(x))$$

Caution

$(f \circ g)(x)$
‎└─ Composition

does not mean

$(f \cdot g)(x)$
‎└─ Multiplication

If $f(x) = 4x$ and $g(x) = 3x + 2$, to find $f \circ g$ and $g \circ f$, we proceed as follows.

$$
\begin{aligned}
(f \circ g)(x) &= f(g(x)) & \qquad (g \circ f)(x) &= g(f(x)) \\
&= f(3x + 2) & &= g(4x) \\
&= 4(3x + 2) & &= 3(4x) + 2 \\
&= 12x + 8 & &= 12x + 2
\end{aligned}
$$

└──────── Different results ────────┘

The different results illustrate that the composition of functions is not commutative. Ususally, we will find that $(f \circ g)(x) \neq (g \circ f)(x)$.

EXAMPLE 3 Let $f(x) = 2x + 1$ and $g(x) = x - 4$. Find:

a. $(f \circ g)(9)$ **b.** $(f \circ g)(x)$ **c.** $(g \circ f)(-2)$

Strategy In part (a), we will find $f(g(9))$. In part (b), we will find $f(g(x))$. In part (c), we will find $g(f(-2))$.

The Language of Algebra

$f(g(x))$

We call these
nested parentheses.

Why To evaluate a composition function written with the circle $\circ$ notation, we rewrite it using nested parentheses: $(f \circ g)(x) = f(g(x))$.

Solution

a. $(f \circ g)(9)$ means $f(g(9))$. In figure (a) on the next page, function g receives the number 9, subtracts 4, and releases the number $g(9) = 5$. Then 5 goes into the f function, which doubles 5 and adds 1. The final result, 11, is the output of the composite function $f \circ g$:

Read as "f of g of 9."
↓

$$
\begin{aligned}
(f \circ g)(9) &= f(g(9)) & &\text{Change from } \circ \text{ notation to nested parentheses.} \\
&= f(5) & &\text{Evaluate: } g(9) = 9 - 4 = 5. \\
&= 2(5) + 1 & &\text{Evaluate } f(5) \text{ using } f(x) = 2x + 1. \\
&= 11
\end{aligned}
$$

Thus, $(f \circ g)(9) = 11$.

b. $(f \circ g)(x)$ means $f(g(x))$. In figure (a) on the next page, function g receives the number x, subtracts 4, and releases the number $x - 4$. Then $x - 4$ goes into the f function, which doubles $x - 4$ and adds 1. The final result, $2x - 7$, is the output of the composite function $f \circ g$.

Read as "f of g of x."
↓

$$
\begin{aligned}
(f \circ g)(x) &= f(g(x)) & &\text{Change from } \circ \text{ notation to nested parentheses.} \\
&= f(x - 4) & &\text{We are given } g(x) = x - 4. \\
&= 2(x - 4) + 1 & &\text{Find } f(x - 4) \text{ using } f(x) = 2x + 1. \\
&= 2x - 8 + 1 \\
&= 2x - 7
\end{aligned}
$$

Thus, $(f \circ g)(x) = 2x - 7$.

Notation

The notation $f \circ g$ can also be read as "f circle g." Remember, it means that the function g is applied first and function f is applied second.

c. $(g \circ f)(-2)$ means $g(f(-2))$. In figure (b) below, function f receives the number -2, doubles it and adds 1, and releases -3 into the g function. Function g subtracts 4 from -3 and outputs a final result of -7. Thus,

Read as "g of f of -2."
$\downarrow$

$$(g \circ f)(-2) = g(f(-2)) \quad \text{Change from } \circ \text{ notation to nested parentheses.}$$
$$= g(-3) \quad \text{Evaluate } f(-2) \text{ using } f(x) = 2x + 1.$$
$$= -3 - 4 \quad \text{Evaluate } g(-3) \text{ using } g(x) = x - 4.$$
$$= -7$$

Thus, $(g \circ f)(-2) = -7$.

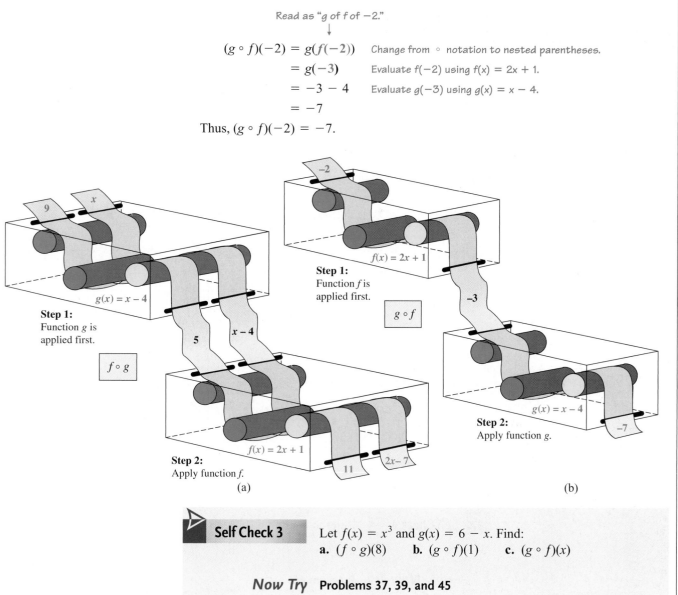

(a)

(b)

Self Check 3   Let $f(x) = x^3$ and $g(x) = 6 - x$. Find:
a. $(f \circ g)(8)$ **b.** $(g \circ f)(1)$ **c.** $(g \circ f)(x)$

Now Try   **Problems 37, 39, and 45**

---

3   **Use Graphs to Evaluate Functions.**

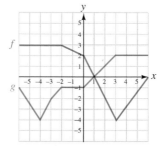

EXAMPLE 4   Refer to the graphs of functions f and g on the left to find each of the following.
a. $(f + g)(-4)$
b. $(f \cdot g)(2)$
c. $(f \circ g)(-3)$

Strategy   We will express the sum, product, and composition functions in terms of the functions from which they are formed.

Why We can evaluate sum, product, and composition functions at a given x-value by evaluating each function from which they are formed at that x-value.

Solution

a. $(f + g)(-4) = f(-4) + g(-4)$

$\qquad = 3 + (-4)$

$\qquad = -1$

b. $(f \cdot g)(2) = f(2) \cdot g(2)$

$\qquad = -2 \cdot 1$

$\qquad = -2$

c. $(f \circ g)(-3) = f(g(-3))$

$\qquad = f(-2)$

$\qquad = 3$

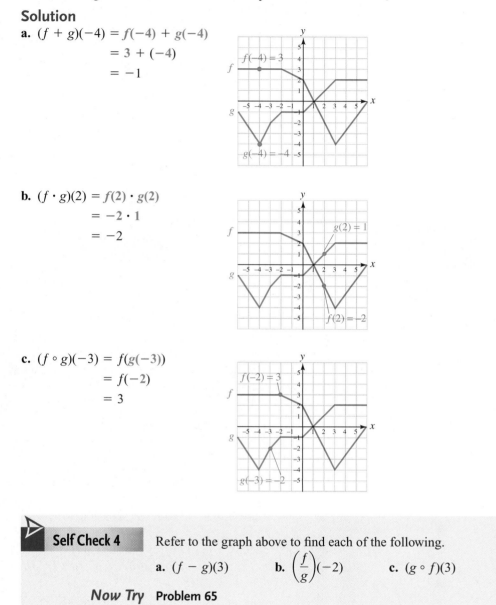

Self Check 4 Refer to the graph above to find each of the following.

a. $(f - g)(3)$ **b.** $\left(\dfrac{f}{g}\right)(-2)$ **c.** $(g \circ f)(3)$

Now Try **Problem 65**

4 **Define the Identity Function.**

The **identity function** is defined by the equation $I(x) = x$. Under this function, the value that is assigned to any real number x is x itself. For example $I(2) = 2$, $I(-3) = -3$, and $I(7.5) = 7.5$. If f is any function, the composition of f with the identity function is just the function f:

$$(f \circ I)(x) = (I \circ f)(x) = f(x)$$

EXAMPLE 5 Let f be any function and let I be the identity function, $I(x) = x$. Show that **a.** $(f \circ I)(x) = f(x)$ **b.** $(I \circ f)(x) = f(x)$

Strategy In part (a), we will find $f(I(x))$ and in part (b), we will find $I(f(x))$.

Why To find a composition function written with the circle $\circ$ notation, we rewrite it using the nested parentheses notation.

Solution

The Language of Algebra
The *identity* function pairs each real number with itself such that each output is *identical* to its corresponding input.

a. $(f \circ I)(x)$ means $f(I(x))$. Because $I(x) = x$, we have

$$(f \circ I)(x) = f(I(x)) = f(x)$$

b. $(I \circ f)(x)$ means $I(f(x))$. Because I passes any number through unchanged, we have

$$(I \circ f)(x) = I(f(x)) = f(x)$$

 Now Try Problems 69 and 71

⑤ **Use Composite Functions to Solve Problems.**

EXAMPLE 6 *Biological Research.* A specimen is stored in refrigeration at a temperature of 15° Fahrenheit. Biologists remove the specimen and warm it at a controlled rate of 3°F per hour. Express its Celsius temperature as a function of the time t since it was removed from refrigeration.

Strategy We will express the Fahrenheit temperature of the specimen as a function of the time t since it was removed from refrigeration. Then we will express the Celsius temperature of the specimen as a function of its Fahrenheit temperature and find the composition of the two functions.

Why The Celsius temperature of the specimen is a function of its Fahrenheit temperature. Its Fahrenheit temperature is a function of the time since it was removed from refrigeration. This chain of dependence suggests that we write a composition of functions.

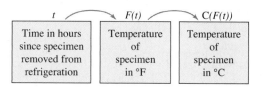

Solution The temperature of the specimen is 15°F when the time $t = 0$. Because it warms at a rate of 3°F per hour, its initial temperature of 15°F increases by $3t$°F in t hours. The Fahrenheit temperature at time t of the specimen is given by the function

$$F(t) = 3t + 15$$

The Celsius temperature C is a function of this Fahrenheit temperature F, given by the function

$$C(F) = \frac{5}{9}(F - 32)$$

To express the specimen's Celsius temperature as a function of *time,* we find the composite function $(C \circ F)(t)$.

$$(C \circ F)(t) = C(F(t))$$

$$= C(3t + 15) \qquad \text{Substitute } 3t + 15 \text{ for } F(t).$$

$$= \frac{5}{9}[(3t + 15) - 32] \qquad \text{Substitute } 3t + 15 \text{ for } F \text{ in } \frac{5}{9}(F - 32).$$

$$= \frac{5}{9}(3t - 17) \qquad \text{Simplify within the brackets.}$$

$$= \frac{15}{9}t - \frac{85}{9} \qquad \text{Distribute the multiplication by } \frac{5}{9}.$$

$$= \frac{5}{3}t - \frac{85}{9}$$

The composite function, $C(t) = \frac{5}{3}t - \frac{85}{9}$, gives the temperature of the specimen in degrees Celsius *t* hours after it is removed from refrigeration.

Now Try Problem 75

ANSWERS TO SELF CHECKS   **1. a.** $(f + g)(x) = 2x^2 + 6x - 2$ **b.** $(f - g)(x) = -2x^2 - 2$
2. a. $(f \cdot g)(x) = 2x^4 - x^2 - 3$ **b.** $(f/g)(x) = \dfrac{2x^2 - 3}{x^2 + 1}$ **3. a.** -8 **b.** 5 **c.** $(g \circ f)(x) = 6 - x^3$
4. a. -6 **b.** -3 **c.** -4

STUDY SET
9.1

VOCABULARY

Fill in the blanks.

1. The _____ of *f* and *g*, denoted as $f + g$, is defined by $(f + g)(x) =$ _____ and the _____ of *f* and *g*, denoted as $f - g$, is defined by $(f - g)(x) =$ _____ .

2. The _____ of *f* and *g*, denoted as $f \cdot g$, is defined by $(f \cdot g)(x) =$ _____ and the _____ of *f* and *g*, denoted as f/g, is defined by $(f/g)(x) =$ _____ .

3. The _____ of the function $f + g$ is the set of real numbers *x* that are in the domain of both *f* and *g*.

4. The _____ function $f \circ g$ is defined by $(f \circ g)(x) =$ _____ .

5. Under the _____ function, the value that is assigned to any real number *x* is *x* itself: $I(x) = x$.

6. When reading the notation $f(g(x))$, we say "*f* ____ *g* ____ *x*."

CONCEPTS

7. Fill in the blanks.
   a. $(f \circ g)(3) = f(\quad)$
   b. To find $f(g(3))$, we first find _____ and then substitute that value for *x* in $f(x)$.

8. a. If $f(x) = 3x + 1$ and $g(x) = 1 - 2x$, find $f(g(3))$ and $g(f(3))$.
   b. Is the composition of functions commutative?

9. Fill in the three blanks in the drawing of the function machines that show how to compute $g(f(-2))$.

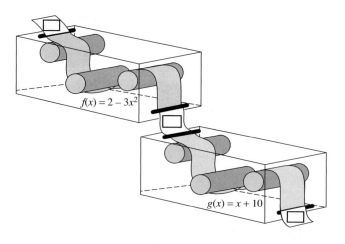

$f(x) = 2 - 3x^2$

$g(x) = x + 10$

10. Complete the table of values for the identity function, $I(x) = x$. Then graph it.

x	$I(x)$
-3	
-2	
-1	
0	
1	
2	
3	

NOTATION

Complete each solution.

11. Let $f(x) = 3x - 1$ and $g(x) = 2x + 3$. Find $f \cdot g$.

$(f \cdot g)(x) = f(x) \cdot$

$\qquad = $ ░░░ $(2x + 3)$

$\qquad = 6x^2 + $ ░ $-$ ░ $- 3$

$(f \cdot g)(x) = 6x^2 + 7x - 3$

12. Let $f(x) = 3x - 1$ and $g(x) = 2x + 3$. Find $f \circ g$.

$(f \circ g)(x) = f($ ░ $)$

$\qquad = f($ ░░ $)$

$\qquad = 3($ ░░ $) - 1$

$\qquad = $ ░ $ + $ ░ $- 1$

$(f \circ g)(x) = 6x + 8$

GUIDED PRACTICE

Let $f(x) = 3x$ and $g(x) = 4x$. Find each function and its domain. See Examples 1 and 2.

13. $f + g$

14. $f - g$

15. $g - f$

16. $g + f$

17. $f \cdot g$

18. f/g

19. g/f

20. $g \cdot f$

Let $f(x) = 2x + 1$ and $g(x) = x - 3$. Find each function and its domain. See Examples 1 and 2.

21. $f + g$

22. $f - g$

23. $g - f$

24. $g + f$

25. $f \cdot g$

26. f/g

27. g/f

28. $g \cdot f$

Let $f(x) = 3x - 2$ and $g(x) = 2x^2 + 1$. Find each function and its domain. See Examples 1 and 2.

29. $f - g$

30. $f + g$

31. f/g

32. $f \cdot g$

Let $f(x) = x^2 - 1$ and $g(x) = x^2 - 4$. Find each function and its domain.

33. $f - g$

34. $f + g$

35. g/f

36. $g \cdot f$

Let $f(x) = 2x + 1$ and $g(x) = x^2 - 1$. Find each of the following. See Example 3.

37. $(f \circ g)(2)$

38. $(g \circ f)(2)$

39. $(g \circ f)(-3)$

40. $(f \circ g)(-3)$

41. $(f \circ g)(0)$

42. $(g \circ f)(0)$

43. $(f \circ g)\left(\dfrac{1}{2}\right)$

44. $(g \circ f)\left(\dfrac{1}{3}\right)$

45. $(f \circ g)(x)$

46. $(g \circ f)(x)$

47. $(g \circ f)(2x)$

48. $(f \circ g)(2x)$

Let $f(x) = 3x - 2$ and $g(x) = x^2 + x$. Find each of the following. See Example 3.

49. $(f \circ g)(4)$

50. $(g \circ f)(4)$

51. $(g \circ f)(-3)$

52. $(f \circ g)(-3)$

53. $(g \circ f)(0)$

54. $(f \circ g)(0)$

55. $(g \circ f)(x)$

56. $(f \circ g)(x)$

Let $f(x) = \frac{1}{x}$ and $g(x) = \frac{1}{x^2}$. Find each of the following.

57. $(f \circ g)(4)$ 58. $(f \circ g)(6)$

59. $(g \circ f)\left(\frac{1}{3}\right)$ 60. $(g \circ f)\left(\frac{1}{10}\right)$

61. $(g \circ f)(8x)$ 62. $(f \circ g)(5x)$

63. If $f(x) = x + 1$ and $g(x) = 2x - 5$, show that $(f \circ g)(x) \neq (g \circ f)(x)$.

64. If $f(x) = x^2 + 1$ and $g(x) = 3x^2 - 2$, show that $(f \circ g)(x) \neq (g \circ f)(x)$.

See Example 4.

65. Refer to graphs.
   a. $(f + g)(-5)$
   b. $(f - g)(3)$
   c. $(f \cdot g)(-3)$
   d. $(f/g)(0)$
   e. $(f \circ g)(3)$
   f. $(g \circ f)(2)$

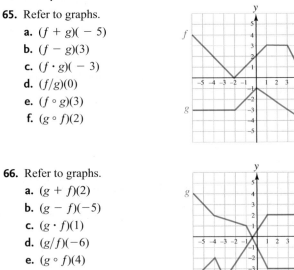

66. Refer to graphs.
   a. $(g + f)(2)$
   b. $(g - f)(-5)$
   c. $(g \cdot f)(1)$
   d. $(g/f)(-6)$
   e. $(g \circ f)(4)$
   f. $(f \circ g)(6)$

67. Use the tables of values for functions f and g to find each of the following.

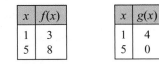

x	$f(x)$
1	3
5	8

x	$g(x)$
1	4
5	0

   a. $(f + g)(1)$ b. $(f - g)(5)$
   c. $(f \cdot g)(1)$ d. $(g/f)(5)$

68. Use the table of values for functions f and g to find each of the following.

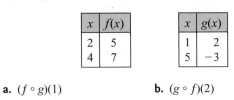

x	$f(x)$
2	5
4	7

x	$g(x)$
1	2
5	-3

   a. $(f \circ g)(1)$ b. $(g \circ f)(2)$

Let $f(x) = 3x + 1$ and find each composition. See Example 5.

69. $(f \circ I)(x)$ 70. $(I \circ f)(x)$

Let $f(x) = -2x - 5$ and find each composition. See Example 5.

71. $(I \circ f)(x)$ 72. $(f \circ I)(x)$

APPLICATIONS

73. **SAT SCORES**   Refer to the following illustration. The graph of function m gives the average score on the mathematics portion of the SAT college entrance exam, the graph of function v gives the average score on the verbal portion, and x represents the number of years since 1990. (In 2006, the SAT Test was renamed the SAT Reasoning Test and it now includes three parts: mathematics, critical reading, and writing.)

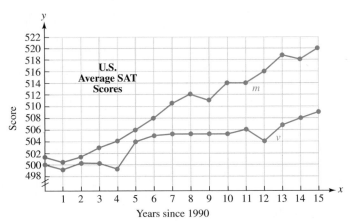

Source: *The World Almanac*, 1999 and Infoplease.com

   a. Find $(m + v)(3)$ and explain what information about SAT scores it gives.

   b. Find $(m - v)(6)$ and explain what information about SAT scores it gives.

   c. Find: $(m + v)(15)$

   d. Find: $(m - v)(15)$

74. BACHELOR'S DEGREES   Refer to the following illustration. The graph of function m gives the actual and projected number of bachelor's degrees awarded to men, and the graph of function w gives the actual and projected number of bachelor's degrees awarded to women.

**Actual and Projected Numbers
for Bachelor's Degrees Awarded in the U.S.**

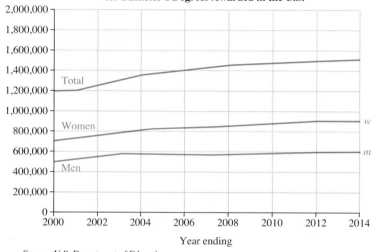

Source: U.S. Department of Education

a. Estimate $(w + m)(2000)$ and explain what information about bachelor's degrees it gives.

b. Estimate $(w - m)(2000)$ and explain what information about bachelor's degrees it gives.

c. Estimate: $(w + m)(2012)$

d. Estimate: $(w - m)(2012)$

75. METALLURGY   A molten alloy must be cooled slowly to control crystallization. When removed from the furnace, its temperature is 2,700°F, and it will be cooled at 200° per hour. Write a composition function that expresses the Celsius temperature as a function of the number of hours t since cooling began. (*Hint*: $C = \frac{5}{9}(F - 32)$.)

76. WEATHER FORECASTING   A high-pressure area promises increasingly warmer weather for the next 48 hours. The temperature is now 34° Celsius and is expected to rise 1° every 6 hours. Write a composition function that expresses the Fahrenheit temperature as a function of the number of hours from now. (*Hint*: $F = \frac{9}{5}C + 32$.)

77. VACATION MILEAGE COSTS

a. Use the following graphs to determine the cost of the gasoline consumed if a family drove 500 miles on a summer vacation.

b. Write a composition function that expresses the cost of the gasoline consumed on the vacation as a function of the miles driven.

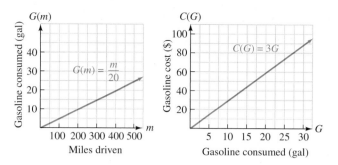

78. HALLOWEEN COSTUMES   The tables on the back of a pattern package can be used to determine the number of yards of material needed to make a rabbit costume for a child.

a. How many yards of material are needed if the child's chest measures 29 inches?

b. In this exercise, one quantity is a function of a second quantity that depends, in turn, on a third quantity. Explain this dependence.

PATTERN 9810   Simplicity

Costumes have front zipper, long raglan sleeves, elastic sleeve and leg casings, hood. Fabrics: Fleece or suede

BODY MEASUREMENTS

Chest (in.)	21	22	23	25	26	27	$28\frac{1}{2}$	29	30
Pattern Size	2	3	4	6	7	8	10	11	12

YARDAGE NEEDED

Pattern Size	2-4	6-8	10-12
Yards	$2\frac{5}{8}$	$3\frac{3}{8}$	$3\frac{3}{4}$

WRITING

79. Exercise 77 illustrates a chain of dependence between the cost of the gasoline, the gasoline consumed, and the miles driven. Describe another chain of dependence that could be represented by a composition function.

80. In this section, what operations are performed on functions? Give an example of each.

81. Write out in words how to say each of the following:

$$(f \circ g)(2) \qquad g(f(-8))$$

82. If $Y_1 = f(x)$ and $Y_2 = g(x)$, explain how to use the following tables to find $g(f(2))$.

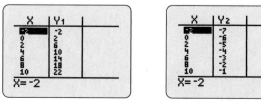

REVIEW

Simplify each complex fraction.

83. $\dfrac{\dfrac{ac - ad - c + d}{a^3 - 1}}{\dfrac{c^2 - 2cd + d^2}{a^2 + a + 1}}$

84. $\dfrac{2 + \dfrac{1}{x^2 - 1}}{1 + \dfrac{1}{x - 1}}$

CHALLENGE PROBLEMS

Fill in the blanks.

85. If $f(x) = x^2$ and $g(x) = \underline{\quad}$, then $(f \circ g)(x) = 4x^2 + 20x + 25$.

86. If $f(x) = \sqrt{3x}$ and $g(x) = \underline{\quad}$, then $(g \circ f)(x) = 9x^2 + 7$.

Refer to the following graphs of functions f and g.

87. Graph the sum function $f + g$ on the given coordinate system.

88. Graph the difference function $f - g$ on the given coordinate system.

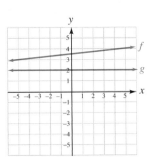

SECTION 9.2
Inverse Functions

Objectives

1 Determine whether a function is a one-to-one function.

2 Use the horizontal line test to determine whether a function is one-to-one.

3 Find the equation of the inverse of a function.

4 Find the composition of a function and its inverse.

5 Graph a function and its inverse.

In the previous section, we created new functions from given functions by using the operations of arithmetic and composition. Another way to create new functions is to find the *inverse of a function*.

1 **Determine Whether a Function Is a One-to-One Function.**

In figure (a) below, the arrow diagram defines a function f. If we reverse the arrows as shown in figure (b), we obtain a new correspondence where the range of f becomes the domain of the new correspondence, and the domain of f becomes the range. The new correspondence is a function because to each member of the domain, there corresponds exactly one member of the range. We call this new correspondence the **inverse** of f, or f inverse.

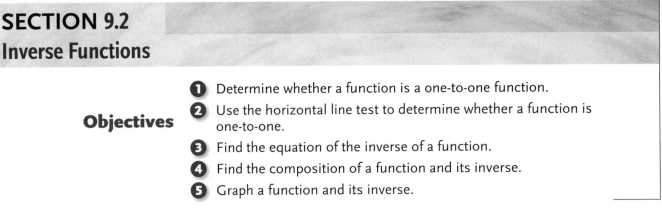

This reversing process does not always produce a function. For example, if we reverse the arrows in function g defined by the diagram in figure (a) below, the resulting correspondence is not a function. This is because to the number 2 in the domain, there corresponds two members of the range: 8 and 4.

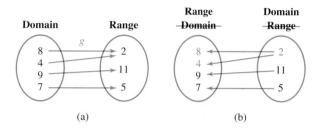

(a)   (b)

The question that arises is, "What must be true of an original function to guarantee that the reversing process produces a function?" The answer is: *the original function must be one-to-one.*

We have seen that in a function, each input determines exactly one output. For some functions, different inputs determine different outputs, as in figure (a). For other functions, different inputs might determine the *same* output, as in figure (b). When a function has the property that different inputs determine different outputs, as in figure (a), we say the function is *one-to-one.*

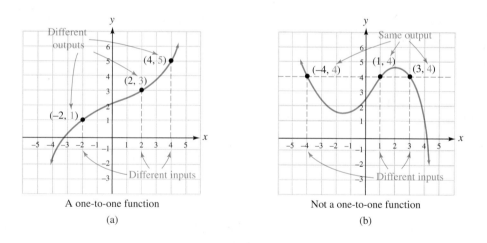

A one-to-one function   Not a one-to-one function
(a)   (b)

One-to-One Functions   A function is called a **one-to-one function** if different inputs determine different outputs.

EXAMPLE 1   Determine whether each function is one-to-one. **a.** $f(x) = x^2$
b. $f(x) = x^3$

Strategy   We will determine whether different inputs have different outputs.

Why   If different inputs have different outputs, the function is one-to-one. If different inputs have the same output, the function is not one-to-one.

Solution

a. Since two different inputs, -3 and 3, have the same output 9, $f(x) = x^2$ is not a one-to-one function.

$f(-3) = (-3)^2 = 9$ and $f(3) = 3^2 = 9$

x	$f(x)$
-3	9
3	9

The output 9 does not correspond to exactly one input.

b. Since different numbers have different cubes, each input of $f(x) = x^3$ determines a different output. This function is one-to-one.

Self Check 1   Determine whether each function is one-to-one. If not, find an output that corresponds to more than one input.

   a. $f(x) = 2x + 3$ **b.** $f(x) = x^2$

Now Try **Problems 19 and 21**

② Use the Horizontal Line Test to Determine Whether a Function Is One-to-One.

To determine whether a function is one-to-one, it is often easier to view its graph rather than its defining equation. If two (or more) points on the graph of a function have the same y-coordinate, the function is not one-to-one. This observation suggests the following *horizontal line test.*

The Horizontal Line Test

A function is one-to-one if each horizontal line that intersects its graph does so exactly once.

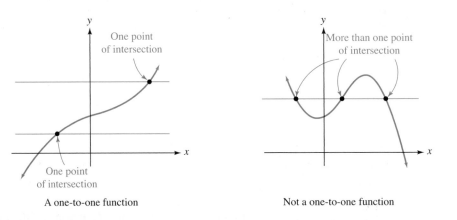

A one-to-one function   Not a one-to-one function

EXAMPLE 2   Use the horizontal line test to determine whether the following graphs of functions represent one-to-one functions.

Strategy   We will draw horizontal lines through the graph of the function and see how many times each line intersects the graph.

Why   If each horizontal line intersects the graph of the function exactly once, the graph represents a one-to-one function. If any horizontal line intersects the graph of the function more than once, the graph does not represent a one-to-one function.

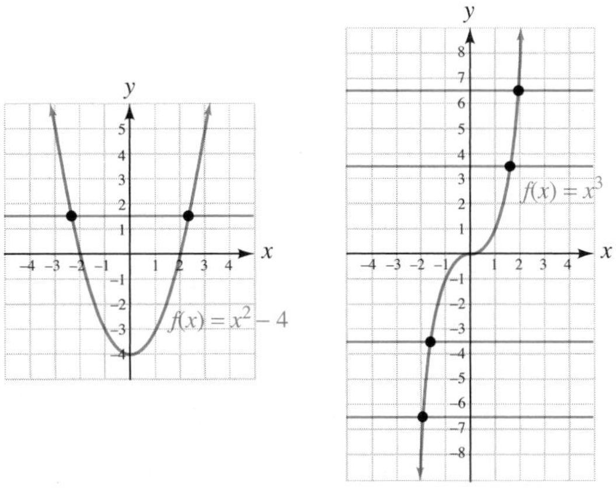

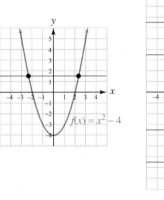

Success Tip

Recall that we use the vertical line test to determine whether a graph represents a function. We use the horizontal line test to determine whether the function that is graphed is one-to-one.

Solution

a. Because we can draw a horizontal line that intersects the graph of the function shown in figure (a) twice, the graph does not represent a one-to-one function.

b. Because every horizontal line that intersects the graph of the function in figure (b) does so exactly once, the graph represents a one-to-one function.

(a) (b)

Self Check 2 Determine whether the following graphs represent one-to-one functions.

a.

b.

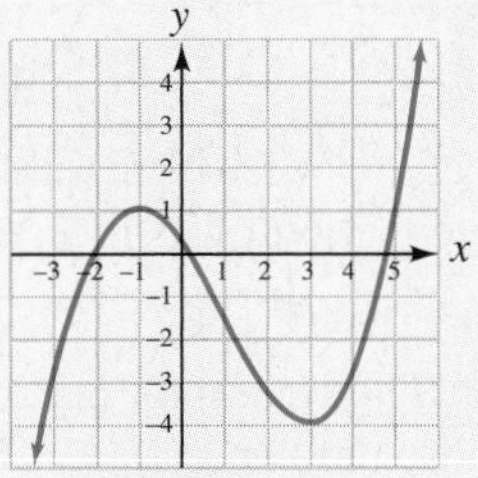

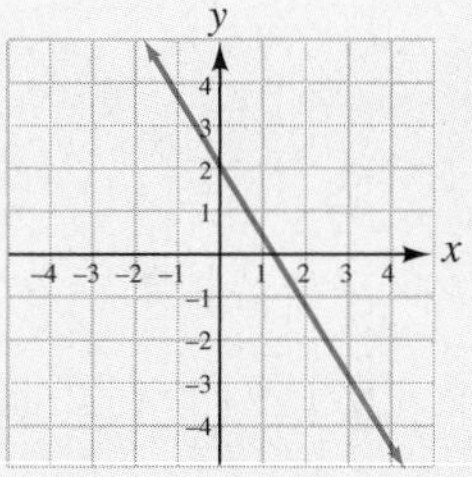

Now Try **Problems 27 and 29**

3 Find the Equation of the Inverse of a Function.

If f is the one-to-one function defined by the arrow diagram in figure (a), it turns the number 1 into 10, 2 into 20, and 3 into 30. The ordered pairs that define f can be listed in a table. Since the inverse of f must turn 10 back into 1, 20 back into 2, and 30 back into 3, it consists of the ordered pairs shown in the table in figure (b).

Function f

x	y	(x, y)
1	10	(1, 10)
2	20	(2, 20)
3	30	(3, 30)

f inverse

x	y	(x, y)
10	1	(10, 1)
20	2	(20, 2)
30	3	(30, 3)

(a) (b)

The x- and y-coordinates are interchanged.

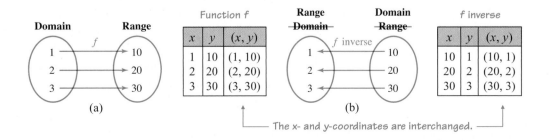

We note that the domain of f and the range of its inverse is $\{1, 2, 3\}$. The range of f and the domain of its inverse is $\{10, 20, 30\}$.

This example suggests that to form the inverse of a function f, we simply interchange the coordinates of each ordered pair that determines f. When the inverse of a function is also a function, we call it **f inverse** and denote it with the symbol f^{-1}. The symbol $f^{-1}(x)$ is read as "the inverse of $f(x)$" or "f inverse of x."

The Inverse of a Function

If f is a one-to-one function consisting of ordered pairs of the form (x, y), the **inverse of f**, denoted f^{-1}, is the one-to-one function consisting of all ordered pairs of the form (y, x).

When a one-to-one function is defined by an equation, we use the following method to find the equation of its inverse.

Finding the Equation of the Inverse of a Function

If a function is one-to-one, we find its inverse as follows:

1. If the function is written using function notation, replace $f(x)$ with y.
2. Interchange the variables x and y.
3. Solve the resulting equation for y.
4. Substitute $f^{-1}(x)$ for y.

EXAMPLE 3   Determine whether each function is one-to-one. If so, find the equation of its inverse.   **a.** $f(x) = 4x + 2$ **b.** $f(x) = x^3$

Strategy   We will determine whether each function is one-to-one. If it is, we can find the equation of its inverse by replacing $f(x)$ with y, interchanging x and y, and solving for y.

Why   The reason for interchanging the variables is this: If a one-to-one function takes an input x into an output y, by definition, its inverse function has the reverse effect.

Solution

a. We recognize $f(x) = 4x + 2$ as a linear function whose graph is a straight line with slope 4 and y-intercept $(0, 2)$. Since such a graph would pass the horizontal line test, we conclude that f is one-to-one.

To find the inverse, we proceed as follows:

$$f(x) = 4x + 2$$
$$y = 4x + 2 \qquad \text{Replace f(x) with y.}$$
$$x = 4y + 2 \qquad \text{Interchange the variables x and y.}$$
$$x - 2 = 4y \qquad \text{To isolate the term 4y, subtract 2 from both sides.}$$
$$\frac{x - 2}{4} = y \qquad \text{To solve for y, divide both sides by 4.}$$
$$y = \frac{x - 2}{4} \qquad \text{Write the equation with y on the left side.}$$

To denote that this equation is the inverse of function f, we replace y with $f^{-1}(x)$.

$$f^{-1}(x) = \frac{x - 2}{4}$$

b. In Example 2, we used the horizontal line test to determine that $f(x) = x^3$ is a one-to-one function. See figure (b) in Example 2.

To find its inverse, we proceed as follows:

Caution
Only one-to-one functions have inverse functions.

$$f(x) = x^3$$
$$y = x^3 \qquad \text{Replace } f(x) \text{ with } y.$$
$$x = y^3 \qquad \text{Interchange the variables } x \text{ and } y.$$
$$\sqrt[3]{x} = y \qquad \text{To solve for } y, \text{ take the cube root of both sides.}$$
$$y = \sqrt[3]{x} \qquad \text{Write the equation with } y \text{ on the left side}$$

Replacing y with $f^{-1}(x)$, we have

$$f^{-1}(x) = \sqrt[3]{x}$$

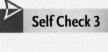

Self Check 3   Determine whether each function is one-to-one. If it is, find the equation of its inverse.   **a.** $f(x) = -5x - 3$
  b. $f(x) = x^5$

Now Try   **Problems 35 and 45**

4   **Find the Composition of a Function and Its Inverse.**

To emphasize a relationship between a function and its inverse, we substitute some number x, such as $x = 3$, into the function $f(x) = 4x + 2$ of Example 3(a). The corresponding value of y that is produced is

$$f(3) = 4(3) + 2 = 14 \qquad f \text{ determines the ordered pair (3, 14).}$$

If we substitute 14 into the inverse function, $f^{-1}(x) = \frac{x-2}{4}$, the corresponding value of y that is produced is

$$f^{-1}(14) = \frac{14 - 2}{4} = 3 \qquad f^{-1} \text{ determines the ordered pair (14, 3).}$$

Thus, the function f turns 3 into 14, and the inverse function f^{-1} turns 14 back into 3.

In general, the composition of a function and its inverse function is the identity function, $I(x) = x$, such that any input x has the output x. This fact can be stated symbolically as follows.

The Composition of Inverse Functions

For any one-to-one function f and its inverse, f^{-1},

$$(f \circ f^{-1})(x) = x \quad \text{and} \quad (f^{-1} \circ f)(x) = x$$

We can use this property to determine whether two functions are inverses.

EXAMPLE 4   Show that $f(x) = 4x + 2$ and $f^{-1}(x) = \frac{x-2}{4}$ are inverses.

Strategy   We will find the composition of $f(x)$ and $f^{-1}(x)$ in both directions and show that the result is x.

Why   Only when the result of the composition is x in both directions are the functions inverses.

Solution   To show that $f(x) = 4x + 2$ and $f^{-1}(x) = \frac{x - 2}{4}$ are inverses, we must show that for each composition, an input of x gives an output of x.

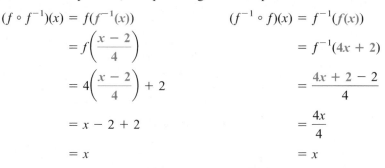

$$(f \circ f^{-1})(x) = f(f^{-1}(x)) \qquad\qquad (f^{-1} \circ f)(x) = f^{-1}(f(x))$$

$$= f\!\left(\frac{x - 2}{4}\right) \qquad\qquad = f^{-1}(4x + 2)$$

$$= 4\!\left(\frac{x - 2}{4}\right) + 2 \qquad\qquad = \frac{4x + 2 - 2}{4}$$

$$= x - 2 + 2 \qquad\qquad = \frac{4x}{4}$$

$$= x \qquad\qquad = x$$

Because $(f \circ f^{-1})(x) = x$ and $(f^{-1} \circ f)(x) = x$, the functions are inverses.

Self Check 4   Show that $f(x) = x - 4$ and $g(x) = x + 4$ are inverses.

Now Try   **Problem 55**

---

5   **Graph a Function and Its Inverse.**

Success Tip
Recall that the line $y = x$ passes through points whose x- and y-coordinates are equal: $(-1, -1)$, $(0, 0)$, $(1, 1)$, $(2, 2)$, and so on.

If a point (a, b) is on the graph of function f, it follows that the point (b, a) is on the graph of f^{-1}, and vice versa. There is a geometric relationship between a pair of points whose coordinates are interchanged. For example, in the graph, we see that the line segment between $(1, 3)$ and $(3, 1)$ is perpendicular to and cut in half by the line $y = x$. We say that $(1, 3)$ and $(3, 1)$ are mirror images of each other with respect to $y = x$.

Since each point on the graph of f^{-1} is a mirror image of a point on the graph of f, and vice versa, the graphs of f and f^{-1} must be mirror images of each other with respect to $y = x$.

EXAMPLE 5   Find the equation of the inverse of $f(x) = -\frac{3}{2}x + 3$. Then graph f and its inverse on one coordinate system.

Strategy   We will determine whether the function has an inverse. If so, we will replace $f(x)$ with y, interchange x and y, and solve for y to obtain the equation of the inverse.

Why   The reason for interchanging the variables is this: If a one-to-one function takes an input x into an output y, by definition, its inverse function has the reverse effect.

Solution   Since $f(x) = -\frac{3}{2}x + 3$ is a linear function, it is one-to-one and has an inverse. To find the inverse function, we replace $f(x)$ with y, and interchange x and y to obtain

$$x = -\frac{3}{2}y + 3$$

Then we solve for y to get

$$x - 3 = -\frac{3}{2}y \qquad \text{Subtract 3 from both sides.}$$

$$-\frac{2}{3}x + 2 = y \qquad \text{To isolate } y, \text{ multiply both sides by } -\frac{2}{3}.$$

When we replace y with $f^{-1}(x)$, we have $f^{-1}(x) = -\frac{2}{3}x + 2$.

To graph f and f^{-1}, we construct tables of values and plot points. Because the functions are inverses of each other, their graphs are mirror images about the line $y = x$

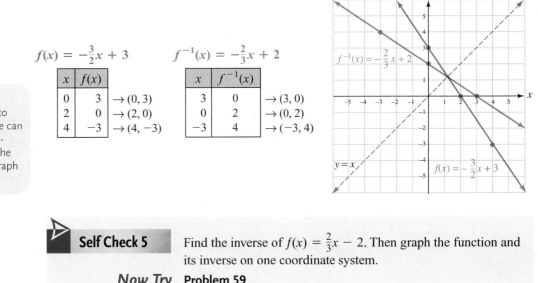

$$f(x) = -\frac{3}{2}x + 3 \qquad f^{-1}(x) = -\frac{2}{3}x + 2$$

x	$f(x)$	
0	3	→ (0, 3)
2	0	→ (2, 0)
4	−3	→ (4, −3)

x	$f^{-1}(x)$	
3	0	→ (3, 0)
0	2	→ (0, 2)
−3	4	→ (−3, 4)

Success Tip

To graph f^{-1}, we don't need to construct a table of values. We can simply interchange the coordinates of the ordered pairs in the table for f and use them to graph f^{-1}.

Self Check 5 Find the inverse of $f(x) = \frac{2}{3}x - 2$. Then graph the function and its inverse on one coordinate system.

Now Try Problem 59

Using Your Calculator *Graphing the Inverse of a Function*

We can use a graphing calculator to check the result found in Example 5. First, we enter $f(x) = -\frac{3}{2}x + 3$ and then enter what we believe to be the inverse function, $f^{-1}(x) = -\frac{2}{3}x + 2$, as well as the equation $y = x$. See figure (a). Before graphing, we adjust the display so that the graphing grid will be composed of squares. The axis of symmetry is then at a 45° angle to the positive x-axis.

In figure (b), it appears that the two graphs are symmetric about the line $y = x$. Although it is not definitive, this visual check does help to validate the result of Example 5.

        (a)   (b)

EXAMPLE 6 Graph the inverse of function *f* shown in figure (a).

Strategy We will find the coordinates of several points on curve *f* in figure (a). After interchanging the coordinates of these points, we will plot them as shown in figure (b).

Why The reason for interchanging the coordinates is this: If (a, b) is a point on the graph of a one-to-one function, then the point (b, a) is on the graph of its inverse.

Solution In figure (a), we see that the points $(-5, -3)$, $(-2, -1)$, $(0, 2)$, $(3, 3)$, $(5, 4)$, and $(7, 5)$ lie on the graph of function *f*. To graph the inverse, we interchange their coordinates, and plot them, as shown in figure (b). Then we graph the line $y = x$ and use symmetry to draw a smooth curve through those points to get the graph of f^{-1}.

> **The Language of Algebra**
> We can also say that the graphs of f and f^{-1} are *reflections* of each other about the line $y = x$, or they are *symmetric about* $y = x$.

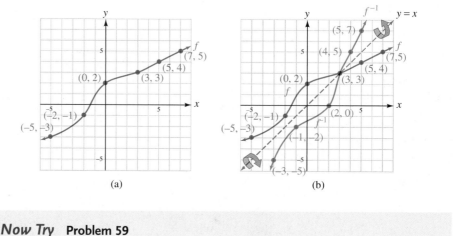

(a)   (b)

Now Try Problem 59

ANSWER TO SELF CHECKS   **1. a.** Yes   **b.** No, $(-1, 1), (1, 1)$ **2. a.** No   **b.** Yes

3. a. $f^{-1}(x) = \dfrac{-x - 3}{5}$ **b.** $f^{-1}(x) = \sqrt[5]{x}$ **5.**

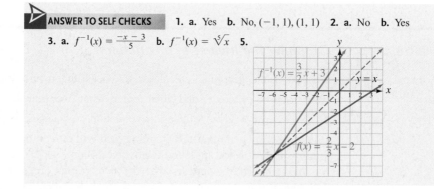

STUDY SET
9.2

VOCABULARY

Fill in the blanks.

1. A function is called a _____ function if different inputs determine different outputs.

2. The _____ line test can be used to determine whether the graph of a function represents a one-to-one function.

3. The functions f and f^{-1} are _____.

4. The graphs of a function and its inverse are mirror _____ of each other with respect to $y = x$. We also say that their graphs are _____ with respect to the line $y = x$.

CONCEPTS

Fill in the blanks.

5. If any horizontal line that intersects the graph of a function does so more than once, the function is not _____.

6. To find the inverse of the function $f(x) = 2x - 3$, we begin by replacing $f(x)$ with y, and then we _____ x and y.

7. If f is a one-to-one function, the domain of f is the _____ of f^{-1}, and the range of f is the _____ of f^{-1}.

8. If a function turns an input of 2 into an output of 5, the inverse function will turn an input of 5 into the output __.

9. If f is a one-to-one function, and if $f(1) = 6$, then $f^{-1}(6) = $ __?

10. If the point $(9, -4)$ is on the graph of the one-to-one function f, then the point (__ , __) is on the graph of f^{-1}.

11. a. Is the correspondence defined by the arrow diagram a one-to-one function?

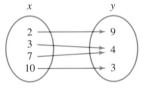

    b. Is the correspondence defined by the table a one-to-one function?

x	$f(x)$
-2	4
-1	1
0	0
2	4
3	9

12. Is the inverse of a one-to-one function always a function?

13. Use the table of values of the one-to-one function f to complete a table of values for f^{-1}.

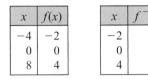

14. Redraw the graph of function f. Then graph f^{-1} and the axis of symmetry on the same coordinate system.

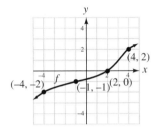

NOTATION

Complete each solution.

15. Find the inverse of $f(x) = 2x - 3$.

$$\boxed{} = 2x - 3$$
$$x = \boxed{} - 3$$
$$x + \boxed{} = 2y$$
$$\frac{x + 3}{2} = \boxed{}$$

The inverse of $f(x) = 2x - 3$ is $\boxed{\phantom{f^{-1}}}(x) = \dfrac{x + 3}{2}$.

16. Find the inverse of $f(x) = \sqrt[3]{x} + 2$.

$$\boxed{} = \sqrt[3]{x} + 2$$
$$x = \sqrt[3]{\boxed{}} + 2$$
$$x - \boxed{} = \sqrt[3]{y}$$
$$(x - 2)^3 = \boxed{}$$

The inverse of $f(x) = \sqrt[3]{x} + 2$ is $\boxed{}(x) = (x - 2)^3$.

17. The symbol f^{-1} is read as "the _____ of f" or "f _____."

18. Explain the difference in the meaning of the -1 in the notation $f^{-1}(x)$ as compared with x^{-1}.

GUIDED PRACTICE

Determine whether each function is one-to-one. See Example 1.

19. $f(x) = 2x$

20. $f(x) = |x|$

21. $f(x) = x^4$

22. $f(x) = x^3 + 1$

23. $f(x) = -x^2 + 3x$

24. $f(x) = \dfrac{2}{3}x + 8$

25. $\{(1, 1), (2, 1), (3, 1), (4, 1)\}$

26. $\{(3, 2), (2, 1), (1,0)\}$

Each graph represents a function. Use the horizontal line test to determine whether the function is one-to-one. See Example 2.

27.

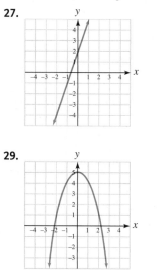

28.

29.

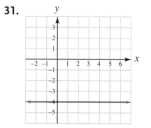

30.

31.

32.

33.

34.

Each of the following functions is one-to-one. Find the inverse of each function and express it using $f^{-1}(x)$ notation. See Example 3.

35. $f(x) = 2x + 4$

36. $f(x) = 5x - 1$

37. $f(x) = \dfrac{x}{5} + \dfrac{4}{5}$

38. $f(x) = \dfrac{x}{3} - \dfrac{1}{3}$

39. $f(x) = \dfrac{x - 4}{5}$

40. $f(x) = \dfrac{2x + 6}{3}$

41. $f(x) = \dfrac{2}{x - 3}$

42. $f(x) = \dfrac{3}{x + 1}$

43. $f(x) = \dfrac{4}{x}$

44. $f(x) = \dfrac{1}{x}$

45. $f(x) = x^3 + 8$

46. $f(x) = x^3 - 4$

47. $f(x) = \sqrt[3]{x}$

48. $f(x) = \sqrt[3]{x - 5}$

49. $f(x) = (x + 10)^3$

50. $f(x) = (x - 9)^3$

51. $f(x) = 2x^3 - 3$

52. $f(x) = \dfrac{3}{x^3} - 1$

53. $f(x) = \dfrac{x^7}{2}$

54. $f(x) = \dfrac{x^9}{4}$

Show that each pair of functions are inverses. See Example 4.

55. $f(x) = 2x + 9,\ f^{-1}(x) = \dfrac{x - 9}{2}$

56. $f(x) = 5x - 1,\ f^{-1}(x) = \dfrac{x + 1}{5}$

57. $f(x) = \dfrac{2}{x - 3},\ f^{-1}(x) = \dfrac{2}{x} + 3$

58. $f(x) = \sqrt[3]{x - 6},\ f^{-1}(x) = x^3 + 6$

Find the inverse of each function. Then graph the function and its inverse on one coordinate system. Show the line of symmetry on the graph. See Examples 5 and 6.

59. $f(x) = 2x$

60. $f(x) = -3x$

61. $f(x) = 4x + 3$

62. $f(x) = \dfrac{x}{3} + \dfrac{1}{3}$

63. $f(x) = -\dfrac{2}{3}x + 3$

64. $f(x) = -\dfrac{1}{3}x + \dfrac{4}{3}$

65. $f(x) = x^3$

66. $f(x) = x^3 + 1$

67. $f(x) = x^2 - 1\ (x \geq 0)$

68. $f(x) = x^2 + 1\ (x \geq 0)$

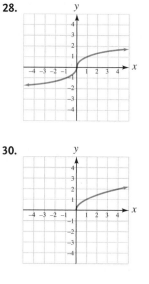
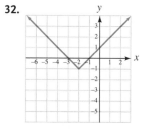
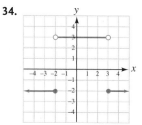

APPLICATIONS

69. INTERPERSONAL RELATIONSHIPS   Feelings of anxiety in a relationship can increase or decrease, depending on what is going on in the relationship. The graph shows how a person's anxiety might vary as a relationship develops over time.

a. Is this the graph of a function? Is its inverse a function?

b. Does each anxiety level correspond to exactly one point in time? Use the dashed lined labeled *Maximum threshold* to explain.

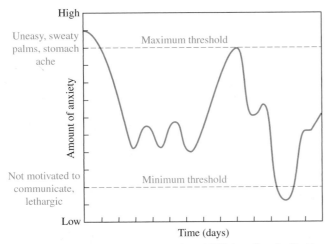

Source: Gudykunst, *Building Bridges: Interpersonal Skills for a Changing World* (Houghton Mifflin, 1994)

70. LIGHTING LEVELS   The ability of the eye to see detail increases as the level of illumination increases. This relationship can be modeled by a function E, whose graph is shown here.

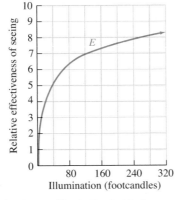

a. From the graph, determine $E(240)$.

b. Is function E one-to-one? Does E have an inverse?

c. If the effectiveness of seeing in an office is 7, what is the illumination in the office? How can this question be asked using inverse function notation?

WRITING

71. In your own words, what is a one-to-one function?

72. Two functions are graphed on the square grid on the right along with the line $y = x$. Explain why the functions cannot be inverses of each other.

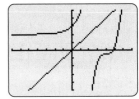

73. Explain how the graph of a one-to-one function can be used to draw the graph of its inverse function.

74. a. Explain the purpose of the vertical line test.

b. Explain the purpose of the horizontal line test.

75. In the illustration, a function f and its inverse f^{-1} have been graphed on the same coordinate system. Explain what concept can be demonstrated by folding the graph paper on the dashed line.

76. Write in words how to read the notation.

a. $f^{-1}(x) = \dfrac{1}{2}x - 3$

b. $(f \circ f^{-1})(x) = x$

REVIEW

Simplify. Write the result in a + bi form.

77. $3 - \sqrt{-64}$

78. $(2 - 3i) + (4 + 5i)$

79. $(3 + 4i)(2 - 3i)$

80. $\dfrac{6 + 7i}{3 - 4i}$

81. $(6 - 8i)^2$

82. i^{100}

CHALLENGE PROBLEMS

83. Find the inverse of $f(x) = \dfrac{x + 1}{x - 1}$.

84. Using the functions of Exercise 83, show that $(f \circ f^{-1})(x) = x$ and $(f^{-1} \circ f)(x) = x$.

85. A table of values for a function f is shown in figure (a). A table of values for f^{-1} is shown in figure (b). Use the tables to find $f^{-1}(f(4))$ and $f(f^{-1}(2))$.

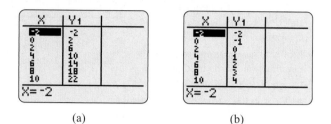

    (a)   (b)

86. a. The graph of a one-to-one function lies entirely in quadrant I. In what quadrant does the graph of its inverse lie?

b. The graph of a one-to-one function lies entirely in quadrant II. In what quadrant does the graph of its inverse lie?

c. The graph of a one-to-one function lies entirely in quadrant III. In what quadrant does the graph of its inverse lie?

d. The graph of a one-to-one function lies entirely in quadrant IV. In what quadrant does the graph of its inverse lie?

SECTION 9.3
Exponential Functions

Objectives

① Simplify expressions containing irrational exponents.

② Define exponential functions.

③ Graph exponential functions.

④ Use exponential functions in applications involving growth or decay.

The graph in figure (a) below shows the balance in a bank account in which $10,000 was invested in 2006 at 9%, compounded monthly. The graph shows that in the year 2016, the value of the account will be approximately $25,000, and in the year 2036, the value will be approximately $147,000. The rapidly rising red curve is the graph of a function called an *exponential function.*

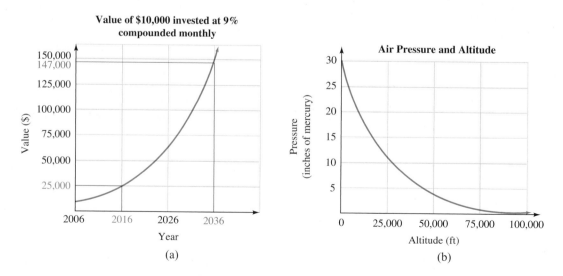

If you have ever climbed a high mountain or gone up in an airplane that does not have a pressurized cabin, you have probably felt the effects of low air pressure. The graph in figure

(b) on the previous page shows the how atmospheric pressure decreases with increasing altitude. The rapidly falling red curve is also the graph of an *exponential function*.

Exponential functions are used to model many other situations, such as population growth, the spread of an epidemic, the temperature of a heated object as it cools, and radioactive decay. Before we can discuss exponential functions in more detail, we must define irrational exponents.

 Simplify Expressions Containing Irrational Exponents.

We have discussed expressions of the form b^x, where x is a rational number.

$8^{1/2}$ means "the square root of 8."

$5^{1/3}$ means "the cube root of 5."

$3^{-2/5} = \dfrac{1}{3^{2/5}}$ means "the reciprocal of the fifth root of 3^2."

To give meaning to b^x when x is an irrational number, we consider the expression

$5^{\sqrt{2}}$ where $\sqrt{2}$ is the irrational number $1.414213562\ldots$

We can successively approximate $5^{\sqrt{2}}$ by the following rational powers:

$$5^{1.4}, \quad 5^{1.41}, \quad 5^{1.414}, \quad 5^{1.4142}, \quad 5^{1.41421}, \ldots$$

Using concepts from advanced mathematics, it can be shown that there is exactly one number that these powers approach. We define $5^{\sqrt{2}}$ to be that number. This process can be used to approximate $5^{\sqrt{2}}$ to as many decimal places as desired.

Any other positive irrational exponent can be defined in the same manner, and negative irrational exponents can be defined using reciprocals. Thus, if b is positive, b^x has meaning for any real number x.

We can use a calculator to obtain a very good approximation of an exponential expression with an irrational exponent.

Using Your Calculator | *Evaluating Exponential Expressions*

To find the value of $5^{\sqrt{2}}$ with a scientific calculator, we enter these numbers and press these keys:

$5\ \boxed{y^x}\ 2\ \boxed{\sqrt{}}\ =$ $\boxed{9.738517742}$

With a graphing calculator, we enter these numbers and press these keys:

$5\ \boxed{\wedge}\ \boxed{\text{2nd}}\ \boxed{\sqrt{}}\ 2\ \boxed{)}\ \boxed{\text{ENTER}}$ $\boxed{\begin{array}{l} 5\wedge\sqrt{}(2) \\ \qquad 9.738517742 \end{array}}$

It can be shown that all of the familiar rules of exponents are also true for irrational exponents.

EXAMPLE 1   Simplify: **a.** $\left(5^{\sqrt{2}}\right)^{\sqrt{2}}$ **b.** $b^{\sqrt{3}} \cdot b^{\sqrt{12}}$

Strategy   We will use the power rule $(x^m)^n = x^{mn}$ to simplify part (a) and the product rule $x^m x^n = x^{m+n}$ to simplify part (b).

Why   These rules for exponents hold true for irrational exponents.

Solution

a. $\left(5^{\sqrt{2}}\right)^{\sqrt{2}} = 5^{\sqrt{2}\sqrt{2}}$ Keep the base and multiply the exponents.

$= 5^2$ Multiply: $\sqrt{2}\sqrt{2} = \sqrt{4} = 2$.

$= 25$

b. $b^{\sqrt{3}} \cdot b^{\sqrt{12}} = b^{\sqrt{3}+\sqrt{12}}$ Keep the base and add the exponents.

$= b^{\sqrt{3}+2\sqrt{3}}$ Simplify: $\sqrt{12} = \sqrt{4}\sqrt{3} = 2\sqrt{3}$.

$= b^{3\sqrt{3}}$ Combine like radicals. $\sqrt{3} + 2\sqrt{3} = 3\sqrt{3}$.

 Self Check 1   Simplify: **a.** $\left(3^{\sqrt{2}}\right)^{\sqrt{8}}$ **b.** $b^{\sqrt{2}} \cdot b^{\sqrt{18}}$

Now Try   **Problems 15 and 17**

2   Define Exponential Functions.

If $b > 0$ and $b \neq 1$, the function $f(x) = b^x$ is called an **exponential function**. Since x can be any real number, its domain is the set of real numbers. This is the interval $(-\infty, \infty)$.

Because b is positive, the value of $f(x)$ is positive, and the range is the set of positive numbers. This is the interval $(0, \infty)$.

Since $b \neq 1$, an exponential function cannot be the constant function $f(x) = 1^x$, in which $f(x) = 1$ for every real number x.

Exponential Functions	An **exponential function with base b** is defined by the equations $f(x) = b^x \quad$ or $\quad y = b^x$ where $b > 0$, $b \neq 1$, and x is a real number. The domain of $f(x) = b^x$ is the interval $(-\infty, \infty)$, and the range is the interval $(0, \infty)$.

3   Graph Exponential Functions.

Since the domain and range of $f(x) = b^x$ are sets of real numbers, we can graph exponential functions on a rectangular coordinate system.

EXAMPLE 2   Graph: $f(x) = 2^x$

Strategy   We will graph the function by creating a table of function values and plotting the corresponding ordered pairs.

Why   After drawing a smooth curve though the plotted points, we will have the graph.

Solution   To graph $f(x) = 2^x$, we choose several values for x and find the corresponding values of $f(x)$. If x is -1, we have

$f(x) = 2^x$

$f(-1) = 2^{-1}$ Substitute -1 for x.

$= \dfrac{1}{2}$

Notation

We have previously graphed the linear function $f(x) = 2x$ and the squaring function $f(x) = x^2$. For the exponential function $f(x) = 2^x$, note that the variable is in the exponent.

The point $\left(-1, \frac{1}{2}\right)$ is on the graph of $f(x) = 2^x$. In a similar way, we find the corresponding values of $f(x)$ for x values of 0, 1, 2, 3, and 4 and list them in a table. Then we plot the ordered pairs and draw a smooth curve through them.

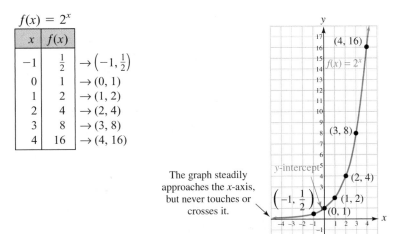

$f(x) = 2^x$

x	$f(x)$	
-1	$\frac{1}{2}$	$\rightarrow \left(-1, \frac{1}{2}\right)$
0	1	$\rightarrow (0, 1)$
1	2	$\rightarrow (1, 2)$
2	4	$\rightarrow (2, 4)$
3	8	$\rightarrow (3, 8)$
4	16	$\rightarrow (4, 16)$

The graph steadily approaches the *x*-axis, but never touches or crosses it.

The Language of Algebra
We have encountered the word *asymptote* earlier, when we graphed rational functions. Recall that an asymptote is not part of the graph. It is a line that the graph approaches and, in this case, never touches.

From the graph, we can see that the domain of $f(x) = 2^x$ is the interval $(-\infty, \infty)$ and the range is the interval $(0, \infty)$. Since the graph passes the horizontal line test, the function is one-to-one.

Note that as x decreases, the values of $f(x)$ decrease and approach 0. Thus, the *x*-axis is a horizontal asymptote of the graph. The graph does not have an *x*-intercept, the *y*-intercept is (0, 1), and the graph passes through the point (1, 2).

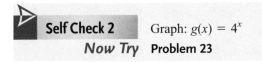

Self Check 2   Graph: $g(x) = 4^x$

Now Try **Problem 23**

EXAMPLE 3   Graph: $f(x) = \left(\frac{1}{3}\right)^x$

Strategy   We will graph the function by creating a table of function values and plotting the corresponding ordered pairs.

Why   After drawing a smooth curve through the plotted points, we will have the graph.

Solution   If $x = -2$, we have

$$f(x) = \left(\frac{1}{3}\right)^x$$

$$f(-2) = \left(\frac{1}{3}\right)^{-2}$$

$$= \left(\frac{3}{1}\right)^2 \qquad \text{Recall: } \left(\frac{x}{y}\right)^{-n} = \left(\frac{y}{x}\right)^n.$$

$$= 9$$

The point $(-2, 9)$ is on the graph of $f(x) = \left(\frac{1}{3}\right)^x$. In a similar way, we find the corresponding values of $f(x)$ for other x-values and list them in a table.

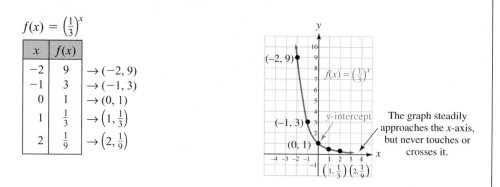

$$f(x) = \left(\frac{1}{3}\right)^x$$

x	$f(x)$	
-2	9	$\rightarrow (-2, 9)$
-1	3	$\rightarrow (-1, 3)$
0	1	$\rightarrow (0, 1)$
1	$\frac{1}{3}$	$\rightarrow \left(1, \frac{1}{3}\right)$
2	$\frac{1}{9}$	$\rightarrow \left(2, \frac{1}{9}\right)$

The graph steadily approaches the x-axis, but never touches or crosses it.

From the graph, we can verify that the domain of $f(x) = \left(\frac{1}{3}\right)^x$ is the interval $(-\infty, \infty)$ and the range is the interval $(0, \infty)$. Since the graph passes the horizontal line test, the function is one-to-one.

Note that as x increases, the values of $f(x)$ decrease and approach 0. Thus, the x-axis is a horizontal asymptote of the graph. The graph does not have an x-intercept, the y-intercept is $(0, 1)$, and the graph passes through the point $\left(1, \frac{1}{3}\right)$.

> **Self Check 3**   Graph: $g(x) = \left(\frac{1}{2}\right)^x$
>
> **Now Try**   **Problem 27**

Examples 2 and 3 illustrate the following properties of exponential functions.

Properties of Exponential Functions	The domain of the exponential function $f(x) = b^x$ is the interval $(-\infty, \infty)$.
	The range is the interval $(0, \infty)$.
	The graph has a y-intercept of $(0, 1)$.
	The x-axis is an asymptote of the graph.
	The graph of $f(x) = b^x$ passes through the point $(1, b)$.

In Example 2 (where $b = 2$), the values of y increase as the values of x increase. Since the graph rises as we move to the right, we call the function an *increasing function.* When $b > 1$, the larger the value of b, the steeper the curve.

In Example 3 $\left(\text{where } b = \frac{1}{3}\right)$, the values of y decrease as the values of x increase. Since the graph drops as we move to the right, we call the function a *decreasing function.* When $0 < b < 1$, the smaller the value of b, the steeper the curve.

In general, the following is true.

Increasing and Decreasing Functions

If $b > 1$, then $f(x) = b^x$ is an **increasing function.**

If $0 < b < 1$, then $f(x) = b^x$ is a **decreasing function.**

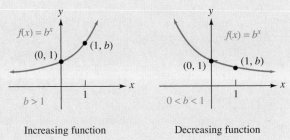

Increasing function   Decreasing function

An exponential function with base b is either increasing (for $b > 1$) or decreasing ($0 < b < 1$). Since different real numbers x determine different values of b^x, exponential functions are one-to-one.

Using Your Calculator

Graphing Exponential Functions

To use a graphing calculator to graph $f(x) = \left(\frac{2}{3}\right)^x$ and $g(x) = \left(\frac{3}{2}\right)^x$, we enter the right sides of the equations after the symbols $Y_1 =$ and $Y_2 =$. The screen will show the following equations.

$$Y_1 = (2/3)^\wedge X$$
$$Y_2 = (3/2)^\wedge X$$

If we press GRAPH, we will obtain the display shown.

We note that the graph of $f(x) = \left(\frac{2}{3}\right)^x$ passes through $(0, 1)$. Since $\frac{2}{3} < 1$, the function is decreasing. The graph of $g(x) = \left(\frac{3}{2}\right)^x$ also passes through $(0, 1)$. Since $\frac{3}{2} > 1$, the function is increasing. Since both graphs pass the horizontal line test, each function is one-to-one.

The graphs of many exponential functions are translations of basic graphs.

EXAMPLE 4   Graph each function by using a translation:
   a. $g(x) = 2^x - 4$ **b.** $g(x) = \left(\frac{1}{3}\right)^{x+3}$

Strategy   We will graph $g(x) = 2^x - 4$ by translating the graph of $f(x) = 2^x$ downward 4 units. We will graph $g(x) = \left(\frac{1}{3}\right)^{x+3}$ by translating the graph of $f(x) = \left(\frac{1}{3}\right)^x$ to the left 3 units.

Why   The subtraction of 4 in $g(x) = 2^x - 4$ causes a vertical shift of the graph of the base-2 exponential function 4 units downward. The addition of 3 to x in $g(x) = \left(\frac{1}{3}\right)^{x+3}$ causes a horizontal shift of the graph of the base-$\frac{1}{3}$ exponential function 3 units to the left.

Solution

a. The graph of $g(x) = 2^x - 4$ will be the same shape as the graph of $f(x) = 2^x$, except it is shifted 4 units downward.

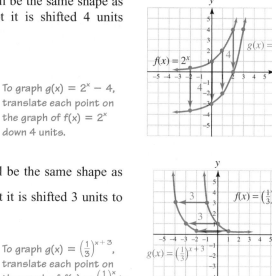

$g(x) = 2^x - 4$

$f(x) = 2^x$

To graph $g(x) = 2^x - 4$, translate each point on the graph of $f(x) = 2^x$ down 4 units.

b. The graph of $g(x) = \left(\frac{1}{3}\right)^{x+3}$ will be the same shape as the graph of $f(x) = \left(\frac{1}{3}\right)^x$, except it is shifted 3 units to the left.

$f(x) = \left(\frac{1}{3}\right)^x$

$g(x) = \left(\frac{1}{3}\right)^{x+3}$

To graph $g(x) = \left(\frac{1}{3}\right)^{x+3}$, translate each point on the graph of $f(x) = \left(\frac{1}{3}\right)^x$ to the left 3 units.

> **Self Check 4**   Graph each function by using a translation:
>
> **a.** $g(x) = \left(\frac{1}{4}\right)^x + 2$ **b.** $g(x) = 4^{x-3}$
>
> *Now Try*   **Problems 31 and 33**

---

④   **Use Exponential Functions in Applications Involving Growth or Decay.**

EXAMPLE 5   *Professional Baseball Salaries.* The exponential function $s(t) = 170{,}000(1.12)^t$ approximates the average salary of a major league baseball player, where t is the number of years after 1980 and $0 \leq t \leq 25$. Sketch the graph of the function. (Source: Baseball Almanac)

Strategy   We will graph the function by creating a table of function values and plotting the corresponding ordered pairs.

Why   After drawing a smooth curve though the plotted points, we will have the graph.

Solution   The function values for $t = 0$ and $t = 5$ are computed as follows:

$t = 0$ (*the year 1980*)

$s(0) = 170{,}000(1.12)^0$

$\quad\quad = 170{,}000(1)$ Any nonzero number to the 0 power is 1.

$\quad\quad = 170{,}000$

$t = 5$ (*the year 1985*)

$s(5) = 170{,}000(1.12)^5$

$\quad\quad \approx 299{,}598$ Use a calculator.

The Language of Algebra
The word *exponential* is used in many settings to describe rapid growth. For example, we hear that the processing power of computers is growing *exponentially*.

To approximate $s(5)$, use the keystrokes 170000 $\boxed{\times}$ 1.12 $\boxed{y^x}$ 5 $\boxed{=}$ on a scientific calculator and 170000 $\boxed{\times}$ 1.12 $\boxed{\wedge}$ 5 $\boxed{\text{ENTER}}$ on a graphing calculator.

In a similar way, we find the corresponding values of $s(t)$ for t-values of 10, 15, 20, and 25 and list them in a table. Then we plot the ordered pairs and draw a smooth curve through them to get the graph.

t	$s(t)$
0	170,000
5	299,598
10	527,994
15	930,506
20	1,639,870
25	2,890,011

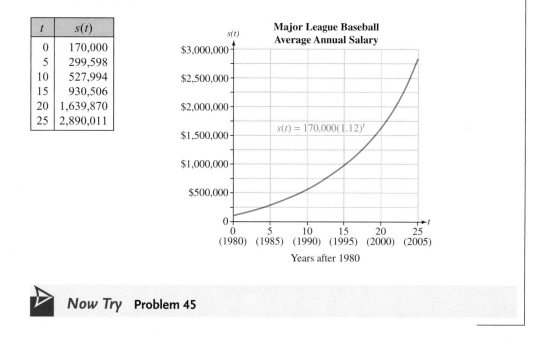

Now Try   **Problem 45**

Using Your Calculator

Graphing Exponential Functions

To use a graphing calculator to graph the exponential function $s(t) = 170,000(1.12)^t$, we enter the right side of the equation after the symbol $Y_1 =$ and replace the variable t with x. The display will show the equation

$$Y_1 = 170000(1.12\wedge X)$$

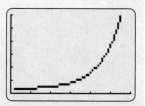

With window settings $[0, 30]$ for x and Xscale $= 5$ and $[0, 3000000]$ for y and Yscale $= 500000$, we obtain the display shown when we press $\boxed{\text{GRAPH}}$.

Exponential functions can be used to calculate compound interest. The following discussion gives some insight into why this is so.

If an amount of money P, called the **principal,** is deposited in an account paying an annual interest rate r, we can find the amount A in the account at the end of t years by using the formula

$$A = P + Prt \quad \text{or} \quad A = P(1 + rt)$$

The Language of Algebra
The following words indicate the number of times that interest is paid by a financial institution in one year.

Annually: 1 time
Semiannually: 2 times
Quarterly: 4 times
Monthly: 12 times
Daily: 365 times

Suppose that we deposit $500 in such an account that pays interest every 6 months. Then $P = 500$, and after 6 months $\left(\frac{1}{2}\text{ year}\right)$, the amount in the account will be

$$A = 500(1 + rt)$$

$$= 500\left(1 + r \cdot \frac{1}{2}\right) \quad \text{Substitute } \tfrac{1}{2} \text{ for } t.$$

$$= 500\left(1 + \frac{r}{2}\right)$$

The account will begin the second 6-month period with a value of $\$500\left(1 + \frac{r}{2}\right)$. After the second 6-month period, the amount in the account will be

$$A = P(1 + rt)$$

$$A = \left[500\left(1 + \frac{r}{2}\right)\right]\left(1 + r \cdot \frac{1}{2}\right) \quad \text{Substitute } 500\left(1 + \tfrac{r}{2}\right) \text{ for } P \text{ and } \tfrac{1}{2} \text{ for } t.$$

$$= 500\left(1 + \frac{r}{2}\right)\left(1 + \frac{r}{2}\right)$$

$$= 500\left(1 + \frac{r}{2}\right)^2$$

At the end of a third 6-month period, the amount in the account will be

$$A = 500\left(1 + \frac{r}{2}\right)^3$$

In this discussion, the earned interest is deposited back in the account and also earns interest, and we say that the account is earning **compound interest.** The preceding example suggests the following formula for compound interest.

Formula for Compound Interest	If $\$P$ is deposited in an account and interest is paid k times a year at an annual rate r, the amount A in the account after t years is given by $$A = P\left(1 + \frac{r}{k}\right)^{kt}$$

EXAMPLE 6   *Educational Savings Plan.*   To save for college, parents of a newborn child invest $12,000 in a mutual fund at 10% interest, compounded quarterly.

a. Find a function for the amount in the account after t years.

b. If the quarterly interest paid is continually reinvested, how much money will be in the account when the child is 18 years old?

Strategy   To write a function for the amount in the account after t years, we will substitute the given values for P, r, and k into the compound interest formula.

Why   The resulting equation will involve only two variables, A and t. Then we can write that equation using function notation.

Solution

a. When we substitute 12,000 for P, 0.10 for r, and 4 for k in the formula for compound interest, the resulting formula involves only two variables, A and t.

$$A = P\left(1 + \frac{r}{k}\right)^{kt}$$

$$A = 12{,}000\left(1 + \frac{0.10}{4}\right)^{4t} \quad \text{Since the interest is compounded quarterly, } k = 4.$$
$$\text{Express } r = 10\% \text{ as a decimal.}$$

Since the value of A depends on the value of t, we can express this relationship using function notation.

$$A(t) = 12{,}000\left(1 + \frac{0.10}{4}\right)^{4t}$$

$$A(t) = 12{,}000(1 + 0.025)^{4t} \quad \text{Evaluate: } \tfrac{0.10}{4} = 0.025.$$

$$A(t) = 12{,}000(1.025)^{4t} \quad \text{This exponential function has a base of 1.025.}$$

b. To find how much money will be in the account when the child is 18 years old, we need to find $A(18)$.

$$A(\mathbf{18}) = 12{,}000(1.025)^{4(18)} \quad \text{Substitute 18 for } t.$$
$$= 12{,}000(1.025)^{72}$$
$$= 12{,}000(1.025)^{72}$$
$$\approx 71{,}006.74 \quad \text{Use a scientific calculator and press these keys:}$$
$$\text{12000 } \boxed{\times} \text{ 1.025 } \boxed{y^x} \text{ 72 } \boxed{=} \text{ .}$$

When the child is 18 years old, the account will contain $71,006.74.

> **Self Check 6** How much money would be in the account after 18 years if the parents initially invested $20,000?
>
> **Now Try** Problem 49

Using Your Calculator

Solving Investment Problems

Suppose $1 is deposited in an account earning 6% annual interest, compounded monthly. To use a graphing calculator to estimate how much will be in the account in 100 years, we can substitute 1 for P, 0.06 for r, and 12 for k in the formula and simplify.

$$A = P\left(1 + \frac{r}{k}\right)^{kt} = 1\left(1 + \frac{0.06}{12}\right)^{12t} = (1.005)^{12t}$$

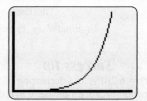

We now graph the function $A(t) = (1.005)^{12t}$ using window settings of $[0, 120]$ and $[0, 400]$ with Xscale = 1 and Yscale = 1 to obtain the graph shown. We can then trace and zoom to estimate that $1 grows to be approximately $397 in 100 years. From the graph, we can see that the money grows slowly in the early years and rapidly in the later years.

11. Match each situation to the exponential graph that best models it.

   a. The number of cell phone subscribers in the world over the past 5 years

   b. The level of caffeine in the bloodstream after drinking a cup of coffee

   c. The amount of money in a bank account earning interest compounded quarterly

   d. The number of rabbits in a population with a high birth rate

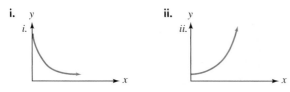

12. What formula is used to determine the amount of money in a savings account earning compound interest?

NOTATION

13. For an exponential function of the form $f(x) = b^x$, what are the restrictions on b?

14. In $A(t) = 16{,}000\left(1 + \frac{0.05}{365}\right)^{365t}$, what is the base and what is the exponent?

GUIDED PRACTICE

Simplify each expression. Write answers using positive exponents. See Example 1.

15. $\left(2^{\sqrt{3}}\right)^{\sqrt{3}}$

16. $\left(3^{\sqrt{5}}\right)^{\sqrt{5}}$

17. $7^{\sqrt{3}}\,7^{\sqrt{12}}$

18. $3^{\sqrt{2}}\,3^{\sqrt{18}}$

19. $\dfrac{3^{2\sqrt{7}}}{3^{\sqrt{7}}}$

20. $\dfrac{5^{6\sqrt{2}}}{5^{4\sqrt{2}}}$

21. $5^{-\sqrt{5}}$

22. $4^{-\sqrt{5}}$

Graph each function. See Examples 2 and 3.

23. $f(x) = 3^x$

24. $f(x) = 6^x$

25. $f(x) = 5^x$

26. $f(x) = 7^x$

27. $f(x) = \left(\dfrac{1}{4}\right)^x$

28. $f(x) = \left(\dfrac{1}{5}\right)^x$

29. $f(x) = \left(\dfrac{1}{6}\right)^x$

30. $f(x) = \left(\dfrac{1}{8}\right)^x$

Graph each function by plotting points or using a translation. See Example 4.

31. $g(x) = 3^x - 2$

32. $g(x) = 2^x + 1$

33. $g(x) = 2^{x+1}$

34. $g(x) = 3^{x-1}$

35. $g(x) = 4^{x-1} + 2$

36. $g(x) = 4^{x+1} - 2$

37. $g(x) = -2^x$

38. $g(x) = -3^x$

Use a graphing calculator to graph each function. Determine whether the function is an increasing or a decreasing function. See Using Your Calculator: Graphing Exponential Functions.

39. $f(x) = \dfrac{1}{2}(3^{x/2})$

40. $f(x) = -3(2^{x/3})$

41. $f(x) = 2(3^{-x/2})$

42. $f(x) = -\dfrac{1}{4}(2^{-x/2})$

APPLICATIONS

43. WORLD POPULATION   See the following graph.

   a. Estimate when the world's population reached $\frac{1}{2}$ billion and when it reached 1 billion.

   b. Estimate the world's population in the year 2000.

   c. What type of function does it appear could be used to model the world's population growth?

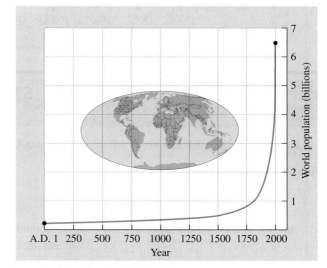

Source: United Nations Population Division

44. GLOBAL WARMING The following graph from the United States Environmental Protection Agency shows the projected sea level changes due to anticipated global warming.

a. What type of function does it appear could be used to model the sea level change?

b. When were the earliest instrumental records of sea level change made?

c. For the year 2100, what is the upper-end projection for sea level change? What is the lower-level projection?

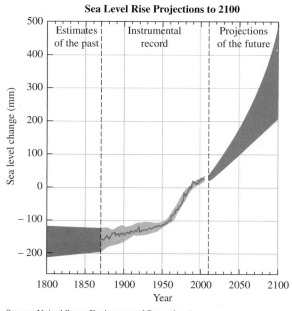

Sea Level Rise Projections to 2100

Source: United States Environmental Protection Agency

45. COMPUTER VIRUSES Suppose the number of computers infected by the spread of a virus through an e-mail is described by the exponential function $c(t) = 5(1.03)^t$, where t is the number of minutes since the first infected e-mail was opened. Graph the function.

46. SALVAGE VALUE A small business purchased a computer for $5,000. It is expected that its value each year will be 75% of its value the preceding year. The value (in dollars) of the computer, t years after its purchase, is given by the exponential function $v(t) = 5,000(0.75)^t$. Graph the function.

47. DIVING *Bottom time* is the time a scuba diver spends descending plus the actual time spent at a certain depth. Graph the bottom time limits given in the table.

Bottom time limits			
Depth (ft)	Bottom time (min)	Depth (ft)	Bottom time (min)
30	no limit	80	40
35	310	90	30
40	200	100	25
50	100	110	20
60	60	120	15
70	50	130	10

48. VALUE OF A CAR The graph shows how the value of the average car depreciates as a percent of its original value over a 10-year period. It also shows the yearly maintenance costs as a percent of the car's value.

a. When is the car worth half of its purchase price?

b. When is the car worth a quarter of its purchase price?

c. When do the average yearly maintenance costs surpass the value of the car?

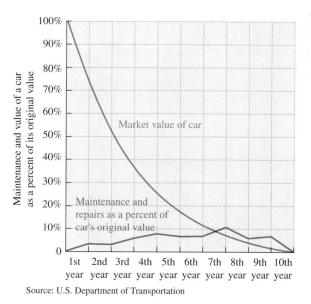

Source: U.S. Department of Transportation

In Exercises 49–54, assume that there are no deposits or withdrawals.

49. COMPOUND INTEREST   An initial deposit of $10,000 earns 8% interest, compounded quarterly. How much will be in the account after 10 years?

50. COMPOUND INTEREST   An initial deposit of $10,000 earns 8% interest, compounded monthly. How much will be in the account after 10 years?

51. COMPARING INTEREST RATES   How much more interest could $1,000 earn in 5 years, compounded quarterly, if the annual interest rate were $5\frac{1}{2}\%$ instead of 5%?

52. COMPARING SAVINGS PLANS   Which institution in the ads provides the better investment?

Fidelity Savings & Loan
Earn 5.25%
compounded monthly

Union Trust
Money Market Account
paying 5.35%
compounded annually

53. COMPOUND INTEREST   If $1 had been invested on July 4, 1776, at 5% interest, compounded annually, what would it be worth on July 4, 2076?

54. FREQUENCY OF COMPOUNDING $10,000 is invested in each of two accounts, both paying 6% annual interest. In the first account, interest compounds quarterly, and in the second account, interest compounds daily. Find the difference between the accounts after 20 years.

55. GUITARS   The frets on the neck of a guitar are placed so that pressing a string against them determines the strings' vibrating length. The exponential function $f(n) = 650(0.94)^n$ gives the vibrating length (in millimeters) of a string on a certain guitar for the fret number n. Find the length of the vibrating string when a guitarist holds down a string at the 7th fret.

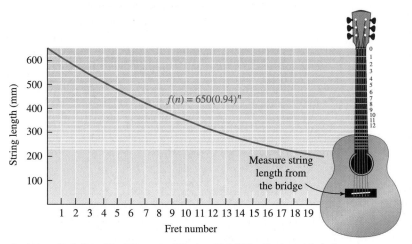

$f(n) = 650(0.94)^n$

Measure string length from the bridge

56. BACTERIAL CULTURES   A colony of 6 million bacteria was determined to be growing in the culture medium shown in illustration (a). If the population P of bacteria after t hours is given by the function $P(t) = 6,000,000(2.3)^t$, find the population in the culture later in the day using the information given in illustration (b).

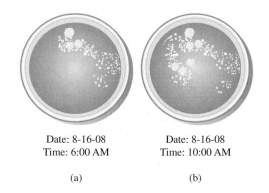

Date: 8-16-08   Date: 8-16-08
Time: 6:00 AM   Time: 10:00 AM

(a)   (b)

57. RADIOACTIVE DECAY   Five hundred grams of a radio-active material decays according to the formula $A = 500\left(\frac{2}{3}\right)^t$, where t is measured in years. Find the amount present in 10 years. Round to the nearest one-tenth of a gram.

58. DISCHARGING A BATTERY   The charge remaining in a battery decreases as the battery discharges. The charge C (in coulombs) after t days is given by the function $C(t) = 0.0003(0.7)^t$. Find the charge after 5 days.

59. POPULATION GROWTH   The population of North Rivers is decreasing exponentially according to the formula $P = 3,745(0.93)^t$, where t is measured in years from the present date. Find the population in 6 years, 9 months.

60. THE LOUISIANA PURCHASE   In 1803, the United States negotiated the Louisiana Purchase with France. The country doubled its territory by adding 827,000 square miles of land for $15 million. If the land appreciated at the rate of 6% each year, what would one square mile of land be worth in 2005?

WRITING

61. If world population is increasing exponentially, why is there cause for concern?

62. How do the graphs of $f(x) = 3^x$ and $g(x) = \left(\frac{1}{3}\right)^x$ differ? How are they similar?

63. A snowball rolling downhill grows *exponentially* with time. Explain what this means. Sketch a simple graph that models the situation.

64. Explain why the graph of $f(x) = 3^x$ gets closer and closer to the x-axis as the values of x decrease. Does the graph ever cross the x-axis? Explain why or why not.

65. Describe the graphs of $f(x) = x^2$ and $g(x) = 2^x$ in words.

66. Write a paragraph explaining the concept that is illustrated in the following graph.

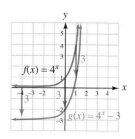

REVIEW

In Exercises 67–70, refer to the illustration below in which lines r and s are parallel.

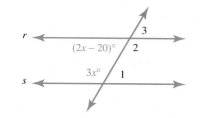

67. Find x.

68. Find the measure of $\angle 1$.

69. Find the measure of $\angle 2$.

70. Find the measure of $\angle 3$.

CHALLENGE PROBLEMS

71. In the definition of the exponential function, b could not be negative. Why?

72. Graph $f(x) = 3^x$. Then use the graph to estimate the value of $3^{1.5}$.

73. Graph $y = x^{1/2}$ and $y = \left(\frac{1}{2}\right)^x$ on the same set of coordinate axes. Estimate the coordinates of any point(s) that the graphs have in common.

74. Find the value of b that would cause the graph of $f(x) = b^x$ to look like the graph below.

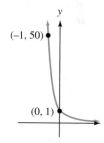

SECTION 9.4
Base-*e* Exponential Functions

Objectives

❶ Define *e* and identify the formula for exponential growth/decay.

❷ Define the natural exponential function.

❸ Graph the natural exponential function.

❹ Use base-*e* exponential functions in applications involving growth or decay.

Any positive number can be used as a base of an exponential function. However, some bases are used more often than others. An exponential function that has many applications is one whose base is an irrational number represented by the letter *e*. We will now discuss one way of arriving at this important number *e*.

❶ **Define *e* and Identify the Formula for Exponential Growth/Decay.**

If a bank pays interest twice a year, we say that interest is compounded semiannually. If it pays interest four times a year, we say that interest is compounded quarterly. If it pays interest continuously (infinitely many times in a year), we say that interest is compounded continuously.

To develop the formula for continuous compound interest, we start with the formula

$$A = P\left(1 + \frac{r}{k}\right)^{kt}$$ This is the formula for compound interest: *r* is the annual rate and *k* is the number of times per year interest is paid.

Notation
Swiss born Leonhard Euler (1707–1783) is said to have published more than any mathematician in history. He had a great influence on the notation that we use today. Through his work, the symbol *e* came into common use.

and let $rn = k$. Since *r* and *k* are positive numbers, so is *n*.

$$A = P\left(1 + \frac{r}{rn}\right)^{rnt}$$

We can then simplify the fraction $\frac{r}{rn}$ and use the commutative property of multiplication to change the order of the exponents.

$$A = P\left(1 + \frac{1}{n}\right)^{nrt}$$

Finally, we can use a property of exponents to write the formula as

(1) $$A = P\left[\left(1 + \frac{1}{n}\right)^{n}\right]^{rt}$$ Use the property $a^{mn} = (a^m)^n$.

n	$\left(1 + \frac{1}{n}\right)^n$
1	2
2	2.25
4	2.44140625 …
12	2.61303529 …
365	2.71456748 …
1,000	2.71692393 …
100,000	2.71826830 …
1,000,000	2.71828137 …

To find the value of $\left(1 + \frac{1}{n}\right)^n$, we evaluate it for several values of *n*, as shown in the table. The results suggest that as *n* gets larger, the value of $\left(1 + \frac{1}{n}\right)^n$ approaches the number 2.71828 . . . . This number is called *e*, which has the following value.

$$e = \mathbf{2.718281828459} \ldots$$

Like π, the number *e* is irrational. Its decimal representation is nonterminating and non-repeating. Rounded to four decimal places, $e \approx 2.7183$.

If we replace $\left(1 + \frac{1}{n}\right)^n$ in Equation 1 with e, we will get the formula for exponential growth.

$$A = P\left[\left(1 + \frac{1}{n}\right)^n\right]^{rt}$$

$$A = Pe^{rt} \qquad \text{Substitute } e \text{ for } \left(1 + \frac{1}{n}\right)^n.$$

Formula for Exponential Growth/Decay

If a quantity P increases or decreases at an annual rate r, compounded continuously, the amount A after t years is given by

$$A = Pe^{rt}$$

If time is measured in years, r is called the **annual growth rate.** If r is negative, the growth represents a decrease.

For a given quantity P, say 10,000, and a given rate r, say 5%, we can write the formula for exponential growth using function notation:

$$A(t) = 10,000e^{0.05t}$$

2 **Define the Natural Exponential Function.**

Of all possible bases for an exponential function, e is the most convenient for problems involving growth or decay. Since these situations occur often in natural settings, we call $f(x) = e^x$ the *natural exponential function.*

The Natural Exponential Function

The function defined by $f(x) = e^x$ is the **natural exponential function** (or the **base-e exponential function**) where $e = 2.71828. \ldots$. The domain of $f(x) = e^x$ is the interval $(-\infty, \infty)$. The range is the interval $(0, \infty)$.

The $\boxed{e^x}$ key on a calculator is used to evaluate the natural exponential function.

Using Your Calculator

The Natural Exponential Function Key

To compute the amount to which $12,000 will grow if invested for 18 years at 10% annual interest, compounded continuously, we substitute 12,000 for P, 0.10 for r, and 18 for t in the formula for continuous compound interest and simplify.

$$A = Pe^{rt} = 12,000e^{0.10(18)} = 12,000e^{1.8} \qquad \text{Write 10\% as 0.10.}$$

To evaluate this expression using a scientific calculator, we enter

1.8 $\boxed{e^x}$ $\boxed{\times}$ 12000 $\boxed{=}$ $\boxed{72595.76957}$

Using a graphing calculator, we enter

12000 $\boxed{\times}$ $\boxed{\text{2nd}}$ $\boxed{e^x}$ 1.8 $\boxed{)}$ $\boxed{\text{ENTER}}$ $\boxed{\begin{array}{l} 12000*e^\wedge(1.8) \\ \qquad\qquad 72595.76957 \end{array}}$

After 18 years, the account will contain $72,595.77. This is $1,589.03 more than the result in Example 6 in the Section 9.3, where interest was compounded quarterly.

EXAMPLE 1   *Investing.* If $25,000 accumulates interest at an annual rate of 8%, compounded continuously, find the balance in the account in 50 years.

Strategy   We will substitute 25,000 for P, 0.08 for r, and 50 for t in the formula $A = Pe^{rt}$ and calculate the value of A.

Why   The words *compounded continuously* indicate that we should use the exponential growth/decay formula.

Solution

$$A = Pe^{rt} \qquad \text{This is the formula for continuous compound interest.}$$

$$A = 25{,}000e^{0.08(50)} \qquad \text{Write 8\% as 0.08.}$$

$$= 25{,}000e^{4}$$

$$\approx 1{,}364{,}953.75 \qquad \text{Use a calculator.}$$

In 50 years, the balance will be $1,364,953.75—more than a million dollars.

Self Check 1   Find the balance in 60 years.

Now Try   Problems 19 and 37

③ Graph the Natural Exponential Function.

To graph $f(x) = e^{x}$, we construct a table of function values by choosing several values for x and finding the corresponding values of $f(x)$. For example, $x = -2$, we have

$$f(x) = e^{x}$$

$$f(-2) = e^{-2}$$

$$= 0.135335283 \ldots \qquad \text{Use a calculator.}$$

$$\approx 0.1 \qquad\qquad \text{Round to the nearest tenth.}$$

We enter $(-2, 0.1)$ in the table. Similarly, we find $f(-1)$, $f(0)$, $f(1)$, and $f(2)$, enter each result in the table, and plot the ordered pairs. We draw a smooth curve through the points to get the graph.

From the graph, we can verify that the domain of $f(x) = e^{x}$ is the interval $(-\infty, \infty)$ and the range is the interval $(0, \infty)$. Since the graph passes the horizontal line test, the function is one-to-one.

Note that as x decreases, the values of $f(x)$ decrease and approach 0. Thus, the x-axis is a horizontal asymptote of the graph. The graph does not have an x-intercept, the y-intercept is $(0, 1)$, and the graph passes through the point $(1, e)$.

$f(x) = e^{x}$

x	$f(x)$	
-2	$\frac{1}{e^{2}} \approx 0.1$	$\to (-2, 0.1)$
-1	$\frac{1}{e^{1}} \approx 0.4$	$\to (-1, 0.4)$
0	$e^{0} = 1$	$\to (0, 1)$
1	$e^{1} = e$	$\to (1, e)$
2	$e^{2} \approx 7.4$	$\to (2, 7.4)$

↑

The outputs can be found using the $\boxed{e^{x}}$ key on a calculator. Some are rounded to the nearest tenth to make point-plotting easier.

The graph steadily approaches the x-axis, but never touches or crosses it.

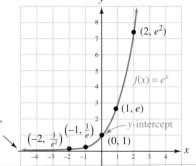

To graph more complicated natural exponential functions, point-plotting can be tedious. In such cases, we will use a graphing calculator.

Using Your Calculator   *Graphing Exponential Functions*

The figure shows the calculator graph of $f(x) = 3e^{-x/2}$. To graph this function, we enter the right side of the equation after the symbol $Y_1 =$. The display will show the equation

$$Y_1 = 3(e\char`\^(-X/2))$$

The graphs of many functions are translations of the natural exponential function.

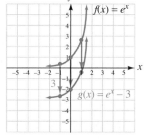
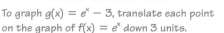
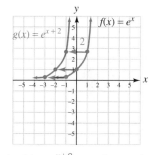

To graph $g(x) = e^x - 3$, translate each point on the graph of $f(x) = e^x$ down 3 units.

To graph $g(x) = e^{x+2}$, translate each point on the graph of $f(x) = e^x$ to the left 2 units.

We can illustrate the effects of vertical and horizontal translations of the natural exponential function by using a graphing calculator.

Using Your Calculator   *Translations of the Natural Exponential Function*

Figure (a) shows the calculator graphs of $f(x) = e^x$, $g(x) = e^x + 5$, and $h(x) = e^x - 3$. To graph them, we enter the right sides of the equations after $Y_1 =$, $Y_2 =$, and $Y_3 =$. The display will show:

$$Y_1 = e\char`\^(X) \qquad Y_2 = e\char`\^(X) + 5 \qquad Y_3 = e\char`\^(X) - 3$$

The graph of $g(x) = e^x + 5$ is 5 units above the graph of $f(x) = e^x$ and the graph of $h(x) = e^x - 3$ is 3 units below the graph of $f(x) = e^x$.

Figure (b) shows the calculator graphs of $f(x) = e^x$, $g(x) = e^{x+5}$, and $h(x) = e^{x-3}$. The graph of $g(x) = e^{x+5}$ is 5 units to the left of the graph of $f(x) = e^x$ and the graph of $h(x) = e^{x-3}$ is 3 units to the right of the graph of $f(x) = e^x$.

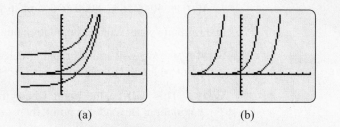

(a)   (b)

4 **Use Base-*e* Exponential Functions in Applications Involving Growth or Decay.**

An equation based on the natural exponential function provides a model for **population growth.** In the **Malthusian model for population growth,** the future population of a colony is related to the present population by the formula $A = Pe^{rt}$. (Note that this is the same formula as for continuous compound interest.)

EXAMPLE 2   *City Planning.*   The population of a city is currently 15,000, but economic conditions are causing the population to decrease 3% each year. If this trend continues, find the population in 30 years.

Strategy   We will substitute 15,000 for P, -0.03 for r, and 30 for t in the formula $A = Pe^{rt}$ and calculate the value of A.

Why   Since the population is decreasing 3% each year, the annual growth rate is -3%, or -0.03.

Solution

<table>
<tr><td>$A = Pe^{rt}$</td><td>This is the model for population growth/decay.</td></tr>
<tr><td>$A = 15{,}000e^{-0.03(30)}$</td><td>Substitute.</td></tr>
<tr><td>$= 15{,}000e^{-0.9}$</td><td></td></tr>
<tr><td>$\approx 6{,}099$</td><td>Use a calculator and round to the nearest whole number.</td></tr>
</table>

In 30 years, the expected population will be 6,099.

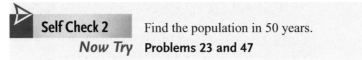

Self Check 2   Find the population in 50 years.

     Now Try   **Problems 23 and 47**

> **Success Tip**
>
> For quantities that are decreasing, remember to enter a negative value for r, the annual rate, in the formula $A = Pe^{rt}$.

The English economist Thomas Robert Malthus (1766–1834) was a pioneer in studying population. He believed that poverty and starvation were unavoidable because the human population tends to grow exponentially but the food supply tends to grow linearly.

EXAMPLE 3   *Food Shortages.*   Suppose that a country with a population of 1,000 people is growing exponentially according to the function

$$P(t) = 1{,}000e^{0.02t} \qquad \text{\em The annual growth rate is 2\% = 0.02.}$$

where t is in years. Furthermore, assume that the food supply F, measured in adequate food per day per person, is growing linearly according to the function

$$F(t) = 30.625t + 2{,}000 \quad (t \text{ is time in years})$$

In how many years will the population outstrip the food supply?

Strategy   We will use a graphing calculator to graph these two functions and find the point where the graphs intersect.

Why   This will be the point where the food supply is exactly adequate to feed the population. Beyond this point, there will be a shortage of food.

Solution   We can use a graphing calculator with window settings of [0, 100] for *x* and [0, 10,000] for *y*. After graphing the functions, we obtain figure (a). The food supply is modeled by the straight line, and the population is modeled by the curved line. If we trace, as in figure (b), we can find the point where the two graphs intersect. From the graph, we can see that the food supply will be adequate for about 71 years. At that time, the population of approximately 4,200 people will begin to have problems.

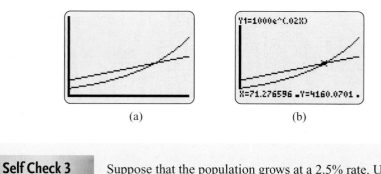

(a)   (b)

Self Check 3   Suppose that the population grows at a 2.5% rate. Use a graphing calculator to determine for how many years the food supply will be adequate.

Now Try   **Problem 35**

EXAMPLE 4   ***Baking.***   A mother takes a cake out of the oven and sets it on a rack to cool. The function $T(t) = 68 + 220e^{-0.18t}$ gives the cake's temperature in degrees Fahrenheit after it has cooled for *t* minutes. If her children will be home from school in 20 minutes, will the cake have cooled enough for the children to eat it? (Assume that 80°F, or cooler, would be a comfortable eating temperature.)

Strategy   We will substitute 20 for *t* in the function $T(t) = 68 + 220e^{-0.18t}$.

Why   The variable *t* represents the number of minutes the cake has cooled.

Solution   When the children arrive home, the cake will have cooled for 20 minutes. To find the temperature of the cake at that time, we need to find $T(20)$.

$$T(t) = 68 + 220e^{-0.18t}$$
$$T(20) = 68 + 220e^{-0.18(20)} \qquad \text{Substitute 20 for } t.$$
$$= 68 + 220e^{-3.6}$$
$$\approx 74.0 \qquad \text{Use a calculator.}$$

When the children return home, the temperature of the cake will be about 74°, and it can be eaten.

Now Try   **Problem 55**

**ANSWERS TO SELF CHECKS**   **1.** $3,037,760.44   **2.** 3,347   **3.** About 51 years

STUDY SET
9.4

VOCABULARY

Refer to the graph of $f(x) = e^x$.

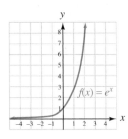

$f(x) = e^x$

1. What is the name of the function $f(x) = e^x$?

2. What is the domain of the function?
3. What is the range of the function?
4. **a.** What is the y-intercept of the graph?
   b. What is the x-intercept of the graph?
5. Is the function one-to-one?
6. What is an asymptote of the graph?
7. Is f an increasing or a decreasing function?
8. The graph passes through the point $(1, y)$. What is y?

CONCEPTS

Fill in the blanks.

9. In _____ compound interest, the number of compoundings is infinitely large.
10. The formula for exponential growth/decay is $A = \quad e^{\quad}$.
11. To two decimal places, the value of e is _____ .
12. If n gets larger and larger, the value of $\left(1 + \frac{1}{n}\right)^n$ approaches the value of _____ .
13. Graph each irrational number on the number line: $\left\{\pi, e, \sqrt{2}\right\}$.

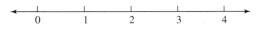

14. Complete the table of values. Round to the nearest hundredth.

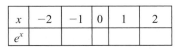

x	-2	-1	0	1	2
e^x					

15. The function $f(x) = e^x$ is graphed and the TRACE feature is used, as shown here. What is the y-coordinate of the point on the graph having an x-coordinate of 1? What is the symbol that represents this number?

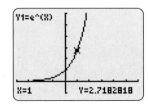

16. The illustration shows a table of values for $f(x) = e^x$. As x decreases, what happens to the values of $f(x)$ listed in the Y_1 column? Will the value of $f(x)$ ever be 0 or negative?

X	Y₁
0	1
-1	.36788
-2	.13534
-3	.04979
-4	.01832
-5	.00674
-6	.00248

X=0

NOTATION

Find A using the formula $A = Pe^{rt}$ given the following values of P, r, and t. Round to the nearest tenth.

17. $P = 1,000$, $r = 0.09$, and $t = 10$

$$A = \underline{\quad} e^{(0.09)(\quad)}$$
$$= 1,000e^{\underline{\quad}}$$
$$\approx \underline{\quad} \qquad \text{Use a calculator.}$$

18. $P = 50,000$, $r = -0.12$, and $t = 50$

$$A = 50,000e^{(\quad)(50)}$$
$$= 50,000e^{\underline{\quad}}$$
$$\approx \underline{\quad} \qquad \text{Use a calculator.}$$

GUIDED PRACTICE

Find A using the formula $A = Pe^{rt}$ given the following values of P, r, and t. Round to the nearest hundredth. See Example 1.

19. $P = 5,000$, $r = 8\%$, $t = 20$ years
20. $P = 15,000$, $r = 6\%$, $t = 40$ years
21. $P = 20,000$, $r = 10.5\%$, $t = 50$ years
22. $P = 25,000$, $r = 6.5\%$, $t = 100$ years

Find A using the formula $A = Pe^{rt}$ given the following values of P, r, and t. Round to the nearest hundredth. **See Example 2.**

23. $P = 15,895$, $r = -2\%$, $t = 16$ years

24. $P = 33,999$, $r = -4\%$, $t = 21$ years

25. $P = 565$, $r = -0.5\%$, $t = 8$ years

26. $P = 110$, $r = -0.25\%$, $t = 9$ years

Graph each function. **See Objective 3.**

27. $f(x) = e^x$ **28.** $f(x) = -e^x$

29. $f(x) = e^x + 1$ **30.** $f(x) = e^x - 2$

31. $y = e^{x+3}$ **32.** $y = e^{x-5}$

33. $f(x) = 2e^x$ **34.** $f(x) = \dfrac{1}{2}e^x$

APPLICATIONS

In Exercises 35 and 36, use a graphing calculator to solve each problem.

35. THE MALTHUSIAN MODEL   In Example 3, suppose that better farming methods changed the function for food growth to $F(t) = 31t + 2,000$. How long would the food supply be adequate?

36. THE MALTHUSIAN MODEL   In Example 3, suppose that a birth-control program changed the function for population growth to $P(t) = 1,000e^{0.01t}$. How long would the food supply be adequate?

In Exercises 37–42, assume that there are no deposits or withdrawals.

37. CONTINUOUS COMPOUND INTEREST   An initial investment of $5,000 earns 8.2% interest, compounded continuously. What will the investment be worth in 12 years?

38. CONTINUOUS COMPOUND INTEREST   An initial investment of $2,000 earns 8% interest, compounded continuously. What will the investment be worth in 15 years?

39. COMPARISON OF COMPOUNDING METHODS   An initial deposit of $5,000 grows at an annual rate of 8.5% for 5 years. Compare the final balances resulting from annual compounding and continuous compounding.

40. COMPARISON OF COMPOUNDING METHODS   An initial deposit of $30,000 grows at an annual rate of 8% for 20 years. Compare the final balances resulting from annual compounding and continuous compounding.

41. DETERMINING THE INITIAL DEPOSIT   An account now contains $11,180 and has been accumulating interest at 7% annual interest, compounded continuously, for 7 years. Find the initial deposit.

42. DETERMINING THE PREVIOUS BALANCE   An account now contains $3,610 and has been accumulating interest at 8% annual interest, compounded continuously. How much was in the account 4 years ago?

43. POPULATION OF THE UNITED STATES   Graph the U.S. census population figures shown in the table (in millions). What type of function does it appear could be used to model the population?

Year	Population	Year	Population
1790	3.9	1900	76.0
1800	5.3	1910	92.2
1810	7.2	1920	106.0
1820	9.6	1930	123.2
1830	12.9	1940	132.1
1840	17.0	1950	151.3
1850	23.1	1960	179.3
1860	31.4	1970	203.3
1870	38.5	1980	226.5
1880	50.1	1990	248.7
1890	62.9	2000	281.4

44. OZONE CONCENTRATIONS   A *Dobson* unit is the most basic measure used in ozone research. Roughly 300 Dobson units are equivalent to the height of 2 pennies stacked on top of each other. Suppose the ozone layer thickness (in Dobsons) over a certain city is modeled by the function $A(t) = 300e^{-0.0011t}$, where t is the number of years after 1990. Estimate how thick the ozone layer will be in 2015.

45. POPULATION OF THE UNITED STATES   The exponential function $A(t) = 123e^{0.0117t}$ approximates the population of the United States (in millions), where t is the number of years after 1930. Use the function to estimate the U.S. population for these important dates:

- 1937 The Golden Gate Bridge is completed
- 1941 The United States enters World War II
- 1955 Rosa Parks refuses to give up her seat on a Montgomery, Alabama, bus
- 1969 Astronaut Neil Armstrong walks on the moon
- 1974 President Nixon resigns
- 1986 The *Challenger* space shuttle explodes
- 1997 *The Simpsons* becomes the longest running cartoon television series in history

46. WORLD POPULATION GROWTH   The population of Earth is approximately 6.6 billion people and is growing at an annual rate of 1.167%. Use the exponential growth model to find the world population in 30 years.

47. HIGHS AND LOWS Liberia, located on the west coast of Africa, has one of the greatest population growth rates in the world. Bulgaria, in southeastern Europe, has one of the smallest. Use an exponential growth/decay model to complete the table.

Country	Population 2007	Annual growth rate	Estimated population 2020
Liberia	3,195,931	4.84%	
Bulgaria	7,322,858	−0.84%	

Source: CIA World Factbook

48. DISINFECTANTS The exponential function $A(t) = 2,000,000e^{-0.588t}$ approximates the number of germs on a table top, t minutes after disinfectant was sprayed on it. Estimate the germ count on the table 5 minutes after it is sprayed.

49. from Campus to Careers
    Social Worker

Social workers often use occupational test results when counseling their clients about employment options. The "learning curve" below shows that as a factory trainee assembled more chairs, the assembly time per chair generally decreased. If company standards required an average assembly time of 10 minutes or less, how many chairs did the trainee have to assemble before meeting company standards?

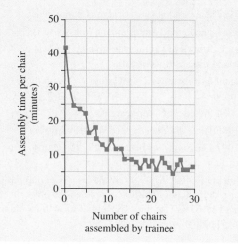

Number of chairs assembled by trainee

50. ANTS Shortly after an explorer ant discovers a food source, a recruitment process begins in which numerous additional ants travel to the source. The number of ants at the source grows exponentially according to the function $a(t) = 1.36\left(\frac{e}{2.5}\right)^t$, where t is the number of minutes since the explorer discovered the food. How many ants will be at the source in 40 minutes?

51. EPIDEMICS The spread of hoof-and-mouth disease through a herd of cattle can be modeled by the function

$$P(t) = 2e^{0.27t} \quad (t \text{ is in days})$$

If a rancher does not quickly treat the two cows that now have the disease, how many cattle will have the disease in 12 days?

52. OCEANOGRAPHY The width w (in millimeters) of successive growth spirals of the sea shell *Catapulus voluto,* shown below, is given by the exponential function

$$w(n) = 1.54e^{0.503n}$$

where n is the spiral number. Find the width, to the nearest tenth of a millimeter, of the sixth spiral.

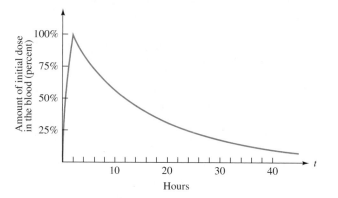

53. HALF-LIFE OF A DRUG The quantity of a prescription drug in the bloodstream of a patient t hours after it is administered can be modeled by an exponential function. (See the graph.) Determine the time it takes to eliminate half of the initial dose from the body from the graph.

54. MEDICINE   The concentration of a certain prescription drug in an organ after t minutes is modeled by the function

$$f(t) = 0.08(1 - e^{-0.1t})$$

where $f(t)$ is the concentration at time t. Find the concentration of the drug at 30 minutes.

55. SKYDIVING   Before the parachute opens, a skydiver's velocity in meters per second is modeled by the function

$$f(t) = 50(1 - e^{-0.2t})$$

where $f(t)$ is the velocity at time t. Find the velocity after 20 seconds of free fall.

56. FREE FALL   After t seconds a certain falling object has a velocity $f(t)$ modeled by the function $f(t) = 50(1 - e^{-0.3t})$. Which is falling faster after 2 seconds—the object or the skydiver in Exercise 55?

WRITING

57. Explain why the graph of $y = e^x - 5$ is five units below the graph of $y = e^x$.

58. A feature article in a newspaper stated that the sport of snowboarding was growing *exponentially*. Explain what the author of the article meant by that.

59. As of 2007, the population growth rate for Russia was -0.37% annually. What are some of the consequences for a country that has a negative population growth?

60. What is e?

REVIEW

Simplify each expression. Assume that all variables represent positive numbers.

61. $\sqrt{240x^5}$

62. $\sqrt[3]{-125x^5y^4}$

63. $4\sqrt{48y^3} - 3y\sqrt{12y}$

64. $\sqrt[4]{48z^5} + \sqrt[4]{768z^5}$

CHALLENGE PROBLEMS

65. Without using a calculator, determine whether the statement $e^e > e^3$ is true or false. Explain your reasoning.

66. Graph the function defined by the equation

$$f(x) = \frac{e^x + e^{-x}}{2}$$

from $x = -2$ to $x = 2$. The graph will look like a parabola, but it is not. The graph, called a **catenary,** is important in the design of power distribution networks, because it represents the shape of a uniform flexible cable whose ends are suspended from the same height. The function is called the **hyperbolic cosine function.**

67. If $e^{t+5} = ke^t$, find k.

68. If $e^{5t} = k^t$, find k.

SECTION 9.5
Logarithmic Functions

Objectives

 1 Define logarithm.

2 Write logarithmic equations as exponential equations.

3 Write exponential equations as logarithmic equations.

4 Evaluate logarithmic expressions.

5 Graph logarithmic functions.

6 Use logarithmic functions in applications.

In this section, we will discuss inverses of exponential functions. These functions are called *logarithmic functions,* and they can be used to solve problems from fields such as electronics, seismology (the study of earthquakes), and business.

1 **Define Logarithm.**

The graph of the exponential function $f(x) = 2^x$ is shown in red on the next page. Since it passes the horizontal line test, it is a one-to-one function and has an inverse. To graph f^{-1}, we interchange the coordinates of the ordered pairs in the table, plot those points, and draw a smooth curve through them, as shown in blue. As expected, the graphs of f and f^{-1} are symmetric with respect to the line $y = x$.

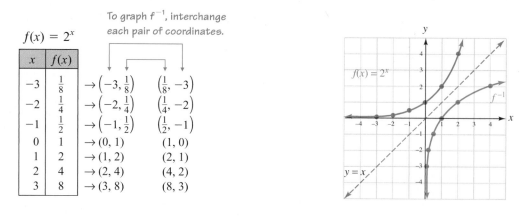

To graph f^{-1}, interchange each pair of coordinates.

$f(x) = 2^x$

x	$f(x)$
-3	$\frac{1}{8}$
-2	$\frac{1}{4}$
-1	$\frac{1}{2}$
0	1
1	2
2	4
3	8

$\rightarrow \left(-3, \frac{1}{8}\right)$ $\left(\frac{1}{8}, -3\right)$
$\rightarrow \left(-2, \frac{1}{4}\right)$ $\left(\frac{1}{4}, -2\right)$
$\rightarrow \left(-1, \frac{1}{2}\right)$ $\left(\frac{1}{2}, -1\right)$
$\rightarrow (0, 1)$ $(1, 0)$
$\rightarrow (1, 2)$ $(2, 1)$
$\rightarrow (2, 4)$ $(4, 2)$
$\rightarrow (3, 8)$ $(8, 3)$

To write an equation for the inverse of $f(x) = 2^x$, we proceed as follows:

$$f(x) = 2^x$$
$$y = 2^x \qquad \text{Replace } f(x) \text{ with } y.$$
$$x = 2^y \qquad \text{Interchange the variables } x \text{ and } y.$$

We cannot solve the equation for y because we have not discussed methods for solving equations with a variable in an exponent. However, we can translate the relationship $x = 2^y$ into words:

$$y = \text{ the power to which we raise 2 to get } x$$

If we substitute $f^{-1}(x)$ for y, we see that

$$f^{-1}(x) = \text{ the power to which we raise 2 to get } x$$

If we define the symbol $\log_2 x$ to mean *the power to which we raise 2 to get x*, we can write the equation for the inverse as

$$f^{-1}(x) = \log_2 x \qquad \text{Read } \log_2 x \text{ as "the logarithm, base 2, of x" or "log, base 2, of x."}$$

We have found that the inverse of the exponential function $f(x) = 2^x$ is $f^{-1}(x) = \log_2 x$. To find the inverse of exponential functions with other bases, such as $f(x) = 3^x$ and $f(x) = 10^x$, we define logarithm in the following way.

Definition of Logarithm

For all positive numbers b, where $b \neq 1$, and all positive numbers x,

$$y = \log_b x \quad \text{is equivalent to} \quad x = b^y$$

Success Tip
Here are examples of inverses of other exponential functions:

$$f(x) = 3^x \qquad f^{-1}(x) = \log_3 x$$
$$g(x) = 10^x \qquad g^{-1}(x) = \log_{10} x$$

This definition guarantees that any pair (x, y) that satisfies the logarithmic equation $y = \log_b x$ also satisfies the exponential equation $x = b^y$. Because of this relationship, a statement written in logarithmic form can be written in an equivalent exponential form, and vice versa. The following diagram will help you remember the respective positions of the exponent and base in each form.

Exponent

$$y = \log_b x \qquad x = b^y$$

Base

2 Write Logarithmic Equations as Exponential Equations.

The following table shows several pairs of equivalent equations.

> **Success Tip**
> Study the relationship between exponential and logarithmic statements carefully. Your success with the material in the rest of this chapter depends greatly on your understanding of this definition.

Logarithmic equation	*Exponential equation*
$\log_2 8 = 3$	$2^3 = 8$
$\log_3 81 = 4$	$3^4 = 81$
$\log_4 4 = 1$	$4^1 = 4$
$\log_5 \frac{1}{125} = -3$	$5^{-3} = \frac{1}{125}$

EXAMPLE 1 Write each logarithmic equation as an exponential equation:

a. $\log_4 64 = 3$ **b.** $\log_7 \sqrt{7} = \frac{1}{2}$ **c.** $\log_6 \frac{1}{36} = -2$

Strategy To write an equivalent exponential equation, we will determine which number will serve as the base and which will serve as the exponent.

Why We can then use the definition of logarithm to move from one form to the other: $\log_b x = y$ is equivalent to $x = b^y$.

Solution

Exponent

a. $\log_4 64 = 3$ is equivalent to $4^3 = 64$.

Base

Exponent

b. $\log_7 \sqrt{7} = \frac{1}{2}$ is equivalent to $7^{1/2} = \sqrt{7}$.

Base

Exponent

c. $\log_6 \frac{1}{36} = -2$ is equivalent to $6^{-2} = \frac{1}{36}$.

Base

Self Check 1 Write $\log_2 128 = 7$ as an exponential equation.

Now Try **Problem 21**

3 Write Exponential Equations as Logarithmic Equations.

EXAMPLE 2 Write each exponential equation as a logarithmic equation:

a. $8^0 = 1$ **b.** $6^{1/3} = \sqrt[3]{6}$ **c.** $\left(\frac{1}{4}\right)^2 = \frac{1}{16}$

Strategy   To write an equivalent logarithmic equation, we will determine which number will serve as the base and where we will place the exponent.

Why   We can then use the definition of logarithm to move from one form to the other: $x = b^y$ is equivalent to $\log_b x = y$.

Solution

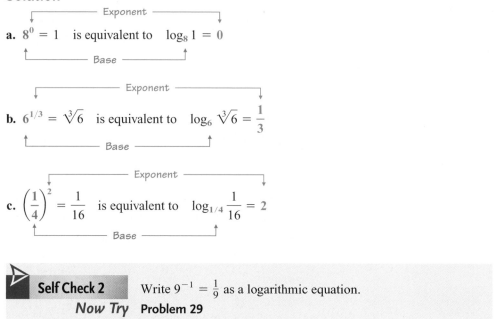

a. $8^0 = 1$ is equivalent to $\log_8 1 = 0$

b. $6^{1/3} = \sqrt[3]{6}$ is equivalent to $\log_6 \sqrt[3]{6} = \dfrac{1}{3}$

c. $\left(\dfrac{1}{4}\right)^2 = \dfrac{1}{16}$ is equivalent to $\log_{1/4} \dfrac{1}{16} = 2$

> ▷ **Self Check 2**   Write $9^{-1} = \frac{1}{9}$ as a logarithmic equation.
>
> *Now Try*   **Problem 29**

Certain logarithmic equations can be solved by writing them as exponential equations.

EXAMPLE 3   Solve each equation for x:

a. $\log_x 25 = 2$ **b.** $\log_3 x = -3$ **c.** $\log_{1/2} \dfrac{1}{16} = x$

Strategy   To solve each logarithmic equation, we will instead write and solve an equivalent exponential equation.

Why   The resulting exponential equation is easier to solve because the variable term is often isolated on one side.

Solution

a. Since $\log_x 25 = 2$ is equivalent to $x^2 = 25$, we can solve $x^2 = 25$ to find x.

$$x^2 = 25$$
$$x = \pm\sqrt{25} \quad \text{Use the square root property.}$$
$$x = \pm 5$$

In the expression $\log_x 25$, the base of the logarithm is x. Because the base must be positive, we discard -5 and we have

$$x = 5$$

To check the solution of 5, we verify that $\log_5 25 = 2$.

b. Since $\log_3 x = -3$ is equivalent to $3^{-3} = x$, we can solve $3^{-3} = x$ to find x.

$$3^{-3} = x$$

$$\frac{1}{3^3} = x$$

$$x = \frac{1}{27}$$

To check the solution of $\frac{1}{27}$, we verify that $\log_3 \frac{1}{27} = -3$.

c. Since $\log_{1/2} \frac{1}{16} = x$ is equivalent to $\left(\frac{1}{2}\right)^x = \frac{1}{16}$, we can solve $\left(\frac{1}{2}\right)^x = \frac{1}{16}$ to find x.

$$\left(\frac{1}{2}\right)^x = \frac{1}{16}$$

$$\left(\frac{1}{2}\right)^x = \left(\frac{1}{2}\right)^4 \qquad \text{Write } \tfrac{1}{16} \text{ as a power of } \tfrac{1}{2} \text{ to match the bases: } \tfrac{1}{2} \cdot \tfrac{1}{2} \cdot \tfrac{1}{2} \cdot \tfrac{1}{2} = \tfrac{1}{16}.$$

$$x = 4 \qquad \text{Since the bases are the same, and since exponential functions are one-to-one, the exponents must be equal.}$$

To check the solution of 4, we verify that $\log_{1/2} \frac{1}{16} = 4$.

> **Success Tip**
> To solve this equation, we note that if the bases are equal, the exponents must be equal.
>
>
> $$\left(\frac{1}{2}\right)^x = \left(\frac{1}{2}\right)^4$$

> **Self Check 3**   Solve each equation for x:
> **a.** $\log_x 49 = 2$ **b.** $\log_{1/3} x = 2$ **c.** $\log_6 216 = x$
>
> ***Now Try***   **Problems 37, 39, and 41**

4   **Evaluate Logarithmic Expressions.**

In the previous examples, we have seen that the logarithm of a number is an exponent. In fact,

$\log_b x$ *is the exponent to which* b *is raised to get* x.

Translating this statement into symbols, we have

$$b^{\log_b x} = x$$

EXAMPLE 4   Evaluate each logarithmic expression:

a. $\log_8 64$ **b.** $\log_3 \frac{1}{3}$ **c.** $\log_4 2$

Strategy   After identifying the base, we will ask "To what power must the base be raised to get the other number?"

Why   That power is the value of the logarithmic expression.

Solution

a. $\log_8 64 = 2$ Ask: "To what power must we raise 8 to get 64?"
Since $8^2 = 64$, the answer is the 2nd power.

b. $\log_3 \frac{1}{3} = -1$ Ask: "To what power must we raise 3 to get $\frac{1}{3}$?"
Since $3^{-1} = \frac{1}{3}$, the answer is the −1 power.

c. $\log_4 2 = \dfrac{1}{2}$ Ask: "To what power must we raise 4 to get 2?"

Since $\sqrt{4} = 4^{1/2} = 2$, the answer is the $\frac{1}{2}$ power.

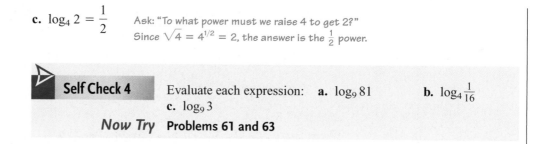

Self Check 4   Evaluate each expression: **a.** $\log_9 81$ **b.** $\log_4 \frac{1}{16}$

c. $\log_9 3$

Now Try   **Problems 61 and 63**

The Language of Algebra
London professor Henry Briggs (1561–1630) and Scottish lord John Napier (1550–1617) are credited with developing the concept of common logarithms. Their tables of logarithms were useful tools at that time for those performing large calculations.

For computational purposes and in many applications, we will use base-10 logarithms (also called **common logarithms**). When the base b is not indicated in the notation $\log x$, we assume that $b = 10$:

$$\log x \quad \text{means} \quad \log_{10} x$$

The table below shows several pairs of equivalent statements involving base-10 logarithms.

Logarithmic form	*Exponential form*	
$\log 100 = 2$	$10^2 = 100$	Read $\log 100$ as "log of 100."
$\log \dfrac{1}{10} = -1$	$10^{-1} = \dfrac{1}{10}$	
$\log 1 = 0$	$10^0 = 1$	

In general, we have

$$\log_{10} 10^x = x$$

EXAMPLE 5   Evaluate each logarithmic expression, if possible:

a. $\log 1{,}000$ **b.** $\log \dfrac{1}{100}$ **c.** $\log 10$ **d.** $\log(-10)$

Strategy   After identifying the base, we will ask "To what power must 10 be raised to get the other number?"

Why   That power is the value of the logarithmic expression.

Solution

a. $\log 1{,}000 = 3$ Ask: "To what power must we raise 10 to get 1,000?"

Since $10^3 = 1{,}000$, the answer is: the 3rd power.

b. $\log \dfrac{1}{100} = -2$ Ask: "To what power must we raise 10 to get $\frac{1}{100}$?"

Since $10^{-2} = \frac{1}{100}$, the answer is: the -2 power.

c. $\log 10 = 1$ Ask: "To what power must we raise 10 to get 10?"

Since $10^1 = 10$, the answer is: the 1st power.

d. To find $\log(-10)$, we must find a power of 10 such that $10^? = -10$. There is no such number. Thus, $\log(-10)$ is undefined.

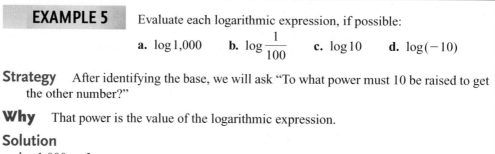

Self Check 5   Evaluate each expression: **a.** $\log 10{,}000$

b. $\log \dfrac{1}{1{,}000}$ **c.** $\log 0$

Now Try   **Problems 65 and 67**

Many logarithmic expressions cannot be evaluated by inspection. For example, to find log 2.34, we ask, "To what power must we raise 10 to get 2.34?" This answer isn't obvious. In such cases, we use a calculator.

Using Your Calculator | **Evaluating Logarithms**

To find log 2.34 with a scientific calculator we enter

2.34 [LOG]   .369215857

On some calculators, the 10^x key also serves as the [LOG] key when [2nd] or [SHIFT] is pressed. This is because $f(x) = 10^x$ and $f(x) = \log x$ are inverses.

To use a graphing calculator, we enter

[LOG] 2.34 [)] [ENTER]

log(2.34)
                                                        .369215857

To four decimal places, log 2.34 = 0.3692. This means, $10^{0.3692} \approx 2.34$.

If we attempt to evaluate logarithmic expressions such as log 0, or the logarithm of a negative number, such as log(−5), an error message like the following will be displayed.

[Error]

ERR:DOMAIN
1:QUIT
2:Go to

ERR:NONREAL ANS
1:QUIT
2:Go to

EXAMPLE 6   Solve $\log x = 0.3568$ and round to four decimal places.

Strategy   To solve this logarithmic equation, we will instead write and solve an equivalent exponential equation.

Why   The resulting exponential equation is easier to solve because the variable term is isolated on one side.

Solution   The equation $\log x = 0.3568$ is equivalent to $10^{0.3568} = x$. Since we cannot determine $10^{0.3568}$ by inspection, we will use a calculator to find an approximate solution. We enter

10 y^x .3568 =

The display reads 2.274049951. To four decimal places,

$x = 2.2740$

If your calculator has a 10^x key, enter .3568 and press it to get the same result. The solution is 2.2740. To check, use your calculator to verify that log 2.2740 ≈ 0.3568.

Self Check 6   Solve $\log x = 1.87737$ and round to four decimal places.

Now Try   Problem 77

5 **Graph Logarithmic Functions.**

Because an exponential function defined by $f(x) = b^x$ is one-to-one, it has an inverse function that is defined by $x = b^y$. When we write $x = b^y$ in the equivalent form $y = \log_b x$, the result is called a *logarithmic function*.

Logarithmic Functions	If $b > 0$ and $b \neq 1$, the **logarithmic function with base b** is defined by the equations
	$$f(x) = \log_b x \quad \text{or} \quad y = \log_b x$$
	The domain of $f(x) = \log_b x$ is the interval $(0, \infty)$ and the range is the interval $(-\infty, \infty)$.

Since every logarithmic function is the inverse of a one-to-one exponential function, logarithmic functions are one-to-one.

We can plot points to graph logarithmic functions. For example, to graph $f(x) = \log_2 x$, we construct a table of function values, plot the resulting ordered pairs, and draw a smooth curve through the points to get the graph, as shown in figure (a). To graph $f(x) = \log_{1/2} x$, we use the same method, as shown in figure (b).

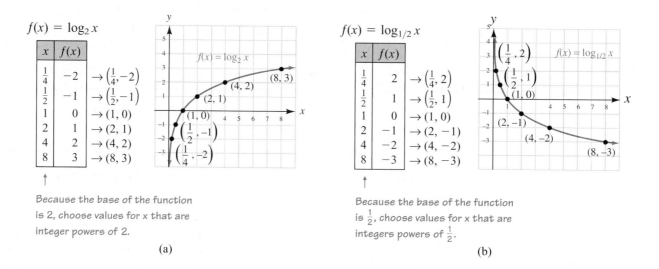

$f(x) = \log_2 x$

x	$f(x)$	
$\frac{1}{4}$	-2	$\rightarrow \left(\frac{1}{4}, -2\right)$
$\frac{1}{2}$	-1	$\rightarrow \left(\frac{1}{2}, -1\right)$
1	0	$\rightarrow (1, 0)$
2	1	$\rightarrow (2, 1)$
4	2	$\rightarrow (4, 2)$
8	3	$\rightarrow (8, 3)$

↑

Because the base of the function is 2, choose values for x that are integer powers of 2.

(a)

$f(x) = \log_{1/2} x$

x	$f(x)$	
$\frac{1}{4}$	2	$\rightarrow \left(\frac{1}{4}, 2\right)$
$\frac{1}{2}$	1	$\rightarrow \left(\frac{1}{2}, 1\right)$
1	0	$\rightarrow (1, 0)$
2	-1	$\rightarrow (2, -1)$
4	-2	$\rightarrow (4, -2)$
8	-3	$\rightarrow (8, -3)$

↑

Because the base of the function is $\frac{1}{2}$, choose values for x that are integers powers of $\frac{1}{2}$.

(b)

The graphs of all logarithmic functions are similar to those shown below. If $b > 1$, the logarithmic function is increasing, as in figure (a). If $0 < b < 1$, the logarithmic function is decreasing, as in figure (b).

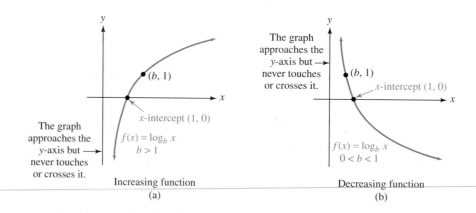

The graph approaches the y-axis but never touches or crosses it.

$(b, 1)$

x-intercept $(1, 0)$

$f(x) = \log_b x$
$b > 1$

Increasing function
(a)

The graph approaches the y-axis but never touches or crosses it.

$(b, 1)$

x-intercept $(1, 0)$

$f(x) = \log_b x$
$0 < b < 1$

Decreasing function
(b)

Properties of Logarithmic Functions	The graph of $f(x) = \log_b x$ (or $y = \log_b x$) has the following properties.

1. It passes through the point $(1, 0)$.
2. It passes through the point $(b, 1)$.
3. The y-axis (the line $x = 0$) is an asymptote.
4. The domain is the interval $(0, \infty)$ and the range is the interval $(-\infty, \infty)$.

The exponential and logarithmic functions are inverses of each other, so their graphs have symmetry about the line $y = x$. The graphs of $f(x) = \log_b x$ and $g(x) = b^x$ are shown in figure (a) when $b > 1$ and in figure (b) when $0 < b < 1$.

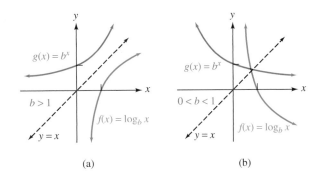

(a)   (b)

The graphs of many functions involving logarithms are translations of the basic logarithmic graphs.

EXAMPLE 7   Graph each function by using a translation:
   a. $g(x) = 3 + \log_2 x$ **b.** $g(x) = \log_{1/2}(x - 1)$

Strategy   We will graph $g(x) = 3 + \log_2 x$ by translating the graph of $f(x) = \log_2 x$ upward 3 units. We will graph $g(x) = \log_{1/2}(x - 1)$ by translating the graph of $f(x) = \log_{1/2} x$ to the right 1 unit.

Why   The addition of 3 in $g(x) = 3 + \log_2 x$ causes a vertical shift of the graph of the base-2 logarithmic function 3 units upward. The subtraction of 1 from x in $g(x) = \log_{1/2}(x - 1)$ causes a horizontal shift of the graph of the base-$\frac{1}{2}$ logarithmic function 1 unit to the right.

Solution

a. The graph of $g(x) = 3 + \log_2 x$ will be the same shape as the graph of $f(x) = \log_2 x$, except that it is shifted 3 units upward.

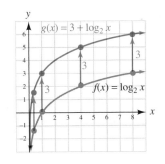

To graph $g(x) = 3 + \log_2 x$, translate each point on the graph of $f(x) = \log_2 x$ up 3 units.

Notation

Parentheses are used to write the function in Example 7b, because $g(x) = \log_{1/2} x - 1$ could be interpreted as

$$g(x) = \log_{1/2}(x - 1)$$

or as

$$g(x) = (\log_{1/2} x) - 1$$

b. The graph of $g(x) = \log_{1/2}(x - 1)$ will be the same shape as the graph of $f(x) = \log_{1/2} x$, except it is shifted 1 unit to the right.

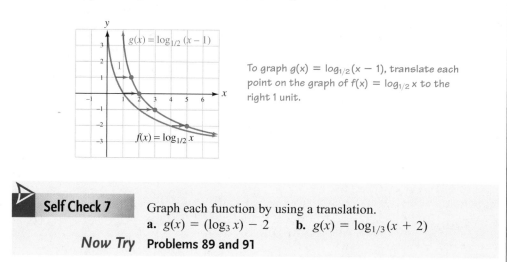

To graph $g(x) = \log_{1/2}(x - 1)$, translate each point on the graph of $f(x) = \log_{1/2} x$ to the right 1 unit.

 Self Check 7   Graph each function by using a translation.

a. $g(x) = (\log_3 x) - 2$ **b.** $g(x) = \log_{1/3}(x + 2)$

Now Try   **Problems 89 and 91**

To graph more complicated logarithmic functions, a graphing calculator is a useful tool.

Using Your Calculator

Graphing Logarithmic Functions

To use a calculator to graph the logarithmic function $f(x) = -2 + \log_{10} \frac{x}{2}$, we enter the right side of the equation after the symbol $Y_1 =$. The display will show the equation

$$Y_1 = -2 + \log(X/2)$$

If we use window settings of $[-1, 5]$ for x and $[-4, 1]$ for y and press the $\boxed{\text{GRAPH}}$ key, we will obtain the graph shown.

---

⑥ Use Logarithmic Functions in Applications.

Logarithmic functions, like exponential functions, can be used to model certain types of growth and decay. Common logarithms are used in electrical engineering to express the voltage gain (or loss) of an electronic device such as an amplifier. The unit of gain (or loss), called the **decibel,** is defined by a logarithmic relation.

Decibel Voltage Gain

If E_O is the output voltage of a device and E_I is the input voltage, the decibel voltage gain of the device (db gain) is given by

$$\text{db gain} = 20 \log \frac{E_O}{E_I}$$

| EXAMPLE 8 | **db Gain.** If the input to an amplifier is 0.5 volt and the output is 40 volts, find the decibel voltage gain of the amplifier. |

Strategy We will substitute into the formula for db gain and evaluate the right side using a calculator.

Why We can use this formula to find the db gain because we are given the input voltage E_I and the output voltage E_O.

Solution We can find the decibel voltage gain by substituting 0.5 for E_I and 40 for E_O into the formula for db gain:

$$\text{db gain} = 20 \log \frac{E_O}{E_I}$$

$$\text{db gain} = 20 \log \frac{40}{0.5}$$

$$= 20 \log 80 \qquad \text{Divide: } \frac{40}{0.5} = 80.$$

$$\approx 38 \qquad \text{Use a calculator: } 20 \log 80 \text{ means } 20 \cdot \log 80.$$

The amplifier provides a 38-decibel voltage gain.

Self Check 8 If the input to an amplifier is 0.6 volt and the output is 40 volts, find the decibel voltage gain of the amplifier.

Now Try Problem 97

In seismology, common logarithms are used to measure the intensity of earthquakes on the **Richter scale.** The intensity of an earthquake is given by the following logarithmic function.

Richter Scale

If R is the intensity of an earthquake, A is the amplitude (measured in micrometers) of the ground motion, and P is the period (the time of one oscillation of the Earth's surface measured in seconds), then

$$R = \log \frac{A}{P}$$

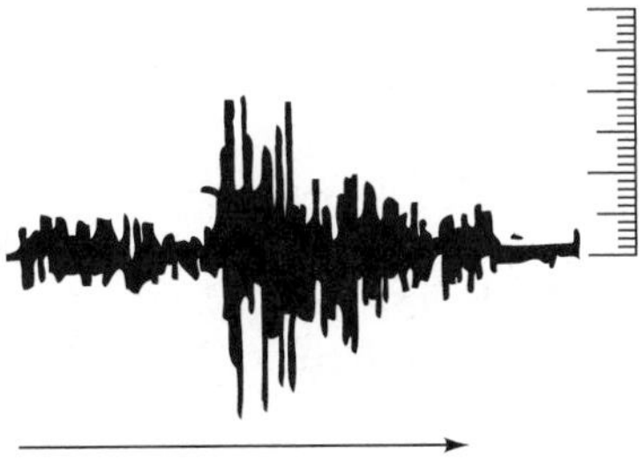

| EXAMPLE 9 | **Earthquakes.** Find the measure on the Richter scale of an earthquake with an amplitude of 5,000 micrometers (0.5 centimeter) and a period of 0.1 second. |

Strategy We will substitute into the formula for intensity of an earthquake and evaluate the right side using a calculator.

Why We can use this formula to find the intensity of the earthquake because we are given the amplitude A and the period P.

Time / Amplitude

Solution   We substitute 5,000 for A and 0.1 for P in the Richter scale formula and proceed as follows:

$$R = \log \frac{A}{P}$$

$$R = \log \frac{5,000}{0.1}$$

$= \log 50,000$ *Divide:* $\frac{5,000}{0.1} = 50,000$.

≈ 4.698970004 *Use a calculator.*

The earthquake measures about 4.7 on the Richter scale.

> **The Language of Algebra**
> The Richter scale was developed in 1935 by Charles F. Richter of the California Institute of Technology.

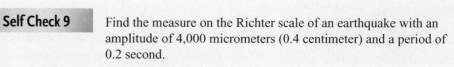

Self Check 9   Find the measure on the Richter scale of an earthquake with an amplitude of 4,000 micrometers (0.4 centimeter) and a period of 0.2 second.

Now Try   Problem 101

ANSWERS TO SELF CHECKS   **1.** $2^7 = 128$ **2.** $\log_9 \frac{1}{9} = -1$ **3. a.** 7   **b.** $\frac{1}{9}$ **c.** 3

4. a. 2   **b.** -2 **c.** $\frac{1}{2}$ **5. a.** 4   **b.** -3,   **c.** Undefined   **6.** 75.3998

7. a.   **b.**   **8.** About 36 db   **9.** 4.3

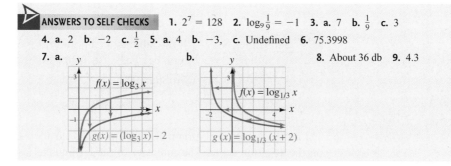

STUDY SET
9.5

VOCABULARY

Refer to the graph of $f(x) = \log_4 x$.

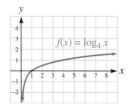

1. What type of function is $f(x) = \log_4 x$?

2. What is the domain of the function?

3. What is the range of the function?

4. a. What is the y-intercept of the graph?

    b. What is the x-intercept of the graph?

5. Is f a one-to-one function?

6. What is an asymptote of the graph?

7. Is f an increasing or a decreasing function?

8. The graph passes through the point $(4, y)$. What is y?

CONCEPTS

Fill in the blanks.

9. The equation $y = \log_b x$ is equivalent to the exponential equation ____ = ____ .

10. $\log_b x$ is the _____ to which b is raised to get x.

11. The functions $f(x) = \log_{10} x$ and $f(x) = 10^x$ are _____ functions.

12. The inverse of an exponential function is called a _____ function.

Complete the table of values, where possible.

13. $f(x) = \log x$

x	f(x)
100	
$\frac{1}{100}$	

14. $f(x) = \log_5 x$

x	f(x)
25	
$\frac{1}{25}$	

15. $f(x) = \log_6 x$

Input	Output
6	
−6	
0	

16. $f(x) = \log_8 x$

Input	Output
8	
−8	
0	

17. a. Use a calculator to complete the table of values for $f(x) = \log x$. Round to the nearest hundredth.

x	f(x)
0.5	
1	
2	
4	
6	
8	
10	

b. Graph $f(x) = \log x$. Note that the units on the x- and y-axes are different.

18. For each function, determine its inverse, $f^{-1}(x)$.

a. $f(x) = 10^x$ **b.** $f(x) = 3^x$

c. $f(x) = \log x$ **d.** $f(x) = \log_2 x$

NOTATION

Fill in the blanks.

19. a. $\log x = \log \underline{} x$ **b.** $\log_{10} 10^x = \underline{}$

20. a. We read $\log_5 25$ as "log,   5,   25."

     b. We read $\log x$ as "   of x."

GUIDED PRACTICE

Write each logarithmic equation as an exponential equation. See Example 1.

21. $\log_3 81 = 4$ **22.** $\log_7 7 = 1$

23. $\log_{10} 10 = 1$ **24.** $\log_{10} 100 = 2$

25. $\log_4 \frac{1}{64} = -3$ **26.** $\log_6 \frac{1}{36} = -2$

27. $\log_5 \sqrt{5} = \frac{1}{2}$ **28.** $\log_8 \sqrt[3]{8} = \frac{1}{3}$

Write each exponential equation as a logarithmic equation. See Example 2.

29. $8^2 = 64$ **30.** $10^3 = 1{,}000$

31. $4^{-2} = \frac{1}{16}$ **32.** $3^{-4} = \frac{1}{81}$

33. $\left(\frac{1}{2}\right)^{-5} = 32$ **34.** $\left(\frac{1}{3}\right)^{-3} = 27$

35. $x^y = z$ **36.** $m^n = p$

Solve for x. See Example 3.

37. $\log_x 81 = 2$ **38.** $\log_x 9 = 2$

39. $\log_8 x = 2$ **40.** $\log_7 x = 0$

41. $\log_5 125 = x$ **42.** $\log_4 16 = x$

43. $\log_5 x = -2$ **44.** $\log_3 x = -4$

45. $\log_{36} x = -\frac{1}{2}$ **46.** $\log_{27} x = -\frac{1}{3}$

47. $\log_x 0.01 = -2$ **48.** $\log_x 0.001 = -3$

49. $\log_{27} 9 = x$ **50.** $\log_{12} x = 0$

51. $\log_x 5^3 = 3$ **52.** $\log_x 5 = 1$

53. $\log_{100} x = \frac{3}{2}$ **54.** $\log_x \frac{1}{1{,}000} = -\frac{3}{2}$

55. $\log_x \frac{1}{64} = -3$ **56.** $\log_x \frac{1}{100} = -2$

57. $\log_8 x = 0$ **58.** $\log_4 8 = x$

59. $\log_x \frac{\sqrt{3}}{3} = \frac{1}{2}$ **60.** $\log_x \frac{9}{4} = 2$

Evaluate each logarithmic expression. **See Examples 4 and 5.**

61. $\log_2 8$

62. $\log_3 9$

63. $\log_4 16$

64. $\log_6 216$

65. $\log 1,000,000$

66. $\log 100,000$

67. $\log \dfrac{1}{10}$

68. $\log \dfrac{1}{10,000}$

69. $\log_{1/2} \dfrac{1}{32}$

70. $\log_{1/3} \dfrac{1}{81}$

71. $\log_9 3$

72. $\log_{125} 5$

Use a calculator to find each value. Give answers to four decimal places. **See Using Your Calculator: Evaluating Logarithms.**

73. $\log 3.25$

74. $\log 0.57$

75. $\log 0.00467$

76. $\log 375.876$

Use a calculator to solve each equation. Round answers to four decimal places. **See Example 6.**

77. $\log x = 3.7813$

78. $\log x = 2.8945$

79. $\log x = -0.7630$

80. $\log x = -1.3587$

81. $\log x = -0.5$

82. $\log x = -0.926$

83. $\log x = -1.71$

84. $\log x = 1.4023$

Graph each function. Determine whether each function is an increasing or a decreasing function. **See Objective 5.**

85. $f(x) = \log_3 x$

86. $f(x) = \log_{1/3} x$

87. $y = \log_{1/2} x$

88. $y = \log_4 x$

Graph each function by plotting points or by using a translation. (The basic logarithmic functions graphed in Exercises 85–88 will be helpful.) **See Example 7.**

89. $f(x) = 3 + \log_3 x$

90. $f(x) = (\log_{1/3} x) - 1$

91. $y = \log_{1/2}(x - 2)$

92. $y = \log_4(x + 2)$

Graph each pair of inverse functions on the same coordinate system. Draw the axis of symmetry. **See Objective 1.**

93. $f(x) = 6^x$
$f^{-1}(x) = \log_6 x$

94. $f(x) = 3^x$
$f^{-1}(x) = \log_3 x$

95. $f(x) = 5^x$
$f^{-1}(x) = \log_5 x$

96. $f(x) = 8^x$
$f^{-1}(x) = \log_8 x$

APPLICATIONS

97. INPUT VOLTAGE   Find the db gain of an amplifier if the input voltage is 0.71 volt when the output voltage is 20 volts.

98. OUTPUT VOLTAGE   Find the db gain of an amplifier if the output voltage is 2.8 volts when the input voltage is 0.05 volt.

99. db GAIN   Find the db gain of the amplifier shown below.

100. db GAIN   An amplifier produces an output of 80 volts when driven by an input of 0.12 volt. Find the amplifier's db gain.

101. THE RICHTER SCALE   An earthquake has amplitude of 5,000 micrometers and a period of 0.2 second. Find its measure on the Richter scale.

102. EARTHQUAKES   Find the period of an earthquake with amplitude of 80,000 micrometers that measures 6 on the Richter scale.

103. EARTHQUAKES   An earthquake with a period of $\frac{1}{4}$ second measures 4 on the Richter scale. Find its amplitude.

104. EARTHQUAKES   In 1985, Mexico City experienced an earthquake of magnitude 8.1 on the Richter scale. In 1989, the San Francisco Bay area was rocked by an earthquake measuring 7.1. By what factor must the amplitude of an earthquake change to increase its severity by 1 point on the Richter scale? (Assume that the period remains constant.)

105. CHILDREN'S HEIGHT   The function $h(A) = 29 + 48.8 \log(A + 1)$ gives the percent of the adult height a male child A years old has attained. If a boy is 9 years old, what percent of his adult height will he have reached?

106. DEPRECIATION   In business, equipment is often depreciated using the double declining-balance method. In this method, a piece of equipment with a life expectancy of N years, costing $\$C$, will depreciate to a value of $\$V$ in n years, where n is given by the formula

$$n = \frac{\log V - \log C}{\log\left(1 - \frac{2}{N}\right)}$$

A computer that cost $\$37,000$ has a life expectancy of 5 years. If it has depreciated to a value of $\$8,000$, how old is it?

107. INVESTING   If $\$P$ is invested at the end of each year in an annuity earning annual interest at a rate r, the amount in the account will be $\$A$ after n years, where

$$n = \frac{\log\left[\dfrac{Ar}{P} + 1\right]}{\log(1 + r)}$$

If $\$1,000$ is invested each year in an annuity earning 12% annual interest, how long will it take for the account to be worth $\$20,000$?

108. GROWTH OF MONEY   If $\$5,000$ is invested each year in an annuity earning 8% annual interest, how long will it take for the account to be worth $\$50,000$? (See Exercise 107.)

WRITING

109. Explain the mathematical relationship between $f(x) = \log x$ and $g(x) = 10^x$.

110. Explain why it is impossible to find the logarithm of a negative number.

111. A table of solutions for $f(x) = \log x$ is shown below. As x decreases and gets close to 0, what happens to the values of $f(x)$?

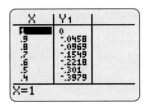

112. What question should be asked when evaluating the expression $\log_4 16$?

REVIEW

Solve each equation.

113. $\sqrt[3]{6x + 4} = 4$

114. $\sqrt{3x + 4} = \sqrt{7x + 2}$

115. $\sqrt{a + 1} - 1 = 3a$

116. $3 - \sqrt{t - 3} = \sqrt{t}$

CHALLENGE PROBLEMS

117. Without graphing, determine the domain of the function $f(x) = \log_5 (x^2 - 1)$. Express the result in interval notation.

118. Evaluate: $\log_6 (\log_5 (\log_4 1{,}024))$

SECTION 9.6
Base-*e* Logarithmic Functions

Objectives

1 Define base-*e* logarithms.

2 Evaluate natural logarithmic expressions.

3 Graph the natural logarithmic function.

4 Use natural logarithmic functions in applications.

We have seen the importance of e in modeling the growth and decay of natural events. Just as $f(x) = e^x$ is called the natural exponential function, its inverse, the base-*e* logarithmic function, is called the *natural logarithmic function*. Natural logarithmic functions have many applications. They play a very important role in advanced mathematics courses, such as calculus.

1 **Define Base-*e* Logarithms.**

Of all possible bases for a logarithmic function, e is the most convenient for problems involving growth or decay. Since these situations occur often in natural settings, base-*e* logarithms are called **natural logarithms** or **Napierian logarithms** after John Napier (1550–1617). They are usually written as $\ln x$ rather than $\log_e x$:

$\ln x$ means $\log_e x$ Read ln x letter-by-letter as "ℓ … n … of x."

In general, the logarithm of a number is an exponent. For natural logarithms,

ln x is the exponent to which e is raised to get x.

Translating this statement into symbols, we have

$$e^{\ln x} = x$$

> **Caution**
> Because of the font used to print the natural log of x, some students initially misread the notation as ln x. In handwriting, ln x should look like

2 Evaluate Natural Logarithmic Expressions.

EXAMPLE 1   Evaluate each natural logarithmic expression:

$$\textbf{a.} \quad \ln e \qquad \textbf{b.} \quad \ln\frac{1}{e^2} \qquad \textbf{c.} \quad \ln 1 \qquad \textbf{d.} \quad \ln\sqrt{e}$$

Strategy   Since the base is e in each case, we will ask "To what power must e be raised to get the given number?"

Why   That power is the value of the logarithmic expression.

Solution

a. $\ln e = 1$ Ask: "To what power must we raise e to get e?"
Since $e^1 = e$, the answer is: the 1st power.

b. $\ln\dfrac{1}{e^2} = -2$ Ask: "To what power must we raise e to get $\frac{1}{e^2}$?"
Since $e^{-2} = \frac{1}{e^2}$, the answer is: the -2 power.

c. $\ln 1 = 0$ Ask: "To what power must we raise e to get 1?"
Since $e^0 = 1$, the answer is: the 0 power.

d. $\ln\sqrt{e} = \dfrac{1}{2}$ Ask: "To what power must we raise e to get $\sqrt{e}$?"
Since $e^{1/2} = \sqrt{e}$, the answer is: the $\frac{1}{2}$ power.

Self Check 1   Evaluate each expression:

$$\textbf{a.} \quad \ln e^3 \qquad \textbf{b.} \quad \ln\frac{1}{e} \qquad \textbf{c.} \quad \ln\sqrt[3]{e}$$

Now Try   Problems 15, 19, and 21

Many natural logarithmic expressions are not as easy to evaluate as those in the previous example. For example, to find ln 2.34, we ask, "To what power must we raise e to get 2.34?" The answer isn't obvious. In such cases, we use a calculator.

Using Your Calculator   *Evaluating Base-e (Natural) Logarithms*

To find ln 2.34 with a scientific calculator, we enter

2.34 $\boxed{\text{LN}}$ $\boxed{.850150929}$

On some calculators, the $\boxed{e^x}$ key also serves as the $\boxed{\text{LN}}$ key when $\boxed{\text{2nd}}$ or $\boxed{\text{SHIFT}}$ is pressed. This is because $f(x) = e^x$ and $g(x) = \ln x$ are inverses.

To use a graphing calculator, we enter

$\boxed{\text{LN}}$ 2.34 $\boxed{)}$ $\boxed{\text{ENTER}}$ $\boxed{\begin{array}{l}\text{ln(2.34)}\\\quad\text{.8501509294}\end{array}}$

To four decimal places, $\boxed{\ln}$ 2.34 = 0.8502. This means that $e^{0.8502} \approx 2.34$.

If we attempt to evaluate logarithmic expressions such as ln 0, or the logarithm of a negative number, such as ln (−5), then one of the following error statements will be are displayed.

$\boxed{\text{Error}}$ $\boxed{\begin{array}{l}\text{ERR:DOMAIN}\\\text{1:QUIT}\\\text{2:Go to}\end{array}}$ $\boxed{\begin{array}{l}\text{ERR:NONREAL ANS}\\\text{1:QUIT}\\\text{2:Go to}\end{array}}$

Certain natural logarithmic equations can be solved by writing them as natural exponential equations.

EXAMPLE 2   Solve each equation: **a.** $\ln x = 1.335$ and **b.** $\ln x = -5.5$. Give each result to four decimal places.

Strategy   To solve this logarithmic equation, we will instead write and solve an equivalent exponential equation.

Why   The resulting exponential equation is easier to solve because the variable term is isolated on one side.

Solution

a. Since the base of the natural logarithmic function is e, the logarithmic equation $\ln x = 1.335$ is equivalent to exponential equation $e^{1.335} = x$. To use a scientific calculator to find x, enter:

1.335 e^x

The display will read 3.799995946. To four decimal places,

$$x = 3.8000$$

The solution is 3.8000. To check, use your calculator to verify that $\ln 3.8000 \approx 1.335$.

b. The equation $\ln x = -5.5$ is equivalent to $e^{-5.5} = x$. To use a scientific calculator to find x, enter:

5.5 +/− e^x

The display will read 0.004086771. To four decimal places,

$$x = 0.0041$$

The solution is 0.0041. To check, use your calculator to verify that $\ln 0.0041 \approx -5.5$.

Self Check 2   Solve each equation. Give each result to four decimal places.
a. $\ln x = 1.9344$ **b.** $-3 = \ln x$

Now Try   **Problems 35 and 39**

---

3   **Graph the Natural Logarithmic Function.**

Because the natural exponential function defined by $f(x) = e^x$ is one-to-one, it has an inverse function that is defined by $x = e^y$. When we write $x = e^y$ in the equivalent form $y = \ln x$, the result is called the *natural logarithmic function*.

The Natural Logarithmic Function	The **natural logarithmic function** with base e is defined by the equations $f(x) = \ln x$ or $y = \ln x$, where $\ln x = \log_e x$. The domain of $f(x) = \ln x$ is the interval $(0, \infty)$, and the range is the interval $(-\infty, \infty)$.

Since the natural logarithmic function is the inverse of the one-to-one natural exponential function, the natural logarithmic function is one-to-one.

To graph $f(x) = \ln x$, we can construct a table of function values, plot the resulting ordered pairs, and draw a smooth curve through the points to get the graph shown in figure (a). Figure (b) shows the calculator graph of $f(x) = \ln x$.

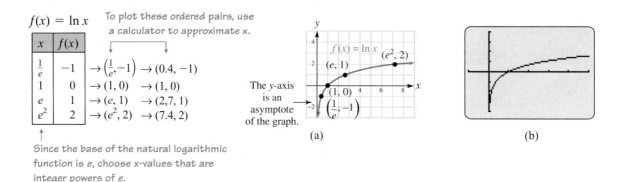

$f(x) = \ln x$

To plot these ordered pairs, use a calculator to approximate x.

x	$f(x)$
$\frac{1}{e}$	-1
1	0
e	1
e^2	2

$\to \left(\frac{1}{e}, -1\right) \to (0.4, -1)$
$\to (1, 0) \to (1, 0)$
$\to (e, 1) \to (2.7, 1)$
$\to (e^2, 2) \to (7.4, 2)$

Since the base of the natural logarithmic function is e, choose x-values that are integer powers of e.

The y-axis is an asymptote of the graph.

(a)   (b)

The natural exponential function and the natural logarithm function are inverse functions. The figure shows that their graphs are symmetric to the line $y = x$.

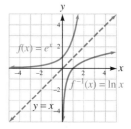

Using Your Calculator

Graphing Base-e Logarithmic Functions

Many graphs of logarithmic functions involve translations of the graph of $f(x) = \ln x$. For example, the figure below shows calculator graphs of the functions $f(x) = \ln x$, $g(x) = (\ln x) + 2$, and $h(x) = (\ln x) - 3$.

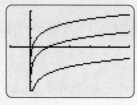

The graph of $g(x) = (\ln x) + 2$ is 2 units above the graph of $f(x) = \ln x$.

The graph of $h(x) = (\ln x) - 3$ is 3 units below the graph of $f(x) = \ln x$.

The next figure shows the calculator graph of the functions $f(x) = \ln x$, $g(x) = \ln (x - 2)$, and $h(x) = \ln (x + 3)$.

The graph of $h(x) = \ln (x + 3)$ is 3 units to the left of the graph of $f(x) = \ln x$.

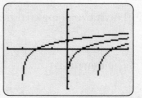

The graph of $g(x) = \ln (x - 2)$ is 2 units to the right of the graph of $f(x) = \ln x$.

4 **Use Natural Logarithmic Functions in Applications.**

If a population grows exponentially at a certain annual rate, the time required for the population to double is called the **doubling time.** It is given by the following formula.

Formula for Doubling Time	If r is the annual rate, compounded continuously, and t is the time required for a population to double, then $$t = \frac{\ln 2}{r}$$

EXAMPLE 3   ***Doubling Time.***   The population of the Earth is growing at the approximate rate of 1.17% per year. If this rate continues, how long will it take for the population to double?

Strategy   We will substitute 1.17% for r in the formula for doubling time and evaluate the right side using a calculator.

Why   We can use this formula because we are given the annual rate of continuous compounding.

Solution   Since the population is growing at the rate of 1.17% per year, we substitute 0.0117 for r in the formula for doubling time and simplify.

$$t = \frac{\ln 2}{r}$$

$$t = \frac{\ln 2}{0.0117}$$

$$\approx 59.24334877 \quad \text{Use a calculator. Find ln 2 first, then divide the result by 0.0117.}$$

The population of the Earth will double in about 59 years.

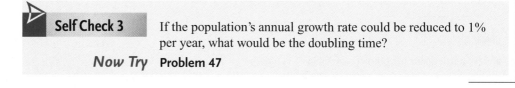

Self Check 3   If the population's annual growth rate could be reduced to 1% per year, what would be the doubling time?

Now Try   **Problem 47**

EXAMPLE 4   ***Doubling Time.***   How long will it take $1,000 to double at an annual rate of 8%, compounded continuously?

Strategy   We will substitute 8% for r in the formula for doubling time and evaluate the right side using a calculator. In this case, the information that the original amount is $1,000 is unnecessary information.

Why   We can use this formula because we are given the annual rate of continuous compounding.

Solution   We substitute 0.08 for r and proceed as follows:

$$t = \frac{\ln 2}{r}$$ *This is the formula for doubling time.*

$$t = \frac{\ln 2}{0.08}$$

$$\approx 8.664339757$$ *Use a calculator. Find ln 2 first, then divide the result by 0.08.*

It will take about $8\frac{2}{3}$ years for the money to double.

Self Check 4   How long will it take at 9%, compounded continuously?

Now Try   Problem 50

ANSWERS TO SELF CHECKS   **1. a.** 3   **b.** -1 **c.** $\frac{1}{3}$ **2. a.** 6.9199   **b.** 0.0498   **3.** About 46 years
4. About 7.7 years

STUDY SET
9.6

VOCABULARY

Fill in the blanks.

1. $f(x) = \ln x$ is called the _____ logarithmic function. The base
is [ ].

2. If a population grows exponentially at a certain annual rate,
the time required for the population to double is called the
_____ _____.

CONCEPTS

3. a. Use a calculator to complete the table of values for
$f(x) = \ln x$. Round to the nearest hundredth.

x	$f(x)$
0.5	
1	
2	
4	
6	
8	
10	

b. Graph $f(x) = \ln x$. Note that the units on the x- and y-axes are
different.

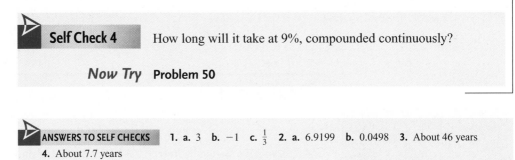

4. What is the inverse of the natural logarithmic function
$f(x) = \ln x$?

Fill in the blanks.

5. The domain of the function $f(x) = \ln x$ is the interval [ ]
and the range of the function is the interval [ ].

6. The graph of $f(x) = \ln x$ has the x-intercept ([ ] , 0). The y-axis
is an _____ of the graph.

7. To find $\ln e^2$, we ask, "To what power must we raise [ ] to get
e^2?" Since the answer is the 2nd power, $\ln e^2 =$ [ ].

8. The logarithmic equation $\ln x = 1.5318$ is equivalent to the exponential equation $ = $.

9. The illustration shows the graph of $f(x) = \ln x$, as well as a vertical translation of that graph. Using the notation $g(x)$ for the translation, write the defining equation for the function.

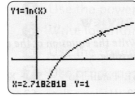

10. In the illustration, $f(x) = \ln x$ was graphed, and the TRACE feature was used. What is the x-coordinate of the point on the graph having a y-coordinate of 1? What is the name given this number?

NOTATION

Fill in the blanks.

11. We read $\ln x$ letter-by-letter as " $\ldots$ $\ldots$ of x."

12. a. $\ln 2$ means $\log\ 2$.

    b. $\log 2$ means $\log\ 2$.

13. To evaluate a base-10 logarithm with a calculator, use the key. To evaluate the base-e logarithm, use the key.

14. If a population grows exponentially at a rate r, the time it will take the population to double is given by the formula $t = $.

GUIDED PRACTICE

Evaluate each expression without using a calculator. See Example 1.

15. $\ln e^5$ **16.** $\ln e^2$

17. $\ln e^6$ **18.** $\ln e^4$

19. $\ln \dfrac{1}{e}$ **20.** $\ln \dfrac{1}{e^3}$

21. $\ln \sqrt[4]{e}$ **22.** $\ln \sqrt[5]{e}$

23. $\ln \sqrt[3]{e^2}$ **24.** $\ln \sqrt[4]{e^3}$

25. $\ln e^{-7}$ **26.** $\ln e^{-10}$

Use a calculator to evaluate each expression, if possible. Express all answers to four decimal places. See Using Your Calculator: Evaluating Base-e Logarithms.

27. $\ln 35.15$ **28.** $\ln 0.675$

29. $\ln 0.00465$ **30.** $\ln 378.96$

31. $\ln 1.72$ **32.** $\ln 2.7$

33. $\ln (-0.1)$ **34.** $\ln (-10)$

Solve each equation. Express all answers to four decimal places. See Example 2.

35. $\ln x = 1.4023$ **36.** $\ln x = 2.6490$

37. $\ln x = 4.24$ **38.** $\ln x = 0.926$

39. $\ln x = -3.71$ **40.** $\ln x = -0.28$

41. $1.001 = \ln x$ **42.** $\ln x = -0.001$

Use a graphing calculator to graph each function. See Objective 2. See Using Your Calculator: Base-e Logarithmic Functions.

43. $f(x) = \ln \left(\dfrac{1}{2}x\right)$ **44.** $f(x) = \ln x^2$

45. $f(x) = \ln (-x)$ **46.** $f(x) = \ln (3x)$

APPLICATIONS

Use a calculator to solve each problem.

47. THE PEACH STATE Chattahoochee County, Georgia, grew by 13.2% between 2005 and 2006, making it the fastest-growing county in the United States at that time. If the growth rate remains constant, how long will it take for the population of the county to double?

Chattahoochee County, Georgia

48. THE SILVER STATE Nevada is the one of the fastest growing states in the United States. Determine the number of years it would take for the population of each of Nevada's four largest cities to double in size using the 2005 data below.

Rank	City	Population 2005	Annual Rate of Increase
1.	Las Vegas	569,838	3.7%
2.	Henderson	241,134	4.8%
3.	Reno	206,735	3.8%
4.	North Las Vegas	180,219	9.2%

49. THE NORTH STAR STATE Minnesota's population increased by 40,362 persons, or 0.8%, between July 2005 and July 2006. If the growth rate remains constant, how long will it take for the population of the state to double?

50. DOUBLING MONEY How long will it take $1,000 to double if it is invested at an annual rate of 5% compounded continuously?

Solution

a. By property 1, $\log_5 1 = 0$, because $5^0 = 1$.

b. By property 2, $\log_3 3 = 1$, because $3^1 = 3$.

c. By property 3, $\ln e^3 = 3$, because $e^3 = e^3$.

d. By property 4, $6^{\log_6 7} = 7$, because $\log_6 7$ is the power to which 6 is raised to get 7.

 Self Check 1   Simplify: **a.** $\log_4 1$ **b.** $\log_4 4$ **c.** $\log_2 2^4$
d. $5^{\log_5 2}$

Now Try   Problems 19, 21, 23, and 27

---

2   **Use the Product Rule for Logarithms.**

The next property of logarithms is related to the product rule for exponents: $x^m \cdot x^n = x^{m+n}$.

The Product Rule for Logarithms	The logarithm of a product is equal to the sum of the logarithms. For all positive real numbers M, N, and b, where $b \ne 1$, $$\log_b MN = \log_b M + \log_b N$$

EXAMPLE 2   Write each expression as a sum of logarithms. Then simplify, if possible.   **a.** $\log_2(2 \cdot 7)$ **b.** $\log 100x$ **c.** $\log_5 125yz$

Strategy   In each case, we will use the product rule for logarithms.

Why   We use the product rule because each of the logarithmic expressions has the form $\log_b MN$.

Solution

a. $\log_2(2 \cdot 7) = \log_2 2 + \log_2 7$ The log of a product is the sum of the logs.

$\qquad\qquad\quad = 1 + \log_2 7$ Simplify: By property 2, $\log_2 2 = 1$.

b. Recall that $100x$ means $100 \cdot x$.

$\log 100x = \log 100 + \log x$ The log of a product is the sum of the logs.

$\qquad\quad = 2 + \log x$ Simplify: By property 3, $\log 100 = \log 10^2 = 2$.

c. We can write $125yz$ as $(125y)z$.

$\log_5 125yz = \log_5(125y)z$ Group the first two factors together.

$\qquad\quad = \log_5(125y) + \log_5 z$ The log of a product is the sum of the logs.

$\qquad\quad = \log_5 125 + \log_5 y + \log_5 z$ The log of a product is the sum of the logs.

$\qquad\quad = 3 + \log_5 y + \log_5 z$ Simplify: By property 3, $\log_5 125 = \log_5 5^3 = 3$.

Notation
To avoid any confusion, we can use parentheses when writing the logarithm of a product:

$\log 100x = \log(100x)$

 Self Check 2   Write each expression as the sum of logarithms. Then simplify, if possible.   **a.** $\log_3(3 \cdot 4)$ **b.** $\log 1{,}000y$
c. $\log_5 25cd$

Now Try   Problems 31 and 35

Caution

As we apply properties of logarithms to rewrite expressions, we assume that all variables represent positive numbers.

| **PROOF** | To prove the product rule for logarithms, we let $x = \log_b M$, $y = \log_b N$, and use the definition of logarithm to write each equation in exponential form. |

$$M = b^x \quad \text{and} \quad N = b^y$$

Then $MN = b^x b^y$, and a property of exponents gives

$$MN = b^{x+y} \quad \textit{Keep the base and add the exponents: } b^x b^y = b^{x+y}.$$

We write this exponential equation in logarithmic form as

$$\log_b MN = x + y$$

Substituting the values of x and y completes the proof.

$$\log_b MN = \log_b M + \log_b N$$

Caution

The log of a sum does not equal the sum of the logs. The log of a difference does not equal the difference of the logs.

Caution   By the product rule, the logarithm of a *product* is equal to the *sum* of the logarithms. The logarithm of a sum or a difference usually does not simplify. In general,

$$\log_b (M + N) \neq \log_b M + \log_b N \quad \text{and} \quad \log_b (M - N) \neq \log_b M - \log_b N$$

For example,

$$\log_2 (2 + 7) \neq \log_2 2 + \log_2 7 \quad \text{and} \quad \log (100 - y) \neq \log 100 - \log y$$

Using Your Calculator   *Verifying Properties of Logarithms*

We can use a calculator to illustrate the product rule for logarithms by showing that

$$\ln (3.7 \cdot 15.9) = \ln 3.7 + \ln 15.9$$

We calculate the left and right sides of the equation separately and compare the results. To use a scientific calculator to find $\ln (3.7 \cdot 15.9)$, we enter

3.7 $\boxed{\times}$ 15.9 $\boxed{=}$ $\boxed{\text{LN}}$ $\boxed{4.074651929}$

To find $\ln 3.7 + \ln 15.9$, we enter

3.7 $\boxed{\text{LN}}$ $\boxed{+}$ 15.9 $\boxed{\text{LN}}$ $\boxed{=}$ $\boxed{4.074651929}$

Since the left and right sides are equal, the equation $\ln (3.7 \cdot 15.9) = \ln 3.7 + \ln 15.9$ is true.

 Use the Quotient Rule for Logarithms.

The next property of logarithms is related to the quotient rule for exponents: $\frac{x^m}{x^n} = x^{m-n}$.

The Quotient Rule for Logarithms

The logarithm of a quotient is equal to the difference of the logarithms. For all positive real numbers M, N, and b, where $b \neq 1$,

$$\log_b \frac{M}{N} = \log_b M - \log_b N$$

The proof of the quotient rule for logarithms is similar to the proof for the product rule for logarithms.

EXAMPLE 3   Write each expression as a difference of logarithms. Then simplify, if possible.   **a.** $\ln \dfrac{10}{7}$ **b.** $\log_4 \dfrac{x}{64}$

Strategy   In both cases, we will apply the quotient rule for logarithms.

Why   We use the quotient rule because each of the logarithmic expressions has the form $\log_b \dfrac{M}{N}$.

Solution

a. $\ln \dfrac{10}{7} = \ln 10 - \ln 7$ The log of a quotient is the difference of the logs.

b. $\log_4 \dfrac{x}{64} = \log_4 x - \log_4 64$ The log of a quotient is the difference of the logs.

$\qquad\qquad = \log_4 x - 3$ Simplify: $\log_4 64 = \log_4 4^3 = 3$.

▷ **Self Check 3**   Write each expression as a difference of logarithms. Then simplify, if possible.

a. $\log_6 \dfrac{6}{5}$ **b.** $\ln \dfrac{y}{100}$

Now Try   **Problem 39**

Caution   By the quotient rule, the logarithm of a *quotient* is equal to the *difference* of the logarithms. The logarithm of a quotient is not the quotient of the logarithms:

$$\log_b \frac{M}{N} \neq \frac{\log_b M}{\log_b N}$$

For example,

$$\ln \frac{10}{7} \neq \frac{\ln 10}{\ln 7} \quad \text{and} \quad \log_4 \frac{x}{64} \neq \frac{\log_4 x}{\log_4 64}$$

In the next example, the product and quotient rules for logarithms are used in combination to rewrite an expression.

EXAMPLE 4   Write $\log \dfrac{xy}{10z}$ as the sum and/or difference of logarithms of a single quantity. Then simplify, if possible.

Strategy   We will use the quotient rule for logarithms and then the product rule.

Why   We use the quotient rule because $\log \dfrac{xy}{10z}$ has the form $\log_b \dfrac{M}{N}$. We later use the product rule because the numerator and denominator of $\dfrac{xy}{10z}$ contain products.

Solution   We begin by applying the quotient rule for logarithms.

$$\log \frac{xy}{10z} = \log xy - \log 10z$$ The log of a quotient is the difference of the logs.

$$= \log x + \log y - (\log 10 + \log z)$$ The log of a product is the sum of the logs.

Write parentheses here so that the sum, $\log 10 + \log z$, is subtracted.

$$= \log x + \log y - \log 10 - \log z$$ Change the sign of each term of $\log 10 + \log z$ and drop the parentheses.

$$= \log x + \log y - 1 - \log z$$ Simplify: $\log 10 = 1$.

> **Self Check 4**   Write $\log_b \dfrac{x}{yz}$ as the sum and/or difference of logarithms of a single quantity. Then simplify, if possible.
>
> ***Now Try***   **Problem 45**

❹ Use the Power Rule for Logarithms.

The next property of logarithms is related to the power rule for exponents: $(x^m)^n = x^{mn}$.

The Power Rule for Logarithms	The logarithm of a power is equal to the power times the logarithm.
	For all real positive numbers M and b, where $b \neq 1$, and any real number p,
	$$\log_b M^p = p \log_b M$$

> **EXAMPLE 5**   Write each logarithm without an exponent or a square root:
>
> **a.** $\log_5 6^2$ **b.** $\log \sqrt{10}$

Strategy   In each case, we will use the power rule for logarithms.

Why   We use the power rule because $\log_5 6^2$ has the form $\log_b M^p$, as will $\log \sqrt{10}$ if we write $\sqrt{10}$ as $10^{1/2}$.

Solution

a. $\log_5 6^2 = 2 \log_5 6$ The log of a power is equal to the power times the log.

b. $\log \sqrt{10} = \log(10)^{1/2}$ Write $\sqrt{10}$ using a fractional exponent: $\sqrt{10} = (10)^{1/2}$.

$$= \frac{1}{2} \log 10$$ The log of a power is equal to the power times the log.

$$= \frac{1}{2}$$ Simplify: $\log 10 = 1$.

> **Self Check 5**   Write each logarithm without an exponent or a cube root:
>
> **a.** $\ln x^4$ **b.** $\log_2 \sqrt[3]{3}$
>
> ***Now Try***   **Problems 51 and 53**

PROOF   To prove the power rule, we let $x = \log_b M$, write the expression in exponential form, and raise both sides to the pth power:

$$M = b^x$$

$$(M)^p = (b^x)^p \qquad \text{Raise both sides to the pth power.}$$

$$M^p = b^{px} \qquad \text{Keep the base and multiply the exponents.}$$

Using the definition of logarithms gives

$$\log_b M^p = px$$

Substituting the value for x completes the proof.

$$\log_b M^p = p \log_b M$$

EXAMPLE 6   Write each logarithm as the sum and/or difference of logarithms of a single quantity: **a.** $\log_b x^2 y^3 z$ **b.** $\ln \dfrac{y^3 \sqrt{x}}{z}$

Strategy   In part (a), we will use the product rule and the power rules for logarithms. In part (b), we will use the quotient rule, the product rule, and the power rule for logarithms.

Why   In part (a), we first use the product rule because the expression has the form $\log_b MN$. In part (b), we first use the quotient rule because the expression has the form $\log_b \dfrac{M}{N}$.

Solution
a. We recognize that $\log_b x^2 y^3 z$ is the logarithm of a product.

$$\log_b x^2 y^3 z = \log_b x^2 + \log_b y^3 + \log_b z \qquad \text{The log of a product is the sum of the logs.}$$

$$= 2 \log_b x + 3\log_b y + \log_b z \qquad \text{The log of a power is the power times the log.}$$

The Language of Algebra
In Examples 2, 3, 4, and 6, we use properties of logarithms to *expand* logarithmic expressions.

b. The expression $\ln \dfrac{y^3 \sqrt{x}}{z}$ is the logarithm of a quotient.

$$\ln \frac{y^3 \sqrt{x}}{z} = \ln y^3 \sqrt{x} - \ln z \qquad \text{The log of a quotient is the difference of the logs.}$$

$$= \ln y^3 + \ln \sqrt{x} - \ln z \qquad \text{The log of a product is the sum of the logs.}$$

$$= \ln y^3 + \ln x^{1/2} - \ln z \qquad \text{Write } \sqrt{x} \text{ as } x^{1/2}.$$

$$= 3 \ln y + \frac{1}{2} \ln x - \ln z \qquad \text{The log of a power is the power times the log.}$$

▷ **Self Check 6**   Expand: $\log \sqrt[4]{\dfrac{x^3 y}{z}}$

Now Try Problems 59 and 61

Write Logarithmic Expressions as a Single Logarithm.

EXAMPLE 7 Write each logarithmic expression as one logarithm:

a. $3 \log_5 x + \frac{1}{2} \log_5 y$ **b.** $\frac{1}{2} \log_b (x - 2) - \log_b y + 3 \log_b z$

Strategy In part (a), we will use the power rule and product rule for logarithms in reverse. In part (b), we will use the power rule, the quotient rule, and the product rule for logarithms in reverse.

Why We use the power rule because we see expressions of the form $p \log_b M$. The $+$ symbol between logarithmic terms suggests that we use the product rule and the $-$ symbol between such terms suggests that we use the quotient rule.

Solution

a. We begin by using the power rule on both terms of the expression.

$$3 \log_5 x + \frac{1}{2} \log_5 y = \log_5 x^3 + \log_5 y^{1/2} \qquad \text{A power times a log is the log of the power.}$$

$$= \log_5 (x^3 \cdot y^{1/2}) \qquad \text{The sum of two logs is the log of the product.}$$

$$= \log_5 x^3 y^{1/2}$$

$$= \log_5 x^3 \sqrt{y} \qquad \text{Write } y^{1/2} \text{ as } \sqrt{y}.$$

b. The first and third terms of this expression can be rewritten using the power rule of logarithms.

> **The Language of Algebra**
> In these examples, we use properties of logarithms to *condense* the given expression into a single logarithmic expression. To *condense* means to make more compact. Summer school is a *condensed* version of the regular semester.

$$\frac{1}{2} \log_b (x - 2) - \log_b y + 3 \log_b z$$

$$= \log_b (x - 2)^{1/2} - \log_b y + \log_b z^3 \qquad \text{A power times a log is the log of the power.}$$

$$= \log_b \frac{(x - 2)^{1/2}}{y} + \log_b z^3 \qquad \text{The difference of two logs is the log of the quotient.}$$

$$= \log_b \frac{\sqrt{x - 2}}{y} + \log_b z^3 \qquad \text{Write } (x - 2)^{1/2} \text{ as } \sqrt{x - 2}.$$

$$= \log_b \left(\frac{\sqrt{x - 2}}{y} \cdot z^3 \right) \qquad \text{The sum of two logs is the log of the product.}$$

$$= \log_b \frac{z^3 \sqrt{x - 2}}{y}$$

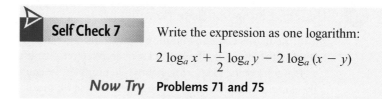

Self Check 7 Write the expression as one logarithm:

$$2 \log_a x + \frac{1}{2} \log_a y - 2 \log_a (x - y)$$

Now Try **Problems 71 and 75**

The properties of logarithms can be used when working with numerical values.

EXAMPLE 8   Find approximations for each logarithm given that $\log 2 \approx 0.3010$ and $\log 3 \approx 0.4771$:   **a.** $\log 6$ **b.** $\log 18$

Strategy   We will express 6 and 18 using factors of 2 and 3 and then use properties of logarithms to simplify each resulting expression.

Why   We express 6 and 18 using factors of 2 and 3 because we are given values of $\log 2$ and $\log 3$.

Solution

a. $\log 6 = \log (2 \cdot 3)$ Write 6 using the factors 2 and 3.

$\qquad = \log 2 + \log 3$ The log of a product is the sum of the logs.

$\qquad \approx 0.3010 + 0.4771$ Substitute the value of each logarithm.

$\qquad \approx 0.7781$

b. $\log 18 = \log (2 \cdot 3^2)$ Write 18 using the factors 2 and 3.

$\qquad = \log 2 + \log 3^2$ The log of a product is the sum of the logs.

$\qquad = \log 2 + 2 \log 3$ The log of a power is the power times the log.

$\qquad \approx 0.3010 + 2(0.4771)$ Substitute the value of each logarithm.

$\qquad \approx 1.2552$

 Self Check 8   Find approximations for each logarithm.   **a.** $\log 1.5$ **b.** $\log 0.75$

Now Try   **Problems 79 and 81**

We summarize the properties of logarithms as follows.

Properties of Logarithms

If b, M, and N are positive real numbers, $b \neq 1$, and p is any real number,

1. $\log_b 1 = 0$ **2.** $\log_b b = 1$

3. $\log_b b^x = x$ **4.** $b^{\log_b x} = x$

5. $\log_b MN = \log_b M + \log_b N$ **6.** $\log_b \dfrac{M}{N} = \log_b M - \log_b N$

7. $\log_b M^p = p \log_b M$

6   **Use the Change-of-Base Formula.**

Most calculators can find common logarithms and natural logarithms. If we need to find a logarithm with some other base, we use a conversion formula.

If we know the base-a logarithm of a number, we can find its logarithm to some other base b by using a formula called the **change-of-base formula.**

Change-of-Base Formula

For any logarithmic bases a and b, and any positive real number x,

$$\log_b x = \frac{\log_a x}{\log_a b}$$

We can use any positive number other than 1 for base b in the change-of-base formula. However, we usually use 10 or e because of the capabilities of a standard calculator.

EXAMPLE 9   Find: $\log_3 5$

Strategy   To evaluate this base-3 logarithm, we will substitute into the change-of-base formula.

Why   We assume that the reader does not have a calculator that evaluates base-3 logarithms (at least not directly). Thus, the only alternative is to change the base.

Solution   To find $\log_3 5$, we substitute 3 for b, 10 for a, and 5 for x in the change-of-base formula and simplify:

$$\log_b x = \frac{\log_a x}{\log_a b}$$

$$\log_3 5 = \frac{\log_{10} 5}{\log_{10} 3} \qquad \text{Substitute: } b = 3, x = 5, \text{ and } a = 10.$$

$$\approx 1.464973521 \qquad \text{Use a scientific calculator and enter } 5 \boxed{\log} \div 3 \boxed{\log} \boxed{=}.$$

To four decimal places, $\log_3 5 = 1.4650$.

We can also use the natural logarithm function (base e) in the change-of-base formula to find a base-3 logarithm.

$$\log_b x = \frac{\log_a x}{\log_a b}$$

$$\log_3 5 = \frac{\log_e 5}{\log_e 3} \qquad \text{Substitute: } b = 3, x = 5, \text{ and } a = e.$$

$$\log_3 5 = \frac{\ln 5}{\ln 3} \qquad \text{Write } \log_e 5 \text{ as ln 5 and } \log_e 3 \text{ as ln 3.}$$

$$\approx 1.464973521 \qquad \text{Use a calculator.}$$

We obtain the same result.

> **Caution**
> Don't misapply the quotient rule: $\frac{\log_{10} 5}{\log_{10} 3}$ means $\log_{10} 5 \div \log_{10} 3$. It is the expression $\log_{10} \frac{5}{3}$ that means $\log_{10} 5 - \log_{10} 3$.

> **Caution**
> Wait until the final calculation has been made to round. Don't round any values when performing intermediate calculations. That could make the final result incorrect because of a build-up of rounding error.

Self Check 9   Find $\log_5 3$ to four decimal places.

Now Try   Problem 87

PROOF   To prove the change-of-base formula, we begin with the equation $\log_b x = y$.

$$y = \log_b x$$

$$x = b^y \qquad \text{Change the equation from logarithmic to exponential form.}$$

$$\log_a x = \log_a b^y \qquad \text{Take the base-}a \text{ logarithm of both sides.}$$

$$\log_a x = y \log_a b \qquad \text{The log of a power is the power times the log.}$$

$$y = \frac{\log_a x}{\log_a b} \qquad \text{Divide both sides by } \log_a b.$$

$$\log_b x = \frac{\log_a x}{\log_a b} \qquad \text{Refer to the first equation and substitute } \log_b x \text{ for } y.$$

7 **Use Properties of Logarithms to Solve Application Problems.**

In chemistry, common logarithms are used to express the acidity of solutions. The more acidic a solution, the greater the concentration of hydrogen ions. This concentration is indicated by the *pH scale,* or *hydrogen ion index.* The pH of a solution is defined as follows.

pH of a Solution	If $[H^+]$ is the hydrogen ion concentration in gram-ions per liter, then $$pH = -\log[H^+]$$

EXAMPLE 10   *pH Meters.*   One of the most accurate ways to measure pH is with a probe and meter. What reading should the meter give for pure water if water has a hydrogen ion concentration $[H^+]$ of approximately 10^{-7} gram-ions per liter?

Strategy   We will substitute into the formula for pH and use the power rule for logarithms to simplify the right side.

Why   After substituting 10^{-7} for H^+ in $-\log[H^+]$, the resulting expression will have the form $\log_b M^p$.

Solution   Since pure water has approximately 10^{-7} gram-ions per liter, its pH is

$$pH = -\log[H^+] \qquad \text{This is the formula for pH.}$$
$$pH = -\log 10^{-7}$$
$$\quad = -(-7)\log 10 \qquad \text{The log of a power is the power times the log.}$$
$$\quad = -(-7)\cdot 1 \qquad \text{Simplify: } \log 10 = 1.$$
$$\quad = 7$$

The meter should give a reading of 7.

▷ *Now Try*   **Problem 99**

EXAMPLE 11   *Hydrogen Ion Concentration.*   Find the hydrogen ion concentration of seawater if its pH is 8.5.

Strategy   To find the hydrogen ion concentration, we will substitute 8.5 for pH in the formula $pH = -\log[H^+]$ and solve the resulting equation for $[H^+]$.

Why   After substituting for pH, the resulting logarithmic equation can be solved by solving an equivalent exponential equation.

Solution
$$\mathbf{pH} = -\log[H^+] \qquad \text{This is the formula for pH.}$$
$$8.5 = -\log[H^+] \qquad \text{Substitute 8.5 for pH.}$$
$$-8.5 = \log[H^+] \qquad \text{Multiply both sides by } -1.$$
$$[H^+] = 10^{-8.5} \qquad \text{Write the equation in the equivalent exponential form.}$$

We can use a calculator to find that

$$[H^+] \approx 3.2 \times 10^{-9} \text{ gram-ions per liter}$$

 Now Try Problem 101

STUDY SET
9.7

VOCABULARY

Fill in the blanks.

1. The expression $\log_3 4x$ is the logarithm of a _____.
2. The expression $\log_2 \frac{5}{x}$ is the logarithm of a _____.
3. The expression $\log 4^x$ is the logarithm of a _____.
4. In the expression $\log_5 4$, the number 5 is the _____ of the logarithm.

CONCEPTS

Fill in the blanks.

5. $\log_b 1 =$
6. $\log_b b =$
7. $\log_b MN = \log_b \quad + \log_b$
8. $b^{\log_b x} =$
9. $\log_b \frac{M}{N} = \log_b M \quad \log_b N$
10. $\log_b M^p = p \log_b$
11. $\log_b b^x =$
12. $\log_b (A + B) \quad \log_b A + \log_b B$
13. $\log_b \frac{M}{N} \quad \frac{\log_b M}{\log_b N}$
14. $\log_b AB \quad \log_b A + \log_b B$
15. $\log_b x = \dfrac{\log_a x}{}$
16. $pH =$

NOTATION

Complete each solution.

17. $\log_b rst = \log_b (\quad)t$
$\quad = \log_b (rs) + \log_b$
$\quad = \log_b \quad + \log_b \quad + \log_b t$

18. $\log \dfrac{r}{st} = \log r - \log (\quad)$
$\quad = \log r - (\log \quad + \log t)$
$\quad = \log r - \log s$

GUIDED PRACTICE

In this Study Set, assume that all variables represent positive numbers and $b \neq 1$.

Evaluate each expression. See Example 1.

19. $\log_6 1$
20. $\log_9 9$
21. $\log_4 4^7$
22. $\ln e^8$
23. $5^{\log_5 10}$
24. $8^{\log_8 10}$
25. $\log_5 5^2$
26. $\log_4 4^2$
27. $\ln e$
28. $\log_7 1$
29. $\log_3 3^7$
30. $5^{\log_5 8}$

Write each logarithm as a sum. Then simplify, if possible. See Example 2.

31. $\log_2 (4 \cdot 5)$
32. $\log_3 (27 \cdot 5)$
33. $\log 25y$
34. $\log xy$
35. $\log 100pq$
36. $\log 1{,}000rs$
37. $\log 5xyz$
38. $\log 10abc$

Write each logarithm as a difference. Then simplify, if possible.
See Example 3.

39. $\log \dfrac{100}{9}$

40. $\ln \dfrac{27}{e}$

41. $\log_6 \dfrac{x}{36}$

42. $\log_8 \dfrac{y}{8}$

Write each logarithm as the sum and/or difference of logarithms of a single quantity. Then simplify, if possible. See Example 4.

43. $\log \dfrac{7c}{2}$

44. $\log \dfrac{9t}{4}$

45. $\log \dfrac{10x}{y}$

46. $\log_2 \dfrac{ab}{4}$

47. $\ln \dfrac{exy}{z}$

48. $\ln \dfrac{5p}{e}$

49. $\log_8 \dfrac{1}{8m}$

50. $\log_6 \dfrac{1}{36r}$

Write each logarithm without an exponent or a radical symbol.
Then simplify, if possible. See Example 5.

51. $\ln y^7$

52. $\ln z^9$

53. $\log \sqrt{5}$

54. $\log \sqrt[3]{7}$

55. $\log e^{-3}$

56. $\log e^{-1}$

57. $\log_7 \left(\sqrt[5]{100}\right)^3$

58. $\log_2 \left(\sqrt{10}\right)^5$

Write each logarithm as the sum and/or difference of logarithms of a single quantity. Then simplify, if possible. See Example 6.

59. $\log xyz^2$

60. $\log 4xz^2$

61. $\log_2 \dfrac{2\sqrt[3]{x}}{y}$

62. $\log_3 \dfrac{\sqrt[4]{x}}{yz}$

63. $\log x^3y^2$

64. $\log xy^2z^3$

65. $\log_b \sqrt{xy}$

66. $\log_b x^3\sqrt{y}$

67. $\log_a \dfrac{\sqrt[3]{x}}{\sqrt[4]{yz}}$

68. $\log_b \sqrt[4]{\dfrac{x^3y^2}{z^4}}$

69. $\ln x\sqrt{z}$

70. $\ln \sqrt{xy}$

Write each logarithmic expression as one logarithm. See Example 7.

71. $\log_2 (x + 1) + 9 \log_2 x$

72. $2 \log x + \dfrac{1}{2} \log y$

73. $\log_3 x + \log_3 (x + 2) - \log_3 8$

74. $-2 \log x - 3 \log y + \log z$

75. $-3\log_b x - 2 \log_b y + \dfrac{1}{2} \log_b z$

76. $3 \log_b (x + 1) - 2 \log_b (x + 2) + \log_b x$

77. $\ln \left(\dfrac{x}{z} + x\right) - \ln \left(\dfrac{y}{z} + y\right)$

78. $\ln (xy + y^2) - \ln (xz + yz) + \ln z$

Assume that $\log_b 4 = 0.6021$, $\log_b 7 = 0.8451$, and $\log_b 9 = 0.9542$. Use these values to evaluate each logarithm. See Example 8.

79. $\log_b 28$

80. $\log_b \dfrac{7}{4}$

81. $\log_b \dfrac{4}{63}$

82. $\log_b 36$

83. $\log_b \dfrac{63}{4}$

84. $\log_b 2.25$

85. $\log_b 64$

86. $\log_b 49$

Use the change-of-base formula to find each logarithm to four decimal places. See Example 9.

87. $\log_3 7$

88. $\log_7 3$

89. $\log_{1/3} 3$

90. $\log_{1/2} 6$

91. $\log_3 8$

92. $\log_5 10$

93. $\log_{\sqrt{2}} \sqrt{5}$

94. $\log_\pi e$

Use a calculator to verify that each equation is true. See Using Your Calculator: Verifying Properties of Logarithms.

95. $\log (2.5 \cdot 3.7) = \log 2.5 + \log 3.7$

96. $\ln (2.25)^4 = 4 \ln 2.25$

97. $\ln \dfrac{11.3}{6.1} = \ln 11.3 - \ln 6.1$

98. $\log \sqrt{24.3} = \dfrac{1}{2} \log 24.3$

APPLICATIONS

99. pH OF A SOLUTION   Find the pH of a solution with a hydrogen ion concentration of 1.7×10^{-5} gram-ions per liter.

100. pH OF PICKLES   The hydrogen ion concentration of sour pickles is 6.31×10^{-4}. Find the pH.

101. HYDROGEN ION CONCENTRATION   Find the hydrogen ion concentration of a saturated solution of calcium hydroxide whose pH is 13.2.

102. AQUARIUMS   To test for safe pH levels in a freshwater aquarium, a test strip is compared with the scale shown below. Find the corresponding range in the hydrogen ion concentration.

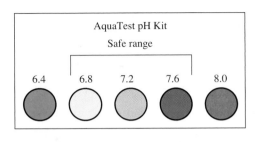

WRITING

103. Explain the difference between a logarithm of a product and the product of logarithms.

104. How can the LOG key on a calculator be used to find $\log_2 7$?

Explain why each statement is false.

105. $\log xy = (\log x)(\log y)$

106. $\log ab = \log a + 1$

107. $\log_b (A - B) = \dfrac{\log_b A}{\log_b B}$

108. $\dfrac{\log_b A}{\log_b B} = \log_b A - \log_b B$

REVIEW

Consider the line that passes through P(−2, 3) and Q(4, −4).

109. Find the slope of line PQ.

110. Find the distance between P and Q.

111. Find the midpoint of line segment PQ.

112. Write the equation in slope–intercept form of line PQ.

CHALLENGE PROBLEMS

113. Explain why $e^{\ln x} = x$.

114. If $\log_b 3x = 1 + \log_b x$, find b.

115. Show that $\log_{b^2} x = \dfrac{1}{2} \log_b x$.

116. Show that $e^{x \ln a} = a^x$.

SECTION 9.8
Exponential and Logarithmic Equations

Objectives

  1 Solve exponential equations.

2 Solve logarithmic equations.

3 Solve radioactive decay problems.

4 Solve population growth problems.

An **exponential equation** contains a variable in one of its exponents. Some examples of exponential equations are

$$3^{x+1} = 81, \qquad 6^{x-3} = 2^x, \qquad \text{and} \qquad e^{0.9t} = 8$$

A **logarithmic equation** is an equation with a logarithmic expression that contains a variable. Some examples of logarithmic equations are

$$\log 5x = 3, \qquad \log (3x + 2) = \log (2x - 3), \qquad \text{and} \qquad \log_2 7 - \log_2 x = 5$$

In this section, we will learn how to solve exponential and logarithmic equations.

1   **Solve Exponential Equations.**

If both sides of an exponential equation can be expressed as a power of the same base, we can use the following property to solve it.

Exponent Property of Equality	If two exponential expressions with the same base are equal, their exponents are equal. For any real number b, where $b \neq -1$, 0, or 1, $$b^x = b^y \quad \text{is equivalent to} \quad x = y$$

EXAMPLE 1   Solve: $3^{x+1} = 81$

Strategy   We will express the right side of the equation as a power of 3.

Why   We can then use the exponent property of equality to set the exponents equal and solve for x.

Solution

$$3^{x+1} = 81 \qquad \text{This is the equation to solve.}$$

$$3^{x+1} = 3^4 \qquad \text{Write 81 as a power of 3: } 81 = 3^4.$$

$$x + 1 = 4 \qquad \text{If two exponential expressions with the same base are equal, their exponents are equal.}$$

$$x = 3$$

The solution is 3 and the solution set is $\{3\}$. To check this result, we substitute 3 for x in the original equation.

$$\textbf{\textit{Check:}} \quad 3^{x+1} = 81$$
$$3^{3+1} \stackrel{?}{=} 81$$
$$3^4 \stackrel{?}{=} 81$$
$$81 = 81 \qquad \text{True}$$

Self Check 1   Solve: $5^{3x-4} = 25$

Now Try   **Problem 21**

EXAMPLE 2   Solve: $2^{x^2+2x} = \frac{1}{2}$

Strategy   We will express the right side of the equation as a power of 2.

Why   We can then use the exponent property of equality to set the exponents equal and solve for x.

Solution

$$2^{x^2+2x} = \frac{1}{2} \qquad \text{This is the equation to solve.}$$

$$2^{x^2+2x} = 2^{-1} \qquad \text{Write } \frac{1}{2} \text{ as a power of 2: } \frac{1}{2} = 2^{-1}.$$

$$x^2 + 2x = -1 \qquad \text{If two exponential expressions with the same base are equal, their exponents are equal.}$$

$$x^2 + 2x + 1 = 0 \qquad \text{Add 1 to both sides.}$$

$$(x + 1)(x + 1) = 0 \qquad \text{Factor the trinomial.}$$

$$x + 1 = 0 \quad \text{or} \quad x + 1 = 0 \qquad \text{Set each factor equal to 0.}$$

$$x = -1 \qquad \qquad x = -1$$

We see that the two solutions are the same. Thus, -1 is a repeated solution and the solution set is $\{-1\}$. Verify that -1 satisfies the original equation.

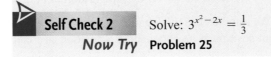

Self Check 2   Solve: $3^{x^2-2x} = \frac{1}{3}$

Now Try   Problem 25

Using Your Calculator

Solving Exponential Equations Graphically

To use a graphing calculator to approximate the solutions of $2^{x^2+2x} = \frac{1}{2}$ (see Example 2), we can subtract $\frac{1}{2}$ from both sides of the equation to get

$$2^{x^2+2x} - \frac{1}{2} = 0$$

and graph the corresponding function

$$f(x) = 2^{x^2+2x} - \frac{1}{2}$$

as shown in figure (a).

The solutions of $2^{x^2+2x} - \frac{1}{2} = 0$ are the x-coordinates of the x-intercepts of the graph of $f(x) = 2^{x^2+2x} - \frac{1}{2}$. Using the ZERO feature, we see in figure (a) that the graph has only one x-intercept, $(-1, 0)$. Therefore, -1 is the only solution of $2^{x^2+2x} - \frac{1}{2} = 0$.

We can also solve $2^{x^2+2x} = \frac{1}{2}$ using the INTERSECT feature found on most graphing calculators. After graphing $Y_1 = 2^{x^2+2x}$ and $Y_2 = \frac{1}{2}$, we select INTERSECT, which approximates the coordinates of the point of intersection of the two graphs. From the display shown in figure (b), we can conclude that the solution is -1. Verify this by checking.

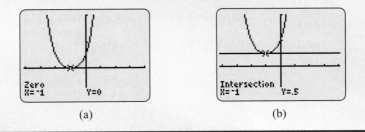

(a)   (b)

When it is difficult or impossible to write each side of an exponential equation as a power of the same base, we can often use the following property of logarithms to solve the equation.

Logarithm Property of Equality

If two positive numbers are equal, the logarithms base-b of the numbers are equal. For any positive number b, where $b \neq 1$, and positive numbers x and y,

$$\log_b x = \log_b y \quad \text{is equivalent to} \quad x = y$$

EXAMPLE 3   Solve: $3^x = 5$

Strategy   We will take the base-10 logarithm of both sides of the equation.

Why   We can then use the logarithm property of equality to move the variable x from its current position as an exponent to a position as a factor.

Solution   Unlike Example 1, where we solved $3^{x+1} = 81$, it is not possible to write each side of $3^x = 5$ as a power of the same base 3. Instead, we use the logarithm property of equality and *take the logarithm of each side* to solve the equation. Although any base logarithm can be chosen, the computations with a calculator are usually simplest if we use a common or natural logarithm.

$$3^x = 5 \qquad \text{This is the equation to solve.}$$

$$\log 3^x = \log 5 \qquad \text{Take the common logarithm of each side.}$$

$$x \log 3 = \log 5 \qquad \begin{array}{l} \text{The log of a power is the power times the log: } \log 3^x = x \log 3. \\ \text{Note that the variable } x \text{ is now a factor of } x \log 3. \end{array}$$

$$x = \frac{\log 5}{\log 3} \qquad \text{To isolate } x \text{, divide both sides by log 3. This is the exact solution.}$$

$$x \approx 1.464973521 \qquad \text{Use a calculator.}$$

The exact solution is $\dfrac{\log 5}{\log 3}$. To four decimal places, the solution is 1.4650.

We can also take the natural logarithm of each side of the equation to solve for x.

$$3^x = 5$$

$$\ln 3^x = \ln 5 \qquad \text{Take the natural logarithm of each side.}$$

$$x \ln 3 = \ln 5 \qquad \text{Use the power rule of logarithms: } \ln 3^x = x \ln 3.$$

$$x = \frac{\ln 5}{\ln 3} \qquad \text{To isolate } x \text{, divide both sides by ln 3.}$$

$$x \approx 1.464973521 \qquad \text{Use a calculator.}$$

The result is the same using the natural logarithm. To check the approximate solution, we substitute 1.4650 for x in 3^x and see if $3^{1.4650}$ is close to 5.

$$\textbf{\textit{Check:}} \qquad 3^x = 5$$

$$3^{1.4650} \stackrel{?}{=} 5$$

$$5.000145454 \approx 5 \qquad \text{Use a calculator: Enter 3 } \boxed{y^x} \text{ 1.4650 } \boxed{=}.$$

Self Check 3   Solve $5^x = 4$ and give the answer to four decimal places.

Now Try   Problem 29

EXAMPLE 4   Solve: $6^{x-3} = 2^x$

Strategy   We will take the common logarithm of both sides of the equation.

Why   We can then move the expression $x - 3$ from its current position as an exponent to a position as a factor.

Solution

$$6^{x-3} = 2^x$$ This is the equation to solve.

$$\log 6^{x-3} = \log 2^x$$ Take the common logarithm of each side.

$$(x - 3) \log 6 = x \log 2$$ The log of a power is the power times the log. Note that the expression $x - 3$ is now a factor of $(x - 3) \log 6$.

$$x \log 6 - 3 \log 6 = x \log 2$$ Distribute the multiplication by $\log 6$.

$$x \log 6 - x \log 2 = 3 \log 6$$ On both sides, add $3 \log 6$ and subtract $x \log 2$.

$$x(\log 6 - \log 2) = 3 \log 6$$ Factor out x on the left side.

$$x = \frac{3 \log 6}{\log 6 - \log 2}$$ To isolate x, divide both sides by $\log 6 - \log 2$.

$$x \approx 4.892789261$$ Use a calculator.

The Language of Algebra

$\frac{3 \log 6}{\log 6 - \log 2}$ is the *exact* solution of $6^{x-3} = 2^x$. An *approximate* solution is 4.8928.

To four decimal places, the solution is 4.8928. To check the approximate solution, we substitute 4.8928 for each x in $6^{x-3} = 2^x$. The resulting values on the left and right sides of the equation should be approximately equal.

Self Check 4   Solve: $5^{x-2} = 3^x$

Now Try   **Problem 33**

EXAMPLE 5   Solve: $e^{0.9t} = 10$

Strategy   We will take the natural logarithm of both sides of the equation.

Why   We can then move the expression $0.9t$ from its current position as an exponent to a position as a factor.

Solution   The exponential expression on the left side has base e. In such cases, the computations are easier when we take the natural logarithm of each side.

$$e^{0.9t} = 10$$ This is the equation to solve.

$$\ln e^{0.9t} = \ln 10$$ Take the natural logarithm of each side.

$$0.9t \ln e = \ln 10$$ Use the power rule of logarithms: $\ln e^{0.9t} = 0.9t \ln e$. Note that the expression $9t$ is now a factor of $0.9t \ln e$.

$$0.9t \cdot 1 = \ln 10$$ Simplify: $\ln e = 1$.

$$0.9t = \ln 10$$

$$t = \frac{\ln 10}{0.9}$$ To isolate t, divide both sides by 0.9.

$$t \approx 2.558427881$$ Use a calculator.

To four decimal places, the solution is 2.5584. Verify this using a check.

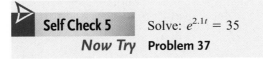

Self Check 5   Solve: $e^{2.1t} = 35$

Now Try   **Problem 37**

2 **Solve Logarithmic Equations.**

A **logarithmic equation** is an equation containing a variable in a logarithmic expression. We can solve many logarithmic equations using properties of logarithms.

EXAMPLE 6 Solve: $\log 5x = 3$

Strategy Recall that $\log 5x = \log_{10} 5x$. To solve $\log 5x = 3$, we will instead write and solve an equivalent base-10 exponential equation.

Why The resulting exponential equation is easier to solve because the variable term is isolated on one side.

Solution

$\log 5x = 3$	This is the equation to solve.
$10^3 = 5x$	Write the equivalent base-10 exponential equation.
$1,000 = 5x$	Simplify: $10^3 = 1,000$.
$200 = x$	To isolate x, divide both sides by 5.

The solution is 200 and the solution set is $\{200\}$.

> **Caution**
> Always check your solutions to a logarithmic equation.

Check:
$$\log 5x = 3$$
$$\log 5(\mathbf{200}) \overset{?}{=} 3 \quad \text{Substitute 200 for x.}$$
$$\log 1,000 \overset{?}{=} 3 \quad \text{Multiply } 5(200) = 1,000.$$
$$3 = 3 \quad \text{Evaluate: } \log 1,000 = \log 10^3 = 3.$$

▷ **Self Check 6** Solve: $\log_2 (x - 3) = -1$

Now Try **Problem 41**

EXAMPLE 7 Solve: $\log(3x + 2) = \log(2x - 3)$

Strategy We will use the logarithmic property of equality to see that $3x + 2 = 2x - 3$.

Why We can use the logarithm property of equality because the given equation, $\log(3x + 2) = \log(2x - 3)$, has the form $\log_b x = \log_b y$.

Solution

$\log(3x + 2) = \log(2x - 3)$	This is the equation to solve.
$3x + 2 = 2x - 3$	If the logarithms of two numbers are equal, the numbers are equal.
$x + 2 = -3$	Subtract 2x from both sides.
$x = -5$	Subtract 2 from both sides.

> **Caution**
> Don't make this error of trying to "distribute" log:
>
> $\log(3x + 2)$
>
> The notation **log** is not a number, it is the name of a function and cannot be distributed.

Check:
$$\log(3x + 2) = \log(2x - 3)$$
$$\log[3(-5) + 2] \overset{?}{=} \log[2(-5) - 3] \quad \text{Substitute } -5 \text{ for x.}$$
$$\log(-13) \overset{?}{=} \log(-13) \quad \begin{array}{l}\text{Evaluate within brackets.}\\ \text{Recall that } \log(-13) \text{ is undefined.}\end{array}$$

Since the logarithm of a negative number does not exist, the proposed solution of -5 must be discarded. This equation has no solutions. Its solution set is $\varnothing$.

> ▷ **Self Check 7**   Solve: $\log(5x + 14) = \log(7x - 2)$
>
>   ***Now Try*** **Problem 49**

EXAMPLE 8   Solve: $\log x + \log(x - 3) = 1$

Strategy   We will use the product rule for logarithms in reverse: The sum of two logarithms is equal to the logarithm of a product. Then we will write and solve an equivalent exponential equation.

Why   We use the product rule for logarithms because the left side of the equation, $\log x + \log(x - 3)$, has the form $\log_b M + \log_b N$.

Solution

$$\log x + \log(x - 3) = 1 \qquad \text{This is the equation to solve.}$$

$$\log x(x - 3) = 1 \qquad \text{On the left side, use the product rule for logarithms.}$$

$$\log_{10} x(x - 3) = 1 \qquad \text{The base of the logarithm is 10.}$$

$$x(x - 3) = 10^1 \qquad \text{Write an equivalent base-10 exponential equation.}$$

$$x^2 - 3x - 10 = 0 \qquad \text{Distribute the multiplication by x, and then subtract 10 from both sides.}$$

$$(x + 2)(x - 5) = 0 \qquad \text{Factor the trinomial.}$$

$$x + 2 = 0 \quad \text{or} \quad x - 5 = 0 \qquad \text{Set each factor equal to 0.}$$

$$x = -2 \mid \qquad x = 5$$

Check: The number -2 is not a solution because it does not satisfy the equation (a negative number does not have a logarithm). We will check the other result, 5.

> **Caution**
> The proposed solutions of a logarithmic equation must be checked to see whether they produce undefined logarithms in the original equation.

$$\log x + \log(x - 3) = 1$$

$$\log 5 + \log(5 - 3) \overset{?}{=} 1 \qquad \text{Substitute 5 for x.}$$

$$\log 5 + \log 2 \overset{?}{=} 1$$

$$\log 10 \overset{?}{=} 1 \qquad \text{Use the product rule of logarithms:}$$
$$\qquad\qquad\qquad\qquad \log 5 + \log 2 = \log(5 \cdot 2) = \log 10.$$

$$1 = 1 \qquad \text{Evaluate: } \log 10 = 1.$$

Since 5 satisfies the equation, it is the solution.

> ▷ **Self Check 8**   Solve: $\log x + \log(x + 3) = 1$
>
>   ***Now Try*** **Problem 53**

Using Your Calculator   ***Solving Logarithmic Equations Graphically***

To use a graphing calculator to approximate the solutions of the logarithmic equation $\log x + \log(x - 3) = 1$ (see Example 8), we can subtract 1 from both sides of the equation to get

$$\log x + \log(x - 3) - 1 = 0$$

and graph the corresponding function

$$f(x) = \log x + \log (x - 3) - 1$$

as shown in figure (a). Since the solution of the equation is the x-value that makes $f(x) = 0$, the solution is the x-coordinate of the x-intercept of the graph. We can use the ZERO feature to find that this x-value is 5.

We can also solve $\log x + \log (x - 3) = 1$ using the INTERSECT feature. After graphing $Y_1 = \log x + \log (x - 3)$ and $Y_2 = 1$, we select INTERSECT, which approximates the coordinates of the point of intersection of the two graphs. From the display shown in figure (b), we can conclude that the solution is 5.

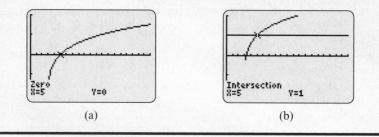

(a)   (b)

EXAMPLE 9   Solve: $\log_2 7 - \log_2 x = 5$

Strategy   We will use the quotient rule for logarithms in reverse: The difference of two logarithms is equal to the logarithm of a quotient. Then we will write and solve an equivalent exponential equation.

Why   We use the quotient rule for logarithms because the left side of the equation, $\log_2 7 - \log_2 x$, has the form $\log_b M - \log_b N$.

Solution

$$\log_2 7 - \log_2 x = 5 \qquad \text{This is the equation to solve.}$$

$$\log_2 \frac{7}{x} = 5 \qquad \text{On the left side, use the quotient rule for logarithms.}$$

$$\frac{7}{x} = 2^5 \qquad \text{Write an equivalent base-2 exponential equation.}$$

$$\frac{7}{x} = 32 \qquad \text{Evaluate: } 2^5 = 32.$$

$$7 = 32x \qquad \text{To clear the equation of the fraction, multiply both sides by } x.$$

$$\frac{7}{32} = x \qquad \text{To isolate } x, \text{ divide both sides by 32.}$$

The solution is $\frac{7}{32}$. Verify that it satisfies the original equation.

Self Check 9   Solve: $\log_2 9 - \log_2 x = 4$
Now Try   **Problem 57**

3   **Solve Radioactive Decay Problems.**

Experiments have determined the time it takes for half of a sample of a radioactive material to decompose. This time is a constant, called the material's **half-life.**

When living organisms die, the oxygen–carbon dioxide cycle common to all living things ceases, and carbon-14, a radioactive isotope with a half-life of 5,700 years, is no longer absorbed. By measuring the amount of carbon-14 present in an ancient object, archaeologists can estimate the object's age by using the radioactive decay formula.

Radioactive Decay Formula	If A is the amount of radioactive material present at time t, A_0 was the amount present at $t = 0$, and h is the material's half-life, then $$A = A_0 2^{-t/h}$$

EXAMPLE 10   *Carbon-14 Dating.*   How old is a piece of wood that retains only one-third of its original carbon-14 content?

Strategy   If A_0 is the original carbon-14 content, then today's content $A = \frac{1}{3}A_0$. We will substitute $\frac{A_0}{3}$ for A and 5,700 for h in the radioactive decay formula and solve for t.

Why   The value of t is the estimated age of the piece of wood.

Solution   To find the time t when $A = \frac{1}{3}A_0$, we substitute $\frac{A_0}{3}$ for A and 5,700 for h in the radioactive decay formula and solve for t:

$$A = A_0 2^{-t/h}$$ This is the radioactive decay model.

$$\frac{A_0}{3} = A_0 2^{-t/5,700}$$ The half-life of carbon-14 is 5,700 years.

$$1 = 3(2^{-t/5,700})$$ Divide both sides by A_0 and multiply both sides by 3.

$$\log 1 = \log 3(2^{-t/5,700})$$ Take the common logarithm of both sides.

$$0 = \log 3 + \log 2^{-t/5,700}$$ $\log 1 = 0$, and use the product rule for logarithms.

$$-\log 3 = -\frac{t}{5,700}\log 2$$ Subtract log 3 from both sides and use the power rule of logarithms.

$$5,700\left(\frac{\log 3}{\log 2}\right) = t$$ Multiply both sides by $-\frac{5,700}{\log 2}$.

$$t \approx 9,034.286254$$ Use a calculator.

> **Notation**
> The initial amount of radioactive material is represented by A_0, and it is read as "A sub 0."

The piece of wood is approximately 9,000 years old.

▷ **Self Check 10**   How old is a piece of wood that retains 25% of its original carbon-14 content?

Now Try   **Problem 97**

4   **Solve Population Growth Problems.**

When there is sufficient food and space available, populations of living organisms tend to increase exponentially according to the following growth model.

Exponetial Growth Model

If P is the population at some time t, P_0 is the initial population at $t = 0$, and k depends on the rate of growth, then

$$P = P_0e^{kt}$$

Notation

The initial population of bacteria is represented by P_0, and it is read as "P sub 0."

EXAMPLE 11

Population Growth. The bacteria in a laboratory culture increased from an initial population of 500 to 1,500 in 3 hours. How long will it take for the population to reach 10,000?

Strategy We will substitute 500 for P_0, 1,500 for P, and 3 for t into the exponential growth model and solve for k:

Why Once we know the value of k, we can substitute 10,000 for P, 500 for P_0, and the value of k into the exponential growth model and solve for the time t.

Solution

$P = P_0e^{kt}$ This is the population growth formula.

$1{,}500 = 500(e^{k3})$ Substitute 1,500 for P, 500 for P_0, and 3 for t.

$3 = e^{3k}$ Divide both sides by 500.

$3k = \ln 3$ Write the equivalent base-e logarithmic equation.

$k = \frac{\ln 3}{3}$ Divide both sides by 3.

To find when the population will reach 10,000, we substitute 10,000 for P, 500 for P_0, and $\frac{\ln 3}{3}$ for k in the equation $P = P_0e^{kt}$ and solve for t:

$P = P_0e^{kt}$

$10{,}000 = 500e^{[(\ln 3)/3]t}$

$20 = e^{[(\ln 3)/3]t}$ Divide both sides by 500.

$\left(\frac{\ln 3}{3}\right)t = \ln 20$ Write the equivalent base-e logarithmic equation.

$t = \frac{3 \ln 20}{\ln 3}$ Multiply both sides by $\frac{3}{\ln 3}$.

≈ 8.180499084 Use a calculator.

The culture will reach 10,000 bacteria in about 8 hours.

Self Check 11 How long will it take the population to reach 20,000?

Now Try **Problem 109**

ANSWERS TO SELF CHECKS **1.** 2 **2.** A repeated solution of 1 **3.** 0.8614 **4.** $\frac{2 \log 5}{\log 5 - \log 3} \approx 6.3013$ **5.** $\frac{\ln 35}{2.1} \approx 1.6930$ **6.** $\frac{7}{2}$ **7.** 8 **8.** 2 **9.** $\frac{9}{16}$ **10.** About 11,400 years **11.** About 10 hours

STUDY SET
9.8

VOCABULARY

Fill in the blanks.

1. An equation with a variable in its exponent, such as $3^{2x} = 8$, is called a(n) _____ equation.

2. An equation with a logarithmic expression that contains a variable, such as $\log_5 (2x - 3) = \log_5 (x + 4)$, is a(n) _____ equation.

CONCEPTS

Fill in the blanks.

3. a. The exponent property of equality: If two exponential expressions with the same base are equal, their exponents are _____.

$b^x = b^y$ is equivalent to _____ .

b. The logarithm property of equality: If the logarithms base-b of two numbers are equal, the numbers are _____.

$\log_b x = \log_b y$ is equivalent to _____ .

4. The right side of the exponential equation $5^{x-3} = 125$ can be written as a power of _____.

5. If $6^{4x} = 6^{-2}$, then $4x =$ _____ .

6. a. Write the equivalent base-10 exponential equation for $\log (x + 1) = 2$.

b. Write the equivalent base-e exponential equation for $\ln (x + 1) = 2$.

Fill in the blanks.

7. To solve $5^x = 2$, we can take the _____ of both sides of the equation to get $\log 5^x = \log 2$.

8. The power rule for logarithms provides a way of moving the variable x from its position as an _____ to a position as a factor of $x \log 5$.

9. If the power rule for logarithms is used on the left side of the equation $\log 7^x = 12$, the resulting equation is _____ $\log 7 = 12$.

10. If $e^{x+2} = 4$, then $\ln e^{x+2} =$ _____ .

11. Perform a check to determine whether -2 is a solution of $5^{2x+3} = \frac{1}{5}$.

12. Perform a check to determine whether 4 is a solution of $\log_5 (x + 1) = 2$.

13. Use a calculator to determine whether 2.5646 is an approximate solution of $2^{2x+1} = 70$.

14. How do we solve $x \ln 3 = \ln 5$ for x?

15. a. Find $\dfrac{\log 8}{\log 5}$. Round to four decimal places.

b. Find $\dfrac{2 \ln 12}{\ln 9}$. Round to four decimal places.

16. Does $\dfrac{\log 7}{\log 3} = \log 7 - \log 3$?

17. Complete each formula.

a. Radioactive decay: $A =$ _____ .

b. Population growth: $P =$ _____ .

18. Use the graphs to estimate the solution of each equation.

a. $2^x = 3^{-x+3}$

b. $3 \log (x - 1) = 2 \log x$

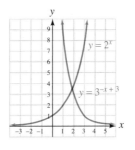

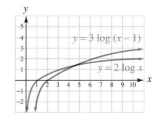

NOTATION

Complete each solution.

19. Solve: $2^x = 7$.

$$\boxed{} \, 2^x = \log 7$$
$$x \boxed{} = \log 7$$
$$x = \frac{\log 7}{\log 2}$$
$$x \approx$$

20. Solve: $\log_2 (2x - 3) = \log_2 (x + 4)$.

$$\boxed{} = x + 4$$
$$x = 7$$

GUIDED PRACTICE

Solve each equation. See Example 1.

21. $6^{x-2} = 36$

22. $3^{x+1} = 27$

23. $5^{4x} = \dfrac{1}{125}$

24. $8^{-2x+1} = \dfrac{1}{64}$

Solve each equation. See Example 2.

25. $2^{x^2 - 2x} = 8$ **26.** $3^{x^2 - 3x} = 81$

27. $3^{x^2 + 4x} = \dfrac{1}{81}$ **28.** $7^{x^2 + 3x} = \dfrac{1}{49}$

Solve each equation. Give answers to four decimal places. See Example 3.

29. $4^x = 5$ **30.** $7^x = 12$

31. $13^{x-1} = 2$ **32.** $5^{x+1} = 3$

Solve each equation. Give answers to four decimal places. See Example 4.

33. $2^{x+1} = 3^x$ **34.** $6^x = 7^{x-4}$

35. $5^{x-3} = 3^{2x}$ **36.** $8^{3x} = 9^{x+1}$

Solve each equation. Give answers to four decimal places. See Example 5.

37. $e^{2.9x} = 4.5$ **38.** $e^{3.3t} = 9.1$

39. $e^{-0.2t} = 14.2$ **40.** $e^{-0.7x} = 6.2$

Solve each equation. See Example 6.

41. $\log 2x = 4$ **42.** $\log 5x = 4$

43. $\log_3 (x - 3) = 2$ **44.** $\log_4 (2x - 1) = 3$

45. $\log (7 - x) = 2$ **46.** $\log (2 - x) = 3$

47. $\log \dfrac{1}{8} x = -2$ **48.** $\log \dfrac{1}{5} x = -3$

Solve each equation. See Example 7.

49. $\log (3 - 2x) = \log (x + 24)$

50. $\log (3x + 5) = \log (2x + 6)$

51. $\ln (3x + 1) = \ln (x + 7)$

52. $\ln (x^2 + 4x) = \ln (x^2 + 16)$

Solve each equation. See Example 8.

53. $\log x + \log (x - 48) = 2$

54. $\log x + \log (x + 9) = 1$

55. $\log_5 (4x - 1) + \log_5 x = 1$

56. $\log_2 (x - 7) + \log_2 x = 3$

Solve each equation. See Example 9.

57. $\log 5 - \log x = 1$ **58.** $\log 11 - \log x = 2$

59. $\log_3 4x - \log_3 7 = 2$ **60.** $\log_2 5x - \log_2 3 = 4$

TRY IT YOURSELF

Solve each equation. Give approximate solutions to four decimal places.

61. $\log 2x = \log 4$ **62.** $\log 3x = \log 9$

63. $\ln x = 1$ **64.** $\ln x = 5$

65. $7^{x^2} = 10$ **66.** $8^{x^2} = 11$

67. $\log (x + 90) + \log x = 3$

68. $\log (x - 90) + \log x = 3$

69. $3^{x-6} = 81$ **70.** $5^{x+4} = 125$

71. $\log \dfrac{4x + 1}{2x + 9} = 0$ **72.** $\log \dfrac{2 - 5x}{2(x + 8)} = 0$

73. $15 = 9^{x+1}$ **74.** $29 = 5^{x-6}$

75. $\log x^2 = 2$ **76.** $\log x^3 = 3$

77. $\log (x - 6) - \log (x - 2) = \log \dfrac{5}{x}$

78. $\log (3 - 2x) - \log (x + 9) = 0$

79. $\log_3 x = \log_3 \left(\dfrac{1}{x} \right) + 4$

80. $\log_5 (7 + x) + \log_5 (8 - x) - \log_5 2 = 2$

81. $2 \log_2 x = 3 + \log_2 (x - 2)$

82. $2 \log_3 x - \log_3 (x - 4) = 2 + \log_3 2$

83. $\log (7y + 1) = 2 \log (y + 3) - \log 2$

84. $2 \log (y + 2) = \log (y + 2) - \log 12$

85. $e^{3x} = 9$ **86.** $e^{4x} = 60$

87. $\dfrac{\log (5x + 6)}{2} = \log x$ **88.** $\dfrac{1}{2} \log (4x + 5) = \log x$

Use a graphing calculator to solve each equation. If an answer is not exact, round to the nearest tenth. See Using Your Calculator: Solving Exponential Equations Graphically *or* Solving Logarithmic Equations Graphically.

89. $2^{x+1} = 7$ **90.** $3^{x-1} = 2^x$

91. $3^x - 10 = 3^{-x}$ **92.** $2^x - 8 = 5 + 2^{-x}$

93. $\log x + \log (x - 15) = 2$

94. $\log x + \log (x + 3) = 1$

95. $\ln (2x + 5) - \ln 3 = \ln (x - 1)$

96. $2 \log (x^2 + 4x) = 1$

APPLICATIONS

97. TRITIUM DECAY   The half-life of tritium is 12.4 years. How long will it take for 25% of a sample of tritium to decompose?

98. RADIOACTIVE DECAY   In 2 years, 20% of a radioactive element decays. Find its half-life.

99. THORIUM DECAY   An isotope of thorium, written as ^{227}Th, has a half-life of 18.4 days. How long will it take for 80% of the sample to decompose?

100. LEAD DECAY   An isotope of lead, written as ^{201}Pb, has a half-life of 8.4 hours. How many hours ago was there 30% more of the substance?

101. CARBON-14 DATING   A bone fragment analyzed by archaeologists contains 60% of the carbon-14 that it is assumed to have had initially. How old is it?

102. CARBON-14 DATING   Only 10% of the carbon-14 in a small wooden bowl remains. How old is the bowl?

103. COMPOUND INTEREST   If $500 is deposited in an account paying 8.5% annual interest, compounded semiannually, how long will it take for the account to increase to $800?

104. CONTINUOUS COMPOUND INTEREST   In Exercise 103, how long will it take if the interest is compounded continuously?

105. COMPOUND INTEREST   If $1,300 is deposited in a savings account paying 9% interest, compounded quarterly, how long will it take the account to increase to $2,100?

106. COMPOUND INTEREST   A sum of $5,000 deposited in an account grows to $7,000 in 5 years. Assuming annual compounding, what interest rate is being paid?

107. RULE OF SEVENTY   A rule of thumb for finding how long it takes an investment to double is called the **rule of seventy.** To apply the rule, divide 70 by the interest rate written as a percent. At 5%, an investment takes $\frac{70}{5} = 14$ years to double. At 7%, it takes $\frac{70}{7} = 10$ years. Explain why this formula works.

108. BACTERIAL GROWTH   A bacterial culture grows according to the function

$$P(t) = P_0 a^t$$

If it takes 5 days for the culture to triple in size, how long will it take to double in size?

109. RODENT CONTROL   The rodent population in a city is currently estimated at 30,000. If it is expected to double every 5 years, when will the population reach 1 million?

110. POPULATION GROWTH   The population of a city is expected to triple every 15 years. When can the city planners expect the present population of 140 persons to double?

111. BACTERIA CULTURE   A bacteria culture doubles in size every 24 hours. By how much will it have increased in 36 hours?

112. OCEANOGRAPHY   The intensity I of a light a distance x meters beneath the surface of a lake decreases exponentially. Use the illustration to find the depth at which the intensity will be 20%.

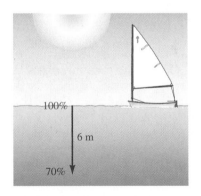

113. NEWTON'S LAW OF COOLING   Water initially at 100°C is left to cool in a room at temperature 60°C. After 3 minutes, the water temperature is 90°. The water temperature T is a function of time t given by the following formula. Find k.

$$T = 60 + 40e^{kt}$$

114. NEWTON'S LAW OF COOLING   Refer to Exercise 113 and find the time for the water temperature to reach 70°C.

WRITING

115. Explain how to solve the equation $2^{x+1} = 31$.

116. Explain how to solve the equation $2^{x+1} = 32$.

117. Write a justification for each step of the solution.

$$15^x = 9 \qquad \text{This is the equation to solve.}$$

$$\log 15^x = \log 9 \qquad \underline{\hspace{3cm}}.$$

$$x \log 15 = \log 9 \qquad \underline{\hspace{3cm}}.$$

$$x = \frac{\log 9}{\log 15} \qquad \underline{\hspace{3cm}}.$$

118. What is meant by the term *half-life*?

REVIEW

119. Find the length of leg AC.

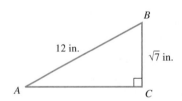

120. The amount of medicine a patient should take is often proportional to his or her weight. If a patient weighing 83 kilograms needs 150 milligrams of medicine, how much will be needed by a person weighing 99.6 kilograms?

CHALLENGE PROBLEMS

121. Without solving the following equation, find the values of x that cannot be a solution:

$$\log (x - 3) - \log (x^2 + 2) = 0$$

122. Solve: $x^{\log x} = 10,000$

123. Solve: $\dfrac{\log_2 (6x - 8)}{\log_2 x} = 2$

124. Solve: $\dfrac{\log (3x - 4)}{\log x} = 2$

CHAPTER 9
Summary & Review

SECTION 9.1 Algebra and Composition of Functions

DEFINITIONS AND CONCEPTS	EXAMPLES

Just as it is possible to perform arithmetic operations on real numbers, it is possible to perform those operations on functions.

The **sum, difference, product,** and **quotient functions** are defined as:

$$(f + g)(x) = f(x) + g(x)$$
$$(f - g)(x) = f(x) - g(x)$$
$$(f \cdot g)(x) = f(x)g(x)$$
$$(f/g)(x) = \frac{f(x)}{g(x)}, \quad \text{with } g(x) \neq 0$$

Let $f(x) = 2x + 1$ and $g(x) = x^2$.

$$(f + g)(x) = f(x) + g(x) \qquad (f - g)(x) = f(x) - g(x)$$
$$= 2x + 1 + x^2 \qquad\qquad = 2x + 1 - x^2$$
$$= x^2 + 2x + 1 \qquad\qquad = -x^2 + 2x + 1$$

$$(f \cdot g)(x) = f(x) \cdot g(x) \qquad (f/g)(x) = \frac{f(x)}{g(x)}$$

$$= (2x + 1)x^2 \qquad\qquad = \frac{2x + 1}{x^2}$$

$$= 2x^3 + x^2$$

Often one quantity is a function of a second quantity that depends, in turn, on a third quantity. Such chains of dependence can be modeled by a **composition of functions.**

Composition of functions:

$$(f \circ g)(x) = f(g(x))$$

Let $f(x) = 4x - 9$ and $g(x) = x^3$. Find $(f \circ g)(2)$ and $(f \circ g)(x)$.

$$(f \circ g)(2) = f(g(2)) \qquad \text{Change to nested parentheses notation.}$$
$$= f(8) \qquad \text{Evaluate: } g(2) = 2^3 = 8.$$
$$= 4(8) - 9 \qquad \text{Evaluate } f(8) \text{ using } f(x) = 4x - 9.$$
$$= 23$$

$$(f \circ g)(x) = f(g(x)) = f(x^3) = 4x^3 - 9$$

REVIEW EXERCISES

Let $f(x) = 2x$ and $g(x) = x + 1$. Find each function and its domain.

1. $f + g$

2. $f - g$

3. $f \cdot g$

4. f/g

Let $f(x) = x^2 + 2$ and $g(x) = 2x + 1$. Find each of the following.

5. $(f \circ g)(-1)$

6. $(g \circ f)(0)$

7. $(f \circ g)(x)$

8. $(g \circ f)(x)$

9. Use the graphs of functions f and g to find each of the following.

    a. $(f + g)(2)$

    b. $(f \cdot g)(-4)$

    c. $(f \circ g)(4)$

    d. $(g \circ f)(6)$

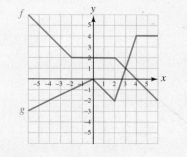

10. MILEAGE COSTS The function $f(m) = \frac{m}{8}$ gives the number of gallons of fuel consumed if a bus travels m miles. The function $C(f) = 3.25f$ gives the cost (in dollars) of f gallons of fuel. Write a composition function that expresses the cost of the fuel consumed as a function of the number of miles driven.

SECTION 9.2   Inverse Functions

DEFINITIONS AND CONCEPTS	EXAMPLES
A function is called a **one-to-one function** if different inputs determine different outputs.	The function $f(x) = 3x - 5$ is a one-to-one function because different inputs have different outputs. Since two different inputs, -2 and 2, have the same output 16, the function $f(x) = x^4$ is not one-to-one.
Horizontal line test: A function is one-to-one if every horizontal line intersects the graph of the function at most once.	The function $f(x) = |x + 1|$ is a not one-to-one function because we can draw a horizontal line that intersects its graph twice.
To find the inverse of a function, replace $f(x)$ with y, interchange the variables x and y, solve for y, and replace y with $f^{-1}(x)$.	To find the inverse of the one-to-one function $f(x) = 2x + 1$, we proceed as follows: $$f(x) = 2x + 1$$ $$y = 2x + 1 \quad \text{Replace } f(x) \text{ with } y.$$ $$x = 2y + 1 \quad \text{Interchange the variables } x \text{ and } y.$$ $$\frac{x - 1}{2} = y \quad \text{Solve for } y.$$ $$f^{-1}(x) = \frac{x - 1}{2} \quad \text{Replace } y \text{ with } f^{-1}(x).$$
If a point (a, b) is on the graph of function f, it follows that the point (b, a) is on the graph of f^{-1}, and vice versa. The graph of a function and its inverse are **symmetric about the line $y = x$**.	The graphs of $f(x) = 2x + 1$ and $f^{-1}(x) = \frac{x-1}{2}$ are symmetric about the line $y = x$ as shown in the illustration.
For any one-to-one function f and its inverse, f^{-1}, $$(f \circ f^{-1})(x) = x \quad \text{and} \quad (f^{-1} \circ f)(x) = x$$	The composition of $f(x) = 2x + 1$ and its inverse $f^{-1}(x) = \frac{x-1}{2}$ is the identity function $f(x) = x$. $$(f \circ f^{-1})(x) = f(f^{-1}(x)) = f\left(\frac{x-1}{2}\right) = 2\left(\frac{x-1}{2}\right) + 1 = x - 1 + 1 = x$$ $$(f^{-1} \circ f)(x) = f^{-1}(f(x)) = f^{-1}(2x + 1) = \frac{2x + 1 - 1}{2} = \frac{2x}{2} = x$$

REVIEW EXERCISES

In Exercises 11–16, determine whether the function is one-to-one.

11. $f(x) = x^2 + 3$

12. $f(x) = \dfrac{1}{3}x - 8$

13. $\{(3, 4), (5, 10), (10, -1), (6, 6)\}$

14.

x	f(x)
0	-5
2	10
4	-5
6	15

15.

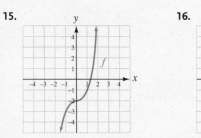

16.

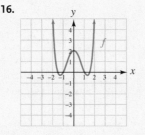

17. Use the table of values of the one-to-one function f to complete a table of values for f^{-1}.

x	f(x)
-6	-6
-1	-3
7	12
20	3

x	$f^{-1}(x)$
-6	
-3	
12	
3	

18. Given the graph of function f, graph f^{-1} on the same coordinate axes. Label the axis of symmetry.

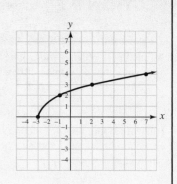

Find the inverse of each function.

19. $f(x) = 6x - 3$

20. $f(x) = \dfrac{4}{x - 1}$

21. $f(x) = (x + 2)^3$

22. $f(x) = \dfrac{x}{6} - \dfrac{1}{6}$

23. Find the inverse of $f(x) = \sqrt[3]{x - 1}$. Then graph the function and its inverse on one coordinate system. Show the axis of symmetry on the graph.

24. Use composition to show that $f(x) = 5 - 4x$ and $f^{-1}(x) = -\dfrac{x - 5}{4}$ are inverse functions.

SECTION 9.3   Exponential Functions

DEFINITIONS AND CONCEPTS	EXAMPLES
An **exponential function** with base b is defined by the equation $$f(x) = b^x, \text{ with } b > 0, b \neq 1$$ Properties of an exponential function $f(x) = b^x$: The **domain** is the interval $(-\infty, \infty)$. The **range** is the interval $(0, \infty)$. Its graph has a **y-intercept** of $(0, 1)$. The x-axis is an **asymptote** of its graph. The graph **passes through** the point $(1, b)$. If $b > 1$, then $f(x) = b^x$ is an **increasing function**. If $0 < b < 1$, then $f(x) = b^x$ is a **decreasing function.**	The graphs of $f(x) = 2^x$ and $g(x) = \left(\dfrac{1}{2}\right)^x$ are shown below. 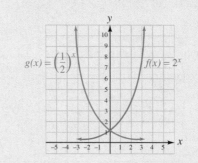 Since the base 2 is greater than 1, the function $f(x) = 2^x$ is an increasing function. Since the base $\dfrac{1}{2}$ is such that $0 < \dfrac{1}{2} < 1$, the function $g(x) = \left(\dfrac{1}{2}\right)^x$ is a decreasing function.

SECTION 9.3   Exponential Functions–*continued*

DEFINITIONS AND CONCEPTS	EXAMPLES

Exponential functions are used to model many situations, such as population **growth,** the spread of an epidemic, the temperature of a heated object as it cools, and radioactive **decay.**

Exponential functions are suitable models for describing **compound interest:**

If P is the deposit, and interest is paid k times a year at an annual rate r, the amount A in the account after t years is given by

$$A = P\left(1 + \frac{r}{k}\right)^{kt}$$

If \$15,000 is deposited in an account paying an annual interest rate of 7.5%, compounded monthly, how much will be in the account in 60 years?

$$A(t) = 15{,}000\left(1 + \frac{0.075}{12}\right)^{12t}$$ To write the formula in function notation, substitute for P, r, and k.

$$A(60) = 15{,}000\left(1 + \frac{0.075}{12}\right)^{12(60)}$$ Substitute 60 for t.

$$= 15{,}000\left(1 + \frac{0.075}{12}\right)^{720}$$

$$\approx 1{,}331{,}479.52$$ Use a calculator.

In 60 years, the account will contain \$1,331,479.52.

REVIEW EXERCISES

Use properties of exponents to simplify each expression.

25. $5^{\sqrt{6}} \cdot 5^{3\sqrt{6}}$

26. $\left(2^{\sqrt{14}}\right)^{\sqrt{2}}$

Graph each function and give the domain and the range. Label the y-intercept.

27. $f(x) = 3^x$

28. $f(x) = \left(\frac{1}{3}\right)^x$

29. $f(x) = \left(\frac{1}{2}\right)^x - 2$

30. $f(x) = 3^{x-1}$

31. In Exercise 30, what is the asymptote of the graph of $f(x) = 3^{x-1}$?

32. COAL PRODUCTION   The table gives the number of tons of coal produced in the United States for the years 1800–1920. Graph the data. What type of function does it appear could be used to model coal production over this period?

Year	Tons	Year	Tons
1800	108,000	1870	40,429,000
1810	178,000	1880	79,407,000
1820	881,000	1890	157,771,000
1830	1,334,000	1900	269,684,000
1840	2,474,000	1910	501,596,000
1850	8,356,000	1920	658,265,000
1860	20,041,000		

Source: *World Book Encyclopedia*

33. COMPOUND INTEREST   How much will \$10,500 become if it earns 9% annual interest, compounded quarterly, for 60 years?

34. DEPRECIATION   The value (in dollars) of a certain model car is given by the function $V(t) = 12{,}000\left(10^{-0.155t}\right)$, where t is the number of years from the present. Find the value of the car in 5 years.

SECTION 9.4 Base-*e* Exponential Functions

DEFINITIONS AND CONCEPTS	EXAMPLES
Of all possible bases for an exponential function, *e* is the most convenient for problems involving growth or decay. $e = 2.718281828459 \ldots$ The function defined by $f(x) = e^x$ is the **natural exponential function.**	From the graph, we see that the domain of the natural exponential function is $(-\infty, \infty)$ and the range is $(0, \infty)$. The *x*-axis is an asymptote of the graph.

Graph showing $f(x) = e^x$ with points $(-1, \frac{1}{e})$, $(0, 1)$, $(1, e)$, and $(2, e^2)$.

Exponential growth/decay: If a quantity increases or decreases at an annual rate *r*, **compounded continuously,** the amount *A* after *t* years is given by $A = Pe^{rt}$ If *r* is negative, the amount decreases.	If \$30,000 accumulates interest at an annual rate of 9%, compounded continuously, find the amount in the account after 25 years. $A = Pe^{rt}$ This is the formula for continuous compound interest. $= 30{,}000e^{0.09 \cdot 25}$ Substitute 30,000 for P, 0.09 for r, and 25 for t. $= 30{,}000e^{2.25}$ $\approx 284{,}632.08$ Use a calculator. In 30 years, the account will contain \$284,632.08. Suppose the population of a city of 50,000 people is decreasing exponentially according to the function $P(t) = 50{,}000e^{-0.003t}$, where *t* is measured in years from the present date. Find the expected population of the city in 20 years. $P(t) = 50{,}000e^{-0.003t}$ Since r is negative, this is the exponential decay model. $P(20) = 50{,}000e^{-0.003(20)}$ Substituite 20 for t. $= 50{,}000e^{-0.06}$ $\approx 47{,}088$ Use a calculator. After 20 years, the expected population will be about 47,088 people.

REVIEW EXERCISES

Graph each function, and give the domain and the range.

35. $f(x) = e^x + 1$

36. $f(x) = e^{x-3}$

37. INTEREST COMPOUNDED CONTINUOUSLY   If \$10,500 accumulates interest at an annual rate of 9%, compounded continuously, how much will be in the account in 60 years?

38. THE GRAND CANYON STATE   In 2006, Arizona ended Nevada's 19-year reign as the nation's fastest growing state. The population of Arizona at the time was 6,166,318 with an annual growth rate of 3.6%. Predict the population of Arizona in 2016, assuming the growth rate remains the same.

39. MORTGAGE RATES   There was the housing boom in the 1980s as the baby boomers (those born from 1946–1964) bought their homes. The average annual interest rate in percent on a 30-year fixed-rate home mortgage for the years 1980–1996 can be approximated by the function $r(t) = 13.9e^{-0.035t}$, where *t* is the number of years since 1980. To the nearest hundredth of a percent, what does this model predict was the 30-year fixed rate in 1980? In 1985? In 1990?

40. MEDICAL TESTS   A radioactive dye is injected into a patient as part of a test to detect heart disease. The amount of dye remaining in his bloodstream *t* hours after the injection is given by the function $f(t) = 10e^{-0.27t}$. How can you determine from the function that the amount of dye in the bloodstream is decreasing?

SECTION 9.5   Logarithmic Functions

DEFINITIONS AND CONCEPTS	EXAMPLES
Definition of logarithm: If $b > 0$, $b \neq 1$, and x is positive, then $\quad y = \log_b x \quad$ means $\quad x = b^y$	*Logarithmic form* $\qquad$ *Exponential form* $\quad \log_5 125 = 3 \quad$ is equivalent to $\quad 5^3 = 125$ $\quad \log_2 \dfrac{1}{8} = -3 \quad$ is equivalent to $\quad 2^{-3} = \dfrac{1}{8}$
$\log_b x$ is the exponent to which b is raised to get x.	To evaluate $\log_4 16$ we ask: "To what power must we raise 4 to get 16?" Since $4^2 = 16$, the answer is: the 2nd power. Thus, $\quad \log_4 16 = 2$
For computational purposes and in many applications, we use base-10 logarithms, called **common logarithms.** $\quad \log x \quad$ means $\quad \log_{10} x$	$\quad \log \dfrac{1}{1,000} = -3 \quad$ because $\quad 10^{-3} = \dfrac{1}{1,000}$
If $b > 0$ and $b \neq 1$, the **logarithmic function with base b** is defined by $f(x) = \log_b x$. The domain is $(0, \infty)$ and the range is $(-\infty, \infty)$. If $b > 1$, then $f(x) = \log_b x$ is an increasing function. If $0 < b < 1$, then $f(x) = \log_b x$ is a decreasing function.	The graph of the logarithmic function $f(x) = \log_2 x$. From the graph, we see that $f(x) = \log_2 x$ is an increasing function.
The exponential function $f(x) = b^x$ and the logarithmic function $f(x) = \log_b x$ are inverses of each other.	$f(x) = 3^x$ and $f^{-1}(x) = \log_3 x$ are inverses of each other. Their graphs are symmetric about the line $y = x$.
Logarithmic functions, like exponential functions, can be used to **model** certain types of growth and decay. *Decibel voltage gain:* $\quad \text{db gain} = 20 \log \dfrac{E_O}{E_I}$ *The Richter scale:* $\quad R = \log \dfrac{A}{P}$	If the input to an amplifier is 0.4 volt and the output is 30 volts, find the decibel voltage gain. $\quad \text{db gain} = 20 \log \dfrac{E_O}{E_I}$ $\qquad\qquad = 20 \log \dfrac{30}{0.4} \qquad$ Substitute 30 for E_O and 0.4 for E_I. $\qquad\qquad \approx 37.50122527 \qquad$ Use a calculator. The db gain is about 38 decibels.

REVIEW EXERCISES

41. Give the domain and range of $f(x) = \log x$.

42. Explain why a student got the following message when she used a calculator to evaluate log 0.

Error

43. Write the statement $\log_4 64 = 3$ in exponential form.

44. Write the statement $7^{-1} = \frac{1}{7}$ in logarithmic form.

Evaluate, if possible.

45. $\log_3 9$

46. $\log_9 \frac{1}{81}$

47. $\log_{1/2} 1$

48. $\log_5 (-25)$

49. $\log_6 \sqrt{6}$

50. $\log 1{,}000$

Solve for x.

51. $\log_2 x = 5$

52. $\log_3 x = -4$

53. $\log_x 16 = 2$

54. $\log_x \frac{1}{100} = -2$

55. $\log_9 3 = x$

56. $\log_{27} 3 = x$

Use a calculator to find the value of x to four decimal places.

57. $\log 4.51 = x$

58. $\log x = 1.43$

Graph each function and its inverse on the same coordinate system. Draw the axis of symmetry.

59. $f(x) = \log_4 x$ and $g(x) = 4^x$

60. $f(x) = \log_{1/3} x$ and $g(x) = \left(\frac{1}{3}\right)^x$

Graph each function. Label the x-intercept.

61. $f(x) = \log(x - 2)$

62. $f(x) = 3 + \log x$

63. ELECTRICAL ENGINEERING Find the db gain of an amplifier with an output of 18 volts and an input of 0.04 volt.

64. EARTHQUAKES An earthquake had a period of 0.3 second and an amplitude of 7,500 micrometers. Find its measure on the Richter scale.

SECTION 9.6   Base-*e* Logarithmic Functions

DEFINITIONS AND CONCEPTS	EXAMPLES
Of all possible bases for a logarithmic function, e is the most convenient for problems involving growth or decay. Since these situations occur often in natural settings, base-e logarithms are called **natural logarithms**: $\ln x$ means $\log_e x$	$\ln 5.7$ means $\log_e 5.7$
$\ln x$ is the exponent to which e is raised to get x.	To evaluate $\ln \frac{1}{e^4}$ we ask: "To what power must we raise e to get $\frac{1}{e^4}$?" Since $e^{-4} = \frac{1}{e^4}$, the answer is: the -4th power. Thus, $\ln \frac{1}{e^4} = -4$
The **natural logarithmic function** with base e is defined by $f(x) = \ln x$ The domain is the interval $(0, \infty)$ and the range is the interval $(-\infty, \infty)$.	The graph of the natural logarithmic function $f(x) = \ln x$ From the graph, we see that $f(x) = \ln x$ is an increasing function. The y-axis is an asymptote of the graph.

SECTION 9.6   Base-*e* Logarithmic Functions—*continued*

DEFINITIONS AND CONCEPTS	EXAMPLES
The natural exponential function $f(x) = e^x$ and the natural logarithmic function $f^{-1}(x) = \ln x$ are **inverses** of each other.	The graphs are symmetric about the line $y = x$.
If a population grows exponentially at a certain annual rate r, the time required for the population to double is called the **doubling time.** It is given by the formula: $$t = \frac{\ln 2}{r}$$	The population of a town is growing at a rate of 3% per year. If this rate continues, how long will it take the population to double? We substitute 0.03 for r and use a calculator to perform the computation. $$t = \frac{\ln 2}{r} = \frac{\ln 2}{0.03} \approx 23.10490602$$ The population will double in about 23.1 years.

REVIEW EXERCISES

Evaluate each expression, if possible. Do not use a calculator.

65. $\ln e$

66. $\ln e^2$

67. $\ln \frac{1}{e^3}$

68. $\ln \sqrt{e}$

69. $\ln (-e)$

70. $\ln 0$

71. $\ln 1$

72. $\ln e^{-7}$

Use a calculator to evaluate each expression. Express all answers to four decimal places.

73. $\ln 452$

74. $\ln 0.85$

Solve each equation. Express all answers to four decimal places.

75. $\ln x = 2.336$

76. $\ln x = -8.8$

77. Explain the difference between the functions $f(x) = \log x$ and $g(x) = \ln x$.

78. What function is the inverse of $f(x) = \ln x$?

Graph each function.

79. $f(x) = 1 + \ln x$

80. $f(x) = \ln (x + 1)$

81. POPULATION GROWTH   How long will it take the population of Mexico to double if the growth rate is currently about 1.153%?

82. BOTANY   The height (in inches) of a certain plant is approximated by the function $H(a) = 13 + 20.03 \ln a$, where a is its age in years. How tall will it be when it is 19 years old?

SECTION 9.7 Properties of Logarithms

DEFINITIONS AND CONCEPTS	EXAMPLES
Properties of logarithms: If M, N, and b are positive real numbers, $b \neq 1$ **1.** $\log_b 1 = 0$ **2.** $\log_b b = 1$ **3.** $\log_b b^x = x$ **4.** $b^{\log_b x} = x$ **5.** *Product rule for logarithms:* $$\log_b MN = \log_b M + \log_b N$$ **6.** *Quotient rule for logarithms:* $$\log_b \frac{M}{N} = \log_b M - \log_b N$$ **7.** *Power rule for logarithms:* $$\log_b M^P = p \log_b M$$	Apply a property of logarithms and then simplify, if possible. **1.** $\log_3 1 = 0$ **2.** $\log_7 7 = 1$ **3.** $\log_5 5^3 = 3$ **4.** $9^{\log_9 10} = 10$ **5.** $\log_2(6 \cdot 8) = \log_2 6 + \log_2 8$ $\qquad\qquad\quad = \log_2 6 + 3$ **6.** $\log_2 \dfrac{8}{6} = \log_2 8 - \log_2 6$ $\qquad\quad = 3 - \log_2 6$ **7.** $\log_2 7^3 = 3 \log_2 7$
Properties of logarithms can be used to **expand** logarithmic expressions.	Write $\log_3 (x^2 y^3)$ as the sum and/or difference of logarithms of a single quantity. $\log_3 (x^2 y^3) = \log_3 x^2 + \log_3 y^3$ The log of a product is the sum of the logs. $\qquad\qquad\;\; = 2 \log_3 x + 3 \log_3 y$ The log of a power is the power times the log.
Properties of logarithms can be used to **condense** certain logarithmic expressions.	Write $3 \ln x - \frac{1}{2} \ln y$ as a single logarithm. $3 \ln x - \dfrac{1}{2} \ln y = \ln x^3 - \ln y^{1/2}$ A power times a log is the log of the power. $\qquad\qquad = \ln \dfrac{x^3}{y^{1/2}}$ The difference of two logs is the log of the quotient. $\qquad\qquad = \ln \dfrac{x^3}{\sqrt{y}}$ Write $y^{1/2}$ as $\sqrt{y}$.
If we need to find a logarithm with some base other than 10 or e, we can use a conversion formula. **Change-of-base formula:** $$\log_b x = \frac{\log_a x}{\log_a b}$$	Find $\log_7 6$ to four decimal places. $$\log_7 6 = \frac{\log 6}{\log 7} \approx 0.920782221$$ To four decimal places, $\log_7 6 = 0.9208$. To check, verify that $7^{0.9208}$ is approximately 6.
In chemistry, common logarithms are used to express the acidity of solutions. **pH scale:** $\text{pH} = -\log[H^+]$	Find the pH of a liquid with a hydrogen ion concentration of 10^{-8} gram-ions per liter. $\text{pH} = -\log[H^+]$ $\quad\;\; = -\log 10^{-8}$ Substitute 10^{-8} for $[H^+]$. $\quad\;\; = -(-8) \log 10$ The log of a power is the power times the log. $\quad\;\; = 8$ Simplify: $\log 10 = 1$.

REVIEW EXERCISES

Simplify each expression.

83. $\log_2 1$

84. $\log_9 9$

85. $\log 10^3$

86. $7^{\log_7 4}$

Write each logarithm as the sum and/or difference of logarithms of a single quantity. Then simplify, if possible.

87. $\log_3 27x$

88. $\log \frac{100}{x}$

89. $\log_5 \sqrt{27}$

90. $\log_b 10ab$

Write each logarithm as the sum and/or difference of logarithms of a single quantity.

91. $\log_b \frac{x^2 y^3}{z}$

92. $\ln \sqrt{\frac{x}{yz^2}}$

Write each logarithmic expression as one logarithm.

93. $3 \log_2 x - 5 \log_2 y + 7 \log_2 z$

94. $-3 \log_b y - 7 \log_b z + \frac{1}{2} \log_b (x + 2)$

Assume that $\log_b 5 = 1.1609$ and $\log_b 8 = 1.5000$ and find each value to four decimal places.

95. $\log_b 40$

96. $\log_b 64$

97. Find $\log_5 17$ to four decimal places.

98. pH OF GRAPEFRUIT   The pH of grapefruit juice is about 3.1. Find its hydrogen ion concentration.

SECTION 9.8   Exponential and Logarithmic Equations

DEFINITIONS AND CONCEPTS	EXAMPLES
An **exponential equation** contains a variable in one of its exponents. If both sides of an exponential equation can be expressed as a power of the same base, we can use the following property to solve it: $\qquad b^x = b^y$ is equivalent to $x = y$	Solve: $3^{x+2} = 27$ We express the right side of the equation as a power of 3. $\quad 3^{x+2} = 3^3$ Write 27 as 3^3. $\quad x + 2 = 3$ If two exponential expressions with the same base are equal, their exponents are equal. $\qquad x = 1$ The solution is 1. Check it in the original equation.
When it is difficult to write each side of an exponential equation as a power of the same base, **take the logarithm of each side.**	Solve $4^x = 7$ and give the answer to four decimal places. We take the base-10 logarithm of both sides of the equation. $\quad \log 4^x = \log 7$ $\quad x \log 4 = \log 7$ The log of a power is the power times the log. $\quad x = \dfrac{\log 7}{\log 4}$ To isolate x, divide both sides by log 4. $\quad x \approx 1.4037$ Use a calculator. To four decimal places, the solution is 1.4037. To check the approximate solution, we substitute 1.4037 for x in $4^x = 7$ and use a calculator to evaluate the left side.

19. Graph $f(x) = e^x$. Label the y-intercept and the asymptote of the graph.

20. POPULATION GROWTH As of July 2007, the population of India was estimated to be 1,129,866,154, with an annual growth rate of 1.606%. If the growth rate remains the same, how large will the population be in 30 years?

21. Write the statement $\log_6 \frac{1}{36} = -2$ in exponential form.

22. a. What are the domain and range of the function $f(x) = \log x$?
b. What is the inverse of $f(x) = \log x$?

Evaluate each logarithmic expression, if possible.

23. $\log_5 25$

24. $\log_9 \frac{1}{81}$

25. $\log(-100)$

26. $\ln \frac{1}{e^6}$

27. $\log_4 2$

28. $\log_{1/3} 1$

Solve for x.

29. $\log_x 32 = 5$

30. $\log_8 x = \frac{4}{3}$

31. $\log_3 x = -3$

32. $\ln x = 1$

Graph each function.

33. $f(x) = -\log_3 x$

34. $f(x) = \ln x$

35. CHEMISTRY pH Find the pH of a solution with a hydrogen ion concentration of 3.7×10^{-7}. (*Hint:* pH $= -\log[H^+]$.)

36. ELECTRONICS Find the db gain of an amplifier when $E_O = 60$ volts and $E_I = 0.3$ volt. *Hint:* db gain $= 20 \log \frac{E_O}{E_I}$.

37. Use a calculator to find x to four decimal places: $\log x = -1.06$

38. Use the change-of-base formula to find $\log_7 3$ to four decimal places.

39. Write the expression $\log_b a^2 bc^3$ as the sum and/or difference of logarithms of a single quantity. Then simplify, if possible.

40. Write the expression $\frac{1}{2} \ln(a + 2) + \ln b - 3 \ln c$ as a logarithm of a single quantity.

Solve each equation. Give approximate answers to four decimal places.

41. $5^x = 3$

42. $3^{x-1} = 27$

43. $\ln(5x + 2) = \ln(2x + 5)$

44. $\log x + \log(x - 9) = 1$

45. The illustration shows the graphs of $y = \frac{1}{2} \ln(x - 1)$ and $y = \ln 2$ and the approximate coordinates of their point of intersection. Estimate the solution of the logarithmic equation $\frac{1}{2} \ln(x - 1) = \ln 2$. Then check the result.

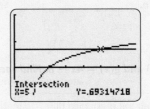

46. INSECTS The number of insects attracted to a bright light is currently 5. If the number is expected to quadruple every 6 minutes, how long will it take for the number to reach 500?

GROUP PROJECT

THE NUMBER *e*

Overview: In this activity, you will use a calculator to find progressively more accurate approximations of *e*.

Instructions: Form groups of two students. Each student will need a scientific calculator.

Begin by finding an approximation of *e* using the $\boxed{e^x}$ key on your calculator. Copy the table shown on the next page, and write the number displayed on the calculator screen at the top of the table.

The value of *e* can be calculated to any degree of accuracy by adding the terms of the following pattern:

$$e = 1 + 1 + \frac{1}{2} + \frac{1}{2 \cdot 3} + \frac{1}{2 \cdot 3 \cdot 4} + \frac{1}{2 \cdot 3 \cdot 4 \cdot 5} + \cdots$$

The more terms that are added, the closer the sum will be to *e*.

You are to add as many terms as necessary until you obtain a sum that matches the value of e ,given by the $\boxed{e^x}$ key on your calculator. Work together as a team. One member of the group should compute the fractional form of the term to be added. (See the middle column of the table.) The other member should take that information and calculate the cumulative sum. (See the right column of the table.)

How many terms must be added so that the cumulative sum approximation and the $\boxed{e^x}$ key approximation match in each decimal place?

Approximation of e found using the $\boxed{e^x}$ key: $e \approx$ _____

Number of terms in the sum	Term (Expressed as a fraction)	Cumulative sum (An approximation of e)
1	1	1
2	1	2
3	$\frac{1}{2}$	2.5
4	$\frac{1}{2 \cdot 3} = \frac{1}{6}$	2.666666667
$\vdots$	$\vdots$	$\vdots$

CHAPTER 10

Conic Sections; More Graphing

© Rich LaSalle/Getty Images

10.1 The Circle and the Parabola

10.2 The Ellipse

10.3 The Hyperbola

10.4 Solving Nonlinear Systems of Equations

CHAPTER SUMMARY AND REVIEW

CHAPTER TEST

Group Project

from **Campus to Careers**
Traffic Engineer

Traffic engineers design roads, streets, and highways for the safe and efficient movement of people and goods. They use traffic flow formulas to determine what kinds of roads are needed and then find economical ways to construct and operate them. During the planning stages, traffic engineers make detailed drawings and graphs of the project. Because highway and street construction is often publicly funded, they make budgets, submit bid proposals, and perform cost analysis studies to make sure that highway tax money is spent wisely.

In **Problem 85** of **Study Set 10.1,** you will design two sections of a freeway that are to be joined with a curve that is one-quarter of a circle.

JOB TITLE:
Traffic Engineer

EDUCATION:
A bachelor's degree in civil engineering is required.

JOB OUTLOOK:
Good; It is expected to increase 9% to 17% through 2014.

ANNUAL EARNINGS:
The median salary in 2007 was $95,300.

FOR MORE INFORMATION:
http://careers.stateuniversity.com

Preparing for Your Next Math Course

Before moving on to a new mathematics course, it's worthwhile to take some time to reflect on your effort and performance in this course.

Now Try This

As this course draws to a close, here are some questions to ask yourself.

1. How was my attendance?
2. Was I organized? Did I have the right materials?
3. Did I follow a regular schedule?
4. Did I pay attention in class and take good notes?
5. Did I spend the appropriate amount of time on homework?
6. How did I prepare for tests? Did I have a test-taking strategy?
7. Was I part of a study group? If not, why not? If so, was it worthwhile?
8. Did I ever seek extra help from a tutor or from my instructor?
9. In what topics was I the strongest? In what topics was I the weakest?
10. If I had it to do over, would I do anything differently?

SECTION 10.1
The Circle and the Parabola

Objectives

❶ Identify conic sections and some of their applications.

❷ Graph equations of circles written in standard form.

❸ Write the equation of a circle, given its center and radius.

❹ Convert the general form of the equation of a circle to standard form.

❺ Solve problems involving circles.

❻ Convert the general form of the equation of a parabola to standard form to graph it.

We have previously graphed first-degree equations in two variables such as $y = 3x + 8$ and $4x - 3y = 12$. Their graphs are lines. In this section, we will graph second-degree equations in two variables such as $x^2 + y^2 = 25$ and $x = -3y^2 - 12y - 13$. The graphs of these equations are *conic sections*.

❶ **Identify Conic Sections and Some of Their Applications.**

The curves formed by the intersection of a plane with an infinite right-circular cone are called **conic sections.** Those curves have four basic shapes, called **circles, parabolas, ellipses,** and **hyperbolas,** as shown on the next page.

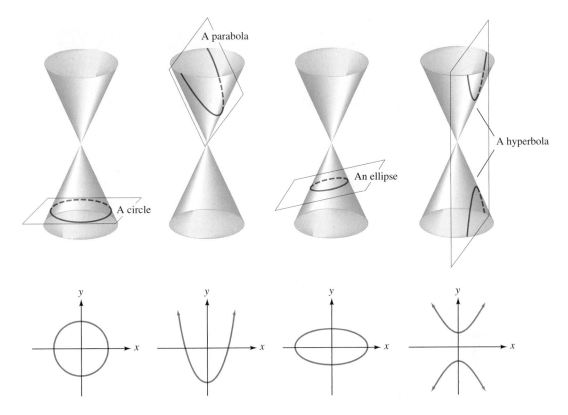

Conic sections have many applications. For example, everyone is familiar with circular wheels and gears, pizza cutters, and hula hoops.

Parabolas can be rotated to generate dish-shaped surfaces called **paraboloids.** Any light or sound placed at the **focus** of a paraboloid is reflected outward in parallel paths. This property makes parabolic surfaces ideal for flashlight and headlight reflectors. It also makes parabolic surfaces good antennas, because signals captured by such antennas are concentrated at the focus. Parabolic mirrors are capable of concentrating the rays of the sun at a single point, thereby generating tremendous heat. This property is used in the design of solar furnaces.

Any object thrown upward and outward travels in a parabolic path. An example of this is a stream of water flowing from a drinking fountain. In architecture, many arches are parabolic in shape, because this gives them strength. Cables that support suspension bridges hang in the shape of a parabola.

Radar dish

Stream of water

Support cables

Ellipses have optical and acoustical properties that are useful in architecture and engineering. Many arches are portions of an ellipse, because the shape is pleasing to the eye. The planets and many comets have elliptical orbits. Certain gears have elliptical shapes to provide nonuniform motion.

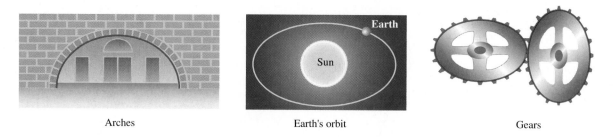

Arches   Earth's orbit   Gears

Hyperbolas serve as the basis of a navigational system known as LORAN (LOng RAnge Navigation). They are also used to find the source of a distress signal, are the basis for the design of hypoid gears, and describe the orbits of some comets.

A sonic shock wave created by a jet aircraft has the shape of a cone. In level flight, the sound wave intersects the ground as one branch of a hyperbola, as shown below. People in different places along the curve on the ground hear and feel the sonic boom at the same time.

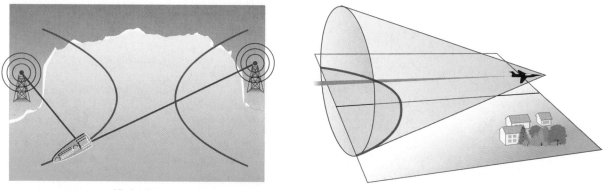

Navigation   Sonic boom

② Graph Equations of Circles Written in Standard Form.

Every conic section can be represented by a second-degree equation in x and y. To find the equation of a circle, we use the following definition.

Definition of a Circle	A **circle** is the set of all points in a plane that are a fixed distance from a fixed point called its **center.** The fixed distance is called the **radius** of the circle.

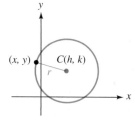

If we let (h, k) be the center of a circle and (x, y) be some point on a circle that is graphed on a rectangular coordinate system, the distance from (h, k) to (x, y) is the radius r of the circle. We can use the distance formula to find r.

$$r = \sqrt{(x - h)^2 + (y - k)^2}$$

We can square both sides to eliminate the radical and obtain

$$r^2 = (x - h)^2 + (y - k)^2$$

This result is called the *standard form of the equation of a circle* with radius r and center at (h, k).

Equation of a Circle	The **standard form of the equation of a circle** with radius r and center at (h, k) is $$(x - h)^2 + (y - k)^2 = r^2$$

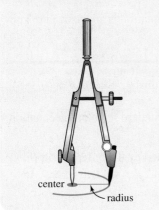

EXAMPLE 1   Find the center and the radius of each circle and then graph it:

a. $(x - 4)^2 + (y - 1)^2 = 9$ b. $x^2 + y^2 = 25$

c. $(x + 3)^2 + y^2 = 12$

Strategy   We will compare each equation to the standard form of the equation of a circle, $(x - h)^2 + (y - k)^2 = r^2$, and identify h, k, and r.

Why   The center of the circle is the point with coordinates (h, k) and the radius of the circle is r.

Solution

a. The color highlighting shows how to compare the given equation to the standard form to find h, k, and r.

$$(x - 4)^2 + (y - 1)^2 = 9$$
$$\uparrow \qquad \uparrow \qquad \uparrow$$
$$(x - h)^2 + (y - k)^2 = r^2$$

$h = 4$, $k = 1$, and $r^2 = 9$. Since the radius of a circle must be positive, $r = 3$.

The center of the circle is $(h, k) = (4, 1)$ and the radius is 3.

To plot four points on the circle, we move up, down, left, and right 3 units from the center, as shown in figure (a). Then we draw a circle through the points to get the graph of $(x - 4)^2 + (y - 1)^2 = 9$, as shown in figure (b).

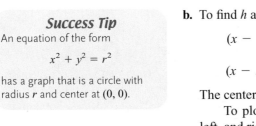

(a)   (b)

b. To find h and k, we will write $x^2 + y^2 = 25$ in the following way:

$$(x - 0)^2 + (y - 0)^2 = 25$$
$$\uparrow \qquad \uparrow \qquad \uparrow$$
$$(x - h)^2 + (y - k)^2 = r^2$$

$h = 0$, $k = 0$, and $r^2 = 25$. Since the radius must be positive, $r = 5$.

The center of the circle is at $(0, 0)$ and the radius is 5.

To plot four points on the circle, we move up, down, left, and right 5 units from the center. Then we draw a circle through the points to get the graph of $x^2 + y^2 = 25$, as shown.

c. To find h, we will write $x + 3$ as $x - (-3)$.

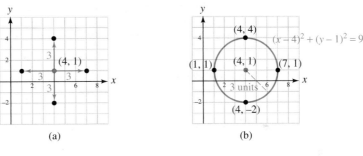

Standard form requires
a minus symbol here.

$$[x - (-3)]^2 + (y - 0)^2 = 12$$
$$\uparrow \qquad \uparrow \qquad \uparrow$$
$$(x - h)^2 + (y - k)^2 = r^2$$

$h = -3$, $k = 0$, and $r^2 = 12$.

Since $r^2 = 12$, we have

$$r = \pm\sqrt{12} = \pm 2\sqrt{3}$$ Use the square root property.

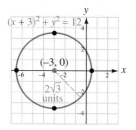

Since the radius can't be negative, $r = 2\sqrt{3}$. The center of the circle is at $(-3, 0)$ and the radius is $2\sqrt{3}$.

To plot four points on the circle, we move up, down, left, and right $2\sqrt{3} \approx 3.5$ units from the center. We then draw a circle through the points to get the graph of $(x + 3)^2 + y^2 = 12$, as shown on the left.

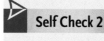 **Self Check 1**   Find the center and the radius of each circle and then graph it:

a. $(x - 3)^2 + (y + 4)^2 = 4$ **b.** $x^2 + y^2 = 8$

Now Try   **Problems 15, 19, and 21**

3   **Write the Equation of a Circle, Given its Center and Radius.**

Because a circle is determined by its center and radius, that information is all we need to know to write its equation.

EXAMPLE 2   Write the equation of the circle with radius 9 and center at $(6, -5)$.

Strategy   We substitute 9 for r, 6 for h, and -5 for k in the standard form of the equation of a circle, $(x - h)^2 + (y - k)^2 = r^2$.

Why   When writing the standard form, the center is represented by the ordered pair (h, k) and the radius as r.

Solution

$$(x - h)^2 + (y - k)^2 = r^2$$
$$(x - 6)^2 + [y - (-5)]^2 = 9^2 \quad \text{Substitute 6 for } h, \ -5 \text{ for } k, \text{ and 9 for } r.$$
$$(x - 6)^2 + (y + 5)^2 = 9^2 \quad \text{Write } y - (-5) \text{ as } y + 5.$$

If we express 9^2 as 81, we have

$$(x - 6)^2 + (y + 5)^2 = 81$$

> ### Notation
> Standard form can be written
>
> $$(x - 6)^2 + (y + 5)^2 = 9^2$$
> or
> $$(x - 6)^2 + (y + 5)^2 = 81$$

  Self Check 2   Write the equation of the circle with radius 10 and center at $(-7, 1)$.

Now Try   **Problems 23, 27, and 31**

4   **Convert the General Form of the Equation of a Circle to Standard Form.**

In Example 2, the result was written in standard form: $(x - 6)^2 + (y + 5)^2 = 81$. If we square $x - 6$ and $y + 5$, we obtain a different form for the equation of the circle.

$$(x - 6)^2 + (y + 5)^2 = 9^2$$
$$x^2 - 12x + 36 + y^2 + 10y + 25 = 81 \quad \text{Square each binomial.}$$
$$x^2 - 12x + y^2 + 10y - 20 = 0 \quad \text{Subtract 81 from both sides. Combine like terms.}$$
$$x^2 + y^2 - 12x + 10y - 20 = 0 \quad \text{Rearrange the terms, writing the squared terms first.}$$

> ### Success Tip
> This example illustrates an important fact: The equation of a circle contains both x^2 and y^2 terms on the same side of the equation with equal coefficients.

This result is written in the *general form of the equation of a circle.*

Equation of a Circle	The **general form of the equation of a circle** is

$$x^2 + y^2 + Dx + Ey + F = 0$$

We can convert from the general form to the standard form of the equation of a circle by completing the square.

EXAMPLE 3   Write the equation $x^2 + y^2 - 4x + 2y - 11 = 0$ in standard form and graph it.

Strategy   We will rearrange the terms to write the equation in the form $x^2 - 4x + y^2 + 2y = 11$ and complete the square on x and y.

Why   Standard form contains the expressions $(x - h)^2$ and $(y - k)^2$. We can obtain a perfect-square trinomial that factors as $(x - 2)^2$ by completing the square on $x^2 - 4x$. We can complete the square on $y^2 + 2y$ to obtain an expression of the form $(y + 1)^2$.

Solution   To write the equation in standard form, we complete the square twice.

$$x^2 + y^2 - 4x + 2y - 11 = 0$$

$$x^2 - 4x \quad + y^2 + 2y \qquad = 11 \quad \text{\small Write the x-terms together, the y-terms}$$
$$\text{\small together, and add 11 to both sides.}$$

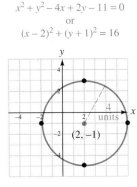

$$x^2 + y^2 - 4x + 2y - 11 = 0$$
$$\text{or}$$
$$(x - 2)^2 + (y + 1)^2 = 16$$

To complete the square on $x^2 - 4x$, we note that $\frac{1}{2}(-4) = -2$ and $(-2)^2 = $ **4**. To complete the square on $y^2 + 2y$, we note that $\frac{1}{2}(2) = 1$ and $1^2 = $ **1**. We add **4** and **1** to both sides of the equation.

$$x^2 - 4x + 4 + y^2 + 2y + 1 = 11 + 4 + 1$$

$$(x - 2)^2 + (y + 1)^2 = 16 \qquad \text{\small Factor } x^2 - 4x + 4 \text{ and } y^2 + 2y + 1.$$

The equation can also be written as $(x - 2)^2 + (y + 1)^2 = 4^2$.

We can determine the circle's center and radius by comparing this equation to the standard form of the equation of a circle, $(x - h)^2 + (y - k)^2 = r^2$. We see that $h = 2$, $k = -1$, and $r = 4$. We can use the center, $(h, k) = (2, -1)$ and the radius $r = 4$, to graph the circle as shown on the left.

Self Check 3   Write the equation $x^2 + y^2 + 12x - 6y - 4 = 0$ in standard form and graph it.

Now Try   **Problem 35**

Using Your Calculator	*Graphing Circles*

Since the graphs of circles fail the vertical line test, their equations do not represent functions. It is more difficult to use a graphing calculator to graph equations that are not functions. For example, to graph the circle described by $(x - 1)^2 + (y - 2)^2 = 4$, we must split the equation into two functions and graph each one separately. We begin by solving the equation for y.

$$(x - 1)^2 + (y - 2)^2 = 4$$

$$(y - 2)^2 = 4 - (x - 1)^2 \qquad \text{\small Subtract } (x - 1)^2 \text{ from both sides.}$$

$$y - 2 = \pm\sqrt{4 - (x - 1)^2} \qquad \text{\small Use the square root property.}$$

$$y = 2 \pm \sqrt{4 - (x - 1)^2} \qquad \text{\small Add 2 to both sides.}$$

This equation defines two functions. If we graph

$$y = 2 + \sqrt{4 - (x - 1)^2} \quad \text{and} \quad y = 2 - \sqrt{4 - (x - 1)^2}$$

we get the distorted circle shown in figure (a). To get a better circle, we can use the graphing calculator's square window feature, which gives an equal unit distance on both the x- and y-axes. (Press ZOOM, 5, ENTER.) Using this feature, we get the circle shown in figure (b). Sometimes the two arcs will not connect because of approximations made by the calculator at each endpoint.

The graph of
$y = 2 + \sqrt{4 - (x - 1)^2}$
is the top half of the circle.

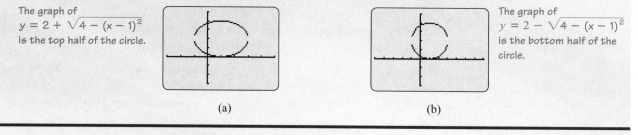

The graph of
$y = 2 - \sqrt{4 - (x - 1)^2}$
is the bottom half of the circle.

(a)   (b)

5 **Solve Problems Involving Circles.**

EXAMPLE 4   *Radio Translators.*   The broadcast area of a television station is bounded by the circle $x^2 + y^2 = 3{,}600$, where x and y are measured in miles. A translator station picks up the signal and retransmits it from the center of a circular area bounded by

$$(x + 30)^2 + (y - 40)^2 = 1{,}600$$

Find the location of the translator and the greatest distance from the main transmitter that the signal can be received.

Strategy   Refer to the figure below. We will find two distances: the distance from the TV station transmitter to the translator and the distance from the translator to the outer edge of its coverage.

Why   The greatest distance of reception from the main transmitter is the sum of those two distances.

Solution   The coverage of the TV station is bounded by $x^2 + y^2 = 60^2$, a circle centered at the origin with a radius of 60 miles, as shown in yellow in the figure. Because the translator is at the center of the circle $(x + 30)^2 + (y - 40)^2 = 1{,}600$, it is located at $(-30, 40)$, a point 30 miles west and 40 miles north of the TV station. The radius of the translator's coverage is $\sqrt{1{,}600}$, or 40 miles.

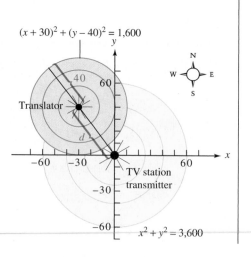

As shown in the figure, the greatest distance of reception is the sum of d, the distance from the translator to the television station, and 40 miles, the radius of the translator's coverage.

To find d, we use the distance formula to find the distance between the origin, $(x_1, y_1) = (0, 0)$, and $(x_2, y_2) = (-30, 40)$.

$$d = \sqrt{(x_2 - x_1)^2 + (y_2 - y_1)^2} \qquad \text{The distance formula was introduced in Section 7.6.}$$

$$d = \sqrt{(-30 - 0)^2 + (40 - 0)^2}$$

$$d = \sqrt{(-30)^2 + 40^2}$$

$$= \sqrt{900 + 1{,}600}$$

$$= \sqrt{2{,}500}$$

$$= 50$$

The translator is located 50 miles from the television station, and it broadcasts the signal 40 miles. The greatest reception distance from the main transmitter signal is, therefore, $50 + 40$, or 90 miles.

 Now Try Problem 83

6 **Convert the General Form of the Equation of a Parabola to Standard Form to Graph It.**

Another type of conic section is the parabola.

| **Definition of a Parabola** | A **parabola** is the set of all points in a plane that are equidistant from a fixed point, called the **focus,** and a fixed line, called the **directrix.** |

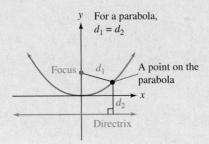

We have previously discussed parabolas whose graphs open upward or downward. Parabolas can also open to the right and to the left, but they do not define functions because their graphs fail the vertical line test.

The two general forms of the equation of a parabola are similar.

| **Equation of a Parabola** | The **general forms of the equation of a parabola** are:

1. $y = ax^2 + bx + c$ The graph opens upward if $a > 0$ and downward if $a < 0$.
2. $x = ay^2 + by + c$ The graph opens to the right if $a > 0$ and to the left if $a < 0$. |

Recall from Chapter 8 that equations written in the standard form $y = a(x - h)^2 + k$ represent parabolas with vertex at (h, k) and axis of symmetry $x = h$. They open upward when $a > 0$ and downward when $a < 0$.

EXAMPLE 5   Write $y = -2x^2 + 12x - 15$ in standard form and graph it.

Strategy   We will complete the square on x to write the equation in standard form, $y = a(x - h)^2 + k$.

Why   Standard form contains the expression $(x - h)^2$. We can obtain a perfect-square trinomial that factors into that form by completing the square on x.

Solution   Because the equation is not in standard form, the coordinates of the vertex are not obvious. To write the equation in standard form, we complete the square on x.

$y = -2x^2 + 12x - 15$

$y = -2(x^2 - 6x \quad\;) - 15$ Factor out -2 from $-2x^2 + 12x$.

This step adds $-2 \cdot 9$ Add 18 to counteract
or -18 to this side.   the addition of -18.

$y = -2(x^2 - 6x + 9) - 15 + 18$ Complete the square on $x^2 - 6x$.

$y = -2(x - 3)^2 + 3$ Factor $x^2 - 6x + 9$ and combine like terms.

This equation is written in the form $y = a(x - h)^2 + k$, where $a = -2$, $h = 3$, and $k = 3$. Thus, the graph of the equation is a parabola that opens downward with vertex at $(3, 3)$ and an axis of symmetry $x = 3$. We can construct a table of solutions and use symmetry to plot several points on the parabola. Then we draw a smooth curve through the points to get the graph of $y = -2x^2 + 12x - 15$, as shown below.

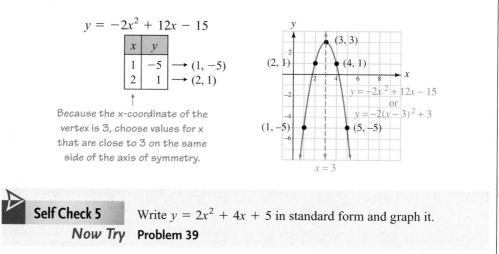

$y = -2x^2 + 12x - 15$

x	y
1	-5	$\rightarrow (1, -5)$
2	1	$\rightarrow (2, 1)$

Because the x-coordinate of the vertex is 3, choose values for x that are close to 3 on the same side of the axis of symmetry.

Self Check 5   Write $y = 2x^2 + 4x + 5$ in standard form and graph it.

Now Try   Problem 39

The *standard form* for the equation of a parabola that opens to the right or left is similar to $y = a(x - h)^2 + k$, except that the variables, x and y, exchange positions as do the constants, h and k.

Success Tip

Recall that we can find the x-coordinate of the vertex using

$$x = -\frac{b}{2a} = -\frac{12}{2(-2)} = 3$$

To find the y-coordinate, substitute:

$$y = -2(3)^2 + 12(3) - 15$$
$$= 3$$

The vertex is at $(3, 3)$.

Standard Form of the Equation of a Parabola

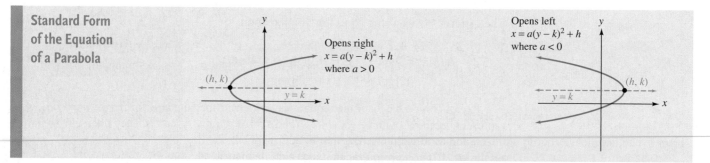

Opens right
$x = a(y - k)^2 + h$
where $a > 0$

Opens left
$x = a(y - k)^2 + h$
where $a < 0$

EXAMPLE 6   Graph: $x = \frac{1}{2}y^2$

Strategy   We will compare the equation to the standard form of the equation of a parabola to find a, h, and k.

Why   Once we know these values, we can locate the vertex of the graph. We also know whether the parabola will open to the left or to the right.

Solution   This equation is written in the form $x = a(y - k)^2 + h$, where $a = \frac{1}{2}$, $k = 0$, and $h = 0$. The graph of the equation is a parabola that opens to the right with vertex at $(0, 0)$ and an axis of symmetry $y = 0$.

To construct a table of solutions, we choose values of y and find their corresponding values of x. For example, if $y = 1$, we have

$$x = \frac{1}{2}y^2$$

$$x = \frac{1}{2}(1)^2 \quad \text{Substitute 1 for } y.$$

$$x = \frac{1}{2}$$

The point $\left(\frac{1}{2}, 1\right)$ is on the parabola.

We plot the ordered pairs from the table and use symmetry to plot three more points on the parabola. Then we draw a smooth curve through the points to get the graph of $x = \frac{1}{2}y^2$, as shown below.

$x = \frac{1}{2}y^2$

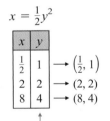

x	y	
$\frac{1}{2}$	1	$\longrightarrow \left(\frac{1}{2}, 1\right)$
2	2	$\longrightarrow (2, 2)$
8	4	$\longrightarrow (8, 4)$

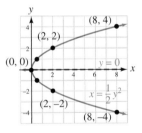

Because the y-coordinate of the vertex is 0, choose values for y that are close to 0 on the same side of the axis of symmetry.

Self Check 6   Graph: $x = -\frac{2}{3}y^2$

Now Try   **Problem 43**

EXAMPLE 7   Write $x = -3y^2 - 12y - 13$ in standard form and graph it.

Strategy   We will complete the square on y to write the equation in standard form, $x = a(y - k)^2 + h$.

Why   Standard form contains the expression $(y - k)^2$. We can obtain a perfect-square trinomial that factors into that form by completing the square on y.

Solution   To write the equation in standard form, we complete the square.

$$x = -3y^2 - 12y - 13$$
$$x = -3(y^2 + 4y \quad\quad) - 13 \qquad \text{Factor out } -3 \text{ from } -3y^2 - 12y.$$
$$x = -3(y^2 + 4y + 4) - 13 + 12 \qquad \text{Complete the square on } y^2 + 4y. \text{ Then add 12 to the right side to counteract } -3 \cdot 4 = -12.$$
$$x = -3(y + 2)^2 - 1 \qquad \text{Factor } y^2 + 4y + 4 \text{ and combine like terms.}$$

This equation is in the standard form $x = a(y - k)^2 + h$, where $a = -3$, $k = -2$, and $h = -1$. The graph of the equation is a parabola that opens to the left with vertex at $(-1, -2)$ and an axis of symmetry $y = -2$.

We can construct a table of solutions and use symmetry to plot several points on the parabola. Then we draw a smooth curve through the points to get the graph of $x = -3y^2 - 12y - 13$, as shown below.

$$x = -3y^2 - 12y - 13$$
$$\text{or}$$
$$x = -3(y + 2)^2 - 1$$

x	y	
-4	-1	$\rightarrow (-4, -1)$
-13	0	$\rightarrow (-13, 0)$

↑ Choose values for y, and find the corresponding x-values.

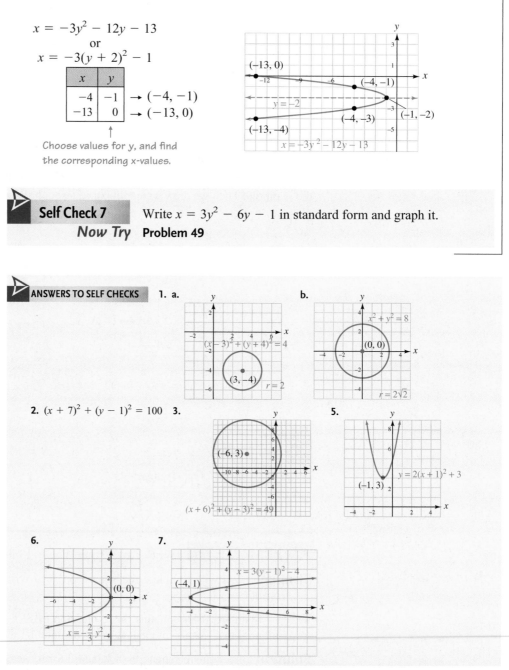

Self Check 7 Write $x = 3y^2 - 6y - 1$ in standard form and graph it.

Now Try Problem 49

ANSWERS TO SELF CHECKS

1. a. $(x - 3)^2 + (y + 4)^2 = 4$ b. $x^2 + y^2 = 8$

2. $(x + 7)^2 + (y - 1)^2 = 100$ 3.

5. $y = 2(x + 1)^2 + 3$

6. $x = -\frac{2}{3}y^2$

7. $x = 3(y - 1)^2 - 4$

STUDY SET
10.1

VOCABULARY

Fill in the blanks.

1. The curves formed by the intersection of a plane with an infinite right-circular cone are called _____ _____.

2. Give the name of each curve shown below.

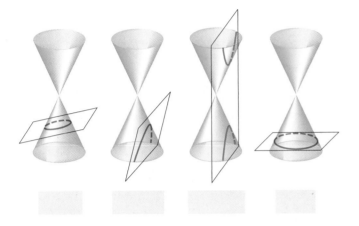

3. A _____ is the set of all points in a plane that are a fixed distance from a fixed point called its center. The fixed distance is called the _____.

4. A parabola is the set of all points in a plane that are equidistant from a fixed point and a fixed _____.

CONCEPTS

5. **a.** Write the standard form of the equation of a circle.

   b. Write the standard form of the equation of a circle with the center at the origin.

6. **a.** Find the center and the radius of the circle graphed on the right.

   b. Write the equation of the circle.

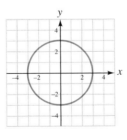

7. **a.** Find the center and the radius of the circle graphed on the right.

   b. Write the equation of the circle.

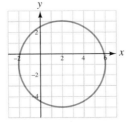

8. Fill in the blanks. To complete the square on $x^2 + 2x$ and on $y^2 - 6y$, what numbers must be added to each side of the equation?

$$x^2 + 2x + y^2 - 6y = 2$$

$$x^2 + 2x + \boxed{} + y^2 - 6y + \boxed{} = 2 + \boxed{} + \boxed{}$$

9. **a.** What is the standard form of the equation of a parabola opening upward or downward?

   b. What is the standard form of the equation of a parabola opening to the right or left?

10. Fill in the blanks.

    a. To complete the square on the right side, what should be factored from the first two terms?

    $$x = 4y^2 + 16y + 9$$
    $$x = \boxed{}(y^2 + 4y) + 9$$

    b. To complete the square on $y^2 + 4y$, what should be added within the parentheses, and what should be subtracted outside the parentheses?

    $$x = 4(y^2 + 4y + \boxed{}) + 9 - \boxed{}$$

11. Determine whether the graph of each equation is a circle or a parabola.

    a. $x^2 + y^2 - 6x + 8y - 10 = 0$

    b. $y^2 - 2x + 3y - 9 = 0$

    c. $x^2 + 5x - y = 0$

    d. $x^2 + 12x + y^2 = 0$

12. Draw a parabola using the given facts.

    • Opens right
    • Passes through $(-2, 1)$
    • Vertex $(-3, 2)$
    • x-intercept $(1, 0)$

NOTATION

13. Find h, k, and r: $(x - 6)^2 + (y + 2)^2 = 9$

14. **a.** Find a, h, and k: $y = 6(x - 5)^2 - 9$

    b. Find a, h, and k: $x = -3(y + 2)^2 + 1$

GUIDED PRACTICE

Find the center and radius of each circle and graph it. **See Example 1.**

15. $x^2 + y^2 = 9$

16. $x^2 + y^2 = 16$

17. $x^2 + (y + 3)^2 = 1$

18. $(x + 4)^2 + y^2 = 1$

19. $(x + 3)^2 + (y - 1)^2 = 16$

20. $(x - 1)^2 + (y + 4)^2 = 9$

21. $x^2 + y^2 = 6$

22. $x^2 + y^2 = 10$

Write the equation of a circle in standard form with the following properties. See Example 2.

23. Center at the origin; radius 1

24. Center at the origin; radius 4

25. Center at (6, 8); radius 5

26. Center at (5, 3); radius 2

27. Center at (−2, 6); radius 12

28. Center at (5, −4); radius 6

29. Center at (0, 0); radius $\dfrac{1}{4}$

30. Center at (0, 0); radius $\dfrac{1}{3}$

31. Center at $\left(\dfrac{2}{3}, -\dfrac{7}{8}\right)$; radius $\sqrt{2}$

32. Center at (−0.7, −0.2); radius $\sqrt{11}$

33. Center at the origin; diameter $4\sqrt{2}$

34. Center at the origin; diameter $8\sqrt{3}$

Write each equation of a circle in standard form and graph it. Give the coordinates of its center and give the radius. See Example 3.

35. $x^2 + y^2 - 2x + 4y = -1$

36. $x^2 + y^2 + 6x - 4y = -12$

37. $x^2 + y^2 + 4x + 2y = 4$

38. $x^2 + y^2 + 8x + 2y = -13$

Write each equation of a parabola in standard form and graph it. Give the coordinates of the vertex. See Example 5.

39. $y = 2x^2 - 4x + 5$

40. $y = x^2 + 4x + 5$

41. $y = -x^2 - 2x + 3$

42. $y = -2x^2 - 4x$

Graph each equation of a parabola. Give the coordinates of the vertex. See Example 6.

43. $x = y^2$

44. $x = 2y^2$

45. $x = 2(y + 1)^2 + 3$

46. $x = 3(y - 2)^2 - 1$

Write each equation of a parabola in standard form and graph it. Give the coordinates of the vertex. See Example 7.

47. $x = y^2 - 2y + 5$

48. $x = y^2 + 6y + 8$

49. $x = -3y^2 + 18y - 25$

50. $x = -2y^2 + 4y + 1$

Use a graphing calculator to graph each equation. (Hint: *Solve for y and graph two functions.*) *See Using Your Calculator: Graphing Circles.*

51. $x^2 + y^2 = 7$

52. $x^2 + y^2 = 5$

53. $(x + 1)^2 + y^2 = 16$

54. $x^2 + (y - 2)^2 = 4$

Use a graphing calculator to graph each equation. (Hint: *Solve for y and graph two functions when necessary.*)

55. $x = 2y^2$

56. $x = y^2 - 4$

57. $x^2 - 2x + y = 6$

58. $x = -2(y - 1)^2 + 2$

TRY IT YOURSELF

Write each equation in standard form, if it is not already so, and graph it. If the graph is a circle, give the coordinates of its center and its radius. If the graph is a parabola, give the coordinates of its vertex.

59. $x = y^2 - 6y + 4$

60. $x = y^2 - 8y + 13$

61. $(x - 2)^2 + y^2 = 25$

62. $x^2 + (y - 3)^2 = 25$

63. $x^2 + y^2 - 6x + 8y + 18 = 0$

64. $x^2 + y^2 - 4x + 4y - 3 = 0$

65. $y = 4x^2 - 16x + 17$

66. $y = 4x^2 - 32x + 63$

67. $(x - 1)^2 + (y - 3)^2 = 15$

68. $(x + 1)^2 + (y + 1)^2 = 8$

69. $x = -y^2 + 1$

70. $x = -y^2 - 5$

71. $(x - 2)^2 + (y - 4)^2 = 36$

72. $(x - 3)^2 + (y - 2)^2 = 36$

73. $x = -\dfrac{1}{4}y^2$

74. $x = 4y^2$

75. $x = -6(y - 1)^2 + 3$

76. $x = -6(y + 1)^2 - 4$

77. $x^2 + y^2 + 2x - 8 = 0$

78. $x^2 + y^2 - 4y = 12$

79. $x = \dfrac{1}{2}y^2 + 2y$

80. $x = -\dfrac{1}{3}y^2 - 2y$

81. $y = -4(x + 5)^2 + 5$

82. $y = -4(x - 4)^2 - 4$

APPLICATIONS

83. BROADCAST RANGES Radio stations applying for licensing may not use the same frequency if their broadcast areas overlap. One station's coverage is bounded by $x^2 + y^2 - 8x - 20y + 16 = 0$, and the other's by $x^2 + y^2 + 2x + 4y - 11 = 0$. May they be licensed for the same frequency?

84. MESHING GEARS For design purposes, the large gear is described by the circle $x^2 + y^2 = 16$. The smaller gear is a circle centered at $(7, 0)$ and tangent to the larger circle. Find the equation of the smaller gear.

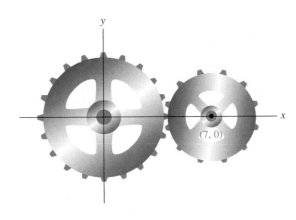

(7, 0)

85. from Campus to Careers
Traffic Engineer

Suppose you are a traffic engineer and you are designing two sections of a new freeway so that they join with a curve that is one-quarter of a circle, as shown. The equation of the circle is $x^2 + y^2 - 10x - 12y + 52 = 0$, where distances are measured in miles.

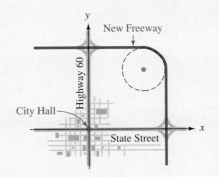

© Rich LaSalle/Getty Images

a. How far from City Hall will the new freeway intersect State Street?

b. How far from City Hall will the new freeway intersect Highway 60?

86. WALKWAYS The walkway shown is bounded by the two circles $x^2 + y^2 = 2,500$ and $(x - 10)^2 + y^2 = 900$, measured in feet. Find the largest and the smallest width of the walkway.

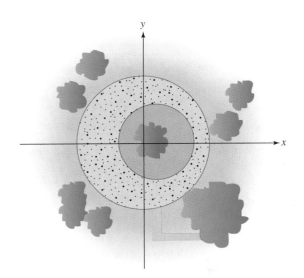

87. PROJECTILES The cannonball in the illustration follows the parabolic path $y = 30x - x^2$. How far short of the castle does it land?

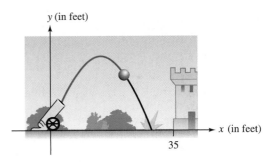

88. PROJECTILES In Exercise 87, how high does the cannonball get?

89. COMETS If the orbit of the comet is approximated by the equation $2y^2 - 9x = 18$, how far is it from the sun at the vertex V of the orbit? Distances are measured in astronomical units (AU).

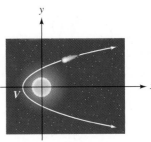

90. SATELLITE ANTENNAS   The cross section of the satellite antenna in the illustration is a parabola given by the equation $y = \frac{1}{16}x^2$, with distances measured in feet. If the dish is 8 feet wide, how deep is it?

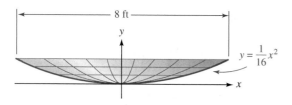

WRITING

91. Explain how to decide from its equation whether the graph of a parabola opens up, down, right, or left.

92. From the equation of a circle, explain how to determine the radius and the coordinates of the center.

93. On the day of an election, the following warning was posted in front of a school. Explain what it means.

> *No electioneering within a 1,000-foot radius of this polling place.*

94. What is meant by the *turning radius* of a truck?

REVIEW

Solve each equation.

95. $|3x - 4| = 11$

96. $\left|\dfrac{4 - 3x}{5}\right| = 12$

97. $|3x + 4| = |5x - 2|$

98. $|6 - 4x| = |x + 2|$

CHALLENGE PROBLEMS

99. Could the intersection of a plane with an infinite right-circular cone as shown on page 939 be a single point? If so, draw a picture that illustrates this.

100. Under what conditions will the graph of $x = a(y - k)^2 + h$ have no y-intercepts?

101. Write the equation of a circle with a diameter whose endpoints are at $(-2, -6)$ and $(8, 10)$.

102. Write the equation of a circle with a diameter whose endpoints are at $(-5, 4)$ and $(7, -3)$.

SECTION 10.2
The Ellipse

Objectives

① Define an ellipse.

② Graph ellipses centered at the origin.

③ Graph ellipses centered at (h, k).

④ Solve problems involving ellipses.

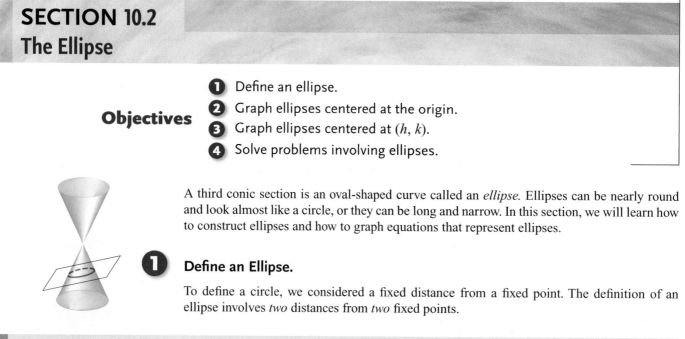

A third conic section is an oval-shaped curve called an *ellipse*. Ellipses can be nearly round and look almost like a circle, or they can be long and narrow. In this section, we will learn how to construct ellipses and how to graph equations that represent ellipses.

①   **Define an Ellipse.**

To define a circle, we considered a fixed distance from a fixed point. The definition of an ellipse involves *two* distances from *two* fixed points.

Definition of an Ellipse	An **ellipse** is the set of all points in a plane for which the sum of the distances from two fixed points is a constant.

The figure on the next page illustrates that any point on an ellipse is a constant distance $d_1 + d_2$ from two fixed points, each of which is called a **focus**. Midway between the **foci** is the **center** of the ellipse.

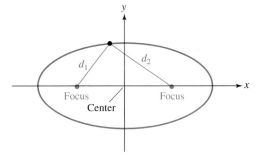

The Language of Algebra

The word *foci* (pronounced foe-sigh) is plural for the word *focus*. In the illustration on the right, the foci are labeled using subscript notation. One focus is F_1 and the other is F_2.

We can construct an ellipse by placing two thumbtacks fairly close together to serve as foci. We then tie each end of a piece of string to a thumbtack, catch the loop with the point of a pencil, and (keeping the string taut) draw the ellipse.

 Graph Ellipses Centered at the Origin.

The definition of an ellipse can be used to develop the standard equation of an ellipse. To learn more about the derivation, see Problem 70 in the Challenge Problem section of the Study Set.

Equation of an Ellipse Centered at the Origin	The **standard form of the equation of an ellipse** that is symmetric with respect to both axes and centered at $(0, 0)$ is $$\frac{x^2}{a^2} + \frac{y^2}{b^2} = 1 \quad \text{where } a > 0 \text{ and } b > 0$$

To graph an ellipse centered at the origin, it is helpful to know the intercepts of the graph. To find the x-intercepts of the graph of

$$\frac{x^2}{a^2} + \frac{y^2}{b^2} = 1$$

we let $y = 0$ and solve for x.

$$\frac{x^2}{a^2} + \frac{0^2}{b^2} = 1 \qquad \text{Substitute 0 for y.}$$

$$\frac{x^2}{a^2} + 0 = 1 \qquad \text{Simplify: } \frac{0^2}{b^2} = 0.$$

$$x^2 = a^2 \qquad \text{Simplify and multiply both sides by } a^2.$$

$$x = \pm a \qquad \text{Use the square root property.}$$

The x-intercepts are $(a, 0)$ and $(-a, 0)$.

To find the y-intercepts of the graph, we can let $x = 0$ and solve for y.

$$\frac{0^2}{a^2} + \frac{y^2}{b^2} = 1 \qquad \text{Substitute 0 for x.}$$

$$0 + \frac{y^2}{b^2} = 1 \qquad \text{Simplify: } \frac{0^2}{a^2} = 0.$$

$$y^2 = b^2 \qquad \text{Simplify and multiply both sides by } b^2.$$

$$y = \pm b \qquad \text{Use the square root property.}$$

The y-intercepts are $(0, b)$ and $(0, -b)$.

In general, we have the following results.

| **The Intercepts of an Ellipse** | The graph of $\frac{x^2}{a^2} + \frac{y^2}{b^2} = 1$ is an ellipse, centered at the origin, with x-intercepts $(a, 0)$ and $(-a, 0)$ and y-intercepts $(0, b)$ and $(0, -b)$. |

For $\frac{x^2}{a^2} + \frac{y^2}{b^2} = 1$, if $a > b$, the ellipse is horizontal, as shown in figure (a). If $b > a$, the ellipse is vertical, as shown in figure (b). The points V_1 and V_2 are called the **vertices** of the ellipse. The line segment joining the vertices is called the **major axis,** and its midpoint is called the **center** of the ellipse. The line segment whose endpoints are on the ellipse and that is perpendicular to the major axis at the center is called the **minor axis** of the ellipse.

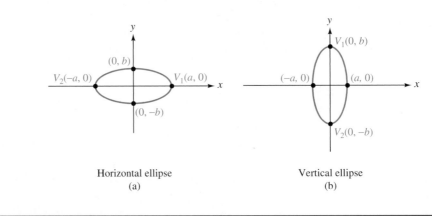

Horizontal ellipse
(a)

Vertical ellipse
(b)

EXAMPLE 1   Graph: $\dfrac{x^2}{36} + \dfrac{y^2}{9} = 1$

Strategy   This equation is in standard $\frac{x^2}{a^2} + \frac{y^2}{b^2} = 1$ form. We will identify a and b.

Why   Once we know a and b, we can determine the intercepts of the graph of the ellipse.

Solution   The color highlighting shows how to compare the given equation to the standard form to find a and b.

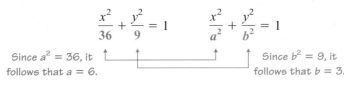

$$\frac{x^2}{36} + \frac{y^2}{9} = 1 \qquad \frac{x^2}{a^2} + \frac{y^2}{b^2} = 1$$

Since $a^2 = 36$, it follows that $a = 6$.   Since $b^2 = 9$, it follows that $b = 3$.

The x-intercepts are $(a, 0)$ and $(-a, 0)$, or $(6, 0)$ and $(-6, 0)$. The y-intercepts are $(0, b)$ and $(0, -b)$, or $(0, 3)$ and $(0, -3)$. Using these four points as a guide, we draw an oval-shaped curve through them, as shown in figure (a). The result is a horizontal ellipse.

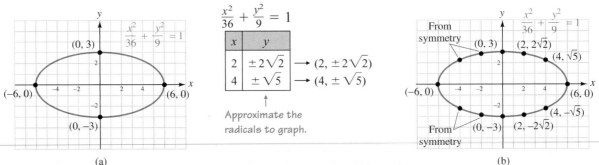

x	y	
2	$\pm 2\sqrt{2}$	$\rightarrow (2, \pm 2\sqrt{2})$
4	$\pm\sqrt{5}$	$\rightarrow (4, \pm\sqrt{5})$

Approximate the radicals to graph.

(a)

(b)

To increase the accuracy of the graph, we can find additional ordered pairs that satisfy the equation and plot them. For example, if $x = 2$, we have

$$\frac{2^2}{36} + \frac{y^2}{9} = 1 \qquad \text{Substitute 2 for } x \text{ in the equation of the ellipse.}$$

$$36\left(\frac{4}{36} + \frac{y^2}{9}\right) = 36(1) \qquad \text{To clear the fractions, multiply both sides by the LCD, 36.}$$

$$4 + 4y^2 = 36 \qquad \text{Distribute the multiplication by 36 and simplify.}$$

$$y^2 = 8 \qquad \text{Subtract 4 from both sides and divide both sides by 4.}$$

$$y = \pm\sqrt{8} \qquad \text{Use the square root property.}$$

$$y = \pm 2\sqrt{2} \qquad \text{Simplify the radical.}$$

Since two values of y, $2\sqrt{2}$ and $-2\sqrt{2}$, correspond to the x-value 2, we have found two points on the ellipse: $\left(2, 2\sqrt{2}\right)$ and $\left(2, -2\sqrt{2}\right)$.

In a similar way, we can find the corresponding values of y for the x-value 4. In figure (b) we record these ordered pairs in a table, plot them, use symmetry with respect to the y-axis to plot four other points, and draw the graph of the ellipse.

Self Check 1   Graph: $\frac{x^2}{49} + \frac{y^2}{25} = 1$

Now Try   **Problem 17**

EXAMPLE 2   Graph: $16x^2 + y^2 = 16$

Strategy   We will write the equation in standard $\frac{x^2}{a^2} + \frac{y^2}{b^2} = 1$ form.

Why   When the equation is in standard form, we will be able to identify the center and the intercepts of the graph of the ellipse.

Solution   The given equation is not in standard form. To write it in standard form with 1 on the right side, we divide both sides by 16.

$$16x^2 + y^2 = 16$$

$$\frac{16x^2}{16} + \frac{y^2}{16} = \frac{16}{16} \qquad \text{Divide both sides by 16.}$$

$$\frac{x^2}{1} + \frac{y^2}{16} = 1 \qquad \text{Simplify: } \frac{16x^2}{16} = x^2 = \frac{x^2}{1} \text{ and } \frac{16}{16} = 1.$$

Success Tip

Although the term $\frac{16x^2}{16}$ simplifies to x^2, we write it as the fraction $\frac{x^2}{1}$ so that it has the form $\frac{x^2}{a^2}$.

To determine a and b, we can write the equation in the form

$$\frac{x^2}{1^2} + \frac{y^2}{4^2} = 1 \qquad \text{To find } a, \text{ write 1 as } 1^2. \text{ To find } b, \text{ write 16 as } 4^2.$$

Since a^2 (the denominator of x^2) is 1^2, it follows that $a = 1$, and since b^2 (the denominator of y^2) is 4^2, it follows that $b = 4$. Thus, the x-intercepts of the graph are $(1, 0)$ and $(-1, 0)$ and the y-intercepts are $(0, 4)$ and $(0, -4)$. We use these four points as guides to sketch the graph of the ellipse, as shown. The result is a vertical ellipse.

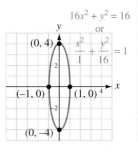

> **Self Check 2** Graph: $9x^2 + y^2 = 9$
>
> **Now Try** Problem 21

3 Graph Ellipses Centered at (h, k).

Not all ellipses are centered at the origin. As with the graphs of circles and parabolas, the graph of an ellipse can be translated horizontally and vertically.

The Equation of an Ellipse Centered at (h, k)

The **standard form of the equation of a horizontal or vertical ellipse** centered at (h, k) is

$$\frac{(x - h)^2}{a^2} + \frac{(y - k)^2}{b^2} = 1 \quad \text{where } a > 0 \text{ and } b > 0$$

For a horizontal ellipse, a is the distance from the center to a vertex. For a vertical ellipse, b is the distance from the center to a vertex.

EXAMPLE 3 Graph: $\dfrac{(x - 2)^2}{16} + \dfrac{(y + 3)^2}{25} = 1$

Strategy The equation is in standard $\dfrac{(x - h)^2}{a^2} + \dfrac{(y - k)^2}{b^2} = 1$ form. We will identify h, k, a, and b.

Why If we know h, k, a, and b, we can graph the ellipse.

Solution To determine h, k, a, and b, we write the equation in the form

$$\frac{(x - 2)^2}{4^2} + \frac{[y - (-3)]^2}{5^2} = 1 \qquad \begin{array}{l}\text{To find } k \text{, write } y + 3 \text{ as } y - (-3).\\ \text{To find } a \text{, write } 16 \text{ as } 4^2. \text{ To find } b \text{, write } 25 \text{ as } 5^2.\end{array}$$

We find the center of the ellipse in the same way we would find the center of a circle, by examining $(x - 2)^2$ and $(y + 3)^2$. Since $h = 2$ and $k = -3$, this is the equation of an ellipse centered at $(h, k) = (2, -3)$. From the denominators, 4^2 and 5^2, we find that $a = 4$ and $b = 5$. Because $b > a$, it is a vertical ellipse.

We first plot the center, as shown below. Since b is the distance from the center to a vertex for a vertical ellipse, we can locate the vertices by counting 5 units above and 5 units below the center. The vertices are the points $(2, 2)$ and $(2, -8)$.

To locate two more points on the ellipse, we use the fact that a is 4 and count 4 units to the left and to the right of the center. We see that the points $(-2, -3)$ and $(6, -3)$ are also on the graph.

Using these four points as guides, we draw the graph shown below.

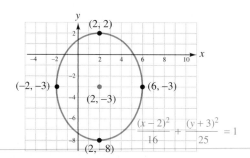

Self Check 3   Graph: $\dfrac{(x-1)^2}{9} + \dfrac{(y+2)^2}{16} = 1$

Now Try   Problem 25

Using Your Calculator

Graphing Ellipses

To use a graphing calculator to graph the equation from Example 3,

$$\frac{(x-2)^2}{16} + \frac{(y+3)^2}{25} = 1$$

we clear the equation of fractions and solve for y.

$$25(x-2)^2 + 16(y+3)^2 = 400 \qquad \text{Multiply both sides by 400.}$$

$$16(y+3)^2 = 400 - 25(x-2)^2 \qquad \begin{array}{l}\text{Subtract } 25(x-2)^2 \text{ from} \\ \text{both sides.}\end{array}$$

$$(y+3)^2 = \frac{400 - 25(x-2)^2}{16} \qquad \text{Divide both sides by 16.}$$

$$y + 3 = \pm\frac{\sqrt{400 - 25(x-2)^2}}{4} \qquad \text{Use the square root property.}$$

$$y = -3 \pm \frac{\sqrt{400 - 25(x-2)^2}}{4} \qquad \text{Subtract 3 from both sides.}$$

The previous equation represents two functions. On a calculator, we can graph them in a square window to get the ellipse shown below.

$$y = -3 + \frac{\sqrt{400 - 25(x-2)^2}}{4} \qquad \text{and} \qquad y = -3 - \frac{\sqrt{400 - 25(x-2)^2}}{4}$$

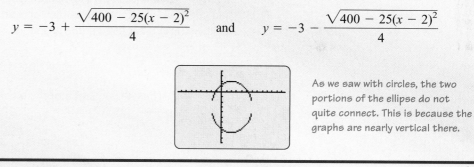

As we saw with circles, the two portions of the ellipse do not quite connect. This is because the graphs are nearly vertical there.

EXAMPLE 4   Graph: $4(x-2)^2 + 9(y-1)^2 = 36$

Strategy   We will write the equation in standard $\dfrac{(x-h)^2}{a^2} + \dfrac{(y-k)^2}{b^2} = 1$ form. Then we will identify h, k, a, and b.

Why   If we know h, k, a, and b, we can graph the ellipse.

Solution   This equation is not in standard form. To write it in standard form with 1 on the right side, we divide both sides by 36.

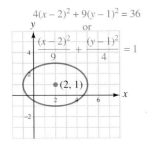

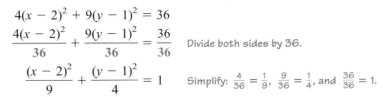

$$4(x - 2)^2 + 9(y - 1)^2 = 36$$

$$\frac{4(x - 2)^2}{36} + \frac{9(y - 1)^2}{36} = \frac{36}{36} \qquad \text{Divide both sides by 36.}$$

$$\frac{(x - 2)^2}{9} + \frac{(y - 1)^2}{4} = 1 \qquad \text{Simplify: } \frac{4}{36} = \frac{1}{9}, \frac{9}{36} = \frac{1}{4}, \text{ and } \frac{36}{36} = 1.$$

This is the standard form of the equation of a horizontal ellipse, centered at (2, 1), with $a = 3$ and $b = 2$. The graph of the ellipse is shown in the margin.

> ▷ **Self Check 4**   Graph: $12(x - 1)^2 + 3(y + 1)^2 = 48$
>
> **Now Try**   **Problem 29**

❹   **Solve Problems Involving Ellipses.**

EXAMPLE 5   *Landscape Design.* A landscape architect is designing an elliptical pool that will fit in the center of a 20-by-30-foot rectangular garden, leaving 5 feet of clearance on all sides, as shown in the illustration below. Find the equation of the ellipse.

Strategy   We will establish a coordinate system with its origin at the center of the garden. Then we will determine the x- and y-intercepts of the edge of the pool.

Why   If we know the x- and y-intercepts of the graph of the edge of the elliptical pool, we can use that information to write its equation.

Solution   We place the rectangular garden in the coordinate system shown below. To maintain 5 feet of clearance at the ends of the ellipse, the x-intercepts must be the points (10, 0) and (−10, 0). Similarly, the y-intercepts are the points (0, 5) and (0, −5).

Since the ellipse is centered at the origin, its equation has the form

$$\frac{x^2}{a^2} + \frac{y^2}{b^2} = 1$$

with $a = 10$ and $b = 5$. Thus, the equation of the boundary of the pool is

$$\frac{x^2}{100} + \frac{y^2}{25} = 1$$

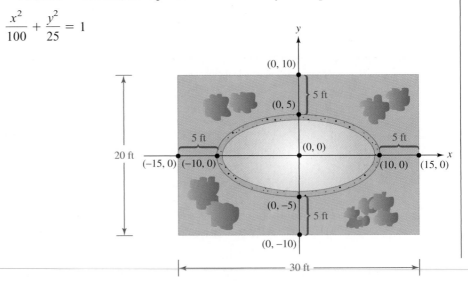

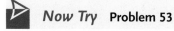

Now Try Problem 53

Ellipses, like parabolas, have reflective properties that are used in many practical applications. For example, any light or sound originating at one focus of an ellipse is reflected by the interior of the figure to the other focus.

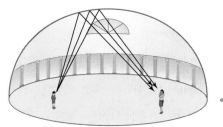

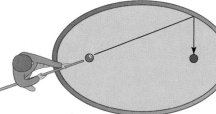

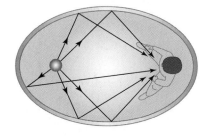

Whispering galleries

In an elliptical dome, even the slightest whisper made by a person standing at one focus can be heard by a person standing at the other focus.

Elliptical billiards tables

When a ball is shot from one focus, it will rebound off the side of the table into a pocket located at the other focus.

Treatment for kidney stones

The patient is positioned in an elliptical tank of water so that the kidney stone is at one focus. High-intensity sound waves generated at another focus are reflected to the stone to shatter it.

ANSWERS TO SELF CHECKS

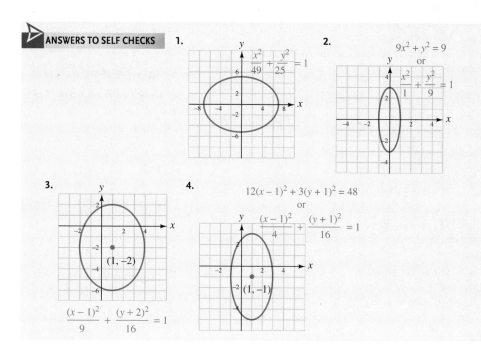

1. $\dfrac{x^2}{49} + \dfrac{y^2}{25} = 1$

2. $9x^2 + y^2 = 9$ or $\dfrac{x^2}{1} + \dfrac{y^2}{9} = 1$

3. $\dfrac{(x-1)^2}{9} + \dfrac{(y+2)^2}{16} = 1$

4. $12(x-1)^2 + 3(y+1)^2 = 48$ or $\dfrac{(x-1)^2}{4} + \dfrac{(y+1)^2}{16} = 1$

STUDY SET
10.2

VOCABULARY

Fill in the blanks.

1. The curve graphed below is an _____.

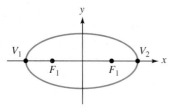

2. An _____ is the set of all points in a plane for which the sum of the distances from two fixed points is a constant.

3. In the graph above, F_1 and F_2 are the _____ of the ellipse. Each one is called a _____ of the ellipse.

4. In the graph above, V_1 and V_2 are the _____ of the ellipse. Each one is called a _____ of the ellipse.

5. The line segment joining the vertices of an ellipse is called the _____ axis of the ellipse.

6. The midpoint of the major axis of an ellipse is the _____ of the ellipse.

CONCEPTS

7. Write the standard form of the equation of an ellipse centered at the origin and symmetric to both axes.

8. Write the standard form of the equation of a horizontal or vertical ellipse centered at (h, k).

9. Find the x- and the y-intercepts of the graph of $\frac{x^2}{a^2} + \frac{y^2}{b^2} = 1$.

10. a. Find the center of the ellipse graphed on the right. What are a and b?

   b. Is the ellipse horizontal or vertical?

   c. Find the equation of the ellipse.

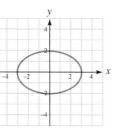

11. a. Find the center of the ellipse graphed on the right. What are a and b?

   b. Is the ellipse horizontal or vertical?

   c. Find the equation of the ellipse.

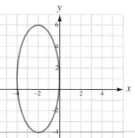

12. Find two points on the graph of $\frac{x^2}{16} + \frac{y^2}{4} = 1$ by letting $x = 2$ and finding the corresponding values of y.

13. Divide both sides of the equation by 64 and write the equation in standard form:

$$4(x - 1)^2 + 64(y + 5)^2 = 64$$

14. Determine whether the graph of each equation is a circle, a parabola, or an ellipse.

   a. $x = y^2 - 2y + 10$

   b. $\frac{x^2}{49} + \frac{y^2}{64} = 1$

   c. $(x - 3)^2 + (y + 4)^2 = 25$

   d. $2(x - 1)^2 + 8(y + 5)^2 = 32$

NOTATION

15. Find h, k, a, and b: $\frac{(x + 8)^2}{100} + \frac{(y - 6)^2}{144} = 1$

16. Write each denominator in the equation $\frac{x^2}{81} + \frac{y^2}{49} = 1$ as the square of a number.

GUIDED PRACTICE

Graph each equation. See Example 1.

17. $\frac{x^2}{25} + \frac{y^2}{4} = 1$ **18.** $\frac{x^2}{16} + \frac{y^2}{9} = 1$

19. $\frac{x^2}{4} + \frac{y^2}{9} = 1$ **20.** $\frac{x^2}{16} + \frac{y^2}{25} = 1$

Graph each equation. See Example 2.

21. $x^2 + 9y^2 = 9$ **22.** $25x^2 + 9y^2 = 225$

23. $16x^2 + 4y^2 = 64$ **24.** $4x^2 + 9y^2 = 36$

Graph each equation. See Example 3.

25. $\frac{(x - 2)^2}{9} + \frac{(y - 1)^2}{4} = 1$ **26.** $\frac{(x - 1)^2}{9} + \frac{(y - 3)^2}{4} = 1$

27. $\frac{(x + 2)^2}{64} + \frac{(y - 2)^2}{100} = 1$ **28.** $\frac{(x - 6)^2}{36} + \frac{(y + 6)^2}{144} = 1$

Graph each equation. See Example 4.

29. $(x + 1)^2 + 4(y + 2)^2 = 4$

30. $25(x + 1)^2 + 9y^2 = 225$

31. $16(x - 2)^2 + 4(y + 4)^2 = 256$

32. $4(x - 2)^2 + 9(y - 4)^2 = 144$

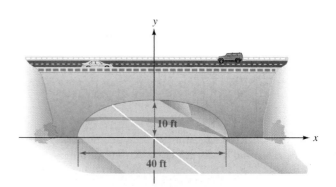

54. DESIGNING AN UNDERPASS   The arch of an underpass is a part of an ellipse. Find the equation of the ellipse.

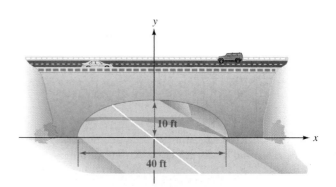

10 ft

40 ft

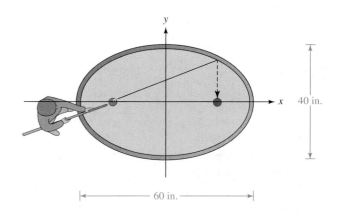

55. CALCULATING CLEARANCE   Find the height of the elliptical arch in Exercise 54 at a point 10 feet from the center of the roadway that passes under the arch.

56. POOL TABLES   Find the equation of the outer edge of the elliptical pool table shown below.

40 in.

60 in.

57. AREA OF AN ELLIPSE   The area A bounded by the following ellipse is given by $A = \pi ab$.

$$\frac{x^2}{a^2} + \frac{y^2}{b^2} = 1$$

Find the area bounded by the ellipse described by $9x^2 + 16y^2 = 144$.

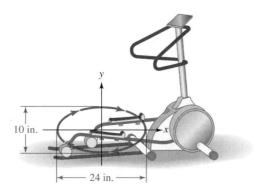

Use a graphing calculator to graph each equation. **See Using Your Calculator: Graphing Ellipses.**

33. $\dfrac{x^2}{9} + \dfrac{y^2}{4} = 1$ **34.** $x^2 + 16y^2 = 16$

35. $\dfrac{x^2}{4} + \dfrac{(y-1)^2}{9} = 1$ **36.** $\dfrac{(x+1)^2}{9} + \dfrac{(y-2)^2}{4} = 1$

TRY IT YOURSELF

Write each equation in standard form, if it is not already so, and graph it. The problems include equations that describe circles, parabolas, and ellipses.

37. $(x+1)^2 + (y-2)^2 = 16$ **38.** $(x-3)^2 + (y+1)^2 = 25$

39. $\dfrac{x^2}{16} + \dfrac{y^2}{1} = 1$ **40.** $\dfrac{x^2}{1} + \dfrac{y^2}{9} = 1$

41. $x = \dfrac{1}{2}(y-1)^2 - 2$ **42.** $x = -\dfrac{1}{2}(y+4)^2 + 5$

43. $x^2 + y^2 - 25 = 0$ **44.** $x^2 = 36 - y^2$

45. $x^2 = 100 - 4y^2$ **46.** $x^2 = 36 - 4y^2$

47. $y = -3x^2 - 24x - 43$ **48.** $y = 5x^2 - 60x + 173$

49. $x^2 + y^2 - 2x + 4y - 4 = 0$

50. $x^2 + y^2 + 4x + 6y + 9 = 0$

51. $9(x-1)^2 + 4(y+2)^2 = 36$

52. $16(x-5)^2 + 25(y-4)^2 = 400$

APPLICATIONS

53. FITNESS EQUIPMENT   With elliptical cross-training equipment, the feet move through the natural elliptical pattern that one experiences when walking, jogging, or running. Write the equation of the elliptical pattern shown below.

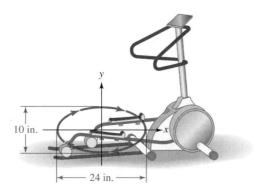

10 in.

24 in.

58. AREA OF A TRACK The elliptical track shown in the figure is bounded by the ellipses $4x^2 + 9y^2 = 576$ and $9x^2 + 25y^2 = 900$. Find the area of the track. (See Exercise 57.)

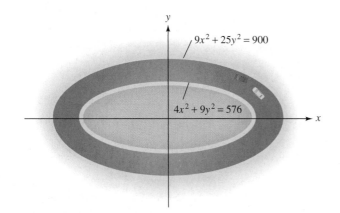

WRITING

59. What is an ellipse?

60. Explain the difference between the focus of an ellipse and the vertex of an ellipse.

61. Compare the graphs of $\frac{x^2}{81} + \frac{y^2}{64} = 1$ and $\frac{x^2}{64} + \frac{y^2}{81} = 1$. Do they have any similarities?

62. What are the reflective properties of an ellipse?

REVIEW

Find each product.

63. $3x^{-2}y^2(4x^2 + 3y^{-2})$

64. $(2a^{-2} - b^{-2})(2a^{-2} + b^{-2})$

Simplify each expression.

65. $\dfrac{x^{-2} + y^{-2}}{x^{-2} - y^{-2}}$

66. $\dfrac{2x^{-3} - 2y^{-3}}{4x^{-3} + 4y^{-3}}$

CHALLENGE PROBLEMS

67. What happens to the graph of the equation $\frac{x^2}{a^2} + \frac{y^2}{b^2} = 1$ when $a = b$?

68. Graph: $9x^2 + 4y^2 = 1$

69. Write the equation $9x^2 + 4y^2 - 18x + 16y = 11$ in the standard form of the equation of an ellipse.

70. Let the foci of an ellipse be $(c, 0)$ and $(-c, 0)$. Suppose that the sum of the distances from any point (x, y) on the ellipse to the two foci is the constant $2a$. Show that the equation for the ellipse is

$$\frac{x^2}{a^2} + \frac{y^2}{a^2 - c^2} = 1$$

Then let $b^2 = a^2 - c^2$ to obtain the standard form of the equation of an ellipse.

SECTION 10.3
The Hyperbola

Objectives

1. Define a hyperbola.
2. Graph hyperbolas centered at the origin.
3. Graph hyperbolas centered at (h, k).
4. Graph equations of the form $xy = k$.
5. Solve problems involving hyperbolas.

The final conic section that we will discuss, the *hyperbola,* is a curve that has two branches. In this section, we will learn how to graph equations that represent hyperbolas.

1 **Define a Hyperbola.**

Ellipses and hyperbolas have completely different shapes, but their definitions are similar. Instead of the *sum* of distances, the definition of a hyperbola involves a *difference* of distances.

Definition of a Hyperbola	A **hyperbola** is the set of all points in a plane for which the difference of the distances from two fixed points is a constant.

The figure below illustrates that any point P on the hyperbola is a constant distance $d_1 - d_2$ from two fixed points, each of which is called a **focus.** Midway between the **foci** is the **center** of the hyperbola.

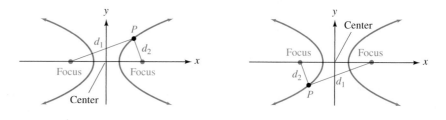

2 **Graph Hyperbolas Centered at the Origin.**

The graph of the equation

$$\frac{x^2}{25} - \frac{y^2}{9} = 1$$

is a hyperbola. To graph the equation, we make a table of solutions that satisfy the equation, plot each point, and join them with a smooth curve.

Caution
Although the two branches of a hyperbola look like parabolas, they are not parabolas.

$$\frac{x^2}{25} - \frac{y^2}{9} = 1$$

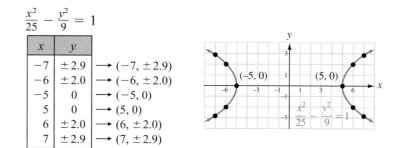

x	y	
-7	± 2.9	$\longrightarrow (-7, \pm 2.9)$
-6	± 2.0	$\longrightarrow (-6, \pm 2.0)$
-5	0	$\longrightarrow (-5, 0)$
5	0	$\longrightarrow (5, 0)$
6	± 2.0	$\longrightarrow (6, \pm 2.0)$
7	± 2.9	$\longrightarrow (7, \pm 2.9)$

This graph is centered at the origin and intersects the x-axis at $(5, 0)$ and $(-5, 0)$. We also note that the graph does not intersect the y-axis.

It is possible to draw a hyperbola without plotting points. For example, if we want to graph the hyperbola with an equation of

$$\frac{x^2}{a^2} - \frac{y^2}{b^2} = 1$$

we first find the x- and y-intercepts. To find the x-intercepts, we let $y = 0$ and solve for x:

$$\frac{x^2}{a^2} - \frac{0^2}{b^2} = 1$$
$$x^2 = a^2$$
$$x = \pm a \quad \text{Use the square root property.}$$

The hyperbola crosses the x-axis at the points $V_1(a, 0)$ and $V_2(-a, 0)$, called the **vertices** of the hyperbola.

To attempt to find the y-intercepts, we let $x = 0$ and solve for y:

$$\frac{0^2}{a^2} - \frac{y^2}{b^2} = 1$$

$$y^2 = -b^2$$

$$y = \pm\sqrt{-b^2}$$

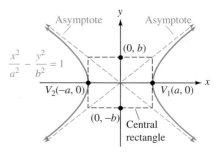

Since b^2 is always positive, $\sqrt{-b^2}$ is an imaginary number. This means that the hyperbola does not intersect the y-axis.

If we construct a rectangle, called the **central rectangle,** whose sides pass horizontally through $\pm b$ on the y-axis and vertically through $\pm a$ on the x-axis, the extended diagonals of the rectangle will be **asymptotes** of the hyperbola. As the hyperbola gets farther away from the origin, its branches get closer and closer to the asymptotes. The asymptotes are not part of the hyperbola, but they serve as a guide when drawing its graph. Since the slopes of the diagonals are $\frac{b}{a}$ and $-\frac{b}{a}$, the equations of the asymptotes are

The Language of Algebra
The central rectangle is also called the *fundamental rectangle.*

$$y = \frac{b}{a}x \quad \text{and} \quad y = -\frac{b}{a}x$$

Standard Form of the Equation of a Hyperbola Centered at the Origin and Intersecting the x-Axis

Any equation that can be written in the form

$$\frac{x^2}{a^2} - \frac{y^2}{b^2} = 1$$

has a graph that is a hyperbola centered at the origin. The x-intercepts are the vertices $V_1(a, 0)$ and $V_2(-a, 0)$. There are no y-intercepts.

The asymptotes of the hyperbola are the extended diagonals of the central rectangle, and their equations are $y = \frac{b}{a}x$ and $y = -\frac{b}{a}x$.

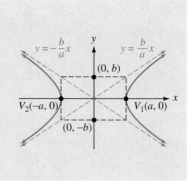

The branches of the hyperbola in previous discussions open to the left and to the right. It is possible for hyperbolas to have different orientations with respect to the x- and y-axes. For example, the branches of a hyperbola can open upward and downward. In that case, the following equation applies.

Standard Form of the Equation of a Hyperbola Centered at the Origin and Intersecting the y-Axis

Any equation that can be written in the form

$$\frac{y^2}{a^2} - \frac{x^2}{b^2} = 1$$

has a graph that is a hyperbola centered at the origin. The y-intercepts are the vertices $V_1(0, a)$ and $V_2(0, -a)$. There are no x-intercepts.

The asymptotes of the hyperbola are the extended diagonals of the central rectangle, and their equations are $y = \frac{a}{b}x$ and $y = -\frac{a}{b}x$.

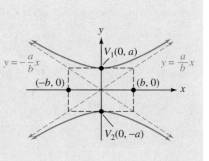

EXAMPLE 1   Graph: $\dfrac{x^2}{9} - \dfrac{y^2}{16} = 1$

Strategy   This equation is in standard $\dfrac{x^2}{a^2} - \dfrac{y^2}{b^2} = 1$ form. We will identify a and b.

Why   We can use a and b to find the vertices of the graph of the hyperbola and the location of the central rectangle.

Solution   The color highlighting shows how to compare the given equation with the standard form to find a and b.

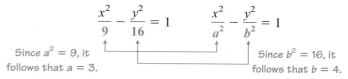

$$\dfrac{x^2}{9} - \dfrac{y^2}{16} = 1 \qquad \dfrac{x^2}{a^2} - \dfrac{y^2}{b^2} = 1$$

Since $a^2 = 9$, it follows that $a = 3$.   Since $b^2 = 16$, it follows that $b = 4$.

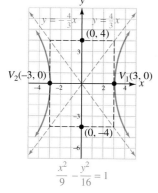

$$\dfrac{x^2}{9} - \dfrac{y^2}{16} = 1$$

This is the standard form of the equation of a hyperbola, centered at the origin, that opens left and right. The x-intercepts are $(a, 0)$ and $(-a, 0)$, or $(3, 0)$ and $(-3, 0)$. They are also the vertices of the hyperbola.

To construct the central rectangle, we use the values of $a = 3$ and $b = 4$. The rectangle passes through $(3, 0)$ and $(-3, 0)$ on the x-axis, and $(0, 4)$ and $(0, -4)$ on the y-axis. We draw extended diagonal dashed lines through the rectangle to obtain the asymptotes and write their equations: $y = \dfrac{4}{3}x$ and $y = -\dfrac{4}{3}x$. Then we draw a smooth curve through each vertex that gets close to the asymptotes.

Self Check 1   Graph: $\dfrac{x^2}{25} - \dfrac{y^2}{4} = 1$
Now Try   **Problem 17**

EXAMPLE 2   Graph: $9y^2 - 4x^2 = 36$

Strategy   We will write the equation in standard $\dfrac{y^2}{a^2} - \dfrac{x^2}{b^2} = 1$ form.

Why   When the equation is in standard form, we will be able to identify the center and the vertices of the graph of the hyperbola and the location of the central rectangle.

Solution   To write the equation in standard form, we divide both sides by 36.

$$9y^2 - 4x^2 = 36$$

$$\dfrac{9y^2}{36} - \dfrac{4x^2}{36} = \dfrac{36}{36} \qquad \text{To get a 1 on the right side, divide both sides by 36.}$$

$$\dfrac{y^2}{4} - \dfrac{x^2}{9} = 1 \qquad \text{Simplify each fraction.}$$

This is the standard form of the equation of a hyperbola, centered at the origin, that opens up and down. The color highlighting shows how we compare the resulting equation to the standard form to find a and b.

> **Success Tip**
> The positive variable term in the standard form equation determines whether a hyperbola is vertical or horizontal. In this example, the positive variable term involves y, so the hyperbola is vertical.
>
>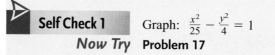
>
> $$\dfrac{y^2}{4} - \dfrac{x^2}{9} = 1$$

$$\dfrac{y^2}{4} - \dfrac{x^2}{9} = 1 \qquad \dfrac{y^2}{a^2} - \dfrac{x^2}{b^2} = 1$$

Since $a^2 = 4$, it follows that $a = 2$.   Since $b^2 = 9$, it follows that $b = 3$.

The y-intercepts are $(0, a)$ and $(0, -a)$, or $(0, 2)$ and $(0, -2)$. They are also the vertices of the hyperbola.

Since $a = 2$ and $b = 3$, the central rectangle passes through $(0, 2)$ and $(0, -2)$, and $(3, 0)$ and $(-3, 0)$. We draw its extended diagonals and sketch the hyperbola.

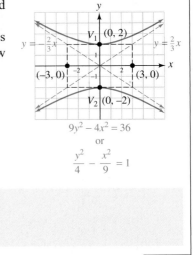

$$9y^2 - 4x^2 = 36$$
or
$$\frac{y^2}{4} - \frac{x^2}{9} = 1$$

> ▷ **Self Check 2**   Graph: $16y^2 - x^2 = 16$
> *Now Try*   **Problem 21**

We can determine whether an equation, when graphed, will be a circle, a parabola, an ellipse, or a hyperbola by examining its variable terms.

$x^2 + y^2 = 16$	With the variable terms on the same side of the equation, we see that the coefficients of the squared terms are the same. The graph is a circle.
$4x^2 + 9y^2 = 144$	With the variable terms on the same side of the equation, we see that the coefficients of the squared terms are different, but have the same sign. The graph is an ellipse.
$4x^2 - 9y^2 = 144$	With the variable terms on the same side of the equation, we see that the coefficients of the squared terms have different signs. The graph is a hyperbola.
$x = y^2 + y - 16$	Since one variable is squared and the other is not, the graph is a parabola.

Using Your Calculator   *Graphing Hyperbolas*

To graph $\frac{x^2}{9} - \frac{y^2}{16} = 1$ from Example 1 using a graphing calculator, we follow the same procedure that we used for circles and ellipses. To write the equation as two functions, we solve for y to get $y = \pm\frac{\sqrt{16x^2 - 144}}{3}$. Then we graph the following two functions in a square window setting to get the graph of the hyperbola shown below.

$$y = \frac{\sqrt{16x^2 - 144}}{3} \quad \text{and} \quad y = -\frac{\sqrt{16x^2 - 144}}{3}$$

Graph Hyperbolas Centered at (h, k)**.**

If a hyperbola is centered at a point with coordinates (h, k), the following equations apply.

Standard Form of the Equation of a Hyperbola Centered at (h, k)	Any equation that can be written in the form $$\frac{(x - h)^2}{a^2} - \frac{(y - k)^2}{b^2} = 1$$ is a hyperbola that has its center at (h, k) and opens left and right. Any equation of the form $$\frac{(y - k)^2}{a^2} - \frac{(x - h)^2}{b^2} = 1$$ is a hyperbola that has its center at (h, k) and opens up and down.

EXAMPLE 3   Graph:

$$\textbf{a.} \quad \frac{(x - 3)^2}{16} - \frac{(y + 1)^2}{4} = 1 \qquad \textbf{b.} \quad \frac{(y - 2)^2}{9} - \frac{(x - 1)^2}{9} = 1$$

Strategy   We will write each equation in a form that makes it easy to identify h, k, a, and b.

Why   If we know h, k, a, and b, we can graph the hyperbola and the central rectangle.

Solution

a. We can write the given equation as

$$\frac{(x - 3)^2}{4^2} - \frac{[y - (-1)]^2}{2^2} = 1$$

To find k, write $y + 1$ as $y - (-1)$.
To find a, write 16 as 4^2. To find b, write 4 as 2^2.

Because the term involving x is positive, the hyperbola opens left and right. We find the center by examining $(x - 3)^2$ and $[y - (-1)]^2$. Since $h = 3$ and $k = -1$, the hyperbola is centered at $(h, k) = (3, -1)$. From the denominators, 4^2 and 2^2, we find that $a = 4$ and $b = 2$. Thus, its vertices are located 4 units to the right and left of the center, at $(7, -1)$ and $(-1, -1)$. Since $b = 2$, we can count 2 units above and below the center to locate points $(3, 1)$ and $(3, -3)$. With these four points, we can draw the central rectangle along with its extended diagonals (the asymptotes). We can then sketch the hyperbola, as shown.

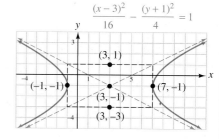

b. We can write the given equation as

$$\frac{(y - 2)^2}{3^2} - \frac{(x - 1)^2}{3^2} = 1$$

Because the term involving y is positive, the hyperbola opens up and down. We find its center by examining $(y - 2)^2$ and $(x - 1)^2$. Since $k = 2$ and $h = 1$, the hyperbola is centered at $(h, k) = (1, 2)$. From the denominators, 3^2 and 3^2, we find that $a = 3$ and $b = 3$, and we use that information to draw the central rectangle and its extended diagonals (the asymptotes), as shown.

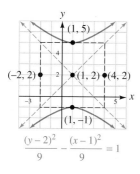

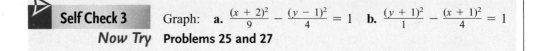

Self Check 3   Graph: **a.** $\dfrac{(x + 2)^2}{9} - \dfrac{(y - 1)^2}{4} = 1$ **b.** $\dfrac{(y + 1)^2}{1} - \dfrac{(x + 1)^2}{4} = 1$

Now Try   **Problems 25 and 27**

4   Graph Equations of the Form $xy = k$.

There is a special type of hyperbola (also centered at the origin) that does not intersect either the x- or the y-axis. These hyperbolas have equations of the form $xy = k$, where $k \neq 0$.

EXAMPLE 4   Graph: $xy = -8$

Strategy   We will make a table of solutions, plot the points, and connect the points with a smooth curve.

Why   Since this equation cannot be written in standard form, we cannot use the methods used in the previous examples.

Solution   To make a table of solutions, we can solve the equation $xy = -8$ for y:

$$y = \dfrac{-8}{x}$$

Then we choose several values for x, find the corresponding values of y, and record the results in the table below. We plot the ordered pairs and join them with a smooth curve to obtain the graph of the hyperbola.

The Language of Algebra
The asymptotes of this hyperbola are the x- and y-axes. A hyperbola for which the asymptotes are perpendicular is called a **rectangular** hyperbola.

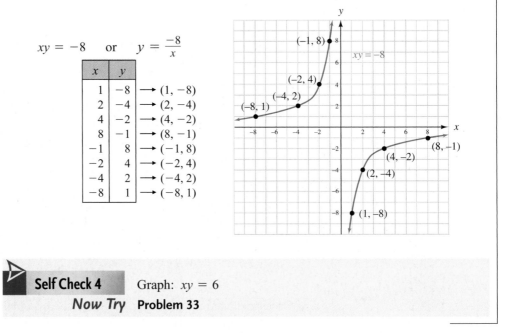

$$xy = -8 \quad \text{or} \quad y = \dfrac{-8}{x}$$

x	y	
1	-8	→ $(1, -8)$
2	-4	→ $(2, -4)$
4	-2	→ $(4, -2)$
8	-1	→ $(8, -1)$
-1	8	→ $(-1, 8)$
-2	4	→ $(-2, 4)$
-4	2	→ $(-4, 2)$
-8	1	→ $(-8, 1)$

Self Check 4   Graph: $xy = 6$

Now Try   **Problem 33**

The result in Example 4 illustrates the following general equation.

| Equations of Hyperbolas of the Form $xy = k$ | Any equation of the form $xy = k$, where $k \neq 0$, has a graph that is a **hyperbola,** which does not intersect either the x- or y-axis. |

5 **Solve Problems Involving Hyperbolas.**

EXAMPLE 5   *Atomic Structure.*   In an experiment that led to the discovery of the atomic structure of matter, Lord Rutherford (1871–1937) shot high-energy alpha particles toward a thin sheet of gold. Many of them were reflected, and Rutherford showed the existence of the nucleus of a gold atom. An alpha particle is repelled by the nucleus at the origin; it travels along the hyperbolic path given by $4x^2 - y^2 = 16$. How close does the particle come to the nucleus?

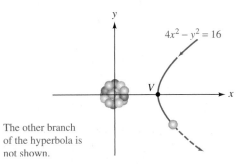

The other branch of the hyperbola is not shown.

Strategy   We will write the equation in standard form and find the coordinates of point V.

Why   The distance from the origin to point V is the closest the particle comes to the nucleus.

Solution   To find the distance from the nucleus at the origin, we must find the coordinates of the vertex V. To do so, we write the equation of the particle's path in standard form:

$$4x^2 - y^2 = 16$$

$$\frac{4x^2}{16} - \frac{y^2}{16} = \frac{16}{16} \qquad \text{Divide both sides by 16.}$$

$$\frac{x^2}{4} - \frac{y^2}{16} = 1 \qquad \text{Simplify.}$$

$$\frac{x^2}{2^2} - \frac{y^2}{4^2} = 1 \qquad \text{To determine } a \text{ and } b, \text{ write 4 as } 2^2 \text{ and 16 as } 4^2.$$

This equation is in the form

$$\frac{x^2}{a^2} - \frac{y^2}{b^2} = 1$$

with $a = 2$. Thus, the vertex of the path is $(2, 0)$. The particle is never closer than 2 units from the nucleus.

Now Try   **Problem 61**

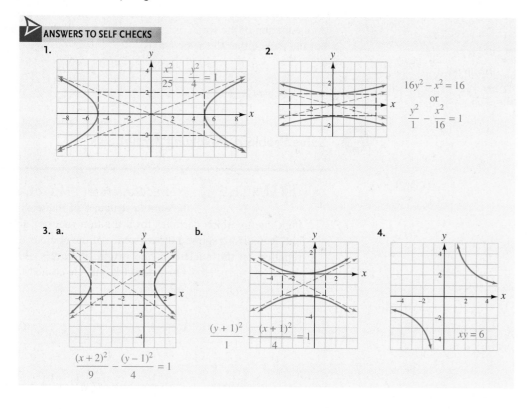

3. a. $\dfrac{(x+2)^2}{9} - \dfrac{(y-1)^2}{4} = 1$

b. $\dfrac{(y+1)^2}{1} - \dfrac{(x+1)^2}{4} = 1$

STUDY SET
10.3

VOCABULARY

Fill in the blanks.

1. The two-branch curve graphed on the right is a _____.

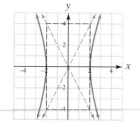

2. A _____ is the set of all points in a plane for which the difference of the distances from two fixed points is a constant.

3. In the graph above, V_1 and V_2 are the _____ of the hyperbola.

4. In the graph above, the figure drawn using dashed black lines is called the _____ _____.

5. The extended _____ of the central rectangle are asymptotes of the hyperbola.

6. To write $9x^2 - 4y^2 = 36$ in _____ form, we divide both sides by 36.

CONCEPTS

7. Write the standard form of the equation of a hyperbola centered at the origin that opens left and right.

8. Write the standard form of the equation of a hyperbola centered at (h, k) that opens up and down.

9. Write the standard form of the equation of a hyperbola centered at (h, k) that opens left and right.

10. a. Find the center of the hyperbola graphed below. What are a and b?

b. Find the x-intercepts of the graph. What are the y-intercepts of the graph?

c. Find the equation of the hyperbola.

d. Find the equations of the asymptotes.

11. a. Find the center of the hyperbola graphed on the right. What are a and b?

    b. Find the equation of the hyperbola.

12. a. Fill in the blank: An equation of the form $xy = k$, where $k \neq 0$, has a graph that is a _____ that does not intersect either the x-axis or the y-axis.

    b. Complete the table of solutions for $xy = 10$.

x	y
-2	
5	

13. Divide both sides of the equation by 100 and write the equation in standard form:

$$100(x + 1)^2 - 25(y - 5)^2 = 100$$

14. Determine whether the graph of the equation will be a circle, a parabola, an ellipse, or a hyperbola.

    a. $x^2 + y^2 = 10$ **b.** $9y^2 - 16x^2 = 144$

    c. $x = y^2 - 3y + 6$ **d.** $4x^2 + 25y^2 = 100$

NOTATION

15. Find h, k, a, and b: $\dfrac{(x - 5)^2}{25} - \dfrac{(y + 11)^2}{36} = 1$

16. Write each denominator in the equation $\dfrac{x^2}{36} - \dfrac{y^2}{81} = 1$ as the square of a number.

GUIDED PRACTICE

Graph each hyperbola. See Example 1.

17. $\dfrac{x^2}{9} - \dfrac{y^2}{4} = 1$ **18.** $\dfrac{x^2}{4} - \dfrac{y^2}{4} = 1$

19. $\dfrac{y^2}{4} - \dfrac{x^2}{9} = 1$ **20.** $\dfrac{y^2}{4} - \dfrac{x^2}{64} = 1$

Graph each hyperbola. See Example 2.

21. $y^2 - 4x^2 = 16$ **22.** $9y^2 - 25x^2 = 225$

23. $25x^2 - y^2 = 25$ **24.** $9x^2 - 4y^2 = 36$

Graph each hyperbola. See Example 3.

25. $\dfrac{(x - 2)^2}{9} - \dfrac{y^2}{16} = 1$ **26.** $\dfrac{(x + 2)^2}{16} - \dfrac{(y - 3)^2}{25} = 1$

27. $\dfrac{(y + 1)^2}{1} - \dfrac{(x - 2)^2}{4} = 1$ **28.** $\dfrac{(y - 2)^2}{4} - \dfrac{(x + 1)^2}{1} = 1$

29. $\dfrac{(x + 1)^2}{9} - \dfrac{(y + 1)^2}{9} = 1$ **30.** $\dfrac{(x - 2)^2}{16} - \dfrac{(y - 1)^2}{16} = 1$

31. $\dfrac{(y - 3)^2}{25} - \dfrac{x^2}{25} = 1$ **32.** $\dfrac{(y - 1)^2}{9} - \dfrac{x^2}{9} = 1$

Graph each equation. See Example 4.

33. $xy = 8$ **34.** $xy = 4$

35. $xy = -10$ **36.** $xy = -12$

Use a graphing calculator to graph each equation. See Using Your Calculator: Graphing Hyperbolas.

37. $\dfrac{x^2}{9} - \dfrac{y^2}{4} = 1$ **38.** $y^2 - 16x^2 = 16$

39. $\dfrac{x^2}{4} - \dfrac{(y - 1)^2}{9} = 1$ **40.** $\dfrac{(y + 1)^2}{9} - \dfrac{(x - 2)^2}{4} = 1$

TRY IT YOURSELF

Write each equation in standard form, if it is not already so, and graph it. The problems include equations that describe circles, parabolas, ellipses, and hyperbolas.

41. $(x + 1)^2 + (y - 2)^2 = 16$ **42.** $(x - 3)^2 + (y + 4)^2 = 1$

43. $9x^2 - 49y^2 = 441$ **44.** $25y^2 - 16x^2 = 400$

45. $4(x + 1)^2 + 9(y + 1)^2 = 36$ **46.** $16x^2 + 25(y - 3)^2 = 400$

47. $4(x + 3)^2 - (y - 1)^2 = 4$ **48.** $(x + 5)^2 - 16y^2 = 16$

49. $xy = -6$ **50.** $xy = 10$

51. $x = \dfrac{1}{2}(y - 1)^2 - 2$ **52.** $x = -\dfrac{1}{4}(y - 3)^2 + 2$

53. $\dfrac{y^2}{25} - \dfrac{(x - 2)^2}{4} = 1$ **54.** $\dfrac{y^2}{36} - \dfrac{(x + 2)^2}{4} = 1$

55. $y = -x^2 + 6x - 4$ **56.** $y = x^2 - 2x + 5$

57. $\dfrac{x^2}{1} + \dfrac{y^2}{36} = 1$ **58.** $\dfrac{x^2}{4} + \dfrac{y^2}{16} = 1$

59. $x^2 + y^2 + 4x - 6y - 23 = 0$

60. $x^2 + y^2 + 8x - 2y - 8 = 0$

APPLICATIONS

61. ALPHA PARTICLES The particle in the illustration on the next page approaches the nucleus at the origin along the path $9y^2 - x^2 = 81$ in the coordinate system shown. How close does the particle come to the nucleus?

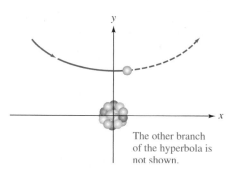

The other branch
of the hyperbola is
not shown.

62. LORAN By determining the difference of the distances between the ship in the illustration and two radio transmitters, the LORAN navigation system places the ship on the hyperbola $x^2 - 4y^2 = 576$ in the coordinate system shown. If the ship is 5 miles out to sea, find its coordinates.

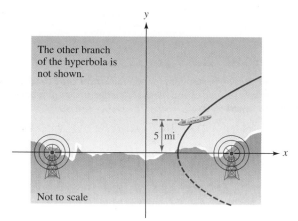

The other branch
of the hyperbola is
not shown.

5 mi

Not to scale

63. SONIC BOOM The position of a sonic boom caused by the faster-than-sound aircraft is one branch of the hyperbola $y^2 - x^2 = 25$ in the coordinate system shown. How wide is the hyperbola 5 miles from its vertex?

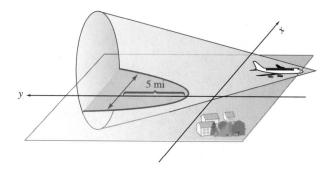

5 mi

64. FLUIDS See the illustration in the next column. Two glass plates in contact at the left, and separated by about 5 millimeters on the right, are dipped in beet juice, which rises by capillary action to form a hyperbola. The hyperbola is modeled by an equation of the form $xy = k$. If the curve passes through the point (12, 2), what is k?

WRITING

65. What is a hyperbola?

66. Compare the graphs of $\frac{x^2}{81} - \frac{y^2}{64} = 1$ and $\frac{y^2}{81} - \frac{x^2}{64} = 1$. Do they have any similarities?

67. Explain how to determine the dimensions of the central rectangle that is associated with the graph of

$$\frac{x^2}{36} - \frac{y^2}{25} = 1$$

68. Explain why the graph of the following hyperbola has no y-intercept.

$$\frac{x^2}{a^2} - \frac{y^2}{b^2} = 1$$

REVIEW

Find each value of x.

69. $\log_8 x = 2$

70. $\log_{25} x = \frac{1}{2}$

71. $\log_{1/2} \frac{1}{8} = x$

72. $\log_{12} x = 0$

73. $\log_x \frac{9}{4} = 2$

74. $\log_6 216 = x$

75. $\log_x 1,000 = 3$

76. $\log_2 \sqrt{2} = x$

CHALLENGE PROBLEMS

77. Write the equation $x^2 - y^2 - 2x + 4y = 12$ in standard form to show that it describes a hyperbola.

78. Write the equation $x^2 - 4y^2 + 2x - 8y = 7$ in standard form to show that it describes a hyperbola.

79. Write the equation $36x^2 - 25y^2 - 72x - 100y = 964$ in standard form to show that it describes a hyperbola.

80. Write an equation of a hyperbola whose graph has the following characteristics:
- vertices $(\pm 1, 0)$
- equations of asymptotes: $y = \pm 5x$

81. Graph: $16x^2 - 25y^2 = 1$

82. Show that the equations of the extended diagonals of the fundamental rectangle of the hyperbola

$$\frac{x^2}{a^2} - \frac{y^2}{b^2} = 1 \quad \text{are} \quad y = \frac{b}{a}x \quad \text{and} \quad y = -\frac{b}{a}x$$

SECTION 10.4
Solving Nonlinear Systems of Equations

Objectives

1. Solve systems by graphing.
2. Solve systems by substitution.
3. Solve systems by elimination (addition).

In Chapter 3, we discussed how to solve systems of linear equations by the graphing, substitution, and elimination methods. In this section, we will use these methods to solve systems where at least one of the equations is nonlinear.

  Solve Systems by Graphing.

A solution of a **nonlinear system of equations** is an ordered pair of real numbers that satisfies all of the equations in the system. The **solution set of a nonlinear system** is the set of all such ordered pairs. One way to solve a system of two equations in two variables is to graph the equations on the same rectangular coordinate system.

| **EXAMPLE 1** | Solve $\begin{cases} x^2 + y^2 = 25 \\ 2x + y = 10 \end{cases}$ by graphing. |

Strategy We will graph both equations on the same coordinate system.

Why If the equations are graphed on the same coordinate system, we can see whether they have any common solutions.

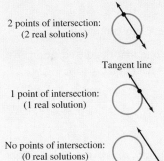
Solution The graph of $x^2 + y^2 = 25$ is a circle with center at the origin and radius of 5. The graph of $2x + y = 10$ is a line. Depending on whether the line is a **secant** (intersecting the circle at two points) or a **tangent** (intersecting the circle at one point) or does not intersect the circle at all, there are two, one, or no solutions to the system, respectively.

After graphing the circle and the line, it appears that the points of intersection are $(5, 0)$ and $(3, 4)$. To verify that they are solutions of the system, we need to check each one.

Check:	**For (5, 0)**		**For (3, 4)**	
	$2x + y = 10$	$x^2 + y^2 = 25$	$2x + y = 10$	$x^2 + y^2 = 25$
	$2(5) + 0 \overset{?}{=} 10$	$5^2 + 0^2 \overset{?}{=} 25$	$2(3) + 4 \overset{?}{=} 10$	$3^2 + 4^2 \overset{?}{=} 25$
	$10 = 10$	$25 = 25$	$10 = 10$	$25 = 25$
	True	True	True	True

The ordered pair $(5, 0)$ satisfies both equations of the system, and so does $(3, 4)$. Thus, there are two solutions, $(5, 0)$ and $(3, 4)$, and the solution set is $\{(5, 0), (3, 4)\}$.

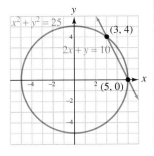

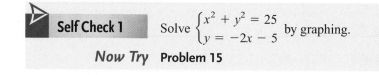

Self Check 1   Solve $\begin{cases} x^2 + y^2 = 25 \\ y = -2x - 5 \end{cases}$ by graphing.

Now Try   **Problem 15**

Using Your Calculator

Solving Systems of Equations

To solve Example 1 with a graphing calculator, we graph the circle and the line on one set of coordinate axes. See figure (a). We then trace to find the coordinates of the intersection points of the graphs. See figures (b) and (c).

We can zoom for better results.

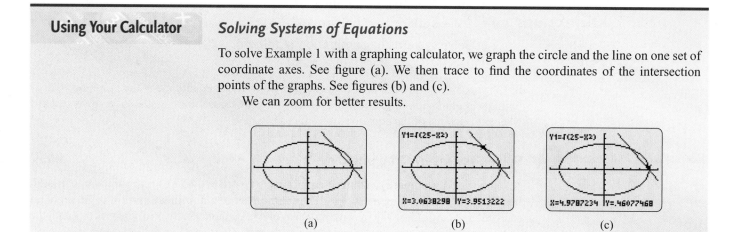

(a)   (b)   (c)

---

2 **Solve Systems by Substitution.**

When solving a system by graphing, it is often difficult to determine the coordinates of the intersection points. A more precise algebraic method called the **substitution method** can be used to solve certain systems involving nonlinear equations.

EXAMPLE 2   Solve $\begin{cases} x^2 + y^2 = 2 \\ 2x - y = 1 \end{cases}$ by substitution.

Strategy   We will solve the second equation for y and substitute the result for y in the first equation.

Why   We can solve the resulting equation for x and then back substitute to find y.

Solution   This system has one second-degree equation and one first-degree equation. We can solve this type of system by substitution. Solving the linear equation for y gives

$$2x - y = 1$$
$$-y = -2x + 1 \quad \text{Subtract 2x from both sides.}$$
$$y = 2x - 1 \quad \text{Multiply both sides by } -1. \text{ We call this the substitution equation.}$$

Because y and $2x - 1$ are equal, we can substitute $2x - 1$ for y in the first equation of the system.

$$y = \boxed{2x - 1} \qquad\qquad x^2 + y^2 = 2$$

Success Tip
With this method, the objective is to use an appropriate substitution to obtain *one* equation in *one* variable.

Then we solve the resulting quadratic equation for x.

$$x^2 + y^2 = 2$$
$$x^2 + (2x - 1)^2 = 2 \quad \text{Substitute 2x − 1 for y.}$$
$$x^2 + 4x^2 - 4x + 1 = 2 \quad \text{Use a special-product rule to find } (2x - 1)^2.$$

$$5x^2 - 4x - 1 = 0 \qquad \text{To get 0 on the right side, subtract 2 from both sides and then combine like terms.}$$

$$(5x + 1)(x - 1) = 0 \qquad \text{Factor.}$$

$$5x + 1 = 0 \quad \text{or} \quad x - 1 = 0 \qquad \text{Set each factor equal to 0.}$$

$$x = -\frac{1}{5} \qquad\qquad x = 1$$

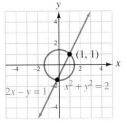

If we substitute $-\frac{1}{5}$ for x in the equation $y = 2x - 1$, we get $y = -\frac{7}{5}$. If we substitute 1 for x in $y = 2x - 1$, we get $y = 1$. Thus, the system has two solutions, $\left(-\frac{1}{5}, -\frac{7}{5}\right)$ and $(1, 1)$. Verify that each ordered pair satisfies both equations of the original system.

The graph in the margin confirms that the system has two solutions, and that one of them is $(1, 1)$. However, it would be virtually impossible to determine from the graph that the coordinates of the second point of intersection are $\left(-\frac{1}{5}, -\frac{7}{5}\right)$.

Self Check 2   Solve $\begin{cases} x^2 + y^2 = 10 \\ y = x + 2 \end{cases}$ by substitution.

Now Try   **Problem 23**

EXAMPLE 3   Solve: $\begin{cases} 4x^2 + 9y^2 = 5 \\ y = x^2 \end{cases}$

Strategy   Since $y = x^2$, we will substitute y for x^2 in the first equation.

Why   This will give an equation in one variable that we can solve for y. We can then find x by back substitution.

Solution   We can solve this system by substitution.

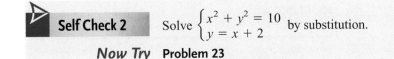

When we substitute y for x^2 in the first equation, the result is a quadratic equation in y.

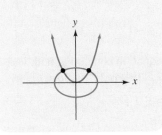

$$4x^2 + 9y^2 = 5$$
$$4y + 9y^2 = 5 \qquad \text{Substitute } y \text{ for } x^2.$$
$$9y^2 + 4y - 5 = 0 \qquad \text{To get 0 on the right side, subtract 5 from both sides.}$$
$$(9y - 5)(y + 1) = 0 \qquad \text{Factor } 9y^2 + 4y - 5.$$
$$9y - 5 = 0 \quad \text{or} \quad y + 1 = 0 \qquad \text{Set each factor equal to 0.}$$
$$y = \frac{5}{9} \qquad\qquad y = -1$$

Since $y = x^2$, the values of x are found by solving the equations

$$x^2 = \frac{5}{9} \quad \text{or} \quad x^2 = -1$$

Because $x^2 = -1$ has no real solutions, this possibility is discarded. The solutions of $x^2 = \frac{5}{9}$ are

$$x = \sqrt{\frac{5}{9}} = \frac{\sqrt{5}}{\sqrt{9}} = \frac{\sqrt{5}}{3} \quad \text{or} \quad x = -\sqrt{\frac{5}{9}} = -\frac{\sqrt{5}}{\sqrt{9}} = -\frac{\sqrt{5}}{3}$$

Thus, the solutions of the system are

$$\left(\frac{\sqrt{5}}{3}, \frac{5}{9}\right) \quad \text{and} \quad \left(-\frac{\sqrt{5}}{3}, \frac{5}{9}\right)$$

> **Self Check 3** Solve: $\begin{cases} x^2 + y^2 = 20 \\ y = x^2 \end{cases}$
>
> **Now Try** Problem 27

③ Solve Systems by Elimination (Addition).

Another method for solving nonlinear system of equations is the **elimination** or **addition method.** With this method, we combine the equations in a way that will eliminate the terms of one of the variables.

EXAMPLE 4 Solve: $\begin{cases} 3x^2 + 2y^2 = 36 \\ 4x^2 - y^2 = 4 \end{cases}$

Strategy We will multiply both sides of the second equation by 2 and add the result to the first equation.

Why This will eliminate the y^2-terms and produce an equation that we can solve for x.

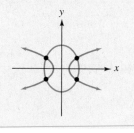

Solution To solve this system of two second-degree equations, we can use either the substitution or the elimination method. We will use the elimination method because the y^2-terms can be eliminated by multiplying the second equation by 2 and adding it to the first equation.

$$\begin{cases} 3x^2 + 2y^2 = 36 \\ 4x^2 - y^2 = 4 \end{cases} \xrightarrow[\text{Multiply by 2}]{\text{Unchanged}} \begin{cases} 3x^2 + 2y^2 = 36 \\ 8x^2 - 2y^2 = 8 \end{cases}$$

We add the two equations on the right to eliminate y^2 and solve the resulting equation for x:

$$11x^2 = 44$$
$$x^2 = 4$$
$$x = 2 \quad \text{or} \quad x = -2$$

To find y, we can substitute 2 for x and then -2 for x into any equation containing both variables. It appears that the computations will be simplest if we use $3x^2 + 2y^2 = 36$.

For x = 2	*For x = −2*
$3x^2 + 2y^2 = 36$	$3x^2 + 2y^2 = 36$
$3(2)^2 + 2y^2 = 36$	$3(-2)^2 + 2y^2 = 36$
$12 + 2y^2 = 36$	$12 + 2y^2 = 36$
$2y^2 = 24$	$2y^2 = 24$
$y^2 = 12$	$y^2 = 12$
$y = \sqrt{12}$ or $y = -\sqrt{12}$	$y = \sqrt{12}$ or $y = -\sqrt{12}$
$y = 2\sqrt{3}$	$y = -2\sqrt{3}$	$y = 2\sqrt{3}$	$y = -2\sqrt{3}$

The four solutions of this system are

$$\left(2, 2\sqrt{3}\right), \qquad \left(2, -2\sqrt{3}\right), \qquad \left(-2, 2\sqrt{3}\right), \qquad \text{and} \qquad \left(-2, -2\sqrt{3}\right)$$

Self Check 4   Solve: $\begin{cases} x^2 + 4y^2 = 16 \\ x^2 - y^2 = 1 \end{cases}$

Now Try   **Problem 31**

ANSWERS TO SELF CHECKS   **1.** $(-4, 3), (0, -5)$

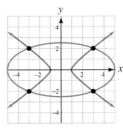

$x^2 + y^2 = 25$
$(-4, 3)$
$(0, -5)$
$y = -2x - 5$

2. $(1, 3), (-3, -1)$
3. $(2, 4), (-2, 4)$
4. $\left(2, \sqrt{3}\right),$
$\left(2, -\sqrt{3}\right),$
$\left(-2, \sqrt{3}\right),$
$\left(-2, -\sqrt{3}\right)$

STUDY SET
10.4

VOCABULARY

Fill in the blanks.

1. $\begin{cases} 4x^2 + 6y^2 = 24 \\ 9x^2 - y^2 = 9 \end{cases}$ is a _____ of two nonlinear equations.

2. The graph of $2x + y = 10$ is a _____ and the graph of $x^2 + y^2 = 25$ is a _____.

3. When solving a system by graphing, it is often difficult to determine the coordinates of the points of _____ of the graphs.

4. Two algebraic methods for solving systems of nonlinear equations are the _____ method and the _____ method.

5. A _____ is a line that intersects a circle at two points.

6. A _____ is a line that intersects a circle at one point.

CONCEPTS

7. a. A line can intersect an ellipse in at most _____ points.
 b. An ellipse can intersect a parabola in at most _____ points.
 c. An ellipse can intersect a circle in at most _____ points.
 d. A hyperbola can intersect a circle in at most _____ points.

8. Determine whether $(1, -1)$ is a solution of the system $\begin{cases} 2x + y - 1 = 0 \\ x^2 - y^2 = 3 \end{cases}$.

9. Find the solutions of the system $\begin{cases} x^2 + 4y^2 = 25 \\ x^2 - 2y^2 = 1 \end{cases}$ that is graphed on the right.

10. Find a substitution equation that can be used to solve the system $\begin{cases} x^2 + y^2 = 9 \\ 2x - y = 3 \end{cases}$.

11. Consider the system $\begin{cases} 6x^2 + y^2 = 9 \\ 3x^2 + 4y^2 = 36 \end{cases}$.
 a. If the y^2-terms are to be eliminated, by what should the first equation be multiplied?
 b. If the x^2-terms are to be eliminated, by what should the second equation be multiplied?

12. Suppose you begin to solve the system $\begin{cases} x^2 + y^2 = 10 \\ 4x^2 + y^2 = 13 \end{cases}$ and find that x is ± 1. Use the first equation to find the corresponding y-values for $x = 1$ and $x = -1$. State the solutions as ordered pairs.

NOTATION

Complete each solution to solve the system.

13. Solve: $\begin{cases} x^2 + y^2 = 5 \\ y = 2x \end{cases}$

$x^2 + y^2 = 5$ *This is the first equation.*

$x^2 + ()^2 = 5$

$x^2 + 4x^2 = $

$x^2 = 5$

$x^2 = $

$x = 1 \quad$ or $\quad x = -1$

If $x = 1$, then $y = 2() = 2$.   *Use the second equation.*

If $x = -1$, then $y = 2() = -2$.

The solutions are $(1, 2)$ and $\left(-1, \right)$.

14. Solve: $\begin{cases} y = x^2 + 2 \\ y = -x^2 + 4 \end{cases}$

$2y = $ *Add the equations.*

$y = $

If $y = 3$, then

$ = x^2 + 2$ *This is the first equation.*

$1 = x^2$

$1 = x$

The solutions are

$\left(1, \right)$ and $\left(, 3\right)$

GUIDED PRACTICE

Solve each system of equations by graphing. See Example 1.

15. $\begin{cases} x^2 + y^2 = 9 \\ y - x = 3 \end{cases}$

16. $\begin{cases} x^2 + y^2 = 16 \\ y - x = -4 \end{cases}$

17. $\begin{cases} 9x^2 + 16y^2 = 144 \\ 9x^2 - 16y^2 = 144 \end{cases}$

18. $\begin{cases} x^2 + 9y^2 = 9 \\ 9y^2 - x^2 = 9 \end{cases}$

19. $\begin{cases} y = x^2 - 4x \\ x^2 + y = 0 \end{cases}$

20. $\begin{cases} x^2 - y = 0 \\ y = -x^2 + 4x \end{cases}$

21. $\begin{cases} x^2 + 4y^2 = 4 \\ x = 2y^2 - 2 \end{cases}$

22. $\begin{cases} 4x^2 + y^2 = 4 \\ y = 2x^2 - 2 \end{cases}$

Solve each system of equations by substitution for real values of x and y. See Examples 2 and 3.

23. $\begin{cases} x^2 + y^2 = 5 \\ x + y = 3 \end{cases}$

24. $\begin{cases} x^2 - x - y = 2 \\ 4x - 3y = 0 \end{cases}$

25. $\begin{cases} y = x^2 + 6x + 7 \\ 2x + y = -5 \end{cases}$

26. $\begin{cases} 2x + y = 1 \\ x^2 + y = 4 \end{cases}$

27. $\begin{cases} x^2 + y^2 = 13 \\ y = x^2 - 1 \end{cases}$

28. $\begin{cases} x^2 + y^2 = 10 \\ y = 3x^2 \end{cases}$

29. $\begin{cases} x^2 + y^2 = 30 \\ y = x^2 \end{cases}$

30. $\begin{cases} x^2 + y^2 = 20 \\ y = x^2 \end{cases}$

Solve each system of equations by elimination for real values of x and y. See Example 4.

31. $\begin{cases} x^2 + y^2 = 20 \\ x^2 - y^2 = -12 \end{cases}$

32. $\begin{cases} x^2 + y^2 = 13 \\ x^2 - y^2 = 5 \end{cases}$

33. $\begin{cases} 9x^2 - 7y^2 = 81 \\ x^2 + y^2 = 9 \end{cases}$

34. $\begin{cases} x^2 + y^2 = 25 \\ 2x^2 - 3y^2 = 5 \end{cases}$

35. $\begin{cases} 2x^2 + y^2 = 6 \\ x^2 - y^2 = 3 \end{cases}$

36. $\begin{cases} x^2 + y^2 = 36 \\ 49x^2 + 36y^2 = 1{,}764 \end{cases}$

37. $\begin{cases} x^2 - y^2 = -5 \\ 3x^2 + 2y^2 = 30 \end{cases}$

38. $\begin{cases} 6x^2 + 8y^2 = 182 \\ 8x^2 - 3y^2 = 24 \end{cases}$

Solve each system. See Using Your Calculator: Solving Systems of Equations.

39. $\begin{cases} x^2 - 6x - y = -5 \\ x^2 - 6x + y = -5 \end{cases}$

40. $\begin{cases} x^2 - y^2 = -5 \\ 3x^2 + 2y^2 = 30 \end{cases}$

TRY IT YOURSELF

Solve each system of equations for real values of x and y.

41. $\begin{cases} 2x^2 - 3y^2 = 5 \\ 3x^2 + 4y^2 = 16 \end{cases}$

42. $\begin{cases} 2x^2 - y^2 + 2 = 0 \\ 3x^2 - 2y^2 + 5 = 0 \end{cases}$

43. $\begin{cases} y = x^2 - 4 \\ x^2 - y^2 = -16 \end{cases}$

44. $\begin{cases} y - x = 0 \\ 4x^2 + y^2 = 10 \end{cases}$

45. $\begin{cases} 3y^2 = xy \\ 2x^2 + xy - 84 = 0 \end{cases}$

46. $\begin{cases} x^2 + y^2 = 10 \\ 2x^2 - 3y^2 = 5 \end{cases}$

47. $\begin{cases} y^2 = 40 - x^2 \\ y = x^2 - 10 \end{cases}$

48. $\begin{cases} 25x^2 + 9y^2 = 225 \\ 5x + 3y = 15 \end{cases}$

49. $\begin{cases} 3x - y = -3 \\ 25y^2 - 9x^2 = 225 \end{cases}$ **50.** $\begin{cases} x - 2y = 2 \\ 9x^2 - 4y^2 = 36 \end{cases}$

51. $\begin{cases} x^2 - y = 0 \\ x^2 - 4x + y = 0 \end{cases}$ **52.** $\begin{cases} xy = -\dfrac{9}{2} \\ 3x + 2y = 6 \end{cases}$

53. $\begin{cases} x^2 - 2y^2 = 6 \\ x^2 + 2y^2 = 2 \end{cases}$ **54.** $\begin{cases} x^2 + 9y^2 = 1 \\ x^2 - 9y^2 = 3 \end{cases}$

55. $\begin{cases} y = x^2 - 4 \\ 6x - y = 13 \end{cases}$ **56.** $\begin{cases} y = x + 1 \\ x^2 - y^2 = 1 \end{cases}$

57. $\begin{cases} x^2 + y^2 = 4 \\ 9x^2 + y^2 = 9 \end{cases}$

58. $\begin{cases} 2x^2 - 6y^2 + 3 = 0 \\ 4x^2 + 3y^2 = 4 \end{cases}$

59. $\begin{cases} xy = \dfrac{1}{6} \\ y + x = 5xy \end{cases}$ **60.** $\begin{cases} xy = \dfrac{1}{12} \\ y + x = 7xy \end{cases}$

61. $\begin{cases} x^2 = 4 - y \\ y = x^2 + 2 \end{cases}$ **62.** $\begin{cases} 3x + 2y = 10 \\ y = x^2 - 5 \end{cases}$

63. $\begin{cases} x^2 - y^2 = 4 \\ x + y = 4 \end{cases}$ **64.** $\begin{cases} x - y = -1 \\ y^2 - 4x = 0 \end{cases}$

APPLICATIONS

Use a nonlinear system of equations to solve each problem.

65. INTEGER PROBLEM   The product of two integers is 32, and their sum is 12. Find the integers.

66. NUMBER PROBLEM   The sum of the squares of two numbers is 221, and the sum of the numbers is 9. Find the numbers.

67. ARCHERY   See the illustration. An arrow shot from the base of a hill follows the parabolic path $y = -\frac{1}{6}x^2 + 2x$, with distances measured in meters. The inclined hill has a slope of $\frac{1}{3}$ and can therefore be modeled by the equation $y = \frac{1}{3}x$. Find the coordinates of the point of impact of the arrow and then its distance from the archer.

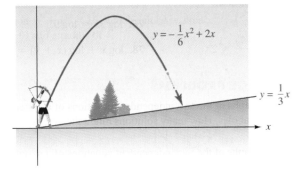

68. GEOMETRY   The area of a rectangle is 63 square centimeters, and its perimeter is 32 centimeters. Find the dimensions of the rectangle.

69. FENCING PASTURES   The rectangular pasture shown here is to be fenced in along a riverbank. If 260 feet of fencing is to enclose an area of 8,000 square feet, find the dimensions of the pasture.

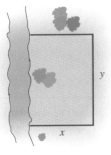

70. DRIVING RATES   Jim drove 306 miles. Jim's brother made the same trip at a speed 17 mph slower than Jim did and required an extra $1\frac{1}{2}$ hours. What was Jim's rate and time?

71. INVESTING   Grant receives $225 annual income from one investment. Jeff invested $500 more than Grant, but at an annual rate of 1% less. Jeff's annual income is $240. What are the amount and rate of Grant's investment?

72. INVESTING   Carol receives $67.50 annual income from one investment. John invested $150 more than Carol at an annual rate of $1\frac{1}{2}$% more. John's annual income is $94.50. What are the amount and rate of Carol's investment?   (*Hint:* There are two answers.)

WRITING

73. a. Describe the benefits of the graphical method for solving a system of equations.

 b. Describe the drawbacks of the graphical method.

Exploring Careers

Ultimately, your choice of career will determine the math course(s) that you need to take after Intermediate Algebra. Before the end of this term, it would be wise to have at least a general idea of your career goals.

HOW DO YOU DECIDE?:   Seek the advice of a counselor, visit your school's career center, search the Internet, or read books that will help you discover your interests and possible related careers.

ONCE YOU'VE DECIDED: Talk to your counselor and consult the appropriate college catalogs to develop a long-term plan that will put you on the correct educational path.

Now Try This

1. Do you have a career goal in mind? If so, what is it?
2. Take at least two personality tests and two career-choice tests. A list of tests offered online can be found at http://academic.cengage.com/math/tussy/.
3. Visit a counselor to discuss which classes you should take during your next term and beyond. Make a list of classes that your counselor suggests that you take.

SECTION 11.1
The Binomial Theorem

Objectives

1. Raise binomials to powers.
2. Use Pascal's triangle to expand binomials.
3. Use factorial notation.
4. Use the binomial theorem to expand binomials.
5. Find a specific term of a binomial expansion.

We have discussed how to raise binomials to positive-integer powers. For example, we have seen that

The Language of Algebra
Recall that two-term polynomial expressions such as $a + b$ and $3u - 2v$ are called *binomials*.

$$(a + b)^2 = a^2 + 2ab + b^2$$

and that $(a + b)^3 = (a + b)(a + b)^2$

$$= (a + b)(a^2 + 2ab + b^2)$$
$$= a^3 + 2a^2b + ab^2 + a^2b + 2ab^2 + b^3$$
$$= a^3 + 3a^2b + 3ab^2 + b^3$$

In this section, we will learn how to raise binomials to positive-integer powers without performing the multiplications.

1 Raise Binomials to Powers.

To see how to raise binomials to nonnegative-integer powers, we consider the following binomial expansions of $a + b$.

The Language of Algebra
When we expand a power of a binomial, the result is called a *binomial expansion*. For powers greater than or equal to 2, an expansion has more terms than the original binomial.

$(a + b)^0 =$	1	1 term
$(a + b)^1 =$	$a + b$	2 terms
$(a + b)^2 =$	$a^2 + 2ab + b^2$	3 terms
$(a + b)^3 =$	$a^3 + 3a^2b + 3ab^2 + b^3$	4 terms
$(a + b)^4 =$	$a^4 + 4a^3b + 6a^2b^2 + 4ab^3 + b^4$	5 terms
$(a + b)^5 =$	$a^5 + 5a^4b + 10a^3b^2 + 10a^2b^3 + 5ab^4 + b^5$	6 terms
$(a + b)^6 =$	$a^6 + 6a^5b + 15a^4b^2 + 20a^3b^3 + 15a^2b^4 + 6ab^5 + b^6$	7 terms

Several patterns appear in these expansions:

1. Each expansion has one more term than the power of the binomial.

2. For each term of an expansion, the sum of the exponents on a and b is equal to the exponent of the binomial being expanded. For example, in the expansion of $(a + b)^5$, the sum of the exponents in each term is 5:

$$(a + b)^5 = a^5 \quad + \quad \overset{4+1=5}{5a^4b} \quad + \quad \overset{3+2=5}{10a^3b^2} \quad + \quad \overset{2+3=5}{10a^2b^3} \quad + \quad \overset{1+4=5}{5ab^4} \quad + \quad b^5$$

3. The first term in each expansion is a, raised to the power of the binomial, and the last term in each expansion is b, raised to the power of the binomial.

4. The exponents on a decrease by one in each successive term, ending with $a^0 = 1$ in the last term. The exponents on b, beginning with $b^0 = 1$ in the first term, increase by one in each successive term. For example, the expansion of $(a + b)^4$ could be written as

$$a^4b^0 + 4a^3b^1 + 6a^2b^2 + 4a^1b^3 + a^0b^4$$

Thus, the variables have the pattern

$$a^n, \quad a^{n-1}b, \quad a^{n-2}b^2, \quad \dots, \quad ab^{n-1}, \quad b^n$$

5. The coefficients of each expansion begin with 1, increase through some values, and then decrease through those same values, back to 1.

2 Use Pascal's Triangle to Expand Binomials.

To see another pattern, we write the coefficients of each expansion of $a + b$ in a triangular array:

The Language of Algebra
This array of numbers is named *Pascal's triangle* in honor of the French mathematician Blaise Pascal (1623–1662).

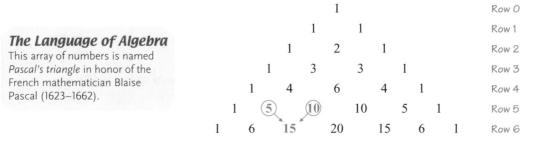

					1							Row 0
				1		1						Row 1
			1		2		1					Row 2
		1		3		3		1				Row 3
	1		4		6		4		1			Row 4
1		5		10		10		5		1		Row 5
1		6		15		20		15		6	1	Row 6

STUDY SET
11.1

VOCABULARY

Fill in the blanks.

1. The two-term polynomial expression $a + b$ is called a _____.

2. $a^4 + 4a^3b + 6a^2b^2 + 4ab^3 + b^4$ is the binomial _____ of $(a + b)^4$.

3. We can use the _____ theorem to raise binomials to positive-integer powers without doing the actual multiplication.

4. The array of numbers that gives the coefficients of the terms of a binomial expansion is called _____ triangle.

5. $n!$ (read as "n _____") is the product of consecutively _____ natural numbers from n to 1.

6. In the expansion $a^3 - 3a^2b + 3ab^2 - b^3$, the signs _____ between $+$ and $-$.

CONCEPTS

Fill in the blanks.

7. The binomial expansion of $(m + n)^6$ has ___ more term than the power of the binomial.

8. For each term of the expansion of $(a + b)^8$, the sum of the exponents of a and b is ___.

9. The first term of the expansion of $(r + s)^{20}$ is r ___ and the last term is s ___.

10. In the expansion of $(m - n)^{15}$, the exponents on m _____ and the exponents on n _____.

11. The coefficients of the terms of the expansion of $(c + d)^{20}$ begin with ___, increase through some values, and then decrease through those same values, back to ___.

12. Complete Pascal's Triangle:

```
                1
            1       1
         1      2
      1             3     1
   1         6          4     1
1      5        10    10      5     1
  1          15           15     6    1
    7     21      35         21     7    1
1    8    28     56     70     56         8
```

13. $n \cdot ($ ___ $-$ ___ $)! = n!$ **14.** $8! = 8 \cdot$ ___ $!$

15. $0! =$ ___

16. According to the binomial theorem, the third term of the expansion of $(a + b)^n$ is $\dfrac{!}{\,!(n - 2)!}a^{n-2}b$.

17. The coefficient of the fourth term of the expansion of $(a + b)^9$ is $9!$ divided by $3!($ ___ $-$ ___ $)!$.

18. The exponent on a in the fourth term of the expansion of $(a + b)^6$ is ___ and the exponent on b is ___.

19. The exponent on a in the fifth term of the expansion of $(a + b)^6$ is ___ and the exponent on b is ___.

20. The expansion of $(a - b)^4$ is

$$a^4 \quad 4a^3b \quad 6a^2b^2 \quad 4ab^3 \quad b^4$$

21. $(x + y)^3$

$$= x \quad + \quad \frac{!}{1!(3 - 1)!}x^2 \quad + \quad \frac{3!}{\,!(3 - 2)!}xy \quad + y$$

22. Fill in the blanks.

a. The $(r + 1)$st term of the expansion of $(a + b)^n$ is $\dfrac{n!}{r!(n - \ )!}a^{\ -r}b$.

b. To use this formula to find the 6th term of the expansion of $\left(m + \dfrac{n}{2}\right)^8$, we note that $r =$ ___ , $n =$ ___ , $a =$ ___ , and $b =$ ___ .

NOTATION

Fill in the blanks.

23. $n! = n($ ___ $-$ ___ $)(n - 2) \cdot \cdots \cdot 3 \cdot 2 \cdot 1$

24. The symbol $5!$ is read as "____ _____" and it means $5 \cdot$ ___ $\cdot$ ___ $\cdot$ ___ $\cdot$ ___ .

GUIDED PRACTICE

Use Pascal's triangle to expand each binomial. **See Examples 1 and 2.**

25. $(a + b)^3$

26. $(m + p)^4$

27. $(m - p)^5$

28. $(a - b)^3$

Evaluate each expression. **See Examples 3 and 4.**

29. $3!$ **30.** $7!$

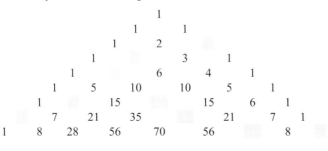

```
                1
              1
          1       2
       1      3
    1      4      6
  1     5     10
1   6    15    20
1  7   21    35
1  8  28   56   70
```

31. $5!$

32. $6!$

33. $3! + 4!$

34. $4! + 4!$

35. $3!(4!)$

36. $2!(3!)$

37. $8(7!)$

38. $4!(5)$

39. $\dfrac{49!}{47!}$

40. $\dfrac{101!}{100!}$

41. $\dfrac{9!}{11!}$

42. $\dfrac{13!}{10!}$

43. $\dfrac{9!}{7!0!}$

44. $\dfrac{7!}{5!0!}$

45. $\dfrac{5!}{1!(5-1)!}$

46. $\dfrac{15!}{14!(15-14)!}$

47. $\dfrac{5!}{3!(5-3)!}$

48. $\dfrac{6!}{4!(6-4)!}$

49. $\dfrac{5!(8-5)!}{4! \cdot 7!}$

50. $\dfrac{6! \cdot 7!}{(8-3)!(7-4)!}$

51. $\dfrac{7!}{5!(7-5)!}$

52. $\dfrac{8!}{6!(8-6)!}$

Use a calculator to evaluate each expression. See Using Your Calculator: Factorials.

53. $11!$

54. $13!$

55. $20!$

56. $55!$

Use the binomial theorem to expand each expression. See Examples 5 and 6.

57. $(m + n)^4$

58. $(a - b)^4$

59. $(c - d)^5$

60. $(c + d)^5$

61. $(a - b)^9$

62. $(a + b)^7$

63. $(s + t)^6$

64. $(s - t)^6$

Use the binomial theorem to expand each expression. See Example 7.

65. $(2x + y)^3$

66. $(x + 2y)^3$

67. $(2t - 3)^5$

68. $(2b + 1)^4$

69. $(5m - 2n)^4$

70. $(2m + 3n)^5$

71. $\left(\dfrac{x}{3} + \dfrac{y}{2}\right)^3$

72. $\left(\dfrac{x}{2} - \dfrac{y}{3}\right)^3$

73. $\left(\dfrac{x}{3} - \dfrac{y}{2}\right)^4$

74. $\left(\dfrac{x}{2} + \dfrac{y}{3}\right)^4$

75. $(c^2 - d^2)^5$

76. $(u^2 - v^3)^5$

Find the indicated term of each binomial expansion. See Examples 8 and 9.

77. $(x + y)^8$; 3rd term

78. $(x + y)^9$; 7th term

79. $(r + s)^6$; 5th term

80. $(r + s)^7$; 5th term

81. $(x - 1)^{13}$; 3rd term

82. $(x - 1)^{10}$; 5th term

83. $(x - 3y)^4$; 2nd term

84. $(3x - y)^5$; 3rd term

85. $(2x - 3y)^5$; 5th term

86. $(3x - 2y)^4$; 2nd term

87. $\left(\dfrac{c}{2} - \dfrac{d}{3}\right)^4$; 2nd term

88. $\left(\dfrac{c}{3} + \dfrac{d}{2}\right)^5$; 4th term

89. $(2t - 5)^7$; 4th term

90. $(2t - 3)^6$; 6th term

91. $(a^2 + b^2)^6$; 2nd term

92. $(a^2 + b^2)^7$; 6th term

WRITING

93. Describe how to construct Pascal's triangle.

94. Explain why the signs alternate in the expansion of $(x - y)^9$.

95. Explain why the third term of the expansion of $(m + 3n)^9$ could not be $324m^7n^3$.

96. Using your own words, write a definition of $n!$.

REVIEW

Assume that x, y, z, and b represent positive numbers. Use the properties of logarithms to write each expression as the logarithm of a single quantity.

97. $2\log x + \dfrac{1}{2}\log y$

98. $-2\log x - 3\log y + \log z$

99. $\ln(xy + y^2) - \ln(xz + yz) + \ln z$

100. $\log_2(x + 1) - \log_2 x$

CHALLENGE PROBLEMS

101. Find the constant term in the expansion of $\left(x + \dfrac{1}{x}\right)^{10}$.

102. Find the coefficient of a^5 in the expansion of $\left(a - \dfrac{1}{a}\right)^9$.

103. a. If we applied the pattern of the coefficients to the coefficient of the first term in a binomial expansion, the coefficient would be $\dfrac{n!}{0!(n-0)!}$. Show that this expression is 1.

   b. If we applied the pattern of the coefficients to the coefficient of the last term in a binomial expansion, the coefficient would be $\dfrac{n!}{n!(n-n)!}$. Show that this expression is 1.

104. Expand $(i - 1)^7$, where $i = \sqrt{-1}$.

SECTION 11.2
Arithmetic Sequences and Series

Objectives

1. Find terms of a sequence given the general term.
2. Find terms of an arithmetic sequence by identifying the first term and the common difference.
3. Find arithmetic means.
4. Find the sum of the first n terms of an arithmetic sequence.
5. Solve application problems involving arithmetic sequences.
6. Use summation notation.

The word *sequence* is used in everyday conversation when referring to an ordered list. For example, a history instructor might discuss the sequence of events that led up to the sinking of the *Titanic.* In mathematics, a **sequence** is a list of numbers written in a specific order.

1   Find Terms of a Sequence Given the General Term.

Each number in a sequence is called a **term** of the sequence. **Finite sequences** contain a finite number of terms and **infinite sequences** contain infinitely many terms. Two examples of sequences are:

> **Finite sequence:** 1, 5, 9, 13, 17, 21, 25
>
> **Infinite sequence:** 3, 6, 9, 12, 15, . . .   The . . . indicates that the sequence goes on forever.

Sequences are defined formally using the terminology of functions.

Finite and Infinite Sequences

A **finite sequence** is a function whose domain is the set of natural numbers $\{1, 2, 3, 4, \ldots, n\}$ for some natural number n.

An **infinite sequence** is a function whose domain is the set of natural numbers: $\{1, 2, 3, 4, \ldots\}$.

Instead of using $f(x)$ notation, we use a_n (read as "a sub n") notation to write the value of a sequence at a number n. For the infinite sequence introduced earlier, we have:

1st term	2nd term	3rd term	4th term	5th term
3,	6,	9,	12,	15, . . .
↑	↑	↑	↑	↑
a_1	a_2	a_3	a_4	a_5

To specifically describe all the terms of a sequence, we can write a formula for a_n, called the **general term** of the sequence. For the sequence 3, 6, 9, 12, 15, . . . , we note that $a_1 = 3 \cdot 1$, $a_2 = 3 \cdot 2$, $a_3 = 3 \cdot 3$, and so on. In general, the nth term of the sequence is found by multiplying n by 3.

$$a_n = 3n \quad \text{Read } a_n \text{ as "}a \text{ sub } n\text{."}$$

We can use this formula to find any term of the sequence. For example, to find the 12th term, we substitute 12 for n.

$$a_{12} = 3(12) = 36$$

EXAMPLE 1 Given an infinite sequence with $a_n = 2n - 3$, find each of the following: **a.** the first four terms **b.** a_{50}

Strategy We will substitute 1, 2, 3, 4, and 50 for n in the formula that defines the sequence.

Why To find the first term of the sequence, we let $n = 1$. To find the second term, let $n = 2$, and so on.

Solution

a. $a_1 = 2(1) - 3 = -1$ Substitute 1 for n. $\qquad a_2 = 2(2) - 3 = 1$ Substitute 2 for n.

$a_3 = 2(3) - 3 = 3$ Substitute 3 for n. $\qquad a_4 = 2(4) - 3 = 5$ Substitute 4 for n.

The first four terms of the sequence are -1, 1, 3, and 5.

b. To find a_{50}, the 50th term of the sequence, we let $n = 50$ in the formula for the nth term:

$$a_{50} = 2(50) - 3 = 97$$

Self Check 1 Given an infinite sequence with $a_n = 3n + 5$, find each of the following: **a.** the first three terms **b.** a_{100}

Now Try Problem 17

EXAMPLE 2 Find the first four terms of the sequence whose general term is $a_n = \dfrac{(-1)^n}{2^n}$.

Strategy We will substitute 1, 2, 3, and 4 for n in the formula that defines the sequence.

Why To find the first term of the sequence, we let $n = 1$. To find the second term, let $n = 2$, and so on.

Solution

Success Tip.

The factor $(-1)^n$ in $a_n = \frac{(-1)^n}{2^n}$ causes the signs of the terms to alternate between positive (when n is even) and negative (when n is odd).

$a_1 = \dfrac{(-1)^1}{2^1} = -\dfrac{1}{2}$ $\qquad (-1)^1 = -1 \qquad$ $a_2 = \dfrac{(-1)^2}{2^2} = \dfrac{1}{4}$ $\qquad (-1)^2 = 1$

$a_3 = \dfrac{(-1)^3}{2^3} = \dfrac{-1}{8} = -\dfrac{1}{8}$ $\quad (-1)^3 = -1 \qquad$ $a_4 = \dfrac{(-1)^4}{2^4} = \dfrac{1}{16}$ $\qquad (-1)^4 = 1$

The first four terms of the sequence are $-\dfrac{1}{2}, \dfrac{1}{4}, -\dfrac{1}{8},$ and $\dfrac{1}{16}$.

Self Check 2 Find the first four terms of the sequence whose general term is $a_n = \dfrac{(-1)^n}{n}$.

Now Try Problem 25

Find Terms of an Arithmetic Sequence by Identifying the First Term and the Common Difference.

A sequence where each term is found by adding the same number to the previous term is called an *arithmetic sequence*. Two examples are

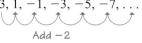

5, 12, 19, 26, 33, 40   This is a finite arithmetic sequence where each term

Add 7   is found by adding 7 to the previous term.

3, 1, −1, −3, −5, −7, . . .   This is an infinite arithmetic sequence where each term

Add −2   is found by adding −2 to the previous term.

Arithmetic Sequence

An **arithmetic sequence** is a sequence of the form

$$a_1, \quad a_1 + d, \quad a_1 + 2d, \quad a_1 + 3d, \quad \ldots, \quad a_1 + (n-1)d, \ldots$$

where a_1 is the **first term** and d is the **common difference**. The nth term is given by

$$a_n = a_1 + (n-1)d$$

We note that the second term of an arithmetic sequence has an addend of $1d$, the third term has an addend of $2d$, the fourth term has an addend of $3d$, and the nth term has an addend of $(n-1)d$. We also note that the *difference between any two consecutive terms in an arithmetic sequence is d*.

EXAMPLE 3   An arithmetic sequence has a first term 5 and a common difference 4. Write the first five terms of the sequence and find the 25th term.

Strategy   To find the first five terms, we will write the first term and add 4 to each successive term until we produce five terms. To find the 25th term, we will substitute 5 for a_1, 4 for d, and 25 for n in the formula $a_n = a_1 + (n-1)d$.

Why   The same number is added to each term of an arithmetic sequence to get the next term. However, successively adding 4 to find the 25th term would be time consuming. Using the formula is faster.

Solution   Since the first term is 5 and the common difference is 4, the first five terms are

5, 9, 13, 17, 21

Add 4

Since the first term is $a_1 = 5$ and the common difference is $d = 4$, the arithmetic sequence is defined by the formula

$$a_n = 5 + (n-1)4 \qquad \text{In } a_n = a_1 + (n-1)d, \text{ substitute 5 for } a_1 \text{ and 4 for } d.$$

To find the 25th term, we substitute 25 for n and simplify.

$$a_{25} = 5 + (25-1)4$$
$$= 5 + (24)4$$
$$= 101$$

The 25th term is 101.

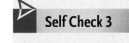

Self Check 3   Write the first five terms of an arithmetic sequence with a first term of 10 and a common difference of 8. Then find the 30th term.

Now Try   **Problem 29**

EXAMPLE 4   The first three terms of an arithmetic sequence are 3, 8, and 13. Find the 100th term.

Strategy   We can use the first three terms to find the common difference. Then we will know the first term of the sequence and the common difference.

Why   Once we know the first term and the common difference, we can use the formula $a_n = a_1 + (n-1)d$ to find the 100th term by letting $n = 100$.

Solution   The common difference d is the difference between any two successive terms. Since $a_1 = 3$ and $a_2 = 8$, we can find d using subtraction.

$$d = a_2 - a_1 = 8 - 3 = 5 \qquad \text{Also note that } a_3 - a_2 = 13 - 8 = 5.$$

To find 100th term, we substitute 3 for a_1, 5 for d, and 100 for n in the formula for the nth term.

$$a_n = a_1 + (n-1)d$$
$$a_{100} = 3 + (100-1)5$$
$$= 3 + (99)5$$
$$= 498$$

> **Success Tip**
> The common difference d of an arithmetic sequence is defined to be
> $$d = a_{n+1} - a_n$$

Self Check 4   The first three terms of an arithmetic sequence are -3, 6, and 15. Find the 99th term.

Now Try   **Problem 37**

EXAMPLE 5   The first term of an arithmetic sequence is 12 and the 50th term is 3,099. Write the first six terms of the sequence.

Strategy   We will find the common difference by substituting 3,099 for a_n, 12 for a_1, and 50 for n in the formula $a_n = a_1 + (n-1)d$.

Why   Once we know the first term and the common difference, we can successively add the common difference to each term to produce the first six terms.

Solution   Since the 50th term of the sequence is 3,099, we substitute 3,099 for a_{50}, 12 for a_1, and 50 for n in the formula $a_n = a_1 + (n-1)d$ and solve for d.

$$a_{50} = a_1 + (n-1)d$$
$$3{,}099 = 12 + (50-1)d \qquad \text{Substitute 3,099 for } a_{50}, \text{ 12 for } a_1, \text{ and 50 for } n.$$
$$3{,}099 = 12 + 49d \qquad \text{Simplify.}$$
$$3{,}087 = 49d \qquad \text{Subtract 12 from both sides.}$$
$$63 = d \qquad \text{Divide both sides by 49.}$$

Since the first term is 12 and the common difference is 63, the first six terms are

12, 75, 138, 201, 264, 327   *Add 63 to a term to get the next term.*

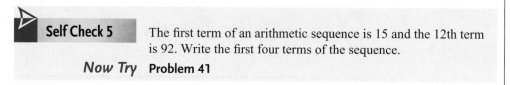

Self Check 5   The first term of an arithmetic sequence is 15 and the 12th term is 92. Write the first four terms of the sequence.

Now Try   **Problem 41**

③ **Find Arithmetic Means.**

If numbers are inserted between two numbers *a* and *b* to form an arithmetic sequence, the inserted numbers are called **arithmetic means** between *a* and *b*. If a single number is inserted, it is called **the arithmetic mean** between *a* and *b*.

EXAMPLE 6   Insert two arithmetic means between 6 and 27.

Strategy   Because two arithmetic means are to be inserted between 6 and 27, we will consider a sequence of four terms, with a first term of 6 and a fourth term of 27. We will then use $a_n = a_1 + (n - 1)d$ to find the common difference *d*.

Why   Once we know the first term and the common difference, we can add the common difference to find the two unknown terms.

Solution   The first term is $a_1 = 6$ and the fourth term is $a_4 = 27$. We must find the common difference so that the terms

$$6, \quad 6 + d, \quad 6 + 2d, \quad 27$$
$$\uparrow \qquad \uparrow \qquad \quad \uparrow \qquad \quad \uparrow$$
$$a_1 \qquad a_2 \qquad \; a_3 \qquad \; a_4$$

form an arithmetic sequence. To find the common difference *d*, we substitute 6 for a_1, 4 for *n*, and 27 for a_4 in the formula for the 4th term:

$a_4 = a_1 + (n - 1)d$ *This gives the 4th term of any arithmetic sequence.*

$27 = 6 + (4 - 1)d$ *Substitute.*

$27 = 6 + 3d$ *Simplify.*

$21 = 3d$ *Subtract 6 from both sides.*

$7 = d$ *Divide both sides by 3.*

To find the two arithmetic means between 6 and 27, we add the common difference 7, as shown:

$6 + d = 6 + 7$ or $6 + 2d = 6 + 2(7)$

$\qquad = 13$ *This is a_2.* $\qquad = 6 + 14$

$\qquad\qquad\qquad\qquad\qquad\qquad\qquad = 20$ *This is a_3.*

Two arithmetic means between 6 and 27 are 13 and 20.

Self Check 6   Insert two arithmetic means between 8 and 44.

Now Try   **Problem 45**

4 **Find the Sum of the First *n* Terms of an Arithmetic Sequence.**

To develop a formula for finding the sum of the first *n* terms of an arithmetic sequence, we let S_n represent the sum of the first *n* terms of an arithmetic sequence:

$$S_n = \quad a_1 \quad + \quad [a_1 + d] \quad + \quad [a_1 + 2d] \quad + \cdots + [a_1 + (n - 1)d]$$

We write the same sum again, but in reverse order:

$$S_n = [a_1 + (n - 1)d] + [a_1 + (n - 2)d] + [a_1 + (n - 3)d] + \cdots + \quad a_1$$

Adding these equations together, term by term, we get

$$2S_n = [2a_1 + (n - 1)d] + [2a_1 + (n - 1)d] + [2a_1 + (n - 1)d] + \cdots + [2a_1 + (n - 1)d]$$

Because there are *n* equal terms on the right side of the preceding equation, we can write

(1) $\quad 2S_n = n[2a_1 + (n - 1)d]$

(2) $\quad 2S_n = n[a_1 + a_1 + (n - 1)d] \qquad$ Write $2a_1$ as $a_1 + a_1$.

$\qquad 2S_n = n(a_1 + a_n) \qquad\qquad\qquad$ Substitute a_n for $a_1 + (n - 1)d$.

$\qquad S_n = \dfrac{n(a_1 + a_n)}{2} \qquad\qquad\quad$ Divide both sides by 2.

This reasoning establishes the following formula.

Sum of the First *n* Terms of an Arithmetic Sequence

The sum of the first *n* terms of an arithmetic sequence is given by the formula

$$S_n = \frac{n(a_1 + a_n)}{2}$$

where a_1 is the first term, a_n is the *n*th (or last) term, and *n* is the number of terms in the sequence.

EXAMPLE 7 Find the sum of the first 40 terms of the arithmetic sequence: 4, 10, 16, . . .

Strategy We know the first term is 4 and we can find the common difference *d*. We will substitute these values into the formula $a_n = a_1 + (n - 1)d$ to find the last term to be added, a_{40}.

Why To use the formula $S_n = \dfrac{n(a_1 + a_n)}{2}$ to find the sum of the first 40 terms, we need to know the first term, a_1, and the last term, a_{40}.

Solution We can substitute 4 for a_1, 40 for *n*, and $10 - 4 = 6$ for *d* into $a_n = a_1 + (n - 1)d$ to get $a_{40} = 4 + (40 - 1)6 = 238$. We then substitute these values into the formula for S_{40}:

$$S_n = \frac{n(a_1 + a_{40})}{2}$$

$$S_{40} = \frac{40(4 + 238)}{2} \qquad \text{Substitute: } a_1 = 4, n = 40, \text{ and } a_{40} = 238.$$

$$= 20(242)$$

$$= 4{,}840$$

The sum of the first 40 terms is 4,840.

Success Tip

An alternate form of the summation formula for arithmetic sequences can be obtained from Equation (2) above by combining like terms and dividing both sides by 2.

$$S_n = \frac{n[2a_1 + (n - 1)d]}{2}$$

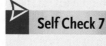

> **Self Check 7**   Find the sum of the first 50 terms of the arithmetic sequence: 3, 8, 13, . . .
>
> *Now Try*   **Problem 53**

⑤   **Solve Application Problems Involving Arithmetic Sequences.**

EXAMPLE 8   *Halftime Performances.*   Each row of a formation formed by the members of a college marching band has one more person in it than the previous row. If 4 people are in the front row and 21 are in the 18th (and last) row, how many band members are there?

Strategy   To find the number of band members, we will write an arithmetic sequence to model the situation and find the sum of its terms.

Why   We can use an arithmetic sequence to model this situation because each row has one more person in it than the previous one. Thus, the common difference is 1.

Solution   When we list the number of band members in each row of the formation, we get the arithmetic sequence 4, 5, 6, . . . , 21, where $a_1 = 4$, $d = 1$, $n = 18$, and $a_{18} = 21$. We can use the formula $S_n = \frac{n(a_1 + a_n)}{2}$ to find the sum of the terms of the sequence.

$$S_{18} = \frac{18(4 + 21)}{2} = \frac{18(25)}{2} = 225$$

There are 225 members of the marching band.

> **Self Check 8**   How many band members would it take to form a 10-row formation, if the first row has 5 people in it, the second row has 7 people, the third row has 9 people, and so on?
>
> *Now Try*   **Problem 95**

⑥   **Use Summation Notation.**

When the commas between the terms of a sequence are replaced with + signs, we call the sum a **series.** The sum of the terms of an arithmetic sequence is called an **arithmetic series.** Some examples are

$$4 + 8 + 12 + 16 + 20 + 24$$ *Since this series has a limited number of terms, it is a finite arithmetic series.*

$$5 + 8 + 11 + 14 + 17 + \cdots$$ *Since this series has infinitely many terms, it is an infinite arithmetic series.*

When the general term of a sequence is known, we can use a special notation to write a series. This notation, called **summation notation,** involves the Greek letter Σ (sigma). The expression

$$\sum_{k=1}^{4} 3k$$ *Read as "the summation of 3k as k runs from 1 to 4."*

designates the sum of all terms obtained if we successively substitute the numbers 1, 2, 3, and 4 for k, called the **index of the summation.** Thus, we have

$$k = 1 \quad k = 2 \quad k = 3 \quad k = 4$$
$$\downarrow \qquad \downarrow \qquad \downarrow \qquad \downarrow$$

$$\sum_{k=1}^{4} 3k = 3(1) + 3(2) + 3(3) + 3(4)$$

$$= 3 + 6 + 9 + 12$$

$$= 30$$

EXAMPLE 9   Write the series associated with each summation and find the sum:

a. $\displaystyle\sum_{k=1}^{3} (2k + 1)$ **b.** $\displaystyle\sum_{k=2}^{8} k^2$

Strategy   In part (a), we will substitute 1, 2, and 3 for k and add the resulting numbers. In part (b), we will substitute 2, 3, 4, 5, 6, 7, and 8 for k and add the resulting numbers.

Why   Think of k as a counter that begins with the number written at the bottom of the notation and successively increases by 1 until it reaches the number written at the top.

Solution

a. We substitute the integers 1, 2, and 3 for k and find the sum.

$$\sum_{k=1}^{3} (2k + 1) = [2(1) + 1] + [2(2) + 1] + [2(3) + 1]$$

$$= 3 + 5 + 7$$

$$= 15$$

b. We substitute the integers from 2 to 8 for k and find the sum.

$$\sum_{k=2}^{8} k^2 = 2^2 + 3^2 + 4^2 + 5^2 + 6^2 + 7^2 + 8^2$$

$$= 4 + 9 + 16 + 25 + 36 + 49 + 64$$

$$= 203$$

▷ **Self Check 9**   Find the sum: $\displaystyle\sum_{k=1}^{4} (2k^2 - 2)$

Now Try   Problems 57 and 61

▷ **ANSWERS TO SELF CHECKS**   **1. a.** 8, 11, 14 **b.** 305   **2.** $-1, \frac{1}{2}, -\frac{1}{3}, \frac{1}{4}$ **3.** 10, 18, 26, 34, 42; 242
4. 879   **5.** 15, 22, 29, 36   **6.** 20, 32   **7.** 6,275   **8.** 140   **9.** 52

STUDY SET
11.2

VOCABULARY

Fill in the blanks.

1. A _____ is a function whose domain is the set of natural numbers.

2. A sequence with an unlimited number of terms is called a(n) _____ sequence. A sequence with a specific number of terms is called a(n) _____ sequence.

3. Each term of a(n) _____ sequence is found by adding the same number to the previous term.

4. 5, 15, 25, 35, 45, 55, . . . is an example of a(n) _____ sequence. The first _____ is 5 and the common _____ is 10.

5. If a single number is inserted between a and b to form an arithmetic sequence, the number is called the arithmetic _____ between a and b.

6. The sum of the terms of an arithmetic sequence is called an arithmetic _____.

CONCEPTS

7. Write the first three terms of an arithmetic sequence if $a_1 = 1$ and $d = 6$.

8. Given the arithmetic sequence 4, 7, 10, 13, 16, 19, . . . , find a_5 and d.

9. **a.** Write the formula for a_n, the general term of an arithmetic sequence.

   b. Write the formula for S_n, the sum of the first n terms of an arithmetic sequence.

10. An infinite arithmetic sequence is of the form

$$a_1, a_1 + d, \quad\quad , a_1 + 3d, \quad\quad , \ldots$$

NOTATION

Fill in the blanks.

11. The notation a_n represents the _____ term of a sequence.

12. To find the common difference of an arithmetic sequence, we use the formula $d = a_{} - a_{}$.

13. The symbol Σ is the Greek letter _____.

14. In the symbol $\sum_{k=1}^{5} (2k - 5)$, k is called the _____ of summation.

15. We read $\sum_{k=1}^{10} 3k$ as "the _____ of $3k$ as k _____ from 1 to 10."

16. $\sum_{k=1}^{5} k = + + + + $

GUIDED PRACTICE

Write the first five terms of each sequence and then find the specified term. See Example 1.

17. $a_n = 4n - 1$, a_{40}

18. $a_n = 5n - 3$, a_{25}

19. $a_n = -3n + 1$, a_{30}

20. $a_n = -6n + 2$, a_{15}

21. $a_n = -n^2$, a_{20}

22. $a_n = -n^3$, a_{10}

23. $a_n = \dfrac{n-1}{n}$, a_{12} 24. $a_n = \dfrac{n+1}{2n}$, a_{100}

Write the first four terms of each sequence. See Example 2.

25. $a_n = \dfrac{(-1)^n}{3^n}$ 26. $a_n = \dfrac{(-1)^n}{4^n}$

27. $a_n = (-1)^n(n + 6)$ 28. $a_n = (-1)^n(7n)$

Write the first five terms of each arithmetic sequence with the given properties and find the specified term. See Example 3.

29. First term: 3, common difference: 2; find the 10th term.

30. First term: -2, common difference 3; find the 20th term.

31. First term -5, common difference: -3; find the 15th term.

32. First term: 8, common difference: -5; find the 25th term.

33. First term: 7, common difference: 12; find the 30th term.

34. First term: -1, common difference: 4; find the 55th term.

35. First term: -7; common difference: -2; find the 15th term.

36. First term: 8, common difference -3; find the 25th term.

The first three terms of an arithmetic sequence are shown below. Find the specified term. See Example 4.

37. 1, 4, 7, . . . ; 30th term

38. 2, 6, 10, . . . ; 28th term

39. $-5, -1, 3, \ldots$; 17th term

40. $-7, -1, 5, \ldots$; 15th term

Write the first five terms of the arithmetic sequence with the following properties. See Example 5.

41. The first term is 5 and the fifth term is 29.

42. The first term is 4 and the sixth term is 39.

43. The first term is -4 and the sixth term is -39.

44. The first term is -5 and the fifth term is -37.

Insert the given number of arithmetic means between the numbers. **See Example 6.**

45. Two arithmetic means between 2 and 11

46. Four arithmetic means between 5 and 25

47. Four arithmetic means between 10 and 20

48. Three arithmetic means between 20 and 80

49. Three arithmetic means between 20 and 30

50. Two arithmetic means between 10 and 19

51. One arithmetic mean between -4.5 and 7

52. One arithmetic mean between -6.5 and 8.5

For each arithmetic sequence, find the sum of the specified number of terms. **See Example 7.**

53. The first 35 terms of $5, 9, 13, \ldots$

54. The first 50 terms of $7, 12, 17, \ldots$

55. The first 40 terms of $-5, -1, 3, \ldots$

56. The first 25 terms of $2, -3, -8, \ldots$

Write the series associated with each summation. **See Example 9.**

57. $\displaystyle\sum_{k=1}^{4} (3k)$

58. $\displaystyle\sum_{k=1}^{4} (k - 9)$

59. $\displaystyle\sum_{k=2}^{4} k^2$

60. $\displaystyle\sum_{k=3}^{5} (-2k)$

Find each sum. **See Example 9.**

61. $\displaystyle\sum_{k=1}^{4} (6k)$

62. $\displaystyle\sum_{k=2}^{5} (3k)$

63. $\displaystyle\sum_{k=3}^{4} k^3$

64. $\displaystyle\sum_{k=2}^{4} (-k^2)$

65. $\displaystyle\sum_{k=3}^{4} (k^2 + 3)$

66. $\displaystyle\sum_{k=2}^{6} (k^2 + 1)$

67. $\displaystyle\sum_{k=4}^{4} (2k + 4)$

68. $\displaystyle\sum_{k=3}^{3} (3k^2 - 7)$

69. $\displaystyle\sum_{k=2}^{5} (5k)$

70. $\displaystyle\sum_{k=2}^{5} (3k - 5)$

71. $\displaystyle\sum_{k=4}^{6} (4k - 1)$

72. $\displaystyle\sum_{k=3}^{5} (k^3)$

TRY IT YOURSELF

73. Find the common difference of the arithmetic sequence with a first term of 40 if its 44th term is 556.

74. Find the first term of the arithmetic sequence with a common difference of -5 if its 23rd term is -625.

75. Find the sum of the first 12 terms of the arithmetic sequence if its second term is 7 and its third term is 12.

76. Find the sum of the first 16 terms of the arithmetic sequence if its second term is 5 and its fourth term is 9.

77. Find the first five terms of the arithmetic sequence if the common difference is 7 and the sixth term is -83.

78. Find the first five terms of the arithmetic sequence if the common difference is 3 and the seventh term is 12.

79. Find the first six terms of the arithmetic sequence if the common difference is -3 and the ninth term is 10.

80. Find the first six terms of the arithmetic sequence if the common difference is -5 and the tenth term is -27.

81. The first three terms of an arithmetic sequence are 5, 12, and 19. Find the 200th term.

82. The first three terms of an arithmetic sequence are 10, 14, and 18. Find the 500th term.

83. Find the sum of the first 50 natural numbers.

84. Find the sum of the first 100 natural numbers.

85. Find the 37th term of the arithmetic sequence with a second term of -4 and a third term of -9.

86. Find the 40th term of the arithmetic sequence with a second term of 6 and a fourth term of 16.

87. Find the first term of the arithmetic sequence with a common difference of 11 if its 27th term is 263.

88. Find the common difference of the arithmetic sequence with a first term of -164 if its 36th term is -24.

89. Find the 15th term of the arithmetic sequence $\frac{1}{2}, \frac{1}{4}, 0$

90. Find the 14th term of the arithmetic sequence $\frac{2}{3}, \frac{1}{2}, \frac{1}{3}, \ldots$

91. Find the sum of the first 50 odd natural numbers.

92. Find the sum of the first 50 even natural numbers.

APPLICATIONS

93. SAVING MONEY Yasmeen puts $60 into a safety deposit box. After each succeeding month, she puts $50 more in the box. Write the first six terms of an arithmetic sequence that gives the monthly amounts in her savings, and find her savings after 10 years.

94. INSTALLMENT LOANS Maria borrowed $10,000, interest-free, from her mother. She agreed to pay back the loan in monthly installments of $275. Write the first six terms of an arithmetic sequence that shows the balance due after each month, and find the balance due after 17 months.

95. DESIGNING PATIOS   Refer to the illustration. Each row of bricks in a triangular patio floor is to have one more brick than the previous row, ending with the longest row of 150 bricks. How many bricks will be needed?

96. LOGGING   Logs are stacked so that the bottom row has 30 logs, the next row has 29 logs, the next row has 28 logs, and so on.

a. If there are 20 rows in the stack, how many logs are in the top row?

b. How many logs are in the stack?

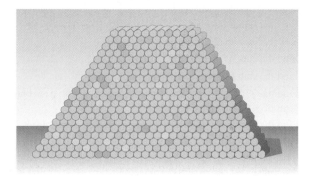

97. HOLIDAY SONGS   A popular song of European origin lists the gifts received from someone's "true love" over a 12-day span: a partridge in a pear tree, two turtle doves, three French hens, four calling birds, five gold rings, six geese a-laying, seven swans a-swimming, eight maids a-milking, nine ladies dancing, ten lords a-leaping, eleven pipers piping, and twelve drummers drumming. Use a formula from this section to determine the total number of gifts in this list.

98. INTERIOR ANGLES   The sums of the angles of several polygons are given in the table in the next column. Assuming that the pattern continues, complete the table.

Figure	Number of sides	Sum of angles
Triangle	3	180°
Quadrilateral	4	360°
Pentagon	5	540°
Hexagon	6	720°
Octagon	8	
Dodecagon	12	

WRITING

99. Explain why 1, 4, 8, 13, 19, 26, . . . is not an arithmetic sequence.

100. What is the difference between a sequence and a series?

101. What is the difference between a_n and S_n?

102. How is the symbol Σ used in this section?

REVIEW

Assume that x, y, z, and b represent positive numbers. Use the properties of logarithms to write each expression in terms of the logarithms of x, y, and z.

103. $\log_2 \dfrac{2x}{y}$

104. $\ln x\sqrt{z}$

105. $\log x^3 y^2$

106. $\log x^3 y^{1/2}$

CHALLENGE PROBLEMS

Write the summation notation for each sum.

107. Write the summation notation for

$$1 + 4 + 9 + 16 + 25$$

108. Write the summation notation for $3 + 4 + 5 + 6$ without using $k = 1$ in your answer.

109. For what value of x will $x - 2$, $2x + 4$, and $5x - 8$, in that order, form an arithmetic sequence?

110. For what value of x will the arithmetic mean of $x + 4$ and $x + 8$ be 5?

SECTION 11.3
Geometric Sequences and Series

Objectives

1. Find terms of a geometric sequence by identifying the first term and the common ratio.
2. Find geometric means.
3. Find the sum of the first n terms of a geometric sequence.
4. Define and use infinite geometric series.
5. Solve application problems involving geometric sequences.

We have seen that the same number is added to each term of an arithmetic sequence to get the next term. In this section, we will consider another type of sequence where we multiply each term by the same number to get the next term. This type of sequence is called a *geometric sequence.* Two examples are

$2, 8, 32, 128, \ldots$ This is an infinite geometric sequence where each term is found by multiplying the previous term by 4.

Multiply by 4

$27, 9, 3, 1, \dfrac{1}{3}, \dfrac{1}{9}$ This is a finite geometric sequence where each term is found by multiplying the previous term by $\frac{1}{3}$.

Multiply by $\frac{1}{3}$

 Find Terms of a Geometric Sequence by Identifying the First Term and the Common Ratio.

Each term of a geometric sequence is found by multiplying the previous term by the same number.

Geometric Sequence	A **geometric sequence** is a sequence of the form $$a_1, \ a_1r, \ a_1r^2, \ a_1r^3, \ldots, \ a_1r^{n-1}, \ldots$$ where a_1 is the **first term** and r is the **common ratio.** The nth term is given by $$a_n = a_1 r^{n-1}$$

We note that the second term of a geometric sequence has a factor of r^1, the third term has a factor of r^2, the fourth term has a factor of r^3, and the nth term has a factor of r^{n-1}. We also note that *r is the quotient obtained when any term is divided by the previous term.*

EXAMPLE 1   A geometric sequence has a first term 5 and a common ratio 3.
a. Write the first five terms of the sequence.   **b.** Find the 9th term.

Strategy   In part (a), we will write the first term and successively multiply each term by 3 until we produce five terms. In part (b), we will substitute 5 for a_1, 3 for r, and 9 for n in the formula for the nth term of a geometric sequence $a_n = a_1 r^{n-1}$.

Why   To find the terms of a geometric sequence, we multiply each term by the same number to get the next term. To answer part (b), successively multiplying by 3 to find the 9th term would be time consuming. Using the formula is faster.

Solution

a. Because the first term is $a_1 = 5$ and the common ratio is $r = 3$, the first five terms are

$$5, \quad 5(3), \quad 5(3^2), \quad 5(3^3), \quad 5(3^4)$$

Each term is found by multiplying the previous term by 3.

$$\uparrow a_1 \quad \uparrow a_2 \quad \uparrow a_3 \quad \uparrow a_4 \quad \uparrow a_5$$

or

$$5, 15, 45, 135, 405$$

b. The nth term is $a_n = a_1 r^{n-1}$ with $a_1 = 5$ and $r = 3$. Because we want the ninth term, we let $n = 9$:

$$a_n = a_1 r^{n-1}$$
$$a_9 = 5(3)^{9-1}$$
$$= 5(3)^8$$
$$= 5(6{,}561)$$
$$= 32{,}805$$

Self Check 1   A geometric sequence has a first term 3 and a common ratio 4.
a. Write the first four terms.
b. Find the 8th term.

Now Try **Problem 15**

EXAMPLE 2   The first three terms of a geometric sequence are 16, 4, and 1. Find the 7th term.

Strategy   We can use the first three terms to find the common ratio. Then we will know the first term and the common ratio.

Why   Once we know the first term, a_1, and the common ratio, r, we can use the formula $a_n = a_1 r^{n-1}$ to find the seventh term by letting $n = 7$.

Solution   The common ratio r is the ratio between any two successive terms. Since $a_1 = 16$ and $a_2 = 4$, we can find r as follows:

> **Success Tip**
> The common ratio r of a geometric sequence is defined to be
> $$r = \frac{a_{n+1}}{a_n}$$

$$r = \frac{a_2}{a_1} = \frac{4}{16} = \frac{1}{4}$$ Also note that $\frac{a_3}{a_2} = \frac{1}{4}$.

To find the seventh term, we substitute 16 for a_1, $\frac{1}{4}$ for r, and 7 for n in the formula for the nth term and simplify:

$$a_n = a_1 r^{n-1}$$ This is the formula for the nth term of a geometric sequence.

$$a_7 = 16\left(\frac{1}{4}\right)^{7-1}$$ Substitute 16 for a_1, 7 for n, and $\frac{1}{4}$ for r.

$$= 16\left(\frac{1}{4}\right)^6$$

$$= 16\left(\frac{1}{4{,}096}\right)$$

$$= \frac{1}{256}$$

Self Check 2   The first three terms of a geometric sequence are 25, 5, and 1. Find the 7th term.

Now Try   **Problem 19**

EXAMPLE 3   Find the first five terms of the geometric sequence with a first term of 2, a third term of 32, and a common ratio that is positive.

Strategy   We will substitute $a_1 = 2$, $a_3 = 32$, and $n = 3$ into the formula for the nth term of a geometric sequence $a_n = a_1 r^{n-1}$ and solve for r.

Why   Once we know the common ratio, we can successively multiply each term by the common ratio to produce the first five terms.

Solution   We will substitute 3 for n, 2 for a_1, 32 for a_3 and solve for r.

$a_n = a_1 r^{n-1}$	This is the formula for the nth term of a geometric sequence.
$a_3 = 2r^{3-1}$	Substitute 3 for n and 2 for a_1.
$32 = 2r^2$	Substitute 32 for a_3 and simplify.
$16 = r^2$	Divide both sides by 2.
$\pm 4 = r$	Use the square root property.

Since r is given to be positive, $r = 4$. The first five terms are produced by multiplying by the common ratio:

$$2,\ 2 \cdot 4,\ 2 \cdot 4^2,\ 2 \cdot 4^3,\ 2 \cdot 4^4$$

or

$$2,\ 8,\ 32,\ 128,\ 512$$

Self Check 3   Find the first five terms of the geometric sequence with a first term of -2, a fourth term of -54.

Now Try   **Problem 23**

 Find Geometric Means.

If numbers are inserted between two numbers a and b to form a geometric sequence, the inserted numbers are called **geometric means** between a and b. If a single number is inserted, that number is called **the geometric mean** between a and b.

EXAMPLE 4   Insert two geometric means between 7 and 1,512.

Strategy   Because two geometric means are to be inserted between 7 and 1,512, we will consider a sequence of four terms, with a first term of 7 and a fourth term of 1,512. We will then use $a_n = a_1 r^{n-1}$ to find the common ratio r.

Why   Once we know the first term and the common ratio, we can multiply by the common ratio to find the two unknown terms.

Solution   In this example, the first term is $a_1 = 7$, and the fourth term is $a_4 = 1,512$. To find the common ratio r so that the terms

$$7, \quad 7r, \quad 7r^2, \quad 1,512$$
$$\uparrow \qquad \uparrow \qquad \uparrow \qquad \uparrow$$
$$a_1 \qquad a_2 \qquad a_3 \qquad a_4$$

form a geometric sequence, we substitute 4 for n and 7 for a_1 in the formula for the nth term of a geometric sequence and solve for r.

$$a_n = a_1 r^{n-1} \qquad \text{This is the formula for the } n\text{th term of a geometric sequence.}$$
$$a_4 = 7r^{4-1} \qquad \text{Substitute 4 for } n \text{ and 7 for } a_1.$$
$$1,512 = 7r^3 \qquad \text{Substitute 1,512 for } a_4.$$
$$216 = r^3 \qquad \text{Divide both sides by 7.}$$
$$6 = r \qquad \text{Take the cube root of both sides.}$$

To find the two geometric means between 7 and 1,512, we multiply by the common ratio 6, as shown:

$$7r = 7(6) = 42 \quad \text{and} \quad 7r^2 = 7(6)^2 = 7(36) = 252$$

The numbers 7, 42, 252, and 1,512 are the first four terms of a geometric sequence.

Self Check 4   Insert three positive geometric means between 1 and 16.

Now Try   **Problem 27**

EXAMPLE 5   Find the geometric mean between 2 and 20.

Strategy   Because one geometric mean is to be inserted between 2 and 20, we will consider a sequence of three terms, with a first term of 2 and a third term of 20. We will then use $a_n = a_1 r^{n-1}$ to find the common ratio r.

Why   Once we know the first term and the common ratio, we can multiply by the common ratio to find the unknown term.

Solution   We want to find the middle term of the three-termed geometric sequence

$$2, \quad 2r, \quad 20$$
$$\uparrow \qquad \uparrow \qquad \uparrow$$
$$a_1 \qquad a_2 \qquad a_3$$

with $a_1 = 2$, $a_3 = 20$, and $n = 3$. To find r, we substitute these values into the formula for the nth term of a geometric sequence:

$$a_n = a_1 r^{n-1} \qquad \text{This is the formula for the } n\text{th term of a geometric sequence.}$$
$$a_3 = 2r^{3-1} \qquad \text{Substitute 3 for } n \text{ and 2 for } a_1.$$

$$20 = 2r^2 \qquad \text{Substitute 20 for } a_3.$$
$$10 = r^2 \qquad \text{Divide both sides by 2.}$$
$$\pm \sqrt{10} = r \qquad \text{Use the square root property.}$$

Because r can be either $\sqrt{10}$ or $-\sqrt{10}$, there are two values for the geometric mean. They are

$$2r = 2\sqrt{10} \quad \text{and} \quad 2r = -2\sqrt{10}$$

The sets of numbers 2, $2\sqrt{10}$, 20 and 2, $-2\sqrt{10}$, 20 both form geometric sequences. The common ratio of the first sequence is $\sqrt{10}$, and the common ratio of the second sequence is $-\sqrt{10}$.

> **Self Check 5**   Find the positive geometric mean between 2 and 200.
>
> **Now Try**   **Problem 31**

3   **Find the Sum of the First *n* Terms of a Geometric Sequence.**

There is a formula that gives the sum of the first n terms of a geometric sequence. To develop this formula, we let S_n represent the sum of the first n terms of a geometric sequence.

(1) $S_n = a_1 + a_1 r + a_1 r^2 + a_1 r^3 + \cdots + a_1 r^{n-1}$

We multiply both sides of Equation 1 by r to get

(2) $S_n r = \qquad a_1 r + a_1 r^2 + a_1 r^3 + \cdots + a_1 r^{n-1} + a_1 r^n$

We now subtract Equation 2 from Equation 1 and solve for S_n:

$$S_n - S_n r = a_1 - a_1 r^n$$
$$S_n(1 - r) = a_1 - a_1 r^n \qquad \text{Factor out } S_n \text{ from the left side.}$$
$$S_n = \frac{a_1 - a_1 r^n}{1 - r} \qquad \text{Divide both sides by } 1 - r.$$

This reasoning establishes the following formula.

Success Tip

If the common factor of a_1 in the numerator of $\frac{a_1 - a_1 r^n}{1 - r}$ is factored out, the formula can be written in a different form:

$$S_n = \frac{a_1(1 - r^n)}{1 - r}$$

Sum of the First *n* Terms of a Geometric Sequence

The sum of the first n terms of a geometric sequence is given by the formula

$$S_n = \frac{a_1 - a_1 r^n}{1 - r} \quad \text{or} \quad S_n = \frac{a_1(1 - r^n)}{1 - r} \quad \text{where } r \neq 1$$

where S_n is the sum, a_1 is the first term, r is the common ratio, and n is the number of terms.

EXAMPLE 6   Find the sum of the first six terms of the geometric sequence: 250, 50, 10, . . .

Strategy   We will find the common ratio r.

Why   If we know the first term and the common ratio, we can use the formula $S_n = \frac{a_1 - a_1 r^n}{1 - r}$ to find the sum of the first six terms by letting $n = 6$.

Solution   The common ratio r is the ratio between any two successive terms. Since $a_1 = 250$ and $a_2 = 50$, we can find r as follows:

$$r = \frac{a_2}{a_1} = \frac{50}{250} = \frac{1}{5} \quad \text{Also note that } \frac{a_3}{a_2} = \frac{10}{50} = \frac{1}{5}.$$

In this sequence, $a_1 = 250$, $r = \frac{1}{5}$, and $n = 6$. We substitute these values into the formula for the sum of the first n terms of a geometric sequence and simplify:

$$S_n = \frac{a_1 - a_1 r^n}{1 - r}$$

$$S_6 = \frac{250 - 250\left(\frac{1}{5}\right)^6}{1 - \frac{1}{5}}$$

$$= \frac{250 - 250\left(\frac{1}{15{,}625}\right)}{\frac{4}{5}}$$

$$= \left(250 - \frac{250}{15{,}625}\right) \cdot \frac{5}{4} \quad \text{Multiply the numerator by the reciprocal of the denominator.}$$

$$= 312.48 \quad \text{Use a calculator.}$$

The sum of the first six terms is 312.48.

    Self Check 6   Find the sum of the first five terms of the geometric sequence 100, 20, 4, . . .

Now Try   **Problem 35**

4   **Define and Use Infinite Geometric Series.**

When we add the terms of a geometric sequence, we form a **geometric series.** If we form the sum of the terms of an infinite geometric sequence, we get a series called an **infinite geometric series.** For example, if the common ratio r is 3, we have

Infinite geometric sequence   *Infinite geometric series*

2, 6, 18, 54, 162, . . .   2 + 6 + 18 + 54 + 162 + · · ·

As the number of terms of this series gets larger, the value of the series gets larger. We can see that this is true by forming some **partial sums.**

The first partial sum of the series is $S_1 = 2$.

The second partial sum of the series is $S_2 = 2 + 6 = 8$.

The third partial sum of the series is $S_3 = 2 + 6 + 18 = 26$.

The fourth partial sum of the series is $S_4 = 2 + 6 + 18 + 54 = 80$.

The Language of Algebra
The word *partial* means only a part, not total. Have you ever seen a *partial* eclipse of the moon?

We can see that as the number of terms gets infinitely large, the value of this series gets infinitely large.

The values of some infinite geometric series get closer and closer to a specific number as the number of terms approaches infinity. One such series is

$$\frac{3}{2} + \frac{3}{4} + \frac{3}{8} + \frac{3}{16} + \frac{3}{32} + \frac{3}{64} + \cdots \qquad \textit{Here, } r = \frac{1}{2}.$$

To see that this is true, we form some partial sums.

The first partial sum is $S_1 = \dfrac{3}{2} = \mathbf{1.5}$

The second partial sum is $S_2 = \dfrac{3}{2} + \dfrac{3}{4} = \dfrac{9}{4} = \mathbf{2.25}$

The third partial sum is $S_3 = \dfrac{3}{2} + \dfrac{3}{4} + \dfrac{3}{8} = \dfrac{21}{8} = \mathbf{2.625}$

The fourth partial sum is $S_4 = \dfrac{3}{2} + \dfrac{3}{4} + \dfrac{3}{8} + \dfrac{3}{16} = \dfrac{45}{16} = \mathbf{2.8125}$

The fifth partial sum is $S_5 = \dfrac{3}{2} + \dfrac{3}{4} + \dfrac{3}{8} + \dfrac{3}{16} + \dfrac{3}{32} = \dfrac{93}{32} = \mathbf{2.90625}$

The sixth partial sum is $S_6 = \dfrac{3}{2} + \dfrac{3}{4} + \dfrac{3}{8} + \dfrac{3}{16} + \dfrac{3}{32} + \dfrac{3}{64} = \dfrac{189}{64} = \mathbf{2.953125}$

As the number of terms in this series gets larger, the values of the partial sums approach the number 3. We say that 3 is the **limit** of S_n as n approaches infinity, and we say that 3 is the **sum of the infinite geometric series.**

To develop a formula for finding the sum of an infinite geometric series, we consider the formula that gives the sum of the first n terms.

$$S_n = \frac{a_1 - a_1 r^n}{1 - r} \qquad \text{where } r \neq 1$$

If $|r| < 1$ and a_1 is constant, the term $a_1 r^n$ in the above formula approaches 0 as n becomes very large. For example,

$$a_1 \left(\frac{1}{2}\right)^1 = \frac{1}{2} a_1, \qquad a_1 \left(\frac{1}{2}\right)^2 = \frac{1}{4} a_1, \qquad a_1 \left(\frac{1}{2}\right)^3 = \frac{1}{8} a_1$$

and so on. Thus, when n is very large, the value of $a_1 r^n$ is negligible, and the term $a_1 r^n$ in the above formula can be ignored. This reasoning justifies the following formula.

Sum of the Terms of an Infinite Geometric Series	If a_1 is the first term and r is the common ratio of an infinite geometric sequence, and $|r| < 1$, the sum of the terms of the corresponding series is given by $$S = \frac{a_1}{1 - r}$$

EXAMPLE 7 Find the sum of the terms of the infinite geometric series $125 + 25 + 5 + \cdots$

Strategy We will identify a_1 and find r.

Why To use the formula $S = \frac{a_1}{1 - r}$ to find the sum, we need to know the first term, a_1, and the common ratio, r.

Notation
The sum of the terms of an infinite geometric sequence is also denoted S_∞.

Solution In this geometric series, $a_1 = 125$ and $r = \frac{25}{125} = \frac{1}{5}$. Since $|r| = \left|\frac{1}{5}\right| = \frac{1}{5} < 1$, we can find the sum of the series. We do this by substituting 125 for a_1 and $\frac{1}{5}$ for r in the formula $S = \frac{a_1}{1-r}$ and simplifying:

$$S = \frac{a_1}{1-r} = \frac{125}{1 - \dfrac{1}{5}} = \frac{125}{\dfrac{4}{5}} = 125 \cdot \frac{5}{4} = \frac{625}{4}$$

The sum of the series $125 + 25 + 5 + \cdots$ is $\frac{625}{4} = 156.25$.

Self Check 7 Find the sum of the infinite geometric series:
$100 + 20 + 4 + \cdots$

Now Try Problem 43

EXAMPLE 8 Find the sum of the infinite geometric series:
$64 + (-4) + \frac{1}{4} + \cdots$

Strategy We will identify a_1 and find r.

Notation
This infinite geometric series could also be written:
$$64 - 4 + \frac{1}{4} - \cdots$$

Why To use the formula $S = \frac{a_1}{1-r}$ to find the sum, we need to know the first term, a_1, and the common ratio, r.

Solution In this geometric series, $a_1 = 64$ and $r = \frac{-4}{64} = -\frac{1}{16}$. Since $|r| = \left|-\frac{1}{16}\right| = \frac{1}{16} < 1$, we can find the sum of the series. We substitute 64 for a_1 and $-\frac{1}{16}$ for r in the formula $S = \frac{a_1}{1-r}$ and simplify:

Caution
If $|r| \geq 1$ for an infinite geometric sequence, the sum of the terms of the sequence does not exist.

$$S = \frac{a_1}{1-r} = \frac{64}{1 - \left(-\dfrac{1}{16}\right)} = \frac{64}{\dfrac{17}{16}} = 64 \cdot \frac{16}{17} = \frac{1{,}024}{17}$$

The sum of the geometric series $64 + (-4) + \frac{1}{4} + \cdots$ is $\frac{1{,}024}{17}$.

Self Check 8 Find the sum of the infinite geometric series:
$81 + (-27) + 9 + \ldots$

Now Try Problem 51

EXAMPLE 9 Change $0.\overline{8}$ to a common fraction.

Strategy First, we will show that the decimal $0.888\ldots$ can be represented by an infinite geometric series whose terms are fractions. Then we will identify a_1, find the common ratio r, and find the sum.

Why When we use the formula $S = \frac{a_1}{1-r}$ to find the sum, the result will be the required common fraction.

Solution   The decimal $0.\overline{8}$ can be written as the infinite geometric series

$$0.\overline{8} = 0.888\ldots$$
$$= 0.8 + 0.08 + 0.008 + \cdots$$
$$= \frac{8}{10} + \frac{8}{100} + \frac{8}{1,000} + \cdots$$

Here, $a_1 = \frac{8}{10}$ and $r = \frac{1}{10}$. Because $|r| = \left|\frac{1}{10}\right| = \frac{1}{10} < 1$, we can find the sum as follows:

$$S = \frac{a_1}{1 - r} = \frac{\dfrac{8}{10}}{1 - \dfrac{1}{10}} = \frac{\dfrac{8}{10}}{\dfrac{9}{10}} = \frac{8}{10} \cdot \frac{10}{9} = \frac{8}{9}$$

Thus, $0.\overline{8} = \frac{8}{9}$. Long division will verify that $\frac{8}{9} = 0.888\ldots$.

Self Check 9   Change $0.\overline{6}$ to a common fraction.

Now Try   **Problem 55**

---

5   **Solve Application Problems Involving Geometric Sequences.**

EXAMPLE 10   *Inheritances.*   A father decides to give his son part of his inheritance early. Each year, on the son's birthday, the father will pay the son 15% of what remains in the inheritance fund. If the fund initially begins with $100,000, how much money will be left in the fund after 20 years of payments?

Strategy   We will model the facts of the problem using a geometric sequence with a first term of 100,000 and a common ratio of 0.85.

Why   One of the terms of the sequence will represent the amount of money left in the inheritance fund after 20 years of payments.

Solution   If 15% of the money in the inheritance fund is given to the son each year, 85% of that amount remains after each payment. To find the amount of money that remains in the fund after a payment is made, we multiply the amount that was in the fund by 0.85. Over the years, the amounts of money that are left in the fund after a payment form a geometric sequence.

Amount of money remaining in the inheritance fund

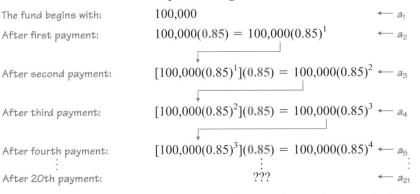

The amount of money remaining in the inheritance fund after 20 years is represented by the 21st term of a geometric sequence, where $a_1 = 100{,}000$, $r = 0.85$, and $n = 21$.

$$a_n = a_1 r^{n-1} \qquad \text{This is the formula for the nth term.}$$

$$a_{21} = a_1 r^{21-1} \qquad \text{Substitute 21 for n.}$$

$$= 100{,}000(0.85)^{21-1} \qquad \text{Substitute 100,000 for } a_1 \text{ and 0.85 for } r.$$

$$= 100{,}000(0.85)^{20}$$

$$\approx 3{,}876 \qquad \text{Use a calculator. Round to the nearest dollar.}$$

In 20 years, approximately $3,876 of the inheritance fund will be left.

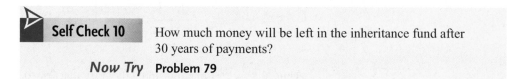

Self Check 10   How much money will be left in the inheritance fund after 30 years of payments?

Now Try   **Problem 79**

EXAMPLE 11   *Testing Steel.*   One way to measure the hardness of a steel anvil is to drop a ball bearing onto the face of the anvil. The bearing should rebound $\frac{4}{5}$ of the distance from which it was dropped. If a bearing is dropped from a height of 10 inches onto a hard forged steel anvil, and if it could bounce forever, what total distance would the bearing travel?

Strategy   We will show that the facts in the problem can be modeled by an infinite geometric sequence.

Why   The sum of the terms of the infinite geometric sequence will give the total distance the bearing will travel.

Solution   The total distance the ball bearing travels is the sum of two motions, falling and rebounding. The bearing falls 10 inches, then rebounds $\frac{4}{5} \cdot 10 = 8$ inches, and falls 8 inches, and rebounds $\frac{4}{5} \cdot 8 = \frac{32}{5}$ inches, and falls $\frac{32}{5}$ inches, and rebounds $\frac{4}{5} \cdot \frac{32}{5} = \frac{128}{25}$ inches, and so on.

The distance the ball falls is given by the sum

$$10 + 8 + \frac{32}{5} + \frac{128}{25} + \cdots \qquad \text{This is an infinite geometric series with } a_1 = 10 \text{ and } r = \frac{4}{5}.$$

The distance the ball rebounds is given by the sum

$$8 + \frac{32}{5} + \frac{128}{25} + \cdots \qquad \text{This is an infinite geometric series with } a_1 = 8 \text{ and } r = \frac{4}{5}.$$

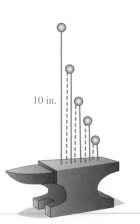

10 in.

Since each of these is an infinite geometric series with $|r| < 1$, we can use the formula $S = \dfrac{a_1}{1 - r}$ to find each sum.

$$\text{Falling: } \frac{10}{1 - \dfrac{4}{5}} = \frac{10}{\dfrac{1}{5}} = 50 \text{ in.} \qquad \text{Rebounding: } \frac{8}{1 - \dfrac{4}{5}} = \frac{8}{\dfrac{1}{5}} = 40 \text{ in.}$$

The total distance the bearing travels is 50 inches + 40 inches = 90 inches.

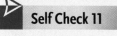

Self Check 11 If a bearing was dropped from a height of 15 inches onto a hard forged steel anvil, and if it could bounce forever, what total distance would it travel?

Now Try **Problem 85**

ANSWERS TO SELF CHECKS **1. a.** 3, 12, 48, 192 **b.** 49,152 **2.** $\frac{1}{625}$ **3.** $-2, -6, -18, -54, -162$
4. 2, 4, 8 **5.** 20 **6.** 124.96 **7.** 125 **8.** $\frac{243}{4}$ **9.** $\frac{2}{3}$ **10.** About \$763 **11.** 135 in.

STUDY SET
11.3

VOCABULARY

Fill in the blanks.

1. Each term of a _____ sequence is found by multiplying the previous term by the same number.

2. 8, 16, 32, 64, 128, . . . is an example of a _____ sequence. The first _____ is 8 and the common _____ is 2.

3. If a single number is inserted between a and b to form a geometric sequence, the number is called the geometric _____ between a and b.

4. The sum of the terms of a geometric sequence is called a geometric _____. The sum of the terms of an infinite geometric sequence is called an _____ geometric series.

CONCEPTS

5. Write the first three terms of a geometric sequence if $a_1 = 16$ and $r = \frac{1}{4}$.

6. Given the geometric sequence 1, 2, 4, 8, 16, 32, . . . , find a_5 and r.

7. Write the formula for a_n, the general term of a geometric sequence.

8. a. Write the formula for S_n, the sum of the first n terms of a geometric sequence.

b. Write the formula for S, the sum of the terms of an infinite geometric sequence, where $|r| < 1$.

9. Which of the following values of r satisfy $|r| < 1$?

a. $r = \frac{2}{3}$ **b.** $r = -3$

c. $r = 6$ **d.** $r = -\frac{1}{5}$

10. Write $0.\overline{7}$ as an infinite geometric series:

$$0.\overline{7} = 0.7 + 0.07 + 0.007 + \cdots$$

$$= \frac{7}{\rule{1cm}{0.4pt}} + \frac{7}{\rule{1cm}{0.4pt}} + \frac{7}{\rule{1cm}{0.4pt}} + \cdots$$

NOTATION

Fill in the blanks.

11. An infinite geometric sequence is of the form

$$a_1, \ a_1r, \ a_1\rule{0.6cm}{0.4pt}, \ a_1r^3, \ \rule{0.8cm}{0.4pt}, \ldots$$

12. The first four terms of the sequence defined by $a_n = 4(3)^{n-1}$ are ___, ___, ___, ___.

13. To find the common ratio of a geometric sequence, we use the formula $r = \dfrac{a_{n+1}}{a}$.

14. S_8 represents the sum of the first _____ terms of a geometric sequence.

GUIDED PRACTICE

Write the first five terms of each geometric sequence with the given properties and then find the specified term. **See Example 1.**

15. First term: 3, common ratio: 2; find the 9th term

16. First term: -2, common ratio: 2; find the 8th term

17. First term: -5, common ratio: $\frac{1}{5}$; find the 8th term

18. First term: 8, common ratio: $\frac{1}{2}$; find the 10th term

Find the specified term of the geometric sequence with the following properties. **See Example 2.**

19. The first three terms are 2, 6, 18; find the 7th term

20. The first three terms are 50, 100, 200; find the 10th term

21. The first three terms are $\frac{1}{2}, -\frac{5}{2}, \frac{25}{2}$; find the 6th term

22. The first three terms are $\frac{1}{4}, -\frac{3}{4}, \frac{9}{4}$; find the 9th term

Write the first five terms of the geometric sequence with the following properties. See Example 3.

23. First term: 2, $r > 0$, third term: 18

24. First term: 2, $r < 0$, third term: 50

25. First term: 3, fourth term: 24

26. First term: -3, fourth term: -192

Find the geometric means to be inserted in each geometric sequence. See Example 4.

27. Insert three positive geometric means between 2 and 162.

28. Insert four geometric means between 3 and 96.

29. Insert four geometric means between -4 and $-12,500$.

30. Insert three geometric means (two positive and one negative) between -64 and $-1,024$.

Find a geometric mean to be inserted in each geometric sequence. See Example 5.

31. Find the geometric mean between 2 and 128.

32. Find the geometric mean between 3 and 243.

33. Find the geometric mean between 10 and 20.

34. Find the geometric mean between 5 and 15.

For each geometric sequence, find the sum of the specified number of terms. See Example 6.

35. The first 6 terms of 2, 6, 18, . . .

36. The first 6 terms of 2, -6, 18, . . .

37. The first 5 terms of 2, -6, 18, . . .

38. The first 5 terms of 2, 6, 18, . . .

39. The first 8 terms of 3, -6, 12, . . .

40. The first 8 terms of 3, 6, 12, . . .

41. The first 7 terms of 3, 6, 12, . . .

42. The first 7 terms of 3, -6, 12, . . .

Find the sum of each infinite geometric series, if possible. See Examples 7 and 8.

43. $8 + 4 + 2 + \cdots$

44. $12 + 6 + 3 + \cdots$

45. $54 + 18 + 6 + \cdots$

46. $45 + 15 + 5 + \cdots$

47. $-\frac{27}{2} + (-9) + (-6) + \cdots$

48. $-112 + (-28) + (-7) + \cdots$

49. $\frac{9}{2} + 6 + 8 + \cdots$

50. $\frac{18}{25} + \frac{6}{5} + 2 + \cdots$

51. $12 + (-6) + 3 + \cdots$

52. $8 + (-4) + 2 + \cdots$

53. $-45 + 15 + (-5) + \cdots$

54. $-54 + 18 + (-6) + \cdots$

Write each decimal in fraction form. Then check the answer by performing long division. See Example 9.

55. $0.\overline{1}$

56. $0.\overline{2}$

57. $0.\overline{3}$

58. $0.\overline{4}$

59. $0.\overline{12}$

60. $0.\overline{21}$

61. $0.\overline{75}$

62. $0.\overline{57}$

TRY IT YOURSELF

63. Find the common ratio of the geometric sequence with a first term -8 and a sixth term $-1,944$.

64. Find the common ratio of the geometric sequence with a first term 12 and a sixth term $\frac{3}{8}$.

65. Write the first five terms of the geometric sequence if its first term is -64, $r < 0$, and its fifth term is -4.

66. Write the first five terms of the geometric sequence if its first term is -64, $r > 0$, and its fifth term is -4.

67. Find a geometric mean, if possible, between -50 and 10.

68. Find a negative geometric mean, if possible, between -25 and -5.

69. Find the 10th term of the geometric sequence with $a_1 = 7$ and $r = 2$.

70. Find the 12th term of the geometric sequence with $a_1 = 64$ and $r = \frac{1}{2}$.

71. Write the first four terms of the geometric sequence if its first term is -64 and its sixth term is -2.

72. Write the first four terms of the geometric sequence if its first term is -81 and its sixth term is $\frac{1}{3}$.

73. Find the sum of the terms of the geometric sequence $3, \frac{3}{4}, \frac{3}{16}, \frac{3}{64}, \cdots$

74. Find the sum of the terms of the geometric sequence $1, -\frac{1}{2}, \frac{1}{4}, -\frac{1}{8}, \cdots$

75. Find the first term of the geometric sequence with a common ratio -3 and an eighth term -81.

76. Find the first term of the geometric sequence with a common ratio 2 and a tenth term 384.

77. Find the sum of the first five terms of the geometric sequence if its first term is 3 and the common ratio is 2.

78. Find the sum of the first five terms of the geometric sequence if its first term is 5 and the common ratio is -6.

APPLICATIONS

Use a calculator to help solve each problem.

79. DECLINING SAVINGS   John has $10,000 in a safety deposit box. Each year, he spends 12% of what is left in the box. How much will be in the box after 15 years?

80. SAVINGS GROWTH   Sally has $5,000 in a savings account earning 12% annual interest. How much will be in her account 10 years from now? (Assume that Sally makes no deposits or withdrawals.)

81. *from Campus to Careers*
   Real Estate Sales Agent

Suppose you are a real estate sales agent and you are working with a client who is considering buying a $250,000 house in Seattle as an investment. If the property values in that area have a track record of increasing at a rate of 8% per year, what will the house be worth 10 years from now?

© Image Source Black/Getty Images

82. BOAT DEPRECIATION A boat that cost $5,000 when new depreciates at a rate of 9% per year. How much will the boat be worth in 5 years?

83. INSCRIBED SQUARES Each inscribed square in the illustration joins the midpoints of the next larger square. The area of the first square, the largest, is 1 square unit. Find the area of the 12th square.

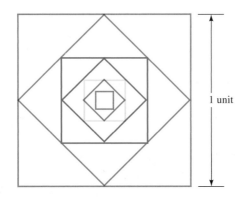

1 unit

84. GENEALOGY The following family tree spans 3 generations and lists 7 people. How many names would be listed in a family tree that spans 10 generations?

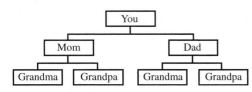

85. BOUNCING BALLS On each bounce, the rubber ball in the illustration rebounds to a height one-half of that from which it fell. Find the total vertical distance the ball travels.

10 m

86. BOUNCING BALLS A golf ball is dropped from a height of 12 feet. On each bounce, it returns to a height that is two-thirds of the distance it fell. Find the total vertical distance the ball travels.

87. PEST CONTROL To reduce the population of a destructive moth, biologists release 1,000 sterilized male moths each day into the environment. If 80% of these moths alive one day survive until the next, then after a long time the population of sterile males is the sum of the infinite geometric series

$$1,000 + 1,000(0.8) + 1,000(0.8)^2 + 1,000(0.8)^3 + \cdots$$

Find the long-term population.

88. PENDULUMS On its first swing to the right, a pendulum swings through an arc of 96 inches. Each successive swing, the pendulum travels $\frac{99}{100}$ as far as on the previous swing. Determine the total distance that the pendulum will travel by the time it comes to rest.

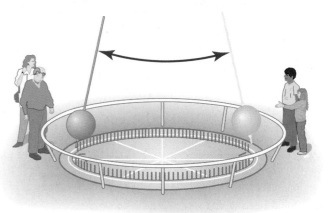

WRITING

89. Describe the real numbers that satisfy $|r| < 1$.

90. Why must the absolute value of the common ratio be less than 1 before an infinite geometric sequence can have a sum?

91. Explain the difference between an arithmetic sequence and a geometric sequence.

92. Why is $1 - \frac{1}{2} + \frac{1}{4} - \frac{1}{8} + \frac{1}{16} - \frac{1}{32} + \cdots$ called an alternating infinite geometric series?

REVIEW

Solve each inequality. Write the solution set using interval notation.

93. $x^2 - 5x - 6 \leq 0$

94. $a^2 - 7a + 12 \geq 0$

95. $\dfrac{x - 4}{x + 3} > 0$

96. $\dfrac{t^2 + t - 20}{t + 2} < 0$

CHALLENGE PROBLEMS

97. If $f(x) = 1 + x + x^2 + x^3 + x^4 + \cdots$, find $f\left(\frac{1}{2}\right)$ and $f\left(-\frac{1}{2}\right)$.

98. Find the sum:

$$\frac{1}{\sqrt{3}} + \frac{1}{3} + \frac{1}{3\sqrt{3}} + \frac{1}{9} + \cdots$$

99. If $a > b > 0$, which is larger: the arithmetic mean between a and b or the geometric mean between a and b?

100. Is there a geometric mean between -5 and 5?

CHAPTER 11
Summary & Review

SECTION 11.1   The Binomial Theorem

DEFINITIONS AND CONCEPTS	EXAMPLES
Pascal's triangle gives the coefficients of the terms of the expansion of $(a + b)^n$.	**Pascal's triangle:** 1 — Row 0 1 1 — Row 1 1 2 1 — Row 2 1 3 3 1 — Row 3 1 4 6 4 1 — Row 4 1 5 10 10 5 1 — Row 5 $(x + y)^5 = 1x^5 + 5x^4y + 10x^3y^2 + 10x^2y^3 + 5xy^4 + 1y^5$
The symbol $n!$ (*n factorial*) is the product of consecutively decreasing natural numbers from n to 1. $n! = n(n - 1)(n - 2) \cdot \ldots \cdot 2 \cdot 1$	Evaluate each expression: $4! = 4 \cdot 3 \cdot 2 \cdot 1 = 24$ $3! \cdot 2! = 3 \cdot 2 \cdot 1 \cdot 2 \cdot 1 = 12$ $\dfrac{6!}{5!} = \dfrac{6 \cdot \overset{1}{\cancel{5!}}}{\underset{1}{\cancel{5!}}} = 6$ $1! = 1$ and $0! = 1$
The **binomial theorem** is usually the best way to expand a binomial. $(a + b)^n = a^n + \dfrac{n!}{1!(n - 1)!}a^{n-1}b +$ $\dfrac{n!}{2!(n - 2)!}a^{n-2}b^2 + \ldots + b^n$	Use the binomial theorem to expand $(p + q)^4$, where $n = 4$. $(p + q)^4 = p^4 + \dfrac{4!}{1!(4 - 1)!}p^3q + \dfrac{4!}{2!(4 - 2)!}p^2q^2 + \dfrac{4!}{3!(4 - 3)!}pq^3 + q^4$ $= p^4 + 4p^3q + 6p^2q^2 + 4pq^3 + q^4$
To find a specific term of an expansion: The $(r + 1)$st term of the expansion of $(a + b)^n$ is $\dfrac{n!}{r!(n - r)!}a^{n-r}b^r$ Remember that r is always 1 less than the number of the term that you are finding.	Find the third term of the expansion of $(a + b)^5$. In the third term of $(a + b)^5$, $n = 5$ and $r = 2$. $\dfrac{5!}{2!(5 - 2)!}a^{5-2}b^2 = \dfrac{5 \cdot 4 \cdot \overset{1}{\cancel{3!}}}{2 \cdot 1 \cdot \underset{1}{\cancel{3!}}}a^3b^2 = 10a^3b^2$

REVIEW EXERCISES

1. Complete Pascal's triangle. List the row that gives the coefficients for the expansion of $(a + b)^5$.

```
                1
              1   1
          1     2     1
       1     3
    4     6     4     1
 1   5          10      5    1
1   6    15    20           6    1
```

2. Consider the expansion of $(a + b)^{12}$.

   a. How many terms does the expansion have?

   b. For each term, what is the sum of the exponents on a and b?

   c. What is the first term? What is the last term?

   d. How do the exponents on a and b change from term to term?

Evaluate each expression.

3. $4! \cdot 3!$

4. $\dfrac{5!}{3!}$

5. $\dfrac{6!}{2!(6 - 2)!}$

6. $\dfrac{12!}{3!(12 - 3)!}$

7. $(n - n)!$

8. $\dfrac{8!}{7!}$

Use the binomial theorem to find each expansion.

9. $(x + y)^5$

10. $(x - y)^9$

11. $(4x - y)^3$

12. $\left(\dfrac{c}{2} + \dfrac{d}{3}\right)^4$

Find the specified term in each expansion.

13. $(x + y)^4$; third term

14. $(x - y)^6$; fourth term

15. $(3x - 4y)^3$; second term

16. $(u^2 - v^3)^5$; fifth term

SECTION 11.2   Arithmetic Sequences and Series

DEFINITIONS AND CONCEPTS	EXAMPLES
A **sequence** is a list of numbers written in a specific order.	**Finite sequence:** 1, 4, 7, 10, 13 **Infinite sequence:** 2, 6, 10, 14, 18, . . .
To specifically describe all the terms of a sequence, we can write a formula for a_n, called the **general term** of the sequence.	Find the first four terms of the sequence described by $a_n = 5n + 1$. We substitute 1, 2, 3, and 4, for n in the formula: $a_1 = 5(1) + 1 = 6 \qquad a_2 = 5(2) + 1 = 11$ $a_3 = 5(3) + 1 = 16 \qquad a_4 = 5(4) + 1 = 21$ The first four terms are 6, 11, 16, and 21.
A sequence where each term is found by adding the same number to the previous term is called an **arithmetic sequence.** An arithmetic sequence has the form $a_1,\ a_1 + d,\ a_1 + 2d,\ \ldots,\ a_1 + (n-1)d, \ldots$ where a_1 is the first term and d is the common difference. The **nth term of an arithmetic sequence** is given by $a_n = a_1 + (n-1)d$. The **common difference d** of an arithmetic sequence is the difference between any two consecutive terms: $d = a_{n+1} - a_n$.	Find the 12th term of the arithmetic sequence $-4, -1, 2, 5, 8, \ldots$. First, we find the common difference d. It is the difference between any two successive terms: $d = a_2 - a_1 = -1 - (-4) = 3$ Then we substitute into the formula for the nth term: $a_n = a_1 + (n-1)d$ $a_{12} = -4 + (12 - 1)3$ In this sequence, $n = 12$, $a_1 = -4$, and $d = 3$. $\phantom{a_{12}} = -4 + (11)3$ $\phantom{a_{12}} = -4 + 33$ $\phantom{a_{12}} = 29$ The 12th term of the sequence is 29.
If numbers are inserted between two given numbers a and b to form an arithmetic sequence, the inserted numbers are **arithmetic means** between a and b.	Find three arithmetic means between -6 and 14. Since there will be five terms, $n = 5$. We also know that $a_1 = -6$ and $a_5 = 14$. We will substitute these values in the formula for the nth term and solve for d. $a_5 = a_1 + (n-1)d$ $14 = -6 + (5 - 1)d$ $14 = -6 + 4d$ $20 = 4d$ $5 = d$ Since the common difference is 5, we successively add 5 to the terms to get the sequence $-6, -1, 4, 9, 14$. Thus, the three arithmetic means are -1, 4, and 9.

SECTION 11.2   Arithmetic Sequences and Series—*continued*

DEFINITIONS AND CONCEPTS	EXAMPLES
When the commas between the terms of a sequence are replaced with + signs, we call the sum a **series**. The **sum of the first *n* terms of an arithmetic sequence** is given by $$S_n = \frac{n(a_1 + a_n)}{2}$$	Find the sum of the first 12 terms of the arithmetic sequence $-4, -1, 2, 5, 8, \ldots$. Here $a_1 = -4$ and $n = 12$. On the previous page, we found that for this sequence $a_{12} = 29$. We substitute these values into the formula for S_n and simplify. $$S_n = \frac{n(a_1 + a_n)}{2}$$ $$S_{12} = \frac{12(-4 + 29)}{2}$$ $$= \frac{300}{2}$$ $$= 150$$ The sum of the first 12 terms is 150.
Summation notation involves the Greek letter sigma Σ. It designates the sum of terms called a **series**.	$$\underset{\substack{\uparrow \\ n=1}}{\sum^{4}} (3k - 2) = \overset{k=1}{[3(1) - 2]} + \overset{k=2}{[3(2) - 2]} + \overset{k=3}{[3(3) - 2]} + \overset{k=4}{[3(4) - 2]}$$ $$= 1 + 4 + 7 + 10$$ $$= 22$$

REVIEW EXERCISES

17. Find the first four terms of the sequence defined by $a_n = 2n - 4$.

18. Find the first five terms of the sequence defined by $a_n = \frac{(-1)^n}{n + 1}$.

19. Find the 50th term of the sequence defined by $a_n = 100 - \frac{n}{2}$.

20. Find the eighth term of an arithmetic sequence whose first term is 7 and whose common difference is 5.

21. Write the first five terms of the arithmetic sequence whose ninth term is 242 and whose seventh term is 212.

22. The first three terms of an arithmetic sequence are 6, -6, and -18. Find the 101st term.

23. Find the common difference of an arithmetic sequence if its 1st term is -515 and the 23rd term is -625.

24. Find two arithmetic means between 8 and 25.

25. Find the sum of the first ten terms of the sequence 9, 6.5, 4, ...

26. Find the sum of the first 28 terms of an arithmetic sequence if the second term is 6 and the sixth term is 22.

Find each sum.

27. $\displaystyle\sum_{k=4}^{6} \frac{1}{2}k$

28. $\displaystyle\sum_{k=2}^{5} 7k^2$

29. $\displaystyle\sum_{k=1}^{4} (3k - 4)$

30. $\displaystyle\sum_{k=10}^{10} 36k$

31. What is the sum of the first 200 natural numbers?

32. SEATING   The illustration shows the first 2 of a total of 30 rows of seats in an amphitheater. The number of seats in each row forms an arithmetic sequence. Find the total number of seats.

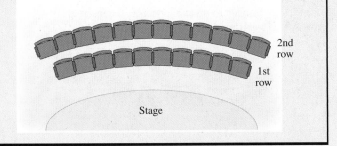

SECTION 11.3   Geometric Sequences and Series

DEFINITIONS AND CONCEPTS	EXAMPLES
Each term of a **geometric sequence** is found by multiplying the previous term by the same number. A geometric sequence has the form: $$a_1, a_1 r, a_1 r^2, a_1 r^3, \ldots, a_1 r^{n-1}, \ldots$$ where a_1 is the first term and r is the common ratio. The ***n*th term of a geometric sequence** is given by $a_n = a_1 r^{n-1}$. The **common ratio** r of a geometric sequence is the quotient obtained when any term is divided by the previous term: $$r = \frac{a_{n+1}}{a_n}$$	Find the 8th term of the geometric sequence 8, 4, 2, ... First, we find the common ratio r. It is the quotient obtained when any term is divided by the previous term: $$r = \frac{a_2}{a_1} = \frac{4}{8} = \frac{1}{2}$$ Then we substitute into the formula for the *n*th term: $$a_n = a_1 r^{n-1}$$ $$a_8 = 8\left(\frac{1}{2}\right)^{8-1} \quad \text{Substitute: } n = 8,\ a_1 = 8, \text{ and } r = \tfrac{1}{2}.$$ $$= 8\left(\frac{1}{2}\right)^7$$ $$= \frac{8}{128} \quad \text{Evaluate: } \left(\tfrac{1}{2}\right)^7 = \tfrac{1}{128}.$$ $$= \frac{1}{16} \quad \text{Simplify the fraction: Remove a common factor of 8 in the numerator and denominator.}$$
If numbers are inserted between a and b to form a geometric sequence, the inserted numbers are **geometric means** between a and b.	Find two geometric means between 6 and 162. Since there will be four terms, $n = 4$. We also know that $a_1 = 6$ and $a_4 = 162$. We will substitute these values in the formula for the *n*th term and solve for r. $$a_n = a_1 r^{n-1}$$ $$162 = 6 r^{4-1}$$ $$27 = r^3 \quad \text{Divide both sides by 6.}$$ $$3 = r \quad \text{Take the cube root of both sides.}$$ Since the common ratio is 3, we successively multiply the terms by 3 to get the sequence 6, 18, 54, 162. Thus, the two geometric means are 18 and 54.
The **sum of the first *n* terms of a geometric sequence** is given by: $$S_n = \frac{a_1 - a_1 r^n}{1 - r} \quad r \neq 1$$	Find the sum of the first five terms of the geometric sequence 2, 6, 18, ... Here $a_1 = 2$, $r = 3$, and $n = 5$. We substitute these values into the formula for S_n and simplify. $$S_n = \frac{a_1 - a_1(r)^n}{1 - r}$$ $$S_5 = \frac{2 - (2)(3)^5}{1 - 3}$$ $$= \frac{2 - 2(243)}{-2} \quad \text{Evaluate: } (3)^5 = 243.$$ $$= \frac{2 - 486}{-2}$$ $$= \frac{-484}{-2}$$ $$= 242$$ The sum of the first five terms is 242.

SECTION 11.3 Geometric Sequences and Series–*continued*

DEFINITIONS AND CONCEPTS	EXAMPLES
If we form the sum of the terms of an infinite geometric sequence, we get a series called an **infinite geometric series.** The sum of an infinite geometric series is given by: $$S = \frac{a_1}{1-r} \qquad \text{where } |r| < 1$$	Find the sum of the infinite geometric series: $12 + 8 + \frac{16}{3} + \cdots$ Here $a_1 = 12$ and $r = \frac{8}{12} = \frac{2}{3}$. (Note that $r < 1$) We substitute these values into the formula for the sum of an infinite series. $$S = \frac{a_1}{1-r} = \frac{12}{1 - \frac{2}{3}} = \frac{12}{\frac{1}{3}} = 12 \cdot \frac{3}{1} = 36$$ The sum is 36.

REVIEW EXERCISES

33. Find the sixth term of a geometric sequence with a first term of $\frac{1}{8}$ and a common ratio of 2.

34. Write the first five terms of the geometric sequence whose fourth term is 3 and whose fifth term is $\frac{3}{2}$.

35. Find the first term of a geometric sequence if it has a common ratio of -3 and the ninth term is 243.

36. Find two geometric means between -6 and 384.

37. Find the sum of the first seven terms of the sequence 162, 54, 18, ...

38. Find the sum of the first eight terms of the sequence: $\frac{1}{8}, -\frac{1}{4}, \frac{1}{2}, \ldots$.

39. FEEDING BIRDS   Tom has 50 pounds of birdseed stored in his garage. Each month, he uses 25% of what is left in the bag to feed the birds in his yard. How much birdseed will be left in 12 months?

40. Find the sum of the infinite geometric series: $25 + 20 + 16 + \cdots$

41. Change the decimal $0.\overline{05}$ to a common fraction.

42. WHAM-O TOYS   Tests have found that 1998 Superballs rebound $\frac{9}{10}$ of the distance from which they are dropped. If a Superball is dropped from a height of 10 feet, and if it could bounce forever, what total distance would it travel?

CHAPTER 11
Test

1. Fill in the blanks.

   a. The array of numbers that gives the coefficients of the terms of a binomial expansion is called _____ triangle.

   b. In the expansion $a^3 - 3a^2b + 3ab^2 - b^3$, the signs _____ between $+$ and $-$.

   c. Each term of an _____ sequence is found by adding the same number to the previous term.

   d. The sum of the terms of an arithmetic sequence is called an arithmetic _____.

   e. Each term of a _____ sequence is found by multiplying the previous term by the same number.

2. Find the first 4 terms of the sequence defined by: $a_n = -6n + 8$

3. Find the first 5 terms of the sequence defined by: $a_n = \frac{(-1)^n}{n^3}$.

4. Evaluate: $\frac{10!}{6!(10 - 6)!}$

5. Expand: $(a - b)^6$.

6. Find the third term in the expansion of $(x^2 + 2y)^4$.

7. Find the tenth term of an arithmetic sequence whose first 3 terms are 3, 10, and 17.

8. Find the sum of the first 12 terms of the sequence: $-2, 3, 8, \ldots$

9. Find two arithmetic means between 2 and 98.

10. Find the common difference of an arithmetic sequence if the second term is $\frac{5}{4}$ and the 17th term is 5.

11. Find the sum of the first 27 terms of an arithmetic sequence if the 4th term is -11 and the 20th term is -75.

12. PLUMBING   Plastic pipe is stacked so that the bottom row has 25 pipes, the next row has 24 pipes, the next row has 23 pipes, and so on until there is 1 pipe at the top of the stack. If a worker removes the top 15 rows of pipe, how many pieces of pipe will be left in the stack?

13. FALLING OBJECTS   If an object is in free fall, the sequence 16, 48, 80, . . . represents the distance in feet that object falls during the 1st second, during the 2nd second, during the 3rd second, and so on. How far will the object fall during the first 10 seconds?

14. Evaluate: $\displaystyle\sum_{k=1}^{3}(2k-3)$

15. Find the seventh term of the geometric sequence whose first 3 terms are $-\frac{1}{9}, -\frac{1}{3}$, and -1.

16. Find the sum of the first 6 terms of the sequence: $\frac{1}{27}, \frac{1}{9}, \frac{1}{3}, \dots$

17. Find the first term of a geometric sequence if the common ratio is $-\frac{2}{3}$ and the fourth term is $-\frac{16}{9}$.

18. Find two geometric means between 3 and 648.

19. Find the sum of infinite geometric series: $9 + 3 + 1 + \cdots$

20. DEPRECIATION   A yacht that cost $1,500,000 when new depreciates at a rate of 8% per year. How much will the yacht be worth in 10 years?

21. PENDULUMS   On its first swing to the right, a pendulum swings through an arc of 60 inches. Each successive swing, the pendulum travels $\frac{97}{100}$ as far as on the previous swing. Determine the total distance the pendulum will travel by the time it comes to rest.

60 in

22. Change the decimal $0.\overline{7}$ to a common fraction.

GROUP PROJECT

THE LANGUAGE OF ALGEBRA

Overview: This activity will help you review for the final exam.

Instructions: Form groups of two students. Match each instruction in column I with the most appropriate item in column II. Each letter in column II is used only once.

Column I

_____ 1. Use the FOIL method.

_____ 2. Apply a rule for exponents to simplify.

_____ 3. Add the rational expressions.

_____ 4. Rationalize the denominator.

_____ 5. Factor completely.

_____ 6. Evaluate the expression for $a = -1$ and $b = -6$.

_____ 7. Express in lowest terms.

_____ 8. Solve for t.

_____ 9. Combine like terms.

_____ 10. Remove parentheses.

_____ 11. Solve the system by graphing.

_____ 12. Find $f(g(x))$.

_____ 13. Solve using the quadratic formula.

_____ 14. Identify the base and the exponent.

_____ 15. Write without a radical symbol.

_____ 16. Write the equation of the line having the given slope and y-intercept.

_____ 17. Solve the inequality.

_____ 18. Complete the square to make a perfect-square trinomial.

_____ 19. Find the slope of the line passing through the given points.

_____ 20. Use a property of logarithms to simplify.

_____ 21. Set each factor equal to zero and solve for x.

_____ 22. State the solution of the compound inequality using interval notation.

_____ 23. Find the inverse function, $h^{-1}(x)$.

_____ 24. Write using scientific notation.

_____ 25. Write the logarithmic statement in exponential form.

_____ 26. Find the sum of the first 6 terms of the sequence.

Column II

a. $2,300,000,000$

b. e^3

c. $f(x) = x^2 + 1$ and $g(x) = 5 - 3x$

d. $-2x(3x^2 - 4x + 8)$

e. $4x - 7 > -3x - 7$

f. $\begin{cases} 2x = y - 5 \\ x + y = -1 \end{cases}$

g. $(x^2 - 5)(x^2 + 3)$

h. $(x + 2)(x - 10) = 0$

i. $\dfrac{x - 1}{2x^2} + \dfrac{x + 1}{8x}$

j. $\sqrt{4x^2}$

k. $\ln 6 + \ln x$

l. $\dfrac{10}{\sqrt{6} - \sqrt{2}}$

m. $2x - 8 + 6y - 14$

n. $(3, -2)$ and $(0, -5)$

o. $x^4 \cdot x^3$

p. $\dfrac{4x^2y}{16xy}$

q. $h(x) = 10^x$

r. $\log_2 8 = 3$

s. $m = \dfrac{2}{3}$ and passes through $(0, 2)$

t. $2, 6, 18, \dots$

u. $3y^3 - 243b^6$

v. $x + 7 \geq 0$ and $-x < -1$

w. $x^2 - 3x - 4 = 0$

x. $Rt = cd + 2t$

y. $-2\pi a^2 - 3b^3$

z. $x^2 + 4x$

CUMULATIVE REVIEW
Chapters 1–11

1. Give the elements of the set $\left\{-\frac{4}{3},\ \pi,\ 5.6,\ \sqrt{2},\ 0,\ -23,\ e,\ 7i\right\}$ that belong to each of the following sets. [Section 1.2]

  a. Whole numbers

  b. Rational numbers

  c. Irrational numbers

  d. Real numbers

2. Solve: $6[x - (2 - x)] = -4(8x + 3)$ [Section 1.5]

3. Solve $A = \frac{1}{2}h(b_1 + b_2)$ for b_2. [Section 1.6]

4. MARTIAL ARTS   Find the measure of each angle of the triangle shown in the illustration. [Section 1.7]

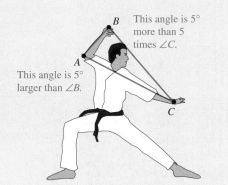

This angle is 5° more than 5 times ∠C.

This angle is 5° larger than ∠B.

5. FINANCIAL PLANNING   Anna has some money to invest. Her financial planner tells her that if she can come up with $3,000 more, she will qualify for an 11% annual interest rate. Otherwise, she will have to invest the money at 7.5% annual interest. The financial planner urges her to invest the larger amount, because the 11% investment would yield twice as much annual income as the 7.5% investment. How much does she originally have on hand to invest? [Section 1.8]

6. BOATING   Use the following graph to determine the average rate of change in the sound level of the engine of a boat in relation to the number of revolutions per minute (rpm) of the engine. [Section 2.3]

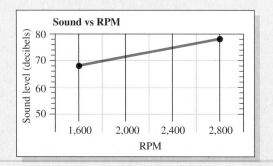

7. Decide whether the graphs of the equations are parallel or perpendicular. [Section 2.3]

  a. $3x - 4y = 12,\ \ y = \frac{3}{4}x - 5$

  b. $y = 3x + 4,\ \ x = -3y + 4$

8. SALVAGE VALUES   A truck was purchased for $28,000. Its salvage value at the end of 6 years is expected to be $7,600. Find the straight-line depreciation equation.

Find an equation of the line with the given properties. Write the equation in slope–intercept form.

9. $m = -2$, passing through $(0, 5)$ [Section 2.4]

10. Passing through $(8, -5)$ and $(-5, 4)$ [Section 2.4]

11. Explain why the graph does not represent a function. [Section 2.5]

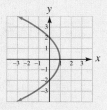

12. If $f(x) = 3x^5 - 2x^2 + 1$, find $f(-1)$ and $f(a)$. [Section 2.5]

13. Use the graph of function h to find each of the following. [Section 2.6]

  a. $h(-3)$

  b. $h(4)$

  c. The value(s) of x for which $h(x) = 1$

  d. The value(s) of x for which $h(x) = 0$

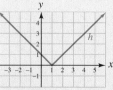

14. The graphs of $y = 4(x - 5) - x - 2$ and $y = -(2x + 6) - 1$ are shown in the illustration. Use the information in the display to solve $4(x - 5) - x - 2 = -(2x + 6) - 1$ graphically. [Section 3.1]

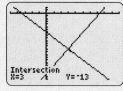

15. Use substitution to solve: $\begin{cases} 2x - y = -21 \\ 4x + 5y = 7 \end{cases}$ [Section 3.2]

16. Use addition to solve: $\begin{cases} 4y + 5x - 7 = 0 \\ \dfrac{10}{7}x - \dfrac{4}{9}y = \dfrac{17}{21} \end{cases}$ [Section 3.2]

17. MIXING COFFEE   How many pounds of regular coffee (selling for $4 per pound) and how many pounds of Brazilian coffee (selling for $11.50 per pound) must be combined to get 40 pounds of a mixture worth $6 per pound? [Section 3.3]

18. Solve: $\begin{cases} b + 2c = 7 - a \\ a + c = 8 - 2b \\ 2a + b + c = 9 \end{cases}$ [Section 3.4]

19. Use matrices to solve the system: $\begin{cases} 2x + y = 1 \\ x + 2y = -4 \end{cases}$

[Section 3.6]

20. Use Cramer's rule to solve: $\begin{cases} 2(x + y) + 1 = 0 \\ 3x + 4y = 0 \end{cases}$

[Section 3.7]

Give the solution in interval notation and graph the solution set.

21. Solve: $4.5x - 1 < -10$ or $6 - 2x \geq 12$
[Section 4.2]

22. Solve: $|5 - 3x| - 14 \leq 0$ [Section 4.3]

23. Graph the solution set: $\begin{cases} 3x - 2y \leq 6 \\ y < -x + 2 \end{cases}$ [Section 4.5]

24. WORK SCHEDULES   A student works two part-time jobs. She earns \$12 an hour for working at the college book store and \$22.50 an hour working graveyard shift at an airport. To save time for study, she limits her work to 25 hours a week. If she enjoys the work at the bookstore more, how many hours can she work at the bookstore and still earn at least \$450 a week? [Section 4.1]

Simplify each expression. Write answers using positive exponents.

25. $(x^2)^5 y^7 y^3 x^{-2} y^0$ **26.** $\left(\dfrac{3x^5 y^2}{6x^5 y^{-2}} \right)^{-4}$

[Section 5.1]   [Section 5.1]

27. Write 173,000,000,000,000 and 0.000000046 in scientific notation. [Section 5.2]

28. Write each number in scientific notation and perform the indicated operations. Give the answer in scientific notation and in standard notation. [Section 5.2]

$$\dfrac{(0.00024)(96,000,000)}{(640,000,000)(0.025)}$$

29. Complete the table of values for the polynomial function $f(x) = -x^3 - x^2 + 6x$ and then graph it. What are the x- and y-intercepts of the graph? [Section 5.3]

x	$f(x)$
-4	
-3	
-2	
-1	
0	
1	
2	
3	

30. Simplify the polynomial: $\dfrac{9}{4}rt^2 - \dfrac{5}{3}rt - \dfrac{1}{2}rt^2 + \dfrac{5}{6}rt$

[Section 5.3]

Perform the indicated operations and simplify, if possible.

31. $(-2x^2 y^3 + 6xy + 5y^2) - (-4x^2 y^3 - 7xy + 2y^2)$
[Section 5.3]

32. $(x - 3y)(x^2 + 3xy + 9y^2)$ [Section 5.4]

33. $(2m^5 - 7)(3m^5 - 1)$ [Section 5.4]

34. $(9ab^2 - 4)^2$ [Section 5.4]

Factor the expression completely.

35. $3x^3 y - 4x^2 y^2 - 6x^2 y + 8xy^2$ [Section 5.5]

36. $b^3 - 4b^2 - 3b + 12$ [Section 5.5]

37. $12y^2 + 23y + 10$ [Section 5.6]

38. $256x^4 y^4 - z^8$ [Section 5.7]

39. $27t^3 + u^3$ [Section 5.7]

40. $a^4 b^2 - 20a^2 b^2 + 64b^2$ [Section 5.8]

41. Solve for λ: $\dfrac{A\lambda}{2} + 1 = 2d + 3\lambda$ [Section 5.5]

42. Solve: $(x + 7)^2 = -2(x + 7) - 1$ [Section 5.9]

43. Solve: $x^3 + 8x^2 = 9x$ [Section 5.9]

44. PAINTING   When it is spread out, a rectangular-shaped painting tarp covers an area of 84 square feet. Its length is 1 foot longer than five times its width. Find its width and length. [Section 5.9]

45. Find the domain of the rational function $f(x) = \dfrac{2x^2 - 3x - 2}{x^2 + 2x - 24}$. Write the domain using interval notation. [Section 6.1]

46. Simplify: $\dfrac{6x^2 + 13x + 6}{6x^2 + 5x - 6}$ [Section 6.1]

Perform the indicated operations and simplify, if possible.

47. $\dfrac{p^3 - q^3}{q^2 - p^2} \cdot \dfrac{q^2 + pq}{p^3 + p^2 q + pq^2}$ [Section 6.2]

48. $\dfrac{2}{a - 2} + \dfrac{3}{a + 2} - \dfrac{a - 1}{a^2 - 4}$ [Section 6.3]

49. $\dfrac{\dfrac{y}{x} - \dfrac{x}{y}}{\dfrac{1}{x} + \dfrac{1}{y}}$ [Section 6.4]

50. $(16x^4 + 3x^2 + 13x + 3) \div (4x + 3)$ [Section 6.5]

51. Solve: $\dfrac{1}{a + 5} = \dfrac{1}{3a + 6} - \dfrac{a + 2}{a^2 + 7a + 10}$ [Section 6.6]

52. Solve $\dfrac{1}{R} = \dfrac{1}{R_1} + \dfrac{1}{R_2} + \dfrac{1}{R_3}$ for R. [Section 6.7]

53. PRINTING PAYCHECKS   It takes a printer 6 hours to print the payroll checks for all of the employees of a large company. A faster printer can print the paychecks in 4 hours. How long will it take the two printers working together to print all of the paychecks? [Section 6.7]

54. CAPTURE-RELEASE METHOD   To estimate the ground squirrel population on his acreage, a farmer trapped, tagged, and then released two dozen squirrels. Two weeks later, the farmer trapped 31 squirrels and noted that 8 were tagged. Use this information to estimate the number of ground squirrels on his acreage. [Section 6.8]

55. LIGHT   The intensity of a light source is inversely proportional to the square of the distance from the source. If the intensity is 18 lumens at a distance of 4 feet, what is the intensity when the distance is 12 feet? [Section 6.8]

56. Graph: $f(x) = \sqrt{x} + 2$. Give the domain and range of the function. [Section 7.1]

57. Evaluate: $\left(\dfrac{25}{49}\right)^{-3/2}$ [Section 7.2]

58. Simplify: $\sqrt{112a^3b^5}$ [Section 7.3]

Simplify each expression. All variables represent positive numbers.

59. $\sqrt{98} + \sqrt{8} - \sqrt{32}$ [Section 7.3]

60. $12\sqrt[3]{648x^4} + 3\sqrt[3]{81x^4}$ [Section 7.3]

61. $\left(2\sqrt{7} + 1\right)\left(\sqrt{7} - 1\right)$ [Section 7.4]

62. $3\left(\sqrt{5x} - \sqrt{3}\right)^2$ [Section 7.4]

Rationalize each denominator.

63. $\dfrac{\sqrt[3]{4}}{\sqrt[3]{b}}$ [Section 7.4]

64. $\dfrac{3t - 1}{\sqrt{3t} + 1}$ [Section 7.4]

Solve each equation.

65. $2x = \sqrt{16x - 12}$ [Section 7.5]

66. $\sqrt[3]{12m + 4} = 4$ [Section 7.5]

67. $\sqrt{x + 3} - \sqrt{3} = \sqrt{x}$ [Section 7.5]

68. CHANGING DIAPERS   The following illustration shows how to put a diaper on a baby. If the diaper is a square with sides 16 inches long, what is the largest waist size that this diaper can wrap around, assuming an overlap of 1 inch to pin the diaper? [Section 7.6]

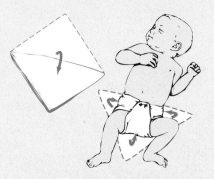

69. Express $\sqrt{-25}$ in terms of i. [Section 7.7]

70. Simplify: i^{42} [Section 7.7]

Write each expression in $a + bi$ form.

71. $(-7 + 9i) - (-2 - 8i)$ [Section 7.7]

72. $\dfrac{2 - 5i}{2 + 5i}$ [Section 7.7]

73. Solve: $t^2 = 24$ [Section 8.1]

74. Solve $m^2 + 10m - 7 = 0$ by completing the square. [Section 8.1]

Solve each equation.

75. $4w^2 + 6w + 1 = 0$ [Section 8.2]

76. $3x^2 - 4x = -2$ [Section 8.2]

77. $2(2x + 1)^2 - 7(2x + 1) + 6 = 0$ [Section 8.3]

78. $x^4 + 19x^2 + 18 = 0$ [Section 8.3]

79. TIRE WEAR   Refer to the graph. [Section 8.4]

   a. What type of function does it appear would model the relationship between the inflation of a tire and the percent of service it gives?

   b. At what percent(s) of inflation will a tire offer only 90% of its possible service?

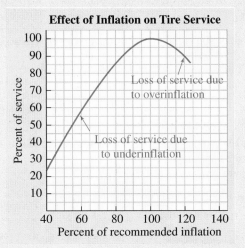

80. Graph $f(x) = -6x^2 - 12x - 8$ using the vertex formula. Then determine the x- and y-intercepts of the graph. Finally, plot several points and complete the graph. [Section 8.4]

81. Solve $x^2 - 8x \leq -15$. Write the solution set in interval notation and graph it. [Section 8.5]

82. If $f(x) = x^2 - 2$ and $g(x) = 2x + 1$, find $(f \circ g)(x)$. [Section 9.1]

83. Find the inverse function of $f(x) = 2x^3 - 1$. [Section 9.2]

84. Graph $f(x) = \left(\dfrac{1}{2}\right)^x$ and give the domain and range of the function. [Section 9.3]

85. Graph $f(x) = e^x$ and its inverse on the same coordinate system. [Section 9.4]

86. POPULATION GROWTH   As of 2007, the population of Mexico was about 109 million and the annual growth rate was 1.153%. If the growth rate remains the same, estimate the population of Mexico in 25 years. [Section 9.4]

Find x.

87. $\log 1{,}000 = x$ [Section 9.5] **88.** $\log_8 64 = x$ [Section 9.5]

89. $\log_3 x = -3$ [Section 9.5] **90.** $\log_x 25 = 2$ [Section 9.5]

91. $\ln e = x$ [Section 9.6] **92.** $\ln \dfrac{1}{e} = x$ [Section 9.6]

93. Find $\ln 0$, if possible. [Section 9.6]

94. Use the properties of logarithms to simplify $\log_6 \dfrac{36}{x^3}$. [Section 9.7]

95. Write the expression $\dfrac{1}{2}\ln x + \ln y - \ln z$ as a single logarithm. [Section 9.7]

96. BACTERIA GROWTH   The bacteria in a laboratory culture increased from an initial population of 200 to 600 in 4 hours. How long will it take the population to reach 8,000? [Section 9.8]

Solve each equation. Round to four decimal places when necessary.

97. $5^{4x} = \dfrac{1}{125}$ [Section 9.8]

98. $2^{x+2} = 3^x$ [Section 9.8]

99. $\log x + \log(x + 9) = 1$ [Section 9.8]

100. $\log_3 x = \log_3\left(\dfrac{1}{x}\right) + 4$ [Section 9.8]

101. Write the equation of the circle that has its center at $(1, -3)$ and a radius of 2. Graph the equation. [Section 10.1]

102. Complete the square to write the equation $y^2 + 4x - 6y = -1$ in $x = a(y - k)^2 + h$ form. Determine the vertex and the axis of symmetry of the graph. Then plot several points and complete the graph. [Section 10.1]

103. Graph: $\dfrac{(x + 1)^2}{4} + \dfrac{(y - 3)^2}{16} = 1$ [Section 10.2]

104. Write the equation in standard form and graph it: $(x - 2)^2 - 9y^2 = 9$ [Section 10.2]

105. Use the binomial theorem to expand $(3a - b)^4$. [Section 11.1]

106. Find the seventh term of the expansion of $(2x - y)^8$. [Section 11.1]

107. Evaluate: $\dfrac{12!}{10!(12 - 10)!}$ [Section 11.1]

108. Find the 20th term of an arithmetic sequence with a first term -11 and a common difference 6. [Section 11.2]

109. Find the sum of the first 20 terms of an arithmetic sequence with a first term 6 and a common difference 3. [Section 11.2]

110. Evaluate: $\displaystyle\sum_{k=3}^{5} (2k + 1)$ [Section 11.2]

111. Find the seventh term of a geometric sequence with a first term $\dfrac{1}{27}$ and a common ratio 3. [Section 11.3]

112. BOAT DEPRECIATION   How much will a $9,000 boat be worth after 9 years if it depreciates 12% per year? [Section 11.3]

113. Find the sum of the first ten terms of the sequence: $\dfrac{1}{64}, \dfrac{1}{32}, \dfrac{1}{16}, \ldots$ [Section 11.3]

114. Find the sum of the infinite series: $9 + 3 + 1 + \cdots$ [Section 11.3]

APPENDIX 1
Roots and Powers

n	n^2	$\sqrt{n}$	n^3	$\sqrt[3]{n}$	n	n^2	$\sqrt{n}$	n^3	$\sqrt[3]{n}$
1	1	1.000	1	1.000	51	2,601	7.141	132,651	3.708
2	4	1.414	8	1.260	52	2,704	7.211	140,608	3.733
3	9	1.732	27	1.442	53	2,809	7.280	148,877	3.756
4	16	2.000	64	1.587	54	2,916	7.348	157,464	3.780
5	25	2.236	125	1.710	55	3,025	7.416	166,375	3.803
6	36	2.449	216	1.817	56	3,136	7.483	175,616	3.826
7	49	2.646	343	1.913	57	3,249	7.550	185,193	3.849
8	64	2.828	512	2.000	58	3,364	7.616	195,112	3.871
9	81	3.000	729	2.080	59	3,481	7.681	205,379	3.893
10	100	3.162	1,000	2.154	60	3,600	7.746	216,000	3.915
11	121	3.317	1,331	2.224	61	3,721	7.810	226,981	3.936
12	144	3.464	1,728	2.289	62	3,844	7.874	238,328	3.958
13	169	3.606	2,197	2.351	63	3,969	7.937	250,047	3.979
14	196	3.742	2,744	2.410	64	4,096	8.000	262,144	4.000
15	225	3.873	3,375	2.466	65	4,225	8.062	274,625	4.021
16	256	4.000	4,096	2.520	66	4,356	8.124	287,496	4.041
17	289	4.123	4,913	2.571	67	4,489	8.185	300,763	4.062
18	324	4.243	5,832	2.621	68	4,624	8.246	314,432	4.082
19	361	4.359	6,859	2.668	69	4,761	8.307	328,509	4.102
20	400	4.472	8,000	2.714	70	4,900	8.367	343,000	4.121
21	441	4.583	9,261	2.759	71	5,041	8.426	357,911	4.141
22	484	4.690	10,648	2.802	72	5,184	8.485	373,248	4.160
23	529	4.796	12,167	2.844	73	5,329	8.544	389,017	4.179
24	576	4.899	13,824	2.884	74	5,476	8.602	405,224	4.198
25	625	5.000	15,625	2.924	75	5,625	8.660	421,875	4.217
26	676	5.099	17,576	2.962	76	5,776	8.718	438,976	4.236
27	729	5.196	19,683	3.000	77	5,929	8.775	456,533	4.254
28	784	5.292	21,952	3.037	78	6,084	8.832	474,552	4.273
29	841	5.385	24,389	3.072	79	6,241	8.888	493,039	4.291
30	900	5.477	27,000	3.107	80	6,400	8.944	512,000	4.309
31	961	5.568	29,791	3.141	81	6,561	9.000	531,441	4.327
32	1,024	5.657	32,768	3.175	82	6,724	9.055	551,368	4.344
33	1,089	5.745	35,937	3.208	83	6,889	9.110	571,787	4.362
34	1,156	5.831	39,304	3.240	84	7,056	9.165	592,704	4.380
35	1,225	5.916	42,875	3.271	85	7,225	9.220	614,125	4.397
36	1,296	6.000	46,656	3.302	86	7,396	9.274	636,056	4.414
37	1,369	6.083	50,653	3.332	87	7,569	9.327	658,503	4.431
38	1,444	6.164	54,872	3.362	88	7,744	9.381	681,472	4.448
39	1,521	6.245	59,319	3.391	89	7,921	9.434	704,969	4.465
40	1,600	6.325	64,000	3.420	90	8,100	9.487	729,000	4.481
41	1,681	6.403	68,921	3.448	91	8,281	9.539	753,571	4.498
42	1,764	6.481	74,088	3.476	92	8,464	9.592	778,688	4.514
43	1,849	6.557	79,507	3.503	93	8,649	9.644	804,357	4.531
44	1,936	6.633	85,184	3.530	94	8,836	9.695	830,584	4.547
45	2,025	6.708	91,125	3.557	95	9,025	9.747	857,375	4.563
46	2,116	6.782	97,336	3.583	96	9,216	9.798	884,736	4.579
47	2,209	6.856	103,823	3.609	97	9,409	9.849	912,673	4.595
48	2,304	6.928	110,592	3.634	98	9,604	9.899	941,192	4.610
49	2,401	7.000	117,649	3.659	99	9,801	9.950	970,299	4.626
50	2,500	7.071	125,000	3.684	100	10,000	10.000	1,000,000	4.642

APPENDIX 2
Answers to Selected Exercises

Study Set Section 1.1 (page 7)

1. variable **3.** equation **5.** addition **7. a.** Expression
b. Equation **c.** Expression **d.** Expression **9.** $b = t - 10$; the
height of the base is 10 ft less than the height of the tower
11. a. $7d = h$ **b.** $t = 2,500 - d$ **13.** $c = 13u + 24$
15. $w = \frac{c}{75}$ **17.** $A = t + 15$ **19.** $c = 12b$ **21.** The amount
(in dollars) that the husband will keep is the quotient of the amount of
the couple's refund and 2, decreased by 75. $h = \frac{t}{2} - 75$ (Answers may
vary) **23.** The number of bottles of water left on his truck is the
difference of 300 and the product of 6 and the number of stops that he
has made. $b = 300 - 6s$ (Answers may vary) **25.** 2, 6, 15
27. 22.44, 21.43, 0 **29. a.** A bar graph **b.** 1 year; $100
c. 2000, $700; 2005, $860 **31.** $r, 2, r, 4$ **33. a.** The rental cost is
the product of 10 and the number of hours it is rented, increased by 20.
b. $C = 10h + 20$

c. 30, 40, 50, 60, 100

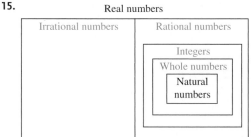

Hours rented

Study Set Section 1.2 (page 20)

1. whole, natural, integers **3.** prime, composite **5.** Irrational
7. inequality **9.** $-8, 4$ **11.** Nonrepeating, irrational
13. Repeating, rational
15.

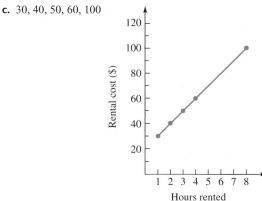

17. a. $12 < 19$ **b.** $-5 \geq -6$ **19.** is less than; is greater than or
equal to **21.** braces
23. $\left\{ \frac{a}{b} \mid a \text{ and } b \text{ are integers, with } b \neq 0. \right\}$
25. a. Rational numbers **b.** Irrational numbers **c.** Real numbers
27. True **29.** False **31.** False **33.** True **35.** 0.6,
terminating **37.** $-0.7\overline{3}$, repeating **39.** $1.2\overline{27}$, repeating
41. 0.016, terminating **43.** 1, 2, 9 **45.** $-3, 0, 1, 2, 9$
47. $\sqrt{3}, \pi$ **49.** 2 **51.** 2 **53.** 9
55.

57.

59.

61.

63.

65.

67. $<$ **69.** $>$ **71.** $>$ **73.** $<$ **75.** 20 **77.** 5.9 **79.** -6
81. $-\frac{9}{4}$ **83.** $3\frac{2}{25} = 3.0800, \frac{77}{50} = 1.5400, \frac{15}{16} = 0.9375,$
$2\frac{5}{8} = 2.6250, \frac{\pi}{4} \approx 0.7854, \sqrt{8} \approx 2.8284$ **85.** π in.
(3.141592654 . . . in.); 3.14 in. **91.** Expression **93.** 2.2, 8.5, 29.1

Study Set Section 1.3 (page 35)

1. sum, difference **3.** reciprocal **5.** squared, cubed
7. parentheses, outermost **9. a.** multiplication **b.** opposite
c. negative, positive **11. a.** Addition first or multiplication first
b. 12; multiplication is performed before addition
13. a. $(-4)^2 = 16$ **b.** $-4^2 = -16$ **15.** -8 **17.** -4.3
19. -13 **21.** $\frac{1}{6}$ **23.** -7 **25.** 0 **27.** $\frac{11}{10}$ **29.** -2
31. -12 **33.** -1.5 **35.** 60 **37.** $-\frac{6}{7}$ **39.** -2 **41.** -14
43. 9 **45.** $\frac{24}{25}$ **47.** 1,296 **49.** 62.41 **51.** -25 **53.** $-\frac{27}{125}$
55. 8 **57.** -9 **59.** $-\frac{3}{4}$ **61.** 0.2 **63.** -17 **65.** 32
67. -8 **69.** -32 **71.** 10,000 **73.** 64 **75.** -8 **77.** 13
79. -114 **81.** -32 **83.** $-\frac{1}{18}$ **85.** $\frac{1}{2}$ **87.** 5 **89.** $-\frac{1}{2}$
91. -24 **93.** -2 **95.** 61 **97.** 1 **99.** 10 **101.** 4

103. There was a net inflow of $390 billion.   **105.** New York, Atlanta, Boise, Omaha, Helena   **107.** 9   **111.** -7 and 3   **113.** $\{\ldots, -2, -1, 0, 1, 2, \ldots\}$ **115.** True

Study Set Section 1.4 (page 47)

1. term   **3.** constant   **5.** commutative, associative   **7.** Like
9. a. $(x + y) + z = x + (y + z)$ **b.** $xy = yx$
c. $r(s + t) = rs + rt$ **11. a.** a, Commutative Property of Multiplication   **b.** $a(bc)$, Associative Property of Multiplication
c. 0, Multiplicative Property of 0 **d.** a, Identity Property of Multiplication   **e.** 1, Multiplicative Inverse Property   **13. a.** 0
b. 1 **c.** $-x$ **d.** $\frac{1}{x}$ **15. a.** 5 **b.** $\frac{1}{5}$
17. Multiplication by -1 **19.** $3x^3, 11x^2, -x, 9; 3, 11, -1, 9$
21. $\frac{11}{12}a^4, -\frac{3}{4}b^2, 25b; \frac{11}{12}, -\frac{3}{4}, 25$ **23.** $7 + 3$ **25.** $3 \cdot 2 + 3d$
27. c **29.** 1 **31.** $(8 + 7) + a$ **33.** $2(x + y)$ **35.** $72m$
37. $-45q$ **39.** $-49x$ **41.** $64ry$ **43.** $81x + 18$
45. $12t - 12$ **47.** $-24 + d$ or $d - 24$ **49.** $2s^2 - 6$
51. $0.7m + 1.4n$ **53.** $9x + 2y$ **55.** $45t^2 - 60t - 15$
57. $4x - 5y + 1$ **59.** $2t + 3$ **61.** $-3y + 6$ **63.** $18x$
65. $-3.1h$ **67.** $0.8x^2$ **69.** $-2x$ **71.** $\frac{9}{10}ab$ **73.** $\frac{14}{15}t$
75. $12ad - 44a$ **77.** $6m + 2t$ **79.** $-2x^2 + 15x$
81. $-6p + 17$ **83.** $8x - 9$ **85.** $17y - 27$ **87.** $-4x + 87$
89. $56b + 6$ **91.** $-2a - A - 3$ **93.** $14cd + 62c$
95. $6.4a^2 + 2.6a + 5.7$ **97.** $-\frac{19}{16}x$ **99.** $14z - 5$
101. $18h^2 - 8h$ **103.** $-3.8y + 38.7$ **105.** 0
107. a. $20(x + 6)$ m^2 **b.** $(20x + 120)$ m^2
c. $20(x + 6) = 20x + 120$; Distributive Property   **113.** $-\frac{7}{8}$
115. 988

Study Set Section 1.5 (page 60)

1. equation   **3.** satisfies   **5.** identity   **7.** c, c, Adding, both
9. a. 3   **b.** 9   **c.** 2   **d.** 18   **11. a.** All real numbers; $\mathbb{R}$
b. No solution; $\varnothing$ **13.** $-2x, 14, 14, -2, -2, -17, -10, \stackrel{?}{=}, 20,$
-17 **15.** Yes   **17.** No   **19.** 6   **21.** $\frac{15}{8}$ **23.** 28   **25.** -30
27. 2.52   **29.** -0.25 **31.** 29   **33.** 7   **35.** 15   **37.** $-\frac{5}{2}$
39. -16 **41.** 18   **43.** -11 **45.** 1   **47.** 1.7   **49.** $\frac{17}{4}$
51. $\frac{21}{5}$ **53.** 0   **55.** -8 **57.** -11 **59.** 4   **61.** 2   **63.** 30
65. $-\frac{1}{2}$ **67.** 63   **69.** 24   **71.** 0   **73.** $\frac{21}{19}$ **75.** 3   **77.** 24
79. All real numbers, $\mathbb{R}$; identity   **81.** No solution, $\varnothing$; contradiction
83. All real numbers, $\mathbb{R}$; identity   **85.** No solution, $\varnothing$; contradiction
87. 13   **89.** 6   **91.** -15 **93.** 1,000   **95.** $\frac{9}{13}$ **97.** No
solution, $\varnothing$; contradiction   **99.** -1.2 **101.** -5 **103.** $\frac{28}{57}$
105. $\frac{5}{2}$ **111. a.** $a + b = b + a$ **b.** $(ab)c = a(bc)$
c. $a(b + c) = ab + ac$ **113. a.** $0 + a = a$ **b.** $1 \cdot a = a$

Study Set Section 1.6 (page 70)

1. formula   **3.** volume   **5. a.** Area; ft^2 **b.** Volume; ft^3
c. Circumference; ft   **d.** Perimeter; ft   **7.** t, t, a
9. $ad, ad, bc, b, b, c, \frac{t - ad}{b}$ **11.** 8 yd   **13.** 37 in.   **15.** 10.2 ft^2
17. 295.84 mi^2 **19.** 23.56 in.   **21.** 15.71 ft   **23.** 102.1 in.^{2}
25. 86.6 ft^2 **27.** 95.08 ft^3 **29.** 808.86 m^3 **31.** $t = \frac{d}{r}$
33. $h = \frac{V}{lw}$ **35.** $h = \frac{3V}{\pi r^2}$ **37.** $W = T - ma$

39. $a = \frac{2h - 96t}{t^2}$ or $a = \frac{2(h - 48t)}{t^2}$ **41.** $b_2 = \frac{2A - b_1 h}{h}$ or
$b_2 = \frac{2A}{h} - b_1$ **43.** $n = \frac{l - a + d}{d}$ or $n = \frac{l - a}{d} + 1$
45. $w = \frac{P - 2h - 2l}{2}$ or $w = \frac{P}{2} - h - l$ **47.** $A = \frac{\lambda}{x + B}$
49. $T_a = \frac{T_f}{1 - F}$ **51.** $d = \frac{l - a}{n - 1}$ **53.** $t = \frac{d_1 - d_2}{v}$
55. $y = \frac{2}{5}x - 4$ **57.** $y = -\frac{4}{3}x - 4$ **59.** $x = \frac{y - b}{m}$
61. $R = \frac{L - 2d - 3.25r}{3.25}$ **63.** $g = \frac{2(s - vt)}{t^2}$ **65.** $x = \frac{y - y_1 + mx_1}{m}$
67. $S = \frac{U + pV - G}{T}$ **69.** $r = \frac{PV}{nt}$ **71.** $R = \frac{E - Ir}{I}$ or $R = \frac{E}{I} - r$
73. $s_3 = 3A - s_1 - s_2$ **75.** $d = \frac{2S - 2an}{n(n - 1)}$ **77.** $h = \frac{3d}{4\pi}$
79. Perimeter; 216 in.   **81.** 1st term: area of bottom flap; 2nd term: area of left and right flaps; 3rd term: area of top flap; 4th term: area of face; 42.5 in.^{2} **83.** $C = \frac{5}{9}(F - 32)$ or $C = \frac{5(F - 32)}{9}$; 432, -179;
58, -89; 17, -66 **85.** $d = \frac{360A}{\pi(r_1^2 - r_2^2)}$; 140, 160
87. $n = \frac{PV}{R(T + 273)}$; 0.008, 0.090   **89.** $n = \frac{C - 6.50}{0.07}$; 621, 1,000,
about 1,692.9 kwh   **91.** $h = \frac{A - 2\pi r^2}{2\pi r}$ **97.** $26r + 132t + 1$
99. $-12a + 101$ **101.** $-0.6pt + 11p$

Study Set Section 1.7 (page 82)

1. acute   **3.** complementary   **5.** right   **7.** angles
9. $d + 15, 2d - 10, 2d + 20, \frac{d}{2} - 10, 2d$
11. a. $5x, 6x, 10(x - 2), 5x$ **b.** $5x + 6x + 10(x - 2) + 5x$
c. $5x + 6x + 10(x - 2) + 5x = 110$ **13.** Cheerios: $689 million;
Frosted Flakes: $250 million   **15.** Pedestal: 154 ft; statue: 151 ft
17. 42 min   **19.** 6 in.   **21.** 20   **23.** 310 mi   **25.** 5,000 shares
of stock funds, 7,000 shares of bond funds   **27.** 35 $12 calculators,
50 $99 calculators   **29.** 30°, 150°   **31. a.** 5.6°   **b.** 1.4°
33. $\angle 1$: 50°; $\angle 2$: 60°; $\angle 3$: 70°   **35.** 50°   **37.** 10°   **39.** 10 ft
41. 156 ft by 312 ft   **43.** 8 ft, 11 ft   **47.** Repeating
49. $\{\ldots, -2, -1, 0, 1, 2, \ldots\}$ **51.** 0

Study Set Section 1.8 (page 96)

1. amount, base   **3.** mean, mode, median   **5.** $\square$ is $\square$% of $\square$?
7. a. 51,824   **b.** 51,824, what, 3,734,536
9. a. $0.055x, 0.07(10,850 - x)$
b. $0.055x + 0.07(10,850 - x) = 1,205$
11. a. $1, 0.15x, x, 1, 0.18x$ **b.** $0.15x + 0.18x = 3,300$
13. a. $7.45p, 50 - p, 8.25(50 - p), 7.75(50)$
b. $7.75p + 8.25(50 - p) = 7.75(50)$ **15. a.** 0.025   **b.** 6%
17. $x = 0.05 \cdot 10.56$ **19.** $32.5 = 0.74x$ **21.** 448 quadrillion Btu
23. 597   **25.** 20%   **27.** $50   **29.** 9.3%   **31.** $-0.7\%, 4.1\%$
33. City: mean 37.2, median 32.5, mode 32; hwy: mean 41, median 40,
mode 40   **35.** 94   **37.** CD: $10,000; Money market: $2,000
39. a. $15,000 at 7%; $30,000 at 10%   **b.** $45,000   **41.** $100,000
43. $\frac{1}{4}$ hr = 15 min   **45.** $\frac{2}{3}$ hr   **47.** 3:30 P.M.   **49.** $1\frac{1}{2}$ hr
51. 10 lb   **53.** 1.8 lb of each   **55.** 4,000 ft^3 of the premium mix;
2,000 ft^3 of sawdust   **57.** 28 lb   **59.** 2 gal   **61.** 10 oz   **67.** 0
69. 8

Chapter 1 Review (page 102)

1. a. $C = 2t + 15$ **b.** $l = \frac{25}{w}$ **c.** $P = u - 3$
2. 180, 195, 210, 225, 240

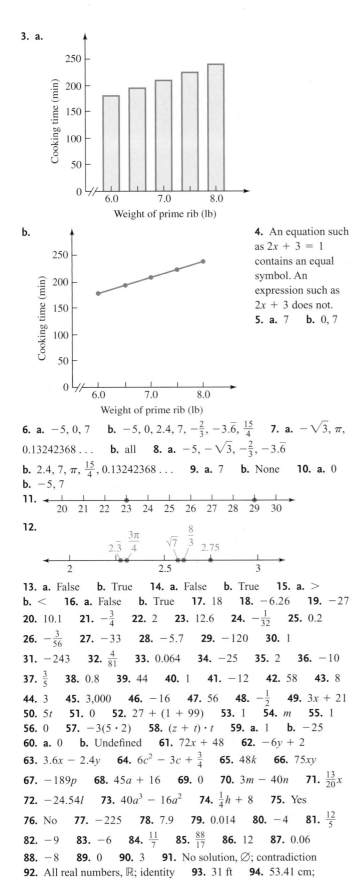

3. a.

b.

4. An equation such as $2x + 3 = 1$ contains an equal symbol. An expression such as $2x + 3$ does not.
5. a. 7 **b.** 0, 7

6. a. $-5, 0, 7$ **b.** $-5, 0, 2.4, 7, -\frac{2}{3}, -3.\overline{6}, \frac{15}{4}$ **7. a.** $-\sqrt{3}, \pi,$
$0.13242368\ldots$ **b.** all **8. a.** $-5, -\sqrt{3}, -\frac{2}{3}, -3.\overline{6}$
b. $2.4, 7, \pi, \frac{15}{4}, 0.13242368\ldots$ **9. a.** 7 **b.** None **10. a.** 0
b. $-5, 7$
11.

12.

13. a. False **b.** True **14. a.** False **b.** True **15. a.** $>$
b. $<$ **16. a.** False **b.** True **17.** 18 **18.** -6.26 **19.** -27
20. 10.1 **21.** $-\frac{3}{4}$ **22.** 2 **23.** 12.6 **24.** $-\frac{1}{32}$ **25.** 0.2
26. $-\frac{3}{56}$ **27.** -33 **28.** -5.7 **29.** -120 **30.** 1
31. -243 **32.** $\frac{4}{81}$ **33.** 0.064 **34.** -25 **35.** 2 **36.** -10
37. $\frac{3}{5}$ **38.** 0.8 **39.** 44 **40.** 1 **41.** -12 **42.** 58 **43.** 8
44. 3 **45.** 3,000 **46.** -16 **47.** 56 **48.** $-\frac{1}{2}$ **49.** $3x + 21$
50. $5t$ **51.** 0 **52.** $27 + (1 + 99)$ **53.** 1 **54.** m **55.** 1
56. 0 **57.** $-3(5 \cdot 2)$ **58.** $(z + t) \cdot t$ **59. a.** 1 **b.** -25
60. a. 0 **b.** Undefined **61.** $72x + 48$ **62.** $-6y + 2$
63. $3.6x - 2.4y$ **64.** $6c^2 - 3c + \frac{3}{4}$ **65.** $48k$ **66.** $75xy$
67. $-189p$ **68.** $45a + 16$ **69.** 0 **70.** $3m - 40n$ **71.** $\frac{13}{20}x$
72. $-24.54l$ **73.** $40a^3 - 16a^2$ **74.** $\frac{1}{4}h + 8$ **75.** Yes
76. No **77.** -225 **78.** 7.9 **79.** 0.014 **80.** -4 **81.** $\frac{12}{5}$
82. -9 **83.** -6 **84.** $\frac{11}{7}$ **85.** $\frac{88}{17}$ **86.** 12 **87.** 0.06
88. -8 **89.** 0 **90.** 3 **91.** No solution, $\varnothing$; contradiction
92. All real numbers, $\mathbb{R}$; identity **93.** 31 ft **94.** 53.41 cm;

226.98 cm^2 **95.** 1,767.15 m^3 **96. a.** 100 in.^{2}
b. $80\pi \approx 251.3$ in.^{3} **97.** $h = \frac{3V}{\pi r^2}$ **98.** $M = \frac{2K - Iw^2}{v_0^2}$
99. $d = \frac{l - a}{n - 1}$ **100.** $y = \frac{9}{5}x - 7$ **101.** O'Hare: 76.2 million;
Atlanta: 84.8 million **102.** $245 - 5c$ **103.** 600 **104.** 42 ft,
45 ft, 48 ft, 51 ft **105.** $50°, 130°$ **106.** 27 in. by 40 in.
107. 120 **108.** 32% **109. a.** 3.4% **b.** 4.0% **110.** 2, 1.5, 0
111. \$18,000 at 10%; \$7,000 at 9% **112.** 2 min after the
photographer leaves **113.** $6\frac{2}{3}$ gal **114.** Mild: 50 lb; robust: 40 lb

Chapter 1 Test (page 113)
1. a. undefined **b.** inequality **c.** like terms **d.** solve
e. addition, equality **2. a.** $s = T + 10$ **b.** $A = \frac{1}{2}bh$
3. a. 200 calories **b.** 1,200 **c.** $3\frac{1}{2}$ **4. a.** $-2, 0, 5$
b. $-2, 0, -3\frac{3}{4}, 9.2, \frac{14}{5}, 5$ **c.** $\pi, -\sqrt{7}$ **d.** All **5. a.** True
b. False **c.** True **d.** True
6.

7.

8. a. False **b.** False **9.** $\frac{4}{15}$ **10.** $\frac{8}{9}$ **11.** -209 **12.** -3
13. 100 mg **14. a.** Commutative Property of Addition
b. Associative Property of Multiplication **c.** Additive Inverse
Property **d.** Multiplicative Identity Property
15. $11.1n^2 - 7.8n - 9.8$ **16.** $90st$ **17.** $-12c + 108$
18. $-\frac{1}{36}xy + 16x$ **19.** 15 **20.** 6 **21.** No solution, $\varnothing$,
contradiction **22.** 12 **23.** Yes **24.** $i = \frac{f(P - L)}{s}$
25. $x_1 = \frac{y_1 + mx - y}{m}$ **26.** 1,018 m^2 **27.** 8 **28.** 25
29. $85°, 85°, 10°$ **30.** 4 cm by 9 cm **31.** 28% **32. a.** 2.2
b. 2 **c.** 2 **33.** \$4,000 **34.** 400 mi **35.** 10 oz **36.** *Skin*
Soother: 3 oz; *Cool Sport:* 5 oz

Study Set Section 2.1 (page 126)
1. ordered **3.** origin **5.** rectangular **7.** midpoint **9.** origin,
right, down **11.** II **13.** Yes **15.** A capital letter
17. $x^2 = x \cdot x$; x_2 represents the x-coordinate of a point.
19-26.

19. Quadrant I **21.** Quadrant
IV **23.** x-axis **25.** y-axis
27. $(2, 4)$ **29.** $(-2.5, -1.5)$
31. $(3, 0)$ **33.** $(0, 0)$

35.

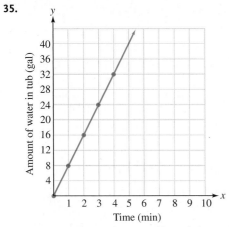

37. a. On the surface   **b.** Diving deeper   **c.** 3 hr   **d.** 500 ft
39. a. $2   **b.** $9   **c.** 3 days   **d.** No. The first-day rental fee is $2. The second day, the cost jumps another $2. The third day, and all subsequent days, the cost jumps $1.

41. $(3, 4)$ **43.** $(9, 12)$ **45.** $\left(\frac{1}{2}, -8\right)$ **47.** $\left(-\frac{3}{2}, \frac{5}{2}\right)$ **49.** $(4, 1)$
51. $(-20, -3)$ **53.** Jonesville (5, B), Easley (1, B), Hodges (2, E), Union (6, C)   **55. a.** $(2, -1)$ **b.** No   **c.** Yes   **57. a.** 6
b. 7 strokes   **c.** 16th   **d.** 18th   **59. a.** 1   **b.** 1   **c.** 1
d. A and C   **e.** Runner 1 was running; Runner 2 was stopped.   **f.** 2
61. a. 75¢   **b.** 90¢   **c.** 3.5 oz   **63. a.** T   **b.** R   **69.** -5
71. 5   **73.** $w = \dfrac{P - 2l}{2}$

Study Set Section 2.2 (page 142)
1. ordered pair   **3.** linear   **5.** y-intercept, x-intercept
7. $(0, 3), (3, 1), (6, -1)$ **9.** $1, {}^1, {}^1$ **11. a.** $(-3, 0); (0, 4)$
b. False   **13.** $y = -4x - 1$ **15. a.** The y-axis   **b.** The x-axis
17. a. Yes   **b.** No   **19. a.** Yes   **b.** No
21. 5, 4, 2

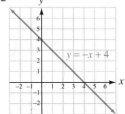

23. $0, -1, -2$

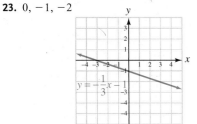

25.   **27.**

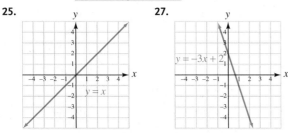

29.   **31.**   **33.**   **35.**   **37.**   **39.**   **41.**   **43.**   **45.**   **47.**   **49.** 1.22   **51.** 4.67   **53.**   **55.**

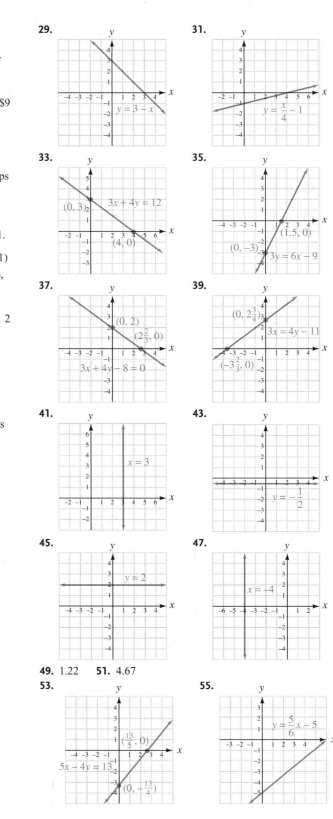

57.

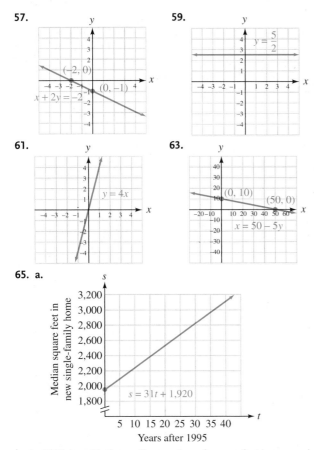

59.

61.

63.

65. a.

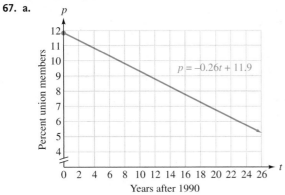

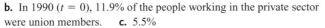

b. In 1995 ($t = 0$), the median number of square feet in a new single-family home was 1,920.   **c.** 2,800

67. a.

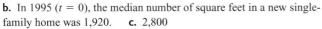

b. In 1990 ($t = 0$), 11.9% of the people working in the private sector were union members.   **c.** 5.5%

69. a.

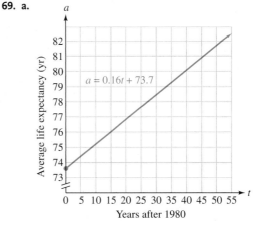

b. The average life expectancy in 1980 ($t = 0$) was 73.7 years.
c. 81.7 yr   **71.** In 8 years ($x = 8$), the computer will have no value. When new ($x = 0$), the computer was worth \$3,000.   **73.** 12.5 yr
75. 200   **81.** $\{11, 13, 17, 19, 23, 29\}$ **83.** III   **85.** $-480s$
87. $-24x - 64$

Study Set Section 2.3 (page 156)

1. ratio, ratio   **3.** rate, change   **5.** change, rise   **7. a.** 200, 40
b. $-160, 16$ **c.** $-160, 16, -10$ **9. a.** -6 **b.** 2   **c.** -3
11. a. l_3; 0   **b.** l_2; undefined   **c.** l_1; 2   **d.** l_4; -3
13. a. $\frac{5}{7}, \frac{4}{7}$; no   **b.** $\frac{4}{7}, -\frac{7}{4}$; yes   **15.** per   **17.** $\frac{6}{7}$ **19.** $-\frac{8}{3}$
21. -3 **23.** $\frac{1}{2}$ **25.** 3   **27.** -1 **29.** $-\frac{1}{3}$ **31.** $\frac{11}{7}$ **33.** 0
35. Undefined   **37.** $\frac{9}{2}$ **39.** -0.5 or $\frac{1}{2}$ **41.** Parallel
43. Perpendicular   **45.** Neither   **47.** Parallel   **49.** Parallel
51. Perpendicular   **53. a.** An increase of 65.5 million units/yr
b. A decrease of 47 million units/yr   **55.** 0.028 °F/yr
57. $\frac{3}{140}, \frac{1}{15}, \frac{1}{20}$; part 2   **59.** $\frac{1}{25}$; 4%   **61.** $\frac{5}{6}$ **63.** 250 ft/min
65. No; they are equally steep. The steepness of each course would be found by finding the slope of the same line.   **71.** 40 lb licorice; 20 lb gumdrops

Study Set Section 2.4 (page 171)

1. slope–intercept   **3. a.** No   **b.** No   **5. a.** -1 **b.** Undefined
7. $-\frac{2}{5}, 3$; $m = -\frac{2}{5}$, $(0, 3)$ **9.** $m = -\frac{2}{3}$; $(-1, 3)$ **11.** Yes
13. $\frac{1}{3}x, 2, 2, 1, \frac{1}{3}, -1$ **15.** $y = 3x + 6$ **17.** $y = -\frac{2}{3}x - \frac{7}{3}$
19. $y = \frac{1}{2}x + 3$ **21.** $y = -\frac{7}{5}x - 1$ **23.** $y = 7x + 54$
25. $y = -9x + 14$ **27.** $y - 7 = 10(x - 1)$
29. $y + 4 = -\frac{2}{3}(x + 2)$ **31.** $y = 5x - 25$
33. $y = -9x - 28.8$ **35.** $y = -\frac{1}{2}x + 11$ **37.** $y = \frac{2}{3}x + \frac{11}{3}$
39. $1, (0, -1)$

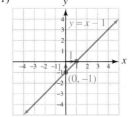

41. $-\frac{5}{4}$, $(0, -3)$

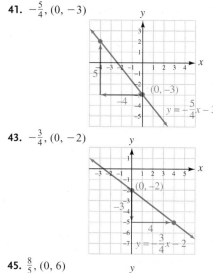

43. $-\frac{3}{4}$, $(0, -2)$

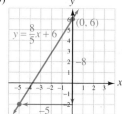

45. $\frac{8}{5}$, $(0, 6)$

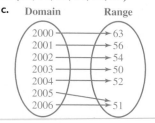

47. $\frac{3}{2}$, $(0, -4)$ **49.** $-\frac{1}{3}$, $\left(0, -\frac{5}{6}\right)$ **51.** Parallel
53. Perpendicular   **55.** Neither   **57.** Perpendicular
59. $y = 4x - 3$ **61.** $y = 3x - \frac{7}{4}$ **63.** $y = -\frac{1}{4}x$
65. $y = -\frac{5}{4}x + 3$ **67.** $y = \frac{7}{3}x - 3$ **69.** $y = \frac{4}{3}x + \frac{7}{3}$
71. $y = -\frac{1}{4}x + \frac{11}{2}$ **73.** $y = \frac{7}{5}x - \frac{28}{5}$ **75.** $y = x$ **77.** $y = 4x$
79. $y = -3x + 6$ **81.** $y = -\frac{1}{3}x + \frac{17}{36}$ **83.** $y = \frac{4}{3}x + 4$
85. $y = \frac{4}{5}x - \frac{26}{5}$ **87. a.** $C = 1.58t + 16$ **b.** C-intercept: price
in 1990 was $16; slope: the yearly increase in cost was $1.58.
c. $63.40   **89. a.** $c = 7.8m + 220$ **b.** About $10\frac{1}{4}$ min
91. a. $B = \frac{1}{100}p - 195$ **b.** 905   **93. a.** $y = 4,000x + 246,000$
b. $346,000   **95.** $y = -765x + 3,295$
97. a. $y = 12.5x + 100$ **b.** The year 2032
99. Not quite: $0.351(-2.799) \neq -1$ **105.** $29,100

Study Set Section 2.5 (page 187)

1. relation, domain, range   **3.** function, variable, dependent
5. linear   **7. a.** $\{(2000, 63), (2001, 56), (2002, 54), (2003, 50),$
$(2004, 52), (2005, 51), (2006, 51)\}$
b. D: $\{2000, 2001, 2002, 2003, 2004, 2005, 2006\}$;
R: $\{50, 51, 52, 54, 56, 63\}$

c.

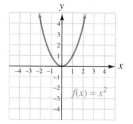

9. a. $(-2, 4)$, $(-2, -4)$ **b.** No;
to the x-value -2, there
corresponds more than one y-value
(4 and -4).   **11.** $f(x)$, y
13. a. of   **b.** of   **15.** $f(8)$
17. D: $\{-5, -1, 7, 8\}$;
R: $\{-11, -6, -1, 3\}$
19. D: $\{-23, 0, 7\}$; R: $\{1, 35\}$

21. Yes   **23.** No; $(4, 2)$, $(4, 4)$, $(4, 6)$ **25.** No; $(3, 4)$, $(3, -4)$ or
$(4, 3)$, $(4, -3)$ **27.** Yes   **29.** Yes   **31.** No; $(-1, 0)$, $(-1, 2)$
33. Yes   **35.** Yes   **37.** No; $(1, 1)$, $(1, -1)$ **39.** Yes   **41.** Yes
43. No; $(1, 1)$, $(1, -1)$ **45.** $9, -3$ **47.** $3, -5$
49. $22, 2$ **51.** $3, 11$ **53.** $4, 9$ **55.** $7, 26$ **57.** $9, 16$
59. $6, 15$ **61.** $4, 4$ **63.** $2, 2$ **65.** $\frac{1}{5}, 1$ **67.** $-2, \frac{2}{5}$
69. $3.7, 1.1, 3.4$ **71.** $-\frac{27}{64}, \frac{1}{216}, \frac{125}{8}$ **73.** $2w, 2w + 2$
75. $3w - 5, 3w - 2$ **77.** 0 **79.** 1 **81. a.** The set of real
numbers   **b.** The set of all real numbers except 4   **83. a.** The set
of real numbers   **b.** The set of all real numbers except $-\frac{1}{2}$

85.   **87.**

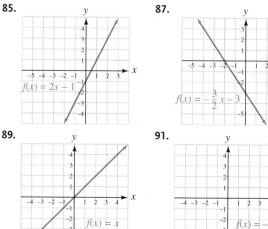

89.   **91.**

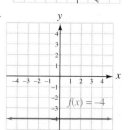

93. No, $(0, 2)$, $(0, -2)$ **95.** Yes   **97.** Yes   **99.** No, $(3, 0)$, $(3, 1)$
101. Between 20°C and 25°C   **103. a.** $I(b) = 4.75b - 125$
b. $397.50   **105. a.** $(200, 25)$, $(200, 90)$, $(200, 105)$ **b.** It
doesn't pass the vertical line test.   **107. a.** $3,372.50; the tax on an
adjusted gross income of $25,000 is $3,372.50.
b. $T(a) = 4,220 + 0.25(a - 30,650)$ **113.** No solution, $\varnothing$;
contradiction

Study Set Section 2.6 (page 201)

1. nonlinear   **3.** nonnegative
5. a. The squaring function   **b.** The cubing function

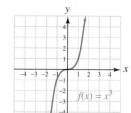

c. The absolute value function   **7.** $(5, 9)$

9. a.

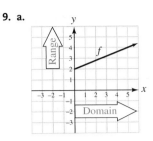

b. D: the set of nonnegative real numbers, R: the set of real numbers greater than or equal to 2
11. 4, 0, −2   **13. a.** 4, left
b. 4, up   **15. a.** −4   **b.** 0
c. 2   **d.** −1   **17. a.** 2   **b.** 4
c. −1, 1, 5   **d.** 2, 4   **19.** D: the set of real numbers, R: the set of real numbers   **21.** D: the set of real numbers, R: the set of real numbers less than or equal to 5
23. D: the set of real numbers, R: the set of real numbers greater than or equal to −4   **25.** D: the set of nonnegative real numbers, R: the set of nonnegative real numbers

27. D: the set of real numbers, R: the set of real numbers greater than or equal to 2

29. D: the set of real numbers, R: the set of real numbers

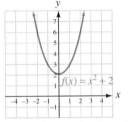

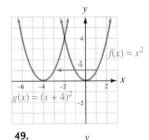

31. D: the set of real numbers, R: the set of nonnegative real numbers

33. D: the set of real numbers, R: the set of nonnegative real numbers

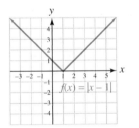

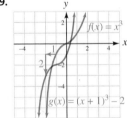

35. D: the set of real numbers, R: the set of real numbers greater than or equal to −2

37. D: the set of real numbers, R: the set of real numbers

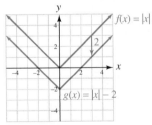

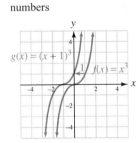

39. D: the set of real numbers, R: the set of all real numbers greater than or equal to −3

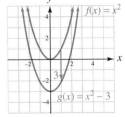

41. D: the set of real numbers, R: the set of real numbers

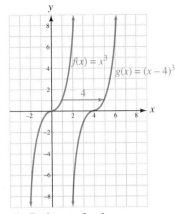

43. D: the set of real numbers, R: the set of real numbers

45. D: the set of real numbers, R: the set of nonnegative real numbers

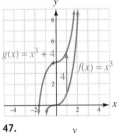

47.

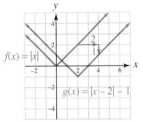

49.

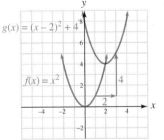

51.

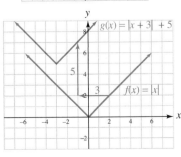

53.

55.

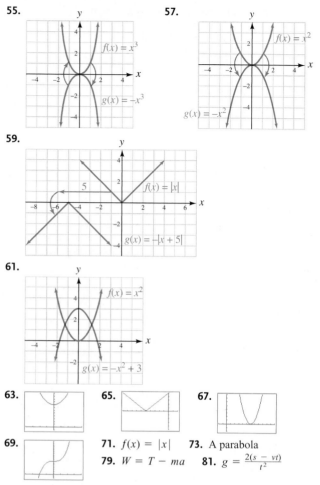

57.

59.

61.

63.   **65.**   **67.**

69.   **71.** $f(x) = |x|$ **73.** A parabola
79. $W = T - ma$ **81.** $g = \frac{2(s - vt)}{t^2}$

Chapter 2 Review (page 206)

1.

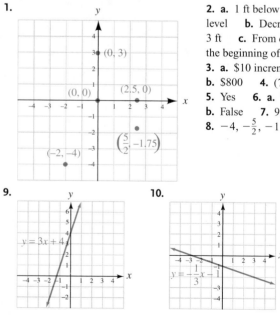

2. a. 1 ft below its normal level   **b.** Decreased by 3 ft   **c.** From day 3 to the beginning of day 4
3. a. $10 increments
b. $800   **4.** $(7, -3)$
5. Yes   **6. a.** True
b. False   **7.** 9, 0, -9
8. $-4, -\frac{5}{2}, -1$

9.   **10.**

11.   **12.**

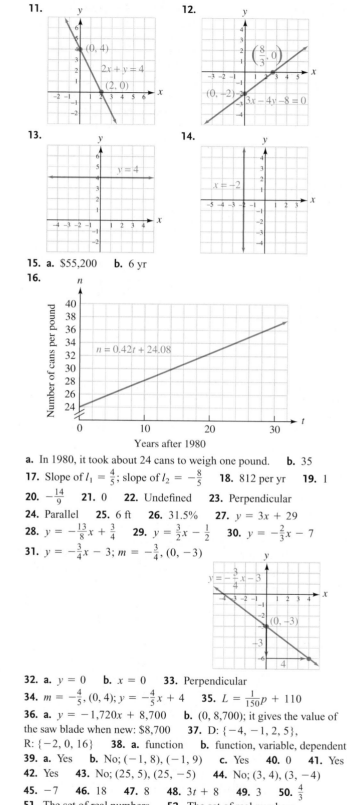

13.   **14.**

15. a. $55,200   **b.** 6 yr
16.

a. In 1980, it took about 24 cans to weigh one pound.   **b.** 35
17. Slope of $l_1 = \frac{4}{5}$; slope of $l_2 = -\frac{8}{5}$ **18.** 812 per yr   **19.** 1
20. $-\frac{14}{9}$ **21.** 0   **22.** Undefined   **23.** Perpendicular
24. Parallel   **25.** 6 ft   **26.** 31.5%   **27.** $y = 3x + 29$
28. $y = -\frac{13}{8}x + \frac{3}{4}$ **29.** $y = \frac{3}{2}x - \frac{1}{2}$ **30.** $y = -\frac{2}{3}x - 7$
31. $y = -\frac{3}{4}x - 3; m = -\frac{3}{4}, (0, -3)$

32. a. $y = 0$ **b.** $x = 0$ **33.** Perpendicular
34. $m = -\frac{4}{5}, (0, 4); y = -\frac{4}{5}x + 4$ **35.** $L = \frac{1}{150}p + 110$
36. a. $y = -1,720x + 8,700$ **b.** $(0, 8,700)$; it gives the value of the saw blade when new: $8,700   **37.** D: $\{-4, -1, 2, 5\}$,
R: $\{-2, 0, 16\}$ **38. a.** function   **b.** function, variable, dependent
39. a. Yes   **b.** No; $(-1, 8), (-1, 9)$ **c.** Yes   **40.** 0   **41.** Yes
42. Yes   **43.** No; $(25, 5), (25, -5)$ **44.** No; $(3, 4), (3, -4)$
45. -7 **46.** 18   **47.** 8   **48.** $3t + 8$ **49.** 3   **50.** $\frac{4}{3}$
51. The set of real numbers   **52.** The set of real numbers
53. The set of all real numbers except 2   **54.** The set of all real numbers except -5 **55.** Function   **56.** Not a function;
$(0, 2), (0, 3)$ **57. a.** $R(t) = 0.02t + 5.05$ **b.** 7.05 milliohms

58.

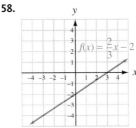

59. a. −4   **b.** 3   **c.** 1
60. a. 4   **b.** 1   **c.** −4, 2
61. D: the set of real numbers, R: the set of real numbers   **62.** D: the set of real numbers, R: the set of real numbers greater than or equal to 1

3. $\left(\frac{5}{2}, -8\right)$ **4.**

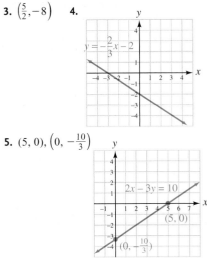

63. D: the set of real numbers, R: the set of nonnegative real numbers

64. a. 6, up   **b.** 6, left

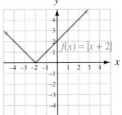

5. $(5, 0), \left(0, -\frac{10}{3}\right)$

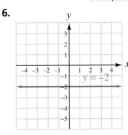

65. D: the set of real numbers, R: the set of real numbers greater than or equal to −3

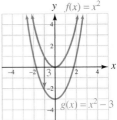

6.

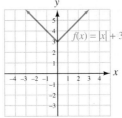

66. D: the set of real numbers, R: the set of nonnegative real numbers

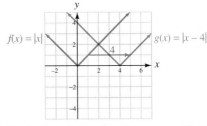

7. −1.5 degree/hr   **8.** $\frac{1}{2}$ **9.** $\frac{2}{3}$ **10. a.** Undefined   **b.** 0
11. $y = 3x + 1$ **12.** $y = 8x + 22$ **13.** $m = -\frac{1}{3}, \left(0, -\frac{3}{2}\right)$
14. Perpendicular   **15.** $y = -\frac{3}{2}x$ **16. a.** $v = -600x + 4{,}000$
b. $(0, 4{,}000)$; it gives the value of the copier when new: $4,000
17. $n = \frac{11}{200}p + 1{,}400$ **18. a.** No; $(1, -8), (1, 6)$ **b.** Yes
c. Yes   **d.** No; $(4, -4), (4, 4)$ **19.** The set of all real numbers except 6   **20.** −20   **21.** 10   **22.** 49   **23.** $\frac{9}{16}$ **24.** $3r + 25$
25. −2   **26.** 2   **27.** Function   **28.** Not a function; $(2, 2), (2, -2)$ **29.** D: the set of real numbers, R: the set of real numbers greater than or equal to 3

67. D: the set of real numbers, R: the set of real numbers

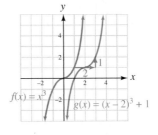

68. D: the set of real numbers, R: the set of real numbers

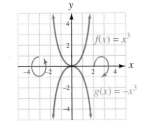

Chapter 2 Test (page 216)

1. a. rectangular   **b.** function   **c.** function, variable, dependent
d. domain, range   **e.** translation   **2. a.** 11 P.M.   **b.** 10 A.M., 7 P.M.   **c.** 4 hr   **d.** 10 A.M. to 3 P.M.; 7 P.M. to 11 P.M.   **e.** 3 P.M.

30.

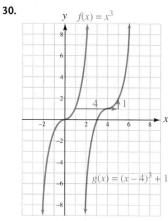

D: the set of real numbers,
R: the set of real numbers

31. D: the set of real numbers, R: the set of nonnegative real numbers
32. $f(x) = -x^2 + 1$

Chapters 1–2 Cumulative Review (page 218)

1. a. 1, 2, 6, 7   **b.** 0, 1, 2, 6, 7   **c.** $-2, 0, 1, 2, \frac{13}{12}, 6, 7$
d. $\sqrt{5}, \pi$ **e.** -2 **f.** $-2, 0, 1, 2, \frac{13}{12}, 6, 7, \sqrt{5}, \pi$ **g.** 2, 7
h. $-2, 0, 2, 6$
2.

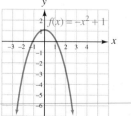

3. -2 **4.** -149 **5.** $\frac{24}{25}$ **6.** 2   **7.** 33   **8.** -5
9. a. Associative property of addition   **b.** Distributive property
c. Commutative property of addition   **d.** Associative property of
multiplication   **10. a.** -6 **b.** -8 **11.** $-28st$ **12.** $15y + 22$
13. $-\frac{13}{12}s$ **14.** $-12x + 101$ **15.** $-\frac{8}{3}$ **16.** -1 **17.** 6
18. 24   **19.** No solution, $\varnothing$ **20.** $\frac{33}{5}$ **21.** $B = \frac{c + Tx}{3y}$
22. $h = \frac{2A}{b_1 + b_2}$ **23.** 164   **24.** 70°, 70°, 40°   **25.** $14,000
26. 39 mph going, 65 mph returning   **27.** $\left(1, \frac{1}{2}\right)$ **28.** Yes
29.

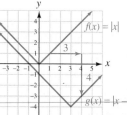

30. $-\frac{5}{6}$ **31.** $y = -\frac{7}{5}x + \frac{11}{5}$
32. $y = -3x - 3$
33. The points (2, 20),
(3, 40), and (6, 60) do not
lie on a straight line.
34. a. $T = \frac{1}{1,000}p + 25$
b. 60   **35.** 0   **36.** 2   **37.** Yes
38. D: $\{-6, 0, 1, 5\}$,
R: $\{-12, 4, 7, 8\}$; no   **39.** No; (4, 2), (4, -2)   **40.** -60
41. 5   **42.** -1 **43.** 3   **44.** $3r^2 + 2$ **45.** The set of all real
numbers except -1 **46.** 6
47. D: the set of real numbers,
R: the set of real numbers less
than or equal to 1

48. D: the set of real numbers,
R: the set of real numbers
greater than or equal to -4

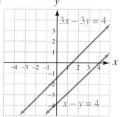

Study Set Section 3.1 (page 231)

1. system   **3.** independent, dependent   **5. a.** True   **b.** False
c. True   **d.** True   **7. a.** No solution; independent
b. Infinitely many solutions; $(-3, 0), (-2, -2), (0, -6)$; consistent
9. 1   **11.** brace   **13.** Yes   **15.** No   **17.** No   **19.** Yes
21. (4, 2)

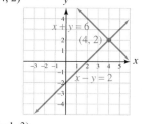

23. $(-1, 3)$

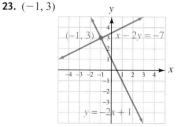

25. No solution, $\varnothing$; inconsistent system

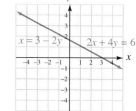

27. $\{(x, y) \mid x = 3 - 2y\}$ or $\{(x, y) \mid x + 2y = 3\}$, infinitely many
solutions; dependent equations

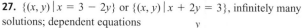

29. $(-3, -3)$

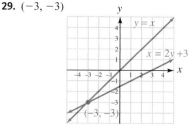

31. $(-2, -1)$

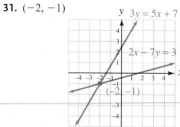

33. 2

35. 3

55. $\left(3, \frac{5}{2}\right)$

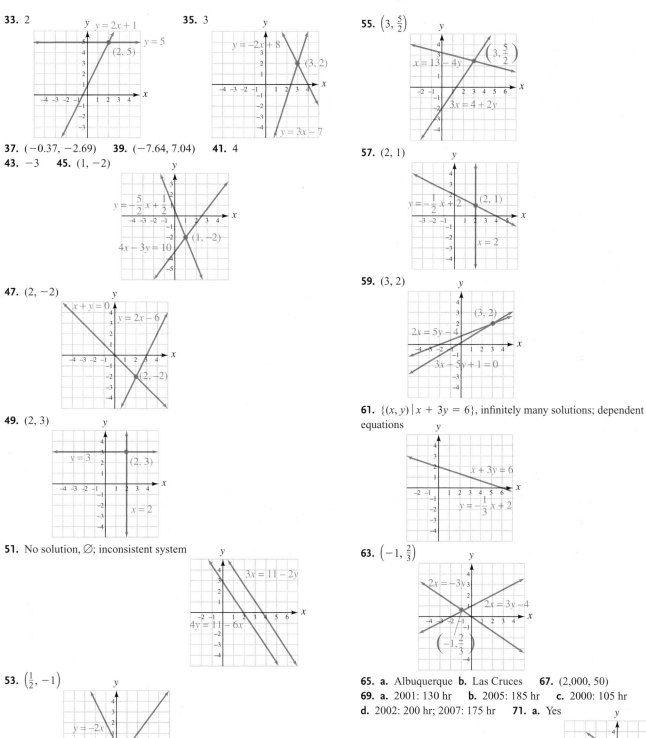

37. $(-0.37, -2.69)$ **39.** $(-7.64, 7.04)$ **41.** 4

43. -3 **45.** $(1, -2)$

57. $(2, 1)$

47. $(2, -2)$

59. $(3, 2)$

49. $(2, 3)$

61. $\{(x, y) \mid x + 3y = 6\}$, infinitely many solutions; dependent equations

51. No solution, $\varnothing$; inconsistent system

63. $\left(-1, \frac{2}{3}\right)$

53. $\left(\frac{1}{2}, -1\right)$

65. a. Albuquerque **b.** Las Cruces   **67.** $(2,000, 50)$
69. a. 2001: 130 hr   **b.** 2005: 185 hr   **c.** 2000: 105 hr
d. 2002: 200 hr; 2007: 175 hr   **71. a.** Yes

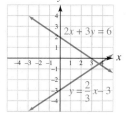

b. No. The ships would have to reach this point at the same time.
77. -3 **79.** 0   **81.** $-t^3 + 2t - 2$
83. D: the set of all real numbers except -2

Study Set Section 3.2 (page 244)

1. standard/general **3.** eliminated **5.** y; second equation
7. a. 3; -4 (answers may vary) **b.** 2; -3 (answers may vary)
9. Elimination method **11. a.** $7x + 4y = 8, x - y = 8$
b. $2x + y = 12, 3x - 9y = 170$ **13.** $(2, 6)$ **15.** $(5, 3)$
17. $(-2, 4)$ **19.** $(4, -7)$ **21.** $(9, 2)$ **23.** $(-4, 0)$
25. $(3, -2)$ **27.** $(12, 4)$ **29.** $(22, 25)$ **31.** $\left(-2, \frac{3}{2}\right)$
33. No solution, $\varnothing$; inconsistent system
35. $\left\{(x, y) \mid 2x - \frac{5}{2} = y\right\}$ or $\left\{(x, y) \mid 4x - 5 = 2y\right\}$, infinitely many
solutions; dependent equations **37.** $\left(5, \frac{3}{2}\right)$ **39.** $\left(\frac{1}{2}, \frac{2}{3}\right)$
41. $\{(x, y) \mid 2x - 4 = y\}$, infinitely many solutions; dependent
equations **43.** $(-20, -30)$ **45.** $(9, -1)$ **47.** $(4, 8)$
49. $(20, -12)$ **51.** No solution, $\varnothing$; inconsistent system
53. $\left(\frac{2}{3}, \frac{3}{2}\right)$ **55.** $\left(\frac{1}{2}, -3\right)$ **57.** $(2, 3)$ **59.** $\left(-\frac{1}{3}, 1\right)$ **67.** $-\frac{5}{2}$
69. $-\frac{8}{5}$ **71.** $\frac{4}{3}$

Study Set Section 3.3 (page 257)

1. parallel **3. a.** $70°, 75°$ **b.** alternate, interior
5. a. $x + c$ **b.** $x - c$ **7. a.** 0.06 **b.** 0.048 **c.** 0.135
9. $0.05x, 0.04y, 25,000, 1,050,$ $\begin{cases} x + y = 25,000 \\ 0.05x + 0.04y = 1,050 \end{cases}$
11. $13.90x, 5.10y, 6, 61.50,$ $\begin{cases} x + y = 6 \\ 13.90x + 5.10y = 61.50 \end{cases}$
13. 750 ohms, 625 ohms **15.** Montana: 406; Idaho: 208
17. Canada: 8; United States: 24 **19.** $75°, 25°$ **21.** $150°, 30°$
23. 16 m by 20 m **25.** 15 sec: \$475; 30 sec: \$800
27. 85 racing bikes, 120 mountain bikes **29.** Rolling Stones: \$264;
Jimmy Buffet: \$132 **31. a.** 400 tires
b.

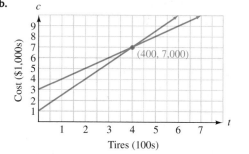

c. The second mold **33. a.** $4,666\frac{2}{3}$ books **b.** The newer press
35. 4,031 **37. a.** 590 units per month **b.** 620 units per month
c. A (smaller loss) **39.** \$3,000 at 10%, \$5,000 at 12%
41. Credit union: \$8,000; money market \$24,000
43. VISA: \$11,800; Robinsons-May: \$4,700 **45.** 525 mph, 75 mph
47. Walking: 6 ft per sec; moving walkway: 2 ft per sec
49. 25 mph, 5 mph **51.** Gummy bears: 45 lb; jelly beans: 15 lb
53. Regular: $14\frac{2}{3}$ lb; kona: $5\frac{1}{3}$ lb **55.** Small flake: 900 lb;
mylar: 1,100 lb **57.** 10%: 8 pints; 40%: 16 pints
59. 148 g of the 0.2%, 37 g of the 0.7% **67.** rational **69.** identity
71. isosceles

Study Set Section 3.4 (page 272)

1. system, standard **3.** triples **5.** dependent **7. a.** No solution
b. No solution **9.** $x + y + 4z = 3, 7x - 2y + 8z = 15,$
$3x + 2y - z = 4$ **11.** Yes **13.** No **15.** $(1, 1, 2)$
17. $(-1, 3, 0)$ **19.** $(3, -1, 3)$ **21.** $(-3, 0, 5)$ **23.** $(7, -6, 3)$

25. $\left(\frac{1}{2}, 2, -1\right)$ **27.** $(40, -20, 5)$ **29.** $\left(\frac{2}{3}, 2, -1\right)$ **31.** $(9, 0, -4)$
33. $(10, -5, 5)$ **35.** No solution, $\varnothing$; inconsistent system
37. Infinitely many solutions; dependent equations
39. $(8, 4, 5)$ **41.** $(12, -9, 4)$ **43.** $(3, 2, 1)$
45. No solution, $\varnothing$; inconsistent system **47.** $\left(\frac{3}{4}, \frac{1}{2}, \frac{1}{3}\right)$
49. No solution, $\varnothing$; inconsistent system **51.** $(2, 4, 8)$
53. Infinitely many solutions; dependent equations **55.** $(2.5, 3, 3.5)$
57. $(2, 6, 9)$ **59. a.** Infinitely many solutions, all lying on the line
running down the binding **b.** 3 parallel planes (shelves); no solution
c. Each pair of planes (cards) intersect; no solution
d. 3 planes (faces of die) intersect at a corner; 1 solution
61. 100, 81, 78
67. **69.**

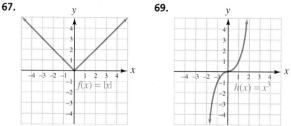

Study Set Section 3.5 (page 279)

1. satisfy **3.** $\begin{cases} x + y + z = 50 \\ 5x + 6y + 7z = 295 \\ 2x + 3y + 4z = 145 \end{cases}$ **5.** $-3 = 4a + 2b + c$

7. 30 large, 50 medium, 100 small **9.** Food A: 2, Food B: 3,
Food C: 1 **11.** 120 coats, 200 shirts, 150 slacks **13.** Young: 85,
Montana: 55, Gannon: 16 **15.** Nitrogen: 78%, oxygen: 21%,
other: 1% **17.** $\angle A = 40°, \angle B = 60°, \angle C = 80°$ **19.** *X-Files:* 201,
Will & Grace: 194, *Seinfeld:* 180 **21.** 6 lb rose petals, 3 lb lavender,
1 lb buckwheat hulls **23.** Nickels: 20, dimes: 40, quarters: 4
25. $y = \frac{1}{2}x^2 - 2x - 1$ **27.** $x^2 + y^2 - 2x - 2y - 2 = 0$
31. Yes **33.** Yes **35.** No; $(1, 2), (1, -2)$ **37.** No; $(4, 2), (4, -2)$

Study Set Section 3.6 (page 291)

1. matrix **3.** rows, columns, by **5.** augmented **7. a.** 2×3
b. 3×4 **9. a.** $x, y, y, 6, 4$ **b.** $x, z, 2z, z, 4, 0, -20$
11. a. Interchange rows 1 and 2 **b.** Multiply row 1 by $\frac{1}{2}$
c. Add row 3 to 6 times row 2
13. $\begin{bmatrix} 1 & 2 & | & 6 \\ 3 & -1 & | & -10 \end{bmatrix}$ **15.** $\begin{cases} x + 6y = 7 \\ y = 4 \end{cases}$ **17.** $\begin{bmatrix} 1 & -4 & | & 4 \\ -3 & 1 & | & -6 \end{bmatrix}$

19. $\begin{bmatrix} 1 & -\frac{1}{3} & | & 2 \\ 1 & -4 & | & 4 \end{bmatrix}$ **21.** $\begin{bmatrix} 3 & 6 & -9 & | & 0 \\ -2 & 2 & -2 & | & 5 \\ 1 & 5 & -2 & | & 1 \end{bmatrix}$

23. $\begin{bmatrix} 3 & 6 & -9 & | & 0 \\ -2 & -1 & 7 & | & 1 \\ -2 & 2 & -2 & | & 5 \end{bmatrix}$ **25.** $(1, 1)$ **27.** $(2, -3)$

29. $(3, 2, 1)$ **31.** $(4, 5, 4)$ **33.** No solution, $\varnothing$; inconsistent system
35. $\{(x, y) \mid x + y = 3\}$, infinitely many solutions; dependent equations
37. $(-2, -1, 1)$ **39.** $(-1, -1)$ **41.** $(0, -3)$
43. No solution, $\varnothing$; inconsistent system **45.** $\{(x, y) \mid 4x - y = 2\}$,
infinitely many solutions; dependent equations
47. $\left(\frac{1}{2}, 1, -2\right)$ **49.** $(0, 1, 3)$ **51.** $(-4, 8, 5)$ **53.** 262,144
55. $22°, 68°$ **57.** $40°, 65°, 75°$ **59.** $76°, 104°$ **65.** $m = \frac{y_2 - y_1}{x_2 - x_1}$,
$(x_2 \neq x_1)$ **67.** $y - y_1 = m(x - x_1)$

Study Set Section 3.7 (page 302)

1. determinant, diagonal   **3.** minor   **5.** *ad, bc*

7. The third column   **9.** $\begin{vmatrix} 1 & 2 & 0 \\ 3 & 1 & -1 \\ 8 & 4 & -1 \end{vmatrix}$ **11.** $(-2, -1, 1)$

13. 6, 30   **15.** 4   **17.** -10 **19.** -170 **21.** 6   **23.** 35
25. 0   **27.** 4   **29.** 26   **31.** -37 **33.** 5   **35.** -79
37. 1   **39.** $(4, 2)$ **41.** $(-5, -8)$ **43.** $\left(5, \frac{3}{2}\right)$ **45.** $(11, 3)$
47. No solution, $\varnothing$; inconsistent system   **49.** $\{(x, y) \mid 5x = 12 - 6y\}$, infinitely many solutions; dependent equations   **51.** $(1, 1, 2)$
53. $(-2, 0, 2)$ **55.** $(3, -2, 1)$ **57.** $(2, -1)$ **59.** $\left(-\frac{1}{2}, -1, -\frac{1}{2}\right)$
61. No solution, $\varnothing$; inconsistent system   **63.** No solution, $\varnothing$; inconsistent system   **65.** $(-2, 3, 1)$ **67.** $\left(-\frac{1}{2}, \frac{1}{3}\right)$
69. Dependent equation; infinitely many solutions   **71.** $-46,811$
73. $-60,527,941$ **75.** $50°, 80°$ **83.** No
85. The graph of function *g* is 2 units below the graph of function *f*.
87. *y*-intercept   **89.** $x; y$

Chapter 3 Review (page 305)

1. Yes   **2.** No   **3. a.** $(1, 3), (2, 1), (4, -3)$ (Answers may vary)
b. $(0, -4), (2, -2), (4, 0)$ (Answers may vary)   **c.** $(3, -1)$
4. The point of intersection is (2001, 50%). In 2001, the percent of time spent viewing broadcast networks and cable networks was the same, 50%.   **5.** $(3, 5)$

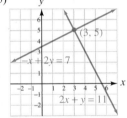

6. $(-2, 3)$

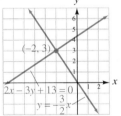

7. $\{(x,y) \mid 3x + 2y = 12\}$, infinitely many solutions; dependent equations

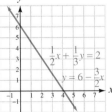

8. No solution, $\varnothing$; inconsistent system

9. 2   **10.** -1 **11.** $(-1, 3)$ **12.** $(-3, -1)$ **13.** $(-5, -8)$
14. $(-1, -1)$ **15.** $(-6, 16)$ **16.** $\left(4, \frac{1}{2}\right)$ **17.** $\left(\frac{1}{4}, -\frac{1}{2}\right)$
18. $\{(x, y) \mid x + 3y = -2\}$, infinitely many solutions; dependent equations   **19.** No solution, $\varnothing$; inconsistent system   **20.** $(3, 4)$
21. $(-1, 0.7)$ (Answers may vary); $\left(-1, \frac{2}{3}\right)$
23. Austin–Houston: 162 mi; Austin–San Antonio: 83 mi
24. 8 mph, 2 mph   **25.** 500 oz of 6%, 250 oz of 18%
26. $4,000 at 6%, $6,000 at 12%
27. Teaspoon: 5 mL; tablespoon: 15 mL   **28.** 17,500 bottles
29. No   **30.** Yes; infinitely many solutions   **31.** $(2, 0, -3)$
32. $(2, -1, 3)$ **33.** $(-1, 1, 3)$ **34.** $(-5, -14, -4)$
35. $(6, -2, 2)$ **36.** $(1, 4, 1)$ **37.** No solution, $\varnothing$; inconsistent system   **38.** Infinitely many solutions; dependent equations
39. Small: 50; medium: 60; large: 40   **40.** 2 cups mix A, 1 cup mix B, 1 cup mix C   **41.** $5,000 at 5%, $7,000 at 6%, and $10,000 at 7%

42. $a = -\frac{1}{16}, b = 2, c = 0$ **43.** $\begin{bmatrix} 5 & 4 & \vdots & 3 \\ 1 & -1 & \vdots & -3 \end{bmatrix}$

44. $\begin{bmatrix} 1 & 2 & 3 & \vdots & 6 \\ 1 & -3 & -1 & \vdots & 4 \\ 6 & 1 & -2 & \vdots & -1 \end{bmatrix}$ **45. a.** $\begin{bmatrix} 1 & 3 & \vdots & -2 \\ 6 & 12 & \vdots & -6 \end{bmatrix}$

b. $\begin{bmatrix} 1 & 2 & \vdots & -1 \\ 1 & 3 & \vdots & -2 \end{bmatrix}$ **c.** $\begin{bmatrix} 0 & -6 & \vdots & 6 \\ 1 & 3 & \vdots & -2 \end{bmatrix}$

46. a. $\begin{bmatrix} 1 & 1 & 0 & \vdots & -1 \\ 2 & -1 & 1 & \vdots & 3 \\ 3 & -1 & -2 & \vdots & 7 \end{bmatrix}$ **b.** $\begin{bmatrix} 2 & -1 & 1 & \vdots & 3 \\ 3 & 3 & 0 & \vdots & -3 \\ 3 & -1 & -2 & \vdots & 7 \end{bmatrix}$

c. $\begin{bmatrix} 0 & -3 & 1 & \vdots & 5 \\ 1 & 1 & 0 & \vdots & -1 \\ 3 & -1 & -2 & \vdots & 7 \end{bmatrix}$ **47.** $(1, -3)$ **48.** $(5, -3, -2)$

49. $\{(x, y) \mid -2x + y = -4\}$, infinitely many solutions; dependent equations   **50.** No solution, $\varnothing$; inconsistent system   **51.** 18
52. 38   **53.** -3 **54.** 28   **55.** $(2, 1)$ **56.** No solution, $\varnothing$; inconsistent system   **57.** $(1, -2, 3)$ **58.** $(-3, 2, 2)$

Chapter 3 Test (page 314)

1. a. system   **b.** rows, columns   **c.** triples   **d.** plane
e. matrix   **2.** $(2, 1)$

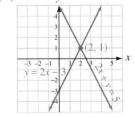

3. It is a solution.   **4.** The point of intersection is (6/06, 44%). That means that Governor Schwarzenegger's job approval and disapproval ratings were the same in June of 2006, approximately 44%.

5. a. Inconsistent system; no solution, $\varnothing$ **b.** Dependent equations; infinitely many solutions   **6.** 3   **7.** (7, 0)   **8.** (2, −3)

9. $\{(x, y) \mid 2x + 3y = -3\}$, infinitely many solutions; dependent equations   **10.** $\left(\frac{1}{3}, 2\right)$ **11.** 55, 70   **12.** 15 gal of 40%, 5 gal of 80%

13. 1,375 impressions   **14.** No   **15.** (3, 2, −1)   **16.** (−6, 2, 5)

17. Children: 60, general admission: 30, seniors: 10

18. $\begin{bmatrix} 1 & 7 & \vdots & -3 \\ 0 & -22 & \vdots & 22 \end{bmatrix}$ **19.** (2, 2)   **20.** No solution, $\varnothing$; inconsistent system   **21.** −2   **22.** 4   **23.** (−3, 3)   **24.** −1

Chapters 1–3 Cumulative Review (page 316)

1.

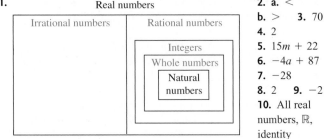

Real numbers

Irrational numbers | Rational numbers

Integers

Whole numbers

Natural numbers

2. a. $<$
b. $>$ **3.** 70
4. 2
5. $15m + 22$
6. $-4a + 87$
7. −28
8. 2   **9.** −2
10. All real numbers, $\mathbb{R}$, identity

11. $B = \frac{\lambda - Ax}{A}$ **12.** $d_2 = d_1 - vt$ **13.** Los Angeles: 6 wk; Las Vegas: 4 wk; Dallas: 7 wk   **14.** $1,190 billion   **15.** 20 mph
16. 5 lb apple slices, 5 lb banana chips

17.

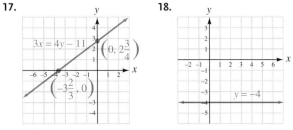

$3x = 4y - 11$ $\left(0, 2\frac{3}{4}\right)$ $\left(-3\frac{2}{3}, 0\right)$

18.

$y = -4$

19. $-\frac{4}{5}$ **20.** Perpendicular   **21.** $y = -3x + 17$
22. $v = 17.5x + 300$ **23.** −105   **24.** $-r^2 - \frac{r}{2}$
25. $f(x) = x^3$ (Answers may vary)   **26.** Yes   **27. a.** 1
b. 2   **28.** No. It does not pass the vertical line test.
29. D: the set of real numbers, R: the set of real numbers greater than or equal to 0

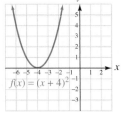

$f(x) = (x + 4)^2$

30. (−2, −3)

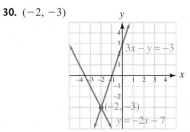

$3x - y = -3$ $(-2, -3)$ $y = -2x - 7$

31. a. An inconsistent system has no solution.   **b.** Dependent equations have the same graph.   **32.** (2, −1)   **33.** 200 of the $67 phones, 160 of the $100 phones
34. (−1, 0, 2)

35. $2,500 in the 12-month CD, $15,000 in the 24-month CD, $12,500 in the 36-month CD   **36.** (8, 3)   **37. a.** 40   **b.** 26   **38.** $\left(\frac{14}{3}, \frac{2}{3}\right)$

Study Set Section 4.1 (page 330)

1. inequality   **3.** interval   **5.** builder, such that
7. $7t - 5 > 4, \frac{x}{2} \le -1$ **9. a.** Yes   **b.** No   **c.** Yes   **d.** No
11. a. $(-\infty, \infty)$ 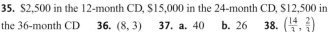   **b.** $\varnothing$

13. $-10, \le, -5, 2, -\infty, x \le 2$ **15.** $<$
17. $(-\infty, 14); \{x \mid x < 14\}$

19. $[-2, \infty); \{x \mid x \ge -2\}$

21. $(-\infty, 1); \{x \mid x < 1\}$

23. $(-3, \infty); \{x \mid x > -3\}$

25. $[-11, \infty); \{x \mid x \ge -11\}$

27. $(-\infty, 2]; \{a \mid a \le 2\}$

29. $(0, \infty); \{x \mid x > 0\}$

31. $(-\infty, 7]; \{t \mid t \le 7\}$

33. $\left[\frac{10}{3}, \infty\right); \left\{x \mid x \ge \frac{10}{3}\right\}$

35. $\left(-\infty, -\frac{16}{7}\right); \left\{x \mid x < -\frac{16}{7}\right\}$

37. $\left(-\infty, \frac{11}{5}\right); \left\{x \mid x < \frac{11}{5}\right\}$

39. $\left[-\frac{18}{13}, \infty\right); \left\{n \mid n \ge -\frac{18}{13}\right\}$

41. $(-\infty, \infty); \mathbb{R}$ **43.** No solution; $\varnothing$

45. $\left[-\frac{2}{5}, \infty\right); \left\{t \mid t \ge -\frac{2}{5}\right\}$

47. $\left(-\infty, \frac{1}{2}\right); \left\{x \mid x < \frac{1}{2}\right\}$

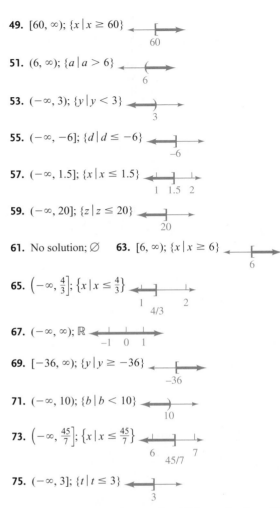
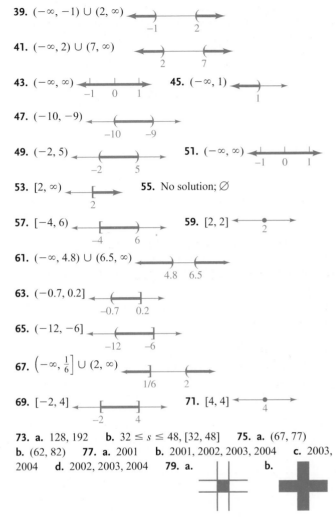

49. $[60, \infty)$; $\{x \,|\, x \geq 60\}$

51. $(6, \infty)$; $\{a \,|\, a > 6\}$

53. $(-\infty, 3)$; $\{y \,|\, y < 3\}$

55. $(-\infty, -6]$; $\{d \,|\, d \leq -6\}$

57. $(-\infty, 1.5]$; $\{x \,|\, x \leq 1.5\}$

59. $(-\infty, 20]$; $\{z \,|\, z \leq 20\}$

61. No solution; $\varnothing$ **63.** $[6, \infty)$; $\{x \,|\, x \geq 6\}$

65. $\left(-\infty, \frac{4}{3}\right]$; $\left\{x \,\middle|\, x \leq \frac{4}{3}\right\}$

67. $(-\infty, \infty)$; $\mathbb{R}$

69. $[-36, \infty)$; $\{y \,|\, y \geq -36\}$

71. $(-\infty, 10)$; $\{b \,|\, b < 10\}$

73. $\left(-\infty, \frac{45}{7}\right]$; $\left\{x \,\middle|\, x \leq \frac{45}{7}\right\}$

75. $(-\infty, 3]$; $\{t \,|\, t \leq 3\}$

77. $(-\infty, 1)$ **79.** $[-4, \infty)$ **81.** Midwest, South   **83.** d is zero
85. 8 hr   **87.** 10 hr   **89.** 11   **97.** 4, 5, 3

Study Set Section 4.2 (page 343)
1. intersection, union   **3.** double   **5. a.** both   **b.** one, both
7. a. intersection   **b.** union   **9. a.** No   **b.** Yes
11. a. $[-2, 1)$ **b.** $[2, 2]$ **c.** $\varnothing$ **13.** union, intersection
15. All real numbers   **17.** $\{4, 6\}$ **19.** $\{-3, 1, 2\}$

21. $\{-3, -1, 0, 1, 2, 4, 6, 8, 10\}$ **23.** $\{-3, 0, 1, 2, 3, 4, 5, 6, 8\}$
25. $(-2, 5]$ **27.** $(2, 3]$

29. $(-\infty, -6]$

31. No solution; $\varnothing$
33. $[1, 4]$

35. $(0.8, 1.1)$

37. $(-\infty, -2] \cup (6, \infty)$

39. $(-\infty, -1) \cup (2, \infty)$

41. $(-\infty, 2) \cup (7, \infty)$

43. $(-\infty, \infty)$ **45.** $(-\infty, 1)$

47. $(-10, -9)$

49. $(-2, 5)$ **51.** $(-\infty, \infty)$

53. $[2, \infty)$ **55.** No solution; $\varnothing$

57. $[-4, 6)$ **59.** $[2, 2]$

61. $(-\infty, 4.8) \cup (6.5, \infty)$

63. $(-0.7, 0.2]$

65. $(-12, -6]$

67. $\left(-\infty, \frac{1}{6}\right] \cup (2, \infty)$

69. $[-2, 4]$ **71.** $[4, 4]$

73. a. 128, 192   **b.** $32 \leq s \leq 48$, $[32, 48]$ **75. a.** $(67, 77)$
b. $(62, 82)$ **77. a.** 2001   **b.** 2001, 2002, 2003, 2004   **c.** 2003,
2004   **d.** 2002, 2003, 2004   **79. a.**   **b.**

87. Mean: 6, median: 6, mode: 6

Study Set Section 4.3 (page 357)
1. absolute value   **3.** isolate   **5.** compound   **7.** compound
9. a. $-2, 2$ **b.** $-1.99, -1, 0, 1, 1.99$ **c.** $-3, -2.01, 2.01, 3$
11. a. $8, -8$ **b.** $x - 3, -(x - 3)$ **13. a.** $x = 8$ or $x = -8$
b. $x \leq -8$ or $x \geq 8$ **c.** $-8 \leq x \leq 8$ **d.** $5x - 1 = x + 3$ or
$5x - 1 = -(x + 3)$ **15. a.** ii   **b.** iii   **c.** i   **17.** $23, -23$
19. $\frac{3}{4}, -\frac{3}{4}$ **21.** $13, -3$ **23.** $\frac{14}{3}, -6$ **25.** $50, -50$
27. $3.1, -6.7$ **29.** No solution; $\varnothing$ **31.** No solution; $\varnothing$
33. $25, -19$ **35.** $7, -\frac{7}{3}$ **37.** $2, -\frac{1}{2}$ **39.** $\frac{32}{7}, -16$ **41.** -10
43. -8 **45.** $\frac{9}{2}$ **47.** $-\frac{1}{2}$ **49.** $-4, \frac{28}{9}$ **51.** $-2, \frac{18}{11}$
53. $0, -2$ **55.** $11, \frac{1}{3}$ **57.** $(-4, 4)$

59. $[-21, 3]$

61. $[-24, 0]$

63. $\left(-\frac{8}{3}, 4\right)$ **65.** No solution; $\varnothing$

67. No solution; $\varnothing$　　**69.** $(-\infty, -3) \cup (3, \infty)$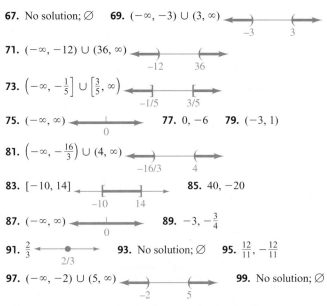

71. $(-\infty, -12) \cup (36, \infty)$

73. $\left(-\infty, -\frac{1}{5}\right] \cup \left[\frac{3}{5}, \infty\right)$

75. $(-\infty, \infty)$　　**77.** $0, -6$　　**79.** $(-3, 1)$

81. $\left(-\infty, -\frac{16}{3}\right) \cup (4, \infty)$

83. $[-10, 14]$　　　　　　**85.** $40, -20$

87. $(-\infty, \infty)$　　**89.** $-3, -\frac{3}{4}$

91. $\frac{2}{3}$ 　　**93.** No solution; $\varnothing$　　**95.** $\frac{12}{11}, -\frac{12}{11}$

97. $(-\infty, -2) \cup (5, \infty)$　　　　**99.** No solution; $\varnothing$

101. $70° \le t \le 86°$　　**103. a.** $|c - 0.6°| \le 0.5°$　　**b.** $[0.1°, 1.1°]$
105. $26.45\%, 24.76\%$　　**111.** $50°, 130°$

Study Set Section 4.4 (page 368)

1. linear, two　　**3.** edge　　**5. a.** Yes　**b.** No　**c.** Yes　**d.** No
7. a. $m = 3; (0, -1)$　　**b.** $(-3, 0), (0 -2)$　　**9. a.** No; dashed
b. Yes; solid　　**c.** Yes; solid　　**d.** No; dashed

11. 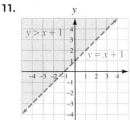　　**13.**

15. 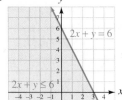　　**17.**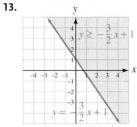

19. 　　**21.**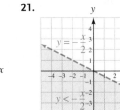

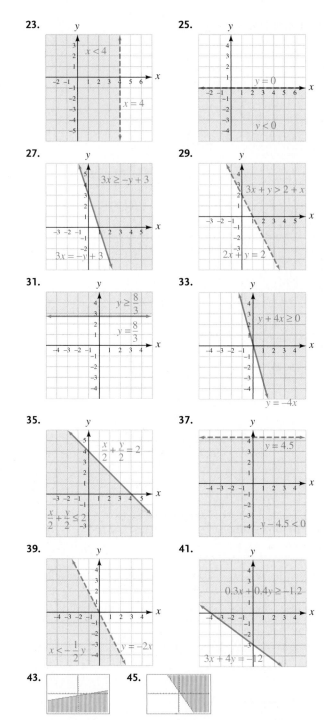

47. a. The Mississippi River　　**b.** The area of the U.S. west of the Mississippi River

49. $4x + 6y \leq 120$; (5, 15), (15, 10), (20, 5)

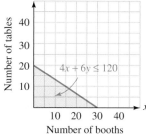

51. $10x + 15y \geq 1,200$; (40, 80), (80, 80), (120, 40)

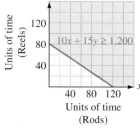

57. Yes   **59.** $(-2, -3)$

Study Set Section 4.5 (page 376)

1. inequalities   **3.** point   **5. a.** Yes   **b.** No   **c.** No   **d.** Yes
7. a. False   **b.** True   **c.** True   **d.** False   **e.** True   **f.** True

9. **11.**

13. **15.**

17. **19.**

21. **23.**

25. **27.**

29.

31. No solution

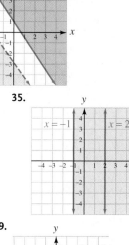

33. **35.**

37. **39.**

41. **43.**

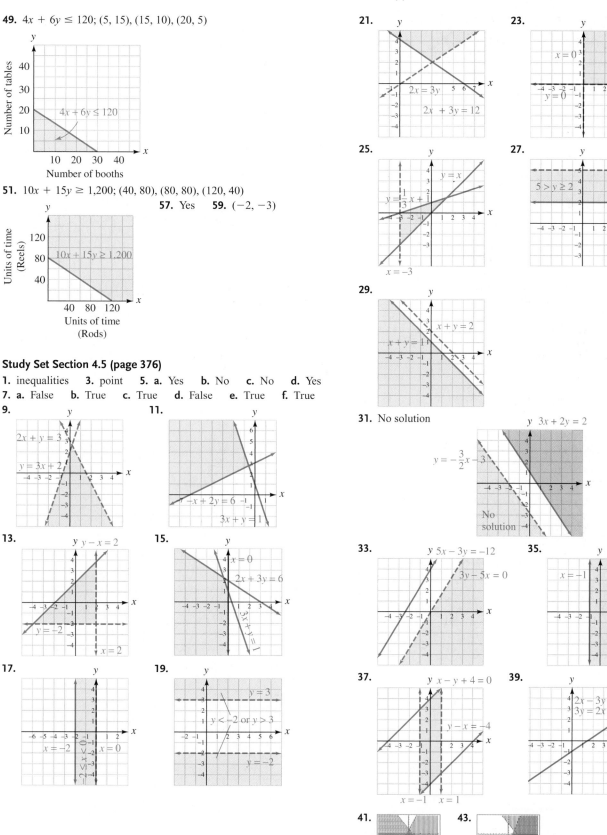

45.

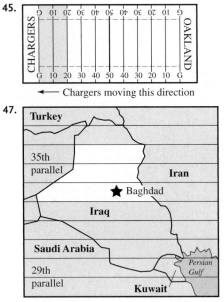

Chargers moving this direction

47.

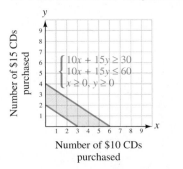

49. 1 $10 CD and 2 $15 CDs, 4 $10 CDs and 1 $15 CD

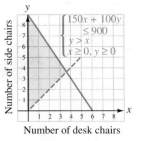

51. 2 desk chairs and 4 side chairs, 1 desk chair and 5 side chairs   **57.** IV   **59.** II

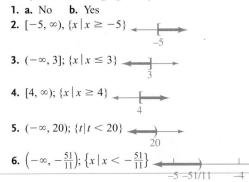

Chapter 4 Review (page 380)

1. a. No   **b.** Yes

2. $[-5, \infty)$, $\{x \mid x \geq -5\}$

3. $(-\infty, 3]$; $\{x \mid x \leq 3\}$

4. $[4, \infty)$; $\{x \mid x \geq 4\}$

5. $(-\infty, 20)$; $\{t \mid t < 20\}$

6. $\left(-\infty, -\frac{51}{11}\right)$; $\{x \mid x < -\frac{51}{11}\}$

7. $(-\infty, \infty)$, $\mathbb{R}$ **8.** No solution; $\varnothing$

9. $20,000 or more   **10.** She needs to receive a score that is greater than 5.2.   **11.** 5 hr   **13.** $\{-3,3\}$ **14.** $\{-6, -5, -3, 0, 3, 6, 8\}$

15. Yes   **16.** No   **17.**

18.   **19.** $[-10, -4)$

20. $(-\infty, -11)$ **21.** No solution; $\varnothing$

22. $[0, 0]$ **23.** $\left(-\frac{1}{3}, 2\right)$

24. $[1, 9]$ **25.** Yes   **26.** No

27. $(-\infty, -5) \cup (4, \infty)$

28. $(-\infty, \infty)$

29. $17 \leq 4x \leq 25$, 4.25 ft $\leq x \leq 6.25$ ft, $[4.25, 6.25]$

30. a. ii, iv   **b.** i, iii   **31.** $2, -2$ **32.** $3, -\frac{11}{3}$ **33.** $\frac{26}{3}, -\frac{10}{3}$

34. No solution; $\varnothing$ **35.** 3   **36.** $\frac{1}{8}, -\frac{19}{8}$ **37.** $\frac{1}{5}, -5$ **38.** $\frac{13}{12}$

39. $[-3, 3]$

40. $(-5, -2)$

41. $\left[-3, \frac{19}{3}\right]$ **42.** No solution; $\varnothing$

43. $(-\infty, -1) \cup (1, \infty)$

44. $(-\infty, -4] \cup \left[\frac{22}{5}, \infty\right)$

45. $\left(-\infty, \frac{4}{3}\right) \cup (4, \infty)$

46. $(-\infty, \infty)$, $\mathbb{R}$

47. Since $|0.04x - 8.8|$ is always greater than or equal to 0 for any real number x, this absolute value inequality has no solution.

48. Since $\left|\frac{3x}{50} + \frac{1}{45}\right|$ is always greater than or equal to 0 for any real number x, this absolute value inequality is true for all real numbers.

49. a. $8, 2$ **b.** $[6, 10]$ **50.** $3, -3$ **51.** No   **52.** Yes

53.   **54.**

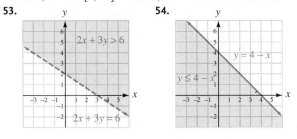

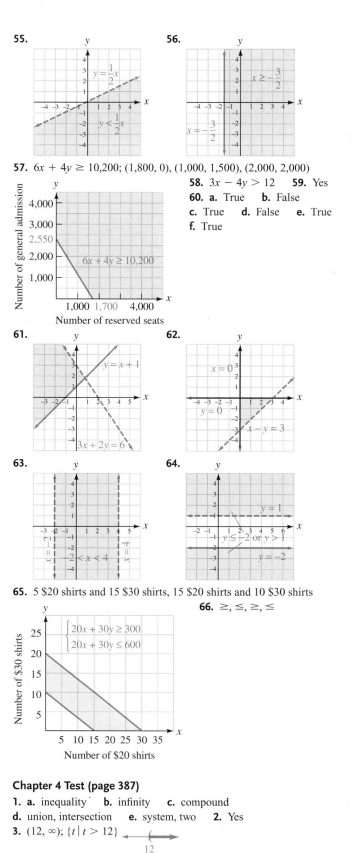

55.

56.

57. $6x + 4y \geq 10,200$; $(1,800, 0)$, $(1,000, 1,500)$, $(2,000, 2,000)$

58. $3x - 4y > 12$ **59.** Yes
60. a. True   **b.** False
c. True   **d.** False   **e.** True
f. True

61.

62.

63.

64.

65. 5 $20 shirts and 15 $30 shirts, 15 $20 shirts and 10 $30 shirts

66. $\geq, \leq, \geq, \leq$

$$\begin{cases} 20x + 30y \geq 300 \\ 20x + 30y \leq 600 \end{cases}$$

Chapter 4 Test (page 387)

1. a. inequality   **b.** infinity   **c.** compound
d. union, intersection   **e.** system, two   **2.** Yes
3. $(12, \infty)$; $\{t \mid t > 12\}$

4. $(-\infty, -5]$; $\{x \mid x \leq -5\}$

5. $(-\infty, -14)$; $\{x \mid x < -14\}$

6. $(-\infty, \infty)$; $\mathbb{R}$

7. More than 78   **8.** 11 hr   **9.** No   **10.** $\{-4, 0, 11\}$
11. $\{-5, -4, 0, 7, 8, 9, 10, 11\}$ **12. a.**

b.   **13.** $\left[1, \frac{9}{4}\right]$

14. $(-\infty, -3) \cup (8, \infty)$

15. $(-2, 16)$ **16.** No solution; $\emptyset$

17. $-5, \frac{23}{3}$ **18.** $4, -4$ **19.** $\frac{8}{9}, -\frac{8}{9}$ **20.** No solution; $\emptyset$
21. 10   **22.** $|x - 0.0625| \leq 0.0015$; $[0.0610, 0.0640]$
23. $[-7, 1]$

24. $(-\infty, -9) \cup (13, \infty)$

25. $(-\infty, 1) \cup (3, \infty)$

26. $\left[\frac{4}{3}, \frac{8}{3}\right]$

27. $(-\infty, \infty)$, $\mathbb{R}$ **28.** $(-6, -3)$

29.

30.

31.

32.

33.

34. (1, 1), (2, 1), (2, 2)

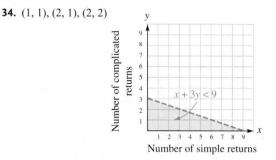

Number of simple returns

35. a. (3, −4) is a solution of inequality 2. **b.** No, it does not lie in the doubly shaded region. **36.** ≤, ≥, ≥, ≤

Chapters 1–4 Cumulative Review (page 390)

1. Rational numbers: terminating and repeating decimals; irrational numbers: nonterminating and nonrepeating decimals

2. 0.125, 0.0625, 0.03125, 0.054125 **3.** 10 **4.** −6

5. −2a + b − 2 **6.** 8t − 9 **7.** 3 **8.** 6 **9.** No solution; ∅

10. −2 **11.** $d = \frac{l - a}{n - 1}$ **12.** 201 ft^2 **13.** $20,000

14. $\frac{1}{4}$ hr **15.** $-\frac{8}{5}$ **16. a.** 26,000 prisoners/yr

b. 1990–1995; 67,200 prisoners/yr **17.** Parallel **18.** $y = \frac{1}{3}x + \frac{11}{3}$

19. a. $y = -0.06x + 8.2$ **b.** 2.8 L/min

20. D: {−12, −6, 5, 8}, R: {−6, 4, 6} **21.** 14 **22.** 3t^2 − t

23.

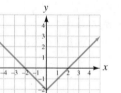

D: the set of real numbers, R: the set of all real numbers greater than or equal to −2

24. Yes

25. (2, 1)

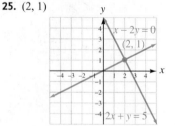

26. (7, 23), in 1907 the percent of U.S. workers in white-collar and farming jobs was the same (23%); (45, 42), in 1945 the percent of U.S. workers in white-collar and blue-collar jobs was the same (42%).

27. (3, 1) **28.** (1, 1)

29. 750 **30.** (−1, −1, 3)

31. 250 $5 tickets, 375 $3 tickets, 125 $2 tickets **32.** (−1, −1)

33. −10 **34.** (3, −2) **35.** (−∞, 11], {x | x ≤ 11}

36. (−3, 3), {x | −3 < x < 3}

37. (−∞, 2) ∪ (7, ∞)

38. $\left[1, \frac{9}{4}\right]$ **39.** 3, $-\frac{3}{2}$ **40.** −5, $-\frac{3}{5}$

41. $\left[-\frac{2}{3}, 2\right]$

42. (−∞, −4) ∪ (1, ∞)

43.

44.

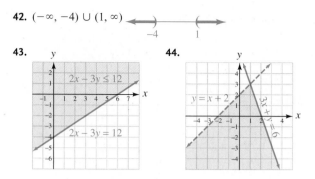

Study Set Section 5.1 (page 405)

1. exponential **3.** factor **5.** power **7. a.** x^{m+n} **b.** x^{mn}

c. $x^n y^n$ **d.** $\frac{x^n}{y^n}$ **e.** 1 **f.** $\frac{1}{x^n}$ **g.** x^{m-n} **h.** $\left(\frac{y}{x}\right)^n$ **i.** $\frac{y^n}{x^m}$

9. multiply **11. a.** 1 **b.** reciprocal **13.** $9, (-2), 11$ **15.** 6x; 3

17. x; 5 **19.** b; 6 **21.** $\frac{n}{4}$; 3 **23.** m − 8; 6 **25.** 10; 0

27. x^5 **29.** y^{12} **31.** $-6t^{23}$ **33.** $a^4 b^5$ **35.** 64 **37.** x^{28}

39. r^{55} **41.** g^{32} **43.** $x^{20} y^4$ **45.** $64m^{21}$ **47.** $\frac{m^{80}}{n^8}$ **49.** $\frac{9a^{32}}{49n^{22}}$

51. 1 **53.** 1 **55.** 60 **57.** −3t **59.** $\frac{1}{25}$ **61.** $-\frac{1}{27}$ **63.** $\frac{8}{x^9}$

65. $-\frac{1}{h}$ **67.** $\frac{1}{m^{10}}$ **69.** $\frac{1}{s^6}$ **71.** a^5 **73.** 49 **75.** $\frac{1}{16}$

77. $-\frac{p^8}{12t^6}$ **79.** m^{12} **81.** $\frac{33}{y^{12}}$ **83.** t^{14} **85.** $\frac{m^2}{6n^9}$ **87.** $\frac{9}{4}$

89. $\frac{t^{120}}{g^{80}}$ **91.** $16a^{20}$ **93.** $\frac{64b^{12}}{27a^9}$ **95.** −8 **97.** $x^{19} y^4$ **99.** x^{73}

101. $25r^{13}$ **103.** $\frac{1}{m^{10}}$ **105.** $\frac{27z^{21}}{64a^{12}b^{12}}$ **107.** $\frac{1}{25}$ **109.** $\frac{1}{9x^3}$

111. $\frac{64d^6}{9}$ **113.** $-\frac{27}{c^{15}d^8}$ **119.** $26^3 \cdot 10^4$; 175,760,000

121. $10^{-2}, 10^{-3}, 10^{-4}, 10^{-5}, 10^{-6}, 10^{-7}, 10^{-8}, 10^{-9}$ **123. a.** x^6 ft^2

b. x^9 ft^3 **129.** $\left(-\infty, -\frac{10}{9}\right]$

Study Set Section 5.2 (page 415)

1. scientific, standard **3.** 10^n, integer **5.** left **7. a.** 60.22 is not between 1 and 10. **b.** 0.6022 is not between 1 and 10.

9. 3.9×10^3 **11.** 7.8×10^{-3} **13.** 1.73×10^{14}

15. 9.6×10^{-6} **17.** 2.03×10^{-9} **19.** 5.016×10^{16}

21. 2.365×10^7 **23.** 9.009×10^{-10} **25.** 3.17×10^{-4}

27. 5.27×10^3 **29.** 3.23×10^7 **31.** 6.0×10^{-4} **33.** 27,000

35. 0.00323 **37.** 796,000,000 **39.** 0.00000035 **41.** 5.23

43. 80,000,000,000,000 **45.** 2.6×10^9 **47.** 1.817×10^{12}

49. 5.005×10^8 **51.** 7.2×10^{-6} **53.** 4.3×10^9

55. 5.1×10^{11} **57.** 3.4×10^{16} **59.** 5.0×10^{-8}

61. 2.25×10^{16}; 22,500,000,000,000,000 **63.** 3.2×10^{-3}; 0.0032

65. 4.005×10^{20}; 400,500,000,000,000,000,000

67. 1.44×10^7; 14,400,000 **69.** 4.2×10^9; 4,200,000,000

71. 4.75×10^{-14}; 0.0000000000000475 **73.** 2,600,000 to 1

75. 1.0×10^{21} **77.** 2.97×10^4 dollars; $29,700

79. 8.5×10^{-28} g **81.** About 9.5×10^{15} m

83. a. 2.5×10^9 sec = 2,500,000,000 sec **b.** About 79 years

85. 1.209×10^8 mi

93. $\left[1, \frac{9}{4}\right]$

95. (−∞, 2) ∪ (7, ∞)

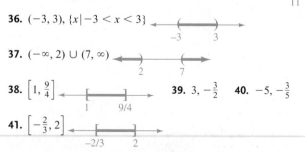

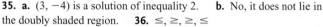

Study Set Section 5.3 (page 429)

1. polynomial   **3.** leading, coefficient, constant   **5.** degree
7. like   **9. a.** No   **b.** Yes   **c.** No   **d.** Yes   **11. a.** ^{6} **b.** ^{5}
13. a. 15   **b.** Approximately 3   **c.** $0, -2, -4$ **d.** D: $(-\infty, \infty)$;
R: $(-\infty, \infty)$ **15.** $3, 3, 3, -27, -29$

17.

Term	Coefficient	Degree
$25x^2$	25	2
$-x$	-1	1
4	4	0
Degree of the polynomial: 2		
Type of polynomial: Trinomial		

19.

Term	Coefficient	Degree
$5a^7b^4$	5	11
$-33a^2b$	-33	3
Degree of the polynomial: 11		
Type of polynomial: Binomial		

21. a. -1 **b.** 95   **23. a.** 6.5   **b.** 11
25. $-42, 0, 12, 6, -6, -12, 0, 42$; D: $(-\infty, \infty)$; R: $(-\infty, \infty)$

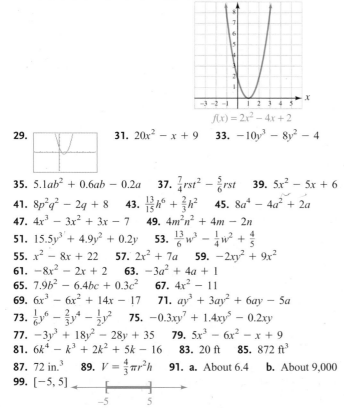

$f(x) = 2x^3 - 3x^2 - 11x + 6$

27. $8, 2, 0, 2, 8$; D: $(-\infty, \infty)$; R: $[0, \infty)$

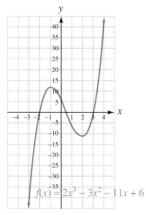

$f(x) = 2x^2 - 4x + 2$

29. **31.** $20x^2 - x + 9$ **33.** $-10y^3 - 8y^2 - 4$

35. $5.1ab^2 + 0.6ab - 0.2a$ **37.** $\frac{7}{4}rst^2 - \frac{5}{6}rst$ **39.** $5x^2 - 5x + 6$
41. $8p^2q^2 - 2q + 8$ **43.** $\frac{13}{15}h^6 + \frac{2}{3}h^2$ **45.** $8a^4 - 4a^2 + 2a$
47. $4x^3 - 3x^2 + 3x - 7$ **49.** $4m^2n^2 + 4m - 2n$
51. $15.5y^3 + 4.9y^2 + 0.2y$ **53.** $\frac{13}{6}w^3 - \frac{1}{4}w^2 + \frac{4}{5}$
55. $x^2 - 8x + 22$ **57.** $2x^2 + 7a$ **59.** $-2xy + 9x^2$
61. $-8x^2 - 2x + 2$ **63.** $-3a^2 + 4a + 1$
65. $7.9b^2 - 6.4bc + 0.3c^2$ **67.** $4x^2 - 11$
69. $6x^3 - 6x^2 + 14x - 17$ **71.** $ay^3 + 3ay^2 + 6ay - 5a$
73. $\frac{1}{6}y^6 - \frac{2}{3}y^4 - \frac{1}{2}y^2$ **75.** $-0.3xy^7 + 1.4xy^5 - 0.2xy$
77. $-3y^3 + 18y^2 - 28y + 35$ **79.** $5x^3 - 6x^2 - x + 9$
81. $6k^4 - k^3 + 2k^2 + 5k - 16$ **83.** 20 ft   **85.** 872 ft^3
87. 72 in.^{3} **89.** $V = \frac{4}{3}\pi r^2 h$ **91. a.** About 6.4   **b.** About 9,000
99. $[-5, 5]$

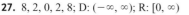

101. $(-1, 9)$

Study Set Section 5.4 (page 444)

1. monomials, binomials   **3.** first, outer, inner, last   **5. a.** factors
b. term   **c.** term   **7.** square, square, $x^2 - y^2$ **9. a.** $6b - 2$
b. $2b$ **c.** $8b^2 - 6b + 1$ **11.** $6a^7$ **13.** $-15a^2b^3$ **15.** $6g^{11}h^{16}$
17. $-40x^3y^{14}z^{12}$ **19.** $30x^{10} + 20x^7$ **21.** $-6d^7 + 6d^6 - 4d^5$
23. $7r^3s^3t + 7rs^5t - 7rs^3t^3$ **25.** $28x^7y^9z^4 - 16x^3y^7z^7$
27. $14t^2 + 17t - 6$ **29.** $6y^6 - 11y^3 + 4$ **31.** $32b^2 + b - \frac{3}{8}$
33. $27b^5c - 9b^4c^2 - 3b^2c + bc^2$ **35.** $6y^3 + 11y^2 + 9y + 2$
37. $x^3 - y^3$ **39.** $6a^3 + 17a^2 + 8a - 6$
41. $2x^4 - 6x^3 - 12x^2 + 58x - 42$ **43.** $18p^4 + 30p^3 - 72p^2$
45. $24m^3y - 20m^2y^2 + 4my^3$ **47.** $4a^2 + 4ab + b^2$
49. $25r^4 + 60r^2 + 36$ **51.** $\frac{1}{16}b^2 - b + 4$
53. $81a^2b^4 - 72ab^2 + 16$ **55.** $25y^2 - 5.76$ **57.** $x^4y^2 - \frac{36}{25}$
59. $36a^2 + 12ab + b^2 + 48a + 8b + 16$
61. $49 - 14x + 14y + x^2 - 2xy + y^2$ **63.** $25 - 4n^2 - 4np - p^2$
65. $27x^3 - 54x^2 + 36x - 8$ **67.** $b^2 - 6b - 5$ **69.** $a^2 - 8a + 3$
71. $48x^2 - 13x + 13$ **73.** $5x^2 - 36x + 7$
75. $9.2127x^2 - 7.7956x - 36.0315$
77. $299.29y^2 - 150.51y + 18.9225$ **79.** $8a^3 - b^3$
81. $9b^2 + 9b - 7$ **83.** $55m^3 + 22m^2n^2 + 15mn^3 + 6n^5$
85. $15s^6t^5u^2$ **87.** $2a^2 + ab - b^2 - 3bc - 2c^2$
89. $a^3 - 3a^2b - ab^2 + 3b^3$ **91.** $16k^2 - 104k + 169$
93. $\frac{1}{4}x^2 + 6x - 64$ **95.** $0.2t^2 - 2.7t + 9$ **97.** $y^6 - 4$
99. $9a^6 + 24a^3c + 16c^2$ **101.** $-6m^5n^2 - 6m^3n^5$
103. $125s^3 - 75s^2t^3 + 15st^6 - t^9$
105. $9d^2 + 6df + f^2 + 12d + 4f + 4$ **107. a.** $x^2 + 2x - 8$
b. $\frac{1}{2}b^2 + \frac{3}{2}b - 5$ **109. a.** $(x + y)(x - y)$ **b.** $x(x - y); x^2 - xy$
c. $y(x - y); xy - y^2$ **d.** They represent the same area.
$(x + y)(x - y) = x^2 - y^2$ **111.** $(4x^2 - 164x + 1,656)$ in.^{2}

117. **119.**

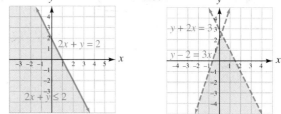

Study Set Section 5.5 (page 456)

1. factored   **3.** greatest common factor   **5.** binomial   **7.** $6xy^2$
9. a. 4   **b.** No   **c.** $5, s$ **11.** $2x$ **13.** $15a^4$ **15.** $8y^2$
17. $12r^3s^4$ **19.** $18xy^2z^6$ **21.** $c - d$ **23.** $8(3x + 2)$
25. $y(3y^2 + 5y - 9)$ **27.** $9a(5a - 1)$ **29.** $5x^2y(3 - 2y)$
31. Prime   **33.** $3a^6b^4c^2(9a^3c^3 - 4a + 10c)$ **35.** $-(a + b)$
37. $-(5xy - y + 4)$ **39.** $-(-x + 2y)$ or $-(2y - x)$
41. $-(-p^2 - p + 17)$ **43.** $-8(a + 2)$ **45.** $-3x(2x + y)$
47. $-6ab(3a - 2b)$ **49.** $-4a^2c^8(2a^2 - 7a + 5c)$
51. $(x + y)(4 + t)$ **53.** $(a - b)(r - s + t)$
55. $(m + n + p)(3 + x)$ **57.** $(x + 7)^2(3x - 2)$
59. $(a + 8)(x + y)$ **61.** $(3 - c)(c + d)$ **63.** $(a + b)(a - 4)$
65. $(t - 3)(t^2 - 7)$ **67.** $(x + y)(x + 1)$ **69.** $(1 - n)(1 - m)$
71. $(y^2 + 3)(y - 4)$ **73.** $(t + v)(s + r)$

75. $2c(2b^3 + 1)(7a^3 - 1)$ **77.** $x(m + n)(p + q)$ **79.** $f = \dfrac{d_1 d_2}{d_2 + d_1}$

81. $a^2 = \dfrac{b^2 x^2}{b^2 - y^2}$ **83.** $y = \dfrac{rx}{b + t}$ **85.** $r = \dfrac{S - a}{S - l}$

87. $-7u^2(9u - 4)$ **89.** $(a^3 - 3)(b^2 + 1)$

91. $9x^7 y^3(5x^3 - 7y^4 + 9x^3 y^7)$ **93.** $t(r^2 + s^2)(a - b)$ **95.** Prime

97. $\frac{1}{5}x^2(3ax^2 + b - 4ax)$ **99.** $(2x + y)(a - b + c)$

101. $5y^6(9y^6 + 6y^4 + 5y^2 - 1)$ **103.** $(a^2 + b)(x - 1)$

105. $(b + c)(a + b + c)$ **107. a.** $\frac{1}{2}b_1 h$ **b.** $\frac{1}{2}b_2 h$

c. $\frac{1}{2}h(b_1 + b_2)$; the formula for the area of a trapezoid

109. $r^2(4 - \pi)$ **117.** $4{,}900$

Study Set Section 5.6 (page 471)

1. perfect   **3.** leading, coefficient, 2, descending   **5. a.** Positive
b. Negative   **c.** Positive   **7.** 9, 6, -9, -6, product, sum   **9.** No
11. $-$ **13.** $+$ **15.** $4, -4, 1$ **17.** $(a + 9)^2$ **19.** $(2y + 7)^2$
21. $(y^2 - 5)^2$ **23.** $(3b^2 - 2c^2)^2$ **25.** $(x + 2)(x + 3)$
27. $(t + 6)(t + 8)$ **29.** $(x - 5)(x - 11)$ **31.** $(y - 8)(y - 9)$
33. $(y^2 - 3)(y^2 - 10)$ **35.** $(x^2 + 5)(x^2 + 10)$
37. $3(x + 7)(x - 3)$ **39.** $15(x + 2)(x + 1)$
41. $2y(x + y)(x - 7y)$ **43.** $3t^2(s - 2t)(s + 8t)$
45. $(5x + 3)(x + 2)$ **47.** $(7a + 5)(a + 1)$ **49.** $(3r - 2)(r + 5)$
51. $(11y - 1)(y + 3)$ **53.** Prime   **55.** Prime
57. $(4a - 7y)(2a - y)$ **59.** $(7g + 2h)(g - 2h)$
61. $-2(9p + 2q)(p - q)$ **63.** $-5(3x - 4y)(2x + y)$
65. $a^2(3b^2 + 4)(3b^2 + 1)$ **67.** $c^8(4b^2 - 3)(2b^2 + 5)$
69. $b^2(2b^2 - 3)(5b^2 - 2)$ **71.** $m^3(3m^2 + 4)(3m^2 - 2)$
73. $(x + a + 1)^2$ **75.** $(a + b + 4)(a + b - 6)$
77. $(7q - 7r + 2)(2q - 2r - 3)$ **79.** $(8s + 8t + 9)(2s + 2t - 3)$
81. $-(a - 8)(a + 4)$ **83.** $(2z + 3)(3z + 4)$ **85.** $(5y - 1)^2$
87. $4h^4(8h - 1)(2h + 1)$ **89.** Prime   **91.** $x^2(b^2 - 7)(b^2 - 5)$
93. $-(3a + 2b)(a - b)$ **95.** $14(4a^2 + 3a - 5)$
97. $(3t + 3w - 2)(2t + 2w + 5)$ **99.** $(3y^3 + 2)(4y^3 + 5)$
101. $(m + n)(6a - 5)(a + 3)$ **103.** $5(2a - 3b)^2$
105. $-3a^2(x - 3)(x - 2)$ **107.** $(5m^4 - 6n)^2$ **109.** $5x - 4$
115. 5   **117.** 6

Study Set Section 5.7 (page 480)

1. squares   **3. a.** 1, 4, 9, 16, 25, 36, 49, 64, 81, 100
b. 1, 8, 27, 64, 125, 216, 343, 512, 729, 1,000   **5. a.** $F - L$
b. $F^2 - FL + L^2$ **c.** $F^2 + FL + L^2$
7. a. $x^2 - 4$ (Answers may vary)   **b.** $(x - 4)^2$ (Answers may vary)
c. $x^2 + 4$ (Answers may vary)   **d.** $x^3 + 8$ (Answers may vary)
e. $(x + 8)^3$ (Answers may vary)   **9.** $(x + 4)(x - 4)$
11. $(3y + 8)(3y - 8)$ **13.** $(12 + c)(12 - c)$
15. $(10m + 1)(10m - 1)$ **17.** $(9a + 7b)(9a - 7b)$ **19.** Prime
21. $(3r^2 + 11s)(3r^2 - 11s)$ **23.** $(4t + 5w^2)(4t - 5w^2)$
25. $(10rs^2 + t^2)(10rs^2 - t^2)$ **27.** $(6x^2 y + 7z^3)(6x^2 y - 7z^3)$
29. $(x^2 + y^2)(x + y)(x - y)$ **31.** $(4a^2 + 9b^2)(2a + 3b)(2a - 3b)$
33. $(x + y + z)(x + y - z)$ **35.** $(r - s + t^2)(r - s - t^2)$
37. $2(x + 12)(x - 12)$ **39.** $3x(x + 9)(x - 9)$
41. $5a(b^2 + 1)(b + 1)(b - 1)$ **43.** $4b(4 + b^2)(2 + b)(2 - b)$
45. $(c + d)(c - d + 1)$ **47.** $(a - b)(a + b + 2)$
49. $(x + 6 + y)(x + 6 - y)$ **51.** $(x - 1 + 3z)(x - 1 - 3z)$
53. $(a + 5)(a^2 - 5a + 25)$ **55.** $(2r + s)(4r^2 - 2rs + s^2)$
57. $(4t^2 - 3v)(16t^4 + 12t^2 v + 9v^2)$
59. $(x - 6y^2)(x^2 + 6xy^2 + 36y^4)$
61. $(a - b + 3)(a^2 - 2ab + b^2 - 3a + 3b + 9)$

63. $(4 - a - b)(16 + 4a + 4b + a^2 + 2ab + b^2)$
65. $(x + 1)(x^2 - x + 1)(x - 1)(x^2 + x + 1)$
67. $(x^2 + y)(x^4 - x^2 y + y^2)(x^2 - y)(x^4 + x^2 y + y^2)$
69. $5(x + 5)(x^2 - 5x + 25)$ **71.** $4x^2(x - 4)(x^2 + 4x + 16)$
73. $(4a - 5b^2)(16a^2 + 20ab^2 + 25b^4)$ **75.** $2b^2(12 + b^2)(12 - b^2)$
77. $(x + y)(x - y + 8)$ **79.** $(x + y)(x^2 - xy + y^2)(x^6 - x^3 y^3 + y^6)$
81. $(12at + 13b^3)(12at - 13b^3)$ **83.** Prime
85. $(9c^2 d^2 + 4t^2)(3cd + 2t)(3cd - 2t)$
87. $2u^2(4v - t)(16v^2 + 4tv + t^2)$ **89.** $(y + 2x - t)(y - 2x + t)$
91. $(x + 10 + 3z)(x + 10 - 3z)$
93. $(c - d + 6)(c^2 - 2cd + d^2 - 6c + 6d + 36)$
95. $\left(\frac{1}{6} + y^2\right)\left(\frac{1}{6} - y^2\right)$
97. $(m + 2)(m^2 - 2m + 4)(m - 2)(m^2 + 2m + 4)$
99. $(a + b)(x + 3)(x^2 - 3x + 9)$
101. $(x^3 - y^4 z^5)(x^6 + x^3 y^4 z^5 + y^8 z^{10})$
103. $\frac{4}{3}\pi(r_1 - r_2)(r_1^2 + r_1 r_2 + r_2^2)$
109.

111. $y = -4$

Study Set Section 5.8 (page 487)

1. prime   **3.** cubes, cubes   **5.** common   **7.** trinomial
9. Multiply the factors of $y^2 z^3(x + 6)(x + 1)$ to see if the product is
$x^2 y^2 z^3 + 7xy^2 z^3 + 6y^2 z^3$.   **11.** $3ab, 2b, 2a$
13. $4bc(a - 5)(a + 6)$ **15.** $-3xy(x + 2 - 4)$
17. $(y + 1)(y - 1)(y - 3)(y^2 + 3y + 9)$
19. $36(x^2 + 1)(x + 1)(x - 1)$ **21.** Prime   **23.** $2(3x - 4)(x - 1)$
25. $y^2(2x + 1)^2$ **27.** $2x^2 y^2 z^2(2 - 13z)$ **29.** $a^2(3a^2 - 8)^2$
31. $6(a - b)(a^2 + ab + b^2)(a + b)(a - b)$
33. $(x - y + 5)[(x - y)^2 - 5(x - y) + 25]$
35. $(2a - 2b + 3)(a - b + 1)$ **37.** $(2x - 9)(3x + 7)$
39. $(x + 1)(x - 1)(x + 4)(x - 4)$ **41.** $(x + 5 + y^4)(x + 5 - y^4)$
43. $(3x - 1 - 5y)(3x - 1 + 5y)$ **45.** $(a + b - 2)(a - b + 2)$
47. $(x - y)^2(a - 1)$ **49.** $\left(\frac{9}{4}x^2 + y^{20}\right)\left(\frac{3}{2}x + y^{10}\right)\left(\frac{3}{2}x - y^{10}\right)$
51. $16(m^8 + 1)(m^4 + 1)(m^2 + 1)(m + 1)(m - 1)$
53. $(3y + 1)(3y^4 + y^3 + 1)$ **55.** $(x - 4)(x + y)(x - y)$
57. $c(c + 2a - b)(c - 2a + b)$ **59.** $(3x + 1)(3x^3 + x^2 + 1)$
61. $(2x + 1)^2$ **63.** Prime   **65.** $4 - 9x^2 = (2 + 3x)(2 - 3x)$ in.^{2}
69. 6   **71.** -13

Study Set Section 5.9 (page 498)

1. quadratic   **3.** standard   **5. a.** Yes   **b.** No   **c.** Yes
d. No   **7. a.** At least one is 0.   **b.** zero, 0, 0
9. 4 is a solution; -5 is not a solution   **11. a.** $10 + 2x$
b. $20 + 2x$ **13.** $y + 2, y - 4, -2$ **15.** $-3, -5$ **17.** $-2, -4$
19. $1, \frac{1}{2}$ **21.** $-\frac{1}{3}, -3$ **23.** $0, -1$ **25.** $0, 2$ **27.** $-4, 4$
29. $-\frac{3}{4}, \frac{3}{4}$ **31.** $\frac{1}{3}, -1$ **33.** $\frac{1}{5}, -2$ **35.** $\frac{1}{4}, -\frac{3}{2}$ **37.** $-3, 3$
39. A repeated solution of -8 **41.** $2, -\frac{5}{2}$ **43.** $1, 2$ **45.** $0, 2, 4$
47. $0, 7, -3$ **49.** $0, 7, -7$ **51.** $6, -6, 1, -1$ **53.** $3, -2, 2$
55. $2.78, 0.72$ **57.** 1   **59.** $2, -\frac{5}{6}$ **61.** $0, -1$ **63.** $1, -4$

65. $-4, 4, 2$ **67.** $0, \frac{25}{6}$ **69.** 0 and a repeated solution of 9

71. $1, -\frac{5}{7}$ **73.** $-1, -2, 2$ **75.** $-\frac{9}{8}, \frac{9}{8}$ **77.** $0, \frac{5}{6}, -7$

79. $11, -6$ **81.** $5, -\frac{7}{2}$ **83.** $5.5, -3$ **85.** $-5, 5, -3, 3$

87. $16, 18$ **89.** Base: 8 ft; height: 4 ft **91.** Length: 16 in.; width: 10 in. **93.** 3 ft **95.** 2 ft **97.** 9 sec **99.** 2 sec

101. 6 m/sec **103.** 4 **111.** 25 ft^2

Chapter 5 Review (page 503)

1. 243 **2.** -32 **3.** -64 **4.** $\frac{4}{9}$ **5.** x^6 **6.** $\frac{m^3}{n^5}$ **7.** $64m^{16}$

8. t^{13} **9.** $9x^4y^6$ **10.** $\frac{x^{16}}{b^4}$ **11.** -10 **12.** $\frac{h^{12}}{5}$ **13.** $\frac{b}{2a}$

14. $\frac{x^5}{5}$ **15.** $\frac{x^6}{9}$ **16.** $\frac{2}{9x}$ **17.** $-c^{10}$ **18.** $\frac{25}{16}$ **19.** $\frac{16}{27}$ **20.** 64

21. s^{73} **22.** $-\frac{b^3c^3}{8a^{21}}$ **23.** 1.93×10^{10} **24.** 2.735×10^{-8}

25. 72,770,000 **26.** 0.0000000083 **27.** 7.6×10^2 sec

28. 1.67248×10^{-18} g **29.** 8.4×10^6 **30.** 1.875×10^{-21}

31. No **32.** Yes **33.** Yes **34.** No **35.** Binomial, 2

36. Monomial, 4 **37.** None of these, 4 **38.** Trinomial, 8

39. 134 in.^3 **40.** -29

41.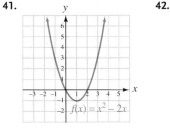

42.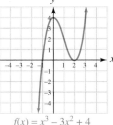

43. $2t^3 - 2t^2 - 5t$ **44.** $\frac{3}{4}ab^2c - \frac{5}{6}abc$ **45.** $3x^2y^3 - 8x^2y + 8y$

46. $19m^2 - 13m - 9$ **47.** $\frac{1}{2}s^6 - \frac{5}{3}s^4$ **48.** $6k^4 - 6k^3 + 9k^2 - 2$

49. $4c^2d^2 + 4cd^2$ **50. a.** -1 **b.** $-2, -1, 1$

c. D: $(-\infty, \infty)$; R: $[-1, \infty)$ **51.** $-4a^3$ **52.** $6x^3y^2z^5$

53. $2x^4y^3 - 8x^2y^7$ **54.** $a^4b + 2a^3b^2 - a^2b^3$

55. $6x^3 - 12x^2 + 4x - 8$ **56.** $25a^2t^2 - 60at + 36$

57. $49c^8d^6 - d^2$ **58.** $15x^4 - 22x^3 + 58x^2 - 40x$

59. $r^3 - 3r^2s - rs^2 + 3s^3$ **60.** $9c^2 - \frac{9}{2}c + \frac{9}{16}$

61. $25 - 10a + 10b + a^2 - 2ab + b^2$

62. $2x^2 + xy - y^2 - 3yz - 2z^2$ **63.** $41a^2 - 12a + 5$

64. $b^2 - b - 7$ **65. a.** $(12x - 2)$ in. **b.** $(8x^2 - 2x)$ in.^{2}

c. $(16x^3 + 12x^2 - 4x)$ in.^{3} **66. a.** $f(x) = x^3 + 3x^2 + 2x$

b. 210 in.^3 **67.** 6 **68.** $3xy^3$ **69.** $4(x^4 + 2)$

70. $\frac{x}{5}(3x^2 - 6x + 1)$ **71.** Prime **72.** $7a^3b(ab + 7)$

73. $5x^2(x + y)(1 - 3x)$ **74.** $9x^2y^3z^2(3xz + 9x^2y^2 - 10z^5)$

75. $-(x + 9)$ **76.** $-(-4r + 7)$ or $-(7 - 4r)$ **77.** $-7(b^3 - 2c)$

78. $-7a^2b^2(a - b)^3(7a^2 - 7ab - 9b^2)$ **79.** $(x + 2)(y + 4)$

80. $(ry - a + 1)(r - 1)$ **81.** $(t^2 + 1)(t - 9)$

82. $(1 - x)(1 - 3z)$ **83.** $m_1 = \frac{mm_2}{m_2 - m}$ **84.** $A = 2\pi r(r + h)$

85. $(x + 5)^2$ **86.** $(7a^3 + 6b^2)^2$ **87.** $(y - 20)(y - 1)$

88. $(z - 5)(z - 6)$ **89.** $-(x + 7)(x - 4)$ **90.** $(a - 8b)(a + 3b)$

91. $(4a - 1)(a - 1)$ **92.** Prime **93.** $y^6(y + 2)(y - 1)$

94. $9st(3r - 2)(r + 4)$ **95.** $(r + s)(6t - 5)(t + 3)$

96. $(v^2 - 7)(v^2 - 6)$ **97.** $(w^4 - 10)(w^4 + 9)$ **98.** $(s + t - 1)^2$

99. $(z + 4)(z - 4)$ **100.** $(xy^2 + 8z^3)(xy^2 - 8z^3)$ **101.** Prime

102. $(c + a + b)(c - a - b)$ **103.** $10m^2(m^2 + 4)(m + 2)(m - 2)$

104. $(m + n)(m - n + 1)$ **105.** $2c(4a^2 + 9b^2)(2a + 3b)(2a - 3b)$

106. $(k + 1 + 3m)(k + 1 - 3m)$ **107.** $(t + 4)(t^2 - 4t + 16)$

108. $(2a - 5b^3)(4a^2 + 10ab^3 + 25b^6)$

109. $4d^4(d + 1)(d^2 - d + 1)$

110. $(b + c + 3)(b^2 + 2bc + c^2 - 3b - 3c + 9)$

111. $4rs(q - 5t)(q + 6t)$ **112.** $(2m + 2n + 3)(m + n - 1)$

113. $(z - 2)(z + x + 2)$ **114.** Prime

115. $(x + 2 + 2p^2)(x + 2 - 2p^2)$ **116.** $(y + 2)(y + 1 + x)$

117. $4c^2(ab + 4)(a^2b^2 - 4ab + 16)$

118. $(a + 3)(a - 3)(a + 2)(a - 2)$ **119.** $(2x + 3)(2x^3 + 3x^2 + 1)$

120. $\frac{\pi}{2}h(r_1 + r_2)(r_1 - r_2)$ **121.** $0, \frac{3}{4}$ **122.** $6, -6$ **123.** $\frac{1}{2}, -\frac{5}{6}$

124. $3, -3, 1, -1$ **125.** $0, -\frac{2}{3}, \frac{4}{5}$ **126.** $-\frac{2}{3}, 7, 0$

127. A repeated solution of -8 **128.** $-7, -1, 1$ **129.** $1, 3$

130. 7 m by 10 m **131.** 5 ft **132.** $-\frac{1}{2}, 1$

Chapter 5 Test (page 512)

1. a. perfect **b.** product **c.** binomials **d.** quadratic

e. difference, cubes **f.** greatest common **2. a.** $x^{10} \text{ ft}^2$

b. $x^{15} \text{ ft}^3$ **3.** $\frac{a^5}{9m^5}$ **4.** $\frac{-8x^6y^9}{125}$ **5.** $-\frac{64}{b^8}$ **6.** $\frac{m^4}{9n^{10}}$

7. $2.9 \times 10^6 = 2,900,000$ **8.** 1.116×10^7 mi

9. a. $3, -4, -3, -\frac{5}{3}; 5$ **b.** $8, -1, 1, -6, 4; 13$ **10.** 110 ft

11. a.

b. D: $(-\infty, \infty)$; R: $(-\infty, \infty)$

c. $-2, 0$ **12. a.** 0 **b.** 2, 6

c. D: $(-\infty, \infty)$; R: $(-\infty, 2]$

13. $2x^2y^3 + 13xy + 3y^2$

14. $2x^5 + \frac{2}{15}x^2$

15. $-12a^4y^{12}z^{12}$

16. $-15a^3b^4 + 10a^3b^5$

17. $6y^3 + 11y^2 + 9y + 2$

18. $0.06d^2 + 1.6d - 6$

19. $36 + 12m - 12n + m^2 - 2mn + n^2$ **20.** $32s^3 - 50st^2$

21. $15t^2 - 21t + 13$ **22.** $c^2 - c + 4$

23. $3abc(4a^2b - abc + 2c^2)$ **24.** $(k + z)(h + b)$

25. $(x - 6)(x + 5)$ **26.** $(y^2 + 9)(y + 3)(y - 3)$ **27.** $(5m - 6n)^2$

28. $(s + 3)(s - 3)(s + 2)(s - 2)$ **29.** $-x^2(7x - 8)(3x + 2)$

30. Prime **31.** $5(x + 5)(x^2 - 5x + 25)$

32. $(4a - 5b^2)(16a^2 + 20ab^2 + 25b^4)$

33. $(x - y + 5)(x - y - 2)$ **34.** $(3b + 2c)(2b - c)$

35. $(a + b)(a - b + 1)$ **36.** $v = \frac{v_1v_3}{v_3 + v_1}$ **37.** $0, 5$ **38.** $-5, \frac{7}{2}$

39. $\frac{3}{4}, -\frac{3}{2}$ **40.** $1, 0, -9$ **41.** $-1, -4, 4$ **42.** 1.5 ft

Study Set Section 6.1 (page 526)

1. rational **3.** simplify **5.** opposites **7. a.** 1 **b.** 0.5

c. 0.25 **d.** D: $(0, \infty)$; R: $(0, \infty)$ **9. a.** $\frac{3y}{7x^2}$ **b.** $\frac{x - 3}{x + 2}$

c. $-\frac{a^3}{a + 9}$ **11. a.** iii **b.** i **c.** iv **d.** ii

13. Yes, Yes, Yes, Yes, Yes

15. 6, 3, 1.5, 1, 0.75, 0.6, 0.5

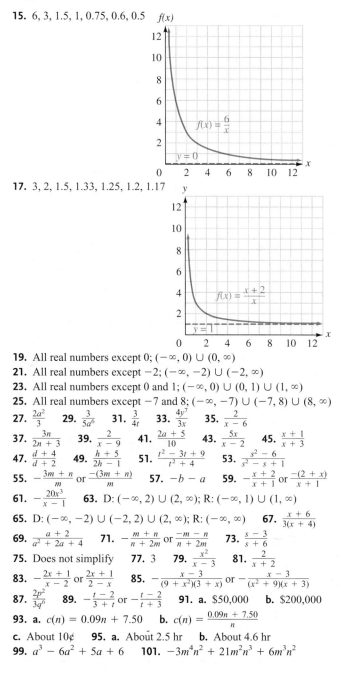

$f(x) = \frac{6}{x}$

$y = 0$

17. 3, 2, 1.5, 1.33, 1.25, 1.2, 1.17

$f(x) = \frac{x+2}{x}$

$y = 1$

19. All real numbers except 0; $(-\infty, 0) \cup (0, \infty)$

21. All real numbers except -2; $(-\infty, -2) \cup (-2, \infty)$

23. All real numbers except 0 and 1; $(-\infty, 0) \cup (0, 1) \cup (1, \infty)$

25. All real numbers except -7 and 8; $(-\infty, -7) \cup (-7, 8) \cup (8, \infty)$

27. $\frac{2a^2}{3}$ **29.** $\frac{3}{5a^6}$ **31.** $\frac{3}{4t}$ **33.** $\frac{4y^7}{3x}$ **35.** $\frac{2}{x-6}$

37. $\frac{3n}{2n+3}$ **39.** $\frac{2}{x-9}$ **41.** $\frac{2a+5}{10}$ **43.** $\frac{5x}{x-2}$ **45.** $\frac{x+1}{x+3}$

47. $\frac{d+4}{d+2}$ **49.** $\frac{h+5}{2h-1}$ **51.** $\frac{t^2-3t+9}{t^2+4}$ **53.** $\frac{s^2-6}{s^2-s+1}$

55. $-\frac{3m+n}{m}$ or $\frac{-(3m+n)}{m}$ **57.** $-b-a$ **59.** $-\frac{x+2}{x+1}$ or $\frac{-(2+x)}{x+1}$

61. $-\frac{20x^3}{x-1}$ **63.** D: $(-\infty, 2) \cup (2, \infty)$; R: $(-\infty, 1) \cup (1, \infty)$

65. D: $(-\infty, -2) \cup (-2, 2) \cup (2, \infty)$; R: $(-\infty, \infty)$ **67.** $\frac{x+6}{3(x+4)}$

69. $\frac{a+2}{a^2+2a+4}$ **71.** $-\frac{m+n}{n+2m}$ or $\frac{-m-n}{n+2m}$ **73.** $\frac{s-3}{s+6}$

75. Does not simplify **77.** 3 **79.** $\frac{x^2}{x-3}$ **81.** $\frac{2}{x+2}$

83. $\frac{-2x+1}{x-2}$ or $\frac{2x+1}{2-x}$ **85.** $-\frac{x-3}{(9+x^2)(3+x)}$ or $-\frac{x-3}{(x^2+9)(x+3)}$

87. $\frac{2p^2}{3q^6}$ **89.** $-\frac{t-2}{3+t}$ or $-\frac{t-2}{t+3}$ **91. a.** \$50,000 **b.** \$200,000

93. a. $c(n) = 0.09n + 7.50$ **b.** $c(n) = \frac{0.09n + 7.50}{n}$

c. About 10¢ **95. a.** About 2.5 hr **b.** About 4.6 hr

99. $a^3 - 6a^2 + 5a + 6$ **101.** $-3m^4n^2 + 21m^2n^3 + 6m^3n^2$

Study Set Section 6.2 (page 538)

1. rational **3.** invert **5.** numerators, denominators, $\frac{AC}{BD}$

7. $(x-5)$, $(x+3)$, x, 5, x **9.** Yes, No, Yes **11.** $\frac{11}{4}$ **13.** $\frac{2}{5}$

15. $\frac{1}{25a^3}$ **17.** $\frac{3x^4}{14y^3}$ **19.** $\frac{y(y+3)}{10}$ **21.** $\frac{(x-2)^2}{x}$ **23.** $\frac{x+1}{9x^2}$

25. $\frac{(t+3)^2}{(t-3)(t+2)}$ **27.** 1 **29.** $\frac{t-1}{t+1}$ **31.** $\frac{3}{2x}$ **33.** $\frac{a-b}{a}$

35. $x+1$ **37.** $2y+16$ or $2(y+8)$ **39.** $-\frac{a^4}{2(a^2+3)}$

41. $-\frac{(x+1)^2(x+2)}{x+2c}$ **43.** $\frac{x^2-6x+9}{x^4+8x^2+16}$

45. $\frac{4m^4-4m^3-11m^2+6m+9}{x^4-2x^2+1}$ **47.** $\frac{5}{6}$ **49.** $\frac{9}{2}$ **51.** $\frac{2y^5}{3x^6}$

53. $\frac{3}{10p^9}$ **55.** $\frac{x^{11}}{x^2+2x+4}$ **57.** $\frac{2x(x-4)}{x+5}$ **59.** $\frac{n+2}{n+1}$

61. $\frac{c+d}{d(25c^2-5cd+d^2)}$ **63.** $\frac{y(y+3)}{y+2}$ **65.** $\frac{2}{x+1}$ **67.** $\frac{a+1}{a-1}$

69. $\frac{3x}{2}$ **71.** $-\frac{(x-3)(x-6)}{(x+2)(x+3)}$ **73.** $\frac{2(x+1)}{x(x+3)}$ **75.** $\frac{q}{p}$ **77.** $\frac{5r^3}{2s^2}$

79. $-10h + 30$ or $-10(h-3)$ **81.** $\frac{x+y}{x-y}$ **83.** $\frac{x+2}{x+1}$

85. $\frac{2x(x-5)}{x+5}$ **87.** $k_1(k_1+2)$, k_2+6 **93.** x^{m+n} **95.** x^ny^n

97. 1 **99.** x^{m-n} **101.** $\frac{y^n}{x^m}$

Study Set Section 6.3 (page 550)

1. denominator **3.** build **5.** numerators, denominator, $A+B, A-B$ **7.** factor, greatest **9. a.** ii

b. Adding or subtracting rational expressions

c. Simplifying a rational expression **11. a.** Twice **b.** Once **13.**

a. $2 \cdot 2 \cdot 2 \cdot 5 \cdot x \cdot x$ **b.** $2x(x-3)$ **c.** $(n+8)(n-8)$

15. $3x-2$, 3, $3x-1$, 3 **17.** $\frac{13}{3x}$ **19.** $\frac{t}{2r}$ **21.** 4 **23.** $\frac{3}{x+3}$

25. $\frac{15q+2p}{pq}$ **27.** $\frac{21a-22b}{6ab}$ **29.** $\frac{8x-2}{(x+2)(x-4)}$ **31.** $\frac{2x^2-30x}{(x+3)(x-3)}$

33. $\frac{x}{x-3}$ **35.** $\frac{9m+2}{m-n}$ **37.** $\frac{4x-7}{x-2}$ **39.** $\frac{7x^2+x}{7x-3}$ **41.** $36x^2y$

43. $x(x+3)(x-3)$ **45.** $(x+3)^2(x^2-3x+9)$

47. $(2x+3)^2(x+1)^2$ **49.** $\frac{41}{30m}$ **51.** $\frac{3a-10b}{4a^2b^2}$ **53.** $\frac{1}{x+1}$

55. $\frac{m-5}{(m+3)(m+5)}$ **57.** $\frac{2x^2+x}{(x+3)(x+2)(x-2)}$ **59.** $\frac{-x+28}{30(x-7)}$

61. 3 **63.** $\frac{-2(2x^2-7x-27)}{x(x+3)(x-3)}$ **65.** $\frac{11x^2+7x-3}{(2x-1)(3x+2)}$ **67.** 0

69. $\frac{14s+58}{(s+3)(s+7)}$ **71.** $\frac{5x-y}{6}$ **73.** $\frac{6}{x-3}$ **75.** $\frac{1}{a-b}$

77. $\frac{2x^2+5x+4}{x+1}$ **79.** $\frac{-x^2+11x+8}{(3x+2)(x+1)(x-3)}$ **81.** $\frac{2}{x+1}$

83. $\frac{16y^2+3}{18y^4}$ **85.** $\frac{3}{(b+2)(b-3)}$ **87.** $\frac{a^2-a}{a-5}$ **89.** $\frac{3(a-1)}{3a-2}$ or $\frac{3a-3}{3a-2}$

91. $\frac{2x}{x+1}$ **93.** $\frac{m-3}{(m+3)(m+1)}$ **95.** $\frac{9p^3}{5}$ **97.** $\frac{6-3d}{5d(d-1)}$

99. $\frac{2s-3t}{t(2s+3t)}$ **101.** $\frac{10r+20}{r}$, $\frac{3t+9}{t}$ **107.** A repeated solution of 3

109. A repeated solution of $0, -1$

Study Set Section 6.4 (page 561)

1. rational, complex **3.** $\frac{t^2}{t^2}$ **5.** $\div$, $\frac{3}{25m}$, m, 3, 3, m, 10 **7. a.** $\div$

b. $6 - k - \frac{5}{k}$; $k^2 - 9$ **9.** $\frac{2a^5}{3}$ **11.** $\frac{5y}{9}$ **13.** $\frac{27y^3}{5x^3}$ **15.** $\frac{20}{49c^6d^4}$

17. $\frac{3-2a}{a^2-3a}$ **19.** $\frac{p^2-1}{3p-1}$ **21.** $y-x$ **23.** $-\frac{1}{a+b}$ **25.** $\frac{2+a}{2a+1}$

27. $\frac{b-1}{b+2}$ **29.** $\frac{y+x}{xy}$ **31.** $\frac{a^3b^2-a^2}{a^2b^3-b^2}$ or $\frac{a^2(ab^2-1)}{b^2(a^2b-1)}$ **33.** $\frac{5z-12}{5z}$

35. $\frac{x-9}{3}$ **37.** $\frac{y+x}{y-x}$ **39.** $\frac{2x+4}{3-4}$ **41.** $\frac{x+2}{x-3}$ **43.** $\frac{1}{c-d}$

45. $\frac{a-1}{a+1}$ **47.** $125b$ **49.** -1 **51.** $\frac{3a+7}{2a}$ **53.** $\frac{xy^2}{y-x}$

55. $\frac{y-2x}{2y+x}$ **57.** $\frac{5x^2y^2}{xy+1}$ **59.** $\frac{1}{2y}$ **61.** $\frac{1}{x-y}$ **63.** $\frac{4}{c+d}$

65. $\frac{k_1k_2}{k_2+k_1}$ **67.** $\frac{4d-1}{3d-1}$ **73.** 8 **75.** 2, -2, 3, -3

Study Set Section 6.5 (page 573)

1. monomial, polynomial, binomial **3.** Divisor, Quotient, Dividend, Remainder **5. a.** term **b.** 9, 9 **c.** 6, 6, 6

7. $(2x-1)(x^2+3x-4) = 2x^3 + 5x^2 - 11x + 4$

9. x^2, $7x$, 28 **11.** $3a^2 + 5 + \frac{6}{3a-2}$

13. $\frac{x^2-x-12}{x-4}$, $x-4\overline{)x^2-x-12}$, $(x^2-x-12) \div (x-4)$

15. $\frac{y}{2x^3}$ **17.** $\frac{3}{4a^2}$ **19.** $2x^4 + 3x$ **21.** $\frac{2x}{3} - \frac{x^2}{6}$ **23.** $2a - \frac{2a^2y}{3}$

25. $-\frac{x^4y^4}{2} + \frac{x^3y^9}{4} - \frac{3}{4xy^2}$ **27.** $x+2$ **29.** $x-3$ **31.** $4x-5$

33. $3x^2 + 4x + 3$ **35.** $t^2 + 2t + 1 + \frac{3}{t+6}$

37. $2x^2 + 5x - 3 + \frac{-8}{3x-4}$ **39.** $a+1$ **41.** $2y+2$

43. $3x^2 - x + 2$ **45.** $4x^3 - 3x^2 + 3x + 1$ **47.** $4a^2 - 2a + 1$

49. $5a^2 - 3a - 4$ **51.** $x^2 + 3x + 4$ **53.** $2x + 3 + \frac{20x - 13}{3x^2 - 7x + 4}$

55. $6y - 12$ **57.** $4a^2 - 3a + \frac{7}{a + 1}$ **59.** $x^4 + x^2 + 4$

61. $\frac{a^2}{5} - \frac{2}{5a^2}$ **63.** $s + 5 + \frac{-10}{2s + 3}$ **65.** $\frac{8m^2}{7n^{10}}$

67. $m - 4 + \frac{m + 3}{m^2 + 1}$ **69.** $y^2 + 4y + 16$ **71.** $a^4 - a^2 + 1$

73. $10x^2z - 2x - \frac{1}{x}$ **75.** $x^2 - 2 + \frac{-x^2 + 7x + 4}{x^3 + 2x + 1}$

77. $2x^2 - x + 1$ **79.** $\frac{5x^2}{9} + \frac{x}{3} - \frac{1}{9}$ **81.** $x^4 - x^2 - 3$

83. $x^2 - 5x + 6$ **85.** $3x - 2, x + 5$ **89.** $8x^2 + 2x + 4$

91. $-2y^3 - y^2 + 10y - 14$

Study Set Section 6.6 (page 582)

1. synthetic   **3.** divisor   **5.** theorem

7. a. $(5x^3 + x + 3) \div (x + 2)$ **b.** $5x^2 - 10x + 21 + \frac{-45}{x + 2}$

9. $6x^3 - x^2 - 17x + 9, x - 8$ **11.** 2, 6, 1, 12, 26, 2, 8

13. $2x + 3$ **15.** $5x - 2$ **17.** $3x - 4$ **19.** $5x + 6$

21. $a^2 - a - 2$ **23.** $3a^2 + 12a + 1$ **25.** $3b^2 + 9b - 4 + \frac{1}{b - 3}$

27. $4t^2 + 8t + 15 + \frac{12}{t - 2}$ **29.** $x^2 - 5x + 6$

31. $3x^2 - 4x - 4 + \frac{-10}{x + 8}$ **33.** $2x^2 - 3x + 12 + \frac{-52}{x + 5}$

35. $x^2 - 2x + 3 + \frac{-3}{x + 10}$ **37.** $7.2x - 0.66 + \frac{0.368}{x - 0.2}$

39. $9x^2 - 513x + 29{,}241 - \frac{1{,}666{,}762}{x + 57}$ **41.** -1 **43.** -37

45. 23   **47.** -1 **49.** 2   **51.** -1 **53.** 18   **55.** 174

57. -8 **59.** 59   **61.** 44   **63.** $\frac{29}{32}$ **65.** Yes   **67.** No

69. $6x^2 - x + 1 + \frac{3}{x + 1}$ **71.** $x - 7 + \frac{28}{x + 2}$

73. $a^4 + a^3 + a^2 + a + 1$ **75.** $-6c^4 - 10c^3 - 2c^2 - 4c + 9$

77. $9a^2 - 21$ **79.** $4x^3 - x + 2 + \frac{6}{x + 3}$ **81.** $3x^2 - x + 2$

83. $2x^2 + 4x + 5$ **85.** $4x^2 - 3x + 6 + \frac{-13}{x + 2}$

87. $8a^2 - 16a - 20$ **93.** 0   **95.** 2

Study Set Section 6.7 (page 592)

1. rational   **3. a.** Yes   **b.** No   **5.** The LCD, $10(y - 5)$

7. $30y, \frac{9}{2y}, 30y, 30y, 30y, 7y, 35$ **9.** 12   **11.** 5   **13.** 4

15. $\frac{5}{3}$ **17.** 5   **19.** $-\frac{2}{3}$ **21.** 1   **23.** 7   **25.** $4, -1$

27. $2, -3$ **29.** $-\frac{1}{3}, 7$ **31.** $-\frac{1}{2}, 6$ **33.** -8 **35.** 6

37. No solution; 9 is extraneous   **39.** No solution; -3 is extraneous

41. $A = LQ + I$ **43.** $r = \frac{E - IR_L}{I}$ **45.** $n_2 = \frac{2\mu_R - n_1{}^2 - n_1}{n_1}$

47. $Q_1 = \frac{PQ_2}{1 + P}$ **49.** $R = \frac{R_1R_2R_3}{R_2R_3 + R_1R_3 + R_1R_2}$ **51.** $r = \frac{eR}{E - e}$

53. 2   **55.** $\frac{1}{2}$ **57.** No solution; 0 is extraneous   **59.** $1, -11$

61. 1, 6   **63.** A repeated solution of 2   **65.** $-\frac{1}{2}$ **67.** 0   **69.** 5

71. 5; 3 is extraneous   **73.** 1   **75.** 6   **77.** 1   **79.** 2

81. $\frac{1}{6}$ **83.** $-\frac{3}{2}, 2$ **85. a.** $\frac{11x + 30}{12x}$ **b.** 6

87. a. $\frac{m^2 - 4m + 2}{(m - 2)(m - 3)}$ **b.** 4   **89. a.** $f = \frac{s_1s_2}{s_1 + s_2}$ **b.** $4\frac{8}{13}$ in.

91. a. $L = \frac{SN - CN}{V - C}$ **b.** 8 yr   **97.** 9.0×10^9 **99.** 4.4×10^{-22}

Study Set Section 6.8 (page 602)

1. work, motion   **3.** x **5.** $\frac{x}{15}, \frac{x}{8}$ **7.** $\frac{12}{x}, \frac{12}{x + 15}$ **9.** $4\frac{5}{9}$ hr

11. $2\frac{6}{11}$ days   **13. a.** $1\frac{7}{8}$ days   **b.** Santos: \$412.50, Mays: \$375

15. $5\frac{5}{6}$ min   **17.** $4\frac{8}{13}$ sec   **19.** $2\frac{4}{13}$ weeks

21. Waiter: 10 min; busboy: 15 min

23. Faster worker: 6 hr; slower worker: 12 hr

25. Experienced plumber: 6 days; apprentice: 12 days

27. $1\frac{1}{2}$ hours   **29.** About 110 sec

31. Going: 40 mph; returning: 30 mph   **33.** Going: 60 mph, returning: 40 mph   **35.** 35 mph and 45 mph   **37.** 150 mph

39. 2 mph   **41.** 3 mph   **45.** $\frac{m^{80}}{n^8}$ **47.** $-\frac{1}{w^2}$ **49.** -4

51. $-x^8y^{10}$

Study Set Section 6.9 (page 616)

1. ratio   **3.** extremes, means, extremes, means   **5.** similar

7. inverse, decreases   **9.** Direct   **11.** Inverse   **13.** 7, 6, 18, 66, 11

15. 3   **17.** 5   **19.** $\frac{21}{5}$ **21.** -4 **23.** 2, -2 **25.** 4, -1

27. $-\frac{5}{2}, -1$ **29.** $0, -3$ **31.** $A = kp^2$ **33.** $z = \frac{k}{t^3}$

35. $C = kxyz$ **37.** $P = \frac{ka^2}{j^3}$ **39.** r varies directly as t.

41. b varies inversely as h.   **43.** U varies jointly as r, the square of s, and t.   **45.** P varies directly as m and inversely as n.   **47.** $\frac{7}{2}$

49. $-5, 2$ **51.** No solution   **53.** $-\frac{1}{2}, 0, 5$ **55.** 6

57. $-\frac{5}{2}, -1, 1$ **59.** -0.1 **61.** $-10, 10$ **63.** 202 mg, 139 mg, 125 mg   **65.** About 2 gal   **67.** Eye: 49.9 in.; seat: 17.6 in.; elbow: 27.8 in.   **69.** 12.5 in.   **71.** 25 ft   **73.** $46\frac{7}{8}$ ft   **75.** 4 cm

77. a. False   **b.** False   **c.** True   **79.** 1,600 ft   **81.** 25 days

83. 12 in.^{3} **85.** \$9,000   **87.** 3 ohms   **89.** 0.275 in.

91. 12.8 lb   **95.** $\frac{13}{6}w^3 - \frac{1}{4}w^2 + \frac{4}{5}$ **97.** $6y^3 + 11y^2 + 9y + 2$

Chapter 6 Review (page 623)

1. 8, 4, 2, 1.33, 1, 0.8, 0.67, 0.57, 0.5 $f(x)$

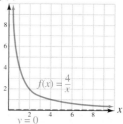

2. a. 1   **b.** 2   **c.** D: $(0, \infty)$; R: $(0, \infty)$ **3.** The domain is the set of all real numbers except -6 and 4: $(-\infty, -6) \cup (-6, 4) \cup (4, \infty)$.

4. $y = 3, x = 0$; D: $(-\infty, 0) \cup (0, \infty)$; R: $(-\infty, 3) \cup (3, \infty)$ **5.** $\frac{12x}{19y^7}$

6. $\frac{x - 7}{x + 7}$ **7.** $\frac{1}{2x^2(x + 2)}$ **8.** $\frac{1}{x^4(x - 6)}$ **9.** $\frac{-a - b}{c + d}$ **10.** $\frac{m + 2n}{2m + n}$

11. $\frac{2x + 1}{3x - 4}$ **12.** -2 **13.** Does not simplify

14. $-\frac{3m - 4}{m + 3}$ or $\frac{4 - 3m}{m + 3}$ **15.** $\frac{2}{49x^2}$ **16.** $-x$ **17.** $\frac{2a - 1}{4a(a + 2)}$

18. $\frac{t - 2}{t}$ **19.** $\frac{h^2 - 4h + 4}{h^6 + 8h^3 + 16}$ **20.** $\frac{a + 6}{(m + 4)(m - 3)}$

21. $\frac{2m - n}{m + n}$ **22.** $\frac{3x(x - 1)}{(x - 3)(x + 1)}$ **23.** $\frac{5y - 3}{x - y}$ **24.** $\frac{1}{c - d}$

25. $-\frac{2}{t - 3}$ **26.** $-\frac{1}{p + 12}$ **27.** $60a^2h^3$ **28.** $ab^2(b - 1)$

29. $(x - 5)(x + 5)(x + 1)$ **30.** $(m^2 + 2m + 4)(m - 2)^2$

31. $\frac{9a + 8}{a + 1}$ **32.** $\frac{40x + 7y^2z}{112z^2}$ **33.** $\frac{4x^2 + 9x + 12}{(x - 4)(x + 3)}$ **34.** $\frac{2a^2 + 8a - 19}{3(a + 2)}$

35. $\frac{14y + 58}{(y + 3)(y + 7)}$ **36.** $\frac{12y + 20}{15y(x - 2)}$ **37.** $\frac{1}{(a + 3)(a + 2)}$

38. $\frac{2a - 3}{2a}$ **39.** $\frac{2bc^3}{7}$ **40.** $\frac{p - 3}{2(p + 2)}$ **41.** $\frac{b + 2a}{2b - a}$

42. $\frac{x - 2}{x + 3}$ **43.** $\frac{x^2y^2}{(x - y)^2(y^2 - x^2)}$ **44.** $\frac{1 + b + d}{1 - b - d}$ **45.** $\frac{2b^2 - 3b + 3}{3b - 2}$

46. $\frac{4r - 8}{2r + 5}$ **47.** $\frac{5h^3}{11k^2}$ **48.** $\frac{1}{2y^3z^{10}}$ **49.** $6a + \frac{16}{3}$

50. $-3x^2y + \frac{3x}{2} + y$ **51.** $b + 4$ **52.** $v^2 - 3v - 10$

53. $x^2 - 2x + 4$ **54.** $2m - 5 + \frac{-4}{4m + 1}$ **55.** $3a - 2 + \frac{-15a + 2}{a^2 + 5}$

56. $m^4 - m^2 + 1 + \frac{2}{m^4 + 2m^2 - 3}$ **57.** $x - 7$ **58.** $m^2 - 3m + 2$

59. $-3n^4 - 2n^3 - n^2 - 2n + 1$ **60.** $4x^2 - 3x + 6 + \frac{-13}{x + 2}$

61. $3a^3 - a + 4 + \frac{6}{a + 1}$ **62.** $x^3 + 3x^2 + 9x + 27 + \frac{82}{x - 3}$

63. 54   **64.** -227 **65.** Yes   **66.** No   **67.** 5   **68.** $\frac{77}{5}$

69. $-2, 3$ **70.** $-1, -2$ **71.** $\frac{3}{2}$ **72.** 0

73. No solution; 3 is extraneous   **74. a.** $\frac{x + 26}{x(x - 4)}$ **b.** $\frac{26}{9}$

75. $b = \frac{Ha}{2a - H}$ **76.** $y^2 = \frac{x^2b^2 - a^2b^2}{a^2}$ **77.** $R = \frac{R_1 R_2}{R_2 + R_1}$

78. $F = \frac{ma}{k}$ **79. a.** $\frac{1}{10}$ of the job per hour

b. $\frac{x}{10}$ of the job is completed   **80.** $14\frac{2}{5}$ hr

81. Experienced electrician: 10 days; apprentice: 15 days

82. 6 min   **83.** 5 mph   **84.** 50 mph   **85.** 5   **86.** $-4, -12$

87. $0, -1$ **88.** $-2, 3$ **89.** 70.4 ft   **90.** 20   **91.** 66 in.   **92.**
$5,460   **93.** 1.25 amps   **94.** 126.72 lb   **95.** Inverse variation

96. 0.2

Chapter 6 Test (page 634)

1. a. rational   **b.** reciprocal   **c.** complex   **d.** extraneous

e. proportion   **3.** $\frac{2}{3xy}$ **4.** $\frac{2}{x - 2}$ **5.** -3 **6.** $\frac{2x + y}{4y}$

7. y

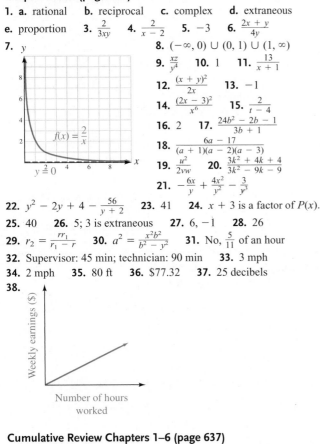

$f(x) = \frac{2}{x}$

$y = 0$

8. $(-\infty, 0) \cup (0, 1) \cup (1, \infty)$

9. $\frac{xz}{y^4}$ **10.** 1   **11.** $\frac{13}{x + 1}$

12. $\frac{(x + y)^2}{2x}$ **13.** -1

14. $\frac{(2x - 3)^2}{x^6}$ **15.** $\frac{2}{t - 4}$

16. 2   **17.** $\frac{24b^2 - 2b - 1}{3b + 1}$

18. $\frac{6a - 17}{(a + 1)(a - 2)(a - 3)}$

19. $\frac{u^2}{2vw}$ **20.** $\frac{3k^2 + 4k + 4}{3k^2 - 9k - 9}$

21. $-\frac{6x}{y} + \frac{4x^2}{y^2} - \frac{3}{y^3}$

22. $y^2 - 2y + 4 - \frac{56}{y + 2}$ **23.** 41   **24.** $x + 3$ is a factor of $P(x)$.

25. 40   **26.** 5; 3 is extraneous   **27.** $6, -1$ **28.** 26

29. $r_2 = \frac{rr_1}{r_1 - r}$ **30.** $a^2 = \frac{x^2b^2}{b^2 - y^2}$ **31.** No, $\frac{5}{11}$ of an hour

32. Supervisor: 45 min; technician: 90 min   **33.** 3 mph

34. 2 mph   **35.** 80 ft   **36.** $77.32   **37.** 25 decibels

38.

Weekly earnings ($)

Number of hours worked

Cumulative Review Chapters 1–6 (page 637)

1. -372 **2.** $22a^3 + 20a$ **3.** -2 **4.** $n = \frac{l - a + d}{d}$

5. 490.9 in.^{2} **6.** 24   **7.** Life expectancy will increase 0.1 year each
year during this period.   **8.** $y = -7x + 54$ **9.** $y = -\frac{11}{6}x - \frac{7}{3}$

10. a. $(-3, 0), (0, -7)$ **b.** $-\frac{7}{3}$ **c.** I   **d.** $y = -\frac{7}{3}x - 7$

e. Yes   **11.** 18   **12.** No; $(1, 1), (1, -1)$

13. D: $(-\infty, \infty)$; R: $[0, \infty)$

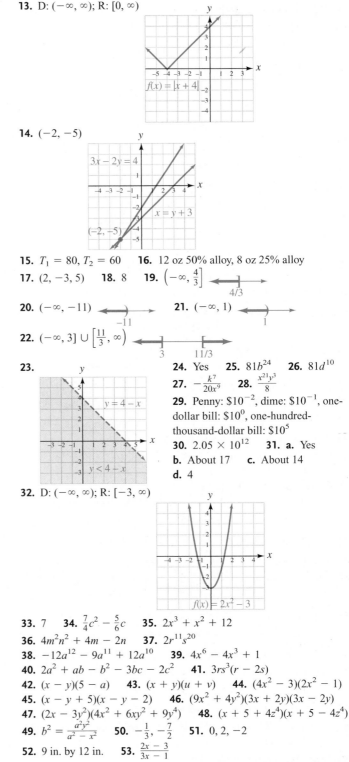

$f(x) = |x + 4|$

14. $(-2, -5)$

$3x - 2y = 4$

$x = y + 3$

$(-2, -5)$

15. $T_1 = 80, T_2 = 60$ **16.** 12 oz 50% alloy, 8 oz 25% alloy

17. $(2, -3, 5)$ **18.** 8   **19.** $\left(-\infty, \frac{4}{3}\right]$

4/3

20. $(-\infty, -11)$

-11

21. $(-\infty, 1)$

1

22. $(-\infty, 3] \cup \left[\frac{11}{3}, \infty\right)$

3   11/3

23.

$y = 4 - x$

$y < 4 - x$

24. Yes   **25.** $81b^{24}$ **26.** $81d^{10}$

27. $-\frac{k^7}{20x^9}$ **28.** $\frac{x^{21}y^3}{8}$

29. Penny: 10^{-2}, dime: 10^{-1}, one-
dollar bill: 10^0, one-hundred-
thousand-dollar bill: 10^5

30. 2.05×10^{12} **31. a.** Yes

b. About 17   **c.** About 14

d. 4

32. D: $(-\infty, \infty)$; R: $[-3, \infty)$

$f(x) = 2x^2 - 3$

33. 7   **34.** $\frac{7}{4}c^2 - \frac{5}{6}c$ **35.** $2x^3 + x^2 + 12$

36. $4m^2n^2 + 4m - 2n$ **37.** $2r^{11}s^{20}$

38. $-12a^{12} - 9a^{11} + 12a^{10}$ **39.** $4x^6 - 4x^3 + 1$

40. $2a^2 + ab - b^2 - 3bc - 2c^2$ **41.** $3rs^3(r - 2s)$

42. $(x - y)(5 - a)$ **43.** $(x + y)(u + v)$ **44.** $(4x^2 - 3)(2x^2 - 1)$

45. $(x - y + 5)(x - y - 2)$ **46.** $(9x^2 + 4y^2)(3x + 2y)(3x - 2y)$

47. $(2x - 3y^2)(4x^2 + 6xy^2 + 9y^4)$ **48.** $(x + 5 + 4z^4)(x + 5 - 4z^4)$

49. $b^2 = \frac{a^2y^2}{a^2 - x^2}$ **50.** $-\frac{1}{3}, -\frac{7}{2}$ **51.** $0, 2, -2$

52. 9 in. by 12 in.   **53.** $\frac{2x - 3}{3x - 1}$

54. All real numbers except 0 and 2: $(-\infty, 0) \cup (0, 2) \cup (2, \infty)$

55. $-\frac{n^6}{n^2 - 2}$ **56.** $-\frac{q}{p}$ **57.** $\frac{4}{x - y}$ **58.** $\frac{c + 9}{c - 5}$ **59.** $\frac{y^2}{3} + 3y$

60. $4x^2 - x - 1$ **61.** -17 **62.** 0   **63.** It will rise sharply.

64. It has dropped.   **65.** 13 cups   **66.** 1 day

Study Set Section 7.1 (page 654)

1. square, cube   **3.** index, radicand   **5.** odd, even   **7.** a
9. two, 5   **11.** x **13.** 3, up   **15. a.** 3   **b.** 0   **c.** 6
d. D: $[2, \infty)$; R: $[0, \infty)$ **17.** $\sqrt{x^2} = |x|$ **19.** $f(x) = \sqrt{x - 5}$
21. 10   **23.** -8 **25.** $\frac{1}{3}$ **27.** 0.5   **29.** Not a real number
31. 11   **33.** 3.4641   **35.** 26.0624   **37.** $2|x|$ **39.** $9h^2$
41. $6|s^3|$ **43.** $12m^4$ **45.** $|y - 1|$ **47.** $a^2 + 3$ **49. a.** 2   **b.** 5
51. a. 2   **b.** -3
53. $0, -1, -2, -3, -4$; D: $[0, \infty)$; R: $(-\infty, 0]$

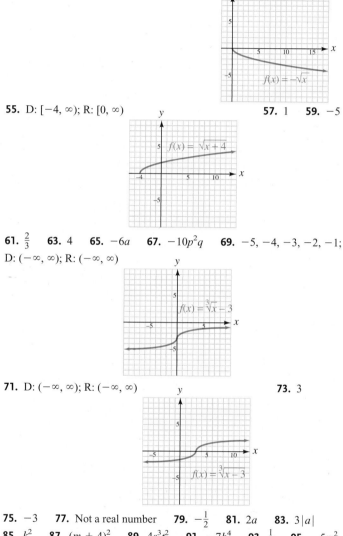

$f(x) = -\sqrt{x}$

55. D: $[-4, \infty)$; R: $[0, \infty)$ **57.** 1   **59.** -5

$f(x) = \sqrt{x + 4}$

61. $\frac{2}{3}$ **63.** 4   **65.** $-6a$ **67.** $-10p^2q$ **69.** $-5, -4, -3, -2, -1$;
D: $(-\infty, \infty)$; R: $(-\infty, \infty)$

$f(x) = \sqrt[3]{x} - 3$

71. D: $(-\infty, \infty)$; R: $(-\infty, \infty)$ **73.** 3

$f(x) = \sqrt[3]{x} - 3$

75. -3 **77.** Not a real number   **79.** $-\frac{1}{2}$ **81.** $2a$ **83.** $3|a|$
85. k^2 **87.** $(m + 4)^2$ **89.** $4s^3t^2$ **91.** $-7b^4$ **93.** $\frac{1}{2}$ **95.** $-5m^2$
97. $2|ab|$ **99.** $x + 2$ **101.** Not a real number   **103.** $|n + 6|$
105. 7.0 in.   **107.** 1,138.5 mm   **109.** About 58.6 beats/min
111. 13.4 ft   **113.** 3.5%   **119.** 1   **121.** $\frac{3(m^2 + 2m - 1)}{(m + 1)(m - 1)}$

Study Set Section 7.2 (page 668)

1. rational (or fractional)   **3.** index, radicand
5. $25^{1/5}, \left(\sqrt[3]{-27}\right)^2, 16^{-3/4}, \left(\sqrt{81}\right)^3, -\left(\frac{9}{64}\right)^{1/2}$
7.

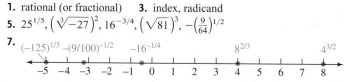

$(-125)^{1/3}$ $-(9/100)^{-1/2}$ $-16^{-1/4}$ $8^{2/3}$ $4^{3/2}$

9. $\sqrt[n]{x}$ **11.** $\frac{1}{x^{m/n}}$ **13.** $100a^4, 10a^2$ **15.** 5   **17.** 3   **19.** 2
21. -6 **23.** -2 **25.** $\frac{1}{2}$ **27.** $2x^2$ **29.** $|x|$ **31.** m^2
33. Not a real number   **35.** $-3n^3$ **37.** $2|x|$ **39.** Not a real number
41. $|x + 1|$ **43.** 216   **45.** 8   **47.** $\frac{1}{36}$ **49.** -32 **51.** $125x^6$
53. $4x^4y^2$ **55.** $27x^3y^6$ **57.** $-\frac{x^4}{16}$ **59.** $(8abc)^{1/5}$
61. $(a^2 - b^2)^{1/3}$ **63.** $\sqrt[4]{6x^3y}$ **65.** $\sqrt{2s^2 - t^2}$ **67.** $\frac{1}{2}$ **69.** $\frac{1}{5}$
71. $\frac{1}{64}$ **73.** $-\frac{1}{100y^2}$ **75.** $\frac{16}{81}$ **77.** $\frac{27y^3}{8}$ **79.** 2   **81.** 243
83. $9^{5/7}$ **85.** $\frac{1}{36}$ **87.** m^6 **89.** $a^{3/4}b^{1/2}$ **91.** 3   **93.** a
95. $y + y^2$ **97.** $x^2 - 1 + x^{3/5}$ **99.** $\sqrt{5}$ **101.** $\sqrt[3]{11}$ **103.** $\sqrt{p}$
105. $\sqrt[5]{xy}$ **107.** $\sqrt[18]{c}$ **109.** $\sqrt[15]{7m}$ **111.** 2.47   **113.** 1.02
115. $5y$ **117.** $-\frac{a^3}{27}$ **119.** p **121.** $-\frac{1}{3x^2}$ **123.** $n^{3/5} - 1$
125. 243   **127.** $\sqrt{5b}$ **129.** $-\frac{1}{4a^2b^4}$ **131.** 736 ft/sec
133. 1.96 units   **135.** 145.8 in. or 12.1 ft   **139.** $2\frac{1}{2}$ hr

Study Set Section 7.3 (page 680)

1. like   **3.** factor, perfect   **5.** $\sqrt[n]{a}\sqrt[n]{b}$, product, roots
7. a. $\sqrt{4 \cdot 5}$ **b.** $\sqrt{4}\sqrt{5}$ **c.** $\sqrt{4 \cdot 5} = \sqrt{4}\sqrt{5}$ **9. a.** $\sqrt{5}$,
$\sqrt[3]{5}$ (Answers may vary); no   **b.** $\sqrt{5}, \sqrt{6}$ (Answers may vary); no
11. $8k^3, 8k^3, 4k$ **13.** $5\sqrt{2}$ **15.** $3\sqrt{5}$ **17.** $2\sqrt[3]{4}$ **19.** $2\sqrt[4]{3}$
21. $5a\sqrt{3}$ **23.** $4\sqrt{2b}$ **25.** $8ab^2\sqrt{2ab}$ **27.** $10\sqrt{3xy}$
29. $-3x^2\sqrt[3]{2}$ **31.** $2x^3y\sqrt[4]{2}$ **33.** $11\sqrt{2}$ **35.** $4a\sqrt{7a}$
37. $-2\sqrt[5]{3a^4}$ **39.** $3x^4y\sqrt[3]{15y}$ **41.** $\frac{\sqrt{11}}{3}$ **43.** $\frac{\sqrt[3]{7}}{4}$ **45.** $\frac{\sqrt[4]{3}}{5}$
47. $\frac{x^2\sqrt[5]{3}}{2}$ **49.** $\frac{z}{4x}$ **51.** $\frac{\sqrt[4]{5x}}{2z}$ **53.** 10   **55.** $7x$ **57.** $2x^2$
59. $3a\sqrt[3]{a}$ **61.** $8\sqrt{7}$ **63.** $5\sqrt[3]{4}$ **65.** $3\sqrt{2}$ **67.** $-4\sqrt{2}$
69. $-\sqrt[3]{4}$ **71.** -10 **73.** $13\sqrt[4]{2}$ **75.** $7\sqrt{5} - 3\sqrt{3}$
77. $10\sqrt{2x}$ **79.** $\sqrt[5]{7a^2}$ **81.** $3\sqrt{2t} + \sqrt{3t}$ **83.** $-11\sqrt[3]{2}$
85. $12\sqrt[3]{a}$ **87.** $4\sqrt[4]{4}$ **89.** $m\sqrt[6]{m^5}$ **91.** $5y^3\sqrt{2y}$
93. $\frac{5n^2\sqrt{5}}{8}$ **95.** $4x\sqrt[5]{xy^2}$ **97.** $2m\sqrt[4]{13n}$ **99.** $\frac{a^2}{4}$ **101.** $2t^2\sqrt[5]{2t}$
103. $3\sqrt[3]{3x}$ **105.** $8\pi\sqrt{5}$ ft²; 56.2 ft²   **107.** $5\sqrt{3}$ amps; 8.7 amps
109. $\left(26\sqrt{5} + 10\sqrt{3}\right)$ in.; 75.5 in.   **115.** $-\frac{15x^5}{y}$
117. $3p + 4 - \frac{5}{2p - 5}$

Study Set Section 7.4 (page 692)

1. FOIL   **3.** irrational   **5.** perfect   **7. a.** $6\sqrt{6}$ **b.** 48
c. Can't be simplified   **d.** $-6\sqrt{6}$ **9.** $\sqrt{6}, 48, 16, 4$ **11.** $3\sqrt{5}$
13. $6\sqrt{2}$ **15.** 18   **17.** $2\sqrt[3]{3}$ **19.** $48ab^2$ **21.** $5a\sqrt[4]{a}$
23. $-20r\sqrt[3]{10s}$ **25.** $x^2(x + 3)$ **27.** $12\sqrt{5} - 15$
29. $8\sqrt{3} + 2\sqrt{14}$ **31.** $-8x\sqrt{10} + 6\sqrt{15x}$ **33.** $8 + 2\sqrt[3]{3}$
35. $-1 - 2\sqrt{2}$ **37.** $3x - 2y$
39. $12\sqrt[3]{2} + 8\sqrt[3]{5} - 18 - 6\sqrt[3]{20}$ **41.** $\sqrt[3]{25z^2} + 3\sqrt[3]{15z} + 2\sqrt[3]{9}$
43. 7   **45.** 12   **47.** 18   **49.** $-16x^2$ **51.** $18r - 12\sqrt{2r} + 4$
53. $3x + 6\sqrt{x} + 3$ **55.** $\frac{\sqrt{14}}{7}$ **57.** $\frac{2\sqrt{6}}{3}$ **59.** $\frac{2\sqrt{6}}{3}$ **61.** $\frac{\sqrt[3]{4}}{2}$
63. $\sqrt[3]{3}$ **65.** $\frac{\sqrt[4]{4}}{2}$ **67.** $\frac{\sqrt{5y}}{y}$ **69.** $\frac{\sqrt{6y}}{y}$ **71.** $\frac{\sqrt[3]{6t}}{3}$ **73.** $\frac{\sqrt[3]{2ab^2}}{b}$
75. $\frac{23\sqrt{2p}}{10p^3}$ **77.** $\frac{7\sqrt{6b}}{12b^2}$ **79.** $\frac{\sqrt[3]{20}}{4}$ **81.** $\frac{\sqrt[3]{36}}{9}$ **83.** $\frac{19\sqrt[3]{25c}}{5c}$
85. $\frac{\sqrt[3]{12r^2}}{2r}$ **87.** $\frac{\sqrt[4]{54t^2}}{3t}$ **89.** $\frac{25\sqrt[3]{2a^3}}{2a}$ **91.** $\frac{3\sqrt{2} - \sqrt{10}}{4}$
93. $\frac{2(\sqrt{x} - 1)}{x - 1}$ or $\frac{2\sqrt{x} - 2}{x - 1}$ **95.** $\frac{9 - 2\sqrt{14}}{5}$ **97.** $\frac{3\sqrt{6} + 4}{2}$

99. $\frac{x - 2\sqrt{xy} + y}{x - y}$ **101.** $\sqrt{2z} + 1$ **103.** $\frac{x - 9}{x(\sqrt{x} - 3)}$

105. $\frac{x - y}{\sqrt{x}(\sqrt{x} - \sqrt{y})}$ **107.** $x\sqrt{14} + \sqrt{2x}$ **109.** $2,000x$

111. $9p^2 + 6p\sqrt{5} + 5$ **113.** $\frac{6m^2\sqrt{2m}}{5}$ **115.** $3n\sqrt[4]{n}$ **117.** $\frac{\sqrt[3]{3x}}{3}$

119. $\frac{\sqrt{2\pi}}{2\pi\sigma}$ **121.** $\frac{\sqrt{2}}{2}$ **129.** $\frac{1}{3}$

Study Set Section 7.5 (page 705)

1. radical   **3.** extraneous   **5.** $^n,^n$ **7. a.** Square both sides.

b. Subtract 3 from both sides.   **c.** Add $\sqrt{2x + 9}$ to both sides.

9. $x - 6\sqrt{x} + 9$ **11.** $6, ^2, ^2, 3x + 3, 33, 11,$ Yes   **13.** 4   **15.** 5

17. 6   **19.** 198   **21.** 7, ∕1   **23.** 14, ∕6   **25.** 3, ∕2

27. 0, ∕1   **29.** $\frac{4}{5}$, no solution   **31.** ∕4, no solution   **33.** 4

35. 2, −1   **37.** 1, −3   **39.** 10   **41.** 85   **43.** −7   **45.** 12

47. 0   **49.** 3   **51.** 3   **53.** 9   **55.** 1   **57.** 4   **59.** 0

61. ∕3, no solution   **63.** 1, 9   **65.** $h = \frac{v^2}{2g}$ **67.** $l = \frac{8T^2}{\pi^2}$

69. $A = P(r + 1)^3$ **71.** $v^2 = c^2\left(1 - \frac{L_A^2}{L_B^2}\right)$ **73.** 16   **75.** −1, 1

77. 2, ∕1   **79.** −16   **81.** 4   **83.** $\frac{5}{2}, \frac{1}{2}$ **85.** $-\frac{1}{2}$ **87.** ∕6, no

solution   **89.** 1   **91.** 4, 3   **93.** ∕8, no solution   **95.** 1

97. 178 ft   **99.** About 488 watts   **101.** 10 lb   **103.** $5

113. 2.5 foot-candles   **115.** 0.41511 in.

Study Set Section 7.6 (page 716)

1. hypotenuse   **3.** Pythagorean   **5.** a^2, b^2, c^2 **7.** $\sqrt{2}$ **9.** $\sqrt{3}$

11. $(x_2 - x_1)^2, (y_2 - y_1)^2$ **13.** 6, 52, 4, 2, 7.21   **15.** 10 ft

17. 17 ft   **19.** 40 ft   **21.** 24 cm   **23.** $x = 2, h = 2\sqrt{2} \approx 2.83$

25. $3.2\sqrt{2}$ ft ≈ 4.53 ft   **27.** $x = \frac{3\sqrt{2}}{2} \approx 2.12, y = \frac{3\sqrt{2}}{2} \approx 2.12$

29. $5\sqrt{2}$ in.   **31.** $x = 5\sqrt{3} \approx 8.66, h = 10$

33. $75\sqrt{3}$ cm ≈ 129.90 cm, 150 cm   **35.** $x = \frac{40\sqrt{3}}{3} \approx 23.09,$

$h = \frac{80\sqrt{3}}{3} \approx 46.19$ **37.** $\frac{55\sqrt{3}}{3}$ mm ≈ 31.75 mm,

$\frac{110\sqrt{3}}{3}$ mm ≈ 63.51 mm   **39.** $x = 50, y = 50\sqrt{3} \approx 86.60$

41. 0.75 ft, $0.75\sqrt{3}$ ft ≈ 1.30 ft   **43.** 5   **45.** 13   **47.** 10

49. $\sqrt{34}$ **51.** $2\sqrt{5}$ **53.** $3\sqrt{10}$ **55. a.** 118.73 m

b. 133.14 m   **57.** $7\sqrt{3}$ cm   **59.** $\left(5\sqrt{2}, 0\right), \left(0, 5\sqrt{2}\right), \left(-5\sqrt{2}, 0\right),$

$\left(0, -5\sqrt{2}\right); (7.07, 0), (0, 7.07), (-7.07, 0), (0, -7.07)$

61. $10\sqrt{3}$ mm ≈ 17.32 mm   **63.** $10\sqrt{181}$ ft ≈ 134.54 ft

65. About 0.13 ft   **67. a.** 21.21 units   **b.** 8.25 units

c. 13.00 units   **69.** Yes   **75.** 7   **77.** 9

Study Set Section 7.7 (page 731)

1. imaginary, power   **3.** real, imaginary   **5. a.** $\sqrt{-1}$ **b.** -1

c. $-i$ **d.** 1   **e.** four   **7.** real, imaginary   **9.** $1 + 8i, 1 + 8i$

11.

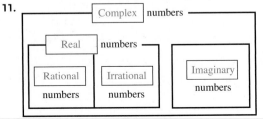

13. $9, 6i, 2, 11$ **15. a.** True   **b.** False   **c.** False   **d.** False

---

17. $3i$ **19.** $\sqrt{7}i$ or $i\sqrt{7}$ **21.** $2\sqrt{6}i$ or $2i\sqrt{6}$

23. $-6\sqrt{2}i$ or $-6i\sqrt{2}$ **25.** $45i$ **27.** $\frac{5}{3}i$ **29. a.** $5 + 0i$

b. $0 + 7i$ **31. a.** $1 + 5i$ **b.** $-3 + 2i\sqrt{2}$ **33. a.** $76 - 3i\sqrt{6}$

b. $-7 + i\sqrt{19}$ **35. a.** $-6 - 3i$ **b.** $3 + i\sqrt{6}$ **37.** $8 - 2i$

39. $15 + 2i$ **41.** $3 - 5i$ **43.** $1 + 3i$ **45.** -6 **47.** $-2\sqrt{6}$

49. $6 - 27i$ **51.** $35 - 28i$ **53.** $6 + 14i$ **55.** $-25 - 25i$

57. $7 + i$ **59.** $12 + 5i$ **61.** $13 - i$ **63.** $3 + 4i$ **65.** 40

67. 65   **69.** $\frac{45}{26} - \frac{9}{26}i$ **71.** $-\frac{77}{65} + \frac{44}{65}i$ **73.** $\frac{14}{17} - \frac{5}{17}i$

75. $-\frac{6}{29} + \frac{43}{29}i$ **77.** $\frac{11}{10} + \frac{13}{10}i$ **79.** $0 - i$ **81.** $4 + 0i$

83. $-2 + 0i$ **85.** $0 - \frac{5}{3}i$ **87.** $0 + \frac{2}{7}i$ **89.** i **91.** $-i$

93. 1   **95.** -1 **97.** $4 - 11i$ **99.** $14 - 8i$ **101.** $-2 - 6i$

103. $-\frac{4}{13} - \frac{6}{13}i$ **105.** $18 + 12i$ **107.** $0 + \frac{4}{5}i$

109. $8 + \sqrt{2}i$ or $8 + i\sqrt{2}$ **111.** $-2 + 7i$ **113.** $-48 - 64i$

115. $\frac{1}{4} - \frac{\sqrt{15}}{4}i$ **117.** $-1 + i$ **123.** 20 mph

Chapter 7 Review (page 734)

1. 7   **2.** -11 **3.** $\frac{15}{7}$ **4.** Not a real number   **5.** $10a^6$

6. $5|x|$ **7.** x^4 **8.** $|x + 2|$ **9.** -3 **10.** -6 **11.** $4x^2y$

12. $\frac{x^3}{5}$ **13.** 2   **14.** -2 **15.** $4x^2|y|$ **16.** $x + 1$ **17.** $-\frac{1}{2}$

18. Not a real number   **19.** Not a real number   **20.** 0   **21.** 13 ft

22. 24 cm^2 **23.** D: $[0, \infty)$; R: $[0, \infty)$

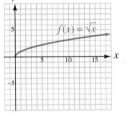

24. D: $(-\infty, \infty)$; R: $(-\infty, \infty)$

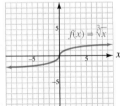

25. D: $[-2, \infty)$; R: $[0, \infty)$

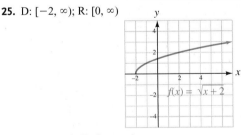

26. D: $(-\infty, \infty)$; R: $(-\infty, \infty)$

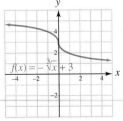

27. $\sqrt{t}$ **28.** $\sqrt[4]{5xy^3}$ **29.** 5   **30.** -6 **31.** Not a real number

32. 1   **33.** $\frac{3}{x}$ **34.** -2 **35.** 5   **36.** $3cd$ **37.** 27   **38.** $\frac{1}{4}$

39. $-16{,}807$ **40.** 10   **41.** $\frac{27}{8}$ **42.** $\frac{1}{3,125}$ **43.** $125x^3y^6$

44. $\frac{1}{4u^4v^2}$ **45.** $5^{3/4}$ **46.** $a^{1/7}$ **47.** k^8 **48.** $3^{2/3}$ **49.** $u - 1$

50. $v + v^2$ **51.** $\sqrt{a}$ **52.** $\sqrt[6]{c}$ **53.** 183 mi   **54.** Two true statements result: $32 = 32$.   **55.** $4\sqrt{5}$ **56.** $3\sqrt[3]{2}$ **57.** $2\sqrt[4]{10}$

58. $-2\sqrt[5]{3}$ **59.** $2x^2\sqrt{2x}$ **60.** $r^4\sqrt[4]{r}$ **61.** $-3j^2\sqrt[3]{jk}$

62. $-2xy\sqrt[3]{2x^2y}$ **63.** $\frac{\sqrt{m}}{12n^6}$ **64.** $\frac{\sqrt{17xy}}{8a^2}$ **65.** $2x$ **66.** $3x^3$

67. $3\sqrt{2}$ **68.** $11\sqrt{5}$ **69.** 0   **70.** $-8a\sqrt[4]{2a}$ **71.** $29x\sqrt{2}$

72. $13x\sqrt[3]{2}$ **73.** $9\sqrt[4]{2t^3} - 8\sqrt[4]{6t^3}$ **74.** $12x^2\sqrt[4]{x} + 5x\sqrt[4]{x}$

76. $\left(6\sqrt{2} + 2\sqrt{10}\right)$ in., 14.8 in.   **77.** 7   **78.** $6\sqrt{10}$ **79.** 32

80. $6\sqrt{10}$ **81.** $3x$ **82.** $x + 1$ **83.** $-2x^3\sqrt[3]{x}$ **84.** 3

85. $42t + 9t\sqrt{21t}$ **86.** $-2x^3y^3\sqrt[4]{2x^2y^2}$ **87.** $3b + 6\sqrt{b} + 3$

88. $\sqrt[3]{9p^2} - \sqrt[3]{6p} - 2\sqrt[3]{4}$ **89.** $\frac{10\sqrt{3}}{3}$ **90.** $\frac{\sqrt{15xy}}{5xy}$ **91.** $\frac{\sqrt[6]{6u^2}}{u^2}$

92. $\frac{\sqrt[4]{27ab^2}}{3b}$ **93.** $2\left(\sqrt{2} + 1\right)$ or $2\sqrt{2} + 2$

94. $\frac{12\sqrt{xz} - 16x - 2z}{z - 16x}$ **95.** $\frac{a - b}{a + \sqrt{ab}}$ **96.** $r = \frac{\sqrt[3]{6\pi^2 V}}{2\pi}$

97. 22   **98.** 16, 9   **99.** $\frac{13}{2}$ **100.** $\frac{9}{16}$ **101.** 2, -4 **102.** 7

103. 1   **104.** $-\frac{3}{2}, 1$ **105.** 8, no solution   **106.** 6, -2

107. $-\frac{1}{2}, 4$ **108.** 2   **109.** $P = \frac{A}{(r + 1)^2}$ **110.** $I = \frac{h^3 b}{12}$

111. 17 ft   **112.** 88 yd   **113.** $7\sqrt{2}$ m ≈ 9.90 m

114. $\frac{15\sqrt{2}}{2}$ yd ≈ 10.61 yd   **115.** Shorter leg: 6 cm, longer leg: $6\sqrt{3}$ cm ≈ 10.39 cm   **116.** $40\sqrt{3}$ ft ≈ 69.28 ft, $20\sqrt{3}$ ft ≈ 34.64 ft   **117.** $x = 5\sqrt{2} \approx 7.07$, $y = 5$

118. $x = 25\sqrt{3} \approx 43.30$, $y = 25$ **119.** 13   **120.** $2\sqrt{2}$

121. $5i$ **122.** $3i\sqrt{2}$ **123.** $-i\sqrt{6}$ **124.** $\frac{3}{8}i$

125. Real, Imaginary   **126. a.** True   **b.** True   **c.** False **d.** False   **127. a.** $3 - 6i$ **b.** $0 - 19i$ **128. a.** $-1 + 7i$ **b.** $0 + i$ **129.** $8 - 2i$ **130.** $3 - 5i$ **131.** $3 + 6i$

132. $22 + 29i$ **133.** $-3\sqrt{3} + 0i$ **134.** $-81 + 0i$ **135.** $4 + i$

136. $0 - \frac{3}{11}i$ **137.** -1 **138.** i

Chapter 7 Test (page 743)

1. a. radical   **b.** imaginary   **c.** extraneous   **d.** isosceles **e.** rationalize   **f.** complex   **2. a.** If $\sqrt[n]{a}$ and $\sqrt[n]{b}$ are real numbers, then $\sqrt[n]{ab} = \sqrt[n]{a}\sqrt[n]{b}$.   **b.** If $\sqrt[n]{a}$ and $\sqrt[n]{b}$ are real numbers, then $\sqrt[n]{\frac{a}{b}} = \frac{\sqrt[n]{a}}{\sqrt[n]{b}}$, $(b \neq 0)$.

3. D: $[1, \infty)$; R: $[0, \infty)$ y

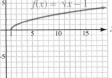

4. 46 ft/sec   **5. a.** -1 **b.** 2   **c.** 1   **d.** D: $(-\infty, \infty)$; R: $(-\infty, \infty)$ **6.** No real number raised to the fourth power is -16.

7. $7x^2$ **8.** -9 **9.** $\frac{1}{216}$ **10.** $\frac{25n^4}{4}$ **11.** $2^{4/3}$ **12.** $a^{1/9}$

13. $|x|$ **14.** $|y - 5|$ **15.** $-4xy^2$ **16.** $\frac{2}{3}a$ **17.** $t + 8$

18. $6xy^2\sqrt{15xy}$ **19.** $2x^5y\sqrt[3]{3}$ **20.** $2\sqrt[4]{2}$ **21.** $2y^2\sqrt{3y}$

22. $14\sqrt[3]{5}$ **23.** $5z^3\sqrt[4]{3z}$ **24.** $-6x\sqrt{y} - 2xy^2$ **25.** $3 - 7\sqrt{6}$

26. $\sqrt[3]{4a^2} + 18\sqrt[3]{2a} + 81$ **27.** $\frac{4\sqrt{10}}{5}$ **28.** $\sqrt{3t} + 1$

29. $\frac{\sqrt[3]{18a^2}}{2a}$ **30.** $\frac{1}{\sqrt{2}(\sqrt{5} - 3)} = \frac{1}{\sqrt{10} - 3\sqrt{2}}$ **31.** $\frac{1}{15}$ **32.** 10

33. $\varnothing$, no solution   **34.** 3, -3 **35.** 2, 3   **36.** 108, no solution

37. -2 **38.** $G = \frac{4\pi^2 r^3}{Mt^2}$ **39.** $x = \frac{8\sqrt{3}}{3}$ cm ≈ 4.62 cm, $h = \frac{16\sqrt{3}}{3}$ cm ≈ 9.24 cm   **40.** $x = \frac{(12.26)\sqrt{2}}{2}$ in. ≈ 8.67 in., $y = \frac{(12.26)\sqrt{2}}{2}$ in. ≈ 8.67 in.   **41.** 25   **42.** 28 in.   **43.** $3i\sqrt{5}$

44. -1 **45.** $-4 + 11i$ **46.** $4 - 7i$ **47.** $75 + 45i$

48. $-46 - 78i$ **49.** $0 - \frac{\sqrt{2}}{2}i$ **50.** $\frac{1}{2} + \frac{1}{2}i$

Study Set Section 8.1 (page 759)

1. quadratic   **3.** square   **5.** $\sqrt{c}, -\sqrt{c}$ **7. a.** 36   **b.** $\frac{25}{4}$

9. a. Subtract 7 from both sides   **b.** Divide both sides by 4

11. a. False   **b.** True   **13.** plus or minus   **15.** $0, -2$

17. $5, -5$ **19.** $-2, -4$ **21.** $2, \frac{1}{2}$ **23.** ± 9 **25.** ± 6

27. $\pm 5\sqrt{2}$ **29.** $\pm\frac{4\sqrt{3}}{3}$ **31.** $-8, -2$ **33.** $0, -8$

35. $-5 \pm \sqrt{3}$ **37.** $2 \pm 2\sqrt{2}$ **39.** $\pm 4i$ **41.** $\pm\frac{9}{2}i$

43. $3 \pm i\sqrt{5}$ **45.** $1 \pm 2i$ **47.** $x^2 + 24x + 144 = (x + 12)^2$

49. $a^2 - 7a + \frac{49}{4} = \left(a - \frac{7}{2}\right)^2$ **51.** $2, -4$ **53.** 2, 6   **55.** 1, -6

57. $-1, 4$ **59.** $-\frac{1}{2}, 1$ **61.** $\frac{3}{4}, -\frac{1}{3}$ **63.** $\frac{6 \pm \sqrt{33}}{3}$

65. $\frac{-5 \pm \sqrt{41}}{4}$ **67.** $-1 \pm i$ **69.** $-4 \pm i\sqrt{2}$ **71.** $2, -\frac{4}{3}$

73. $\frac{3 \pm 2\sqrt{3}}{3}$ **75.** $-4 \pm \sqrt{10}$ **77.** $\pm 2i\sqrt{3}$ **79.** $1 \pm 3\sqrt{2}$

81. $\frac{7 \pm \sqrt{37}}{2}$ **83.** $\pm\sqrt{5}$ **85.** $\frac{-7 \pm \sqrt{29}}{10}$ **87.** $-\frac{1}{2} \pm \frac{\sqrt{11}}{2}i$

89. $\frac{-5 \pm 2\sqrt{6}}{8}$ **91.** $-3, 9$ **93.** $-\frac{1}{4} \pm \frac{\sqrt{11}}{4}i$ **95.** 4.4 sec

97. 3,959 mi   **99.** 1.70 in.   **101.** 0.92 in.   **103.** 2.9 ft, 6.9 ft

109. $2ab^2\sqrt[3]{5}$ **111.** $\frac{2}{5}$

Study Set Section 8.2 (page 770)

1. quadratic   **3. a.** $x^2 + 2x + 5 = 0$ **b.** $3x^2 + 2x - 1 = 0$

5. a. True   **b.** True   **7.** 2, -4 **9. a.** 2; $\frac{-2 + \sqrt{3}}{2}, \frac{-2 - \sqrt{3}}{2}$

b. [number line from -2 to 2 marked at $\frac{-2 - \sqrt{3}}{2}$ and $\frac{-2 + \sqrt{3}}{2}$]   **11. a.** The fraction bar wasn't drawn under both parts of the numerator.

b. A $\pm$ sign wasn't written between $-b$ and the radical.   **13.** 1, 2

15. A repeated solution of -6 **17.** $-\frac{3}{2}, 1$ **19.** $\frac{2}{3}, -\frac{1}{4}$

21. $\frac{1 \pm \sqrt{29}}{2}$; 3.19, -2.19 **23.** $\frac{-5 \pm \sqrt{5}}{10}$; $-0.28, -0.72$

25. $\frac{-3 \pm \sqrt{6}}{3}$; $-0.18, -1.82$ **27.** $\frac{1 \pm 2\sqrt{5}}{2}$; 2.74, -1.74

29. $-\frac{1}{4} \pm \frac{\sqrt{7}}{4}i$ **31.** $\frac{1}{3} \pm \frac{\sqrt{2}}{3}i$ **33.** $1 \pm i$ **35.** $-\frac{1}{2} \pm i$

37. a. $5x^2 - 9x + 2 = 0$ **b.** $16t^2 + 24t - 9 = 0$

39. a. $3x^2 + 2x - 1 = 0$ **b.** $2m^2 - 3m - 2 = 0$ **41.** $\frac{4}{3}, -\frac{2}{5}$

43. $\frac{2}{3} \pm \frac{\sqrt{2}}{3}i$ **45.** $\frac{1}{4}, -\frac{3}{4}$ **47.** $\frac{3 \pm \sqrt{17}}{4}$ **49.** $5 \pm \sqrt{7}$

51. 23, -17 **53.** $\frac{-5 \pm 3\sqrt{5}}{2}$ **55.** $\frac{1}{3} \pm \frac{\sqrt{6}}{3}i$ **57.** $\frac{-3 \pm \sqrt{29}}{10}$

59. $\frac{10 \pm \sqrt{55}}{30}$ **61.** $2 \pm 2i$ **63.** $3 \pm i\sqrt{5}$ **65.** $\frac{-5 \pm \sqrt{17}}{2}$

67. $\frac{9 \pm \sqrt{89}}{2}$ **69.** 1.22, −0.55   **71.** 8.98, −3.98   **73.** 40 ft

75. 0.7, 2.4, 2.5   **77.** 97 ft by 117 ft   **79.** 0.5 mi by 2.5 mi

81. 25 sides   **83.** $4.80 or $5.20   **85.** 4,000   **87.** Late 1976

91. $n^{1/2}$ **93.** $(3b)^{1/4}$ **95.** $\sqrt[3]{t}$ **97.** $\sqrt[4]{3t}$

Study Set Section 8.3 (page 781)

1. discriminant   **3.** conjugates   **5.** rational   **7. a.** x^2 **b.** $\sqrt{x}$
c. $x^{1/3}$ **d.** $\frac{1}{x}$ **e.** $x + 1$ **9.** $4ac$, 5, 6, 24, rational
11. One repeated rational-number solution   **13.** Two imaginary-
number solutions (complex conjugates)   **15.** Two different irrational-
number solutions   **17.** Two different rational-numbers solutions
19. Two different irrational-number solutions   **21.** Two different
rational-numbers solutions   **23.** −1, 1, −4, 4   **25.** 2, −2, 3i, −3i
27. 25, 64   **29.** 1   **31.** −1, 27   **33.** −64, 8   **35.** 0, 2
37. $\frac{1}{4}, \frac{1}{2}$ **39.** $-\frac{1}{3}, \frac{1}{2}$ **41.** $-4, \frac{2}{3}$ **43.** $\frac{5 \pm \sqrt{65}}{2}$ **45.** $\frac{7 \pm \sqrt{105}}{2}$
47. $\frac{9}{4}$ **49.** $-\frac{1}{3}, 1$ **51.** $-i, i, -3i\sqrt{2}, 3i\sqrt{2}$ **53.** $4 \pm i$
55. $\frac{3 \pm \sqrt{57}}{6}$ **57.** 16, 4   **59.** $\pm i\sqrt{2}, \pm 3\sqrt{2}$
61. Repeated solutions of 1 and −1   **63.** $2, -2, i\sqrt{7}, -i\sqrt{7}$
65. $\frac{243}{32}, 1$ **67.** A repeated solution of $-\frac{3}{2}$ **69.** $3 \pm \sqrt{7}$
71. 49, 225   **73.** $1 \pm i$ **75.** No solution   **77.** $1, -1, \sqrt{5}, -\sqrt{5}$
79. $-\frac{5}{7}, 3$ **81.** $-1, -\frac{27}{13}$ **83.** 10.3 min   **85.** 20 ft/sec
87. 32.4 ft   **91.** $x = 3$ **93.** $y = \frac{2}{3}x$

Study Set Section 8.4 (page 795)

1. quadratic, parabola   **3.** axis of symmetry
5. a. (1, 0), (3, 0)   **b.** (0, −3)   **c.** (2, 1)   **d.** $x = 2$
e. Domain: $(-\infty, \infty)$; range: $(-\infty, 1]$
7.

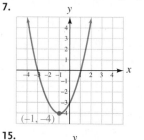

9. a. 2   **b.** 9, 18   **11.** −3, 5
13. $h = -1; f(x) = 2[x - (-1)]^2 + 6$

15.

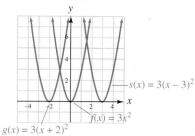

17.

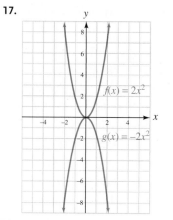

19.

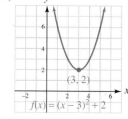

21.

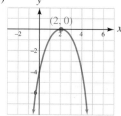

23. (1, 2); $x = 1$; upward   **25.** (−3, −4); $x = -3$; downward
27. (7.5, 8.5); $x = 7.5$; downward   **29.** (0, −4); $x = 0$; upward
31. (3, 2)

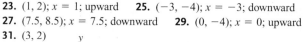

33. (2, 0)

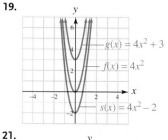

35. $(-3, 4)$

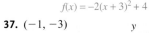
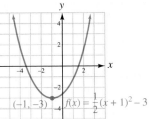

$f(x) = -2(x + 3)^2 + 4$

37. $(-1, -3)$

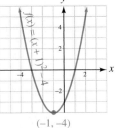

$f(x) = \frac{1}{2}(x + 1)^2 - 3$

39. $(-2, 1); x = -2;$ upward   **41.** $(3, -6); x = 3;$ downward
43. $f(x) = (x + 1)^2 - 4; (-1, -4), x = -1$

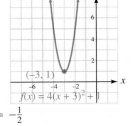

45. $f(x) = 4(x + 3)^2 + 1; (-3, 1), x = -3$

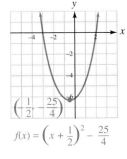

$f(x) = 4(x + 3)^2 + 1$

47. $f(x) = \left(x + \frac{1}{2}\right)^2 - \frac{25}{4}; \left(-\frac{1}{2}, -\frac{25}{4}\right); x = -\frac{1}{2}$

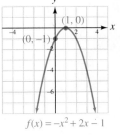

$f(x) = \left(x + \frac{1}{2}\right)^2 - \frac{25}{4}$

49. $f(x) = -4(x - 2)^2 + 6; (2, 6), x = 2$

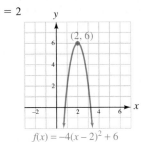

$f(x) = -4(x - 2)^2 + 6$

51. $f(x) = 2(x + 2)^2 - 2; (-2, -2), x = -2$

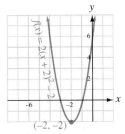

$f(x) = 2(x + 2)^2 - 2$

53. $f(x) = -(x + 4)^2 - 1; (-4, -1), x = -4$

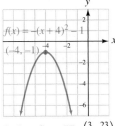

$f(x) = -(x + 4)^2 - 1$

55. $(-1, -6)$ **57.** $\left(\frac{3}{4}, \frac{23}{8}\right)$ **59.** $(-5, 0), (7, 0); (0, -35)$
61. $(0, 0), (2, 0); (0, 0)$
63. $(-2, 0); (-2, 0); (0, 4)$

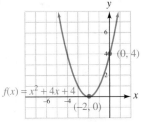

$f(x) = x^2 + 4x + 4$

65. $(1, 0); (1, 0); (0, -1)$

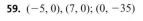

$f(x) = -x^2 + 2x - 1$

67. $(1, -1)$; $(0, 0)$, $(2, 0)$; $(0, 0)$

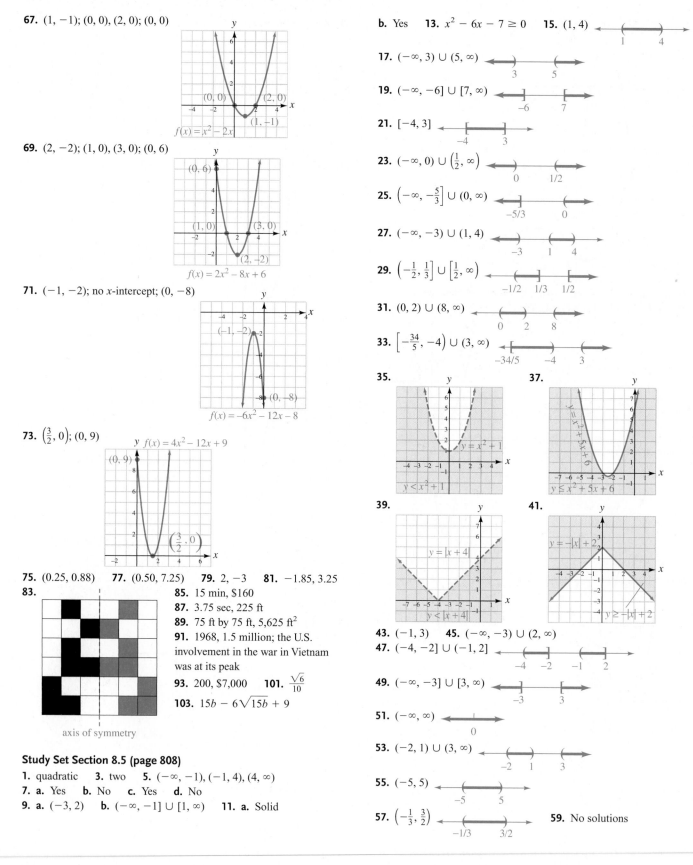

69. $(2, -2)$; $(1, 0)$, $(3, 0)$; $(0, 6)$

71. $(-1, -2)$; no x-intercept; $(0, -8)$

73. $\left(\frac{3}{2}, 0\right)$; $(0, 9)$

75. $(0.25, 0.88)$ **77.** $(0.50, 7.25)$ **79.** $2, -3$ **81.** $-1.85, 3.25$

83.

85. 15 min, $160
87. 3.75 sec, 225 ft
89. 75 ft by 75 ft, 5,625 ft^2
91. 1968, 1.5 million; the U.S. involvement in the war in Vietnam was at its peak
93. 200, $7,000   **101.** $\frac{\sqrt{6}}{10}$
103. $15b - 6\sqrt{15b} + 9$

axis of symmetry

Study Set Section 8.5 (page 808)

1. quadratic   **3.** two   **5.** $(-\infty, -1), (-1, 4), (4, \infty)$
7. a. Yes   **b.** No   **c.** Yes   **d.** No
9. a. $(-3, 2)$ **b.** $(-\infty, -1] \cup [1, \infty)$ **11. a.** Solid

b. Yes   **13.** $x^2 - 6x - 7 \geq 0$ **15.** $(1, 4)$

17. $(-\infty, 3) \cup (5, \infty)$

19. $(-\infty, -6] \cup [7, \infty)$

21. $[-4, 3]$

23. $(-\infty, 0) \cup \left(\frac{1}{2}, \infty\right)$

25. $\left(-\infty, -\frac{5}{3}\right] \cup (0, \infty)$

27. $(-\infty, -3) \cup (1, 4)$

29. $\left(-\frac{1}{2}, \frac{1}{3}\right] \cup \left[\frac{1}{2}, \infty\right)$

31. $(0, 2) \cup (8, \infty)$

33. $\left[-\frac{34}{5}, -4\right) \cup (3, \infty)$

35.   **37.**

39.   **41.**

43. $(-1, 3)$ **45.** $(-\infty, -3) \cup (2, \infty)$
47. $(-4, -2] \cup (-1, 2]$

49. $(-\infty, -3] \cup [3, \infty)$

51. $(-\infty, \infty)$

53. $(-2, 1) \cup (3, \infty)$

55. $(-5, 5)$

57. $\left(-\frac{1}{3}, \frac{3}{2}\right)$ **59.** No solutions

61. $(-1, 4) \cup \left(\frac{23}{2}, \infty\right)$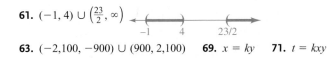

63. $(-2,100, -900) \cup (900, 2,100)$ **69.** $x = ky$ **71.** $t = kxy$

Chapter 8 Review (page 811)

1. $-5, -4$ **2.** $-\frac{1}{3}, -\frac{5}{2}$ **3.** $\pm 2\sqrt{7}$ **4.** $4, -8$ **5.** $\pm 5i$

6. $\pm\frac{7\sqrt{5}}{5}$ **7.** $r = \frac{\sqrt{\pi A}}{\pi}$ **8.** $x^2 - x + \frac{1}{4} = \left(x - \frac{1}{2}\right)^2$

9. $-4, -2$ **10.** $\frac{3 \pm \sqrt{3}}{2}$ **11.** $\frac{6 \pm \sqrt{30}}{6}$ **12.** $1 \pm 2i\sqrt{3}$

13. Because 7 is an odd number and not divisible by 2, the computations involved in completing the square on $x^2 + 7x$ involve fractions. The computations involved in completing the square on $x^2 + 6x$ do not.   **14.** 2 is not a factor of the numerator—it is a term. Only common factors of the numerator and denominator can be removed.   **15. a.** $(2 + 2x)$ ft   **b.** $(6 + 2x)$ ft

16. 6 seconds before midnight   **17.** $\frac{1}{2}, -7$ **18.** $5 \pm \sqrt{7}$

19. $0, 10$ **20.** $\frac{13 \pm \sqrt{163}}{3}$ **21.** $-\frac{3}{4} \pm \frac{\sqrt{15}}{4}i$ **22.** $\frac{2}{3} \pm \frac{\sqrt{2}}{3}i$

23. $\frac{-3 \pm \sqrt{29}}{10}$ **24.** $\frac{3 \pm 3\sqrt{13}}{2}$ **25.** \$24 or \$26

26. Sides: 1.25 in. wide; top/bottom: 2.5 in. wide

27. 0.7 sec, 1.8 sec   **28.** 33 in., 56 in.

29. Two different irrational-number solutions

30. Two imaginary-number solutions that are complex conjugates

31. One repeated solution, a rational number   **32.** Two different rational-number solutions   **33.** $1, 144$ **34.** $8, -27$

35. $i, -i, \frac{\sqrt{6}}{3}, -\frac{\sqrt{6}}{3}$ **36.** $1, -\frac{8}{5}$ **37.** $4 \pm i$ **38.** Repeated solutions of -1 and 1 **39.** A repeated solution of $-\frac{2}{5}$ **40.** $\frac{1}{32}, 2$

41. About 81 min   **42.** 30 mph   **43.** 30.8 million   **44.** h, k, x

45.

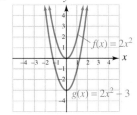

46.

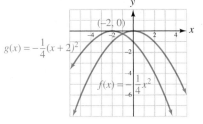

47. $(1, 4), x = 1$

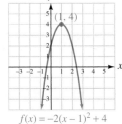

48. $f(x) = 4(x + 2)^2 - 7; (-2, -7), x = -2$

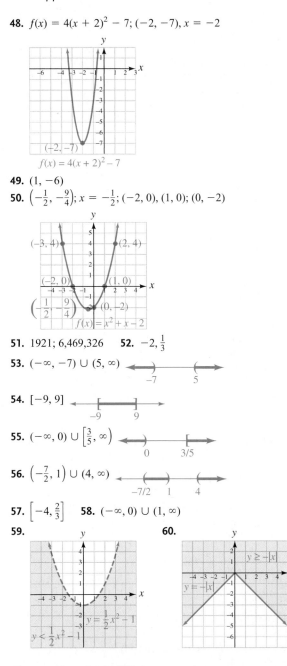

$f(x) = 4(x + 2)^2 - 7$

49. $(1, -6)$

50. $\left(-\frac{1}{2}, -\frac{9}{4}\right); x = -\frac{1}{2}; (-2, 0), (1, 0); (0, -2)$

51. $1921; 6,469,326$ **52.** $-2, \frac{1}{3}$

53. $(-\infty, -7) \cup (5, \infty)$

54. $[-9, 9]$

55. $(-\infty, 0) \cup \left[\frac{3}{5}, \infty\right)$

56. $\left(-\frac{7}{2}, 1\right) \cup (4, \infty)$

57. $\left[-4, \frac{2}{3}\right]$ **58.** $(-\infty, 0) \cup (1, \infty)$

59.

60.

Chapter 8 Test (page 818)

1. a. quadratic   **b.** completed, square   **c.** vertex   **d.** rational
e. nonlinear   **2.** $\pm 3\sqrt{7} \approx 7.94$ **3.** $-7 \pm 5\sqrt{2}$ **4.** $\pm 2i$

5. $x^2 + 11x + \frac{121}{4} = \left(x + \frac{11}{2}\right)^2$ **6.** $\frac{3}{2}, \frac{5}{2}$ **7.** $\frac{-1 \pm \sqrt{2}}{2}$

8. $1 \pm \sqrt{5}$ **9.** $2 \pm 3i$ **10.** $-5, -3$ **11.** $1, \frac{1}{4}$

12. $-1, \frac{1}{3}$ **13.** $2, -2, i\sqrt{3}, -i\sqrt{3}$ **14.** $-\frac{4}{5}, \frac{4}{7}$

15. $2 \pm \sqrt{10}$ **16.** $-8, -\frac{1}{125}$ **17.** $c = \frac{\sqrt{Em}}{m}$

18. a. Two different imaginary-number solutions that are complex conjugates   **b.** One repeated solution, a rational number

19. 4.5 ft by 1,502 ft   **20.** About 46 min   **21.** 20 in.   **22.** iii

23. $(1, -2), x = 1$

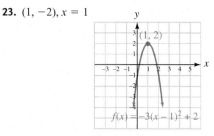

$f(x) = -3(x - 1)^2 + 2$

24. $f(x) = 5(x + 1)^2 - 6; (-1, -6), x = -1$

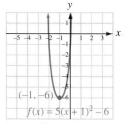

$f(x) = 5(x + 1)^2 - 6$

25. $\left(-\frac{1}{4}, -\frac{9}{8}\right), x = -\frac{1}{4}, (-1, 0), \left(\frac{1}{2}, 0\right); (0, -1)$

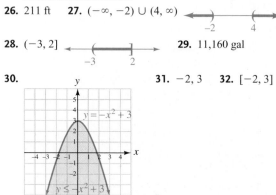

$f(x) = 2x^2 + x - 1$

26. 211 ft   **27.** $(-\infty, -2) \cup (4, \infty)$

28. $(-3, 2]$ **29.** 11,160 gal

30.

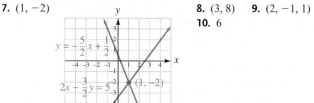

$y \le -x^2 + 3$

31. $-2, 3$ **32.** $[-2, 3]$

Cumulative Review Chapters 1–8 (page 821)

1. -3 **2.** 6 L   **3.** $y = 3x + 2$ **4.** $y = -\frac{2}{3}x - 2$

5. Supply: a decrease of 15,000 nurses per year; demand: an increase of 55,000 nurses per year   **6. a.** Domain: [0, 24]   **b.** 1.5

c. At noon, the low tide mark was -2.5 m.   **d.** 0, 2, 9, 17

7. $(1, -2)$ **8.** $(3, 8)$ **9.** $(2, -1, 1)$ **10.** 6

$y = -\frac{5}{2}x + \frac{1}{2}$

$2x - \frac{3}{2}y = 5$ $(1, -2)$

11. $\left(-\infty, -\frac{10}{9}\right)$ $-10/9$

12. $\left[1, \frac{9}{4}\right]$ $1 \quad 9/4$

13. $(-\infty, -10] \cup [15, \infty)$ $-10 \quad 15$

14.

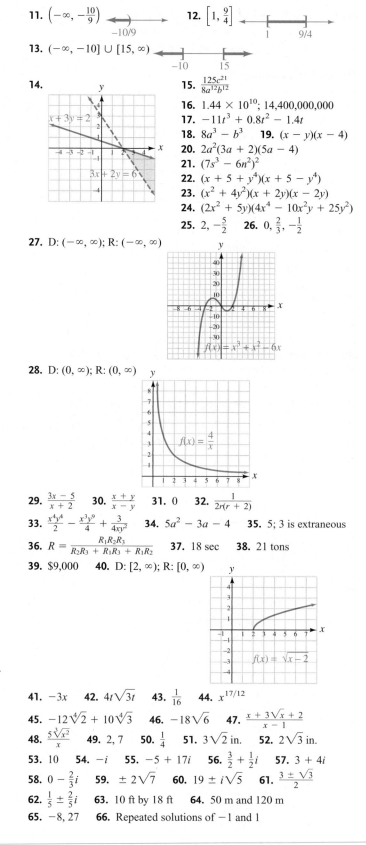

$x + 3y = 2$

$3x + 2y = 6$

15. $\frac{125c^{21}}{8a^{12}b^{12}}$

16. 1.44×10^{10}; 14,400,000,000

17. $-11t^3 + 0.8t^2 - 1.4t$

18. $8a^3 - b^3$ **19.** $(x - y)(x - 4)$

20. $2a^2(3a + 2)(5a - 4)$

21. $(7s^3 - 6n^2)^2$

22. $(x + 5 + y^4)(x + 5 - y^4)$

23. $(x^2 + 4y^2)(x + 2y)(x - 2y)$

24. $(2x^2 + 5y)(4x^4 - 10x^2y + 25y^2)$

25. $2, -\frac{5}{2}$ **26.** $0, \frac{2}{3}, -\frac{1}{2}$

27. D: $(-\infty, \infty)$; R: $(-\infty, \infty)$

$f(x) = x^3 + x^2 - 6x$

28. D: $(0, \infty)$; R: $(0, \infty)$

$f(x) = \frac{4}{x}$

29. $\frac{3x - 5}{x + 2}$ **30.** $\frac{x + y}{x - y}$ **31.** 0   **32.** $\frac{1}{2r(r + 2)}$

33. $\frac{x^4y^4}{2} - \frac{x^3y^9}{4} + \frac{3}{4xy^2}$ **34.** $5a^2 - 3a - 4$ **35.** 5; 3 is extraneous

36. $R = \frac{R_1R_2R_3}{R_2R_3 + R_1R_3 + R_1R_2}$ **37.** 18 sec   **38.** 21 tons

39. \$9,000   **40.** D: $[2, \infty)$; R: $[0, \infty)$

$f(x) = \sqrt{x - 2}$

41. $-3x$ **42.** $4t\sqrt{3t}$ **43.** $\frac{1}{16}$ **44.** $x^{17/12}$

45. $-12\sqrt[4]{2} + 10\sqrt[4]{3}$ **46.** $-18\sqrt{6}$ **47.** $\frac{x + 3\sqrt{x} + 2}{x - 1}$

48. $\frac{5\sqrt[3]{x^2}}{x}$ **49.** 2, 7   **50.** $\frac{1}{4}$ **51.** $3\sqrt{2}$ in.   **52.** $2\sqrt{3}$ in.

53. 10   **54.** $-i$ **55.** $-5 + 17i$ **56.** $\frac{3}{2} + \frac{1}{2}i$ **57.** $3 + 4i$

58. $0 - \frac{2}{3}i$ **59.** $\pm 2\sqrt{7}$ **60.** $19 \pm i\sqrt{5}$ **61.** $\frac{3 \pm \sqrt{3}}{2}$

62. $\frac{1}{5} \pm \frac{2}{5}i$ **63.** 10 ft by 18 ft   **64.** 50 m and 120 m

65. $-8, 27$ **66.** Repeated solutions of -1 and 1

67. $(-2, 4)$, $x = -2$; $(-4, 0)$, $(0, 0)$; $(0, 0)$

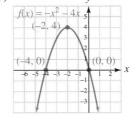

68. $(-9, 9)$

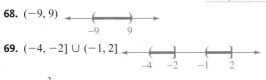

69. $(-4, -2] \cup (-1, 2]$

70. a. $-\frac{3}{4}$ **b.** No solution

Study Set Section 9.1 (page 834)

1. sum, $f(x) + g(x)$, difference, $f(x) - g(x)$ **3.** domain
5. identity **7. a.** $g(3)$ **b.** $g(3)$
9.

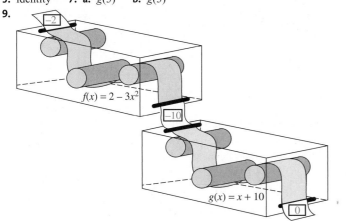

11. $g(x)$, $(3x - 1)$, $9x$, $2x$ **13.** $(f + g)(x) = 7x$, $(-\infty, \infty)$
15. $(g - f)(x) = x$, $(-\infty, \infty)$ **17.** $(f \cdot g)(x) = 12x^2$, $(-\infty, \infty)$
19. $(g/f)(x) = \frac{4}{3}$, $(-\infty, 0) \cup (0, \infty)$
21. $(f + g)(x) = 3x - 2$, $(-\infty, \infty)$
23. $(g - f)(x) = -x - 4$, $(-\infty, \infty)$
25. $(f \cdot g)(x) = 2x^2 - 5x - 3$, $(-\infty, \infty)$
27. $(g/f)(x) = \frac{x-3}{2x+1}$, $\left(-\infty, -\frac{1}{2}\right) \cup \left(-\frac{1}{2}, \infty\right)$
29. $(f - g)(x) = -2x^2 + 3x - 3$, $(-\infty, \infty)$
31. $(f/g)(x) = \frac{3x-2}{2x^2+1}$; $(-\infty, \infty)$ **33.** $(f - g)(x) = 3$, $(-\infty, \infty)$
35. $(g/f)(x) = \frac{x^2-4}{x^2-1}$, $(-\infty, -1) \cup (-1, 1) \cup (1, \infty)$ **37.** 7
39. 24 **41.** -1 **43.** $-\frac{1}{2}$ **45.** $(f \circ g)(x) = 2x^2 - 1$
47. $(g \circ f)(2x) = 16x^2 + 8x$ **49.** 58 **51.** 110 **53.** 2
55. $(g \circ f)(x) = 9x^2 - 9x + 2$ **57.** 16 **59.** $\frac{1}{9}$
61. $(g \circ f)(8x) = 64x^2$ **65. a.** 0 **b.** 6 **c.** -3 **d.** -2
e. 1 **f.** -3 **67. a.** 7 **b.** 8 **c.** 12 **d.** 0 **69.** $3x + 1$
71. $-2x - 5$ **73. a.** 1,003; in 1993, the average combined score on
the SAT was 1,003. **b.** 3; in 1996, the average difference in the math
and verbal scores was 3. **c.** 1028 **d.** 11
75. $C(t) = \frac{5}{9}(2,668 - 200t)$ **77. a.** About \$75
b. $C(m) = \frac{3m}{20} = 0.15m$ **83.** $\frac{1}{c-d}$

Study Set Section 9.2 (page 847)
1. one-to-one **3.** inverses **5.** one-to-one **7.** range, domain
9. 1 **11. a.** No **b.** No **13.** $-4, 0, 8$ **15.** y, $2y$, 3, y, f^{-1}
17. inverse, inverse **19.** Yes **21.** No **23.** No **25.** No
27. One-to-one **29.** Not one-to-one **31.** Not one-to-one
33. One-to-one **35.** $f^{-1}(x) = \frac{x-4}{2}$ **37.** $f^{-1}(x) = 5x - 4$
39. $f^{-1}(x) = 5x + 4$ **41.** $f^{-1}(x) = \frac{2}{x} + 3$ **43.** $f^{-1}(x) = \frac{4}{x}$
45. $f^{-1}(x) = \sqrt[3]{x - 8}$ **47.** $f^{-1}(x) = x^3$ **49.** $f^{-1}(x) = \sqrt[3]{x} - 10$
51. $f^{-1}(x) = \sqrt[3]{\frac{x+3}{2}}$ **53.** $f^{-1}(x) = \sqrt[7]{2x}$
59. $f^{-1}(x) = \frac{1}{2}x$

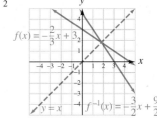

61. $f^{-1}(x) = \frac{x-3}{4}$

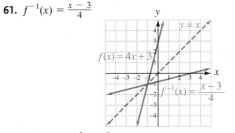

63. $f^{-1}(x) = -\frac{3}{2}x + \frac{9}{2}$

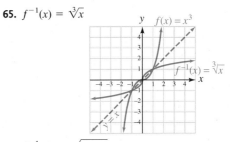

65. $f^{-1}(x) = \sqrt[3]{x}$

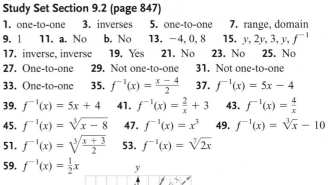

67. $f^{-1}(x) = \sqrt{x + 1}$

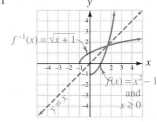

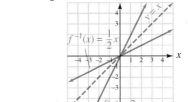
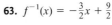

69. a. Yes; no   **b.** No. Twice during this period, the person's anxiety level was at the maximum threshold value.   **77.** $3 - 8i$
79. $18 - i$ **81.** $-28 - 96i$

Study Set Section 9.3 (page 861)

1. Exponential   **3.** $(0, \infty)$ **5.** Yes   **7.** Increasing   **9. a.** $\frac{1}{9}$
b. $\frac{1}{16}$ **c.** 25   **11. a.** ii   **b.** i   **c.** ii   **d.** ii   **13.** $b > 0, b \neq 1$
15. 8   **17.** $7^{3\sqrt{3}}$ **19.** $3^{\sqrt{7}}$ **21.** $\frac{1}{5\sqrt{5}}$

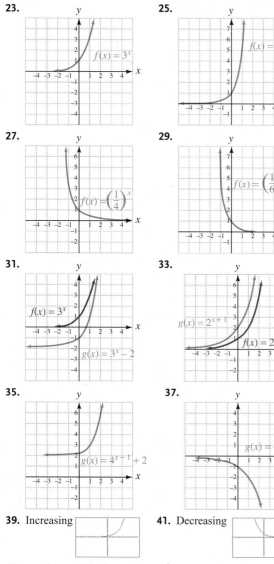

23.

25.

27.

29.

31.

33.

35.

37.

39. Increasing

41. Decreasing

43. a. About 1500, about 1825   **b.** About 6.5 billion

c. Exponential   **45.**

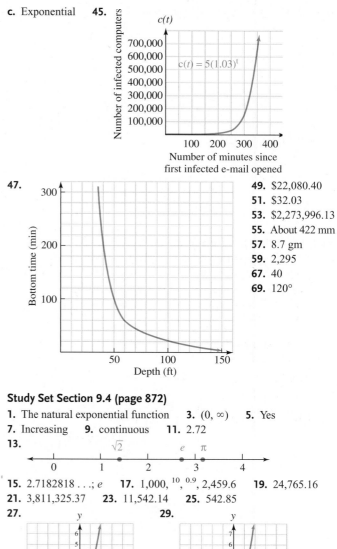

47.

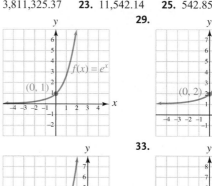

49. $22,080.40
51. $32.03
53. $2,273,996.13
55. About 422 mm
57. 8.7 gm
59. 2,295
67. 40
69. 120°

Study Set Section 9.4 (page 872)

1. The natural exponential function   **3.** $(0, \infty)$ **5.** Yes
7. Increasing   **9.** continuous   **11.** 2.72
13.

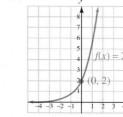

15. 2.7182818 . . .; e **17.** 1,000, ^{10}, $^{0.9}$, 2,459.6   **19.** 24,765.16
21. 3,811,325.37   **23.** 11,542.14   **25.** 542.85
27.

29.

31.

33.

35. About 72 yr   **37.** $13,375.68   **39.** $7,518.28 from annual compounding, $7,647.95 from continuous compounding
41. $6,849.16

43.

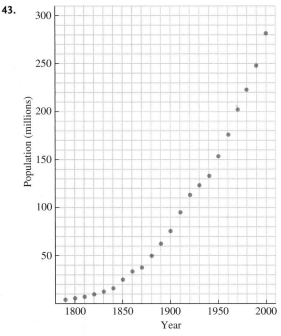

45. 133 million, 140 million, 165 million, 194 million, 206 million, 237 million, 269 million **47.** 5,995,915; 6,565,316 **49.** 14
51. 51 **53.** 12 hr **55.** 49 meters per second **61.** $4x^2\sqrt{15x}$
63. $10y\sqrt{3y}$

Study Set Section 9.5 (page 886)

1. Logarithmic **3.** $(-\infty, \infty)$ **5.** Yes **7.** Increasing **9.** x, b^y
11. inverse **13.** 2, −2 **15.** 1, none, none
17. a. −0.30, 0, 0.30, 0.60, 0.78, 0.90, 1
b.

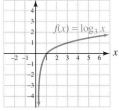

19. a. 10 **b.** x
21. $3^4 = 81$
23. $10^1 = 10$
25. $4^{-3} = \frac{1}{64}$
27. $5^{1/2} = \sqrt{5}$
29. $\log_8 64 = 2$
31. $\log_4 \frac{1}{16} = -2$
33. $\log_{1/2} 32 = -5$
35. $\log_x z = y$ **37.** 9
39. 64 **41.** 3 **43.** $\frac{1}{25}$
45. $\frac{1}{6}$ **47.** 10 **49.** $\frac{2}{3}$

51. 5 **53.** 1,000 **55.** 4 **57.** 1 **59.** $\frac{1}{3}$ **61.** 3 **63.** 2
65. 6 **67.** −1 **69.** 5 **71.** $\frac{1}{2}$ **73.** 0.5119 **75.** −2.3307
77. 6,043.6597 **79.** 0.1726 **81.** 0.3162 **83.** 0.0195
85. Increasing

87. Decreasing

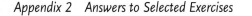

89.   **91.**

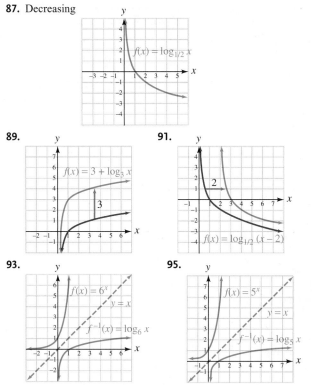

93.   **95.**

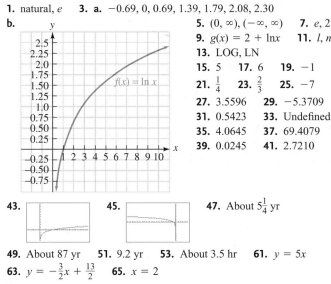

97. 29 db **99.** 49.5 db **101.** 4.4 **103.** 2,500 micrometers
105. 77.8% **107.** 10.8 yr **113.** 10 **115.** 0; $-\frac{5}{9}$ does not check

Study Set Section 9.6 (page 894)

1. natural, e **3. a.** −0.69, 0, 0.69, 1.39, 1.79, 2.08, 2.30
b.

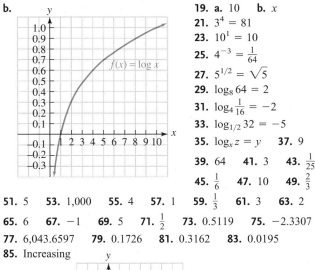

5. $(0, \infty), (-\infty, \infty)$ **7.** $e, 2$
9. $g(x) = 2 + \ln x$ **11.** l, n
13. LOG, LN
15. 5 **17.** 6 **19.** −1
21. $\frac{1}{4}$ **23.** $\frac{2}{3}$ **25.** −7
27. 3.5596 **29.** −5.3709
31. 0.5423 **33.** Undefined
35. 4.0645 **37.** 69.4079
39. 0.0245 **41.** 2.7210

43.   **45.**   **47.** About $5\frac{1}{4}$ yr

49. About 87 yr **51.** 9.2 yr **53.** About 3.5 hr **61.** $y = 5x$
63. $y = -\frac{3}{2}x + \frac{13}{2}$ **65.** $x = 2$

Study Set Section 9.7 (page 907)

1. product **3.** power **5.** 0 **7.** M, N **9.** − **11.** x
13. ≠ **15.** $\log_a b$ **17.** rs, t, r, s **19.** 0 **21.** 7 **23.** 10
25. 2 **27.** 1 **29.** 7 **31.** $2 + \log_2 5$ **33.** $\log 25 + \log y$
35. $2 + \log p + \log q$ **37.** $\log 5 + \log x + \log y + \log z$
39. $2 - \log 9$ **41.** $\log_6 x - 2$ **43.** $\log 7 + \log c - \log 2$

45. $1 + \log x - \log y$ **47.** $1 + \ln x + \ln y - \ln z$
49. $-1 - \log_8 m$ **51.** $7 \ln y$ **53.** $\frac{1}{2} \log 5$ **55.** $-3 \log e$
57. $\frac{3}{5} \log_7 100$ **59.** $\log x + \log y + 2 \log z$
61. $1 + \frac{1}{3} \log_2 x - \log_2 y$ **63.** $3 \log x + 2 \log y$
65. $\frac{1}{2}(\log_b x + \log_b y)$ **67.** $\frac{1}{3} \log_a x - \frac{1}{4} \log_a y - \frac{1}{4} \log_a z$
69. $\ln x + \frac{1}{2} \ln z$ **71.** $\log_2 x^9(x + 1)$ **73.** $\log_3 \frac{x(x + 2)}{8}$
75. $\log_b \frac{\sqrt{z}}{x^3 y^2}$ **77.** $\ln \frac{\frac{x}{z} + x}{\frac{y}{z} + y} = \ln \frac{x}{y}$ **79.** 1.4472 **81.** -1.1972
83. 1.1972 **85.** 1.8063 **87.** 1.7712 **89.** -1.0000
91. 1.8928 **93.** 2.3219 **99.** About 4.8
101. 6.3×10^{-14} gram-ions per liter **109.** $-\frac{7}{6}$ **111.** $\left(1, -\frac{1}{2}\right)$

Study Set Section 9.8 (page 919)
1. exponential **3. a.** equal, $x = y$ **b.** equal, $x = y$ **5.** -2
7. logarithm **9.** x **11.** It is a solution. **13.** Yes
15. a. 1.2920 **b.** 2.2619 **17. a.** $A_0 2^{-t/h}$ **b.** $P_0 e^{kt}$
19. log, log 2, 2.8074 **21.** 4 **23.** $-\frac{3}{4} = -0.75$ **25.** 3, -1
27. A repeated solution of 2 **29.** 1.1610 **31.** 1.2702
33. 1.7095 **35.** -8.2144 **37.** 0.5186 **39.** -13.2662
41. 5,000 **43.** 12 **45.** -93 **47.** 0.08 **49.** -7 **51.** 3
53. 50 **55.** $\frac{5}{4} = 1.25$ **57.** 0.5 **59.** 15.75 **61.** 2
63. $e \approx 2.7183$ **65.** ± 1.0878 **67.** 10 **69.** 10 **71.** 4
73. 0.2325 **75.** 10, -10 **77.** 10 **79.** 9
81. A repeated solution of 4 **83.** 1, 7 **85.** 0.7324 **87.** 6
89. 1.8 **91.** 2.1 **93.** 20 **95.** 8 **97.** 5.1 yr **99.** 42.7 days
101. About 4,200 yr **103.** 5.6 yr **105.** 5.4 yr
107. Because $\ln 2 \approx 0.7$ **109.** 25.3 yr **111.** 2.828 times larger
113. $\frac{1}{3} \ln 0.75 \approx -0.0959$ **119.** $\sqrt{137}$ in.

Chapter 9 Review (page 922)
1. $(f + g)(x) = 3x + 1, (-\infty, \infty)$
2. $(f - g)(x) = x - 1, (-\infty, \infty)$
3. $(f \cdot g)(x) = 2x^2 + 2x, (-\infty, \infty)$
4. $(f/g)(x) = \frac{2x}{x + 1}, (-\infty, -1) \cup (-1, \infty)$ **5.** 3 **6.** 5
7. $(f \circ g)(x) = 4x^2 + 4x + 3$ **8.** $(g \circ f)(x) = 2x^2 + 5$
9. a. 0 **b.** -8 **c.** 0 **d.** -1 **10.** $C(m) = \frac{3.25m}{8}$
11. No **12.** Yes **13.** Yes **14.** No **15.** Yes **16.** No
17. $-6, -1, 7, 20$ **18.**

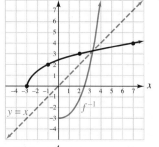

19. $f^{-1}(x) = \frac{x + 3}{6}$ **20.** $f^{-1}(x) = \frac{4}{x} + 1$ **21.** $f^{-1}(x) = \sqrt[3]{x} - 2$
22. $f^{-1}(x) = 6x + 1$

23. $f^{-1}(x) = x^3 + 1$

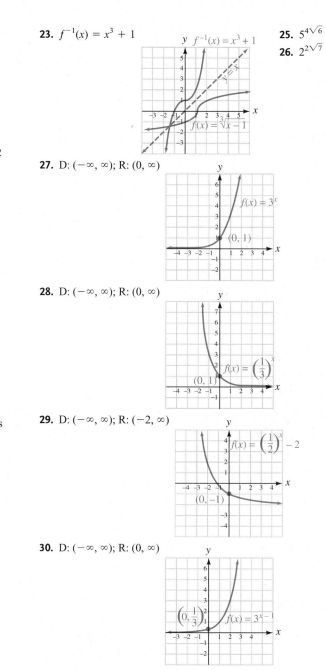

25. $5^{4\sqrt{6}}$
26. $2^{2\sqrt{7}}$

27. D: $(-\infty, \infty)$; R: $(0, \infty)$

28. D: $(-\infty, \infty)$; R: $(0, \infty)$

29. D: $(-\infty, \infty)$; R: $(-2, \infty)$

30. D: $(-\infty, \infty)$; R: $(0, \infty)$

31. The x-axis ($y = 0$)

32. An exponential function

33. $2,189,703.45
34. About $2,015

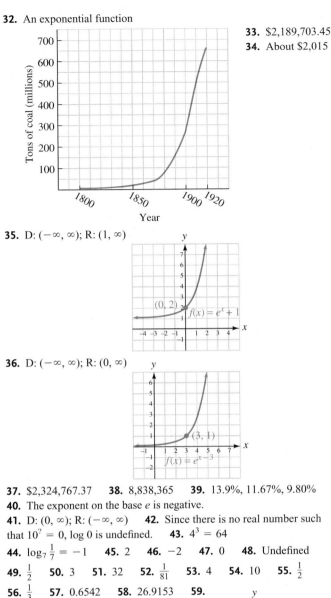

35. D: $(-\infty, \infty)$; R: $(1, \infty)$

36. D: $(-\infty, \infty)$; R: $(0, \infty)$

37. $2,324,767.37 **38.** 8,838,365 **39.** 13.9%, 11.67%, 9.80%
40. The exponent on the base e is negative.
41. D: $(0, \infty)$; R: $(-\infty, \infty)$ **42.** Since there is no real number such that $10^? = 0$, log 0 is undefined. **43.** $4^3 = 64$
44. $\log_7 \frac{1}{7} = -1$ **45.** 2 **46.** -2 **47.** 0 **48.** Undefined
49. $\frac{1}{2}$ **50.** 3 **51.** 32 **52.** $\frac{1}{81}$ **53.** 4 **54.** 10 **55.** $\frac{1}{2}$
56. $\frac{1}{3}$ **57.** 0.6542 **58.** 26.9153 **59.**

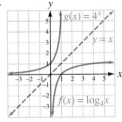

60.

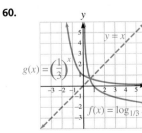

61.

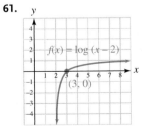

62.

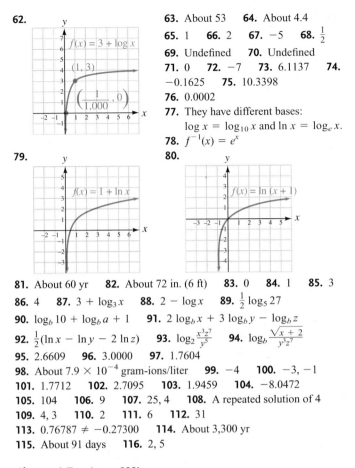

63. About 53 **64.** About 4.4
65. 1 **66.** 2 **67.** -5 **68.** $\frac{1}{2}$
69. Undefined **70.** Undefined
71. 0 **72.** -7 **73.** 6.1137 **74.** -0.1625 **75.** 10.3398
76. 0.0002
77. They have different bases: $\log x = \log_{10} x$ and $\ln x = \log_e x$.
78. $f^{-1}(x) = e^x$
79.

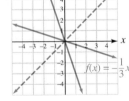

80.

81. About 60 yr **82.** About 72 in. (6 ft) **83.** 0 **84.** 1 **85.** 3
86. 4 **87.** $3 + \log_3 x$ **88.** $2 - \log x$ **89.** $\frac{1}{2} \log_5 27$
90. $\log_b 10 + \log_b a + 1$ **91.** $2 \log_b x + 3 \log_b y - \log_b z$
92. $\frac{1}{2}(\ln x - \ln y - 2 \ln z)$ **93.** $\log_2 \frac{x^3 z^7}{y^5}$ **94.** $\log_b \frac{\sqrt{x+2}}{y^3 z^7}$
95. 2.6609 **96.** 3.0000 **97.** 1.7604
98. About 7.9×10^{-4} gram-ions/liter **99.** -4 **100.** $-3, -1$
101. 1.7712 **102.** 2.7095 **103.** 1.9459 **104.** -8.0472
105. 104 **106.** 9 **107.** 25, 4 **108.** A repeated solution of 4
109. 4, 3 **110.** 2 **111.** 6 **112.** 31
113. $0.76787 \neq -0.27300$ **114.** About 3,300 yr
115. About 91 days **116.** 2, 5

Chapter 9 Test (page 933)
1. a. composite **b.** natural **c.** continuous **d.** inverse
e. logarithmic **2. a.** f composed with g **b.** g of f of eight
3. $(f + g)(x) = 4x^2 - 2x + 11, (-\infty, \infty)$
4. $(g/f)(x) = \frac{4x^2 - 3x + 2}{x + 9}, (-\infty, -9) \cup (-9, \infty)$ **5.** 76
6. $32x^2 - 128x + 131$ **7. a.** -16 **b.** 17 **8. a.** 0 **b.** 3
c. 1 **d.** -2 **e.** -5 **9.** Yes **10.** No
11. $f^{-1}(x) = -3x$

12. $f^{-1}(x) = \sqrt[3]{x} + 15$ **14. a.** Yes **b.** Yes
c. 80; when the temperature of the tire tread is 260°, the vehicle is traveling 80 mph

15. D: $(-\infty, \infty)$; R: $(1, \infty)$

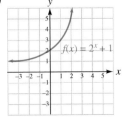

16. D: $(-\infty, \infty)$; R: $(0, \infty)$

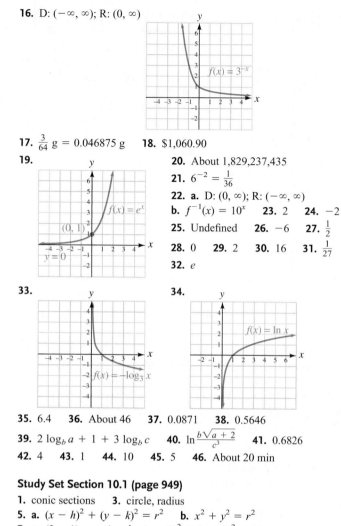

$f(x) = 3^{-x}$

17. $\frac{3}{64}$ g $= 0.046875$ g   **18.** \$1,060.90

19.

$f(x) = e^x$

$(0, 1)$

$y = 0$

20. About 1,829,237,435

21. $6^{-2} = \frac{1}{36}$

22. a. D: $(0, \infty)$; R: $(-\infty, \infty)$
b. $f^{-1}(x) = 10^x$ **23.** 2   **24.** -2

25. Undefined   **26.** -6 **27.** $\frac{1}{2}$

28. 0   **29.** 2   **30.** 16   **31.** $\frac{1}{27}$

32. e

33.

$f(x) = -\log_3 x$

34.

$f(x) = \ln x$

35. 6.4   **36.** About 46   **37.** 0.0871   **38.** 0.5646

39. $2 \log_b a + 1 + 3 \log_b c$ **40.** $\ln \frac{b\sqrt{a+2}}{c^3}$ **41.** 0.6826

42. 4   **43.** 1   **44.** 10   **45.** 5   **46.** About 20 min

Study Set Section 10.1 (page 949)

1. conic sections   **3.** circle, radius
5. a. $(x - h)^2 + (y - k)^2 = r^2$ **b.** $x^2 + y^2 = r^2$
7. a. $(2, -1)$; $r = 4$ **b.** $(x - 2)^2 + (y + 1)^2 = 16$
9. a. $y = a(x - h)^2 + k$ **b.** $x = a(y - k)^2 + h$
11. a. Circle   **b.** Parabola   **c.** Parabola   **d.** Circle
13. 6, -2, 3   **15.** $(0, 0)$, $r = 3$

$x^2 + y^2 = 9$

$(0, 0)$

17. $(0, -3)$, $r = 1$

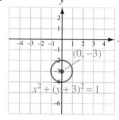

$(0, -3)$

$x^2 + (y + 3)^2 = 1$

19. $(-3, 1)$, $r = 4$

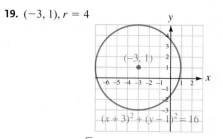

$(-3, 1)$

$(x + 3)^2 + (y - 1)^2 = 16$

21. $(0, 0)$, $r = \sqrt{6} \approx 2.4$

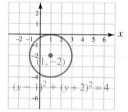

$(0, 0)$

$x^2 + y^2 = 6$

23. $x^2 + y^2 = 1$ **25.** $(x - 6)^2 + (y - 8)^2 = 25$
27. $(x + 2)^2 + (y - 6)^2 = 144$ **29.** $x^2 + y^2 = \frac{1}{16}$
31. $\left(x - \frac{2}{3}\right)^2 + \left(y + \frac{7}{8}\right)^2 = 2$ **33.** $x^2 + y^2 = 8$
35. $(x - 1)^2 + (y + 2)^2 = 4$; $(1, -2)$, $r = 2$

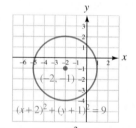

$(1, -2)$

$(x - 1)^2 + (y + 2)^2 = 4$

37. $(x + 2)^2 + (y + 1)^2 = 9$; $(-2, -1)$, $r = 3$

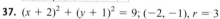

$(-2, -1)$

$(x + 2)^2 + (y + 1)^2 = 9$

39. $y = 2(x - 1)^2 + 3$; vertex: $(1, 3)$

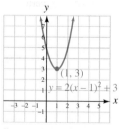

$(1, 3)$

$y = 2(x - 1)^2 + 3$

41. $y = -(x + 1)^2 + 4$; vertex: $(-1, 4)$

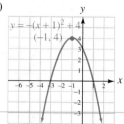

$y = -(x + 1)^2 + 4$

$(-1, 4)$

43. Vertex: (0, 0)

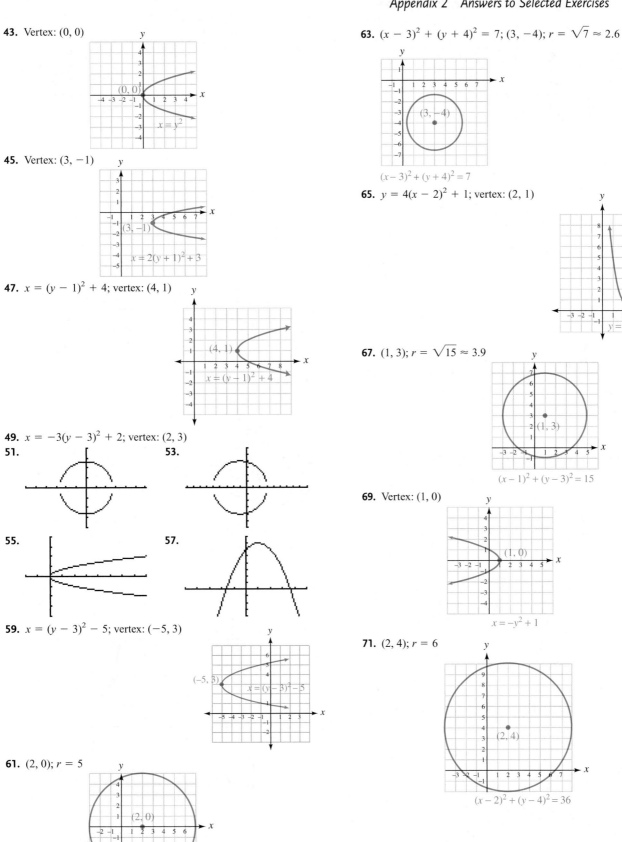

45. Vertex: (3, −1)

47. $x = (y - 1)^2 + 4$; vertex: (4, 1)

49. $x = -3(y - 3)^2 + 2$; vertex: (2, 3)

51.

53.

55.

57.

59. $x = (y - 3)^2 - 5$; vertex: (−5, 3)

61. (2, 0); $r = 5$

63. $(x - 3)^2 + (y + 4)^2 = 7$; (3, −4); $r = \sqrt{7} \approx 2.6$

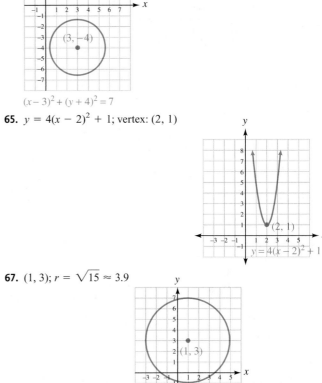

65. $y = 4(x - 2)^2 + 1$; vertex: (2, 1)

67. (1, 3); $r = \sqrt{15} \approx 3.9$

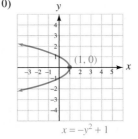

69. Vertex: (1, 0)

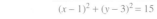

71. (2, 4); $r = 6$

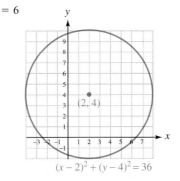

73. Vertex: $(0, 0)$

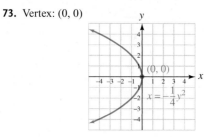

$$x = -\frac{1}{4}y^2$$

75. Vertex: $(3, 1)$

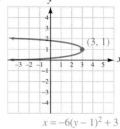

$$x = -6(y - 1)^2 + 3$$

77. $(x + 1)^2 + y^2 = 9$; $(-1, 0)$; $r = 3$

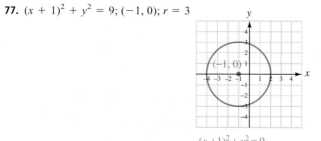

$$(x + 1)^2 + y^2 = 9$$

79. $x = \frac{1}{2}(y + 2)^2 - 2$; vertex: $(-2, -2)$

$$x = \frac{1}{2}(y + 2)^2 - 2$$

81. Vertex: $(-5, 5)$

$$y = -4(x + 5)^2 + 5$$

83. No
85. a. 8 mi
b. 9 mi   **87.** 5 ft
89. 2 AU
95. $5, -\frac{7}{3}$
97. $3, -\frac{1}{4}$

Study Set Section 10.2 (page 960)

1. ellipse   **3.** foci, focus   **5.** major   **7.** $\frac{x^2}{a^2} + \frac{y^2}{b^2} = 1$
9. x-intercepts: $(a, 0)$, $(-a, 0)$; y-intercepts: $(0, b)$, $(0, -b)$
11. a. $(-2, 1)$; $a = 2, b = 5$ **b.** Vertical
c. $\frac{(x + 2)^2}{4} + \frac{(y - 1)^2}{25} = 1$ **13.** $\frac{(x - 1)^2}{16} + \frac{(y + 5)^2}{1} = 1$

15. $h = -8, k = 6, a = 10, b = 12$ **17.**

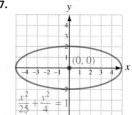

$$\frac{x^2}{25} + \frac{y^2}{4} = 1$$

19.

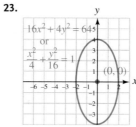

$$\frac{x^2}{4} + \frac{y^2}{9} = 1$$

21.

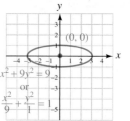

$$x^2 + 9y^2 = 9$$
or
$$\frac{x^2}{9} + \frac{y^2}{1} = 1$$

23.

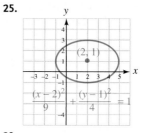

$$16x^2 + 4y^2 = 64$$
or
$$\frac{x^2}{4} + \frac{y^2}{16} = 1$$

25.

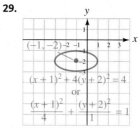

$$\frac{(x - 2)^2}{9} + \frac{(y - 1)^2}{4} = 1$$

27.

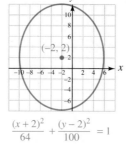

$$\frac{(x + 2)^2}{64} + \frac{(y - 2)^2}{100} = 1$$

29.

$$(x + 1)^2 + 4(y + 2)^2 = 4$$
or
$$\frac{(x + 1)^2}{4} + \frac{(y + 2)^2}{1} = 1$$

31.

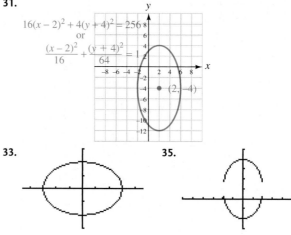

$$16(x - 2)^2 + 4(y + 4)^2 = 256$$
or
$$\frac{(x - 2)^2}{16} + \frac{(y + 4)^2}{64} = 1$$

33.

35.

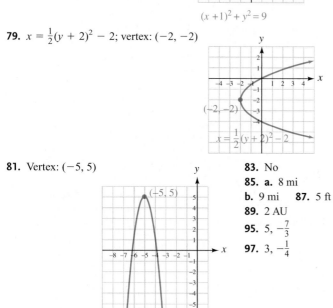

37.

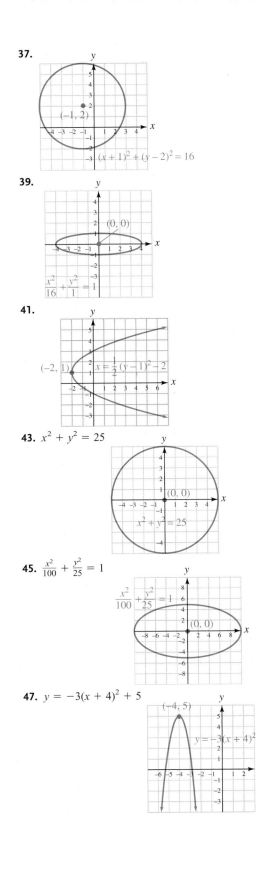

39.

41.

43. $x^2 + y^2 = 25$

45. $\frac{x^2}{100} + \frac{y^2}{25} = 1$

47. $y = -3(x + 4)^2 + 5$

49. $(x - 1)^2 + (y + 2)^2 = 9$

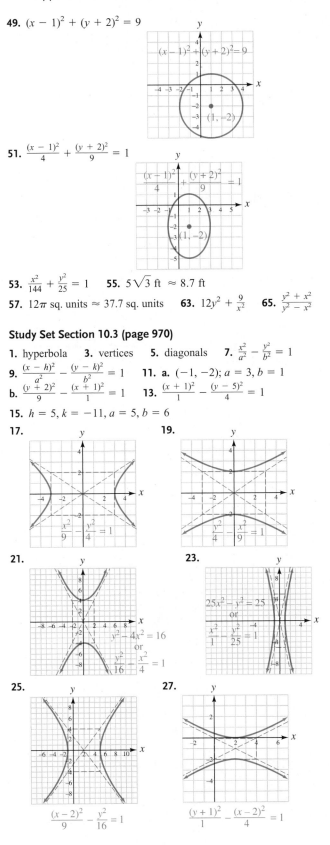

51. $\frac{(x - 1)^2}{4} + \frac{(y + 2)^2}{9} = 1$

53. $\frac{x^2}{144} + \frac{y^2}{25} = 1$ **55.** $5\sqrt{3}$ ft ≈ 8.7 ft

57. 12π sq. units ≈ 37.7 sq. units   **63.** $12y^2 + \frac{9}{x^2}$ **65.** $\frac{y^2 + x^2}{y^2 - x^2}$

Study Set Section 10.3 (page 970)

1. hyperbola   **3.** vertices   **5.** diagonals   **7.** $\frac{x^2}{a^2} - \frac{y^2}{b^2} = 1$

9. $\frac{(x - h)^2}{a^2} - \frac{(y - k)^2}{b^2} = 1$ **11. a.** $(-1, -2)$; $a = 3, b = 1$

b. $\frac{(y + 2)^2}{9} - \frac{(x + 1)^2}{1} = 1$ **13.** $\frac{(x + 1)^2}{1} - \frac{(y - 5)^2}{4} = 1$

15. $h = 5, k = -11, a = 5, b = 6$

17.

19.

21.

23.

25.

27.

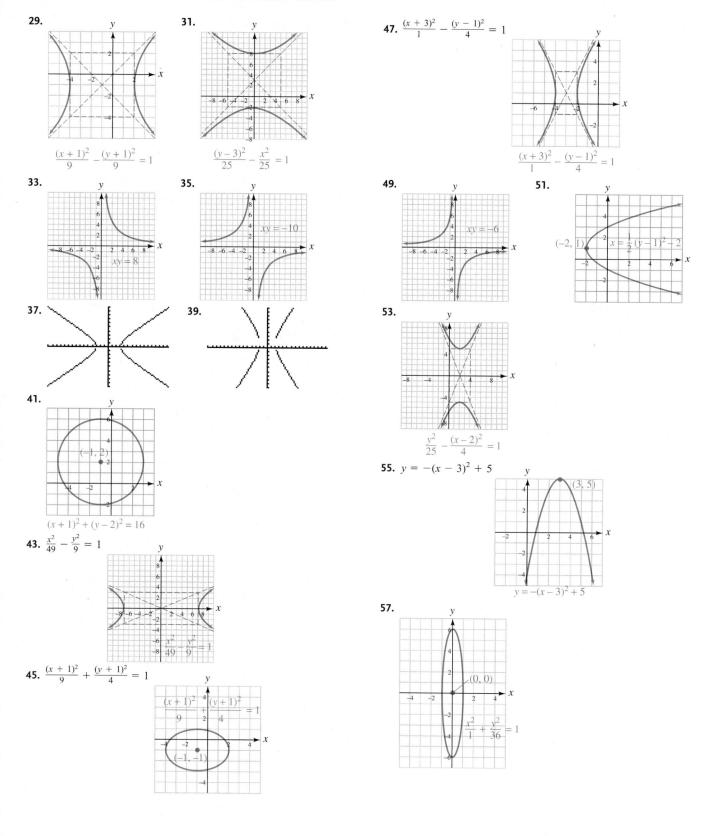

29.

$$\frac{(x+1)^2}{9} - \frac{(y+1)^2}{9} = 1$$

31.

$$\frac{(y-3)^2}{25} - \frac{x^2}{25} = 1$$

33.

$xy = 8$

35.

$xy = -10$

37.

39.

41.

$(-1, 2)$

$(x+1)^2 + (y-2)^2 = 16$

43. $\dfrac{x^2}{49} - \dfrac{y^2}{9} = 1$

$$\frac{x^2}{49} - \frac{y^2}{9} = 1$$

45. $\dfrac{(x+1)^2}{9} + \dfrac{(y+1)^2}{4} = 1$

$$\frac{(x+1)^2}{9} + \frac{(y+1)^2}{4} = 1$$

$(-1, -1)$

47. $\dfrac{(x+3)^2}{1} - \dfrac{(y-1)^2}{4} = 1$

$$\frac{(x+3)^2}{1} - \frac{(y-1)^2}{4} = 1$$

49.

$xy = -6$

51.

$(-2, 1)$ $x = \dfrac{1}{2}(y-1)^2 - 2$

53.

$$\frac{y^2}{25} - \frac{(x-2)^2}{4} = 1$$

55. $y = -(x-3)^2 + 5$

$(3, 5)$

$y = -(x-3)^2 + 5$

57.

$(0, 0)$

$$\frac{x^2}{1} + \frac{y^2}{36} = 1$$

59. $(x + 2)^2 + (y - 3)^2 = 36$

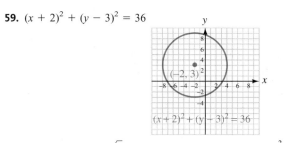

61. 3 units   **63.** $10\sqrt{3}$ miles   **69.** 64   **71.** 3   **73.** $\frac{3}{2}$ **75.** 10

Study Set Section 10.4 (page 977)

1. system   **3.** intersection   **5.** secant   **7. a.** two   **b.** four
c. four   **d.** four   **9.** $(-3, 2), (3, 2), (-3, -2), (3, -2)$
11. a. -4 **b.** -2 **13.** $2x, 5, 5, 1, 1, -1, -2$
15. $(0, 3), (-3, 0)$

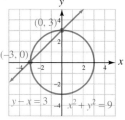

17. $(-4, 0), (4, 0)$

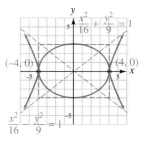

19. $(0, 0), (2, -4)$

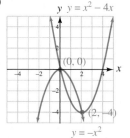

21. $(-2, 0), (0, -1), (0, 1)$

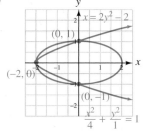

23. $(1, 2), (2, 1)$ **25.** $(-6, 7), (-2, -1)$ **27.** $(-2, 3), (2, 3)$
29. $(\sqrt{5}, 5), (-\sqrt{5}, 5)$ **31.** $(2, 4), (2, -4), (-2, 4), (-2, -4)$
33. $(3, 0), (-3, 0)$ **35.** $(\sqrt{3}, 0), (-\sqrt{3}, 0)$
37. $(-2, 3), (2, 3), (-2, -3), (2, -3)$

39. $(1, 0), (5, 0)$

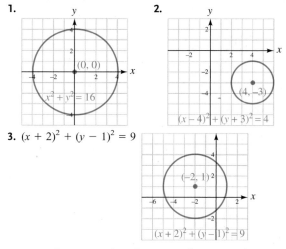

41. $(2, 1), (-2, 1), (2, -1), (-2, -1)$ **43.** $(0, -4), (-3, 5), (3, 5)$
45. $(6, 2), (-6, -2), \left(-\sqrt{42}, 0\right), \left(\sqrt{42}, 0\right)$
47. $\left(-\sqrt{15}, 5\right), \left(\sqrt{15}, 5\right), (-2, -6), (2, -6)$
49. $(0, 3), \left(-\frac{25}{12}, -\frac{13}{4}\right)$ **51.** $(0, 0), (2, 4)$ **53.** No solution, $\varnothing$
55. $(3, 5)$
57. $\left(\frac{\sqrt{10}}{4}, \frac{3\sqrt{6}}{4}\right), \left(\frac{\sqrt{10}}{4}, -\frac{3\sqrt{6}}{4}\right), \left(-\frac{\sqrt{10}}{4}, \frac{3\sqrt{6}}{4}\right), \left(-\frac{\sqrt{10}}{4}, -\frac{3\sqrt{6}}{4}\right)$
59. $\left(\frac{1}{2}, \frac{1}{3}\right), \left(\frac{1}{3}, \frac{1}{2}\right)$ **61.** $(-1, 3), (1, 3)$ **63.** $\left(\frac{5}{2}, \frac{3}{2}\right)$ **65.** 4, 8
67. $\left(10, \frac{10}{3}\right); \frac{10}{3}\sqrt{10}$ m   **69.** 80 ft by 100 ft or 50 ft by 160 ft
71. \$2,500 at 9%   **75.** 2,000   **77.** 7

Chapter 10 Review (page 982)

1.

2.

3. $(x + 2)^2 + (y - 1)^2 = 9$

4. $(x - 9)^2 + (y - 9)^2 = 9^2$ or $(x - 9)^2 + (y - 9)^2 = 81$
5. $(-6, 0); r = 2\sqrt{6}$ **6.** center, radius

7. $(0, 0)$

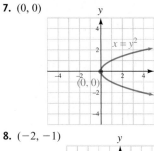

8. $(-2, -1)$

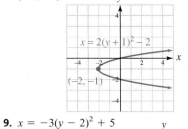

9. $x = -3(y - 2)^2 + 5$

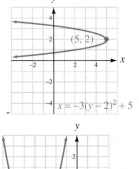

10. $y = (x + 4)^2 - 5$

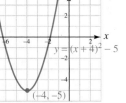

11. $(2, -2), (8, -3)$

12. When $x = 22, y = 0$: $-\frac{5}{121}(22 - 11)^2 + 5 = 0$

13.

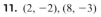
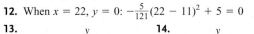

14.

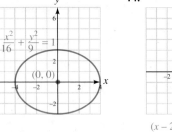

15.

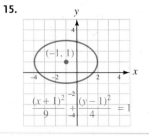

16. $\frac{x^2}{12^2} + \frac{y^2}{1^2} = 1$ **17. a.** Circle
b. Ellipse   **c.** Parabola
d. Ellipse   **18.** $\frac{x^2}{25} + \frac{y^2}{9} = 1$
19. ellipse, focus

20.

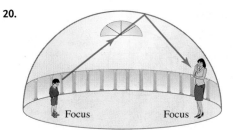

21.

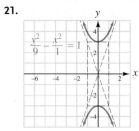

22.

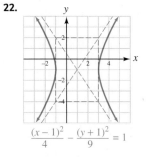

23.

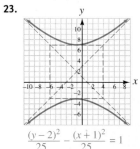

24.

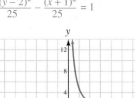

25. 4 units   **26. a.** Ellipse
b. Hyperbola   **c.** Parabola
d. Circle   **27.** Yes
28. $(0, 3), (0, -3)$

29. $(2, 2), (-1, -4)$

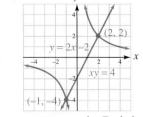

30. a. 2   **b.** 4
c. 4   **d.** 4
31. $(0, 1), (0, -1)$
32. $y = 3x - 1$

33. $(0, -4), (-3, 5), (3, 5)$ **34.** $\left(\sqrt{2}, 0\right), \left(-\sqrt{2}, 0\right)$
35. $(2, 2), \left(-\frac{2}{9}, -\frac{22}{9}\right)$ **36.** $(4, 2), (4, -2), (-4, 2), (-4, -2)$
37. $(2, 3), (2, -3), (-2, 3), (-2, -3)$
38. $\left(2\sqrt{2}, \sqrt{2}\right), \left(-2\sqrt{2}, -\sqrt{2}\right), (1, 4), (-1, -4)$
39. No solution, $\varnothing$ **40.** $(-2, 1)$

Chapter 10 Test (page 986)

1. a. conic   **b.** center, radius   **c.** hyperbola   **d.** nonlinear
e. ellipse   **2.** $(0, 0)$; $r = 10$

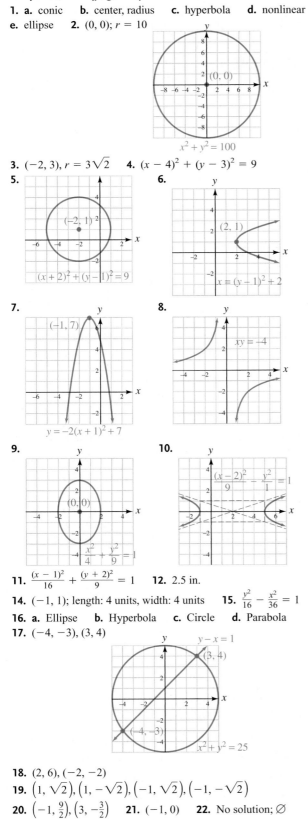

3. $(-2, 3), r = 3\sqrt{2}$ **4.** $(x - 4)^2 + (y - 3)^2 = 9$
5.

$(x + 2)^2 + (y - 1)^2 = 9$

6.

$x = (y - 1)^2 + 2$

7.

$y = -2(x + 1)^2 + 7$

8.

$xy = -4$

9.

$\frac{x^2}{4} + \frac{y^2}{9} = 1$

10.

$\frac{(x - 2)^2}{9} - \frac{y^2}{1} = 1$

11. $\frac{(x - 1)^2}{16} + \frac{(y + 2)^2}{9} = 1$ **12.** 2.5 in.
14. $(-1, 1)$; length: 4 units, width: 4 units   **15.** $\frac{y^2}{16} - \frac{x^2}{36} = 1$
16. a. Ellipse   **b.** Hyperbola   **c.** Circle   **d.** Parabola
17. $(-4, -3), (3, 4)$

$y - x = 1$

$x^2 + y^2 = 25$

18. $(2, 6), (-2, -2)$
19. $\left(1, \sqrt{2}\right), \left(1, -\sqrt{2}\right), \left(-1, \sqrt{2}\right), \left(-1, -\sqrt{2}\right)$
20. $\left(-1, \frac{9}{2}\right), \left(3, -\frac{3}{2}\right)$ **21.** $(-1, 0)$ **22.** No solution; $\varnothing$

Study Set Section 11.1 (page 998)

1. binomial   **3.** binomial   **5.** factorial, decreasing   **7.** one
9. $20, 20$ **11.** $1, 1$ **13.** $n, 1$ **15.** 1 **17.** $9, 3$ **19.** $2, 4$
21. $3, 3, y, 2, 2, 3$ **23.** $n, 1$ **25.** $a^3 + 3a^2b + 3ab^2 + b^3$
27. $m^5 - 5m^4p + 10m^3p^2 - 10m^2p^3 + 5mp^4 - p^5$ **29.** 6
31. 120 **33.** 30 **35.** 144 **37.** $40,320$ **39.** $2,352$ **41.** $\frac{1}{110}$
43. 72 **45.** 5 **47.** 10 **49.** $\frac{1}{168}$ **51.** 21 **53.** $39,916,800$
55. $2.432902008 \times 10^{18}$ **57.** $m^4 + 4m^3n + 6m^2n^2 + 4mn^3 + n^4$
59. $c^5 - 5c^4d + 10c^3d^2 - 10c^2d^3 + 5cd^4 - d^5$
61. $a^9 - 9a^8b + 36a^7b^2 - 84a^6b^3 + 126a^5b^4 - 126a^4b^5 +$
    $84a^3b^6 - 36a^2b^7 + 9ab^8 - b^9$
63. $s^6 + 6s^5t + 15s^4t^2 + 20s^3t^3 + 15s^2t^4 + 6st^5 + t^6$
65. $8x^3 + 12x^2y + 6xy^2 + y^3$
67. $32t^5 - 240t^4 + 720t^3 - 1,080t^2 + 810t - 243$
69. $625m^4 - 1,000m^3n + 600m^2n^2 - 160mn^3 + 16n^4$
71. $\frac{x^3}{27} + \frac{x^2y}{6} + \frac{xy^2}{4} + \frac{y^3}{8}$ **73.** $\frac{x^4}{81} - \frac{2x^3y}{27} + \frac{x^2y^2}{6} - \frac{xy^3}{6} + \frac{y^4}{16}$
75. $c^{10} - 5c^8d^2 + 10c^6d^4 - 10c^4d^6 + 5c^2d^8 - d^{10}$ **77.** $28x^6y^2$
79. $15r^2s^4$ **81.** $78x^{10}$ **83.** $-12x^3y$ **85.** $810xy^4$ **87.** $-\frac{1}{6}c^3d$
89. $-70,000t^4$ **91.** $6a^{10}b^2$ **97.** $\log x^2y^{1/2}$ or $\log x^2\sqrt{y}$ **99.** $\ln y$

Study Set Section 11.2 (page 1008)

1. sequence   **3.** arithmetic   **5.** mean   **7.** 1, 7, 13
9. a. $a_n = a_1 + (n - 1)d$ **b.** $S_n = \frac{n(a_1 + a_n)}{2}$ **11.** nth
13. sigma   **15.** summation, runs   **17.** 3, 7, 11, 15, 19; 159
19. $-2, -5, -8, -11, -14; -89$
21. $-1, -4, -9, -16, -25; -400$ **23.** $0, \frac{1}{2}, \frac{2}{3}, \frac{3}{4}, \frac{11}{12}$
25. $-\frac{1}{3}, \frac{1}{9}, -\frac{1}{27}, \frac{1}{81}$ **27.** $-7, 8, -9, 10$ **29.** 3, 5, 7, 9, 11; 21
31. $-5, -8, -11, -14, -17; -47$ **33.** 7, 19, 31, 43, 53; 355
35. $-7, -9, -11, -13, -15; -35$ **37.** 88   **39.** 59
41. 5, 11, 17, 23, 29   **43.** $-4, -11, -18, -25, -32$ **45.** 5, 8
47. 12, 14, 16, 18   **49.** $\frac{45}{2}, 25, \frac{55}{2}$ **51.** $\frac{5}{4}$ **53.** 2,555
55. 2,920   **57.** $3 + 6 + 9 + 12$ **59.** $4 + 9 + 16$ **61.** 60
63. 91   **65.** 31   **67.** 12   **69.** 70   **71.** 57   **73.** 12   **75.** 354
77. $-118, -111, -104, -97, -90$ **79.** 34, 31, 28, 25, 22, 19
81. 1,398   **83.** 1,275   **85.** -179 **87.** -23 **89.** -3
91. 2,500   **93.** $60, $110, $160, $210, $260, $310; $6,060
95. 11,325   **97.** 78   **103.** $1 + \log_2 x - \log_2 y$
105. $3 \log x + 2 \log y$

Study Set Section 11.3 (page 1021)

1. geometric   **3.** mean   **5.** 16, 4, 1   **7.** $a_n = a_1r^{n-1}$
9. a. Yes   **b.** No   **c.** No   **d.** Yes   **11.** r^2, a_1r^4 **13.** n
15. 3, 6, 12, 24, 48; 768   **17.** $-5, -1, -\frac{1}{5}, -\frac{1}{25}, -\frac{1}{125}; -\frac{1}{15,625}$
19. 1,458   **21.** $-\frac{3,125}{2}$ **23.** 2, 6, 18, 54, 162
25. 3, 6, 12, 24, 48   **27.** 6, 18, 54   **29.** $-20, -100, -500, -2,500$
31. $-16, 16$ **33.** $-10\sqrt{2}, 10\sqrt{2}$ **35.** 728   **37.** 122
39. -255 **41.** 381   **43.** 16   **45.** 81   **47.** $-\frac{81}{2}$ **49.** No sum
51. 8   **53.** $-\frac{135}{4}$ **55.** $\frac{1}{9}$ **57.** $\frac{1}{3}$ **59.** $\frac{4}{33}$ **61.** $\frac{25}{33}$ **63.** 3
65. $-64, 32, -16, 8, -4$ **67.** No geometric mean exists.
69. 3,584   **71.** $-64, -32, -16, -8$ **73.** 4   **75.** $\frac{1}{27}$ **77.** 93
79. $1,469.74$ **81.** About $539,731
83. $\left(\frac{1}{2}\right)^{11} \approx 0.0005$ square unit   **85.** 30 m   **87.** 5,000
93. $[-1, 6]$ **95.** $(-\infty, -3) \cup (4, \infty)$

Chapter 11 Review (page 1024)

1. 1, 3, 1, 1, 10, 15; 1, 5, 10, 10, 5, 1   **2. a.** 13   **b.** 12

c. a^{12}, b^{12} **d.** a: decrease; b: increase   **3.** 144   **4.** 20

5. 15   **6.** 220   **7.** 1   **8.** 8

9. $x^5 + 5x^4y + 10x^3y^2 + 10x^2y^3 + 5xy^4 + y^5$

10. $x^9 - 9x^8y + 36x^7y^2 - 84x^6y^3 + 126x^5y^4 - 126x^4y^5 + 84x^3y^6 - 36x^2y^7 + 9xy^8 - y^9$ **11.** $64x^3 - 48x^2y + 12xy^2 - y^3$

12. $\frac{c^4}{16} + \frac{c^3d}{6} + \frac{c^2d^2}{6} + \frac{2cd^3}{27} + \frac{d^4}{81}$ **13.** $6x^2y^2$ **14.** $-20x^3y^3$

15. $-108x^2y$ **16.** $5u^2v^{12}$ **17.** $-2, 0, 2, 4$

18. $-\frac{1}{2}, \frac{1}{3}, -\frac{1}{4}, \frac{1}{5}, -\frac{1}{6}$ **19.** 75   **20.** 42

21. 122, 137, 152, 167, 182   **22.** $-1,194$ **23.** -5

24. $\frac{41}{3}, \frac{58}{3}$ **25.** $-\frac{45}{2}$ **26.** 1,568   **27.** $\frac{15}{2}$ **28.** 378

29. 14   **30.** 360   **31.** 20,100   **32.** 1,170   **33.** 4

34. 24, 12, 6, 3, $\frac{3}{2}$ **35.** $\frac{1}{27}$ **36.** 24, -96 **37.** $\frac{2,186}{9}$

38. $-\frac{85}{8}$ **39.** About 1.6 lb   **40.** 125   **41.** $\frac{5}{99}$ **42.** 190 ft

Chapter 11 Test (page 1028)

1. a. Pascal's   **b.** alternate   **c.** arithmetic   **d.** series

e. geometric   **2.** 2, $-4, -10, -16$ **3.** $-1, \frac{1}{8}, -\frac{1}{27}, \frac{1}{64}, -\frac{1}{125}$

4. 210   **5.** $a^6 - 6a^5b + 15a^4b^2 - 20a^3b^3 + 15a^2b^4 - 6ab^5 + b^6$

6. $24x^4y^2$ **7.** 66   **8.** 306   **9.** 34, 66   **10.** $\frac{1}{4}$ **11.** $-1,377$

12. 205   **13.** 1,600 ft   **14.** 3   **15.** -81 **16.** $\frac{364}{27}$ **17.** 6

18. 18, 108   **19.** $\frac{27}{2}$ **20.** About \$651,583   **21.** 2,000 in.

22. $\frac{7}{9}$

Group Project (page 1029)

1. g   **2.** o   **3.** i   **4.** l   **5.** u   **6.** y   **7.** p   **8.** x   **9.** m

10. d   **11.** f   **12.** c   **13.** w   **14.** b   **15.** j   **16.** s   **17.** e

18. z   **19.** n   **20.** k   **21.** h   **22.** v   **23.** q   **24.** a

25. r   **26.** t

Cumulative Review Chapters 1–11 (page 1030)

1. a. 0   **b.** $-\frac{4}{3}, 5.6, 0, -23$ **c.** $\pi, \sqrt{2}, e$

d. $-\frac{4}{3}, \pi, 5.6, \sqrt{2}, 0, -23, e$ **2.** 0

3. $b_2 = \frac{2A - b_1h}{h}$ or $b_2 = \frac{2A}{h} - b_1$ **4.** $85°, 80°, 15°$ **5.** \$8,250

6. $\frac{1}{120}$ decibels/rpm   **7. a.** Parallel   **b.** Perpendicular

8. $y = -3,400x + 28,000$ **9.** $y = -2x + 5$

10. $y = -\frac{9}{13}x + \frac{7}{13}$ **11.** It doesn't pass the vertical line test. The graph passes through $(0, 2)$ and $(0, -2)$.   **12.** $-4; 3a^5 - 2a^2 + 1$

13. a. 4   **b.** 3   **c.** 0, 2   **d.** 1   **14.** $(3, -13)$ **15.** $(-7, 7)$

16. $\left(\frac{4}{5}, \frac{3}{4}\right)$ **17.** Regular: $29\frac{1}{3}$ lb; Brazilian: $10\frac{2}{3}$ lb   **18.** $(3, 2, 1)$

19. $(2, -3)$ **20.** $\left(-2, \frac{3}{2}\right)$ **21.** $(-\infty, -2)$ ⟵━━━➤
                                                         -2

22. $\left[-3, \frac{19}{3}\right]$ ⟵━━[━━━━━]━━➤
                       -3 19/3

23.

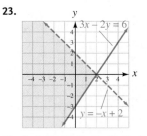

24. 10 hr   **25.** x^8y^{10} **26.** $\frac{16}{y^{16}}$

27. $1.73 \times 10^{14}; 4.6 \times 10^{-8}$

28. $1.44 \times 10^{-3}; 0.00144$

29. $(-3, 0), (0, 0), (2, 0); (0, 0); 24, 0, -8, -6, 0, 4, 0, -18$

30. $\frac{7}{4}rt^2 - \frac{5}{6}rt$ **31.** $2x^2y^3 + 13xy + 3y^2$ **32.** $x^3 - 27y^3$

33. $6m^{10} - 23m^5 + 7$ **34.** $81a^2b^4 - 72ab^2 + 16$

35. $xy(3x - 4y)(x - 2)$ **36.** $(b - 4)(b^2 - 3)$

37. $(3y + 2)(4y + 5)$ **38.** $(16x^2y^2 + z^4)(4xy + z^2)(4xy - z^2)$

39. $(3t + u)(9t^2 - 3tu + u^2)$ **40.** $b^2(a + 4)(a - 4)(a + 2)(a - 2)$

41. $\lambda = \frac{4d - 2}{A - 6}$ **42.** A repeated solution of -8 **43.** $0, 1, -9$

44. Length: 21 ft; width: 4 ft   **45.** $(-\infty, -6) \cup (-6, 4) \cup (4, \infty)$

46. $\frac{3x + 2}{3x - 2}$ **47.** $-\frac{q}{p}$ **48.** $\frac{4a - 1}{(a + 2)(a - 2)}$ **49.** $y - x$

50. $4x^3 - 3x^2 + 3x + 1$ **51.** $-\frac{7}{5}$ **52.** $R = \frac{R_1R_2R_3}{R_2R_3 + R_1R_3 + R_1R_2}$

53. $2\frac{2}{5}$ hr $= 2.4$ hr   **54.** 93   **55.** 2 lumens

56. D: $[0, \infty)$; R: $[2, \infty)$

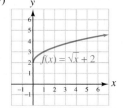

57. $\frac{343}{125}$ **58.** $4ab^2\sqrt{7ab}$ **59.** $5\sqrt{2}$ **60.** $81x\sqrt[3]{3x}$

61. $13 - \sqrt{7}$ **62.** $15x - 6\sqrt{15x} + 9$ **63.** $\frac{\sqrt[3]{4b^2}}{b}$

64. $\sqrt{3t} - 1$ **65.** 1, 3   **66.** 5   **67.** 0   **68.** About $21\frac{1}{2}$ in.

69. $5i$ **70.** -1 **71.** $-5 + 17i$ **72.** $-\frac{21}{29} - \frac{20}{29}i$ **73.** $\pm 2\sqrt{6}$

74. $-5 \pm 4\sqrt{2}$ **75.** $\frac{-3 \pm \sqrt{5}}{4}$ **76.** $\frac{2}{3} \pm \frac{\sqrt{2}}{3}i$ **77.** $\frac{1}{4}, \frac{1}{2}$

78. $-i, i, -3i\sqrt{2}, 3i\sqrt{2}$ **79. a.** A quadratic function

b. At about 85% and 120% of the suggested inflation

80. $(-1, -2)$; $(0, -8)$; no x-intercepts

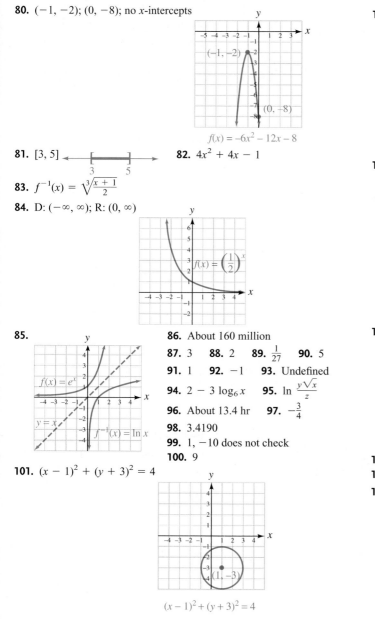

$f(x) = -6x^2 - 12x - 8$

81. $[3, 5]$

82. $4x^2 + 4x - 1$

83. $f^{-1}(x) = \sqrt[3]{\dfrac{x+1}{2}}$

84. D: $(-\infty, \infty)$; R: $(0, \infty)$

$f(x) = \left(\dfrac{1}{2}\right)^x$

85.

$f(x) = e^x$

$y = x$

$f^{-1}(x) = \ln x$

86. About 160 million

87. 3   **88.** 2   **89.** $\dfrac{1}{27}$ **90.** 5

91. 1   **92.** -1 **93.** Undefined

94. $2 - 3\log_6 x$ **95.** $\ln \dfrac{y\sqrt{x}}{z}$

96. About 13.4 hr   **97.** $-\dfrac{3}{4}$

98. 3.4190

99. 1, -10 does not check

100. 9

101. $(x - 1)^2 + (y + 3)^2 = 4$

$(x - 1)^2 + (y + 3)^2 = 4$

102. $x = -\dfrac{1}{4}(y - 3)^2 + 2$; $(2, 3)$

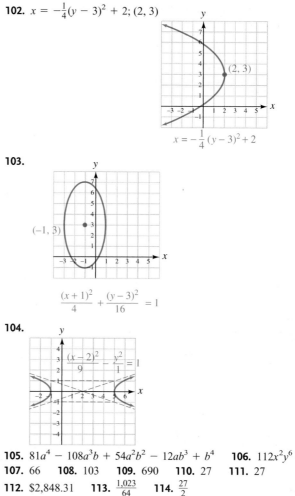

$x = -\dfrac{1}{4}(y - 3)^2 + 2$

103.

$(-1, 3)$

$\dfrac{(x + 1)^2}{4} + \dfrac{(y - 3)^2}{16} = 1$

104.

$\dfrac{(x - 2)^2}{9} - \dfrac{y^2}{1} = 1$

105. $81a^4 - 108a^3b + 54a^2b^2 - 12ab^3 + b^4$ **106.** $112x^2y^6$

107. 66   **108.** 103   **109.** 690   **110.** 27   **111.** 27

112. $2,848.31   **113.** $\dfrac{1{,}023}{64}$ **114.** $\dfrac{27}{2}$

INDEX

Absolute value
  equations, 347–352
  equations with two, 351–352
  functions, 196
  inequalities, 347–352
  of real numbers, 18–20
Acute angles, 79
Addition
  associative property of, 38
  commutative property of, 38
  of complex numbers, 724
  of cubes, 473–480
  of functions, 826–828
  method, 238, 976–977
  of negative numbers, 23
  of polynomials, 426–427
  of positive numbers, 23
  property of equality, 51, 238
  property of inequality, 323
  of radical expressions, 677–679
  of radicals, 678
  of rational expressions, 542–550
  of rational expressions with like
    denominators, 542–543
  of rational expressions with
    unlike denominators, 543–546
  of real numbers, 23
Additive identity, 39
Additive inverse, 18, 40
Algebra, 2
Algebraic expressions, 3
  evaluating, 33–34
  simplifying, using real numbers,
    37–46
Al-Jabr wa'l muquabalah
  (al-Kwarizmi), 2
al-Kwarizmi, 2
Altitude, 712
And, 336–339
Angles
  acute, 79
  base, 80
  complementary, 79
  obtuse, 79
  right, 79
  straight, 79
  supplementary, 79
  vertex, 80
Applications
  involving functions, 186–187
  involving geometric sequences,
    1019–1021
  involving sequences, 1006
  logarithmic functions in, 883–885
  logarithmic properties and,
    905–906
  natural logarithmic functions in,
    892–893
  quadratic equations, 757–758
  slope, 151–153
Applied problems, 138–142
Area, 63–66
Arithmetic mean, 1004
Arrow diagrams, 177
Ascending powers, 418

Associative property
  of addition, 38
  of multiplication, 38
Asymptotes, 964
Axis
  horizontal, 6
  major, 954
  minor, 954
  of symmetry, 785
  vertical, 6
  x, 119
  y, 119

Bar graphs, 6
Base angles, 80
Base-e exponential functions,
  865–870
  decay and, 869–870
  growth and, 869–870
Base-e logarithmic functions,
  888–893
  defining, 888
Bases, 28
  identifying, 394–395
Binomial theorem, 990–998
  expanding binomials with,
    994–996
Binomials, 418
  expanding, 991–998
  multiplication of, 436–437
  raising, to powers, 991
Bounded intervals, 337
Bracket, 322
Break points, systems and, 250–251
Break-even point, 222
Briggs, Henry, 879

Cartesian coordinate systems, 119
Celsius, 69
Center, 940
  of ellipses, 952
  of hyperbolas, 963
  writing equations given, 942
Central rectangle, 964
Change-of-base formulas, 903–904
Circles, 938–948
  center of, 940
  definition of, 940
  equations of, 942
  equations of, in standard form,
    940–944
  general form of equations of,
    942–944
  problems involving, 944–945
  radius of, 940
Closed intervals, 337, 338
Coefficients, 38
  leading, 418, 460
  numerical, 418
Common difference, 1002–1004
Common logarithms, 879
Common ratio, 1011–1013
Commutative property
  of addition, 38
  of multiplication, 38

Complementary angles, 79
Completing squares, 748–759,
    788–790
  method, 775
  solving quadratic equations with,
    752–757
Complex conjugates, 726
Complex fraction
  simplifying, 554–561
  simplifying, using division,
    554–555
  simplifying, using LCD, 555–561
Complex numbers, 721–730
  addition of, 724
  division of, 726–729
  multiplication of, 725–726
  subtraction of, 724
  writing, 723
Composite functions, 829–830
Composite numbers, 11
Composition of functions, 826–834,
    843–844
Compound inequalities, 335–343
  containing "and," 336–339
  containing "or," 340–343
  graphing, 373–374
Compound interest, 858
Computations with scientific
    notation, 412–414
Conditional equations, 59–60
Conic sections, 938–940
Conjugates, complex, 726
Consistent systems, 224
Constant function, 184
Constant terms, 418
Constants, 418
  degrees of nonzero, 419
  of proportionality, 611
  of variation, 611, 613, 615
Contradictions, identifying, 59–60,
    327–328
Coordinates, 17
  Cartesian, 119
  planes, 119
  of points, 118–121
  rectangular, 118–126
  x, 119
  y, 120
Cramer's Rule
  for systems of three equations,
    300–301
  for systems of two equations,
    297–300
Critical numbers, 800
Cube(s)
  difference of, 473–480
  factoring difference of, 477–480
  factoring sum of, 477–480
  perfect, 649
  sum of, 473–480
Cube root function, graphing,
    650–651
Cube roots, 649–650
  notation, 649
Cubic functions, 420

Cubing function, 195
Curve-fitting problems, using
    systems to solve, 278–279

Data tables, 5–6
Decay, 865–866
  base-e exponential functions and,
    869–870
  exponential functions and,
    856–861
  radioactive, 916
Decibels, 883
Decimals
  in equations, 56–59
  repeating, 13
  terminating, 13
Decreasing functions, 855
Degrees
  of nonzero constants, 419
  of terms of polynomials, 419
Denominators
  like, 542–543
  rationalizing, 687–690
  rationalizing, with two terms,
    690–691
  unlike, 543–546
Dependent equations, 225–229
  elimination and, 242–243
  identifying, 269–271
  matrices in identification of,
    289–291
  substitution and, 242–243
Dependent variables, 178
Depreciation, straight-line, 140
Descending powers, 418
Determinants
  array of signs for, 296
  solving systems of equations
    using, 294–301
  values of, 294, 295
Difference, 4, 24
  common, 1002–1004
  of squares, 473–480
Direct variation, 611–613
Discriminant, 773–774
Distance formula, 714–716
Distributive property, 42–44
  extended, 43
  of multiplication over addition,
    42
Division
  of complex numbers, 726–729
  of functions, 826–828
  of monomials by monomials,
    565–566
  of polynomials, 565–573
  of polynomials by monomials,
    566–567
  of polynomials by polynomials,
    567–571
  of polynomials with missing
    terms, 571–572
  properties, 41
  property of equality, 51
  property of inequality, 324

Division (*continued*)
radical expression, 683–693
of rational expressions, 530–538
of real numbers, 25–27
simplifying complex fractions using, 554–555
synthetic, 576–581
with zero, 41
Domains, 177
of functions, 178, 182–183, 193
of rational functions, 518–520
Double linear equalities, 339–340
Double-sign notation, 749
Doubling time, 892

Elementary row operations, 284–285
Elimination, 238–242
inconsistent systems and, 242–243
method, 243
solving systems of linear equations by, 265
solving systems of nonlinear equations by, 976–977
Ellipses, 938, 952–959
center of, 952
defining, 952
equations of, centered at origin, 953
foci of, 952
graphing, 956–958
graphing, centered at origin, 953–955
horizontal, 956
intercepts of, 954
problems involving, 958–959
standard form of equations of, 953, 956
vertical, 956
Empty sets, 60
Endpoints, 125
Equality
addition property of, 51, 238
division property of, 51
exponential property of, 910–911
logarithm property of, 49–60
multiplication property of, 51
properties of, 49–60
subtraction property of, 51
Equations, 4, 50
absolute value, 347–351
of circles in general form, 942–944
of circles in standard form, 940–944
conditional, 59
containing radicals, 697
data tables with, 5–6
decimals in, 56–59
dependent, 225–229, 242–243, 269–271, 289–291
of ellipses centered at origin, 953
equivalent, 51, 765–766
exponential, 876–878, 908–917
forming, 76
fractions in, 56–59
higher-degree polynomial, 493–495
of horizontal lines, 138
independent, 226
of inverse of functions, 841–843
left side of, 50
linear, 49–60, 132–142, 169–171, 224–225, 263, 265–266
logarithmic, 876–878, 908–917

nonlinear systems of, 973–977
ordered pairs as solutions to, 132–133
of parabolas, 945
point-slope form and, 164–167
polynomial, 488
quadratic, 488–493, 762–765, 775, 794–795
in quadratic form, 774–779
radical, 696–704
rational, 584–592
regression, 170
right side of, 50
simplifying expressions to solve, 54–56
solutions in two variables, 133
solving, algebraically, 235–244
solving, by factoring, 488–498
solving, graphically, 229–231
solving problems with, 75–81
solving systems of, using Cramer's Rule, 297–301
solving systems of, using determinants, 294–301
solving systems of, using matrices, 282–291
solving systems of, with missing variable terms, 266–269
solving, with matrices, 286–289
standard form of hyperbolas, 964
standard form of parabola, 946
supply, 145
systems of, 222–231, 235–244
systems of, in three variables, 263–271
systems of three, 275–279
systems of two, 246–257
of vertical lines, 138
with two absolute values, 351–352
writing, of lines, 161–171
Equilateral triangles, 80, 712–714
Equivalent equations, 51
quadratic formula computations and, 765–766
Equivalent expressions, 39, 521
Equivalent inequalities, 324
Equivalent systems, 228
Evaluation
of algebraic expressions, 33–34
of functions using graphs, 831–832
of logarithmic expressions, 878–880
of natural logarithmic expressions, 889–890
of rational functions, 517–518
Exponent(s), 394–405
changing, 401
identifying, 394–395
irrational, 851–852
natural-number, 28, 395
negative, 399–402
positive, 401
power rules for, 395–399
product rules for, 395–399
property of equality, 909
quotient rule for, 402–404
rational, 658–666
rules, 405
for simplifying expressions, 664–665
zero, 399–402
Exponential equations, 908–917
as logarithmic equations, 876–878
solving, 908–912

Exponential functions, 850–861
base-*e*, 865–870
decay and, 856–861, 865–866
defining, 852
graphing, 852–856
growth and, 856–861, 865–866
natural, 866–867
properties of, 854
Exponential growth, 917
Expressions, 3
algebraic, 3, 33–34, 37–46
equivalent, 39, 521
exponents for simplifying, 664–665
logarithmic, 878–880, 902–903
natural logarithmic, 889–890
radical, 642–654, 665–666, 671–679, 727
rational, 516–525, 530–538, 542–550
simplifying, 54–56, 443–444, 658–662, 663–664, 851–852
square roots of, containing variables, 645–646
Extraneous solutions, 589–590

Factor theorem, 580–581
Factorial notation, 993–994
Factorial property, 994
Factoring
difference of two cubes, 477–480
difference of two squares, 474
by grouping, 447–456
grouping method for, trinomials, 468–470
method, 775
out GCF, 449–452
polynomials completely, 454
polynomials using factor theorem, 580–581
polynomials using remainder theorem, 579–580
solving equations by, 488–498
in solving formulas for specified variables, 455–456
for solving quadratic equations, 490
steps for, polynomials, 483
substitution in, of trinomials, 467
sum of two cubes, 477–480
techniques, 483–486
trinomials, 459–470
Factorization, prime, 674–675
Factors, 25, 38
greatest common, 447–456
opposite, 524–525
Fahrenheit, 69
Finite sequences, 1000
First terms, 1002–1004
Focus, 939
of ellipses, 952
of hyperbolas, 963
FOIL method, 436, 548
Formulas
change-of-base, 903–904
containing radicals, 704
distance, 714–716
for exponential growth/decay, 865–866
quadratic, 761–771, 775
radioactive decay, 916
solving, 63–70
solving application problems using, 69–70
solving, for specified variables, 590–592

for specified variables, 455–456
for vertex of parabolas, 791
Fractions
complex, 554–561
in equations, 56–59
negative, 13
properties of, 520
Functions, 176–187
absolute value, 196
addition of, 826–828
applications involving, 186–187
base-*e*, 865–870, 888–893
composite, 829–830
composition of, 826–834
constant, 184
cubic, 420
cubing, 195
decreasing, 855
division of, 826–828
domains of, 178, 182–183, 193
evaluating, with graphs, 831–832
exponential, 850–861
graphs of, 192–201
graphs of inverse, 844–846
identifying, 177–180
identity, 184, 832–833
increasing, 855
inverse, 838–846
linear, 183–184
logarithmic, 874–885
multiplication of, 826–828
natural exponential, 866–867
natural logarithmic, 890–891, 892–893
nonlinear, 193
notation, 180–182
one-to-one, 838–841
operations on, 826–828
polynomial, 417–428
quadratic, 420, 783–795
radical, 642–654
ranges of, 178, 193
rational, 516–525
reflecting graphs of, 200–201
square root, 646–648
squaring, 193
subtraction of, 826–828
translations of graphs of, 197–200

GCF. *See* Greatest common factor
General form, equations of circles in, 942–944
General terms, 1000–1001
Geometric means, 1013–1015
Geometric sequences
application problems involving, 1019–1021
infinite, 1016–1019
sums of terms in, 1015–1016
terms of, 1011–1013
Geometry, 63–70, 79–80
radicals and, 708–716
systems in, 248–249
Graph(s), 120
bar, 6
evaluating functions with, 831–832
of functions, 192–201
of inverse functions, 844–846
line, 6
of nonlinear functions, 193–197
of quadratic functions, 783–795
reading, 122–124
of real numbers, 17
reflections of, 200–201

Graph(s) (*continued*)
 slopes of lines and, 148–149
 step, 123
 translations of, 197–200
Graphical models, 6–7
 ellipses centered at origin,
  953–955
Graphing, 120
 compound inequalities,
  373–374
 cube root functions, 650–651
 dependent equations and,
  225–229
 ellipses, 956–958
 exponential functions, 852–856
 functions, 197
 horizontal lines, 137–138
 hyperbolas, 967–968
 hyperbolas centered at origin,
  963–966
 inconsistent systems and,
  225–229
 intervals, 321–323
 linear equations, 133–137
 linear functions, 183–184
 linear inequalities, 361–353
 linear inequalities with boundary
  through origin, 363–364
 logarithmic functions, 881–883
 method, 224, 227, 243
 natural exponential functions,
  867–868
 natural logarithmic functions,
  890–891
 nonlinear inequalities, 806–808
 paired data, 121–122
 polynomial functions, 421–424
 quadratic functions, 792
 rational functions, 518
 slopes and, 167–168
 solving nonlinear systems by,
  973–974
 solving systems of equations by,
  222–231
 square root function, 646–648
 vertical lines, 137–138
 x-intercepts, 142
Greatest common factor (GCF),
  447–456
 factoring out, 449–452
 finding, in lists of terms, 448
 strategies for finding, 448
Grouping
 factoring by, 447–456
 factoring trinomials by, 468–470
 symbols, 32
Growth, 865–866
 base-*e* exponential functions and,
  869–870
 exponential, 917
 exponential functions and,
  856–861

Half-life, 916
Half-open interval, 338
Higher-degree polynomial
  equations, 493–495
Horizontal axis, 6
Horizontal ellipses, 956
Horizontal lines
 boundary, 364–365
 equations of, 138
 graphing, 137–138
 slope of, 151
 tests, 840–841
Horizontal translations, 198

Hyperbolas, 938, 962–970
 center of, 963
 defining, 962–963
 foci of, 963
 graphing, 967–968
 graphing, centered at origin,
  963–966
 problems involving, 969–970
 standard form of equations of, 964
 vertices of, 963

Identities
 additive, 39
 identifying, 59–60, 327–328
 multiplicative, 39
Identity function, 184
 defining, 832–833
Imaginary numbers, 644, 721
Imaginary part, 723
Inconsistent systems, 225–229
 elimination and, 242–243
 identifying, 269–271
 matrices in identification of,
  289–291
 substitution and, 242–243
Increasing functions, 855
Independent equations, 226
Independent variables, 178
Index, 649, 651
 of summation, 1007
Inequalities
 absolute value, 347–352
 addition property of, 323
 compound, 335–343
 division property of, 324
 double linear, 339–340
 equivalent, 324
 linear, 320–330
 multiplication property of, 323
 nonlinear, 799–808
 properties of, 323–327
 quadratic, 799–808
 rational, 802–806
 satisfying, 361
 solving, graphically, 805–806
 subtraction property of, 323
Inequality symbols, 17–18
 interpreting, 321
Infinite sequences, 1000
 geometric, 1016–1019
Infinity
 negative, 322
 positive, 321
Inputs, 179
Integers, 12
 even, 12
 odd, 12
Intercepts
 determining, 142
 of ellipses, 954
 finding, 135–137
 of lines, 135
 x, 135, 142
 y, 135
Interest, 251–257
 compound, 858
 simple, 91
Intersections, of two sets, 335–336
Intervals
 bounded, 337
 closed, 337, 338
 graphing, 321–323
 half-open, 338
 notation, 321–323
 open, 338
 unbounded, 322

Inverse functions, 838–846
 composition of, 843–844
 equations of, 841–843
 graphs of, 844–846
Inverse variation, 613–614
Inverses
 additive, 40
 multiplicative, 40
Investment problems, 91–93
Irrational exponents, 851–852
Irrational numbers, 14–15
 approximating, 15
Irreducible polynomials, 450
Isosceles triangles, 80, 710–711

Joint variation, 614–615

Law of reflection, 204
LCD. *See* Least common
  denominator
Leading coefficients, 418, 460
Leading terms, 418
Least common denominator (LCD),
  546–549
 simplifying complex fractions
  using, 555–561
Left side, 50
Like terms, combining, 44–46,
  424–425
Line(s)
 boundary, 364–365
 horizontal, 137–138, 151,
  364–365, 840–841
 intercepts of, 135
 number, 11
 parallel, 153–156, 168–169
 perpendicular, 153–156,
  168–169
 point–slope form and, 164–167
 slope, 146–156
 vertical, 137–138, 151, 364–365
 writing equations of, 161–171
Line graphs, 6
Line segments, midpoints of,
  125–126
Linear equations, 49–60, 132–142
 data collections and, 169–171
 defining, 51
 graphing, 133–137
 solving, by elimination, 265
 solving, by graphing, 224–225
 solving, by substitution, 235–238
 solving, in one variable, 56
 solving, with matrices, 287
 standard form of, 135
 systems, 235–244
 systems of three, 265–266
 in three variables, 263
Linear functions, graphing, 183–184
Linear inequalities, 320–330
 applied problems involving,
  365–367
 double, 339–340
 graphing, 361–363
 problems involving systems of,
  375–376
 solution sets of, 323
 solving, 323–327
 solving problems with, 329–330
 systems of, 370–376
 in two variables, 360–367
 with horizontal boundary lines,
  364–365
 with vertical boundary lines,
  364–365
Linear models, 138–142

Logarithm(s)
 common, 879
 defining, 874–875
 logarithmic expressions as single,
  902–903
 Napierian, 888
 natural, 888
 power rule for, 900–901
 product rule for, 897–898
 properties of, 896–906
 property of equality, 910–911
 quotient rule for, 898–900
Logarithmic equations, 876–878,
  908–917
 solving, 913–915
Logarithmic expressions
 evaluating, 878–880
 natural, 889–890
 as single logarithms, 902–903
Logarithmic functions, 874–885
 in applications, 883–885
 base-*e*, 888–893
 exponential equations as,
  876–878
 graphing, 881–883
 natural, 890–893
 properties of, 882
LORAN, 940

Major axis, 954
Mapping diagrams, 177
Mathematical models, 2, 4
Matrices, 282–291
 augmented, 283–284
 defining, 283
 linear equations and, 287
 performing row operations on,
  284–285
 in solving systems of three
  equations, 287–289
 in solving systems of two
  equation, 286–287
 square, 294
Maximum values, 793–794
Mean, 89–91
 arithmetic, 1004
 geometric, 1013–1015
Median, 89–91
Midpoints, 125–126
 formula, 125
Minimum values, 793–794
Minor axis, 954
Missing terms, 571–572
Mixed operations, 537–538,
  549–550
Mixture problems, 94–96, 251–257
Mode, 89–91
Models, 2
 graphical, 6–7
 linear, 138–142
 mathematical, 2
 verbal, 2, 3
Monomials, 418
 division of, by monomials,
  565–566
 division of polynomials by,
  566–567
 multiplication of polynomials by,
  435
Multiplication
 associative property of, 38
 of binomials, 436–437
 commutative property of, 38
 of complex numbers, 725–726
 distributive property of, 42
 of functions, 826–828

Multiplication (*continued*)
   by negative one, 545
   of polynomials, 434–444
   of polynomials by monomials, 435
   property of equality, 51
   property of inequality, 323
   property of zero, 39
   radical expression, 683–693
   of rational expressions, 530–538
   simplifying expressions with, 443–444
   of three polynomials, 439–440
Multiplicative identity, 39
Multiplicative inverse, 40

Napier, John, 879, 888
Napierian logarithms, 888
Natural exponential functions, 866–867
   graphing, 867–868
Natural logarithmic expressions, evaluating, 889–890
Natural logarithmic functions, 890–891
   in applications, 892–893
Natural logarithms, 888
Natural numbers, 11
   even, 659
   exponents, 28, 395
   odd, 659
Negative
   infinity, 322
   reciprocals, 155
Negative exponents, 399–402
   reciprocals, 404–405
Negative fractions, 13
Negative integer rules, 399–402
Negative numbers, 11
   adding, 23
   square roots, 721–722
Negative powers, 404–405
Negative rational exponents, 663–664
Negative square roots, 643
Nonlinear functions, graphs of, 193–197
Nonlinear inequalities, 799–808
   graphing, in two variables, 806–808
Nonlinear systems, 973–977
   solution sets of, 973
   solving, by elimination, 976–977
   solving, by substitution, 974–976
Notation
   cube root, 649
   double-sign, 749
   factorial, 993–994
   function, 180–182
   interval, 321–323
   scientific, 408–414
   set-builder, 14, 321–323
   square root, 643
   standard, 410–411
   summation, 1006–1007
*n*th roots, 651–654
Null sets, 60
Numbers, 11–12
   complex, 721–730
   composite, 11
   critical, 800
   imaginary, 644, 721
   irrational, 14–15
   line, 11
   natural, 11, 659

negative, 11, 23
positive, 11, 23
prime, 11
rational, 12–14
real, 15–16, 18–20, 23–34, 37–46
   in scientific notation, 408–414
   as solutions, 50
   whole, 11
Number-value problems, 78–79
Numerators, 692

Obtuse angles, 79
One-to-one functions, 838–840
   horizontal line test of, 840–841
Open intervals, 338
Operations
   on functions, 826–828
   involving powers, 730
   mixed, 537–538, 549–550
   row, 284–285
Opposites, 12
   quotients of, 525
   of real numbers, 18
Or, 340–343
Order of operations, 30–33
Ordered pairs, 118–121
   as solutions, 132–133
   as solutions of systems, 223–224
Ordered triples, 263–264
Orders, 651
Origins, 119
   of ellipses, 953–955
   graphing hyperbolas centered at, 963–966
   linear inequalities with boundary through, 363–364
Outputs, 179
Overbars, 14

Paired data, graphing, 121–122
Parabola, 194, 785, 938–948
   definition of, 945
   equations of, 945
   formula for vertex of, 791
   standard form of equations of, 946
Paraboloids, 939
Parallel lines
   recognizing, 168–169
   slope of, 153–156
Parenthesis, 321
Partial sums, 1016
Parts
   imaginary, 723
   real, 723
Pascal's triangle, 991–993
Pendulums, 648
Percent
   of decrease, 88
   of increase, 88
   problems, 86–89
   sentences, 86
Perfect cubes, 649
Perfect squares, 644
Perfect-square trinomials, 459–460, 752
Perimeter, 63–66
Perpendicular lines
   recognizing, 168–169
   slope of, 153–156
Plotting, 120
Points
   break, 250–251
   break-even, 222
   coordinates of, 118–121

plotting, 133–135
   slopes of lines and, 149–150
Point–slope form, 164–167
Polynomials, 417–428
   addition of, 426–427
   degrees of, 419
   division of, 565–573
   division of, by monomials, 566–567
   division of, by polynomials, 567–571
   division of, with missing terms, 571–572
   equations, 488
   factoring, completely, 454
   factoring, random, 483
   factoring, using factor theorem, 580–581
   factoring, using remainder theorem, 579–580
   functions, 417–428
   higher degree, equations, 493–495
   multiplication of, 434–444
   multiplication of, by monomials, 435
   multiplication of three, 439–440
   in one variable, 418
   prime, 450
   simplifying, 424–425
   steps for factoring, 483
   subtraction of, 427–428
   in three variables, 418
   in two variables, 418
Population growth, 916–917
Positive exponents, 399–402
Positive infinity, 321
Positive numbers, 11
   adding, 23
Positive square roots, 643
Powers, 394
   ascending, 418
   descending, 418
   negative, 404–405
   operations involving, 730
   of product and quotient, 398
   of radical expressions, 686–687
   raising binomials to, 991
   of rational expressions, 534–535
   of roots, 686
   rules, 395–399, 900–901
Prime factorization, 674–675
Prime polynomials, 450
Principal square root, 30, 643
Problems
   application, 757–758, 1006, 1019–1021
   applied, 138–142, 365–367
   curve-fitting, 278–279
   geometry, 79–80
   interest, 251–257
   investment, 91–92
   involving ellipses, 958–959
   involving hyperbolas, 969–970
   involving systems of linear inequalities, 375–376
   mixture, 94–96, 251–257
   number-value, 78–79, 249–250
   percent, 86–89
   population growth, 916–917
   radioactive decay, 916
   shared-work, 596–599
   solving, 86–96
   solving strategies, 247
   solving, using rational equations, 596–602

solving, with equations, 75–81
   strategies for solving, 75–76
   uniform motion, 93, 251–257, 599–602
   variation, 613
Product, 4, 25
   powers of, 398
   rules, 395–399, 671–674, 683, 897–898
   simplifying, 41–42
   special, 440–443
Properties
   addition, of equality, 51, 238
   addition, of inequality, 323
   additive inverse, 40
   associative, of addition, 38
   associative, of multiplication, 38
   commutative, of addition, 38
   commutative, of multiplication, 38
   distributive, 42–44
   division, 41
   division, of equality, 51
   division, of inequality, 324
   of equality, 49–60
   exponent, of equality, 909
   of exponential functions, 854
   extended distributive, 43
   factorial, 994
   of fractions, 520
   fundamental, 607
   of inequality, 323–327
   logarithm, of equality, 910–911
   of logarithmic functions, 882
   of logarithms, 896–906
   multiplication, of equality, 51
   multiplication, of inequality, 323
   multiplicative inverse, 40
   of one, 39
   of real numbers, 38–40
   subtraction, of equality, 51
   subtraction, of inequality, 323
   of zero, 39
   zero-factor, 488–493
Proportion, 606–616
   constants of, 611
   fundamental properties, 607
   identifying, 606–607
   inverse, 613
   solving, 607–608
   using, to solve problems, 608–609
Pythagorean theorem, 709–710

Quadrants, 119
Quadratic equations, 488–493
   checking solutions of, 756
   completing squares to solve, 752–757
   factoring method for solving, 490
   general, 762
   for solving application problems, 757–758
   solving, graphically, 794–795
   solving problems involving, 779–780
   solving problems with, 495–498
   solving, using quadratic formula, 762–765
   square root property for solving, 749–752
   strategies for solving, 775
Quadratic form, 773–781
   solving equations in, 774–779

Quadratic formula, 761–771, 775
  deriving, 762
  equivalent equations and, 765–766
  for solving application problems, 766–771
Quadratic functions, 420
  defining, 784
  graphing, 792
  graphs of, 783–795
  in standard form, 784, 787
Quadratic inequalities, 799–808
  solving, 799–802
Quotient rule, 402–404
  for logarithms, 898–900
  for radicals, 676
  for simplifying rational expressions, 675–677
Quotients, 26
  of opposites, 525
  powers of, 398
  simplifying, raised to negative powers, 404–405

Radical(s), 643
  addition of, 678
  conversion from, 662–663
  equations containing, 697
  formulas containing, 704
  geometric applications of, 708–716
  product rule for, 672, 683
  quotient rule for, 676
  radical equations with one, 696–701
  radical equations with two, 701–703
  simplifying, 665
  subtraction of, 678
Radical equations
  solving, 696–704
  with one radical, 696–701
  with two radicals, 701–703
Radical expressions, 642–654, 727
  addition of, 677–679
  combining, 671–679
  division of, 683–693
  multiplication of, 683–693
  powers of, 686–687
  prime factorization and, 674–675
  product rule for simplification of, 671–674
  simplified form of, 672
  simplifying, 665–666, 671–679
  subtraction of, 677–679
Radical functions, 642–654, 646
Radical symbols, 30, 643
Radicand, 643
Radioactive decay
  formula, 916
  problems, 916
Radius, 940
  writing equations given, 942
Random polynomials, 483
Range, 177
  of functions, 178
Rate of change, 146–156
  average, 146–148
Rate of work, 596
Rates, 146
  identifying, 606–607
Rational equations
  problem solving using, 596–602
  solving, 584–592
  with extraneous solutions, 589–590

Rational exponents, 658–666
  conversion to, 662–663
  negative, 663–664
Rational expressions
  addition of, 542–550
  addition of, with like denominators, 542–543
  addition of, with unlike denominators, 543–546
  building, 544
  defining, 516–517
  division of, 530–538
  multiplication of, 530–538
  powers of, 534–535
  simplifying, 516–525
  simplifying, with opposite factors, 524–525
  subtraction of, 542–550
  subtraction of, with like denominators, 542–543
  subtraction of, with unlike denominators, 543–546
Rational functions, 516–525
  domains of, 518–520
  evaluating, 517–518
  graphing, 518
Rational inequalities, 802–806
Rational numbers, 12–14
Rationalizing denominators, 687–690
  with two terms, 690–691
Rationalizing numerators, 692
Ratios, 146
  common, 1011–1013
  identifying, 606–607
Real numbers, 18–20
  absolute values of, 18–20
  addition of, 23
  algebraic expressions and, 37–46
  classifying, 15–16
  dividing, 25–27
  graphing, 17
  multiplying, 25–27
  opposites of, 18
  ordering, 17–18
  powers of, 27–30
  properties of, 38–40
  subtraction of, 24
Real part, 723
Reciprocals, 27, 40
  negative, 155, 404–405
Rectangles, central, 964
Rectangular coordinate system, 118–126
Reflections
  of graphs of functions, 200–201
  law of, 204
Regression, 169
  equations, 170
Relations, 177
Remainder theorem, in factoring polynomials, 579–580
Repeating decimals, 13
Richter scale, 884
Right angles, 79
Right side, 50
Right triangles, 80
Row operations, 284–285
Rules
  Cramer's, 297–301
  exponents, 405
  negative integer exponent, 399–402
  power, 395–399
  product, 395–399

quotient, for exponents, 402–404
  special product, 440
  zero exponent, 399–402
Run, 148

Scientific notation, 408–414
  computations with, 412–414
  standard notation conversion and, 410–411
  writing numbers in, 408–414
Secants, 973
Segments, line, 125–126
Sequences, 1000–1007
  application problems involving, 1006
  arithmetic, 1000–1007
  finding, given general terms, 1000–1001
  finite, 1000
  geometric, 1011–1021
  infinite, 1000
  sums of terms in, 1005–1006
  terms of, 1002–1004
Series, 1000–1007
Set-builder notation, 14, 321–323
Sets
  empty, 60
  intersection of, 335–336
  null, 60
  solution, 50, 323, 973
  union of, 335–336
Shared-work problems, 596–599
Similar triangles, 609–611
Simple interest, 91
Simplifying
  algebraic expressions, 37–46
  exponents for, expressions, 664–665
  expressions, 54–56, 443–444, 658–662, 851–852
  polynomials, 424–425
  products, 41–42
  quotients raised to negative powers, 404–405
  radical expressions, 665–666, 671–679
  radicals, 665
  rational expressions, 516–525
Slope–intercept form, 161–163
Slopes
  applications of, 151–153
  graphing and, 167–168
  of horizontal lines, 151
  of lines, 146–156
  parallel lines and, 153–156
  perpendicular lines and, 153–156
  triangles, 148
  of vertical lines, 151
Solutions
  extraneous, 589–590
  infinitely many, 227
  many, 227
  numbers as, 50
  one, 227
  ordered pairs as, 132–133
  sets, 50, 323, 973
  in two variables, 133
Special products, 440–443
  rules, 440
Specified variables, 66–69, 455–456, 590–592
Square(s)
  completing, 748–759, 752–757, 775, 788–790
  difference of, 473–480

perfect, 644
  of square roots, 697
Square matrix, 294
Square root function, graphing, 646–648
Square root property, 748–759
  for solving quadratic equations, 749–752
Square roots
  of expressions containing variables, 645–646
  finding, 642–645
  method, 775
  negative, 643
  of negative numbers, 721–722
  notation, 643
  positive, 643
  principal, 30, 643
  of real numbers, 27–30
  squares of, 697
Squaring function, 193
Standard form, 489
  equations of circles in, 940–944
  of equations of ellipses, 956
  of equations of parabolas, 946
  of quadratic functions, 784, 787
Standard notation, scientific notation and, 410–411
Step graphs, 123
Straight angles, 79
Straight-line depreciation, 140
Subscript notation, 149
Substitution
  in factoring of trinomials, 467
  inconsistent systems and, 242–243
  method, 235, 243, 974–976
  solving equations by, 235–238
Subtraction
  of complex numbers, 724
  of cubes, 473–480
  of functions, 826–828
  of polynomials, 427–428
  property of equality, 51
  property of inequality, 323
  of radical expressions, 677–679
  of radicals, 678
  of rational expressions, 542–550
  of rational expressions with like denominators, 542–543
  of rational expressions with unlike denominators, 543–546
  of real numbers, 24
  of squares, 473–480
Summation notation, 1006–1007
Supplementary angles, 79
Symbols
  grouping, 32
  inequality, 17–18
  radical, 30, 643
Symmetry, 785
Synthetic division, 576–581
  performing, 576–579
Systems
  consistent, 224
  in curve-fitting problems, 278–279
  of equations, 222–231, 235–244
  of equations in three variables, 263–271
  of equations with missing variable terms, 266–269
  equivalent, 228
  for finding break points, 250–251

Systems (*continued*)
   inconsistent, 225–229, 242–243,
      269–271, 289–291
   in interest problems, 251–257
   of linear equations, 235–244, 265
   of linear inequalities, 370–376
   in mixture problems, 251–257
   nonlinear, of equations, 973–977
   in number-value problems,
      249–250
   solving, by graphing, 973–974
   solving, by substitution, 974–976
   solving, of equations using
      Cramer's rule, 297–301
   solving, of equations using
      determinants, 294–301
   solving, of equations using
      matrices, 282–291
   of three equations, 275–279
   of two equations, 246–257
   in uniform motion problems,
      251–257

Tangents, 973
Terminating decimals, 13
Terms, 38
   constant, 418
   degrees of, 419
   finding GCF in lists of, 448
   first, 1002–1004, 1011–1013
   general, 1000–1001
   of geometric sequences,
      1011–1013
   leading, 418
   like, 44–46, 424–425
   missing, 571–572

   of sequences, 1002–1004
   specific, 996–998
   sums of, 1005–1006, 1015–1017
   unlike, 44
Theorem
   binomial, 990–998
   factor, 580–581
   Pythagorean, 709–710
   remainder, 579–580
Translations
   of graphs of functions, 197–200
   horizontal, 198
   vertical, 197
Trial-and-check method, 465
Triangles
   equilateral, 80, 712–714
   isosceles, 80, 710–711
   Pascal's, 991–993
   right, 80
   similar, 609–611
   slope, 148
Trinomials, 418
   factoring, 459–470
   grouping method for factoring,
      468–470
   perfect-square, 459–460, 752
   substitution in factoring, 467

Unbounded intervals, 322
Uniform motion, 93, 251–257,
     599–602
Unions, of two sets, 335–336
Unknowns
   assigning variables to, 246–248
   assigning variables to three,
     275–278

Unlike terms, 44

Values
   maximum, 793–794
   minimum, 793–794
Variables, 3, 50
   assigning, 246–248
   assigning to three unknowns,
     275–278
   Cramer's Rule for equations with
     three, 300
   Cramer's Rule for equations with
     two, 298
   dependent, 178
   expressions containing, 645–646
   graphing nonlinear inequalities in
     two, 806–808
   independent, 178
   isolating, 56
   linear equations in three, 263
   linear inequalities in two,
     360–367
   missing, 266–269
   polynomials in one, 418
   polynomials in three, 418
   polynomials in two, 418
   solutions in two, 133
   solving linear equations in one, 56
   specified, 66–69, 455–456,
     590–592
   systems of equations in three,
     263–271
Variation, 606–616
   combined, 615–616
   constants of, 611, 613, 615
   direct, 611–613

   inverse, 613–614
   joint, 614–615
   problems, 613
Venn Diagrams, 336
Verbal models, 2
Vertex, 790, 954
   formula for, 791
   of hyperbolas, 963
Vertex angles, 80
Vertical axis, 6
Vertical ellipses, 956
Vertical lines
   boundary, 364–365
   equations of, 138
   graphing, 137–138
   slope of, 151
   test, 184–185
Vertical translations, 197
Volume, 63–66

Whole numbers, 11
Work
   amount, 596
   rate, 596
   shared, 596–602

x-axis, 119
x-coordinates, 119
x-intercepts, 135, 142

y-axis, 119
y-coordinates, 120
y-intercepts, 135

Zero exponents, 399–402
Zero-factor property, 488–493